Environmental SCIENCE

A Global Concern

William P. Cunningham
University of Minnesota

Mary Ann Cunningham
Vassar College

Mc Graw Hill Education

ENVIRONMENTAL SCIENCE: A GLOBAL CONCERN, FOURTEENTH EDITION

1 2 3 4 5 6 7 8 9 LWI 21 20 19 18 17

ISBN 978–1259–63115–3
MHID 1–259–63115–X

Chief Product Officer, SVP Products & Markets: *G. Scott Virkler*
Vice President, General Manager, Products & Markets: *Marty Lange*
Vice President, Content Design & Delivery: *Betsy Whalen*
Managing Director: *Thomas Timp*
Brand Manager: *Michael Ivanov, Ph.D.*
Director, Product Development: *Rose M. Koos*
Product Developer: *Jodi Rhomberg*
Marketing Manager: *Noah Evans*
Market Development Manager: *Tamara Hodge*
Digital Product Analyst: *Patrick Diller*
Digital Product Developer: *Joan Weber*
Director, Content Design & Delivery: *Linda Avenarius*
Program Manager: *Lora Neyens*
Content Project Manager: *Sherry Kane/Tammy Juran*
Buyer: *Laura Fuller*
Design: *Tara McDermott*
Content Licensing Specialists: *Carrie Burger/Lorraine Buczek*
Cover Image: *© Georgetta Douwma/Getty Images*
Compositor: *SPi Global*
Printer: *LSC Communications*

All credits appearing on page or at the end of the book are considered to be an extension of the copyright page.

Design Elements: TOC, Glossary, Index: ©imagebroker/Alamy RF; Preface: ©Daryl Leniuk/Getty Images RF;
Author: ©Glow Images/SuperStock RF

Library of Congress Cataloging-in-Publication Data

Names: Cunningham, William P., author. | Cunningham, Mary Ann, author.
Title: Environmental science: a global concern/William P. Cunningham,
   University of Minnesota, Mary Ann Cunningham, Vassar College.
Description: Fourteenth edition. | New York: McGraw-Hill Education, [2017] |
   Audience: Ages: 18+
Identifiers: LCCN 2016040835 | ISBN 9781259631153 (acid-free paper)
Subjects: LCSH: Environmental sciences—Textbooks.
Classification: LCC GE105 .C86 2017 | DDC 304.2—dc23
LC record available at https://lccn.loc.gov/2016040835

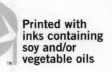

logo applies to the text stock only

mheducation.com/highered

William P. Cunningham

© Tom Finkle

William P. Cunningham is an emeritus professor at the University of Minnesota. In his 38-year career at the university, he taught a variety of biology courses, including Environmental Science, Conservation Biology, Environmental Health, Environmental Ethics, Plant Physiology, and Cell Biology. He is a member of the Academy of Distinguished Teachers, the highest teaching award granted at the University of Minnesota. He was a member of a number of interdisciplinary programs for international students, teachers, and nontraditional students. He also carried out research or taught in Sweden, Norway, Brazil, New Zealand, China, and Indonesia.

Professor Cunningham has participated in a number of governmental and nongovernmental organizations over the past 40 years. He was chair of the Minnesota chapter of the Sierra Club, a member of the Sierra Club national committee on energy policy, vice president of the Friends of the Boundary Waters Canoe Area, chair of the Minnesota governor's task force on energy policy, and a citizen member of the Minnesota Legislative Commission on Energy.

In addition to environmental science textbooks, he edited three editions of the *Environmental Encyclopedia,* published by Thompson-Gale Press. He has also authored or coauthored about 50 scientific articles, mostly in the fields of cell biology and conservation biology, as well as several invited chapters or reports in the areas of energy policy and environmental health. His Ph.D. from the University of Texas was in botany.

Professor Cunningham's hobbies include photography, birding, hiking, gardening, and traveling. He lives in St. Paul, Minnesota, with his wife, Mary. He has three children (one of whom is coauthor of this book) and seven grandchildren.

Both authors have a long-standing interest in the topics in this book. Nearly half the photos in the book were taken on trips to the places they discuss.

Mary Ann Cunningham

© Tom Finkle

Mary Ann Cunningham is an associate professor of geography at Vassar College. A biogeographer with interests in landscape ecology, geographic information systems (GIS), and climate impacts on biodiversity and food production, she teaches environmental science, natural resource conservation, land-use planning, and GIS. Field research methods, statistical methods, and data analysis and visualization are regular components of her teaching. Every aspect of this book is woven into, and informed by, her courses and her students' work. As a scientist and an educator, Mary Ann enjoys teaching and conducting research with both science students and non-science liberal arts students. As a geographer, she likes to engage students with the ways their physical surroundings and social context shape their world experience. In addition to teaching at a liberal arts college, she has taught at community colleges and research universities.

Professor Cunningham has been writing in environmental science for nearly two decades, and she has been coauthor of this book since its seventh edition. She is also coauthor of Principles of Environmental Science (now in its eighth edition) and an editor of the Environmental Encyclopedia (third edition, Thompson-Gale Press). She has published work on pedagogy in cartography, as well as instructional and testing materials in environmental science, and a GIS lab manual that introduces students to spatial and environmental analysis. She has also been a leader in sustainability programs and climate action planning at Vassar.

In addition to environmental science, Professor Cunningham's primary research activities focus on land-cover change, habitat fragmentation, and distributions of bird populations. This work allows her to conduct field studies in the grasslands of the Great Plains, as well as in the woodlands of the Hudson Valley. In her spare time she loves to travel, hike, and watch birds. Professor Cunningham holds a bachelor's degree from Carleton College, a master's degree from the University of Oregon, and a Ph.D. from the University of Minnesota.

Brief Contents

Contents

List of Case Studies

About the Cover

Coral reefs are among the most magnificent biological communities on Earth. They rival tropical rainforests in beauty and diversity. About one quarter of all marine species spend some or all of their life cycle in the shelter of coral reefs, and in some areas, reefs provide three-fourths of the protein in human diets. Reefs protect shorelines from storms, and provide valuable recreation and educational resources. Most corals are in serious trouble because of environmental change, pollution, and physical degradation, which threaten reef communities nearly everywhere. We've already lost about 30 percent of our coral worldwide, and another 60 percent is threatened with extinction. But environmental scientists are working to protect and restore corals around the world. Plans to slow climate change are expanding, as are efforts to reduce marine pollution. And with tools of the growing field of ecosystem restoration (chapter 11), we may be able to regrow reefs, and even breed resistant corals that can repopulate damaged reefs.

Environmental Science: A Search for Solutions

Environmental science focuses on understanding challenges that affect our lives, and on finding solutions to those challenges. Your decision to study environmental science will help you develop the tools to find answers to some of the most important problems facing us today.

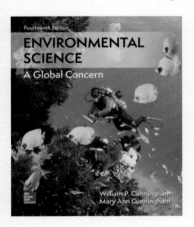

Coral reefs (shown on the cover, and the focus of the opening case study for chapter 11, Biodiversity: Preserving Species) are one of many fascinating systems explored in environmental science. These ecosystems are built on complex, intricately evolved symbiotic and competitive relationships. Energy and nutrients that flow through these systems support countless varieties of organisms— fish, shrimp, crabs, colorful snails and worms, and many others. Species that spend at least part of their life cycle in reefs provide nourishment for hundreds of millions of people. Thus, humans are also part of the reef system.

Like other ecosystems, coral reefs are also affected by factors in the surrounding environment: temperatures, nutrient sources, the sun's energy, and also human-caused pollution, disturbances, and, increasingly, climate warming that results primarily from burning fossil fuels. Researchers warn that we've already lost about one third of our existing reefs, and that another 60 percent are threatened by human activities.

An increasingly frequent consequence of climate warming is coral bleaching, the loss of the vivid colors characteristic of coral reefs. Bleaching was an unusual and mysterious phenomenon only a decade or so ago. Biologists have now shown that when stress occurs in a coral system, due to especially high temperatures, it causes coral polyps to eject the colorful symbiotic algae that give a reef its brilliance—and the energy for survival. Long-lasting bleaching causes death of the corals. A cascading chain of loss follows, as the countless fish, crustaceans, crabs, and other creatures inhabiting the reef lose their food and shelter. Human communities dependent on reef-based ecosystems become part of the cascade as well.

There Are Many Ways to Address Complex Problems

In addition to societal changes aimed at removing the threats to coral reef survival, there are many efforts, as you'll read in chapter 11, to restore these biological marvels. This is part of a larger movement to repair damage to our environment, and to rebuild natural systems. Like many of the issues in environmental science, threats to coral reefs are complex and often global. The global nature of the problem also means that no matter where you live, you can help find answers. The survival of reefs and many other ecosystems depends on strategies such as reducing greenhouse gas emissions, developing renewable energy systems, better practices of pollution prevention, development of sustainable farming systems, and sustainable consumption patterns. These are all topics you can study in this course.

As you will find in the "What Can You Do?" boxes in every chapter, there are numerous practical opportunities to protect and sustain natural resources. It doesn't always take a huge project to do important work for your local environment. Individuals and small groups have many opportunities to make positive change.

All these ideas make environmental science an exciting and important subject. As you read this book, you can discover many ways to engage with the issues and ideas involved in environmental science. Whether you are a biologist, a geologist, a chemist, an economist, a political scientist, a writer, or an artist or poet who can capture our imagination, you can find fruitful and interesting ways to connect with the topics in this book.

We Are Surrounded by Challenges and Progress

All around us are examples of continuing challenges and evidence of progress. Human population growth continues, but it is slowing almost everywhere as women's education and economic opportunity allow for small, well-cared-for families. We remain addicted to fossil fuels, but new energy technologies now provide reliable alternatives in many countries. Solar, wind, biomass, geothermal energy, and conservation could supply all the energy we need, if we chose to invest in them. Water quality and air pollution remain dire problems in many areas, but we have shown that we can dramatically improve water quality, air quality, and environmental health, when we put our minds to it.

Governments around the world are acknowledging the costs of environmental degradation and are taking steps to reduce their environmental impacts. From China to Europe to North America and developing countries, policymakers have plans to restore forests, conserve water, reduce air and water pollution, and develop sustainable energy supplies. Public support for environmental protection has been overwhelmingly enthusiastic. Grants and tax incentives, historically given to polluting industries, are now also supporting more sustainable energy and millions of green jobs.

Businesses everywhere now recognize the opportunities in conservation, recycling, producing nontoxic products, and reducing their ecological footprints. New jobs are being created in environmental fields. Public opinion supports environmental protection because voters see the importance of environmental health for the economy, society, and quality of life.

College and university students are also finding new ways to organize, network, and take action to protect the environment they will inherit (see chapter 25). Ecologist Norman Meyers has said, "The present has a unique position in history. Now, as never before, we have technical, political, and economic resources to solve our global environmental crisis. And if we don't do it now, it may be too late for future generations to do so."

What Sets This Book Apart?

As practicing scientists and educators, we bring to this book decades of experience in the classroom, in the practice of science, and in civic engagement. This experience helps give students a clear sense of what environmental science is and why it matters.

Engaged and active learning

We've given particular attention to learning styles and active learning features in this edition, both in the text and in online **Connect** study materials and supplements. Throughout, the text promotes active, engaged learning practices. In each section heading, **key concepts** identify ideas for students to focus on as they read. **Section reviews** encourage students to check their learning at the end of each main section. These practices of active reading have been shown to improve retention of class topics, as well as higher-order thinking about concepts. **Key terms** at the end of each chapter encourage students to test their understanding. **Critical thinking and discussion questions** and **Data Analysis** exercises push students to explore further the concepts in the text.

A rich collection of online study resources is available on the **Connect** website. **LearnSmart** study resources, practice quizzes, animations, videos, and other resources improve understanding and retention of course material.

The book also engages course material with students' own lives: **What Can You Do?** sections help students identify ways to apply what they are learning to their own lives and communities. **What Do You Think?** readings ask students to critically evaluate their own assessments of a complex problem. We devote a special introduction (**Learning to Learn**) to the ways students can build study habits, take ownership of this course, and practice critical, analytical, and reflective thinking.

Many of these resources are designed as starting points for lectures, discussions in class, essays, lab activities, or projects. Some data analysis exercises involve simple polls of classes, which can be used for graphing and interpretation. Data analysis exercises vary in the kinds of learning and skills involved, and all aim to give students an opportunity to explore data or ideas discussed in the text.

Quantitative reasoning and methods of science

Quantitative reasoning is increasingly recognized as essential in many aspects of education, and this book has greater coverage of this topic, and provides more up-to-date data and graphs, than other books on the market. **Quantitative reasoning** questions in the text push students to evaluate data and graphs they have read about. Attention to statistics, graphing, graph interpretation, and abundant up-to-date data are some of the resources available to help students practice their skills with data interpretation.

Exploring Science readings show how science is done, to demystify the process of answering questions with scientific and quantitative methods. Throughout the text, we emphasize principles and methods of science through discussions of scientific methods, uncertainty and probability, and detailed examination of how scientists observe the world, gather data, and use data to answer relevant questions.

A positive focus on opportunities

Our intent is to empower students to make a difference in their communities by becoming informed, critical thinkers with an awareness of environmental issues and the scientific basis of these issues. Many environmental problems remain severe, but there have been many improvements in recent decades, including cleaner water and cleaner air for most Americans, declining rates of hunger and fertility, and increasing access to education. An entire chapter (chapter 13) focuses on ecological restoration, one of the most important aspects of ecology today. Case studies show examples of real progress, and What Can You Do? sections give students ideas for contributing to solutions. Throughout this text we balance evidence of serious environmental challenges with ideas about what we can do to overcome them.

A balanced presentation for critical thinking

Among the most important practices a student can learn are to think analytically about evidence, to consider uncertainty, and to skeptically evaluate the sources of information. This book offers abundant opportunities to practice the essential skills of critically analyzing evidence, of evaluating contradictory interpretation, and identifying conflicting interests. We ask students to practice critical and reflective thinking in What Do You Think? readings, in end-of-chapter discussion questions, and throughout the text. We present balanced evidence, and we provide the tools for students to discuss and form their own opinions.

An integrated, global perspective

Globalization spotlights the interconnectedness of environmental resources and services, as well as our common interest in how to safeguard them. To remain competitive in a global economy, it is critical that we understand conditions in other countries and cultures. This book provides case studies and topics from regions around the world, with maps and data illustrating global issues. These examples show the integration between environmental conditions at home and abroad.

Google Earth™ placemarks

Our global perspective is supported by placemarks and questions you can explore in Google Earth. This free, online program lets students view detailed satellite images of the earth that aid in understanding the geographical context of topics in the book. Through Connect, students can access placemarks, descriptions, and questions about those places. These stimulate a thoughtful exploration of each site and its surroundings. This interactive geographical exploration is a wonderful tool to give an international perspective on environmental issues.

What's New in This Edition?

Throughout the book, we have used data from LearnSmart online testing and review resources to identify and revise concepts and terms that students find especially challenging. This edition is closely tied to online resources in Connect, which support teaching, studying, and grading. Resources on Connect include figures, animations, movie clips, data analysis exercises, online quizzes, and course management software.

One third of the case studies are new in this edition, and approximately one-third of chapters have new boxed readings. Data, tables, and figures throughout the text have been updated. Live links have also been added to the ebook version of the text. New concepts and developments are added, such as the UN Sustainable Development Goals, emerging post-carbon energy technologies, global population growth, and recent climate data.

Specific changes to chapters

- Learning to Learn (in the Introduction section following the Preface) has been revised and shortened to focus on critical thinking and study habits.
- Chapter 1 opens with a new case study on development challenges in the Kibera settlement in Nairobi, Kenya. The newly released UN Sustainable Development Goals are discussed, along with a revised discussion of the tragedy of the commons and managing the commons, and updated development statistics.

- Chapter 2 has an expanded Data Analysis: Working with Graphs, which gives an overview of different graph types and graph reading.
- Chapter 3 opens with an updated case study on nutrients in Chesapeake Bay. A new Exploring Science reading examines gene editing with CRISPR, an exciting new technology that is changing not only genetics but also ecology and many other scientific disciplines. An updated Data Analysis reflects new data on improving conditions in Chesapeake Bay.
- Chapter 4 opens with a new Exploring Science section on the microbiome, the bacterial ecosystem that inhabits our bodies and keeps us healthy.
- Chapter 6 now includes the "rule of 70," which describes population doubling.
- Chapter 7 has been updated with data on population trends and doubling times. Sections on population growth, poverty, and technology have been updated, as has the "What Do You Think?" section on China relaxing its one-child policy. New graphs now show the responses of total fertility and infant mortality to education.
- Chapter 8 opens with a new case study on the history and risks of PFOA. A new table lists leading causes of global disease burden. Information on Ebola and HIV has been updated, and a new discussion of the Zika virus has been added. The section on Toxicology has been revised, with added information about lead exposures in Flint, Michigan, and other cities, as well as a revised list of persistent organic pollutants, including Bisphenol A. The Exploring Science box on epigenomes has been updated.
- Chapter 9 contains a new section on fishing methods, as well as rates of increase in farmed and wild-caught fisheries. The world status of hunger, as well as food production, was updated, and a new section on climate impacts on food production was added. Data, figures, and tables were updated throughout.
- Chapter 10 has new data on genetically modified crops and on sustainable agriculture.
- Chapter 11 opens with a new case study titled "Restoring Coral Reefs." Table 11.1 updates estimates of known and threatened species. The discussion of climate change impacts on species has been updated. A new Exploring Science reading discusses the challenge of protecting rhinos.
- Chapter 12 opens with a new case study on the effects of palm oil plantations on endangered orangutans. This case study ties to the Exploring Science boxed reading on multinational REDD payments to protect forests and reduce climate change. Forest harvesting methods are discussed in a new section (with a new diagram), and aerial photos of road-building and forest destruction in the Amazon are updated, as are a discussion and figure of cooperative range management in the Malpai borderlands project.

- Chapter 13 contains revised discussions of prairie restoration with bison and with fire, of wetland and stream restoration, and of the challenges of restoring Florida's Everglades.

- Chapter 14 provides a revised boxed reading on whether we should reform the 1872 mining laws, including recent events around copper nickel mining near the Boundary Waters Canoe Area Wilderness in northern Minnesota. The Exploring Science box on rare earth metals is updated, and discussion of the blow-out of the Gold King mine and pollution of Colorado's Animas River has been added, as well as a new section about human-induced earthquakes associated with wastewater disposal, especially from oil and gas drilling. It has a new discussion of the tsunami risks along the Cascadia fault off the Pacific Northwest coast of North America.

- Chapter 15 has been updated to reflect recent climate reports and policy initiatives, including the 2015 Paris Climate Agreement (section 15.6). Extensive attention has been given to climate concepts flagged as challenging by LearnSmart data.

- Chapter 16 begins with a new case study on air pollution in Beijing. Data has been updated on pollution sources and amounts. A new boxed reading discusses the recent Minamata Convention on mercury pollution; the discussion of methane emissions from natural gas wells and other sources is updated; the boxed reading on the London smog of 1952 (formerly case study) highlights evolving efforts at pollution control. Data on benefits of air pollution control are updated.

- Chapter 17 has been updated with data and figures on global water shortages. It includes a new boxed reading on GRACE satellite measurements of groundwater.

- Chapter 18 opens with a new case study on water pollution in the Ganges River. Data and figures on water quality and treatment have been upgraded.

- Chapter 19 opens with a new case study about the end of coal use, followed by a new discussion of declining coal use in the U.S. and in China, and the challenges of clean coal technology. We have updated all data about current energy sources both in the U.S. and worldwide. Extensive revisions have been added about coal use and the need to leave 80% of fossil fuels in the ground. Unstable oil prices and their effects on oil production are discussed, as are fracking, water use, and the effect of abundant new gas supplies on prices and other energy sources. A new map of shale gas deposits in the U.S. is included.

- Chapter 20 opens with a new case study on the renewable energy transition (*Energiewende*) in Germany and continues with extensive discussions of recent developments in renewable energy: improvements in energy efficiency, stable power generation and declining primary energy consumption in Germany, and a new section on low-energy "passive house" standards. Recent advances in solar energy are covered, including opportunities for residential solar, solar gardens, and the challenge to utilities of distributed energy and feed-in tariffs. China's world leadership in producing and installing wind turbines is discussed, and a description and photo of a vertical axis wind turbine is added. The chapter includes added focus on the necessity of renewable energy if we are to contain climate change, as well as a new Exploring Science box titled "Greening Gotham: Can New York Reach Its 80 by 50 goal?" which complements examples from Germany.

- Chapter 22 has updated data on the growth of cities, including a table of the world's largest urban areas.

- Chapter 23 opens with a new case study about British Columbia's carbon tax, an economic strategy for assigning a price to carbon emissions. This is followed by discussions in section 23.5 on emissions markets and carbon taxes. The previous case study on Kiva micro-loans, which continue to be an important development strategy, is now an Exploring Science boxed reading.

- Chapter 24 includes a new discussion of recent developments in the Mercury and Air Toxics Standards and the Clean Power Plan.

- Chapter 25 opens with an updated case study on the People's Climate March in New York and other recent events. A new Exploring Science box discusses "Doing Citizen Science with eBird." The section on sustainable development and sustainability goals has been expanded and updated.

Acknowledgments

We owe a great debt to the hardworking, professional team that has made this the best environmental science text possible. We express special thanks for editorial support to Michael Ivanov, PhD. and Jodi Rhomberg. We are grateful to Lora Neyens, Sherry Kane, Carrie Burger, Lorraine Buczek, and Tara McDermott, for their work in putting the book together, and marketing leadership by Noah Evans. We thank Mike McGee for copyediting and Jerry Marshall for excellent work on photographs.

The following individuals helped write and review learning goal–oriented content for LearnSmart/ for Environmental Science:

College of DuPage, Shamili Ajgaonkar Sandiford

Florida Atlantic University, Jessica Miles

Georgia Southern University, J. Michelle Cawthorn

Northern Arizona University, Sylvester Allred

Roane State Community College, Arthur C. Lee

Rock Valley Community College, Joseph E. Haverly

Rock Valley Community College, Megan M. Pease

State University of New York at Cortland, Noelle J. Relles

University of North Carolina at Chapel Hill, Trent McDowell

University of Wisconsin, Milwaukee, Tristan J. Kloss

University of Wisconsin, Milwaukee, Gina Seegers Szablewski

Input from instructors teaching this course is invaluable to the development of each new edition. Our thanks and gratitude go out to the following individuals who either completed detailed chapter reviews of *Environmental Science, A Global Concern*, fourteenth edition, or provided market feedback for this course.

American University, Priti P. Brahma

Antelope Valley College, Zia Nisani

Arizona Western College, Alyssa Haygood

Aurora University, Carrie Milne-Zelman

Baker College, Sandi B. Gardner

Baylor College, Heidi Marcum

Boston University, Kari L. Lavalli

Bowling Green State University, Daniel M. Pavuk

Bradley University, Sherri J. Morris

Broward College, Elena Cainas

Broward College, Nilo Marin

California Energy Commission, James W. Reede

California State University–East Bay, Gary Li

California State University, Natalie Zayas

Campbellsville University, Ogochukwu Onyiri

Central Carolina Community College, Scott Byington

Central State University, Omokere E. Odje

Clark College, Kathleen Perillo

Clemson University, Scott Brame

College of DuPage, Shamili Ajgaonkar Sandiford

College of Lake County, Kelly S. Cartwright

College of Southern Nevada, Barry Perlmutter

College of the Desert, Tracy Albrecht

College of the Desert, Candice Weber

College of the Desert, Kurt Leuschner

Columbia College, Jill Bessetti

Columbia College, Daniel Pettus

Community College of Baltimore County, Katherine M. Van de Wal

Connecticut College, Jane I. Dawson

Connecticut College, Chad Jones

Connors State College, Stuart H. Woods

Cuesta College, Nancy Jean Mann

Dalton State College, David DesRochers

Dalton State College, Gina M. Kertulis-Tartar

Deanza College, Dennis Gorsuch

East Tennessee State University, Alan Redmond

Eastern Oklahoma State College, Patricia C. Bolin Ratliff

Edison State College, Cheryl Black

Elgin Community College, Mary O'Sullivan

Erie Community College, Gary Poon

Estrella Mountain Community College, Rachel Smith

Farmingdale State College, Paul R. Kramer

Fashion Institute of Technology, Arthur H. Kopelman

Flagler College, Barbara Blonder

Florida State College at Jacksonville, Catherine Hurlbut

Franklin Pierce University, Susan Rolke

Galveston College, James J. Salazar

Gannon University, Amy L. Buechel

Gardner-Webb University, Emma Sandol Johnson

Gateway Community College, Ramon Esponda

Geneva College, Marjory Tobias

Georgia Perimeter College, M. Carmen Hall

Georgia Perimeter College, Michael L. Denniston

Gila Community College, Joseph Shannon

Golden West College, Tom Hersh

Gulf Coast State College, Kelley Hodges

Gulf Coast State College, Linda Mueller Fitzhugh

Holy Family University, Robert E. Cordero

Houston Community College, Yiyan Bai

Hudson Valley Community College, Daniel Capuano

Hudson Valley Community College, Janet Wolkenstein

Illinois Mathematics and Science Academy, C. Robyn Fischer

Illinois State University, Christy N. Bazan

Indiana University of Pennsylvania, Holly J. Travis

Indiana Wesleyan University, Stephen D. Conrad

James Madison University, Mary Handley

James Madison University, Wayne S. Teel

John A. Logan College, Julia Schroeder

Kentucky Community & Technical College System-Big Sandy District, John G. Shiber

Lake Land College, Jeff White

Lane College, Satish Mahajan

Lansing Community College, Lu Anne Clark

Lewis University, Jerry H. Kavouras

Lindenwood University, David M. Knotts

Longwood University, Kelsey N. Scheitlin

Louisiana State University, Jill C. Trepanier

Lynchburg College, David Perault

Marshall University, Terry R. Shank

Menlo College, Neil Marshall

Millersville University of Pennsylvania, Angela Cuthbert

Minneapolis Community and Technical College, Robert R. Ruliffson

McGraw-Hill Connect®
Learn Without Limits

Connect is a teaching and learning platform that is proven to deliver better results for students and instructors.

Connect empowers students by continually adapting to deliver precisely what they need, when they need it, and how they need it, so your class time is more engaging and effective.

73% of instructors who use **Connect** require it; instructor satisfaction **increases** by 28% when **Connect** is required.

Analytics

Connect Insight®

Connect Insight is Connect's new one-of-a-kind visual analytics dashboard—now available for both instructors and students—that provides at-a-glance information regarding student performance, which is immediately actionable. By presenting assignment, assessment, and topical performance results together with a time metric that is easily visible for aggregate or individual results, Connect Insight gives the user the ability to take a just-in-time approach to teaching and learning, which was never before available. Connect Insight presents data that empowers students and helps instructors improve class performance in a way that is efficient and effective.

Connect's Impact on Retention Rates, Pass Rates, and Average Exam Scores

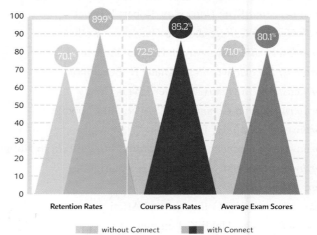

Using **Connect** improves retention rates by **19.8%**, passing rates by **12.7%**, and exam scores by **9.1%**.

Impact on Final Course Grade Distribution

without Connect		with Connect
22.9%	A	31.0%
27.4%	B	34.3%
22.9%	C	18.7%
11.5%	D	6.1%
15.4%	F	9.9%

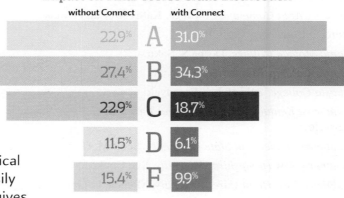

Students can view their results for any **Connect** course.

Mobile

Connect's new, intuitive mobile interface gives students and instructors flexible and convenient, anytime–anywhere access to all components of the Connect platform.

Adaptive

THE **ADAPTIVE**
READING EXPERIENCE
DESIGNED TO TRANSFORM
THE WAY STUDENTS READ

More students earn **A's** and **B's** when they use McGraw-Hill Education **Adaptive** products.

SmartBook®

Proven to help students improve grades and study more efficiently, SmartBook contains the same content within the print book, but actively tailors that content to the needs of the individual. SmartBook's adaptive technology provides precise, personalized instruction on what the student should do next, guiding the student to master and remember key concepts, targeting gaps in knowledge and offering customized feedback, and driving the student toward comprehension and retention of the subject matter. Available on tablets, SmartBook puts learning at the student's fingertips—anywhere, anytime.

Over **8 billion questions** have been answered, making McGraw-Hill Education products more intelligent, reliable, and precise.

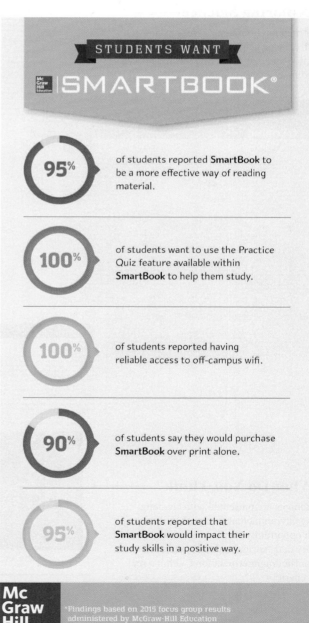

STUDENTS WANT

SMARTBOOK®

95% of students reported **SmartBook** to be a more effective way of reading material.

100% of students want to use the Practice Quiz feature available within **SmartBook** to help them study.

100% of students reported having reliable access to off-campus wifi.

90% of students say they would purchase **SmartBook** over print alone.

95% of students reported that **SmartBook** would impact their study skills in a positive way.

*Findings based on 2015 focus group results administered by McGraw-Hill Education

www.mheducation.com

Key Elements

A global perspective is vital to learning about environmental science.

Case Studies

All chapters open with a real-world case study to help students appreciate and understand how environmental science impacts lives and how scientists study complex issues.

Exploring Science

Current environmental issues exemplify the principles of scientific observation and data-gathering techniques to promote scientific literacy.

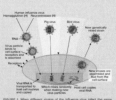

Google Earth

Google Earth interactive satellite imagery gives students a geographical context for global places and topics discussed in the text. Google Earth icons indicate a corresponding exercise in Connect. In these exercises, students will find links to locations mentioned in the text, as well as corresponding assessments that will help them understand environmental topics.

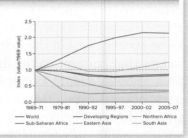

Data Analysis

At the end of every chapter, these exercises give students further opportunities to apply critical-thinking skills and analyze data. These are assigned through Connect in an interactive online environment. Students are asked to analyze data in the form of documents, videos, and animations.

What Do You Think?

Students are presented with challenging environmental studies that offer an opportunity to consider contradictory data, special interest topics, and conflicting interpretations within a real scenario.

Learning Outcomes

Found at the beginning of each chapter, and organized by major headings, these outcomes give students an overview of the key concepts they will need to understand.

Learning Outcomes

After studying this chapter, you should be able to:

9.1 Describe patterns of world hunger and nutritional requirements.

9.2 Identify key food sources, including protein-rich foods.

9.3 Explain new crops and genetic engineering.

9.4 Discuss how policy can affect food resources.

Section Reviews

Section reviews are a series of content-specific questions that appear at the end of each section in the chapter. These questions encourage students to periodically review what they have read and offers an opportunity to check their understanding of key concepts.

Section Review

1. How many people in the world are chronically undernourished? What does chronically undernourished mean?
2. List at least five African countries with high rates of hunger (fig. 9.3; use a world map to help identify countries).
3. What are some of the health risks of overeating? What percentage of adults are overweight in the United States?

Conclusion

This section summarizes the chapter by highlighting key ideas and relating them to one another.

Conclusion

The potential location of biological communities is determined in large part by climate, moisture availability, soil type, geomorphology, and other natural features. Understanding the global distribution of biomes, and knowing the differences in who lives where and why, are essential to the study of global environmental science. Human occupation and use of natural resources are strongly dependent on the biomes found in particular locations. Humans tend to prefer mild climates and the highly productive biological communities found in temperate zones. These biomes also suffer the highest rates of degradation and overuse. ... characteristics that allow ... seasonal tropical forests, ... these adaptations helps ... al in those biomes.

Oceans cover over 70 percent of the earth's surface, yet we know relatively little about them. Some marine biomes, such as coral reefs, can be as biologically diverse and productive as any terrestrial biome. People have always depended on rich, complex ecosystems. In recent times, the rapid growth of human populations, coupled with more powerful ways to harvest resources, has led to extensive destruction of these environments. Still, it is possible for us to protect these living communities. The opening case study of this chapter illustrates how people can work together to protect and even restore the biological communities on which they depend. Perhaps we can find similar solutions in other biologically rich but endangered biomes.

Critical Thinking and Discussion Questions

1. Do people around you worry about hunger? Do you think they should? Why or why not? What factors influence the degree to which people worry about hunger in the world?
2. Global issues such as hunger and food production often seem far too large to think about solving, but it may be that many strategies can help us address chronic hunger. Consider your own skills and interests. Think of at least one skill that could be applied (if you had the time and resources) to helping reduce hunger in your community or elsewhere.
3. Suppose you are a farmer who wants to start a confined animal feeding operation. What conditions make this a good strategy for you, and what factors would you consider in weighing its costs and benefits? What would you say to neighbors who wish to impose restrictions on how you run the operation?
4. Debate the claim that famines are caused more by human actions (or inactions) than by environmental forces. What kinds of evidence would be needed to resolve this debate?
5. Outline arguments you would make to your family and friends for why they should buy shade-grown, fair-trade coffee and cocoa. How much of a premium would you pay for these products? What factors would influence how much you would pay?
6. Given what you know about GMO crops, identify some of the costs and benefits associated with them. Which of the costs and benefits do you find most important? Why?
7. Corn is by far the dominant crop in the United States. In what ways is this a good thing for Americans? How is it a problem? Who are the main beneficiaries of this system?

Critical Thinking and Discussion Questions

Brief scenarios of everyday occurrences or ideas challenge students to apply what they have learned to their lives.

What Can You Do?

This feature gives students realistic steps for applying their knowledge to make a positive difference in our environment.

What Can You Do?

Controlling Pests

Based on the principles of integrated pest management, the U.S. EPA releases helpful guides to pest control. Among their recommendations:

1. *Identify pests, and decide how much pest control is necessary.* Does your lawn really need to be totally weed-free? Could you tolerate some blemished fruits and vegetables? Could you replace sensitive plants with ones less sensitive to pests?
2. *Eliminate pest sources.* Remove from your house or yard any food, water, and habitat that encourages pest growth. Eliminate hiding places or other habitats. Rotate crops in your garden.
3. *Develop a weed-resistant yard.* Pay attention to your soil's pH, nutrients, texture, and organic content. Grow grass or cover varieties suited to your climate. Set realistic goals for weed control.
4. *Use biological controls.* Encourage beneficial insect predators such as birds, bats that eat insects, ladybugs, spiders, centipedes, dragonflies, wasps, and ants.
5. *Use simple manual methods.* Cultivate your garden and handpick weeds and pests from your garden. Set traps to control rats, mice, and some insects. Mulch to reduce weed growth.
6. *Use chemical pesticides carefully.* If you decide that the best solution is chemical, choose the right pesticide product, read safety warnings and handling instructions, buy the amount you need, store the product safely, and dispose of any excess properly.

Source: *Citizen's Guide to Pest Control and Pesticide Safety:* EPA 730-K-95-001

Quantitative Reasoning

Compare the "hot spot" map in figure 11.4 with the biomes map in figure 5.4 . Which of the "hot spots" has the largest number of endemic species? Which has the least? Can you detect any patterns when you compare these two maps?

Quantitative Reasoning

Quantitative reasoning questions in the text push students to evaluate data and graphs they have read about. Attention to statistics, graphing, graph interpretation, and abundant up-to-date data are some of the resources available to help students

Relevant Photos and Instructional Art Support Learning

High-quality photos and realistic illustrations display detailed diagrams, graphs, and real-life situations.

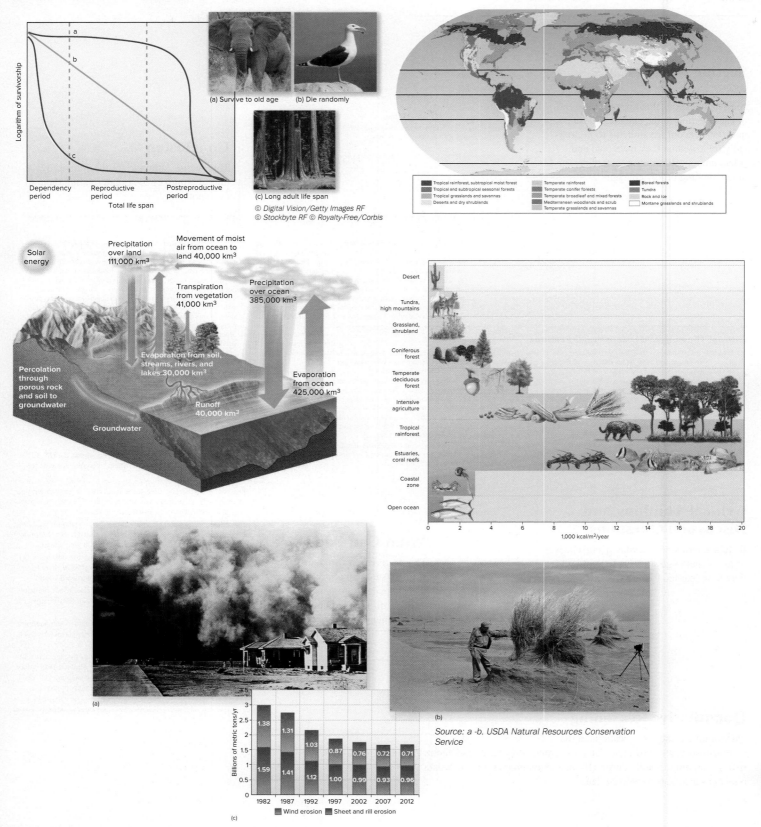

(a) Survive to old age (b) Die randomly

(c) Long adult life span

© Digital Vision/Getty Images RF
© Stockbyte RF © Royalty-Free/Corbis

Source: a -b. USDA Natural Resources Conservation Service

Introduction
Learning to Learn

▲ Learning to learn is a lifelong skill.
© William P. Cunningham

Learning Outcomes

After studying this introduction, you should be able to:

L.1 Form a plan to organize your efforts and become a more effective and efficient student.

L.2 Be prepared to apply critical and reflective thinking in environmental science.

"What kind of world do you want to live in? Demand that your teachers teach you what you need to know to build it."

– Peter Kropotkin

How can I do well in environmental science?

Welcome to environmental science. This is a field that helps you develop the knowledge and skills to understand problems in the world around us, and to help find answers to them. The subject involves a diversity of topics, with connections to basic ecology, natural resources, and policy questions that influence those systems. Topics in this course primarily involve our natural environment, but we also examine our human environment, including the built world of technology and cities, as well as human social or cultural institutions. All of these interrelated aspects of our life affect us, and, in turn, are affected by what we do.

One focus of this chapter is how to organizing your learning process as you study the diverse topics in environmental science. This means being aware and intentional about your study habits. Take time as you read this chapter to consider what you do well as you study, and what you need to do better to be effective with study time. This is another skill set that will serve you well in other contexts.

Another focus of this chapter is critical thinking, that is, assessing how and why we think about things as we do. Critical thinking is one of the most useful skills you can learn in any of your classes, and so it is a focus of this chapter. Many central topics in environmental science are highly contested: What kinds of energy are most important? Where should they come from? What is a resource? How should we manage

FIGURE L.1 How do environmental problems come about? Who made what decisions to get us to this point, and why? Critical thinking helps us evaluate problems and find the solutions we need.
Source: Photo by Eric Vanceonse, U.S. EPA

and conserve water resources? Who should pay the cost of controlling air pollution? Answering these questions requires analysis of evidence. But evidence can depend on when and by whom it was gathered and evaluated. For every opinion there is an equal and opposite opinion. How can you make sense out of this welter of ever-changing information? You need to develop a capacity to think independently, systematically, and skillfully to form your own opinions (fig. L.1). These qualities and abilities can help you in many aspects of life. Throughout this book you will find "What Do You Think?" boxes that invite you to practice your critical and reflective thinking skills.

Thinking about how we think is a practice that applies in ordinary conversation, as well as in media you encounter, and even in textbooks. Finding these patterns in arguments can be fun; it's also important. Paying attention to these sorts of argument strategies is also a good practice in any class you take. These are a few of the logical errors you can watch for:

- *Red herring:* Introducing extraneous information to divert attention from the important point.
- *Ad hominem attacks:* Criticizing the opponent rather than the logic of the argument.
- *Hasty generalization:* Drawing conclusions about all members of a group based on evidence that pertains only to a selected sample.
- *False cause:* Drawing a link between premises and conclusions that depends on some imagined causal connection that does not, in fact, exist.
- *Appeal to ignorance:* Because some facts are in doubt, a conclusion is impossible.
- *Appeal to authority:* It's true because someone says so.
- *Equivocation:* Using words with double meanings to mislead the listener.
- *Slippery slope:* A claim that some event or action will cause some subsequent action.
- *False dichotomy:* Giving either/or alternatives as if they are the only choices.

These skills are important to doing well in this class, and they are part of becoming a responsible and productive environmental citizen. Each of us needs a basis for learning and evaluating scientific principles, as well as some insights into the social, political, and economic systems that impact our global environment. We hope this book and the class you're taking will give you the information you need to reach those goals. As the noted Senegalese conservationist and educator Baba Dioum once said, "In the end, we will conserve only what we love, we will love only what we understand, and we will understand only what we are taught."

L.1 How Can I Get an A in This Class?

- *Making a frank and honest assessment of your strengths and weaknesses will help you do well in this class.*
- *Reading in a purposeful, deliberate manner is an important part of productive learning.*

What do you need to know to succeed in a class on environmental science? This chapter provides an overview of some skills to keep in mind as you begin. As Henry Ford once said, "If you think you can do a thing, or think you can't do a thing, you're right."

One of the first things that will help you do well in this class—and enjoy it—is to understand that science is useful and accessible, if you just take your time with it. To do well in this class, start by identify the ways that science connects with your interests and passions. Most environmental scientists are motivated by a love for something: a fishery biologist might love fishing; a plant pathologist might love gardening; an environmental chemist might be motivated by wanting to improve children's health in the city in which she lives. All these people use the tools of science to help them understand something they get excited about. Finding that angle can help you do better in this class, and it can help you be a better and happier member of your community (fig. L.2).

Another key to success is understanding what "science" is. Basically, science is about trying to figure out how things work. This means examining a question carefully and methodically. It means questioning your own assumptions, as well as the statements you hear from others. Understanding some basic ideas in science can be very empowering: learning to look for evidence and to question your assumptions is a life skill, and building comfort with thinking about numbers can help you budget your groceries, prioritize your schedule, or plan your vacation. Ideas in this book can help you understand the food you eat, the weather you encounter, the policies you hear about in the news—from energy policy to urban development to economics.

FIGURE L.2 Finding the connections between your studies and the community, places, and ideas you care about can make this class more rewarding and fun.
Source: Photo by Gwen Bausmith, U.S. EPA

What are good study habits?

What are your current study skills and habits? Making a frank and honest assessment of your strengths and weaknesses will help you set goals and make plans for achieving them during this class. A good way to start is to examine your study habits. Rate yourself on each of the following study skills and habits on a scale of 1 (excellent) to 5 (needs improvement). If you rate yourself below 3 on any item, think about an action plan to improve that competence or behavior.

- How well do you manage your time (do you tend to run late, or do you complete assignments on time)?
- Do you have a regular study environment where you can focus?
- How effective are you at reading and note-taking (do you remember what you've read; do you take notes regularly)?
- Do you attend class regularly, listen for instructions, and participate actively in class discussions? Do you bring questions to class about the material?
- Do you generally read assigned chapters in the textbook before attending class, or do you wait until the night before the exam?
- How do you handle test anxiety (do you usually feel prepared for exams and quizzes or are you terrified of them? Do you have techniques to reduce anxiety or turn it into positive energy)?
- Do you actively evaluate how you are doing in a course based on feedback from your instructor and then make corrections to improve your effectiveness?
- Do you seek out advice and assistance outside of class from your instructors or teaching assistants?

Procrastination is something almost everyone does, but a few small steps can help you build better habits. If you routinely leave your studying until the last minute, then consider making a study schedule, and keep a written record how much time you spend studying. Schedule time for sleep, meals, exercise, and recreation so that you will be rested and efficient when you do study. Divide your work into reasonable sized segments that you can accomplish on a daily basis. Carry a calendar to keep track of assignments. And find a regular study space in which you can be effective and productive.

How you behave in class and interact with your instructor also can have a big impact on how much you learn and what grade you get. Make an effort to get to know your instructor. Sit near the front of the room where you can see and be seen. Learn to ask questions: This can keep you awake and engaged in class. Practice the skills of good note-taking (table L.1). Attend every class and arrive on time. Don't fold up your papers and prepare to leave until after the class period is over. Arriving late and leaving early says to your instructor that you don't care much about either the class or your grade.

Practice active, purposeful learning. It isn't enough to passively absorb knowledge provided by your instructor and this textbook. You need to actively engage the material in order to really understand it. The more you invest yourself in the material, the easier it will be to comprehend and remember. It is very helpful to have a study buddy with whom you can compare notes and try out ideas (fig. L.3).

It's well known that the best way to learn something is to teach it to someone else. Take turns with your study buddy explaining the material you're studying. You may think you've mastered a topic by quickly skimming the text, but you're likely to find that you have to struggle to give a clear description in your own words. Anticipating possible exam questions and taking turns quizzing each other can be a very good way to prepare for tests.

FIGURE L.3 Cooperative learning, in which you take turns explaining ideas and approaches with a friend, can be one of the best ways to comprehend material.
© BananaStock/JupiterImages RF

How can you use this textbook effectively?

An important part of productive learning is to read assigned material in a purposeful, deliberate manner. Ask yourself questions as you read. What is the main point being made here? How does the evidence presented support the assertions being made? What personal experience have you had or what prior knowledge can you bring to bear on this question? Can you suggest alternative explanations for the phenomena being discussed? A study technique developed by Frances Robinson and called the **SQ3R** method can improve your reading comprehension. It's also helpful to have a study group (fig. L.4). After class and before exams, you can compare notes, identify priorities, and sort out points that are unclear. Try these steps as you read the first few chapters of this book, and see if they improve your recall of the material:

1. *Survey* the entire chapter or section you are about to read, so you can see how it fits together. What are the major headings and subdivisions?

2. *Question* what the main points are likely to be in each of the sections. Which parts look most important or interesting? Where you should invest the most time and effort?

3. *Read* the material, taking brief notes as you go. Read in small segments and stop frequently for reflection and to make notes.

FIGURE L.4 Talking through ideas with your peers is an excellent way to test your knowledge. If you can explain it, then you probably understand the material.
© PhotoDisc/Getty Images RF

4. *Recite*: Stop periodically to recite to yourself what you have just read. Check your comprehension at the end of each major section. Ask yourself: Did I understand what I just read? What are the main points being made here? Summarize the information in your own words to be sure that you really understand and are not just depending on rote memory.

5. *Review:* Once you have completed a section, review the main points to make sure you remember them clearly. Did you miss any important points? Do you understand things differently the second time through? This is a chance to think critically about the material. Do you agree with the conclusions suggested by the authors?

Will this be on the test?

You should develop different study strategies depending on whether you are expected to remember and choose between a multitude of facts and details, or whether you will be asked to write a paragraph summarizing some broad topic. Organize the ideas you're reading and hearing in lecture. This course will probably include a great deal of information, so try to organize for yourself what ideas are most important? What's the big picture? As you read and review, ask yourself what might be some possible test questions in each section. Memorize some benchmark figures: Just a few will help a lot. Pay special attention to ideas, relationships, facts, and figures about which your instructor seemed especially interested. Usually those points are emphasized in class because your teacher thinks they are most important to remember. There is a good chance you'll see those topics again on a test.

Pay special attention to tables, graphs, and diagrams. They were chosen because they illustrate important points, and they are often easy to put on a test. Also pay attention to units. You probably won't be expected to remember all the specific numbers in this book, but you probably should know orders of magnitude. The world population is about 7.3 *billion* people (not thousands, millions, or trillions). It often helps to remember facts and figures if you can relate them to some other familiar example. The United States, for instance, has about 314 million residents. The populations of the European Union is slightly larger, India and China are each more than four times as large. Those general relationships are usually easier to remember and compare than detailed figures.

Section Review

1. What is your strongest learning style?
2. What are the five techniques of SQ3R method for studying?

L.2 THINKING ABOUT THINKING

- Critical thinking is a valuable tool in learning and in life.
- Certain attitudes, skills and approaches are essential for well-reasoned analysis.

Perhaps the most valuable skill you can learn in any of your classes is the ability to think clearly, creatively, and purposefully.

Developing the ability to learn new skills, examine new facts, evaluate new theories, and formulate your own interpretations is essential to keep up in a changing world. In other words, you need to learn how to learn on your own.

Thinking about thinking means pausing to examine you are forming ideas, or how you interpret what you hear and read. A number of approaches can help us evaluate information and make decisions. **Analytical thinking** asks, "How can I break this problem down into its constituent parts?" **Creative thinking** asks, "How might I approach this problem in new and inventive ways?" **Logical thinking** asks, "How can orderly, deductive reasoning help me think clearly?" **Critical thinking** asks, "What am I trying to accomplish here and how will I know when I've succeeded?" **Reflective thinking** asks, "What does it all mean?" As figure L.5 suggests, critical thinking is central in the constellation of thinking skills. Thinking critically can help us discover hidden ideas and means, develop strategies for evaluating reasons and conclusions in arguments, recognize the differences between facts and values, and avoid jumping to conclusions.

How do you tell the news from the noise?

With the explosion of cable channels, blogs, social networks, and e-mail access, most of us are interconnected constantly to a degree unique in history. There are well over 150 million blogs on the Web, and 15,000 new ones are added every day. More than a billion people are linked in social networks. Every day several billion e-mails, tweets, text messages, online videos, and social media postings connect us to one another. As you participate in these networks, you probably already think about the sources of information you are exposed to on a daily basis (fig. L.6).

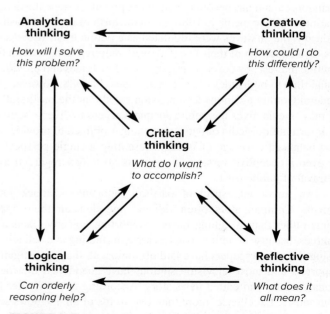

FIGURE L.5 Different approaches to thinking are used to solve different kinds of problems or to study alternate aspects of a single issue.

FIGURE L.6 "There is absolutely no cause for alarm at the nuclear plant!"
Source: © Tribune Media Services. Reprinted with permission.

One of the issues that has emerged with this proliferation of media is partisan journalism—reports that serve one viewpoint, rather than trying to weigh diverse evidence and perspectives. Partisan journalism has become much more prevalent since the deregulation of public media in 1988. From the birth of the broadcasting industry, the airwaves were regulated as a public trust. Broadcasters, as a condition of their licenses, were required to operate in the "public interest" by covering important policy issues and providing equal time to both sides of contested issues. In 1988, however, the Federal Communications Commission ruled that the proliferation of mass media gives the public adequate access to diverse sources of information. Media outlets are no longer obliged to provide fair and balanced coverage of issues. Presenting a single perspective or even a deceptive version of events is no longer regarded as a betrayal of public trust.

An important aspect of partisan reporting is attack journalism. Commentators often ridicule and demean their opponents rather than weighing ideas or reporting objective facts and sources, because shouting matches are entertaining and sell advertising. Most newspapers have laid off almost all their investigative reporters and most television stations have abandoned the traditional written and edited news story. According to the Center for Journalistic Excellence, more than two-thirds of all TV news segments now consist of on-site "stand-up" reports or live interviews in which a single viewpoint is presented as news without any background or perspective.

Part of the reason for the growth of sensationalist media is that real news—topics that affect your community and your environment—often don't make exciting visuals. So they don't make it into TV coverage. Instead, crime, accidents, disasters, lifestyle stories, sports, and weather make up more than 90 percent of the coverage on a typical television news program. An entire day of cable TV news would show, on average, only 1 minute each about the environment and health care, 2 minutes each on science and education, and 4 minutes on art and culture. More than 70 percent of the segments are less than 1 minute long, which allows them to convey lots of emotion but little substance. People who get their news primarily from TV are significantly more fearful and pessimistic than those who get news from print media. And it becomes hard to separate rumor from truth. Evidence and corroboration take a backseat to dogma and passion.

How can you detect bias in blogs, social media, or news reporting? Ask the questions below as you look at media. Also ask these questions as you examine your own work, to avoid falling into these traps.

1. Are speakers discussing facts and rational ideas, or are they resorting to innuendo, name-calling, character assassination, and *ad hominem* (personal) attacks? When people start calling each other Nazi or communist (or both), civil discourse has probably come to an end.

2. What special interests might be involved? Who stands to gain presenting a particular viewpoint? Who is paying for the message?

3. What sources are used as evidence in this communication? How credible are they?

4. Are facts or statistics cited in the presentation? Are they credible? Are citations provided so you can check the sources?

5. If the presentation claims to be fair and balanced, are both sides represented by credible spokespersons, or is one simply a foil set up to make the other side look good?

6. Are the arguments presented based on evidence, or are they purely emotional appeals?

Applying critical thinking

In logic, an argument is made up of one or more introductory statements (called **premises**), and a **conclusion** that supposedly follows logically from the premises. Often in ordinary conversation, different kinds of statements are mixed together, so it is difficult to distinguish between them or to decipher hidden or implied meanings.

We all use critical or reflective thinking at times. Suppose a television commercial tells you that a new breakfast cereal is tasty and good for you. You may be suspicious and ask yourself a few questions. What do they mean by good? Good for whom or what? Does "tasty" simply mean more sugar and salt? Might the sources of this information have other motives in mind besides your health and happiness? Although you may not have been aware of it, you already have been using some of the techniques of critical analysis. Working to expand these skills helps you recognize the ways information and analysis can be distorted, misleading, prejudiced, superficial, unfair, or otherwise defective. Here are some steps in critical thinking:

Identify and evaluate premises and conclusions in an argument. What is the basis for the claims made here? What evidence is presented to support these claims and what conclusions are drawn from this evidence? If the premises and evidence are correct, does it follow that the conclusions are necessarily true?

Acknowledge and clarify uncertainties, vagueness, equivocation, and contradictions. Do the terms used have more than one meaning? If so, are all participants in the argument using the same meanings? Are ambiguity or equivocation deliberate? Can all the claims be true simultaneously?

Distinguish between facts and values. Are claims made that can be tested? (If so, these are statements of fact and should be able to be verified by gathering evidence.) Are claims made about the worth or lack of worth of something? (If so, these are value statements or opinions and probably cannot be verified objectively.) For example, claims of what we *ought* to do to be moral or righteous or to respect nature are generally value statements.

Recognize and assess assumptions. Given the backgrounds and views of the protagonists in this argument, what underlying reasons might there be for the premises, evidence, or conclusions presented? Does anyone have an "axe to grind" or a personal agenda in this issue? What do they think you know, need, want, or believe? Is there a subtext based on race, gender, ethnicity, economics, or some belief system that distorts this discussion?

Distinguish the reliability or unreliability of a source. What makes the experts qualified in this issue? What special knowledge or information do they have? What evidence do they present? How can we determine whether the information offered is accurate, true, or even plausible?

Recognize and understand conceptual frameworks. What are the basic beliefs, attitudes, and values that this person, group, or society holds? What dominating philosophy or ethics control their outlook and actions? How do these beliefs and values affect the way people view themselves and the world around them? If there are conflicting or contradictory beliefs and values, how can these differences be resolved?

As you read this book, you will have many opportunities to practice critical thinking. Every chapter includes facts, figures, opinions, and theories. Are all of them true? Probably not. They were the best information available when this text was written, but scientific knowledge always growing. Data change constantly as does our interpretation of them. Environmental conditions change, evidence improves, and different perspectives and explanations evolve over time.

As you read this book or any book, try to distinguish between statements of fact and opinion. Ask yourself if the premises support the conclusions drawn from them. Although we have tried to present the best available scientific data and to represent the main consensus among environmental scientists, it is always important for you, as a reader, to think for yourself and utilize your critical and reflective thinking skills to find the truth.

Section Review

1. Describe seven attitudes needed for critical thinking.
2. List six steps in critical thinking.

Conclusion

Whether you find environmental science interesting and useful depends largely on your own attitudes and efforts. Developing good study habits, setting realistic goals for yourself, taking the initiative to look for interesting topics, finding an appropriate study space, and working with a study partner can make your study time more efficient and also can improve your final grade.

We all have our own learning styles. You may understand and remember things best if you see them in writing, hear them spoken by someone else, reason them out for yourself, or learn by doing. By determining your preferred style, you can study in the way that is most comfortable and effective for you.

1 Understanding Our Environment

▲ Many of the most important challenges in environmental quality and sustainable development occur in informal settlements like Kibera, in Nairobi, Kenya.
© Eoghan Rice/Trócaire via www.flickr.com/photos/trocaire/7365580178/

Learning Outcomes

After studying this chapter, you should be able to:

1.1 Explain what environmental science is, and how it draws on different kinds of knowledge.

1.2 Identify some early thinkers on environment and resources, and contrast some of their ideas.

1.3 Describe sustainable development and its goals.

1.4 Explain core concepts in sustainable development.

1.5 Identify ways in which ethics and faith might promote sustainability and conservation.

"Working together, we have proven that sustainable development is possible; that reforestation of degraded land is possible; and that exemplary governance is possible when ordinary citizens are informed, sensitized, mobilized and involved in direct action for their environment."

– Wangari Maathai (1940–2011)

Winner of 2004 Nobel Peace Prize

Sustainable Development Goals for Kibera

The central idea of sustainability is that we can improve well-being for poor populations, including reducing severe poverty, while maintaining or improving the environment on which we depend. These goals might seem contradictory, but increasing evidence shows that they can go together. In fact, as our resource consumption and population grow, it is increasingly necessary that they go together. Starting in 2016, the United Nations launched a new program to promote 17 Sustainable Development Goals, including access to education, health care, a safe natural environment, clean water, and other priorities, as well as conserving biodiversity and slowing climate change. Are all these goals possible (fig. 1.1)?

Perhaps the greatest test case of this question is in fast-growing urban settlements of the developing world. One of the largest of these is a slum known as Kibera in Nairobi, Kenya. Every week, some 2,500 people arrive in Nairobi, drawn by hopes for better jobs and education. The city cannot build housing fast enough for this influx. Nor can it provide sanitary sewage, safe water systems, electric power, or other services. New arrivals build

FIGURE 1.1 Sustainable development goals include access to education and electricity to study at night.
© Corrie Wingate Photography/SolarAid via www.flickr.com/photos/solaraid/16748134826/

informal neighborhoods on the margins, using whatever materials are available to construct simple shelters of mud, brick, and tin roofing. Kibera is the largest of about 200 such settlements in Nairobi. These are home to over 2.5 million people, around 60 percent of the city's population (although reliable numbers are hard to come by).

Kibera grew on lowlands along the Nairobi River, in an area prone to flooding that periodically inundates houses and muddy informal streets. Because there is no system for managing sewage or garbage, both end up in the river, often entering homes with flood waters. Much of the time, a fetid odor of decomposing waste fills the air. and plastic shopping bags and other debris fill the corners of roadways and buildings. Occupying degraded outskirts of large cities, neighborhoods like Kibera suffer from the pollution produced by wealthy neighborhoods, and also create their own pollution and health hazards.

The city government has a complicated relationship with Kibera. The settlement provides much-needed housing, and residents contribute labor and consumer markets for growing businesses. But sub-standard housing is an embarrassment for city governments. Impoverished and unemployed populations turn to crime, even while they are the main victims of criminal activity. The city regularly tries to remove this and other slums, replacing them with modern housing, but the new flats are usually too expensive, and insufficient in supply, for the displaced residents.

Similar settlements exist in many of the world's fast-growing urban areas—Rio de Janeiro, Manila, Lagos, Cairo, Mumbai, Delhi, and many others—because global processes drive the growth of these vast slums. Rural population growth reduces access to farmland; forest destruction and soil degradation make traditional lifestyles difficult to maintain. Large landholders expand, displacing rural communities. Climate change threatens crop production. Declining water resources make farming difficult, and farmers are driven to the city.

In striving to enter the middle class, residents of Kibera also increase their environmental impacts. As they succeed, they consume more material goods, more energy, more cars and fuel, and electronics. All of these expand the environmental footprint of residents. On the other hand, the per capita energy and resource consumption of most Kibera residents is vanishingly small compared to consumption of their wealthy neighbors, who may have multiple cars and large houses, many appliances, and rich diets.

The global challenge of sustainable development is to find ways to improve the lives and the environment of people everywhere, including those in Kibera and other informal settlements. Slum residents have energy and ideas and are eager to improve the lives of their children, like people everywhere. Increasingly, global efforts, such as the Sustainable Development Goals, seek to increase the amount and effectiveness of money and development strategies transferred to poor countries. At the Paris climate talks of 2015, for example, the international community renewed its pledge to the Green Climate Fund, aiming for US $100 billion by 2020, to help developing areas produce clean power, and provide electricity to growing populations without increasing greenhouse gas emissions. Investing in renewable energy is one of the most important ways to improve the well-being of the poor without accelerating climate change.

Even without a Green Climate Fund, African countries are poised to dramatically increase investments in renewable energy. By 2015, Kenya produced more than half its electric power production from geothermal, wind, and solar energy sources. The World Bank calculates that Sub-Saharan Africa could produce 170 gigawatts of low-carbon energy. This sustainable development potential exists across much of the world. In the Americas, Uruguay achieved 95 percent renewable electricity due to a decade of policy commitment. Costa Rica, with abundant geothermal energy, is approaching 100 percent renewable energy. Nicaragua is aiming for 90 percent renewable energy by 2020.

Environmental science involves understanding the natural systems we depend on, as well as ways to promote sustainable development without destroying those systems. These are among the most important questions we face today, and you will explore them as you read this book.

1.1 WHAT IS ENVIRONMENTAL SCIENCE?

- *This subject draws on many disciplines, skills, and interests.*
- *We face persistent challenges, as well as progress, in themes such as population growth, climate change, pollution, and biodiversity losses.*
- *Ecological footprints are a way to estimate our impacts.*

Humans have always inhabited two worlds. One is the natural world of plants, animals, soils, air, and water that preceded us by billions of years and of which we are a part. The other is the world of social institutions and artifacts that we create for ourselves using science, technology, and political organization. Both worlds are essential to our lives, but their intersections often cause enduring tensions: More than ever before, we have power to extract and consume resources, produce waste, and modify our world in ways that threaten both our continued existence and that of many organisms with which we share the planet. We also have better access than ever before to new ideas, efficient technologies, and opportunities to cooperate in finding sustainable strategies. To ensure a sustainable future for ourselves and future generations, we need to understand more about how our world works, what we are doing to it, and what we can do to protect and improve it.

Environment (from the French *environner:* to encircle or surround) can be defined as (1) the circumstances or conditions that surround an organism or group of organisms, or (2) the complex of social or cultural conditions that affect an individual or community. Because humans inhabit the natural world, as well as the "built" or technological, social, and cultural world, all constitute important parts of our environment (fig. 1.2).

Environmental science is the systematic study of our environment and our proper place in it. Environmental science is interdisciplinary, integrating natural sciences, social sciences, and humanities in a broad, holistic study of the world around us. Much of environmental science focuses on understanding and resolving problems in our natural environment, such as pollution or lost biodiversity. But

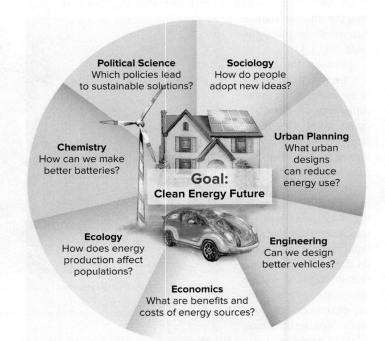

FIGURE 1.2 Many kinds of knowledge contribute to solutions in environmental science. For a goal such as achieving clean and sustainable energy, strategies involve input from many disciplines.

solutions have to do with how we consume resources and dispose of waste. This is why environmental science also includes discussion of policy, population, economics, and urbanization.

As distinguished economist Barbara Ward pointed out, for an increasing number of environmental issues, the difficulty is not to identify remedies. Remedies are now well understood. The problem is to make them socially, economically, and politically acceptable. Foresters know how to plant trees, but not how to establish conditions under which we can agree to let forests grow to maturity. Engineers know how to control pollution, but not how to persuade factories to install the necessary equipment. City planners know how to build housing and design safe drinking water systems, but not how to make them affordable for the poorest members of society. These are complex problems, then, that require input from multiple perspectives.

As you study environmental science, you should aim to do the following:

- understand how natural systems function;
- understand ecological concepts that explain biological diversity;
- understand current environmental challenges, such as pollution and climate change; and
- use critical thinking to envision solutions to these challenges.

Environmental science is about understanding where we live

In this course, you will examine processes in our physical environment, including factors affecting biological diversity, biological productivity, sources of earth resources and energy, and circulation of climate and of water resources, as well as the ways resource use, policy, and practices influence those environmental functions. But as you read, also remember that the systems we discuss are amazing and beautiful. Imagine you are an astronaut returning to Earth after a trip to the moon or Mars. What a relief it would be to come back to this beautiful, bountiful planet after experiencing the hostile, desolate environment of outer space. We live in a remarkably prolific and hospitable world. Compared to the conditions on other planets in our solar system, temperatures on the earth are mild and relatively constant. Plentiful supplies of clean air, fresh water, and fertile soil are regenerated endlessly and spontaneously by geological and biological cycles (discussed in chapters 3 and 4).

Perhaps the most amazing feature of our planet is the rich diversity of life that exists here. Millions of beautiful and intriguing species populate the earth and help sustain a habitable environment (fig. 1.3). This vast multitude of life creates complex, interrelated communities where towering trees and huge animals live together with, and depend upon, tiny life-forms such as viruses, bacteria, and fungi. Together, all these organisms make up delightfully diverse, self-sustaining communities, including dense, moist forests, vast sunny savannas, and richly colorful coral reefs. From time to time, we should pause to remember that, in spite of the challenges and complications of life on earth, we are incredibly lucky to be here. We should ask ourselves: What is our proper place in nature? What *ought* we do and what *can* we do to protect the irreplaceable habitat that produced and supports us?

To really understand our environment, we also need to get outdoors and experience nature, in our backyard, a local park, or somewhere more exotic. As author Ed Abbey said, "It is not enough to fight for the land; it is even more important to enjoy it. While you can. While it is still there. So get out there and mess around with your friends, ramble out yonder and explore the forests, encounter the grizz, climb the mountains. Run the rivers, breathe deep of that yet sweet and lucid air, sit quietly for a while and contemplate the precious stillness, that lovely, mysterious and awesome space. Enjoy yourselves, keep your brain in your head and your head firmly attached to your body, the body active and alive."

FIGURE 1.3 Perhaps the most amazing feature of our planet is its rich diversity of life.
© Royalty-Free/Corbis

What topics will you study in this course?

Throughout this book, you will find both problems that are getting worse and conditions that are improving, as in the major themes listed below. While there is much to be pessimistic about, there are many areas for optimism. Often, one of our biggest challenges is understanding how much worse conditions used to be. If you are interested in finding solutions, the idea that change is possible is a good place to start. In this book, therefore, you will find a mix of bad news and good news. Recognizing where conditions have improved over time also reminds us that the hard work of generations before us has been fruitful. We have inherited an extraordinary natural world, which we hope to pass on to future generations in as good a condition—perhaps even better—than when we arrived.

Population and resource consumption

One of the most widely debated challenges is population growth. With over 7.5 billion humans on earth, we're adding about 80 million more every year. Family sizes have declined almost everywhere, from about five children per family 60 years ago to about two today, but still demographers project a population between 8 and 10 billion by 2050 (fig. 1.4a). The impacts of that many people on our natural resources and ecological systems is a serious concern. All high–birth rate countries are low-income, often war-affected areas. Of the 40 countries with the highest birth rates, all are in Africa except Afghanistan.

On the other hand, population growth has stabilized in nearly all industrialized countries and even in most poor countries where social security and democracy have been established. Over the last 20 years, the average number of children born per woman worldwide has decreased from 6.1 to 2.5 (fig. 1.4b). The UN Population Division predicts that by 2050 all developed countries and 75 percent of the developing world will experience a below-replacement fertility rate of 2.1 children per woman. This prediction suggests that the world population could stabilize sooner and lower than previously estimated.

Rising resource consumption per person is also an urgent concern. Poor populations consume very little energy, food, and other resources, compared to wealthy populations, which consume energy and goods from around the globe. As wealth rises around

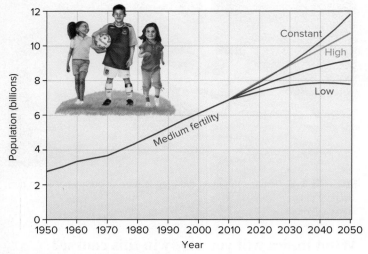

(a) Possible population trends: Where will we be in 2050?

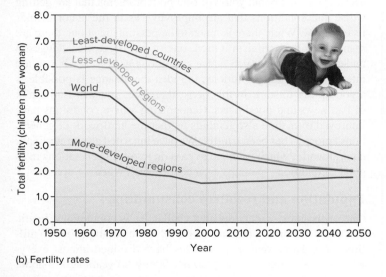

(b) Fertility rates

FIGURE 1.4 Bad news and good news: globally, populations continue to rise, but our rate of growth has plummeted. Nearly half of countries are below the replacement rate of about two children per woman.

the world, people emerging from poverty desire the same high levels of consumption. Thus, both population and consumption rates are persistent questions in environmental science.

Climate change

Burning fossil fuels, making cement, cultivating rice paddies, clearing forests, and other activities release carbon dioxide and other "greenhouse gases" that trap heat in the atmosphere. Over the past 200 years, atmospheric CO_2 concentrations have increased about 35 percent. Climatologists warn that if current trends continue, by 2100 mean global temperatures will probably increase by 2° to 6°C (3.6° to 12.8°F) compared to temperatures in 1900 (fig. 1.5a). This warming is probably responsible for the increasing severity and frequency of droughts, storms, and wildfires in recent years. Melting alpine glaciers and snowfields are depleting

water supplies on which millions of people, in cities such as Los Angeles and Denver, depend. Canadian environment minister David Anderson has said that global climate change is a greater threat than terrorism, because it threatens the homes and livelihoods of billions of people and could trigger worldwide social and economic catastrophe.

Climate change is the most severe and disruptive problem we face, but the need to slow climate change is leading to unprecedented efforts to find global solutions. At the Paris Climate Conference in December 2015, nearly all the world's nations agreed to carbon-reduction commitments. These included pledges for more renewable energy, for replanting forests, for efficiency improvements, and many other strategies. International cooperation is required to meet these pledges, and success is far from certain, but this agreement has given a boost to needed policy strategies (such as carbon fees) and technologies (such as renewable energy). In addition, there are many "co-benefits" to these measures, including reductions in poverty, pollution, and illness.

Hunger

In spite of population growth that added nearly a billion people to the world during the 1990s, the number facing food insecurity and chronic hunger during this period declined by about 40 million. Global food production has more than kept pace with human population growth, but hunger persists in many areas (fig. 1.5b). In a world of food surpluses, the United Nations estimates that some 925 million people are chronically undernourished, often because of drought, floods, displacement from land, or war. Soil scientists report that about two-thirds of all agricultural lands show signs of degradation. Biotechnology and intensive farming techniques, responsible for much of our recent production gains, are too expensive for poor farmers, and they contaminate waterways and deplete soils. Small-scale farms still produce 80 percent of food consumed worldwide, according to the United Nations Development Programme. Can we ensure the sustainability of these farms without further environmental degradation?

Biodiversity loss and conservation efforts

Biologists report that habitat destruction, overexploitation, pollution, and the introduction of exotic organisms are eliminating species at a rate comparable to the great extinction that marked the end of the age of dinosaurs. The UN Environment Programme reports that, over the past century, more than 800 species have disappeared and at least 10,000 species are now considered threatened (fig. 1.5c). This includes about half of all primates and freshwater fish, together with around 10 percent of all plant species. Top predators, including nearly all the big cats in the world, are particularly rare and endangered. At least half of the forests existing before the introduction of agriculture have been cleared, and much of the diverse "old growth" on which many species depend for habitat is rapidly being cut and replaced by ecologically impoverished forest plantations.

Despite ongoing losses, we are also finding ways to conserve resources and use them more sustainably. Restoration ecology

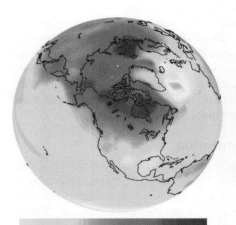

2 2.8 3.6 5 7 9 11 13 15 20°F
Projected winter temperature increase

(a) Climate change

(b) Hunger

(c) Biodiversity

(d) Resource management

FIGURE 1.5 Major environmental themes: (a) Climate change is projected to raise temperatures, especially in northern winter months. (b) Nearly a billion people suffered from chronic hunger in 2010. (c) Many species, including rhinos, are severely threatened, but (d) sustainable resource use can safeguard fisheries and other vital resources.

Source: a: NOAA Geophysical Fluid Dynamics Laboratory; b: © Norbert Schiller/The Image Works; c: © Tom Finkle; d: © William P. Cunningham

(chapter 13) has made strides in species monitoring and recovery. Improved monitoring of fisheries and networks of marine protected areas promote species conservation, as well as human development (fig. 1.5d). Brazil, which has the largest area of tropical rainforests in the world, has reduced forest destruction by nearly two-thirds in the past five years. In addition to protecting endangered species, this is great news in the battle to stabilize our global climate. Nature preserves and protected areas have increased sharply, from about 7 million km^2 in 1990 to over 17.4 million km^2 in 2015. This represents about 11.7 percent of all land area, a dramatic expansion (chapter 12). At the same time, the need for protection has also increased, with rapidly expanding land conversion for agriculture, forestry, mining, and urbanization.

Energy

How we obtain and use energy will play a crucial role in our environmental future. Fossil fuels (oil, coal, and natural gas) presently provide around 80 percent of the energy used in industrialized countries. But acquiring and using these fuels causes air and water pollution, mining damage, shipping accidents, and political conflict. Cleaner renewable energy resources, including solar power, wind, geothermal, and biomass, together with conservation, could give us cleaner, less destructive options if we invest appropriately. Cities and regions everywhere are investing in renewable energy sources in order to protect energy security, employment, and the climate (fig. 1.6a).

Rapidly developing countries have the capacity to make real progress. China leads the world in solar energy, wind turbines, and biogas generation (from agricultural waste), and China is investing in these technologies in other developing regions. Progress in photovoltaic production has helped prices for solar panels in the United States drop by from $20 per watt in the 1980s to less than 50 cents today. In many places, solar and wind are competitive with fossil fuels. The European Union has pledged to get 20 percent of its energy from renewable sources by 2020. Improved permitting, financing, and installation strategies have been almost as important as improved technology. The United Kingdom aims to cut carbon dioxide emissions by 60 percent through energy conservation and a switch to renewables. Denmark and Sweden aim to eliminate most fossil fuel uses by 2050.

Pollution and environmental health

In developing areas, especially China and India, air quality has worsened dramatically in recent years. Over southern Asia, for example, satellite images recently revealed that a 3-km (2-mile)-thick toxic haze of ash, acids, aerosols, dust, and photochemical products regularly covers the entire Indian subcontinent for much of the year. At least 3 million people die each year from diseases triggered by air pollution. The United Nations estimates that, worldwide, more than 2 billion metric tons of air pollutants (not including carbon dioxide or windblown soil) are emitted each year, and many of these pollutants travel worldwide. Mercury, pesticides, perfluorocarbons, and other long-lasting pollutants accumulate in arctic ecosystems and native people after being transported by air currents from industrial regions thousands of kilometers to the south. And on some days, 75 percent of the smog and particulate pollution recorded in California can be traced to Asia.

The good news is that we know how to control air pollution. Metals, dust, even greenhouse gases can be captured before they leave the smoke stack. Most cities in Europe and North America are cleaner and healthier now than they were a half century ago. Clean technology benefits the economy and saves lives. The question is how to ensure that pollution controls are used where they are needed.

Water resources

Water may well be the most critical resource in the twenty-first century. Climate change is reducing irrigation supplies in many farming regions. Over 600 million people (9 percent of us) lack safe drinking water, and 2.4 billion (32 percent) don't have safe sanitation (fig. 1.6b). These figures are considerably better than 25 years ago, but polluted water and inadequate sanitation are estimated to contribute to illness in more than a billion people annually, and to the death of over 5 million children per year. About 40 percent of the world population lives in countries where water demands now exceed supplies, and the UN projects that by 2025 as many as three-fourths of us could live under similar conditions. Water shortages and drought are frequently blamed for displacement of "climate refugees," who lack water for farming or basic subsistence.

The incidence of water-borne infectious diseases has declined in many areas, however. These and other infectious diseases have declined, while life expectancy has nearly doubled, on average (fig. 1.6c). Smallpox has been completely eradicated, and polio has been vanquished except in a few countries. Since 1990, more than 800 million people have gained access to improved water supplies and modern sanitation.

Information and education

Education for girls is now recognized to be the most powerful strategy for slowing population growth and reducing child mortality. In this and many other cases, increasing access to education and information are transforming lives around the world. Rates of illiteracy

(a) Renewable energy

(b) Water resources

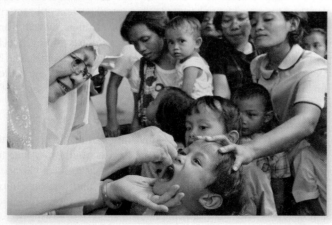

(c) Health care

(d) Education

FIGURE 1.6 Renewable energy (a) is a central theme. Water quality (b) continues to cause illness around the world, but there has been substantial progress in (c) health care, and (d) education.
Source: a: Dennis Schroeder/NREL; b: © Roger A. Clark/Science Source; c: © Dimas Ardian/Getty Images; d: © Christopher S. Collins, Pepperdine University

are falling in many areas, including very poor regions (fig. 1.6d). Because so many environmental issues can be fixed by new ideas, technologies, and strategies, expanding access to knowledge is essential to progress. The increased speed at which information now moves around the world offers unprecedented opportunities for sharing ideas. Developing countries may be able to avoid the mistakes made by industrialized countries and grow with new, efficient, and environmentally sustainable technologies.

Calculating Your Ecological Footprint

Can the earth sustain our current lifestyles? Will there be adequate natural resources for future generations? These questions are among the most important in environmental science today. We depend on our environment for food, water, energy, oxygen, waste disposal, and other life-support systems. For resource use to be sustainable, we cannot consume them faster than nature can replenish them. Degradation of ecological systems ultimately threatens everyone's well-being. Although we may be able to overspend nature's budget temporarily, future generations will have to pay the debts we leave behind.

To calculate your debts, you need a good accounting system. Organizations such as Redefining Progress provide tools to calculate an **ecological footprint,** a measure used to quantify the demands placed on nature by individuals or by nations. Online footprint calculators, such as the WWF Footprint Calculator, or the Redefining Progress calculator, let you assess your own footprint by answering a simple questionnaire about consumption patterns, such as electricity use, shopping, and driving habits.

Footprints are often calculated in terms of global hectares ("gha") of productive capacity, or the global area that would be needed to support one person. Part of the power of this metaphor is that we can visualize a specific area of land—one hectare is an area 100 m x 100 m—and we can use the numbers to compare overall consumption patterns among countries. The term "global hectares" also reminds us that we are always consuming resources from around the world.

According to Redefining Progress, the average world citizen has an ecological footprint equivalent to 2.7 gha, while the biologically productive land available is only 1.8 gha per person. How can this be? We're using nonrenewable resources (such as fossil fuels) to support a lifestyle beyond the productive capacity of our environment. It's like living by borrowing on your credit cards. You can do it for a while, but eventually you have to pay off the deficit. The imbalance is far more pronounced in wealthier countries. The average resident of the United States, for example, lives at a consumption level that requires 7.2 gha of biologically productive land. If everyone in the world were to adopt a North American lifestyle, we'd need about four more planets to support us all.

Like any model, an ecological footprint gives a usefully simplified description of a system. Also like any model, it is built on a number of simplifying assumptions: (1) Various measures of resource consumption and waste flows can be converted into the biologically productive area required to maintain them; (2) different kinds of resource use and dissimilar types of productive land can be standardized into roughly equivalent areas; (3) because these areas stand for mutually exclusive uses, they can be added up to a total—a total representing humanity's demand—that can be compared to the total world area of productive land.

Technological change sometimes can reduce our footprint: For example, world food production has increased about fourfold since 1950, mainly through advances in irrigation, fertilizer use, and higher-yielding crop varieties, rather than through increased croplands. How to sustain this level of production is another question, but this progress shows that land area isn't always an absolute limit. Similarly, switching to renewable energy sources such as wind and solar power can greatly reduce our ecological footprint. Note that in figure 1.7, carbon emissions (from energy consumption) make up about half of the calculated footprint globally. In Germany, which has invested heavily in wind, solar, small-scale hydropower, and public transportation, the ecological footprint is only 4.6 gha per person.

What are the most important steps your community could take to reduce its footprint? Are there things you could do to reduce your personal footprint? Is technological progress most important, or are there policy measures that could be just as important in helping developing areas grow without increasing their ecological footprint?

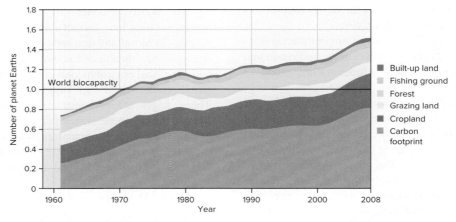

FIGURE 1.7 Humanity's ecological footprint has nearly tripled since 1961, when we began to collect global environmental data.
Source: WWF, 2012.

Section Review

1. Why is population an important question in environmental science? In what ways is population less of a problem than in earlier years?

2. In what ways is pollution still a problem? Has it improved? Why?

3. What is an "ecological footprint"?

1.2 WHERE DO OUR IDEAS ABOUT OUR ENVIRONMENT COME FROM?

- *Utilitarian conservation seeks to protect useful resources.*

- *Wilderness preservation aims to preserve wilderness for aesthetic, intellectual, or philosophical value.*

- *Modern environmental movements have formed to fight pollution, injustice, and poverty.*

Debates about human misuse of nature have a long history. Plato complained in the fourth century B.C.E. that Greece once had been blessed with fertile soil and clothed with abundant forests of fine trees. After the trees were cut to build houses and ships, however, heavy rains washed the soil into the sea, leaving only a rocky "skeleton of a body wasted by disease." Springs and rivers dried up and farming became all but impossible (fig. 1.8). Many classical authors regarded Earth as a living being, vulnerable to aging, illness, and even mortality following the devastation of forest clearing, soil degradation, and other activities.

Some of the earliest *scientific* studies of environmental degradation were carried out in the eighteenth century by French and British colonial administrators. These administrators, some of whom were trained scientists, observed rapid soil loss and drying wells that resulted from intensive colonial production of sugar and other commodities. These early conservationists observed and understood the connection between deforestation, soil erosion, and local climate change, so they recognized that environmental stewardship was an economic necessity. The pioneering British plant physiologist Stephen Hales, for instance, suggested that conserving green plants preserved rainfall. His ideas were put into practice in 1764 on the Caribbean island of Tobago, where about 20 percent of the land was marked as "reserved in wood for rains."

Similarly, Pierre Poivre, an early French governor of Mauritius, an island in the Indian Ocean, was appalled at the environmental and social devastation caused by destruction of wildlife

FIGURE 1.8 Nearly 2,500 years ago, Plato lamented land degradation that denuded the hills of Greece. Have we learned from history's lessons?
© Ken Walsh/Alamy Stock Photo

(such as the flightless dodo) and the felling of ebony forests on the island by early European settlers. In 1769, Poivre declared that one-quarter of the island was to be preserved in forests, particularly on steep mountain slopes and along waterways. Mauritius remains a model for balancing nature and human needs. Its forest reserves shelter a larger percentage of its original flora and fauna than most other human-occupied islands.

Current ideas have followed industrialization

Many of our current ideas about our environment and its resources were articulated by writers and thinkers in the past 150 years. Although many earlier societies had negative impacts on their environments, recent technological innovations have greatly accelerated our impacts. As a consequence of these changes, different approaches have developed for understanding and protecting our environment.

We can divide conservation history and environmental activism into at least four distinct stages: (1) pragmatic resource conservation, (2) moral and aesthetic nature preservation, (3) a growing concern about health and ecological damage caused by pollution, and (4) global environmental citizenship. Each era focused on different problems and each suggested a distinctive set of solutions. These stages are not necessarily mutually exclusive, however. Ideas from all these stages persist today, shaping our ideas and priorities about environmental resources and conservation. But it is useful to associate these ideas with particular stages in history that inspired their widespread adoption.

Stage 1. Resource waste inspired pragmatic, utilitarian conservation

Many historians consider the publication of *Man and Nature* in 1864 by geographer George Perkins Marsh as the wellspring of environmental protection in North America. Marsh, who also was a lawyer, politician, and diplomat, traveled widely around the Mediterranean as part of his diplomatic duties in Turkey and Italy. He read widely in the classics (including Plato) and personally observed the damage caused by the excessive grazing by goats and sheep and by the deforestation of steep hillsides. Alarmed by the wanton destruction and profligate waste of resources still occurring on the American frontier in his lifetime, he warned of its ecological consequences. Largely as a result of his book, national forest reserves were established in the United States in 1873 to protect dwindling timber supplies and endangered watersheds.

Among those influenced by Marsh's warnings were President Theodore Roosevelt (fig. 1.9a) and his chief conservation advisor, Gifford Pinchot (fig. 1.9b). In 1905, Roosevelt, who was the leader of the populist, progressive movement, moved the Forest Service out of the corruption-filled Interior Department into the Department of Agriculture. Pinchot, who was the first native-born professional forester in North America, became the founding head of this new agency. He put resource management on an honest, rational, and scientific basis for the first time in our history. Together with naturalists and activists such as John Muir, William Brewster, and George Bird Grinnell, Roosevelt and Pinchot established the framework of our national forest, park, and wildlife refuge systems, passed game protection laws, and tried to stop some of the most flagrant abuses of the public domain.

The basis of Roosevelt's and Pinchot's policies was pragmatic **utilitarian conservation.** They argued that the forests should be saved "not because they are beautiful or because they shelter wild creatures of the wilderness, but only to provide homes and jobs for people." Resources should be used "for the greatest good, for the greatest number for the longest time." "There has been a fundamental misconception," Pinchot said, "that conservation means nothing but husbanding of resources for future generations. Nothing could be further from the truth. The first principle of conservation is development and use of the natural resources now existing on this continent for the benefit of the people who live here now. There may be just as much waste in neglecting the development and use of certain natural resources as there is in their destruction." This pragmatic approach still can be seen today in the multiple use policies of the Forest Service.

Stage 2. Ethical and aesthetic concerns inspired the preservation movement

John Muir (fig. 1.9c) was a geologist, author, and first president of the Sierra Club. He strenuously opposed Pinchot's utilitarian approach. Muir argued that nature deserves to exist for its own sake, regardless of its usefulness to us. Aesthetic and spiritual values formed the core of his philosophy of nature protection. This outlook has been called **biocentric preservation** because it emphasizes the fundamental right of other organisms to exist and to pursue their own interests. Muir wrote: "The world, we are told, was made for man. A presumption that is totally unsupported by the facts. . . . Nature's object in making animals and plants might possibly be first of all the happiness of each one of them. . . . Why ought man to value himself as more than an infinitely small unit of the one great unit of creation?"

Muir, who was an early explorer and interpreter of the Sierra Nevada Mountains in California, fought long and hard for the establishment of Yosemite and Kings Canyon National Parks. The National Park Service, established in 1916, was first headed by Muir's disciple Stephen Mather and has always been oriented toward preservation of nature in its purest state. It has often been at odds with Pinchot's utilitarian Forest Service.

In 1935, pioneering wildlife ecologist Aldo Leopold (fig. 1.9d) bought a small, worn-out farm in central Wisconsin. A dilapidated chicken shack, the only remaining building, was remodeled into a rustic cabin (fig. 1.10). Working together with his children, Leopold planted thousands of trees in a practical experiment in restoring the health and beauty of the land. Leopold argued for stewardship of the land. He wrote of "the land ethic," by which we should care for the land because it's the right thing to do—as well as the smart thing. "Conservation," he wrote, "is the positive exercise of skill and insight, not merely a negative exercise of abstinence or caution." The shack became a writing refuge and became the main focus of *A Sand County Almanac,* a much beloved collection of essays about our relation with nature. In it, Leopold wrote, "We abuse land because we regard it as a commodity belonging to us. When we see land as a community to which we belong, we may begin to use it with love and respect."

(a) President Teddy Roosevelt

(b) Gifford Pinchot

(c) John Muir

(d) Aldo Leopold

FIGURE 1.9 Some early pioneers of the American conservation movement. President Teddy Roosevelt (a) and his main advisor Gifford Pinchot (b) emphasized pragmatic resource conservation, while John Muir (c) and Aldo Leopold (d) focused on ethical and aesthetic relationships.
Source: a: Library of Congress Prints & Photographs Division [LC-USZ62-77199]; b: Source: Library of Congress Prints & Photographs Division [LC-USZ62-103915]; c: © Bettmann/Corbis; d: © AP/Wide World Photos

FIGURE 1.10 Aldo Leopold's Wisconsin shack, the main location for his *Sand County Almanac,* in which he wrote, "A thing is right when it tends to preserve the integrity, stability, and beauty of the biotic community. It is wrong when it tends otherwise." How might you apply this to your life?
© William P. Cunningham

(a) Rachel Carson

(b) David Brower

(c) Barry Commoner

(d) Wangari Maathai

FIGURE 1.11 Among many distinguished environmental leaders in modern times, (a) Rachel Carson, (b) David Brower, (c) Barry Commoner, and (d) Wangari Maathai stand out for their dedication, innovation, and bravery.
a: © AP/Wide World Photos; b: © Tom Turney/The Brower Fund/Earth Island Institute; c: Source: Photo of Barry Commoner; courtesy University Archives, Rare Book & Manuscript Library, Columbia University in the City of New York; d: © AP/Wide World Photos

Stage 3. Rising pollution levels led to the modern environmental movement

The undesirable effects of pollution probably have been recognized at least as long as those of forest destruction. In 1273, King Edward I of England threatened to hang anyone burning coal in London, because of the acrid smoke it produced. In 1661, the English diarist John Evelyn complained about the noxious air pollution caused by coal fires and factories and suggested that sweet-smelling trees be planted to purify city air. Increasingly dangerous smog attacks in Britain led in 1880 to the formation of a national Fog and Smoke Committee to combat this problem.

The tremendous industrial expansion during and after the Second World War added a new set of concerns to the environmental agenda. *Silent Spring,* written by Rachel Carson (fig. 1.11*a*) and published in 1962, awakened the public to the threats of pollution and toxic chemicals to humans as well as other species. The movement she engendered might be called **environmentalism,** because its concerns are extended to include both environmental resources and pollution. Like many environmentalists, Carson was especially concerned with the ways pollution endangered human health. Other pioneers of this movement were activist David Brower, who fought to protect public lands from industrial destruction (fig. 1.11*b*) and biologist Barry Commoner (fig. 1.11*c*). Brower,

while executive director of the Sierra Club, Friends of the Earth, and the Earth Island Institute, introduced many of the techniques of modern environmentalism, including litigation, intervention in regulatory hearings, book and calendar publishing, and using mass media for publicity campaigns.

Like Rachel Carson, Barry Commoner was principally interested in environmental health—an issue that is especially urgent for low-income, minority, and inner-city residents. Trained as a molecular biologist, Commoner protested the health impacts of nuclear testing and of industrial pollution, hazards revealed by his work as a biologist. Many of today's efforts to curb climate change or reduce biodiversity losses are led by scientists who raise the alarm about environmental problems.

Stage 4. Environmental quality is tied to social progress

Many people today believe that the roots of the environmental movement are elitist—promoting the interests of a wealthy minority who can afford to vacation in wilderness. In fact, most environmental leaders have seen social justice and environmental equity as closely linked. Gifford Pinchot, Teddy Roosevelt, and John Muir all strove to keep land and resources accessible to everyone, at a time when public lands, forests, and waterways were increasingly controlled by industrial interests, especially railroad, mining, and logging companies. The idea of national parks, one of our principal strategies for nature conservation, is to provide public access to natural beauty and outdoor recreation. Aldo Leopold, a founder of the Wilderness Society, promoted ideas of land stewardship among farmers, fishers, and hunters. Robert Marshall, also a founder of the Wilderness Society, campaigned all his life for social and economic justice for low-income groups. Many environmental leaders grew up in working-class families, so their sympathy with social causes is not surprising.

Increasingly, environmental activists are linking environmental quality and social progress on a global scale. Barry Commoner, for example, emphasized that uneven poverty and wealth are a root cause of environmental health risks, of pollution, and of population growth. One of the core concepts of modern environmental thought is **sustainable development,** often defined as "meeting the needs of the present without compromising the ability of future generations to meet their own needs." In other words, we should be able to improve conditions for the world's poorest populations without devastating the environment. This definition was given in a 1987 report of the World Commission on Environment and Development, *Our Common Future,* This report is often called the Brundtland Report, after the chair of the commission, Norwegian prime minister Gro Harlem Brundtland.

The Brundtland Report led to the pivotal 1992 Earth Summit, a United Nations meeting held in Rio de Janeiro, Brazil. The Rio meeting was a pivotal event because it brought together many diverse groups. Environmentalists and politicians from wealthy countries, indigenous people and workers struggling for rights and land, and government representatives from developing countries all came together and became more aware of their common needs. The Rio meeting is largely credited with spreading the idea of sustainable development.

Some of today's leading environmental thinkers come from developing nations, where poverty and environmental degradation plague hundreds of millions of people. Dr. Wangari Maathai of Kenya (1940–2011) was a notable example. In 1977, Dr. Maathai (see fig. 1.11d) founded the Green Belt Movement in her native Kenya as a way of both organizing poor rural women and restoring their environment. Beginning at a small, local scale, this organization has grown to more than 600 grassroots networks across Kenya. They have planted more than 30 million trees, while mobilizing communities for self-determination, justice, equity, poverty reduction, and environmental conservation. Dr. Maathai was elected to the Kenyan Parliament and served as assistant minister for environment and natural resources. Her leadership helped bring democracy and good government to her country. In 2004, she received the Nobel Peace Prize for her work, the first time a Nobel has been awarded for environmental action. In her acceptance speech, she said, "Working together, we have proven that sustainable development is possible; that reforestation of degraded land is possible; and that exemplary governance is possible when ordinary citizens are informed, sensitized, mobilized and involved in direct action for their environment."

Under the leadership of a number of other brilliant and dedicated activists and scientists, the environmental agenda was expanded in the 1960s and 1970s to include issues such as human population growth, atomic weapons testing and atomic power, fossil fuel extraction and use, recycling, air and water pollution, wilderness protection, and a host of other pressing problems that are addressed in this textbook. Environmentalism has become well established on the public agenda since the first national Earth Day in 1970. A majority of Americans now consider themselves environmentalists, although there is considerable variation in what that term means.

Photographs of the earth from space (fig. 1.12) provide a powerful icon for the fourth wave of ecological concern, which might be called **global environmentalism.** These photos remind us how small, fragile, beautiful, and rare our home planet is. We all share a common environment at this global scale. As our attention shifts from questions of preserving particular landscapes or preventing pollution of a specific watershed or airshed, we begin to worry about the life-support systems of the whole planet.

A growing number of Chinese activists are part of this global environmental movement. In 2006, Yu Xiaogang was awarded the Goldman Prize, the world's top honor for environmental protection. Yu was recognized for his work on Yunan's Lashi Lake, where he brought together residents, government officials, and entrepreneurs to protect wetlands, restore fisheries, and improve water quality. He also worked on sustainable development programs, such as women's schools and microcredit loans. His leadership was instrumental in stopping plans for 13 dams on the

FIGURE 1.12 The life-sustaining ecosystems on which we all depend are unique in the universe, as far as we know.
© Stocktrek/age fotostock RF

Nu River (known as the Salween in Thailand and Burma). Another Goldman Prize winner is Dai Qing, who was jailed for her book that revealed the social and environmental costs of the Three Gorges Dam on the Yangzi River.

Section Review

1. Differentiate "conservation" and "preservation." Identify one person associated with each.
2. What was Rachel Carson's *Silent Spring* about? Why?
3. In what ways is environmental quality tied to social progress?

1.3 SUSTAINABLE DEVELOPMENT

- *Poverty affects quality of life in many ways.*
- *Poverty has declined dramatically in recent years, but so has environmental quality.*
- *Sustainable development aims to reduce poverty without damaging environmental resources.*

Policymakers are becoming aware that eliminating poverty and protecting our common environment are inextricably interlinked, because the world's poorest people are both victims and agents of environmental degradation. The poorest people are often forced to meet short-term survival needs at the cost of long-term sustainability. Desperate for croplands to feed their families, and for fuel, many clear forests or cultivate steep hillsides, where soil is rapidly eroded. Others migrate to the crowded shantytowns that surround most major cities in the developing world.

The World Bank estimates that 700 million people (nearly 10 percent of us) live below an international poverty line of (U.S.) $1.90 per day. This is less than a third as many severely impoverished people as there were in 1990, so it represents tremendous progress. Still, the human suffering engendered by this poverty is tragic. The very poor often lack access to an adequate diet, decent housing, basic sanitation, clean water, education, medical care, and other essentials for a humane existence (fig. 1.13). Seventy percent of those people are women and children.

Poverty affects many **quality-of-life indicators** (table 1.1). Infant mortality in the least-developed countries is about 25 times as high as in the most-developed countries. Modern sanitation, essential for controlling disease, is available to only 23 percent of residents in poorer countries. Meanwhile, carbon dioxide emissions (a measure of both energy use and contributions to global warming) are over 30 times greater in rich countries.

Affluence is a goal and a liability

About one-sixth of the world's population live in the richest countries, where the average per capita income is above (U.S.) $41,000 per year. Most of these countries are in North America or Western Europe, but Japan, Singapore, and Australia also fall into this group. Another one-sixth live in the least-developed countries, with per capita income averaging less than $2,500. The remaining two-thirds of the world's population live in middle- or low-income

FIGURE 1.13 People living in extreme poverty generally lack adequate food, housing, medical care, clean water, and safety.
© William P. Cunningham

Table 1.1 Quality-of-Life Indicators

	Least-Developed Countries	Most-Developed Countries
GDP/Person[1]	$2,122	$41,395
Life Expectancy	63 years	81 years
Adult Literacy	58%	99%
Child Labor[2]	21.7	~0
Female Secondary Education	17%	95%
Total Fertility[3]	4.2	1.8
Infant Mortality[4]	55	5
Percent Urban	29.8%	81.9%
Electricity Access	34.2%	99.9%
CO_2/Capita[5]	0.3 tons	11 tons

[1]Annual gross domestic product per person, U.S. dollar equivalent
[2]Percent ages 5–14
[3]Average births/woman
[4]Per 1,000 live births
[5]Metric tons/yr/person
Source: UNDP Human Development Index, 2015

countries, where incomes average over $10,000, and where most people live above the extreme poverty line, by global standards.

Those of us in the richer nations enjoy unprecedented affluence and comfort. To do so, we consume an inordinate share of the world's resources and produce unsustainable amounts of pollution—although much of that pollution occurs far from where we live, in industrial cities and degraded mining regions and farmlands of distant countries. The United States, for example, with about 4.6 percent of the world's population, consumes about 20 percent of all oil, and produces about 15 percent of all carbon dioxide and half of all toxic wastes in the world (fig. 1.14).

FIGURE 1.14 "And may we continue to be worthy of consuming a disproportionate share of this planet's resources."
© Lee Lorenz/The New Yorker Collection/www.cartoonbank.com

FIGURE 1.15 A rapidly growing economy has brought increasing affluence to China that has improved standards of living for many Chinese people, but it also brings environmental and social problems associated with Western lifestyles.
© Justin Guariglia/Corbis

Low-income countries hope to emulate this prosperity. Take the example of China, the most dramatic recent case of development. In the early 1960s, some 300 million Chinese suffered from chronic hunger, and perhaps 50 million starved to death in the worst famine in world history. Since then, China has experienced amazing economic growth and modernization. The national GDP has been increasing at about 5–10 percent per year. Hundreds of millions of people have been lifted out of extreme poverty. Chronic hunger has become uncommon. Average life expectancy has increased from 42 to 76 years, and infant mortality dropped from 150 per 1,000 live births in 1960 to 11 today. Annual per capita GDP has grown from less than (U.S.) $200 per year to more than $12,000 (fig. 1.15).

As a consequence of this growth, pollution has become more severe each year, as demand has exploded for resources, consumer goods, cars, and other luxuries. Most Chinese still consume far less than Americans or Europeans, though. In terms of ecological footprints (What Do You Think?, later in this section), it takes about 9.7 global hectares (gha, or hectares-worth of resources) to support the average American each year. By contrast, the average person in China consumes about 2.1 gha per year. Providing the 1.3 billion Chinese with American standards of consumption would require about 10 billion gha, or almost another entire earth's worth of resources. China's use of coal-powered electricity, one of the most important sources of greenhouse gases and air pollution, has doubled in less than a decade (fig. 1.16).

Historically, some problems such as waste generation and carbon dioxide emissions have tended to increase with economic growth (fig. 1.17). Other problems, such as urban air pollution, tend to rise and then decline. A common explanation for the decline is that increasing wealth leads to better regulation and pollution controls. Some environmental costs can fall steadily with increasing development, as in the case of household sanitation: Communities can pay more to clean and protect water resources as they become wealthier (fig. 1.17). Some of these rules are changing, too. Economic growth has been tied to coal and oil burning ever since the Industrial Revolution. But in recent years, global economic growth has occurred despite falling greenhouse gas emissions.

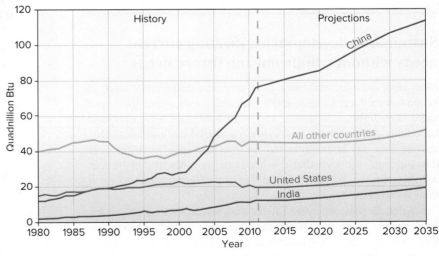

FIGURE 1.16 Coal consumption, most of it used for electricity generation, has fueled much of China's recent growth. Because coal is our primary source of air pollutants and greenhouse gases, projected increases would be disastrous.
Source: US Energy Information Agency 2013

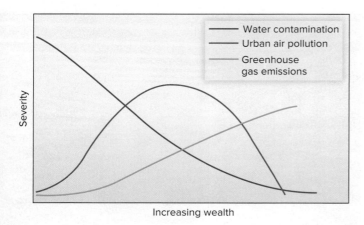

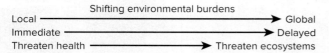

FIGURE 1.17 Environmental indicators show different patterns as incomes rise. Water contamination decreases as people can afford wastewater treatment and drinking water filtration. Local air pollution, on the other hand, often increases as more fuel is burned; eventually, development reaches a point at which people can afford clean air technology. Delayed, distant problems, such as greenhouse gas emissions that lead to global climate change, tend to rise steadily with income because people make decisions based on immediate needs and wants rather than long-term consequences. Thus, we tend to shift environmental burdens from local and immediate to distant and delayed if we can afford to do so.
Source: World Energy Assessment, UNDP

Quantitative Reasoning

Examine figure 1.17. Describe in your own words how increasing wealth affects water contamination, air pollution, and greenhouse gas emissions. Why might greenhouse gas emissions rise? Describe and explain the air pollution curve.

Sustainable development: meeting current needs without compromising future needs

Most poor nations, and most poor families, desperately want to become wealthier, to have more material goods, more food, and more cars, lights, computers, and other amenities. Can everyone have their share without destroying our shared environment? One answer that has been proposed for this dilemma is sustainable development, the idea that the needs of the present can be met without compromising the ability of future generations to meet their own needs.

Another way of saying this is that we depend on nature for food, water, energy, fiber, waste disposal, and other life-support services. We can't deplete resources or create wastes faster than nature can recycle them, if we hope to be here for the long term. Development means improving people's lives. Sustainable development, then, means progress in human well-being that can be extended or prolonged over many generations. To be truly enduring, the benefits of sustainable development must be available to all humans rather than to just the members of a privileged group.

Sustainable development can mean many things. Many forms of sustainable farming, for example, produce food while protecting wildlife habitat, soil, and water resources. Sustainable fisheries have persisted for generations in some areas, without destroying the resource. Even energy production can increase with limited environmental costs, as solar, wind, geothermal, and conservation technologies improve. Often, sustainable development simply means distributing investment to small producers, who circulate money in their local communities. Fair trade organizations, for example, help people in developing countries to grow or make high-value products—often using traditional techniques and designs—that can be sold on world markets for good prices (fig. 1.18). Any time you buy fair trade products, you are probably contributing directly to sustainable development.

FIGURE 1.18 A Mayan woman from Guatemala weaves on a backstrap loom. A member of a women's weaving cooperative, she sells her work to nonprofit organizations in the United States at much higher prices than she would get at the local market.
© *McGraw-Hill Education/Barry Barker, photographer*

Growth can also occur without increased resource consumption. Markets in arts, entertainment, education, services, and leisure time can improve our lives with little environmental cost. These are growth industries whose primary resource is human ingenuity. This is an idea recognized by economists at least since John Stuart Mill in 1857: "It is scarcely necessary to remark that [resource limitation] implies no stationary state of human improvement. There would be just as much scope as ever for all kinds of mental culture and moral and social progress; as much room for improving the art of living and much more likelihood of its being improved when minds cease to be engrossed by the art of getting on."

The UN has identified Sustainable Development Goals

Beginning in 2016, the United Nations initiated a program to promote 17 **Sustainable Development Goals** (SDGs). The goals are ambitious and global, and they include eliminating the most severe poverty and hunger, promoting health, education, and gender equality, providing safe water and clean energy, and preserving biodiversity (see fig. 1.19). This global effort, developed by representatives of the member states of the UN, seeks to coordinate data gathering and reporting, so that countries can monitor their progress, to share resources, and to promote sustainable investment in developing areas.

For each of the 17 goals, organizers identified targets, some quantifiable, some more general. For example, Goal 1, "End poverty," includes targets to eradicate extreme poverty everywhere, and to ensure that all people have rights to basic services,

ownership and inheritance of property, and other economic needs. Goal 7, "Ensure access to affordable, sustainable energy," includes targets of doubling energy efficiency and enhancing international investment in clean energy. Goal 12, "Ensure sustainable consumption and production," calls for cutting per capita food waste in half, as well as phasing out fossil fuel subsidies that encourage wasteful consumption. The UN aims to meet these targets by 2030, in a span of 15 years.

The SDGs also include a number of targets for economic and social equity, and for better governance. To most economists and policymakers, it seems obvious that economic growth is the only way to bring about a long-range transformation to more advanced and productive societies and to provide resources to improve the lot of all people. As former U.S. president John F. Kennedy said, "A rising tide lifts all boats." But the Brundtland report (previous section) emphasized that equity is also essential: Political stability, democracy, and fair access to resources and opportunity are needed to ensure that the poor will get a fair share of the benefits of greater wealth in a society. According to a study released in 2006 by researchers at Yale and Columbia Universities, environmental sustainability tends to occur where there are open political systems and good government.

The Millennium Development Goals were largely successful

These ambitious goals might appear unrealistic, but they build on the remarkable (though not complete) successes of the **Millennium Development Goals**. These eight goals were a 15-year effort, from 2000 to 2015, to improve literacy, health, access to safe

FIGURE 1.19 The United Nations Sustainable Development Goals are intended to improve well-being of the world's poorest people while also protecting biodiversity, natural resources, and climate. These goals follow the largely successful Millennium Development Goals.
Source: UN Development Programme

water, child survival, and other goals. Targets included ending poverty and hunger, universal education, gender equity, child health, maternal health, combating HIV/AIDS, environmental sustainability, and global cooperation in development efforts. While only modest progress was achieved on some goals, UN Secretary General Ban Ki-Moon called that effort "the most successful anti-poverty movement in history." Extreme poverty dropped from nearly half the population of developing countries to just 14 percent in only 15 years. The proportion of undernourished people dropped by almost half, from 23 percent to 13 percent. Primary school enrollment rates climbed from 83 percent to 91 percent in developing countries. Girls gained access to education, employment, and political representation in national parliaments.

At the same time, UN reports stressed that many goals were not met. Some 2.6 billion people gained access to safe drinking water, but over 40 percent of the world's population still lacks access to piped drinking water at home. The proportion of urban populations living in slums fell from 39 percent to 30 percent, but that still represents a large population with inadequate and unsafe housing.

The value of having clearly stated goals, especially with quantifiable targets, is that they help people agree on what to work for. With so many simultaneous problems in developing areas, it can be hard for leaders to know where to focus first. Agreed-upon targets, especially when they are shared and monitored by many countries, can strongly motivate action. International agreement on goals can also help motivate financial and planning assistance, both often badly needed in developing areas.

Could we eliminate acute poverty through aid?

Economist Jeffery Sachs, director of the UN Millennium Development Project, says we could end extreme poverty worldwide by 2025 if the richer countries would donate just 0.7 percent of their national income for development aid in the poorest nations. These funds could be used for universal childhood vaccination against common infectious diseases, access to primary schools for everyone, family planning and maternal health services, safe drinking water and sanitation for all, food supplements for the hungry, and microcredit loans to promote self-employment.

The United States, the world's largest total donor, sets aside only 0.16 percent of its gross domestic product for development aid. That amounts to about 18 cents per citizen per day for both private and government aid to foreign nations. As former Canadian prime minister Jean Chrétien said, "Aid to developing countries isn't charity; it's an investment. It will make us safer, and when standards of living increase in those countries, they'll become customers who will buy tons of stuff from us."

The United Nations Development Programme has estimated that it would take about (U.S.) $135 billion per year to abolish extreme poverty and the worst infectious diseases over the next 20 years. That's a lot of money, but it's not much more than the $120 billion in subsidies and tax breaks the U.S. government gives to oil companies each year. And it's far less than the $1 trillion of global military spending each year (fig. 1.20).

FIGURE 1.20 Every year, military spending equals the total income of half the world's people. The cost of a single large aircraft carrier equals 10 years of human development aid given by all the world's industrialized countries.
Source: Photo by Kyle Gahlau/DVIDS

Many experts propose that if we were to shift one-tenth of that spending to development aid, we would be much safer than we are spending it on the military. The Worldwatch Institute warns that "poverty, disease and environmental decline are the true axis of evil." Terrorist attacks—and the responses they provoke—are symptoms of the underlying sources of global instability, including the dangerous interplay among poverty, hunger, disease, environmental degradation, and rising competition for resources. Unless the world takes action to promote sustainability and equity, Worldwatch suggests, we are unlikely to resolve persistent problems of wars, terrorism, and natural disasters. As writer H. L. Mencken once said, "If you want peace, work for justice."

Section Review

1. List any three quality of life indicators (table 1.1). How do they differ between wealthy and poor countries?
2. Why is affluence a liability? Give an example.
3. Explain the idea of sustainable development.

1.4 CORE CONCEPTS IN SUSTAINABLE DEVELOPMENT

- *"Ecosystem services" is a term for goods, services, and products we rely on; often these are invisible.*
- *Shared resources and ecosystem services can be described as "common property," or as a "commons." Managing common property is a key challenge.*
- *Indigenous peoples often protect biodiversity.*

Some general organizing ideas help us make sense of the problem of sustainable development and of how to think about prosperity in a world of limits. The ideas raised in this section will help you think about issues such as water resource management, ways to preserve water quality, how to encourage greenhouse gas reductions, questions of biodiversity, air quality, and many other natural resource considerations. These issues will also come up in later chapters on economics and policy.

How do we describe resource use?

The natural world supplies the water, food, metals, energy, and other resources we use. Because we use so many kinds of resources, one widely used measure for evaluating resource consumption is the ecological footprint (see What Do You Think?, back in section 1.1, and the Data Analysis section, at the end of this chapter). Our ecological footprint is an index, a number representing a complex array of factors.

Another widely used concept for describing resource use is **throughput,** the amount of material or resources that flow through a system. A household that consumes abundant consumer goods, foods, and energy brings in a great deal of natural resource–based materials; that household also disposes of a great deal of materials. Conversely, a household that consumes very little also tends to produce little waste.

Ecosystem services refers to services or resources provided by environmental systems (fig. 1.21). Provisioning of resources, such as the fuels we burn, may be the most obvious service we require. Supporting services are less obvious until you start

listing them: these include water purification, production of food and atmospheric oxygen by plants, or decomposition of waste by fungi and bacteria. Regulating services include maintenance of temperatures suitable for life by the earth's atmosphere, or carbon capture by green plants, which maintains a stable atmospheric composition. Cultural services include a diverse range of recreational, aesthetic, and other nonmaterial benefits. Usually we rely on these resources without thinking about them. They support all our economic activities in some way, but we don't put a price on them because nature doesn't force us to pay for them.

How can we protect these services over the long term? One of the answers to this basic question was given in an essay entitled **"Tragedy of the Commons,"** published in 1968 in the journal *Science* by ecologist Garret Hardin. In this classic framing of the problem, Hardin argued that population growth leads inevitably to overuse and then destruction of common resources—such as shared pastures, unregulated fisheries, fresh water, land, and clean air. Hardin's essay has influenced our ideas about resource management for decades.

Examples of destroyed commons abound. The North Atlantic cod fishery, once one of the world's greatest fish populations, was functionally destroyed by a free-for-all of unregulated fishing by fleets of many nations. The species is not extinct, but it may never recover to its historic abundance. Air pollution is another familiar example: industries emit pollution from unregulated incinerators and burners, spilling soot, sulfur dioxide, and carbon dioxide into the air, which is then contaminated for all users. Hardin proposed that there are only two ways to avoid this destruction: a system of private property, in which owners protect resources because of self-interest, or coercive regulation by the state.

An alternative perspective to Hardin's framework is strategies for **managing the commons;** that is, for collectively safe-guarding commonly used resources. The importance of common property management was publicized by Elinor Ostrom, who won the 2009 Nobel Prize in Economic Sciences. Ostrom and her colleagues demonstrated that ordinary people have often created rules and institutions for the sustainable management of shared resources. They examined cases of successful long-term management of sustainable fisheries, common forests, common grazing lands, and other resources in communities around the world. They emphasized that not all common properties are well managed, but a great many are.

What conditions can help communities manage their commons over the long term? Many strategies exist, but Ostrom and her colleagues found that some conditions occur frequently in successful cases. Among these are (1) effective and inexpensive monitoring of resource use;

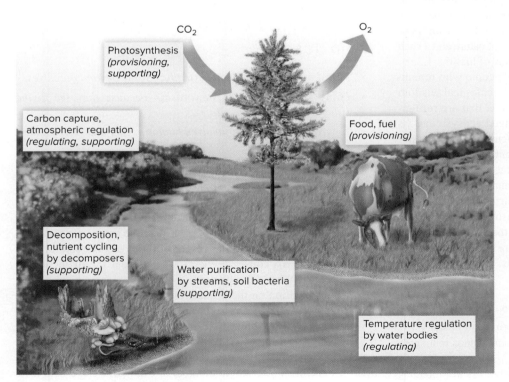

FIGURE 1.21 Ecosystem services we depend on are countless and often invisible.

CO$_2$

O$_2$

Photosynthesis
(provisioning, supporting)

Carbon capture, atmospheric regulation
(regulating, supporting)

Food, fuel
(provisioning)

Decomposition, nutrient cycling by decomposers
(supporting)

Water purification by streams, soil bacteria
(supporting)

Temperature regulation by water bodies
(regulating)

(2) an ability to exclude outsiders, who don't understand rules of use; and (3) frequent face-to-face communications and strong social networks among users, which reduce distrust and promote communication about the state of the resource.

What type of institution is best for managing a global commons, such as climate or biodiversity? Garret Hardin, with many others, has argued that local solutions to climate change are irrelevant as long as countries and international institutions fail to make policy changes. Ostrom and her colleagues argued that these large institutions are often incapable of taking strong or quick action for resource conservation. So it is also important to invest in smaller, local, even individual policy changes, whose effects and ideas may spread contagiously or inform broader improvements in resource management. Both positions are probably correct: Often mixtures of different types of institutions, large and small, may be needed, to contribute simultaneously to solutions at different scales.

Indigenous peoples often protect biodiversity

Development challenges are especially severe for indigenous peoples in both rich and poor countries. Typically, descendants of the original inhabitants of an area taken over by more powerful outsiders, indigenous peoples are distinct from their country's dominant language, culture, religion, and racial communities. Consequently, these groups often generally are the least powerful, most neglected groups in an area. Of the world's nearly 6,000 recognized cultures, 5,000 are indigenous; and these account for only about 10 percent of the total world population. In many countries, traditional caste systems, discriminatory laws, economics, or prejudices repress indigenous people. At least half of the world's 6,000 distinct languages are dying because they are no longer taught to children. When the last elders who still speak the language die, so will the culture that was its origin. Lost with those cultures will be a rich repertoire of knowledge about nature and a keen understanding of a particular environment and way of life (fig. 1.22).

Nonetheless, the 500 million indigenous people who remain in traditional homelands still possess valuable ecological wisdom and are the guardians of little-disturbed habitats that are refuges for rare and endangered species and undamaged ecosystems. As we seek strategies for sustainable development and biodiversity conservation, this knowledge may be an essential resource.

Recognizing native land rights and promoting political pluralism can be among the best ways to safeguard ecological processes and endangered species. A few countries, such as Papua New Guinea, Fiji, Ecuador, Canada, and Australia, acknowledge indigenous title to extensive land areas. As the Kuna Indians of Panama say, "Where there are forests, there are native people, and where there are native people, there are forests."

Section Review

1. Think of five ecosystem services on which you rely.
2. What is the "tragedy of the commons"? What are two ways to avoid it?
3. List any two of the factors that can help communities manage a commons.
4. Why do indigenous people often have an interest in protecting biodiversity?

FIGURE 1.22 Do indigenous people have unique knowledge about nature and inalienable rights to traditional territories?
© McGraw-Hill Education/Barry Barker, photographer

1.5 Environmental Ethics, Faith, and Justice

- *Moral extensionism means extending value beyond ourselves.*
- *Many faiths encourage stewardship because they see divine value in our environment.*
- *Environmental justice involves human rights and environmental justice.*

The ways we interpret environmental issues, or our decisions about what we should or should not do with natural resources, depend partly on our basic worldviews. Perhaps you have a basic ethical assumption that you should be kind to your neighbors, or that you should try to contribute in positive ways to your community. Do you have similar responsibilities to take care of your environment? To conserve energy? To prevent the extinction of rare species? Why? Or why not?

Your position on these questions is partly a matter of **ethics,** or your sense of what is right and wrong. Some of these ideas you learn early in life; some might change over time. Ethical views in society also change over time. In ancient Greece, many philosophers who were concerned with ethics and morality owned slaves. A slave owner could mistreat or even kill a slave with little or no consequences. Today, few societies condone slavery. Most societies now believe it is unethical to treat other humans as property. On the other hand, most societies now consider land, water, forests, and other natural resources as private property. It is the owner's right to conserve or degrade those private resources as they like. Other people (or other organisms) have no legal right to restrict how private property is used or abused. Normally, if you have ancient trees on your property, it is your right to cut and sell them, regardless of your neighbors' opinions. Is this ethical? That depends on your perspective.

Often, our core beliefs are so deeply held that we have difficulty even identifying them. But they can influence how you act, how you spend money, or how you vote. Try to identify some of your core beliefs. What is a basic thing you simply should or should not do? Where does your understanding come from about those actions?

We can extend moral value to people and things

One of the reasons we don't accept slavery now, as the ancient Greeks did, is that most societies believe that all humans have basic rights. The Greeks granted **moral value** (value or worth, based on moral principles) only to adult male citizens within their own community. Women, slaves, and children had few rights and were essentially treated as property. Over time, we have gradually extended our sense of moral value to a wider and wider circle, an idea known as **moral extensionism;** that is, extending moral value to a larger circle of people, organisms, or objects (fig. 1.23). In most countries, women and minorities have basic civil rights, children cannot be treated as property, even domestic pets have some legal protections against cruel treatment. For many people, moral value also extends to domestic livestock (cattle, hogs, poultry), which makes eating meat a fundamentally wrong thing to do. For others, this moral extension ends with pets, or with humans. Some people extend moral value to include forests, biodiversity, inanimate objects, or the earth as a whole.

These philosophical questions aren't simply academic or historical. In 2004, the journal *Science* caused public uproar by publishing a study demonstrating that fish feel pain. Many recreational anglers had long managed to suppress worries that they were causing pain to fish, and the story was so unsettling that it made national headlines and provoked fresh public debates on the ethics of fishing.

How we treat other people, animals, or things can also depend on whether we believe they have **inherent value** (an intrinsic right

Me
Parents
Family
Humanity
Sentient animals
All life
The world

Scope of moral consideration

FIGURE 1.23 Moral extensionism describes an increasing consideration of moral value in other living things—or even nonliving things.

to exist) or **instrumental value** (usefulness to someone). If I hurt you, I owe you an apology. If I borrow your car and crash it into a tree, I don't owe the car an apology, I owe *you* an apology—or reimbursement.

Does this apply to other species? Domestic animals clearly have an instrumental value because they are useful to (or give comfort to) their owners. But some philosophers would say they also have inherent values and interests. By living, breathing, struggling to stay alive, the animal carries on its own life independent of its usefulness to someone else.

Some people believe that even nonliving things also have inherent worth. Rocks, rivers, mountains, landscapes, and certainly the earth itself, have value. These things were in existence before we came along, and we couldn't re-create them if they are altered or destroyed. This philosophical debate became a legal dispute in a historic 1969 court case, when the Sierra Club sued the Disney Corporation on behalf of the trees, rocks, and wildlife of Mineral King Valley in the Sierra Nevada Mountains (fig. 1.24), where Disney wanted to build a ski resort. The Sierra Club argued that it represented the interests of beings that could not speak for themselves in court.

A legal brief entitled *Should Trees Have Standing?,* written for this case by Christopher D. Stone, proposed that organisms as well as ecological systems and processes should have standing (or rights) in court. After all, corporations—such as Disney—are treated as persons and given legal rights even though their "personhood" is only a figment of our imagination. Why shouldn't nature have similar standing? The case went all the way to the Supreme Court but was overturned on a technicality. In the meantime, Disney lost interest in the project and the ski resort was never built. What do you think? Where would you draw the line, regarding what deserves moral standing? Are there ethical limits on what we can do to nature?

Many faiths promote conservation and justice

Ethical and moral values are often rooted in religious traditions, which try to guide us in what is right and wrong to do. With growing public awareness of environmental problems, religious organizations have begun to take stands on environmental concerns. They recognize that some of our most pressing environmental problems don't need technological or scientific solutions; they're not so much a question of what we're able to do, but what we're willing to do. Are we willing to take the steps

FIGURE 1.24 Mineral King Valley at the southern border of Sequoia National Park was the focus of an important environmental law case in 1969. The Disney Corporation wanted to build a ski resort here, but the Sierra Club sued to protect the valley on behalf of the trees, rocks, and native wildlife.
© photo by Laurel Di Silvestro courtesy of the Mineral King Valley Preservation Society

Table 1.2 Principles and Actions for Stewardship in the Ohito Declaration
Spiritual Principles
1. Religious beliefs and traditions call us to care for the earth.
2. For people of faith, maintaining and sustaining environmental life systems is a religious responsibility.
3. Environmental understanding is enhanced when people learn from the example of prophets and of nature itself.
4. People of faith should give more emphasis to a higher quality of life, in preference to a higher standard of living, recognizing that greed and avarice are root causes of environmental degradation and human debasement.
5. People of faith should be involved in the conservation and development process.
Recommended Courses of Action
The Ohito Declaration calls upon religious leaders and communities to
1. emphasize environmental issues within religious teaching: faith should be taught and practiced as if nature mattered.
2. commit themselves to sustainable practices and encourage community use of their land.
3. promote environmental education, especially among youth and children.
4. pursue peacemaking as an essential component of conservation action.
5. take up the challenge of instituting fair trading practices devoid of financial, economic, and political exploitation.

necessary to stop global climate change? Do our values and ethics require us to do so? In what ways might religious views influence our attitudes toward nature?

Environmental scientists have long been concerned about religious perspectives. In 1967, historian Lynn White Jr. published a widely influential paper, "The Historic Roots of Our Ecological Crisis." He argued that Christian societies have often exploited natural resources carelessly because the Bible says that God commanded Adam and Eve to dominate nature: "Be fruitful, and multiply, and replenish the earth and subdue it: and have dominion over the fish of the sea, and over the fowl of the air, and over every living thing that moveth upon the earth" (Genesis 1:28). Since then, many religious scholars have pointed out that God also commanded Adam and Eve to care for the garden they were given, "to till it and keep it" (Genesis 2:15). Furthermore, Noah was commanded to preserve individuals of all living species, so that they would not perish in the great Flood. Passages such as these inspire many Christians to insist that it is our responsibility to act as stewards of nature, and to care for God's creations.

The idea of **stewardship,** or taking care of the resources we are given, inspires many religious leaders to promote conservation. "Creation care" is a term that has become prominent among evangelical Christians in the United States. In 1995, representatives of nine major religions met in Ohito, Japan, to discuss their various traditions' views of environmental stewardship. The resulting document, the Ohito Declaration, outlined common beliefs and responsibilities of these different faiths toward protecting the earth and its life (table 1.2). In recent years, religious organizations have played important roles in nature protection. A coalition of evangelical Christians has been instrumental in promoting stewardship of many aspects of our environment, from rare plants and animals to our global climate.

Religious concern extends beyond our treatment of plants and animals. Pope Francis has repeatedly called for Christians to care for God's creation: "let us become ... channels through which God can water the earth, protect all creation and make justice and peace flourish." Pope John Paul II and Orthodox Patriarch Bartholomew called on countries bordering the Black Sea to stop pollution, saying that "to commit a crime against nature is a sin."

Many religious organizations are also working for integrated justice and environmental goals. Interfaith Power and Light is an organization promoting indigenous interests, clean energy, and greenhouse gas reduction. The Creation Care Network has also launched initiatives against energy inefficiency, mercury pollution, mountaintop removal mining, and destruction of endangered species, in addition to its campaign to combat global warming. For many people, religious beliefs provide the best justification for environmental protection.

Calls for both stewardship and human domination over nature can be found in the writings of most major faiths. The Koran teaches that "each being exists by virtue of the truth and is also owed its due according to nature," a view that extends moral rights and value to all other creatures. Hinduism and Buddhism teach *ahimsa,* or the practice of not harming other living creatures, because all living beings are divinely connected (fig. 1.25).

FIGURE 1.25 Many religions emphasize the divine relationships among humans and the natural world. The Tibetan Buddhist goddess Tara represents compassion for all beings.
© William P. Cunningham

Environmental justice integrates civil rights and environmental protection

People of color in the United States and around the world are subjected to a disproportionately high level of environmental health risks in their neighborhoods and on their jobs. Minorities, who tend to be poorer and more disadvantaged than other residents, work in the dirtiest jobs where they are exposed to toxic chemicals and other hazards. More often than not they also live in urban ghettos, barrios, reservations, and rural poverty pockets that have shockingly high pollution levels and are increasingly the site of unpopular industrial facilities, such as toxic waste dumps, landfills, smelters, refineries, and incinerators. **Environmental justice** combines civil rights with environmental protection to demand a safe, healthy, life-giving environment for everyone.

Among the evidence of environmental injustice is the fact that three out of five African Americans and Hispanics, and nearly half of all Native Americans, Asians, and Pacific Islanders, live in communities with one or more uncontrolled toxic waste sites, incinerators, or major landfills, while fewer than 10 percent of all whites live in these areas. Using zip codes or census tracts as a unit of measurement, researchers found that minorities make up twice as large a population share in communities with these locally unwanted land uses (**LULUs**) as in communities without them. A recent study

using "distance-based" methods found an even greater correlation between race and location of hazardous waste facilities.

Although it is difficult to distinguish between race, class, historical locations of ethnic groups, economic disparities, and other social factors in these disputes, racial origins often seem to play a role in exposure to environmental hazards. Simple correlation doesn't prove causation; still, while poor people in general are more likely to live in polluted neighborhoods than rich people, the discrepancy between the pollution exposure of middle-class blacks and middle-class whites is even greater than the difference between poorer whites and blacks. Where upper-class whites can "vote with their feet" and move out of polluted and dangerous neighborhoods, blacks and other minorities tend to be restricted by color barriers and prejudice (overt or covert) to the less desirable locations (fig. 1.26).

Racial prejudice is a belief that people are inferior merely because of their race. **Environmental racism** is inequitable distribution of environmental hazards based on race. Evidence of environmental racism can be seen in lead poisoning in children. The Federal Agency for Toxic Substances and Disease Registry considers lead poisoning to be the number one environmental health problem for children in the United States. Some 4 million children—many of whom are African American, Latino, Native American, or Asian, and most of whom live in inner-city areas—have dangerously high lead levels in their bodies. This lead is absorbed from old lead-based house paint, contaminated drinking water from lead pipes or lead solder, and soil polluted by industrial effluents and automobile exhaust. The evidence of racism is that at every income level, whether rich or poor, black children are two to three times more likely than whites to suffer from lead poisoning.

Because of their quasi-independent status, most Native American reservations are considered sovereign nations that are not covered

FIGURE 1.26 Poor people and people of color often live in the most dangerous and least desirable places. Here children play next to a chemical refinery in Texas City, Texas.
© Sam Kittner

by state environmental regulations. Court decisions holding that reservations are specifically exempt from hazardous waste storage and disposal regulations have resulted in a land rush of seductive offers from waste disposal companies to Native American reservations for onsite waste dumps, incinerators, and landfills. The short-term economic incentives can be overwhelming for communities with chronic poverty. Nearly every tribe in America has been approached with proposals for some dangerous industry or waste facility.

The practice of targeting poor communities of color in the developing nations for waste disposal or experimentation with risky technologies has been described as **toxic colonialism.** Internationally, the trade in toxic waste has mushroomed in recent years as wealthy countries have become aware of the risks of industrial refuse (fig. 1.27). Poor minority communities at home and abroad are being increasingly targeted as places to dump unwanted wastes. Although a treaty regulating international shipping of toxics was signed by 105 nations in 1989, millions of tons of toxic and hazardous materials continue to move—legally or illegally—from the richer countries to the poorer ones every year. This issue is discussed further in chapter 23.

Section Review

1. Explain the idea of moral extensionism, and give an example.
2. How does *inherent value* differ from *instrumental value*?
3. Why is stewardship important in many faiths?
4. What is environmental justice?

FIGURE 1.27 Much of our waste is exported to developing countries where environmental controls are limited. Here workers in a Chinese village sort electronic waste materials.
© Basel Action Network 2006

Conclusion

We face serious environmental challenges, but there are also many opportunities for improving lives without damaging our shared environment. Many of these challenges are especially visible in the rapidly expanding urban areas in the developing world. Measures of sustainable development show that we continue to face air and water pollution, chronic hunger, water shortages, and other problems. On the other hand, the UN Millennium Development Goals produced substantial progress, and the Sustainable Development Goals aim to build on that progress. We have seen important innovations in transportation, energy sources, food production, and international cooperation for environmental protection. Environmental science is a discipline that uses an understanding of natural systems to seek solutions to environmental problems and to help find solutions—which can draw on knowledge from technological, biological, economic, political, social, and many other fields of study.

Ideas about environmental quality and environmental protection have deep historic roots, and these ideas include several main themes. Utilitarian conservation has been a common incentive; aesthetic preservation also motivates many people to work for conservation. Social progress and a concern for making sure that all people have access to a healthy environment have also been important motivating factors in environmental science and in environmental conservation. Inequitable distribution of resources has been a persistent concern, and programs like the Sustainable Development Goals are intended to address these inequities. Growing consumption of energy, water, land, and other resources makes many questions in environmental science more urgent.

Sustainable development is the idea that tries to bring together the needs for human well-being and resource conservation. The aim is to improve people's lives without reducing resources and opportunities for future generations. A number of core ideas are helpful in seeking strategies for sustainable development, such as ecosystem services, the tragedy of the commons, and strategies for managing common property.

Ethics and faith-based perspectives also provide guidance for resource conservation, because ethical frameworks and religions often promote ideas of fairness and stewardship, which has been a guiding principle for many faith-based groups. Environmental justice is also often rooted in ethical and religious principles, which have often led religious groups to lead the struggle for environmental justice for minority and low-income communities.

Reviewing Key Terms

Can you define the following terms in environmental science?

biocentric preservation 1.2

ecological footprint 1.1

ecosystem services 1.4

environment 1.1

environmentalism 1.2

environmental justice 1.5

environmental racism 1.5

environmental science 1.1

ethics 1.5

global environmentalism 1.2

inherent value 1.5

instrumental value 1.5

LULUs 1.5

managing the commons 1.4

Millennium Development Goals 1.4

moral extensionism 1.5

moral value 1.5

quality-of-life indicators 1.3

stewardship 1.5

sustainable development 1.2

Sustainable Development Goals 1.3

throughput 1.4

Tragedy of the Commons 1.4

toxic colonialism 1.5

utilitarian conservation 1.2

Critical Thinking and Discussion Questions

1. Overall, do environmental and social conditions in Kibera give you hope or fear about the future?

2. What are the underlying assumptions and values of utilitarian conservation and altruistic preservation? Which do you favor?

3. Suppose a beautiful grove of trees near your house is scheduled to be cut down for a civic project such as a swimming pool. Would you support this? Why or why not?

4. What resource uses are most strongly represented in the ecological footprint? What are the advantages and disadvantages of using this assessment?

5. Are there enough resources in the world for 8 or 10 billion people to live decent, secure, happy lives? What do these terms mean to you? Try to imagine what they mean to residents of other countries.

6. Identify several types of ecosystem services. How might their value be accounted for?

Data Analysis

Working with graphs

Graphs are one of the most common and important ways scientists communicate their results. Learning to understand the language of graphs will help you understand ideas in this book. Graphs help us identify trends and understand relationships. We could present a table of numbers, but most of us have difficulty seeing a pattern in

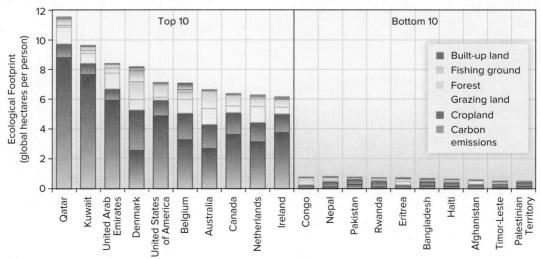

FIGURE 1 Ecological footprint for the 10 highest and 10 lowest countries, in terms of impact per person
Source: WWF Living Planet Report 2012

a large field of numbers. In a graph, we can quickly and easily see trends and relationships.

Below is a graph showing ecological footprints. Often we pass quickly over graphs like these that appear in text, but it's rewarding to investigate them more closely, because their relationships can raise interesting questions. Go to Connect to answer questions about this graph and to demonstrate your understanding of graph reading.

For a full discussion of ecological footprints and the many components that contribute to its calculation, see the WWF Living Planet Report, http://wwf.panda.org/about_our_earth/all_publications/living_planet_report/. To calculate your own footprint, visit the Global Footprint Network http://footprintnetwork.org/.

TO ACCESS ADDITIONAL RESOURCES FOR THIS CHAPTER, PLEASE VISIT CONNECT AT www.connect.mheducation.com.

You will find LearnSmart, an adaptive learning system, Google Earth™ exercises, additional Case Studies, Data Analysis exercises, and an interactive ebook.

2

Principles of Science and Systems

▲ Student interns measure plant growth in experimental plots in the B4Warmed study in northern Minnesota.
© Artur Stefanski

Learning Outcomes

After studying this chapter, you should be able to:

2.1 Describe the scientific method and explain how it works.
2.2 Explain systems and how they're useful in science.
2.3 Evaluate the role of scientific consensus and conflict.

"The ultimate test of a moral society is the kind of world that it leaves to its children."

– Dietrich Bonhoeffer

Forest Responses to Global Warming

How will forests respond to climate change? This is one of the great unknowns in environmental science today. Will northern regions that now support boreal forests shift to another biome—hardwood forest, open savannah, grassland, or something entirely different? With rising emissions of CO_2 and other greenhouse gases, climate models predict that boreal forests will move north by about 480 km (300 mi) within this century. But there's a great deal of uncertainty in this prediction.

How do environmental scientists approach and analyze such complex questions? One strategy is to grow plants in a greenhouse, and test plant responses to different temperature and moisture levels. By changing just one variable at a time, we can get an approximation of responses to environmental change. But this approach misses the complex species interactions that influence plant growth in a real ecosystem, so an alternative approach is to use field tests in which mixtures of plants are grown in natural settings that include competition for resources, predator/prey interactions, natural climatic variations, and other ecological factors.

Professor Peter Reich and his colleagues and student research assistants are now carrying out such a field study in a patch of boreal (northern) forest in Minnesota. Calling this experiment B4Warmed, which stands for Boreal Forest Warming at an Ecotone in Danger, they are artificially raising ambient temperatures in a series of boreal forest plots, to emulate warming climate conditions.

The group established 96 circular experimental plots, each 3 meters (9.8 ft) in diameter (fig. 2.1). Each plot was planted with a mixture of tree species and annual understory plants. The plots were then randomly assigned to one of four treatments. Half the plots are in mature forest, and half are in forest openings. Half are kept 2°C above ambient temperatures, and half are kept 4°C higher than ambient temperatures, using infrared lamps placed around the plots, as well as buried heat cables (fig. 2.2). Control plots (with no temperature manipulations) are also maintained for comparison with treatments.

Seed germination, growth, survival, and biomass accumulation are recorded for ten important boreal and temperate tree species in this experiment. Preliminary results suggest, as might be expected, that boreal species, such as balsam fir and white spruce, which are now at the southern edge of their range in the study area, tend to show negative responses to warming. Temperate broad-leaved species, such as oak and maples, by contrast, tend to have neutral or positive growth and survival responses.

It's too early to know exactly what the long-term effects of climate change will be on the northern forest community. It's predicted that rainfall patterns are likely to change along with ambient temperatures. Rain exclosures were installed on 18 plots in 2012 to evaluate the effects of reduced rain.

One preliminary result from this study that appears to offer good news is that the CO_2 emissions both from forest plants and from the soil are lower than expected at higher temperatures. Apparently both standing vegetation and soil microbes alter their metabolic rates to acclimate to ambient environmental conditions. Thus, the feedback cycles predicted to exacerbate global warming effects may not be as bad as we feared.

This kind of careful, rational, systematic research is the hallmark of modern science. It has given us powerful insights into how our world works. In this chapter, we'll look at how scientists form and answer other questions about our environment.

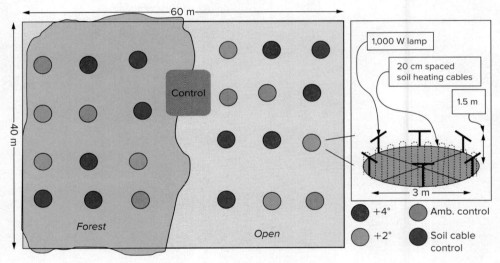

FIGURE 2.1 Experimental design for B4Warmed study.

FIGURE 2.2 A student researcher adjusts the electrical panel that controls heat lamps and heating cables.
© Artur Stefanski

2.1 WHAT IS SCIENCE?

- *Science depends on careful, objective, logical analysis.*
- *Reproducible results are essential in science.*
- *Science depends on orderly testing of hypotheses.*

Science is a process for producing knowledge methodically and logically. Derived from *scire,* "to know" in Latin, science depends on making precise observations of natural phenomena. We develop or test theories (proposed explanations of how a process works) using these observations. "Science" also refers to the cumulative body of knowledge produced by many scientists. Science is valuable because it helps us understand the world and meet practical needs, such as new medicines, new energy sources, or new foods.

Science rests on the assumption that the world is knowable and that we can learn about the world by careful observation (table 2.1). For early philosophers of science, this assumption was a radical departure from religious and philosophical approaches. In the Middle Ages, the ultimate sources of knowledge about how crops grow, how diseases spread, or how the stars move, were religious authorities or cultural traditions. These sources provided many useful insights, but there was no way to test their explanations independently and objectively. The benefit of scientific thinking is that it searches for testable evidence. By testing our ideas with observable evidence, we can evaluate whether our explanations are reasonable or not.

Table 2.1 Basic Principles of Science

1. *Empiricism:* We can learn about the world by careful observation of empirical (real, observable) phenomena; we can expect to understand fundamental processes and natural laws by observation.

2. *Uniformitarianism:* Basic patterns and processes are uniform across time and space; the forces at work today are the same as those that shaped the world in the past, and they will continue to do so in the future.

3. *Parsimony:* When two plausible explanations are equally reasonable, the simpler (more parsimonious) one is preferable. This rule is also known as Ockham's razor, after the English philosopher who proposed it.

4. *Uncertainty:* Knowledge changes as new evidence appears, and explanations (theories) change with new evidence. Theories based on current evidence should be tested on additional evidence, with the understanding that new data may disprove the best theories.

5. *Repeatability:* Tests and experiments should be repeatable; if the same results cannot be reproduced, then the conclusions are probably incorrect.

6. *Proof is elusive:* We rarely expect science to provide absolute proof that a theory is correct, because new evidence may always undermine our current understanding.

7. *Testable questions:* To find out whether a theory is correct, it must be tested; we formulate testable statements (hypotheses) to test theories.

Science depends on skepticism and accuracy

Ideally, scientists are skeptical. They are cautious about accepting proposed explanations until there is substantial evidence to support them. Even then, as we saw in the case study about global warming that opened this chapter, explanations are considered only provisionally true, because there is always a possibility that some additional evidence may appear to disprove them. Scientists also aim to be methodical and unbiased. Because bias and methodical errors are hard to avoid, scientific tests are subject to review by informed peers, who can evaluate results and conclusions (fig. 2.3). The peer review process is an essential part of ensuring that scientists maintain good standards in study design, data collection, and interpretation of results.

Scientists demand **reproducibility** because they are cautious about accepting conclusions. Making an observation or obtaining a result just once doesn't count for much. You have to produce the same result consistently to be sure that your first outcome wasn't a fluke. Even more important, you must be able to describe the conditions of your study so that someone else can reproduce your findings. Repeating studies or tests is known as **replication.**

Science also relies on accuracy and precision. Accuracy is correctness of measurements. Inaccurate data can produce sloppy and misleading conclusions (fig. 2.4). Precision means repeatability of results and level of detail. The classic analogy for repeatability is throwing darts at a dart board. You might throw ten darts and miss the center every time, but if all the darts hit nearly the same spot, they were very precise. Another way to think of precision is levels of detail. Suppose you want to measure how much snow fell last

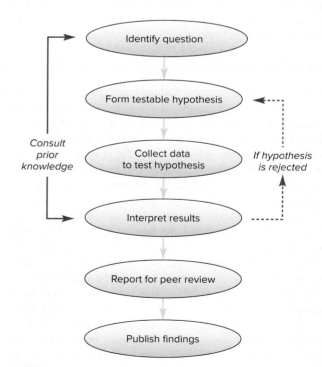

FIGURE 2.3 Ideally, scientific investigation follows a series of logical, orderly steps to formulate and test hypotheses.

FIGURE 2.4 Making careful, accurate measurements and keeping good records are essential in scientific research.
© David L. Hansen, University of Minnesota Agricultural Experiment Station

night, so you take out your ruler, which is marked in centimeters, and you find that the snow is just over 6 cm deep. You cannot tell if it is 6.3 cm or 6.4 cm because the ruler doesn't report that level of detail. If you average several measurements, you might find an average depth of 6.4333 cm. If you report all four decimal places, it will imply that you know more than you really do about the snow depth. If you had a ruler marked in millimeters (one-tenth of a centimeter), you could find a depth of 6.4 cm. Here, the one decimal place would be a **significant figure,** or a level of detail you actually knew. Reporting 6.4333 cm would be inappropriate because the last three digits are not meaningful.

Deductive and inductive reasoning are both useful

Ideally, scientists deduce conclusions from general laws that they know to be true. For example, if we know that massive objects attract each other (because of gravity), then it follows that an apple will fall to the ground when it releases from the tree. This logical reasoning from general to specific is known as **deductive reasoning.** Often, however, we do not know general laws that guide natural systems. We observe, for example, that birds appear and disappear as a year goes by. Through many repeated observations in different places, we can infer that the birds move from place to place. We can develop a general rule that birds migrate seasonally. Reasoning from many observations to produce a general rule is **inductive reasoning.** Although deductive reasoning is more logically sound than inductive reasoning, it only works when our general laws are correct. We often rely on inductive reasoning to understand the world, because we have few immutable laws.

Sometimes it is insight, as much as reasoning, that leads us to an answer. Many people fail to recognize the role that insight, creativity, aesthetics, and luck play in research. Some of our most important discoveries were made not because of superior scientific method and objective detachment, but because the investigators were passionately interested in their topics and pursued hunches that appeared unreasonable to fellow scientists. A good example is Barbara McClintock, the geneticist who discovered that genes in corn can move and recombine spontaneously. Where other corn geneticists saw random patterns of color and kernel size, McClintock's years of experience in corn breeding, and an uncanny ability to recognize patterns, led her to guess that genes could recombine in ways that no one had yet imagined. Her intuitive understanding led to a theory that took other investigators years to accept.

Testable hypotheses and theories are essential tools

Science also depends on orderly testing of hypotheses, a process known as the scientific method. You may already be using the scientific method without being aware of it. Suppose you have a flashlight that doesn't work. The flashlight has several components (switch, bulb, batteries) that could be faulty. If you change all the components at once, your flashlight might work, but a more methodical series of tests will tell you more about what was wrong with the system—knowledge that may be useful the next time you have a faulty flashlight. So you decide to follow the standard scientific steps:

1. *Observe* that your flashlight doesn't light; also, that there are three main components of the lighting system (batteries, bulb, and switch).

2. Propose a **hypothesis,** a testable explanation: "The flashlight doesn't work because the batteries are dead."

3. Develop a *test* of the hypothesis and *predict* the result that would indicate your hypothesis was correct: "I will replace the batteries; the light should then turn on."

4. Gather *data* from your test: After you replaced the batteries, did the light turn on?

5. *Interpret* your results: If the light works now, then your hypothesis was right; if not, then you should formulate a new hypothesis, perhaps that the bulb is faulty, and develop a new test for that hypothesis.

In systems more complex than a flashlight, it is almost always easier to prove a hypothesis wrong than to prove it unquestionably true. This is because we usually test our hypotheses with observations, but there is no way to make every possible observation. The philosopher Ludwig Wittgenstein illustrated this problem as follows: Suppose you saw hundreds of swans, and all were white. These observations might lead you to hypothesize that all swans were white. You could test your hypothesis by viewing thousands of swans, and each observation might support your hypothesis, but you could never be entirely sure that it was correct. On the other hand, if you saw just one black swan, you would know with certainty that your hypothesis was wrong.

As you'll read in later chapters, the elusiveness of absolute proof is a persistent problem in environmental policy and law. You can never absolutely prove that the toxic waste dump up the street is making you sick. The elusiveness of proof often decides environmental liability lawsuits.

When an explanation has been supported by a large number of tests, and when a majority of experts have reached a general consensus that it is a reliable description or explanation, we call it a **scientific theory.** Note that scientists' use of this term is very different from the way the public uses it. To many people, a theory is speculative and unsupported by facts. To a scientist, it means just the opposite: While all explanations are tentative and open to revision and correction, an explanation that counts as a scientific theory is supported by an overwhelming body of data and experience, and it is generally accepted by the scientific community, at least for the present (fig. 2.5).

Understanding probability helps reduce uncertainty

One strategy to improve confidence in the face of uncertainty is to focus on probability. Probability is a measure of how likely something is to occur. Usually, probability estimates are based on a set of previous observations or on standard statistical measures. Probability does not tell you what *will* happen, but it tells you what *is likely* to happen. If you hear on the news that you have a 20 percent chance of catching a cold this winter, that means that 20 of every 100 people are likely to catch a cold. This doesn't mean that you will catch one. In fact, it's more likely that you won't catch a cold than that you will. If you hear that 80 out of every 100 people will catch a cold, you still don't know whether you'll get sick, but there's a much higher chance that you will.

Science often involves probability, so it is important to be familiar with the idea. Sometimes probability has to do with random chance: If you flip a coin, you have a random chance of getting heads or tails. Every time you flip, you have the same 50 percent probability of getting heads. The chance of getting ten heads in a

row is small (in fact, the chance is 1 in 210, or 1 in 1,024), but on any individual flip you have exactly the same 50 percent chance, because this is a random test.

Sometimes probability is weighted by circumstances: Suppose that about 10 percent of the students in this class earn an A each semester. Your likelihood of being in that 10 percent depends a great deal on how much time you spend studying, how many questions you ask in class, and other factors. Sometimes there is a combination of chance and circumstances: The probability that you will catch a cold this winter depends partly on whether you encounter someone who is sick (largely random chance) and whether you take steps to stay healthy (get enough rest, wash your hands frequently, eat a healthy diet, and so on).

Scientists often increase their confidence in a study by comparing results to a random sample or a larger group. Suppose that 40 percent of the students in your class caught a cold last winter. This *seems* like a lot of colds, but is it? One way to decide is to compare to the cold rate in a larger group. You call your state epidemiologist, who took a random sample of the state population last year: She collected 200 names from the telephone book and called each to find out if each got a cold last year. A larger sample, say 2,000 people, would have been more likely to represent the actual statewide cold rate. But a sample of 200 is much better than a sample of 50 or 100. The epidemiologist tells you that in your state as a whole, only 20 percent of people caught a cold.

Now you know that the rate in your class (40 percent) was quite high, and you can investigate possible causes for the difference. Perhaps people in your class got sick because they were short on sleep, because they tended to stay up late studying. You could test whether studying late was a contributing factor by comparing the frequency of colds in two groups: those who study long and late, and those who don't. Suppose it turns out that among the 40 late-night studiers, 30 got colds (a rate of 75 percent). Among the 60 casual studiers, only 10 got colds (17 percent). This difference would give you a good deal of confidence that staying up late contributes to getting sick. (Note, however, that all 40 of the studying group got good grades!)

Statistics can indicate the probability that your results were random

Statistics can help in experimental design as well as in interpreting data (see Exploring Science). Many statistical tests focus on calculating the probability that observed results could have occurred by chance. Often, the degree of confidence we can assign to results depends on sample size as well as the amount of variability between groups.

Ecological tests are often considered significant if there is less than 5 percent probability that the results were achieved by random chance. A probability of less than 1 percent gives still greater confidence in the results.

As you read this book, you will encounter many statistics, including many measures of probability. When you see these numbers, stop and think: Is the probability high enough to worry

FIGURE 2.5 Data collection and repeatable tests support scientific theories. Here students use telemetry to monitor radio-tagged fish.
© *David L. Hansen, University of Minnesota Agricultural Experiment Station*

Why Do Scientists Answer Questions with a Number?

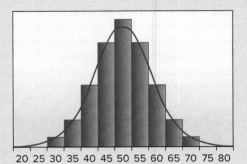

Statistics are numbers that let you evaluate and compare things. "Statistics" is also a field of study that has developed meaningful methods of comparing those numbers. By both definitions, statistics are widely used in environmental sciences, partly because they can give us a useful way to assess patterns in a large population, and partly because the numbers can give us a measure of confidence in our research or observations. Understanding the details of statistical tests can take years of study, but a few basic ideas will give you a good start toward interpreting statistics.

1. **Descriptive statistics help you assess the general state of a group.** In many towns and cities, the air contains dust, or particulate matter, as well as other pollutants. From personal experience you might know your air isn't as clean as you'd like, but you may not know how clean or dirty it is. You could start by collecting daily particulate measurements to find average levels. An averaged value is more useful than a single day's values, because daily values may vary a great deal, but general, long-term conditions affect your general health. Collect a sample every day for a year; then divide the sum by the number of days, to get a **mean** (average) dust level. Suppose you found a mean particulate level of 30 micrograms per cubic meter ($\mu g/m^3$) of air. Is this level high or low? In 1997, the EPA set a standard of 50 $\mu g/m^3$ as a limit for allowable levels of coarse particulates (2.5–10 micrometers in diameter). Higher levels tend to be associated with elevated rates of asthma and other respiratory diseases. Now you know that your town, with an annual average of 30 $\mu g/m^3$, has relatively safe air, after all.

2. **Statistical samples.** Although your town is clean by EPA standards, how does it compare with the rest of the cities in the country? Testing the air in *every* city is probably not possible. You could compare your town's air quality with a **sample,** or subset of cities, however. A large, random sample of cities should represent the general "population" of cities reasonably well. Taking a large sample reduces the effects of outliers (unusually high or low values) that might be included. A random sample minimizes the chance that you're getting only the worst sites, or only a

collection of sites that are close together, which might all have similar conditions. Suppose you get average annual particulate levels from a sample of 50 randomly selected cities. You can draw a frequency distribution, or histogram, to display your results (fig. 1). The mean value of this group is 36.8 $\mu g/m^3$, so by comparison your town (at 30 $\mu g/m^3$) is relatively clean.

Many statistical tests assume that the sample has a normal, or Gaussian, frequency distribution, often described as a bell-shaped curve (fig. 2). In this distribution, the mean is near the center of the range of values, and most values are fairly close to the mean. Large and random samples are more likely to fit this shape than are small and nonrandom samples.

3. **Confidence.** How do you know that the 50 cities you sampled really represent all the cities in the country? You can't ever be completely certain, but you can use estimates, such as confidence limits, to express the reliability of your mean statistic. Depending on the size of your sample (not 10, not 500, but 50) and the amount of variability in the sample data, you can calculate a confidence interval that the mean represents the whole population (all cities). Confidence limits, or confidence intervals, represent the likelihood that your statistics correctly represent the entire population. For the mean of your sample, a confidence interval tells you the probability that your sample is similar to other random samples of the population. A common convention is to compare values with a 95

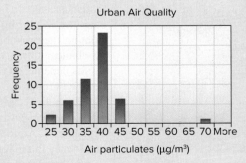

FIGURE 1 Average annual airborne dust levels for 50 cities.

percent confidence level, or a probability of 5 percent or less that your conclusions are misleading. Using statistical software, we can calculate that, for our 50 cities, the mean is 36.8 $\mu g/m^3$, and the confidence interval is 35.0 to 38.6. This suggests that, if you take 1,000 samples from the entire population of cities, 95 percent of those samples ought to be within 2 $\mu g/m^3$ of your mean. This indicates that your mean is reliable and representative.

4. **Is your group unusual?** Once you have described your group of cities, you can compare it with other groups. For example, you might believe that Canadian cities have cleaner air than U.S. cities. You can compare mean air quality levels for the two groups. Then you can calculate confidence intervals for the difference between the means, to see if the difference is meaningful.

5. **Evaluating relationships between variables.** Are respiratory diseases correlated with air pollution? For each city in your sample, you could graph pollution and asthma rates (fig. 3). If the graph looks like a loose cloud of dots, there is no clear relationship. A tight, linear pattern of dots trending upward to the right indicates a strong and positive relationship. You can also use a statistical package to calculate an equation to describe the relationship and, again, confidence intervals for the equation. This is known as a regression equation.

6. **Lies, damned lies, and statistics.** Can you trust a number to represent a complex or large phenomenon? One of the devilish details of representing the world with numbers is that those numbers can be tabulated in many

FIGURE 2 is shown above right.

FIGURE 2 A normal distribution.

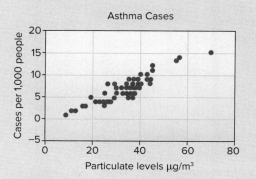

Asthma Cases

(Graph: y-axis "Cases per 1,000 people" ranging from -5 to 20; x-axis "Particulate levels µg/m³" ranging from 0 to 80)

FIGURE 3 A dot plot shows relationships between variables.

ways. If we want to assess the greatest change in air quality statistics, do we report rates of change or the total amount of change? Do we look at change over five years? Twenty-five years? Do we accept numbers selected by the EPA, by the cities themselves, by industries, or by environmental groups? Do we trust that all the data were collected with a level of accuracy and precision that we would accept if we knew the hidden details in the data-gathering process? Like all information, statistics need to be interpreted in terms of who produced them, when, and why. Awareness of some of the standard assumptions behind statistics, such as sampling, confidence, and probability, will help you interpret statistics that you see and hear.

Test your comprehension

1. What is a mean? How would you use one?
2. What is a Gaussian or normal distribution? What shape does it create in a graph?
3. What do statisticians mean by confidence limits?

---

about? How high is it compared to other risks or chances you've read about? What are the conditions that make probability higher or lower? Science involves many other aspects of statistics.

Experimental design can reduce bias

The study of colds and sleep deprivation is an example of an observational experiment, one in which you observe natural events and interpret a causal relationship between the variables. This kind of study is also called a **natural experiment;** that is, a study of events that have already happened. Many scientists depend on natural experiments: A geologist, for instance, might want to study mountain building, or an ecologist might want to learn about how species coevolve, but neither scientist can spend millions of years watching the process happen. Similarly, a toxicologist cannot give people a disease just to see how lethal it is.

Other scientists can use **manipulative experiments,** such as the B4Warmed experiment in the opening case study for this chapter, in which some conditions are deliberately altered and all other variables are held constant (fig. 2.6). In one famous manipulative study, ecologists Edward O. Wilson and Robert MacArthur were interested in how quickly species colonize small islands, depending on distance to the mainland. They fumigated several tiny islands in the Florida Keys, killing all resident insects, spiders, and other invertebrates. They then monitored the islands to learn how quickly ants and spiders recolonized them from the mainland or other islands.

Most manipulative experiments are done in the laboratory, where conditions can be carefully controlled. Suppose you are interested in studying whether lawn chemicals contribute to deformities in tadpoles. You might keep two groups of tadpoles in fish tanks and expose one to chemicals. In the lab, you could ensure that both tanks had identical temperatures, light, food, and oxygen.

FIGURE 2.6 A researcher gathers data from the B4Warmed field experiment in the boreal forest.
© Artur Stefanski

By comparing a treatment (exposed) group and a control (unexposed) group, you have also made this a **controlled study.**

Often there is a risk of experimenter bias. Suppose the researcher sees a tadpole with a small nub that looks like it might become an extra leg. Whether she calls this nub a deformity might depend on whether she knows that the tadpole is in the treatment group or the control group. To avoid this bias, **blind experiments** are often used, in which the researcher doesn't know which group is treated until after the data have been analyzed. In health studies, such as tests of new drugs, **double-blind experiments** are used, in which neither the subject (who receives a drug or a placebo) nor the researcher knows who is in the treatment group and who is in the control group.

In each of these studies there is one **dependent variable** and one, or perhaps more, **independent variables.** The dependent variable, also known as a response variable, is affected by the independent variables. Independent variables are rarely really independent (they are affected by the same environmental conditions as the dependent variable, for example). Many people prefer to call them explanatory variables, because we hope they will explain differences in the dependent variable.

You will encounter many graphs in this book. Working with graphs takes practice, but it is an essential skill that will serve you well in this course and many others. The Data Analysis exercise at the end of this chapter reviews some of the most common types of graphs, and explains how they're used.

Models are an important experimental strategy

Another way to gather information about environmental systems is to use **models.** A model is a simple representation of something. Perhaps you have built a model airplane. The model doesn't have all the elements of a real airplane, but it has the most important ones for your needs. A simple wood or plastic airplane has the proper shape, enough to allow a child to imagine it is flying (fig. 2.7). A more complicated model airplane might have a small gas engine, just enough to let a teenager fly it around for short distances.

Similarly, scientific models vary greatly in complexity, depending on their purposes. Some models are physical models: Engineers test new cars and airplanes in wind tunnels to see how they perform, and biologists often test theories about evolution and genetics using "model organisms" such as fruit flies or rats as a surrogate for humans.

Most models are numeric, though. A model could be a mathematical equation, such as a simple population growth model ($N_t = rN_{(t-1)}$). Here the essential components are number (N) of individuals at time t (N_t), and the model proposes that N_t is equal to the growth rate (r) times the number in the previous time period ($N_{(t-1)}$). This model is a very simplistic representation of population change, but it is useful because it precisely describes a relationship between population size and growth rate. Also, by

FIGURE 2.7 A model uses just the essential elements to represent a complex system.
© F. Schussler/PhotoLink/Getty Images RF

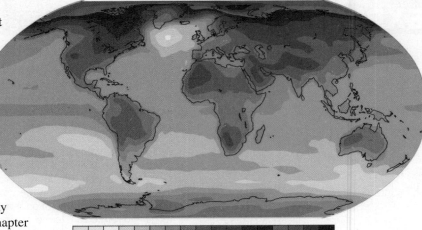

Geographical pattern of surface warming

0 0.5 1 1.5 2 2.5 3 3.5 4 4.5 5 5.5 6 6.5 7 7.5
Temperature change (°C)

FIGURE 2.8 Numerical models, calculated from observed data, can project future scenarios. Here, temperature changes in 2090–2099 are modeled, relative to 1980–1999 temperatures.
Figure SPM.6 from Climate Change 2007: Synthesis Report. Contribution of Working Groups I, II and III to the Fourth Assessment Report of the Intergovernmental Panel on Climate Change [Core Writing Team, Pachauri, R.K. and Reisinger, A. (eds.)]. IPCC, Geneva, Switzerland. Used with permission.

converting the symbols to numbers, we can predict populations over time. For example, if last year's rabbit population was 100, and the growth rate is 1.6 per year, then this year's population will be 160. Next year's population will be 160 × 1.6, or 256. This is a simple model, then, but it can be useful. A more complicated model might account for deaths, immigration, emigration, and other factors.

More complicated mathematical models can be used to describe and calculate more complex processes, such as climate change (fig. 2.8) or economic growth. These models are also useful because they allow the researcher to manipulate variables without actually destroying anything. An economist can experiment with different interest rates to see how they affect economic growth. A climatologist can raise the variables for CO_2 levels and see how quickly the variables for temperatures respond. These models are often called simulation models, because they simulate a complex system. Of course, the results depend on the assumptions built into the models. One model might show temperature rising quickly in response to CO_2; another might show temperature rising more slowly, depending on how evaporation, cloud cover, and other variables are taken into account. Consequently, simulations can produce powerful but controversial results. If multiple models generally agree, though, as in the cases of climate models that agree on generally upward temperature trends, we can have confidence that the overall predictions are reliable. These models are also very useful in laying out and testing our ideas about how a system works.

1. What is science? What are some of its basic principles?

2. Why are widely accepted, well-defended scientific explanations called "theories"?

3. Draw a diagram showing the steps of the scientific method, and explain why each is important.

2.2 SYSTEMS INVOLVE INTERACTIONS

- *A system is a network of interdependent components and processes that together have properties beyond those of individual parts.*

- *Feedbacks are self-regulating mechanisms in which the results of a process affect the process itself.*

- *Homeostasis (the ability to maintain stability) and resilience (the ability to recover from disturbance) are important characteristics of systems.*

The forest ecosystem you examined in the opening case study of this chapter is interesting because it is composed of many interdependent parts. By studying those parts, we can understand how similar ecosystems might function, and why. The idea of **systems,** including ecosystems, is central in environmental science. A system is a network of interdependent components and processes, with materials and energy flowing from one component of the system to another. For example, "ecosystem" is probably a familiar term for you. This simple word stands for a complex assemblage of animals, plants, and their environment, through which materials and energy move.

The idea of systems is useful because it helps us organize our thoughts about the inconceivably complex phenomena around us. For example, an ecosystem might consist of countless animals and plants and their physical surroundings. You yourself are a system consisting of millions of cells, complex organs, and innumerable bits of energy and matter that move through your body. Keeping track of all the elements and their relationships in an ecosystem would probably be an impossible task. But if we step back and think about them in terms of broad functional categories, then we can start to comprehend how it works (fig. 2.9).

We can use some general terms to describe the components of a system. A simple system consists of state variables (also called compartments), which store resources such as energy, matter, or water; and flows, or the pathways by which those resources move from one state variable to another. In figure 2.9, the plant and animals represent state variables. The plant represents many different plant types, all of which are things that store solar energy and create carbohydrates from carbon, water, and sunlight. The rabbit represents many kinds of herbivores, all of which consume plants, then store energy, water, and carbohydrates until they are used, transformed, or consumed by a carnivore. We can describe the flows in terms of herbivory, predation, or photosynthesis, all processes that transfer energy and matter from one state variable to another.

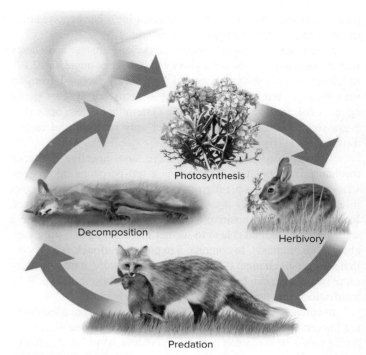

FIGURE 2.9 A system can be described in very simple terms.

It may seem cold and analytical to describe a rabbit or a flower as a state variable, but it is also helpful to do so. When we start discussing natural complexity in the simple terms of systems, we can identify common characteristics. Understanding these characteristics can help us diagnose disturbances or changes in the system: for example, if rabbits become too numerous, herbivory can become too rapid for plants to sustain. Overgrazing can lead to widespread collapse of this system. Let's examine some of the common characteristics we can find in systems.

Systems can be described in terms of their characteristics

Open systems are those that receive inputs from their surroundings and produce outputs that leave the system. Almost all natural systems are open systems. In principle, a **closed system** exchanges no energy or matter with its surroundings, but these are rare. Often we think of pseudo-closed systems, those that exchange only a little energy but no matter with their surroundings. **Throughput** is a term we can use to describe the energy and matter that flow into, through, and out of a system. Larger throughput might expand the size of state variables. For example, you can consider your household economy in terms of throughput. If you have a large income and high expenses, you may have a stable situation (balanced budget). A more frugal person might have an equally stable budget with far less cash flow (or throughput). In a grassland, inputs of energy (sunlight) and matter (carbon dioxide and water) are stored in biomass. If there is lots of water, the biomass storage might increase (in the form of trees). If there's little input, biomass might decrease (grass could become short or sparse). Eventually, stored matter and energy may be exported (by fire, grazing, land clearing). The exported matter and energy can be thought of as throughput.

A grassland is an *open system:* the exchange matter and energy with its surroundings (the atmosphere and soil, for example; fig. 2.10). In theory, a closed system would be entirely isolated from its surroundings, but in fact all natural systems are at least partly open. A fish tank is an example of a system that is less open than a grassland, because it can exist with only sunlight and carbon dioxide inputs (fig. 2.11).

Systems also experience positive and negative feedback mechanisms. A **positive feedback** is a self-perpetuating process. In a grassland, a grass plant grows new leaves, and the more leaves it has, the more energy it can capture for producing more leaves. In other words, in a positive feedback mechanism, increases in a state variable (biomass) lead to further increases in that state variable (more biomass). In contrast, a **negative feedback** is a process that suppresses change. If grass grows very rapidly, it may produce more leaves than can be supported by available soil moisture. With insufficient moisture, the plant begins to die back.

In climate systems (chapter 15), positive and negative feedbacks are important ideas. For example, as warm summers melt ice in the Arctic, newly exposed water surfaces absorb heat, which leads to further melting, which leads to further heat absorption . . . This is positive feedback. In contrast, clouds can have a negative feedback effect (although there are debates on the net effect of clouds). A warming atmosphere can evaporate more water, producing clouds. Clouds block some solar heat, which reduces the evaporation. Thus, clouds can slow the warming process.

Positive and negative feedback mechanisms are also important in understanding population dynamics (chapter 6). For example, more individuals produce more young, which produces more

(a) A simple system

FIGURE 2.10 Environmental scientists often study open systems. Here students at Cedar Creek study the climate-vegetation system, gathering plant samples that grew in carbon dioxide–enriched air pumped from the white poles, but other factors (soil, moisture, sunshine, temperature) are not controlled.
© William P. Cunningham

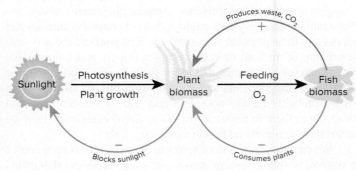

(b) A model of a system

FIGURE 2.11 Systems consist of compartments (also known as state variables) such as fish and plants, and flows of resources, such as nutrients or O_2 (a). Feedback loops (b) enhance or inhibit flows and the growth of compartments.

individuals . . . (a positive feedback). But sometimes environmental limits reduce the number of young that survive to reproduce (a negative feedback). Your body is a system with active negative feedback mechanisms: For example, if you exercise, you become hot, and your skin sweats, which cools your body.

Systems may exhibit stability

Negative feedbacks tend to maintain stability in a system. We often think of systems exhibiting **homeostasis,** or a tendency to remain more or less stable and unchanging. Equilibrium is another term for stability in a system. Your body temperature stays remarkably constant despite dramatic changes in your environment and your activity levels. Changing by just a few degrees is extremely unusual—a fever just 4–6°F above normal is unusual and serious. Natural systems such as grasslands can be fairly stable, too. If the climate isn't too dry or too wet, a grassland tends to remain grassland, although the grass may be dense in some years and sparse in others. Cycles of wet and dry years may be part of the system's normal condition.

Disturbances, events that can destabilize or change the system, might also be normal for the system. There can be many kinds of disturbance in a grassland. Severe drought can set back the community, so that it takes some time to recover. Many grasslands also experience occasional fires, a disturbance that stimulates grass growth (by clearing accumulated litter and recycling nutrients) but destroys trees that might be encroaching on the grassland. Thus, disturbances are often a normal part of natural systems. Sometimes we consider this "dynamic equilibrium," or a tendency for a system to change and then return to normal.

Grassland plots show **resilience,** an ability to recover from disturbance. In fact, studies indicate that species-rich plots may show more resilience than species-poor plots. Sometimes severe disturbance can lead to a **state shift,** in which conditions do not return to "normal." For example, a climate shift that drastically reduced rainfall could lead to a transition from grassland to desert. Plowing up grassland to plant crops is basically a state shift from a complex system to a single-species system.

Emergent properties are another interesting aspect of systems. Sometimes a system is more than the sum of its parts. For example, a tree is more than just a mass of stored carbon. It provides structure to a forest, habitat for other organisms, it shades and cools the ground, and it holds soil in place with its roots. All these functions are emergent properties, or characteristics of the system that are greater than the sum of its parts. An ecosystem can have emergent properties such as spatial structure, complexity, and diversity that individual components of the system could not have. It can also have emergent properties that we appreciate, such as beautiful sights and sounds, beyond its functioning as a system (fig. 2.12). In a similar way, you are a system made up of component parts, but you have many emergent properties, including your ability to think, share ideas with people around you, sing, and dance. These are properties that emerge because you function well as a system.

Section Review

1. Why are systems important in our environment?
2. What are feedback mechanisms?
3. Describe some emergent properties of ecosystems.

FIGURE 2.12 Emergent properties of systems, including beautiful sights and sounds, make them exciting to study.
© William P. Cunningham

2.3 SCIENTIFIC CONSENSUS AND CONFLICT

- *Science is an incremental process in which many people gradually reach a consensus.*
- *Critical thinking helps us evaluate scientific evidence.*
- *Many people misunderstand the role of uncertainty in science.*

The scientific method outlined in figure 2.3 is the process used to carry out individual studies. Larger-scale accumulation of scientific knowledge involves cooperation and contributions from countless people. Good science is rarely carried out by a single individual working in isolation. Instead, a community of scientists collaborates in a cumulative, self-correcting process. You often hear about big breakthroughs and dramatic discoveries that change our understanding overnight, but in reality these changes are usually the culmination of the labor of many people, each working on different aspects of a common problem, each adding small insights to solve the problem. Ideas and information are exchanged, debated, tested, and retested to arrive at **scientific consensus,** or general agreement among informed scholars.

The idea of consensus is important. For those not deeply involved in a subject, the multitude of contradictory results can be bewildering: Are shark populations disappearing, and does it matter? Is climate changing, and how much? Among those who have performed and read many studies, there tends to emerge a general agreement about the state of a problem. Scientific consensus now holds that many shark populations are in danger, though opinions vary on how severe the problem is. Consensus is that global climates are changing, though models differ somewhat on how rapidly they will change under different policy scenarios.

Sometimes new ideas emerge that cause major shifts in scientific consensus. Two centuries ago, geologists explained many earth features in terms of Noah's flood. The best scientists held that the flood created beaches well above modern sea level, scattered boulders erratically across the landscape, and gouged enormous valleys where there is little water now (fig. 2.13). Then the Swiss glaciologist Louis Agassiz and others suggested that the earth had once been much colder and that glaciers had covered large areas. Periodic ice ages better explained changing sea levels, boulders transported far from their source rock, and the great, gouged valleys. This new idea completely altered the way geologists explained their subject. Similarly, the idea of tectonic plate movement, in which continents shift slowly around the earth's surface, revolutionized the ways geologists, biogeographers, ecologists, and others explained the development of the earth and its life-forms.

These great changes in explanatory frameworks were termed **paradigm shifts** by Thomas Kuhn, who studied revolutions in scientific thought. According to Kuhn, paradigm shifts occur when a majority of scientists accept that the old explanation no longer explains new observations very well. The shift is often contentious and political, because whole careers and worldviews, based on one sort of research and explanation, can be undermined by a new model. Sometimes a revolution happens rather quickly. Quantum mechanics and Einstein's theory of relativity, for example, overturned classical physics in only about 30 years. Sometimes a whole generation of scholars has to retire before new paradigms can be accepted.

FIGURE 2.13 Paradigm shifts change the ways we explain our world. Geologists now attribute Yosemite's valleys to glaciers, where once they believed events like Noah's flood carved its walls.
© McGraw-Hill Education/John A. Karachewski, photographer

As you study this book, try to identify some of the paradigms that guide our investigations, explanations, and actions today. This is one of the skills involved in critical thinking, discussed in the introductory chapter of this book.

Detecting pseudoscience relies on independent, critical thinking

Ideally, science should serve the needs of society. Deciding what those needs are, however, is often a matter of politics and economics. Should water be taken from a river for irrigation or left in the river for wildlife habitat? Should we force coal-burning power plants to reduce air pollution in order to lower health costs and respiratory illnesses, or are society and our economy better served by having cheap but dirty energy? These thorny questions are decided by a combination of scientific evidence, economic priorities, political positions, and ethical viewpoints.

On the other hand, in every political debate, lawyers and lobbyists can find scientists who will back either side. Politicians hold up favorable studies, proclaiming them "sound science," while they dismiss others as "junk science." Opposing sides dispute the scientific authority of the study they dislike. What is "sound" science, anyway? If science is often embroiled in politics, does this mean that science is always a political process?

Consider the case of climate change. If you judge only from reports in newspapers or on television, you'd probably conclude that scientific opinion is about equally divided on whether climate change is a threat or not. In fact, the vast majority of scientists working on this issue agree that the earth's climate is being affected by human activities, and that threats to the systems we depend on are serious. In a study of 928 papers published in peer-reviewed scientific journals between 1993 and 2003, not one disagreed with the broad scientific consensus on global warming.

Why, then, is there so much confusion among the public about this issue? Why do politicians continue to assert that the dangers of climate change are uncertain at best, or "the greatest hoax ever perpetrated on the American people," as Senator James Inhofe, chair of the Senate Committee on Environment and Public Works, claims. A part of the confusion lies in the fact that media often present the debate as if it's evenly balanced. The fact that an overwhelming majority of working scientists agree on the issue doesn't make good drama, so some media give equal time to minority viewpoints just to make an interesting fight.

Perhaps a more important source of misinformation comes from conservative foundations and political action funds that finance climate deniers. Between 2002 and 2010, a small group of billionaires donated at least $120 million through secret channels to fund more than 100 think tanks, media outlets, and other groups that promote skepticism about global warming. Some of these organizations sound like legitimate science or grassroots groups but are really only public relations operations. Others are run by individuals who find it rewarding to offer contrarian views. This tactic of spreading doubt and disbelief through innocuous-sounding organizations or seemingly authentic experts isn't limited to the climate change debate. This strategy was pioneered by the tobacco industry to mislead the public about the dangers of smoking. Notably, some of the same individuals, groups, and lobbying firms employed by tobacco companies are now working to spread confusion about climate change.

Given this highly sophisticated battle of "experts," how do you interpret these disputes, and how do you decide whom to trust? The most important strategy is to apply critical thinking as you watch or read the news. What is the agenda of the person making the report? What is the source of their expertise? What economic or political interests do they serve? Do they appeal to your reason or to your emotions? Do they use inflammatory words (such as "junk"), or do they claim that scientific uncertainty makes their opponents' study meaningless? If they use statistics, what is the context for their numbers?

It helps to seek further information as you answer some of these questions. When you watch or read the news, you can look for places where reporting looks incomplete, you can consider sources and ask yourself what unspoken interests might lie behind the story.

Another strategy for deciphering the rhetoric is to remember that there are established standards of scientific work, and to investigate whether an "expert" follows these standards: Is the report peer-reviewed? Do a majority of scholars agree? Are the methods used to produce results well documented?

Harvard's Edward O. Wilson writes, "We will always have contrarians whose sallies are characterized by willful ignorance, selective quotations, disregard for communications with genuine experts, and destructive campaigns to attract the attention of the media rather than scientists. They are the parasite load on scholars who earn success through the slow process of peer review and approval." How can we identify misinformation and questionable claims? The astronomer Carl Sagan proposed a "Baloney Detection Kit" to help identify questionable sources and arguments (table 2.2).

Table 2.2 Questions for Baloney Detection
1. How reliable are the sources of this claim? Is there reason to believe that they might have an agenda to pursue in this case?
2. Have the claims been verified by other sources? What data are presented in support of this opinion?
3. What position does the majority of the scientific community hold in this issue?
4. How does this claim fit with what we know about how the world works? Is this a reasonable assertion or does it contradict established theories?
5. Are the arguments balanced and logical? Have proponents of a particular position considered alternate points of view or only selected supportive evidence for their particular beliefs?
6. What do you know about the sources of funding for a particular position? Are they financed by groups with partisan goals?
7. Where was evidence for competing theories published? Has it undergone impartial peer review or is it only in proprietary publication?

Most scientists have an interest in providing knowledge that is useful, and our ideas of what is useful and important depend partly on our worldviews and priorities. Science is not necessarily political, but it is often used for political aims. The main task of educated citizens is to discern where it is being misused or disregarded for purposes that undermine public interests.

Section Review

1. Why do we say that proof is elusive in science?
2. How can we evaluate the validity of claims about science?
3. What is the role of consensus in science?

Conclusion

Science is a process for producing knowledge methodically and logically. Scientists try to understand the world by making observations and trying to discern patterns and rules that explain those observations. Scientists try to remain cautious and skeptical of conclusions, because we understand that any set of observations is only a sample of all possible observations. In order to make sure we follow a careful and methodical approach, we use the scientific method, which is the step-by-step process of forming a testable question, doing tests, and interpreting results. Scientists use both deductive reasoning (deducing an explanation from general principles) and inductive reasoning (deriving a general rule from observations).

Hypotheses and theories are basic tools of science. A hypothesis is a testable question. A theory is a well-tested explanation that explains observations and that is accepted by the scientific community. Probability is also a key idea: Chance is involved in many events, and circumstances can influence probabilities—such as your chances of getting a cold or of getting an A in this class.

We often use probability to measure uncertainty when we test our hypotheses.

Models and systems are also central ideas. A system is a network of interdependent components and processes. For example, an ecosystem consists of plants, animals, and other components, and energy and nutrients transfer among those components. Systems have general characteristics we can describe, including throughput, feedbacks, homeostasis, resilience, and emergent properties. Often we use models (simplified representations of systems) to describe or manipulate a system. Models vary in complexity, according to their purposes, from a paper airplane to a global circulation model.

Science aims to foster debate and inquiry, but scientific consensus emerges as most experts come to agree on well-supported theoretical explanations. Sometimes new explanations revolutionize science, but scientific consensus helps us identify which ideas and theories are well supported by evidence, and which are not supported.

Reviewing Key Terms

Can you define the following terms in environmental science?

blind experiment 2.1

closed system 2.2

confidence 2.1

controlled study 2.1

deductive reasoning 2.1

dependent variable 2.1

disturbances 2.2

double-blind experiment 2.1

emergent properties 2.2

homeostasis 2.2

hypothesis 2.1

independent variable 2.1

inductive reasoning 2.1

manipulative experiment 2.1

mean 2.1

models 2.1

natural experiment 2.1

negative feedback 2.2

open system 2.2

paradigm shift 2.3

positive feedback 2.2

replication 2.1

reproducibility 2.1

resilience 2.2

sample 2.1

science 2.1

scientific consensus 2.3

scientific theory 2.1

significant figure 2.1

state shift 2.2

statistics 2.1

systems 2.2

throughput 2.2

Critical Thinking and Discussion Questions

1. Explain why scientific issues are or are not influenced by politics. Can scientific questions ever be entirely free of political interest? If you say no, does that mean that all science is merely politics? Why or why not?

2. Review the questions for "baloney detection" in table 2.2, and apply them to an ad on TV. How many of the critiques in this list are easily detected in the commercial?

3. How important is scientific thinking for you, personally? How important do you think it should be? How important is it for society to have thoughtful scientists? How would your life be different without the scientific method?

4. Many people consider science too remote from everyday life and from nonscientists. Do you feel this way? Are there aspects of scientific methods (such as reasoning from observations) that you use?

5. Many scientific studies rely on models for experiments that cannot be done on real systems, such as climate, human health, or economic systems. If assumptions are built into models, then are model-based studies inherently weak? What would increase your confidence in a model-based study?

Data Analysis

Working with Graphs

To understand trends and compare values in environmental science, we need to examine many numbers. Most people find it hard to quickly assess large amounts of data in a table. But if you can make a picture or a diagram of the data, it can be much easier to see patterns, trends, and connections. You will encounter many graphs in this book. Reading graphs takes practice, but it is an essential skill that will serve you well in this course and many others.

A graph is a diagram that shows relationships between different factors. Many graphs show how variables change with respect to one another. A dependent variable is a factor or category whose value "depends on" other factors. An independent variable stands alone and isn't contingent on other factors. For example, in most studies, time changes independently of what you may do to your subject.

In most graphs, independent variables are represented on the X (horizontal) axis. Dependent variables, then, are represented on the Y (vertical) axis. That isn't always true, however, because we often rotate graphs to fit in printed columns. To determine the

actual number for any data point in the graph, simply draw horizontal and vertical lines from that data point to the X and Y axes.

1. Line graphs often show trends over time. For example, figure 1 shows coal consumption by different countries over a 50-year period. Time, on the X-axis, is the independent variable, while the Y-axis represents the amount consumed in quadrillion tons over this period. This shows very clearly that consumption in China has increased very rapidly in recent years, while other countries have been relatively constant. Data for this table was collected around 2012, so amounts to the right of this year are projections, or estimates, of what we expect to happen.

2. Often, it's useful to plot two or more data sets on the same line graph. Figure 2 shows changing populations of Canadian lynx and snowshoe hares, as estimated by the number of pelts brought into Hudson Bay Company trading posts between 1840 and 1930. As you can see, the numbers of Canada lynx fluctuate on about a ten-year cycle that is similar to,

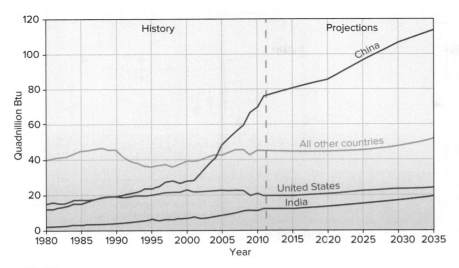

FIGURE 1 Line graph

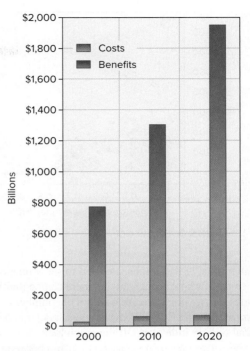

FIGURE 3 Bar graph

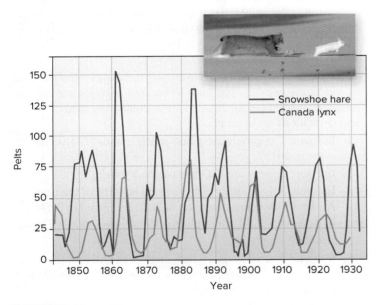

FIGURE 2 Line graph

the Clean Air Act provisions are shown for three years (the data for 2020 are estimates). This gives a powerful understanding of both the relative values for costs and benefits in a single year, as well as the trends in these data by decade.

4. Pie charts compare portions of a whole. For example, the proven oil reserves by country are shown in figure 4. Each country is represented by a slice of the pie, which makes it very easy to compare the relative percentage of the proven global oil supply that each possesses. Giving each country a different color makes it easier to distinguish between them. Note that these data are percentages of the global supply and not absolute numbers.

but slightly out of phase with, the population peaks of snowshoe hares. Although there are some doubts now about how and where these data were collected, this remains a classic example of population dynamics. When prey populations (hares) are abundant, predators (lynx) reproduce more successfully and their population grows. When hare populations crash, so do the lynx, showing an interdependence between predator and prey populations.

3. Bar graphs (or bar charts) are used when one variable is a category and the other is a number. Each category is depicted by a rectangular bar whose length is proportional to the value it represents. The bars can be plotted vertically or horizontally to fit in a column. In figure 3, the costs and benefits of

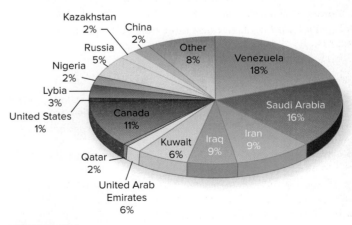

FIGURE 4 Pie chart

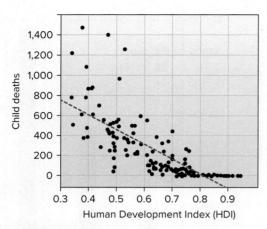

FIGURE 5 Scatter plot

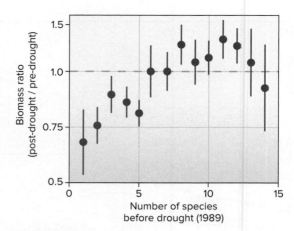

FIGURE 6 Error bars

5. Scatterplots, or dot graphs, are used to show relationships when data are not sequential and there can be multiple Y-axis values for any point. Figure 4 shows the number of child deaths compared to the Human Development Index (a composite of social welfare indicators). If the graph looks like a loose cloud of dots, there may be no clear relationship between variables. A tight, linear pattern of dots trending in one direction probably indicates a strong relationship. You can also use a statistical package to calculate an equation to describe the relationship. This is known as regression analysis, and it can be used to generate a line or curve that best fits the trends in the data set.

6. Error Bars and Confidence Intervals

   Uncertainty is a key idea in science. We can rarely have absolute proof in experimental results, because our conclusions rest on observations, but we often have only a small sample of all possible observations. Because uncertainty is always present, it's useful to describe how much uncertainty you have, relative to what you know. It might seem ironic, but in science, knowing about uncertainty increases our confidence in our conclusions.

   Figure 6 is from a landmark field study by D. Tilman et al. It shows change in biomass within experimental plots containing varying numbers of native prairie plants after a severe drought. Each dot shows the mean ratio of biomass before and after drought. Because more than 200 replicate (repeated

test) plots were used, this study was able to give an estimate of uncertainty. Vertical bars show the standard error of the mean for vegetation plots planted with different numbers of species. Regrowth after a drought was poorer for plots with five or fewer species (blue) than for more diverse plots (red).

Questions:

Look carefully at these graphs to answer these questions. Note that in comparing numbers you must sometimes approximate and give rounded answers (as for figure 1). For comparing general trends, approximations are appropriate.

1. In figure 1, approximately how much coal did China consume in 2012?

2. In figure 2, what are the units on the Y axis?

3. In figure 3, how much higher are the benefits of the Clean Air Provisions expected to be in 2020 compared to 2000?

4. In figure 4, what country has the largest percentage of proven world oil supplies?

5. In figure 5, what is the highest number of child deaths in this time period? What appears to be the lowest? Why might the lowest number be misleading?

6. Each dot in figure 6 shows the average species count for a set of text plots. About how many species are represented by the leftmost dot? By the rightmost dot?

TO ACCESS ADDITIONAL RESOURCES FOR THIS CHAPTER, PLEASE VISIT CONNECT AT www.connect.mheducation.com.

You will find LearnSmart, an adaptive learning system, Google Earth™ exercises, additional Case Studies, Data Analysis exercises, and an interactive ebook.

3

Matter, Energy, and Life

▲ Chesapeake Bay's ecosystem supports fisheries, recreation, and communities. But the estuary is an ecosystem out of balance.
© Aliaksandr Zhamasek/iStock/Getty Images

Learning Outcomes

After studying this chapter, you should be able to:

3.1 Describe matter, elements, and molecules and give simple examples of the role of four major kinds of organic compounds in living cells.

3.2 Define energy and explain how thermodynamics regulates ecosystems.

3.3 Understand how living organisms capture energy and create organic compounds.

3.4 Define species, populations, communities, and ecosystems, and summarize the ecological significance of trophic levels.

3.5 Understand pathways in the water, carbon, nitrogen, sulfur, and phosphorus cycles.

"When one tugs at a single thing in nature, he finds it attached to the rest of the world."

– John Muir

Chesapeake Bay: How Do We Improve on a C?

Each year Chesapeake Bay, the largest estuary in the United States, gets a report card, just as you do at the end of a semester. Like your report card, this one summarizes several key performance measures. Unlike your grades, the bay's grades are based on measures such as water clarity, oxygen levels, the health of sea grass beds, and the condition of the microscopic plankton community. These factors reflect the overall stability of fish and shellfish populations, which are critical to the region's ecosystems and economy. Since record keeping began, the bay's performance has been poor, with scores hovering between 35 and 57 out of 100, and an average grade of C. This is better than the C- or D of past years, but there is considerable room for improvement. The main reason for the low grades? Excessive levels of nitrogen and phosphorus, two common life-supporting elements that have destabilized the ecosystem.

Chesapeake Bay's watershed is a vast and complex system, with over 17,600 km (11,000 mi) of tidal shoreline in six states, and a population of 20 million people. Approximately 100,000 streams and rivers drain into the bay. All these streams carry runoff from forests, farmlands, cities, and suburbs from as far away as New York (fig. 3.1a).

The system has consistently bad grades, but it's clearly worth saving. Even in its impaired state, the bay provides 240 million kg (500 million lb) of seafood every year. It supports fishing and recreational economies worth $33 billion a year. But this is just a fraction of what it should be. The bay once provided abundant harvests of oysters, blue crabs, rockfish, white perch, shad, sturgeon, flounder, eel, menhaden, alewives, and soft-shell clams. Overharvesting, disease, and declining ecosystem productivity have decimated fisheries. Blue crabs are just above population survival levels. The oyster harvest, which was 15 to 20 million bushels per year in the 1890s, has declined to less than 1 percent of that amount. According to the Environmental Protection Agency (EPA), the bay should support more than twice the fish, crabs, and oysters that are there today. Human health is also at risk. After heavy rainfall, people are advised to stay out of the water for 48 hours, to avoid contamination from sewer overflows and urban and agricultural runoff.

The principal problem is simply excessive levels of nitrogen and phosphorus. These two elements are essential for life, because they are part of all cells and biological reactions, but the system is overloaded by excess loads from farm fields, livestock manure, urban streets, suburban lawn fertilizer, the legal discharges of over 3,000 sewage treatment plants, and half a million aging household septic systems. Air pollution from cars, power plants, and factories also introduce nitrogen to the bay (fig. 3.1b; see chapter 16). Sediment is also a key issue: It washes in from fields and streets, smothers eelgrass beds, and blocks sunlight, further reducing photosynthesis in the bay.

Just as too many donuts are bad for you, an excessive diet of nutrients is bad for an estuary. Excess nutrients fertilize superabundant growth of algae, which further blocks sunlight and reduces photosynthesis and oxygen levels in the bay. Lifeless, oxygen-depleted areas result. Fish, oysters, and crabs die off. These algal blooms in nutrient-enriched waters are increasingly common in bays and estuaries worldwide.

Progress has been discouragingly slow for decades, but in 2010 the EPA finally addressed the problem seriously, complying with its charge from Congress (under the Clean Water Act) to protect the bay. Where piecemeal, mostly voluntary efforts by individual states had long failed to improve the Chesapeake's report cards, the EPA brought all neighboring states to the negotiating table. Total maximum daily loads (TMDLs) for nutrients and sediments were established, and states were given freedom to decide how to meet their share of nitrogen reductions. But the EPA has legal authority from the Clean Water Act to enforce reductions. The aim is to cut

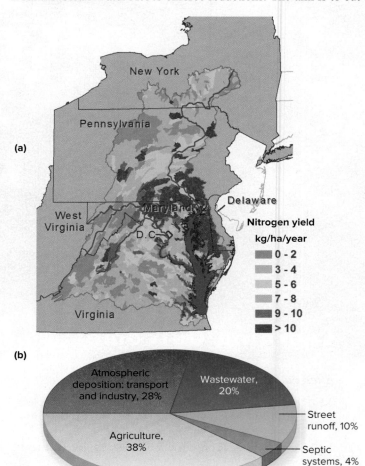

FIGURE 3.1 America's largest and richest estuary, Chesapeake Bay (shown in blue) suffers from pollutants from six states (a), and many sources (b). *Sources: USGS, EPA 2010*

nitrogen levels by 25 percent, phosphorus by 24 percent, and sediment by 20 percent. The nitrogen target of 85 million kg (186 million lb) per year is still 4–5 times greater than would be released by an undisturbed watershed, but it's a huge improvement.

States from Virginia to New York have chosen their own strategies to meet limits. Maryland plans to capture and sell nitrogen and phosphorus from chicken manure. New York promises better urban wastewater treatment. Pennsylvania is strengthening soil conservation efforts to retain nutrients on farmland. Together, over time, these changes may rescue this magnificent estuary.

Chesapeake Bay has long been a symbol of the intractable difficulty of managing large, complex systems. Progress has required better understanding of several issues: the integrated functioning of the uplands and the waterways, the interdependence of the diverse human communities and economies that depend on the bay, and the pathways of nitrogen and phosphorus through an ecosystem.

Environmental scientists have led the way to the EPA's solution with years of ecosystem research and data collection. Through their efforts, and with EPA leadership, Chesapeake Bay could become the largest, and perhaps the most broadly beneficial, ecosystem restoration ever attempted in the United States. In this chapter, we'll examine how these and other elements move through systems, and why they are important. Understanding these basic ideas will help you explain how many different systems function, including Chesapeake Bay's, your local ecosystem's, and even your own body's.

3.1 ELEMENTS OF LIFE

- *From living organisms to ecosystems, life can be understood in terms of the movement of matter and energy.*
- *To understand how matter and energy cycle through living things, we must understand how atoms bond together to form compounds.*
- *Carbon-based ("organic") compounds are the foundation of organisms.*

The accumulation and transfer of energy and nutrients allows living systems, including yourself, to exist. These processes tie together the parts of an ecosystem—or an organism. In this chapter, we'll introduce a number of concepts that are essential to understanding how living things function in their environment. These concepts include fundamental ideas of matter and energy, the ways organisms acquire and use energy, and the nature of chemical elements. We then apply these ideas to feeding relationships among organisms—the ways that energy and nutrients are passed from one living thing to another. In other words, we'll trace components from atoms to elements to compounds to cells to organisms to ecosystems.

Atoms, elements, and compounds

Everything that takes up space and has mass is **matter.** Matter exists in four distinct states, or phases—solid, liquid, gas, and plasma—which vary in energy intensity and the arrangement of particles that make up the substance. Water, for example, can exist as ice (solid), as liquid water, or as water vapor (gas). The fourth phase, plasma, occurs when matter is heated so intensely that electrons are released and particles become ionized (electrically charged). We can observe plasma in the sun, lightning, and very hot flames.

Under ordinary circumstances, matter is neither created nor destroyed; rather, it is recycled over and over again. This idea is known as the principle of **conservation of matter.** The molecules that make up your body may contain atoms that once were part of the body of a dinosaur. Most certainly you contain atoms that were part of many smaller prehistoric organisms. This is because chemical elements are used and reused by living organisms. Matter is transformed and combined in different ways, but it doesn't disappear; everything goes somewhere.

How does this principle apply to environmental science? It explains how components of environmental systems are intricately connected. From Chesapeake Bay to your local ecosystem to your own household, all matter comes from somewhere, and all waste goes somewhere. Pause to consider what you have eaten, used, or bought today. Then think of where those materials will go when you are done with them. You are intricately tied to both the sources and the destinations of everything you use. This is a useful idea for us as residents of a finite world. Ultimately, when we throw away our disposable goods, they don't really go "away," they just go somewhere else, to stay there for a while and then move on.

Matter consists of **elements** (basic substances that cannot be broken down into simpler forms by ordinary chemical reactions), such as carbon or oxygen. Each of the 122 known elements (92 natural, plus 30 created under special conditions) has distinct chemical characteristics. Just four elements—oxygen, carbon, hydrogen, and nitrogen—are responsible for more than 96 percent of the mass of most living organisms. See if you can find these four elements in the periodic table of the elements.

Atoms are the smallest particles that exhibit the characteristics of an element. Atoms are composed of positively charged protons, negatively charged electrons, and electrically neutral neutrons. Protons and neutrons, which have approximately the same mass, are clustered in the nucleus in the center of the atom (fig. 3.2). Electrons, which are tiny in comparison to the other particles, orbit the nucleus at the speed of light.

Each element has a characteristic number of protons per atom, called its **atomic number.** Carbon, for example, has 6 protons (see fig. 3.2), so its atomic number is 6. Each element also has a

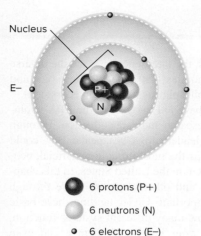

Nucleus

E−

P+

N

● 6 protons (P+)

○ 6 neutrons (N)

◦ 6 electrons (E−)

FIGURE 3.2 As difficult as it may be to imagine when you look at a solid object, all matter is composed of tiny, moving particles, separated by space and held together by energy. It is hard to capture these dynamic relationships in a drawing. This model represents carbon-12, with a nucleus of six protons and six neutrons; the six electrons are represented as a fuzzy cloud of potential locations rather than as individual particles.

characteristic atomic mass, which is the sum of protons and neutrons (each having a mass of about 1 atomic mass unit). Carbon normally has 6 neutrons, as well as its 6 protons, which sum to an atomic mass of 12. However, the number of neutrons can vary slightly. Forms of the same element that differ in atomic mass are called **isotopes.** For example, some carbon atoms have 7 or 8 neutrons. These atoms have a mass of 13 or 14, rather than the usual 12. Similarly, hydrogen (H) is the lightest element, and normally it has just one proton and one electron (and no neutrons) and an atomic mass of 1. A small percentage of hydrogen atoms also have a neutron in the nucleus, which gives those atoms an atomic mass of 2 (one proton + one neutron). We call this isotope deuterium (^2H). An even smaller percentage of natural hydrogen called tritium (^3H) has one proton plus two neutrons. Oxygen atoms can also have one or two extra neutrons, making them the isotopes ^{17}O or ^{18}O, instead of the normal ^{16}O.

This difference is important in environmental science. Oxygen isotopes, for example, tell us about ancient climates. Water (H_2O) containing ^{18}O is slightly more massive than water containing the normal ^{16}O. The higher-mass H_2O evaporates into the air more easily in hot climates, where there is plenty of evaporative energy, than in cold climates. When we examine bubbles of ancient air trapped in ice cores, abundant ^{18}O indicates a relatively warm ancient climate. Lower amounts of ^{18}O indicate a colder climate (chapter 15). Some isotopes are unstable—that is, they spontaneously emit electromagnetic energy or subatomic particles, or both. Radioactive waste and nuclear energy are both environmentally hazardous because they involve unstable isotopes of elements such as uranium and plutonium (chapters 19 and 21).

Chemical bonds hold molecules together

Atoms often join to form **compounds,** or substances composed of different kinds of atoms (fig. 3.3). A pair or group of atoms that can exist as a single unit is known as a **molecule.** Some elements commonly occur as molecules, such as molecular oxygen (O_2) or molecular nitrogen (N_2), and some compounds can exist as molecules, such as glucose ($C_6H_{12}O_6$). In contrast to these molecules, sodium chloride (NaCl, table salt) is a compound that cannot exist as a single pair of atoms. Instead it occurs in a solid mass of Na and Cl atoms or as two ions, Na^+ and Cl^-, dissolved in solution.

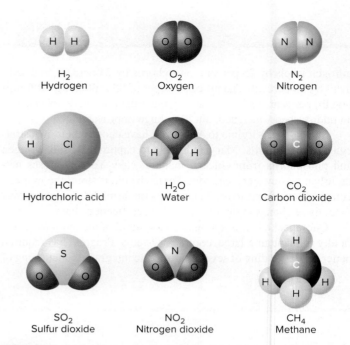

H_2 Hydrogen

O_2 Oxygen

N_2 Nitrogen

HCl Hydrochloric acid

H_2O Water

CO_2 Carbon dioxide

SO_2 Sulfur dioxide

NO_2 Nitrogen dioxide

CH_4 Methane

FIGURE 3.3 These common molecules, with atoms held together by covalent bonds, are important components of the atmosphere or are important pollutants.

Most molecules consist of only a few atoms. But many, such as proteins and nucleic acids, discussed below, can include millions or even billions of atoms.

Electrical attraction holds atoms together to form compounds. When *ions* with opposite charges (such as Na^+ and Cl^-) form a compound, the electrical attraction holding them together is an *ionic bond.* Sometimes neither atom readily gives up an electron to the other, as when two hydrogen atoms meet. In this case, atoms form bonds by *sharing* electrons. Two hydrogen atoms can bond by sharing a pair of electrons—they orbit the two hydrogen nuclei equally and hold the atoms together. Such electron-sharing bonds are known as *covalent bonds.* Carbon (C) can form covalent bonds simultaneously with four other atoms, so carbon can create complex structures such as sugars and proteins. Atoms in covalent bonds do not always share electrons evenly. An important example in environmental science is the covalent bonds in water (H_2O). The oxygen atom attracts the shared electrons more strongly than do the two hydrogen atoms. Consequently, the hydrogen portion of the molecule has a slight positive charge, while the oxygen has a slight negative charge. These charges create a mild attraction between water molecules, making water somewhat cohesive. This fact helps explain some of the remarkable properties of water that we'll discuss in the next section.

When an atom gives up one or more electrons, we say it is *oxidized* (because it is very often oxygen, an abundant and highly reactive element, that takes the electron). When an atom gains electrons, we say it is *reduced.* Oxidation and reduction reactions are necessary for life: Oxidation of sugar and starch molecules, for example, is an important part of how you gain energy from food.

Forming bonds usually releases energy. Breaking bonds generally requires energy. Think of this in burning wood: carbon-rich organic compounds such as cellulose are *broken*, which requires energy; at the same time, oxygen from the air *forms* bonds with carbon from the wood, making CO_2. In a fire, more energy is produced than is consumed, and the net effect is that it feels hot to us.

Unique properties of water

If travelers from another solar system were to visit our lovely, cool, blue planet, they might call it Aqua rather than Terra because of its outstanding feature: the abundance of streams, rivers, lakes, and oceans of liquid water. Our planet is the only place we know where water exists as a liquid in any appreciable quantity. Water covers nearly three-fourths of the earth's surface and moves around constantly via the hydrologic cycle (discussed in chapter 15) that distributes nutrients, replenishes freshwater supplies, and shapes the land. Water makes up 60 to 70 percent of the weight of most living organisms. It fills cells, giving form and support to tissues. Among water's unique, even serendipitous qualities are the following:

1. Water molecules are polar: They have a slight positive charge on one side and a slight negative charge on the other side. Therefore, water readily dissolves polar or ionic substances, including sugars and nutrients, and carries materials to and from cells.

2. Water is the only inorganic substance that normally exists as a liquid at temperatures suitable for life. Most substances exist as either a solid or a gas, with only a very narrow liquid temperature range. Organisms synthesize organic compounds such as oils and alcohols that remain liquid at ambient temperatures and are therefore extremely valuable to life, but the original and predominant liquid in nature is water.

3. Water molecules are cohesive: They hold together tenaciously and create high surface tension (fig 3.4). You have experienced this property if you have ever done a belly flop off a diving board. Water has the highest surface tension of any common, natural liquid. Water also adheres to surfaces. As a result, water is subject to *capillary action:* It can be drawn into small channels. Without capillary action, movement of water and nutrients into groundwater reservoirs and through living organisms might not be possible.

4. Water is unique in that it expands when it crystallizes. Most substances shrink as they change from liquid to solid. Ice floats because it is less dense than liquid water. When temperatures fall below freezing, the surface layers of lakes, rivers, and oceans cool faster and freeze before deeper water. Floating ice then insulates underlying layers, keeping most water bodies liquid (and aquatic organisms alive) throughout the winter in most places. Without this feature, many aquatic systems would freeze solid in winter.

5. Water has a high heat of vaporization: It takes a great deal of heat to convert from liquid to vapor. Consequently, evaporating water is an effective way for organisms to shed excess heat. Many animals pant or sweat to moisten evaporative cooling surfaces. Why do you feel less comfortable on a hot, humid day than on a hot, dry day? Because the water vapor–laden air inhibits the rate of evaporation from your skin, thereby impairing your ability to shed heat.

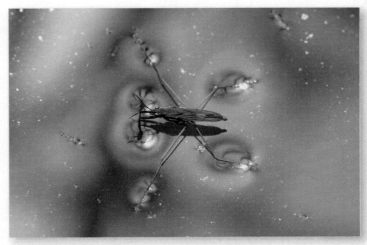

FIGURE 3.4 Surface tension is demonstrated by the resistance of a water surface to penetration, as when it is walked upon by a water strider.
© Nigel Cattlin/Alamy Stock Photo

6. Water also has a high specific heat: A great deal of heat is absorbed before it changes temperature. The slow response of water to temperature change helps moderate global temperatures, keeping the environment warm in winter and cool in summer. This effect is especially noticeable near the ocean, but it is important globally.

All these properties make water a unique and vitally important component of the ecological cycles that transfer matter and energy and that make life on earth possible.

Generally, some energy input (activation energy) is needed to start these reactions. In your fireplace, a match might provide the needed activation energy. In your car, a spark from the battery provides activation energy to initiate the oxidation (burning) of gasoline.

Ions react and bond to form compounds

Atoms frequently gain or lose electrons, acquiring a negative or positive electrical charge. Charged atoms (or combinations of atoms) are called **ions.** Negatively charged ions (with one or more extra electrons) are *anions.* Positively charged ions are *cations.* A hydrogen (H) atom, for example, can give up its sole electron to become a hydrogen ion (H^+). Chlorine (Cl) readily gains electrons, forming chlorine ions (Cl^-).

Substances that readily give up hydrogen ions in water are known as **acids.** Hydrochloric acid, for example, dissociates in water to form H^+ and Cl^- ions. In later chapters, you may read about acid rain (which has an abundance of H^+ ions), acid mine drainage, and many other environmental problems involving acids. In general, acids cause environmental damage because the H^+ ions react readily with living tissues (such as your skin or tissues of fish larvae) and with nonliving substances (such as the limestone on buildings, which erodes under acid rain).

Substances that readily bond with H^+ ions are called **bases** or alkaline substances. Sodium hydroxide (NaOH), for example, releases hydroxide ions (OH^-) that bond with H^+ ions in water. Bases can be highly reactive, so they also cause significant

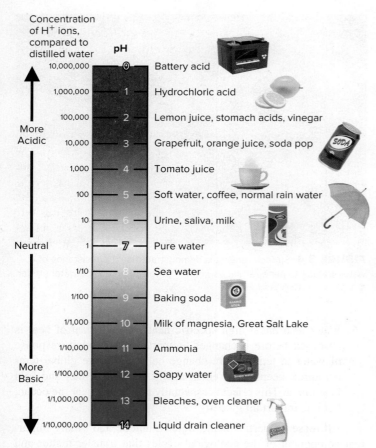

FIGURE 3.5 The pH scale. The numbers represent the negative logarithm of the hydrogen ion concentration in water. Alkaline (basic) solutions have a pH greater than 7. Acids (pH less than 7) have high concentrations of reactive H⁺ ions.

Organic compounds have a carbon backbone

Organisms use some elements in abundance, others in trace amounts, and others not at all. Certain vital substances are concentrated within cells, while others are actively excluded. Carbon is a particularly important element because chains and rings of carbon atoms form the skeletons of **organic compounds,** the material of which biomolecules, and therefore living organisms, are made.

There are four major categories of organic compounds in living things ("bio-organic compounds"): lipids, carbohydrates,

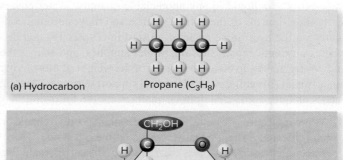

(a) Hydrocarbon   Propane (C_3H_8)

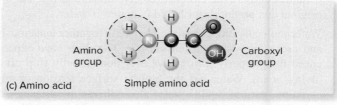

(b) Sugar   Glucose ($C_6H_{12}O_6$)

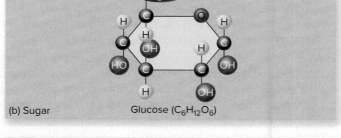

Amino group   Carboxyl group

(c) Amino acid   Simple amino acid

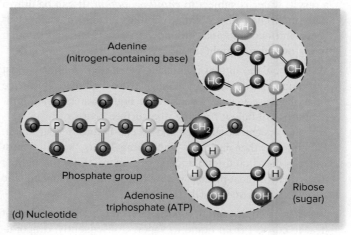

Adenine (nitrogen-containing base)

Phosphate group

Adenosine triphosphate (ATP)

Ribose (sugar)

(d) Nucleotide

FIGURE 3.6 The four major groups of biologically important organic molecules are based on repeating subunits of these carbon-based structures. Basic structures are shown for (a) butyric acid (a building block of lipids) and a hydrocarbon, (b) a simple carbohydrate, (c) a protein, and (d) a nucleotide (a component of nucleic acids).

environmental problems. Acids and bases can also be essential to living things: The acids in your stomach dissolve food, for example, and acids in soil help make nutrients available to growing plants.

We describe the strength of an acid and base by its **pH,** the negative logarithm of its concentration of H⁺ ions (fig. 3.5). Acids have a pH below 7; bases have a pH greater than 7. A solution of exactly pH 7 is "neutral." Because the pH scale is logarithmic, pH 6 represents *ten times* more hydrogen ions in solution than pH 7.

A solution can be neutralized by adding buffers—substances that accept or release hydrogen ions. In the environment, for example, alkaline rock can buffer acidic precipitation, decreasing its acidity. Lakes with acidic bedrock, such as granite, are especially vulnerable to acid rain because they have little buffering capacity.

Quantitative Reasoning

The pH scale shows the availability of reactive hydrogen ions (H⁺) in a liquid. The scale is logarithmic, so milk has 10 times as many H⁺ ions as pure water, for a given volume. How many more H⁺ ions does normal rain have compared to pure water? Soda pop? Vinegar? Is sea water more acidic or more basic than pure water?

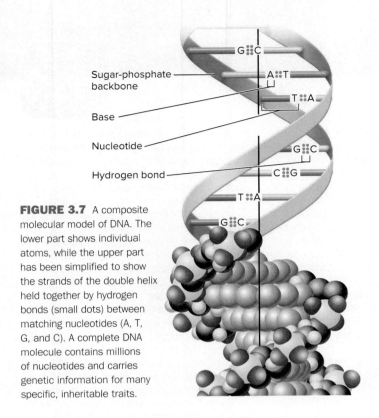

FIGURE 3.7 A composite molecular model of DNA. The lower part shows individual atoms, while the upper part has been simplified to show the strands of the double helix held together by hydrogen bonds (small dots) between matching nucleotides (A, T, G, and C). A complete DNA molecule contains millions of nucleotides and carries genetic information for many specific, inheritable traits.

Labels in figure: Sugar-phosphate backbone; Base; Nucleotide; Hydrogen bond; G∷C; A∷T; T∷A; G∷C; C∷G; T∷A; G∷C

Long chains of DNA bind together to form a stable double helix (fig. 3.7). These chains separate and are duplicated when cells divide, so that genetic information is replicated. Every individual has a unique set of DNA molecules, which create the differences between individuals and between species.

Cells are the fundamental units of life

All living organisms are composed of **cells,** minute compartments within which the processes of life are carried out (fig. 3.8). Microscopic organisms such as bacteria, some algae, and protozoa are composed of single cells. Higher organisms have many cells, usually with many different cell varieties. Your body, for instance, is composed of several trillion cells of about two hundred distinct types. Every cell is surrounded by a thin but dynamic membrane of lipid and protein that receives information about the exterior

proteins, and nucleic acids. Lipids (including fats and oils) store energy for cells, and they provide the core of cell membranes and other structures. Lipids do not readily dissolve in water, and their basic structure is a chain of carbon atoms with attached hydrogen atoms. This structure makes them part of the family of hydrocarbons (fig. 3.6*a*). Carbohydrates (including sugars, starches, and cellulose) also store energy and provide structure to cells. Like lipids, carbohydrates have a basic structure of carbon atoms, but hydroxyl (OH) groups replace half the hydrogen atoms in their basic structure, and they usually consist of long chains of sugars. Glucose (fig. 3.6*b*) is an example of a very simple sugar.

Proteins are composed of chains of subunits called amino acids (fig. 3.6*c*). Folded into complex three-dimensional shapes, proteins provide structure to cells and are used for countless cell functions. Most enzymes, such as those that release energy from lipids and carbohydrates, are proteins. Proteins also help identify disease-causing microbes, make muscles move, transport oxygen to cells, and regulate cell activity.

Nucleotides are complex molecules made of a five-carbon sugar (ribose or deoxyribose), one or more phosphate groups, and an organic nitrogen-containing base called either a purine or pyrimidine (fig. 3.6*d*). Nucleotides serve many functions. They carry information between cells, tissues, and organs. They are sources of energy for cells. They also form long chains called *ribo*nucleic *a*cid (RNA) or **deoxyribonucleic acid (DNA)** that are essential for storing and expressing genetic information.

Just four kinds of nucleotides make up all DNA (these are adenine, guanine, cytosine, and thymine), but there can be billions of these molecules lined up in a very specific sequence.

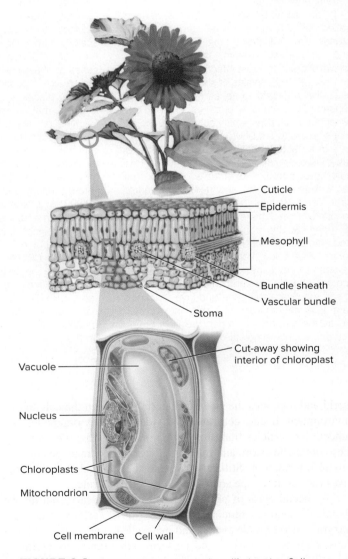

Labels in figure: Cuticle; Epidermis; Mesophyll; Bundle sheath; Vascular bundle; Stoma; Cut-away showing interior of chloroplast; Vacuole; Nucleus; Chloroplasts; Mitochondrion; Cell membrane; Cell wall

FIGURE 3.8 Plant tissues and a single cell's interior. Cell components include a cellulose cell wall, a nucleus, a large empty vacuole, and several chloroplasts, which carry out photosynthesis.

Gene Editing

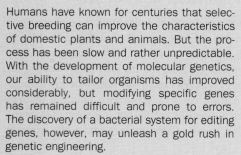

Humans have known for centuries that selective breeding can improve the characteristics of domestic plants and animals. But the process has been slow and rather unpredictable. With the development of molecular genetics, our ability to tailor organisms has improved considerably, but modifying specific genes has remained difficult and prone to errors. The discovery of a bacterial system for editing genes, however, may unleash a gold rush in genetic engineering.

This bacterial gene editing system is called CRISPR (pronounced crisper), which stands for "Clustered regularly-interspaced short palindromic repeats." It's a system that allows bacteria to resist infection by pathogenic viruses. CRISPR consists of short sequences of viral DNA attached to genes (the palindromic repeats) for a group of enzymes, chief among which are nucleases, which act like tiny molecular scissors to cut DNA in specific places.

In a bacterial cell, the viral DNA is translated into "guide" RNA that binds to the associated proteins to form a complex that recognizes, binds to, and destroys invading viral DNA. Several years ago, researchers recognized that this system can be tweaked to recognize and edit genes in higher plants and animals. Rather than using viral DNA as a target, RNA guide molecules are synthesized to bond to any genes you want to edit. When the complex binds to and cuts the target DNA, it can inactivate the gene and give us important information about its role in the cell. Or, as the cell tries to repair the DNA break, you can supply templates for new versions of

the target gene that will replace the original sequences. Think of this as a molecular version of the search and replace function in your word processor.

The tool is already being used in the lab to: make human cells impervious to HIV; correct a mutation that leads to blindness in humans; cure mice of muscular dystrophy, cataracts, and a hereditary liver disease; and improve crops including wheat, rice, soybeans, tomatoes, oranges, and wheat. Libraries of tens of thousands of guide RNAs are now available to target and activate, or silence, specific genes. One of the most exciting features of CRISPR is that it can modify multiple genes at the same time in a single cell. This may make it possible to study and/or treat complex diseases, such as Alzheimer's or Parkinson's, that are regulated by many genes.

In 2015, researchers in China reported that they had successfully used the CRISPR system in human zygotes to modify a gene that causes the blood disorder β thalassemia. The zygotes they used were not viable, but this study triggered a furious debate over the proper use of CRISPR, and whether it's ever OK to manipulate human embryos or gametes. Critics worry about designer babies and mindless-drones engineered for dull or dangerous jobs.

Thus, CRISPR has become the latest example in the on-going debate about genetically modified organisms (GMOs). On one hand, CRISPR makes it easier, quicker, and less expensive to create modified genomes. This will probably make GMOs more abundant. On the other hand, CRISPR edits genes much

more precisely than other tools. Previous methods of creating GMOs generally involved blasting large chunks of DNA into cells. We couldn't be sure exactly what genes were being introduced, where in the genome they'd end up, or exactly how they'd act once introduced.

Using CRISPR, we could edit a single gene—or even a single nucleotide in a gene— very precisely, and we wouldn't need to move DNA between species or families. The results should be more accurate and predictable, perhaps minimizing unintended consequences.

What, you may be wondering, does this have to do with environmental science? One of the big worries about GMOs is that genetic engineering might inadvertently create a superbug that would cause a terrible epidemic or disrupt the balance of nature. This outcome seems less likely given CRISPR's precision. And genetic engineering has been used to control disease vectors, such as *Aedes aegypti* mosquitoes that carry the Zika virus, which may cause microcephly (see chapter 8). The GMO mosquitoes mate with wild relatives and produce non-viable larvae. In Brazilian tests, *Aedes aegypti* larvae numbers dropped by 82 percent in only eight months. Thus, we may be able to control some diseases without blasting ecosystems with powerful pesticides. We also might be able to engineer important plant and animal species to make them resistant to changing climate and water availability.

What do you think? If you were appointed to a regulatory panel commissioned to oversee gene editing, what limits—if any—would you impose on this new technology?

---

world and regulates the flow of materials between the cell and its environment. Inside, cells are subdivided into tiny organelles and subcellular particles that provide the machinery for life. Some of these organelles store and release energy. Others manage and distribute information. Still others create the internal structure that gives the cell its shape and allows it to fulfill its role.

A special class of proteins called **enzymes** carry out all the chemical reactions required to create these various structures. Enzymes also provide energy and materials to carry out cell functions, dispose of wastes, and perform other functions of life at the cellular level. Enzymes are molecular catalysts: they regulate chemical reactions without being used up or inactivated in the process. Like hammers or wrenches, they do their jobs without being

consumed or damaged as they work. There are generally thousands of different kinds of enzymes in every cell, which carry out the many processes on which life depends. Altogether, the multitude of enzymatic reactions performed by an organism is called its **metabolism.**

Section Review

1. Define *atom* and *element*. Are these terms interchangeable?
2. Your body contains vast numbers of carbon atoms. How is it possible that some of these carbon atoms may have been part of the body of a prehistoric creature?
3. What are six characteristics of water that make it so valuable for living organisms and their environment?

3.2 ENERGY

- *Energy occurs in different forms, such as kinetic energy, potential energy, chemical energy, or heat.*
- *The laws of thermodynamics state that energy is neither created nor destroyed, but that energy degrades to lower-intensity forms when used.*

Energy is the ability to do work, such as moving matter over a distance or causing a heat transfer between two objects at different temperatures. Energy can take many different forms. Heat, light, electricity, and chemical energy are examples that we all experience. Here we examine differences between forms of energy.

Energy varies in intensity

The energy contained in moving objects is called **kinetic energy.** A rock rolling down a hill, the wind blowing through the trees, water flowing over a dam (fig. 3.9), or electrons speeding around the nucleus of an atom are all examples of kinetic energy.

Potential energy is stored energy that is latent but available for use. A rock poised at the top of a hill and water stored behind a dam are examples of potential energy. **Chemical energy** is potential energy stored in the chemical bonds of molecules. The energy provided by the food you eat and the gasoline you put into your car are also examples of chemical energy that can be released to do useful work. Energy is often measured in units of heat (calories) or work (joules). One joule (J) is the work done when one kilogram is accelerated at one meter per second per second. One calorie is the amount of energy needed to heat one gram of pure water one degree Celsius. A calorie can also be measured as 4.184 J.

Heat is the energy that can be transferred between objects due to their difference in temperature. When a substance absorbs heat, the kinetic energy of its molecules increases, or it may change state: A solid may become a liquid, or a liquid may become a gas.

FIGURE 3.9 Water stored behind this dam represents potential energy. Water flowing over the dam has kinetic energy, some of which is converted to heat.
© William P. Cunningham

We sense change in heat content as change in temperature (unless the substance changes state).

An object can have a high heat content but a low temperature, such as a lake that freezes slowly in the fall. Other objects, like a burning match, have a high temperature but little heat content. Heat storage in lakes and oceans is essential to moderating climates and maintaining biological communities. Heat absorbed in changing states is also critical. As you will read in chapter 15, evaporation and condensation of water in the atmosphere help distribute heat around the globe.

Energy that is diffused, dispersed, and low in temperature is considered low-quality energy because it is difficult to gather and use for productive purposes. The heat stored in the oceans, for instance, is immense but hard to capture and use, so it is low quality. Conversely, energy that is intense, concentrated, and high in temperature is high-quality energy because of its usefulness in carrying out work. The intense flames of a very hot fire or high-voltage electrical energy are examples of high-quality forms that are valuable to humans. Many of our alternative energy sources (such as wind) are diffuse compared to the higher-quality, more concentrated chemical energy in oil, coal, or gas. This can mean that alternative energy sources are less intense—and less easy to use for work—than oil or gas.

Thermodynamics regulates energy transfers

Atoms and molecules cycle endlessly through organisms and their environment, but energy flows in a one-way path. A constant supply of energy—nearly all of it from the sun—is needed to keep biological processes running. Energy can be used repeatedly as it flows through the system, and it can be stored temporarily in the chemical bonds of organic molecules, but eventually it is released and dissipated.

The study of thermodynamics deals with how energy is transferred in natural processes. More specifically, it deals with the rates of flow and the transformation of energy from one form or quality to another. Thermodynamics is a complex, quantitative discipline, but you don't need a great deal of math to understand some of the broad principles that shape our world and our lives.

The **first law of thermodynamics** states that energy is *conserved;* that is, it is neither created nor destroyed under normal conditions. Energy may be transformed, for example, from the energy in a chemical bond to heat energy, but the total amount does not change.

The **second law of thermodynamics** states that, with each successive energy transfer or transformation in a system, less energy is available to do work. That is, energy is degraded to lower-quality forms, or it dissipates and is lost, as it is used. When you drive a car, for example, the chemical energy of the gas is degraded to kinetic energy and heat, which dissipates, eventually, to space. The second law recognizes that disorder, or **entropy,** tends to increase in all natural systems. Consequently, there is always less *useful* energy available when you finish a process than there was before you started. Because of this loss, everything in the universe tends to fall apart, slow down, and get more disorganized.

How does the second law of thermodynamics apply to organisms and biological systems? Organisms are highly organized, both structurally and metabolically. Constant care and maintenance is required to keep up this organization, and a constant supply of energy is required to maintain these processes. Every time some energy is used by a cell to do work, some of that energy is dissipated or lost as heat. If cellular energy supplies are interrupted or depleted, the result—sooner or later—is death.

Section Review

1. Restate the first and second law of thermodynamics.
2. The oceans store a vast amount of heat, but (except for climate moderation) this huge reservoir of energy is of little use to humans.
3. Explain the difference between high-quality and low-quality energy.

3.3 ENERGY FOR LIFE

- *Nearly all energy for life comes from the sun.*
- *Green plants capture this energy through photosynthesis; plants and animals release this energy through cellular respiration.*

The sun provides energy for nearly all plants and animals on earth. In this section, we examine how organisms capture and use this energy. We also explore an alternative energy source, chemical reactions using elements from the earth's crust.

Extremophiles gain energy without sunlight

Until recently, the deep ocean floor was believed to be essentially lifeless. Cold, dark, subject to crushing pressures, and without any known energy supply, it was a place where scientists thought nothing could survive. Undersea explorations in the 1970s, however, revealed dense colonies of animals—blind shrimp, giant tubeworms, strange crabs, and bizarre clams—clustered around vents called black chimneys, where boiling hot, mineral-laden water bubbles up through cracks in the earth's crust. How do these sunless ecosystems get energy? The answer is **chemosynthesis,** the process in which bacteria use chemical bonds between inorganic elements, such as hydrogen sulfide (H_2S) or hydrogen gas (H_2), to provide energy for synthesis of organic molecules.

Discovering organisms living under the severe conditions of deep-sea hydrothermal vents led to exploration of other sites that seem exceptionally harsh to us. Fascinating organisms have been discovered in hot springs, such as in Yellowstone National Park, in intensely salty lakes, and even in deep rock formations, up to 1,500 m (nearly a mile) deep in Columbia River basalts. Some species are amazingly hardy. The recently described *Pyrolobus fumarii* can withstand temperatures up to 113°C (235°F). Most of these extremophiles are archaea, single-celled organisms that are thought to be the most primitive of all living organisms, and the conditions under which they live are thought to be similar to those in which life first evolved.

FIGURE 3.10 A colony of tube worms and mussels cluster over a cool, deep-sea methane seep in the Gulf of Mexico.
Source: Image courtesy of Gulf of Mexico 2002, NOAA/OER

Deep-sea exploration of areas without thermal vents also has found abundant life (fig. 3.10). We now know that archaea live in oceanic sediments in astonishing numbers. The deepest of these species (they can be 800 m or more below the ocean floor) make methane from gaseous hydrogen (H_2) and carbon dioxide (CO_2), derived from rocks. Other species oxidize methane using sulfur to create hydrogen sulfide (H_2S), which is consumed by bacteria, that serves as a food source for more complex organisms such as tubeworms, crabs, and shrimp.

The vast supply of methane generated by this community could be either a great resource or a terrible threat to us. The total amount of methane made by these microbes is probably greater than all the known reserves of coal, gas, and oil. If we could safely extract the huge supplies of methane hydrate in ocean sediments, it could supply our energy needs for hundreds of years. Of greater immediate importance is that if methane-eating microbes weren't intercepting the methane produced by their neighbors, more than 300 million tons per year of this potent greenhouse gas would probably be bubbling to the surface, and we'd have runaway global warming. Methane-using bacteria can also help clean up pollution. After the Deepwater Horizon oil spill in the Gulf of Mexico in 2010, a deep-sea bloom of methane-metabolizing bacteria apparently consumed most of the methane (natural gas) escaping the spill.

Green plants get energy from the sun. Our sun is a star—a fiery ball of exploding hydrogen gas. Its energy comes from fission of hydrogen atoms, which releases intense ultraviolet energy and nuclear radiation (fig. 3.11), yet life here depends upon this searing energy source.

Solar energy is essential to life for two main reasons. First, the sun provides warmth. Most organisms survive within a relatively narrow temperature range: above 40°C, most biomolecules begin to break down or become distorted and nonfunctional. At low temperatures (near 0°C), some chemical reactions of metabolism

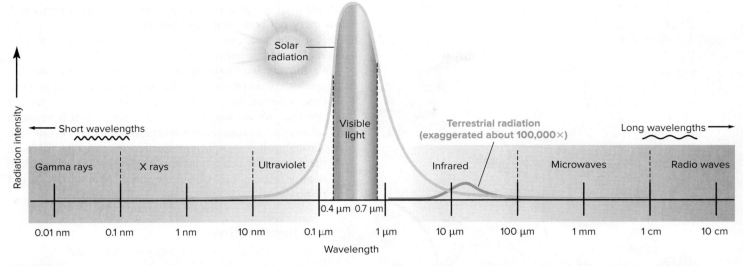

FIGURE 3.11 The electromagnetic spectrum. Our eyes are sensitive to light wavelengths, which make up nearly half the energy that reaches the earth's surface (represented by the area under the curve). Photosynthesizing plants also use the most abundant solar wavelengths. The earth reemits lower-energy, longer wavelengths, mainly the infrared part of the spectrum.

occur too slowly to enable organisms to grow and reproduce. Other planets in our solar system are either too hot or too cold to support life as we know it. The earth's water and atmosphere help to moderate, maintain, and distribute the sun's heat.

Second, nearly all organisms on the earth's surface depend on solar radiation for life-sustaining energy, which is captured by green plants, algae, and some bacteria in a process called **photosynthesis.** Photosynthesis converts radiant energy into high-quality chemical energy in the bonds that hold together organic molecules. Photosynthetic organisms (plants, algae, and bacteria) capture roughly 105 billion metric tons of carbon every year and store it as biomass. About half of this carbon capture is on land; about half is in the ocean.

This photosynthesis is accomplished using particular wavelengths of solar radiation that pass through our earth's atmosphere and reach the surface. About 45 percent of the radiation at the surface is visible, another 45 percent is infrared, and 10 percent is ultraviolet. Photosynthesis chiefly uses the most abundant wavelengths: visible and near infrared. Of the visible wavelengths, photosynthesis uses mainly red and blue light. Most plants reflect green wavelengths, so that is the color they appear to us. Half of the energy plants absorb is used in evaporating water. In the end, only 1 to 2 percent of the sunlight falling on plants is captured by photosynthesis. This small percentage is the energy base for virtually all life in the biosphere.

Photosynthesis captures energy; respiration releases that energy

Photosynthesis occurs in tiny organelles called chloroplasts that reside within plant cells (fig. 3.8). The main key to this process is chlorophyll, a green molecule that can absorb light energy and

use the energy to create high-energy chemicals in compounds that serve as the fuel for all subsequent cellular metabolism. Chlorophyll doesn't do this important job all alone, however. It is assisted by a large group of other lipid, sugar, protein, and nucleotide molecules. Together, these components carry out two interconnected cyclic sets of reactions (fig. 3.12).

Some photosynthetic reactions require energy from sunlight, and some don't. The process begins with a series of *light-dependent reactions*. These use solar energy directly to split water molecules into oxygen (O_2), which is released to the atmosphere, and hydrogen (H). This is the source of all the oxygen in the atmosphere on which all animals, including you, depend for life. Separating the hydrogen atom from its electron produces H^+ and an electron, both of which are used to form mobile, high-energy molecules called adenosine triphosphate (ATP) and nicotinamide adenine dinucleotide phosphate (NADPH).

Light-independent reactions then use the energy stored in ATP and NADPH molecules to create simple carbohydrates and sugar molecules (glucose, $C_6H_{12}O_6$) from carbon atoms (from CO_2) and water (H_2O). Glucose provides the energy and the building blocks for larger, more complex organic molecules. As ATP and NADPH give up some of their chemical energy, they are transformed to adenosine *di*phosphate (ADP) and NADP. These molecules are then reused in another round of light-dependent reactions. In most temperate-zone plants, photosynthesis can be summarized in the following equation:

$$6H_2O + 6CO_2 + \text{solar energy} \xrightarrow[\text{chlorophyll}]{} C_6H_{12}O_6 \text{ (sugar)} + 6O_2$$

We read this equation as "water plus carbon dioxide plus energy produces sugar plus oxygen." The reason the equation uses six water and six carbon dioxide molecules is that it takes six

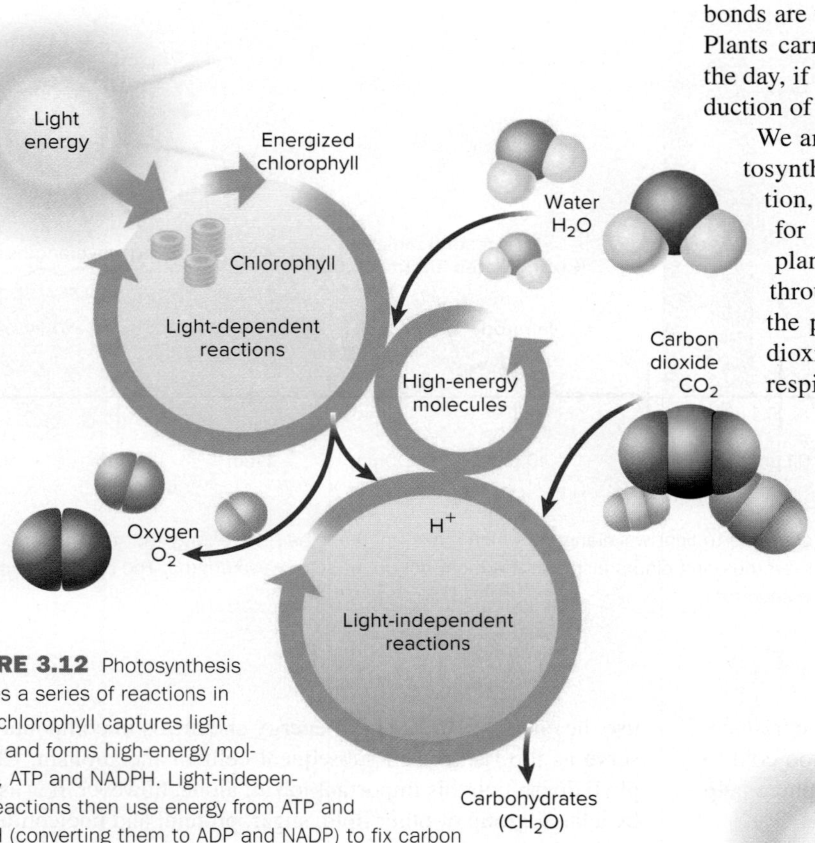

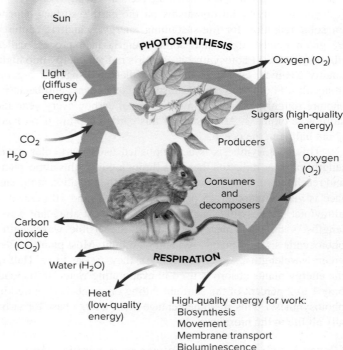

FIGURE 3.12 Photosynthesis involves a series of reactions in which chlorophyll captures light energy and forms high-energy molecules, ATP and NADPH. Light-independent reactions then use energy from ATP and NADPH (converting them to ADP and NADP) to fix carbon from air in organic molecules.

bonds are used to capture, store, and deliver energy within a cell. Plants carry out both photosynthesis and respiration, but during the day, if light, water, and CO_2 are available, they have a net production of O_2 and carbohydrates.

We animals don't have chlorophyll and can't carry out photosynthetic food production. We do perform cellular respiration, however. In fact, this is how we get all our energy for life. We eat plants—or other animals that have eaten plants—and break down the organic molecules in our food through cellular respiration to obtain energy (fig. 3.13). In the process, we also consume oxygen and release carbon dioxide, thus completing the cycle of photosynthesis and respiration.

Section Review

1. What are primary producers? Consumers?
2. What is the source of carbon for green plants? What is one product of photosynthesis?
3. How are photosynthesis and respiration related?

carbon atoms to make glucose (a sugar). The CO_2 in this equation is captured from the air by plant tissues. Thus, you could say that a plant is made primarily from air and water.

What does the plant do with glucose? Because glucose is an energy-rich compound, it serves as the central, primary fuel for all metabolic processes of cells. The energy in its chemical bonds—created by photosynthesis—can be released by other enzymes and used to make other molecules (lipids, proteins, nucleic acids, or other carbohydrates), or it can drive kinetic processes such as the movement of ions across membranes, the transmission of messages, changes in cellular shape or structure, or movement of the cell itself in some cases.

This process of releasing chemical energy, called **cellular respiration,** involves splitting carbon and hydrogen atoms from the sugar molecule and recombining them with oxygen to re-create carbon dioxide and water. The net chemical reaction, then, is the reverse of photosynthesis:

$$C_6H_{12}O_6 + 6O_2 \longrightarrow 6H_2O + 6CO_2 + \text{released energy}$$

Note that in photosynthesis, energy is *captured,* while in respiration, energy is *released.* Similarly, photosynthesis *consumes* water and carbon dioxide to *produce* sugar and oxygen, while respiration does just the opposite. In both sets of reactions, chemical

FIGURE 3.13 Energy exchange in ecosystems. Plants use sunlight, water, and carbon dioxide to produce sugars and other organic molecules. Consumers use oxygen and break down sugars during cellular respiration. Plants also carry out respiration, but during the day, if light, water, and CO_2 are available, they have a net production of O_2 and carbohydrates.

3.4 FROM SPECIES TO ECOSYSTEMS

- *Ecosystems consist of organisms and the systems they depend on.*
- *Food webs are structured by trophic levels.*
- *Higher trophic levels have fewer organisms and less mass than lower trophic levels.*

When we discuss Chesapeake Bay as a complex system (opening case study), we are concerned with rates of photosynthesis, abundance of photosynthesizing algae, and the ways that changes to the bay's chemistry influence population sizes for different species. Numbers of blue crabs, oysters, menhaden, and other species all contribute to our assessment of the system's stability and health. Understanding how nutrients and energy function in a system, and where they come from, and where they go, are essential to understanding **ecology,** the scientific study of relationships between organisms and their environment.

Ecosystems include living and nonliving parts

To biologists, terms like *species*, *population*, and *community* have very particular meanings. In Latin, *species* literally means "kind." In biology, **species** generally refers to all organisms of the same kind that are genetically similar enough to breed in nature and produce live, fertile offspring. There are important exceptions to this definition, and taxonomists increasingly use genetic differences to define species, but for our purposes this is a useful working definition.

A **population** consists of all the members of a species living in a given area at the same time. All the populations living and interacting in a particular area make up a **biological community.** What populations make up the biological community of which you are a part? If you consider all the populations of animals, plants, fungi, and microorganisms in your area, your community is probably large and complex. We'll explore the dynamics of populations and communities more in chapters 4 and 6.

As discussed in chapter 2, systems are networks of interaction among many interdependent factors. Your body, for example, is a very complex, self-regulating system. An ecological system, or **ecosystem,** is composed of a biological community and its physical environment. The environment includes abiotic factors (nonliving components), such as climate, water, minerals, and sunlight, as well as biotic factors, such as organisms and their products (secretions, wastes, and remains) and effects in a given area. It is useful to think about the biological community and its environment together, because energy and matter flow through both. Understanding how those flows work is a major theme in ecology.

For simplicity, we think of ecosystems as distinct ecological units with fairly clear boundaries. If you look at a patch of woods surrounded by farm fields, for instance, a relatively sharp line might separate the two areas, and conditions such as light levels, wind, moisture, and shelter are quite different in the woods than in the fields around them. Because of these variations, distinct populations of plants and animals live in each place. By studying each of these areas, we can make important and interesting discoveries about who lives where and why and about how conditions are established and maintained there.

The division between the fields and woods is not always clear, however. Air, of course, moves freely from one to another, and the runoff after a rainfall may carry soil, leaf litter, and live organisms between the areas. Birds may feed in the field during the day but roost in the woods at night, giving them roles in both places. Are they members of the woodland community or the field community? Is the edge of the woodland ecosystem where the last tree grows, or does it extend to every place that has an influence on the woods?

As you can see, it may be difficult to draw clear boundaries around communities and ecosystems. To some extent, we define these units by what we want to study and how much information we can handle. Thus, an ecosystem might be as large as a whole watershed or as small as a pond or even your own body. The thousands of species of bacteria, fungi, protozoans, and other organisms that live in and on your body make up a complex, interdependent community called the **microbiome.** You keep the other species warm and fed; they help you with digestion, nutrition, and other bodily functions. Some members of your community are harmful, but many are beneficial. You couldn't survive easily without them. We'll discuss your microbiome further in chapter 4.

You, as an ecosystem, have clear boundaries, but you are open in the sense that you take in food, water, energy, and oxygen from your surrounding environment, and you excrete wastes. This is true of most ecosystems, but some are relatively closed; that is, they import and export comparatively little from outside. Others, such as a stream, are in a constant state of flux with materials and even whole organisms coming and going. Because of the second law of thermodynamics, however, every ecosystem must have a constant inflow of energy and a way to dispose of heat. Thus, with regard to energy flow, every ecosystem is open.

Many ecosystems have feedback mechanisms that maintain generally stable structures and functions. A forest tends to remain a forest, for the most part, and to have forest-like conditions if it isn't disturbed by outside forces. Some ecologists suggest that ecosystems—or perhaps all life on the earth—may function as superorganisms, because they maintain stable conditions and can be resilient to change.

Food webs link species of different trophic levels

All ecosystems are based on photosynthesis (or, rarely, chemosynthesis). Organisms that photosynthesize, mainly green plants and algae, are therefore known as **producers.** One of the major properties of an ecosystem is its **productivity,** the amount of **biomass** (biological material) produced in a given area during a given period of time. Photosynthesis is described as *primary productivity* because it is the basis for almost all other growth in an ecosystem. Manufacture of biomass by organisms that eat plants is termed *secondary productivity*. A given ecosystem may have very high total productivity, but if decomposers decompose organic material as rapidly as it is formed, the *net primary productivity* will be low.

Think about what you have eaten today and trace it back to its photosynthetic source. If you have eaten an egg, you can trace it back to a chicken, which probably ate corn. This is an example of a **food chain,** a linked feeding series. Now think about a more complex food chain involving you, a chicken, a corn plant, and a grasshopper. The chicken could eat grasshoppers that had eaten leaves of the corn plant. You also could eat the grasshopper directly—some humans do. Or you could eat corn yourself, making the shortest possible food chain. Humans have several options of where we fit into food chains.

In ecosystems, some consumers feed on a single species, but most consumers have multiple food sources. Similarly, some species are prey to a single kind of predator, but many species in an ecosystem are beset by several types of predators and parasites. In this way, individual food chains become interconnected to form a **food web.** Figure 3.14 shows feeding relationships among some of the larger organisms in a woodland and lake community. If we were to add all the insects, worms, and microscopic organisms that belong in this picture, however, we would have overwhelming complexity. Perhaps you can imagine the challenge ecologists face in trying to quantify and interpret the precise matter and energy transfers that occur in a natural ecosystem!

An organism's feeding status in an ecosystem can be expressed as its **trophic level** (from the Greek *trophe,* "food"). In our first example, the corn plant is at the producer level; it transforms solar energy into chemical energy, producing food molecules. Other organisms in the ecosystem are **consumers** of the chemical energy harnessed by the producers. An organism that eats producers is a primary consumer. An organism that eats primary consumers is a secondary consumer, which may, in turn, be eaten by a tertiary consumer, and so on. Most terrestrial food chains are relatively short (seeds → mouse → owl), but aquatic food chains may be quite long (microscopic algae → copepod → minnow → crayfish → bass → osprey). The length of a food chain also may reflect the physical characteristics of a particular ecosystem. A harsh arctic landscape, with relatively low species diversity, can have a much shorter food chain than a temperate or tropical one (fig. 3.15).

Organisms can be identified both by the trophic level at which they feed and by the *kinds* of food they eat (fig. 3.16). **Herbivores** are plant eaters, **carnivores** are flesh eaters, and **omnivores** eat both plant and animal matter. What are humans? We are natural omnivores, by history and by habit. Tooth structure is an important clue to understanding animal food preferences, and humans are no exception. Our teeth are suited for an omnivorous diet, with a combination of cutting and crushing surfaces that are not highly adapted for one specific kind of food, as are the teeth of a wolf (carnivore) or a horse (herbivore).

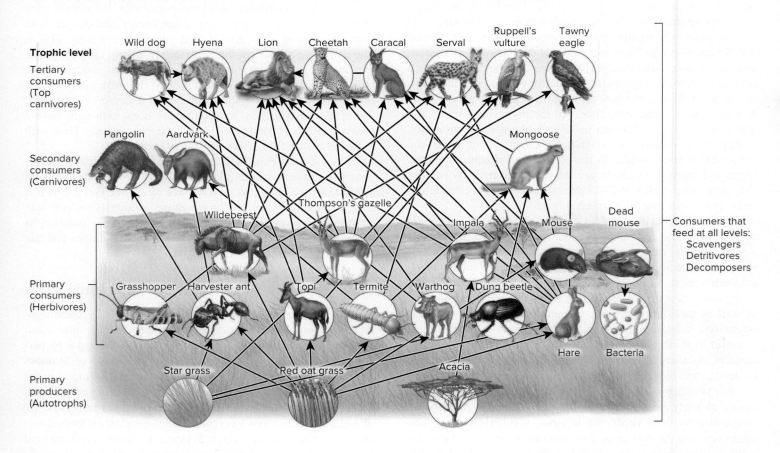

FIGURE 3.14 Each time an organism feeds, it becomes a link in a food chain. In an ecosystem, food chains become interconnected when predators feed on more than one kind of prey, thus forming a food web. The arrows in this diagram and in figure 3.15 indicate the direction in which matter and energy are transferred through feeding relationships. Only a few representative relationships are shown here. What others might you add?

FIGURE 3.15 Harsh environments tend to have shorter food chains than environments with more favorable physical conditions. Compare the arctic food chains depicted here with the longer food chains in the food web in figure 3.14.

One of the most important trophic levels is occupied by the many kinds of organisms that remove and recycle the dead bodies and waste products of others. **Scavengers** such as crows, jackals, and vultures clean up dead carcasses of larger animals. **Detritivores** such as ants and beetles consume litter, debris, and dung, while **decomposer** organisms such as fungi and bacteria complete the final breakdown and recycling of organic materials. It could be argued that these microorganisms are second in importance only to producers, because without their activity, nutrients would remain locked up in the organic compounds of dead organisms and discarded body wastes, rather than being made available to successive generations of organisms.

Ecological pyramids describe trophic levels

If we arrange the organisms according to trophic levels, they generally form a pyramid with a broad base representing primary producers and only a few individuals in the highest trophic levels. This pyramid arrangement is especially true if we look at the energy content of an ecosystem (fig. 3.17).

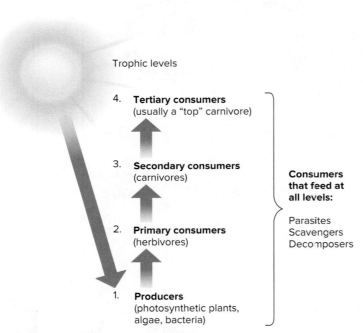

Trophic levels

4. **Tertiary consumers**
(usually a "top" carnivore)

3. **Secondary consumers**
(carnivores)

2. **Primary consumers**
(herbivores)

1. **Producers**
(photosynthetic plants,
algae, bacteria)

**Consumers
that feed at
all levels:**

Parasites
Scavengers
Decomposers

FIGURE 3.16 Organisms in an ecosystem may be identified by how they obtain food for their life processes (producer, herbivore, carnivore, omnivore, scavenger, decomposer, reducer) or by consumer level (producer; primary, secondary, or tertiary consumer) or by trophic level (1st, 2nd, 3rd, 4th).

Quantitative Reasoning

As a rule of thumb, about one-tenth of the energy or biomass consumed is stored at each trophic level. About how many kg of feed should it take to produce 1 kg of chicken meat that we eat? How much more energy should it take to provide you a meal of meat compared to vegetables?

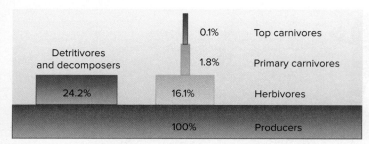

FIGURE 3.17 A classic example of an energy pyramid from Silver Springs, Florida. The numbers in each bar show the percentage of the energy captured in the primary producer level that is incorporated into the biomass of each succeeding level. Detritivores and decomposers feed at every level but are shown attached to the producer bar because this level provides most of their energy.

Why is there so much less energy in each successive level in figure 3.17? Because of the second law of thermodynamics, which says that energy dissipates and degrades as it is reused. Thus, a rabbit consumes a great deal of chemical energy stored in carbohydrates in grass, and much of that energy is transformed to kinetic energy, when the rabbit moves, or to heat, which dissipates to the environment. A fox eats the rabbit, and the same degradation and dissipation happen again. As the fox uses the energy to live, some is lost in heat or movement. A little of the energy it has eaten is stored in chemical bonds in the fox's tissues. From an ecosystem energy perspective, there will always be smaller amounts of energy at successively higher trophic levels. Large top carnivores need a very large pyramid, and a large home range, to support them. A tiger, for example, may require a home range of several hundred square kilometers to survive.

As a broad generalization, only about one-tenth of the energy in one trophic level is represented in the next higher level (fig. 3.18). The amount of energy available is often expressed in biomass. For example, it generally takes about 100 kg of clover to produce 10 kg of rabbit, and 10 kg of rabbit to make 1 kg of fox.

The total number of organisms and the total amount of biomass in each successive trophic level of an ecosystem also may

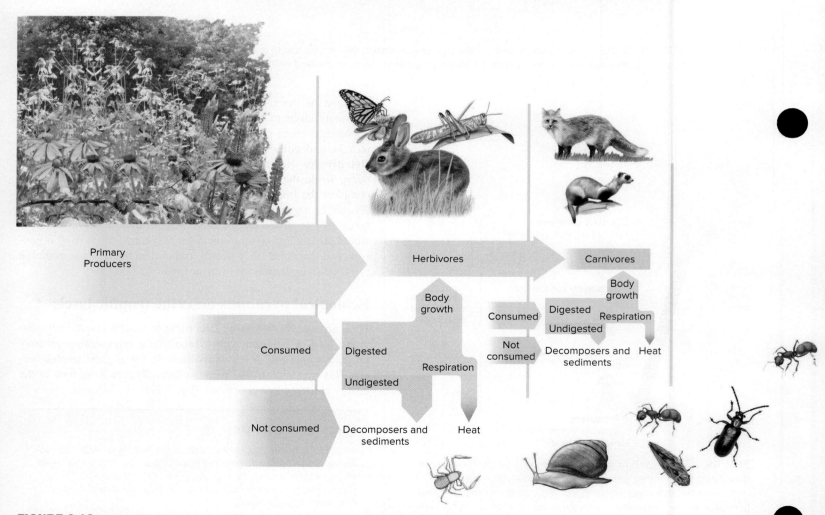

FIGURE 3.18 A biomass pyramid shows that, like energy, biomass storage decreases at higher trophic levels.

form pyramids (fig. 3.19) similar to those describing energy content. The relationship between biomass and numbers is not as dependable as energy, however. The biomass pyramid, for instance, can be inverted by periodic fluctuations in producer populations (for example, low plant and algal biomass present during winter in temperate aquatic ecosystems). The numbers pyramid also can be inverted. One coyote can support numerous tapeworms, for example. Numbers inversion also occurs at the lower trophic levels (for example, one large tree can support thousands of caterpillars).

Section Review

1. Describe the following: producers; consumers; secondary consumers; decomposers.

2. Ecosystems require energy to function. Where does this energy go as it is used? How does the flow of energy conform to the laws of thermodynamics?

3. Why are there generally fewer organisms at the top of the food pyramid than at the bottom?

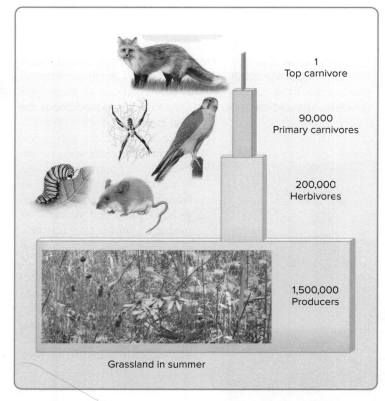

FIGURE 3.19 Usually, smaller organisms are eaten by larger organisms and it takes numerous small organisms to feed one large organism. The classic study represented in this pyramid shows numbers of individuals at each trophic level per 1,000 m^2 of grassland, and reads like this: To support one individual at the top carnivore level, there were 90,000 primary carnivores feeding upon 200,000 herbivores that in turn fed upon 1,500,000 producers.

3.5 MATERIAL CYCLES

- *The water cycle distributes water among atmosphere, biosphere, surface, and groundwater.*
- *Carbon, nitrogen, and phosphorous are among the essential elements that also move through biological, atmospheric, and earth systems (biogeochemical cycles).*

Earth is the only planet in our solar system that provides a suitable environment for life as we know it. Even our nearest planetary neighbors, Mars and Venus, do not meet these requirements. Maintenance of these conditions requires a constant recycling of materials between the biotic (living) and abiotic (nonliving) components of ecosystems.

The hydrologic cycle redistributes water

The path of water through our environment, known as the **hydrologic cycle,** is perhaps the most familiar material cycle, and it is discussed in greater detail in chapter 17. Most of the earth's water is stored in the oceans, but solar energy continually evaporates this water, and winds distribute water vapor around the globe. Water that condenses over land surfaces, in the form of rain, snow, or fog, supports all terrestrial (land-based) ecosystems (fig. 3.20). Living organisms emit the moisture they have consumed through respiration and perspiration. Eventually this moisture reenters the atmosphere or enters lakes and streams, from which it ultimately returns to the ocean again.

As it moves through living things and through the atmosphere, water is responsible for metabolic processes within cells, for maintaining the flows of key nutrients through ecosystems, and for global-scale distribution of heat and energy (chapter 15). Water performs countless services because of its unusual properties. Water is so important that when astronomers look for signs of life on distant planets, traces of water are the key evidence they seek.

Everything about global hydrological processes is awesome in scale. Each year, the sun evaporates approximately 496,000 km^3 of water from the earth's surface. More water evaporates in the tropics than at higher latitudes, and more water evaporates over the oceans than over land. Although the oceans cover about 70 percent of the earth's surface, they account for 86 percent of total evaporation. Ninety percent of the water evaporated from the ocean falls back on the ocean as rain. The remaining 10 percent is carried by prevailing winds over the continents, where it combines with water evaporated from soil, plant surfaces, lakes, streams, and wetlands to provide a total continental precipitation of about 111,000 km^3.

What happens to the surplus water on land—the difference between what falls as precipitation and what evaporates? Some of it is incorporated by plants and animals into biological tissues. A large share of what falls on land seeps into the ground to be stored for a while as soil moisture or groundwater. Water might stay for just a few days or weeks in soil and shallow groundwater. In deep aquifers, water can reside for centuries or millennia. Eventually, all the water makes its way back downhill to the oceans.

FIGURE 3.20 The hydrologic cycle. Most exchange occurs with evaporation from oceans and precipitation back to oceans. About one-tenth of water evaporated from oceans falls over land, is recycled through terrestrial systems, and eventually drains back to oceans in rivers.

The 40,000 km³ carried back to the ocean each year by surface runoff or underground flow is the renewable supply available for human uses and sustaining freshwater-dependent ecosystems.

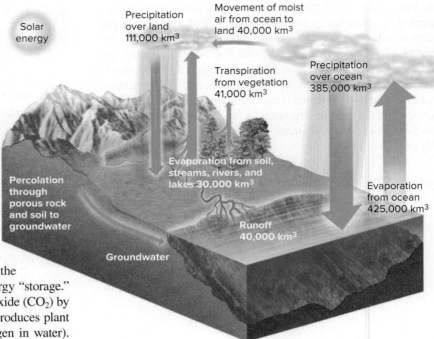

Carbon cycles through earth, air, water, and life

Carbon serves a dual purpose for organisms: (1) it is a structural component of organic molecules, and (2) the energy-holding chemical bonds it forms represent energy "storage." The **carbon cycle** begins with the intake of carbon dioxide (CO_2) by photosynthetic organisms (fig. 3.21). Photosynthesis produces plant cells from carbon in the air (plus hydrogen and oxygen in water). Carbon dioxide is eventually released during respiration, closing the cycle. The carbon cycle is of special interest because biological accumulation and release of carbon is a major factor in climate regulation.

The path followed by an individual carbon atom in this cycle may be quite direct and rapid, depending on how it is used in an organism's body. Imagine for a moment what happens to a simple sugar molecule you swallow in a glass of fruit juice. The sugar molecule is absorbed into your bloodstream, where it is made available to your cells for cellular respiration or for making more complex biomolecules. If it is used in respiration, you may exhale the same carbon atom as CO_2 the same day.

Some forms of carbon cycle extremely slowly. Coal and oil are the compressed, chemically altered remains of plants or microorganisms that lived millions of years ago. Their carbon atoms (and hydrogen, oxygen, nitrogen, sulfur, etc.) are not released until the coal and oil are burned. We dramatically accelerate the movement of this carbon when we use fossil fuels. Enormous amounts of carbon also are locked up as calcium carbonate ($CaCO_3$), used to build shells and skeletons of marine organisms from tiny protozoans to corals. Most of these deposits are at the bottom of the oceans. The world's extensive surface limestone deposits are biologically formed calcium carbonate from ancient oceans, exposed by geological events. The carbon in limestone has been

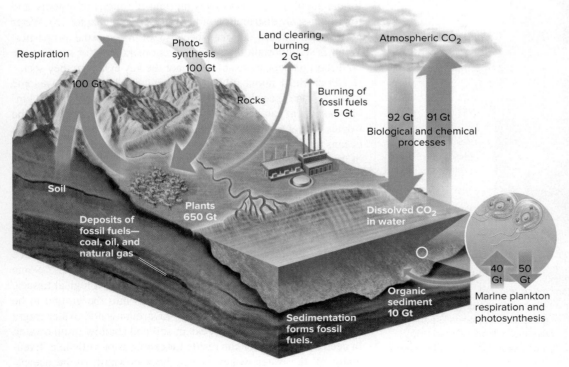

FIGURE 3.21 The carbon cycle. Numbers indicate the approximate exchange of carbon in gigatons (Gt) per year. Natural exchanges are balanced, but human sources produce a net increase of CO_2 in the atmosphere.

locked away for millennia, which is probably the fate of carbon currently being deposited in ocean sediments. Eventually, even the deep ocean deposits are recycled, as they are drawn into deep molten layers and released via volcanic activity. Geologists estimate that every carbon atom on the earth has made about 30 such round trips over the last 4 billion years.

How does tying up so much carbon in the bodies and by-products of organisms affect the biosphere? Favorably. It helps balance CO_2 generation and utilization. Carbon dioxide is one of the so-called greenhouse gases because it absorbs heat radiated from the earth's surface, retaining it instead in the atmosphere (see chapter 15). Photosynthesis, accumulation of organic matter in soils and wetlands, and deposition of $CaCO_3$ remove atmospheric carbon dioxide; therefore, expansive forested areas, such as the boreal forests, and the oceans are very important **carbon sinks** (storage deposits). Cellular respiration and combustion both release CO_2, so they are referred to as carbon sources in the cycle.

Presently, combustion of organic fuels (mainly wood, coal, and petroleum products), removal of standing forests, and soil degradation are releasing huge quantities of CO_2 at rates that surpass the pace of CO_2 removal, a problem discussed in chapters 15 and 16.

Nitrogen occurs in many forms

As in Chesapeake Bay (opening case study), nitrogen often is a key limiting factor in ecosystems. Nitrogen surpluses or shortages can dramatically alter biological communities. This is because nitrogen is an essential component of proteins, enzymes, DNA, and countless other organic compounds (see fig. 3.6). Primary producers incorporate nitrogen into these compounds, and consumers, such as yourself, acquire nitrogen by consuming plants or animals that have eaten plants.

Nitrogen is abundant in our environment—nitrogen gas molecules (N_2) make up 78 percent of our atmosphere. But these nitrogen molecules have strong bonds, and green plants can't use nitrogen from N_2 directly. Plants rely on bacteria living in soils, in plant tissues, or in aquatic systems to capture N_2. These "nitrogen fixing" bacteria have proteins that can break the N_2 bonds. The N is then available to be combined with hydrogen to form NH_3 (ammonia) or NH_4^+ (ammonium; the "+" indicates that the molecule has a positive ionic charge). Bacteria can gain energy by recombining these molecules with oxygen to form NO_2^- (nitrite) or NO_3^- (nitrate).

Of the many nitrogen compounds, only two, nitrate (NO_3^-) and ammonium (NH_4^+) can be used directly by plants. These are the sources of nitrogen for forming amino acids, the building blocks for complex organic compounds such as proteins.

Some plants, such as those in the bean and clover families, have nitrogen-fixing bacteria that reside in their roots. In beans, these bacteria live in nodules on plant roots (fig. 3.22). Having resident bacteria gives these plants built-in fertilizer. In return, the plant gives carbohydrates to the bacteria. This symbiotic relationship also enriches soil, so farmers often rotate legumes, such as beans, with nitrogen-hungry crops such as corn. Using natural fertilizers this way can improve both soil and productivity.

FIGURE 3.22 The roots of this bean plant are covered with bumps called nodules. Each nodule is a mass of root tissue containing many bacteria that help to convert nitrogen in the soil to a form the bean plants can assimilate and use to manufacture amino acids.
© Nigel Cattlin/Alamy Stock Photo

Bacteria in soil or water provide chemical reactions that drive many parts of the complex **nitrogen cycle** (fig. 3.23). Where oxygen is available, bacteria may combine ammonia (NH_3) with oxygen to form nitrous oxide (N_2O), nitric oxide (NO), nitrite (NO_2^-), or nitrate (NO_3^-).

In oxygen-poor conditions, such as in streambed sediments, saturated soils, or wetlands, denitrifying bacteria may remove oxygen from nitrate to form gaseous compounds, especially nitrous oxide or nitrogen gas (table 3.1). Conversion to these gaseous forms is known as denitrification. Denitrification can be important in removing nitrogen from aquatic systems that suffer from eutrophication.

Nitrogen moves through the food web as organisms die, decompose, or are consumed. Decomposers (fungi and bacteria) release ammonia and ammonium ions, which then are available for nitrate formation. Organisms also release proteins when plants shed their leaves, needles, flowers, fruits, and cones; or when animals shed hair, feathers, skin, exoskeletons, pupal cases, and silk, excrement, or urine, all of which are rich in nitrogen. Urinary wastes are especially high in nitrogen because they contain the detoxified wastes of protein metabolism.

Humans now capture a great deal of atmospheric nitrogen, and our activities have more than doubled the amount of nitrogen in circulation. Synthetic ammonia (NH_3) fertilizers are largely responsible. This form of nitrogen is produced from atmospheric nitrogen and natural gas under extreme heat and pressure. The spread of nitrogen-fixing soybeans, now one of the world's dominant crops, has also increased nitrogen capture. Even more important is the burning of fossil fuels in car engines and industrial furnaces. The high heat and pressure of combustion fix N_2 to form ammonia (NH_3) or nitrous oxide (N_2O). Rainfall washes these compounds out of the air and into soils and aquatic ecosystems. Excessive nitrogen fertilization is a dominant pollutant in many destabilized rivers, lakes, and estuaries. In terrestrial systems, nitrogen enrichment encourages the spread of weeds into

FIGURE 3.23 The nitrogen cycle. Nitrogen occurs in many different forms. Human activities have doubled the amount of nitrogen circulating, mainly because of industry, gasoline combustion in transportation, and agricultural fertilizers.

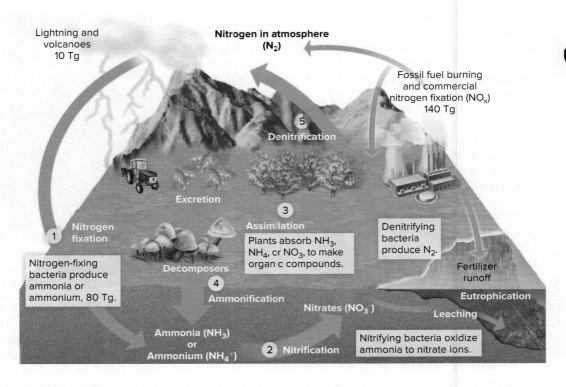

Table 3.1 Common Forms of Nitrogen

Name	Formula	Characteristics
Nitrogen gas	N_2	Makes up 78% of atmosphere; can be fixed by bacteria, cyanobacteria, some marine algae.
Ammonia	NH_3	Gaseous form; corrosive; common agricultural fertilizer that bonds with H^+ to form $NH4^+$.
Ammonium	NH_4^+	Directly usable by plants; positive charge helps adhere to clays in soils.
Nitrite	NO_2^-	Ion with negative electrical charge; a step in *nitrification* process (conversion of $NH4^+$ to $NO3^-$); toxic to plants; usually present temporarily or in low quantities.
Nitrate	NO_3^-	Directly usable by plants; product of *nitrification*.
Organic Nitrogen		Diverse compounds, such as proteins; must be converted to $NH4^+$ for use by plants. Organic to inorganic transition is *mineralization*.
Nitrogen oxides	NO_x	Various combinations such as NO_2 (nitrogen dioxide), NO (nitric oxide), and N_2O (nitrous oxide); fuel combustion in vehicles and industry produces most NO_x; rainfall washes NO_x into soils and waterways. NO_2 and N_2 also result from *denitrification* by bacteria.

ecosystems that historically had low N inputs. In addition, nitrous oxide (N_2O) is an important greenhouse gas.

Phosphorus follows a one-way path

Phosphorus is an essential component of all cells. Compounds containing this element store and release a great deal of energy, so phosphorus-containing compounds, such as ATP, are primary participants in energy-transfer reactions in cells. Phosphorus is also a key component of proteins, enzymes, and tissues. The amount of available phosphorus in an ecosystem can have a dramatic effect on productivity. Low levels of phosphorus can limit plant growth. Abundant phosphorus stimulates lush plant and algal growth, making it a major water pollutant.

Most plants acquire phosphorus from soil, and animals acquire phosphorus by consuming plants or other animals. The ultimate source of phosphorus, though, is rocks. On the scale of centuries, the **phosphorus cycle** (fig. 3.24) is really a one-way path. This is because phosphorus has no atmospheric form in which it can quickly recirculate. Instead, phosphorus travels gradually downstream, as it is leached from rocks and minerals, taken up by the food web, and eventually released into water bodies that deliver it to the ocean. Phosphorus cycles repeatedly through the food web, as inorganic phosphorus is taken up by plants, incorporated into organic molecules, and passed on to consumers. Eventually, though, phosphorus washes down river to the ocean, where it accumulates in ocean sediments. Over geologic time, these deposits can be uplifted and exposed, so they become available to terrestrial life again. The phosphate used for detergents and fertilizers today are mined from exposed ocean sediments millions of

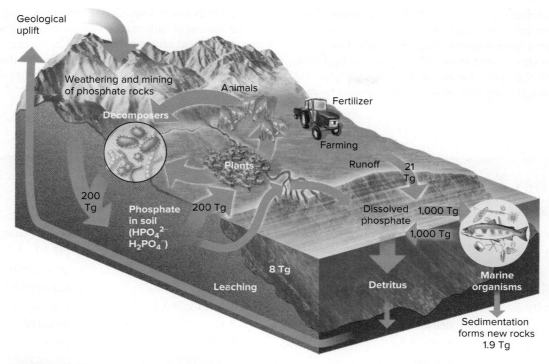

FIGURE 3.24 The phosphorus cycle. Natural movement of phosphorus is slight, involving recycling within ecosystems and some erosion and sedimentation of phosphorus-bearing rock. Use of phosphate (PO_4^{-3}) fertilizers and cleaning agents increases phosphorus in aquatic systems, causing eutrophication. Units are teragrams (Tg) phosphorus per year.

years old. As with nitrogen, we have dramatically accelerated the movement of phosphorus in our environment. The effects of this increased phosphorus on aquatic ecosystems, as in Chesapeake Bay, can include explosive growth of algae and photosynthetic bacteria, or eutrophication.

Phosphorus is essential for agriculture. There are worries about how long global phosphorus supplies will last. Some authorities estimate that we have several hundred years' worth of this essential mineral. Others worry that we may run out much sooner than that.

Section Review

1. What do we mean by carbon-fixation or nitrogen-fixation? Why is it important to humans that carbon and nitrogen be "fixed"?

2. Why is nitrogen important to living things? What type of organism is responsible for making nitrogen available to plants?

3. Describe the relative residence time of water in the ocean, the atmosphere, and in rivers and lakes.

4. In what forms is carbon stored in oceans, the atmosphere, living things, and rocks?

Conclusion

Matter is conserved as it cycles over and over through ecosystems, but energy is always degraded or dissipated as it is transformed or transferred from one place to another. These laws of physics and thermodynamics mean that elements are continuously recycled, but that living systems need a constant supply of external energy to replace that lost to entropy. Some extremophiles living in harsh conditions, such as hot springs or the bottom of the ocean, capture energy from chemical reactions. For most organisms, however, the ultimate source of energy is the sun. Plants capture sunlight through the process of photosynthesis and use the captured energy for metabolic processes and to build biomass (organic material). Herbivores eat plants to obtain energy and nutrients, carnivores

eat herbivores or each other, and decomposers eat the waste products of this food web.

This dependence on solar energy is a fundamental limit for most life on earth. It's estimated that humans now dominate roughly 40 percent of the potential terrestrial net productivity. We directly eat only about 10 percent of that total (mainly because of the thermodynamic limits on energy transfers in food webs), but the crops and livestock that feed, clothe, and house us represent the rest of that photosynthetic output. By dominating nature as we do, we exclude other species.

While energy flows in a complex but ultimately one-way path through nature, materials are endlessly recycled. Four of the major material cycles are water, carbon, nitrogen, and phosphorus. Each of these materials is critically important to living organisms. As humans interfere with these material cycles, we make it easier for some organisms to survive and more difficult for others. Often, we manipulate material cycles for our own short-term gain, but we don't think about the consequences for other species or even for ourselves in the long term. For example, we have accelerated the movement of carbon and nitrogen into the atmosphere, but the resulting global climate change and ecosystem disruption could have disastrous results for us. It's essential that we understand these environmental systems so that we can take them into account as we consider policies for resource use.

Reviewing Key Terms

Can you define the following terms in environmental science?

acids 3.1
atomic number 3.1
atoms 3.1
bases 3.1
biological community 3.4
biomass 3.4
carbon cycle 3.5
carbon sink 3.5
carnivore 3.4
cell 3.1
cellular respiration 3.3
chemical energy 3.2
chemosynthesis 3.3
compound 3.1

conservation of matter 3.1
consumer 3.4
decomposer 3.4
deoxyribonucleic acid
    (DNA) 3.1
detritivore 3.4
ecology 3.4
ecosystem 3.4
elements 3.1
energy 3.2
entropy 3.2
enzymes 3.1
first law of
    thermodynamics 3.2

food chain 3.4
food web 3.4
heat 3.2
herbivore 3.4
hydrologic cycle 3.5
ions 3.1
isotopes 3.1
kinetic energy 3.2
matter 3.1
metabolism 3.1
microbiome 3.4
molecule 3.1
nitrogen cycle 3.5
omnivore 3.4

organic compounds 3.1
pH 3.1
phosphorus cycle 3.5
photosynthesis 3.3
population 3.4
potential energy 3.2
producers 3.4
productivity 3.4
scavenger 3.4
second law of
    thermodynamics 3.2
species 3.4
trophic level 3.4

Critical Thinking and Discussion Questions

1. If your dishwasher detergents contained phosphorus, would you change brands? Would you encourage others to change? Why or why not?

2. Describe one or two practical examples of the laws of thermodynamics in your own life. Do they help explain why you can recycle cans and bottles but not energy? Which law is responsible for the fact that you get hot and sweaty when you exercise?

3. The ecosystem concept revolutionized ecology by introducing holistic systems thinking as opposed to individualistic life history studies. Why was this a conceptual breakthrough?

4. If ecosystems are so difficult to delimit, why is this such a persistent concept? Can you imagine any other ways to define or delimit environmental investigation?

5. Choose one of the material cycles (carbon, nitrogen, phosphorus, or sulfur) and identify the components of the cycle in which you participate. For which of these components would it be easiest to reduce your impacts?

Data Analysis

Inspect the Chesapeake's Report Card

You know that nutrients are an important concern in the Chesapeake Bay watershed in general, but now you can examine the details and see how conditions have changed. Go to http://ecoreportcard.org/report-cards/chesapeake-bay/health and click on "Learn More" to see the 2015 data. This site is maintained by the University of Maryland and the National Oceanic and Atmospheric Administration (NOAA), with the support of many collaborators and data providers. Use this site to examine recent data on chlorophyll, nutrients, and other measures of the bay's health.

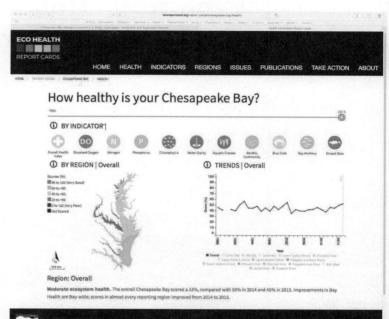

FIGURE 1 The EcoCheck website provides a wealth of water quality data.
Source: Integration and Application Network, University of Maryland Center for Environmental Science (http://ecoreportcard.org/report-cards/chesapeake-bay/health/) Used by permission.

TO ACCESS ADDITIONAL RESOURCES FOR THIS CHAPTER, PLEASE VISIT CONNECT AT www.connect.mheducation.com.

You will find LearnSmart, an adaptive learning system, Google Earth™ exercises, additional Case Studies, Data Analysis exercises, and an interactive ebook.

4 Evolution, Biological Communities, and Species Interactions

▲ The relatively young and barren volcanic islands of the Galápagos, isolated from South America by strong, cold currents and high winds, have developed a remarkable community of unique plants and animals.
© Galen Rowell/Corbis

Learning Outcomes

After studying this chapter, you should be able to:

4.1 Describe how evolution produces species diversity.

4.2 Discuss how species interactions shape biological communities.

4.3 Summarize how community properties affect species and populations.

4.4 Explain why communities are dynamic and change over time.

"When I view all beings not as special creations, but as lineal descendents of some few beings which have lived long before the first bed of the Cambrian system was deposited, they seem to me to become ennobled."

– Charles Darwin

Natural Selection in the Galápagos Islands

The Galápagos Islands are a small archipelago of arid volcanic islands, isolated and remote—nearly 1,000 km from mainland South America. These small, rocky islands lack the profusion of life seen on many tropical islands, yet they are renowned as the place where Charles Darwin revolutionized our understanding of evolution, biodiversity, and biology in general. Why is this?

Charles Darwin (1809–1882) visited the islands in 1835 while serving as ship's naturalist on the HMS *Beagle*. His job was to collect specimens and record observations for general interest. He found there a variety of unusual creatures, most occurring only in the Galápagos and some on just one or two islands. Giant land tortoises fed on tree-size cacti. Unique marine iguanas lived by grazing on algae scraped from rocky shoals. Sea birds were so unafraid of humans that Darwin could pick them off their nests. The islands also had a variety of small brown finches that differed markedly in appearance, food preferences, and habitat. Most finches forage for small seeds, but in the Galápagos there were fruit eaters with thick, parrot-like beaks; seed eaters with heavy, crushing beaks; and insect eaters with thin, probing beaks to catch their prey. The woodpecker finch pecks at tree bark for hidden insects. Lacking the woodpecker's long tongue, however, the finch uses a cactus spine as a tool to extract insects.

Like other naturalists, Darwin was intrigued by the question of how such variety came to be. Most Europeans at the time believed that all living things had existed, unchanged, since a moment of divine creation, just a few thousand years ago. But Darwin had observed fossils of vanished creatures in South America. And he had read the new theories of geologist Charles Lyell (1797–1875), who argued that fossils showed that the world was much older than previously thought, and that species could undergo gradual but profound change over time. After his return to England, Darwin continued to ponder his Galápagos specimens. They seemed to be adapted to particular food resources on the different islands. It seemed likely that these birds were related, but somehow they had been modified to survive under different conditions.

Observing that dog breeders created new varieties of dogs, from dachshunds to Great Danes, by selecting for certain traits, Darwin proposed that "natural selection" could explain the origins of species. Just as dog breeders favored individuals with particular characteristics, such as long legs or short noses, environmental conditions in an area could favor certain characteristics. On an island where only large seeds were available, finches with larger-than-average beaks could have more success in feeding—and reproducing—than smaller-beaked individuals. Thus, competition for limited food resources could explain the prevalence of particular traits in a population. Darwin borrowed the idea of survival of the fittest from Thomas Malthus, a minister whose *Essay on the Principle of Population* (1798) argued that growth of human populations is always held in check by food scarcity (together with war and disease). Only the best competitors are likely to survive, according to Malthus. Darwin proposed that individuals with traits suitable for their environment are the best competitors for scarce resources. As those individuals survive and reproduce, their traits eventually become common in the population.

The explanation of evolution by natural selection has been supported by 150 years of observations and experiments, and Darwin's theory has provided an explanation for countless examples of species variations. In a now-classic study of competition, for example, ecologist David Lack carefully measured the sizes and shapes of finches' beaks on several islands to see how beaks varied with resources. Lack showed not only that beaks vary with resources, but that they specialize further in cases where multiple species compete for those resources. On the islands of Daphne Major and Los Hermanos, two finch species, *Geospiza fuliginosa* and *Geospiza fortis*, had beaks of moderate depth (thickness). But on islands where the two species coexisted, beak sizes shifted to the extremes, a shift that minimized competition for food resources (fig. 4.1). Where three species coexisted, traits again shifted to minimize competition. Thus, competition among species had led to shifts in beak traits. Among Darwin's Galápagos finches,

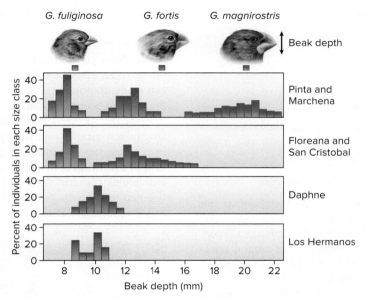

FIGURE 4.1 Two finch species have similar beaks when they occur separately (on the islands of Daphne and Los Hermanos), but beak sizes shift when the species occur together, as they specialize in different feeding strategies. When three finches coexist, feeding strategies and beak sizes are further differentiated. *(After D. Lack, 1947)*

13 modern species are now recognized, probably descended from a few seed-eating ancestors that blew to the islands from South America, where a similar species still exists.

Evolution of species through natural selection is now a cornerstone of biology and its many subfields, from ecology to medicine and health care. Subsequent discoveries have filled in many details. The discovery of DNA in the 1950s, in particular, allowed us to understand how random mutations (changes) in genes can account for the development of the variation in a population on which natural selection acts. In this chapter, we'll look at some of the ways species and communities adapt to their environments. We'll also consider the ways populations interact, and the adaptations that make some species abundant and those that make others rare.

Further Reading

Darwin, Charles. *The Voyage of the* Beagle (1837) and *On the Origin of Species* (1859).

Stix, Gary. 2009. Darwin's living legacy. *Scientific American* 300(1): 38–43.

4.1 EVOLUTION PRODUCES SPECIES DIVERSITY

- *Natural selection acting on spontaneous mutations results in evolution.*
- *All species have environmental tolerance limits.*
- *Taxonomy describes relationships between species.*

What are the mechanisms that promote the great variety of species on earth, and what determines that a species will survive in one environment but not another? In this section, we examine (1) concepts behind the theory of speciation by means of natural selection and adaptation (evolution); (2) the characteristics of species that make some of them weedy and others endangered; and (3) the limitations species face in their environments and implications for their survival. We'll start with the basics: How do species arise?

Natural selection leads to evolution

How does a polar bear stand the long, sunless, extremely cold arctic winter? How does the saguaro cactus survive the blistering temperatures and extreme dryness of the desert? We commonly say that each species is *adapted* to the environment where it lives, but what does that mean? **Adaptation,** the acquisition of traits that allow a species to survive in its environment, is one of the most important concepts in biology.

Adaptation in a species happens over generations, as advantageous traits persist and others die out. This is not the same thing as *acclimation,* which is an individual organism responding to its environment. If you move a houseplant from indoors to outdoors, it may grow a thicker skin or denser pigments to tolerate the increased sunlight. But the increased pigment and thicker skin are not permanent, and the plant's offspring won't inherit those traits. In evolutionary terms, in contrast, *adaptation* affects populations, which consist of many individuals. Genetic traits are passed from generation to generation, and individuals with traits that give an advantage in survival are more likely to reproduce, and their offspring inherit the beneficial traits. Over generations, those traits become common in a population, and the characteristics of the species—for example, beak shape in a finch or neck length in a giraffe—shifts (fig. 4.2). This process of adaptation to environment is central to the theory of evolution. The basic idea of **evolution** is that species change over generations because individuals compete for scarce resources. The process of better-selected individuals passing their traits to the next generation is called **natural selection.**

FIGURE 4.2 Giraffes don't have long necks because they stretch to reach tree-top leaves, but those giraffes that happened to have longer necks got more food and had more offspring, so the trait became fixed in the population.
© Tom Fink'e

Changes in traits occur because of random **mutations** or (changes) in the DNA of individuals. Random changes occur occasionally in any population. Exposure to radiation and toxic materials, and random errors in DNA replication are the main causes of mutations. Those changes in the genetic code can be inherited by offspring if they occur in reproductive cells. Evolutionary change is mostly brought about by many mutations accumulating over time. Only mutations in reproductive cells (gametes) matter; body cell changes—cancers, for example—are not inherited. Most mutations have no effect on fitness, and many have negative effects. During the course of a species' existence—a million or more years—some mutations are thought to have given those individuals an advantage under the **selection pressures** (factors that favor certain traits) in their environment at that time. The result is a species population that differs from those of preceding generations.

All species live within limits

Environmental factors exert selection pressure and influence the fitness of individuals and their offspring. For this reason, species are limited in where they can live. Limiting factors include: (1) physiological stress due to inappropriate levels of some critical environmental factor, such as moisture, light, temperature, pH, or specific nutrients; (2) competition with other species; (3) predation, including parasitism and disease. Any one of these could constrain a species in different circumstances.

An organism's physiology and behavior allow it to survive only in certain environments. Temperature, moisture level, nutrient supply, soil and water chemistry, living space, and other environmental factors must be at appropriate levels for organisms to persist. In 1840, the chemist Justus von Liebig proposed that the single factor in shortest supply relative to demand is the **critical factor** determining where a species lives. The giant saguaro cactus (*Carnegiea gigantea*), which grows in the dry, hot Sonoran Desert of southern Arizona and northern Mexico, offers an example (fig. 4.3). Saguaros are tolerant of extreme heat and drought, intense sunlight, and nutrient-poor soils, but are extremely sensitive to freezing temperatures. A single winter night with temperatures below freezing for 12 or more hours kills growing tips on the branches, preventing further development. Thus, the northern edge of the saguaro's range corresponds to a zone where freezing temperatures last less than half a day at any time.

Ecologist Victor Shelford (1877–1968) later expanded Liebig's principle by stating that each environmental factor has both minimum and maximum levels, called **tolerance limits,** beyond which a particular species cannot survive or is unable to reproduce (fig. 4.4). The single factor closest to these survival limits, Shelford postulated, is the critical factor that limits where a particular organism can live. At one time, ecologists tried to identify unique factors limiting the growth of every plant and animal population. We now know that several factors working together, even in a clear-cut case like the saguaro, usually determine a species' distribution. If you have ever explored the rocky coasts of

FIGURE 4.3 Saguaro cacti, symbolic of the Sonoran desert, are an excellent example of distribution controlled by a critical environmental factor. Extremely sensitive to low temperatures, saguaros are found only where minimum temperatures never dip below freezing for more than a few hours at a time.
© William P. Cunningham

New England or the Pacific Northwest, you have probably noticed that mussels and barnacles grow thickly in the intertidal zone, the place between high and low tides. No one factor decides this pattern. Instead, the distribution of these animals is determined by a combination of temperature extremes, drying time between tides, salt concentrations, competitors, and food availability.

In some species, tolerance limits affect the distribution of young differently than they affect adults. The desert pupfish, for instance, lives in small, isolated populations in warm springs in the northern Sonoran Desert. Adult pupfish can survive temperatures between 0° and 42°C (a remarkably high temperature for a fish) and tolerate an equally wide range of salt concentrations. Eggs and juvenile fish, however, can survive only between 20° and 36°C and are killed by high salt levels. Reproduction, therefore, is limited to a small part of the range of the adult fish.

Sometimes the requirements and tolerances of species are useful **indicators** of specific environmental characteristics. The presence or absence of such species indicates something about the community and the ecosystem as a whole. Lichens and eastern white pine, for example, are indicators of air pollution, because they are extremely sensitive to sulfur dioxide and ozone, respectively. Bull thistle and many other plant weeds grow on disturbed soil but are not eaten by cattle; therefore, a vigorous population of bull thistle or certain other plants in a pasture indicates it is being overgrazed. Similarly, anglers know that trout species require cool, clean, well-oxygenated water; the presence or absence of trout is used as an indicator of water quality.

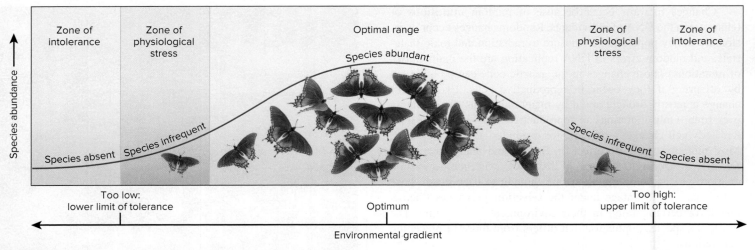

Too low:
lower limit of tolerance

Optimum

Too high:
upper limit of tolerance

Environmental gradient

FIGURE 4.4 Tolerance limits affect species distributions. For every environmental factor, there is an optimal range within which a species lives or reproduces most easily, so abundance is high. The horizontal axis here could represent a factor such as temperature, rainfall, vegetation height, or the availability of some critical resource.

The ecological niche is a species' role and environment

A **habitat** is the place or set of environmental conditions in which a particular organism lives. The term **ecological niche** is a more functional description of both the role played by a species in a biological community and the set of environmental factors that determine its distribution (fig. 4.5). The concept of niche was first defined in 1927 by the British ecologist Charles Elton (1900–1991). To Elton, each species had a role in a community of species, and the niche defined its way of obtaining food, the relationships it had with other species, and the services it provided to its community.

Thirty years later, the American limnologist G. E. Hutchinson (1903–1991) proposed a more physical and biological definition of the niche. Every species, he pointed out, exists within a range of physical and chemical conditions, such as temperature, light levels, acidity, humidity, or salinity. It also exists within a set of biological interactions, such as the presence of predators or prey, or the availability of nutritional resources.

Some species tolerate a wide range of conditions or exploit a wide range of resources. These species are known as **generalists** (fig. 4.6). Many generalists have large geographic ranges, as in the case of the black bear (*Ursus americana*), which is omnivorous, abundant, and ranges across most of North America's forested regions. Some generalists are also "weedy" species or pests, such as rats, cockroaches, or dandelions, because they thrive in a broad variety of environments.

Other species, such as the giant panda (*Ailuropoda melanoleuca*), are specialists and have a narrow ecological niche. Pandas have evolved from carnivorous ancestors (unrelated to bears) to subsist almost entirely on bamboo, a large but low-nutrient grass (fig. 4.7). To acquire enough nutrients, pandas must spend as much as 16 hours a day eating. The panda's slow metabolism,

slow movements, and low reproductive rate help it survive on this highly specialized diet. Its narrow niche now endangers the giant panda, though, as recent destruction of most of its native bamboo forest has reduced its range and its population to the margin of survival.

The giant saguaro is also a specialist: slow-growing and finely adapted to certain climatic conditions, but unable to persist in wetter or cooler environments. Because of their exacting habitat requirements, specialists tend not to tolerate environmental change well. Often, specialists are also **endemic species**—they occur only

FIGURE 4.5 Each of the species in this African savanna has its own ecological niche that determines where and how it lives.
© Tom Finkle

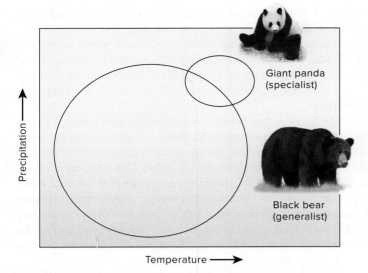

FIGURE 4.6 Generalists, such as the American black bear, tolerate a wide range of environmental conditions. Specialists, in contrast, have a narrower tolerance of environmental conditions.

FIGURE 4.7 The giant panda has evolved from carnivorous, cat-like ancestors to live on a diet composed almost exclusively of bamboo. Adaptions include "thumbs" that help it grasp bamboo leaves, and teeth that help it chew the grass.
© leungchopan/Getty Images RF

in one area (or one type of environment). The giant panda, for example, is endemic to the mountainous bamboo forests of south-western China; the saguaro is endemic to the Sonoran Desert.

In most organisms, genetic traits and instinctive behaviors restrict the ecological niche. But some species have complex social structures that help them expand the range of resources or environments they can use. Elephants, chimpanzees, and dolphins, for example, learn from their social group how to behave and can invent new ways of doing things when presented with novel opportunities or challenges. In effect, they expand their ecological niche by transmitting cultural behavior from one generation to the next.

When two species compete for limited resources, one eventually gains the larger share, while the other finds a different habitat, dies out, or experiences a change in its behavior or physiology so

that competition is minimized. Consequently, as explained by the Russian biologist G. F. Gause (1910–1986), "complete competitors cannot coexist." The general term for this idea is the **principle of competitive exclusion:** no two species can occupy the same ecological niche for long. The species that is more efficient in using available resources will exclude the other. The other species disappears or develops a new niche, exploiting resources differently. This process, in which species evolve to use resources differently, or to use different resources, is known as **resource partitioning** (fig. 4.8). For example, on a gradient from shallow to deep water in a lake, different fish often occur at different depths. Partitioning can allow several species to utilize different parts of the same resource and coexist within a single habitat. A classic example of resource partitioning is that of woodland warblers, studied by ecologist Robert MacArthur. Although several similar warblers species foraged in the

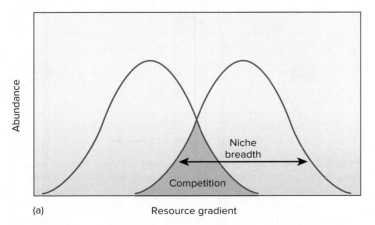

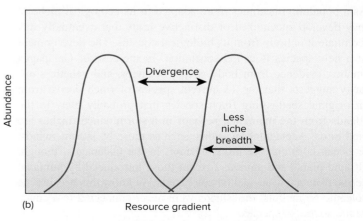

FIGURE 4.8 Competition causes resources partitioning and niche specialization. (a) Where niches of two species overlap along a resource gradient, competition occurs (shaded area). Individuals in this part of the niche have less success producing young. (b) Over time, the traits of the populations diverge, leading to specialization, narrower niche breadth, and less competition between species.

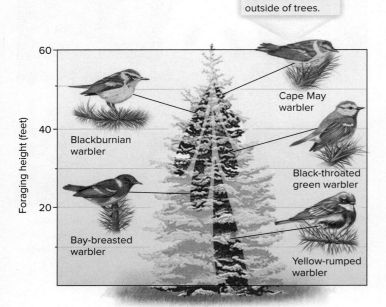

Tends to remain on outside of trees.

Cape May warbler

Blackburnian warbler

Black-throated green warbler

Bay-breasted warbler

Yellow-rumped warbler

FIGURE 4.9 Several species of insect-eating wood warblers occupy the same forests in eastern North America. The competitive exclusion principle predicts that the warblers should partition the resource—insect food—in order to reduce competition. And, in fact, the warblers feed in different parts of the forest.
Source: Original observations by R. H. MacArthur (1958)

same trees, they avoided competition by specializing in different levels of the forest canopy, or in inner and outer branches (fig. 4.9).

Resources can be partitioned in time as well as space. Swallows and insect-eating bats both live by capturing flying insects, but bats hunt for night-flying insects, while swallows hunt during the day. Thus, the two groups have noncompetitive feeding strategies for similar insect prey.

Speciation maintains species diversity

As a population becomes more adapted to its ecological niche, it may develop specialized or distinctive traits that eventually differentiate it entirely from its biological cousins. The development of a new species is called speciation. In the case of Galápagos finches, evidence from body shape, behavior, and genetic similarity suggests that the 13 current species of finch derive from an original seed-eating finch species that probably blew to the islands from the mainland, perhaps in a storm, since finches are land birds. Accidental invasions, such as those by storms, winds, or ocean currents, are probably rare. In the Galápagos, though, all land plants and animals (except those introduced by humans) derive from a few accidental colonizers. We know this because, as volcanic seamounts, the islands were never connected to a continental source of species.

In the Galápagos finches, speciation occurred largely because of **geographic isolation.** The islands are far enough apart that, in many cases, populations were genetically isolated; they couldn't

interbreed with populations on other islands. These isolated populations gradually changed in response to their individual environments, some of which were extremely dry with sparse vegetation, while others were relatively moist with a greater abundance and diversity of food resources. Speciation that occurs when populations are geographically separated is known as **allopatric speciation.**

The barriers that divide subpopulations aren't always physical. For example, two virtually identical tree frogs (*Hyla versicolor* and *H. chrysoscelis*) live in similar habitats of eastern North America but have different mating calls. This difference, which is enough to prevent interbreeding, is known as behavioral isolation. Speciation that occurs within one geographic area is known as **sympatric speciation.** Fern species and other plants sometimes undergo sympatric speciation by doubling or quadrupling the chromosome number of their ancestors, which makes them reproductively incompatible.

Normally, we assume evolution occurs slowly, but some degree of selection can happen in just a few years. A study of the finch species *Geospiza fortis* on the Galápagos island of Daphne Major, for example, showed selection after a two-year drought. As the drought reduced the availability of small seeds on the island, large-billed individuals in the finch population had better success in opening the remaining larger seeds. Within two years, the large-billed trait came to dominate the population. This shift toward one extreme of a trait is known as directional selection (fig. 4.10). In this case, the changes were not dramatic enough to result in speciation. When the drought ended, the population shifted back toward moderate-size beaks, which aided exploitation of a wider range of seeds.

Sometimes environmental conditions can reduce variation in a trait (stabilizing selection), or they can cause traits to diverge to the

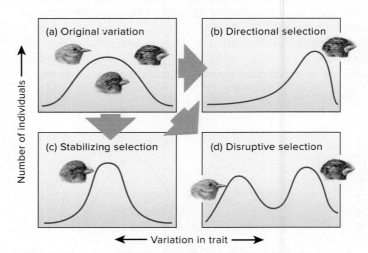

FIGURE 4.10 A species trait, such as beak shape, changes in response to selection pressure. Suppose a population starts with wide variation in beak shape (a). Sometimes selection pressures advantage just one beak type, which becomes dominant (b). Sometimes an intermediate or average type is advantageous (c). Disruptive selection moves characteristics to the extremes of the trait (d). Which selection type plausibly resulted in two distinct beak shapes—narrow in tree finches versus stout in ground finches—among Galápagos finches?

extremes (disruptive selection). Competition can cause disruptive selection, which allows for better partitioning of a resource (see fig. 4.1). Directional selection can be observed in the emergence of antibiotic-resistant bacteria (chapter 8) and of pesticide-tolerant insects (chapter 10). In both cases, chemical compounds (antibiotics or pesticides) create an environmental stressor. A few individuals that happen to have better-than-average tolerance of these compounds tend to survive in a population, while other individuals die off. Resistant survivors produce new generations, leading to a population with resistance to the antibiotic (among bacteria) or to the pesticide (among insects). This resistance, and the loss of effectiveness of antibiotics or pesticides, is a very common—and very costly—problem in public health and in agriculture.

New environmental conditions often lead to speciation, as new opportunities become available—as in the case of finch diversification on the previously unoccupied Galápagos Islands. Geologic time is marked by periods of tremendous diversification that have followed the sudden extinctions of species (chapter 11). The end of the age of dinosaurs, for example, was followed by dramatic diversification of mammals, which expanded to fill newly available niches. The fossil record is one of ever-increasing species diversity, despite several events that wiped out large proportions of species.

We generally believe that species arise slowly, and in fact many have existed unchanged for tens of millions of years (the American alligator has existed unchanged for over 150 million years), but as we saw with Darwin's finches, some organisms evolve swiftly. New flu viruses, for instance, evolve every season (see below, Exploring Science). Fruit flies in Hawaii mutate frequently and can give rise to new species in just a few years. Reproductive isolation has led to the development of new traits in populations of red squirrels in Arizona in just the past 10,000 years or so, since the last glacial period (fig. 4.11).

Many examples from both laboratory experiments and from nature shows evolution is still at work. Geneticists have modified many fruit fly properties—including body size, eye color, growth rate, life span, and feeding behavior—using artificial selection. In one experiment, researchers selected fruit flies with many bristles (stiff, hairlike structures) on their abdomen. In each generation, the flies with the most bristles were allowed to mate. After 86 generations, the number of bristles had quadrupled. In a similar experiment with corn, agronomists chose seeds with the highest oil content to plant and mate. After 90 generations, the average oil content had increased 450 percent.

Similarly, the widespread application of pesticides in agriculture and urban settings has led to the rapid evolution of resistance in more than 500 insect species, just as the extensive use of antibiotics in human medicine and livestock operations has led to antibiotic resistance in many microbes. The Centers for Disease Control and Prevention estimates that 90,000 Americans die every year from hospital-acquired infections, most of which are resistant to one or more antibiotics. We're engaged in a kind of arms race with germs. As quickly as new drugs are invented, microbes become impervious to them. Currently, vancomycin is the drug of last resort. When resistance to it becomes widespread, we may have no protection from infections.

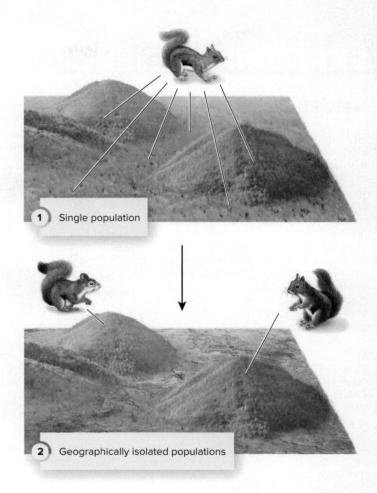

FIGURE 4.11 Geographic barriers can result in allopatric speciation. During cool, moist glacial periods, what is now Arizona was forest-covered, and squirrels could travel and interbreed freely. As the climate warmed and dried, desert replaced forest on the plains. Squirrels were confined to cooler mountaintops, which acted as island refugia, where new, reproductively isolated species gradually evolved.

On the other hand, evolution sometimes works in our favor. We've spread a number of persistent organic pollutants (called POPs), such as pesticides and industrial solvents, throughout our environment. One of the best ways to get rid of them is with microbes that can destroy or convert them to a nontoxic form. It turns out that the best place to look for these species is in the most contaminated sites. The presence of a new food source has stimulated evolution of organisms that can metabolize it. A little artificial selection and genetic modification in the laboratory can turn these species into very useful bioremediation tools.

Taxonomy describes relationships among species

Taxonomy is the study of types of organisms and their relationships. With it you can trace how organisms have descended from common ancestors. Taxonomic relationships among species are displayed like a family tree. Botanists, ecologists, and other

New Flu Vaccines

Why do we need a new flu vaccination every fall? Why can't they make one that lasts for years like the measles/mumps shot we received as infants? The answer is that the flu virus has an alarming ability to mutate rapidly. Our bodies are constantly trying to identify and build defenses against new viruses, while viruses have evolved methods to evolve rapidly and avoid surveillance by our immune system. Understanding the principles of evolution and genetics has made it possible to defend ourselves from the flu—provided we get the vaccines right each year.

Viruses can't replicate by themselves. They have to invade a cell of a higher organism and hijack the cell's biochemical systems. If multiple viruses infect the same cell, their RNA molecules (genes) can be mixed and recombined to create new virus strains. To invade a cell, the virus binds to a receptor on the cell surface (fig. 1). The binding proteins are called hemagglutin (because they also bind to antibodies in our blood). Additionally, the viruses have proteins called neuraminidases on their surface, which play a role in the budding of particles from the cell membrane and modifying sugars on the virus exterior. Influenza has 16 groups of H proteins and 9 groups of N proteins. We identify virus strains by code names, such as H5N1 or H3N2, based on their surface proteins.

Every year, new influenza strains sweep across the world, and because they change their surface proteins, our immune system fails to recognize them. The U.S. Centers for Disease Control and Prevention constantly surveys the flu strains occurring elsewhere to try to guess what varieties are most likely to invade the United States. Vaccines are prepared based on that best guess, but

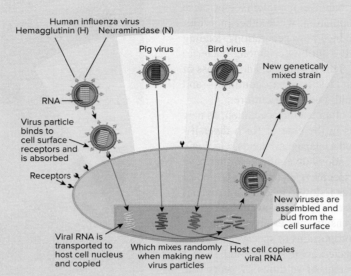

FIGURE 1 When different strains of the influenza virus infect the same cell, their genetic material can intermix to create a new re-assorted variety.

sometimes the best guess is wrong. There can suddenly appear an unknown variety of virus against which we have neither residual immunity nor vaccines. The result is a bad flu season.

An example of the surprises caused by rapid flu evolution occurred in 2009. A virus in the H1N1 family emerged in Mexico, where it infected at least 1,000 people and killed around 150. As it spread into the United States, children were particularly susceptible, while adults, particularly those over 60, often had some degree of immunity. Although that virus wasn't as lethal as first feared, by November 2009 it had infected about 50 million Americans, causing 200,000 hospitalizations and 10,000 deaths.

The H1N1 family is notorious as the source of the worst influenza pandemic (worldwide epidemic) in recorded history. The

1918 Spanish flu killed upward of 50 million people. This family also infects pigs, but it rarely kills them. For years the swine flu viruses seemed to evolve more slowly than human strains, but this picture is changing. Suddenly pig viruses have begun to evolve at a much faster rate and move to humans with increasing frequency. Critics of industrial agriculture charge that pigs increasingly are raised in enormous industrial facilities where diseases can quickly sweep through up to a million crowded animals. Many epidemiologists consider the roughly one billion pigs now raised annually to be laboratories for manufacturing new virus strains.

Pigs also serve as a conduit between humans and other animals. That's because they're susceptible to viruses from many sources. And once inside a cell, viral genes can mix freely to create new, more virulent combinations. The 2009 H1N1, for example, was shown to have genes from at least five different strains: a North America swine flu, North American avian flu, human influenza, and two swine viruses typically found in Asia and Europe. It's thought that the recombination of these various strains occurred in pigs, although we don't know when or where that took place.

So for the time being, we must continue to get a new inoculation annually and hope it protects us against the main flu strains we're likely to encounter in the next flu season. Someday, there may be a universal vaccine that will immunize us against all influenza viruses, but for now, that's just a dream.

For more information, see H. Branswell, 2011, Flu factories, *Scientific American* 304(1): 46–51.

scientists often use the most specific levels of the tree, genus and species, to compose names called **binomials.** Also called scientific or Latin names, binomials identify and describe species using Latin, or Latinized nouns and adjectives, or names of people or places. Scientists communicate about species using these scientific names instead of common names (such as lion, dandelion, or ant), to avoid confusion. A common name can refer to any number

of species in different places, and a single species might have many common names. The binomial *Pinus resinosa,* on the other hand, always is the same tree, whether you call it a red pine, Norway pine, or just pine.

Taxonomy also helps organize specimens and subjects in museum collections and research. You are *Homo sapiens* (human) and eat chips made of *Zea mays* (corn or maize). Both are members

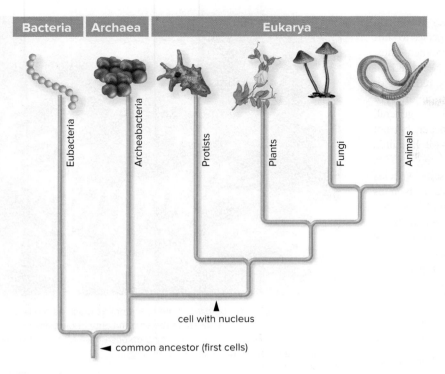

FIGURE 4.12 The six great kingdoms representing all life on earth. The kingdoms are grouped in domains indicating common origins.

of two well-known kingdoms. Scientists, however, recognize six kingdoms (fig. 4.12): animals, plants, fungi (molds and mushrooms), protists (algae, protozoans, slime molds), bacteria (or eubacteria), and archaebacteria (ancient, single-celled organisms that live in harsh environments, such as hot springs). Within these kingdoms are millions of different species, which you will learn more about in chapters 5 and 11.

Section Review

1. Explain how tolerance limits to environmental factors determine distribution of a highly specialized species such as the saguaro cactus.
2. Describe how evolution produces species diversity.
3. Define selective pressure and describe how it affects species.

4.2 SPECIES INTERACTIONS SHAPE BIOLOGICAL COMMUNITIES

- *Competition leads to resource allocation.*
- *Predation is an important type of selective pressure.*
- *Symbiosis benefits both species involved.*

We have learned that adaptation to one's environment, determination of ecological niche, and even speciation is affected not just by bodily limits and behavior, but also by competition and predation. Don't despair. Not all biological interactions are antagonistic, and many involve cooperation or at least benign interactions and tolerance. In some cases, different organisms depend on each other to acquire resources. Now we will look at the interactions within and between species that affect their success and shape biological communities.

Competition leads to resource allocation

Competition is a type of antagonistic relationship within a biological community. Organisms compete for resources that are in limited supply: energy and matter in usable forms, living space, and specific sites to carry out life's activities. Plants compete for growing space to develop root and shoot systems so that they can absorb and process sunlight, water, and nutrients (fig. 4.13). Animals compete for living, nesting, and feeding sites, and also for mates. Competition among members of the same species is called **intraspecific competition,** whereas competition between members of different species is called **interspecific competition.** Recall the competitive exclusion principle as it applies to

FIGURE 4.13 In this tangled Indonesian rainforest, space and light are at a premium. Plants growing beneath the forest canopy have adaptations that help them secure these limited resources. Many understory plants have large leaves to help them catch the scant light that filters down through the canopy. Ferns and bromeliads that grow on tree trunks and branches are epiphytes; they find space and get closer to the sun by perching above ground level. These are just some of the adaptations to life in the dark jungle.
© William P. Cunningham

interspecific competition. Competition shapes a species population and biological community by causing individuals and species to shift their focus from one segment of a resource type to another. Thus, warblers all competing with each other for insect food in New England tend to specialize on different areas of the forest's trees, reducing or avoiding competition. Since the 1950s, there have been hundreds of interspecific competition studies in natural populations. In general, scientists assume that interspecific competition does occur, but not always; and in some groups—carnivores and plants—it has little effect.

In intraspecific competition, members of the same species compete directly with each other for resources. Several avenues exist to reduce competition in a species population. First, the young of the year disperse. Even plants practice dispersal; seeds are carried by wind, water, and passing animals to less crowded conditions away from the parent plants. Second, by exhibiting strong territoriality, many animals force their offspring or trespassing adults out of their vicinity. In this way, territorial species, which include bears, songbirds, ungulates, and even fish, minimize competition between individuals and generations. A third way to reduce intraspecific competition is resource partitioning between generations. The adults and juveniles of these species occupy different ecological niches. For instance, monarch caterpillars munch on milkweed leaves, while metamorphosed butterflies sip nectar. Crabs begin as floating larvae and do not compete with bottom-dwelling adult crabs.

We think of competition among animals as a battle for resources—"nature, red in tooth and claw" is a common description. In fact, many animals avoid fighting if possible, or confront one another with noise and predictable movements. Bighorn sheep and many other ungulates, for example, engage in ritualized combat, with the weaker animal knowing instinctively when to back off. It's worse to be injured than to lose. Instead, competition often is simply about getting to food or habitat first, or being able to use it more efficiently. As we discussed, each species has tolerance limits for nonbiological (abiotic) factors. Studies often show that, when two species compete, the one living in the center of its tolerance limits for a range of resources has an advantage and usually prevails in competition with another species living outside its optimal environmental conditions.

Predation affects species relationships

All organisms need food to live. Producers make their own food, while consumers eat organic matter created by other organisms. As we saw in chapter 3, photosynthetic plants and algae are the producers in most communities. Consumers include herbivores, carnivores, omnivores, parasites, scavengers, detritivores, and decomposers. You may think only carnivores are predators, but ecologically a **predator** is any organism that feeds directly on another living organism, whether or not this kills the prey (fig. 4.14). Herbivores, carnivores, and omnivores, which feed on live prey, are predators, but scavengers, detritivores, and decomposers, which feed on dead things, are not. In this sense, parasites (organisms that feed on a host organism or steal resources

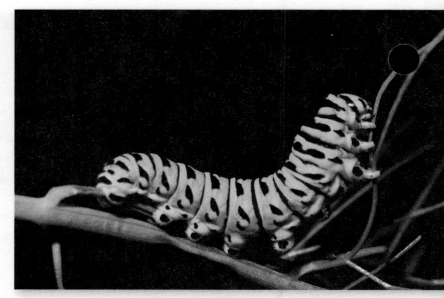

FIGURE 4.14 Insect herbivores are predators as much as are lions and tigers. In fact, insects consume the vast majority of biomass in the world. Complex patterns of predation and defense have often evolved between insect predators and their plant prey.
© Royalty Free/Corbis

from it without necessarily killing it) and even pathogens (disease-causing organisms) can be considered predator organisms. Herbivory is the type of predation practiced by grazing and browsing animals on plants.

Predation is a powerful but complex influence on species populations in communities. It affects (1) all stages in the life cycles of predator and prey species; (2) many specialized food-obtaining mechanisms; and (3) the evolutionary adjustments in behavior and body characteristics that help prey avoid being eaten and help predators more efficiently catch their prey. Predation also interacts with competition. In **predator-mediated competition,** a superior competitor in a habitat builds up a larger population than the competing species; predators take note and increase their hunting pressure on the superior species, reducing its abundance and allowing the weaker competitor to increase its numbers. To test this idea, scientists remove predators from communities of competing species. Often, the superior competitors eliminated other species from the habitat. In a classic example, the ochre starfish (*Pisaster ochraceus*) was removed from Pacific tidal zones, and so its main prey, the common mussel (*Mytilus californicus*), exploded in numbers and crowded out other intertidal species.

Knowing how predators affect prey populations has direct applications to human needs, such as pest control in croplands. The cyclamen mite (*Phytonemus pallidus*), for example, is a pest of California strawberry crops. Its damage to strawberry leaves is reduced by predatory mites (*Typhlodromus* and *Neoseiulus*), which arrive naturally or are introduced into fields. Pesticide spraying to control the cyclamen mite can actually increase the infestation because it also kills the beneficial predatory mites.

An organism's predators, and its position in the food web, can change as individuals mature. In marine ecosystems, crustaceans, mollusks, and worms release eggs directly into the water, where they and hatchling larvae join the floating plankton community (fig. 4.15). Planktonic animals eat each other and are food for larger carnivores, including fish. As prey species mature, their predators change. Barnacle larvae are planktonic and are eaten by small fish, but as adults their hard shells protect them from fish, but not from starfish and predatory snails. Predators often switch prey in the course of their lives. Carnivorous adult frogs usually begin their lives as herbivorous tadpoles. Predators also switch prey when their original prey becomes rare or something else becomes abundant. Many predators have morphologies and behaviors that make them highly adaptable to a changing prey base, but some, like the polar bear, are highly specialized in their prey preferences.

Some adaptations help avoid predation

Predator–prey relationships exert selection pressures that favor evolutionary adaptation. In this world, predators become more efficient at searching and feeding, and prey become more effective at escape and avoidance. Toxic chemicals, body armor, extraordinary speed, and the ability to hide are a few strategies organisms use to protect themselves. Plants have thick bark, spines, thorns, or distasteful and even harmful chemicals in tissues—poison ivy and stinging nettle are examples. Arthropods, amphibians, snakes, and some mammals produce noxious odors or poisonous secretions that cause other species to leave them alone. Animal prey are adept at hiding, fleeing, or fighting back. On the Serengeti Plain of East Africa, the swift Thomson's gazelle and even swifter cheetah are engaged in an arms race of speed, endurance, and quick reactions. The gazelle escapes often because the cheetah lacks stamina, but the cheetah accelerates from 0 to 72 kph in 2 seconds, giving it

the edge in a surprise attack. The response of predator to prey and vice versa, over tens of thousands of years, produces physical and behavioral changes in a process known as **coevolution.** Coevolution can be mutually beneficial: many plants and pollinators have forms and behaviors that benefit each other. A classic case is that of fruit bats, which pollinate and disperse seeds of fruit-bearing tropical plants. The plants provide food for the bats, and get their seeds dispersed in exchange.

Many species with chemical defenses display distinct coloration and patterns to warn away enemies (fig. 4.16). In a neat evolutionary twist, certain species that are harmless resemble poisonous or distasteful ones, gaining protection against predators who remember a bad experience with the actual toxic organism. This is called **Batesian mimicry,** after the English naturalist H. W. Bates (1825–1892), a traveling companion of Alfred Wallace. Many wasps, for example, have bold patterns of black and yellow stripes to warn potential predators (fig. 4.17a). The much rarer longhorn beetle has no stinger but looks and acts much like a wasp, tricking predators into avoiding it (fig. 4.17b). The distasteful monarch and benign viceroy butterflies are a classic case of Batesian mimicry. Another form of mimicry, **Müllerian mimicry** (after the biologist Fritz Müller) involves two unpalatable or dangerous species that look alike. When predators learn to avoid either species, both benefit. Species also display forms, colors, and patterns that help avoid being discovered. Predators also often use camouflage to conceal themselves as they lie in wait for their next meal.

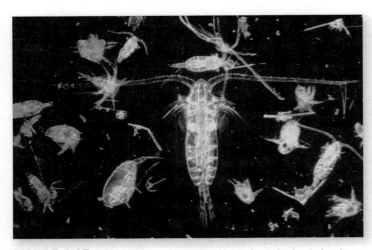

FIGURE 4.15 Microscopic plants and animals form the basic levels of many aquatic food chains and account for a large percentage of total world biomass. Many oceanic plankton are larval forms that have habitats and feeding relationships very different from their adult forms.
© D. P. Wilson/Science Source

FIGURE 4.16 Poison arrow frogs of the family Dendrobatidae display striking patterns and brilliant colors that alert potential predators to the extremely toxic secretions on their skin. Indigenous people in Latin America use the toxin to arm blowgun darts.
© Dirk Ercken/Shutterstock.com

(a)

(b)

FIGURE 4.17 An example of Batesian mimicry. The dangerous wasp (a) has bold yellow and black bands to warn predators away. The longhorn beetle (b) has no poisonous stinger, but it has a similar color and pattern that helps it avoid predators as well.
© Martin Fowler/Alamy RF © Ger Bosma/Moment Open/Getty Images

Symbiosis involves intimate relations among species

In contrast to predation and competition, some interactions between organisms can be nonantagonistic, even beneficial. In such relationships, called **symbiosis,** two or more species live intimately together, with their fates linked. Symbiotic relationships often enhance the survival of one or both partners. In lichens, a fungus and a photosynthetic partner (either an alga or a cyanobacterium) combine tissues to mutual benefit (fig. 4.18a). This association is called **mutualism.** Some ecologists believe that cooperative, mutualistic relationships may be more important in evolution than commonly thought (fig. 4.18b). Survival of the fittest may also mean survival of organisms that can live best together.

Commensalism is a type of symbiosis in which one member clearly benefits and the other apparently is neither benefited nor harmed. Many mosses, bromeliads, and other plants growing on trees in the moist tropics are considered commensals (fig. 4.18c). These epiphytes are watered by rain and obtain nutrients from leaf litter and falling dust, and often they neither help nor hurt the trees

(a) Lichen on a rock

(b) Oxpecker and impala

(c) Bromeliad

FIGURE 4.18 Symbiotic relationships. (a) Lichens represent an obligatory mutualism between a fungus and alga or cyanobacterium. (b) Mutualism between a parasite-eating red-billed oxpecker and parasite-infested impala. (c) Commensalism between a tropical tree and free-loading bromeliad.
© William P. Cunningham © P. de Graaf/Getty Images RF © William P. Cunningham

Say Hello to Your 90 Trillion Little Friends

Have you ever thought of yourself as a biological community or an ecosystem? Researchers estimate that each of us has about 90 trillion bacteria, fungi, protozoans, and other organisms living in or on our bodies. And the viruses inside those commensal species increase our biodiversity by another order of magnitude. The largest group—around 2 kg worth—inhabit your gut, but there are thousands of species living in every orifice, gland, pore, and crevice of your anatomy. Although the 10 trillion or so mammalian cells make up more than 95 percent of the volume of your body, they represent less than 10 percent of all the cell types that occupy that space.

Because most of the other species with which we coexist are microorganisms, we call the collection of cells that inhabit us our microbiome. The species composition of your own microbial community will be very similar to that of other people and pets with whom you live, but each of us has a unique collection of species that may be as distinctive as our fingerprints.

As is the case in other species interactions, these relationships can be mutualistic, symbiotic, commensal, or predatory. We used to think of all microorganisms as germs to be eliminated as quickly and thoroughly as possible. Current research suggests, however, that many of our fellow travelers are beneficial, perhaps even indispensable, to our good health and survival.

Your microbiome is essential, for example, in the digestion and absorption of nutrients. Symbiotic bacteria in your gut supply essential nutrients (important amino acids and short-chain fatty acids), vitamins (such as K and some B varieties), hormones and neurotransmitters (such as serotonin), and a host of other signaling molecules that

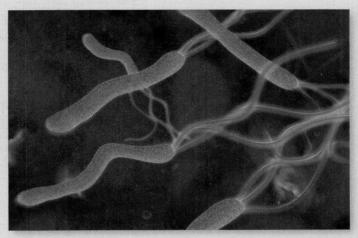

Intestinal bacteria, such as these, help crowd out pathogens, aid in digestion, supply your body with essential nutrients, and may play a role in obesity, diabetes, colitis, allergies, and chronic inflammation, along with a host of other critical diseases.
© ImageBroker/Alamy

communicate with, and modulate, your immune and metabolic systems. They help exclude pathogens by competing with them for living space, or by creating an environment in which harmful species can't grow or prosper.

The inhabitants of different organs can have important roles in specific diseases. Oral bacteria, for example, have been implicated in cardiovascular disease, pancreatic cancer, rheumatoid arthritis, and preterm birth, among other things. Symbionts in the lung have been linked to cystic fibrosis and chronic obstructive pulmonary disease (COPD). And the gut community seems to play a role in obesity, diabetes, colitis, susceptibility to infections, allergies, and other chronic problems. A healthy biome seems to be critical in controlling chronic inflammation that triggers many important long-term diseases.

As is true in many ecosystems, the diversity of your microbiome may play an important role in its stability and resilience.

Having a community rich in good microbes will not only help you resist infection by pathogens but will allow faster recovery after a catastrophic event. People in primitive or rustic societies who eat a wide variety of whole grains, raw fruits and vegetables, and unprocessed meat or dairy products tend to have a much greater species variety than those of us who have a diet full of simple sugars and highly processed foods. Widespread use of antibiotics to treat illnesses, as well as chronic low levels of antimicrobials, preservatives, and stabilizers in our food, toothpaste, soap, and many other consumer products also limits diversity in our symbiotic community.

A growing problem in many places is antibiotic-resistant, hospital-acquired infections. One of the most intractable of these is *Clostridium difficile*, or *C. diff,* which infects 250,000 and kills 14,000 Americans every year. An effective treatment for this superpathogen is a fecal transplant. A sample of the microbiome from a healthy person is implanted either directly through a feeding tube into the patient's stomach or in frozen, encapsulated pellets of feces that are delivered orally. In one trial, 18 of 20 patients who received fecal transplants recovered from *C. diff.*

Similarly, obese mice given fecal transplants from lean mice lose weight, while lean mice that receive samples of gut bacteria from obese mice gain weight. The microbiome may even regulate mood and behavior. When microbes from easygoing, adventurous mice are transplanted into the gut of anxious, timid mice, they become bolder and more adventurous.

So, it may pay to take care of your garden of microbes. If you keep them happy, they may help keep you happy as well.

---

on which they grow. Robins and sparrows that inhabit suburban yards are commensals with humans. **Parasitism,** a form of predation, may also be considered symbiosis because of the dependency of the parasite on its host.

You have many examples of commensalism, symbiosis, and parasitism in and on your own body. Thousands of species live in and on your body. We call this highly diverse community your **microbiome.** You couldn't survive without it (see Exploring Science).

FIGURE 4.19 Coevolution has led to close evolutionary relationships between many species, as in this star orchid and the specially adapted hawk moth that pollinates it. When Charles Darwin saw this orchid, he predicted there must be a moth with a 30-cm proboscis to reach the bottom of the nectary. Nearly a century later, the moth was discovered.

Mutualistic relationships often entail some degree of *coevolution,* where the partners change together. Many moths, for example, are highly adapted to pollinate specific flowering plants (fig. 4.19). Another mutualistic coadaptation is evident between swollen thorn acacias (*Acacia collinsii*) and the ants (*Pseudomyrmex ferruginea*) that tend them in Central and South America. Acacia ant colonies live inside the swollen thorns on the acacia tree branches. Ants feed on nectar that is produced in glands at the leaf bases and also eat special protein-rich structures that are produced on leaflet tips. The acacias thus provide shelter and food for the ants. Although they spend energy to provide these services, the trees are not harmed by the ants. What do the acacias get in return? The ants aggressively defend their territories, driving away herbivorous insects that would feed on the acacias. The ants also trim away vegetation that grows around the tree, reducing competition by other plants for water and nutrients. You can see how mutualism is structuring the biological community in the vicinity of acacias harboring ants, just as competition or predation shapes communities.

Mutualistic relationships can develop quickly. In 2005, the Harvard entomologist E. O. Wilson pieced together evidence to explain a 500-year-old agricultural mystery in the oldest Spanish settlement in the New World, Hispaniola. Using historical accounts and modern research, Dr. Wilson reasoned that mutualism developed between the tropical fire ant (*Solenopsis geminata*), native to the Americas, and a sap-sucking insect that was probably introduced from the Canary Islands in 1516

on a shipment of plantains. The plantains were planted, the sap-suckers were distributed across Hispaniola, and in 1518 a great die-off of crops occurred. Apparently, the native fire ants discovered the foreign sap-sucking insects, consumed their excretions of sugar and protein, and protected them from predators, thus allowing the introduced insect population to explode. The Spanish assumed the fire ants caused the agricultural blight, but a little ecological knowledge would have led them to the real culprit.

Keystone species have disproportionate influence

A **keystone species** plays a critical role in a biological community that is out of proportion to its abundance. Originally, keystone species were thought to be top predators—lions, wolves, tigers—that limited herbivore abundance and reduced the herbivory of plants. Scientists now recognize that less-conspicuous species also play keystone roles. Tropical figs, for example, bear fruit year-round at a low but steady rate. If figs are removed from a forest, many fruit-eating animals (frugivores) would starve in the dry season when other fruits are scarce. In turn, the disappearance of frugivores would affect plants that depend on them for pollination and seed dispersal. It is clear that the effect of a keystone species on communities often ripples across trophic levels.

Keystone functions have been documented for vegetation-clearing elephants, the predatory ochre sea star, and frog-eating salamanders in coastal North Carolina. Even microorganisms can play keystone roles. In many temperate forest ecosystems, groups of fungi that are associated with tree roots (mycorrhizae) facilitate the uptake of essential minerals much like the microbiome in mammals. When fungi are absent, trees grow poorly or not at all. Overall, keystone species seem to be more common in aquatic habitats than in terrestrial ones.

The role of keystone species can be difficult to untangle from other species interactions. Off the northern Pacific coast, a giant brown alga (*Macrocystis pyrifera*) forms dense "kelp forests" (fig. 4.20*a*), which shelter fish and shellfish species from predators, allowing them to become established in the community. It turns out, however, that sea otters (fig. 4.20*b*) eat sea urchins (fig. 4.20*c*) living in the kelp forests; when sea otters are absent, the urchins graze on and eliminate kelp forests. To complicate things, around 1990, killer whales began preying on otters because of the dwindling stocks of seals and sea lions, thereby creating a cascade of effects. Is the kelp, otter, or orca the keystone here? Whatever the case, keystone species exert their influence by changing competitive relationships. In some communities, perhaps we should call it a "keystone set" of organisms.

(b) Sea otters protect kelp ecosystem by preying on urchins

(c) Sea urchins graze on kelp

(a) Kelp shelter fish, seals, and other species

FIGURE 4.20 Sea otters protect kelp forests in the northern Pacific Ocean by eating sea urchins that would otherwise destroy the kelp. But the otters are being eaten by killer whales. Which is the keystone in this community—or is there a keystone set of organisms?
© Gregory Ochocki/Science Source © PhotoLink/PhotoDisc/Getty Images RF © Medioimages/PunchStock RF

Section Review

1. The most intense interactions often occur between individuals of the same species. What concept discussed in this chapter can be used to explain this phenomenon?

2. Explain how predators affect the adaptations of their prey.

3. Describe how competition for a limited quantity of resources occurs in ecosystems.

4.3 COMMUNITY PROPERTIES AFFECT SPECIES AND POPULATIONS

- *Productivity is a measure of biological activity.*

- *Abundance and diversity measure the number and variety of organisms.*

- *Resilience and stability make communities resistant to disturbance.*

The processes and principles that we have studied thus far in this chapter—tolerance limits, species interactions, resource partitioning, evolution, and adaptation—play important roles

in determining the characteristics of populations and species. In this section, we will look at some fundamental properties of biological communities and ecosystems—productivity, diversity, structure, complexity, resilience, and stability—to learn how they are affected by these factors.

Productivity is a measure of biological activity

A community's **primary productivity** is the rate of biomass production, an indication of the rate of solar energy conversion to chemical energy. The energy left after respiration is the net primary production. Photosynthetic rates are regulated by light levels, temperature, moisture, and nutrient availability. Productivity levels vary dramatically among environments (fig. 4.21). Tropical forests, coral reefs, and estuaries (where rivers meet the ocean), for example, have high biological productivity, capturing in biomass around 20,000 kcal of energy per m^2 per year. These environments have abundant light, moisture, and nutrients, and warm temperatures. In deserts, a lack of water limits photosynthesis, and biological productivity may be less than 1/10th that of a tropical forest. On the arctic tundra or in high mountains, low temperatures inhibit plant growth. In the open ocean, a lack of nutrients reduces the ability of algae to make use of plentiful sunshine and water.

Some agricultural crops such as corn (maize) and sugar cane grown under ideal conditions in the tropics approach the productivity levels of tropical forests. Because shallow water ecosystems such as coral reefs, salt marshes, tidal mud flats, and other highly productive aquatic communities are relatively rare compared to the vast extent of open oceans—which often are effectively biological deserts—marine ecosystems are much less productive, on average, than terrestrial ecosystems.

Even in the most photosynthetically active ecosystems, only a small percentage of the available sunlight is captured and used to make energy-rich compounds. Between one-quarter and three-quarters of the light reaching plants is reflected by leaf surfaces. Most of the light absorbed by leaves is converted to heat that is either radiated away or dissipated by evaporation of water. Only

---

Quantitative Reasoning

In figure 4.21, how much *primary productivity* occurs in a desert? How much more is there in a tropical rainforest? Why are the open oceans called the deserts of the sea?

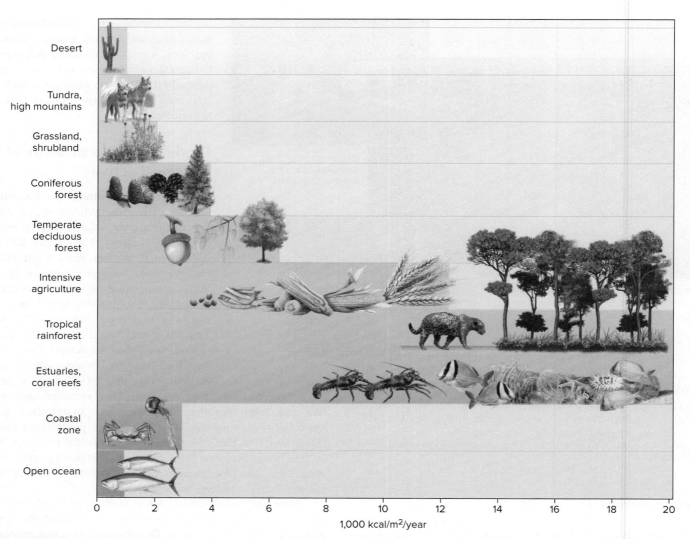

Desert

Tundra,
high mountains

Grassland,
shrubland

Coniferous
forest

Temperate
deciduous
forest

Intensive
agriculture

Tropical
rainforest

Estuaries,
coral reefs

Coastal
zone

Open ocean

0   2   4   6   8   10   12   14   16   18   20

1,000 kcal/m²/year

FIGURE 4.21 Relative biomass accumulation of major world ecosystems. Only plants and some bacteria capture solar energy. Animals consume biomass to build their own bodies.

What Can You Do?

Working Locally for Ecological Diversity

You might think that diversity and complexity of ecological systems are too large or too abstract for you to have any influence. But you can contribute to a complex, resilient, and interesting ecosystem, whether you live in the inner city, a suburb, or a rural area.

- *Keep your cat indoors.* Our lovable domestic cats are also very successful predators. Migratory birds, especially those nesting on the ground, have not evolved defenses against these predators.

- *Plant a butterfly garden.* Use native plants that support a diverse insect population. Native trees with berries or fruit also support birds. (Be sure to avoid non-native invasive species; see chapter 11.) Allow structural diversity (open areas, shrubs, and trees) to support a range of species.

- *Join a local environmental organization.* Often, the best way to be effective is to concentrate your efforts close to home. City parks and neighborhoods support ecological communities, as do farming and rural areas. Join an organization working to maintain ecosystem health; start by looking for environmental clubs at your school, park organizations, a local Audubon chapter, or a local Nature Conservancy branch.

- *Take walks.* The best way to learn about ecological systems in your area is to take walks and practice observing your environment. Go with friends and try to identify some of the species and trophic relationships in your area.

- *Live in town.* Suburban sprawl consumes wildlife habitat and reduces ecosystem complexity by removing many specialized plants and animals. Replacing forests and grasslands with lawns and streets is the surest way to simplify, or eliminate, ecosystems.

0.1 to 0.2 percent of the absorbed energy is used by chloroplasts to synthesize carbohydrates.

In a temperate-climate oak forest, only about half the incident light available on a midsummer day is absorbed by the leaves. Ninety-nine percent of this energy is used to evaporate water. A large oak tree can transpire (evaporate) several thousand liters of water on a warm, dry, sunny day while it makes only a few kilograms of sugars and other energy-rich organic compounds.

Abundance and diversity measure the number and variety of organisms

Abundance is an expression of the total number of organisms in a biological community, while **diversity** is a measure of the number of different species, ecological niches, or genetic variation present. The abundance of a particular species often is inversely related to the total diversity of the community. That is, communities with a very large number of species often have only a few members of any given species in a particular area. As a general rule, diversity decreases but abundance within species increases as we go from the equator toward the poles. The Arctic has vast numbers of insects such as mosquitoes, for example, but only a few species. The tropics, on the other hand, have vast numbers of species—some of which have incredibly bizarre forms and habits—but often only a few individuals of any particular species in a given area.

Consider bird populations. Greenland is home to 56 species of breeding birds, while Colombia, which is only one-fifth the size of Greenland, has 1,395. Why are there so many species in Colombia and so few in Greenland?

Climate and history are important factors. Greenland has such a harsh climate that the need to survive through the winter or escape to milder climates becomes the single most important critical factor that overwhelms all other considerations and severely limits the ability of species to specialize or differentiate into new forms. Furthermore, because Greenland was entirely covered by glaciers until about 10,000 years ago, there has been little time for new species to develop.

Many areas in the tropics, by contrast, have relatively abundant rainfall and warm temperatures year-round so that ecosystems there are highly productive. The year-round dependability of food, moisture, and warmth supports a great exuberance of life and allows a high degree of specialization in physical shape and behavior. Coral reefs are similarly stable, productive, and conducive to proliferation of diverse and amazing life-forms. The enormous abundance of brightly colored and fantastically shaped fish, corals, sponges, and arthropods in the reef community is one of the best examples we have of community diversity.

Productivity is related to abundance and diversity, both of which are dependent on the total resource availability in an ecosystem, as well as the reliability of resources, the adaptations of the member species, and the interactions between species. You shouldn't assume that all communities are perfectly adapted to their environment. A relatively new community that hasn't had time for niche specialization, or a disturbed one where roles such as top predators are missing, may not achieve maximum efficiency of resource use or reach its maximum level of either abundance or diversity.

Community structure is the spatial distribution of organisms

Ecological structure is the spatial distribution or organization of individuals and populations within a community. Individuals in a population can be distributed randomly or in highly regular patterns, or clumped together. In a randomly arranged population, individuals live wherever resources are available (fig. 4.22a). Ordered patterns may be determined by the physical environment but are more often the result of biological competition. For example, competition for nesting space in seabird colonies on the Falkland Islands is often fierce. Each nest tends to be just out of reach of the neighbors sitting on their own nests. Constant squabbling produces a highly regular pattern (fig. 4.22b). Similarly, sagebrush releases toxins from roots and fallen leaves, which inhibit the growth of competitors and create a circle of bare ground around each bush. As neighbors fill in empty spaces up to the limit of this chemical barrier, a regular spacing results.

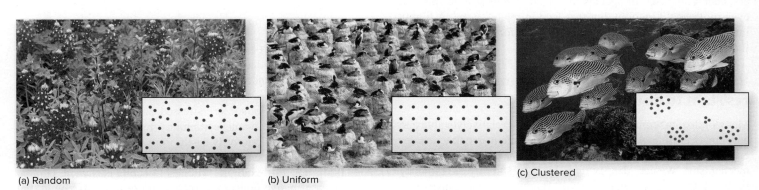

(a) Random (b) Uniform (c) Clustered

FIGURE 4.22 Distribution of members of a population in a given space can be (a) random, (b) uniform, or (c) clustered. The physical environment and biological interactions determine these patterns. The patterns may produce a graininess or patchiness in community structure.
© Royalty-Free/Corbis © Eric and David Hosking/Corbis © image100/PunchStock RF

Some other species cluster together for protection, mutual assistance, reproduction, or access to a particular environmental resource. Dense schools of fish, for instance, cluster closely together in the ocean, increasing their chances of detecting and escaping predators (fig. 4.22c). Similarly, predators, whether sharks, wolves, or humans, often hunt in packs to catch their prey. Both a flock of blackbirds descending on a cornfield, as well as a troop of baboons traveling across the African savanna, band together to avoid predators and to find food more efficiently.

Plants can cluster for protection, as well. A grove of wind-sheared evergreen trees is often found packed tightly together at the crest of a high mountain or along the seashore. The trees offer mutual protection from the wind not only to each other but also to other creatures that find shelter in or under their branches.

Most environments are patchy at some scale. Organisms cluster or disperse according to patchy availability of water, nutrients, or other resources. Distribution in a community can be vertical as well as horizontal. The tropical forest, for instance, has many layers, each with different environmental conditions and combinations of species. Distinct communities of smaller plants, animals, and microbes live at different levels. Similarly, many aquatic communities are stratified into layers based on light penetration in the water, temperature, salinity, pressure, or other factors.

Complexity and connectedness are important ecological indicators

Community complexity and connectedness generally are related to diversity and are important because they help us visualize and understand community functions. **Complexity** in ecological terms refers to the number of species at each trophic level and the number of trophic levels in a community. A diverse community may not be very complex if all its species are clustered in only a few trophic levels and form a relatively simple food chain.

By contrast, a complex, highly interconnected community (fig. 4.23) might have many trophic levels, some of which can be compartmentalized into subdivisions. In tropical rainforests, for instance, the herbivores can be grouped into "guilds" based on the specialized ways they feed on plants. There may be fruit eaters, leaf nibblers, root borers, seed gnawers, and sap suckers, each composed of species of very different size, shape, and even biological kingdom, but that feed in related ways. A highly interconnected community such as this can form a very elaborate food web.

Resilience and stability make communities resistant to disturbance

Many biological communities tend to remain relatively stable and constant over time. An oak forest tends to remain an oak forest, for example, because the species that make it up have self-perpetuating mechanisms. We can identify three kinds of stability or resiliency in ecosystems: *constancy* (lack of fluctuations in composition or functions), *inertia* (resistance to perturbations), and *renewal* (ability to repair damage after disturbance).

FIGURE 4.23 Tropical rainforests are complex structurally and ecologically. Trees form layers, each with a different amount of light and a unique combination of flora and fauna. Many insects, arthropods, birds, and mammals spend their entire life in the canopy. In Brazil's Atlantic Rainforest, a single hectare had 450 tree species and many times that many insects. With so many species, the ecological relationships are complex and highly interconnected.

In 1955, Robert MacArthur, who was then a graduate student at Yale, proposed that the more complex and interconnected a community is, the more stable and resilient it will be in the face of disturbance. If many different species occupy each trophic level, some can fill in if others are stressed or eliminated by external forces, making the whole community resistant to perturbations and able to recover relatively easily from disruptions. This theory has been controversial, however. Some studies support it while others do not. For example, Minnesota ecologist David Tilman, in studies of native prairie and recovering farm fields, found that plots with high diversity were better able to withstand and recover from drought than those with only a few species.

On the other hand, in a diverse and highly specialized ecosystem, removal of a few keystone members can eliminate many other associated species. Eliminating a major tree species from a tropical forest, for example, may destroy pollinators and fruit distributors as well. We might replant the trees, but could we replace the whole web of relationships on which they depend? In this case, diversity has made the forest less resilient rather than more.

Edges and boundaries are the interfaces between adjacent communities

An important aspect of community structure is the boundary between one habitat and its neighbors. We call these relationships **edge effects.** Sometimes the edge of a patch of habitat is relatively sharp and distinct. In moving from a woodland patch into a grassland or cultivated field, you sense a dramatic change from the cool, dark, quiet forest interior to the windy, sunny, warmer, open space of the meadow (fig. 4.24). In other cases, one habitat type intergrades very gradually into another, so there is no distinct border.

Ecologists use the term **ecotone** to refer either to a boundary between adjacent communities or a gradual transition zone between biomes. Often these transition zones have distinctive conditions: The boundary between a forest and an open field, may be sunnier and warmer than the forest interior but cooler and more moist than the center of the field. These edge conditions often support a different combination of plant and animal species than the "core" area of either field or forest.

Most boundaries are readily crossed by organisms. As we saw earlier in this chapter, birds might feed in fields or grasslands but nest in the forest. As they fly back and forth, the birds interconnect the ecosystems by moving energy and material from one to the other. These communities, which exchange of organisms, energy, and material with their surroundings, are known as open communities. In principle, a community that isolated from its neighbors is called a closed community, but truly closed communities are rare.

Depending on how far edge effects extend from the boundary, differently shaped habitat patches may have very dissimilar amounts of interior area (fig. 4.25). In Douglas fir forests of the Pacific Northwest, for example, increased rates of blowdown, decreased humidity, absence of shade-requiring ground cover, and other edge effects can extend as much as 200 m into a forest. A 40-acre block (about 400 m^2) surrounded by clear-cut would have essentially no true core habitat at all.

Many popular game animals, such as white-tailed deer and pheasants are adapted to landscapes that are disturbed and fragmented by human activity. These game species often are most plentiful in boundary zones between different types of habitat. Game managers once were urged to develop as much edge as possible to promote large game populations. Today most wildlife conservationists recognize that the edge effects associated with habitat fragmentation are generally detrimental to species that do not thrive on edges. Often these "interior" species have difficulty

FIGURE 4.24 Complex landscapes include contrasting environments, edges, transition zones, and corridors connecting larger patches. Edges are biologically rich, but core areas are critical for many species.
© Fuse/Getty Images

surviving in landscapes altered and fragmented by human disturbance. Preserving large habitat blocks and linking smaller blocks with migration corridors may be the best ways to protect rare and endangered species (chapter 12).

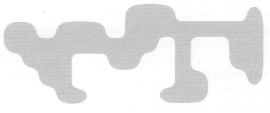

Total area: 50 ha
Core area: 0

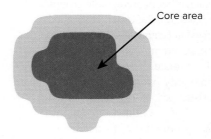

Core area

Total area: 50 ha
Core area: 25 ha

FIGURE 4.25 Shape can be as important as size in small preserves. While these areas are similar in size, no point in the top figure is far enough from the edge to have characteristics of core habitat, while the bottom patch has a significant core.

Section Review

1. Which ecosystems have the greatest biological productivity?
2. What are resilience and stability?
3. What are ecological edge effects?

4.4 COMMUNITIES ARE DYNAMIC AND CHANGE OVER TIME

- *Some biological communities are dependent on periodic disturbance.*
- *Introduced species can cause profound community change.*

If fire sweeps through a biological community, it's destroyed, right? Not so fast. Fire may be good for that community. Up until now we've focused on the day-to-day interactions of organisms with their environments, set in a context of adaptation and selection. In this section, we'll step back and look at more dynamic aspects of communities and how they change over time.

The nature of communities is debated

For several decades, starting in the early 1900s, ecologists argued about the basic nature of communities. Many of the basic debates and ideas still influence how we understand communities, and how and why they change over time. Two ecologists, J. E. B. Warming (1841–1924) in Denmark and Henry Chandler Cowles (1869–1939) in the United States, established the idea that communities develop in a sequence of stages, starting either from bare rock or after a severe disturbance. They worked in sand dunes and watched the changes as plants first took root in bare sand and, with further development, created forest. This example represents constant change, not stability. The community that developed last and lasted the longest was called the **climax community.**

The idea of a climax community was first championed by the biogeographer F. E. Clements (1874–1945). He viewed the process as a relay—species replace each other in predictable groups and in a fixed, regular order. He argued that every landscape has a characteristic climax community, determined mainly by climate. If left undisturbed, this community would mature to a characteristic set of organisms, each performing its optimal functions. A climax community to Clements represented the maximum complexity and stability that was possible. He and others made the analogy that the development of a climax community resembled the maturation of an organism. Both communities and organisms, they argued, begin simply, and then mature through predictable stages until they reach the highest stage of development allowed by genetics or environmental conditions.

This organism-like theory of community was opposed by Clements's contemporary, H. A. Gleason (1882–1975), who saw community history as a less predictable process. He argued that species are individualistic, each establishing in an environment according to its ability to colonize, tolerate the environmental conditions, and reproduce there. This idea allows for myriad temporary associations of plants and animals to form, fall apart, and reconstitute in slightly different forms, depending on environmental conditions and the species in the neighborhood. Imagine a time-lapse movie of a busy airport terminal. Passengers come and go; groups form and dissipate. Patterns and assemblages that seem significant may not mean much a year later. Gleason suggested that we think ecosystems are uniform and stable only because our lifetimes are too short and our geographic scope too limited to understand their actual dynamic nature.

Ecological succession involves changes in community composition

Although there are different explanations for it, change in ecological communities is probably an idea that's familiar to you. A general term for this change is ecological succession, the transition from one community to another. Once you understand likely transitions, you can see the history of an area by observing what is growing there. For example, in humid environments, abandoned farm fields are gradually overtaken by shrubs, then by sun-loving trees, then by mature forest canopy trees. During succession, different types of organisms occupy a site and change the environmental conditions.

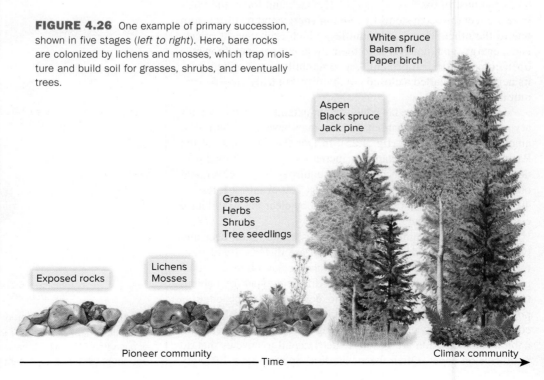

FIGURE 4.26 One example of primary succession, shown in five stages (*left to right*). Here, bare rocks are colonized by lichens and mosses, which trap moisture and build soil for grasses, shrubs, and eventually trees.

White spruce
Balsam fir
Paper birch

Aspen
Black spruce
Jack pine

Grasses
Herbs
Shrubs
Tree seedlings

Lichens
Mosses

Exposed rocks

Pioneer community ——— Time ——— Climax community

In **primary succession**, land that is bare of soil—a sandbar, mudslide, rock face, volcanic flow—is colonized by living organisms where none lived before (fig. 4.26). When an existing community is disturbed, a new one develops from the biological legacy of the old in a process called **secondary succession.** In both kinds of succession, organisms change the environment by modifying soil, light levels, food supplies, and microclimate. This change permits new species to colonize and eventually replace the previous species, a process known as ecological development or facilitation.

In primary succession on land, the first colonists are hardy **pioneer species,** often microbes, mosses, and lichens that can withstand a harsh environment with few resources. When they die, the bodies of pioneer species create patches of organic matter. Organics and other debris accumulates in pockets and crevices, creating soil where seeds lodge and grow. As succession proceeds, the community becomes more diverse, and species compete for space and sunlight. Pioneers disappear as the environment favors new colonizers that have competitive abilities more suited to the new environment.

You can see secondary succession all around you, in abandoned farm fields, in clear-cut forests, and in disturbed urban areas. Soil and possibly plant roots and seeds are present. On bare soil, plants that live one or two years (annuals and biennials) do relatively well. Their light seeds travel far on the wind, and their seedlings tolerate full sun and extreme heat. When they die, they lay down organic material that improves the soil's fertility and shelters other seedlings. Soon, long-lived and deep-rooted perennial grasses, herbs, shrubs, and trees take hold, building up the soil's organic matter and increasing its ability to store moisture. Forest species that cannot survive bare, dry, sunny ground eventually find ample food, a diverse community structure, and shelter from drying winds and low humidity.

Generalists figure prominently in early succession. Over thousands of years, however, competition should decrease as niches proliferate and specialists arise. In theory, long periods of community development lead to greater community complexity, high nutrient conservation and recycling, stable productivity, and great resistance to disturbance—an ideal state to be in when the slings and arrows of misfortune arrive.

Appropriate disturbances can benefit some communities

Disturbances are plentiful on earth: landslides, mudslides, hailstorms, earthquakes, hurricanes, tornadoes, tidal waves, wildfires, and volcanoes, to name just the obvious. A **disturbance** is any force that disrupts the established patterns of species diversity and abundance, community structure, or community properties. Animals can cause disturbance. African elephants rip out small trees, trample shrubs, and tear down tree limbs as they forage and move about, opening up forest communities and creating savannas. People also cause disturbances with agriculture, forestry, new roads and cities, and construction projects for dams and pipelines. It is customary in ecology to distinguish between natural disturbances and human-caused (or anthropogenic) disturbances, but a subtle point of clarification is needed.

Aboriginal populations have disturbed and continue to disturb communities around the world, setting fire to grasslands and savannas, practicing slash-and-burn agriculture in forests, and so on. Because their populations often are or were relatively small, the disturbances usually are patchy and limited in scale in forests, or restricted to quickly passing wildfires in grasslands, savannas, or woodlands, which are comprised of species already adapted to fire. Sometimes, disturbances can have profound, long-lasting effects. When aboriginal people reached Australia, 50,000 years ago, it's thought they set fires and eliminated plant and animal species to produce the dry, desert landscape we see there now.

The disturbances caused by modern people may be long-lasting, as well. In the Kingston Plains of Michigan's Upper Peninsula, clear-cut logging followed by repeated human-set fires from 1880 to 1900 caused a change in basic ecological conditions such that the white pine forest has never regenerated (fig. 4.27). Anthropogenic climate change may have even more global, long-lasting effects on our biosphere (see chapter 15).

Ecologists generally find that disturbance benefits many species, much as predation does, because it sets back supreme competitors and allows less-competitive species to persist. In northern temperate forests, maples (especially the sugar maple) are more prolific seeders and more shade tolerant at different stages of growth than nearly any other tree species. Given decades of succession, maples outcompete other trees for a place in the forest canopy. Most species of oak, hickory, and other light-requiring trees diminish in abundance, as do species of forest herbs. The dense shade of maples basically starves

FIGURE 4.27 These "stump barrens" in Michigan's Upper Peninsula were created over a century ago when clear-cutting of dense white pine forest was followed by repeated burning. The stumps are left from the original forest, which has not grown back in more than 100 years.
© William P. Cunningham

other species for light. When windstorms, tornadoes, wildfires, or ice storms hit a maple forest, trees are toppled, branches broken, and light again reaches the forest floor and stimulates seedlings of oaks and hickories, as well as forest herbs. Breaking the grip of a supercompetitor is the helpful role disturbances often play.

Some landscapes never reach a stable climax in a traditional sense because they are characterized by periodic disturbances and are made up of **disturbance-adapted species;** that is, species that tolerate or even require occasional disturbances. Many species tolerate fires, for example. Plant roots might survive a fire, or tree bark might resist fire, and seeds sprout quickly in the full sunshine after a fire. Grasslands, the chaparral scrubland of California and the Mediterranean region, savannas, and some kinds of coniferous forests are shaped and maintained by periodic fires that have long been a part of their history (fig. 4.28). Often, the dominant plant species in these communities depend on fire to suppress competitors, to prepare the ground for seeds to germinate, or to pop open cones or split thick seed coats and release seeds. Without fire, community structure would be quite different.

FIGURE 4.28 This lodgepole pine forest in Yellowstone National Park was once thought to be a climax forest, but we now know that this forest must be constantly renewed by periodic fire. It is an example of an equilibrium, or disclimax, community.
© William P. Cunningham

People taking an organismal view of such communities believe that disturbance is harmful. In the early 1900s, this view merged with the desire to protect timber supplies from ubiquitous wildfires, and to store water behind dams while also controlling floods. Fire suppression and flood control became the central policies in American natural resource management (along with predator control) for most of the twentieth century. Recently, new concepts about natural disturbances are entering land management discussions and bringing change to land management policies. Grasslands and some forests are now considered "fire-adapted," and fires are allowed to burn in them if weather conditions are appropriate. Floods also are seen as crucial for maintaining floodplain and river health. Policymakers and managers increasingly consider ecological information when deciding on new dams and levee construction projects.

From another view, disturbance resets the successional clock that always operates in every community. Even though all seems chaotic after a disturbance, it may be that preserving species diversity by allowing in natural disturbances (or judiciously applied human disturbances) actually ensures stability over the long run, just as diverse prairies managed with fire recover after drought. In time, community structure and productivity get back to normal, species diversity is preserved, and nature seems to reach its dynamic balance.

Introduced species can cause profound community change

Succession requires the continual introduction of new community members and the disappearance of previously existing species. New species move in as conditions become suitable; others die or move out as the community changes. New species also can be introduced after a stable community already has become established. Some cannot compete with existing species and fail to become established. Others are able to fit into and become part of the community, defining new ecological niches. If, however, an introduced species preys upon or outcompetes one or more populations that are native to the community, the entire nature of the community can be altered.

Human introductions of Eurasian plants and animals to non-Eurasian communities often have been disastrous to native species because of competition or overpredation. Oceanic islands offer classic examples of devastation caused by rats, goats, cats, and pigs liberated from sailing ships. All these animals are prolific, quickly developing large populations. Goats are efficient, nonspecific herbivores; they eat nearly everything vegetational, from grasses and herbs to seedlings and shrubs. In addition, their sharp hooves are hard on plants rooted in thin island soils. Rats and pigs are opportunistic omnivores, eating the eggs and nestlings of seabirds that tend to nest in large, densely packed colonies, and digging up sea turtle eggs. Cats prey upon nestlings of both ground- and tree-nesting birds. Native island species are particularly vulnerable because they have not evolved under circumstances that required them to have defensive adaptations to these predators.

FIGURE 4.29 Mongooses were released in Hawaii in an effort to control rats. The mongooses are active during the day, however, while the rats are night creatures, so they ignored each other. Instead, the mongooses attacked defenseless native birds and became as great a problem as the rats.
© Denis-Huot/hemis.fr/Getty Images

Sometimes we introduce new species in an attempt to solve problems created by previous introductions but end up making the situation worse. In Hawaii and on several Caribbean islands, for instance, mongooses were imported to help control rats that had escaped from ships and were destroying indigenous birds and devastating plantations (fig. 4.29). The mongooses were diurnal (active in the day), however, and rats are nocturnal, so they tended to ignore each other. Instead, the mongooses also killed native birds and further threatened endangered species. Our lessons from this and similar introductions have a new technological twist. Some of the ethical questions currently surrounding the release of genetically engineered organisms are based on concerns that they are novel organisms, and we might not be able to predict how they will interact with other species in natural ecosystems—let alone how they might respond to natural selective forces. It is argued that we can't predict either their behavior or their evolution.

Section Review

1. Describe the process of succession that occurs after a forest fire destroys an existing biological community.
2. Discuss the dangers posed to existing community members when new species are introduced into ecosystems.

Conclusion

Evolution is one of the key organizing principles of biology. It explains how species diversity originates, and how organisms are able to live in highly specialized ecological niches. Natural selection, in which beneficial traits are passed from survivors in one generation to their progeny, is the mechanism by which evolution occurs. Species interactions—competition, predation, symbiosis, and coevolution—are important factors in natural selection. The unique set of organisms and environmental conditions in an ecological community give rise to important properties, such as productivity, abundance, diversity, structure, complexity, connectedness, resilience, and succession. Human introduction of new species, as well as the removal of existing ones, can cause profound changes in biological communities and can compromise the life-supporting ecological services on which we all depend. Understanding these community ecology principles is a vital step in becoming an educated environmental citizen.

Reviewing Key Terms

Can you define the following terms in environmental science?

abundance 4.3	disturbance 4.4	habitat 4.1	parasitism 4.2	resource
adaptation 4.1	disturbance-adapted	indicators 4.1	pioneer species 4.4	partitioning 4.1
allopatric	species 4.4	interspecific	predator 4.2	secondary
speciation 4.1	diversity 4.3	competition 4.2	predator-mediated	succession 4.4
Batesian mimicry 4.2	ecological niche 4.1	intraspecific	competition 4.2	selection pressures 4.1
binomials 4.1	ecotone 4.3	competition 4.2	primary	symbiosis 4.2
climax community 4.4	edge effects 4.3	keystone species 4.2	productivity 4.3	sympatric
coevolution 4.2	endemic species 4.1	Müllerian mimicry 4.2	primary	speciation 4.1
commensalism 4.2	evolution 4.1	mutation 4.1	succession 4.4	tolerance limits 4.1
complexity 4.3	generalist 4.1	mutualism 4.2	principle of competitive	
critical factor 4.1	geographic isolation 4.1	natural selection 4.1	exclusion 4.1	

Critical Thinking and Discussion Questions

1. The concepts of natural selection and evolution are central to how most biologists understand and interpret the world, and yet the theory of evolution is contrary to the beliefs of many religious groups. Why do you think this theory is so important to science and so strongly opposed by others? What evidence would be required to convince opponents of evolution?

2. What is the difference between saying that a duck has webbed feet because it needs them to swim and saying that a duck is able to swim because it has webbed feet?

3. The concept of keystone species is controversial among ecologists because most organisms are highly interdependent. If each of the trophic levels is dependent on all the others, how can we say one is most important? Choose an ecosystem with which you are familiar and decide whether it has a keystone species or keystone set.

4. Some scientists look at the boundary between two biological communities and see a sharp dividing line. Others looking at the same boundary see a gradual transition with much intermixing of species and many interactions between communities. Why are there such different interpretations of the same landscape?

5. The absence of certain lichens is used as an indicator of air pollution in remote areas such as national parks. How can we be sure that air pollution is really responsible? What evidence would be convincing?

6. We tend to regard generalists or "weedy" species as less interesting and less valuable than rare and highly specialized endemic species. What values or assumptions underlie this attitude?

Data Analysis

Species Competition

In a classic experiment on competition between species for a common food source, the Russian microbiologist G. F. Gause grew populations of different species of ciliated protozoans separately and together in an artificial culture medium. He counted the number of cells of each species and plotted the total volume of each population. The organisms were *Paramecium caudatum* and its close relative *Paramecium aurelia*. He plotted the aggregate volume of cells rather than the total number in each population, because *P. caudatum* is much larger than *P. aurelia* (this size difference allowed him to distinguish between them in a mixed culture). The graphs below show the experimental results. As we mentioned earlier in the text, this was one of the first experimental demonstrations of the principle of competitive exclusion. After studying these graphs, answer the following questions.

1. How do you read these graphs? What is shown in the top and bottom panels?
2. How did the total volume of the two species compare after 14 days of separate growth?
3. If *P. caudatum* is roughly twice as large as *P. aurelia,* how did the total number of cells compare after 14 days of separate growth?
4. How did the total volume of the two species compare after 24 days of growth in a mixed population?

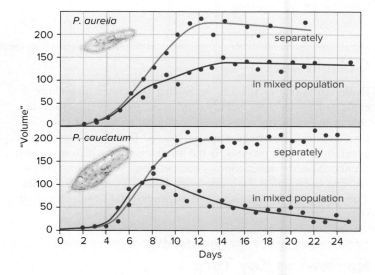

5. Which of the two species is the more successful competitor in this experiment?
6. Does the larger species always win in competition for food? Why not?

Source: Gause, Georgyi Frantsevitch. 1934. The Struggle for Existence. Dover Publications, 1971 reprint of original text.

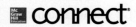

5

Biomes
Global Patterns of Life

▲ Kenya's Greenbelt Movement has planted millions of trees and inspired other groups to plant billions more.
© Claudiad/Vetta/Getty Images

Learning Outcomes

After studying this chapter, you should be able to:

5.1 Recognize the characteristics of major terrestrial biomes and factors that determine their distribution.

5.2 Understand how and why marine environments vary with depth and distance from shore.

5.3 Compare the characteristics and biological importance of major freshwater ecosystems.

5.4 Summarize the overall patterns of human disturbance of world biomes.

"What is the use of a house if you haven't got a tolerable planet to put it on?"

– Henry David Thoreau

Spreading Green Across Kenya

Our environment provides all of us with food, fuel, and shelter, but the world's poorest people often depend most directly on their environment—and they suffer most from a degraded environment. In remote areas of rural Kenya, subsistence farmers depend on local forests, soils, and groundwater for fuel, food, and water. What can these villagers do when overuse or careless management degrades these resources? This question challenges rural communities throughout the developing world.

When the first humans were evolving in Africa, the area that we now call Kenya was a mostly a biologically rich mixture of forests and grasslands. The country still has valuable natural resources, but millions of people live in severe poverty. Growing populations of farmers and herders depend on dwindling forests and degraded soils. Many forests were cleared decades ago for farming, and remaining woodlands are decimated by people gathering fuel and building materials, as well as by farming and grazing. As the forests disappear, the land becomes dry, soils wash away, and women must travel farther in search of fuelwood. Because women traditionally have the responsibility of gathering wood and water, and because they have little economic or political power, women and their children suffer most directly from forest losses. Environmental degradation causes economic instability, and families further exploit remaining forests, causing increasing environmental degradation, in a downward spiral of poverty.

FIGURE 5.1 Dr. Wangari Maathai has worked to restore trees, communities, and peace.
© AP/Wide World Photos

Stories of environmental and social degradation like that in Kenya can be found in developing areas worldwide. But Kenyans also have found a strategy to combat the combined problems of social, economic, and environmental devastation. The Greenbelt Movement, initiated by the environmental leader Dr. Wangari Maathai, (1940–2011, fig. 5.1) is working to teach communities to help themselves by growing and planting trees. Starting with the women and expanding to include their families, the movement is mobilizing people to help themselves. In the process, the Greenbelt Movement is teaching peaceful political involvement and local community development.

Dr. Maathai started out working on both environmental issues and women's empowerment. A native of Nyeri, Kenya, she studied in the United States and Germany in the 1960s and 1970s, and earned a PhD from the University of Nairobi. Dr. Maathai taught at the University of Nairobi, worked with the United Nations Environment Programme (UNEP), based in Nairobi, and eventually became chair of the National Council of Women of Kenya (NCWK).

According to Dr. Maathai, women in the villages told her they suffered from the loss of trees, so she suggested they plant new trees. But the women said they didn't know how to plant trees. And so this is where the Greenbelt Movement began, helping villagers create nurseries, grow seedlings, and plant trees. In 1977, Dr. Maathai and her colleagues from the NCWK celebrated World Environment Day by planting trees, the first of what would eventually become an international greenbelt movement.

Dr. Maathai's experience in multiple fields—like that of many environmental scientists—helped her see the deep ties between the powerlessness of poor women and environmental conditions that made their lives difficult. Dr. Maathai understood that many rural women had to walk miles every day for wood or water, in addition to tending to their farms and families. Making matters worse, many poor women depended on small farm plots with eroding, worn-out soils, to feed their families.

The tree-planting work started small and grew slowly, but it has endured, and community-based reforestation has grown and spread broad roots across Kenya. The Greenbelt Movement has trained people from around the world, and it now has branches in other countries in Africa, Asia, and South America. The program supports community networks that care for over 6,000 tree nurseries. The movement also promotes peace, education, and civic leadership, and recently it has expanded its vision to include climate mitigation through forest conservation. Thousands of community members have planted more than 40 million trees on degraded and eroding lands, in school yards and church yards, on farms, and in cities and villages. Goals of environmental quality and social justice remain a very long way off, but the movement has restored thousands of hectares of land, and it has brought hope to millions. In 2004, Dr. Maathai received the Nobel Peace Prize for her work on promoting peace through environmental stewardship and social justice.

Tree planting is a powerful act of hope. Planting a tree is an investment in the future, empowering people and showing the world that we care about those who will follow after we are gone. Expanding tree cover in once-forested lands helps nurture soils, biodiversity, and communities. The Greenbelt Movement shows that we have many choices other than simply watching while our environment deteriorates.

Finding ways to live sustainably within the limits of our resource bases, without damaging the life-support systems of ecosystems, is a preeminent challenge of environmental science. Sometimes, as this case study shows, ecological knowledge and local action can lead to positive effects on a global scale. We'll examine these and related issues in this chapter.

5.1 TERRESTRIAL BIOMES

- *Characteristics of biological communities vary with temperature, precipitation, and latitude.*
- *Hot, humid regions generally have greater biological productivity than cold or dry regions.*
- *We use climate graphs to describe and compare precipitation and temperature in different biomes.*

The Greenbelt Movement aims to restore components of an expansive biological community. To understand what that community should be like, it is helpful for us to identify some of the general types of communities with similar climate conditions, growth patterns, and vegetation types. We call these broad types of biological communities **biomes.** Understanding the global distribution of biomes, and knowing the differences in what grows where and why, is essential to the study of global environmental science. Biological productivity—and ecosystem resilience—varies greatly from one biome to another. Human use of biomes depends largely on those levels of productivity. Our ability to restore ecosystems and nature's ability to restore itself depend largely on biome conditions. Clear-cut forests can regrow relatively quickly in Kenya, but very slowly in Siberia, where logging is currently expanding. Some grasslands rejuvenate rapidly after grazing, and some are very slow to recover. Why these differences? The sections that follow seek to answer this question.

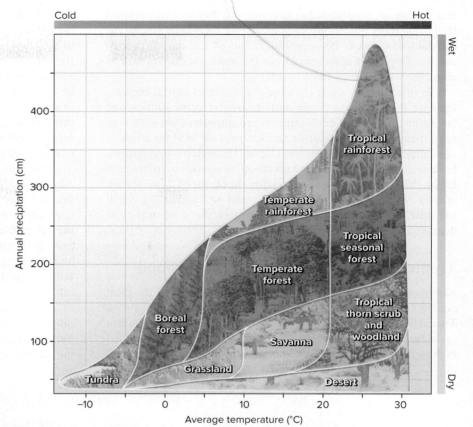

FIGURE 5.2 Temperature and precipitation conditions predict the general type of biome most likely to occur in a place. (Human disturbance, soil types, and other factors can also help determine the biological community that occurs in a local area.)

Temperature and precipitation are the most important determinants in biome distribution on land (fig. 5.2). If we know the general temperature range and precipitation level, we can predict what kind of biological community is likely to occur there in the absence of human disturbance.

Temperature and precipitation change dramatically with elevation. In mountainous regions, temperatures are cooler and precipitation is usually greater at high elevations. **Vertical zonation** occurs as vegetation types change rapidly from warm and dry to cold and wet as you go up a mountain. A 100-km transect from California's Central Valley up to Mt. Whitney, for example, crosses as many vegetation zones as you would find on a journey from southern California to northern Canada (fig. 5.3).

Because temperatures are cooler at high latitudes (away from the equator), temperature-controlled biomes often occur in latitudinal bands. For example, a band of boreal (northern) forests crosses Canada and Siberia, tropical forests occur near the equator,

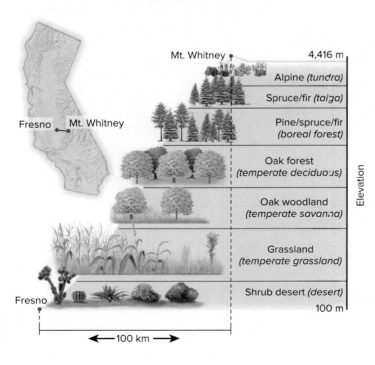

FIGURE 5.3 Vegetation changes with elevation because temperatures are lower and precipitation is greater high on a mountainside. A 100-km transect from Fresno, California, to Mt. Whitney (California's highest point) crosses vegetation zones similar to about seven different biome types.

and expansive grasslands lie near—or just beyond—the tropics (fig. 5.4). Many biomes are even named for their latitudes. Tropical rainforests occur between the Tropic of Cancer (23° north) and the Tropic of Capricorn (23° south); arctic tundra lies near or above the Arctic Circle (66.6° north).

In this chapter, we'll examine the major terrestrial biomes, then we'll investigate ocean and freshwater communities and environments. Ocean environments are important because they cover two-thirds of the earth's surface, provide food for much of humanity, and help regulate our climate through photosynthesis. Freshwater systems have tremendous influence on environmental health, biodiversity, and water quality. In chapter 12, we'll look at how we use these communities; and in chapter 13, we'll see how we preserve, manage, and restore them when they're degraded.

Tropical moist forests have rain year-round

The humid tropical regions of the world support one of the most complex and biologically rich biome types in the world (fig. 5.5). There are several kinds of moist tropical forests; all have abundant moisture and uniform temperatures. Cool **cloud forests** are found high in the mountains where fog and mist keep vegetation wet all the time. **Tropical rainforests** occur where rainfall is abundant—more than 200 cm (80 in.) per year—and temperatures are warm to hot year-round. For aid in reading the climate graphs in these figures, see the Data Analysis box at the end of this chapter.

The soil of both of these tropical moist forest types tends to be old, thin, acidic, and nutrient-poor, yet an enormous number of species can be present. For example, it has been estimated that there are millions of insect species in the canopy of tropical rainforests! And it is believed that one-half to two-thirds of all species of terrestrial plants and insects live in tropical forests.

The nutrient cycles of these forests also are distinctive. Almost all (90 percent) of the nutrients in the system are contained in the bodies of the living organisms. This is a striking contrast to temperate forests, where nutrients are held within the soil and made available for new plant growth. The luxuriant growth in tropical rainforests depends on rapid decomposition and recycling of dead organic material. Leaves and branches that fall to the forest floor decay and are almost immediately incorporated back into living biomass.

When the forest is removed for logging, agriculture, and mineral extraction, the thin soil cannot support continued cropping and cannot resist erosion from the abundant rains. And if the cleared area is too extensive, it cannot be repopulated by the

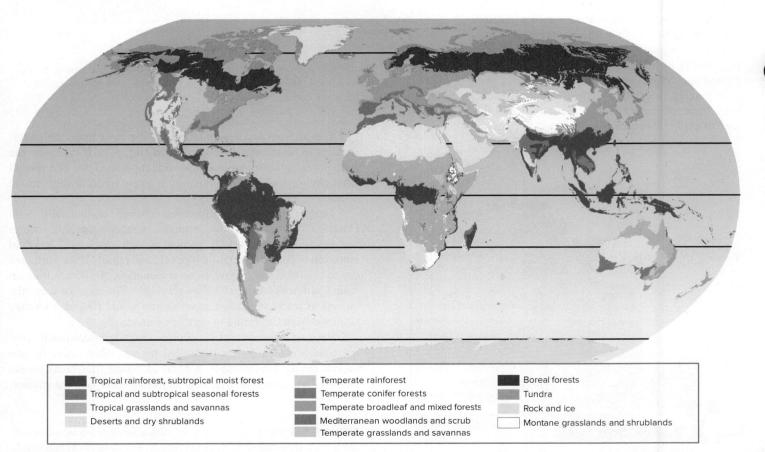

FIGURE 5.4 Major world biomes. Compare this map to figure 5.2 for generalized temperature and moisture conditions that control biome distribution. Also compare it to the satellite image of biological productivity (fig. 5.13).

Legend:
- Tropical rainforest, subtropical moist forest
- Tropical and subtropical seasonal forests
- Tropical grasslands and savannas
- Deserts and dry shrublands
- Temperate rainforest
- Temperate conifer forests
- Temperate broadleaf and mixed forests
- Mediterranean woodlands and scrub
- Temperate grasslands and savannas
- Boreal forests
- Tundra
- Rock and ice
- Montane grasslands and shrublands

How Do We Describe Climate Regions?

Differences between climate regions shape our experience of the world. It makes a difference in your daily life if you live in a hot, humid area, where your air conditioning costs are high, or if you live in a cold, dry region where you pay for heating much of the year. Perhaps you have to follow special water conservation rules because demand for water is greater than rainfall can supply. Climate regions also shape the plants and animals that live around you. Does the climate in your area produce abundant, dense vegetation? Vegetation that is low or sparse? Do plants and animals have special adaptations to survive in local conditions? You probably observe biome conditions every day, whether or not you realize it.

Comparing regions is a more specialized task. Climatologists, geographers, biologists, and others often want to explain differences in biodiversity, or in water resources, for example. To describe and compare regional climate

conditions, they use climographs, graphs that plot precipitation and temperature variation over a year (fig. 1). If you take a few minutes to make sure you understand these graphs, they will help you remember and explain differences in the biomes discussed in this chapter.

Each climograph shows long-term averages for each month. Plotting 12 months gives a picture of the year. Temperature (°C) is plotted on the left axis. Precipitation (mm of rain or snow) uses the right axis. Yellow areas indicate how much of the year evaporation exceeds precipitation—that is, when conditions are dry. Dry conditions could occur because it's hot and evaporation is rapid or simply because there's little precipitation. In yellow months, little moisture is available for plant growth. Blue areas show when precipitation exceeds evaporation, leaving moisture available for plants to thrive. Examine these graphs to answer these questions.

1. What are the maximum and minimum temperatures in each location?
2. What do these temperatures correspond to in Fahrenheit? (Hint: Look at the conversion table in the back of your book.)
3. Which area has the wettest climate? Which is driest?
4. How do the maximum and minimum monthly rainfalls in San Diego and Belém compare? Note that the axis scale changes for Belém, because there's not enough room to plot such high numbers.
5. What would a climate graph look like where you live? Try sketching one out, then compare it to a graph for a biome similar to yours in this chapter.
6. Which of these graphs is probably most similar to that in Kenya, the subject of the opening case study?

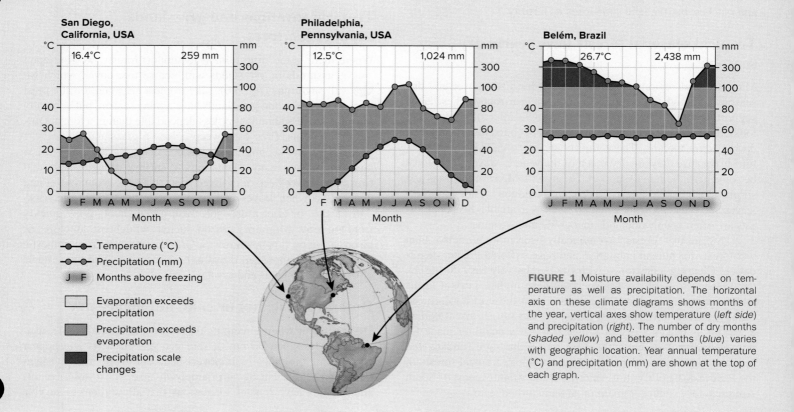

Temperature (°C)
Precipitation (mm)
J F Months above freezing

Evaporation exceeds precipitation

Precipitation exceeds evaporation

Precipitation scale changes

FIGURE 1 Moisture availability depends on temperature as well as precipitation. The horizontal axis on these climate diagrams shows months of the year, vertical axes show temperature (*left side*) and precipitation (*right*). The number of dry months (*shaded yellow*) and better months (*blue*) varies with geographic location. Year annual temperature (°C) and precipitation (mm) are shown at the top of each graph.

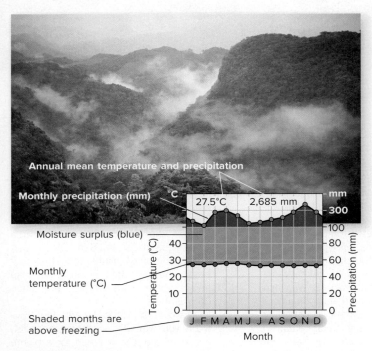

FIGURE 5.5 Tropical rainforests have luxuriant and diverse plant growth. Heavy rainfall in most months, shown in the climate graph, supports this growth.
© Adalberto Rios Szalay/Sexto Sol/Getty Images RF

FIGURE 5.6 Tropical savannas and grasslands experience annual drought and rainy seasons and year-round warm temperatures. Thorny acacias and abundant grazers thrive in this savanna. Yellow areas show moisture deficit.
© Tom Finkle

rainforest community. Rapid deforestation is occurring in many tropical areas as people move into the forests to establish farms and ranches, but the land soon loses its fertility.

Tropical seasonal forests have yearly dry seasons

Many tropical regions are characterized by distinct wet and dry seasons, although temperatures remain hot year-round. These areas support **tropical seasonal forests:** drought-tolerant forests that look brown and dormant in the dry season but burst into vivid green during rainy months. These forests are often called dry tropical forests because they are dry much of the year; however, there must be some periodic rain to support tree growth. Many of the trees and shrubs in a seasonal forest are drought-deciduous: They lose their leaves and cease growing when no water is available. Seasonal forests are often open woodlands that grade into savannas.

Tropical dry forests have typically been more attractive than wet forests for human habitation and have suffered greater degradation. Clearing a dry forest with fire is relatively easy during the dry season. Soils of dry forests often have higher nutrient levels and are more agriculturally productive than those of a rainforest. Finally, having fewer insects, parasites, and fungal diseases than a wet forest makes a dry or seasonal forest a healthier place for humans to live. Consequently, these forests are highly endangered in many places. Less than 1 percent of the dry tropical forests of the Pacific coast of Central America or the Atlantic coast of South America, for instance, remain in an undisturbed state.

Tropical savannas and grasslands support few trees

Where there is too little rainfall to support forests, we find open **grasslands** or grasslands with sparse tree cover, which we call **savannas** (fig. 5.6). Like tropical seasonal forests, most tropical savannas and grasslands have a rainy season, but generally the rains are less abundant or less dependable than in a forest. During dry seasons, fires can sweep across a grassland, killing off young trees and keeping the landscape open. Savanna and grassland plants have many adaptations to survive drought, heat, and fires. Many have deep, long-lived roots that seek groundwater and that persist when leaves and stems above the ground die back. After a fire, or after a drought, fresh green shoots grow quickly from the roots. Migratory grazers such as wildebeest, antelope, or bison thrive on this new growth. Grazing pressure from domestic livestock is an important threat to both the plants and the animals of tropical grasslands and savannas.

Deserts can be hot or cold, but all are dry

Deserts occur where precipitation is rare and unpredictable, usually with less than 30 cm of rain per year. Deserts can appear barren and biologically impoverished, but while their vegetation is sparse, it can be surprisingly diverse. Most desert plants and animals have special adaptations that allow them to survive long droughts and extreme heat, and many can survive extreme cold. Adaptations to these conditions include water-storing leaves and stems, thick epidermal layers to reduce water loss, and salt tolerance. As in other dry environments, many desert plants are drought-deciduous. Most also bloom and set seed quickly when a spring rain does fall.

Warm, dry, high-pressure climate conditions (chapter 15) create desert regions at about 30° latitude north and south (see fig. 5.4). Extensive deserts occur in the continental interiors (far from oceans, which evaporate the moisture for most precipitation) of North America, Central Asia, Africa, and Australia (fig. 5.7). The rain shadow of the Andes produces the world's driest desert in coastal Chile. Deserts can also be cold. Antarctica is a desert. Some inland valleys apparently get almost no precipitation at all.

Like plants, animals in deserts are specially adapted. Many are nocturnal, spending their days in burrows to avoid the sun's heat and desiccation. Pocket mice, kangaroo rats, and gerbils can get most of their moisture from seeds and plants. Desert rodents also have highly concentrated urine and nearly dry feces that allow them to eliminate body waste without losing precious moisture.

Deserts are more vulnerable than you might imagine. Sparse, slow-growing vegetation is quickly damaged by off-road vehicles. Desert soils recover slowly. Tracks left by army tanks practicing in California deserts during World War II can still be seen today.

Deserts are also vulnerable to overgrazing. In Africa's vast Sahel (the southern edge of the Sahara Desert), livestock are destroying much of the plant cover. Bare, dry soil becomes drifting sand, and restabilization is extremely difficult. Without plant roots and organic matter, the soil loses its ability to retain what rain does fall, and the land becomes progressively drier and more bare. Similar depletion of dryland vegetation is happening in many desert areas, including Central Asia, India, and the American Southwest and Plains states.

Temperate grasslands have rich soils

As in tropical latitudes, temperate (midlatitude) grasslands occur where there is enough rain to support abundant grass but not enough for forests (fig. 5.8). Most grasslands are a complex, diverse mix of grasses and flowering herbaceous plants, generally known as forbs. Myriad flowering forbs make a grassland colorful and lovely in summer. In dry grasslands, vegetation may be less than a meter tall. In more humid areas, grasses can exceed 2 m. Where scattered trees occur in a grassland, we call it a savanna.

Deep roots help plants in temperate grasslands and savannas survive drought, fire, and extreme heat and cold. These roots, together with an annual winter accumulation of dead leaves on the surface, produce thick, organic-rich soils in temperate grasslands. Because of this rich soil, many grasslands have been converted to farmland. The legendary tallgrass prairies of the central United States and Canada have been almost completely replaced by corn, soybeans, wheat, and other crops. Most remaining

FIGURE 5.7 Deserts generally receive less than 300 mm (30 cm) of precipitation per year. Hot deserts, as in the American Southwest, endure year-round drought and extreme heat in summer.
© William P. Cunningham

FIGURE 5.8 Grasslands occur at midlatitudes on all continents. Kept open by extreme temperatures, dry conditions, and periodic fires, grasslands can have surprisingly high plant and animal diversity.
© Mary Ann Cunningham

grasslands in this region are too dry to support agriculture, and the greatest threat to them is overgrazing. Excessive grazing eventually kills even deep-rooted plants. As ground cover dies off, soil erosion results, and unpalatable weeds, such as cheatgrass or leafy spurge, spread.

Temperate shrublands have summer drought

Many dry environments support sparse shrubs and trees that are adapted to survive drought, in addition to grasses. These mixed environments can be highly variable. They can also be very rich biologically. Such conditions are often described as Mediterranean (where the hot season coincides with the dry season, producing hot, dry summers and cool, moist winters). Evergreen shrubs with small, leathery, sclerophyllous (hard, waxy) leaves form dense thickets. Scrub oaks, drought-resistant pines, or other small trees often cluster in sheltered valleys. Periodic fires burn fiercely in this fuel-rich plant assemblage and are a major factor in plant succession. Annual spring flowers often bloom profusely, especially after fires. In California, this landscape is called **chaparral**, Spanish for "thicket." These areas are inhabited by drought-tolerant animals such as jackrabbits, kangaroo rats, mule deer, chipmunks, lizards, and many bird species. Very similar landscapes are found along the Mediterranean coast, as well as southwestern Australia, central Chile, and South Africa. Although this biome doesn't cover a very large total area, it contains a high number of unique species and is often considered a "hot spot" for biodiversity. It also is highly desired for human habitation, often leading to conflicts with rare and endangered plant and animal species.

Areas that are drier year-round, such as the African Sahel, northern Mexico, or the American Intermountain West (or Great Basin), tend to have a more sparse, open shrubland, characterized by sagebrush (*Artemisia* sp.), chamiso (*Adenostoma* sp.), or saltbush (*Atriplex* sp.). Some typical animals of this biome in America are a wide variety of snakes and lizards, rodents, birds, antelope, and mountain sheep.

Temperate forests can be evergreen or deciduous

Temperate, or midlatitude, forests occupy a wide range of precipitation conditions but occur mainly between about 30° and 55° latitude (see fig. 5.3). In general, we can group these forests by tree type, which can be broadleaf **deciduous** (losing leaves seasonally) or evergreen **coniferous** (cone-bearing).

Deciduous Forests

Broadleaf forests occur throughout the world where rainfall is plentiful. In midlatitudes, these forests are deciduous and lose their leaves in winter. The loss of green chlorophyll pigments can produce brilliant colors in these forests in autumn (fig. 5.9). At lower latitudes, broadleaf forests may be evergreen or drought-deciduous. Southern live oaks, for example, are broadleaf evergreen trees.

FIGURE 5.9 Temperate deciduous forests have year-round precipitation and winters near or below freezing.
© William P. Cunningham

Although these forests have a dense canopy in summer, they have a diverse understory that blooms in spring, before the trees leaf out. Spring ephemeral (short-lived) plants produce lovely flowers, and vernal (springtime) pools support amphibians and insects. These forests also shelter a great diversity of songbirds.

North American deciduous forests once covered most of what is now the eastern half of the United States and southern Canada. Most of western Europe was once deciduous forest but was cleared a thousand years ago. When European settlers first came to North America, they quickly settled and cut most of the eastern deciduous forests for firewood, lumber, and industrial uses, as well as to clear farmland. Many of those regions have now returned to deciduous forest, though the dominant species have changed.

Deciduous forests can regrow quickly because they occupy moist, moderate climates. But most of these forests have been occupied so long that human impacts are extensive, and most native species are at least somewhat threatened. The greatest threat to broadleaf deciduous forests is in eastern Siberia, where deforestation is proceeding rapidly. Siberia may have the highest deforestation rate in the world. As forests disappear, so do Siberian tigers, bears, cranes, and a host of other endangered species.

Coniferous Forests

Coniferous forests grow in a wide range of temperature and moisture conditions. They often occur where moisture is limited: In

Boreal forests occur at high latitudes

Because conifers can survive winter cold, they tend to dominate the **boreal forest,** or northern forests, that lie between about 50° and 60° north (fig. 5.11). Mountainous areas at lower latitudes may also have many characteristics and species of the boreal forest. Dominant trees are pines, hemlocks, spruce, cedar, and fir. Some deciduous trees are also present, such as maples, birch, aspen, and alder. These forests are slow-growing because of the cold temperatures and a short frost-free growing season, but they are still an expansive resource. In Siberia, Canada, and the western United States, large regional economies depend on boreal forests.

The extreme, ragged edge of the boreal forest, where forest gradually gives way to open tundra, is known by its Russian name, **taiga.** Here the extreme cold and short summers limit the growth rate of trees. A 10 cm diameter tree may be over 200 years old in the far north.

Tundra can freeze in any month

Where temperatures are below freezing most of the year, only small, hardy vegetation can survive. **Tundra,** a treeless landscape that occurs at high latitudes or on mountaintops, has a growing season of only two to three months and may have frost any month of the year. Some people consider tundra a variant of grasslands

FIGURE 5.10 Temperate rainforests have abundant but often seasonal precipitation that supports magnificent trees and luxuriant understory vegetation. Often these forests experience dry summers.

© Mark Karrass/Corbis RF

cold climates, moisture is unavailable (frozen) in winter; hot climates may have seasonal drought; sandy soils hold little moisture, and they are often occupied by conifers. Thin, waxy leaves (needles) help these trees reduce moisture loss. Coniferous forests provide most wood products in North America. Dominant wood production regions include the southern Atlantic and Gulf Coast states, the mountain West, and the Pacific Northwest (northern California to Alaska), but coniferous forests support forestry in many regions.

Remaining fragments of ancient temperate rainforests are important areas of biodiversity. Recent battles over old-growth conservation (chapter 12) focus mainly on these areas. As with deciduous forests, Siberian forests are especially vulnerable to old-growth logging. The rate of this clearing, and its environmental effects, remain largely unknown.

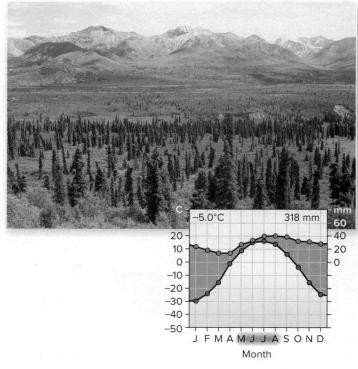

FIGURE 5.11 Boreal forests have moderate precipitation but are often moist because temperatures are cold most of the year. Cold-tolerant and drought-tolerant conifers dominate boreal forests and taiga, the forest fringe.

© William P. Cunningham

because it has no trees; others consider it a very cold desert because water is unavailable (frozen) most of the year.

Arctic tundra covers vast areas of the far north, as in northern Canada and Siberia. This biome has low productivity because it has a short growing season (fig. 5.12). During midsummer, however, 24-hour sunshine supports a burst of plant growth and an explosion of insect life. Tens of millions of waterfowl, shorebirds, terns, and songbirds migrate to the Arctic every year to feast on the abundant invertebrate and plant life and to raise their young on the brief bounty. These birds then migrate to wintering grounds, where they may be eaten by local predators—effectively they carry energy and protein from high latitudes to low latitudes. Arctic tundra is essential for global biodiversity, especially for birds.

Alpine tundra, occurring on or near mountaintops, has environmental conditions and vegetation similar to arctic tundra. These areas have a short, intense growing season. Often one sees a splendid profusion of flowers in alpine tundra; everything must flower at once in order to produce seeds in a few weeks before the snow comes again. Many alpine tundra plants also have deep

pigmentation and leathery leaves to protect against the strong ultraviolet light in the thin mountain atmosphere.

Compared to other biomes, tundra has relatively low diversity. Dwarf shrubs such as willows, sedges, grasses, mosses, and lichens tend to dominate the vegetation. Migratory musk oxen, caribou, or alpine mountain sheep and mountain goats can live on the vegetation because they move frequently to new pastures.

Because these environments are too cold for most human activities, they are not as badly threatened as other biomes. There are important problems, however. Global climate change may be altering the balance of some tundra ecosystems, and air pollution from distant cities tends to accumulate at high latitudes (chapter 15). In eastern Canada, coastal tundra is being badly overgrazed and degraded by overabundant populations of snow geese, whose numbers have exploded due to winter grazing on the rice fields of Arkansas and Louisiana. Oil and gas drilling—and associated truck traffic—threatens tundra in Alaska and Siberia. Clearly, this remote biome is not independent of human activities at lower latitudes.

Section Review

1. A grassland biome occupies much of the center of North America. Why is this, in terms of environmental factors?
2. What is taiga and where is it found? Why might taiga be slower to recover from logging than southern forests?
3. Why are tropical moist forests often less suited for agriculture and human occupation than tropical deciduous forests?
4. Find out the annual temperature and precipitation conditions where you live (fig. 5.2). Which biome type do you occupy?

5.2 MARINE ECOSYSTEMS

- *Marine ecosystems vary mainly with depth, temperature, and salinity.*
- *Coral reefs and estuaries are among the world's most productive and diverse ecosystems.*

The biological communities in oceans and seas are less understood than terrestrial biomes, but they can be just as diverse and complex. Oceans cover nearly three-fourths of the earth's surface, and they contribute in important, although often unrecognized, ways to terrestrial ecosystems. Like land-based systems, most marine communities depend on photosynthetic organisms. Often, it is algae or tiny, free-floating photosynthetic plants (**phytoplankton**) that support a marine food web, rather than the trees and grasses we see on land. In oceans, photosynthetic activity tends to be greatest near coastlines, where nitrogen, phosphorus, and other nutrients wash offshore and fertilize primary producers. Ocean currents also contribute to the distribution of biological productivity, because they transport nutrients and phytoplankton far from shore (fig. 5.13).

As plankton, algae, fish, and other organisms die, they sink toward the ocean floor. Deep-ocean ecosystems, consisting of crabs, filter-feeding organisms, strange phosphorescent fish, and many other life-forms, often rely on this "marine snow" as a

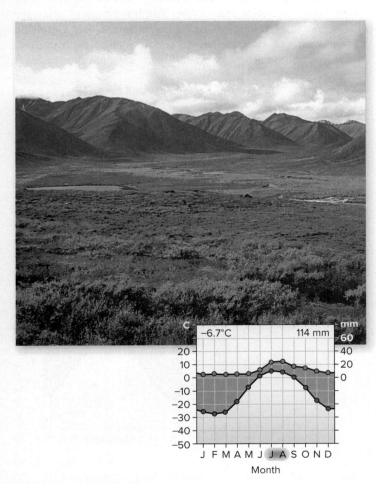

FIGURE 5.12 This landscape in Canada's Northwest Territories has both alpine and arctic tundra. Plant diversity is relatively low, and frost can occur even in summer.
© Mary Ann Cunningham

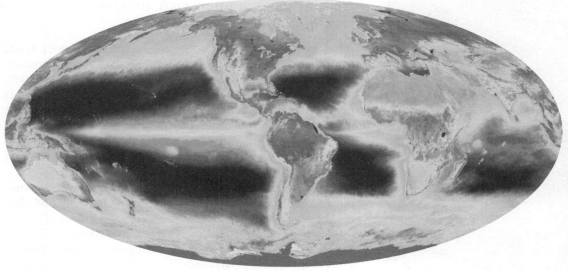

FIGURE 5.13 Satellite measurements of chlorophyll levels in the oceans and on land. Dark green to blue land areas have high biological productivity. Dark blue oceans have little chlorophyll and are biologically impoverished. Light green to yellow ocean zones are biologically rich.
Source: Courtesy of SeaWiFS/NASA

primary nutrient source. Surface communities also depend on this material. Upwelling currents circulate nutrients from the ocean floor back to the surface. Along the coasts of South America, Africa, and Europe, these currents support rich fisheries.

Vertical stratification (differing conditions from upper to lower layers) is a key feature of aquatic ecosystems, mainly because light and temperatures decrease rapidly with depth. Photosynthesis can only occur in the upper sunlit layers (the photic zone, often reaching about 20 m deep). Well-lit, near-surface communities such as coral reefs and estuaries are among the world's most biologically productive environments. Photosynthetic algae and phytoplankton provide the basis of much of the marine food web. Deeper communities gain energy from sources other than photosynthesis, such as detritus drifting down from the surface. Temperature also decreases with depth. Deep-ocean species often grow slowly, in part because metabolism slows in cold conditions. Temperature also affects the amount of oxygen that can be absorbed in water. Cold water holds abundant oxygen, so productivity is often high in cold oceans, as in the North Atlantic, North Pacific, and Antarctic.

Ocean systems can be described by depth and proximity to shore (fig. 5.14). In general, **benthic** communities occur on the bottom, and **pelagic** (from "sea" in Greek) zones comprise the water column. The epipelagic zone (*epi* = on top) has photosynthetic organisms. Below this are the mesopelagic (*meso* = medium), and bathypelagic (*bathos* = deep) zones. The deepest layers are the abyssal zone (to 4,000 m) and hadal zone (deeper than 6,000 m). Shorelines are known as littoral zones, and the area exposed by low tides is known as the intertidal zone. Often, there is a broad, relatively shallow region along a continent's coast, which may reach a few kilometers or hundreds of kilometers from shore. This undersea area is the continental shelf.

Depth controls light penetration and temperature

The open ocean is often referred to as a biological desert because it has relatively low productivity. But like terrestrial deserts, the open ocean has areas of rich productivity and diversity. Fish and plankton abound in regions such as the equatorial Pacific

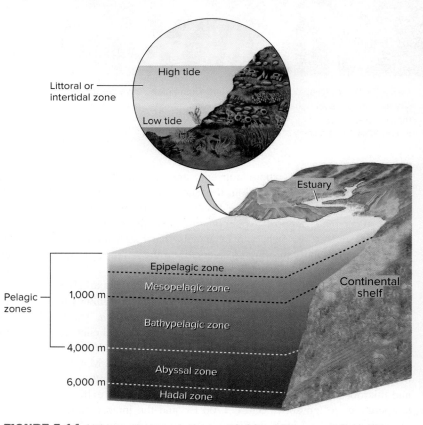

FIGURE 5.14 Light penetrates only the top 10–20 m of the ocean. Below this level, temperatures drop and pressure increases. Nearshore environments include the intertidal zone and estuaries.

FIGURE 5.15 Deep-ocean thermal vent communities have great diversity and are unusual because they rely on chemosynthesis, not photosynthesis, for energy.
Source: NOAA. Image courtesy of C. Van Dover

and Antarctic oceans, where nutrients are distributed by currents. Another notable exception, the Sargasso Sea in the western Atlantic, is known for its free-floating mats of brown algae. These algae mats support a phenomenal diversity of animals, including sea turtles, fish, and even eels that hatch amid the algae, then eventually migrate up rivers along the Atlantic coasts of North America and Europe.

Deep-sea thermal vent communities are another remarkable type of marine system (fig. 5.15) that was completely unknown until 1977 explorations with the deep-sea submarine *Alvin*. These communities are based on microbes that capture chemical energy, mainly by metabolizing sulfur compounds released from thermal vents—jets of hot water and minerals on the ocean floor. Magma below the ocean crust heats these vents. Tube worms, mussels, and microbes on these vents are adapted to survive both extreme temperatures, often above 350°C (700°F), and the intense water pressure at depths of 7,000 m (20,000 ft) or more. Oceanographers have discovered thousands of different types of organisms, most of them microscopic, in these communities (chapter 3).

Coastal zones support rich, diverse communities

As in the open ocean, shoreline communities vary with depth, light, nutrient concentrations, and temperature. Some shoreline communities, such as estuaries, have high biological productivity and diversity because they are enriched by nutrients washing off from the land. But nutrient loading can be excessive. Around the world, more than 200 "dead zones" occur in coastal areas where excess nutrients stimulate bacterial growth that consumes almost all oxygen in the water and excludes most other life. We'll discuss this problem further in chapter 18.

Coral reefs are among the best-known marine ecosystems because of their extraordinary biological productivity and their diverse and beautiful organisms. Reefs are aggregations of minute colonial animals (coral polyps) that live symbiotically with photosynthetic algae. Calcium-rich coral skeletons build up to make reefs, atolls, and islands (fig. 5.16a). Reefs protect shorelines and shelter countless species of fish, worms, crustaceans, and other life-forms. Reef-building corals live where water is shallow and clear enough for sunlight to reach the photosynthetic algae. They need warm (but not too warm) water, and can't survive where high nutrient concentrations or runoff from the land create dense layers of algae, fungi, or sediment. Coral reefs also are among the most endangered biomes in the world. Destructive fishing practices can damage or destroy coral communities. In addition, polluted urban runoff, trash, sewage and industrial effluent, sediment from agriculture, and unsustainable forestry are smothering coral reefs along coastlines that have high human populations. Introduced pathogens and predators also threaten many reefs. Perhaps the greatest threat to reefs is global warming. Elevated water temperatures cause **coral bleaching,** in which corals expel their algal partner and then die. The third UNESCO Conference on Oceans, Coasts, and Islands in 2006 reported that one-third of all coral reefs have been destroyed, and that 60 percent are now degraded and probably will be dead by 2030.

The value of an intact reef in a tourist economy can be upward of $1 million per square kilometer. The costs of conserving these same reefs in a marine-protected area would be just $775 per square kilometer per year, the UN Environment Programme estimates. Of the approximately 30 million small-scale fishers in the developing world, most are dependent to a greater or lesser extent on coral reefs. In the Philippines, the UN estimates that more than 1 million fishers depend directly on coral reefs for their livelihoods. We'll discuss reef restoration efforts further in chapter 13.

Sea-grass beds, or eel-grass beds, often occupy shallow, warm, sandy areas near coral reefs. Like reefs, these communities support a rich diversity of grazers, from snails and worms to turtles and manatees. Also like reefs, these environments are easily smothered by sediment originating from onshore agriculture and development.

Mangroves are a special variety of trees that grow in salt water (fig. 5.16b). They occur along calm, shallow, tropical coastlines around the world. Mangrove forests or swamps help stabilize shorelines, and they are also critical nurseries for fish, shrimp, and

(a) Coral reefs

(b) Mangroves

(c) Estuary and salt marsh

(d) Tide pool

FIGURE 5.16 Coastal environments support incredible diversity and help stabilize shorelines. Coral reefs (a), mangroves (b), and estuaries (c) also provide critical nurseries for marine ecosystems. Tide pools (d) also shelter highly specialized organisms.
(a) © Glen Allison/Getty Images RF (b) © Robert Garvey/Corbis (c) © Andrew Martinez/Science Source (d) © Cary Kalscheuer/Shutterstock.com

other commercial species. Like coral reefs, mangroves line tropical and subtropical coastlines, where they are vulnerable to development, sedimentation, and overuse. Unlike reefs, mangroves provide commercial timber, and they can be clear-cut to make room for aquaculture (fish farming) and other activities. Ironically, mangroves provide the protected spawning beds for most of the fish and shrimp farmed in these ponds. As mangroves become increasingly threatened in tropical countries, villages relying on fishing for income and sustenance are seeing reduced catches and falling income.

Estuaries are bays where rivers empty into the sea, mixing fresh water with salt water. **Salt marshes,** shallow wetlands flooded regularly or occasionally with seawater, occur on shallow coastlines, including estuaries (fig. 5.16c). Usually calm, warm, and nutrient-rich, estuaries and salt marshes are biologically

diverse and productive. Rivers provide nutrients and sediments; and a muddy bottom supports emergent plants (whose leaves emerge above the water surface), as well as the young forms of crustaceans, such as crabs and shrimp, and mollusks, such as clams and oysters. Nearly two-thirds of all marine fish and shellfish rely on estuaries and saline wetlands for spawning and juvenile development.

Estuaries near major American cities once supported an enormous wealth of seafood. Oyster beds and clam banks in the waters adjacent to New York, Boston, and Baltimore provided free and easy food to early residents. Sewage and other contaminants long ago eliminated most of these resources, however. Recently, major efforts have been made to revive Chesapeake Bay, America's largest and most productive estuary. These efforts have shown some success, but many challenges remain (chapter 3).

In contrast to the shallow, calm conditions of estuaries, coral reefs, and mangroves, there are violent, wave-blasted shorelines that support fascinating life-forms in **tide pools.** Tide pools are depressions in a rocky shoreline that are flooded at high tide but retain some water at low tide. These areas remain rocky where wave action prevents most plant growth or sediment (mud) accumulation. Extreme conditions—with frigid flooding at high tide, and hot desiccating sunshine at low tide—make life impossible for most species. But the specialized animals and plants that do occur in this rocky intertidal zone are astonishingly diverse and beautiful (fig. 5.16*d*).

Barrier islands are low, narrow, sandy islands that form parallel to a coastline (fig. 5.17). They occur where the continental shelf is shallow and rivers or coastal currents provide a steady source of sediments. They protect brackish (moderately salty), inshore lagoons and salt marshes from storms, waves, and tides. One of the world's most extensive sets of barrier islands lines the Atlantic coast from New England to Florida, as well as along the Gulf Coast of Texas. Composed of sand that is constantly reshaped by wind and waves, these islands can be formed or removed by a single violent storm. Because they are mostly beach, barrier islands are also popular places for real estate development. About 20 percent of the barrier island surface in the United States has been developed. Barrier islands are also critical to preserving coastal shorelines, settlements, estuaries, and wetlands.

Human occupation often destroys the value that attracts us there in the first place. Barrier islands and beaches are dynamic environments, and sand is hard to keep in place. Wind and wave erosion is a constant threat to beach developments. Walking or driving vehicles over dune grass destroys the stabilizing vegetative cover and accelerates, or triggers, erosion. Cutting roads through the dunes further destabilizes these islands, making them increasingly vulnerable to storm damage. When Hurricane Katrina hit the

FIGURE 5.18 Coastal beaches and barrier islands are extremely vulnerable to erosion, especially when shoreline plant communities are damaged. Protecting property on naturally shifting coastlines is difficult and costly.
© David L. Ryan/The Boston Glove via Getty Images

U.S. Gulf Coast in 2005, it caused at least $200 billion in property damage and displaced 4 million people. Thousands of homes were destroyed (fig. 5.18), particularly on low-lying barrier islands. Similarly, when Superstorm Sandy washed over New York and New Jersey in 2012, most of the $50 billion in damage occurred on the heavily built-up but low-lying beaches and barrier islands of the region.

Because of these problems, we spend billions of dollars each year building protective walls and barriers, pumping sand onto beaches from offshore, and moving sand from one beach area to another. Much of this expense is borne by the public. Some planners question whether we should allow rebuilding on barrier islands, especially after they've been destroyed multiple times.

Section Review

1. How do physical conditions change with depth in marine environments?
2. Describe four different coastal ecosystems.
3. What is coral bleaching?

5.3 FRESHWATER ECOSYSTEMS

- *Freshwater systems vary according to depth and light penetration, which control size and types of vegetation.*
- *Shallow wetlands can be highly productive and diverse.*

Freshwater environments are often small, but they are disproportionately important in biodiversity. Wetlands support many species that can live nowhere else, including amphibians,

FIGURE 5.17 A barrier island, Assateague, along the Maryland–Virginia coast. Grasses cover and protect dunes, which keep ocean waves from disturbing the bay, salt marshes, and coast at right. Roads cut through the dunes expose them to erosion.
Source: USGS

fish, aquatic plants, and many insects and birds. Upland ecosystems (drier areas around wetlands) often rely on these wetland communities. Insects, birds, and frogs, for example, are eaten by predators in grassland and forest ecosystems. In deserts or dry grasslands, isolated pools and streams support astonishing concentrations of biodiversity, including streamside shrubs and trees that support birds, insects, and small mammals.

Temperature and light vary with depth in lakes

Freshwater lakes, like marine environments, have distinct vertical zones (fig. 5.19). Near the surface, a subcommunity of plankton—mainly microscopic plants, animals, and protists (single-celled organisms such as amoebae)—floats freely in the water column. Insects such as water striders and mosquitoes also live at the air-water interface. Fish move through the water column, sometimes near the surface and sometimes at depth.

Finally, the bottom, or *benthos,* is occupied by a variety of snails, burrowing worms, fish, and other organisms. These make up the benthic community. Oxygen levels are lowest in the benthic environment, mainly because there is little mixing to introduce oxygen to this zone. Anaerobic bacteria (not using oxygen) may live in low-oxygen sediments. In the littoral zone, emergent plants such as cattails and rushes grow in the bottom sediment. These plants create important functional links between layers of an aquatic ecosystem, and they may provide the greatest primary productivity to the system.

Lakes, unless they are shallow, have a warmer upper layer that is mixed by wind and warmed by the sun. This layer is the *epilimnion.* Below the epilimnion is the *hypolimnion* (*hypo* = below), a colder, deeper layer that is not mixed. If you have gone swimming in a moderately deep lake, you may have discovered the sharp temperature boundary, known as the **thermocline,** between these layers. Below this boundary, the water is much colder. This boundary is also called the *mesolimnion.*

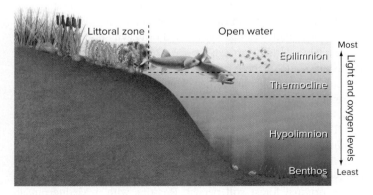

FIGURE 5.19 The layers of a deep lake are determined mainly by gradients of light, oxygen, and temperature. The epilimnion is affected by surface mixing from wind and thermal convections, while mixing between the hypolimnion and epilimnion is inhibited by a sharp temperature and density difference at the thermocline.

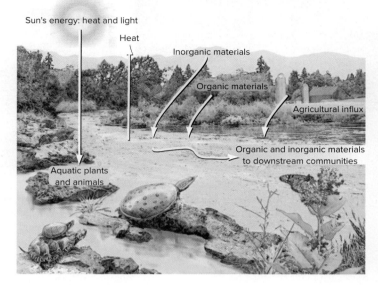

FIGURE 5.20 The character of freshwater ecosystems is greatly influenced by the immediately surrounding terrestrial ecosystems, and even by ecosystems far upstream or far uphill from a particular site.

Local conditions that affect the characteristics of an aquatic community include (1) availability (or excess) of nutrients, such as nitrates and phosphates; (2) suspended matter, such as silt, that affects light penetration; (3) depth; (4) temperature; (5) currents; (6) bottom characteristics, such as muddy, sandy, or rocky floor; (7) internal currents; and (8) connections to, or isolation from, other aquatic and terrestrial systems (fig. 5.20).

Wetlands are shallow and productive

Wetlands are shallow ecosystems in which the land surface is saturated or submerged at least part of the year. Wetlands have vegetation that is adapted to grow under saturated conditions. These legal definitions are important, because although wetlands make up only a small part of most countries, they are disproportionately important in conservation debates and are the focus of continual legal disputes around the world and in North America. Beyond these basic descriptions, defining wetlands is a matter of hot debate. How often must a wetland be saturated, and for how long? How large must it be to deserve legal protection? Answers can vary, depending on political, as well as ecological, concerns.

These relatively small systems support rich biodiversity, and they are essential for both breeding and migrating birds. Although wetlands occupy less than 5 percent of the land in the United States, the Fish and Wildlife Service estimates that one-third of all endangered species spend at least part of their lives in wetlands. Wetlands retain storm water and reduce flooding by slowing the rate at which rainfall reaches river systems. Floodwater storage is worth $3 billion to $4 billion per year in the United States. As

(a) Swamp, or wooded wetland

(b) Marsh

(c) Coastal saltmarsh

FIGURE 5.21 Wetlands provide irreplaceable ecological services, including water filtration, water storage and flood reduction, and habitat. Forested wetlands (a) are often called swamps; marshes (b) have no trees; coastal saltmarshes (c) are tidal and have rich diversity.

(a) © S. Solum/PhotoLink/Getty Images RF (b) © C. McIntyre/PhotoLink/Getty Images RF (c) © William P. Cunningham

water stands in wetlands, it also seeps into the ground, replenishing groundwater supplies. Wetlands filter, and even purify, urban and farm runoff, as bacteria and plants take up nutrients and contaminants in water. They are also in great demand for filling and development. They are often near cities or farms, where land is valuable, and once drained, wetlands are easily converted to more lucrative uses.

Wetlands are generally described by their vegetation. **Swamps,** also known as wooded wetlands, are wetlands with trees (fig. 5.21*a*). **Marshes** are wetlands without trees (fig. 5.21*b*). **Bogs** are areas of saturated ground, and usually the ground is composed of deep layers of accumulated, undecayed vegetation known as peat. **Fens** are similar to bogs except that they are mainly fed by groundwater, so that they have mineral-rich water and specially adapted plant species. Bogs are fed mainly by precipitation. Swamps and marshes have high biological productivity. Bogs and fens, which are often nutrient-poor, have low biological productivity. They may have unusual and interesting species, though, such as sundews and pitcher plants, which are adapted to capture nutrients from insects rather than from soil.

The water in marshes and swamps usually is shallow enough to allow full penetration of sunlight and seasonal warming (fig. 5.21*c*). These mild conditions favor great photosynthetic activity, resulting in high productivity at all trophic levels. In short, life is abundant and varied. Wetlands are major breeding, nesting, and migration staging areas for waterfowl and shorebirds.

Wetlands may gradually convert to terrestrial communities as they fill with sediment, and as vegetation gradually fills in toward the center. Often, this process is accelerated by increased sediment loads from urban development, farms, and roads. Wetland losses are one of the areas of greatest concern to biologists.

Section Review

1. Describe four different kinds of wetlands.
2. Why are wetlands sites of high biodiversity and productivity?

5.4 HUMAN DISTURBANCE

- *Temperate forests and grasslands—where humans live and farm—are the world's most extensively disturbed environments.*
- *Temperate wetlands are severely depleted.*
- *Until recently, tropical rainforests were relatively undisturbed.*

Humans have become dominant organisms over most of the earth, damaging or disturbing more than half of the world's terrestrial ecosystems to some extent. By some estimates, humans preempt about 40 percent of the net terrestrial primary productivity of the biosphere, either by consuming it directly, by interfering with its production or use, or by altering the species composition or physical processes of human-dominated ecosystems. Conversion of natural habitat to human uses is the largest single cause of biodiversity losses.

Human disturbance has displaced many of the world's natural biomes, as shown in the purple areas in fig. 5.22. This map, developed initially by researchers from the environmental group Conservation International, indicates that agricultural or heavily populated regions have largely been transformed: Europe, China, and the American Great Plains are some of these extensively impacted areas. Environments difficult for humans to inhabit, such as tundra, boreal forest, deserts, or tropical rainforest regions (compare to fig. 5.4) are less disturbed. Data from this study are shown in table 5.1.

Temperate broadleaf forests are the most completely human-dominated of any major biome. The climate and soils that support such forests are especially congenial for human occupation. In eastern North America or most of Europe, for example, only remnants of the original forest still persist. Regions with a Mediterranean climate generally are highly desired for human habitation. Because these landscapes also have high levels of biodiversity, conflicts between human preferences and biological values frequently occur.

Temperate grasslands, temperate rainforests, tropical dry forests, and many islands also have been highly disturbed by human activities. If you have traveled through the American cornbelt states,

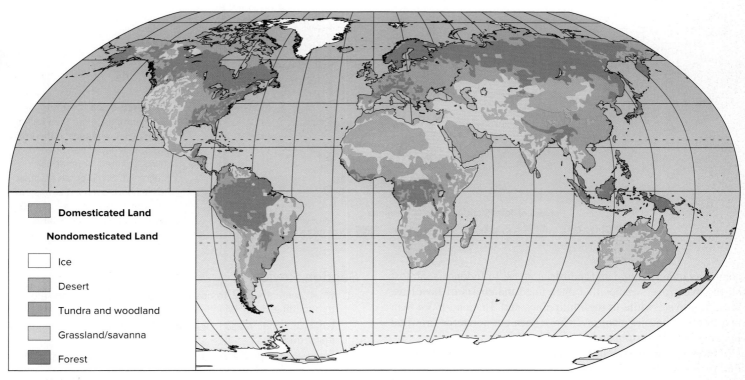

FIGURE 5.22 Domesticated land has replaced much of the earth's original land cover.
Source: United Nations Environment Programme, Global Environment Outlook.

such as Iowa or Illinois, you have seen how thoroughly former prairies have been converted to farmlands. Intensive cultivation of this land exposes the soil to erosion and fertility losses (chapter 9).

Islands are highly vulnerable to disruption. Because of their isolation, islands often have high proportions of unusual species that occur nowhere else. Many islands, such as Madagascar, Haiti, and Java, have lost more than 99 percent of their original land cover, as well as many of their original species.

Tundra and arctic deserts are the least disturbed biomes in the world. Because of harsh climates and unproductive soils, these biomes have relatively little human settlement or conversion to farms. Cool temperate conifer forests also generally are lightly populated, and large areas remain in a relatively natural state. Recent expansion of forest harvesting in Canada and Siberia may threaten the integrity of this biome, however. Similarly, tropical moist forests, with poor soils and heavy rains, historically were hard to farm or log. Large expanses of tropical moist forests still remain in the Amazon and Congo basins. But in other areas of the tropics, such as West Africa, Madagascar, Southeast Asia, and increasingly Brazil and Malaysia, forests are being cleared for timber or converted to palm oil and soybean farming (chapter 12).

Wetlands have suffered severe losses wherever human settlement is extensive. About half of all original wetlands in the United States have been drained, filled, or lost to coastal erosion over the past 250 years. In the prairie states, most small wetlands and seasonally flooded marshes, once the main source of the continent's ducks and important reservoirs for reducing spring floods, have been drained and converted to croplands. Iowa, for example, is estimated to have lost 99 percent of its presettlement wetlands (fig. 5.23). California has lost 90 percent of the extensive marshes and deltas that once stretched across its central valley.

Similar wetland losses have occurred globally. In New Zealand, over 90 percent of natural wetlands have been destroyed since European settlement. In Portugal, some 70 percent of freshwater wetlands and 60 percent of estuarine habitats have been

Table 5.1 Human Disturbance	
Biome	**% Human Dominated**
Temperate broadleaf forests	81.9
Chaparral	67.8
Temperate grasslands	40.4
Temperate rainforests	46.1
Tropical dry forests	45.9
Mixed mountain systems	25.6
Mixed island systems	41.8
Cold deserts/semideserts	8.5
Warm deserts/semideserts	12.2
Moist tropical forests	24.9
Tropical grasslands	4.7
Temperate coniferous forests	11.8
Tundra and arctic desert	0.3

Note: Where undisturbed and human-dominated areas do not add up to 100 percent, the difference represents partially disturbed lands.

Source: Hannah, Lee, et al., "Human Disturbance and Natural Habitat: A Biome Level Analysis of a Global Data Set," in *Biodiversity and Conservation, 1995, 4:128–55.*

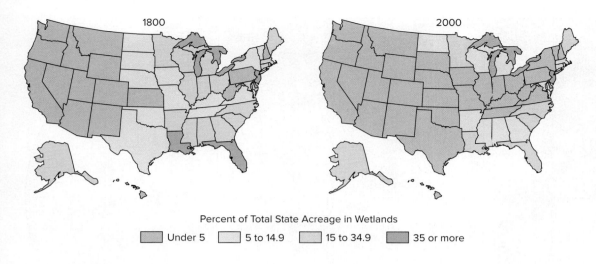

Percent of Total State Acreage in Wetlands

Under 5   5 to 14.9   15 to 34.9   35 or more

FIGURE 5.23 Over the past two centuries, more than half of the original wetlands in the lower 48 states have been drained, filled, polluted, or otherwise degraded. Some of the greatest losses have been in Midwestern farming states, where up to 99 percent of all wetlands have been lost.

converted to agriculture and industrial areas. In Indonesia, almost all the mangrove swamps that once lined the coasts of Java have been destroyed, while in the Philippines and Thailand more than two-thirds of coastal mangroves have been cut down for firewood or conversion to shrimp and fish ponds.

Slowing this destruction, or even reversing it, is a challenge that we will discuss in chapter 13.

Section Review

1. What percentage of temperate grasslands and forests are disturbed (table 5.1)? Why?
2. What is one reason for the clearing of tropical coastal mangroves?
3. How have temperate wetlands in the United States been lost?

Conclusion

The potential location of biological communities is determined in large part by climate, moisture availability, soil type, geomorphology, and other natural features. Understanding the global distribution of biomes, and knowing the differences in who lives where and why, are essential to the study of global environmental science. Human occupation and use of natural resources are strongly dependent on the biomes found in particular locations. Humans tend to prefer mild climates and the highly productive biological communities found in temperate zones. These biomes also suffer the highest rates of degradation and overuse.

Plants and animals have evolved characteristics that allow them to live in particular biomes, such as seasonal tropical forests, alpine tundra, or chaparral. Recognizing these adaptations helps you understand limiting factors for survival in those biomes.

Oceans cover over 70 percent of the earth's surface, yet we know relatively little about them. Some marine biomes, such as coral reefs, can be as biologically diverse and productive as any terrestrial biome. People have always depended on rich, complex ecosystems. In recent times, the rapid growth of human populations, coupled with more powerful ways to harvest resources, has led to extensive destruction of these environments. Still, it is possible for us to protect these living communities. The opening case study of this chapter illustrates how people can work together to protect and even restore the biological communities on which they depend. Perhaps we can find similar solutions in other biologically rich but endangered biomes.

Reviewing Key Terms

Can you define the following terms in environmental science?

barrier islands 5.2	conifers 5.1	grasslands 5.1	swamp 5.3	tropical seasonal
benthic 5.2	coral bleaching 5.2	mangroves 5.2	taiga 5.1	forest 5.1
biome 5.1	coral reefs 5.2	marsh 5.3	temperate	wetlands 5.3
bog 5.3	deciduous 5.1	pelagic 5.2	rainforest 5.1	
boreal forest 5.1	desert 5.1	phytoplankton 5.2	thermocline 5.3	
chaparral 5.1	estuary 5.2	salt marsh 5.2	tide pool 5.2	
cloud forests 5.1	fen 5.3	savannas 5.1	tropical rainforests 5.1	

Critical Thinking and Discussion Questions

1. What physical and biological factors are most important in shaping the biome in which you live? How have those factors changed in the past 100 or 1,000 years?

2. Fire is a common component of forest biomes. As more of us build homes in these areas, can we protect people from natural disturbances? How?

3. Often, humans work to preserve biomes that are visually attractive. Are there biomes that might be lost because they are not attractive? Is this a problem?

4. Disney World in Florida wants to expand onto a wetland. It has offered to buy and preserve a large nature preserve in a different area to make up for the wetland it is destroying. Is that reasonable? Why or why not?

5. Suppose further that the wetland being destroyed in question 4 and its replacement area both contain several endangered species (but different ones). How would you compare different species against each other? Should we preserve animals before we preserve plant or insect species?

6. Historically, barrier islands have been hard to protect because links between them and inshore ecosystems are poorly recognized. What kinds of information would help a community distant from the coast commit to preserving a barrier island?

Data Analysis

Reading Climate Graphs

As climate conditions change, interest in drought, rainfall, and temperature conditions are becoming more important to farmers, producers and users of hydroelectric power, policymakers, and others. The National Drought Mitigation Center at the University of Nebraska is a resource for monitoring climate conditions for the United States. Examine its website (http://drought.unl.edu/DroughtBasics/WhatisClimatology/ClimographsforSelectedUSCities.aspx) and go to Connect to find questions and demonstrate your understanding of these graphs.

Climographs for Selected U.S. Cities

TO ACCESS ADDITIONAL RESOURCES FOR THIS CHAPTER, PLEASE VISIT CONNECT AT
www.connect.mheducation.com

You will find LearnSmart, an adaptive learning system, Google Earth™ exercises, additional Case Studies, Data Analysis exercises, and an interactive ebook.

6

Population Biology

▲ A bluefin tuna, the largest and most expensive commercially harvested tuna, is disentangled from a net.
Source: NOAA

Learning Outcomes

After studying this chapter, you should be able to:

6.1 Describe the dynamics of population growth.

6.2 Compare and contrast the factors that regulate population growth.

6.3 Identify some applications of population dynamics in conservation biology.

"Nature teaches more than she preaches."

– John Burroughs

Are We Fishing to Extinction?

The most expensive tuna ever sold, a 222 kg (489 lb) bluefin tuna, was auctioned in Tokyo in January 2013 for a record-breaking $1.8 million. This one fish, the auspicious first sale of the new year at the Tsukiji fish market, cost a jaw-dropping $7,456 per kg ($3,389 per lb). The price was extreme because the first fish of the year is thought to bring good luck, but plummeting numbers of bluefins and rising demand for sushi and sashimi also helped to push up the bids. The world was watching this sale, because bluefin tuna has been the subject of bitter disputes. On the one side, biologists warn that overfishing has cut populations by as much as 92 percent and is driving the species toward extinction. On the other side, the fishing industry and traders in Japanese sushi are unwilling to sacrifice the enormous profits it brings. Just months before this sale, under pressure from tuna-fishing nations, Atlantic bluefin tuna were denied international endangered species designation. The tuna-fishing industry claimed, despite warnings of its own biologists, that reduced catches were unnecessary to protect populations.

Population biology, the science of modeling changes in species abundance, is key to understanding this controversy. The bluefin tuna is a large, fast, wide-ranging fish. It can live for 50 years, but it matures slowly for a fish—some populations take 8 years or more to reach spawning age. The number of young in a year can be enormous, but that number depends on the availability of spawning-age fish and other factors. Biologists use those numbers to calculate the likely rate of decline in population sizes, the estimated rate of recovery from reduced fishing pressure, or the amount of fishing that the species can safely sustain. In the bluefin's case, spawning-age fish are declining fast, and population models indicate that the species is heading for a crash.

The International Commission for the Conservation of Atlantic Tunas (ICCAT) is in charge of protecting Atlantic tuna, marlin, swordfish, and other species by setting sustainable catch limits. Ideally, ICCAT uses population models to calculate a sustainable catch rate that maintains a stable spawning-age population. But ICCAT data show that Atlantic spawning stock has dropped by at least 75 percent from pre-1950 levels. Despite this decline, allowable catch limits remain high. A sustainable catch would be 8,500 tons or so of Atlantic bluefin tuna per year, but ICCAT has maintained limits 2 to 3 times this high. Moreover, ICCAT member states exceed their legal limits every year.

To make matters worse, unreported illegal catches by ICCAT member states are extremely high. Fishing is a notoriously hard industry to monitor. In the free-for-all on the high seas, where enforcement is weak or impossible, where individual nations subsidize fishing fleets, and where so much money is at stake, it's hard to be completely honest—especially if you don't trust the honesty of your competitors. ICCAT estimates that its records represent just half of actual catches in some years (fig. 6.1). According to the U.S. National Marine Fisheries Service, comparable problems of overfishing are occurring in nearly all the other top-predator fisheries, including marlin, swordfish, and albacore tuna. However, some populations, including Atlantic bigeye and yellowfin tuna, are not currently overfished.

Some tuna-catching nations may see it as being in their best interest to liquidate the species for short-term profits. Others just want to protect the interests of their own fishing fleets in the face of international competition. Because the species belongs to no individual nation, countries have strong profit incentives to catch the last fish before someone else does. Thus, the self-policing ICCAT structure has so far failed to conserve the Atlantic tuna. In 2009, Monaco petitioned for endangered species designation for the east-Atlantic population, but in 2010 that listing was denied on the grounds that ICCAT was already in charge of conserving the species. Just months later, ICCAT declined to substantially reduce allowable catches, although the organization did promise more thorough monitoring in the future.

The science of population biology provides the tools and insights needed to model sustainable catch rates and to identify unsustainable harvests. In this chapter, we'll examine the main concepts of population biology and the uses of these concepts in environmental science.

To find out which fish are best to eat, see the Monterey Bay Aquarium (www.montereybayaquarium.org/cr/seafoodwatch.aspx). You can also see data from ICCAT here: www.iccat.int/en/.

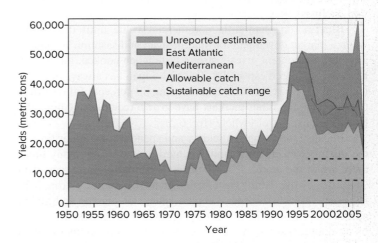

FIGURE 6.1 The bluefin tuna catch in the Atlantic since 1950. Note the differences between allowable, sustainable, and actual catch estimates. *Source: Data from ICCAT.*

- *Unrestrained exponential growth can lead to boom and bust cycles.*
- *Logistic growth slows as populations increase.*
- *Carrying capacity is the number of individuals that live in a given area for an extended period of time.*

Conserving the bluefin tuna depends on a good understanding of how populations grow and decline. Bluefin tuna are top predators, big and fast enough to catch and eat almost anything they encounter. They can grow to 4.6 m (over 15 feet) in length and weigh up to 680 kg (about 1,500 lb). They can dive to depths of 1,000 m (3,280 ft) and swim more than 40 miles (64 km) per hour. Bluefins migrate thousands of kilometers annually. Atlantic bluefins spawned in the Mediterranean travel across the Atlantic to feed along the East Coast of North America. A smaller population spawns on the northern slope of the Gulf of Mexico, then travels north to mix with the Mediterranean population in rich foraging areas in the open ocean. The Southern tuna population spawns between Java and northwestern Australia and ranges throughout the Southern Ocean.

Most bluefin tuna are caught on long-lines, a method that also catches fish, turtles, sharks and many non-target species. This technique is so effective that it's capable of depleting valuable marine resources if it isn't regulated. We'll discuss fishing techniques in chapter 9.

A 500-pound (226-kg) female bluefin can lay 30 million eggs in a single spawning season, but only few of those 30 million offspring will survive to maturity. Most are eaten by predators long before they are old enough to reproduce. Ordinarily, the enormous number of eggs produced by a population of adult females makes up for this high mortality among young tuna. But we're now upsetting that balance by specifically targeting fish of reproductive age, which are the largest. Thus, the number of offspring in the population declines every year.

Bluefin tuna had little commercial value until the 1960s, when a market developed for bluefin sushi and sashimi. Its unusually high fat content gives a strong taste when cooked, but its raw flesh is considered especially flavorful. Japan has always been the leader in the raw fish market, consuming 80 percent of the world's bluefin tuna, but other markets have grown recently in China, the United States, and elsewhere (fig. 6.2).

Because they range so widely and spend most of their time in the open ocean far from land, it's hard to know how many tunas there are in the world. ICCAT depends primarily on harvest data submitted by the fishing fleet itself. But as fish numbers drop, fishing boats just work harder, keeping catch statistics up and giving a false sense of confidence about population dynamics. There can be no doubt, though, that in recent years both the number of tuna caught and the average size of those fish have been dropping rapidly. Recapture studies suggest that the Southern tuna population has dropped by at least 90 percent in the past 30 years. The species is listed as critically endangered on the IUCN (International Union for Conservation of Nature) Red List of Threatened Species.

FIGURE 6.2 Frozen tuna at auction at Tsukiji market. Japan consumes 80 percent of the world's bluefin tuna as sushi and sashimi.
© PunchStock RF

We can describe growth symbolically

Describing the general pattern of population growth is easiest if we can reduce it to a few general factors. Ecologists find it most efficient and simplest to use symbolic terms such as N, r, and t to refer to these factors. At first, this symbolic form might seem hard to inter pret, but as you become familiar with the terms, you'll probably find them quicker to follow than longer text versions.

Here are some examples to show how you can describe population change. Consider cockroaches, a species capable of reproducing very rapidly. Figure 6.3 shows how quickly this population can grow. If there are no predators and food is abundant, then this depends mainly on two factors: the number you start with, and the rate of reproduction. Start with 2 cockroaches, one male and one female, and suppose they can lay eggs and increase to about 20 cockroaches in the course of three months. You can describe the rate of growth (r) per adult in one three-month period like this: $r = 20$ per 2 adults, or 10/adult, or "$r = 10$." If nothing limits population growth, numbers will continue to increase at this rate of $r = 10$ for each three-month time step. You can call each of these time steps (t). The starting point, before population growth begins, is "time 0" (t_0). The first time step is called t_1, the second time step is t_2, and so on. If $r = 10$, and the population (N) starts at 2 cockroaches, then the numbers will increase like this:

time	N	rate (r)	$r \times N$
t_1	2	10	$10 \times 2 = 20$
t_2	20	10	$10 \times 20 = 200$
t_3	200	10	$10 \times 200 = 2,000$
t_4	2,000	10	$10 \times 2,000 = 20,000$

This is a very rapid rate of increase, from 2 to 20,000 in four time steps (fig. 6.3). It's a very simplified explanation of growth, but it's fairly easy to follow. This rate is described as a "geometric" rate of increase. Look carefully at the numbers above, and you might notice that the population at t_2 is $2 \times 10 \times 10$, and the population at t_3 is $2 \times 10 \times 10 \times 10$. Another way to say this is that the population at t_2 is 2×10^2, and at t_3 the population is 2×10^3. In fact, the population at any given time is equal to the starting number (2) times the rate (10) raised to the exponent of the number of time steps (10^t). The short way to express the geometric rate of increase is below. Stop here and make sure you understand the terms N, r, and t:

$$N_t = N_0 r^t$$

Exponential growth involves continuous change

The example in the previous section takes growth one time step at a time, but really cockroaches can reproduce continuously if they live in a warm, humid environment. You can describe continuous change using the same terms, r, N, and t, plus the added term delta (d), for *change* (fig. 6.4).

You can read the equation in fig. 6.4 like this: the change in N (dN) per change in time (dt) equals rate of increase (r) times the population size (N). This equation is a model, a very simplified description of the dynamic process of population growth. Models like this are convenient because you can use them to describe many different growth trends, just by changing the "r" term. If $r > 0$, then dN increases over time. If $r < 0$, then dN is negative, and the population is declining. If $r = 0$, then dN is 0 (no change), and the population is stable.

This particular model describes an **exponential growth** rate. An exponential growth rate has a J-shaped curve, as in the upward parts of the curve in figure 6.5. This growth rate is characteristic of many species that grow rapidly when food is available, including moose and many prey species.

Doubling times and the rule of 70

Often, the easiest way to grasp the effect of rapidly growing populations is to calculate the doubling time (the time it takes to

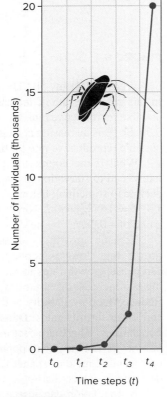

FIGURE 6.3 Population increase with a constant growth rate.

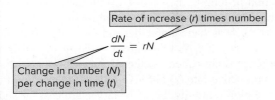

FIGURE 6.4 Exponential growth.

reach twice the previous size). There's a shortcut called the rule of 70 for this computation. If you divide 70 by the rate of growth in percentage, you'll get the approximate doubling time. For example, a population growing 2 percent per year will double in about 35 years. Or a population growing at 10 percent per year will double in 7 years. This rule makes it easy to convert one common measure of growth (percent per year) to another (time to doubling).

Exponential growth leads to crashes

A population can grow at an exponential rate this fast only if nothing limits its growth. Usually there are many factors that reduce the rate of increase. Individuals die, they might mature slowly, they may fail to reproduce. A population that has few or no predators (as in the case of invasive species, chapter 11), though, can grow at an exponential rate, at least for a while.

But all environments have a limited capacity to provide food and other resources for a particular species. **Carrying capacity** is the term for the number or biomass of a species that can be supported in a certain area without depleting resources. Eventually, a rapidly growing population reaches and overshoots this carrying capacity (fig. 6.5). Shortages of food or other resources eventually lead to a **population crash,** or rapid dieback. Once below the carrying capacity, the population may rise again, leading to boom and bust cycles. These oscillations can eventually lower the environmental carrying capacity for an entire food web.

In the case of the bluefin tuna (opening case study), we might say that the population of tuna fishers grew too fast and overshot the carrying capacity of the bluefin resource. The subsequent collapse would appear inevitable to a population biologist.

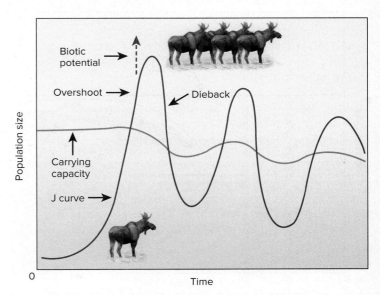

FIGURE 6.5 J curve, or exponential growth curve, with overshoot of carrying capacity. Exponential growth in an unrestrained population (*left side of curve*) leads to a population crash and oscillations below former levels. After the overshoot, carrying capacity may be reduced because of damage to the resources of the habitat. Moose on Isle Royale in Lake Superior may be exhibiting this growth pattern in response to their changing environment.

Logistic growth slows with population increase

Sometimes growth rates slow down as the population approaches carrying capacity—as resources become scarce, for example. In symbolic terms, the rate of change (dN/dt) depends on how close population size (N) is to the carrying capacity (K).

For example, suppose you have an area that can support 100 wolves. Let's say that 20 years ago, there were only 50 wolves, so there was abundant space and prey. The 50 wolves were healthy, many pups survived each year, and the population grew rapidly. Now the population has risen to 90. This number is close to the maximum 100 that the environment can support before the wolves begin to deplete their prey. Now, with less food per wolf, fewer cubs are surviving to adulthood, and the rate of increase has slowed. This slowing rate of growth makes an S-shaped curve, or a "sigmoidal" curve (fig. 6.6). This S-shaped growth pattern is also called **logistic growth,** because the curve is shaped like a logistic function used in math.

You can describe the general case of this growth by modifying the basic exponential equation with a feedback term—a term that can dampen the exponential growth of N (fig. 6.7).

If you are patient, you can see interesting patterns in this equation. Look first at the $\frac{N}{K}$ part. For the wolf example, K is 100 wolves, the maximum that can be supported. If N is 100, then $\frac{N}{K} = \frac{100}{100}$, which is 1. So $1 - \frac{N}{K} = 1 - 1$, which is 0. As a consequence, the right side of the equation is equal to 0 ($rN \times 0 = 0$), so $\frac{dN}{dt} = 0$. So there is no change in N if N is equal to the carrying capacity. Try working out the following examples on paper as you read, so you can see how N changes the equation.

$$\frac{dN}{dt} = rN\left(1 - \frac{N}{K}\right)$$

> Population size as a proportion of carrying capacity

FIGURE 6.7 Logistic growth.

What if N is only 50? Then $\frac{N}{K} = \frac{50}{100} = \frac{1}{2}$. So $1 - \frac{N}{K} = 1 - \frac{1}{2} = \frac{1}{2}$. In this case, the rate of increase is $\frac{1}{2}rN$, or half of the maximum possible reproductive rate. If $N = 10$, then $1 - \frac{N}{K} = 1 - \frac{10}{100}$, which is 0.90. So $\frac{dN}{dt}$ is increasing at a rate 90 percent as fast as the maximum possible reproductive rate for that species.

What if the population grows to 120? Overpopulation will likely lead to starvation or low birth rates, and the population will decline to something below 100 again. In terms of the model, now $1 - \frac{N}{K} = 1 - \frac{120}{100} = 0.2$. Now the rate of change, $\frac{dN}{dt}$, is *declining* at a rate of $-0.2rN$.

Logistic growth is **density-dependent,** meaning that the growth rate slows as population density increases. Many factors can cause growth rates to slow as density increases: overcrowding can increase disease rates, stress, and predation, for example. Reduced food availability or disease can lead to smaller body size and lower fertility rates. **Density-independent** factors also affect populations. These are factors that influence population growth but that do not vary with population density. Often, these are abiotic (nonliving) disturbances, such as drought or fire or habitat destruction, which disrupt an ecosystem.

A population can lose a portion of its numbers every year, but that portion depends on r, N, and K, among other factors. A sustainable harvest is possible, as in a tuna fishery, if the number caught is within that sustainable proportion. A "maximum sustained yield" is the highest number that can be regularly captured.

Population biologists have often been very successful in using growth rates to identify a sustainable yield. In North America, game laws restrict hunting of ducks, deer, fish, and other game species, and most hunters and fishers now understand and defend those limits. Acceptable harvest levels are set by population biologists, who have studied reproductive rates, carrying capacities, and population size of each species, in order to determine a sustainable yield. (In some cases, r is now too rapid for K: See What Do You Think? later in section 6.2.) Similarly, the Pacific salmon fishery, from California to Alaska, is carefully monitored and is considered a healthy and sustainable fishery.

Species respond to limits differently: *r*- and *K*-selected species

Which is more successful for increasing a population, rapid reproduction or long-term survival within the carrying capacity? Different species place their bets on different strategies. Some organisms, such as dandelions, depend on a high rate of reproduction and growth (rN) to secure a place in the environment. These organisms are called ***r*-selected species** because they have a high reproductive rate (r) but give little or no care to offspring, which have high mortality. Seeds or larvae are cast far and wide, and there is always a chance that some will survive and prosper.

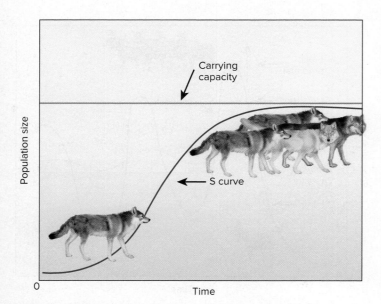

FIGURE 6.6 S curve, or logistic growth curve, describes a population's changing number over time in response to feedback from the environment or its own population density. Over the long run, a conservative and predictable population dynamic may win the race over an exponential population dynamic. Species with this growth pattern tend to be *K*-selected.

If there are no predators or diseases to control their population, these abundantly reproducing species can overshoot carrying capacity and experience population crashes, but as long as vast quantities of young are produced, a few will survive. Other organisms reproduce more conservatively—with longer generation times, late sexual maturity, and fewer young. These are referred to as *K*-selected species, because their growth slows as the carrying capacity (*K*) of their environment is approached.

Many species blend exponential (*r*-selected) and logistic (*K*-selected) growth characteristics. Still, it's useful to contrast the advantages and disadvantages of organisms at the extremes of the continuum. It also helps if we view differences in terms of "strategies" of adaptation and the "logic" of different reproductive modes (table 6.1).

Organisms with *r*-selected, or exponential, growth patterns tend to occupy low trophic levels in their ecosystems (see chapter 3) or they are successional pioneers. These species, which generally have wide tolerance limits for environmental factors, and thus can occupy many different niches and habitats, are the ones we often describe as "weedy." They tend to occupy disturbed or new environments, grow rapidly, mature early, and produce many offspring with excellent dispersal abilities. As individual parents, they do little to care for their offspring or protect them from predation. They invest their energy in producing huge numbers of young and count on some surviving to adulthood.

A female clam, for example, can release up to 1 million eggs in her lifetime. The vast majority of young clams die before reaching maturity, but a few survive and the species persists. Many marine invertebrates, parasites, insects, rodents, and annual plants follow this reproductive strategy. Also included in this group are most invasive and pioneer organisms, weeds, and pests.

Table 6.1 Reproductive Strategies	
***r*-Selected Species**	***K*-Selected Species**
1. Short life	1. Long life
2. Rapid growth	2. Slower growth
3. Early maturity	3. Late maturity
4. Many small offspring	4. Few, large offspring
5. Little parental care or protection	5. High parental care or protection
6. Little investment in individual offspring	6. High investment in individual offspring
7. Adapted to unstable environment	7. Adapted to stable environment
8. Pioneers, colonizers	8. Later stages of succession
9. Niche generalists	9. Niche specialists
10. Prey	10. Predators
11. Regulated mainly by extrinsic factors	11. Regulated mainly by intrinsic factors
12. Low trophic level	12. High trophic level

So-called *K*-selected organisms are usually larger, live long lives, mature slowly, produce few offspring in each generation, and have few natural predators. Elephants, for example, are not reproductively mature until they are 18 to 20 years old. In youth and adolescence, a young elephant belongs to an extended family that cares for it, protects it, and teaches it how to behave. A female elephant normally conceives only once every 4 or 5 years. The gestation period is about 18 months; thus, an elephant herd doesn't produce many babies in any year. Because elephants have few enemies and live a long life (60 or 70 years), this low reproductive rate produces enough elephants to keep the population stable, given good environmental conditions and no poachers.

When you consider the species you recognize from around the world, can you designate them as *r*- or *K*-selected species? What strategies seem to be operating for ants, bald eagles, cheetahs, clams, dandelions, giraffes, or sharks?

Section Review

1. What factors caused the collapse of bluefin tuna populations?
2. Define *exponential growth* and *logistic growth*.
3. Explain these terms: *r, N, t, dN/dt*.

6.2 Factors That Regulate Population Growth

- *Births, immigration, deaths, and emigration impact population size.*
- *Biotic factors are often intrinsic to an organism, while abiotic factors are always extrinsic.*
- *Density-dependent effects can be dramatic.*

By adding carrying capacity, we complicated our first simple population model, and we made it more realistic. To complicate it still further, we can consider the four factors that contribute to *r*, or rate of growth. These factors are **B**irths, **I**mmigration from other areas, **D**eaths, and **E**migration to other areas. More specifically, rate of growth is equal to Births + Immigration − Deaths − Emigration. In a detailed population model, populations receive immigrants and lose individuals to emigration. Number of births might rise more rapidly than number of deaths. Models of human populations (see chapter 7), as well as animal populations, involve detailed calculations of the four BIDE factors.

Too Many Deer?

A century ago, few Americans had ever seen a wild deer. Uncontrolled hunting and habitat destruction had reduced the deer population to about 500,000 animals nationwide. Some states had no deer at all. To protect the remaining deer, laws were passed in the 1920s and 1930s to restrict hunting, and the main deer predators—wolves and mountain lions—were exterminated throughout most of their former range.

As Americans have moved from rural areas to urban centers, forests have regrown, and with no natural predators, deer populations have undergone explosive growth. Maturing at age two, a female deer can give birth to twin fawns every year for a decade or more. Increasing more than 20 percent annually, a deer population can double in just three years, an excellent example of irruptive, exponential growth.

Wildlife biologists estimate that the contiguous 48 states now have a population of more than 30 million white-tailed deer (*Odocoileus virginianus*), probably triple the number present in pre-Columbian times. Some

White-tailed deer (odocoileus virginianus) can become emaciated and sick when they exceed their environment's carrying capacity.
© Royalty-Free/Corbis

areas have as many as 200 deer per square mile (518/km^2). At this density, woodland plant diversity is generally reduced to a few species that deer won't eat. Most deer, in such conditions, suffer from malnourishment, and many die every year of disease and starvation. Other species are diminished as well. Many small mammals and ground-dwelling birds begin to disappear when deer populations reach 25 animals per square mile. At 50 deer per square mile, most ecosystems are seriously impoverished.

The social costs of large deer populations are high. In Pennsylvania alone, where deer numbers are now about 500 times greater than a century ago, deer destroy about $70 million worth of crops and $75 million worth of trees annually. In 2013, some 115,000 collisions with motor vehicles caused nearly $400 million in property damage. Deer help spread Lyme disease, and in some states chronic wasting disease is found in wild deer herds. Some of the most heated criticisms of current deer management policies are in the suburbs. Deer love to browse on the flowers, young trees, and ornamental bushes in suburban yards. Disputes arise between those who love to watch deer and their neighbors who want to defend their gardens.

In remote forest areas, many states have extended hunting seasons, increased the bag limit to four or more animals, and encouraged hunters to shoot does (females) as well as bucks (males). Some hunters criticize these changes because they believe that fewer deer will make it harder to hunt successfully and less likely that they'll find a trophy buck. Others, however, argue that a healthier herd and a more diverse ecosystem is better for all concerned.

In urban areas, increased sport hunting usually isn't acceptable. Wildlife biologists argue that the only practical way to reduce deer herds is culling by professional sharpshooters. Animal rights activists protest lethal control methods as cruel and inhumane. They call instead for fertility controls, the reintroduction of predators, such as wolves and mountain lions, or trap and transfer programs. Birth control works in captive populations but is expensive and impractical with wild animals. Trapping is expensive, and few places are willing to take surplus animals, which often die after relocation.

This case shows that carrying capacity can be more complex than simply the maximum number of organisms an ecosystem can support. While it may be possible for 200 deer to survive in a square mile, the ecological carrying capacity—the population that can be sustained without damage to the ecosystem and to other species—is usually considerably lower. There's also an ethical carrying capacity, if we don't want to see animals suffer from malnutrition, disease, or starvation. There may also be a cultural carrying capacity, if we consider the tolerable rate of depredation on crops and lawns or an acceptable number of motor vehicle collisions.

If you were a wildlife biologist charged with managing the deer herd in your state, how would you reconcile the different interests in this issue? What sources of information or ideas shape views for and against population control in deer? What methods would you suggest to reach the optimal population size? What social or ecological indicators would you look for to gauge whether deer populations are excessive or have reached an appropriate level?

The two terms that make population grow—births and immigration—should be relatively easy to imagine. Birth rates are different for different species (house flies versus elephants, for example), and a birth rate can decline if there are food shortages or if crowding leads to stress, as noted earlier. Of the two negative terms, deaths and emigration, the emigration idea simply means that sometimes individuals leave the population. Deaths, on the other hand, can have some interesting patterns.

Mortality, or death rate, is the portion of the population that dies in any given time period. Some of mortality is determined by environmental factors, and some of it is determined by an organism's physiology, or its natural life span. Life spans vary enormously. Some microorganisms live whole life cycles in a few hours or even minutes. Bristlecone pine trees in the mountains of California, on the other hand, have life spans up to 4,600 years.

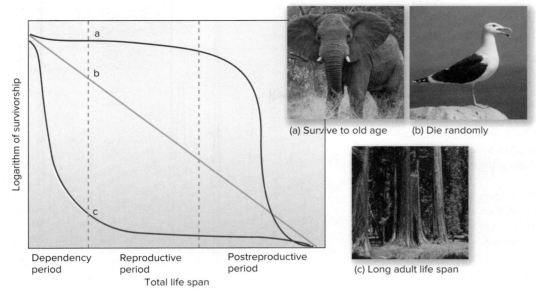

(a) Survive to old age (b) Die randomly

(c) Long adult life span

FIGURE 6.8 Three basic types of survivorship curves for organisms with different life histories. Curve (a) represents organisms such as humans or elephants, which tend to live out the full physiological life span if they survive early growth. Curve (b) represents organisms such as sea gulls, which have a fairly constant mortality at all age levels. Curve (c) represents such organisms as clams and redwood trees, which have a high mortality rate early in life but live a full life if they reach adulthood.

© Digital Vision/Getty Images RF © Stockbyte RF © Royalty-Free/Corbis

Survivorship curves show life histories

Different organisms have different characteristic rates of growth, maturity, and survival over time. Some organisms have just a few offspring, most of which survive to reproductive age (fig. 6.8, curve a). Survivorship declines sharply in the older, postreproductive phase. This pattern is followed by many larger mammals, such as whales, bears, and elephants (and many human populations). Juvenile survival tends to be fairly high, in part because parents invest considerable energy in tending to one or two young at a time. Adult survival is generally high because these organisms have few predators. Some very small organisms also have this type of survivorship curve. Among predatory protozoa, for example, a large proportion survives to what, for a microorganism, is a mature age.

Other organisms are about equally likely to die at any time of their lifespan (fig. 6.8, curve b). Here, the probability of death is unrelated to age, once infancy is past. Sea gulls, mice, rabbits, and other organisms face risks that affect all ages, such as predation, disease, or accidents. Mortality rates can be more or less constant with age, and the survivorship curve can be described as a straight line.

Quantitative Reasoning

In figure 6.8, curve (a) represents animals such as elephants. At what point in the curve does the number of elephants start to decline? What factors might explain the shape of this curve? (Hint: Refer to table 6.1.) For redwood trees, curve (c), at what point does the frequency of survivors change sharply? Explain this survivorship curve in words. Which of these survivorship patterns best describes humans?

Many organisms produce tremendous numbers of offspring, only a few of which survive to reproductive age (fig. 6.8, curve c). Many tree species, fish, clams, crabs, and other invertebrate species produce hundreds of thousands, or even millions, of seeds or eggs. The tremendous number ensures that predators miss at least a few individuals, which are then able to survive to maturity. Those that do survive to adulthood, however, have a very high chance of living nearly the maximum life span for the species.

Intrinsic and extrinsic factors are important

So far, we have seen that differing patterns of natality, mortality, life span, and longevity can produce quite different rates of population growth. The patterns of survivorship and age structure created by these interacting factors not only show us how a population is growing but also can indicate what general role that species plays in its ecosystem. They also reveal a good deal about how that species is likely to respond to disasters or resource bonanzas in its environment. But what factors *regulate* natality, mortality, and the other components of population growth? In this section, we will look at some of the mechanisms that determine how a population grows.

Factors that regulate population growth, primarily by affecting natality or mortality, can be classified in different ways. They can be *intrinsic* (operating within individual organisms or between organisms in the same species) or *extrinsic* (imposed from outside the population). Factors can also be either **biotic** (caused by living organisms) or **abiotic** (caused by nonliving components of the environment). Finally, the regulatory factors can act in a *density-dependent* manner (effects are stronger, or a higher proportion of the population is affected as population density increases) or *density-independent* manner (the effect is the same, or a constant proportion of the population is affected regardless of population density).

In general, biotic regulatory factors tend to be density-dependent, while abiotic factors tend to be density-independent. There has been much discussion about which of these factors is most important in regulating population dynamics. In fact, it probably depends on the particular species involved, its tolerance levels, the stage of growth and development of the organisms involved, the specific ecosystem in which they live, and the way combinations of factors interact. In most cases, density-dependent and density-independent factors probably exert simultaneous influences. Depending on whether regulatory factors are regular and predictable or irregular and unpredictable, species will develop different strategies for coping with them.

Some population factors are density-independent; others are density-dependent

In general, the factors that affect natality or mortality independently of population density tend to be abiotic components of the ecosystem. Often, weather (conditions at a particular time) or climate (average weather conditions over a longer period) are among the most important of these factors. Extreme cold or even moderate cold at the wrong time of year, high heat, drought, excess rain, severe storms, and geologic hazards—such as volcanic eruptions, landslides, and floods—can have devastating impacts on particular populations.

Abiotic factors can have beneficial effects as well, as anyone who has seen the desert bloom after a rainfall can attest. Fire is a powerful shaper of many biomes. Grasslands, savannas, and some montane and boreal forests often are dominated—even created—by periodic fires. Some species, such as jack pine and Kirtland's warblers, are so adapted to periodic disturbances in the environment that they cannot survive without them.

In a sense, these density-independent factors don't necessarily regulate population *per se*, because regulation implies a homeostatic feedback that increases or decreases as density fluctuates. By definition, these factors operate without regard to the number of organisms involved. They may have such a strong impact on a population, however, that they completely overwhelm the influence of any other factor and determine how many individuals make up a particular population at any given time.

Density-dependent mechanisms tend to reduce population size by decreasing natality or increasing mortality as the population size increases. Most of them are the results of interactions *between* populations of a community (especially predation), but some of them are based on interactions *within* a population.

Interspecific Interactions Occur Between Species

As we discussed in chapter 4, a predator feeds on—and usually kills—its prey species. While the relationship is one-sided with respect to a particular pair of organisms, the prey species as a whole may benefit from the predation. For instance, the moose that gets eaten by wolves doesn't benefit individually, but the moose *population* is strengthened because the wolves tend to kill old or sick members of the herd. Their predation helps prevent population overshoot, so the remaining moose are stronger and healthier.

Sometimes predator and prey populations oscillate in a sort of synchrony with each other, as illustrated in figure 6.9, which shows the number of furs brought into Hudson Bay Company trading posts in Canada between 1840 and 1930. As you can see, the numbers of Canada lynx fluctuate on about a ten-year cycle that is similar to, but slightly out of phase with, the population peaks of snowshoe hares. Although there are some doubts now about how and where these data were collected, this remains a classic example of population dynamics. When prey populations (hares) are abundant, predators (lynx) reproduce more successfully and their population grows. When hare populations crash, so do the lynx. This predator–prey oscillation is known as the Lotka–Volterra model, after the scientists who first described it mathematically.

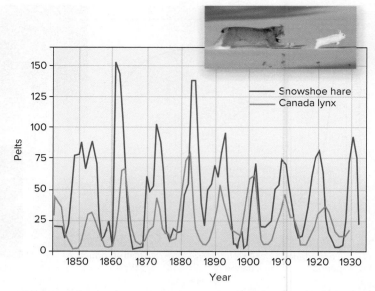

FIGURE 6.9 Ten-year oscillations in the populations of snowshoe hare and lynx in Canada suggest a close linkage of predator and prey, but that might not tell the whole story. These data are based on the number of pelts received by the Hudson Bay Company each year, making fur traders unwitting accomplices in later scientific research.
© Alan & Sandy Carey/Science Source

Not all interspecific interactions are harmful to one of the species involved. Mutualism and commensalism, for instance, are interspecific interactions that are beneficial or neutral in terms of population growth (see chapter 4).

Intraspecific Interactions Occur Within a Species

Intraspecific ("within a species") competition is also common. This competition is especially obvious when populations are dense, and individuals struggle for food, sunlight, or other resources. The stronger, quicker, more aggressive, or luckier members get a larger share, while others get less. If the population exceeds the limits for key resources—that is, if the it exceeds the carrying capacity of its environment—the population is likely to decline.

Establishing separate territories is one principal way many animal species control access to environmental resources (fig. 6.10). The individual, pair, or group that holds the territory will drive off rivals, if possible, by threats, displays of superior features (colors, size, dancing ability), or fighting equipment (teeth, claws, horns, antlers). Members of the opposite sex are attracted to individuals that are able to seize and defend the largest share of the resources. From a selective point of view, these successful individuals presumably are superior members of the population and the ones best able to produce offspring that will survive.

Stress and Crowding Can Affect Reproduction

Stress and crowding also are density-dependent population control factors. When population densities get very high, organisms often exhibit symptoms of what is called stress shock or

stress-related diseases—physical, psychological, and/or behavioral changes that are thought to result from the stress of crowding and competition. Crowding-related changes in reproduction are not well understood. Lower reproduction could result from stress. Alternatively, they may simply result from malnutrition, infectious disease, or other more obvious mechanisms.

Some of the best evidence for the existence of stress-related disease comes from experiments in which laboratory animals, usually rats or mice, are grown in very high densities with plenty of food and water but very little living space. A variety of symptoms are reported, including reduced fertility, low resistance to infectious diseases, and pathological behavior. Dominant animals seem to be affected least by crowding, while subordinate animals—the ones presumably subjected to the most stress in intraspecific interactions—seem to be the most severely affected.

Density-dependent effects can be dramatic

The desert locust, *Schistocerca gregarius*, has been called the world's most destructive insect. Throughout recorded human history, locust plagues have periodically swarmed out of deserts and into settled areas. Their impact on human lives has often been so disruptive that records of plagues have taken on religious significance and made their way into sacred and historical texts.

Locusts usually are solitary creatures resembling ordinary grasshoppers. Every few decades, however, when rain comes to the desert and vegetation flourishes, locusts reproduce rapidly until the ground seems to be crawling with bugs. High population densities and stress bring ominous changes in these normally innocuous insects. They stop reproducing, grow longer wings, group together in enormous swarms, and begin to move across the desert. Dense clouds of insects darken the sky, moving as much as 100 km per day. Locusts may be small, but they can eat their own body weight in vegetation every day. A single swarm can cover 1,200 km² and contain 50 to 100 billion individuals. The swarm can strip pastures, denude trees, and destroy crops in a matter of hours, consuming as much food in a day as 500,000 people would need for a year. Eventually, having exhausted their food supply and migrated far from the desert where conditions favor reproduction, the locusts die and aren't seen again for decades.

Huge areas of crops and rangeland in northern Africa, the Middle East, and Asia are within the reach of the desert locust. This small insect, with its voracious appetite, can affect the livelihood of at least one-tenth of the world's population. During quiet periods, called recessions, African locusts are confined to the Sahara Desert, but when conditions are right, swarms invade countries as far away as Spain, Russia, and India. Swarms are even reported to have crossed the Atlantic Ocean from Africa to the Caribbean.

Unusually heavy rains in the Sahara in 2004 created the conditions for a locust explosion. Four generations bred in rapid succession, and swarms of insects moved out of the desert. Twenty-eight countries in Africa and the Mediterranean area were afflicted. Crop losses reached 100 percent in some places, and food supplies for millions of people were threatened. Officials at the United Nations warned that we could be headed toward another

FIGURE 6.10 Individuals within a species often compete for territory and resources. This is called intraspecific competition. Resource scarcity can induce stress and affect reproductive success.
© Art Wolfe/Iconica/Getty Images

great plague. Hundreds of thousands of hectares of land were treated with pesticides, but millions of dollars of crop damage were reported anyway.

This case study illustrates the power of exponential growth and the disruptive potential of a boom-and-bust life cycle. Stress, population density, migration, and intraspecific interactions all play a role in this story. Although desert conditions usually keep locust numbers under control, the locust's biotic potential for reproduction is a serious worry for residents of many countries.

Section Review

1. Describe three major types of survivorship patterns, and explain what they show about a species' role in its ecosystem.
2. Explain how biotic and abiotic factors affect population growth.
3. What are the main interspecific population regulatory interactions? How do they work?

6.3 POPULATION SIZE AND CONSERVATION

- *Small and isolated populations are prone to extinction, as predicted by the theory of island biogeography.*
- *Loss of genetic diversity can reduce a population's chances of survival.*
- *Gene flow within a metapopulation can maintain genetic diversity.*

Survival of the bluefin tuna (opening case study) depends on its rate of reproduction, growth, and mortality (that is, fishing catch). These are all questions of population dynamics and growth discussed in sections 6.1 and 6.2. Fishery biologists are also concerned about tuna numbers because population size and density can also influence survival. Small or isolated populations of any species can undergo catastrophic declines due to environmental change, loss of genetic diversity, or random accidents. Below a critical size, a small population may not be viable in the long term. This is why understanding population numbers has been an important goal for

How Do You Count Tuna?

Population data are necessary for understanding population stability. Collecting these data is hard in any population, but it's especially difficult when the species live far out in the open ocean, migrate widely, and are rarely seen except by fishing boats that catch them.

One technique often used for population studies is catch, tag, release, and recapture. The proportion of tagged individuals you recapture compared to the total number caught can give you an estimate of the entire population. But if you recapture only a tiny number of tagged specimens out of a catch of millions or billions, the accuracy of your estimate may be very poor.

Because of these difficulties, most data regarding ocean fish populations come from commercial fishing records. A decline in average size, in number of adults, or in size at spawning age indicates that a population is being overfished. The bluefin tuna has shown evidence of all these effects. At Tokyo's Tsukiji fish market, the average weight of a bluefin has fallen since the 1980s from 100–160 kg

An observer measures a big-eye tuna.
Source: NOAA

to just 50 kg today. The proportion of fish younger than 1 year has increased, and the proportion of larger, older fish has fallen sharply.

Another, newer approach is to attach satellite tracking tags, small, plastic-coated rods inserted into the fish's side just below a fin. Electronic tags can record factors such as the location from satellite readings or water

pressure (a measure of depth in the water). Smart tags can be designed to float to the surface if they are released from the fish, and to transmit recorded data to a satellite and then to the researcher's computer, or tags can be returned if fish are caught.

Electronic tagging studies have revealed the astonishing distances tuna travel, where they go, their preferred feeding grounds, and their fidelity to their home spawning grounds. Ideally, this information can be used to designate no-fishing sanctuaries in spawning grounds, as well as provide information on basic population biology. Mapping tag locations also has shown that the western Atlantic and Mediterranean bluefin populations are distinct and that they prefer different types of spawning conditions.

Information such as this gives conservation science a new tool for trying to save a species and the ecosystem that depends on it.

For more information, see Barbara A. Block, *et al.*, 2005. Electronic tagging and population structure of Atlantic bluefin tuna. *Nature* 434: 1121–1127.

---

biologists studying tuna (see the Exploring Sciencebox below) or any other species with declining numbers. In this section, we examine how population size can influence survival.

Small island populations are vulnerable

It has often been observed that small, remote islands have lower biodiversity than larger islands. Why is this? In a classic 1967 study, R. H. MacArthur and E. O. Wilson proposed an explanation they described as the equilibrium theory of **island biogeography.** Their explanation was that diversity (the number of species) in isolated islands depends on rates of colonization and extinction. The rate of colonization depends on proximity of the island to another source of species, such as another island or a continent. If new species can readily reach the island, they may establish new populations there (fig. 6.11). Extinction rates depend on the size of the island and of the population that it can support. Small islands can support only small populations, which are more likely to go extinct due to

natural disasters, diseases, or demographic factors such as imbalance between sexes. Larger islands are more likely to sustain larger populations, and thus extinction rates tend to be lower on large islands. The loss of island species, such as birds in Hawai'i, is one of the greatest concerns for population biologists today.

Island biogeographical patterns have been observed in many places. In the Caribbean, for instance, Cuba is 100 times as large as Monserrat and has about 10 times as many amphibian species. Similarly, in a study on the California Channel Islands, Jared Diamond observed that for island bird species with fewer than 10 breeding pairs, 39 percent of the populations went extinct over an 80-year period. In contrast, for populations with 10–100 breeding pairs, only 10 percent went extinct. No species with more than 1,000 pairs disappeared over this time (fig. 6.12). Island biogeographical patterns, with more species in larger and less remote islands, have since been observed in many different types of "islands," such as isolated patches of forest, or isolated wetlands. Often, greater habitat diversity on larger "islands" helps explain this pattern.

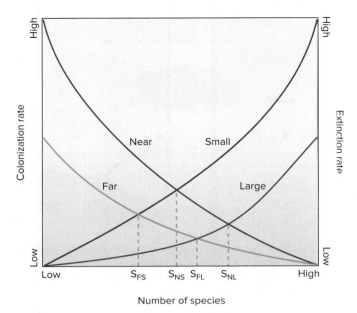

FIGURE 6.11 The theory of island biogeography predicts that the number of species in an island depends on colonization and extinction rates. An island that is small and far from other land masses, for example, should have fewer species (S_{FS}) than a large and near island (S_{NL}).
Source: Modified from The Theory of Island Biogeography, 1967, Princeton University Press.

Genetic diversity may help a population survive

Genetic diversity plays an important role in the survival or extinction of small, isolated populations. In large populations, genetic variation tends to remain stable, provided environmental conditions remain stable. The ratio of genetic traits tends to remain the same from one generation to the next. For example, a large population of Darwin's finches (chapter 3) might maintain a steady proportion of large, medium, and short beaks. This stability in genetic diversity is known as the Hardy–Weinberg principle, named after the two scientists who first described it. A shift in traits could occur, though, if there is environmental change. For example, a shift to a drier climate might make larger beaks advantageous, because they can crack heavier-shelled seeds. In that case, the dominant traits in a population could shift. We call the gradual changes in gene frequencies due to random events **genetic drift.** The presence of genetic variety in a population can be important: If environmental conditions shift, suitable traits are likely to be present that help the species survive.

In small or isolated populations, the pool of traits is smaller. Genetic drift can occur because unusual traits in a few individuals can become dominant in a small population. Suppose a large finch population has an even mixture of large, medium, and small beaks, but a storm blows a small group of four finches to a remote island. If three of those four finches happen to have small beaks, then small beaks are likely to dominate as the birds reproduce and populate the new island. Traits of the resulting population would differ from those of the original population, because of random chance in a small group.

The reduction in genetic diversity in a small population is sometimes called a **founder effect,** when it results from a small colonizing population. It may also be referred to as a **demographic**

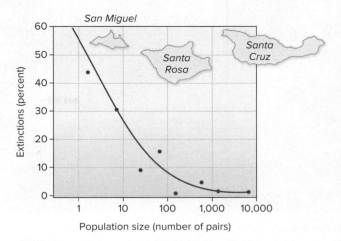

FIGURE 6.12 Extinction rates of bird species on the California Channel Islands as a function of population size over 80 years.
Source: H. L. Jones and J. Diamond, "Short-term-base studies of turnover in breeding bird populations on the California coast island," in Condor, vol. 78:526–549, 1976.

bottleneck, when just a few members of a species survive a catastrophic event (fig. 6.13).

Sometimes the loss of genetic diversity can limit adaptability, reproduction, and species survival. Any deleterious genes present in the founders will be overrepresented in subsequent generations. Inbreeding, the mating of closely related individuals, also makes expression of rare or recessive genes more likely. Cheetahs appear to have undergone a demographic bottleneck sometime in the not-too-distant past. All the male cheetahs alive today appear to be nearly genetically identical, suggesting that they all share a single male ancestor (fig. 6.14). This lack of diversity is thought to be responsible for an extremely low fertility rate, a high abundance of abnormal sperm, and a low survival rate for offspring. All these factors threaten the survival of the species.

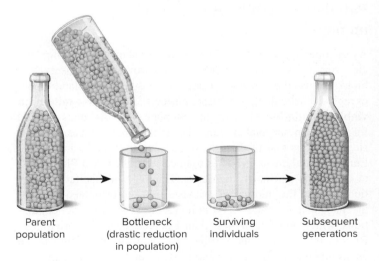

FIGURE 6.13 Genetic drift resulting from a bottleneck effect. The parent population contains roughly equal numbers of blue and yellow individuals. Only a few individuals make it through a "bottleneck" that reduces the population, such as an environmental disturbance. If those few individuals are blue, then subsequent generations will be mostly blue.

FIGURE 6.14 Sometime in the past, cheetahs underwent a severe population crash. Now, all male cheetahs alive today are nearly genetically identical, and deformed sperm, low fertility levels, and low infant survival are common in the species.
© McGraw-Hill Education/Barry Barker, photographer

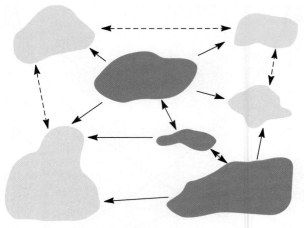

FIGURE 6.15 A metapopulation is composed of several local populations linked by regular (*solid arrows*) or occasional (*dashed lines*) gene flows. Source populations (*dark*) provide excess individuals, which emigrate to and colonize sink habitats (*light*).

In other cases, species seem not to be harmed by inbreeding or lack of genetic diversity. The northern elephant seal, for example, was reduced by overharvesting a century ago, to fewer than 100 individuals. Today, there are more than 150,000 of these enormous animals along the Pacific coast of Mexico and California. No marine mammal is known to have come closer to extinction and then made such a remarkable recovery. All northern elephant seals today appear to be essentially genetically identical and yet they seem to have no apparent problems. Although interpretations of their situation are controversial, in highly selected populations, where only the most fit individuals reproduce, or in which there are few deleterious genes, inbreeding and a high degree of genetic identity may not be such a negative factor.

Population viability can depend on population size

As the opening case study for this chapter shows, predicting species survival can be difficult and controversial. Conservation biologists use the concepts of population dynamics, island biogeography, genetic drift, and founder effects to determine **minimum viable population size,** or the number of individuals needed for long-term survival of rare and endangered species. A classic example is that of the grizzly bear in North America. Before European settlement, grizzlies roamed from the Great Plains west to California and north to Alaska. Hunting and habitat destruction reduced the number of grizzlies from an estimated 100,000 in 1800 to fewer than 1,200 animals in six separate subpopulations, which now occupy less than 1 percent of the historic range. Biologists have calculated that for some sub-populations, the environmental carrying capacities is fewer than 100 bears. But genetic models predicts that an isolated population of 100 bears is unlikely to persist for more than a few generations. Even the 600 bears now in Yellowstone National Park will be susceptible to genetic problems if completely isolated. Interestingly, computer models suggest that translocating only two unrelated bears into

small populations every generation (about ten years) could greatly increase population viability.

For mobile organisms, separated populations can have gene exchange if suitable corridors or migration routes exist. A **metapopulation** is a collection of populations that have regular or occasional gene flow between geographically separate units (fig. 6.15). For example, the northern spotted owl is an endangered species occupying remnant patches of the once-continuous forest of the Pacific Northwest. Historically, there has been interaction among forest areas, so that the many sub-populations are considered a metapopulation.

In a metapopulation, some habitat areas may be better than others at supporting reproductive sub-populations. A "source" habitat, where birth rates are higher than death rates, produces surplus individuals that can migrate to new locations within a metapopulation. In "sink" habitats, on the other hand, mortality exceeds birth rates. For spotted owls, sink habitats might be those too small to support sufficient prey (various small rodents) to raise young successfully, or it might be fragmented habitats where spotted owls must compete with barred owls, which thrive in mixed environments. The species would disappear from sink habitats if it were not periodically replenished from a source population. Recent studies of a metapopulation model for spotted owls predict just such a problem for this species in the Pacific Northwest. If many of these endangered owls end up in habitat patches too small to support reproduction, the entire population may decline. In this case and many others, understanding population biology and population dynamics is critical to conserving the species.

Section Review

1. What is island biogeography and how does it explain population survival?

2. Why does genetic diversity tend to persist in large populations, but gradually drift or shift in small populations?

3. Define the following: *metapopulation, genetic drift, demographic bottleneck.*

Conclusion

Given optimum conditions, populations of many organisms can grow exponentially; that is, they can expand at a constant rate per unit of time. This biotic potential can produce enormous populations that far surpass the carrying capacity of the environment if left unchecked. Obviously, no population grows at this rate forever. Sooner or later, predation, disease, starvation, or some other factor will cause the population to crash. Not all species follow this boom-and-bust pattern, however. Most top predators have intrinsic factors that limit their reproduction and prevent overpopulation. Logistic growth is a general term for growth that slows as the carrying capacity is approached.

Generally, population numbers are determined by rates of births and immigration minus deaths and emigration (BIDE). Many factors influence these rates, including overharvesting, habitat destruction, predator elimination, and the introduction of exotic species. Population dynamics are an important part of conservation biology.

The size and genetic diversity of a population can influence its long-term likelihood of survival. Principles such as island biogeography, genetic drift, demographic bottlenecks, and metapopulation interactions are critical in understanding species viability and in planning for endangered species protection.

Reviewing Key Terms

Can you define the following terms in environmental science?

abiotic 6.2

biotic 6.2

carrying capacity 6.1

demographic bottleneck 6.3

density-dependent 6.1

density-independent 6.1

exponential growth 6.1

founder effect 6.3

genetic drift 6.3

island biogeography 6.3

K-selected species 6.1

logistic growth 6.1

metapopulation 6.3

minimum viable population
   size 6.3

population crash 6.1

r-selected species 6.1

stress-related diseases 6.2

Critical Thinking and Discussion Questions

1. Compare the advantages and disadvantages to a species that result from exponential or logistic growth. Why do you think hares have evolved to reproduce as rapidly as possible, while lynx appear to have intrinsic or social growth limits?

2. Are humans subject to environmental resistance in the same sense that other organisms are? How would you decide whether a particular factor that limits human population growth is ecological or social?

3. What are advantages and disadvantages in living longer or reproducing more quickly? Why hasn't evolution selected for the most advantageous combination of characteristics so that all organisms will be more or less alike?

4. Why do abiotic factors that influence population growth tend to be density-independent, while biotic factors that regulate population growth tend to be density-dependent?

5. Some people consider stress and crowding studies of laboratory animals highly applicable in understanding human behavior. Other people question the cross-species transferability of these results. What considerations would be important in interpreting these experiments?

6. What implications (if any) for human population control might we draw from our knowledge of basic biological population dynamics?

Data Analysis

Experimenting with Population Growth

The previous data analysis lets you work through an example of population growth by hand, which is an important strategy for understanding the equations you've seen in this chapter.

Now try experimenting with more growth rates in an Excel "model." What value of *r* makes the graph extremely steep? What value makes it flat? Can you model a declining population?

Go to Connect to find the Excel spreadsheet that lets you experiment with different growth rates, as well as questions to demonstrate your understanding.

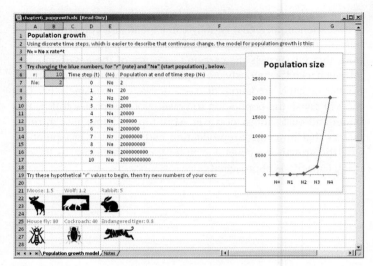

Source: Microsoft

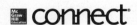

7

Human Populations

▲ Rio de Janeiro, Brazil is one of the world's fast growing cities, where improved standards of living, women's rights, education for girls and new visions of what life could hold for a modern family have slowed population growth without specific government family planning policies.
© sfmthd/iStock RF

Learning Outcomes

After studying this chapter, you should be able to:

7.1 Trace the history of human population growth.

7.2 Summarize different perspectives on population growth.

7.3 Analyze some of the factors that determine population growth.

7.4 Describe how a demographic transition can lead to stable population size.

> *"For every complex problem there is an answer that is clear, simple, and wrong."*
>
> – H. L. Mencken

Population Stabilization in Brazil

Can TV soap operas help control population growth? They are not a usual strategy, but sometimes the answer is yes. In Brazil, a mix of economic growth, female empowerment, urbanization, and the widespread popularity of television and the images it portrays of modern life have all contributed to one of the most abrupt birth rate declines over the past few decades of any country in the world.

The largest country in South America, Brazil has enormous regional differences in geography, race, culture, and environmental conditions, ranging from the lush tropical forests of the Amazon basin to the immense urban agglomerations of Rio de Janeiro and Sao Paulo. Brazil has also seen some of the most dramatic economic growth and urbanization rates in the world. It has the sixth largest population in the world (209 million in 2016), and the sixth largest economy.

This swift economic progress has brought a dramatically improved standard of living for most Brazilians. In 1960, the average wage in Brazil was less than $2,000 (U.S.) per year, only 19 percent of Brazilian households had electricity, TV was rare, the average education for females was two years, the typical woman had 6.2 children, and the annual population growth rate was 3.0 percent. Fifty years later, the average wage had risen fivefold, 95 percent of households had both electricity and TV, the average education for females was 8.6 years (a

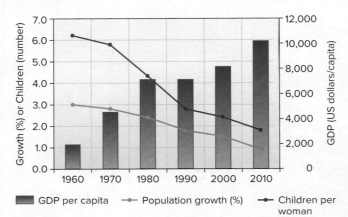

FIGURE 7.1 The number of children per woman and the population growth have dropped sharply in Brazil over the past 50 years as per capita income and standard of living have risen.
Source: UNDP

year more than for males), the fertility rate (number of children per woman in a lifetime) was 1.8, and the growth rate was 0.9 percent (fig. 7.1). Now, 99 percent of urban residents have improved drinking water and 87 percent have improved sanitation. Infant mortality has fallen from 204 per 1,000 children in 1960 to 19 per 1,000 today. Knowing that your children are likely to survive to adulthood makes a big difference in how many you choose to have.

The industrialization that fueled this rapid economic growth also brought urbanization as people moved from rural areas to the cities in search of jobs. Currently, 87 percent of Brazilians live in urban areas, where smaller families are an economic advantage. Apartments are small. With both parents working, having a big family with lots of children is a burden. And with better educational opportunities for children, it makes sense to focus your resources on one or two children who can find better jobs and advance economically.

Although Brazilian culture still emphasizes machismo, as in most of Latin America, women are gaining respect and freedom. Women are now more likely than ever before to have paying jobs outside the home and have more voice in how their wages are used. In 2010, Dilma Rousseff was elected the first woman president in Brazil's history. Before her election, Rousseff was minister of energy and chief of staff for President Luiz Inacio Lula da Silva. In those positions, Rousseff gained public popularity when she pushed a program to extend electricity to remote rural areas and to the favelas (slums) that crowd the hills around major cities.

What does this have to do with soap operas? As people move to the city, they have both more access to TV and more leisure time. Daytime soap operas offer a view of what modern life might be like. Followed avidly by a majority of the population, the actors, plots, and situations in the telenovelas are often widely discussed and admired by the public. In general, the programs show affluent, modern, and small families with lots of material possessions. Women appear as powerful executives and business owners, with successful careers and considerable personal freedom. This image has changed aspirations for many women. The desire to have large families, as their mothers or grandmothers did, is no longer popular among young women.

No official government policy in Brazil has ever promoted family planning. The country is overwhelmingly Roman Catholic (with the largest Catholic population in the world), but most women quietly chose to ignore church teaching about birth control. Abortion is illegal (except in rare cases), but both birth control and morning-after pills are widely available over the counter. Female sterilization is one of the most widely used contraceptive techniques. A high proportion of births are Cesarean, and the doctor can discreetly and cheaply perform a tubal ligation at the same time.

It appears that Brazil has stabilized its population without direct intervention in family planning. Economic growth, better educational opportunities for girls, women's rights, and a vision of what life could hold for a modern family have brought about this change spontaneously.

This case study introduces several important themes of this chapter. What's the best way to achieve sustainable population sizes? How do birth rates affect resource consumption and our impacts on our environment? What is the relative importance of population growth and affluence, which allows us to consume more resources per person? These and other questions about what drives human population dynamics are central themes in environmental science.

7.1 POPULATION GROWTH

- *Human populations have grown rapidly in the past two centuries.*
- *Some people worry that overpopulation will exhaust our natural resources and result in starvation, disease, crime, and misery.*
- *Others argue that our environmental impacts depend on how much we consume and the technology we use to produce goods and services, as well as total population size.*

Every second, on average, four children are born somewhere on the earth. In that same second, one person dies. This difference between births and deaths means a net gain of roughly three more humans per second in the world's population. In 2016, the total world population was at least 7.4 billion people and was growing at 1.08 percent per year. This means we are now adding nearly 80 million more people per year, and if this rate persists, our global population will double in about 66 years. Already humans are among the most numerous and widely distributed of vertebrate species on the earth. We also have a greater global environmental impact than any other species. For families, the birth of a child is a joyous event (fig. 7.2). But collectively, what is the effect of our growing population? The answers to this question depend on our estimates of the causes of change, rates of change, and ideas about resource consumption per person.

Many people worry that overpopulation is causing resource depletion and environmental degradation that threaten the ecological life-support systems on which we all depend. Recall from chapter 6 that carrying capacity is the maximum population size that can be supported for a particular species in a specific environment without depleting resources. Are there enough resources for 8 or 10 billion people to live a good and healthy life? Are we on track to overshoot the Earth's carrying capacity?

Alternatively, can human ingenuity, technology, and enterprise can extend the world carrying capacity and allow us to overcome environmental limits? A larger population means a larger workforce, more inventions, more ideas about better solutions. Along with every new mouth comes a pair of hands. Can continued economic and technological growth provide the stability to reduce population growth voluntarily?

Another question is whether we can address social justice concerns. Countries with the highest current growth rates are also some of the poorest in the world. But the richest countries use the most resources. Many argue that there are sufficient resources for everyone, and the root cause of environmental degradation is inequitable distribution of wealth and power, rather than population size. Fostering democracy, empowering women and minorities, and improving the standard of living of the world's poorest people are what are really needed, from this point of view. A narrow focus on population growth can foster racism and blame the poor for their own poverty, while ignoring the deeper social and economic forces at work.

Will human populations continue to grow at present rates? What would that growth imply for environmental quality and for human life? These are among the most central and pressing questions in environmental science. In this chapter, we examine causes of population growth, as well as the ways populations are measured and described. We also examine factors that slow growth rates and stabilize populations, such as family planning, and the many considerations that influence decisions about family size.

Human populations grew slowly until relatively recently

For most of our history, humans lived by hunting and gathering, and our population was probably not more than a few million. The invention of agriculture and the domestication of animals around 10,000 years ago allowed our numbers to grow. With a larger and more secure food supply, the population reached perhaps 50 million people by 5000 B.C.E. Still, for thousands of years, the number of humans increased slowly. Archaeological evidence and historical descriptions suggest that only about 300 million people were living at the time of Christ (table 7.1). It took all of human history to reach 1 billion people in 1804, but little more than 150 years to reach 3 billion in 1960. To go from 5 to 6 billion took only 12 years.

Demographers often describe this kind of growth in terms of doubling times (see table 7.1). Over time, the growth rate has accelerated, and the time needed to double the population size has decreased from several hundred years to just a few decades. Population growth was most rapid in the early 1960s, when the global population doubling time fell to 33 years or less. The doubling time is now lengthening again, as birth rates slow around the world. Demographers expect that by 2050 the global population will reach about 9.7 billion, with a doubling time of 155 years or more. However, most analysts hope that the population will not actually double again. While we use doubling time as a relative measure of growth rates, the population is expected to stabilize by about 2100 at around 11 billion.

Until the Middle Ages, human populations were held in check by diseases and poor nutrition, which caused high child mortality and short life expectancy for those who reached adulthood.

FIGURE 7.2 The number of children a family chooses to have is determined by many factors. In urban industrialized countries like Brazil, most families now want only one or two children.
© PhotoDisc/Getty Images RF

Table 7.1	World Population Growth and Doubling Times	
Date	Population	Doubling Time
5000 B.C.E.	50 million	?
800 B.C.E.	100 million	4,200 years
200 B.C.E.	200 million	600 years
1200 C.E.	400 million	1,400 years
1700 C.E.	800 million	500 years
1900 C.E.	1,600 million	200 years
1965 C.E.	3,200 million	33 years
2004 C.E.	6,400 million	58 years
2050 C.E. (estimate)	9,710 million	155 years

Source: United Nations Population Division, U.S. Census Bureau

There is evidence that many early societies regulated their population size through cultural taboos and practices such as abstinence and infanticide. War and famine also caused episodes of population decline. Among the most destructive of natural population controls were bubonic plagues (or Black Death) that periodically swept across Europe between 1348 and 1650 (fig. 7.3). It has been estimated that during the worst plague years (1348–1350) at least one-third of the European population perished. Notice, however, that this didn't retard population growth for very long. In 1650, at the end of the last great plague, there were about 600 million people in the world.

As you can see in figure 7.3, human populations began to increase rapidly after 1600 C.E. Many factors contributed to this rapid growth. Increased sailing and navigating skills stimulated commerce and communication between nations. Trade increased access to resources, including more food. Agricultural developments, better sources of power, and better health care and hygiene also played a role.

We have had an exponential or J curve pattern of growth for centuries. Recently, however, the growth rate has begun flattening, and turning into an S-shaped curve (see section 7.3). If we do not shift to an S-shaped curve, will we overshoot the carrying capacity of our environment and experience a catastrophic dieback similar to those described in chapter 6? Or we will reach equilibrium soon enough and at a size that can be sustained over the long term? No one knows, of course, but in the sections that follow we analyze factors that influence growth rates.

Section Review

1. At what point in history did the world population pass its first billion? What is the total population now?
2. What factors contribute to rapid population growth?
3. What factors help control population growth?

7.2 PERSPECTIVES ON POPULATION GROWTH

- *Technology can increase the carrying capacity for humans.*
- *Technology can also cause environmental harm.*
- *Larger populations could bring benefits if we can find ways to live sustainably.*

As with many topics in environmental science, people have widely differing opinions about population and resources. Some believe that population growth is the ultimate cause of poverty and

FIGURE 7.3 Human population levels through history. Since about 1000 C.E., our population curve has assumed a J shape. Are we on the upward slope of a population overshoot? Will we be able to adjust our population growth to an S curve? Or can we just continue the present trend indefinitely?

environmental degradation. Others argue that poverty, environmental degradation, and overpopulation are all merely symptoms of deeper social and political factors. The worldview we choose to believe will profoundly affect our approach to population issues. In this section, we will examine some of the major figures and their arguments in this debate.

Does population growth cause poverty, or does poverty cause growth?

Since the time of the Industrial Revolution, when the population growth began accelerating rapidly, scholars have argued about the causes and consequences of population growth. One of the most influential explanations today was proposed in 1798 by Thomas Malthus (1766–1834), in *An Essay on the Principle of Population*. Malthus argued that populations tend to increase at exponential rates, with an accelerating increase over time. Food production, he proposed, either remains stable or increases only slowly. Eventually, human populations should therefore outstrip their food supply and collapse into starvation, crime, and misery. Population growth, Malthus said, slows only when disease or famine reduces population size, or when constraining social conditions—late marriage, insufficient resources, celibacy, and "moral restraint"—compel a population to reduce birth rates.

In Malthusian terms, then, population growth causes resource depletion and poverty (fig. 7.4*a*). At that time, most economists believed that high fertility was good for the economy, increasing gross domestic output. Malthus converted them to believing that population growth was an economic danger, because per capita output actually fell with rapidly increasing population, and because large populations deplete resources rapidly.

Several decades later, the economist Karl Marx (1818–1883) presented an opposing view. Marx argued that exploitation of workers—such as low wages and job insecurity—causes poverty, which in turn contributes to population growth and other social ills. Poverty can lead to high birth rates, for example, when people in insecure or desperate conditions lack either opportunities (employment or education) or social restraints that discourage early and frequent childbearing. In these conditions, large families can exacerbate insecurity. Others since Marx have argued that the wealthy consume more resources, and cause more resource depletion, than the poor do. Both population growth and exploitation, then, cause resource depletion (fig. 7.4*b*). Reversing poverty, population growth, and environmental degradation, according to this perspective, requires the elimination of exploitation and oppression of the poor. As Mohandas Gandhi stated, "There is enough for everyone's need, but not enough for anyone's greed."

Marx and Malthus developed their theories about human population growth when views about the world, technology, and society were much different than they are today. But these different views continue inform competing approaches to family planning and population growth. If high birth rates are the driving force, causing poverty and extreme resource consumption, then increasing access to birth control, or even coercive reduction in family size, is the most important strategy to reduce both growth and poverty.

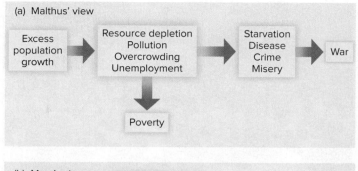

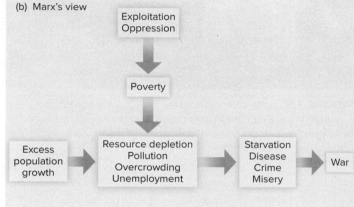

FIGURE 7.4 (a) Thomas Malthus argued that excess population growth is the ultimate cause of many other social and environmental problems. (b) Karl Marx argued that oppression and exploitation are the real causes of poverty and environmental degradation. Population growth in this view is a symptom or result of other problems, not the source.

If poverty contributes to population growth, on the other hand, then the most important task is to reduce poverty and improve opportunity. This is the path that has been followed in Brazil and many other countries. In most European countries, population growth has fallen to below replacement levels, as education and employment opportunities have improved and people have chosen to have smaller families (see section 7.4).

Both perspectives may be partly right. And while there are multiple interpretations of the best strategy for reducing populations, the question still remains whether, or how soon, we will exceed the earth's carrying capacity. The mathematical biologist Joel Cohen, of Rockefeller University, reviewed published estimates of the maximum human population size the planet can sustain. The estimates he found, spanning 300 years of thinking, converged on a median value of 10–12 billion. We are more than 7 billion strong today, and growing, an alarming prospect for some (fig. 7.5). Cornell University entomologist David Pimentel, for example, has said: "By 2100, if current trends continue, twelve billion miserable humans will suffer a difficult life on Earth."

Would you agree with Pimentel or with Ghandi (above)? If environmental quality and resource consumption are the ultimate concern, then is population growth the most important problem to address, or are there other drivers of resource consumption that deserve attention?

FIGURE 7.5 Is the world overcrowded already, or are people a resource? In large part, the answer depends on the kinds of resources we use and how we use them. It also depends on democracy, equity, and justice in our social systems.
© William P. Cunningham

Technology can increase carrying capacity for humans

Technological optimists, those who argue that innovation will allow us to overcome limits, argue that resources do not necessarily constrain human populations as tightly as other species. As resources become scarce, we tend to invent alternative resources or strategies. They note that Malthus was wrong 200 years ago in his predictions of famine and disaster because he failed to account for scientific and technical progress.

For example, even though urban populations have grown dramatically since Malthus' time, we have invented systems of trade, transportation, sanitation, and building that make cities healthier and safer than in the 1790s. And despite having nearly ten times the global population of Malthus' time, we now have have more and better food resources per capita than ever before. According to the UN Food and Agriculture Organization (FAO), the average per capita daily food consumption across the whole world has grown from 2,400 calories per day in 1970, to 2,900 calories in 2015. This is well above the 2,200 calories considered necessary for an average adult. Even poorer developing countries saw a rise, from an average of 2,100 calories per day in 1970 to 2,700. In 2015. In that same period, the world population went from 3.7 to more than 7 billion people. It's important to recognize that food security is not evenly distributed. In 2015, the UN FAO estimated that there were at least 795 million undernourished people despite the abundance of food produced. However, hunger and famines are generally attributable to politics and economics, rather than underproduction as described by Malthus (see section 9.1).

The burst of world population growth that began 200 years ago was stimulated by scientific and industrial revolutions. Progress in agricultural productivity, engineering, information technology, commerce, medicine, sanitation, and other achievements of modern life have made it possible to support thousands of times as many people per unit area as was possible 10,000 years ago. Economist Stephen Moore of the Cato Institute in Washington, D.C., regards this achievement as "a real tribute to human ingenuity and our ability to innovate." There is no reason, he argues, to think that our ability to find technological solutions to our problems will diminish in the future.

Technology can increase our environmental impacts as well as improving our survival rate. Our environmental impacts result from a combination of population size and how much we consume per person. A summary for this explanation of impacts is **I = PAT:** Environmental impacts (I) equal the product of our population size (P) times affluence (A) and the technology (T) used to produce the goods and services we consume (fig. 7.6).

While increased standards of living in Brazil, for example, have helped stabilize population, they also bring about higher technological impacts. A family living an affluent lifestyle that depends on high levels of energy and material consumption, and that produces excessive amounts of pollution, could cause greater environmental damage than a whole village of hunters and gatherers or subsistence farmers.

If the billions of people in Asia, Africa, and Latin America use conventional, resource-intensive technology to reach the standard of living now enjoyed by rich people in North America or Europe, the environmental effects will be disastrous. Already, growing wealth in China, whose middle class is now estimated at about 300 million, or nearly the entire population of the United States, is straining global resources and has made China the world's largest emitter of CO_2. There are now more millionaires in China than in all of Europe, and China has passed the United States in annual automobile production. As you'll recall from the discussion of

Impact = Population × Affluence × Technology

FIGURE 7.6 Environmental impacts of population growth (I) are the product of population size (P), times affluence (A), times the technology (T) used to create wealth.

ecological footprints in chapter 1, it would take about four additional Earths to support everyone if the entire world were living an American lifestyle—using current technology.

But China has also become a global leader in renewable energy, public transportation, and electric vehicles. Ideally, all of us will soon be using nonpolluting, renewable energy and material sources. If we can extend the benefits of environmentally friendly technology to the poorer people of the world, then everyone can enjoy the benefits of a better standard of living without the environmental costs historically associated with development and growth.

Population growth could bring benefits

Large populations of United States, China, and other regions consume tremendous resources, but they also represent gigantic economic engines and sources of innovation. More people mean larger markets, more workers, and efficiencies of scale in mass production of goods. If future economies are based largely on services, such as education, information, or data management, then larger populations mean larger markets for those services. Moreover, adding people boosts human ingenuity and intelligence that will create new resources by finding new materials and discovering new ways of doing things.

Economist Julian Simon (1932–1998), a champion of this rosy view of human history, believed that people are the "ultimate resource" and that no evidence suggests that pollution, crime, unemployment, crowding, the loss of species, or any other resource limitations will worsen with population growth. In a famous bet in 1980, Simon challenged Paul Ehrlich, author of *The Population Bomb,* to pick five commodities that would become more expensive by the end of the decade. The resources Ehrlich chose, copper, chromium, nickel, tin and tungsten, actually declined in price, by an average of 50 percent. Simon, the technological optimist, won the bet. It has been argued that the 10 year span was too short, or that different resources might have performed differently, but it remains true that resource substitution has made many formerly critical resources cheap or obsolete. Leaders of developing countries insist that, instead of being obsessed with population growth, we should focus on the reducing consumption of the world's resources by people in richer countries (see fig. 7.18).

Section Review

1. Who was Thomas Malthus, and why was he worried about population growth?
2. How could technology increase human carrying capacity?
3. What is the I = PAT formula, and what does it mean?

7.3 MANY FACTORS DETERMINE POPULATION GROWTH

- *Fertility rates have fallen rapidly in many countries in recent years.*
- *Life expectancies have risen dramatically in many countries in recent years.*
- *Many factors influence our desire to have children.*

The term **demography** is derived from the Greek words *demos* (people) and *graphos* (to write or to measure). Demography encompasses vital statistics about people, such as births, deaths, and where they live, as well as total population size. In this section, we'll survey the ways human populations are measured and described, and discuss demographic factors that contribute to population growth.

How many of us are there?

There were at least 7.4 billion people in the world in 2016, but that is only an educated guess. Even in this age of information technology and communication, counting the number of people in the world is like shooting at a moving target. People continue to be born and die every moment. Some countries have never even taken a census, or the censuses that have been done may not be accurate. Governments may overstate or understate their populations to make their countries appear larger and more important or smaller and more stable than they really are. Some individuals, especially if they are homeless, refugees, or illegal aliens, may not want to be counted or identified.

We really live in two very different demographic worlds. One is old, rich, and relatively stable. The other is young, poor, and growing rapidly. Most people in Asia, Africa, and Latin America inhabit the latter demographic world (fig. 7.7). These countries represent 80 percent of the world population but more than 90 percent of all projected growth (fig. 7.8).

FIGURE 7.7 We live in two demographic worlds. One is rich, technologically advanced, and has an elderly population that is growing slowly, if at all. The other is poor, crowded, underdeveloped, and growing rapidly.
© Frans Lemmens/Getty Images

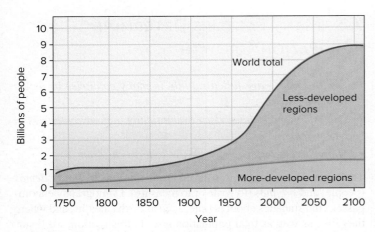

FIGURE 7.8 Estimated human population growth, 1750–2100. In less-developed and more-developed regions. Almost all growth projected for the twenty-first century is in the less-developed countries.
Source: UN Population Division, 2005

Table 7.2	The World's Largest Countries		
2016		**2050***	
Country	Population (millions)	Country	Population (millions)
China	1,367	India	1,692
India	1,252	China	1,295
United States	321	United States	403
Indonesia	256	Nigeria	390
Brazil	209	Indonesia	293
Pakistan	189	Pakistan	275
Nigeria	182	Brazil	223
Bangladesh	162	Bangladesh	194
Russia	143	Philippines	155
Japan	127	Dem. Rep. of Congo	149

*2050 = Estimate
Source: UN Population Division, 2016

The highest population growth rates occur in a few "hot spots," such as sub-Saharan Africa and the Middle East, where economics, politics, religion, and civil unrest keep birth rates high and contraceptive use low. In Niger, for example, annual population growth is currently 3.9 percent. Less than 10 percent of all couples use any form of birth control, women average 7.6 children each, and nearly half the population is less than 15 years old. Even faster growth is occurring in Qatar, where the population doubling time is only 7.3 years. Obviously, a small country with limited resources (except oil) and almost no fresh water can't sustain that high growth rate indefinitely.

Some countries in the developing world have experienced extremely rapid growth rates and are expected to increase dramatically by the middle of the twenty-first century. Table 7.2 shows the ten largest countries in the world, arranged by their estimated size in 2013 and projected size in 2050. Note that, although China was the most populous country throughout the twentieth century, India is expected to pass China in about 2022. Nigeria, which had only 33 million residents in 1950, is forecast to have around 390 million in 2050. Ethiopia, with about 18 million people 50 years ago, is likely to grow nearly eightfold over a century. In many of these countries, rapid population growth is a serious problem. Bangladesh, about the size of Iowa, is already overcrowded at 162 million people. If rising sea levels flood one-third of the country by 2050, as some climatologists predict, adding even more people could be disastrous.

The other demographic world is made up of the richer countries of North America, western Europe, Japan, Australia, and New Zealand. This world is wealthy, old, and mostly shrinking. Italy, Germany, Hungary, and Japan, for example, all have negative growth rates. The average age in these countries is now 40, and the average life expectancy of their residents is expected to exceed 90 by 2050. With many couples choosing to have either one child or none, the populations of these countries are expected to decline significantly over the next century. Japan, which has

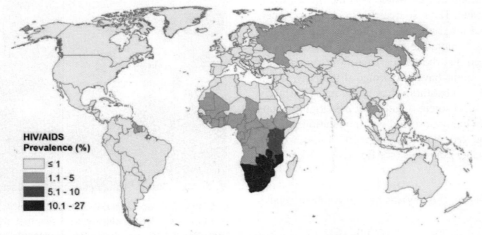

FIGURE 7.9 Proportion of adult population living with HIV/AIDS.
Sources: World Health Organization and UNICEF

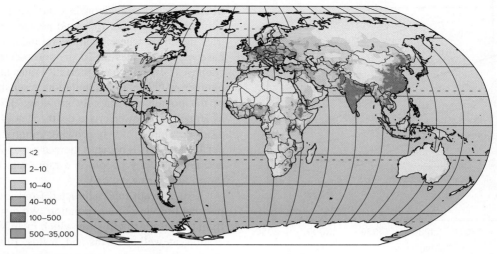

FIGURE 7.10 Population density in persons per square kilometer.
Source: World Bank

Legend:
- <2
- 2–10
- 10–40
- 40–100
- 100–500
- 500–35,000

127 million residents now, is expected to shrink to about 100 million by 2050. Europe, which now makes up about 12 percent of the world population, will constitute less than 7 percent in 50 years, if current trends continue. Even the United States and Canada would have nearly stable populations if immigration were stopped.

Birth rates and life expectancies can decline as a result of political or economic instability. Between about 1990 and 2006, Russia, for instance, declined steadily as death rates soared and birth rates fell to 1.2 children per woman. These changes were generally attributed to a collapsing economy, hyperinflation, crime, and corruption. Since then the birth rate has recovered to 1.6, life expectancy has increased, and the population is approximately stable.

Epidemics can also diminish populations, at least temporarily. The most serious recent case is that of HIV/AIDS, which has affected many African countries especially severely. According to UNICEF, AIDS is the number one cause of death for adolescents in Africa. In South Africa, Zimbabwe, Botswana, and Zambia, for example, over 10 percent of the adult population has AIDS or is HIV positive (fig. 7.9). Botswana, where 23 percent of adults are living with HIV/AIDS, has seen life expectancy decline from 64 to a low of 49 years in 2002; access to antiretroviral treatment has helped survival rates increase since then. The world's highest rate is in Swaziland, where 26.5 percent of adults are living with HIV/AIDS.

The net effect of these variations in growth include high population numbers and densities in China, South Asia, and Europe (figure 7.10). Many areas of high population density are in the fertile river valleys of the Nile, Ganges, Yellow, Yangtze, and Rhine Rivers, and the well-watered coastal plains of India, China, and Europe. Lower populations occur where resources are limited, as in the high Arctic, or the deserts of Australia, North Africa, and Central Asia.

Fertility rates are falling in many countries

The most common demographic statistic of fertility is usually the **crude birth rate,** the number of births in a year per thousand persons. It is statistically "crude" in the sense that it is not adjusted for population characteristics such as the number of women of reproductive age.

The **total fertility rate** is the number of children born to an average woman in a population during her entire reproductive life. Upper-class women in seventeenth- and eighteenth-century England, whose babies were given to wet nurses immediately after birth and who were expected to produce as many children as possible, sometimes had 25 or 30 pregnancies. The highest recorded total fertility rates for working-class people is among some Anabaptist agricultural groups in North America, who have averaged up to 12 children per woman. In most tribal or traditional societies, food shortages, health problems, and cultural practices limit total fertility to about six or seven children per woman even without modern methods of birth control.

Zero population growth (ZPG) occurs when births plus immigration in a population just equal deaths plus emigration. It takes several generations of replacement level fertility (where people just replace themselves) to reach ZPG. Where infant mortality rates are high, the replacement level may be five or more children per couple. In the more highly developed countries, however, this rate is usually about 2.1 children per couple, because some people are infertile, have children who do not survive, or choose not to have children.

As has been the case in Brazil, fertility rates have declined dramatically in every region of the world except Africa over the past 50 years (fig. 7.11). In the 1960s, total fertility rates above 6 were common in many countries. The average family in Mexico in 1975, for instance, had seven children. By 2010, however, the average Mexican woman had only 2.3 children. Similarly, in Iran, total fertility fell from 6.5 in 1975 to 2.04 in 2010. According to the World Health Organization, 100 out of the world's 220 countries are now at or below a replacement rate of 2.1 children per couple, and by 2050 all but a few of the least-developed countries are expected to have reached that milestone. For many of these countries, population growth will continue for a generation because they have such a large number of young people.

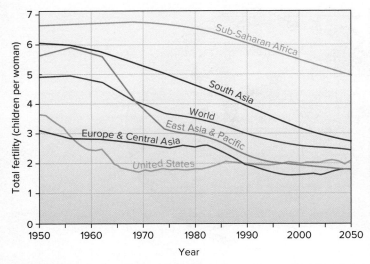

FIGURE 7.11 Total fertility rates for the whole world have fallen by more than half over the past 50 years. Much of this dramatic change has occurred in China and India. Progress has lagged in sub-Saharan Africa, but by 2050 the world average should be approaching the replacement rate of 2.1 children per woman of reproductive age.
Source: World Bank, 2012

Brazil, for example, now has a fertility rate of only 1.8 children per woman. But 26 percent of its population is under 14 years. Those children will mature and start to have families before their parents and grandparents die, so the population will continue to grow for a few decades. Demographers call this **population momentum.**

Some of the greatest fertility reduction has been in East Asia. China's one-child-per-family policy, for example, decreased the fertility rate from 6.5 in 1970 to 1.3 in 2015 (see What Do You Think? later in this section). But as a result of selective abortions for girls (who are less valued in Chinese society), China now reports that 119 boys are being born for every 100 girls. Normal ratios would be about 105 boys to 100 girls. If this imbalance persists, there will be a shortage of brides in another generation. Interestingly, Macao, which is culturally similar to mainland China but hasn't shared its one-child policy, now has a total average fertility rate of only 0.9 and the lowest birth rate in the world.

Although the world as a whole still has an average fertility rate of 2.6, growth rates are now lower than at any time since World War II. If fertility declines like those in Brazil and China were to occur everywhere in the world, global population could begin to decline by 2050, and might be below 6 billion by 2150. Most of Eastern Europe now has fertility levels of 1.2 children per woman. Interestingly, Spain and Italy, although predominantly Roman Catholic like Brazil, have below replacement fertility rates.

Mortality offsets births

A traveler to a foreign country once asked a local resident, "What's the death rate around here?" "Oh, the same as anywhere," was the reply, "about one per person." In demographics, however, **crude death rates** (or crude mortality rates) are expressed in terms of the number of deaths per thousand persons in any given year. Countries in Africa where health care and sanitation are limited may have mortality rates of 20 or more per 1,000 people. Wealthier countries generally have mortality rates around 10 per 1,000. The number of deaths in a population is sensitive to the age structure of the population. Rapidly growing, developing countries such as Brazil have lower crude death rates (4 per 1,000) than do the more-developed, slowly growing countries, such as Denmark (12 per 1,000). This is because there are proportionately more children and fewer elderly people in a rapidly growing country than in a more slowly growing one.

Crude death rate subtracted from crude birth rate gives the **natural increase** of a population. We distinguish natural increase from the **total growth rate,** which includes immigration and emigration, as well as births and deaths. Both of these growth rates are usually expressed as a percentage (number per hundred people) rather than per thousand. A useful rule of thumb is that if you divide 70 by the annual percentage growth, you will get the approximate doubling time in years. Niger, for example, which is growing 3.4 percent per year, is doubling its population every 20 years. The United States, which has a natural increase rate of 0.6 percent per year, would double, without immigration, in 116.7 years. Belgium and Sweden, with natural increase rates of 0.1 percent, are doubling in about 700 years. Ukraine, on the other hand, with a growth rate of -0.2 percent, will lose about 20 percent of its population in the next 50 years. The world growth rate is now 1.08 percent, which means that the population will double in about 66 years if this rate persists.

Life span and life expectancy describe our potential longevity

Life span is the oldest age to which a species is known to survive. Among humans, the oldest living person whose birth was documented was Jeanne Louise Calment of Arles, France, who was 122 years old at her death in 1997. The aging process is still a medical mystery, but it appears that cells in our bodies have a limited ability to repair damage and produce new components. At some point, they simply wear out, and we fall victim to disease, degeneration, accidents, or senility.

Obviously, most people don't reach the maximum age, so usually we examine **life expectancy,** the average age that a newborn infant can expect to attain. This is another way of expressing the average age at death. For most of human history, the average life expectancy in most societies has probably been about 30 years. A few people lived much longer, but many died in early childhood, and most people seem to have died at around 30.

Declining mortality, not rising fertility, is the primary cause of most population growth in the past 300 years. Crude death rates began falling in western Europe during the late 1700s. Most of this advance in survivorship came long before the advent of modern medicine and is due primarily to better food and better sanitation.

China's One-Child Policy

When the People's Republic of China was founded in 1949, it had about 540 million residents, and official government policy encouraged large families. The Republic's First Chairman, Mao Zedong, proclaimed, "Of all things in the world, people are the most precious." He thought that more workers would mean greater output, increasing national wealth, and higher prestige for the country. This optimistic outlook was challenged, however, in the 1960s, when a series of disastrous government policies triggered massive famines and resulted in at least 30 million deaths.

When Deng Xiaoping became Chairman in 1978, he reversed many of Mao's policies by decollectivizing farms, encouraging private enterprise, and discouraging large families. Deng recognized that with an annual growth rate of 2.5 percent, China's population, which had already reached 975 million, would double in only 28 years. China might have nearly 2 billion residents now if that growth had continued. Feeding, housing, educating, and employing all those people would put a severe strain on China's already limited resources.

Deng introduced a successful—but also controversial—one-child-per-family policy. Rural families and ethnic minorities were supposedly exempt from this rule, but local authorities often were capricious and tyrannical in applying sanctions. Ordinary families were punished harshly for having unauthorized children, while government officials and other powerful individuals could have more. There were many reports of bribery, forced abortions, coerced sterilizations, and even infanticide as a result of this policy.

China's one-child-per-family policy, promoted in this billboard, has been remarkably successful in reducing birth rates, but it has had some controversial social effects.
© Alain Le Grasmeur/Corbis

Another result of China's one-child policy is called the 4:2:1 problem. That is, there are now often four grandparents and two parents doting on a single child. Social scientists often refer to this highly spoiled generation as "little emperors."

The Chinese government is beginning to worry about a "birth dearth." If the population declines, will there be enough consumers to keep the economy growing? Will there be enough workers, soldiers, farmers, scientists, inventors, and other productive individuals to keep society functioning in the future? The one-child policy was officially abandoned in 2015. All couples are now allowed to have two children if they want and can afford them. But many demographers doubt that birth rates will change dramatically. They argue that industrialization, urbanization, and economic growth might have brought about a demographic transition without any Draconian policies. Look at neighboring countries, they say. The fertility rate in Singapore is 0.8; Macau is 0.9; Taiwan is 1.11; South Korea is 1.25; and Japan is 1.4; all without policies to restrict child-bearing. If China needed to decrease population growth, they might have depended more on education, women's rights, and economic development as in Brazil (see opening case study for this chapter).

Nevertheless, it's important that Chinese population has declined. China's population in 2016 was about 1.367 billion, or about half a billion less than it might be given its trajectory in 1978. Its annual growth rate is now 0.5 percent, or about 30 percent less than the 0.7 percent annual growth in the United States. China is already the largest contributor to global warning and is driving up world prices for many commodities with its rapidly growing middle class. Think what the effects would be if there were another half a billion Chinese today.

China has also been much more successful in controlling population growth than India. At about the same time that Deng introduced his one-child plan, India, under Indira Gandhi, started a program of compulsory sterilization in an effort to reduce population growth. This cruel policy caused so much public outrage and opposition that the federal government decided to delegate family planning to individual states. Some states have been highly successful in their family planning efforts, while others have not. The net effect, however, is that India is expected to grow to nearly 1.7 billion by 2050, while China is expected to reach zero population growth by 2030.

What do you think? Are there ways that China might have reduced population growth while still respecting human rights? Is the rapid reduction in Chinese population growth worth the social disruption and abuses that it caused? If you were in charge of family planning in China, what policies would you pursue?

The twentieth century has seen a global transformation in human health unmatched in history. By 2015, the average life expectancy was 71.2 years, up from 56.4 fifty years earlier. The greatest progress has been in developing countries (table 7.3).

For example, in 1900 the average Indian man or woman could expect to live about 23 years. A century later, although India had an annual per capita income of only $3,500 (U.S.), the average life expectancy for both men and women had nearly tripled and

Table 7.3	Life Expectancy at Birth for Selected Countries in 1900 and 2015			
	1900		**2015**	
Country	**Males**	**Females**	**Males**	**Females**
India	23	23	66	69
China	39	42	71	77
Russia	31	33	65	76
United States	46	48	76	81
Sweden	57	60	80	84
Japan	42	44	80	87

Source: Data from Population Reference Bureau, 2015

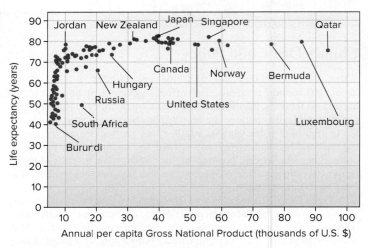

FIGURE 7.12 As incomes rise, so does life expectancy—up to about (U.S.) $10,000. Above that amount, the curve levels off. Some countries, such as South Africa and Russia, have far lower life expectancies than their GDP would suggest. Jordan, on the other hand, which has only one-fifth the per capita GDP of the United States, has a the same life expectancy.
Source: CIA Factbook, 2009

was very close to that of countries with ten times its income level. As in Europe centuries earlier, longer lives were due primarily to better nutrition, improved sanitation, clean water, and education, rather than to miracle drugs or high-tech medicine.

Although the gains were not as great for the already industrialized countries, residents of the United States, Sweden, and Japan, for example, now live about half-again as long as they did at the beginning of the twentieth century, and they can expect to enjoy much of that life in relatively good health. The Disability Adjusted Life Years (DALYs, a measure of disease burden that combines premature death with loss of healthy life resulting from illness or disability) that someone living in Japan can expect is now 74.5 years, compared with only 64.5 DALYs two decades ago.

Large discrepancies exist within countries. In the United States, for example, while the nationwide average life expectancy is 77.5 years, Asian American women in Bergen County, New Jersey, can expect to live 91 years, while Native American men on the Pine Ridge Reservation in South Dakota are reported to typically live only to 48. Two-thirds of African countries have life expectancies greater than that on Pine Ridge.

Women almost always have higher life expectancies than men. Worldwide, the average difference between sexes is three years, but in Russia the difference between men and women is 14 years. Is this because women are biologically superior to men, and thus live longer? Or is it simply that men are generally employed in more hazardous occupations and often engage in more dangerous behaviors, such as drinking, smoking, or reckless driving?

As figure 7.12 shows, life expectancy tends to increase with annual income up to about (U.S.) $10,000 per person. Beyond that level—which is generally enough for adequate food, shelter, and sanitation for most people—life expectancies level out at about 75 years for men and 85 for women.

Some demographers believe that life expectancy is approaching a plateau, while others predict that advances in biology and medicine might make it possible to live 150 years or more. If our average age at death approaches 100 years, as some expect, society will be profoundly affected. In 1970, the median age in the United States was 30. By 2100, the median age could be over 60.

If workers continue to retire at 65, nearly half of the population could be unemployed, and retirees might be facing 35 or 40 years of retirement. We may need to find new ways to structure and finance our lives.

Living longer has demographic implications

A population that is growing rapidly by natural increase has more young people than does a stationary population. One way to show these differences is to graph age classes in a histogram, as shown in figure 7.13. In Niger, which is growing at a rate of 3.4 percent per year, 49 percent of the population is in the prereproductive category (below age 15). Even if total fertility rates were to fall abruptly, the total number of births, and population size, would continue to grow for some years as these young people enter reproductive age.

By contrast, a country with a stable population, like Sweden, has nearly the same number in each age cohort. A population that has recently entered a lower growth rate pattern, such as Singapore, has a bulge in the age classes for the last high-birth-rate generation. Notice that there are more females than males in the older age group in Sweden because of differences in longevity between the sexes.

The greatest challenge to slowing population growth may be that policymakers are not sure how to keep economies growing when populations shrink. Contrary to Malthus' prediction that growing populations deplete resources (section 7.2), economists today worry about collapse if the number of consumers declines. Economists also worry about **dependency ratios,** or the number of nonworking compared to working individuals in a population. In Mexico, for example, each working person supports a high

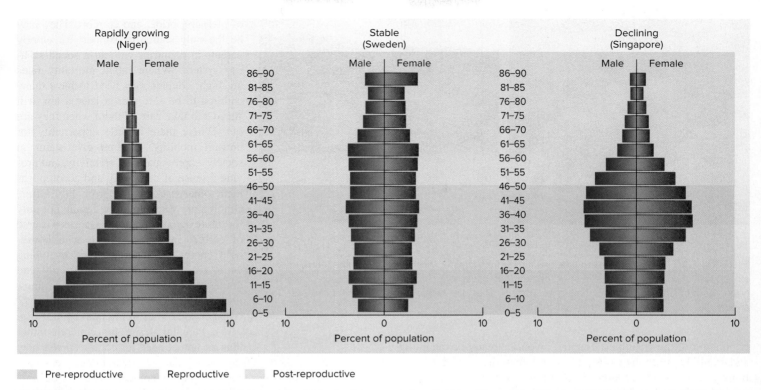

Rapidly growing (Niger)	Stable (Sweden)	Declining (Singapore)

Pre-reproductive   Reproductive   Post-reproductive

FIGURE 7.13 The shape of each age-class histogram is distinctive for a population that is rapidly growing (Niger), stable (Sweden), or declining (Singapore). Horizontal bars represent the percentage of the country's population in consecutive age classes (0–5 yrs, 6–10 yrs, etc.).
Source: U.S. Census Bureau, 2003

number of children. In the United States, by contrast, a declining working population is now supporting an ever larger number of retired persons and there are debates about how well the social security system will hold up. This changing age structure and shifting dependency ratio are occurring worldwide (fig. 7.14). The UN predicts that by 2050 there will be two older persons for every child in the world. Many countries are rethinking their population policies and beginning to offer incentives for marriage and child-bearing.

Emigration and immigration are important demographic factors

Humans are highly mobile, so emigration and immigration play a larger role in human population dynamics than they do in those of many species. Currently, about 800,000 people immigrate legally to the United States each year, and many more enter illegally. Western Europe is particularly stressed by millions of refugees from economic chaos and wars in the Middle East and former socialist states. The United Nations High Commission on Refugees reported in 2016 that 60 million people had been forced from their homes by religious, political, or economic forces. About half had crossed international borders, while the rest were displaced in their own countries. This crisis may become immeasurably worse in coming decades as climate change makes conditions unbearable for billions of people.

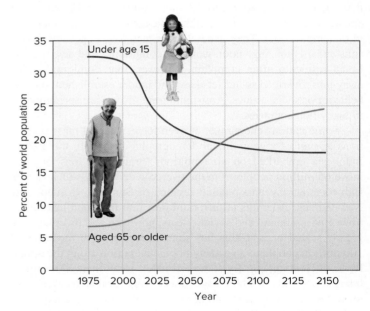

FIGURE 7.14 By the mid-twenty-first century, children under age 15 will make up a smaller percentage of world population, while people over age 65 will comprise an increasingly larger share of the population.

Many countries are trying to limit immigration, but the more-developed regions are expected to gain at least 2 million new residents per year for the next 50 years. Without migration, the

(a)   (b)

FIGURE 7.15 In rural areas with little mechanized agriculture, (a) children are needed to tend live-stock, care for younger children, and help parents with household chores. Where agriculture is mechanized, (b) rural families view children just as urban families do—helpful, but not critical to survival. This affects the decision about how many children to have.
© William P. Cunningham © Neil Beer/Getty Images RF

population of the wealthiest countries would already be declining and would be more than 126 million less than the current 1.2 billion by 2050. In 2008, nearly 45.5 million U.S. residents (15.2 percent of the total population) classified themselves as Hispanic or Latino. They now constitute the largest U.S. minority.

Immigration is a controversial issue in many countries. "Guest workers" often perform heavy, dangerous, or disagreeable work that citizens are unwilling to do. Many migrants and alien workers are of a different racial or ethnic background than the majority in their new home. They generally are paid low wages and given substandard housing, poor working conditions, and few rights. Local residents often complain, however, that immigrants take away jobs, overload social services, and ignore established rules of behavior or social values. Anti-immigrant groups are springing up in many rich countries.

Some nations encourage, or even force, internal mass migrations as part of a geopolitical demographic policy. In the 1970s, Indonesia embarked on an ambitious "transmigration" plan to move 65 million people from the overcrowded islands of Java and Bali to relatively unpopulated regions of Sumatra, Borneo, and New Guinea. Attempts to turn rainforest into farmland had disastrous environmental and social effects, however, and this plan was greatly scaled back. China has announced a plan to move up to 100 million people to a sparsely populated region along the Amur River in Heilongjiang. By some estimates, more than 250 million internal migrants in China have moved from rural areas to the cities to look for work, and another quarter of a billion are expected to do so in the next few decades.

Many factors increase our desire for children

Factors that increase people's desires to have babies are called **pronatalist pressures.** Raising a family may be the most enjoyable and rewarding part of many people's lives. Children can be a source of pleasure, pride, and comfort. They may be the only source of support for elderly parents in countries without a social security system. Where infant mortality rates are high, couples may need to have many children to be sure that at least a few will survive to take care of them when they are old. Where there is little opportunity for upward mobility, children give status in society, express parental creativity, and provide a sense of continuity and accomplishment otherwise missing from life. Often, children are valuable to the family not only for future income, but even more as a source of current income and help with household chores. In much of the developing world, children as young as 6 years old tend domestic animals and younger siblings, fetch water, gather firewood, and help grow crops or sell things in the marketplace (fig. 7.15). Parental desire for children rather than an unmet need for contraceptives may be the most important factor in population growth in many cases.

Society also has a need to replace members who die or become incapacitated. This need often is codified in cultural or religious values that encourage bearing and raising children. In some societies, families with few or no children are looked upon with pity or contempt. The idea of deliberately controlling fertility may be shocking, even taboo. Women who are pregnant or have small children are given special status and protection. Boys frequently are more valued than girls because they carry on the family name and are expected to support their parents in old age. Couples may have more children than they really want in an attempt to produce a son.

Male pride often is linked to having as many children as possible. In Niger and Cameroon, for example, men, on average, want 12.6 and 11.2 children, respectively. Women in these countries consider the ideal family size to be only about one-half that desired by their husbands. Even though a woman might desire fewer children, however, she may have few choices and little control over her own fertility. In many societies, a woman has no status outside of her role as wife and mother. Without children, she has no source of support.

Other factors discourage reproduction

In more highly developed countries, many pressures tend to reduce fertility. Improving opportunities for women is among the most important. Higher education and personal freedom for women tend to provide more alternatives to childbearing, and they are less likely to stay home and have many children. Globally, changes in birth rates show that when they have a choice, most women choose to have two or three children (see fig. 7.11), rather than the historical average of six or more. This decline results partly from later marriage and delayed childbearing when women have other opportunities. When women earn a salary, their income becomes an important part of the family budget. Thus, education

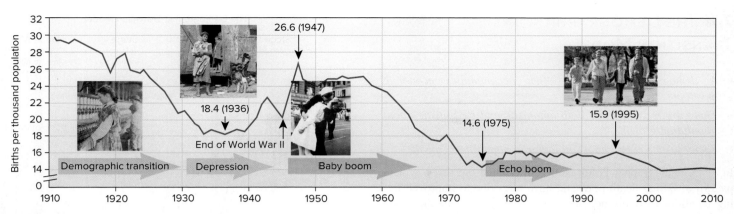

FIGURE 7.16 Birth rates in the United States, 1910–2000. The falling birth rate from 1910 to 1929 represents a demographic transition from an agricultural to an industrial society. The baby boom following World War II lasted from 1945 to 1965. A much smaller "echo boom" occurred around 1980 when the baby boomers started to reproduce.

Sources: Data from the Population Reference Bureau and the U.S. Bureau of the Census; Photos: (1) Library of Congress Prints and Photographs Division LC-DIG-nclc-01366; (2) USDA Photograph Archives; (3) National Archives and Records Administration (NWDNS-80-G-377094); (4) © Stockbyte/Getty Images RF

and socioeconomic status are usually inversely related to fertility in most countries. In very poor countries, rising incomes sometimes produce a temporary increase in fertility, as women become healthier and families are able to support more children. That increase is typically brief, however.

Birthrates in the U.S. have responded to a variety of historical events, but in general they fell between 1910 and 2010 (fig. 7.16). The period between 1910 and 1930 was a time of industrialization and urbanization. Women were getting more education than ever before and entering the workforce. As was the case in Brazil half a century later, birth rates fell. Births increased at the beginning of World War II (as is often the case in wartime). For reasons that are unclear, a higher percentage of boys are usually born during war years.

At the end of the war, there was a "baby boom" as couples were reunited and women were encouraged to leave work outside the home. This high birth rate persisted through the times of prosperity and optimism of the 1950s, but began to fall in the 1960s. Part of this decline was caused by the small number of babies born in the 1930s. This meant fewer young adults to give birth in the 1960s. As in Brazil more recently, part of this was due to changed perceptions of the ideal family size. Whereas in the 1950s women typically wanted four children or more, in the 1970s the norm dropped to one or two (or no) children. A small "echo boom" occurred in the 1980s as people born in the 1960s began to have babies, but changing economics and attitudes seem to have permanently altered our view of ideal family size in the United States.

Could we have a birth dearth?

Most European countries now have birth rates of 1.5 or fewer children per woman, well below the replacement level of 2.1. Italy, Russia, Austria, Germany, Greece, and Spain are experiencing negative rates of natural population change. Families in Japan, Singapore, and Taiwan have fewer than 1.3 children and are also facing a "child shock" as fertility rates have fallen. There are concerns in these countries that declining populations will be unable

to maintain military strength (lack of soldiers), economic power (lack of workers and of consumers), and social systems (not enough workers and taxpayers) if low birth rates persist or are not balanced by immigration. Some analysts now emphasize that the United States and Northern Europe are fortunate to have an influx of immigrants, who bring youth, energy, and consumers to shore up the economy.

Economist Ben Wattenberg warns that this "birth dearth" might seriously erode the powers of Western democracies in world affairs. He points out that Europe and North America accounted for 22 percent of the world's population in 1950. By the 1980s, this number had fallen to 15 percent, and by the year 2030, Europe and North America probably will make up only 9 percent of the world's population. Germany, Hungary, Denmark, and Russia now offer incentives to encourage women to bear children. Japan offers financial support to new parents, and Singapore provides a dating service to encourage marriages among the upper classes as a way of increasing population.

On the other hand, because Europeans and North Americans consume so many more resources per capita than most other people in the world, a reduction in the population of these countries will do more to spare the environment than would a reduction in population almost anywhere else.

An additional reason birth rates have been falling in many industrialized countries may be that toxins and endocrine hormone disrupters in our environment interfere with sperm production. Sperm numbers and quality (fertilization ability) appear to have fallen by about half over the past 50 years in a number of countries. Widespread synthetic chemicals, such as perfluorocarbons, phthalates, and dioxins that disrupt endocrine hormone functions may be responsible for some of this decline. We'll discuss this further in chapter 8.

Section Review

1. What is the replacement rate of births?
2. What is the relation between income and life expectancies?
3. What factors increase or decrease our desire for children?

7.4 THE DEMOGRAPHIC TRANSITION MODEL

- *A demographic transition often accompanies economic development.*
- *There are reasons to be both optimistic and pessimistic about population growth.*
- *Social justice is an important factor in population growth.*

In 1945, demographer Frank Notestein pointed out that a typical pattern of falling death rates and birth rates due to improved living conditions usually accompanies economic development. He called this pattern the **demographic transition** from high birth and death rates to lower birth and death rates. Figure 7.17 shows an idealized model of a demographic transition. This model is often used to explain connections between population growth and economic development.

Economic and social development influence birth and death rates

Stage I in figure 7.17 represents the conditions in a premodern society. Food shortages, malnutrition, lack of sanitation and medicine, accidents, and other hazards generally keep death rates in such a society around 35 per 1,000 people. Birth rates are correspondingly high to keep population densities relatively constant. As economic development brings better jobs, medical care, sanitation, and a generally improved standard of living in Stage II, death rates often fall very rapidly. Birth rates may actually rise at first as more money and better nutrition allow people to have the children they always wanted. Eventually, in a mature industrial economy (Stage III), birth rates fall as people see that all their children are more likely to survive and that the whole family benefits from concentrating more resources on fewer children. Note that population continues to grow rapidly during this stage because of population momentum (baby boomers reaching reproductive age). Depending on how long it takes to complete the transition, the population may go through one or more rounds of doubling before coming into balance again.

Stage IV in figure 7.17 represents conditions in developed countries, where the transition is complete and both birth rates and death rates are low, often a third or less than those in the predevelopment era. The population comes into a new equilibrium in this phase, but at a much larger size than before. Most of the countries of northern and western Europe went through a demographic transition in the nineteenth or early twentieth century similar to the curves shown in this figure.

Many of the most rapidly growing countries in the world, such as Kenya, Yemen, Libya, and Jordan, now are in Stage I of this demographic transition. Their death rates have fallen close to the rates of the fully developed countries, but birth rates have not fallen correspondingly. In fact, both their birth rates and total population are higher than those in most European countries when industrialization began 300 years ago. The large disparity between birth and death rates means that many developing countries now are growing at 3 to 4 percent per year. Such high growth rates in developing countries

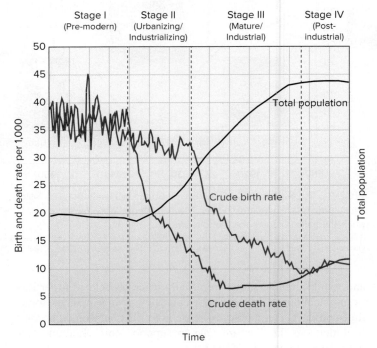

FIGURE 7.17 Theoretical birth, death, and population growth rates in a demographic transition accompanying economic and social development. In a predevelopment society, birth and death rates are both high, and total population remains relatively stable. During development, death rates tend to fall first, followed in a generation or two by falling birth rates. Total population grows rapidly until both birth and death rates stabilize in a fully developed society.

could boost the total world population to 9 billion or more before the end of the twenty-first century. This raises what may be the two most important questions in this entire chapter: Why are birth rates not yet falling in these countries, and what can be done about it?

There are reasons to be optimistic about population

Four conditions are necessary for a demographic transition to occur: (1) improved standard of living, (2) increased confidence that children will survive to maturity, (3) improved social status of women, and (4) increased availability and use of birth control. As the example of Brazil in the opening case study for this chapter shows, these conditions can be met, even in relatively poor countries.

Some demographers claim that a demographic transition already is in progress in most developing nations. Problems in taking censuses and a normal lag between falling death and birth rates may hide this for a time, but the world population should stabilize sometime in the next century. Some evidence supports this view. As we mentioned earlier in this chapter, fertility rates have fallen dramatically nearly everywhere in the world over the past half century. The U.N. Population Division now reports that half the world's countries are at or below the replacement rate of 2.1 children per couple.

Some countries have had remarkable success in population control. In Thailand, Indonesia, Colombia, and Iran, for instance, total fertility dropped by more than half in 20 years. Morocco, the Dominican Republic, Jamaica, Peru, and Mexico all have seen fertility rates fall between 30 percent and 40 percent in a single generation. In all these countries, the following factors contribute to stabilizing populations:

- Growing prosperity and social reforms that accompany development reduce the need and desire for large families in most countries.

- Technology is available to bring advances to the developing world much more rapidly than was the case a century ago, and the rate of technology transfer is much faster than it was when Europe and North America were developing.

- Less-developed countries have historic patterns to follow. They can benefit from our mistakes and chart a course to stability more quickly than they might otherwise do.

- Modern communications (especially television) have caused a revolution of rising expectations that act as a stimulus to spur change and development.

Many people remain pessimistic about population growth

Economist Lester Brown takes a more pessimistic view. He warns that many of the poorer countries of the world appear to be caught in a "demographic trap" that prevents them from escaping from the middle phase of the demographic transition. Their populations are now growing so rapidly that human demands exceed the sustainable yield of local forests, grasslands, croplands, or water resources. High resource demands threaten to destroy the resource base, causing environmental deterioration, economic decline, and political instability. Can these countries escape from this trap and modernize?

Many people argue that the only way to break out of the demographic trap is to immediately and drastically reduce population growth by whatever means are necessary. They argue strongly for birth control education and bold national policies to encourage lower birth rates. Some agree with Malthus that helping the poor will simply increase their reproductive success and further threaten the resources on which we all depend. Author Garret Hardin described this view as lifeboat ethics. "Each rich nation," he wrote, "amounts to a lifeboat full of comparatively rich people. The poor of the world are in other much more crowded lifeboats. Continuously, so to speak, the poor fall out of their lifeboats and swim for a while, hoping to be admitted to a rich lifeboat, or in some other way to benefit from the goodies on board. . . . We cannot risk the safety of all the passengers by helping others in need. What happens if you share space in a lifeboat? The boat is swamped and everyone drowns. Complete justice, complete catastrophe." How would you respond to Professor Hardin?

Social justice is an important consideration

A third view is that **social justice** (a fair share of social benefits for everyone) is the key to successful demographic transitions. The world has enough resources for everyone, but inequitable social, political, and economic systems cause uneven distributions

of those resources. As a consequence, hunger, poverty, violence, environmental degradation, and rapid population growth continue. In this view, it is justice and access, not resources, that are limited. Although overpopulation exacerbates other problems, a narrow focus on this factor alone encourages racism and hatred of the poor. A solution for all these problems is to reduce poverty and inequity, not to blame the victims. Small nations and minorities often regard calls for population control as a form of genocide. Figure 7.18 expresses the opinion of many people in less-developed countries about the relationship between resources and population.

An important part of this view is that many of the rich countries are, or were, colonial powers, while the poor, rapidly growing countries were colonies. The wealth that paid for progress and security for developed countries was often extracted from colonies, which now suffer from exhausted resources, exploding populations, and chaotic political systems. Some of the world's poorest countries, such as India, Ethiopia, Mozambique, and Haiti,

FIGURE 7.18 Population growth, resulting from high birth rates, has important impacts on resource consumption. But small, wealthy families consume far more resources per person, usually sourced from around the globe. Many people debate which impact is more important, but ultimately there are good reasons for reducing both family size and consumption rates.
© Andre Babiak/Alamy © Duncan Vere Green/Alamy

had rich resources and adequate food supplies before they were impoverished by colonialism. Those of us who now enjoy abundance may need to help the poorer countries not only as a matter of justice but because we all share the same environment.

In addition to considering the rights of fellow humans, we should also consider those of other species. Rather than ask what is the maximum number of humans that the world can possibly support, perhaps we should think about the needs of other creatures. As we convert natural landscapes into agricultural or industrial areas, species are crowded out that may have just as much right to exist as we do. What do you think would be the optimum number of people to provide a fair and decent life for all humans while causing the minimum impact on nonhuman neighbors?

Child health affects fertility

Opportunities for education and paying jobs are critical factors in fertility rates (fig. 7.19). Child survival also is crucial in stabilizing population. When infant and child mortality rates are high, as they are in much of the developing world, parents tend to have high numbers of children to ensure that some will survive to adulthood. There has never been a sustained drop in birth rates that was not first preceded by a sustained drop in infant and child mortality. One of the most important distinctions in our demographically divided world is the high infant mortality rates in the less-developed countries. Better nutrition, improved health care, simple oral rehydration therapy, and immunization against infectious diseases (see chapter 8) have brought about dramatic reductions in child mortality rates, which have been accompanied in most regions by falling birth rates. It has been estimated that saving 5 million children each year from easily preventable communicable diseases would avoid 20 or 30 million extra births.

Increasing family income does not always translate into better welfare for children, because in many cultures men control most financial assets. Often, the best way to improve child survival is to ensure the rights of mothers. Land reform, political rights, opportunities to earn an independent income, and the improved health status of women often are better indicators of total fertility and family welfare than rising GNP.

Family planning gives us choices

Family planning allows couples to determine the number and spacing of their children. It doesn't necessarily mean fewer children—people may use family planning to have the maximum number of children possible—but it does imply that the parents will control their reproductive lives and make rational, conscious decisions about how many children they will have and when those children will be born, rather than leaving it to chance. As the desire for smaller families becomes more common, birth control becomes an essential part of family planning in most cases. In this context, **birth control** usually means any method used to reduce births, including abstinence, delayed marriage, contraception, methods that prevent the implantation of embryos, and induced abortions. As the opening case study in this chapter shows, there are many ways to encourage family planning.

Humans have always regulated fertility. Evidence suggests that people in every culture and every historic period have used a variety of techniques to control population size. Studies of hunting and gathering people, such as the !Kung or San of the Kalahari Desert in southwest Africa, indicate that our early ancestors had stable population densities, not because they killed each other or starved to death regularly, but because they controlled fertility.

For instance, San women breast-feed children for three or four years. When calories are limited, lactation depletes body fat stores and suppresses ovulation. Coupled with taboos against intercourse while breast-feeding, this is an effective way of spacing children. Other ancient techniques to control population size include abstinence, folk medicines, abortion, and infanticide. We may find some or all of these techniques unpleasant or morally unacceptable, but we shouldn't assume that other people are too ignorant or too primitive to make decisions about fertility.

Modern medicine gives us many more options for controlling fertility than were available to our ancestors. More than 100 new

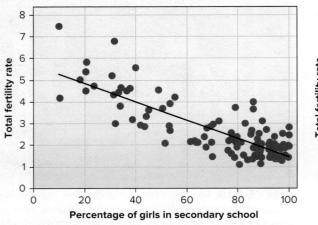 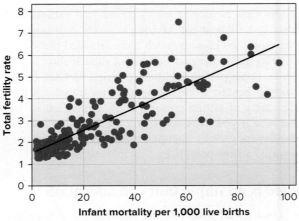

FIGURE 7.19 Total fertility declines as girl's education increases and infant mortality decreases.
Source: World Bank, 2015

contraceptive methods are now being studied, and some appear to have great promise. Nearly all are biologically based (e.g., hormonal) rather than mechanical (e.g., condom, IUD). Recently, the U.S. Food and Drug Administration approved five new birth control products. Four of these use various methods to administer female hormones that prevent pregnancy.

Other methods are years away from use, but take a new direction entirely. Vaccines for women are being developed that will prepare the immune system to reject the hormone chorionic gonadotropin, which maintains the uterine lining and allows egg implant, or that will cause an immune reaction against sperm. Injections for men are focused on reducing sperm production, and have proven effective in mice. Without a doubt, the contemporary couple has access to many more birth control options than their grandparents had.

The choices we make determine our future

Because there's often a lag between the time when a society reaches replacement birth rate and the end of population growth, we are deciding now what the world will look like in a hundred years. How many people will be in the world a century from now? Most demographers believe that world population will stabilize sometime during the twenty-first century at a total of 9–12 billion. The United Nations Population Division projects four population scenarios (fig. 7.20). The optimistic (low) projection suggests that world population might stabilize by about 2030 and then drop back below current levels. This doesn't seem likely. The medium projection shows a population of about 9 billion in 40 years, while the high projection would reach 12 billion by midcentury.

Which of these scenarios will we follow? As you have seen in this chapter, population growth is a complex subject. Stabilizing or reducing human populations will require substantial changes from business as usual. An encouraging sign is that worldwide contraceptive use has increased sharply in recent years. About half of the world's married couples used some family planning techniques in 2000, compared with only 10 percent 30 years earlier, but another 100 million couples say they want, but do not have access to, family planning. If given a choice, people prefer smaller families.

Successful family planning programs often require significant societal changes. Among the most important of these are (1) improved social, educational, and economic status for women (birth control and women's rights are often linked); (2) improved status for children (fewer children are born if they are not needed as a cheap labor source); (3) acceptance of calculated choice as a valid element in life in general and in fertility in particular (the belief that we have no control over our lives discourages a sense of responsibility); (4) social security and political stability that give

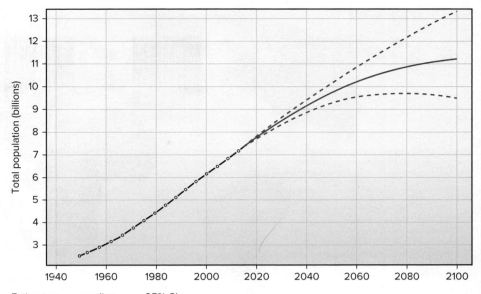

Estimates: —— median - - 95% CI

FIGURE 7.20 Population projections for different growth scenarios. Recent progress in family planning and economic development have led to significantly reduced estimates compared to a few years ago. The median projection is 8.9 billion in 2050, compared to previous estimates of over 10 billion for that date.

From World Population Prospects The 2015 Revision Key Findings and Advance Tables (c) 2015 United Nations. Reprinted with the permission of the United Nations.

people the means and the confidence to plan for the future; and (5) the knowledge, availability, and use of effective and acceptable means of birth control.

The current world average fertility rate of 2.6 births per woman is less than half what it was 50 years ago. If similar progress could be sustained for the next half century, fertility rates could fall to the replacement rate of 2.1 children per woman. Whether this scenario comes true or not depends on choices that all of us make.

Already, nearly half the world population lives in countries where the total fertility rate is at or close to the replacement rate (fig. 7.21). The example of Brazil gives us hope that with rising standards of living, democracy, and social justice, population growth will spontaneously slow without harsh government intervention. However, increasing wealth creates worries that consumption supported by destructive technologies will be unsustainable. The trade-off between population size and affluence may still create unacceptable environmental conditions. Furthermore, as figure 7.21 shows, there are countries, especially in Africa, where wars, corruption, colonial history, religious tensions, and other factors have prevented economic and social development while perpetuating high population growth. Can we overcome all these problems and create a more humane, sustainable world?

Section Review

1. How can prosperity and urbanization affect birth rates?
2. What's the role of women's rights in population growth?
3. Why is Brazil a useful model in demographics?

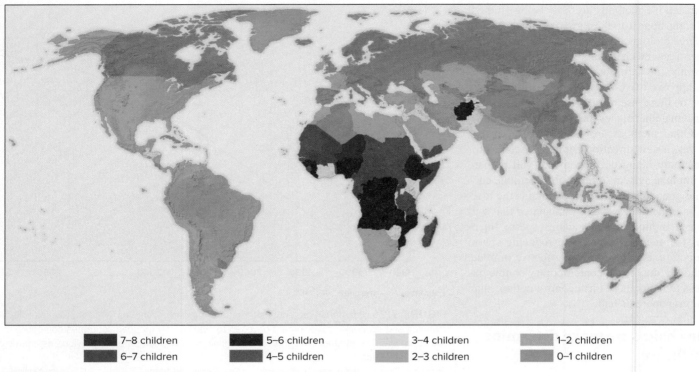

■	7–8 children	■	5–6 children	▨	3–4 children	▨	1–2 children
■	6–7 children	■	4–5 children	▨	2–3 children	▨	0–1 children

FIGURE 7.21 Fertility rates by country. The replacement rate is 2.1 children per woman.

Conclusion

A few decades ago, we were warned that a human population explosion was about to engulf the world. Exponential population growth was seen as a cause or corollary to nearly every important environmental problem. Some people still warn that the total number of humans might grow to 30 or 40 billion by the end of this century. Birth rates have fallen, however, almost everywhere, and most demographers now believe that we will reach an equilibrium around 9 billion people in about 2050. Some claim that if we promote equality, democracy, human development, and modern family planning techniques, population might even decline to below its current level of 7 billion in the next 50 years. How we should carry out family planning and birth control remains a controversial issue. Should we focus on political and economic reforms and

hope that a demographic transition will naturally follow, or should we take more direct action (or any action) to reduce births?

Whether our planet can support 9 billion—or even 7.4 billion—people on a long-term basis remains a vital question. If all those people try to live at a level of material comfort and affluence now enjoyed by residents of the wealthiest nations, using the old, polluting, inefficient technology that we now employ, the answer is almost certainly that even 7 billion people is too many in the long run. If we find more sustainable ways to live, however, it may be that 9 billion people could live happy, comfortable, productive lives. If we don't find new ways to live, we probably face a crisis no matter what happens to our population size. We'll discuss pollution problems, energy sources, and sustainability in subsequent chapters of this book.

Reviewing Key Terms

Can you define the following terms in environmental science?

birth control 7.4

crude birth rate 7.3

crude death rate 7.3

demographic transition 7.4

demography 7.3

dependency ratio 7.3

family planning 7.4

I = PAT 7.2

life expectancy 7.3

natural increase 7.3

population momentum 7.3

pronatalist pressures 7.3

social justice 7.4

total fertility rate 7.3

total growth rate 7.3

zero population growth
(ZPG) 7.3

Critical Thinking and Discussion Questions

1. What do you think is the optimum human population? The maximum human population? Are the numbers different? If so, why?

2. Some people argue that technology can provide solutions for environmental problems; others believe that a "technological fix" will make our problems worse. What personal experiences or worldviews do you think might underlie these positions?

3. Karl Marx called Thomas Malthus a "shameless sycophant of the ruling classes." Why would the landed gentry of the eighteenth century be concerned about population growth of the lower classes? Are there comparable class struggles today?

4. Try to imagine yourself in the position of a person your age in a developing world country. What family planning choices and pressures would you face? How would you choose among your options?

5. Some demographers claim that population growth has already begun to slow; others dispute this claim. How would you evaluate the competing claims of these two camps? Is this an issue of uncertain facts or differing beliefs? What sources of evidence would you accept as valid?

6. What role do race, ethnicity, and culture play in our immigration and population policies? How can we distinguish between prejudice and selfishness, on one hand, and valid concerns about limits to growth, on the other?

Data Analysis

Population Change over Time

Brazil's population trends have shifted dramatically in recent years. Is Brazil unusual in this change? Take a look at population size and trends in different regions, and at some of the factors that influence growth rates. Gapminder.org is a rich source of data on global population, health, and development, including animated graphs showing change over time. Go to Connect to find a link to Gapminder graphs, and answer questions about what they tell you.

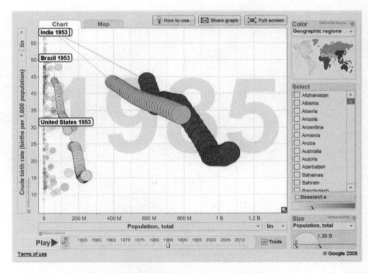

Source: http://www.gapminder.org/world

8 Environmental Health and Toxicology

▲ How dangerous are new, synthetic chemicals like those in non-stick pans? Do we need to worry?
© Paul Bradbury/OJO Images/Getty Images

Learning Outcomes

After studying this chapter, you should be able to:

8.1 Describe health and disease and how the global disease burden is now changing.

8.2 Summarize the principles of toxicology.

8.3 Discuss the movement, distribution, and fate of toxins in the environment.

8.4 Explain ways we evaluate toxicity and risk.

8.5 Relate how we establish health policy.

"To wish to become well is a part of becoming well."

– Seneca

PFOA: Miracle or Menace?

Perfluorocarbons (PFCs) have been called miracle compounds. Recalling DuPont's slogan of "better living through chemistry," these slippery, heat-resistant, durable compounds occur in countless products (fig. 8.1), refrigerants, propellants, water- and stain-repellent coatings, non-stick cookware, electrical insulation, airplane and computer parts, fire-fighting foam, cosmetics, cleaning solvents, and hydraulic fluids.

There's also a dark side to these technological wonders. Because they're so persistent, PFCs are now found throughout the world, even in the most remote and seemingly pristine sites. They accumulate in birds, mammals, fish, and humans. Perfluorooctanoic acid (PFOA), the starting compound for most PFCs, now occurs in the blood of most people in industrialized nations at 2 to 8 parts per billion (ppb). In Korea, average levels are about 60 ppb. In the United States, nearly everyone is thought to have PFOA levels in their blood about ten times higher than the EPA (Environmental Protection Agency) limits suggested for drinking water.

These compounds were not regulated in the U.S. until recently because they were not legally designated as hazardous. There's a common misperception that if a chemical is dangerous, it will be regulated to protect public health. Under the 1976 Toxic Substances Control Act, however, the EPA can investigate a chemical only when it is proven harmful. Because long-lasting environmental exposure is hard to pin to any particular compound, proving cause and effect is extraordinarily difficult. This standard, therefore, largely eliminates public oversight of new chemicals. The EPA has restricted only five synthetic chemicals in the last 40 years, out of an estimated 60,000 or more on the market.

The landmark case that brought PFOA into the public spotlight began in 1999, when a farmer in West Virginia suspected that water draining from a DuPont waste dump was sickening and killing his dairy cows. The unlined waste dump turned out to have 7,000 tons of PFOA residues from Teflon production in a DuPont factory in nearby Parkersburg, West Virginia. The farmer sued DuPont, and the court ordered the company to turn over all documentation related to PFOA.

Surprisingly, the records included confidential medical and health studies, some dating back half a century, documenting a variety of illnesses associated with PFOA. Major manufacturers, such as DuPont and 3M, had documented associations of PFOA exposure with birth defects and cancer in lab animals in the 1960s. They also knew that factory workers had high concentrations of PFOA in their blood, and that workers, as well as residents near industrial facilities and waste disposal sites, suffered from a variety of cancers, as well as kidney, liver, and skin ailments.

As in many environmental health issues, irresponsible waste disposal made up much of the problem. DuPont dumped PFOA and other waste materials into public sewers, waterways, and open waste dumps. These expedient methods avoided the much higher costs of incinerating waste or sending it to secure, contained landfills.

FIGURE 8.1 It's estimated that 20 percent of the PFOA in American blood comes from anti-grease coatings in pizza and popcorn containers.
© CREATISTA/Shutterstock.com

It quickly became apparent that DuPont had long recognized the adverse health effects of PFOA. They also knew that the chemicals were widespread in the environment and were present in local water supplies at unsafe concentrations. The company had been searching for an alternative to PFOA, and had announced in 1993 that they had a compound that appeared to be less toxic, and that stayed in the body for a much shorter time. Nevertheless, the company decided against shifting away from PFOA, which was worth an estimated $1 billion a year in profits.

In 2005, DuPont agreed to a $16.5 million settlement with the EPA, although the company didn't admit any guilt. This was the largest civil penalty the EPA had obtained at that time. The next year, the U.S. government reached an agreement with the eight largest fluorochemical companies in the country to eliminate PFOA and related compounds by 2015. The European Union has ordered a similar phase-out, but companies in Asia and other regions continue to manufacture and release these compounds. The 3M Company, which had been a major supplier of PFOA, ceased production in 2000, and DuPont stopped production and use in 2013. PFOA continues to turn up in groundwater in industrial areas, however, as more public health offices look for it.

This case study demonstrates a number of themes in environmental health. We have been tremendously successful in producing an amazing array of synthetic chemicals with highly useful properties. But we sometimes unwittingly set in motion a tsunami of unintended consequences when we release those materials into the environment. Monitoring new compounds and understanding their effects is extremely difficult. Regulating them is a challenge even when we have an active EPA and laws designed to protect public safety. On the other hand, relatively high exposure is sometimes needed to show up as a health risk. So how dangerous is low-level but widespread exposure to the thousands of new materials we've released? This is a central question of environmental toxicology.

8.1 ENVIRONMENTAL HEALTH

- *The global disease burden is changing. Infectious diseases are decreasing while chronic conditions, such as heart disease, stroke, and diabetes, are increasing.*

- *Emergent diseases are appearing with increasing frequency as we travel more, move into new habitats, and share our germs with others around the world.*

- *Conservation medicine combines ecology with health care.*

What is health? The World Health Organization (WHO) defines **health** as a state of complete physical, mental, and social well-being, not merely the absence of disease or infirmity. By that definition, we all are ill to some extent. Likewise, we all can improve our health to live happier, longer, more productive, and more satisfying lives if we think about what we do.

What is disease? A **disease** is an abnormal change in the body's condition that impairs important physical or psychological functions. Diet and nutrition, infectious agents, toxic substances, genetics, trauma, and stress all play roles in **morbidity** (illness) and **mortality** (death). **Environmental health** focuses on external factors that cause disease, including elements of the natural, social, cultural, and technological worlds in which we live. Figure 8.2 shows some major environmental disease agents, as well as the media through which we encounter them. Ever since the publication of Rachel Carson's *Silent Spring* in 1962, the discharge, movement, fate, and effects of synthetic chemical toxins have been a special focus of environmental health. Later in this chapter, we'll study these topics in detail. First, however, let's look at some of the major causes of illness worldwide.

The global disease burden is changing

Although there are many worries about pollution, environmental toxins, emerging diseases, and other health threats, we should take a moment to consider the remarkable progress we've made over the past hundred years in controlling many terrible diseases. Epidemics that once killed millions of people have been reduced or eliminated. Smallpox was completely wiped out in 1977. Polio has been eliminated everywhere in the world except for a few places in Afghanistan, Pakistan, Laos, and Madagascar. Guinea worms, river blindness, and yaws appear to be on their way to elimination. Epidemics of typhoid fever, cholera, and yellow fever that regularly decimated populations a century ago are now rarely encountered in developed countries. AIDS, which once was an immediate death sentence, has become a highly treatable disease. The average HIV-positive person in the United States now lives 24 years after diagnosis if treated faithfully with modern medicines.

One of the best indicators of health gains is a worldwide increase in child survival. Fifty years ago in South Asia, for example, nearly one child in four died before their fifth birthday. Currently, the global five-year survival rate is over 95 percent

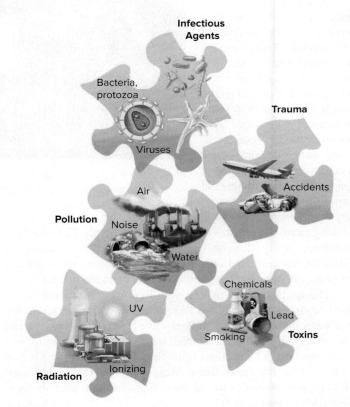

FIGURE 8.2 Major sources of environmental health risks.

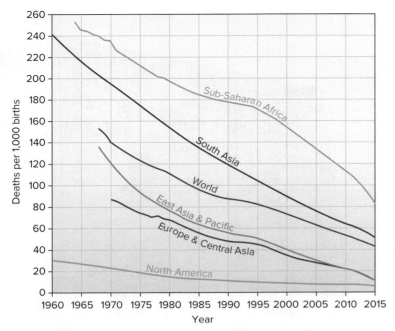

FIGURE 8.3 Child mortality has fallen dramatically over the past 50 years and is expected to continue this decline in the future. East Asia has reached a level matching that of Europe and Central Asia, while sub-Saharan Africa has lagged behind—mainly because of wars, poverty, and poor health.

Source: Data from the World Health Organization, 2012

(fig. 8.3). In another couple of decades, if current trends continue, the world under-5 mortality rate should approach that in the United States of less than 1 percent. Remarkably, East Asia and the Pacific region already have nearly as low a child mortality rate as the richer countries in Europe or North America, despite having a far lower per capita income. Much of this progress is due simply to better nutrition, improved sanitation, and clean water, which keep children healthy. According to the WHO, half of all gains in child survival are brought about by increases in women's education. Only sub-Saharan Africa has lagged in this health revolution, for a number of reasons that include political instability and poverty.

As chapter 7 shows, improved child survival has important demographic implications, including lowered birth rates and longer life expectancies. The fact that average global life expectancies have more than doubled over the past century from 30 to 64 years means many more productive years of life for most of us. but it also means that many more people are reaching old age.

The decrease in many infectious illnesses, along with an aging population, is producing a shift in global disease burden. Many people now live for years with chronic illness or disability. According to the WHO, chronic diseases now account for nearly three-fourths of the 56.5 million total deaths worldwide each year and about half of the global disease burden. To account for the costs of chronic illness, health agencies calculate disease burden for a population in terms of **disability-adjusted life years (DALYs).** This statistic is the sum of overall years lost to disability

or illness plus years lost to early deaths in a population. For example, a teenager permanently paralyzed by a traffic accident will have many years of suffering, health care costs, and lost potential. Similarly, many years of expected life are lost when a child dies of neonatal tetanus. A senior citizen who has a stroke has lost far fewer years to disease or disability.

Major causes of death have changed dramatically in recent decades. Chronic conditions, such as cardiovascular disease, cancer, and strokes were once common only in wealthy countries. These conditions are now among the leading causes of death in lower and middle income countries. Although the traditional killers in developing countries—infections, maternal and perinatal (birth) complications, and nutritional deficiencies—still take a terrible toll in the lowest income areas, diseases such as depression, diabetes, heart attacks, and stroke that once occurred mainly in rich countries are rapidly becoming increasingly common everywhere.

Look at the changes in the relative ranking of major causes of disease burden in table 8.1 Notice how many once-common diseases have been reduced in many cases, while chronic disorders related to lifestyle, age, and affluence are becoming more common. Much of this change in disease burden is occurring in the poorer parts of the world, where people are rapidly adopting the habits and diet of the richer countries. The largest source of unintentional injuries are traffic accidents, while the main sources of intentional injuries are self-harm and interpersonal violence.

Table 8.1 Leading Causes of Global Disease Burden		
	1990	**2012**
Lower respiratory infections	1	4
Diarrheal diseases	2	6
Neonatal conditions	3	1
Ischemic heart disease	4	3
Stroke	5	5
COPD (Chronic Obstructive Pulmonary Disease)	6	8
Malaria	7	13
Tuberculosis	8	15
Protein-energy malnutrition	9	19
Neonatal encephalopathy	10	29
Back and neck pain	11	13
Unintentional injuries	12	2
Congenital anomalies	13	14
Iron anemia	14	15
Depression	15	10
Diabetes	21	12
HIV/AIDS	33	9
Intentional injuries	20	11

Source: Data from World Health Organization, 2016

Taking disability as well as death into account in our assessment of disease burden reveals the increasing role of mental health as a worldwide problem. WHO projections suggest that psychiatric and neurological conditions could increase their share of the global burden from the current 10 percent to 15 percent of the total load by 2020. Again, this isn't just a problem of the developed world. Depression is expected to be the second largest cause of all years lived with disability worldwide, as well as the cause of 1.4 percent of all deaths. For women in both developing and developed regions, depression is the leading cause of disease burden, while suicide, which often is the result of untreated depression, is the fourth largest cause of female deaths.

Pollution contributes to many of the health problems in table 8.1, and has become a leading cause of death and disability in many developing countries. The World Health Organization estimates that smog and toxic agents in outdoor air contribute to at least 3.7 million deaths per year from cardiovascular problems and cancer, particularly in urban areas. China, alone, may suffer 40 percent of those deaths.

Another 4.2 million people died from particulates exposure in indoor air from smoky stoves and cooking fires. Furthermore, about 1 million died from contaminants in soil and water. And nearly a million died from diseases linked to poor sanitation. At least 1.1 billion people smoke today, and this number is expected to increase by at least 50 percent, especially in developing countries. If current patterns persist, about 500 million people alive today will eventually be killed by tobacco. Dr. Gro Harlem Brundtland, former director-general of the WHO, observed that reducing tobacco use and air pollution could save billions of lives.

Infectious and emergent diseases still kill millions of people

Although the ills of modern life are becoming more important almost everywhere in the world, communicable diseases still kill millions of people, especially children (fig. 8.4). Humans are afflicted by a wide variety of pathogens (disease-causing organisms), including viruses, bacteria, protozoans (single-celled animals), parasitic worms, and flukes. Pandemics (worldwide epidemics) have changed the course of history. In the mid-fourteenth century, the "Black Death" (bubonic plague) swept out of Asia and may have killed half the people in Europe. When European explorers and colonists reached the Americas in the late fifteenth and early sixteenth centuries, they brought with them diseases, such as smallpox, measles, cholera, and yellow fever, that killed up to 90 percent of the native population in many areas. The largest loss of life in a pandemic in the past century was in the great influenza pandemic of 1918. Epidemiologists now estimate that at least one-third of all humans living at the time were infected, and that 50 to 100 million died. Businesses, schools, churches, and sports and entertainment events were shut down for months.

We haven't had a pandemic as deadly since the 1918 flu, but epidemiologists warn that new contagious diseases test our defenses every day. An example is the 2014 Ebola outbreak in West Africa, which shows how rapidly infectious diseases can spread, as well as the dangers of lax oversight. Ebola hemorrhagic

FIGURE 8.4 At least six million children die every year from easily preventable diseases. This billboard in Guatemala encourages parents to have their children vaccinated against polio, diphtheria, TB, tetanus, pertussis (whooping cough), and scarlet fever.
© William P. Cunningham

fever is caused by a filovirus. It is highly fatal; in some human outbreaks, 90 percent of patients died. The 2014 epidemic started in rural Guinea but quickly spread into neighboring Sierra Leon and Liberia. It wasn't until the contagion reached the capitols of Freetown and Monrovia, with populations of millions, that officials recognized the threat to the whole world if the disease continued to spiral out of control. Health authorities warned that there could be millions of deaths if the spread wasn't stopped.

It's estimated that Ebola infected 20,000 people in West Africa, and that at least 8,000 died (fig. 8.5). Both infected patients and health workers carried the disease to Europe and the United States. A massive investment from wealthy countries, coupled with improved local infrastructure brought the outbreak mostly under control by 2015, although sporadic cases continue to occur, and health officials worry about additional outbreaks in areas with poor health care infrastructure.

A more widespread virus outbreak came to world attention in 2015. The Zika virus, which was first discovered in 1947 in Uganda, was long thought to cause only a mild fever and rash that dissipate in a few days. Most people who contract the disease never know they had it. The virus, which is transmitted by several mosquito species of the genus *Aedes,* moved slowly across tropical Africa and Asia. In 2013, however, the disease jumped to French Polynesia and infected two-thirds of the population in just seven months. It seemed to be both more virulent and more dangerous, causing 73 cases of Guillain–Barré syndrome, a life threatening paralysis. In 2015, the virus reached Brazil, and a new, frightening symptom was reported. Mothers who had Zika fever while pregnant were having babies with microcephaly, abnormally small

FIGURE 8.5 Ebola workers need complete protection from bodily contact, which is difficult in a hot, humid climate. Protective gear also alienates locals, who tend to be suspicious of strangers and authority.
© Kenzo Tribouillard/AFP/Getty Images

heads and damaged brains. The virus had acquired the ability to cross the blood-brain barrier and infect the central nervous system.

The frequency of microcephaly when mothers have the virus during their first trimester is thought to be between 1 and 13 percent. In 2016 the World Health Organization (WHO) declared Zika fever a Public Health Emergency of International Concern, a rarely used designation. By then, the disease was present in 70 countries, including most of the tropical areas of the Americas and the Caribbean. Hundreds of U.S. citizens who traveled to tropical areas returned with Zika fever. Mosquito vectors for the virus already existed across much of the southern U.S., increasing the likelihood that the disease would

spread rapidly across a broad swath of the country. There's evidence that the disease can be spread sexually between humans as well.

The most common carrier for the virus is the mosquito *Aedes egypti*, which also carries related viruses that cause dengue and chikungunya fever. These mosquitoes, which can breed in water containers as small as a bottle cap, are highly adapted to cities, especially where slum conditions exist and where there's much litter and standing water. Since most mosquitoes travel only a few hundred meters in their lifetime, living in these conditions is an important risk factor. If you do travel to places where *Aedes* are prevalent, the best protection is to cover up and use repellent. Experiments are underway to genetically modify mosquitoes so they can't transmit the disease or to cause certain species to simply die out. Some people, however, worry about the unintended consequences of tampering with nature on a global scale.

The U.S. Centers for Disease Control and Prevention warn women who are pregnant, or might become pregnant, to avoid travel to countries where Zika is prevalent. Some governments, such as Brazil, Colombia, El Salvador, and Jamaica, advise women to simply avoid getting pregnant, but that's a complicated recommendation.

According to the WHO, 2 billion people are currently at risk worldwide and 5 million babies are born annually in parts of the Americas with the Zika virus. In 2016, the Obama Administration requested $1.9 billion in emergency funding to fight this epidemic, but Congress failed to act, a big disappointment for states facing this growing menace.

Emergent diseases are those not previously known or that have been absent for at least 20 years. Recent outbreaks of the Ebola and Zika virus are good examples. There have been hundreds of outbreaks of other emergent diseases over the past two decades (fig. 8.6), including cholera, which had been absent from South America for more

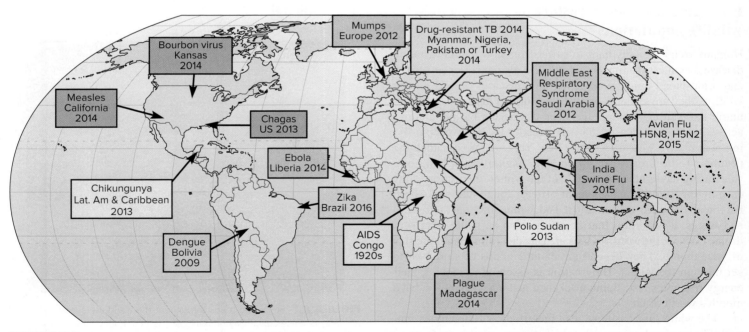

FIGURE 8.6 Start dates of some recent outbreaks of highly lethal infectious diseases. Why are super-contagious organisms emerging in so many different places?
Source: Data from U.S. Centers for Disease Control and Prevention

than a century, but reemerged in Peru in 1992. Some other examples include a new drug-resistant form of tuberculosis, now spreading in Russia; dengue fever, which is spreading through Southeast Asia and the Caribbean; malaria, which is reappearing in places where it had been nearly eradicated; and a new human lymphotropic virus (HTLV), which is thought to have jumped from monkeys into people in Cameroon who handled or ate bushmeat. These HTLV strains are now thought to infect 25 million people.

The largest recent death toll from an emergent disease is for HIV/AIDS. Although virtually unknown 35 years ago, acquired immune-deficiency syndrome is estimated to have killed at least 34 million people so far. The WHO estimates that 37 million people are now infected with HIV, but that half of them are unaware of their condition. Because of improved treatment, the number of deaths is declining, to 1.5 million or fewer per year. Although two-thirds of all current HIV infections are now in sub-Saharan Africa, the disease is spreading rapidly in South and East Asia. However, a greater focus on this disease has brought progress. New infection rates have fallen by 50 percent or more in 25 countries—13 of them in sub-Saharan Africa. Half of all the reductions in HIV infections in the past two years have been among children.

Still, the costs of this disease are terrible. As chapter 7 points out, without AIDS the life expectancy in Swaziland would be about 65 years. With AIDS, the average life expectancy is now only about 33 years. Worldwide more than 14 million children—the equivalent of all children under age five in America—have lost one or both parents to AIDS. The economic costs of treating patients and lost productivity from premature deaths resulting from this disease are estimated to be at least $35 billion per year, or about one-tenth of the total GDP of sub-Saharan Africa.

Emerging diseases devastate wildlife populations

Humans aren't the only ones to suffer from new and devastating diseases. Domestic animals and wildlife also experience sudden and widespread epidemics. When new diseases spread among wildlife, they are sometimes called **ecological diseases.** A particularly devastating example has been destroying bat populations across North America in the past decade. In 2006, people living near a cave west of Albany, New York, reported something peculiar: Little brown bats (*Myotis lucifugus*) were flying outside during daylight in the middle of the winter. Inspection of the cave by the wildlife management officers found numerous dead bats near the cave mouth. Most had white fuzz on their faces and wings, a condition that is now known as white-nose syndrome (WNS). Little brown bats are tiny creatures, about the size of your thumb. They depend on about 2 grams of stored fat to get through the winter. Hibernation is essential to making their energy resources last. Being awakened just once can cost a bat a month's worth of fat.

The white fuzz has now been identified as filamentous fungus (*Geomyces destructans*), which thrives in the cool, moist conditions where bats hibernate. We don't know where the fungus came from, but, true to its name, the pathogen is highly destructive. Occurring in only a few states a decade ago, the disease has now been found in almost every state and province in eastern North America. Biologists estimate that nearly 6 million bats already have died from this disease. It isn't known how the pathogen spreads. Perhaps it moves from animal to animal through physical contact. It's also possible that humans introduce fungal spores on their shoes and clothing when they go from one cave to another.

In some areas, bat populations have fallen by 95 percent or more. So far, seven species of bat are known to be susceptible to this plague, and it has been detected in five additional species. Some researchers fear that bats could be extinct in 20 years in the eastern United States. Losing these important species, which consume insects and pollinate plants, would have devastating ecological consequences.

An even more widespread and lethal epidemic is currently sweeping through amphibians worldwide. A disease called Chytrid fungus (or Chytridiomycosis) is causing dramatic losses or even extinctions of frogs and toads throughout the world (fig. 8.7). A fungus called *Batrachochytrium dendrobatidis* causes the disease. It was first recognized in 1993 in dead and dying frogs in Queensland, Australia, and now seems to be spreading rapidly, perhaps because the fungus has become more virulent or amphibians are more susceptible due to environmental change. Most of the world's approximately 6,000 amphibian species appear to be susceptible to the disease, and around 2,000 species have declined or become extinct in their native habitats as part of this global epidemic. African clawed frogs (*Xenopus* sp.), which are resistant to fungal infections and thus may be carriers of the disease, are a possible vector for this disease. Widespread use of these frogs for research and pregnancy testing may have contributed to the rapid spread of the pathogen.

FIGURE 8.7 Frogs and toads throughout the world are succumbing to a deadly disease called Chytridiomycosis. Is this a newly virulent fungal disease, or are amphibians more susceptible because of other environmental stresses?
© Rosalie Kreulen/Shutterstock.com

Temperatures above 28°C (82°F) kill the fungus, and treating frogs with warm water can cure the disease in some species. Topical application of the drug chloramphenicol also has successfully cured some frogs. And certain skin bacteria seem to confer immunity to fungal infections. In some places, refuges have been established in which frogs can be maintained under antiseptic conditions until a cure is found. It's hoped that survivors can eventually be reintroduced back to their native habitat and species will be preserved.

There have been many other reports of new diseases emerging in wildlife. A new form of leukemia, for example, is ripping through softshell clams along the U.S. Atlantic Coast. Tuberculosis is killing African elephants; a measles-like disease is killing Pacific dolphins; and Ebola (mentioned earlier) is thought to have killed about one-third of all the gorillas in Africa.

Surprisingly, research is showing that land-based parasites are killing marine mammals. The most studied of these pathogens is *Toxoplasma gondii,* a single-celled protozoan that infects cats and other mammals. About one-quarter of the U.S. human population carries this parasite with few ill effects, although pregnant women are warned against cleaning cat boxes because Toxoplasmosis can cause birth defects. But around the world, sea life from California sea otters, to Mediterranean dolphins, to endangered Hawaiian monk seals are dying of *T. gondii* infections. Cats (but not humans) shed *T. gondii* oosysts in their feces, which wash down rivers and into the ocean.

One thing that emergent diseases in humans and ecological diseases in natural communities have in common is environmental change that stresses biological systems and upsets normal ecological relationships. We cut down forests and drain wetlands, destroying habitat for native species. Invasive organisms and diseases are accidentally or intentionally introduced into new areas where they can grow explosively. Increasing incursion into former wilderness is spurred by human population growth and ecotourism. In 1950, only about 3 million people a year flew on commercial jets; by 2000, more than 300 million did. Diseases can spread around the globe in mere days as people pass through international travel hubs.

We are coming to recognize that the delicate ecological balances that we value so highly—and disrupt so frequently—are important to our own health. **Conservation medicine** is an emerging discipline that attempts to understand how our environmental changes threaten our own health, as well as that of the natural communities on which we depend for ecological services. Although it is still small, this new field is gaining recognition from mainstream funding sources such as the World Bank, the World Health Organization, and the U.S. National Institutes of Health.

Resistance to drugs, antibiotics, and pesticides is increasing

Malaria, the most deadly of all insect-borne illnesses, is an example of the return of a disease that once was thought to be nearly vanquished. With the advent of modern medicines and pesticides, including DDT that kills the mosquito hosts of malaria, the disease was nearly wiped out in many places in the mid-twentieth century.

But then it came roaring back as the protozoan parasite that causes the disease became resistant to most drugs and the mosquitoes that transmit it developed resistance to insecticides. Anti-malarial campaigns in India and Sri Lanka, for instance, reduced new infections from millions per year to only a few thousand in the 1950s and 1960s, largely by relying on DDT and the inexpensive anti-malarial drug Chloroquine. By the 1980s, however, malaria incidence was back nearly to pre-DDT levels.

The reason for the resurgence of malaria was overuse of a few drugs and pesticides. Massive exposure to DDT killed off all but a few tolerant mosquito strains, and against these mosquitoes DDT had no effect. Similarly, the drug Chloroquine eliminated all but a few resistant malaria protozoans, which then multiplied freely and abundantly. Current efforts, rather than attempting to completely eliminate this disease, are to control its spread by eliminating wetlands, insecticide-treated bed nets, and selective indoor spraying.

Overexposure now threatens one of our most important health care tools, antibiotics. In recent years, health workers have become increasingly alarmed about the rapid spread of methicillin-resistant *Staphylococcus aureus* (MRSA). Staphylococcus (or Staph) is very common. Most people have at least some of these bacteria. They are a common cause of sore throats and skin infections, but are usually easily controlled. This new strain is resistant to penicillin and related antibiotics (such as methicillin) and can cause deadly infections, especially in people with weak immune systems.

MRSA occurs most frequently in hospitals, nursing homes, correctional facilities, and other places where high levels of antibiotics are used and people are in close contact. It's generally spread through direct skin contact. School locker rooms, gymnasiums, and contact sports also are sources of infections. Several U.S. states have closed schools as a result of MRSA contamination. It's estimated that in the past decade at least 100,000 MRSA infections in the United States resulted in about 19,000 deaths. A much worse situation is reported in China, where about half of the 5 million annual Staph infections are thought to be methicillin-resistant.

How does this resistance develop? Part of the answer is natural selection and the ability of many organisms to evolve rapidly. Another factor is the human tendency to use control measures carelessly. When we discovered that DDT and other insecticides could control mosquito populations, we spread them everywhere. This not only harmed wildlife and beneficial insects, but it created selective pressures that lead to evolution. Many pests and pathogens are exposed to low, chronic doses, allowing those with natural resistance to survive and spread their genes through the population (fig. 8.8). Because of natural selection, surviving microorganisms and their vectors, which are insensitive to almost all our weapons against them, thrive and multiply.

As discussed in chapter 9, one of the most important causes of antibiotic resistance in pathogens is now factory farming. Raising huge numbers of cattle, hogs, and poultry in densely packed barns and feedlots helps spread diseases. Confined animals are dosed constantly with antibiotics and steroid hormones to keep them disease-free and to make them gain weight faster. More than

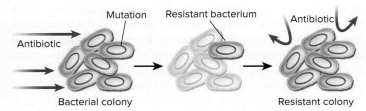

(a) Mutation and selection create drug-resistant strains

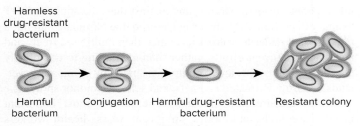

(b) Conjugation transfers drug resistance from one strain to another

FIGURE 8.8 How microbes acquire antibiotic resistance. (a) Random mutations make a few cells resistant. When challenged by antibiotics, only those cells survive to give rise to a resistant colony. (b) Sexual reproduction (conjugation) or plasmid transfer moves genes from one strain or species to another.

half of all antibiotics used in the United States each year is fed to livestock. A significant amount of these antibiotics and hormones are excreted in urine and feces, which are spread, untreated, on the land or discharged into surface water where they contribute further to the evolution of supervirulent pathogens.

At least half of the 100 million antibiotic doses prescribed for humans every year in the United States are unnecessary or are the wrong ones. Furthermore, many people who start a course of antibiotic treatment fail to carry it out for the time prescribed. For your own health and that of the people around you, if you are taking an antibiotic, follow your doctor's orders and don't stop taking the medicine as soon as you start feeling better.

What would better health cost?

The heaviest burden of illness is borne by the poorest people, who can afford neither a healthy environment nor adequate health care. Women in sub-Saharan Africa, for example, suffer six times the disease burden per 1,000 persons as women in most European countries. The WHO estimates that 90 percent of all disease burden occurs in developing countries, where less than one-tenth of all health-care dollars are spent. The group Médecins Sans Frontiéres (MSF, or Doctors Without Borders) calls this the 10/90 gap. While wealthy nations spend billions to treat cosmetic problems in humans or the ailments of their pets, billions of people are sick or dying from treatable infections and parasitic diseases to which little attention is paid. To counter this trend, benefactors, such as the Bill and Melinda Gates Foundation, have pledged $200 million for medical aid to developing countries to help fight AIDS, TB, and malaria.

Dr. Jeffrey Sachs of the Columbia University Earth Institute says that disease is as much a cause as a consequence of poverty and political unrest, yet the world's richest countries now spend just $1 per person per year on global health. He predicts that raising our commitment to about $25 billion annually (about 0.1 percent of the annual GDP of the 20 richest countries) not only would save about 8 million lives each year, but would boost the world economy by billions of dollars.

Section Review

1. What is PFOA and how might you be exposed to it?
2. What are emergent diseases? Give a few examples, and describe their cause and effects.
3. How does antibiotic resistance arise? How do hospitals and feedlots contribute to this phenomenon?

8.2 TOXICOLOGY

- *Many toxic substances are harmful in extremely dilute concentrations. In some cases, billionths, or even trillionths, of a gram can cause irreversible damage.*
- *Many toxic substances, allergens, carcinogens and other dangerous substances can be found in our homes in a wide variety of products.*
- *Chemicals that disrupt normal hormone functions can be especially hazardous, because they can interfere with the growth, development, and physiology of animals—including humans'—at very low doses.*

Toxicology is the study of **toxic substances** (poisons) and their effects, particularly on living organisms. Because many substances are known to be poisonous (whether plant, animal, or microbial), toxicology is a broad field, drawing from biochemistry, histology, pharmacology, pathology, and many other disciplines. Toxic agents damage or kill living organisms because they react with cellular components to disrupt metabolic functions. Because of this reactivity, many are harmful even in extremely diluted concentrations. In some cases, billionths, or even trillionths, of a gram can cause irreversible damage. Toxic substances produced naturally are generally known as toxins. Those made by human activities are generally called toxicants.

All toxic substances are hazardous, but not all hazardous materials are toxic. Some substances, for example, are dangerous because they're flammable, explosive, acidic, caustic, irritants, or sensitizers. Many of these materials must be handled carefully in large doses or high concentrations, but can be rendered relatively innocuous by dilution, neutralization, or other physical treatments. They don't react with cellular components in ways that make them poisonous at low concentrations.

Environmental toxicology, or ecotoxicology, focuses on the effects of toxic substances on living organisms, and on the way these substances interact and are transformed as they move through populations and ecosystems. In aquatic systems, special attention is devoted to the ways pollutants act at the interface of sediment and water, or of water and organisms, or at the water/air interface. In terrestrial environments, the emphasis tends to be on the effects of metals on the soil fauna community and population characteristics.

The U.S. Environmental Protection Agency is responsible for monitoring 275 substances regulated by the Comprehensive Environmental Response, Compensation, and Liability Act (CERCLA), commonly known as the Superfund Act. In 2011, there were 1,280 Superfund sites on the National Priorities List in the United States, and 62 additional sites have been proposed for cleanup. More than 11 million people live within 1 mile (1.2 km) of these sites.

How do toxic substances affect us?

Allergens are substances that activate the immune system. Some allergens act directly as **antigens;** that is, they are recognized as foreign by white blood cells and stimulate the production of specific antibodies (proteins that recognize and bind to foreign cells or chemicals). Other allergens act indirectly by binding to and changing the chemistry of foreign materials so they become antigenic and cause an immune response.

Formaldehyde is a good example of a widely used chemical that is a powerful sensitizer of the immune system. It is directly allergenic and can also trigger reactions to other substances. Widely used in plastics, wood products, insulation, glue, and fabrics, formaldehyde concentrations in indoor air can be thousands of times higher than in normal outdoor air. Some people suffer from what is called **sick building syndrome:** headaches, allergies, chronic fatigue, and other symptoms caused by poorly vented indoor air contaminated by mold spores, carbon monoxide, nitrogen oxides, formaldehyde, and other toxic substances released from carpets, insulation, plastics, building materials, and other sources (fig. 8.9). The Environmental Protection Agency estimates that poor indoor air quality may cost the United States $60 billion a year in absenteeism and reduced productivity.

Immune system depressants are pollutants that suppress the immune system rather than activate it. Little is known

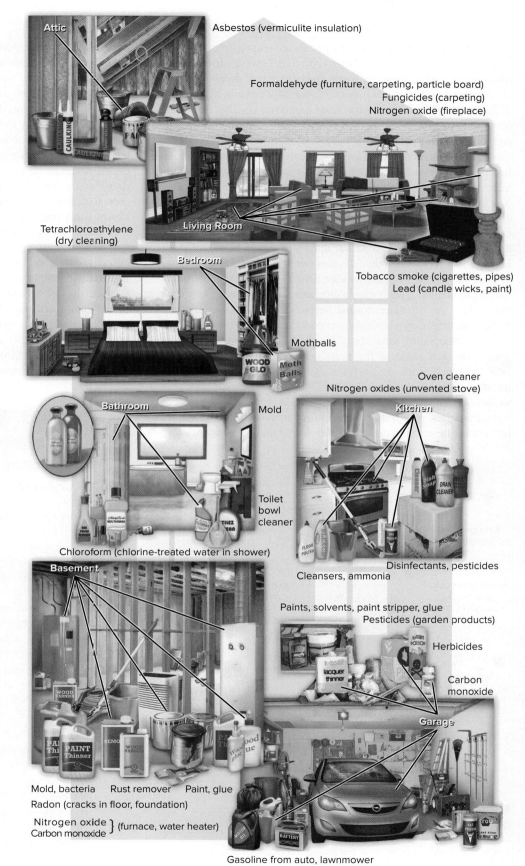

FIGURE 8.9 Some sources of toxic and hazardous substances in a typical home.

about how this occurs or which chemicals are responsible. Immune system failure is thought to have played a role, however, in the widespread deaths of seals in the North Atlantic and of dolphins in the Mediterranean. These dead animals generally contain high levels of pesticide residues, polychlorinated biphenyls (PCBs), and other contaminants that are suspected of damaging the immune system and making it susceptible to a variety of opportunistic infections.

Endocrine disrupters are chemicals that disrupt hormone functions. Hormones are chemicals that living organisms use to regulate the development and function of tissues and organs (fig. 8.10). You undoubtedly have heard about sex hormones and their powerful effects on how we look, behave, and develop, but these are only one example of the many regulatory hormones that rule our lives. Insulin is a hormone that controls our uptake of sugars from food, and adrenaline is a hormone that elevates our breathing and circulation in moments of stress. The hormone thyroxine helps regulate many functions, including heart rate, temperature, and growth. Chemicals that disrupt any of these normal hormone functions can severely damage an organism's health or reproduction.

As the opening case study for this chapter shows, we now know that some of the most insidious effects of persistent chemicals such as perfluorocarbons (PFCs) are that they interfere with normal growth, development, and physiology of a variety of animals—including humans—at very low doses. For some materials, picogram concentrations (trillionths of a gram per liter) may be enough to cause developmental abnormalities in sensitive organisms. Although the official EPA limit for PFOA is 0.4 ppb, Philippe Grandjean of the Harvard School of Public Health suggests that a "safe" level should be only 0.001 ppb or 1 part per trillion. In 2015, more than 200 scientists from 38 countries signed the Madrid Statement, which calls for the regulation of all PFCs due to their tendency to disrupt endocrine hormone functions.

Because these chemicals often cause sexual dysfunction (reproductive health problems in females or the feminization of males, for example), these chemicals are sometimes called environmental estrogens or androgens. They are just as likely, however, to disrupt other important regulatory functions as they are to obstruct sex hormones. Some researchers suggest that endocrine disrupters may be contributing to our obesity crisis.

Neurotoxins are a special class of metabolic poisons that specifically attack nerve cells (neurons). The nervous system is so important in regulating body activities that disruption of its activities is especially fast-acting and devastating. Different types of neurotoxins act in different ways. Heavy metals such as lead and mercury kill nerve cells and cause permanent neurological damage. Anesthetics (ether, chloroform, halothane, etc.) and chlorinated hydrocarbons (DDT, Dieldrin, Aldrin) disrupt nerve cell membranes necessary for nerve action. Organophosphates (Malathion, Parathion) and

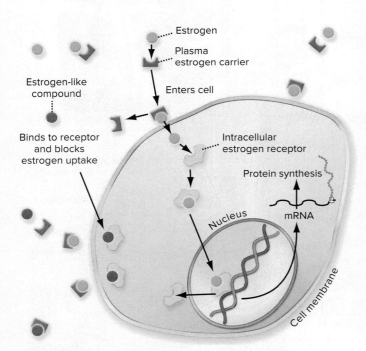

FIGURE 8.10 Steroid hormone action. Plasma hormone carriers deliver regulatory molecules to the cell surface, where they cross the cell membrane. Intracellular carriers deliver hormones to the nucleus, where they bind to and regulate the expression of DNA. Estrogen-like compounds bind to receptors and either block the uptake of endogenous hormones or act as a substitute hormone to disrupt gene expression.

What Can You Do?

Tips for Staying Healthy

- Eat a balanced diet with plenty of fresh fruits, vegetables, legumes, and whole grains. Wash fruits and vegetables carefully; they may well have come from a country where pesticide and sanitation laws are lax.

- Use unsaturated oils such as olive or canola oil rather than hydrogenated or semisolid fats such as margarine.

- Cook meats and other foods at temperatures high enough to kill pathogens; clean utensils and cutting surfaces; store food properly.

- Wash your hands frequently. You transfer more germs from hand to mouth than by any other means of transmission.

- When you have a cold or flu, don't demand antibiotics from your doctor—they aren't effective against viruses.

- If you're taking antibiotics, continue for the entire time prescribed—quitting as soon as you feel well is an ideal way to select for antibiotic-resistant germs.

- Practice safe sex.

- Don't smoke, and avoid smoky places.

- If you drink, do so in moderation. Never drive when your reflexes or judgment are impaired.

- Exercise regularly: walk, swim, jog, dance, garden. Do something you enjoy that burns calories and maintains flexibility.

- Get enough sleep. Practice meditation, prayer, or some other form of stress reduction.

- Make a list of friends and family who make you feel more alive and happy. Spend time with one of them at least once a week.

carbamates (carbaryl, zeneb, maneb) inhibit acetylcholinesterase, an enzyme that regulates signal transmission between nerve cells and the tissues or organs they innervate (for example, muscle). Most neurotoxins are both extremely toxic and fast-acting.

Mutagens are agents that damage or alter genetic material (DNA) in cells. Chemicals and radiation can both act as mutagens. Damage to DNA can lead to birth defects if it occurs during embryonic or fetal growth. Later in life, genetic damage may trigger neoplastic (tumor) growth. When damage occurs in reproductive cells, the results can be passed on to future generations. Cells have repair mechanisms to detect and restore damaged genetic material, but some changes may be hidden, and the repair process itself can be flawed. It is generally accepted that there is no "safe" threshold for exposure to mutagens. Any exposure has some possibility of causing damage.

Teratogens cause birth defects. These are chemicals or other factors that cause abnormalities during embryonic growth and development. Some compounds that are not otherwise harmful can cause tragic problems in these sensitive stages of life. Perhaps the most prevalent teratogen in the world is alcohol. Drinking during pregnancy can lead to **fetal alcohol syndrome**—a cluster of symptoms that includes craniofacial abnormalities, developmental delays, behavioral problems, and mental defects that last throughout a child's life. Even one alcoholic drink a day during pregnancy has been associated with decreased birth weight.

Similarly, by some estimates, 300,000 to 600,000 children born every year in the United States are exposed in the womb to unsafe levels of mercury. The effects are subtle, but include reduced intelligence, attention deficit, and behavioral problems. The total cost of these effects is estimated to be $8.7 billion per year.

Carcinogens are substances that cause **cancer,** invasive, out-of-control cell growth that results in malignant tumors. Cancer rates rose in most industrialized countries during the twentieth century, and cancer is now the second leading cause of death in the United States, killing 589,000 people in 2015. According to the American Cancer Society, 1 in 2 males and 1 in 3 females in the United States will have some form of cancer in their lifetime. Some authors blame this cancer increase on toxic synthetic chemicals in our environment and diet. Others argue that it is attributable mainly to lifestyle (smoking, sunbathing, alcohol) or simply living longer. The U.S. EPA estimates that 200 million U.S. residents live in areas where the combined lifetime cancer risk from environmental carcinogens exceeds 1 in 100,000, or ten times the risk normally considered acceptable.

How does diet influence health?

Diet also has an important effect on health. For instance, there is a strong correlation between cardiovascular disease and the amount of salt and animal fat in one's diet. There's controversy about the role of high-fructose corn syrup, refined sugar, red meat, and other components of our food.

Fruits, vegetables, whole grains, complex carbohydrates, and dietary fiber (plant cell walls) often have beneficial health effects. Certain dietary components seem to have anticancer effects—these components include pectins; vitamins A, C, and E;

substances produced in cruciferous vegetables (cabbage, broccoli, cauliflower, Brussels sprouts); and selenium, which we get from plants. Many of these substances act on the epigenetic system, which we'll discuss later in this chapter.

Eating too much food is a significant dietary health factor in developed countries and among the well-to-do everywhere. Sixty percent of all U.S. adults are now considered overweight, and the worldwide total of obese or overweight people is estimated to be over 1 billion. Every year in the United States, 300,000 deaths are linked to obesity.

The U.S. Centers for Disease Control and Prevention in Atlanta warn that one in three U.S. children will become diabetic unless many more people start eating less and exercising more. The odds are worse for Black and Hispanic children: Nearly half of them are likely to develop the disease. And among the Pima tribe of Arizona, nearly 80 percent of all adults are diabetic. More information about food and its health effects is available in chapter 9.

Section Review

1. What are endocrine disrupters and why are they dangerous?
2. What are teratogens and mutagens?
3. How does diet influence health?

8.3 THE MOVEMENT, DISTRIBUTION, AND FATE OF TOXIC SUBSTANCES

- *Solubility and mobility determine where and when chemicals move through the environment and in our bodies.*
- *Persistent materials, such as heavy metals and some organic compounds, can accumulate and concentrate in food webs so they reach toxic concentrations in top predators.*
- *Synergistic interactions can increase the effects of combined toxic substances far beyond what any single substance might cause alone.*

There are many sources of toxic and hazardous chemicals in the environment and many factors related to each chemical itself—its route or method of exposure, its persistence in the environment, and the characteristics of the target organism (table 8.2)—that determine the dangerousness of the chemical. We can think of both individuals and an ecosystem as sets of interacting compartments between which chemicals move, based on molecular size, solubility, stability, and reactivity (fig. 8.11). The dose (amount), route of entry, timing of exposure, and sensitivity of the organism all play important roles in determining toxicity. In this section, we will consider some of these characteristics and how they affect environmental health.

Compounds dissolve either in water or in fat

Solubility is one of the most important characteristics in determining how, where, and when a toxic material will move through the environment or through the body. Chemicals can be divided into two major groups: those that dissolve more readily in water and

Table 8.2 Factors in Environmental Toxicity

Factors Related to the Toxic Agent

1. Chemical composition and reactivity

2. Physical characteristics (such as solubility, state)

3. Presence of impurities or contaminants

4. Stability and storage characteristics of toxic agent

5. Availability of vehicle (such as solvent) to carry agent

6. Movement of agent through environment and into cells

Factors Related to Exposure

1. Dose (concentration and volume of exposure)

2. Route, rate, and site of exposure

3. Duration and frequency of exposure

4. Time of exposure (time of day, season, year)

Factors Related to Organism

1. Resistance to uptake, storage, or cell permeability of agent

2. Ability to metabolize, inactivate, sequester, or eliminate agent

3. Tendency to activate or alter nontoxic substances so they become toxic

4. Concurrent infections or physical or chemical stress

5. Species and genetic characteristics of organism

6. Nutritional status of subject

7. Age, sex, body weight, immunological status, and maturity

Source: U.S. Department of Health and Human Services, 1995

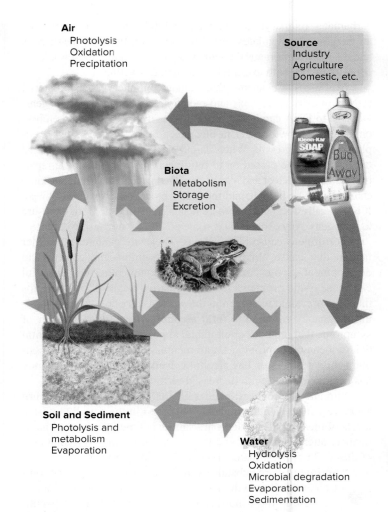

Air
Photolysis
Oxidation
Precipitation

Source
Industry
Agriculture
Domestic, etc.

Biota
Metabolism
Storage
Excretion

Soil and Sediment
Photolysis and metabolism
Evaporation

Water
Hydrolysis
Oxidation
Microbial degradation
Evaporation
Sedimentation

FIGURE 8.11 Living organisms are exposed to chemical compounds in water, soils, and the atmosphere. Commercial products move freely in the environment, increasing the likelihood of exposure.

those that dissolve more readily in oils or fats. Water-soluble compounds move rapidly and widely through the environment because water is everywhere around us. Water-soluble substances also tend to have ready access to most cells in the body, because aqueous solutions bathe all our cells. Molecules that are oil- or fat-soluble (usually these are organic, carbon-based molecules) generally need a carrier to move through the environment and into, and within, the body. Once inside the body, however, oil-soluble compounds penetrate readily into tissues and cells, because the membranes that enclose cells are themselves made of similar oil-soluble chemicals. Once they get inside cells, oil-soluble materials are likely to be accumulated and stored in lipid deposits, where they may be protected from metabolic breakdown and persist for many years.

How do these dangerous substances enter our bodies? There are many routes, although some cause damage more frequently than others (fig. 8.12). Airborne toxic particles generally cause more ill health than any other exposure source. We breathe far more air every day than the volume of food we eat or water we drink. Furthermore, the cellular lining of our lungs is designed to exchange gases very efficiently, but this also means foreign substances move easily from our lungs into our blood stream. Epidemiologists estimate that millions of people die each year from diseases caused or exacerbated by air pollution (see chapter 16).

Food, water, and skin contact also can expose us to a wide variety of toxic substances. The greatest exposures to many of these toxicants are found in industrial settings (including industrial agriculture), where workers may encounter doses thousands of times higher than would be found anywhere else. In the opening case study for this chapter, workers in factories that synthesize perfluorocarbon products had much higher exposure to these chemicals and a greater incidence of cancers and other diseases than the general public. The European Agency for Safety and Health at Work warns that 32 million people (20 percent of all employees) in the European Union are exposed to unacceptable levels of carcinogens and other toxic substances in their workplace.

Condition of the organism and timing of exposure also have strong influences on toxicity. Healthy adults, for example, may be relatively insensitive to doses that would be very dangerous for young children or for someone already weakened by disease. Similarly, exposure to a toxic substance may be very dangerous at certain stages of developmental or metabolic cycles, but may be innocuous at other times. A single dose of the notorious teratogen thalidomide, for example, taken in the third week of pregnancy (a time when many women

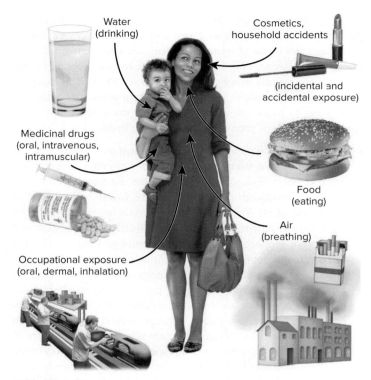

FIGURE 8.12 Routes of exposure to toxic and hazardous environmental factors.

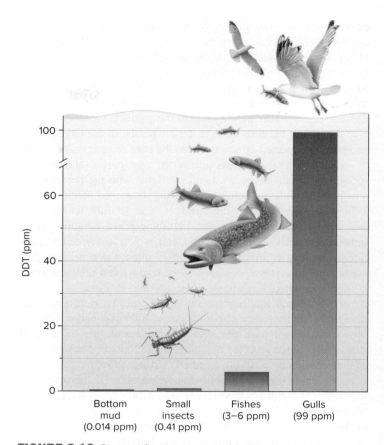

FIGURE 8.13 Biomagnification in a Lake Michigan food chain. The DDT tissue concentration in gulls, a tertiary consumer, was about 240 times that in the small insects sharing the same environment.

aren't aware they're pregnant) can cause severe abnormalities in fetal limb development. A complication in measuring toxicity is that great differences in sensitivity exist between species. Thalidomide was tested on a number of laboratory animals without any deleterious effects. Unfortunately, however, it is a powerful teratogen in humans.

Bioaccumulation and biomagnification increase concentrations of chemicals

Cells selectively absorb and store (accumulate) a great variety of molecules. This process of storing substances in living tissues is known as **bioaccumulation.** Bioaccumulation allows cells to selectively take up and store nutrients and essential minerals. At the same time, though, cells can absorb and store harmful substances. Toxicants that are dilute in the environment, such as pesticides or metals, can reach dangerous levels inside cells and tissues through bioaccumulation.

This effect is magnified through food webs. **Biomagnification** (elevated concentrations of harmful chemicals in the cells of organisms at higher trophic levels) occurs when a predator consumes prey containing pesticides, heavy metals, or other substances. All the toxic substances selectively stored by the cells of the prey are then accumulated in the predator's tissues. One of the first dramatic demonstrations of biomagnification involved the pesticide DDT in aquatic ecosystems (fig. 8.13). Phytoplankton and bacteria take up DDT from water or sediments. Their predators—zooplankton and invertebrates—collect and retain the DDT from prey organisms, building up higher concentrations as they feed, and so on up

the food chain. The cells and tissues of fish and fish-eating birds can contain DDT concentrations hundreds or thousands of times greater than concentrations in the tissues of bacteria and phytoplankton (see horizontal axis, fig. 8.13).

In the 1960s, bioaccumulation of DDT nearly wiped out a number of fish-eating bird species in North America. Inexpensive and poisonous to nearly all insect pests, DDT was sprayed on fields from airplanes, on beaches and wetlands, and used widely for household pest control starting in the late 1940s. Within 15 years it was found that extremely high concentrations of DDT in the tissues of peregrine falcons, brown pelicans, bald eagles, and other birds were interfering with the birds' ability to produce eggs and healthy chicks. Populations plummeted and disappeared from many areas. New controls on the use of DDT were largely responsible for rescuing these species from extinction.

Persistence makes some materials a greater threat

Some chemical compounds are very unstable and degrade (decompose to less harmful compounds) rapidly under most environmental conditions. When substances degrade rapidly, their concentrations and toxicity decline quickly after they are released in the environment. Most modern herbicides and pesticides, for instance, are designed to break down in a few days and lose their toxicity. Other substances are more persistent and last for years or even centuries in the environment without breaking down or losing toxicity. Metals, such as lead or mercury, PVC plastics, chlorinated hydrocarbon pesticides, and asbestos are valuable because they do not degrade rapidly. This stability makes the products valuable, but it is also a problem because these materials persist in the environment and have unexpected effects far from the sites of their original use.

Lead, for example, was one of the first persistent pollutants recognized by toxicologists. Lead interferes with the nervous system both in adults and children, causing decreased mental ability, kidney damage, anemia, miscarriages, and fertility losses. Lead was widely used in house paint, plumbing pipes, ceramic glazes, bullets, pesticides, and in tetraethyl lead, which was added to gasoline to improve combustion. Lead in gasoline was especially dangerous, because it was designed to be burned and released to the atmosphere, where it is widely distributed and easily taken into lung tissue. Developmental impacts on young children were especially disastrous, but they were also widespread and not concentrated near obvious sources. It took many years for epidemiologists to prove the dangers of leaded gasoline and other lead products. Manufacturers of lead-containing products disputed evidence of harm.

The epidemiologist Phillip Landrigan showed that even low lead levels in children's blood correlated with reduced IQ. Lead was banned from gasoline in 1975 and from paint in 1978. This was one of the most successful public health measures ever. By 2005, there was an 88 percent drop in American children's blood levels but the Centers for Disease Control estimate that 535,000 American children, mostly poor minorities living in large cities still have dangerous levels of lead in their bodies. In addition to links to cancer, genital deformities, reproductive problems, and diminished IQ, lead exposure can cause behavioral problems such as ADHD, aggression, and impulsivity. Recently, some epidemiologists have suggested that the decrease in violent crime over the past half-century tracks very closely with the decline of lead in the environment.

In 2016, there was a great concern about children's exposure to lead in Flint, Michigan. To save a few million dollars, the city had switched its source of drinking water from Lake Huron to the Flint River, a long-time dumping ground for industry that once operated in the city. The corrosive river water leached lead out of old plumbing. But Flint's children are hardly the only ones facing this danger. In Flint, 4.9 percent of the children tested have elevated lead levels. In New York State, it's 6.7 percent. In Pennsylvania, it's 8.5 percent. And in parts of Detroit, it's 20 percent. Baltimore, Maryland, Sebring, Ohio, and even Washington, DC have similar problems.

POPs are an especially serious problem

One of the most serious environmental health challenges in recent decades has been a group of **persistent organic pollutants (POPs)** that have become extremely widespread. These compounds, including DDT, PFOA, and related compounds, are dangerous because they are persistent (they don't degrade rapidly) and organic (carbon-based molecules readily taken in by cells, where they can cause significant damage). They have also been used abundantly and globally for decades. POPs are found now in the environment and in the tissues and blood of humans and animals everywhere from the tropics to the Arctic. They often accumulate in food webs and reach toxic concentrations in long-living top predators such as humans, sharks, raptors, swordfish, and bears. In recent years, global treaties have begun to phase out some of these, but many remain in widespread use. POPs of greatest current concern include the following:

- Perfluorooctanoic acid (PFOA, also known as C8) described in the opening case study for this chapter can cause cancer, birth defects, heart disease, and weaken the immune system. It's no longer produced in the U.S., but it's found in the blood of 99 percent of all Americans—even those who weren't yet born when it was used to make hundreds of nonstick, stain-resistant, and waterproof products. Cooking in Teflon-coated pans is thought to be a major source of these chemicals for many Americans. Heating some nonstick cooking pans above 500°F (260°C) can release enough PFOA to kill pet birds. Some alternatives to Teflon cookware are cast iron, stainless steel, and ceramic-coated pans. Grease-resistant coatings in food packaging, such as popcorn bags and pizza boxes, are also important sources of contamination. Exposure may be especially dangerous to women and girls, who may be 100 times more sensitive than men to these chemicals. PFOA can be passed from mother to unborn child in the womb, and can pass from mothers to breast-fed babies.

- Bisphenol A (BPA), a key ingredient of both polycarbonate plastics and epoxy resins, is one of the world's most widely used chemical compounds. In 2015, global production was about 5.4 million metric tons. BPA is used in items ranging from baby bottles, automotive accessories, eyeglasses, water pipes, the linings of cans and bottles, and dental sealants. It's estimated that 95 percent of American adults have measurable amounts of BPA in their bodies. The most likely source of contamination is from food and beverage containers. BPA can leach out, especially when plastic is heated, washed with harsh detergents, scratched, or exposed to acidic compounds. BPA has been linked to myriad health effects including mammary and prostate cancer, genital defects in males, early onset of puberty in females, obesity, and even behavior problems such as attention-deficit hyperactivity disorder. The U.S. Food and Drug administration has banned BPA in baby bottles and packaging for infant formula, but use in other food and water containers is still allowed.

- Phthalates (pronounced *thalates*) are found in cosmetics, deodorants, and many plastics (such as soft polyvinyl chloride, PVC) used for food packaging, children's toys, and medical devices. Some members of this chemical family are known to be toxic to laboratory animals, causing kidney and liver damage and possibly some cancers. In addition, many phthalates act as endocrine hormone disrupters and have been linked to reproductive abnormalities and decreased fertility in humans. A correlation has been found between phthalate levels in urine and low sperm numbers and decreased sperm motility in men. Nearly everyone in the United States has phthalates in their body at levels reported to cause these problems. While not yet conclusive, these results could help explain a 50-year decline in semen quality in most industrialized countries.

- Perchlorate is a waterborne contaminant left over from propellants and rocket fuels. About 12,000 sites in the United States were used by the military for live munition testing and are contaminated with perchlorate. Polluted water used to irrigate crops such as alfalfa and lettuce has introduced the chemical into the human food chain. Tests of cow's milk and human breast milk detected perchlorate in nearly every sample from throughout the United States. Perchlorate can interfere with iodine uptake in the thyroid gland, disrupting adult metabolism and childhood development.

- Atrazine is the second most widely used herbicide in America. More than 60 million pounds of this compound are applied per year, mainly on corn and cereal grains, but also on golf courses, sugarcane, and Christmas trees. It has long been known to disrupt endocrine hormone functions in mammals, resulting in spontaneous abortions, low birth weights, and neurological disorders. Studies of families in corn-producing areas in the American Midwest have found higher rates of developmental defects among infants and certain cancers in families with elevated atrazine levels in their drinking water. University of California professor Tyrone Hayes has shown that atrazine levels as low as 0.1 ppb (30 times less than the EPA maximum contaminant level) causes severe reproductive effects in amphibians, including abnormal gonadal development and hermaphroditism. Atrazine now is found in rain and surface waters nearly everywhere in the United States at levels that could cause abnormal development in frogs. In 2003, the European Union withdrew regulatory approval for this herbicide, and several countries banned its use altogether. Some toxicologists have suggested a similar rule in the United States.

Every one of us has dozens, if not hundreds, of persistent toxicants in our body. This accumulation is called our **body burden.** We acquire it from our air, water, diet, and surroundings. Many of these substances are present in parts per billion, or even parts per trillion. We don't know how dangerous this persistent burden is, but its presence is a matter of concern. If we're anything like the frogs that Tyrone Hayes studies, this accumulated dose of poisons may be a serious problem. Further discussion of POPs can be found in chapter 10.

Synergistic interactions can increase toxicity

Some materials interfere with each other, reducing their effects or breaking them down: For instance, vitamins E and A seem to reduce your body's response to some carcinogens. Other combinations of compounds make each other's impacts worse. This is called a **synergistic effect,** an interaction in which one substance exacerbates the effects of another. For example, occupational exposure to asbestos increases lung cancer rates 20-fold; smoking also increases lung cancer rates by the same amount. Asbestos workers who also smoke, however, have a 400-fold increase in cancer rates. How many other toxic chemicals are we exposed to that are below threshold limits individually but combine to give toxic results?

A fundamental concept in toxicology is that every material can be poisonous under some conditions, but most chemicals have some safe level or threshold below which their effects are undetectable or insignificant. Each of us consumes lethal doses of many chemicals over the course of a lifetime. One hundred cups of strong coffee, for instance, contain a lethal dose of caffeine. Similarly, 100 aspirin tablets, or 10 kilograms (22 lb) of spinach or rhubarb, or a liter of alcohol would be deadly if consumed all at once. Taken in small doses, however, most toxicants can be broken down or excreted before they do much harm. Furthermore, damage they cause can sometimes be repaired, and the repair can make you stronger and more resilient in the future.

Our bodies degrade and excrete toxic substances

Most organisms have enzymes that break down waste products and environmental poisons to reduce their toxicity. In mammals, most of these enzymes are located in the liver, the primary site of detoxification of both natural wastes and introduced poisons. Sometimes these reactions can produce more harmful compounds, however. Benzopyrene, for example, is not toxic in its original form, but our bodies break it down into substances that can cause cancer.

We also reduce the effects of waste products and environmental toxins by eliminating them from our body through excretion. Volatile molecules, such as carbon dioxide, hydrogen cyanide, and ketones, are excreted via breathing. Some excess salts and other substances are excreted in sweat. Primarily, however, excretion is a function of the kidneys, which can eliminate significant amounts of soluble materials through urine formation. Accumulation of toxins in the urine can damage this vital system, however, and the kidneys and bladder often are subjected to harmful levels of toxic compounds. In the same way, the stomach, intestine, and colon often suffer damage from materials concentrated in the digestive system and may be afflicted by diseases and tumors.

Our bodies also have mechanisms to repair damage. Our natural defenses include the ability of cells to repair damaged proteins and DNA. Similarly, tissues and organs that are exposed regularly to physical wear and tear or to toxic or hazardous materials often have mechanisms for damage repair. You have probably seen how your skin can repair damage rapidly. In the same way you continuously replace and repair tissues in your liver, lungs, the linings of your stomach and intestines, and other tissues. Each time tissues are damaged and reproduced, however, there is a chance that some cells will lose normal growth controls and run amok, creating a tumor. Thus, any agent that irritates tissues, such as smoking (which damages lungs) or alcohol (which damages the liver), is likely to be carcinogenic. And tissues with high cell-replacement rates are among the most likely to develop cancers.

Section Review

1. What is biomagnification, and why is it important in environmental health?

2. How do the physical and chemical characteristics of materials affect their movement, persistence, distribution, and fate in the environment?

3. Give three examples of persistent organic pollutants and describe their main effects.

8.4 TOXICITY AND RISK ASSESSMENT

- *A fundamental principal of toxicology is that the dose makes the poison. Almost everything is toxic at some level. The amount, timing, and method of exposure determines your response.*

- *We fear some risks out of proportion to their actual danger, while we ignore other hazards despite their potential effects.*

- *Substances or factors that affect gene expression or disrupt normal growth, development, or metabolism can have serious effects even at very low doses.*

Almost 500 years ago, the Swiss scientist Paracelsus said "the dose makes the poison." He meant by this that almost everything is toxic at some level. This remains the most basic principle of toxicology. Table salt, for example, is necessary at low levels, but if you ate a kilogram of salt all at once, it would make you very sick. A similar amount injected into your bloodstream would kill you. Thus, the delivery method and rate, as well as the dose, matter in determining toxicity.

Substances vary dramatically in their degree of toxicity. Some are so poisonous that a single drop on your skin can kill you. Others require massive amounts injected directly into the blood to be lethal. Measuring and comparing the toxicity of different materials is difficult, because species differ in sensitivity to any substance, and individuals within a species respond differently to a given exposure. In this section, we will look at the ways we measure degrees of toxicity.

Dose-response curves show toxicity in lab animals

Normally, we measure toxicity by exposing a population of laboratory animals to measured doses, to find the statistically most frequent response to a given dose. This procedure is expensive, time-consuming, and often painful and debilitating to the animals being tested. It commonly takes hundreds—or even thousands—of animals, several years of hard work by scientists, and hundreds of thousands of dollars to thoroughly test the effects of a toxin at very low doses. More humane toxicity tests using computer simulations of model reactions, cell cultures, and other substitutes for whole living animals are being developed. However, tests on living animals, which are more complex systems than cell cultures, are considered more reliable, so most public policies about environmental or occupational health hazards are based on live animal tests.

In addition to humanitarian concerns, a challenge for toxicologists and policymakers is that members of a population vary in their response to a toxic substance (fig. 8.14). Some individuals are very sensitive and respond at very low doses. Some respond only at very high doses. Most fall somewhere in the middle, forming a bell-shaped curve. The question for regulators and politicians is whether we should set pollution limits that will protect everyone, including the most sensitive people, or only aim to protect the average person. It might cost billions of extra dollars to protect a

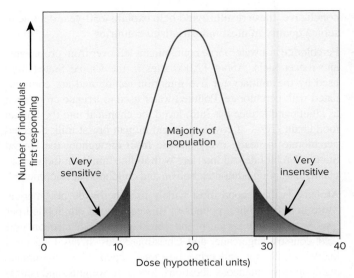

FIGURE 8.14 Probable variations in sensitivity in a population. Some members of a population may be very sensitive to a given substance, while others are much less sensitive. The majority of the population falls somewhere between the two extremes.

very small number of individuals at the extreme end of the curve. Is that a good use of resources?

Cumulative responses are usually depicted in a dose-response curve (fig. 8.15). This graph is like the histogram bell curve of figure 8.14, but it shows the percentage of the population on the vertical axis, rather than the number of individuals. The percentage of organisms responding to low, medium, or high doses thus add up to 100 percent on the graph. A convenient way to describe the toxicity of a chemical is to determine the dose to which 50 percent of the test population is sensitive. In the case of a lethal dose (LD), this is called the **LD50**. In figure 8.15,

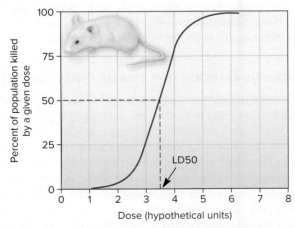

FIGURE 8.15 A dose-response curve shows the cumulative population response to increasing doses of a toxic substance. The LD50 is the dose that is lethal to half the population.

for example, a dose of 3.5 units is lethal for 50 percent of the population.

Dose-response curves are often used to compare toxicity. One substance might be lethal to half the population at 3.5 units, while another might have an LD50 of only 1.5. This would mean the second substance is more than twice as dangerous as the first.

Unrelated species can react very differently to the same compound, not only because body sizes vary but also because of differences in physiology and metabolism. Even closely related species can have very dissimilar reactions to a particular substance. Guinea pigs, for instance, are nearly 5,000 times more sensitive to some dioxins than are hamsters. Of 226 chemicals found to be carcinogenic in either rats or mice, 95 caused cancer in one species but not the other. These variations make it difficult to estimate the risks for humans, because we don't consider it ethical to perform dangerous experiments in which we deliberately expose people to toxins.

Even within a single species there can be variations in responses between different genetic lines. A current controversy in determining the effects of environmental estrogens concerns the type of rats used for toxicology studies. Standard toxicology protocols call for a sturdy strain called the Sprague-Dawley rat. It turns out, however, that these animals, which were bred to grow fast and breed prolifically in lab conditions, are thousands of times less sensitive to endocrine disrupters than ordinary rats. Industry reports that declare certain chemicals to be harmless based on Sprague-Dawley rats are highly suspect.

There is a wide range of toxicity

It is useful to group materials according to their relative toxicity. A moderately toxic substance takes about one gram per kilogram of body weight (about two ounces for an average human) to make a lethal dose. Very toxic materials take about one-tenth that amount, while extremely toxic substances take one-hundredth as much (only a few drops) to kill most people. Supertoxic chemicals are extremely potent; for some, a few micrograms (millionths of a gram—an amount invisible to the naked eye) make a lethal dose. These materials are not all synthetic. One of the most toxic chemicals known, for instance, is ricin, a protein found in castor bean seeds. It is so toxic that 0.3 billionths of a gram given intravenously will generally kill a mouse. If aspirin were this toxic, a single tablet, divided evenly, could kill 1 million people.

Many carcinogens, mutagens, and teratogens are dangerous at levels far below their direct toxic effect because they cause abnormal cell growth, which results in a kind of biological amplification. A single cell, perhaps altered by a single molecular event, can multiply into millions of tumor cells or an entire organism. Just as there are different levels of direct toxicity, however, there are different degrees of carcinogenicity, mutagenicity, and teratogenicity. Methanesulfonic acid, for instance, is highly carcinogenic, while the sweetener saccharin is a suspected carcinogen whose effects may be vanishingly small.

Acute and chronic doses and effects differ

Some toxic effects are sudden and severe, causing immediate impairment or death. We call these **acute effects.** An organism may be able to recover from acute effects, however. Often, if the individual experiencing an acute reaction survives this immediate crisis, the effects are reversible. **Chronic effects,** on the other hand, are long-lasting or even permanent. A chronic effect can result from a single dose of a very toxic substance, or it can be the result of a continuous or repeated sublethal exposure.

Exposure can also be acute or chronic. For example, frequent consumption of alcohol over many years (chronic exposure) can gradually degrade your liver, eventually causing long-lasting damage or even death. Consuming an extreme amount of alcohol all at once (acute exposure) can cause immediate and severe impairment or death. Often, it's difficult to identify or assess health risks of chronic exposures because other factors, such as aging or normal diseases, act simultaneously with the factor under study.

Figure 8.16 shows three possible results from low doses of a toxin. Curve (*a*) shows a baseline level of response in the population, even at a zero dose of the toxin. This suggests that some other factor in the environment also causes this response. Curve (*b*) shows a straight-line relationship from the highest doses to zero exposure. Many carcinogens and mutagens show this kind of response. Any exposure to such agents, no matter how small, carries some risks. Curve (*c*) shows a threshold for the response where some minimal dose is necessary before any effect can be observed. This generally suggests the presence of some defense mechanism that either prevents the toxin from reaching its target in an active form or repairs the damage that it causes. Low levels of exposure to the toxin in question may have no deleterious effects, and it might not be necessary to try to keep exposures to zero.

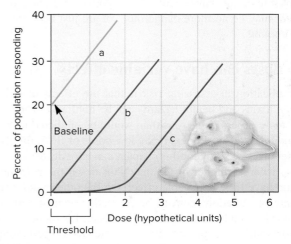

FIGURE 8.16 Three possible dose-response curves at low doses. (a) Some individuals respond, even at zero dose, indicating that some other factor must be involved. (b) Response is linear down to the lowest possible dose. (c) Threshold must be passed before any response is seen.

Which, if any, environmental health hazards have thresholds is an important but difficult question. The 1958 Delaney Clause to the U.S. Food and Drug Act forbids the addition of *any* amount of known carcinogens to food and drugs, based on the assumption that any exposure to these substances is an unacceptable risk. This standard was replaced in 1996 by a *no reasonable harm* requirement, defined as less than one cancer for every million people exposed over a lifetime. This change was supported by a report from the National Academy of Sciences, which concluded that synthetic chemicals in our diet are unlikely to represent an appreciable cancer risk. We will discuss risk analysis in the next section.

Detectable levels aren't always dangerous

You may have seen or heard dire warnings about toxic materials detected in samples of air, water, or food. A typical headline announced recently that 23 pesticides were found in 16 food samples. What does that mean? Is any amount of dangerous materials unacceptable? Not necessarily. We have noted that the dose makes the poison. Toxicity depends on what is there, how much is there, how it is delivered, and other factors.

Toxic pollutants may seem to be more widespread now than in the past, and this is surely a valid perception for many substances. The daily reports we hear of new materials found in new places, however, are also due in part to our more sensitive measuring techniques. Twenty years ago, parts per million was generally the limit of detection for most chemicals. Anything below that amount was often reported as zero or absent rather than more accurately stated as undetected. A decade ago, new machines and techniques were developed to measure parts per billion. Suddenly chemicals were found where none had been suspected. Now we can detect parts per trillion or even parts per quadrillion in some cases. Increasingly sophisticated measuring capabilities may lead us to believe that toxic materials have become more prevalent. In fact, our environment may be no more dangerous; we are just better at finding trace amounts.

Low doses can have variable effects

A complication in assessing risk is that the effects of low doses of some toxics and health hazards can be nonlinear. They may be either more or less dangerous than would be predicted from exposure to higher doses. For example, low doses of BPA, discussed in the opening case study for this chapter, can have devastating health effects, whereas higher doses may shut down the response system and have little noticeable effect. On the other hand, very low amounts of radiation seem to be protective against certain cancers, whereas higher doses are carcinogenic. It's thought now that very low radiation exposure may stimulate DNA repair along with enzymes that destroy free radicals (atoms with unpaired, reactive electrons in their outer shells). Activating these repair mechanisms may defend us from other, unrelated hazards. These nonlinear effects are called hormesis.

Another complication is that some substances can have long-lasting effects on genetic expression. For example, researchers found that exposure of pregnant rats to certain chemicals can have effects, not only on the exposed rats, but on their daughters and granddaughters. A single dose given on a specific day in pregnancy can be expressed several generations later, even if those offspring have never been exposed to the chemical (see the Exploring Science box later in this section).

Some symptoms can be erroneous

Sometimes outbreaks of apparent health problems have a psychological rather than physical basis. When your neighbors talk convincingly about being sick, you may begin to feel ill, too. Some physicians regard the so-called "wind turbine syndrome," for example, as a psychogenic condition with no physical basis. People who are opposed to wind turbines often claim the noise and shadow flicker caused by moving turbine blades make them sick, but there is no medical evidence of cause and effect. More than 155 different symptoms ranging from sleep loss, headaches, dizziness, nausea, muscle pain, and tinnitus to depression and suicidal thoughts have been blamed on wind turbines. Many stresses, including having large machines you don't like near your property may result in such symptoms, but there's little evidence that the motion of the blades by themselves can cause so many illnesses.

Risk perception isn't always rational

Risk is the possibility of suffering harm or loss. **Risk assessment** is the scientific process of estimating the threat that particular hazards pose to human health. This process includes risk identification, dose response assessment, exposure appraisal, and risk characterization. In hazard identification, scientists evaluate all available information to estimate the likelihood that a chemical will cause a certain effect in humans. The best evidence comes from human studies, such as physician case reports. Animal studies are also used to assess health risks. Risk assessment for identified toxicity hazards (for example, lead) includes collection and analysis of site data, development of exposure and risk calculations, and preparation of human health and ecological impact reports. Exposure assessment is the estimation or determination of the magnitude, frequency, duration, and route of exposure to a possible toxicant. Toxicity assessment weighs all available evidence and estimates the potential for adverse health effects to occur.

A number of factors influence how we perceive relative risks associated with different situations, such as the following:

- People with social, political, or economic interests—including environmentalists—tend to downplay certain risks and emphasize others that suit their own agendas. We also tend to tolerate risks that we choose—such as driving, smoking, or overeating—while objecting to risks we cannot control, such as exposure to wind turbines or pesticides.

- Most people have difficulty understanding and believing probabilities. We feel that there must be patterns and connections in events, even though statistical theory says otherwise. If the coin turned up heads last time, we feel certain it will turn up tails the next. In the same way, it is difficult to understand the meaning of a 1-in-10,000 risk of being poisoned by a chemical.

- Our personal experiences are often misleading. When we have not personally experienced a bad outcome, we feel it is more rare and unlikely to occur than it actually may be. Furthermore, we tend to rationalize the things we want to do and exaggerate risks we dislike (fig. 8.17).

- We have an exaggerated view of our own abilities to control our fate. We generally consider ourselves above-average drivers, safer than most when using appliances or power tools, and less likely than others to suffer medical problems, such as heart attacks. People often feel they can avoid hazards because they are wiser or luckier than others.

- News media give us a biased perspective on the frequency of certain kinds of health hazards, overreporting some accidents or diseases while downplaying or underreporting others. Sensational, gory, or especially frightful causes of death, like murders, plane crashes, fires, or terrible accidents, occupy a disproportionate amount of attention in the public media. Heart diseases, cancer, and stroke kill nearly 15 times as many people in the United States as do accidents and 75 times as many people as do homicides, but the emphasis placed by the media on accidents and homicides is nearly inversely proportional to their relative frequency compared to either cardiovascular disease or cancer. This gives us an inaccurate picture of the real risks to which we are exposed.

- We tend to have an irrational fear or distrust of certain technologies or activities that leads us to overestimate their dangers. Nuclear power, for instance, is viewed as very risky, while coal-burning power plants seem to be familiar and relatively benign; in fact, coal mining, shipping, and combustion cause an estimated 10,000 deaths each year in the United States. An old, familiar technology seems safer and more acceptable than does a new, unknown one (see Data Analysis at the end of this chapter).

FIGURE 8.17 How dangerous is trick skating? Many parents regard this as extremely risky, while many students—especially males—believe the risks are acceptable. Perhaps the more important question is whether the benefits outweigh the risks.
© image100/PunchStock RF

Risk acceptance depends on many factors

How much risk is acceptable? What are we willing to sacrifice to avoid exposure to certain risks? Most people will tolerate a higher probability of occurrence of an event if the harm caused by that event is low. Conversely, harm of greater severity is acceptable only at low levels of frequency. A 1-in-10,000 chance of being killed might be of more concern to you than a 1-in-100 chance of being injured. For most people, a 1-in-100,000 chance of dying from some event or some factor is a threshold for changing what we do. That is, if the chance of death is less than 1 in 100,000, we are not likely to be worried enough to change our ways. If the risk is greater, we will probably do something about it. The Environmental Protection Agency generally assumes that a risk of 1 in 1 million is acceptable for most environmental hazards. Critics of this policy ask: Acceptable to whom?

For activities that we enjoy or find profitable, we are often willing to accept far greater risks than this general threshold. Conversely, for risks that benefit someone else, we demand far higher protection. For instance, your chance of dying in a motor vehicle accident in any given year is about 1 in 5,000, but that doesn't deter many people from riding in automobiles. Your lifetime chance of dying from lung cancer if you smoke one pack of cigarettes per day is about 1 in 4. By comparison, the risk from drinking water with the EPA limit of trichloroethylene is about 1 in 10 million. Strangely, many people demand water with zero levels of trichloroethylene, while continuing to smoke cigarettes.

Table 8.3 lists lifetime odds of dying from a few leading diseases and accidents. These are statistical averages, of course, and there clearly are differences in where one lives or how one behaves that affect the danger level of these activities. Although the average lifetime chance of dying in an automobile accident is 1 in 100, there clearly are things you can do—like wearing a seat belt, following safety rules, and avoiding risky situations—that improve your odds. Still, it is interesting how we readily accept some risks while shunning others.

Our perception of relative risks is strongly affected by whether risks are known or unknown, whether we feel in control of the outcome, and how dreadful the results are. Risks that are unknown or unpredictable and results that are particularly gruesome or disgusting seem far worse than those that are familiar and socially acceptable.

The Epigenome

Could your diet, behavior, or environment affect the lives of your children or grandchildren? For a century or more, scientists assumed that the genes you receive from your parents irreversibly fix your destiny and that factors such as stress, habits, exposure to toxins, or parenting should have no effect on future generations.

Now, however, some startling discoveries are making us reexamine those assumptions. Scientists are finding that a complex set of chemical markers and genetic switches—called the epigenome—consisting of DNA, RNA, and their associated proteins and other small molecules, regulate gene function in ways that can affect multiple generations. Understanding how this system works not only helps us see how many environmental factors affect health, but it may become useful in treating a variety of diseases.

One of the most striking experiments on the epigenome was carried out a decade ago by researchers at Duke University. They were studying the affects of diet on a strain of mice carrying an agouti gene that makes them obese, yellow, and prone to cancer and diabetes. Starting just before conception, mother agouti mice were fed a diet rich in B vitamins (folic acid and B_{12}). Amazingly, this simple dietary change resulted in baby mice that were sleek, brown, and healthy. Eating a special diet by mothers had somehow turned off the agouti gene in the offspring.

We know now that B vitamins, as well as some vegetables, such as onions, garlic, and beets, are methyl donors—that is, they can add a carbon atom and three hydrogens to proteins and nucleic acids. Attaching an extra methyl group can switch genes either on or off by changing the way proteins and nucleic acids read and translate the DNA. Similarly, acetylating DNA (adding an acetyl group: CH_3CO) can also either stimulate or inhibit gene expression.

These reactions work not only directly on genes themselves, but also on a huge set of what we once thought was useless, or "junk," DNA in chromosomes, as well as a large amount of protein that once seemed to be merely packing material. We now know that both this extra DNA and the protein around which genes are wrapped play vital roles in controlling gene expression. And methylating or acetylating these proteins or nucleic acids also can have profound effects on heredity.

More remarkable is that changes in the epigenome can carry through multiple generations. In 2004, Michael Skinner, a geneticist at Washington State University, was studying the effects of exposure to a commonly used fungicide on rats. He found that male rats exposed in utero had lower sperm counts later in life. It took only a single exposure to cause this effect. Amazingly, the effect lasted for at least four generations even

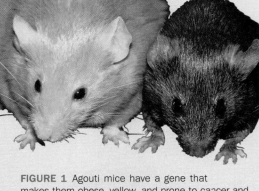

FIGURE 1 Agouti mice have a gene that makes them obese, yellow, and prone to cancer and diabetes. If a mother agouti mouse (*left*) is given B vitamins during pregnancy, the gene is turned off and its baby (*right*) is sleek, brown, and healthy. Amazingly, this genetic change lasts for several generations before the gene resumes its deleterious effects.
© Dana Dolinoy, University of Michigan

though those subsequent offspring were never exposed to the fungicide.

The way a mother rodent nurtures her young also can cause changes in methylation patterns in their babies' brains that are quite similar to the prenatal vitamins and nutrients that affected the agouti gene. It's thought that licking and grooming activates serotonin receptors that turn on genes to reduce stress responses, resulting in profound brain changes. In another study, rats given extra attention, diet, and mental stimulation (toys) did better at memory tests than did environmentally deprived controls. Altered methylation patterns in the hippocampus—the part of the brain that controls memory—were detected in both

these cases. And subsequent generations maintained this methylation.

Epigenetic effects have also been found in humans. One of the most compelling studies involved the comparison of two centuries of health records, climate, and food supply in a remote village in northern Sweden. The village of Overkalix was so isolated that when bad weather caused crop failures, famine struck everyone. In good years, on the other hand, there was plenty of food and people stuffed themselves. A remarkable pattern emerged. When other social factors were factored in, grandfathers who were preteens during lean years had grandsons who lived an amazing 32 years longer than those whose grandfathers were able to gorge themselves as preteens. Similarly, women whose mothers had access to a rich diet while they were pregnant were much more likely to have daughters and granddaughters with health problems and shortened lives.

In another surprising human health study, researchers found in a long-term study of couples in Bristol, England, that fathers who started smoking before they were 11 years old (just as they were starting puberty and sperm formation was beginning) were much more likely than nonsmokers to have sons and grandsons who were overweight and lived significantly shortened lives. Both these results are attributed to epigenetic effects.

A wide variety of factors can cause epigenetic changes. Smoking, for example, leaves a host of persistent methylation markers in your DNA. So does exposure to a number of pesticides, toxics, drugs, and stressors. At the same time, it is possible to help prevent deleterious methylations by getting plenty of polyphenols in green tea and deeply colored fruit, B vitamins, and garlic, onions, turmeric, and other healthy foods. Not surprisingly, epigenetic changes are implicated in many cancers, including colon, prostate, breast, and blood cancers. This may explain many confusing cases in which our environment seems to have long-lasting effects on health and development that can't be explained by ordinary metabolic effects.

Unlike mutations, epigenetic changes aren't permanent. Eventually, the epigenome returns to normal if the exposure wasn't repeated. This makes epigenetic changes candidates for drugs. Currently, the Food and Drug Administration has approved two drugs, Vidaza and Dacogen, that inhibit methylation and are used to treat a precursor to leukemia. Another drug, Zolinza, which enhances acetylation, is approved to treat another form of leukemia. Dozens of other drugs that may treat a variety of diseases, including rheumatoid arthritis, neurodegenerative diseases, and diabetes, are under development.

So your diet, behavior, and environment can have a much stronger impact on both your health and that of your descendants than we previously understood. What you ate, drank, smoked, or did last night may have profound effects on future generations.

Studies of public risk perception show that most people react more to emotion than to statistics. We go to great lengths to avoid some dangers while gladly accepting others. Factors that are involuntary, unfamiliar, undetectable to those exposed, catastrophic, or that have delayed effects or are a threat to future generations are especially feared, whereas those that are voluntary, familiar, detectable, or immediate cause less anxiety. Even though the actual number of deaths from automobile accidents, smoking, or alcohol, for instance, are thousands of times greater than those from pesticides, nuclear energy, or genetic engineering, the latter preoccupy us far more than the former.

Section Review

1. Why is there a threshold in response to some toxic substances?
2. What is an LD50?
3. What is the difference between acute and chronic doses and effects?
4. What is the difference between acute and chronic toxicity?

Table 8.3 Lifetime Chances of Dying in the United States	
Source	**Odds (1 in x)**
Heart disease	6
Cancer	7
Chronic respiratory disease	28
Stroke	29
Intentional self-harm	100
Unintentional poisoning	109
Motor vehicle accident	112
Falls	144
Firearm assault	358
Pedestrian accidents	704
Motorcycle accident	911
Drowning	1,113
Fires	1,442
Bicycle accidents	4,535
Firearm accidental discharge	6,699
Storms	6,778
Bee, wasp, or hornet stings	55,764
Lightning strikes	164,968
Dog bites	144,899
Floods	589,896

Source: U.S. National Safety Council, 2015

8.5 Establishing Health Policy

- *It's difficult to evaluate multiple risks and benefits simultaneously.*
- *In setting health standards, we need to consider combined exposures, different sensitivities, and differences between chronic and acute exposures.*

Risk management combines principles of environmental health and toxicology together with regulatory decisions based on socioeconomic, technical, and political considerations (fig. 8.18). Establishing policies to protect the public is difficult, because we face many sources of harm simultaneously, often without being aware of them. It is difficult to separate the effects of all these different hazards and evaluate their risks accurately. In spite of often vague and contradictory data, public policymakers still must make decisions.

A long-running, contentious debate surrounds the Endocrine Disrupter Screening Program. In 1996, Congress ordered the EPA to start testing 87,000 chemicals for their ability to disrupt endocrine hormone functions that regulate almost every aspect of reproduction, growth, development, and the functioning of our bodies. Twenty years later, out of a subset of 52 compounds, a smaller group of 18 had been identified for further testing, but no clear determinations of endocrine disruptors had been made. Some toxicologists argue that the exposure times may be too short, or the test animals may have been exposed to other chemicals besides the ones being studied. As we mentioned earlier in this chapter, an albino rat breed called the Sprague-Dawley is stipulated in these tests. These rats were originally bred to be resistant to arsenic trioxide pesticides, and they may be unnaturally resistant to endocrine disrupters as well. Meanwhile,

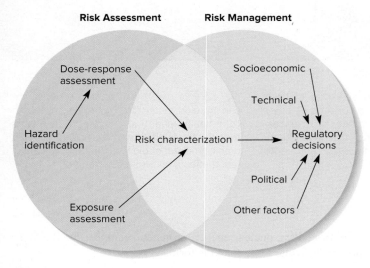

Risk Assessment   Risk Management

Dose-response assessment

Hazard identification

Risk characterization

Exposure assessment

Socioeconomic

Technical

Regulatory decisions

Political

Other factors

FIGURE 8.18 Risk assessment organizes and analyzes data to determine relative risk. Risk management sets priorities and evaluates relevant factors to make regulatory decisions.

the chemical industry disputes the need for any endocrine testing at all. As you can see, establishing public policy isn't simple.

In setting standards for environmental toxins, we need to consider (1) the combined effects of exposure to many different sources of damage, (2) different sensitivities of members of the population, and (3) the effects of chronic as well as acute exposures. Because absolute proof is so elusive, as in the endocrine disrupter study, some European countries have adopted a "precautionary principle" that requires manufacturers to show that substances are safe, rather than for regulators to prove they are dangerous. This is a "better safe than sorry" approach. Critics of the precautionary principle argue that it's impossible to prove a substance is always safe, and it is costly to try to prevent all possible harm from all possible substances (fig. 8.19).

FIGURE 8.19 "Do you want to stop reading those ingredients while we're trying to eat?"
© Richard Guindon. Reprinted with permission.

On the other hand, proponents of the precautionary principle argue that modern industry exposes us to countless new chemicals, and we need to stop releasing them into the environment without understanding their effects. Each challenge to our cells by toxic substances is a stress on our bodies. Although each individual stress may not be life-threatening, the cumulative effects of all the environmental stresses, both natural and human-caused, to which we are exposed may seriously shorten or restrict our lives. Furthermore, some individuals in any population are more susceptible to those stresses than others. Should we set pollution standards so that no one is adversely affected, even the most sensitive individuals, or should the acceptable level of risk be based on the average member of the population?

Finally, policy decisions about hazardous and toxic materials also need to be based on information about how such materials affect the plants, animals, and other organisms that define and maintain our environment. In some cases, pollution can harm or destroy whole ecosystems with devastating effects on the life-supporting cycles on which we depend. In other cases, only the most sensitive species are threatened. Table 8.4 shows the Environmental Protection Agency's assessment of relative risks to human welfare. This ranking reflects a concern that our exclusive focus on reducing pollution to protect human health has neglected risks to natural ecological systems. While there have been many benefits from a case-by-case approach in which we evaluate the health risks of individual chemicals, we have often missed broader ecological problems that may be of greater ultimate importance.

Section Review

1. Why has it been so difficult to regulate endocrine disrupters?
2. What is the "precautionary principle"?

Table 8.4 Relative Risks to Human Welfare
Relatively High-Risk Problems
Habitat alteration and destruction
Species extinction and loss of biological diversity
Stratospheric ozone depletion
Global climate change
Relatively Medium-Risk Problems
Herbicides/pesticides
Toxics and pollutants in surface waters
Acid deposition
Airborne toxics
Relatively Low-Risk Problems
Oil spills
Groundwater pollution
Radionuclides
Thermal pollution

Source: Environmental Protection Agency

Conclusion

We have made marvelous progress in reducing some of the worst diseases that have long plagued humans. Smallpox is the first major disease to have been completely eliminated. Guinea worms and polio are nearly eradicated worldwide; bubonic plague, typhoid fever, cholera, yellow fever, tuberculosis, mumps, and other highly communicable diseases are rarely encountered in advanced countries. Childhood mortality has decreased 90 percent globally, and people almost everywhere are living twice as long, on average, as they did a century ago.

But the technological innovations and affluence that have diminished many terrible diseases have also introduced new risks. Chronic conditions, such as cardiovascular disease, cancer, depression, dementia, diabetes, as well as traffic accidents and other risks that once were confined to richer countries have now become leading health problems nearly everywhere. Part of this change is that we no longer die at an early age of infectious disease, so we live long enough to develop the infirmities of old age. Another factor is that affluent lifestyles, lack of exercise, and unhealthy diets aggravate these chronic conditions.

New emergent diseases are appearing at an increasing rate. With increased international travel, diseases can spread around the globe in a few days. Epidemiologists warn that the next deadly epidemic may be only a plane ride away. In addition, modern industry is introducing thousands of new chemical substances every year, most of which aren't studied thoroughly for health effects. Endocrine disrupters, neurotoxins, carcinogens, mutagens, teratogens, and other toxins, sometimes even at very low levels, can have tragic outcomes. PFOA's role in a wide variety of chronic health effects is an example of how materials we have introduced can have unintended consequences. Many other industrial chemicals could be having similar harmful effects.

Determining what levels of environmental health risk are acceptable is difficult. We are exposed to many different health threats simultaneously. Furthermore, people consider some dangers tolerable, but dread others—especially those that are new, involuntary, and difficult to detect and whose effects are unknown to science. The situation is complicated by the fact that news media give us a biased perspective on some hazards, while our personal experiences and our sense of our own abilities are often misleading.

There are many steps that each of us can take to protect our health. Eating a healthy diet, exercising regularly, drinking in moderation, driving prudently, and practicing safe sex are among the most important.

Reviewing Key Terms

Can you define the following terms in environmental science?

acute effects 8.4	conservation medicine 8.1	fetal alcohol syndrome 8.2	risk 8.4
allergens 8.2	disability-adjusted life	health 8.1	risk assessment 8.4
antigens 8.2	years (DALYs) 8.1	LD50 8.4	sick building
bioaccumulation 8.3	disease 8.1	morbidity 8.1	syndrome 8.2
biomagnification 8.3	ecological diseases 8.1	mortality 8.1	synergistic effect 8.3
body burden 8.3	emergent diseases 8.1	mutagens 8.2	teratogens 8.2
cancer 8.2	endocrine disrupters 8.2	neurotoxins 8.2	toxic substances 8.1
carcinogens 8.2	environmental health 8.1	persistent organic pollutants	
chronic effects 8.4	epigenetic effects 8.4	(POPs) 8.3	

Critical Thinking and Discussion Questions

1. What consequences (positive or negative) do you think might result from defining health as a state of complete physical, mental, and social well-being? Who might favor or oppose such a definition?

2. Do rich countries bear any responsibilities if the developing world adopts unhealthy lifestyles or diets? What could (or should) we do about it?

3. Why do we spend more money on heart or cancer research than on childhood illnesses?

4. What are the premises in the discussion of assessing risk? Could conflicting conclusions be drawn from the facts presented in this section? What is your perception of risk from your environment?

5. Should pollution levels be set to protect the average person in the population or the most sensitive? Why not have zero exposure to all hazards?

6. What level of risk is acceptable to you? Are there some things for which you would accept more risk than others?

7. Some of the mechanisms that help repair wounds or fight off infection can result in cancer when we're old. Why hasn't evolution eliminated these processes? Hint: Do conditions of postreproductive age affect natural selection?

8. What changes could you make in your lifestyle to lessen your risks from the diseases and health risks in table 8.1? What would have the greatest impact on your future well-being?

9. Why has it been so difficult to regulate endocrine disrupters? How should we act in the face of uncertainty?

10. Why should or shouldn't we adopt a "precautionary principle"? What different world views are represented in this debate?

Data Analysis

How Do We Evaluate Risk and Fear?

A central question in environmental health is how we perceive different risks around us. When we evaluate environmental hazards, how do we assess known factors, uncertain risks, and the unfamiliarity of new factors we encounter? Which considerations weigh most heavily in our decisions and our actions as we try to avoid environmental risks?

The graph shown next offers one set of answers to this question, using the aggregate responses of many people to risk. Go to Connect to find further discussion and a set of questions regarding risk, uncertainty, and fear.

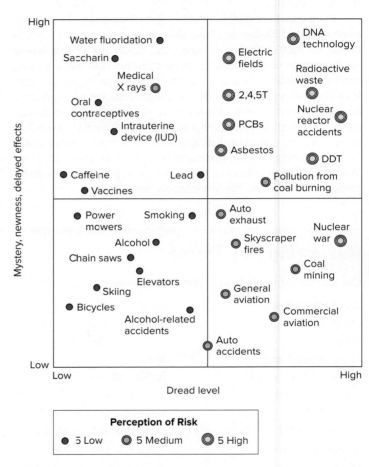

Public perception of risk, depending on the familiarity, apparent potential for harm, and personal control over the risk.

Source: Data from Slovic, Paul, 1987. "Perception of Risk," Science, 236 (4799): 286–290.

TO ACCESS ADDITIONAL RESOURCES FOR THIS CHAPTER, PLEASE VISIT CONNECT AT www.connect.mheducation.com.

You will find LearnSmart, an adaptive learning system, Google Earth™ exercises, additional Case Studies, Data Analysis exercises, and an interactive ebook.

9

Food and Hunger

▲Fresh, local, and organic foods can be hard for college students to find on campus.
© Ron Levin/Getty Images RF

Learning Outcomes

After studying this chapter, you should be able to:

9.1 Describe patterns of world hunger and nutritional requirements.

9.2 Identify key food sources, including protein-rich foods.

9.3 Explain new crops and genetic engineering.

9.4 Discuss how policy can affect food resources.

"It ain't the things we know that cause all the trouble; it's the things we think we know that ain't so."

– Will Rogers

Becoming a Locavore in the Dining Hall

Many people care about good food, and sometimes a factor that makes food satisfying is knowing that it's supporting the local farm economy or that it's sustainably grown. But if you're a university or college student, and if you eat most of your meals in a cafeteria, how much control can you have over where your food comes from? Most cafeteria food seems to originate in a large freezer truck at the loading dock behind the dining hall. Any farm behind those boxes of frozen fries and hamburgers is far away and hard to imagine. If you eat from a cafeteria, even if you have time to care about sustainable or healthy foods, you might feel you have very little power to do anything about it.

At a growing number of colleges and universities, students are speaking up about becoming local eaters—one term for the idea is **"locavore."** One of these schools is Vassar College, in New York's Hudson River Valley. Vassar students have started asking the dining service to provide local foods because they're concerned about the local farm economy, because they want pesticide-free and hormone-free foods, and because they're worried about the environmental costs of foods that travel thousands of miles from the farm to the table.

As more students have pushed for local meals, they have empowered food service managers to take the time to work with local producers. Maureen King, the director of campus dining, and her colleague Ken Oldehoff have found local sources of tomato sauce, salsa, squash, fruit, milk, yogurt, fresh produce, juice, desserts, soups, and other foods (fig. 9.1). Local cider and milk are available in the cafeteria line, along with soft drinks. Fresh apples and other fruit are promoted in the fall. King and Oldehoff have initiated pie bake-offs, pumpkin-carving events, sauce challenges, and even a Local Foods Week that allows students to eat entirely locally all week.

The menu includes more squash, tomatoes, and beets than the standard cafeteria food service truck delivers, but college chefs enjoy using fresh foods, and students keep asking for more local products. There are still plenty of nonlocal alternatives, such as fresh strawberries and melons in midwinter, and, of course, coffee (from Mexico). But for several years now this coffee has been shade-grown, organic,

FIGURE 9.1 Ken Oldehoff and his colleagues have been creative and persistent in bringing local foods to college students.
© Simon Craven, 2005, Vassar College

fair-trade coffee—and it doesn't cost much more at the checkout line than the previous nonorganic coffees did.

Local sourcing isn't always easy. Vassar is located in a region of rapidly growing suburbs, with a struggling farm economy, and Oldehoff and King work constantly on finding new relationships with growers. Sometimes suppliers give up and go out of business; other times, new opportunities suddenly appear. Ordering everything from a single national distributor is usually easier and cheaper than ordering from a miscellaneous group of local growers. But every purchase makes a difference to farmers, and sometimes the college's commitment helps keep local businesses solvent.

What inspires Maureen King and Ken Oldehoff to make this extra effort? They are concerned about serving healthy food, including lots of vegetables and organic food when possible. They like knowing their suppliers and knowing how food was handled. They also live in the community, and they are happy to put dollars into local pockets, and to help protect Hudson River Valley's historic agrarian landscape.

King and Oldehoff also worry about the carbon footprint of the college, so they want to minimize the amount of food they buy from the far side of the continent. If a portion of the college's chicken or tomato sauce can be grown just down the road, then why not try to buy them locally?

These are good justifications, but local purchasing wouldn't go anywhere without student interest. It's not necessary for every student to be an organic vegan locavore, but if a few are willing to stand up and show they care, their voices can shift college policy. Many other institutions have exclusive contracts with national suppliers, which restrict purchasing from local suppliers. Partly because of student action, and partly because of the persistence of King and Oldehoff, Vassar's contracts allow for local purchases. The college has also discovered that good environmental citizenship makes good press and generates good feelings about the institution. As Ken Oldehoff notes, committing to local food is one way student activists can make real change.

Local eating is not just a local concern. There are important connections between local consumption and global patterns of food availability, hunger, and nutrition. We can learn a great deal about global food issues by thinking more carefully about what it would take to become a locavore. In this chapter, we'll think about those connections. We'll look at some of the different kinds of foods we eat; at how much we eat in wealthier countries, compared to poorer ones; and at how food production differs from one area to another. As you read, think about how these global issues help explain why Ken Oldehoff and Maureen King work as hard as they do to procure local foods, and why the students keep asking for them.

9.1 WORLD FOOD AND NUTRITION

- *Food production has grown faster than the population, but chronic hunger remains widespread.*
- *Famines are political as well as environmental.*
- *Overeating is a growing world problem.*
- *A balanced diet is essential for health.*

Despite repeated predictions that runaway population growth would lead to terrible famines (chapter 7), world food supplies have more than kept up with increasing human numbers over the past two centuries. Although population growth slowed to an average 1.08 percent per year during that time, world food production increased an average of 2.2 percent per year. Food availability has improved because of increasing use of irrigation and fertilizers, improved crop varieties, and better distribution systems. Globally, we consume an average of nearly 3,000 kcal per day, well above the 2,200 kcal considered necessary for a healthy and productive life (fig. 9.2*a*). Residents of industrialized countries consume over 3,500 kcal every day. For most of the world, daily protein consumption is also growing, often beyond the 40–60 g per day recommended for a healthy and productive life (fig. 9.2*b*).

The UN Food and Agriculture Organization (FAO) expects world food supplies to continue to increase faster than population growth. In some countries, such as the United States, the problem has long been what to do with surplus food. High production leads to low prices, which make farm profits chronically low. Farmers in these countries are paid billions of dollars per year to keep marginal land out of production. Studies indicate that we still have room to expand farmland and increase production to feed 9 billion people in a few decades, by sacrificing forests and water resources. But is this a good idea? The main question may not be whether we can produce enough food. More important concerns may be how can we improve access to food, how do we provide the right kinds of food, and what are the environmental costs of food production systems?

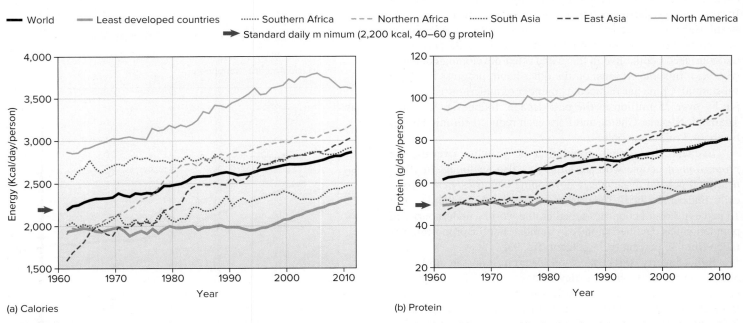

(a) Calories

(b) Protein

FIGURE 9.2 Changes in dietary energy (local) and protein consumption in selected regions. North America and other developed regions consume more calories and protein than are needed.
Source: Data from Food and Agriculture Organizations (FAO), 2015

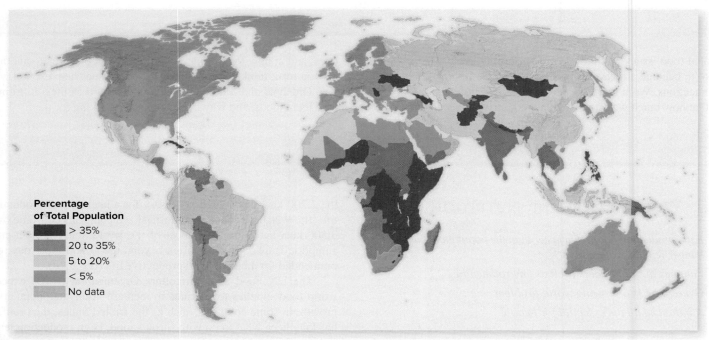

FIGURE 9.3 Hunger rates worldwide. The most severe and chronic hunger is in developing regions, especially sub-Saharan Africa.
Source: Food and Agriculture Organization of the United Nations.

Millions of people are still chronically hungry

Despite bountiful production, some 800 million people in the world today are considered **chronically undernourished,** getting less than the minimum 2,200 kcal per day. This number represents a tragic and persistent problem. It also represents a decline from around 1 billion chronically hungry 25 years ago. In some regions, however, hunger remains widespread, especially in sub-Saharan Africa and in South Asia—which includes India, the second most populous country on earth. Over 95 percent of malnourished people live in these and other developing countries (fig. 9.3). In some regions, on the other hand, progress has been substantial. China has reduced its number of undernourished people by over 75 million in the past decade. Indonesia, Vietnam, Thailand, Nigeria, Ghana, and Peru each reduced chronic hunger by about 3 million people.

Because the global population is growing overall, the *proportion* of people who are hungry is also declining. In 1960, nearly 60 percent of people in developing countries were chronically undernourished. Today that proportion has fallen to just 15 percent. For all countries together, the world's undernourished population has declined slightly in number, but the proportion has fallen from 37 percent to 11 percent (fig. 9.4).

Still, poverty threatens **food security,** or the ability to obtain sufficient food on a day-to-day basis. The 1.5 billion people in the world who live on less than $1 per day often can't afford to buy the food they need and lack the resources to grow it for themselves. Food security is a concern at multiple scales. In the poorest countries, hunger may affect nearly everyone. In other countries, although the average food availability may be satisfactory, some individual communities or families may not have enough to eat. And within families, males often get both the largest share and the most nutritious food, while women and children—who need food most—all too often get the poorest diet. At least 6 million children under 5 years old die every year

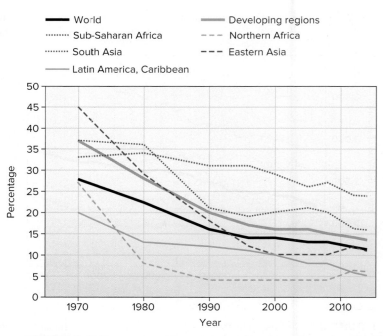

FIGURE 9.4 Changes in numbers and rates of malnourishment by region.
Source: Data from UN Food and Agriculture Organization, 2015

(one every 5 seconds) from hunger and malnutrition. Providing a healthy diet might eliminate as much as 60 percent of all premature deaths worldwide.

People who are hungry get stuck in a cycle of poverty. Nobel Prize–winning economist Robert Fogel has estimated that in 1790 about 20 percent of the population of England and France was effectively excluded from the labor force because they were too weak and hungry to work. Improved nutrition, he calculates, accounted for about half of all European economic growth during the nineteenth century. Because many developing countries are as poor now (in relative terms) as Britain and France were in 1790, his analysis suggests that reducing hunger could yield more than $120 billion in economic growth, produced by longer, healthier, more productive lives for several hundred million people.

Recognizing the role of women in food production is an important step toward food security for all. Throughout the developing world, women do 50 to 70 percent of all farm work but control only a tiny fraction of the land and rarely have access to capital or developmental aid. In Nigeria, for example, home gardens occupy only 2 percent of all cropland but provide half the food families eat. Making land, credit, education, and access to markets available to women could contribute greatly to family nutrition.

Famines usually have political and social causes

Chronic hunger and malnutrition can be silent and often invisible, affecting individuals, families, and communities on an ongoing basis. **Famines,** on the other hand, are characterized by large-scale food shortages, massive starvation, social disruption, and economic chaos. Starving people, especially those uprooted from their farms and villages, may be forced to eat their seed grain and slaughter their breeding stock in a desperate attempt to keep themselves and their families alive. Even if better conditions return, they often have sacrificed their productive capacity or lost their land, and recovery will be slow and difficult. Famines characteristically involve mass migrations, because starving people travel to refugee camps in search of food and medical care (fig. 9.5).

In 2014, the FAO reported that 51 million people in 36 countries (most in sub-Saharan or East Africa) needed emergency food aid. What causes these emergencies? Droughts, earthquakes, severe storms, and other natural disasters are frequently the immediate trigger, but politics and economics are often equally important. Bad weather, insect outbreaks, and other environmental factors cause crop failures and create food shortages. But the Nobel Prize–winning work of Harvard economist Amartya K. Sen has shown that these factors are normally present, and people usually have adaptations to get through hard times if they aren't thwarted by inept or corrupt governments and greedy elites. Policies that favor the elite, however, can make it impossible for poor people to stay on the land and grow food. War and political oppression can drive people from their farms. Professor Sen points out that armed conflict and political oppression almost always are at the root of famine. No democratic country with a relatively free press, he says, has ever had a major famine.

FIGURE 9.5 Children wait for their daily ration of porridge at a feeding station in Somalia. When people are driven from their homes by hunger or war, social systems collapse, diseases spread rapidly, and the situation quickly becomes desperate.
© Norbert Schiller/The Image Works

The aid policies of rich countries often don't help as much as we hope. Despite our best intentions, aid often serves as a way to get rid of surplus commodities rather than stabilize local food production in recipient countries. Even emergency food aid has ambiguous effects. Herding people into feeding camps can badly destabilize communities, and crowding and lack of sanitation in the camps exposes people to diseases. There are no jobs in the refugee camps, so people can't support themselves. Corruption and violence can occur at food-dispensing centers, where aid recipients are vulnerable to theft, extortion, and sexual violence from aid providers. Having left their land and tools behind, people may have difficulty returning to their farms when conditions return to normal.

Overeating is a growing world problem

Although hunger persists, world food supplies are increasing. This is good news, but the downside is increasing overweight and obese populations. In the United States, and increasingly in developing countries, highly processed foods rich in sugars and fats have become a large part of daily diets. Some 64 percent of adult Americans are overweight, up from 40 percent only a decade ago. About one-third of us are seriously overweight, or **obese.** Obesity is quantified in terms of the body mass index (BMI), calculated as weight/height^2. For example, a person weighing 100 kg and standing 2 m tall (220 lb and 6 ft 6 inches) would have a body mass of (100 kg/4m^2) or 25 kg/m^2. Health officials consider a BMI greater than 25 kg/m^2 overweight; over 30 kg/m^2 is considered obese.

Globally, nearly 2 billion adults (18 and older) are overweight, according to a the World Health Organization. This number represents 39 percent of the world's adult population. More than twice as many people are overweight than underweight (870 million). About 13 percent of adults are obese (BMI greater than 30 kg/m^2). This trend is no longer limited to richer countries. Obesity is spreading around the world as Western diets and lifestyles are increasingly adopted in the developing world (fig. 9.6).

Being overweight substantially increases the risk of hypertension, diabetes, heart attacks, stroke, gallbladder disease, osteoarthritis, respiratory problems, and some cancers. In the United States, about 400,000 people die from illnesses related to obesity every year. This number is approaching the number of deaths related to smoking (435,000 annually). Weight-related illnesses and disabilities are now a serious strain on health care systems and health care budgets worldwide.

Growing rates of obesity result partly from increased consumption of oily and sugary foods and soft drinks, and partly from lifestyles that involve less walking, less physical work, and more leisure than previous generations had. Changing these factors can be hard. Just walking to work regularly can be enough to keep weight down, but many of our daily routines are built around sitting still, at a desk or in a car. Many of our social activities, and our traditional holiday meals, focus on rich foods with gravies and sauces, or sweets. We are probably biologically adapted to prefer these energy-rich foods, which were rare and valuable for our ancestors. Today, it can take special effort to cut back on them.

Another cause is the economic necessity for food producers to increase profits. When we already have plenty to eat, and when food prices are low, food processors struggle constantly to ensure continuous growth in production and profits. Manufacturers can achieve better profits with "value added" products: Instead of selling plain oatmeal at 50 cents a pound, a manufacturer might convert oats into flavored, sweetened, instant microwavable oatmeal for $2.50 a pound. Better yet, processing the oats into sweetened, toasted oat flakes might bring $5 a pound. Increases in sugar and

fat content, as well as constant exposure to advertising, encourages us to consume more than we might really need.

Paradoxically, food insecurity and poverty can also contribute to obesity. In one study, more than half the women who reported not having enough to eat were overweight, compared with one-third of the food-secure women. Lack of good-quality food may contribute to a craving for carbohydrates in people with a poor diet. A lack of time for cooking, and limited access to healthy food choices along with ready availability of fast-food snacks and calorie-laden soft drinks, also lead to dangerous dietary imbalances.

We need the right kinds of food

Michael Pollan, who writes about food issues at the University of California, Berkeley, says that plain, simple food is what our bodies are adapted to. Products made of manufactured foodlike substances that your grandmother wouldn't recognize probably are not good for you. Pollan sums up the answer to health and obesity problems this way: "Eat food. Not too much. Mostly plants."

Generally, eating a good variety of foods provides the range of nutrients you need. In general, it's best to have whole grains and vegetables, with only sparing servings of meat, dairy, fats, and sweets. Based on observations of the health effects of Mediterranean diets, as well as a long-term study of 140,000 U.S. health professionals, Dr. Walter Willett and Dr. Meir Stampfer of Harvard University have recommended a dietary pyramid that minimizes red meat and starchy food such as white rice, white bread, potatoes, and pasta (fig. 9.7). Nuts, legumes (beans, peas, and lentils), fruits, vegetables, and whole grain foods form the basis of this diet. The base of this Harvard pyramid is regular, moderate exercise.

Food-insecure people often can't afford the protein, fruits, and vegetables that would ensure a balanced diet. Starchy foods like maize (corn), polished rice, and manioc (tapioca) form the bulk of the diet for poor populations, especially in developing countries. Even if they get enough calories, they may lack sufficient protein, vitamins, and trace minerals. **Malnourishment** is a term for nutritional imbalance caused by a lack of specific dietary components or an inability to absorb or utilize essential nutrients.

The FAO estimates that perhaps 3 billion people (nearly half the world population) suffer from vitamin, mineral, or protein deficiencies. Effects can include devastating illnesses and deaths, as well as slowed mental and physical development. These problems bring an incalculable loss of human potential.

Anemia (low hemoglobin levels in the blood, usually caused by dietary iron deficiency) is the most common nutritional problem in the world. More than 2 billion people suffer from iron deficiencies, especially women and children. The problem is most severe in India, where 80 percent of all pregnant women may be anemic. Anemia increases the risk of maternal deaths from hemorrhage in childbirth and affects childhood

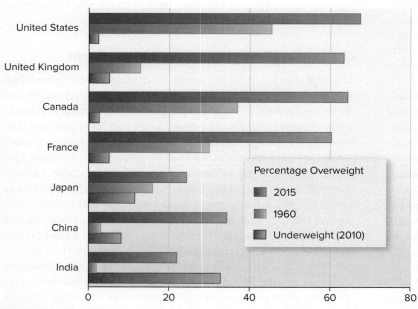

FIGURE 9.6 While over 800 million people are chronically undernourished, more than twice as many are at risk from eating too much. (Latest available data are shown.) *Source: Data from World Health Organization 2015*

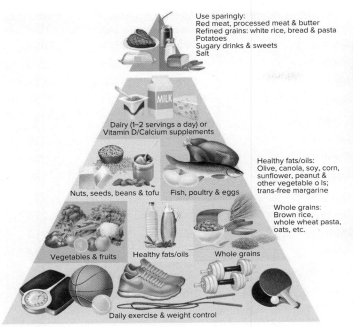

Use sparingly:
Red meat, processed meat & butter
Refined grains: white rice, bread & pasta
Potatoes
Sugary drinks & sweets
Salt

Dairy (1–2 servings a day) or
Vitamin D/Calcium supplements

Healthy fats/oils:
Olive, canola, soy, corn,
sunflower, peanut &
other vegetable oils;
trans-free margarine

Nuts, seeds, beans & tofu   Fish, poultry & eggs

Whole grains:
Brown rice,
whole wheat pasta,
oats, etc.

Vegetables & fruits   Healthy fats/oils   Whole grains

Daily exercise & weight control

FIGURE 9.7 The Harvard food pyramid emphasizes fruits, vegetables, and whole grains as the basis of a healthy diet. Red meat, white rice, pasta, and potatoes should be used sparingly.
Source: Data from Willett and Stampfer, 2002

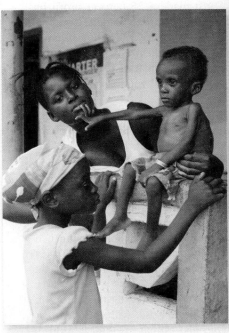
(a) Marasmus

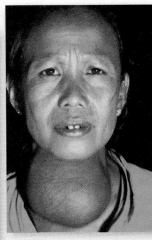
(b) Goiter

FIGURE 9.8 Dietary deficiencies can cause serious illness. (*a*) Marasmus results from protein and calorie deficiency and gives children a wizened look and dry, flaky skin. (*b*) Goiter, a swelling of the thyroid gland, results from an iodine deficiency.
(a) © Burger/Phanie/age fotostock; (b) © Lester Bergman/Corbis

development. Red meat, eggs, legumes, and green vegetables all are good sources of dietary iron.

Protein deficiency can cause conditions such as kwashiorkor and marasmus. **Kwashiorkor** is a West African word meaning "displaced child." (A young child is displaced—and deprived of nutritious breast milk—when a new baby is born.) This condition most often occurs in young children who subsist mainly on cheap starchy foods. Children with kwashiorkor often have puffy, discolored skin and a bloated belly. **Marasmus** (from the Greek "to waste away") is caused by shortages of both calories and protein. A child suffering from severe marasmus is generally thin and shriveled (figure 9.8*a*). Children with these deficiencies have low resistance to infections and may suffer lifelong impacts on mental and physical development.

Iodine is essential for regulating metabolism and brain development. Chronic iodine deficiency causes goiter (a swollen thyroid gland, figure 9.8*b*), stunted growth, and mental impairment. The FAO estimates that 740 million people—mainly in South and Southeast Asia—suffer from iodine deficiency and that 177 million children have stunted growth and development. Adding a few pennies' worth of iodine to table salt has nearly eliminated this problem in developed countries.

Vitamin A deficiencies affect 100–140 million children at any given time. At least 350,000 go blind every year from the effects of this vitamin shortage. Folic acid, found in dark green, leafy vegetables, is essential for early fetal development. Ensuring access to leafy greens can be one of the cheapest ways of providing essential vitamins.

High prices remain a global problem

If we have surplus production, why are high food prices still a problem for struggling families? Nonindustrialized farming economies such as India have also seen long-term price declines, yet impoverished populations still suffer acutely with shorter-term increases in prices for cooking oil, wheat, or other staples. Price changes are inconvenient for food-secure populations. For impoverished families in developing regions, where more than half of the household income may be used to buy food, high prices can be devastating. Why do food prices rise despite global abundance?

Floods, droughts, and storms often trigger spikes in food prices. And droughts and heat waves associated with climate change are reducing supplies in many areas (chapter 15). But a growing factor is that food prices are often driven by global commodity prices. Increasingly, food is traded as an investment commodity, with bankers and investors buying and selling shares in multinational food companies. Traders in Chicago, London, and Tokyo purchase volumes of grain, sugar, coffee, or other commodities simply to make a profit on the trade. This trade can easily inflate prices paid by poor villagers in developing areas. Often, global trade involves speculation and trading in futures: I might promise to pay you $4 a bushel for next summer's corn crop, even though the planting season hasn't even started yet, just so I can reserve the crop and settle the price now. But if there's drought in the spring and the year's production looks poor, someone who really needs the corn might pay me $5 a bushel for the same crop that's not yet in the ground. I just made a 25 percent profit on a future corn crop, and my shareholders are delighted. Consumers somewhere else will cover the higher costs. Trading in commodities and futures, then, can drive food prices, even though the exchanges are far removed from the actual food that a farmer plants and a consumer eats. And expected future shortages can drive up prices today.

To complicate matters further, food prices are driven by non-food demands for crops. In 2007–2008, United States corn prices jumped from around $2 a bushel to over $5 a bushel when the U.S. Congress promised to subsidize corn-based ethanol fuel and to require that ethanol be sold at gas stations nationwide. In that year, future speculation for ethanol drove up corn prices, and wheat and other grains followed in the excitement. Because of the ethanol boom, many small bakers and pasta makers couldn't afford wheat and were driven out of business, and U.S. consumers were pinched as food prices rose throughout the grocery store. Federal policies requiring the use of corn to produce ethanol continue to inflate corn demand, and prices, in many markets.

The same process occurred in 2008–2010 after the European Union passed new rules requiring biofuel use, with the idea that these fuels would be sustainable and climate neutral. Europe's biofuels are produced largely from palm oil, a tropical oily fruit grown mainly in Malaysia and Indonesia. European biofuel rules produced a boom in global palm oil demand. Unfortunately, palm oil is also a cooking staple for poor families across Asia, for whom a doubling of oil prices can be devastating. In developing countries across the globe, riots broke out over rising cooking oil prices, which were driven by well-meaning European legislation for Malaysian biofuel. The palm oil boom is also driving accelerated deforestation and wetland drainage across Malaysia, Indonesia, Ecuador, Colombia, and other palm-oil-producing regions, leading to further social and environmental repercussions (chapter 12).

Section Review

1. How many people in the world are chronically undernourished? What does chronically undernourished mean?

2. List at least five African countries with high rates of hunger (fig. 9.3; use a world map to help identify countries).

3. What are some of the health risks of overeating? What percentage of adults are overweight in the United States?

9.2 KEY FOOD SOURCES

- *Rice, wheat, and a few other crops provide most food.*
- *Meat and fish give excellent protein but consume resources.*
- *Antibiotic overuse is a serious concern in meat production.*

Of the thousands of edible plants and animals in the world, only about a dozen types of seeds and grains, three root crops, 20 or so common fruits and vegetables, six mammals, and two domestic fowl make up most of the food that humans eat (table 9.1). The three crops on which humanity depends for the majority of its

Table 9.1 Key Global Food Sources			
Crop	1965*	1990	2014
Sugar cane	531	1,053	1,911
Maize	227	483	1,018
Rice	254	519	741
Wheat	264	592	716
Milk	358	524	716
Potatoes	271	267	376
Vegetables	66	140	280
Cassava	86	152	277
Soybeans	32	108	276
Meat	72	158	272
Barley	93	144	178
Sweet potatoes	108	123	103
Dry beans, pulses	31	41	44

*Production in million metric tons
Source: Data from UN FAO, 2015

nutrients and calories are wheat, rice, and maize (called corn in the United States). Together, over 2 billion metric tons of these three grains are grown each year. Wheat and rice are especially important as the staple foods for most of the 5.5 billion people in the developing countries of the world. These two grass species supply around 60 percent of the calories consumed directly by humans. As Table 9.1 shows, production of all major crops has increased in the past 50 years.

Dominant crops often depend on local climates. Potatoes, barley, oats, and rye grow well in cool climates, and these are staples in mountainous regions and high latitudes (northern Europe, north Asia). Cassava, sweet potatoes, and other roots and tubers grow well in warm, wet areas and are staples in Amazonia, Africa, Melanesia, and the South Pacific. Sorghum and millet are drought-resistant, and they are staples in the dry regions of Africa.

Fruits, vegetables, and vegetable oils are usually the most important sources of vitamins, minerals, dietary fiber, and complex carbohydrates. In the United States, however, grains make up a far larger part of agricultural production. Corn is by far the most abundant crop, followed by soybeans and wheat (fig. 9.9). Of these three, only wheat is primarily consumed directly by humans. Corn and soy are processed into products such as fuel, livestock feed, or high-fructose corn syrup.

Rising meat production has costs and benefits

Dramatic increases in corn and soy production have led to rising meat consumption worldwide. In developing countries, meat consumption has risen from just 10 kg per person per year in the 1960s to over 26 kg today (fig. 9.10). In the same interval, meat consumption in the United States has risen from 90 kg to 136 kg per person per year. Meat is a concentrated, high-value source of protein, iron, fats, and other nutrients that

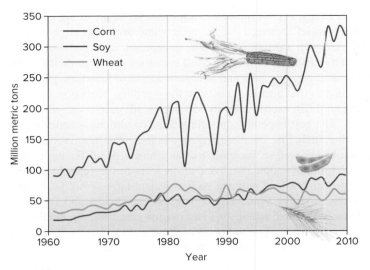

FIGURE 9.9 United States production of the three dominant crops, corn, soybeans, and wheat.
Source: Data from USDA and UN FAO, 2015

FIGURE 9.10 Meat and dairy consumption have quadrupled in the past 40 years, and China represents about 40 percent of that increased demand.
© William P. Cunningham

give us the energy to lead productive lives. Dairy products are also a key protein source: globally we consume more than twice as much dairy as meat. But dairy production per capita has declined slightly while global meat production has doubled in the past 45 years.

Meat is a good indicator of wealth, because it is expensive to produce, in terms of the resources needed to grow an animal (fig. 9.11). As discussed in chapter 3, herbivores use most of the energy they consume for moving and growing; only a portion of energy consumed is stored for consumption by carnivores. A beef steer consumes over 8 kg of grain to produce just 1 kg of beef. Pigs, being smaller, are more efficient. Just 3 kg of pig feed are needed to produce 1 kg of pork. Chickens and herbivorous fish (such as catfish) are still more efficient.

Globally, over one-third of cereals (some 660 million metric tons) are used as livestock feed each year. We could feed about eight times as many people by eating that cereal directly rather than converting it to meat. What differences do you suppose it would make if we did so?

A number of technological and breeding innovations have made this increased production possible. One of the most important is the **confined animal feeding operation (CAFO),** where animals are housed and fed—mainly on soy and corn—for rapid growth (fig. 9.12). These operations dominate livestock raising in the United States, Europe, and increasingly in China. Animals are housed in giant enclosures, with up to 10,000 hogs or a million chickens in an enormous barn complex, or 100,000 cattle in a feedlot. These systems require specially prepared mixes of corn, soy, and animal protein that maximize animals' growth rate. New breeds of livestock have been developed that produce meat rapidly, rather than simply getting fat. The turnaround time is getting shorter, too. A U.S. chicken producer can turn baby chicks into chicken nuggets after just eight weeks of growth. Steers reach full size by just 18 months of age.

Constant use of antibiotics, which are often mixed in daily feed, is also necessary for growing animals in such high densities and with unnaturally rich diets. Over 11 million kg of antibiotics are added to animal feed annually in the United States, about eight times as much as is used in human therapy. Nearly 90 percent of U.S. hogs receive antibiotics in their feed.

Because modern meat production is based on energy-intensive farming practices (see chapter 10), meat is also an energy-intensive product. It takes about 16 times as much fossil fuel energy to produce a kilogram of beef as it takes to produce a kilogram of vegetables or rice. The UN Food and Agriculture Organization estimates that livestock produce 20 percent of the world's greenhouse gases, more than is produced by transportation. In fact, by some estimates Americans could cut energy consumption more if we gave up just one-fifth of our meat consumption than if all of us were to drive a hybrid-electric Prius.

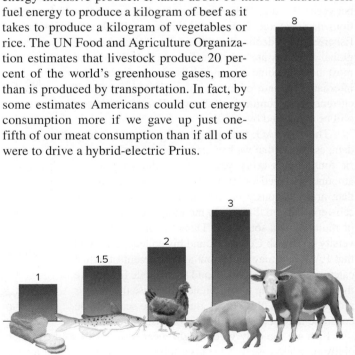

FIGURE 9.11 Number of kilograms of grain needed to produce 1 kg of bread or 1 kg live weight gain.

FIGURE 9.12 Concentrated feeding operations fatten animals quickly and efficiently, but create enormous amounts of waste and expose livestock to unhealthy living conditions.
Source: Photo by Tim McCabe, USDA Natural Resources Conservation Service

Seafood is our only commercial wild-caught protein source

Seafood is the main animal protein source for about 1.5 billion people in developing countries. Most of those people eat mainly locally caught fish. In wealthier countries, consumers eat fish from everywhere in the world, caught by huge, industrial-scale fishing ships. In the 1950s, the fishing industry developed freezer technology that allowed oceangoing factory ships to harvest the world's oceans and bring the catch home in huge volumes of frozen fish. Annual catches of ocean fish rose by about 4 percent annually between 1950 and 1988. Then catches began to fall, as fish populations declined and disappeared. Since the 1980s, all major marine fisheries have declined dramatically. An estimated 90 percent of global fisheries are now overfished or depleted, and most have become commercially unsustainable. An international team of marine biologists warns that if current trends continue, all the world's major fisheries will be exhausted by 2050.

The UN FAO, the only global source of fishery data, estimates that we harvest around 95 million metric tons of fish every year, roughly the same as the amount produced on fish farms (fig. 9.13). Fishery data are notoriously uncertain, however, as they are self-reported and difficult to monitor. A detailed study of multiple data sources by Daniel Pauly, of the University of British Columbia, and his colleagues, found that FAO data may be as low as 50 percent of actual catches. Illegal, unreported, and small-scale fishing are thought to make up the difference. Because the trade is global and difficult to track, much of the seafood you consume may be caught illegally or unsustainably.

Fish are the only wild-caught meat source still sold commercially on a large scale. Because nobody owns fish populations in the open ocean, fishers compete to deplete the population before competitors do. Even in exclusive national waters, illegal fishing is common. Rising numbers of boats, with increasingly efficient technology, exploit the dwindling resource. Ships as big as ocean liners travel thousands of kilometers and drag nets large enough to scoop up a dozen jumbo jets, sweeping a large patch of ocean clean of fish in a few hours. Most countries subsidize their fishing fleets to preserve jobs and to ensure access to fisheries. The FAO estimates that operating costs for the 4 million boats now harvesting wild fish exceed sales by $50 billion per year.

The best hope for controlling these conditions is better monitoring, including satellite imagery and increased international regulation, as well as increasing establishment of no-fishing zones, which protect populations and allow a species to reproduce. Several Pacific island nations, such as Palau, have set up no-fishing zones in recent years, and both President Obama and President George W. Bush established vast marine sanctuaries in the Pacific (see Chapter 6). As monitoring of these areas improves, it is hoped that they will help stabilize global fish populations.

Most commercial fishing operates on an industrial scale

Commercial fishing uses many techniques, depending on the target species. The most common of these are trawling, seining, and longlining. **Trawling** involves dragging a large net, often more than 50 m wide and 10 m tall. Bottom trawlers drag nets along the seafloor to catch shrimp or bottom-dwelling fish, such as halibut or sole (fig 9.14). This method is efficient, but the heavy nets destroy sensitive habitats, often this can include spawning grounds necessary for fish reproduction. Some critics say trawling is like harvesting forest mushrooms with a bulldozer; it's cheap, quick, and effective, but destroys the resource for future use. Trawling also produces tremendous amounts of "bycatch," unwanted marine life, which is usually thrown back dead or dying. In the North Pacific shrimp fishery, for

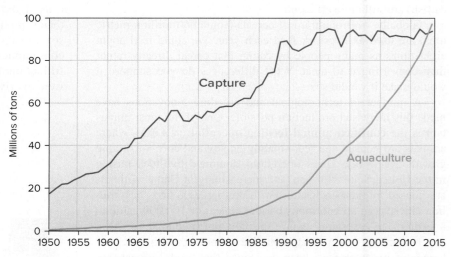

FIGURE 9.13 Estimates of the amount of fish produced globally by fishing and aquaculture. Despite continuously increasing fishing efforts, catches have not increased since the 1990s. Global fisheries kept production up by pursuing ever smaller, more remote species. Aquaculture has recently become an important substitute.
Source: Data source UN FAO FishStatJ 2015

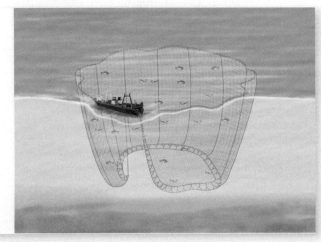

FIGURE 9.14 Trawlers drag a weighted net, often along the bottom, to capture everything in their path. Purse seines encircle a school of fish, then draw the "purse" closed.

example, as much as 98 percent of the catch from trawlers can be discarded as unwanted bycatch, producing over four million tons of waste seafood per year. Increasingly, however, bycatch may be kept to use as feed in fish farms (see below). There are reports that the Chinese shrimp trawl fleet discards very little bycatch, but rather uses it for aquaculture or human food. More damaging, however, are dredges, which use a heavy iron frame, often with rakes attached, to dredge up shellfish and other bottom-dwelling animals, which are then caught in a net attached to the dredge.

Purse seining uses a huge net to surround a school of fish. The bottom of the net is then pulled closed, something like a drawstring purse. Floats prevent fish from escaping out the top. This method is used to catch schooling fish, such as tuna, sardines, herring, and mackerel. It also is used to catch species that gather to spawn in large groups, such as squid. Tuna seiners often set their nets on pods of dolphins, which frequently indicate schools of tuna. The seiners are supposed to allow the dolphins to escape before bringing in the nets, but they often don't do so, and dolphin mortality can be high. Turtles, sharks, sea lions, and other species may also be swept up in giant purse seines, which can be a mile in circumference.

Longlining uses a central fishing line that can be up to 50 miles (80 km) long. This line is strung with shorter lines with multiple baited hooks every couple of meters. A single longline might have 25,000 hooks. Longlines can be set near the surface to catch open-ocean fish, such as tuna or swordfish, or anchored to the bottom to catch bottom-dwellers, such as cod or halibut. Shallow-set lines often catch dolphins, sharks, seabirds, or turtles that are attracted to the bait. By sinking lines deeper or using circular hooks, fishers can reduce unintended catch.

Aquaculture produces about half our seafood

Aquaculture (growing aquatic species in net pens or tanks) can involve farmed marine species, such as salmon or tuna, or freshwater fish in ponds. Many freshwater fish are herbivores, such as carp, so they can be fed on grain or plant material. Most ocean fish, however, are carnivorous and require wild-caught fish

for feed. About one-third of all wild-caught fish is used as food for fish in these operations. Because farmed carnivorous species such as salmon, sea bass, and tuna consume so much wild-caught fish, they also threaten wild fish populations and the seabirds and organisms that depend on them.

In salt-water (marine) aquaculture, net pens are anchored in nearshore areas, where open water circulates through, removing feces and uneaten food particles. A drawback of these systems is that they introduce diseases, with many fish crowded in a small area, as well as pollution and antibiotics into surrounding ecosystems (fig. 9.15). Farmed shrimp are usually produced in ponds built on former mangrove forests and wetlands. These environments provide essential nurseries for marine ecosystems. Studies have shown that the costs of lost ecosystem services can greatly outweigh the value of shrimp produced.

FIGURE 9.15 Fish farms produce about half our seafood, but most rely on wild-caught feed, and pollution and disease are common concerns.
© WaterFame/Alamy Stock Photo

FIGURE 9.16 This state-of-the-art lagoon was built to store manure from a hog farm. Odors, airborne dispersal of ammonia and bacteria, and overflows after storms are risks of open lagoons, but more thorough waste treatment is more expensive.
Source: Photo by Jeff Vanuga, USDA Natural Resources Conservation Service

Aquaculture in land-based ponds or warehouses can eliminate many of these problems, especially when raising herbivorous fish, such as catfish, carp, or tilapia, which consume plant-based feed. These species consume less feed per pound of meat than do carnivorous species, such as farmed salmon. In China, for example, most fish are raised in ponds or rice paddies. One ecologically balanced system uses four carp species that feed at different levels of the food chain. The grass carp, as its name implies, feeds largely on vegetation, while the common carp feeds on detritus that settles to the bottom. Silver carp and bighead carp are filter feeders that consume phytoplankton and zooplankton, respectively. Agricultural wastes such as manure, dead silkworms, and rice straw fertilize ponds and encourage phytoplankton growth. These integrated polyculture systems typically boost fish yields per hectare by 50 percent or more compared with single-crop farming.

Antibiotics are overused in intensive production

In any system that confines many animals in a small space, infections and diseases are likely. Fish farms, beef feedlots, large dairy herds, and pig and poultry operations all rely on antibiotics to keep their stock alive and healthy. In confined situations, bacteria thrive in the manure in the feedlots, or liquid wastes in manure storage lagoons (holding tanks) around hog farms, and in airborne dust around feedlots and manure tanks, which usually are uncontained and adjacent to barns (fig. 9.16). Antibiotics also make livestock put on weight faster. For both reasons, antibiotics are often mixed with daily feed, keeping animals on a constant, low dose.

Constant exposure to antibiotics produces strains of bacteria that are resistant to these drugs, however. This is a simple case of evolution: most of a population is killed off by an antibiotic, but a few resistant individuals are unaffected, and these strains grow and thrive. Antibiotic resistance is increasingly common around livestock operations, as well as in hospitals. This process is gradually rendering our standard antibiotics useless for human health care. The Centers for

Disease Control and Prevention reports that 23,000 Americans die every year from resistant infections (chapter 8). The next time you are prescribed an antibiotic by your doctor, you might ask whether she or he worries about antibiotic resistance, and think about how you would feel if your prescription were ineffectual against your illness.

Because antibiotic resistance is such an important public health issue, several European countries have phased out the use of antibiotics for livestock, except in emergencies. Better monitoring of animal health is a necessary part of this phase-out. Denmark, one of the world's largest pork exporters, outlawed regular feeding of antibiotics in 2000. Since the ban, Denmark's ham production has grown, not declined, and the price is about the same per pound as American ham. Despite vigorous industry opposition, American health agencies are striving to impose similar restrictions.

Alternative systems are also expanding

Despite large-scale development of these production systems, there is also increasing awareness of alternative food systems. In addition to these dominant forms, there is also growing interest in alternative agriculture that can reduce our dependence on oil, antibiotics, and other environmental costs of food production. Like the students described in the opening case study, many people are interested in supporting sustainable food production (What Do You Think?). Organic and sustainable foods are not just vegetables and fruits: meat, eggs, and dairy can be produced sustainably, too. Grass-fed beef, for example, can be an efficient way to convert solar energy into protein. Rotational grazing, using small, easily moved electric fences to concentrate grazing in one area of a field at a time, can invigorate pastures, distribute manure, and keep livestock healthy (fig. 9.17).

As more consumers express preferences for alternative foods and sustainable production practices, the availability of these options is increasing. Between 1994, when the U.S. Department of Agriculture began keeping records on farmers' markets, and 2012, the number of registered farmers' markets grew from 1,700 to more than 8,000.

FIGURE 9.17 Rotational grazing is one strategy for meat production with less reliance on energy, water, and other resources. Here, an electric fence contains cattle in one part of a pasture, while another part recovers for several weeks.
Source: USDA Natural Resources Conservation Service

Shade-Grown Coffee and Cocoa

Are your purchases of coffee and chocolate contributing to the protection or destruction of tropical forests? Coffee and cocoa are examples of food products grown exclusively in developing countries but consumed almost entirely in wealthy nations (vanilla and bananas are some other examples). Coffee grows in cool, mountain areas of the tropics, while cocoa is native to the warm, moist lowlands. Both are small trees of the forest understory, adapted to low light levels.

Until a few decades ago, most of the world's coffee and cocoa were grown under a canopy of large forest trees. Recently, however, new varieties of both crops have been developed that can be grown in full sun. Yields for sun-grown crops are higher because more coffee or cocoa trees can be crowded into these fields, and they capture more solar energy than in a shaded plantation.

There are costs, however, in this new technology. Sun-grown trees die earlier from the stress and diseases common in these fields. Furthermore, ornithologists have found that the number of bird species can be cut in half in full-sun plantations, and the number of individual birds may be reduced by 90 percent. Shade-grown coffee and cocoa generally require fewer pesticides (or sometimes none) because the birds and insects residing in the forest canopy eat many of the pests. Shade-grown plantations also need less chemical fertilizer because many of the plants in these complex forests add nutrients to the soil. In addition, shade-grown crops rarely need to be irrigated because heavy leaf fall protects the soil, while forest cover reduces evaporation.

Currently, about 40 percent of the world's coffee and cocoa plantations have been converted to full-sun varieties and another 25 percent are in the process of converting. Traditional techniques for coffee and

Cocoa pods grow directly on the trunk and large branches of cocoa trees.

© William P. Cunningham

cocoa production are worth preserving. Thirteen of the world's 25 biodiversity hot spots occur in coffee or cocoa regions. If all 20 million ha (49 million acres) of coffee and cocoa plantations in these areas are converted to monocultures, an incalculable number of species will be lost.

The Brazilian state of Bahia is a good example of both the ecological importance of these crops and how they might help preserve forest species. At one time, Brazil produced much of the world's cocoa, but in the early 1900s the crop was introduced into West Africa. Now Côte d'Ivoire alone grows more than 40 percent of the world total, and the value of Brazil's harvest has dropped by 90 percent. Côte d'Ivoire is aided in this competition by a labor system that reportedly includes widespread child slavery. Even adult workers in Côte d'Ivoire get only about $165 per year compared to a minimum wage of $850 per year in Brazil. As African cocoa production ratchets up, Brazilian landowners are converting their plantations to pastures or other crops.

The area of Bahia where cocoa was once king is part of Brazil's Atlantic forest, one of the most threatened forest biomes in the world. Only 8 percent of this forest remains undisturbed. Although cocoa plantations don't represent the full diversity of intact forests, they protect a surprisingly large sample of what once was there. And shade-grown cocoa can provide an economic rationale for preserving that biodiversity. Brazilian cocoa will probably never compete with that from other areas for lowest cost. There is room in the market, however, for specialty products. If consumers were willing to pay a small premium for organic, fair-trade, shade-grown chocolate and coffee, this might provide the incentive needed to preserve biodiversity. Wouldn't you like to know that your chocolate or coffee wasn't grown with child slavery and is helping protect plants and animal species that might otherwise go extinct?

In some cases, greater information and trade are helping bolster markets for alternative foods, too. Although at least 3,000 species of plants have been used for food at one time or another, most of the world's food now comes from only 16 species. There is considerable interest in expanding this number and developing new varieties. For example, the winged bean is not widely used, but it could provide an excellent food source (fig. 9.18). This perennial plant grows well in hot climates, and the entire plant is edible (pods, mature seeds, shoots, flowers, leaves, and tuberous roots), it is resistant to diseases, and it enriches the soil. Another promising crop is triticale, a hybrid between wheat (*Triticum*) and rye (*Secale*) that grows in light, sandy, infertile soil. It is drought-resistant, has nutritious seeds, and is being tested for salt tolerance for growth in saline soils or irrigation with seawater. Some traditional crop varieties grown by Native Americans, such as tepary beans, amaranth, and Sonoran panicgrass, are being collected by seed conservator Gary Nabhan both as a form of cultural revival for native people and as possible food crops for harsh environments.

Food systems are vulnerable to climate change

Most of the world's major crops have narrow temperature ranges for optimal growth. It's increasingly clear that a warming climate will reduce yields in many major crops. Crops suffer from heat in two major ways. First, physiologically they produce seeds and plant biomass best within temperature ranges of a few degrees. Second, plants transpire (release) moisture more in hot conditions. This moisture loss helps cool plants, but it also increases water demand. Increased irrigation is therefore likely to become important with a warming climate. Historically, heat waves and droughts have reduced a country's grain production by about 10 percent. Increasing frequency and severity of heat and drought are likely to have still stronger effects.

Vulnerability is greatest in low-income regions, where small-scale farms produce most basic foods and family income. In this context, people are especially vulnerable to loss of crop productivity with rising temperatures, to increased costs of irrigation, or to

FIGURE 9.18 Winged beans bear fruit year-round in tropical climates and are resistant to many diseases that prohibit growing other bean species. Whole pods can be eaten when they are green, or dried beans can be stored for later use. It is a good protein source in a vegetarian diet.
© William P. Cunningham

increasing frequency of drought (fig. 9.19). But wealthy regions also stand to suffer, because they often depend on just a few major crops, such as corn and soybeans. When crop varieties are optimized for high yields, not for drought tolerance, climate change threatens production, and income for growers.

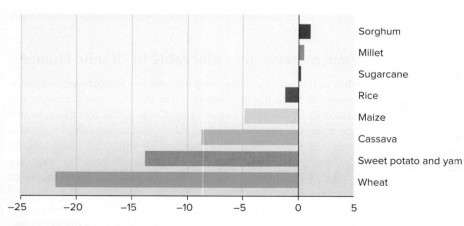

FIGURE 9.19 Expected changes in crop yields in Africa by 2050 with climate change, as compared to expected yields if the climate were not changing. Higher temperatures reduce plant growth in many crops, as well as increasing water demands.
Source: UN Environment Programme GEO5

Section Review

1. Why does meat use more inputs than plant foods?
2. List several benefits and costs of fish and shrimp farming.
3. Why are antibiotics used in livestock production?
4. What is rotational grazing? What are its benefits?

9.3 THE GREEN REVOLUTION AND GENETIC ENGINEERING

- *The green revolution increased production through breeding, fertilizer, and irrigation.*
- *Genetically modified crops have genes transferred from other organisms.*

The FAO predicts that 70 percent of growth in future agricultural production will come from higher yields and new crop varieties, because expanding arable lands is not a reasonable option in many areas. Development of more-intensive farming methods, therefore, is a matter of global interest. How will we accomplish these changes? Since the 1950s, our main improvements in farm production have come from breeding hybrid varieties of a few well-known, widely used species. Yield increases often have been spectacular. A century ago, corn in the United States yielded an average of about 25 bushels per acre. In 2015, because of high-yield varieties, more fertilizer, and other factors, average yields were more than 160 bushels per acre, and peak yields were over 350 bushels per acre. It is these kinds of increases that have allowed us to send food aid overseas and to develop value-added markets for major crops, especially livestock feed.

Starting in the 1950s and 1960s, agricultural research stations began to breed tropical wheat and rice varieties that would provide food for growing populations in developing countries. Cross-breeding plants with desired traits created new, highly productive hybrids known as "miracle" varieties. The first of these was a dwarf, high-yielding wheat developed by Norman Borlaug, who received a Nobel Peace Prize for his work, at a research center in Mexico (fig. 9.20). At about the same time, the International Rice Research Institute in the Philippines developed dwarf rice strains with three or four times the production of varieties in use at the time.

Green revolution crops are high responders

The dramatic increases obtained as these new varieties spread around the world has been called the **green revolution.** The success of these methods is one of the main reasons world food supplies have more than kept pace with the growing human population over the past few decades. Miracle varieties were spread around the world as U.S. and European aid programs helped developing countries adopt new methods and seeds. The green revolution replaced

FIGURE 9.20 Semi-dwarf wheat (*right*), bred by Norman Borlaug, has shorter, stiffer stems and is less likely to lodge (fall over) when wet than its conventional cousin (*left*). This "miracle" wheat responds better to water and fertilizer, and has played a vital role in feeding a growing human population.
© William P. Cunningham

traditional crop varieties and growing methods throughout the developed world, and nearly half of all farmers in the developing world were using green revolution seeds, fertilizers, and pesticides by the 1990s.

Most green revolution breeds really are "high responders." They yield more than other varieties if and only they have steady inputs of fertilizer, water, and pest control (fig. 9.21). Without irrigation and fertilizer, these high responders may yield less than traditional varieties. New methods and inputs are also expensive. Poor farmers who can't afford hybrid seeds, fertilizer, machinery, fuel, and irrigation are put at a disadvantage, compared to wealthier farmers who can afford these inputs. Thus, the green revolution is credited with feeding the world, but it is also often accused of driving poorer farmers off their land, as rising land values and falling commodity prices squeeze them from both sides.

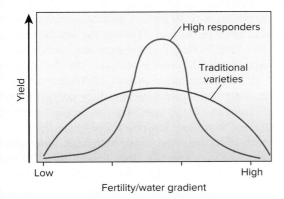

FIGURE 9.21 Green revolution miracle crops are really high responders, meaning that they have excellent yields under optimum conditions. For poor farmers who can't afford the fertilizer and water needed by high responders, traditional varieties may produce better yields.

Recent studies by the FAO and the UN Environment Programme, aimed at finding less expensive methods for low-income farmers, have found that diversified, low-input farming can provide superior food security in poor regions. Soil can be enriched with nitrogen-fixing plants and manure. Rotating crops and interplanting crops reduces pest dispersal. Some data suggest that if farmers are taught better soil management, they can meet or exceed conventional production by improving fertility without costly inputs (see section 10.4).

Genetic engineering moves DNA among species

Genetic engineering involves removing genetic material from one organism and splicing it into the chromosomes of another (fig. 9.22). This technology introduces entirely new traits, at a much faster rate compared to cross-breeding methods. It is now possible to build entirely new genes by borrowing bits of DNA from completely unrelated species, or even synthesizing artificial DNA sequences to create desired characteristics in **genetically modified organisms (GMOs).**

Genetically modified (GM) crops might provide dramatic benefits. Research is under way to improve yields and create crops that resist drought, frost, or diseases. Other strains are being developed to tolerate salty, waterlogged, or nutrient-poor soils. These would allow degraded or marginal farmland to become productive. All of these could be important for reducing hunger in developing countries. Rice researchers have spent decades working on varieties such as "Golden Rice," enriched with vitamin A by inserting genes from daffodils, soil bacteria, or corn. This yellow-colored rice is proposed as a nutrition supplement for very poor populations in developing countries. Vandana Shiva, an Indian food activist, has argued that this rice will be more expensive and less nutritious than traditional vegetables, greens, and other sources of mixed nutrients. Proponents of Golden Rice believe that it could provide essential nutrients to children in developing areas.

Efforts are under way to develop crops such as bananas and potatoes containing oral vaccines that can be grown in developing countries where refrigeration and sterile needles are unavailable. Plants have been engineered to make industrial oils and plastics. Animals, too, are being genetically modified to grow faster, gain weight on less food, and produce pharmaceuticals such as insulin in their milk. It may soon be possible to create animals with human cell–recognition factors that could serve as organ donors.

Most GMOs have been engineered for pest resistance or herbicide tolerance

By far, the dominant gene transfers involve either pest resistance or herbicide tolerance. The dominant pest resistance gene comes from a natural type of soil bacteria, *Bacillus thuringiensis* (Bt). The Bt gene produces toxins lethal to Lepidoptera (butterfly family) and Coleoptera (beetle family). The genes for some of these toxins have been transferred into crops such as maize (to protect against European cut worms), potatoes (to fight potato beetles), and cotton (for protection against boll weevils). This allows farmers to reduce insecticide spraying. Arizona cotton farmers, for

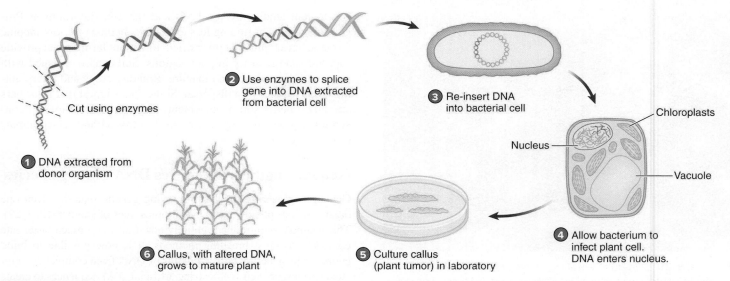

FIGURE 9.22 One method of gene transfer, using an infectious, tumor-forming bacterium such as agrobacterium. Genes with desired characteristics are cut out of donor DNA and spliced into bacterial DNA using special enzymes. The bacteria then infect plant cells and carry altered DNA into cells' nuclei. The cells multiply, forming a tumor, or callus, which can grow into a mature plant.

example, report reducing their use of chemical insecticides by 75 percent on Bt cotton fields.

Constant exposure to Bt toxin has led to the emergence of Bt tolerance in insect pests. Tolerance often develops when a pesticide kills off most of a population, leaving just a few individuals whose genetic traits allow them to survive the pesticide. These individuals then prosper and multiply, unaffected by the pesticide. Declining effectiveness of this natural pesticide—one of the few available to organic growers—is a significant worry for organic food producers, who often rely on natural Bt applications to combat insects. To protect the Bt trait, farmers are often required to plant a part of every field in non-Bt crops that will act as a refuge for nonresistant pests. The hope is that these "refuges" will maintain a nontolerant population of pests.

There also is a concern about the effects on nontarget species. In laboratory tests, about half of a group of monarch butterfly caterpillars died after being fed on plants dusted with pollen from Bt corn. Under field conditions, however, it has been difficult to demonstrate harm to butterflies.

The other major transgenic crops are engineered to tolerate herbicides. These crops are unaffected when fields are sprayed to kill weeds. These crops make up about three-quarters of all genetically engineered acreage. The two main products in this category are Monsanto's "Roundup Ready" crops—so-called because they tolerate Monsanto's best-selling herbicide, Roundup (glyphosate)—and AgrEvo's "Liberty Link" crops, which resist that company's Liberty (glufosinate) herbicide. Because crops with these genes can grow in spite of high herbicide doses, farmers can spray fields heavily to exterminate weeds. This practice allows for conservation tillage and leaving more crop residue on fields to protect topsoil from erosion, both good ideas, but it may also mean an increase in herbicide use.

Like antibiotics, heavily-used herbicides also produce herbicide-tolerant pests (see chapter 10). In 2014, the US Environmental Protection Agency permitted the use of Enlist Duo, a combination of glyphosate and 2,4-D, two widely used herbicides developed by Dow AgroSciences. The new combined herbicide was designed in combination with new GM varieties of corn and soybeans, designed to tolerate both chemicals. As with Roundup Ready crops, the gene modification is designed to support expanded use of the herbicide, as previous weed-control chemicals became impotent.

Safety of GMOs is widely debated

GM crops have been introduced to the world's farmers even more rapidly than green revolution crops were in the 1960s and 1970s. A decade after their introduction in 1996, GM varieties were planted on 400 million hectares (1 billion acres) of farmland. Three years later, that number had doubled to 800 million ha (2 billion acres). This represents just over half of the world's 1.5 billion ha of cultivated land. The United States accounted for 56 percent of that acreage, followed by Argentina with 19 percent. In 2005, China approved GM rice for commercial production. This was the first GM cereal grain approved for direct human consumption and could move China into the forefront of GM crop production. The first GM animals developed for human consumption are GM Atlantic salmon, which grow much faster than normal because they contain growth hormone genes from an oceanic pout. The "enviropig," meanwhile, is being engineered to produce low-phosphorus manure, which should reduce impacts of concentrated hog operations on water quality.

Federal monitoring agencies have decided that GM crops are "substantially equivalent" to non-GM varieties. Consumer groups question this judgment, arguing that we don't fully understand the potential effects of transferred genetic material. California and a number of other states have considered labeling laws to inform consumers about foods containing GM products, but industry groups have worked hard to prevent these policies from becoming law.

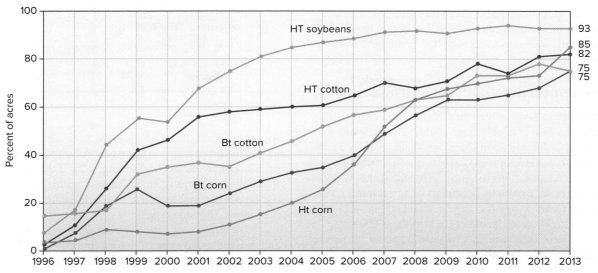

FIGURE 9.23 Growth of genetically engineered corn, cotton, and soy in the United States. HT (herbicide tolerant) varieties mainly tolerate glyphosate (Monsanto's Roundup); Bt varieties contain bacterial (*Bacillus thuringiensis*) proteins that kill insects.
Source: USDA Economic Research Service, 2011

Ultimately, GM foods may be safe for consumers but risky for producers. Farmers producing GM crops must commit to more and more costly herbicides, pesticides, and seeds to stay in business. Although many consumers are wary of GM foods, you have almost certainly eaten them. Over 70 percent of U.S. corn has Bt traits or herbicide tolerance, or both. Nearly 95 percent of soybeans and 80 percent of cotton are modified for herbicide tolerance (fig. 9.23). It has been estimated that over 60 percent of all processed food in America contains GM ingredients.

Section Review

1. What is the "green revolution," and why is it important?
2. What are genetically modified organisms?
3. Why are most U.S. corn and soybeans GM varieties?

9.4 FOOD PRODUCTION POLICIES

- *Government support promotes farm production and innovation.*
- *Food production is an important economic question.*

Much of the increase in food production over the past 50 years has been fueled by government support for agricultural education, research, and development projects that support irrigation systems, transportation networks, crop insurance, and direct subsidies. The World Bank estimates that rich countries pay their own farmers $350 billion per year, or nearly six times as much as all developmental aid to poor countries. A typical cow in Europe enjoys annual subsidies three times the average yearly income for most African farmers.

Agricultural subsidies can make a critical difference for farmers, but they are a concern globally. Subsidies allow American farmers to sell their products overseas at as much as 20 percent below the actual cost of production. Government payments cover losses and costs of production to the American farmers, but when the cheap commodities are sent to developing countries, they undercut local farmers. Often, local farmers are driven out of business, and local food production capacity is lost. Food aid functions similarly, providing free food that destabilizes the market for local producers. The FAO argues that ending distorting financial support in the richer countries would have a far more positive impact on local food supplies and livelihoods in the developing world than any aid program.

Powerful political and economic interests protect agricultural assistance in many countries. Over the past decade, the United States, for example, has spent $143 billion in farm support. This aid is distributed unevenly. According to the Environmental Working Group, 72 percent of all aid goes to the top 10 percent of recipients. One giant rice-farming operation in Arkansas, for example, received $38 million over a five-year period. Aid also is concentrated geographically. Just 5 percent (22) of the nation's 435 congressional districts collect more than 50 percent of all agricultural payments.

Most of this aid has been in direct payments for each bushel of targeted commodities, mainly corn, wheat, soybeans, rice, and cotton, as well as special subsidies for milk, sugar, and peanuts. Proponents insist that crop supports preserve family farms; critics claim that the biggest recipients are corporations that don't need the aid. In 2012, Congress voted to replace direct payments with insurance subsidies, by which the public pays insurance premiums on crops. Insurance reimburses farmers when crops fail because of weather or other risks, remain targeted at dominant commodity crops and are most beneficial for large producers.

An additional effect of these market interventions is to encourage the oil- and sugar-rich diets that lead to the spreading obesity epidemic. Subsidies help ensure that these processed foods are cheaper and more readily available than fresh fruits, vegetables, and whole grains. Many food policy analysts argue that we should support more vegetables and nutrient-rich foods and fewer commodity crops. Public attention to farm policy could help move us toward such policies.

Is genetic engineering about food production?

A leading argument in favor of genetically modified crops is that the world needs the best available, high-yielding varieties to feed the 9 billion people who will soon share our planet. This is a serious concern, but some development experts question whether the GM crops we are adopting are the most efficient strategy. Most GM seed companies are owned by producers of agricultural chemicals: Monsanto, Dow, Syngenta, Bayer, and DuPont—and the world's largest pesticide manufacturers are also the world's largest seed producers. Their seeds are specially designed to maximize production with prescribed combinations of chemicals from their parent companies.

Genetically modified corn, the dominant crop produced in the United States, grown on nearly 100 million acres, or nearly a quarter of U.S. cropland. Nearly all U.S. corn is used for ethanol production and for livestock feed (fig. 9.24). About 10 percent is used for food additives for industrial food production, such as high-fructose corn syrup, glucose, and dextrose. Soybeans the second most important crop in the United States, are used mainly for animal feed, followed by industrial oils, edible oils and meal, and biofuel. Cotton is used for a variety of oil and feed products, in addition to textiles, but only the oil from cotton seeds is used for direct human consumption.

Using these crops for livestock feed has greatly reduced the price of meat. This is true especially in industrialized nations, but also in China and other emerging economies. If we are to continue to expand protein consumption around the world, then GM

Quantitative Reasoning

Figure 9.23 shows the adoption of genetically modified crops. What was the rate of use in soybeans when you were 5 years old? In corn? How much change has there been since then?

soy, corn, and other crops may be necessary. On the other hand, increasing proportions of the world consume more protein and calories than may be healthy (see fig. 9.2). If the concern is to supply 9 billion people with a sufficient and healthy diet while sustaining our land and water systems, then producing meat is less efficient than producing food crops.

Like green revolution varieties, genetically modified seeds and pesticides are expensive for farmers in the developing world. Some food activists claim the GM revolution mainly benefits the corporations that sell seeds and pesticides. Specialized seeds and pesticides are expensive for farmers, who nonetheless must buy them in order to compete with other farmers in the market. When inputs are expensive, wealthy farms have important advantages over poor farmers, especially poor farmers in developing countries, where financial supports are tenuous. Experience in India has shown that the costs of production with GM seeds drives farmers into chronic debt. Waves of farm losses and farmer suicides have swept across India as a result of these debt burdens.

Farm policies can also protect the land

Every year, millions of tons of topsoil and agricultural chemicals wash from U.S. farm fields into rivers, lakes, and, eventually, the ocean. Farmers know that erosion both impoverishes their land and pollutes water, but they're caught in a bind. For every $1 the U.S. government pays farmers to conserve soil and manage nutrients, it pays $7 to support row-crop commodities that cause erosion and destroy soil quality through intensive cultivation and chemical use. The USDA estimates that if federal subsidies didn't promote these commodities, farmers would shift 2.5 million ha (6 million acres) of row crops into pasture, hay, and other crops that minimize erosion and build soil quality.

The United States tries to reduce soil erosion and overproduction of crops with the Conservation Reserve Program (CRP), which pays farmers to keep roughly 12 million ha (30 million acres) of highly erodible land out of production. The USDA reports that CRP lands prevent the annual loss of 450 million tons of soil every year, protect 270,000 km (170,000 miles) of streams, and store 48 million tons of carbon per year. Keeping land enrolled in this soil conservation program is vulnerable to political and economic shifts, however, and the amount of land enrolled changes every year.

Land enrolled in the CRP has been declining, as farmers find it more profitable to plant corn for ethanol, and as farm policy commits less money to supporting land retirement. Between 2007, when new ethanol policies were introduced, and 2012, over a

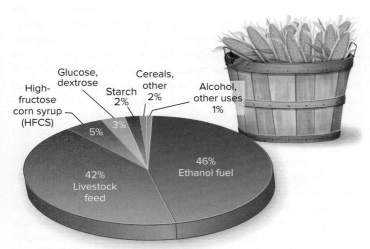

FIGURE 9.24 Little of the U.S. corn crop, our dominant agricultural product, is used directly for food. (Data are rounded to the nearest whole number.)
Source: USDA ERS

quarter of CRP lands were withdrawn from the program; in addition, tens of thousands of acres of uncropped grasslands were converted to corn and soy production. Many agronomists, though, say we should have more CRP land, not less. They argue that the United States could gradually shift payments from production subsidies and crop insurance to conservation programs that would truly support family farms while protecting the environment.

Will novel varieties of crops feed the world, or will they lead to a greater consolidation of corporate wealth and economic inequality? Can higher yields allow poor farmers in developing countries to stop using marginal land and avoid cutting down forests to expand farmland? Would it be more effective and sustainable to help poor farmers develop fish ponds or regenerative farming techniques than to sell them new patented seeds? We may need all the tools we can get, including GM foods, less meat-intensive diets, more land conversion, and other approaches. Many people argue that we should take a better-safe-than-sorry "precautionary approach," and err on the side of safety. Our assessment of GM varieties may also depend on whether we are primarily concerned about human health, environmental health, economic stability of farm economies, or other factors. Debates on all these strategies seem likely to continue for years to come.

Section Review

1. How is money spent supporting agriculture?
2. What are the primary uses of corn in the United States?
3. How can farm policy protect the land?

Conclusion

World food supplies have increased dramatically over the past half century. Despite the fact that the human population has nearly tripled in that time, food production has increased even faster, and we now grow more than enough food for everyone. Because of uneven distribution of food resources, however, there are still some 800 million people who don't have enough to eat on a daily basis, and hunger-related diseases remain widespread. Severe famines continue to occur, although most result more from political and social causes (or a combination of political and environmental conditions) than from environmental causes alone.

While hunger persists in many areas, over a billion people consume more food than is healthy on a daily basis. Epidemics of weight-related illnesses are spreading to developing countries as they adopt the diets and lifestyles of wealthier nations. Obesity is a health risk because it can cause or complicate heart conditions, diabetes, hypertension, and other diseases. In the United States, the death rate from illnesses related to obesity is approaching the death rate associated with smoking. Getting the right nutrients is also important. Many preventable diseases are caused by vitamin deficiencies.

Most of the world's food comes from a few major crops, including grains, vegetables, wheat, rice, corn, and potatoes. In the United States, just three crops—corn, soybeans, and wheat—are principal farm commodities. Corn and soybeans are mainly used for livestock feed and fuel, rather than food. Increasing use of these crops in confined feeding operations has dramatically increased meat production. For this and other reasons, global consumption of protein-rich meat and dairy products has climbed in the past 40 years. Protein gives us the energy to work and study, but raising animals takes a great deal of energy and food, so meat production can be environmentally expensive. However, there are sustainable food alternatives, such as rotational grazing, moderating meat consumption, and eating locally grown foods.

Most increases in food production in recent generations result from "green revolution" varieties of grains, which grow rapidly in response to fertilizer use and irrigation. More recent innovations have focused on genetically modified varieties. Some of these are being developed for improved characteristics, such as vitamin production or tolerance of salty soils. The majority of genetically modified crops are designed to tolerate herbicides, in order to improve competition with weeds.

Meeting the needs of the world's growing population will require a combination of strategies, from new crop varieties to political stabilization in war-torn countries. We can produce enough food for all. How we damage or sustain our environment while doing so is the subject of chapter 10.

Reviewing Key Terms

Can you define the following terms in environmental science?

anemia 9.1
aquaculture 9.2
chronically undernourished 9.1

confined animal feeding operation (CAFO) 9.2
famines 9.1
food security 9.1

genetically modified organisms (GMOs) 9.3
green revolution 9.3
kwashiorkor 9.1

locavore 9.1
malnourishment 9.1
marasmus 9.1
obese 9.1

Critical Thinking and Discussion Questions

1. Do people around you worry about hunger? Do you think they should? Why or why not? What factors influence the degree to which people worry about hunger in the world?

2. Global issues such as hunger and food production often seem far too large to think about solving, but it may be that many strategies can help us address chronic hunger. Consider your own skills and interests. Think of at least one skill that could be applied (if you had the time and resources) to helping reduce hunger in your community or elsewhere.

3. Suppose you are a farmer who wants to start a confined animal feeding operation. What conditions make this a good strategy for you, and what factors would you consider in weighing its costs and benefits? What would you say to neighbors who wish to impose restrictions on how you run the operation?

4. Debate the claim that famines are caused more by human actions (or inactions) than by environmental forces. What kinds of evidence would be needed to resolve this debate?

5. Outline arguments you would make to your family and friends for why they should buy shade-grown, fair-trade coffee and cocoa. How much of a premium would you pay for these products? What factors would influence how much you would pay?

6. Given what you know about GMO crops, identify some of the costs and benefits associated with them. Which of the costs and benefits do you find most important? Why?

7. Corn is by far the dominant crop in the United States. In what ways is this a good thing for Americans? How is it a problem? Who are the main beneficiaries of this system?

Data Analysis

Graphing Relative Values

There are many ways to describe trends in an important subject such as world hunger. One approach is to show the total number or proportion of the population. Another approach is to compare values to a standardized index value (shown here), which compares all years to 1969, when reliable statistics were first gathered by the UN Food and Agriculture Organization (FAO). What different kinds of information do these graphs give? Go to Connect to examine graphs of hunger rates, and to demonstrate your understanding of the data.

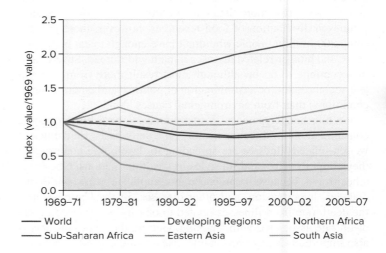

Farming: Conventional and Sustainable Practices

▲ Soybean harvestvest in the Brazilian state of Mato Grosso, whose name means "dense forest."
© Paulo Fridman/Corbis

Learning Outcomes

After studying this chapter, you should be able to:

10.1 Describe the components of soils.

10.2 Explain the ways we use, abuse, and conserve soils.

10.3 Discuss our principal pesticides and their environmental effects.

10.4 Describe several methods of organic and sustainable agriculture.

"We abuse the land because we regard it as a commodity belonging to us. When we see land as a community to which we belong, we may begin to use it with love and respect."

– Aldo Leopold

Farming the Cerrado

A soybean boom is sweeping across South America. Inexpensive land, new crop varieties, growing global markets for soy, and government policies favoring agricultural expansion have made South America the fastest-growing agricultural area in the world. The center of this rapid expansion is the Cerrado, a vast area of grassland and tropical forest stretching from Bolivia and Paraguay across the center of Brazil almost to the Atlantic Ocean (fig. 10.1). Biologically, this rolling expanse of grasslands and tropical woodland is the richest savanna in the world, with at least 130,000 plant and animal species, many of which are threatened by agricultural expansion.

Until recently, the Cerrado was thought to be unsuitable for cultivation. This region, 2 million km^2 in extent, or roughly the size of the American Midwest, has red iron-rich soils that are acidic and poor in essential plant nutrients. The warm, humid climate harbors destructive pests and pathogens. For hundreds of years, the Cerrado was mainly cattle country with poor-quality pastures producing low livestock yields.

In recent decades, however, Brazil has developed new varieties of soybeans specially adapted for the soils and climate of the Cerrado. With applications of lime (calcium carbonate) and phosphorus, new varieties can quadruple yields of soybeans, maize, cotton, and other crops. Increasingly, these new varieties are genetically modified, mainly for pesticide tolerance. Until about 40 years ago, soybeans were a minor crop in Brazil. Since 1975, however, the total area planted with soy has doubled about every four years, reaching more than 33 million ha (82 million acres), an area nearly the size of Montana, in 2015. Over half the Cerrado has now been converted to vast crop fields.

Since 2007, Brazil has been the world's top soy exporter, shipping over 50 million metric tons per year, or about 10 percent more than the United States. With two crops per year, cheap land, low labor costs, favorable tax rates, and yields per hectare equal to those in the American Midwest, Brazilian farmers can produce soybeans for less than half the cost in North America. Agricultural economists predict that by 2020 the global soy crop will be double the current 160 million metric tons per year, and that South America could be responsible for most of that growth. In addition to soy, Brazil now leads the world in exports of beef, maize, oranges, and coffee. This dramatic increase in South American agriculture helps answer the question of how the world's growing human population can be fed.

But people don't eat these soybeans. Most soy is feed for pigs, chicken, and cattle. A major factor in Brazil's current soy expansion is rising income in China. With more money to spend, the Chinese can afford to feed soy to livestock, especially pigs, chickens, and fish. Meat consumption has grown rapidly, although it's still a fraction of what Americans eat. China now imports about 30 million metric tons of soy annually. About half of that comes from Brazil. Production of industrial oils—for lubricants, plastics, and other materials—and biodiesel also consume much of the soy produced here.

All this growth in production has environmental effects of global importance. One of the world's richest biological areas is retreating steadily as croplands expand. This region is also critically important for storing carbon, and forest clearing contributes to climate change. Most of this loss is occurring in the "arc of destruction" between the Cerrado and the Amazon (see fig. 10.1).

Drought also appears to follow land conversion, as agriculture expands into the Amazon rainforest. Cleared land dries rapidly and stores little moisture, and satellite images show increasing amounts of dead and drought-stressed trees in the region. A drying climate threatens agriculture in the region, as well as the Amazon rainforest, one of the world's most important areas for carbon storage.

Conflicts over land also have increased. Small family farms are being gobbled up; farmworkers, displaced by mechanization, have migrated either to the big cities or to frontier forest areas, where they clear still more land. Ongoing conflicts between poor farmers and big

FIGURE 10.1 Brazil's Cerrado, 2 million ha of savanna (grassland) and open woodland, is the site of the world's fastest-growing soybean production. Cattle ranchers and agricultural workers, displaced by mechanized crop production, are moving northward into the "arc of destruction" at the edge of the Amazon rainforest, where the continent's highest rate of forest clearing is occurring.

landowners have led to violent confrontations. The Landless Workers Movement claims that 1,237 rural workers died in Brazil between 1985 and 2000 as a result of assassinations and clashes over land rights. Tens of thousands of landless farmers and displaced families have been forced to live in squatter camps or to migrate to Brazil's growing cities.

Rapid growth of soy production in Brazil has both positive and negative aspects. On one hand, more high-protein food is now available to feed the world. The Cerrado represents one of the world's last opportunities to open a large area of new, highly productive cropland. On the other hand, the rapid expansion of agriculture in Brazil is destroying biodiversity and creating social conflict.

How do we weigh the costs and benefits of land conversion and expanding industrial agriculture? What factors make this expansion possible? What sustainable approaches are available to help negotiate environmental and social priorities? In this chapter, we explore these questions and consider the nature and the future of agricultural production.

10.1 WHAT IS SOIL?

- *Soils are composed of mineral grains (sand, silt, and clay), organic matter, organisms, water, and air.*
- *These affect soil fertility, erodibility, and other factors.*
- *Soils vary in texture, color, and horizon development.*

Agriculture has dramatically changed our environment, altering patterns of vegetation, soils, and water resources worldwide. The story of Brazil's Cerrado involves the conversion of millions of hectares of tropical savanna and rainforest to crop fields and pasture. This is one recent example of agricultural land conversion, but humans have been converting land to agriculture for thousands of years. Some of these agricultural landscapes are ecologically sustainable and have lasted for centuries or millennia. Others have depleted soil and water resources in just a few decades. What are the differences between farming practices that are sustainable and those that are unsustainable? What aspects of our current farming practices degrade the resources we depend on, and in what ways can farming help to restore and rebuild environmental quality?

As you have read in the opening case study, farm expansion has changed the landscape, the environment, and the economy of central Brazil. These changes are driven by financial investments and technological innovations from North American and European corporations. They are supported by rapidly expanding markets in Asia and Europe. But another essential factor has been the development of new ways to modify the region's nutrient-poor, acidic tropical soils. We will begin this chapter by exploring what soils are made of and how they differ from one place to another.

Soils are complex ecosystems

Is soil a renewable resource, or is it a finite resource that we are depleting? It's both. Over time, soil is renewable, because it develops gradually through weathering of bedrock and through the accumulation of organic matter, such as decayed leaves and plant roots. But these processes are extremely slow. Building a few

millimeters of soil can take anything from a few years (in a healthy grassland) to a few thousand years (in a desert or tundra).

Under the best circumstances, topsoil accumulates at about 1 mm per year. This soil growth involves careful management that prevents erosion and allows organic material to accumulate in the soil. But most farming techniques deplete soil. Plowing exposes bare soil to erosion by wind or water, and annual harvests remove organic material such as leaves and roots. Severe erosion can carry away 25 mm or more of soil per year, far more than the 1 mm that can accumulate under the best of conditions.

Soil is a marvelous, complex substance. It is a combination of weathered rocks, plant debris, living fungi, and bacteria, an entire ecosystem that is hidden to most of us. In general, soil has six components:

1. *Sand and gravel* (mineral particles from bedrock, either in place or moved from elsewhere, as in windblown sand)

2. *Silts and clays* (extremely small mineral particles; clays are sticky and hold water because of their flat surfaces and ionic charges)

3. *Dead organic material* (decaying plant matter that stores nutrients and gives soils a black or brown color)

4. *Soil fauna and flora* (living organisms, including soil bacteria, worms, fungi, roots of plants, and insects, that recycle organic compounds and nutrients)

5. *Water* (moisture from rainfall or groundwater, essential for soil fauna and plants)

6. *Air* (tiny pockets of air help soil bacteria and other organisms survive)

Variations in these components produce almost infinite variety in the world's soils. Abundant clays make soil sticky and wet. Abundant organic material and sand make the soil soft and easy to dig. Sandy soils drain quickly, often depriving plants of moisture. Silt particles are larger than clays and smaller than sand (fig. 10.2), so they aren't sticky and soggy, and they don't drain too quickly. Thus, silty soils are ideal for growing crops, but they are also light

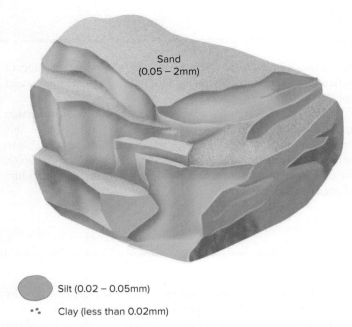

Sand
(0.05 – 2mm)

Silt (0.02 – 0.05mm)

Clay (less than 0.02mm)

FIGURE 10.2 Relative size of sand, silt, and clay particles (magnified about 100 times)

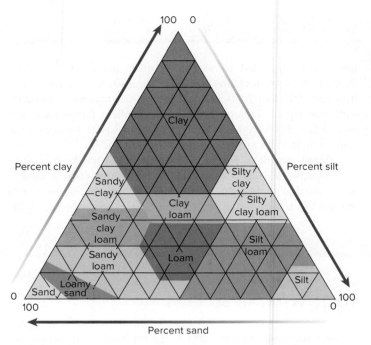

FIGURE 10.3 A soil's texture depends on its proportions of sand, clay, and silt particles. Read the graph by following lines across, up, or down from the axes. For example, "loam" has about 50–75 percent sand, 8–30 percent clay, and 18–50 percent silt. Loamy soils have the best texture for most crops, with enough sand to be loose and workable, yet enough silt and clay to retain water and nutrients.

and blow away easily when exposed to wind. Soils with abundant soil fauna quickly decay dead leaves and roots, making nutrients available for new plant growth. Compacted soils have few air spaces, making soil fauna and plants grow poorly.

Soil texture—the amount of sand, silt, and clay in the soil—is one of the most important characteristics of soils. Texture helps determine whether rainfall drains away quickly or ponds up and drowns plants. Loam soils are usually considered best for farming because they have a mixture of clay, silt, and sand (fig. 10.3).

You can see some of the differences among soils just by looking at them. Reddish soils, including most tropical soils, often are colored by iron-rich, rust-colored clays, which store few nutrients for plants. Deep black soils, on the other hand, are rich in organic material, which is moist, rich in nutrients, and keeps soil soft and workable. Most Brazilian tropical soils are deeply weathered red clays. With frequent rainfall and year-round warm weather, organic material decays quickly and is taken up by living plants or washed away with rainfall. Red, iron-rich, clay soils result. These reddish clays hold few nutrients and little moisture for growing fields of soybeans (fig. 10.4b).

In contrast, the rich, black soils of the Corn Belt of the central United States have abundant organic matter and a good mix of sand, silt, and clay. These soils tend to hold enough moisture for crops without becoming waterlogged, and they tend to be rich in nutrients. Acidic tropical Brazilian soils can be improved by adding lime (calcium carbonate, as in limestone), which improves the soil's ability to retain nutrients applied in fertilizer. Liming vast areas was not economical until recently, but expanding markets for livestock feed in Asia and Europe now make it cost-effective for Brazilian farmers to apply lime to their fields. This is one of the innovations that has allowed recent expansion of Brazilian soy production.

(a)   (b)

FIGURE 10.4 A temperate grassland soil (a) has a thick, black organic layer. Tropical rainforest soils (b) have little organic matter and are composed mostly of nutrient-poor, deeply weathered iron-rich clays. Each of these profiles is about 1 m deep.
Source: a. USDA Natural Resources Conservation Service; b. Source: Photo by Hari Eswaran, USDA Natural Resources Conservation Service

Healthy soil fauna can determine soil fertility

Soil bacteria, algae, and fungi decompose and recycle leaf litter and other organic debris. These microorganisms thus make nutrients available to plants, and also help give soils structure and loose texture (fig. 10.5). The abundance of these organisms can be astonishing. One gram of soil can contain hundreds of soil bacteria and 20 m of tiny strands of fungal material. A cubic meter of soil can contain more than 10 kg of bacteria and fungal biomass. Tiny worms and nematodes process organic matter and create air spaces as they burrow through soil. Slightly larger insects, mites, spiders, and earthworms further loosen and aerate the soil. The sweet aroma of freshly turned soil is caused by actinomycetes, bacteria that grow in fungus-like strands and give us the antibiotics streptomycin and tetracycline. These organisms mostly stay near the surface, often within the top few centimeters. The roots of plants can reach deeper, however, allowing moisture, nutrients, and organic acids to help break down rocks farther down and begin forming new soil.

Many plant species grow best with the help of particular species of soil fungi in relationships called **mycorrhizal symbiosis.** In this relationship, the mycorrhizal fungus (a fungus growing on and around plant roots) provides water and nutrients to the plant, while the plant provides organic compounds to the fungus. Plants growing with their fungal partners often grow better than those growing alone.

The health of the soil ecosystem depends on a variety of factors, such as slope, rainfall, and frequency of disturbance. Too much rain washes away nutrients and organic matter, but soil fauna cannot survive with too little rain. In extreme cold, soil fauna recycle nutrients extremely slowly; in extreme heat, they may work so fast that leaf litter on the forest floor is taken up by plants in just weeks or months—so that the soil retains little organic matter. Frequent disturbance prevents the development of a healthy soil ecosystem, as does steep topography that allows rain to wash away soils. In the United States, the best farming soils tend to occur where the climate is not too wet or dry, on glacial silt deposits, such as those in the upper Midwest, and on silt- and clay-rich flood deposits, like those along the Mississippi River.

Most soil fauna occur in the uppermost layers of a soil, where they consume leaf litter. This layer is known as the "O" (organic) horizon. Just below the O horizon is a layer of mixed organic and mineral soil material, the "A" horizon (fig. 10.6). These layers are known as **topsoil,** or surface soil.

The B horizon, or **subsoil,** tends to be richer in clays than the A; the B horizon is below most organic activity. The B layer accumulates clays that seep downward from the A horizon with rainwater that percolates through the soil. If you dig a hole, you may be able to tell where the B horizon begins, because there the soil tends to become slightly more sticky and cohesive. If you squeeze

FIGURE 10.5 Soil ecosystems include countless organisms that consume and decompose organic material, aerate soil, and distribute nutrients through the soil.

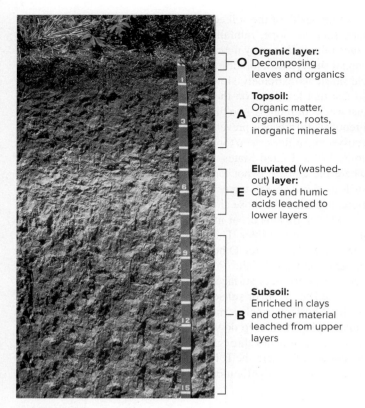

FIGURE 10.6 Soil profile showing possible soil horizons. The actual number, composition, and thickness of these layers varies in different soil types.
© Soil & Land Resources Division, University of Idaho

Organic layer:
O Decomposing leaves and organics

Topsoil:
A Organic matter, organisms, roots, inorganic minerals

Eluviated (washed-out) layer:
E Clays and humic acids leached to lower layers

Subsoil:
B Enriched in clays and other material leached from upper layers

FIGURE 10.7 In many areas, soil or climate constraints limit agricultural production. These hungry goats in Sudan feed on a solitary *Acacia* shrub.
© Eye Ubiquitous/Newscom

a handful of B-horizon soil, it should hold its shape better than a handful of A-horizon soil.

Sometimes an E (eluviated, or washed-out) layer lies between the A and B horizons. An E layer is loose and light-colored because most of its clays and organic material have been washed down to the B horizon. The C horizon, below the subsoil, is mainly decomposed rock fragments. Parent materials underlie the C layer. Parent material is the sand, windblown silt, bedrock, or other mineral material on which the soil is built. About 70 percent of the parent material in the United States was transported to its present site by glaciers, wind, and water, and is not related to the bedrock formations below it.

Your food comes mostly from the A horizon

Ideal farming soils have a thick, organic-rich A horizon. The soils that support the Corn Belt farm states of the Midwest have a rich, black A horizon that can be more than 2 meters thick (although a century of farming has washed much of this soil away and down the Mississippi River). The A horizon in most soils is less than half a meter thick. Desert soils, with slow rates of organic activity, might have almost no O or A horizons (fig. 10.7).

Because topsoil is so important to our survival, we identify soils largely in terms of the thickness and composition of their upper layers. Soils vary endlessly in depth, color, and composition, but for simplicity we can describe a few general groups. The U.S. Department of Agriculture classifies the soils into 11 soil orders

(table 10.1). In the Farm Belt of the United States, the dominant soils are mollisols (*mollic,* "soft"; *sol,* "soil"). These soils have a thick, organic-rich A horizon that developed from the deep, dense roots of prairie grasses that covered the region until about 150 years ago (see fig. 10.4a). Another group that is important for farming is alfisols. Alfisols have a slightly thinner A horizon than mollisols do, and slightly less organic matter. Alfisols develop in deciduous forests where leaf litter is abundant. In contrast, the aridisols (*arid,* "dry") of the desert Southwest have little organic matter, and they often contain accumulations of mineral salts. Mollisols and alfisols dominate most of the farming regions of the United States.

Section Review

1. What are six major components of soil?
2. How are soil organisms important for good soils?
3. Why are Brazilian soils often red?
4. How fast (in mm per year) are soils understood to build under ideal conditions?

Table 10.1	Soil orders defined by the USDA
Soil Order	**Characteristics**
Mollisol	organic-rich, black grassland soil
Alfisol	organic-rich, deciduous forest soil
Spodosol	leached by acidic litter, as from pine needles
Histosol	peaty, wet, largely organic matter
Aridisol	desert soils, may have salt deposits
Ultisol	old, stable, weathered soil
Oxisol	tropical, deeply weathered, red soil
Vertisol	clay-rich, sticky soil
Andisol	volcanic material; mineral-rich
Inceptisol	weak horizon development, as with recently deposited material
Entisol	little/no horizon development, as with sand or eroded material

10.2 How Do We Use, Abuse, and Conserve Soils?

- *Arable land is limited, and expansion has environmental costs.*
- *Crops need many inputs, including water, fertilizer, and fuel.*
- *Wind and water erosion threatens soil fertility and abundance.*
- *Farming techniques, windbreaks, and other strategies can control erosion.*

Much of the world's land is too steep, soggy, salty, cold, or dry for farming. Only about 11 percent of the earth's land area (1.5 billion ha out of 13.4 billion ha of land area) is currently in crop production. In theory, up to four times as much land could potentially be converted to cropland, although that would mean destroying wetlands and forest lands that stabilize our climate and water resources.

Despite those costs, conversion to farmland is continuing in many areas, as global trade in agricultural commodities, such as corn, soy, and palm oil, becomes increasingly lucrative. Brazil's expansion of soy farming into the Cerrado (opening case study) is one of the most dramatic cases of land conversion, but ancient forests and grasslands are also turning into farmland in many parts of the developing world. The ecological costs of these land conversions are hard to calculate. Farmers can easily count the cash income from the farm products they sell, but it is never easy to calculate the value of biodiversity, clean water, and other ecological services of a forest or grassland, compared to the value of crops.

Arable land is unevenly distributed

The best agricultural lands occur where the climate is moderate—not too cold or too dry—and where thick, fertile soils are found. Take a look at the global map in the back of your book. What regions do you think of as the best agricultural areas? The poorest? Much of the United States, Europe, and Canada are fortunate to have temperate climates, abundant water, and high soil fertility. These produce good crop yields that contribute to high standards of living. Other parts of the world, although rich in land area, lack suitable soil, level land, or climates to sustain good agricultural productivity.

In developed countries, 95 percent of recent agricultural growth in the past century has come from a combination of improved crop varieties and increased use of fertilizers, pesticides, and irrigation. Conversion of new land to crop fields has contributed relatively little to increased production. In fact, less land is being cultivated now than 100 years ago in North America or 600 years ago in Europe. Productivity per unit of land has increased, and some marginal land has been retired. Careful management is important for preserving the remaining farmland.

Soil losses threaten farm productivity

Agriculture both causes soil degradation and suffers from it. Soil loss by erosion is the primary concern (fig. 10.8). Every year, erosion makes about 3 million ha of cropland unusable worldwide. Another 4 million ha are converted to nonagricultural uses, such

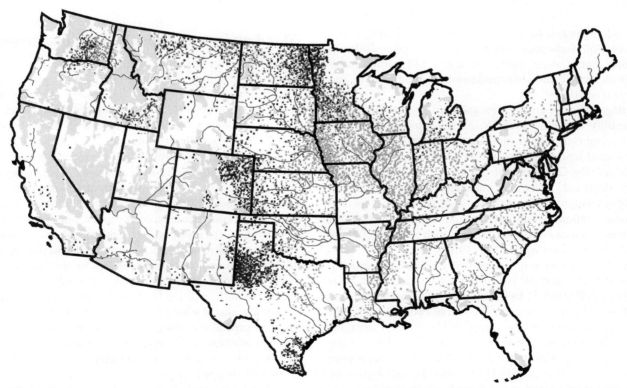

FIGURE 10.8 More than 43.7 million ha (108 million acres) in the United States are subjected to excess erosion by wind (red) or water (blue) each year. Each dot represents 200,000 tons of average annual soil loss.
Source: USDA Natural Resource Conservation Service

(a)

(b)

(c)

FIGURE 10.9 Disastrous erosion during the Dust Bowl years (a) led to national erosion control efforts that have reduced, but not eliminated, soil loss (b). Nationally, wind and water erosion have declined but continue to degrade farmland (c).

Source: a–b. USDA Natural Resources Conservation Service

as urban land, highways, factories, or reservoirs, according to the International Soil Reference and Information Center (ISRIC). In the United States alone, we've lost about 140 million ha of farmland in the past 30 years to urbanization, soil degradation, and other factors (fig. 10.9).

Land degradation is usually slow and hard to observe directly. Usually, fertility declines gradually, as soil washes and blows away, salts accumulate, and organic matter is lost. About 20 percent of vegetated land in Africa and Asia is degraded enough to reduce productivity; 25 percent of lands in Central America and Mexico are degraded. Wind and water erosion are the primary causes of degradation. Additional causes of degradation include chemical deterioration (mainly salt accumulation from salt-laden irrigation water) and physical deterioration, such as compaction by heavy machinery or waterlogging (fig. 10.10).

As a consequence of soil loss, as well as population growth, the amount of arable land per person worldwide has shrunk from about 0.38 ha in 1970 to 0.21 ha in 2010. Consider that a hectare is an area 100 m × 100 m, or roughly the size of two football fields. On average, about five people are supported by that land area. By 2050, the arable land per person will decline to 0.15 ha. In the United States, farmland has fallen from 0.7 to 0.45 ha per person in the past 30 years, according to USDA data. To feed a growing population on declining land area, we are likely to need improvements in production methods, reduced consumption of protein (chapter 9), and improved soil management.

Wind and water cause widespread erosion

When water washes away a thin layer of soil, we call it **sheet erosion.** When little rivulets of running water gather together and cut small channels in the soil, the process is called **rill erosion** (fig. 10.11*a*). When rills enlarge to form bigger channels or ravines that are too large to be removed by normal tillage operations, we call the process **gully erosion** (fig. 10.11*b*). Streambank erosion refers to the washing away of soil from the banks of established streams, creeks, or rivers, often as a result of removing streamside trees and brush or letting livestock trample streambanks.

Sheet and rill erosion are responsible for most soil loss on agricultural land. These types or erosion can be subtle, even invisible, but they can transport large amounts of soil. A farm field can lose 20 metric tons of soil per hectare during winter and spring runoff in rills so small they are erased by the first spring cultivation. That represents a loss of only a few millimeters of soil over the whole surface of the field, hardly an obvious change. But the maximum rate at which soils can rebuild, under the best conditions, is generally considered to be about 1 mm per year (largely through addition of organic matter) or about 11 metric tons per hectare (5 U.S. tons/acre). If soils rebuild slowly and erode rapidly, eventually they will be gone.

Wind can equal or exceed water in erosive force, especially in a dry climate. When plant cover and surface litter are removed from the land, wind lifts loose soil particles and sweeps them away.

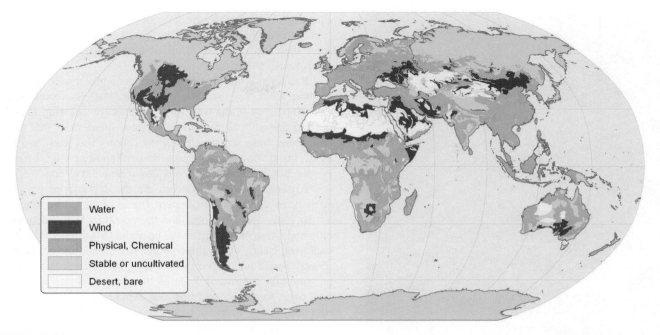

FIGURE 10.10 Global causes of soil erosion and degradation. Globally, 62 percent of eroded land is mainly affected by water; 20 percent is mainly affected by wind.

Source: ISRIC Global Assessment of Human-Induced Soil Degradation, 2008

In extreme conditions, windblown dunes encroach on useful land and cover roads and buildings (fig. 10.11c). Over the past 30 years, China has lost 93,000 km² (about the size of Indiana) to **desertification,** or conversion of productive land to desert. Advancing dunes from the Gobi Desert are now only 160 km (100 mi) from Beijing.

Some of the highest erosion rates in the world occur in the United States and Canada. The U.S. Department of Agriculture reports that 69 million hectares (170 million acres) of U.S.

farmland and range are eroding at rates that are reducing long-term productivity. Intensive farming practices are largely responsible for this situation. Row crops, such as corn and soybeans, leave soil exposed for much of the growing season (fig. 10.12). Deep plowing and heavy herbicide applications create fields that are weed-free but extremely vulnerable to erosion.

Since the Dust Bowl of the 1920s and 1930s, the USDA has been encouraging farmers to plow along contours rather than up and down hills, to avoid planting steep slopes, to leave windbreaks, and

(a) Sheet and rill erosion

(b) Gullying

(c) Wind erosion and desertification

FIGURE 10.11 Land degradation affects more than 1 billion ha yearly, or about two-thirds of all global cropland. Water erosion (a) and gullying (b) account for about half that total. Wind erosion affects a nearly equal area (c).

Source: (a) Photo by Lynn Betts, USDA Natural Resources Conservation Service; (b) Source: Photo by Jeff Vanuga, USDA Natural Resources Conservation Service; (c) © Royalty-Free/Corbis

FIGURE 10.12 Annual row crops leave soil bare and exposed to erosion by wind and water.
© William P. Cunningham

to protect grass-lined waterways in fields. These efforts have greatly reduced erosion, but as farm machinery gets larger and as prices rise for corn and soybeans, soil conservation measures are being abandoned in many areas. Farm policies, especially price supports for ethanol, encourage farmers to plow through grass-lined waterways, to disregard contours, and to pull out windbreaks and fencerows to accommodate the large machines and to get every last square meter into production. Pressed by economic conditions, many farmers have also abandoned traditional practices such as rotating crops or letting fields rest and go fallow from time to time.

Continuous monoculture cropping can increase soil loss tenfold over other farming practices. A soil study in Iowa showed that a three-year rotation of corn, wheat, and clover lost an average of about 6 metric tons per hectare. By comparison, continuous wheat production on the same land caused soil losses of more than 20 tons per hectare. Continuous corn cropping resulted in 40 tons per hectare, seven times as much soil loss as the rotation with wheat and clover.

Desertification affects arid land soils

According to the United Nations, about one-third of the earth's surface and the livelihoods of at least one billion people are threatened by desertification (conversion of productive lands to desert), which contributes to food insecurity, famine, and poverty. Former UN secretary general Kofi Annan called this a "creeping

catastrophe" that creates millions of environmental refugees every year. Forced by economic circumstances to overcultivate and overgraze their land, poor people often are both the agents and the victims of desertification.

Rangelands and pastures, which generally are too dry for cultivation, are highly susceptible to desertification. According to the UN, 80 percent of the world's grasslands are suffering from overgrazing and soil degradation, and three-quarters of that area has undergone some degree of desertification. The world's 3 billion domestic grazing animals provide livelihood and food for many people, but they can have severe environmental effects.

Areas of particular concern are the dry regions of southern Africa and northern and western China. Arid lands, where rains are sporadic and infrequent and the economy is based mainly on crop and livestock raising, make up about two-thirds of the African continent. Nearly 400 million people live around the edges of these deserts. Rapid population growth and poverty create unsustainable pressures on the fragile soils of these areas. Stripping trees and land cover for fodder and firewood exposes the soil to erosion and triggers climate changes that spread desertification, which now affects nearly three-quarters of the arable land in Africa. The fringes of the two great African deserts, the Sahara and the Kalahari, are particularly vulnerable. Much of northern China, similarly, has little rainfall, a growing population, and increasing land degradation from overgrazing. Finding ways to reduce pressure and rebuild soils is one of the important tasks in stabilizing food security for these regions.

Irrigation is needed but can damage soils

Agriculture accounts for the largest single share of global water use. About two-thirds of all fresh water withdrawn from rivers, lakes, and groundwater supplies is used for irrigation (chapter 17). Roughly 15 percent of all cropland is irrigated worldwide. Some regions are water rich and can readily afford to irrigate farmland; other regions are water poor and risk depleting or contaminating precious water resources through irrigation (fig. 10.13).

The efficiency of irrigation water use is low in most countries. Inefficiency starts with water delivery: In some places, evaporation and seepage from unlined canals remove 80 percent of water withdrawn for irrigation before it even reaches the field. Water is also lost to evaporation when farmers irrigate by flooding fields with water. Flood irrigation is cheap in terms of technology, and it can remove salts from desert soils, but much of the water simply evaporates in hot summer sun. Farmers often over-irrigate because water prices are low or because they lack the technology to deliver just the amount needed. Many farmers in the United States and Canada install drains under fields as well as irrigation systems, so they can both add and remove water.

Excessive water often results in **waterlogging.** Waterlogged soil is saturated with water, and plant roots die from lack of oxygen. **Salinization,** in which salts accumulate in the soil, occurs mainly when soils in dry climates are irrigated with salt-laden water. As the water evaporates, it leaves behind a salty crust on the soil surface that can be lethal to plants. Flushing with excess water can wash away this salt accumulation, but the result is even more

FIGURE 10.13 Downward-facing sprinklers on this center-pivot irrigation system deliver water more efficiently than upward facing sprinklers.
© Royalty-Free/Corbis

saline water for downstream users. Extreme salinity is an important problem—for example, downstream of the fruit and vegetable growers of California's Central Valley and Imperial Valley.

Plants need nutrients, but not too much

In addition to water, sunshine, and carbon dioxide, plants need small amounts of nitrogen, potassium, phosphorus, calcium, magnesium, and other nutrients. Shortages of nitrogen, potassium, and phosphorus, in particular, can limit plant growth. Adding these elements in fertilizer stimulates growth and increases crop yields.

Much of the doubling in worldwide crop production in recent decades has involved increased use of inorganic fertilizer, especially ammonia (NH_3). This form of nitrogen fertilizer, developed at the end of World War II by explosives manufacturers, is produced from natural gas (CH_4) and atmospheric nitrogen (N_2). Natural gas provides both the heat and the hydrogen atoms needed to produce ammonia. Since it began in 1947, production has risen to about 150 million metric tons per year. Applications of fertilizer have risen from 20 kg per hectare in 1950 to nearly 100 kg per hectare, on average, worldwide.

Fertilizer doesn't need to derive from this energy-intensive process. Manure from livestock and good soil management can produce yields as high as those from commercial fertilizers. Nitrogen-fixing crops such as beans or lentils, with nitrogen-fixing bacteria living in their roots, are especially valuable for making nitrogen available as a plant nutrient (chapter 9). Interplanting or rotating beans with nitrogen-hungry corn and wheat are traditional ways of increasing nitrogen availability.

In some regions, there is still considerable potential for increasing world food supply by increasing fertilizer. Southern Africa, for instance, uses an average of only 19 kg of fertilizer per hectare, much less than the world average. It has been estimated that the developing world could triple its crop production by raising fertilizer use to the world average.

In many farming regions, though, farmers apply more fertilizer than plants can take up. They do this because they lack information about crop needs, because they fear underfertilizing, because fertilizers are relatively inexpensive, and because in most countries there is nobody whose job it is to prevent overuse. In the United States, for example, the EPA has responsibility for reducing water pollution and has worked to discourage overuse of fertilizer. North American farmers use about 60 kg of nitrogen fertilizer per hectare, whereas European farmers use about 130 kg, and Chinese farmers use nearly 200 kg. Yields with these different application rates are not significantly different, however.

Excess nutrients badly impair water quality. In the American Corn Belt, groundwater is often unsafe because of high nitrate levels. Young children are especially sensitive to the presence of nitrates. Using nitrate-contaminated water to mix infant formula can cause "blue baby syndrome," which can be fatal for newborns. The main environmental consequence of excess fertilizers is eutrophication (excessive growth of algae, followed by decay and oxygen depletion). Rivers carry farmland soil and fertilizer into marine ecosystems around the world. Scores of major "dead zones," bays and seas depleted of oxygen and life, now occur around the world. Urban inputs are also important contributors to dead zones.

The biggest dead zone in the United States is that in the Gulf of Mexico. Each summer, spring rains and snowmelt wash soil and nutrients down the Mississippi River, from Iowa, Illinois, and other farm belt states. Once these nutrients reach the Gulf, eutrophication depletes most marine life in areas as large as 57,000 km^2. Commercial and sport fisheries are severely depleted (chapter 18). Chesapeake bay is another important U.S. water body that has been badly affected by farmland fertilizer and soil runoff (chapter 3).

Conventional farming uses abundant fossil fuels

Farming in industrialized countries is energy-intensive. Reliance on fossil fuels began in the 1920s with the adoption of tractors, and energy use increased sharply after World War II when nitrogen fertilizer made from natural gas became available. Reliance on diesel and gasoline to run tractors, combines, and other machinery has continued to grow in recent decades. Agricultural analyst David Pimentel of Cornell University has calculated the many energy inputs, from fertilizer and pesticides to transportation and irrigation. His estimate amounts to an equivalent of 800 liters of oil (5 barrels of oil) per hectare of corn produced in the United States. A third of this energy is used in producing nitrogen fertilizer. Inputs for machinery and fuel make up another third; herbicides, irrigation, and other fertilizers make up the rest.

After crops leave the farm, additional energy is used in food processing, distribution, storage, and cooking. It has been estimated that the average food item in the American diet travels 2,400 km between the farm that grew it and the person who consumes it. The energy required for this complex processing and distribution system may be five times as much as that used directly in farming.

We can conserve and even rebuild soils

With careful husbandry, soil is a renewable resource that can be replenished and renewed indefinitely. Many sustainable farming practices focus on building soil nutrients. Some rice paddies in

Southeast Asia, for instance, have been farmed continuously for a thousand years without any apparent loss of fertility. The rice-growing cultures that depend on these fields have developed management practices that return organic material to the paddy and carefully nurture the soil (see also the Exploring Science Ancient *Terra Preta* Shows How to Build Soils box at the end of this section).

American agriculture causes far more erosion than is sustainable. But conditions were still worse a few generations ago, before USDA soil conservation programs were established. In a study of one Wisconsin watershed, erosion rates were 90 percent less in 1975–1993 than they were in the 1930s. Ground cover, irrigation, and tillage systems are the most important elements in soil conservation.

Erosion control and reduced tillage are the main strategies for keeping soil on fields. These practices are well understood, but their use varies a great deal. In some cases, high crop values encourage farmers to disregard soil conservation measures. In other cases, farmers practice careful soil management in order to preserve the long-term viability of their farms.

Contours and ground cover reduce runoff

Water runs downhill. The faster it runs, the more soil it carries off the fields. A bare field with a 5 percent slope loses eight times as much soil to erosion as a field with a 1 percent slope. **Contour plowing**—plowing across the hill rather than up and down—is one of the main strategies for controlling soil loss and water runoff. Contour plowing is often combined with strip farming, or planting different kinds of crops in alternating strips along the land contours (fig. 10.14*a*). The ridges created by cultivation also trap water and allow it to seep into the soil.

Terracing involves shaping the land to create level shelves of earth to hold water and soil. The edges of the terrace are planted with soil-anchoring plant species. This is an expensive procedure, requiring either much hand labor or expensive machinery, but it makes it possible to farm very steep hillsides. Rice terraces in Asia create beautiful landscapes as well as highly productive and sustainable agroecosystems (fig. 10.14*b*).

Annual row crops such as corn or beans generally cause the highest erosion rates because they leave soil bare for much of the year (table 10.2). On many steep lands or loose, highly

(a)

(b)

FIGURE 10.14 Contour plowing (a) and terracing, as in these Balinese rice paddies (b), are both strategies to control erosion on farmed hillsides.

Source: (a) Photo by Lynn Betts, USDA Natural Resources Conservation Service; (b) © William P. Cunningham

Table 10.2 Soil Cover and Soil Erosion		
Cropping System	**Average Annual Soil Loss (Tons/Hectare)**	**Percent Rainfall Runoff**
Bare soil (no crop)	41.0	30
Continuous corn	19.7	29
Continuous wheat	10.1	23
Rotation: corn, wheat, clover	2.7	14
Continuous bluegrass	0.3	12

Source: Based on 14 years' data from Missouri Experiment Station, Columbia, Missouri

EXPLORING SCIENCE

Ancient *Terra Preta* Shows How to Build Soils

Although it's ecologically rich, the Amazon rainforest is largely unsuitable for agriculture because of its red, acidic, nutrient-poor soils. But in many parts of the Amazon there are patches of dark, moist, nutrient-rich soils. These patches have long puzzled scientists. Locally known as *terra preta de Indio,* or "dark earth of the Indians," these patches of soil aren't associated with any particular environmental conditions or vegetation. Instead, the presence of bone fragments and pottery pieces hint that they may have a human origin.

Remote sensing surveys show that these dark earth patches, while usually rather small individually, collectively occupy somewhere between 1 and 10 percent of the Amazon. At the upper estimate, this would be

Soils enriched by charcoal centuries ago *(left)* still remain darker and more fertile than the usual weathered, red Amazonian soils *(right).*

Courtesy Dr. Bruno Glaser, University Bayreuth, Deutschland

about twice the size of Britain. Archaeologists now believe that these fertile soils once supported an extensive civilization of farms, fields, and even large cities in the Amazon basin for 1,000 years or more. After Europeans arrived in the sixteenth century, diseases decimated the indigenous population and cities were abandoned, but in many places the *terra preta* remains highly fertile 500 years later.

It's now believed the dark soils were created by native people who deliberately worked charcoal, human and animal manure, food waste, and plant debris into their gardens and fields. In some areas, these black soils, laced with bits of pottery, reach two meters (6 feet) in depth. Much of the dark color seems to come from charcoal that

has been added to the soil. Charcoal also improves the retention of nutrients, water, and other organic matter. Contrary to what scientists expected, the charcoal also seems to be beneficial for the soil-building activities of microorganisms, fungi, and other soil organisms. In short, what seems like a fairly simple practice of soil husbandry has turned extremely poor soils into highly productive gardens. Crops such as bananas, papaya, and mango are as much as three times more productive in *terra preta* than on nearby fields. And although most Amazonian soils need to be fallow for eight to ten years to rebuild nutrients after being farmed, these dark soils can recover after only six months or so.

Native people probably produced charcoal by burning biomass in low-temperature fires, in which fuel is allowed to smolder slowly in an oxygen-poor environment. Modern charcoal makers do this in an enclosed kiln. Some soil scientists are now advocating the use of charcoal, which they call "biochar," to help promote growth. But it turns out that charcoal can have another important benefit. When organic material is burned in an open fire or simply allowed to decompose in the open air, the carbon it contains is converted to CO_2 that contributes to global warming. Charcoal that is turned into the soil, on the other hand, can sequester carbon in the soil for centuries. Some of the Amazonian *terra preta* has five to ten times as much carbon as nearby soils. There's now an international movement to encourage biochar production and use, both to increase food production and to store carbon.

The use of charcoal as a soil amendment wasn't limited to the Amazon. Other places in South America, Africa, and Asia also have had similar soil management traditions, although soil scientists have only recently come to appreciate the benefits of this practice. At recent UN conventions on world food supplies, desertification, and global climate change, there have been discussions of global programs to make and distribute charcoal as a way to combat a whole series of environmental problems. It seems that the rediscovery of ancient methods may improve our soil management today.

---

erodible soils, the best way to keep soil in place is to plant perennial cover, such as grasses, rather than annual plants, which must be harvested every year. Establishing forests, orchards, grassland, or crops such as tea or coffee can minimize the need for regular cultivation. **Cover crops** such as rye, alfalfa, or clover can also be planted after harvest to hold and protect the soil. These cover crops can be plowed under at planting time to provide green manure. Many also fix nitrogen and enrich the soil while the land is idle.

In some cases, interplanting of two different crops in the same field not only protects the soil but also is a more efficient use of the land, providing double harvests. Native Americans and pioneer farmers planted beans or pumpkins between the corn rows. The beans provided nitrogen needed by the corn, pumpkins crowded out weeds, and both crops provided foods that nutritionally balance corn. Traditional swidden (slash-and-burn) cultivators in Africa and South America often plant as many as 20 different crops together in small plots. The crops mature at different times so there is always something to eat, and the soil is never exposed to erosion for very long.

Reduced tillage leaves crop residue

Leaving crop residues on the land after harvest often is the easiest way to protect soil from erosion. Plant stalks and roots left on a field cover the surface and break the erosive power of wind and water; this residue also reduces evaporation and soil temperature in hot climates and protects soil organisms that help aerate and rebuild soil. Residue can increase water infiltration by 99 percent, reduce runoff by 99 percent, and reduce erosion by 98 percent.

There are several methods of reduced tillage. Some farmers cultivate less frequently. Others use a chisel plow, with a row of curved chisel-like blades, which creates ridges on which seeds can be planted. This method leaves up to 75 percent of plant debris on the surface between the rows, helping to prevent erosion. A coulter, a sharp disc like a pizza cutter, can be used to slice through the soil, opening up a furrow or slot just wide enough to insert seeds. In "no-till" planting, seeds are drilled into the ground directly through mulch and ground cover. This allows a cover crop to be interseeded with a subsequent crop (fig. 10.15).

Farmers who use these conservation tillage techniques often depend heavily on pesticides (insecticides, fungicides, and herbicides) to control insects and weeds. Increased use of toxic agricultural chemicals is a matter of great concern. Heavy use of pesticides is not, however, a necessary corollary of soil conservation. It is possible to combat pests and diseases with integrated pest management that combines crop rotation, trap crops, natural repellents, and biological controls.

Section Review

1. What are four kinds of erosion?
2. How can excessive fertilizer and irrigation be a problem?
3. What are some strategies for reducing soil erosion?

10.3 PESTS AND PESTICIDES

- *Modern agriculture relies on herbicides, insecticides, and other chemical pest controls.*
- *Persistence in the environment is one reason DDT was so widely used and was so damaging.*
- *Pesticide use is growing steadily, with neonicotinoids and organophosphates among the most common.*

Every ecosystem has producers and consumers. In an agricultural system, we try to simplify the ecosystem to just one type of producer (the crop plant) and one type of consumer (humans). Other consumers that might compete with us, such as crop-eating insects or fungi, we consider "pests" that need to be controlled. Farmers and the agriculture industry spend a great deal of time and money on controlling crop pests, especially weeds, which compete with crops for space and sunlight, and insects that eat crop plants.

Pest control is an ancient practice. People in every culture have known that salt, smoke, and insect-repelling plants can keep away bothersome organisms and preserve food. The Sumerians controlled insects and mites with sulfur 5,000 years ago. Greeks and Romans used oil sprays, ash and sulfur ointments, lime, and other natural materials to protect themselves, their livestock, and their crops from a variety of pests. Cultures everywhere have used fermentation to preserve foods in acids, alcohol, or brine (salt).

The Romans controlled crop pests by burning fields and by rotating crops. They also employed cover crops to reduce weeds. The Chinese developed plant-derived insecticides and introduced predatory ants in orchards to control caterpillars 1,200 years ago.

Pesticide is a general term for a chemical that kills pests, but sometimes we also consider chemicals that drive pests away to be pesticides. Some pest-control compounds kill a wide range of living things and are called **biocides** (fig. 10.16). Chemicals such as ethylene dibromide, which are used to protect stored grain or to sterilize soils before planting strawberries, are biocides. In addition, there are chemicals aimed at particular groups of pests. **Herbicides** are chemicals that kill plants; **insecticides** kill insects; and **fungicides** kill fungi.

Synthetic (artificially made) chemical pesticides have been one of the main innovations of modern agricultural production. Our use of pesticides has increased dramatically in recent years, although pesticides receive relatively little public attention in most areas. Our current food system relies heavily on synthetic chemicals to control pests. These compounds have brought many

FIGURE 10.15 No-till planting involves drilling seeds through debris from last year's crops. Here, soybeans grow through corn mulch. Debris keeps weeds down, reduces wind and water erosion, and keeps moisture in the soil.
© William P. Cunningham

FIGURE 10.16 Broad-spectrum toxins can eliminate pests quickly and efficiently, but what are the long-term costs to us and to our environment?
© William P. Cunningham

FIGURE 10.17 Before we realized the toxicity of DDT, it was sprayed freely on people to control insects, as shown here at Jones Beach, New York, in 1948.
© Bettmann/Corbis

benefits, but they also bring environmental health problems. Here we review some of the main types of pesticides, how they work, and some alternative strategies.

Modern pesticides provide benefits but also create health risks

The era of synthetic organic pesticides began in 1939 when Swiss chemist Paul Müller discovered the powerful insecticidal properties of dichloro-diphenyl-trichloroethane (DDT). Inexpensive, stable, easily applied, and highly effective, this compound seemed ideal for crop protection and disease prevention. DDT is remarkably lethal to a wide variety of insects but relatively nontoxic to mammals. Mass production of DDT started during World War II, when allied armies used it to protect troops from insect-borne diseases. In less than a decade, manufacture of the compound soared from a few kilograms to thousands of metric tons per year. It was sprayed on crops and houses, dusted on people and livestock, and used to combat insects nearly everywhere (fig. 10.17).

By the 1960s, however, evidence began to accumulate that indiscriminate use of DDT and other long-lasting industrial toxins was having unexpected effects on wildlife. Peregrine falcons, bald eagles, brown pelicans, and other birds at the top of the food chain were disappearing from former territories in eastern North America. Studies revealed that eggshells were thinning in these species as DDT and its breakdown products were concentrated through food chains until they reached endocrine hormone-disrupting levels in top predators (see fig. 8.13). In 1962, biologist Rachel Carson published *Silent Spring,* warning that persistent organic pollutants, such as DDT, pose a threat to wildlife and perhaps to humans. DDT was banned for most uses in developed countries in the late 1960s, but it continues to be used in developing countries

and remains the most prevalent contaminant on food imported to the United States.

Since the 1940s, many new synthetic pesticides have been invented. Many of them, like DDT, have proven to have unintended consequences on nontarget species. Assessing the relative costs and benefits of using these compounds continues to be a contentious topic, especially when unexpected complications arise, such as increasing pest resistance or damage to beneficial insects.

Information on pesticide use is often poorly reported, but the U.S. EPA estimates that world usage of conventional pesticides amounts to some 5.7 billion pounds (2.6 million metric tons) per year of active ingredients. In addition "inert" ingredients are added to pesticides as carriers, stabilizers, and emulsifiers.

Roughly 80 percent of all conventional pesticides applied in the United States are used in agriculture or food storage and shipping. Some 90 million ha of crops in the United States—including 96 percent of all corn and about 95 percent of soybeans—are treated with herbicides every year. In addition, 25 million ha of agricultural fields and 7 million ha of parks, lawns, golf courses, and other lands are treated with insecticides and fungicides. Cotton probably has the highest rate of insecticide application of any crop.

It's important to note that most of us participate in pesticide use. Household uses in homes and gardens account for the fastest-rising sector, about 14 percent of total use, according to the most recent available EPA estimates. Three-quarters of all American homes use some type of pesticide, amounting to 20 million applications per year. Often, people use much larger quantities of chemicals in their homes, yards, or gardens than farmers would use to eradicate the same pests in their fields. Storage and accessibility of toxins in homes also can be a problem. Children's exposure to toxins in their home may be of greater concern than pesticide residues in food. Health effects of these compounds are discussed in chapter 8.

Global use of pesticides is also hard to evaluate, but according to the UN Food and Agriculture Organization, global imports of pesticides have risen about 60-fold since data collection began in 1962 (fig. 10.18). Approximately 20 percent of global pesticide use is in the United States, according to the U.S. EPA.

Organophosphates and chlorinated hydrocarbons are dominant pesticides

Pesticides vary in their chemical structure and composition. Most are organic compounds (carbon-based, such as DDT), with active components including phosphate, chlorine, or other compounds. Some are toxic metals (such as arsenic), and halogen fumigants

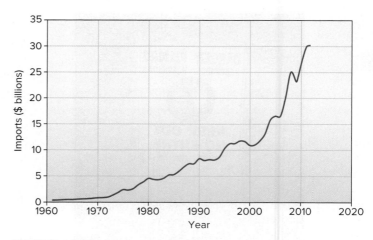

FIGURE 10.18 Value of global trade in pesticides (imports). *Source: Data from the UN Food and Agriculture Organization, 2012*

What Do You Think?

Organic Farming in the City

Farming is remote for most of us, with some 85 percent of Americans living in cities. We eat foods grown far away, processed in anonymous factories, delivered by national grocery store chains. But a growing movement has been reclaiming food production for city-dwellers. Urban farming, urban gardening, and community gardens are just a few of the names and strategies people in cities are using to bring some of their food closer to home.

One of the leading examples is Growing Power, an organization formed in Milwaukee, Wisconsin, by the former basketball player Will Allen, who received a MacArthur "genius" award for his work. Like many older industrial cities, Milwaukee has seen declining population, housing values, incomes, and economic opportunity for decades. Low-income or unemployed minority groups make up an increasing proportion of the city. Young people have few jobs, few training resources, and little food security.

Difficult conditions can make fertile ground for a movement promoting self-sustaining food production. Will Allen brought together unemployed teenagers and other community members, starting on just a 2-acre (0.8 hectare) plot of farmland. Allen's organization teaches kids and their parents to improve the soil with compost and mulch; to grow vegetables, tilapia, chickens, and other foods; and to manage a business and sell food. Growing Power serves kids by teaching them skills and providing internships and paid employment. The organization serves the community by providing a positive focus that brings people together. Who doesn't like fresh food grown by friends? The organization also serves the city by providing wholesome food resources that support the health and food security of low-income neighborhoods.

One of the first steps that members of Growing Power took was to form a land trust, an organization that could take long-term control of the land they work. This stability allows them to invest in the soil and in greenhouses, in projects, and plans. Another step they have taken is to provide workshops that spread their philosophy and techniques nationwide. Growing Power gives people access to fresh food, teaches kids about nurturing the land, and most important, it invests in the next generation of citizens of Milwaukee and other cities.

Urban farming and gardening movements are growing rapidly and have rich potential. One study found that East Lansing, Michigan, could produce 75 percent of its own vegetables on 4,800 acres (2,000 ha) of unbuilt land. Motivations for urban farming include, but are not limited to, issues of food security, community stability, youth employment, improving environmental quality for kids, and fun. Creative and enthusiastic projects abound, from Brooklyn to Detroit to Portland, Oregon, and many places between. Do you think community gardens or urban farming would be useful in areas where you live? What do you think it would take to support these efforts in your area?

Urban farming helps young people and their communities grow stronger. These girls are selling produce from Capuchin Soup Kitchen/Earthworks Urban Farm in Detroit, Michigan.
© *The Capuchin Soup Kitchen-Earthworks Urban Farm, Detroit, MI*

(gases such as bromine). Finally, there are natural organic compounds and biological agents.

Organophosphates, *organic* (carbon-based) compounds containing biologically reactive *phosphate*, are among the most heavily used synthetic pesticides. Glyphosate, the single most heavily used herbicide in the United States, is also known by the trade name Roundup. Glyphosate is applied to 90 percent of U.S. soybeans, as well as to other crops. "Roundup-ready" soybeans and corn—varieties genetically modified to tolerate glyphosate while other plants in the field are destroyed—are the most commonly planted genetically modified crops (chapter 9), and these tolerant varieties are one of the factors that make expanding soy production cost-effective in Brazil (opening case study). These "Roundup-ready" varieties have helped glyphosate surpass atrazine (an herbicide used most abundantly on corn) as the most-used herbicide (fig. 10.19).

Other organophosphates attack the nervous systems of animals, including humans. Parathion, malathion, dichlorvos, and other organophosphates were developed as an outgrowth of nerve gas research during World War II. These compounds can be extremely lethal. Because they break down quickly, usually in just a few days, they are less persistent in the environment than other pesticides. These compounds are very dangerous for workers, however, who are often sent into fields too soon after they have been sprayed (fig. 10.20).

Chlorinated hydrocarbons, also called organochlorines, are persistent and highly toxic to sensitive organisms. Atrazine was the most heavily used herbicide in the United States until the recent increase in glyphosate use (see fig. 10.19). Atrazine is applied to 96 percent of the corn crop in the United States to control weeds in cornfields (fig. 10.21). This widespread use has resulted in concerns about contamination of water supplies. One study of Midwestern Corn Belt states found atrazine in 30 percent of community wells and 60 percent of private wells sampled. This is a worry because atrazine has been linked to sexual abnormalities and population crashes in frogs. Some studies find that atrazine is associated with higher cancer incidence and lower sperm

FIGURE 10.20 The United Farm Workers of America claims that 300,000 farmworkers in the United States suffer from pesticide-related illnesses each year. The World Health Organization estimates that globally 25 million people suffer from pesticide poisoning and 20,000 die each year from improper use or storage of pesticides.
© Charles Smith/Corbis RF

counts in humans. Because of its persistence and uncertain health effects, atrazine was banned in Europe in 2003.

Among the hundreds of other organochlorines are DDT, chlordane, aldrin, dieldrin, toxaphene, and paradichlorobenzene (mothballs). Toxaphene is extremely toxic for fish and can kill goldfish at five parts per billion (5 µg/liter). This group also includes the herbicide 2,4-D, a widely used lawn chemical that selectively suppresses broad-leaf flowering plants, such as dandelions.

Chlorinated hydrocarbons can persist in the soil for decades, and they are stored in fatty tissues of organisms, so they become concentrated through food chains. DDT, which was inexpensive and widely used in the 1950s, has been banned in most developed countries, but it is still produced in the United States and is used in many developing countries.

Neonicotinoid pesticides, developed in the 1990s, are now the most widely used group of insecticides globally. Used to control insects, these compounds are powerful enough that minute concentrations can transfer to pollen and fruits. Insects pollinating these plants, notably honey bees, seem to be easily impaired by these compounds. Global disappearances of honey bee colonies, known as "colony collapse disorder" are widely blamed on neonicotinoid pesticides. Birds and aquatic organisms also appear

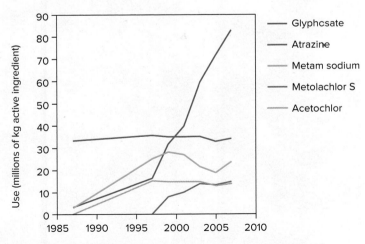

FIGURE 10.19 Usage of the top five pesticides in the United States. All are herbicides applied to soy, corn, or wheat, or to lawns, except metam sodium, a soil fumigant used mainly on ground crops such as carrots, potatoes, peppers, and strawberries.
Source: USDA, 2009

Quantitative Reasoning

What is the change in dollar value of pesticide imports globally, from 1960 to 2010 (fig. 10.18)? What is the percentage change, rounded to the nearest 5 percent, from 1980 to 2010? What is the change in number of pesticide-resistant pest varieties from 1960 to 2010 (fig. 10.25)? The percentage change from 1980 to 2010? Which has risen more steeply?

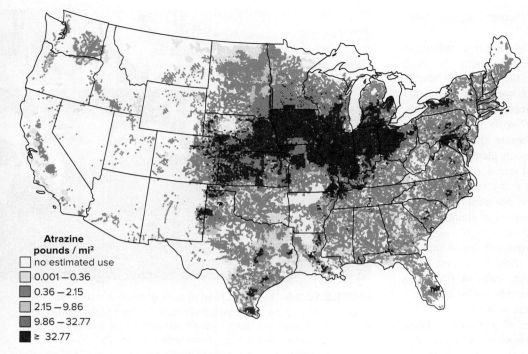

FIGURE 10.21 Atrazine herbicide use, average pounds per square mile of farmland.
Source: USDA

Atrazine
pounds / mi²

- no estimated use
- 0.001 – 0.36
- 0.36 – 2.15
- 2.15 – 9.86
- 9.86 – 32.77
- ≥ 32.77

to be affected by minute amounts of neonicotinoid pesticides. Studies by the American Bird Conservancy have found that the EPA has done relatively little rigorous testing on these compounds, despite their dominance in the insecticide market. Because they are new, highly toxic, slow to break down, little understood, and little regulated, these pesticides worry fruit and vegetable farmers who rely on insects for pollination (fig. 10.22). Because of concerns over pollinators, the European Union imposed a trial ban on neonicotinoid pesticides in 2013. The same week, the USDA published position papers saying that diseases, not pesticides, were the primary cause of colony collapse disorder.

Fumigants are gases composed of small molecules, such as carbon tetrachloride, ethylene dibromide, and methyl bromide. These gases can be injected into soils or used to fumigate enclosed spaces. Fumigants are used to control fungus in strawberry fields and other low-growing crops, as well as to prevent decay or rodent and insect infestations in stored grain. Because these compounds are extremely dangerous for workers who apply them, many have been restricted or banned altogether in some areas.

Inorganic pesticides include compounds of toxic elements such as arsenic, sulfur, copper, and mercury. These broad-spectrum poisons are generally highly toxic and indestructible, remaining in the environment forever. They generally act as nerve toxins. Historically, arsenic powder was a primary pesticide applied to apples and other orchard crops, and traces remain in soil and groundwater in many agricultural areas.

Natural organic pesticides, or "botanicals," generally are extracted from plants. Some important examples are nicotine and nicotinoid alkaloids from tobacco, and pyrethrum, a complex of chemicals extracted from the daisy-like *Chrysanthemum*

cinerariaefolium (fig. 10.23). These compounds also include turpentine, phenols, and other aromatic oils from conifers. All are toxic to insects, and many prevent wood decay.

Microbial agents and **biological controls** are living organisms or toxins derived from them that are used in place of pesticides. The natural soil bacterium *Bacillus thuringiensis* is one of the chief pest-control agents allowed in organic farming. This bacterium kills caterpillars and beetles by producing a toxin that ruptures the digestive tract lining when eaten. Parasitic wasps such as the tiny *Trichogramma* genus attack moth caterpillars and eggs, while lacewings and ladybugs are predators that control aphids.

Pesticides have profound environmental effects

Although we depend on pesticides for most of our food production, and for other purposes such as biofuel production, widespread use of these compounds brings a number of environmental and health risks. The most common risk is exposure of nontarget organisms. Many

FIGURE 10.22 Honeybees, essential for much of our food production, have been disappearing in droves, in what is called Colony Collapse Disorder. Neonicotinoid pesticides are widely thought to be responsible.
© *Michael Maloney/San Francisco Chronicle/Corbis*

FIGURE 10.23 Chrysanthemum flowers are a source of pyrethrum, a natural insecticide.
© William P. Cunningham

FIGURE 10.24 This machine sprays insecticide on orchard trees—and everything nearby. Up to 90 percent of pesticides applied in this fashion never reach target organisms.
© Joe Munroe/Science Source

pesticides are sprayed broadly and destroy beneficial insects—those that pollinate crops or prey on pests—as well as pest species (fig. 10.24). The loss of insect diversity has been a growing problem in agricultural regions: At least a third of the crops we eat rely on pollinators, such as bees and other invertebrates, to reproduce.

Colony collapse disorder in honeybees has received particular attention in recent years. Many crops, including squash, tomatoes, peppers, apples, and other fruit, rely on bees for pollination. The California almond crop, for example, is worth $1.6 billion annually and is entirely dependent on bees for pollination. Honeybee colonies have been dying worldwide, sometimes with half or more disappearing in an area in a single year. While there are many possible explanations, including disease and other stresses, neonicotinoid insecticides are one of the chief suspects.

Pest resurgence, or the rebound of resistant populations, is another important problem in overuse of pesticides. This process occurs when a few resistant individuals survive pesticide treatments and those resistant individuals propagate a new pesticide-resistant population. The Worldwatch Institute reports that at least 1,000 insect pest species and another 550 or so weeds and plant pathogens worldwide have developed chemical resistance. Of the 25 most serious insect pests in California, three-quarters or more are resistant to one or more insecticides. Cornell University entomologist David Pimentel reports that a larger percentage of crops are lost now to insects, diseases, and weeds than in 1944, despite the continuing increase in the use of pest controls (fig. 10.25).

As resistant pests evolve, there is an ever-increasing need for newer, better pesticides—this is called the **pesticide treadmill.** Glyphosate (Roundup), the dominant herbicide used in the United States and one of the primary herbicides in Brazil, Australia, and elsewhere, is no longer effective against a variety of superweeds.

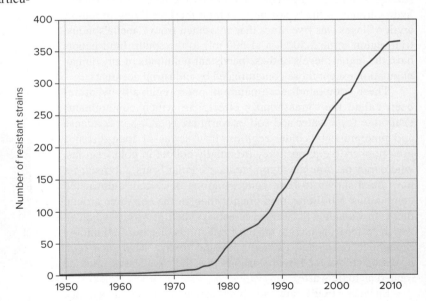

FIGURE 10.25 Pesticide-resistant varieties have increased, resulting in a "pesticide treadmill," in which farmers must use increasingly expensive combinations of chemicals to suppress weeds, disease, and pests.
Source: Data from US EPA: Weed Society of America http://www.weedscience.com

More and more, farmers are advised to mix tanks of various pesticides—metachlor, Flexstar, Gramoxone, diuron, and other combinations—to keep down increasingly aggressive weeds such as pigweed and rye grass. At the same time, ever-larger amounts of glyphosate are needed to combat resistant weeds. In 2010, the U.S. Supreme Court reversed a ban on genetically modified glyphosate-tolerant alfalfa, a decision that crop scientists expect will increase pesticide-tolerant weeds on the 22 million acres of alfalfa grown in the United States. Increasing reliance on glyphosate and other herbicides is sure to increase environmental exposure, with uncertain effects on human health and ecosystems.

POPs accumulate in remote places

Many pesticides break down into less-harmful components several days or weeks after application. Certain compounds, such as DDT and other chlorinated hydrocarbons, are both effective and dangerous because they don't break down easily. **Persistent organic pollutants (POPs)** is a collective term for these chemicals, which are stable (or persistent), easily absorbed into fatty tissues (because they are fat-soluble organic compounds), and highly toxic (see chapter 8).

Because they persist for years, even decades in some cases, and move freely through air, water, and soil, they often show up far from the point of original application. Some of these compounds have been discovered far from any possible source and long after they most likely were used. Because they are easily absorbed by living cells, many chlorinated hydrocarbons are biomagnified (see fig. 8.13) and stored in the bodies of predators at the top of the food web—such as porpoises, whales, polar bears, trout, eagles, ospreys, and humans. In a study of human pesticide uptake and storage, Canadian researchers found that the level of chlorinated hydrocarbons in the breast milk of Inuit mothers living in remote arctic villages was five times that of women from Canada's industrial region some 2,500 km (1,600 mi) to the south. Inuit people have the highest levels of these persistent pollutants of any human population except those contaminated by industrial accidents.

These compounds accumulate in polar regions by what has been called the "grasshopper effect," in which contaminants evaporate from water and soil in warm areas and then condense and precipitate in colder regions. In a series of long-distance grasshopper-like jumps, they eventually collect in polar regions, where they accumulate in top predators. Polar bears, for instance, have been shown to have concentrations of certain chlorinated compounds 3 billion times greater than in the seawater around them. In Canada's St. Lawrence estuary, beluga (white whales), which suffer from a wide range of infectious diseases and tumors thought to be related to environmental toxins, have such high levels of chlorinated hydrocarbons that their carcasses must be treated as toxic waste.

Because POPs are so long-lasting and so dangerous, 127 countries agreed in 2001 to a global ban on the worst of them, including aldrin, chlordane, dieldrin, DDT, endrin, hexachlorobenzene, neptachlor, mirex, toxaphene, polychlorinated biphenyls (PCBs), dioxins, and furans. Most of this "dirty dozen"

had been banned or severely restricted in developed countries for years. However, their production has continued. Between 1994 and 1996, U.S. ports shipped more than 100,000 tons of POPs each year. Most of this was sent to developing countries where regulations were lax. Ironically, many of these pesticides returned to the United States on bananas and other imported crops. According to the 2001 POPs treaty, eight of the dirty dozen were banned immediately; PCBs, dioxins, and furans are being phased out; and use of DDT, still allowed for limited uses such as controlling malaria, must be publicly registered in order to permit monitoring. The POPs treaty has been hailed as a triumph for environmental health and international cooperation. Unfortunately, other compounds—perhaps just as toxic—have been introduced to replace POPs.

Pesticides often impair human health

Pesticide effects on human health can be divided into two categories: (1) acute effects, including poisoning and illnesses caused by relatively high doses and accidental exposures, and (2) chronic effects suspected to include cancer, birth defects, immunological problems, endometriosis, neurological problems, Parkinson's disease, and other chronic degenerative diseases.

The World Health Organization (WHO) estimates that 25 million people suffer pesticide poisoning and at least 20,000 die each year. At least two-thirds of this illness and death results from occupational exposures in developing countries where people use pesticides without proper warnings or protective clothing (fig. 10.26).

FIGURE 10.26 Handling pesticides requires protective clothing and an effective respirator. Pesticide applicators in tropical countries, however, often can't afford these safeguards or can't bear to wear them because of the heat.
Source: Photo by Tim McCabe, USDA Natural Resources Conservation Service

Table 10.2	The 12 Most Contaminated Fruits and Vegetables
Rank	Food
1.	Strawberries
2.	Bell peppers
3.	Spinach
4.	Cherries (U.S.)
5.	Peaches
6.	Cantaloupe (Mexican)
7.	Celery
8.	Apples
9.	Apricots
10.	Green beans
11.	Grapes (Chilean)
12.	Cucumbers

Source: Environmental Working Group, 2002

A tragic example of occupational pesticide exposure is found among workers in the Latin American flower industry. Fueled by the year-round demand in North America for fresh vegetables, fruits, and flowers, a booming export trade has developed in countries such as Guatemala, Colombia, Chile, and Ecuador. To meet demands in North American markets for perfect flowers, table grapes, and other produce, growers use high levels of pesticides, often spraying daily with fungicides, insecticides, and herbicides. Working in warm, poorly ventilated greenhouses with little protective clothing, the workers—70 to 80 percent of whom are women—find it hard to avoid pesticide contact. Almost two-thirds of nearly 9,000 workers surveyed in Colombia experienced blurred vision, nausea, headaches, conjunctivitis, rashes, and asthma. Although harder to document, serious chronic effects such as stillbirths, miscarriages, and neurological problems were also reported.

Pesticide use can expose consumers to agricultural chemicals. In studies of a wide range of foods collected by the USDA, the State of California, and the Consumers Union between 1994 and 2000, 73 percent of conventionally grown food had residue from at least one pesticide and were six times as likely as organic foods to contain multiple pesticide residues. Only 23 percent of the organic samples of the same groups had any residues. Using these data, the Environmental Working Group has assembled a list of the fruits and vegetables most commonly contaminated with pesticides (table 10.2).

Section Review

1. What is the difference between biocides, herbicides, insecticides, and fungicides?
2. Why was DDT considered a "magic bullet"? Why was it listed among the "dirty dozen" persistent organic pollutants (POPs)?
3. Name several of our dominant pesticides.

10.4 ORGANIC AND SUSTAINABLE AGRICULTURE

- *Proponents of conventional and sustainable farming disagree on strategies for feeding the world.*
- *"Organic" and other food labels can have multiple meanings.*
- *Sustainable strategies include IPM, crop rotation, intercropping, soil management, and other strategies.*

Industrial-scale farming produces most food in the United States and other wealthy regions, but it is important to remember that in most of the world, most food continues to be produced by smaller farms. According to the UN Development Programme, in 2015 500 million family farms produced 80 percent of the food consumed worldwide. Most of these cannot afford to use many pesticides, so they are largely organic. These farms continue to feed the world despite dramatic expansion of corporate-scale producers, as described in the opening case study of this chapter.

Traditional methods require substantial labor inputs, however, in planting, weeding, and harvesting diverse crops. Where rural farming populations are declining, as they are in many urbanizing or industrialized countries, many farmers are seeking a compromise: They want to grow crops organically, protecting soil, water quality, and biodiversity, while still producing enough income to live a comfortable life. Is this compromise possible?

Global-scale trade and production like that in the Brazilian Cerrado is unlikely to work with sustainable methods, but at more moderate scales farmers are increasingly developing, or rediscovering, a multitude of strategies, to produce sustainably and to minimize use of pesticides (fig. 10.27). Planting nitrogen-fixing cover crops to avoid fertilizers, using crop rotation to minimize pesticides, strategic water management, mixed cropping, and use of perennial or tree crops are a few of these methods.

FIGURE 10.27 The USDA finds more pesticides in commercial strawberries than in any other fruit. Organically grown berries like these are safer to eat.
© William P. Cunningham

In general, soils stay healthier with these strategies than with chemical-intensive conventional farming. A long-running study in Switzerland has found that average yields on organic plots were 20 percent less than on adjacent fields farmed by conventional methods. But costs were so much lower that farmers' net income was higher with organic crops. Energy use was 56 percent less per unit of yield in organic farming than for conventional approaches. Beneficial root fungi were 40 percent higher, earthworms were three times as abundant, and spiders and other pest-eating predators were doubled in the organic plots. The organic farmers and their families reported better health and greater satisfaction than did their neighbors who used conventional farming methods.

Can sustainable practices feed the world's growing population?

The potential of sustainable farming is hotly debated. High-responding varieties and pesticides have been the foundation of commercial agriculture for over 50 years. The green revolution, and more recently the introduction of genetically modified crops, have produced staggering increases in the volume of harvests worldwide. Proponents of these input-intensive methods argue that sustainable farming is too inefficient and labor intensive to feed large populations.

Proponents of organic production argue that conventional farming has less to do with feeding people than with supporting industrial production, since after half a century of growth in commercial farming, most of the world's population continues to be fed by the small, traditional farms. Commercial crops, such as corn, soybeans, cotton, and sorghum, are used for livestock feed, which feed people indirectly but inefficiently, for biofuels, or for industrial uses (see chapter 9). Sustainable producers say their methods are more productive over time, because conventional practices degrade soils, water, ecosystems, and farm communities. The rapid spread of industrial farming, according to many food activists, serves mainly the multinational agrochemical corporations, which must constantly expand their global markets.

In 2011, the UN Commission on Human Rights and the UN Food and Agriculture Organization (FAO) both weighed in on the matter. Their studies indicate that if the aim is to provide food for impoverished regions, then developing countries should promote innovations in low-cost, low-input, sustainable soil-building and water-preserving methods. Data indicate that costs of irrigation, pesticides, fuel, and newly developed seed varieties have risen at least three times as fast as farm income in poor areas. These costs lead to high debt and widespread farm failures in poor regions. Debt drives many farmers off their land and into the already overcrowded cities of the developing world. In contrast, the FAO reports that areas that have invested in conservation-based farming innovations have increased yields at a rate similar to that of green revolution or genetically modified crops, while offering more sustainable food security in low-income regions.

Currently, less than 1 percent of all American farmland is devoted to organic farming, but the market for organic products may stimulate more conversion to this approach in the future. Organic food is much more popular in Europe than in North America. Liechtenstein is probably the leader among industrialized nations with 18 percent of its modest land area in certified organic agriculture. Sweden is second with 11 percent of its land in organic production. Much of the developing world, where people can afford few fertilizers or pesticides, is effectively organic.

What does "organic" mean?

In general, organic food is grown without artificial pesticides and with only natural fertilizers, such as manure. Legal definitions of the term, though, are more exact and often more controversial. According to USDA rules, products labeled "100 percent organic" must be produced without hormones, antibiotics, pesticides, synthetic fertilizers, or genetic modification. "Organic" means that at least 95 percent of the ingredients must be organic. "Made with organic ingredients" means it must contain at least 70 percent organic contents. Products containing less than 70 percent organic ingredients can list them individually. Organic animals must be raised on organic feed, given access to the outdoors, given no steroidal growth hormones, and treated with antibiotics only to fight diseases.

Walmart has become the top seller of organic products in the United States, a step that has done much to move organic products into the mainstream. However, much of the organic food, cotton, and other products we buy from Walmart comes from overseas. Often, producers in China, India, Mexico, Thailand, and elsewhere have weak oversight compared to that in the United States. More than 2,000 farms in China and India are certified "organic," but how can we be sure what that means? With the market for organic food generating $11 billion per year, it's likely that some farmers and marketers try to pass off foods grown with pesticides as more valuable organic produce. Industrial-scale organic agriculture can also be hard on soils: It often depends on frequent cultivation for weed control, and the constant mechanical disturbance can destroy soil texture and soil microbial communities.

Many who endorse the concept of organic food are disappointed that legal definitions in the United States allow for partial organics and for unsustainable production methods. The term *organic* is also hard to evaluate clearly when you can buy organic intercontinental grapes in which thousands of calories of jet and diesel fuel were consumed to transport every calorie of food energy from Chile to your supermarket, or when you can buy processed snack foods labeled "organic." Many farmers have declined to pay for organic certification because they regard the term as too broad to be meaningful. Often, farmers describe their operations instead as "sustainable" or "natural." Consumers can seek out local foods, both to support growers in their area and because it can be easier to find out about organic practices from local producers (fig. 10.28). Supporting local producers and farmers' markets also benefits the local community and economy.

Strategic management can reduce pests

Organic farming and sustainable farming use a multitude of practices to control pests. In many cases, improved management programs can cut pesticide use by 50 to 90 percent without reducing crop production or creating new diseases. Some of these techniques

FIGURE 10.28 Your local farmers' market is a good source of locally grown and organic produce.
© William P. Cunningham

are relatively simple and save money while maintaining disease control and yielding crops with just as high quality and quantity as we get with current methods (see What Can You Do? further below). In this section, we will examine crop management, biological controls, and integrated pest management systems that could substitute for current pest-control methods.

What Can You Do?

Controlling Pests

Based on the principles of integrated pest management, the U.S. EPA releases helpful guides to pest control. Among their recommendations:

1. *Identify pests, and decide how much pest control is necessary.* Does your lawn really need to be totally weed-free? Could you tolerate some blemished fruits and vegetables? Could you replace sensitive plants with ones less sensitive to pests?

2. *Eliminate pest sources.* Remove from your house or yard any food, water, and habitat that encourages pest growth. Eliminate hiding places or other habitats. Rotate crops in your garden.

3. *Develop a weed-resistant yard.* Pay attention to your soil's pH, nutrients, texture, and organic content. Grow grass or cover varieties suited to your climate. Set realistic goals for weed control.

4. *Use biological controls.* Encourage beneficial insect predators such as birds, bats that eat insects, ladybugs, spiders, centipedes, dragonflies, wasps, and ants.

5. *Use simple manual methods.* Cultivate your garden and handpick weeds and pests from your garden. Set traps to control rats, mice, and some insects. Mulch to reduce weed growth.

6. *Use chemical pesticides carefully.* If you decide that the best solution is chemical, choose the right pesticide product, read safety warnings and handling instructions, buy the amount you need, store the product safely, and dispose of any excess properly.

Source: *Citizen's Guide to Pest Control and Pesticide Safety:* EPA 730-K-95-001

Crop rotation involves growing a different crop in a field each year in a two- to six-year cycle. Most pests are specific to one crop, so rotation keeps pest populations from increasing from year to year: For example, a three-year soybean/corn/hay rotation is effective and economical protection against white-fringed weevils. Mechanical cultivation keeps weeds down, but it also increases erosion. Flooding fields before planting or burning crop residues and replanting with a cover crop can suppress both weeds and insect pests. Habitat diversification, such as restoring windbreaks, hedgerows, and ground cover on watercourses, provides habitat for insect predators, such as birds, and also reduces erosion. Adjusting the timing of planting or cultivation can help avoid pest outbreaks. Switching from vast monoculture fields to mixed polyculture (many crops grown together) makes it more difficult for pests to multiply beyond control.

Useful organisms can help us control pests

Biological controls such as predators (wasps, ladybugs, praying mantises; fig. 10.29) or pathogens (viruses, bacteria, fungi) can control many pests more cheaply and safely than broad-spectrum, synthetic chemicals. *Bacillus thuringiensis* or Bt, for example, is a naturally occurring bacterium that kills the larvae of lepidopteran (butterfly and moth) species but is generally harmless to mammals. A number of important insect pests such as tomato hornworm, corn rootworm, cabbage loopers, and others can be controlled by spraying the Bt bacterium on crops. Larger species are effective as well. Ducks, chickens, and geese, among other species, are used to rid fields of both insect pests and weeds. These biological organisms are self-reproducing and consume a wide variety of invertebrates. A few mantises or ladybugs released in your garden in the spring will keep producing offspring and protect your fruits and vegetables against a multitude of pests for the whole growing season.

Herbivorous insects have been used to control weeds. For example, the prickly pear cactus was introduced to Australia about 150 years ago as an ornamental plant. This hardy cactus escaped from gardens and found an ideal home in the dry soils of the outback. It quickly established huge, dense stands that dominated 25 million ha (more than 60 million acres) of grazing land.

FIGURE 10.29 The praying mantis looks ferocious and is an effective predator against garden pests, but it is harmless to humans. They can even make interesting and useful pets.
© Millard H. Sharp/Science Source

FIGURE 10.30 Neem trees provide oil that is used for medicinal purposes, and the leaves, seeds, and bark provide a natural insecticide.
© Dinodia Photos/Alamy Stock Photo

A natural predator from South America, the cactoblastis moth, was introduced into Australia in 1935 to combat the prickly pear. Within a few years, cactoblastis larvae had eaten so much prickly pear that the cactus has become rare and is no longer economically significant.

Some plants make natural pesticides and insect repellents. The neem tree (*Azadirachta indica*) is native to India but is now grown in many tropical countries (fig. 10.30). The leaves, bark, roots, and flowers all contain compounds that repel insects and can be used to combat a number of crop pests and diseases. Another approach is to use hormones that upset development or use sex attractants to bait traps containing toxic pesticides. To protect human health, many municipalities control mosquitoes with these techniques rather than aerial spraying of insecticides. Briquettes saturated with insect juvenile hormone are scattered in wetlands where mosquitoes breed. The presence of even minute amounts of this hormone prevent larvae from ever turning into biting adults (fig. 10.31).

Genetics and bioengineering can also help in our war against pests. Traditional farmers have long known to save seeds of disease-resistant crop plants or to breed livestock that tolerate pests well. Modern genetic methods have enhanced this process, especially by transferring Bt bacterial genes to corn, soy, and other crops. Heavy reliance on the Bt gene may dilute its effectiveness, however (chapter 9).

IPM uses a combination of techniques

Integrated pest management (IPM) is a flexible, ecologically based strategy that is applied at specific times and aimed at specific crops and pests. It often uses mechanical cultivation and techniques such as vacuuming bugs off crops as an alternative to chemical application (fig. 10.32). IPM doesn't give up chemical pest controls entirely but instead tries to use minimal amounts, only as a last resort, and avoids broad-spectrum, ecologically disruptive products. IPM relies on preventive practices that encourage growth and diversity of beneficial organisms and enhance plant defenses and vigor. Successful IPM requires careful monitoring of pest

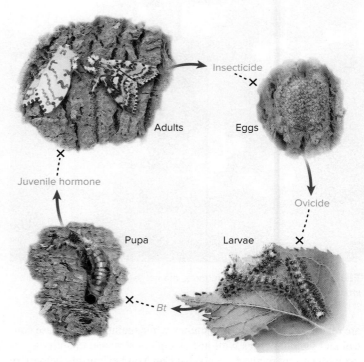

FIGURE 10.31 Different strategies can be used to control pests at various stages of their life cycles. *Bacillus thuringiensis* (*Bt*) kills caterpillars when they eat leaves with these bacteria on the surface. Releasing juvenile hormone in the environment prevents maturation of pupae. Predators attack at all stages.

populations to determine economic thresholds, the point at which potential economic damage justifies pest-control expenditures, and the precise time, type, and method of pesticide application.

Trap crops, small areas planted a week or two earlier than the main crop, are also useful. These plots mature before the rest of the field and attract pests away from other plants. The trap crop then is sprayed heavily with pesticides so that no pests are likely to escape. The trap crop is then destroyed, and the rest of the field should be mostly free of both pests and pesticides.

IPM programs are used on a variety of crops. Massachusetts apple growers who use IPM have cut pesticide use by 43 percent

FIGURE 10.32 This machine, nicknamed the "salad vac," vacuums bugs off crops as an alternative to treating them with toxic chemicals.
© Tanimura and Antle, Inc.

in the past ten years while maintaining per-acre yields of marketable fruit equal to that of farmers who use conventional techniques. Some of the most dramatic IPM success stories come from the developing world. In Brazil, pesticide use on soybeans has been reduced up to 90 percent with IPM, while in Costa Rica, use of IPM on banana plantations has eliminated pesticides altogether in one region. In Africa, mealybugs were destroying up to 60 percent of the cassava crop (the staple food for 200 million people) before IPM was introduced in 1982. A tiny wasp that destroys mealybug eggs was discovered and now controls this pest in over 65 million ha (160 million acres) in 13 countries.

In Indonesia, rice farmers offer a successful IPM model for staple crops. There, brown planthoppers had developed resistance to virtually every insecticide and threatened the country's hard-won self-sufficiency in rice. Researchers found that farmers were spraying their fields habitually—sometimes up to three times a week—regardless of whether fields were infested. In 1986, President Suharto banned 56 of 57 pesticides previously used in Indonesia and declared a crash program to educate farmers about IPM and the dangers of pesticide use. By allowing natural predators to combat pests and spraying only when absolutely necessary with chemicals specific for planthoppers, Indonesian farmers using IPM raised yields and cut pesticide costs by 75 percent. Only two years after its initiation, the program was declared a success. It has been extended throughout the whole country. Because nearly half the people in the world depend on rice as their staple crop, this example could have important implications elsewhere (fig. 10.33).

Although IPM can be a good alternative to chemical pesticides, it also presents environmental risks in the form of exotic organisms. Wildlife biologist George Boettner of the University of Massachusetts reported in 2000 that biological controls of gypsy moths, which attack fruit trees and ornamental plants, have also decimated populations of native North American moths. *Compsilura* flies, introduced in 1905 to control the gypsy moths, have a voracious appetite for other moth caterpillars as well. One of the largest North American moths, the Cecropia moth (*Hyalophora cecropia*), with a 15-cm wingspan, was once ubiquitous in the eastern United States but is now rare in regions where *Compsilura* flies were released.

Low-input agriculture aids farmers and their land

In contrast to the trend toward industrialization and dependence on chemical fertilizers, pesticides, antibiotics, and artificial growth factors common in conventional agriculture, some farmers are going back to a more natural, agroecological farming style. Finding that they can't—or don't want to—compete with factory farms, these producers are making money and staying in farming by returning to small-scale, low-input agriculture. Brent and Regina Beidler, for example, run one of nearly 200 small, organic dairies in Vermont (fig. 10.34). With just 40 cows, the Biedlers move the cattle every 12 hours, after each milking, to a different paddock. This rotation, with intensive grazing in small paddocks, method mimics the movement of herds of natural grazers. Travelling together in tight herds, the cows are forced to eat unpalatable plants as well as the sweet grasses, so that pastures stay healthy and diverse. With time to regenerate between visits, vegetation in the paddocks can remain healthy and diverse, and soil compaction is minimized.

Alternative Pest-Control Strategies

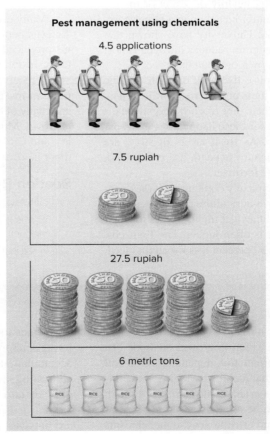
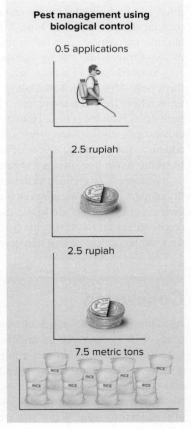

	Pest management using chemicals	Pest management using biological control
Number of times insecticide used in rice season	4.5 applications	0.5 applications
Cost to farmers per hectare	7.5 rupiah	2.5 rupiah
Cost to government per hectare	27.5 rupiah	2.5 rupiah
Rice yield per hectare	6 metric tons	7.5 metric tons

FIGURE 10.33 Indonesia has one of the world's most successful integrated pest management (IPM) programs. Switching from toxic chemicals to natural pest predators has saved money while also increasing rice production.
Source: Tolba, et al., World Environment, 1972–1992, p. 307, Chapman & Hall, 1992 United Nations Environment Programme

FIGURE 10.34 Brent Beidler, of Randolph Center, VT, keeps his cows and his pasture healthy with intensive rotational grazing, without chemical pesticides or fertilizers.
© Melanie Stetson Freeman/The Christian Science Monitor via Getty Images

Like other organic dairies, the Beidlers use no chemical fertilizers or pesticides on pastures. Also like other organic farmers, the Beidlers chose to go organic because it made sense for sustainability of the land, as well as for the health of the cows and the quality of the milk produced. In an organic operation, antibiotics are used only to fight diseases. Cows are kept healthy on a diet of mainly grass and hay, with less grain and antibiotics than in a larger commercial dairy. Studies at Iowa State University have shown that animals on pasture grass rather than grain reduces nitrogen runoff by two-thirds while cutting erosion by more than half.

Similarly, the Franzens, who raise livestock on their organic farm near Alta Vista, Iowa, allow their pigs to roam in lush pastures where they can supplement their diet of corn and soybeans with grasses and legumes. Housing for these happy hogs is in spacious, open-ended hoop structures. As fresh layers of straw are added to the bedding, layers of manure beneath are composted, breaking down into odorless organic fertilizer.

Low-input farms such as these typically don't turn out the quantity of meat or milk that their intensive-agriculture neighbors do, but their production costs are lower and they get higher prices for their crops, so the all-important net gain is often higher. The Franzens, for example, calculate that they pay 30 percent less for animal feed, 70 percent less for veterinary bills, and half as much for buildings and equipment than neighboring confinement operations.

Consumers' choices play an important role

Preserving small-scale, family farms also helps preserve rural culture. As Marty Strange of the Center for Rural Affairs in Nebraska asks, "Which is better for the enrollment in rural schools, the membership of rural churches, and the fellowship of rural communities—two farms milking 1,000 cows each or twenty farms milking 100 cows each?" Family farms help keep rural towns alive by purchasing machinery at the local implement dealer, gasoline at the neighborhood filling station, and groceries at the community grocery store.

These are the arguments that lead many people to shift at least part of their diets to local foods. Eating locally can help sustain local businesses and farm communities. Most profits from conventional foods, in contrast, go to a tiny number of giant food corporations: The top three or four corporations in each commodity group typically control 60 to 80 percent of the U.S. market. Where conventional foods were shipped an average of 2,400 km (1,500 mi) to markets, the average food item at a farmers' market traveled only 72 km (45 miles). Food from local, small-scale farms also often involves less energy for fertilizer, fuel for shipping, and plastic food packaging.

Many co-ops carry food that is locally grown and processed. An even better way to know where your food comes from and how it's produced is to join a community-supported agriculture (CSA) farm. In return for an annual contribution to a local CSA farm, you'll receive a weekly "share" of whatever the farm produces. CSA farms generally practice organic or low-input agriculture, and many of them invite members to visit and learn how their food is grown. Much of America's most fertile land is around major cities, and CSAs and farmers' markets are one way to help preserve these landscapes around metropolitan areas.

Section Review

1. List several arguments for and against sustainable farming.
2. Why is it controversial that Walmart is the largest seller of organic products?
3. What is IPM, and how is it used in pest control?

Conclusion

Agriculture leads to some of our most dramatic environmental changes. It is therefore an area in which improved methods can hold potential for progress. Farm production depends first on soil quality. Soils are complex systems that include biological and mineral components, and soils can be enriched and built up through careful management. Soils can also be eroded and degraded rapidly and irrevocably. Water and wind erosion are the mechanisms damaging most of the world's farming soils. Soil degradation is causing the continuing loss of farmland, even while populations dependent on that farmland grow.

Water for irrigation and energy are two other key resources for agriculture. Irrigation is often necessary, but it can cause salt accumulation or waterlogging in soils. Energy use, in fertilizing, cultivating, harvesting, irrigating, and other activities, continues to grow on farms in the developed world.

Pesticides, an important part of production on modern farms, are increasing dramatically in use. They bring both benefits and environmental costs. In particular, nontarget organisms are often harmed by pesticides, and extensive use often causes the resurgence of pest populations as pests develop immunity to chemicals. Our most abundantly used agricultural chemicals are organophosphates, including glyphosate, and organochlorines, such as atrazine. Glyphosate and atrazine are applied to nearly all soy and corn produced in the United States. Global consumption of these and similar agricultural chemicals continues to grow, but household use is the fastest-growing sector of pesticide application and now makes up about 14 percent of total use.

Alternative strategies for pest control include crop rotation, biological controls, mechanical cultivation, and other methods.

Integrated pest management is a flexible, ecologically based approach that involves monitoring pest populations and using small, targeted applications of pesticides. This approach can dramatically reduce pesticide use.

Other sustainable agriculture practices include soil conservation by terracing, by leaving crop residue on the soil, and by reduced frequency of tilling. These practices can save money for farmers and improve soil fertility. You can help support environmentally sustainable farming practices by buying sustainably and locally produced foods.

Reviewing Key Terms

Can you define the following terms in environmental science?

biocide 10.3

biological controls 10.3

chlorinated hydrocarbons 10.3

contour plowing 10.2

cover crops 10.2

desertification 10.2

fumigants 10.3

fungicide 10.3

gully erosion 10.2

herbicide 10.3

inorganic pesticides 10.3

insecticide 10.3

integrated pest management (IPM) 10.4

microbial agents 10.3

mycorrhizal symbiosis 10.1

natural organic pesticides 10.3

neonicotinoid 10.3

organophosphates 10.3

persistent organic pollutants (POPs) 10.3

pest resurgence 10.3

pesticide 10.3

pesticide treadmill 10.3

rill erosion 10.2

salinization 10.2

sheet erosion 10.2

subsoil 10.1

terracing 10.2

topsoil 10.1

waterlogging 10.2

Critical Thinking and Discussion Questions

1. As you consider the expansion of soybean farming and grazing in Brazil, what are the costs and what are the benefits of these changes? How would you weigh these costs and benefits for Brazilians? If you were a U.S. ambassador to Brazil, how would you advise Brazilians on their farming and grazing policies, and what factors would shape your advice?

2. The discoverer of DDT, Paul Müller, received a Nobel Prize for his work. Would you have given him this prize?

3. Are there steps you could take to minimize your exposure to pesticides, either in things you buy or in your household? What would influence your decision to use household pesticides or not to use them?

4. What criteria should be used to determine whether farmers should use ecologically sound techniques? How would your response differ if you were a farmer, a farmer's neighbor, someone downstream of a farm, or someone far from farming regions?

5. Should we try to increase food production on existing farmland, or should we sacrifice other lands to increase farming areas? Why?

6. Some rice paddies in Southeast Asia have been cultivated continuously for a thousand years or more without losing fertility. Could we, and should we, adapt these techniques to our own country? Why or why not?

7. *Terra preta* soils were a conundrum for soil scientists for decades. What expectations about tropical soils did these black soils violate? What would it take to make similar investments in soils today?

Data Analysis

Graphing Changes in Pesticide Use

The National Agricultural Statistics Service (NASS) keeps records of pesticide use in the United States, and you can access those records by going to www.pestmanagement.info/nass/app_usage.cfm. This data source gives insights into changes in agricultural chemical use.

Go to Connect to graph and examine pesticide data in Excel, and to test your understanding of pesticide use.

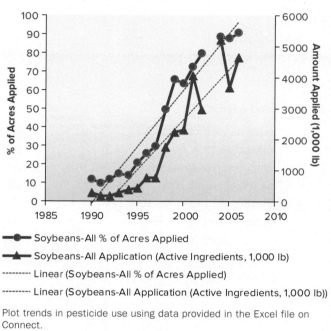

- Soybeans-All % of Acres Applied
- Soybeans-All Application (Active Ingredients, 1,000 lb)
- ------- Linear (Soybeans-All % of Acres Applied)
- ------- Linear (Soybeans-All Application (Active Ingredients, 1,000 lb))

Plot trends in pesticide use using data provided in the Excel file on Connect.

Source: Data from USDA National Agricultural Statistics Service (NASS)

11

Biodiversity
Preserving Species

▲ Juvenile staghorn coral colonies are grown in a nursery before being used for reef restoration.
© Stephen Frink/Corbis Documentary/Getty Images

Learning Outcomes

After studying this chapter, you should be able to:

11.1 Discuss biodiversity and the species concept, as well as some of the ways we benefit from biodiversity.

11.2 Characterize the threats to biodiversity.

11.3 Evaluate endangered species management.

11.4 Assess captive breeding and species survival plans.

"The first rule of intelligent tinkering is to save all the pieces."

– Aldo Leopold

Restoring Coral Reefs

Coral reefs are among the richest biological communities on Earth. They're the marine equivalent of tropical rainforests in diversity, productivity, and complexity. It's estimated that one quarter of all marine species spend some or all of their life cycle in the shelter of coral reefs. Globally, at least 17 percent of the protein we eat comes from coral reef systems. In some coastal areas, that number can be as high as 70 percent. Reefs serve as nurseries and food sources for important commercial species, such as tuna, and act as shelter for important ecological species, such as sharks. Reefs protect shorelines from storms, and are valuable recreation attractions for tourists.

But reefs are in serious trouble. According to recent surveys, we've already lost about 30 percent of coral worldwide, and another 60 percent of this valuable natural resource is threatened by climate change, destructive fishing methods, coral mining, sediment runoff, pollution, and other human-caused stressors. Some researchers warn that if current trends continue there won't be any viable coral reefs anywhere in the world by the end of this century.

Reefs are really colonies of tiny invertebrate animals embedded in calcium carbonate shells cemented together to create branches, digits, brackets, heads, and reefs. Individual animals are called polyps. which have minute fan-shaped tentacles to collect zooplankton and nutrients from the water. There can be thousands of polyps on a single coral branch. Nutrients are sparse in the clear, tropical waters where corals live, so reef-building corals form symbiotic relationships with microscopic algae, called zooxanthellae. Photosynthesis by the algae provide as much as 90 percent of the energy the corals need to grow and survive. Consequently, most corals need clear water and abundant sunlight.

One of the most visible and dangerous signs of reef damage occurs when water temperatures get too high. Under these conditions, the symbiotic algae produce toxic by-products that cause the host corals to expel them in a process called bleaching. This doesn't kill the corals immediately, but if they don't reacquire new symbionts, the coral will starve to death, leaving only stark, white carbonate skeletons. Entire reef ecosystems, starved of their primary producers, die off after a bleaching event. Reef bleaching events have become increasingly common around the world as global warming raises seawater temperatures. In 2016, the hottest year on record at that point, over 90 percent of Australia's Great Barrier Reef was affected by bleaching. In one-third of the areas surveyed, between 60 and 100 percent of corals were bleached.

Climate warming is a global risk to reefs, but scientists, volunteers, and community activists are working to protect and restore coral reef systems around the world. Many of these projects are aimed at reducing pollution and destructive human impacts. In Hawaii, large, barge-mounted vacuum cleaners hoover up invasive algae that are smothering reefs. In Palau, the government, together with international advisors, is training community organizations on how to protect priority marine and coastal areas. In Indonesia, conservation organizations are working with indigenous groups to stop destructive harvest techniques, such as cyanide and dynamite fishing. In the U.S. Virgin Islands, officials are working to reduce sediment, sewage, and pollution runoff from the land. And in Australia, divers are removing or killing crown of thorn urchins that destroy corals.

Some of the most exciting projects are studying ways to regrow—and even improve—corals. Some branched corals, such as staghorn and elkhorn, which are among the most threatened of all species, can grow and reproduce through fragmentation. If a branch breaks off and conditions are favorable, it can reattach to the rock substrate and begin to grow a new colony. Researchers are taking advantage of this feature by harvesting coral fragments and growing them in underwater nurseries (see fig. 11.1) until they're large enough to be relocated to suitable areas. Dozens of these nurseries are now in operation worldwide, and tens of thousands of baby corals have been transplanted to damaged or depleted reefs. Practitioners have found that it's best to create clustered colonies of different coral species so they can protect and support each other.

Some restorers are looking for corals with special characteristics to increase the success in restoration efforts. In Ofu lagoon in American Samoa, for example, corals have been found that can survive much warmer water than most corals can tolerate. If studies can unlock the secret of this unusual heat resistance, it could be valuable in restoration efforts. At this point, most coral reefs in the world have bleached, and many have recovered. What different

FIGURE 11.1 Fragments of staghorn and elkhorn coral can be cultivated in nurseries and used to replenish damaged reef systems.

© ZUMA Press Inc/Alamy Stock Photo

environmental or biological conditions favor recovery? Similarly, an interesting example of natural selection has been discovered in the Miami ship channel, where a rare colony of corals was discovered growing in highly polluted, turbid, unsanitary water. If studies can pinpoint why these specimens are so tough, it might help improve other colonies.

Other scientists aren't waiting for nature to produce resistant coral strains. In Australia and Hawaii, projects are working on "assisted evolution." By growing corals in conditions similar to those we expect to see in the future (warm temperatures, acidic water), they're attempting to select for advantageous genes. Breeding experiments are assisted by using species that reproduce continuously

rather than only once a year, as most corals do. Interestingly, there's evidence that some coral adaptation to adverse conditions may be due to **epigenetic effects.** That is, environmental factors might alter gene expression, potentially affecting, for example, survival in warmer water, in ways that are passed to offspring (see section 8.4). This response could provide added resiliency to coral reefs.

While we probably can't save all reefs, these efforts suggest that we may be able to protect and restore some of this wonderful resource. In this chapter, we'll look at other threats to endangered species, as well as the reasons for protecting biodiversity and habitat. We'll also discuss the politics of endangered species protection and the challenges of recovery projects.

11.1 BIODIVERSITY AND THE SPECIES CONCEPT

- *Biodiversity includes genetic, species, and ecological variety.*
- *A few "hot spots," mostly in the tropics, are a high conservation priority because they have both extremely high biodiversity and a high risk of disruption by humans.*
- *We benefit from other organisms in many ways.*

From the driest desert to the dripping rainforests, from the highest mountain peaks to the deepest ocean trenches, life on earth occurs in a marvelous spectrum of sizes, colors, shapes, life cycles, and interrelationships. Think for a moment how remarkable, varied, abundant, and important are the other living creatures with whom we share this planet (fig. 11.2). How will our lives be impoverished if this biological diversity diminishes?

What is biodiversity?

Previous chapters of this book have described some of the fascinating varieties of organisms and complex ecological relationships that give the biosphere its unique, productive characteristics. Three kinds of **biodiversity** are essential to preserve these ecological systems: (1) *genetic diversity* is a measure of the variety of different versions of the same genes within individual species; (2) *species diversity* describes the number of different kinds of organisms within individual communities or ecosystems; and (3) *ecological diversity* assesses the richness and complexity of a biological community, including the number of niches, trophic levels, and ecological processes that capture energy, sustain food webs, and recycle materials within this system.

Within species diversity, we can distinguish between *species richness* (the total number of species in a community) and *species*

evenness (the relative abundance of individuals within each species). Often, species evenness is considered more informative. Species richness could be high, for example, in a lawn that was 99 percent of one grass species and just a few individuals of nine other species. The total species count would be 10, but most people would not consider the lawn very diverse. A meadow containing even representation of 11 species would appear more diverse, and it would function differently, perhaps producing a greater variety of food sources or habitat for birds and butterflies.

FIGURE 11.2 This coral reef has both high abundance of some species and high diversity of different genera. What will be lost if this biologically rich community is destroyed?
© Georgette Douwma/Getty Images RF

Species are defined in different ways

As you can see, the species concept is fundamental in defining biodiversity, but what exactly do we mean by that term? When Carolus Linnaeus, the great Swedish taxonomist, created our system of scientific nomenclature in 1735, species classification was based entirely on the physical appearance of adult organisms. In recent years, taxonomists have introduced other characteristics as a means of differentiating species. As noted in Chapter 3, species have often been defined in terms of the ability to breed and produce fertile offspring in nature. A general term for this is *reproductive isolation*—that is, organisms may be unable to breed because of physical characteristics, location, habitat, or even differing courtship behaviors. This definition has some serious weaknesses, especially among plants and protists, many of which either reproduce asexually or regularly make fertile hybrids.

Another definition favored by many evolutionary biologists is the *phylogenetic species concept* (*PSC*), which emphasizes the branching (or cladistic) relationships among species or higher taxa, regardless of whether organisms can interbreed successfully.

A third definition, favored by some conservation biologists, is the *evolutionary species concept* (*ESC*), which defines species in evolutionary and historic terms rather than by reproductive potential. The advantage of this definition is that it recognizes that there can be several "evolutionarily significant" populations within a genetically related group of organisms. Unfortunately, we rarely have enough information about a population to judge what its evolutionary importance or fate may be. Ecologists Paul Ehrlich and Gretchen Daily calculated that, on average, there are 220 evolutionarily significant populations per species. This calculation could mean that there are up to 10 billion different populations in total. Deciding which ones we should protect becomes an even more daunting prospect.

Molecular techniques are rewriting taxonomy

Increasingly, DNA sequencing and other molecular techniques are giving us insights into taxonomic and evolutionary relationships. Every individual has a unique hereditary complement called the *genome*. The genome is made up of the millions or billions of nucleotides in its DNA (see Chapter 3). The specific sequence of these nucleotides spells out the structure of all the proteins that make up the cellular composition and machinery of every organism. As you know from modern court cases and paternity suits, we can use that DNA sequence to identify individuals with a high degree of certainty. Now this very precise technology is being applied to identify species in nature.

Because only a small amount of tissue is needed for DNA analysis, species classification—or even the identity of individual animals—can be based on samples such as feathers, fur, or feces when it's impossible to capture living creatures. For example, DNA analysis has shown that whale meat for sale in Japanese markets was from protected species. Sampling of hair from scratching pads has allowed genetic analysis of lynx and bears in North America without causing them the trauma of being captured. Similarly, a new tiger subspecies (*Tigris panthera jacksoni*) was detected in Southeast Asia based on blood, skin, and fur samples from zoo and museum specimens, although no members of this subspecies are known to still exist in the wild (fig. 11.3).

FIGURE 11.3 DNA analysis revealed a new tiger subspecies (*T. panthera jacksoni*) in Malaysia. This technology has become essential in conservation biology.
© Royalty-Free/Corbis

This new technology can help resolve taxonomic uncertainties in conservation. In some cases, an apparently widespread and low-risk species may, in reality, comprise a complex of distinct species, some rare or endangered. Such is the case for a unique New Zealand reptile, the tuatara. Genetic marker studies revealed two distinct species, one of which needed additional protection. Similar studies have shown that the northern spotted owl (*Strix occidentalis caurina*) is a genetically distinct subspecies from its close relatives, the California spotted owl (*S. occidentalis occidentalis*) and the Mexican spotted owl (*S. occidentalis lucida*), and therefore deserves continued protection.

On the other hand, in some cases genetic analysis shows that a protected population is closely related to another much more abundant one. For example, the colonial pocket gopher from Georgia is genetically identical to the common pocket gopher and probably doesn't deserve endangered status. The California gnatcatcher (*Polioptila californica californica*), which lives in the coastal sage scrub between Los Angeles and the Mexican border, was listed as a threatened species in 1993, and thousands of hectares of land worth billions of dollars were put off-limits for development. Genetic studies showed, however, that this population is indistinguishable from the black-tailed flycatcher (*Polioptila californica pontilis*), which is abundant in adjacent areas of Mexico.

In some cases, molecular taxonomy is causing a revision of the basic phylogenetic ideas of how we think evolution proceeded. Studies of corals and other cnidarians (jellyfish and sea anemones), for example, show that they share more genes with primates than do worms and insects. This evidence suggests a branching of the family tree very early in evolution rather than a single sequence from lower to higher animals.

How many species are there?

At the end of the great exploration era of the nineteenth century, some scientists confidently declared that every important kind of living thing on earth would soon be found and named. Most of those explorations focused on charismatic species such as birds and mammals. Recent studies of less conspicuous organisms, such as insects and fungi, suggest that millions of new species and varieties remain to be studied scientifically.

The 1.7 million species presently known (table 11.1) probably represent only a small fraction of the total number that exist. Based on the rate of new discoveries by research expeditions—especially in the tropics—taxonomists estimate that there may be somewhere between 3 million and 50 million different species alive today. In fact, some taxonomists estimate that there are 30 million species of tropical insects alone. The upper limits for these estimates assume a high degree of ecological specialization among tropical insects. A recent study in New Guinea, however, found that 51 plant species were host to 900 species of herbivorous insects. This evidence would suggest no more than 4 to 6 million insect species worldwide.

About 65 percent of all known species are invertebrates (insects, sponges, clams, worms, and other species without backbones). This group probably makes up the vast majority of organisms yet to be discovered. What constitutes a species in bacteria and viruses is even less certain than for other organisms, but some genetic studies suggest there could be 1 trillion different kinds of microbes.

The numbers of endangered species shown in table 11.1 are those officially listed by the International Union for Conservation of Nature and Natural Resources (IUCN). This represents only a small fraction of those actually at risk. Out of more than a million insect species, it's likely that far more than 1,046 are at risk.

Table 11.1	Current Estimates of Known and Threatened Species by Taxonomic Group		
	Known	**Endangered**	**Percentage**
Mammals	5,515	1,197	21.7
Birds	10,424	1,375	13.2
Reptiles	10,272	944	9.2
Amphibians	7,448	2,271	30.5
Fishes	33,200	1,275	3.8
Insects	1,000,000	1,046	0.01
Molluscs	85,000	1,950	2.3
Crustaceans	47,000	728	1.5
Corals	2,175	237	10.9
Arachnids	102,248	164	0.16
Other animals	66,825	76	0.11
Mosses	16,236	75	0.46
Ferns and allies	12,000	197	1.6
Gymnosperms	1,052	400	38.0
Flowering plants	268,000	10,551	3.9
Green & Red Algae	10,356	9	0.07
Lichens	17,000	7	0.04
Mushrooms	31,496	22	0.07
Brown Algae	3,127	6	0.02
Total	**1,729,374**	**23,530**	**1.3**

Source: Data from IUCN Red List, 2015

Hot spots have exceptionally high biodiversity

Of all the world's currently identified species, only 10 to 15 percent live in North America and Europe. The greatest concentration of different organisms tends to be in the tropics, especially in tropical rainforests and coral reefs. Norman Myers, Russell Mittermeier, and others have identified **biodiversity hot spots** that contain at least 1,500 *endemics* (species that occur nowhere else) and have lost at least 70 percent of their habitat owing to, for example, deforestation or invasive species. Using plants and land-based vertebrates as indicators, they have proposed 34 hot spots that are a high priority for conservation because they have both high biodiversity and a high risk of disruption by human activities (fig. 11.4). Although these hot spots occupy only 1.4 percent of the world's land area, they house three-quarters of the world's most threatened mammals, birds, and amphibians. The hot spots also account for about half of all known higher plant species and 42 percent of all terrestrial vertebrate species. The hottest of these hot spots tend to be tropical islands, such as Madagascar, Indonesia, and the Philippines, where geographic isolation has resulted in large numbers of unique plants and animals. Special climatic conditions, such as those found in South Africa, California, and the Mediterranean Basin, also produce highly distinctive flora and fauna.

Some areas with high biodiversity—such as Amazonia, New Guinea, and the Congo basin—aren't included in this hot spot map because most of their land area is relatively undisturbed. Other groups prefer different criteria for identifying important conservation areas. Aquatic biologists, for example, point out that coral reefs, estuaries, and marine shoals host some of the most diverse wildlife communities in the world, and warn that freshwater species are more highly endangered than terrestrial ones. Other scientists worry that the hot spot approach neglects many rare species and major groups that live in less biologically rich areas (cold spots). Nearly half of all terrestrial vertebrates, after all, aren't represented in Myers's hot spots. Focusing on a few hot spots also doesn't recognize the importance of certain species and ecosystems to human beings. Wetlands, for instance, may contain just a few, common plant species, but they perform valuable ecological services, such as filtering water,

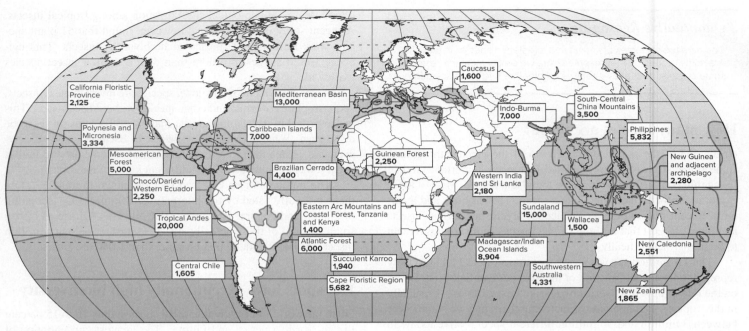

FIGURE 11.4 Biodiversity "hot spots," identified by Conservation International, tend to be in tropical or Mediterranean climates and on islands, coastlines, or mountains where many habitats exist and physical barriers encourage speciation. Numbers indicate endemic species.
Source: Conservation International, 2005

regulating floods, and serving as nurseries for fish. Some conservationists argue that we should concentrate on saving important biological communities or landscapes rather than rare species.

Quantitative Reasoning

Compare the "hot spot" map in figure 11.4 with the biomes map in figure 5.4. Which of the "hot spots" has the largest number of endemic species? Which has the least? Can you detect any patterns when you compare these two maps?

Anthropologists point out that many of the regions with high biodiversity are also home to high cultural diversity as well. It isn't a precise correlation; some countries, like Madagascar, New Zealand, and Cuba, with a high percentage of endemic species, have only a few cultural groups. Often, however, the varied habitat and high biological productivity of places like Indonesia, New Guinea, and the Philippines that allow extensive species specialization also have fostered great cultural variety. By preserving some of the 7,200 recognized language groups in the world—more than half of which are projected to disappear in this century—we might also protect some of the natural settings in which those cultures evolved.

We benefit from biodiversity in many ways

We benefit from other organisms in many ways, and seemingly obscure and insignificant organisms can play irreplaceable roles in ecological systems, or they may be a source of genes or drugs that someday may be indispensable.

Many wild plant species could make important contributions to human food supplies, either as new crops or as a source of genetic material to provide disease resistance or other desirable traits to current domestic crops. Ecologist Norman Myers estimates that as many as 80,000 edible wild plant species could be utilized by humans. Villagers in Indonesia, for instance, are thought to use some 4,000 native plant and animal species for food, medicine, and other valuable products. Few of these species have been explored for possible domestication or more widespread cultivation. A 1975 study by the National Academy of Science (U.S.) found that Indonesia has 250 edible fruits, only 43 of which have been cultivated widely (fig. 11.5).

More than half of all modern medicines are either derived from or modeled on natural compounds from wild species (table 11.2). The United Nations Development Programme estimates the value of pharmaceutical products derived from developing world plants, animals, and microbes to be more than $30 billion per year. Indigenous communities that have protected and nurtured the biodiversity on which these products are based are rarely acknowledged—much less compensated—for the resources extracted from them. Many consider this expropriation "biopiracy," or theft of living things, and there are calls for royalties to be paid for folk knowledge and natural assets.

Consider the success story of vinblastine and vincristine. These anticancer alkaloids are derived from the Madagascar periwinkle (*Catharanthus roseus*) (fig. 11.6). They inhibit the growth of cancer cells and are very effective in treating certain kinds of cancer. Before these drugs were introduced, childhood leukemias were invariably fatal. Now the remission rate for some childhood leukemias

FIGURE 11.5 Mangosteens have been called the world's best-tasting fruit, but they are practically unknown beyond the tropical countries where they grow naturally. There may be thousands of other traditional crops and wild food resources that could be equally valuable but are threatened by extinction.
© William P. Cunningham

FIGURE 11.6 The rosy periwinkle from Madagascar provides anticancer drugs that now make childhood leukemias and Hodgkin's disease highly remissible.
© William P. Cunningham

is 99 percent. Hodgkin's disease was 98 percent fatal a few years ago, but is now only 40 percent fatal, thanks to these compounds. The total value of the periwinkle crop is roughly $150 million to $300 million per year, although Madagascar gets little of those profits.

Pharmaceutical companies are actively prospecting for useful products in many tropical countries. Merck, the world's largest biomedical company, paid $1.4 million to the Instituto Nacional de Biodiversidad (INBIO) of Costa Rica for plant, insect, and microbe samples to be screened for medicinal applications. INBIO, a public/private collaboration, trained native people as practical "parataxonomists" to locate and catalog all the native flora and fauna—between 500,000 and 1 million species—in Costa Rica. This effort may be both a good model for scientific information gathering and a way for developing countries to share in the profits from their native resources.

The UN Convention on Biodiversity calls for a more equitable sharing of the gains from exploiting nature between rich and poor nations. Bioprospectors who discover useful genes or biomolecules in native species will be required to share profits with the countries where those species originate. This is not only a question of fairness; it also provides an incentive to poor nations to protect their natural heritage.

Table 11.2 Some Natural Medicinal Products		
Product	**Source**	**Use**
Penicillin	Fungus	Antibiotic
Bacitracin	Bacterium	Antibiotic
Tetracycline	Bacterium	Antibiotic
Erythromycin	Bacterium	Antibiotic
Digitalis	Foxglove	Heart stimulant
Quinine	Chincona bark	Malaria treatment
Diosgenin	Mexican yam	Birth-control drug
Cortisone	Mexican yam	Anti-inflammation treatment
Cytarabine	Sponge	Leukemia cure
Vinblastine, vincristine	Periwinkle plant	Anticancer drugs
Reserpine	Rauwolfia	Hypertension drug
Bee venom	Bee	Arthritis relief
Allantoin	Blowfly larva	Wound healer
Morphine	Poppy	Analgesic

Biodiversity provides ecological services and brings us many aesthetic and cultural benefits

Human life depends on ecological services provided by other organisms. Bacteria and soil microbes form soil, dispose of waste, purify water, and recycle nutrients. Trees moderate our climate, and plants in healthy ecosystems capture water, build soil, and turn sunlight into nutrients. The total value of these ecological services is at least $33 trillion per year, or about half the total world GNP.

There has been a great deal of controversy about the role of biodiversity in ecosystem stability. It seems intuitively obvious that having more kinds of organisms would make a community better able to withstand or recover from disturbance, but few empirical studies show an unequivocal relationship. A famous long-term study by David Tilman and his associates of diversity and stability in native prairie species found that more diverse plots were more resilient, or more able to recover after a disturbance or drought. Other experiments, however, have found less clear correlation between diversity and stability.

Because we don't fully understand the complex interrelationships among organisms, we often are surprised and dismayed at the effects of removing seemingly insignificant members of biological communities. For instance, wild insects provide a valuable but often unrecognized service in suppressing pests and disease-carrying organisms. It is estimated that 95 percent of the potential pests and disease-carrying organisms in the world are controlled by other species, which keep us safe by preying on or competing with potential pests. Many unsuccessful efforts to control pests with synthetic chemicals (Chapter 10) have shown that biodiversity provides essential pest-control services.

Millions of people enjoy hunting, fishing, camping, hiking, wildlife watching, and other outdoor activities based on nature. These activities keep us healthy by providing invigorating physical exercise. Contact with nature also can be psychologically and emotionally restorative. In some cultures, nature carries spiritual connotations, and a particular species or landscape may be inextricably linked to a sense of identity and meaning. Many moral philosophies and religious traditions hold that we have an ethical responsibility to care for creation and to save "all the pieces" as far as we are able.

Nature appreciation is economically important. The outdoor recreation industry estimates that Americans spend $730 billion annually on nature-based sports (fig. 11.7). This compares to $81 billion spent each year on new automobiles. Forty percent of all adults enjoy

FIGURE 11.7 Birdwatching and other wildlife observation contribute more than $29 million each year to the U.S. economy.
© William P. Cunningham

wildlife, including 39 million who hunt or fish and 76 million who watch, feed, or photograph wildlife. Ecotourism (tourism focused on experiencing wild places or viewing wildlife) can be a good form of sustainable economic development, provided that tourists don't overwhelm the places and cultures they visit.

For many people, the value of wildlife goes beyond the opportunity to hunt, photograph, or simply view a particular species. They argue that **existence value,** based on simply knowing that a species exists, is reason enough to protect and preserve it. We contribute to programs to save bald eagles, redwood trees, whooping cranes, whales, and a host of other rare and endangered organisms because we like to know they still exist somewhere, even if we may never have an opportunity to see them.

Section Review

1. Which group in table 11.1 has the most known species? Which has the highest proportion of endangered species?
2. What are the criteria for defining a biodiversity hot spot?
3. Describe four benefits we obtain from biodiversity.

11.2 WHAT THREATENS BIODIVERSITY?

- *Extinction rates are far higher now than in the past.*
- *Habitat destruction, climate change, invasive species, pollution, human population growth, and overharvesting are the greatest threats to biodiversity.*
- *Islands and specialized habitats are particularly susceptible to invasive species.*

Extinction, the elimination of a species, is a normal process of the natural world. Species die out and are replaced by others, often their own descendants, as part of evolutionary change. In undisturbed ecosystems, the rate of extinction appears to be about one species lost every decade. In this century, however, human impacts on populations and ecosystems have accelerated that rate, causing hundreds or perhaps even thousands of species, subspecies, and varieties to become extinct every year. If present trends continue, we may destroy *millions* of kinds of plants, animals, and microbes in the next few decades. In this section, we will look at some ways we threaten biodiversity.

Extinction is a natural process

Studies of the fossil record suggest that more than 99 percent of all species that ever existed are now extinct. Most of those species were gone long before humans came on the scene. Species arise through processes of mutation and natural selection, and they disappear the same way (chapter 4). Often, new forms replace their own parents. The tiny *Hypohippus,* for instance, has been replaced by the much larger modern horse, but most of its genes probably still survive in its distant offspring.

Periodically, mass extinctions have wiped out vast numbers of species and even whole families (fig. 11.8). The best studied of these

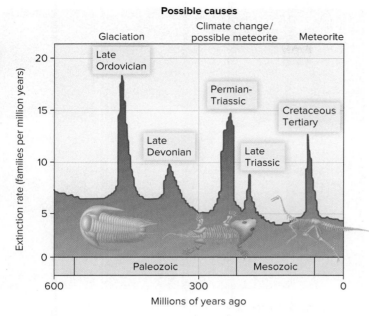

Possible causes

FIGURE 11.8 Major mass extinctions through history. We may be in a sixth mass extinction now, caused by human activities.

events occurred at the end of the Cretaceous period, when dinosaurs disappeared along with at least 50 percent of existing genera and 15 percent of marine animal families. An even greater disaster occurred at the end of the Permian period about 250 million years ago, when 95 percent of all marine species and nearly half of all plant and animal families died out over a period of about 10,000 years—a short time by geological standards. Current theories suggest that some of these catastrophes were caused by climate change, perhaps triggered by volcanic eruptions or large asteroid strikes.

We are accelerating extinction rates

The rate at which species are disappearing appears to have increased dramatically over the last 150 years. It appears that between 1600 and 1850 c.e., human activities were responsible for the extermination of two or three species per decade. By some estimates, we are now losing species at hundreds or even thousands of times natural rates. If present trends continue, the United Nations Environment Programme warns, half of all primates and one-quarter of all bird and invertebrate species could be extinct in the next 50 years. The eminent biologist E. O. Wilson says the impending biodiversity crash could be more abrupt than any previous mass extinction. Some biologists call this the sixth mass extinction, but note that this time it's not asteroids or volcanoes, but human impacts, that are responsible.

Accurate predictions of biodiversity losses are difficult when many species probably haven't yet been identified. Most predictions of human-caused extinction are based on assumptions about habitat area and species abundance: For example, Wilson calculates that if you cut down 90 percent of a forest, you'll eliminate at least 50 percent of the species originally present. In some of

the best-studied biological communities, however, this rule seems not to be reliable. More than 90 percent of Costa Rica's dry seasonal forest, for instance, has been converted to pasture land, yet entomologist Dan Janzen reports that no more than 10 percent of the original flora and fauna appear to have been permanently lost. Wilson and others respond that remnants of the native species may be hanging on temporarily, but that in the long run they're doomed without sufficient habitat.

Still, it's clear that habitat is being destroyed in many places, and that numerous species are less abundant than they once were. Shouldn't we try to protect and preserve as much as we can? Wilson summarizes human threats to biodiversity with the acronym **HIPPO,** which stands for Habitat destruction, Invasive species, Pollution, Population growth (of humans), and Overharvesting. Let's look in more detail at each of these issues.

Habitat destruction is the principal HIPPO factor

Of the five HIPPO factors, habitat loss is often the most important extinction threat. Clear-cutting of forests and plowing of grasslands are especially obvious examples: About half of all the primary tropical forest in the world has been felled for timber or to create farmland, and roughly as much area of former prairie and open woodlands is used for grazing or crops.

One of the most famous and contentious endangered species controversies is the case of the northern spotted owls. Before European settlement, most of the Pacific Northwest between the Cascade Mountains and the coast was covered by a dense, structurally complex forest ideal for spotted owls. Fragmentation by clear-cutting (fig. 11.9) destroys the old-growth characteristics required by many species. As you can see in this photo, the amount of forest edge has increased greatly, while the core areas deep in the forest are mostly eliminated. This is good for species that favor edges, but bad for those, such as spotted owls,

FIGURE 11.9 Habitat degradation is a leading cause of biodiversity loss. Forest fragmentation, such as these clear-cut patches, destroys old-growth characteristics on which species, such as the northern spotted owl, depend.
© William P. Cunningham

that require core habitat. To protect owls, harvesting of old-growth forests in the Pacific Northwest has been greatly diminished. Locals who depend on the forest industry charge that environmentalists have placed owl survival ahead of jobs. Environmentalists, on the other hand, point out that many jobs were lost to mechanization and export of whole logs to foreign countries. A better source of jobs, they claim, are modern industries that don't depend on habitat destruction.

Other kinds of resource extraction, such as mining, dam building, and fishing also destroy habitat. Surface mining, for example, strips off the land covering along with everything growing on it. Especially in mountaintop removal mining, waste can bury valleys and streams with toxic material (chapter 14). The building of dams submerges vital stream habitat under deep reservoirs and eliminates food sources and breeding habitat for some aquatic species. Marine habitats are destroyed by bottom trawling, a fishing practice in which heavy nets are dragged across the ocean floor, scooping up every living thing and crushing the seafloor to lifeless rubble (chapter 9).

Preserving small, scattered areas of habitat often isn't sufficient to maintain a complete species collection. Large mammals, like tigers or wolves, need large expanses of contiguous range relatively free of human incursion. Even species that occupy less space individually suffer when habitat is fragmented into small, isolated pieces. If the intervening areas create a barrier to migration, isolated populations become susceptible to environmental catastrophes such as bad weather or disease epidemics. They also can become inbred and vulnerable to genetic flaws.

Climate change, caused by our releases of greenhouse gases (see chapter 15), is one of the most important human causes of habitat loss. Biologists have observed dozens of species whose migration patterns or behaviors have been altered by changes in temperature, the timing of spring seasons, rainfall, or other climate factors. At least a dozen bird species in Britain have moved north as the climate has warmed; some bird species in Costa Rica have moved from lowland forests to higher mountain slopes. In North America and Europe, 39 butterfly species have shifted their range up to 200 km northward over the past 27 years.

Climate change affects some species more severely than others. Species that disperse easily to new habitat, and those that tolerate a wide range of environments, may simply shift their territories as we enter new climate regimes. Species with narrow habitat or climate requirements, or those that cannot easily disperse, may disappear as their current environment becomes inhospitable. Many species, such as migratory birds, are exquisitely adapted to day length, seasonal changes, the angles in the sky of the sun and stars, and other environmental signals. If they have to move long distances to new territories, those signals may become unreliable.

Polar bears (*Ursus maritimus*) show several of these different vulnerabilities. These bears are highly adapted to hunt seals on sea ice of the Arctic Ocean (fig. 11.10). But rapidly disappearing and thinning sea ice makes it hard for polar bears to hunt and to feed their young. In 2008, the United States declared polar bears endangered because of their rapidly declining population. The

FIGURE 11.10 Polar bears are threatened by rapidly shrinking sea ice, on which they depend for hunting. By mid-century, the Arctic Ocean could be largely free of summer ice, and polar bears may be limited to a small area around northern Greenland and northeastern Canada.
© Vadim Balakin/Getty Images RF

situation has only become worse since then. In 2012, the summer sea ice covered only 24 percent of the Arctic Ocean, down from 50 percent in the 1970s. As a consequence, bears have to swim farther to reach the ice. In some places, bears can no longer reach the summer ice because it is too far from their denning areas on shore. It's difficult to get accurate data on these scarce, wide-ranging, and remote animals, but for the polar bear subpopulations for which we have good data, eight are thought to be declining, three are stable, and just one is increasing. If these trends continue, polar bears may go extinct throughout much of their present range by mid-century.

Alpine species are especially likely to lose habitat as the climate warms. In mountains throughout the world, the treeline (the elevation above which trees don't grow) is shifting toward higher elevations. Alpine tundra environments and high-elevation plants and animals are shrinking rapidly. Although there are few documented examples of species extermination so far, there are numerous examples of diminishing populations as habitat shrinks and adverse climate conditions stress both plants and animals. In 2013, the wolverine (*Gulo gulo*), was proposed for listing as an endangered species. This largest member of the weasel family is dependent on cold climates and deep snow, and its numbers are declining rapidly as its habitat disappears.

Northern forest species also face habitat loss with climate change. The moose (*Alces alces*), once a plentiful and iconic indicator of the boreal forest across northern states, cannot tolerate hot summers. In Minnesota, moose numbers dropped from 8,840 in 2006 to only 3,450 in 2016. Part of the problem is that the climate is just too warm for moose. Furthermore, brain parasites carried by white-tailed deer, which move into moose territory as forests are logged, are lethal for moose. Moose also suffer from ticks, which survive warm winters in greater numbers.

Marine species also are showing the effects of habitat loss and climate change. As the opening case study for this chapter shows, coral reefs—among the richest biological communities on the planet—are dying at alarming rates. Marine scientists warn that there may not be any healthy reef systems on the planet by the end of this century if current trends continue. Pollution, destructive fishing techniques, diseases, and rising temperatures all play a role in this terrible loss. Increasing CO_2 levels make the ocean more acidic, which disrupts calcium chemistry and prevents the coral polyps from creating protective casings. Reefs are expected to stop growing and to begin to disintegrate around the globe when atmospheric CO_2 reaches 560 parts per million—double preindustrial levels—which will probably occur by the end of this century if we don't make immediate changes.

Invasive species displace resident species

Invasive species, the second HIPPO factor, are species that are especially able to colonize new territory. Humans have probably always transported organisms into new habitats, but the rate of movement has increased sharply in recent years with the huge increase in speed and volume of our travel by land, water, and air.

Sometimes we deliberately transplant species we find attractive or useful. Some species hitch a ride in ship ballast water, in the wood of packing crates, inside suitcases or shipping containers, in the soil of potted plants, even on people's shoes. Sometimes we don't move species physically, but merely change habitat in ways that allow them to expand into new territories.

Over the past 300 years, an estimated 50,000 non-native species have become established in the United States. Many of these migrants don't thrive, and we don't notice them. By some estimates, only about 10 percent of introduced species become major problems. Those that do thrive, however, crowd out native species, or alter habitats, and may undermine ecosystem function. We call organisms that readily invade new territory **invasive species.** Nonnative invasives often are especially successful because they are free of the predators, diseases, or resource limitations that controlled their populations in their native habitat (fig. 11.11).

Many ecologists regard invasive species as second only to habitat disruption in terms of threats to rare and endangered species. Others argue that we shouldn't judge species merely on their origins. Merely being native doesn't necessarily give organisms greater evolutionary fitness or positive ecological properties. Perhaps it is unfair to use sinister-sounding, value-laden terms like "exotic" or "alien" for a species just because it doesn't have local origins. On the other hand, many ecologists do view non-native

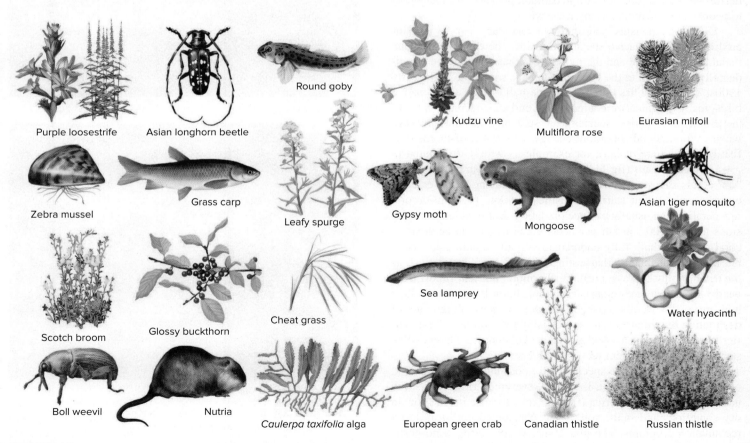

FIGURE 11.11 A few of the approximately 50,000 invasive species in North America. Do you recognize any that occur where you live? What others can you think of?

invasives as truly harmful. They often out-compete other species, including rare and prized native species. Invasive vines can kill trees and bury houses; invasive water weeds clog boats and reduce water quality and lake-side property values. Invasive weeds threaten crops and farm income. It's estimated that invasives cost the United States at least $138 billion annually.

Non-native invasive species are particularly destructive on islands where indigenous species have evolved without defenses against predators or competitors. For example, the New Guinea brown tree snake (*Boiga irregularis*) was introduced onto the Pacific Island of Guam sometime during or shortly after World War II (fig. 11.12a). This large, venomous snake quickly spread throughout the island, eating all manner of birds, lizards, small mammals, and bats. The snake exterminated 12 native bird species. Biodiversity on Guam has declined significantly, as have ecological functions: The decimated bird and bat populations include important pollinators in Guam's forests.

Some of our most destructive invasive species are feral pets or livestock. Cats, rats, mice, dogs, goats, pigs, and domesticated plants that either escape captivity or are deliberately released can consume or crowd out native species. In the United States, domestic and feral house cats are estimated to kill at least 2.4 billion birds and 12.3 billion small mammals every year. Feral pigs are a major ecosystem threat in many U.S. forests. Rooting up plants, destroying vegetation, and hard to capture, pigs are also dangerous to people who encounter them in the wild.

Disease organisms and their vectors can also be considered destructive invasive species. Consider the example of avian malaria. This disease and the mosquitoes that spread it were accidentally introduced to the Hawaiian Islands sometime in the early 1900s. The disease has killed about one-half of the islands' native birds, many of which were already rare and endangered.

Sometimes native species can suddenly become invasive when environmental conditions change. The crown-of-thorns starfish (*Ancanthaster planci*), for example, is widely distributed in tropical oceans (fig 11.12b). The large starfish preys on hard corals, digesting the living polyps and leaving a dead calcareous skeleton. When starfish densities are low (1–30/ha), coral regeneration can usually balance predation. But population explosions up to 1,000 starfish per hectare can destroy large areas of coral reef. We don't fully understand why these outbreaks occur, but it's thought that overharvesting of starfish predators, such as the triton mollusk, plays a roll. Furthermore, nutrient runoff from nearby land areas may nourish phytoplankton blooms and allow high survival rates of starfish larvae. Observations of high rates of destruction by crown-of-thorns starfish on Australia's Great Barrier Reef in the 1960s caused a great deal of concern. Divers killed starfish by injecting them with sodium bisulfate.

Often, we introduce species with good intentions. Tamarisk, or salt cedar (*Tamarix sp*), is a large, evergreen flowering shrub native to Eurasia and Africa (fig 11.12c). Because it can grow in dry conditions and saline soil, it was considered useful in desert reclamation programs. The plants spread quickly by wind-borne or water-borne seeds and became widely established in the Southwestern United States, creating dense thickets along rivers and

(a) Brown tree snake

(b) Crown-of-thorns starfish

(c) Tamarisk or salt cedar

FIGURE 11.12 Three invasive species that have reduced or eliminated indigenous species: (a) brown tree snake from Guam, (b) crown-of-thorns starfish, (c) tamarisk or salt cedar.

(a) © John Mitchell/Science Source; (b) © Jeff Rotman/Getty Images; (c) © William P. Cunningham

streams. Worried that tamarisk was using excessive amounts of water and crowding out fragile native species, the U.S. government spent millions of dollars on failed eradication programs. Subsequent research, however, has shown that the tamarisk uses water at about the same rate as native shrub species, and that it can help stabilize riverbanks and provide habitat for endangered species, such as the southwestern willow flycatcher.

Some ecologists argue that we shouldn't condemn and try to eradicate all new arrivals without first examining their ecological roles. Many new species have a neutral effect, they point out, and some prove to be beneficial. We're concerned about the loss of honeybees, for example, but they're relative newcomers introduced by European settlers. And some native species can have very deleterious impacts. Consider the pine bark beetles now killing thousands of square kilometers of conifer forest in the American West. They have apparently been here for a very long time, but have only recently become virulent as warmer winters allow them to survive and spread explosively. The language we use to describe immigrants: aliens, invasives, exotics, and non-natives have connotations of xenophobia. Why should origin or date of arrival give some species preference over others, regardless of their function in ecosystems? And as global climate change disrupts biomes, it could be that weedy, invasive species may be all that survives in some landscapes.

Pollution and population are direct human impacts

After habitat loss and invasive species, two important HIPPO factors are pollution and growing human populations. We have long known that pollutants can have disastrous effects on biodiversity. Pesticide-linked declines of predators, such as eagles, osprey, falcons, and pelicans, were well documented in the 1970s. Declining populations of marine mammals, alligators, fish, and other wildlife alert us to the connection between pollution and health. This connection has led to a new discipline of conservation medicine (chapter 8). Mysterious, widespread deaths of thousands of seals on both sides of the Atlantic in recent years are thought to be linked to an accumulation of persistent chlorinated hydrocarbons, such as DDT, PCBs, and dioxins, in fat, causing weakened immune systems that make animals vulnerable to infections. Similarly, the mortality of Pacific sea lions, beluga whales in the St. Lawrence estuary, and striped dolphins in the Mediterranean are thought to be caused by accumulation of toxic pollutants.

Lead and mercury poisoning are another major cause of mortality for many species of wildlife. Bottom-feeding waterfowl, such as ducks, swans, and cranes, ingest spent shotgun pellets that fall into lakes and marshes (fig. 11.13). They store the pellets, instead of stones, in their gizzards, and the lead slowly accumulates in their blood and other tissues. The U.S. Fish and Wildlife Service estimates that 3,000 metric tons of lead shot are deposited annually in wetlands and that between 2 and 3 million waterfowl die each year from lead poisoning. There are several non-toxic alternatives for lead shot including steel, tungsteniron, and bismuth-tin alloys.

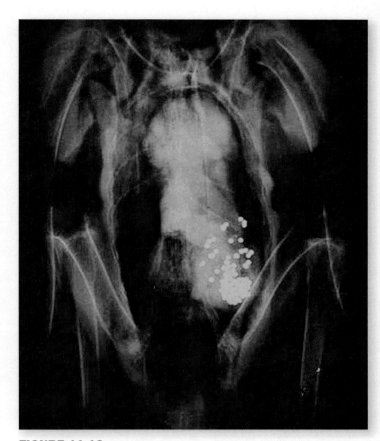

FIGURE 11.13 A bald eagle's stomach contents includes lead shot, which was consumed along with its prey. Fishing weights and shot remain a major cause of lead poisoning in aquatic and fish-eating birds.
© Elliott Jacobson, University of Florida

Mercury levels are elevated principally by burning coal, the fuel for many industrial processes and about 40 percent of global electricity production. Mining and waste incineration can be another source of this neurotoxin. In 1990, about 220 tons of mercury per year were emitted by these sources in the United States. Since the EPA started regulating smokestack emissions, that number has fallen to about 50 tons per year. Mercury enters the food web when aquatic bacteria convert it to methylmercury (CH_3Hg). This compound accumulates and is magnified as it passes from one trophic level to another. Birds and mammals that eat fish may accumulate toxic levels of this stable metal. Mercury poisoning is cumulative. It damages the brain and other internal organs. At high concentrations, harmful effects include reduced reproduction, slower growth and development, abnormal behavior, and death. For fish-eating species, such as loons, mercury poisoning is a serious threat. As we'll see in chapter 16, newly established limits in both the United States and China are expected to significantly reduce mercury emissions from coal-fired power plants.

Human population growth is a threat to biodiversity because everyone has some consumption needs—although some people consume more than others. As our numbers grow, we will need to harvest more timber, catch more fish, plow more land for agriculture, dig up more fossil fuels and minerals, build more houses, and use more water. All of these demands impact wild species. If our population grows to 8 to 10 billion as current projections predict, our impacts will certainly increase. The human population growth curve is leveling off (chapter 7), but it remains unclear whether we can reduce global inequality and provide a tolerable life for all humans while also preserving healthy natural ecosystems and a high level of biodiversity.

Overharvesting results when there is a market for wild species

Overharvesting, the last of the HIPPO factors, is responsible for the depletion or extinction of many species. A classic example is the extermination of the American passenger pigeon (*Ectopistes migratorius*), probably the world's most abundant bird 200 years ago. Although it inhabited only eastern North America, its population is estimated to have been 3 to 5 billion birds (fig. 11.14), possibly one-quarter of all birds in North America. In 1830, John James Audubon saw a single flock of birds estimated to be ten miles wide and hundreds of miles long, and thought to contain perhaps a billion birds. In spite of this vast abundance, market hunters caused the entire population to crash in only about

FIGURE 11.14 A pair of stuffed passenger pigeons (*Ectopistes migratorius*). The last member of this species died in the Cincinnati Zoo in 1914.

Courtesy of Bell Museum, University of MN. Photo taken by Mary Ann Cunningham.

20 years, between 1870 and 1890. They were able to eliminate so many birds so fast because of a ready abundance of firearms after the Civil War, as well as growing urban markets for wild bird meat. Rapid clearing of once-vast eastern forests for farmland also accelerated the decline. The last known wild passenger pigeon was shot in 1900, and the last existing passenger pigeon, a female named Martha, died in 1914 in the Cincinnati Zoo.

At about the same time that passenger pigeons were being extirpated, the American bison, or buffalo (*Bison bison*) was being hunted nearly to extinction on the Great Plains. In 1850, some 60 million bison roamed the western plains. Many were killed merely for sport or for their hides or tongues, leaving millions of carcasses to rot. Some of the bison's destruction was carried out by the U.S. Army so that native peoples who depended on bison for food, clothing, and shelter would be bereft of this resource and could then be forced onto reservations. By 1900, there were only about 150 wild bison left and another 250 in captivity.

More recently, overharvesting has reduced many of the world's fish populations to levels that may barely sustain themselves. A huge increase in the size and efficiency of fishing fleets in recent years is responsible. Of the world's 17 principal fishing zones, 13 are now reported to be commercially exhausted or in steep decline. At least three-quarters of all commercial oceanic species are overharvested. Fisheries biologists estimate that only 10 percent of the top predators, such as swordfish, marlin, tuna, and shark, remain in the Atlantic Ocean. Removal of these top predators from the ocean could have catastrophic ecological effects. You can avoid adding to this overharvest by eating only abundant, sustainably harvested varieties (What Can You Do?). Facebook initiated a campaign to ban longline fishing that threatens sea birds, turtles, and marine mammals. You could start or join similar efforts to raise awareness and change buying habits.

Overharvesting is often illegal and involves endangered species

Despite international bans on trade in products from endangered species, the smuggling of furs, hides, horns, live specimens, and folk medicines amounts to millions of dollars each year.

Developing countries in Asia, Africa, and Latin America with the richest biodiversity in the world are the main sources of wild animals and animal products. Europe, North America, and some of the wealthy Asian countries are the principal buyers. Japan, Taiwan, and Hong Kong buy three-quarters of all cat and snake skins, for instance; European countries buy a similar percentage of live birds. The United States imports 99 percent of all live cacti and 75 percent of all orchids sold each year, many of them rare and collected in the wild.

The profits to be made in wildlife smuggling are enormous. Tiger or leopard fur coats can bring (U.S.) $100,000 in Asia or Europe. And elephants are being slaughtered by the thousands every year for the illicit ivory trade (fig. 11.15). It's estimated that Africa has lost 90 percent of its original elephant population in just the past 50 years. Most of this trade is driven by rising wealth in Asia, where the desire for ivory carvings is intense. The enormous

Don't Buy Endangered Species Products

You probably are not shopping for a fur coat from an endangered tiger, but there might be other ways you are supporting unsustainable harvest and trade in wildlife species. To be a sustainable consumer, you need to learn about the sources of what you buy. Many plant and animal products are farm-raised, not taken from wild populations. But some commercial products are harvested in unsustainable ways. Here are a few products about which you should inquire before you buy:

Seafood includes many top predators that grow slowly and reproduce only when many years old. Despite efforts to manage many fisheries, the following have been severely, sometimes tragically, depleted:

- Top predators: swordfish, marlin, shark, bluefin tuna, albacore ("white") tuna.
- Groundfish and deepwater fish: orange roughy, Atlantic cod, haddock, pollack (source of most fish sticks, artificial crab, generic fish products), yellowtail flounder, monkfish.
- Other species, especially shrimp, yellowfin tuna, and wild sea scallops, are often harvested with methods that destroy other species or habitats.
- Farm-raised species such as shrimp and salmon can be contaminated with PCBs, pesticides, and antibiotics used in their rearing. In addition, aquaculture operations often destroy coastal habitat, pollute surface waters, and deplete wild fish stocks to stock ponds and provide fish meal.

Pets and plants are often collected from wild populations, frequently depleting wild populations:

- Aquarium fish are usually collected in the wild, often using destructive practices that damage reef habitat and other fish.
- Reptiles: snakes and turtles, especially rare varieties, are often collected in the wild.
- Plants: orchids and cacti are the best-known, but not the only, group collected in the wild.

Herbal products such as wild ginseng and wild echinacea (purple coneflower) may be poached from the wild. Ask about the sources of these products.

Do buy some of these sustainably harvested products:

- Shade-grown (or organic) coffee, nuts, and other sustainably harvested forest products.
- Pets from the Humane Society, which works to protect stray animals.
- Organic cotton, linen, and other fabrics.
- Fish products that have relatively little environmental impact or have fairly stable populations: farm-raised catfish or tilapia, wild-caught salmon, mackerel, Pacific pollack, dolphinfish (mahimahi), squids, crabs, crayfish, and mussels.
- Wild freshwater fish like bass, sunfish, pike, catfish, and carp, which are usually better managed than most ocean fish.

---

prices paid for black market ivory finance terrorists, despots, and criminal syndicates around the world. Similarly, rhino horn, valued in Asia because it supposedly enhances virility and allegedly cures a variety of diseases, has risen in value to as much as $300,000 for a single horn. As a consequence, the rhino population has plummeted from approximately around 1 million in 1900 to only about 3,000 in 1980. Recently, however, breeding programs and better anti-poaching efforts have led to dramatic recovery of rhino populations in some African parks. (See the Exploring Sciencebox later in this chapter.)

Plants also are threatened by overharvesting. Wild ginseng has been nearly eliminated in many areas because of the Asian demand for the roots, which are used as an aphrodisiac and folk medicine. In the American Southwest and Mexico, cactus "rustlers" steal cacti by the ton to sell for yard ornaments. With prices ranging as high as $1,000 for rare specimens, it's not surprising that many cacti are now endangered.

Island ecosystems are especially vulnerable to invasive species

New Zealand is a prime example of the damage that can be done by invasive species in island ecosystems. Having evolved for thousands of years without predators, New Zealand's flora and fauna

are particularly susceptible to the introduction of alien organisms. Originally home to more than 3,000 endemic species, including flightless birds such as the kiwi and giant moas, New Zealand has lost at least 40 percent of its native flora and fauna since humans first landed there 1,000 years ago. More than 25,000 plant species have been introduced to New Zealand, and at least 200 of these have become pests that can create major ecological and economic problems. Many animal introductions (both intentional and accidental) also have become major threats to native species. Cats, rats, mice, deer, dogs, goats, pigs, and cattle accompanying human settlers consume native vegetation and eat or displace native wildlife.

One of the most notorious invasive species is the Australian brush-tailed possum, *Trichosurus vulpecula*. This small, furry marsupial was introduced to New Zealand in 1837 to establish a fur trade. In Australia, where their population is held in check by dingoes, fires, diseases, and inhospitable vegetation, possums are rare and endangered. Freed from these constraints in New Zealand, however, possum populations exploded. Now at least 70 million possums chomp their way through about 7 million tons of vegetation per year in their new home. They destroy habitat needed by indigenous New Zealand species, and also eat eggs, nestlings, and even adult birds of species that lack instincts to avoid predators.

FIGURE 11.15 Poachers kill thousands of elephants every year for their ivory. It's estimated that Africa has lost 90 percent of its original elephant population in just the past 50 years.
© Tom Finkle

Several dozen of New Zealand's offshore islands have been declared nature sanctuaries. Efforts are being made to eliminate invasive pests and to restore endangered species and native ecosystems.

Section Review

1. How do current rates of extinction compare to historic rates?
2. Explain HIPPO and its meaning for conserving biodiversity.
3. Why are islands especially sensitive to invasive species?

11.3 ENDANGERED SPECIES MANAGEMENT

- *The Endangered Species Act is one of the most important environmental laws ever passed in the United States.*
- *Protecting keystone, indicator, umbrella, or flagship species can be important beyond their mere survival.*
- *Endangered species protection is controversial.*

Over the years, we have gradually become aware of the harm we have done—and continue to do—to wildlife and biological resources. Slowly we are adopting national legislation and international treaties to protect these irreplaceable assets. Parks, wildlife refuges, nature preserves, zoos, and restoration programs have been established to protect nature and rebuild depleted populations. There has been encouraging progress in this area, but much remains to be done. Most people favor pollution control or protection of favored species such as whales or gorillas, but surveys show that few understand what biological diversity is or why it is important.

Hunting and fishing laws have been effective

In 1874, a bill was introduced in the U.S. Congress to protect the American bison, whose numbers were already falling to dangerously low levels. This initiative failed, however, because most legislators believed that all wildlife—and nature in general—was so abundant and prolific that it could never be depleted by human activity.

By the 1890s, most states had enacted some hunting and fishing restrictions. These laws protected game species, such as deer, ducks, and fish, that were useful to humans. They did not conserve wildlife for its own sake. The wildlife regulations and refuges established since that time have been remarkably successful for many species. In 1900, there were an estimated half million white-tailed deer in the United States; now there are some 14 million—more in some places than the environment can support. Wild turkeys and wood ducks were nearly all gone 50 years ago. Restoring habitat, planting food crops, transplanting breeding stock, building shelters or houses, protecting these birds during breeding season, and other conservation measures have made these beautiful and interesting birds common once again.

At the beginning of the twentieth century, the beautiful plumes from snowy egrets and other shorebirds were highly valued for women's hats (fig. 11.16). Market hunters shot so many birds that the species survival was doubtful. Florida's Pelican Island was established in 1903 as the first wildlife refuge in America to protect birds. The Migratory Bird Act of 1918 extended this protection. Now, egret populations have rebounded.

In 2006, a survey of global fisheries projected that all fish and seafood species would collapse by mid-century. However, in 2013 both Norway and Russia reported that strict enforcement of fishing quotas for North Atlantic cod have resulted in rebounding populations. Similarly, in the northern Pacific, carefully regulated stocks of pollack and salmon have remained strong. This progress has led many scientists to say the future for overfished ocean species is a bit less bleak than it seemed. Still, much of the ocean remains a free-for-all, and many developing countries lack the funds and technology to protect their territorial waters.

The Endangered Species Act is a powerful tool for biodiversity protection

Where earlier regulations had been focused almost exclusively on "game" animals, the Endangered Species Act (ESA) seeks to identify all endangered species and populations and to save as much biodiversity as possible, regardless of its usefulness to humans. As defined by the ESA, **endangered species** are those considered to be in imminent danger of extinction, whereas **threatened species** are those that are likely to become endangered—at least locally—in

FIGURE 11.16 At the beginning of the twentieth century, so many snowy egrets and other shore birds were killed for their plumes that the species seemed doomed to extinction. The Migratory Bird Act of 1918 protected these beautiful birds, and populations have rebounded.
© Corinne Lamontagne/Getty Images

Currently, the United States has 1,372 species on its endangered and threatened species lists and some 386 candidate species waiting to be considered. Worldwide, the International Union for Conservation of Nature and Natural Resources (IUCN) lists 17,741 endangered and threatened species (table 11.1).

Listing of endangered species favors those species that we find interesting or useful. In the United States, invertebrates make up about 75 percent of known species but only 9 percent of those listed for protection. The IUCN Red List describes 21.7 percent of mammals as threatened or endangered; only 0.01 percent of insects are listed as threatened. This is inequitable in two ways. First, there are probably far more endangered insect species than this, even among those we have identified. Furthermore, it's extremely rare to find a new mammal species, whereas the million known insect species may represent only a small fraction of the total insect species on earth.

Listing of new species in the United States has been very slow, sometimes taking decades from the first petition to final determination. Limited funding, political pressures, listing moratoria, and changing administrative policies have created long delays. A few years ago, as a result of lawsuits brought by conservationists, the U.S. Fish and Wildlife Service agreed to either list or reject more than 262 species that have been waiting for protection—some for as long as 30 years. In addition, another 550 candidate species are now being evaluated that had been waiting for consideration.

When Congress passed the original ESA, it probably intended to protect only a few charismatic species, such as raptors and big game animals. Sheltering obscure species such as the Delhi Sands flower-loving fly, the Coachella Valley fringe-toed lizard, Mrs. Furbisher's lousewort, or the orange-footed pimple-back mussel most likely never occurred to those who voted for the bill. This raises some interesting ethical questions about the rights and values of seemingly minor species. Although uncelebrated, these species may play important ecological roles. Protecting them usually preserves habitat and a host of unlisted species.

Recovery plans rebuild populations of endangered species

Once a species is officially listed as endangered, the Fish and Wildlife Service is required to prepare a recovery plan detailing how populations will be rebuilt to sustainable levels. It usually takes years to reach agreement on specific recovery plans. Reaching agreement is difficult for many reasons. Local politicians often resist setting aside land for conservation; recovery is costly; resource industries such as oil, logging, or mining industries lobby hard against conservation plans. The total cost of recovery plans for all currently listed species is estimated to be nearly $5 billion. (For comparison, this is about the cost of two modern Stealth bomber jets.)

The United States currently spends about $150 million per year on endangered species protection and recovery. About half that amount is spent on a dozen charismatic species, like the California condor, the Florida panther, and the grizzly bear, each of which receives around $13 million per year. By contrast, the 137 endangered invertebrates and 532 endangered plants get less than $5 million per

the foreseeable future. Gray wolves, brown (or grizzly) bears, sea otters, and a number of native orchids and other rare plants are a few of the species considered to be locally threatened, even though they remain abundant in other parts of their former range. **Vulnerable species** are naturally rare or have been locally depleted by human activities to a level that puts them at risk. Many of these are candidates for future listing. For vertebrates, protected categories include species, subspecies, and local races or ecotypes.

The ESA regulates a wide range of activities involving endangered species, including "taking" (harassing, harming, pursuing, hunting, shooting, trapping, killing, capturing, or collecting) either accidentally or on purpose; importing into or exporting out of the United States; possessing, selling, transporting, or shipping; and selling or offering for sale any endangered species. Prohibitions apply to live organisms, body parts, and products made from endangered species. Violators of the ESA are subject to fines up to $50,000 and a year's imprisonment. Vehicles and equipment used in violations may be subject to forfeiture. In 1995, the Supreme Court ruled that critical habitat—habitat essential for a species' survival—must be protected, whether on public or private land.

year altogether. Our funding priorities often are based more on emotion and politics than on biology. A variety of terms are used for rare or endangered species thought to merit special attention:

- **Keystone species** are those with major effects on ecological functions and whose elimination would affect many other members of the biological community; examples are prairie dogs (*Cynomys ludovicianus*) and bison (*Bison bison*).

- **Indicator species** are those tied to specific biotic communities or successional stages or a set of environmental conditions. They can be reliably found under certain conditions but not others; an example is brook trout (*Salvelinus fontinalis*).

- **Umbrella species** require large blocks of relatively undisturbed habitat to maintain viable populations. Saving this habitat also benefits other species. Examples of umbrella species are the northern spotted owl (*Strix occidentalis caurina*), the grizzly bear (*Ursus arctos horribilis*), and the African white rhinoceros (*Ceratotherium simum*) discussed in the Exploring Science box in section 11.2.

- **Flagship species** are especially interesting or attractive organisms to which people react emotionally. These species can motivate the public to preserve biodiversity and contribute to conservation; examples are the giant panda (*Ailuropoda melanoleuca*) and killer whales (*Orcinus orca*).

Some recovery plans have been remarkably successful. The American alligator was listed as endangered in 1967 because hunting (for meat, skins, and sport) and habitat destruction had reduced populations to precarious levels. Protection has been so effective that the species is now plentiful throughout its entire southern range. Florida alone estimates that it has at least 1 million alligators.

Sometimes restoring a single species can bring benefits to an entire ecosystem, especially when that species plays a keystone role in the community. Alligators, for example, dig out swimming holes, or wallows, that become dry-season refuges for fish and other aquatic species. American bison are being used in prairie restoration projects to reestablish the health and diversity of grassland ecosystems (see chapter 13).

Before European settlement, the gray wolf (*Canis lupis*) inhabited most of North America south of about 20° latitude (fig. 11.17). By the 1950s, trapping and poisoning eliminated wolves from most of their territory. Only a few hundred wolves survived in northeastern Minnesota and on Isle Royale in Michigan. With the passage of the ESA in 1966, wolves were classified as endangered in the lower 48 states except for Minnesota, where they were listed as threatened. Under protected status, wolf populations rebounded, and in 2012 the species was declared recovered and delisted in the western Great Lakes and the Northern Rockies. Several states have reinstituted hunting seasons for wolves, but the Fish and Wildlife Service continues to monitor populations.

Some other successful recovery programs include bald eagles, peregrine falcons, and whooping cranes. Forty years ago, due mainly to DDT poisoning, only 417 nesting pairs of bald eagles (*Haliaeetus leucocephalis*) remained in the contiguous United States. By 2007, the population had rebounded to more than 9,800 nesting pairs, and the birds were removed from the endangered species list. This

FIGURE 11.17 By the 1950s, poisoning and trapping had eliminated the gray wolf from nearly all the lower 48 United States. Protection and recovery programs have helped reestablish populations in the western Great Lakes states and the northern Rocky Mountains.
© Thomas Kokta/Photolibrary/Getty Images

doesn't mean that eagles are unprotected. Killing, selling, or otherwise harming eagles, their nests, or their eggs is still prohibited. In addition to eagles and falcons, 29 other species, including mammals, fish, reptiles, birds, plants, and even one insect (the Tinian monarch), have been removed or downgraded from the endangered species list.

An important test of the ESA occurred in 1978 in Tennessee where construction of the Tellico Dam threatened a tiny fish called the snail darter. Environmental groups sued to stop the dam, and the case went all the way to the Supreme Court, which ruled that federal agencies were not exempt from the law. Still, opponents of the law have repeatedly tried to require that economic costs and benefits be incorporated into endangered species planning or have tried to eliminate the law altogether.

One of the most costly recovery programs ever may be required for Columbia River salmon and steelhead, endangered by dams that block their migration to the sea. Opening floodgates to allow young fish to run downriver and adults to return to spawning grounds would have high economic costs to barge traffic, farmers, and electric rate payers who have come to depend on abundant water and cheap electricity. On the other hand, commercial and sport fishing for salmon was once worth $1 billion per year and employed about 60,000 people directly or indirectly. There can be economic benefits, as well as costs, in restoring species and their ecosystems.

Private land is vital for species protection

Eighty percent of the habitat for more than half of all listed species in the United States is on private land. The Supreme Court has ruled that destroying habitat is as harmful to endangered species as directly taking (killing) them. Many landowners, however, resist restrictions on how they use their own property to protect what they perceive as insignificant or worthless organisms. This is especially true when the land has potential for economic development. If property is worth millions of dollars as the site of a housing development or shopping center, most owners don't want to be told they have to leave it undisturbed to protect some rare organism. Landowners may be tempted to "shoot, shovel, and shut up," if they discover endangered species on their property. Many feel they should be compensated for lost value caused by ESA regulations.

To reduce these tensions, the Fish and Wildlife Service negotiates agreements called **habitat conservation plans (HCPs)** with private landowners. Under these plans, landowners are allowed to harvest resources or build on part of their land as long as the species benefits overall. In return for improving habitat in some areas, funding conservation research, removing predators and competitors, or taking other steps that benefit the endangered species, developers are allowed to destroy habitat or even "take" endangered organisms.

Scientists and environmentalists often are critical of HCPs, claiming these plans often are based more on politics than biology, and that the potential benefits are frequently overstated. Defenders argue that by making the ESA more landowner-friendly, HCPs benefit wildlife in the long run.

Among the more controversial proposals for HCPs are the so-called Safe Harbor and No-Surprises policies. Under the Safe Harbor clause, any increase in an animal's population resulting from a property owner's voluntary good stewardship would not increase their responsibility or affect future land-use decisions. As long as the property owner complies with the terms of the agreement, he or she can make any use of the property. The No-Surprises provision says that the property owner won't be faced with new requirements or regulations after entering into an HCP. Scientists warn that change, uncertainty, dynamics, and flux are characteristic of all ecosystems. We can't say that natural catastrophes or environmental events won't make it necessary to modify conservation plans in the future.

Endangered species protection is controversial

The U.S. ESA officially expired in 1992. Since then, Congress has debated many alternative proposals, ranging from outright elimination to substantial strengthening of the act. Perhaps no other environmental issue divides Americans more strongly than the ESA. In the western United States, where traditions of individual liberty and freedom are strong and the federal government is viewed with considerable suspicion and hostility, the ESA has been described as a plot to take away private property and to limit industrial activity on public lands (fig. 11.18). Many people believe that the law puts the welfare of plants and animals above economic interests. Farmers, loggers, miners, ranchers, developers, and other ESA opponents repeatedly have tried to scuttle the law or greatly reduce its power. Ecologists, on the other hand, see the ESA as essential to protecting nature and maintaining the viability of the planet. Many regard it as the single most effective law in their arsenal and want it enhanced and improved.

Conservationists also have criticisms of our current endangered species protection. Perhaps chief of these is the focus on individual organisms. Protecting a keystone or umbrella species, such as wolves or elephants, can benefit entire ecological communities, but often we spend millions of dollars attempting to save a single charismatic species when those funds might have done more good, ecologically, by

"DAMN SPOTTED OWL!"

TIMBER MINING CO.

©1990 HERBLOCK

FIGURE 11.18 Endangered species often serve as a barometer for the health of an entire ecosystem and as surrogate protector for myriad less well-known creatures.
"DAMN SPOTTED OWL" A 1990 Herblock Cartoon, copyright by The Herb Block Foundation.

protecting a functional—if less unique—community or biome. Perhaps it would be better to try to preserve representative samples of many different kinds of biological communities and ecological services (even if those communities are missing a few of their historic members) than to save a few rare species just because they are rare.

Gap analysis promotes regional planning

An alternative strategy for species protection is to identify and preserve whole ecosystems that support maximum biological diversity, rather than battle for the rarest species one at a time. By focusing on a few high-profile populations already reduced to only a few individuals, we risk spending most of our conservation funds on species that may be genetically doomed no matter what we do. Furthermore, by concentrating on individual species we spend millions of dollars to breed plants or animals in captivity that have no natural habitat where they can be released. While flagship species such as mountain gorillas or Indian tigers are

What Can You Do?

You Can Help Preserve Biodiversity

If you live in an urban area, as most Americans do, you may not think you have much influence on wildlife. But there are important ways you can help conserve biodiversity.

- Protect or restore native habitat. Environmental organizations or nature preserves near where you live may have volunteer opportunities to remove invasive species, gather native seeds, or replant native vegetation.

- Plant local, native species in your garden. Exotic nursery plants often escape and threaten native ecosystems.

- Don't transport firewood from one region to another. It may carry diseases and insects.

- Follow legislation and management plans for natural areas you value. Lobby or write letters supporting funding and biodiversity-friendly policies.

- Help control invasive species. Never release non-native animals (fish, leaches, turtles, etc.) or vegetation into waterways or sewers. If you boat, wash your boat and trailer when moving from one lake or river to another.

- Don't discard worms in the woods. You probably think of earthworms as beneficial for soils—and they are, in the proper place—but many northern deciduous biomes evolved without them. Worms discarded by anglers are now causing severe habitat destruction in many places.

- Keep your cat indoors. House cats are major predators of woodland birds and other native animals. House cats in the United States kill billions of birds, small mammals, and lizards every year.

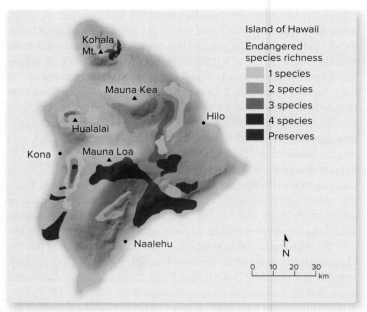

FIGURE 11.19 An example of the biodiversity maps produced by J. Michael Scott and the U.S. Fish and Wildlife Service. Notice that few of the areas of endangered species richness are protected in preserves, which were selected more for scenery or recreation than for biology.

reproducing well in zoos and wild animal parks, the ecosystems they formerly inhabited have largely disappeared.

A leader of this new form of conservation is J. Michael Scott, who was project leader of the California condor recovery program in the mid-1980s and had previously spent ten years working on endangered species in Hawaii. In making maps of endangered species, Scott discovered that even Hawaii, where more than 50 percent of the land is federally owned, has many vegetation types completely outside of natural preserves (fig. 11.19). The gaps between protected areas may contain more endangered species than are preserved within them.

This observation has led to an approach called **gap analysis** in which conservationists and wildlife managers look for unprotected landscapes that are rich in species. Computers and geographic information systems (GIS) make it possible to store, manage, retrieve, and analyze vast amounts of data and create detailed, high-resolution maps relatively easily. This broad-scale, holistic approach seems likely to save more species than a piecemeal approach.

Conservation biologist R. E. Grumbine suggests four management principles for protecting biodiversity in a large-scale, long-range approach:

1. Protect enough habitat for viable populations of all native species in a given region.

2. Manage at regional scales large enough to accommodate natural disturbances (fire, wind, climate change, and so on).

3. Plan over a period of centuries so that species and ecosystems may continue to evolve.

4. Allow for human use and occupancy at levels that do not result in significant ecological degradation.

International treaties improve protection

The 1975 Convention on International Trade in Endangered Species (CITES) was a significant step toward worldwide protection of endangered flora and fauna. It regulated trade in living specimens and products derived from listed species, but it has not been foolproof. Smugglers remove species from countries where they are threatened or endangered, and falsify documents to make it appear that they have come from areas where the species are still common. Investigations and enforcement are especially difficult in developing countries, where wildlife is disappearing most rapidly. Still, eliminating markets for endangered wildlife is an effective way to stop poaching. Appendix I of CITES lists 700 species threatened with extinction by international trade.

Shark survival provides a good example of the need for international cooperation in endangered species protection. Every year, about 100 million sharks and rays are killed, primarily to provide fins to be made into shark fin soup, considered a delicacy in some Asian countries. Depletion of these wide-ranging key predators has risks for the health of entire ocean ecosystems. Marine scientists estimate that global shark populations have declined by as much as 70 to 80 percent, and that about one-third of all shark species are currently threatened with extinction.

Several Pacific Island nations, including Palau, French Polynesia, American Samoa, and the Cook Islands, have banned shark fishing in their territorial waters, but much shark fishing takes place far offshore and is difficult to control. We need joint action to monitor both fisheries and markets to protect these important species.

Section Review

1. Define endangered species and threatened species. Give an example of each.
2. Why are keystone, indicator, umbrella, and flagship species important?
3. What is gap analysis, and how is it related to ecosystem management and the design of nature preserves?

11.4 CAPTIVE BREEDING AND SPECIES SURVIVAL PLANS

- *Zoos and captive breeding are important in protecting biodiversity.*
- *There are limits to how many species we can preserve in captivity.*
- *We need to protect rare species in the wild.*

Breeding programs in zoos and botanical gardens are one way to attempt to save severely threatened species. Institutions like the Missouri Botanical Garden and the Bronx Zoo's Wildlife Conservation Society sponsor conservation and research programs. Botanical gardens, such as Kew Gardens in England, and research stations, such as the International Rice Institute in the Philippines, are repositories for rare and endangered plant species, some of which have ceased to exist in the wild. Valuable genetic traits are preserved in these collections, and in some cases, plants with unique cultural or ecological significance may be reintroduced into native habitats after being cultivated for decades or even centuries in these gardens and seed banks.

Zoos can help preserve wildlife

Until fairly recently, zoos depended primarily on wild-caught animals for most of their collections. This was a serious drain on wild populations, because up to 80 percent of the animals caught died from the trauma of capture and shipping. With better understanding of reproductive biology and better breeding facilities, most mammals in North American zoos now are produced by captive breeding programs.

Some zoos now participate in breeding programs, which can reintroduce endangered species to the wild. The California condor is one of the best-known cases of successful captive breeding. In 1986, only nine of these birds existed in their native habitat. Fearing the loss of these last condors, biologists captured them and brought them to the San Diego and Los Angeles zoos, which had begun breeding programs in the 1970s. By 2016, the population had reached 421 birds, including 228 reintroduced to the wild.

Hawaii's endemic nene (*Nesochen sandvicensis*) also has been successfully bred in captivity and reintroduced into the wild. When Captain Cook arrived in the Hawaiian Islands in 1778, there were probably 25,000 of these land-dwelling geese. By the 1950s, however, habitat destruction and invasive predators had reduced the population to fewer than 30 birds. Today, there are about 800 wild nene, and at least 1,000 in captivity, but there are worries about inbreeding that could reduce the fitness of the species (fig. 11.20).

One of the most successful captive breeding programs for mammals is that of the white rhino (*Ceratotherium simum*) in southern Africa. Although they once ranged widely across Africa, these huge animals were considered extinct in the south until a remnant herd was found in Natal, South Africa, in 1895. Today, there are an estimated 17,500 southern white rhinos in Africa, mainly in national parks and private game ranches. (See the Exploring Science Protecting Rhinos box below.) In 2015, a U.S. hunter paid $350,000 to shoot a male that wildlife managers said was too old to reproduce and was a threat to younger males. The money was used to support parks and further breeding programs. Animal rights activists condemned this "hunt" as legal execution.

Breeding programs have limitations, however. Some species, such as bats, whales, and many reptiles rarely reproduce in captivity, and zoo stock for these animals still comes mainly from the wild. We will never be able to protect the complete spectrum of biological variety in zoos. According to one estimate, if all the space in U.S. zoos were used for captive breeding, only about 100 species of large mammals could be maintained on a long-term basis.

These limitations lead to what is sometimes called the "Noah question": How many species can or should we save? How much

FIGURE 11.20 Nearly extirpated in the 1950s, the land-dwelling nene of Hawaii has been successfully restored by captive breeding programs. From fewer than 30 birds half a century ago, the wild population has grown to more than 800 birds.
© William P. Cunningham

Protecting Rhinos

Rhinos are a prime example of both wildlife overharvesting and successful restoration. Rhinoceros (meaning nose horn) once ranged across most of Africa, as well as much of south and southeast Asia. Next to elephants, rhinos are the largest land animals still in existence. A large male can weigh as much as two metric tons and be up to 4 m (12 feet) long. It's estimated that in the nineteenth century, there may have been a million rhinos in Africa. By the 1960s, poaching had reduced the population to less than 100,000, and by 1980 there were only about 3,000 animals left. More than 99 percent of the species had been exterminated. Of the seven or eight subspecies that once existed in Africa, only the southern black and southern white still remain in viable populations, and those animals are almost exclusively in parks or preserves where they can be guarded day and night.

There are only two species of rhinoceros. The white or square-lipped rhinoceros (*Ceratotherium simum*) is the larger of the two. It has a wide mouth used for grazing and is the most social of rhino species. The black or hook-lipped species (*Diceros bicornis*) is specialized for eating leaves and twigs from thorny bushes. In the 1960s, there were less than 50 southern white rhinos left in Africa. But today, with successful breeding programs in many countries, coupled with a rigorous anti-poaching program, the population has been restored to about 17,500 animals across the continent. There also are about 5,000 southern black rhinos, mainly in Malawi, Zambia, and Botswana.

Although they're irascible and dangerous, the rhinos' poor eyesight and lack of fear of predators make them relatively easy to hunt. Early in the twentieth century, the main demand for rhino horns was for handles of Jambiya, or traditional daggers carried by Middle Eastern men. Powdered rhino horn was recommended in Chinese medicine for a variety of ailments, but it was removed from traditional pharmacopeia in the 1990s and is rarely used now. The current demand for rhino horn comes primarily from Vietnam, where it is regarded as a cure for cancer, an aphrodisiac, a party drug, a status symbol, and an antidote for hangovers.

The current price for rhino horn is as much as $60,000 per pound (or $132,000 per kg), which makes it more valuable by weight than gold, diamonds, or cocaine. A single large horn can bring $300,000, an enormous sum for a would-be poacher. The $20 billion annual traffic in rhino horn and elephant tusks now attracts heavily armed criminal gangs and terrorist groups. Conservation officers are generally outgunned and ill-prepared to confront highly organized and lethal poachers.

Will rhinos continue to exist outside of zoos in rich countries? Even with highly militarized protection in parks and preserves, currently at least two rhinos are killed every day in Africa. But breeding programs have proven to be highly successful. If the global trade in rhino horns can be controlled, the species could be reintroduced in many more areas. Some scientists suggest that a solution to this problem could be to raise rhinos in captivity, harvest their horns humanely, and sell them in a regulated, legal market.

Saving rhinos makes sense both ecologically and economically. Rhinos can be considered to be both flagship and umbrella species. Preserving their habitat will benefit many other less popular species. And wildlife managers estimate that having a viable population of rhinos in South Africa's Kruger National Park brings in at least $4 million annually from tourists who want to see the "big five" game animals (elephants, rhinos, lions, leopards, and buffalo).

These southern white rhinos are protected in Kenya's Lake Nakuru National Park, but poaching by organized crime rings remains a constant threat.
© William P. Cunningham

are we willing to invest to protect the slimy, smelly, crawly things? Would you favor preserving disease organisms, parasites, and vermin, or should we use our limited resources to protect only beautiful, interesting, or seemingly useful organisms?

Even given adequate area and habitat conditions to perpetuate a given species, continued inbreeding of a small population in captivity can lead to the same kinds of fertility and infant survival problems described earlier for wild populations. To reduce genetic problems, zoos often exchange animals or ship individuals long distances to be bred. It sometimes turns out, however, that zoos far distant from each other unknowingly obtained their animals from the same source. Computer databases operated by the International Species Information System, located at the Minnesota Zoo, now keep track of the genealogy of many species. For some species, this system can tell the complete reproductive history of every animal of that species in every zoo in the world. Comprehensive species survival plans based on this genealogy help match breeding pairs and project resource needs.

The ultimate problem with captive breeding, however, is that natural habitat may disappear while we are busy conserving the species itself.

We need to save rare species in the wild

As long as umbrella species—large species such as rhinos and apes—persist in their native habitat, many other species survive as well. Renowned zoologist George Schaller says that ultimately "zoos need to get out of their own walls and put more effort into saving the animals in the wild." An interesting application of this principle is a partnership between the Minnesota Zoo and the Ujung Kulon National Park in Indonesia, home to the world's few remaining Javanese rhinos. Rather than try to capture rhinos and move them to Minnesota, the zoo is helping to protect them in their native habitat by providing patrol boats, radios, housing, training, and salaries for Indonesian guards (fig. 11.21). There are no plans to bring any rhinos to Minnesota, and chances are very slight that any of us will ever see one, but we can gain satisfaction in knowing that, at least for now, a few Javanese rhinos still exist in the wild.

Section Review

1. How many southern white rhinos are there now compared to a century ago?
2. Why aren't zoos a good place to preserve whales, bats, and some reptiles?
3. Describe two examples of successful captive breeding programs.

FIGURE 11.21 The *KM Minnesota* anchored in Tamanjaya Bay in west Java. Funds raised by the Minnesota Zoo paid for local construction of this boat, which allows wardens to patrol Ujung Kulon National Park and protect rare Javanese rhinos from poachers.
Source: Courtesy of Dr. Ronald Tilson, Minnesota Zoological Garden

Conclusion

Biodiversity provides food, fiber, medicines, clean water, and many other products and services we depend upon every day. Yet nearly one-third of native species in the United States are at risk of disappearing. The acronym HIPPO provides a summary of the major classes of threats to biodiversity. Often, these threats are multiple and overlapping: For example, habitat loss, invasive species introduction, and human population growth often proceed together.

The Endangered Species Act has been one of the most powerful tools we have for environmental protection. Increasingly, it tries to support cooperative solutions between landowners and wildlife managers. Because of its effectiveness, the act itself is endangered; opponents have succeeded in limiting its scope and have sought to eliminate it altogether. Still, the act remains a cornerstone of our most basic environmental protections. It has given

new hope for the survival of numerous species that were on the brink of extinction—less than 1 percent of species listed under the ESA have gone extinct since 1973, whereas 10 percent of candidate species still waiting to be listed have suffered that fate.

For some species, though, protection and recovery programs are difficult when the critical habitat on which they depend has largely been degraded or destroyed. In light of the serious threats facing our environment today—including pollution, habitat destruction, invasive species, and global climate change—we prioritize which species we will protect, and how we will protect them. It's clear that we need to be concerned about the other organisms on which we depend for a host of ecological services, and with which we share this planet. In the next two chapters, we'll look at programs that work to protect and restore whole communities and landscapes.

Reviewing Key Terms

Can you define the following terms in environmental science?

biodiversity hot spots 11.1
biodiversity 11.1
endangered species 11.3
existence value 11.1
extinction 11.2

flagship species 11.3
gap analysis 11.3
habitat conservation plans (HCP) 11.3

HIPPO 11.2
indicator species 11.3
invasive species 11.2
keystone species 11.3

overharvesting 11.2
threatened species 11.3
umbrella species 11.3
vulnerable species 11.3

Critical Thinking and Discussion Questions

1. Many ecologists would like to move away from protecting individual endangered species to concentrate on protecting whole communities or ecosystems. Others fear that the public will respond to and support only glamorous "flagship" species such as gorillas, tigers, or otters. If you were designing conservation strategy, where would you put your emphasis?

2. Put yourself in the place of a fishing industry worker. If you continue to catch many species, they will quickly become economically extinct, if not completely exterminated. On the other hand, there are few jobs in your village, and welfare will barely keep you alive. What would you do?

3. Only a few hundred grizzly bears remain in the contiguous United States, but populations are healthy in Canada and Alaska. Should we spend millions of dollars for grizzly recovery and management programs in Yellowstone National Park and adjacent wilderness areas?

4. How could people have believed a century ago that nature is so vast and fertile that human actions could never have a lasting impact on wildlife populations? Are there similar examples of denial or misjudgment occurring now?

5. In the past, mass extinctions have allowed for new growth, including the evolution of our own species. Should we assume that another mass extinction would be a bad thing? Could it possibly be beneficial to us? To the world?

6. Some captive breeding programs in zoos are so successful that they often produce surplus animals that cannot be released into the wild because no native habitat remains. Plans to euthanize surplus animals raise storms of protests from animal lovers. What would you do if you were in charge of the zoo?

7. Compare the estimated numbers of known and threatened species in table 11.1. Are some groups overrepresented? Are we simply more interested in some organisms, or are we really a greater threat to some species?

8. Domestic and feral house cats are estimated to kill 1 billion birds and small mammals in the United States annually. In 2005, a bill was introduced in the Wisconsin legislature to declare an open hunting season year-round on cats that roam out of their owners' yards. Would you support such a measure? Why or why not? What other measures (if any) would you propose to control feline predation?

9. How many species can we afford to save? How much should we spend to protect, slimy, smelly, crawly things?

10. Is it okay to kill one species (say, barred owls) to protect another species (say, spotted owls)?

Data Analysis

Confidence Limits in the Breeding Bird Survey

A central principle of science is the recognition that all knowledge involves uncertainty. No study can observe every possible event in the universe, so there is always missing information. Scientists try to define the limits of their uncertainty, in order to allow a realistic assessment of their results. A corollary of this principle is that the more data we have, the less uncertainty we have. More data increase our confidence that our observations represent the range of possible observations.

One of the most detailed records of wildlife population trends in North America is the Breeding Bird Survey (BBS). Every June, volunteers survey more than 4,000 established 25-mile routes. The accumulated data from thousands of routes, over more than 40 years, indicate population trends, telling which populations are increasing, decreasing, or expanding into new territory. To examine a sample of BBS data, go to Connect, where you can explore the data and explain the importance of uncertainty in data.

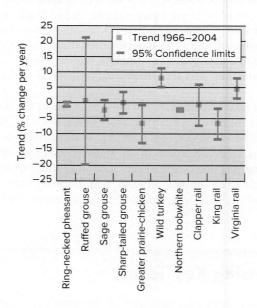

12

Biodiversity
Preserving Landscapes

▲ Orangutans are among the most critically endangered of all the great apes. Over the past 20 years, about 90 percent of their rainforest habitat in Borneo and Sumatra has been destroyed by logging and conversion to palm oil plantations.
© g-miner/iStock/Getty Images RF

Learning Outcomes

After studying this chapter, you should be able to:

12.1 Discuss the types and uses of world forests.
12.2 Describe the location and state of grazing lands around the world.
12.3 Summarize the types and locations of nature preserves.

"If we destroy the land, God may forgive us, but our children will not."

– Togiak Elder

Palm Oil and Endangered Species

Are your donuts, toothpaste, or shampoo killing critically endangered orangutans and tigers in Sumatra and Borneo? How could that be possible, you may wonder. The link is in rapidly expanding Indonesian palm plantations, which are destroying the habitat of rare species, such as orangutans, tigers, rhinos, and elephants. What were once some of the most highly productive and biologically diverse lowland rainforests in the world are rapidly being converted into palm mono-cultures that have no room for endangered species of plants and animals.

In Indonesian, the word *Orang* means person or people, and *utan* means of the forest. Orangutans are among the closest and most charismatic of our primate relatives, sharing at least 97 percent of our genes. They're also among the most critically endangered of all the great apes. It's estimated that between 1,000 and 5,000 of these shy forest giants are killed every year by loggers or poachers. Today, only about 6,000 orangutans are left in Sumatra and about 50,000 in Borneo. The United Nations warns that unless current practices change, there may be no wild orangutans outside protected areas in a few decades.

Palm oil is the most widely used vegetable oil in the world, and together Indonesia and Malaysia currently produce nearly 90 percent of the global supply. You probably have eaten or used more palm oil than you're aware. At least half of all the packaged foods in your local supermarket, along with a wide range of detergents, soaps, cosmetics, and other products, are made with this oil. And palm oil consumption is currently growing faster than that of any other food item.

Fifteen years ago, Indonesia had about 2.5 million ha (6 million acres) of palm plantations. Currently, that area has nearly doubled to more than 11 million ha (27 million acres), now producing around 35 million metric tons of palm oil annually (about 60 percent of the world total). Indonesia's palm operations are projected to double again by 2030. As agribusiness companies slash, burn, and bulldoze the forest into neat rows of palm trees, they're not only eliminating biodiversity but also emitting greenhouse gases and destroying the livelihoods of indigenous and traditional people.

Indonesia, a nation of nearly 17,000 islands lying along the equator between Southeast Asia and Australia, has the third largest area of rainforest in the world, and is one of the most biodiverse

FIGURE 12.1 Over the past 15 years, the area devoted to palm plantations in Indonesia has more than quadrupled to 11 million ha (27 million acres) and now produces about 60 percent of the world supply of this valuable oil. This rapid growth has destroyed habitat and displaced many critically endangered species.
© KhunJompol/iStock/Getty Images RF

hot spots on the planet (see fig 11.4). It also has the highest rate of deforestation of any country, along with the world's third highest greenhouse gas emissions. In addition, the expansion of palm oil plantations is a driving force in both forest destruction and climate-changing gas releases.

Deforestation usually starts with the logging of valuable tropical hardwoods. Logging slash is burned to clear the land for planting (and in many cases, fires cover up illegal logging), and finally, vast areas are planted in sterile monotony (fig 12.1).

Habitat destruction drives out wildlife, while a network of logging roads makes it possible for poachers to enter inaccessible areas. Because mother orangutans are intensely loyal to their babies, poachers will usually kill the mothers so they can then take the babies and sell them on the pet and zoo market.

Oil palms are highly profitable. A single hectare (2.47 acres) of palms can yield 30 metric tons of oil per year—this is as much as ten times the amount of other oilseed crops (fig. 6.1). Palm oil is now Indonesia's third-largest import, bringing in $18 billion annually.

At the 2014 UN Climate Summit in New York, Indonesia promised to stop producing palm oil from newly deforested land. Most of the largest palm oil companies, including Wilmar, Astra Agro Lestari, Cargill, Golden Agri-Resources, and Musim Mas joined in this pledge. And many of the world's largest food and cosmetic companies—including McDonald's, Nestlé, General Mills, Kraft, and Procter & Gamble—vowed to stop using palm oil from unsustainable sources.

It was thought that it wouldn't be difficult to fulfill this promise because Indonesia already has more than enough degraded land to provide all the planned plantation expansion for the next 20 years. However, when Wilmar International, the world's largest palm oil trader created an online platform to trace sustainability—this list included the names and locations of refineries and palm oil plantations in its supply chain—the Indonesian Minister for Economic Affairs complained this would harm small palm oil suppliers. And when Golden Agri tried to convert an area designated for oil plantations in Indonesian Borneo into a conservation forest, local officials threatened to revoke the concession and give it to a competitor who promised to plant oil palms.

Protecting biodiversity and preserving landscapes often are both interconnected and controversial. In this chapter, we'll look at many examples of this important topic.

12.1 WORLD FORESTS

- *Forests provide habitat for biodiversity, valuable resources for humans, and essential ecological services.*
- *Tropical forests are particularly species rich and are threatened by human activities.*
- *Temperate forests also are at risk.*

Forests and woodlands occupy some 4 billion ha (roughly 15 million mi^2), or about 29 percent of the world's land surface (fig. 12.2). Grasslands (pastures and rangelands) cover almost as much area. Together these ecosystems supply many essential resources, such as lumber, paper pulp, and grazing lands for livestock. They also provide vital ecological services, including regulating climate, controlling water runoff, providing wildlife habitat, and purifying air and water. Forests and grasslands also have scenic, cultural, and historic values that deserve protection. These biomes are also among the most heavily disturbed (chapter 5), because they are places in which people prefer to live and work.

As the opening case study for this chapter shows, balancing competing land uses and needs can be complicated. Many conservation debates have concerned protection or use of forests, prairies, and rangelands. This chapter examines the ways we use and abuse these biological communities, as well as some of the ways we can protect them and conserve their resources. We discuss forests first, followed by grasslands, and then strategies for conservation, restoration, and preservation. Chapter 13 focuses on restoration of damaged or degraded ecosystems.

Boreal and tropical forests are most abundant

Forests are widely distributed, but the largest remaining areas are in the tropical humid regions, especially Brazil, western Africa, and Southeast Asia, and in the boreal (northern) forests of Russia, Canada, and the United States (see fig. 12.2 and fig. 12.3). There are differing definitions of what a "forest" is. The UN Food and Agriculture Organization (FAO) defines **forest** as any area where trees cover more than 10 percent of the land. Open wooded areas, with tree cover between about 10 and 20 percent of the ground, are often considered **savannas.** In **closed-canopy forests,** tree crowns overlap to cover most of the ground.

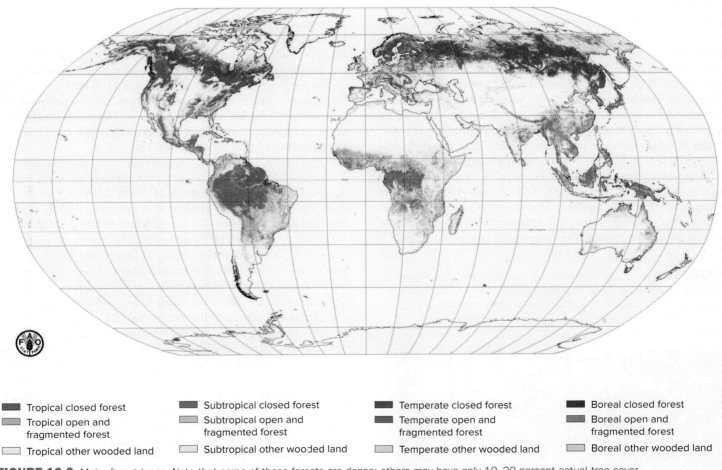

Tropical closed forest
Tropical open and fragmented forest
Tropical other wooded land

Subtropical closed forest
Subtropical open and fragmented forest
Subtropical other wooded land

Temperate closed forest
Temperate open and fragmented forest
Temperate other wooded land

Boreal closed forest
Boreal open and fragmented forest
Boreal other wooded land

FIGURE 12.2 Major forest types. Note that some of these forests are dense; others may have only 10–20 percent actual tree cover.
Source: UN Food and Agriculture Organization, 2002

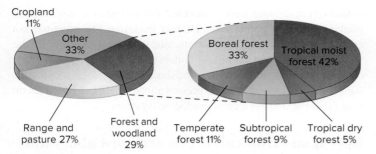

FIGURE 12.3 World land use and forest types. The "other" category includes tundra, desert, wetlands, and urban areas.
Source: UN Food and Agriculture Organization (FAO)

The largest tropical forests are in South America, which has about 22 percent of the world's forest area and by far the most extensive area of undisturbed tropical rainforest. Africa and Southeast Asia also have large areas of tropical forest that are highly important biologically, but both continents are suffering from rapid deforestation. North America and Eurasia have vast areas of relatively unaltered boreal forest. Although many of these forests are harvested regularly, both continents have a net increase in forest area and biomass because of replanting and natural regeneration.

Among the forests of greatest ecological importance are the ancient forests, such as the tropical rainforests described in the opening case study for this chapter, that are home to much of the world's biodiversity, ecological services, and indigenous human cultures. Sometimes called frontier, old-growth, or virgin forests, these are areas large enough and free enough from human modification that native species can live out a natural life cycle, and ecological relationships play out in a relatively normal fashion. A similar term used by the FAO is **primary forests,** which are "composed primarily of native species in which there are no clearly visible indications of human activity and ecological processes are not significantly disturbed."

This doesn't mean that all trees in a primary forest need be enormous or thousands of years old (fig. 12.4). In some biomes, such as lodgepole pine forests of the Rocky Mountains or Canada's boreal forests, most trees live only a century or so before being killed by disease or some natural disturbance. The successional processes (Chapter 4)

FIGURE 12.4 A tropical rainforest in Queensland, Australia. Primary, or old-growth, forests, such as this, aren't necessarily composed entirely of huge, old trees. Instead, they have trees of many sizes and species that contribute to complex ecological cycles and relationships.
© Digital Vision/PunchStock RF

as trees die and are replaced create structural complexity and a diversity of sizes and ages important for specialists, such as the northern spotted owl. Nor does it mean that humans have never been present. Where human occupation entails relatively little impact, a forest may be inhabited for millennia while still retaining its primary characteristics. Even forests that have been logged or converted to cropland often can revert to natural conditions if left alone long enough.

Even though forests still cover about half the area they once did worldwide, only one-quarter of those forests retain old-growth features. The largest remaining areas of old-growth forest are in Russia, Canada, Brazil, Indonesia, and Papua New Guinea. Together, these five countries account for more than three-quarters of all relatively undisturbed forests in the world. In general, remoteness rather than laws protects those forests. Although official data describe only about one-fifth of Russian old-growth forest as threatened, rapid deforestation—both legal and illegal—especially in the Russian Far East, probably puts a much greater area at risk.

Forests provide many valuable products

Wood is generally considered the most important forest product. There is hardly any industry that does not use wood or wood products somewhere in its manufacturing and marketing processes. Think about the amount of printed material, cardboard boxes, and other paper products that each of us in developed countries handles, stores, and disposes of in a single day. Total annual world wood consumption is about 4 billion m^3. This is more than steel and plastic consumption combined. International trade in wood and wood products amounts to more than $100 billion each year. Developed countries consume about 80 percent of wood products, but they produce less than 50 percent of industrial wood. Less-developed countries, mainly in the tropics, produce more than half of all industrial wood but use only 20 percent.

Paper pulp, the fastest-growing type of forest product, accounts for nearly a fifth of all wood consumption. Most of the world's paper is used in the wealthier countries of North America, Europe, and Asia. Global demand for paper is also increasing rapidly as more countries become wealthy. The United States, Russia, and Canada are the largest producers of both paper pulp and industrial wood (lumber and panels). Much industrial logging in Europe and North America now occurs on managed plantations, rather than in untouched old-growth forest. However, paper production is increasingly blamed for deforestation in Southeast Asia, West Africa, and other regions.

Fuelwood accounts for nearly half of global wood use. Roughly one-third of the world's population depends on firewood or charcoal as their principal source of heating and cooking fuel (fig. 12.5). Demand for fuelwood is depleting forests in some developing areas, especially around growing cities. An estimated 1.5 billion people have less fuelwood than they need, and experts expect shortages to worsen as poor urban areas grow. In some countries, firewood harvesting is a major cause of deforestation, but foresters argue that biomass energy could be produced sustainably in most developing countries, with careful management.

Approximately one-quarter of the world's forests are managed for wood production. Ideally, forest management involves scientific planning for sustainable harvests, with particular attention paid

FIGURE 12.5 Firewood accounts for almost half of all wood harvested worldwide and is the main energy source for one-third of all humans.
© William P. Cunningham

- **Group Selection** and corridor harvesting (fig. 12.6c) call for the harvest of mature trees or the thinning of intermediate size trees at relatively short intervals. The remaining groves or rows of mature trees support natural regeneration and shelter new growth, while the open spaces and corridors provide light and room for new growth.
- **Single tree selective harvesting** (fig. 12.6d) calls for the removal of individual mature trees, leaving the majority of trees on a site standing. Trees selected for harvesting can be those with disease or defective growth patterns, or they may be species regarded as of less value, leaving more valuable species to prosper from less competition.

Reforestation for wood production, however, is not the same as maintaining a natural forest. Usually, production is most efficient in large plantations of a single-species, in intensive cropping called **monoculture agroforestry.** Cultivating a single species can maximize growth and simplify harvesting (fig. 12.7). However, a monoculture forest plantation generally supports little biodiversity and provides few ecological services, such as soil erosion control or clean water production. The United States Forest Service has stated that water resources are the most valuable product for U.S. forests, especially in arid Western states.

to forest regeneration. In temperate regions, according to the FAO, more land is being replanted or allowed to regenerate naturally than is being permanently deforested. Increased sourcing from tropical regions explains much of this forest regeneration.

The choice between forest harvest methods is a complex one, involving a wide range of considerations. Foresters evaluate soil, slope, and water conditions; wildlife habitat; recreational opportunities; and ecological principles in making decisions. Every harvest is different, because procedures and equipment to be used must be determined specifically for each site.

- **Clearcutting** (fig. 12.6a) entails removal of all trees in a stand. It can be a valuable tool for managing some species, such as lodgepole pine, that don't grow well in the shade of other trees. And it can be a rapid and economical technique, but it generally destroys wildlife habitat and exposes soil to erosion and invasive species.
- **Shelterwood** or seed tree systems (fig. 12.6b) are designed to remove a majority of mature trees but leave a protective overstory standing to shelter new growth. In some systems, the sheltering trees are selected to provide a seed source for regeneration.

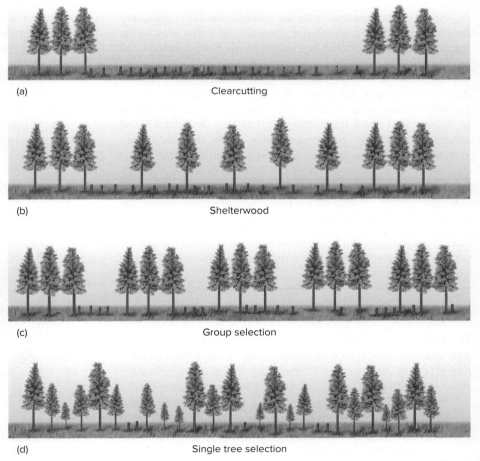

(a) Clearcutting

(b) Shelterwood

(c) Group selection

(d) Single tree selection

FIGURE 12.6 Different harvest methods are appropriate for specific landscapes, species combinations, management objectives, and climate.

FIGURE 12.7 Monoculture forestry, such as this Wisconsin tree farm, produces valuable timber and pulpwood, but has little biodiversity.
© William P. Cunningham

This was an area larger than the entire United States. At least half that forest has already been cleared or degraded. Every day, logging and burning destroy about 30,000 acres (11,000 ha) of tropical forest, while farming, grazing, or conversion to monoculture plantations degrades a roughly equal area. This amounts to an area about the size of Austria each year. It also represents a serious reduction of nature's capacity to store carbon we release by other means. In addition to their value in carbon storage, forests provide important ecological services, such as generating oxygen, storing water, and protecting biodiversity.

If current rates of destruction continue, no primary forest will be left in many countries outside of parks and nature reserves by the end of this century. In its 2015 forest survey, the FAO reported that Africa, South America, and Oceania continue to have a net loss in forest area, while Asia, Europe, and North America have either gained forest or remained stable because of reforestation programs (fig. 12.8). The world's highest current rate of forest loss is in Burundi, which is losing 9 percent of its forest annually.

Not only is tropical deforestation a tragic loss of biological diversity, it also represents a declining livelihood for the millions of people who depend on forests for part or all of their sustenance. Furthermore, approximately 1.7 billion metric tons of carbon are released annually due to deforestation and land use changes, mostly in the tropics. This amounts to about 20 percent of all anthropogenic carbon emissions, or more than the emissions from all forms of transportation combined. Halting forest destruction and soil degradation would help significantly to avoid global climate change (see chapter 15). But climate change also threatens forests. During severe droughts in 2005 and 2010, the Amazon rainforest lost billions of trees. If severe droughts continue, the forest, which now absorbs about one-quarter of all anthropogenic CO_2, could become a carbon source rather than a sink.

As the opening case study for this chapter shows, Indonesia is now thought to have the highest rate of deforestation in the world.

Some of the countries with the most successful reforestation programs are in Asia. China, for instance, cut down most of its forests 1,000 years ago and has suffered centuries of erosion and terrible floods as a consequence. Recently, however, timber cutting in the headwaters of major rivers has been outlawed, and a massive reforestation project has begun. Since 1990, China has planted 50 billion trees, mainly in Xinjiang Province, to stop the spread of deserts. Korea and Japan also have had very successful forest restoration programs. After being almost totally denuded during World War II, both countries are now about 70 percent forested.

Quantitative Reasoning

Looking at figure 12.8, roughly what percentage of world forests was gained or lost during the period shown? If there are about 4 billion ha of forest in the world, how much area was lost per year between 2005 and 2010? How could Europe be shown to have so much forest area?

Tropical forests are especially threatened

Tropical forests are among the richest and most diverse terrestrial systems. This is especially true of the moist forests (rainforests) of the Amazon and Congo River basins and Southeast Asia. Although these forests now occupy less than 10 percent of the earth's land surface, they are thought to contain more than two-thirds of all higher plant biomass and at least half of all the plant, animal, and microbial species in the world.

A century ago, an estimated 12.5 million km^2 (nearly 5 million mi^2) of tropical lands were covered by primary forest.

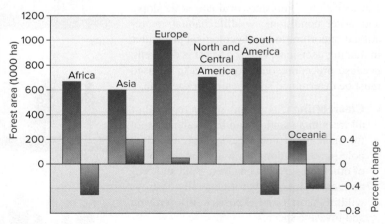

FIGURE 12.8 Annual net change in forest area, 2000–2005. The largest annual net deforestation rate in the world is in Africa. Large-scale tree planting in Asia and Europe has produced a net gain in forest area.
Source: Data from FAO, 2010

1975   1989   2015

FIGURE 12.9 Forest destruction in Rondonia, Brazil, between 1975 and 2015. Construction of logging roads creates a feather-like pattern that opens forests to settlement by farmers.

Source: a–c: USGS Earth Resources Observation and Science (EROS) Center

At the beginning of the twentieth century, at least 84 percent of Indonesia's total land was forested. Between 1990 and 2010, the country lost at least 24 million ha of forest (59 million acres), much of it for illegal palm oil plantations, and only 52 percent of the island nation was forested. Despite the 2014 pledge to reduce forest clearing, Indonesia continues to lose an estimated 840,000 ha per year, surpassing deforestation in Brazil.

Causes of Deforestation

There are many causes of deforestation. In Africa, forest clearing by subsistence farmers is responsible for about two-thirds of the forest destruction, but large-scale commercial logging also takes a toll. In Latin America, the largest single cause of deforestation is the expansion of soy farming and cattle ranching. Loggers start the process by cutting roads into the forest (fig. 12.9) to harvest valuable hardwoods, such as mahogany or cocobolo. Logging roads allow subsistence farmers to move into the forest, which they clear for homesteads. Large-scale landowners, ranchers, and soy farmers (see chapter 9) then move in and remove most remaining forest. In recent years, Brazil reports that rates of deforestation have declined by about half. Primary forest lost in earlier decades, though, remains in soy fields and pasture.

Fires destroy about 350 million ha (1,350 mi^2) of forest every year. Some fires are set intentionally to cover up illegal logging or land clearing. Others are started by

natural causes. The greatest fire hazard in the world, according to the FAO, is in sub-Saharan Africa, which accounts for about half the global total. Uncontrolled fires tend to be worst in countries with corrupt or ineffective governments and high levels of poverty, civil unrest, and internal refugees. And as global warming brings drought and insect infestations to many parts of the world, there's a worry that forest fires may increase catastrophically.

Forest Protection

What can be done to encourage forest protection? A program called Reducing Emissions from Deforestation and Forest Degradation in Developing Countries (REDD) is offering a mechanism by which richer countries can offer aid to poorer countries in exchange for progress in reducing deforestation, replanting trees, and stopping the drying of peatlands (see the Exploring Science box below).

About 12.5 percent of all world forests are now in some form of protected status, but the effectiveness of that protection varies greatly (fig. 12.10). Central America claims that nearly 55 percent of its forests are protected. Costa Rica is attempting not only to

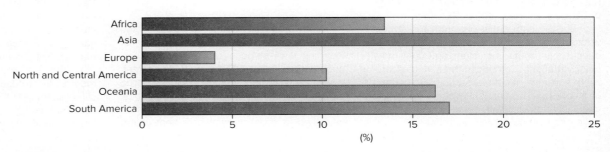

FIGURE 12.10 Percentage of forest area designated for conservation. About 12.5 percent of all world forests are now in some form of protected status, but the effectiveness of that protection varies greatly.

Source: FAO Global Forest Resources Assessment 2010 (Main Report, p. 60)

EXPLORING SCIENCE

Protecting Forests to Prevent Climate Change

One of the most promising agreements reached in the 2015 Paris Climate Change talks was a pledge by the 195 participating governments to reduce deforestation, restore degraded forests and substantially increase global reforestation. Cutting emissions through forest conservation is arguably one of the simplest and most cost-effective ways to address climate change. Experts estimate that aggressive forest management could offset roughly half of the globe's current carbon emissions over the next decade.

How to pay for forest protection remains a question, however. In 2010, participating nations approved a mechanism called REDD (Reducing Emissions from Deforestation and Forest Degradation in Developing Countries). This strategy would allow developed countries to compensate developing nations for protecting forests. One of the first agreements under REDD was between Norway and Indonesia.

Norway has promised up to (U.S.) $1 billion in aid, providing that Indonesia makes acceptable progress toward sustainable forest management. As the opening case study for this chapter shows, Indonesia has the third-largest area of tropical rainforest on the globe (after Brazil and the Democratic Republic of Congo), as well as some of the highest biological diversity in the world. And palm oil plantation expansion is one of the leading causes of forest destruction in Indonesia.

Together, deforestation, land-use change, and the drying, decomposition, and burning of peatlands cause about 80 percent of Indonesia's current greenhouse gas emissions, which are currently the third highest in the world. Deep peatlands, where water-logged soils prevent biomass decomposition can contain more than 28 times as much carbon as mineral soil. The draining and burning of a hectare of peatland can release 15,000 tons of CO_2.

But the high emissions from forest conversion means that Indonesia can make deeper cuts in CO_2 emissions and do it more quickly than most other countries. And according to government estimates, up to 80 percent of Indonesia's logging is illegal, so bringing it under control also will increase national revenue and help build civic institutions.

REDD could result in a major transfer of financial assistance from wealthy countries to poor ones. It's estimated that it will take about (U.S.) $30 billion per year to fund this program. But it offers a chance to save one of the world's most precious ecosystems. Forests would no longer be viewed merely as timber waiting to be harvested or land awaiting clearance for agriculture.

Logging valuable hardwoods is generally the first step in tropical forest destruction. Although loggers may take only one or two large trees per hectare, the damage caused by extracting logs exposes the forest to invasive species, poachers, and fires.
© Digital Vision/PunchStock RF

Many problems need to be solved for the Norway/Indonesia partnership to work. For one thing, it will be necessary to calculate how much carbon is stored in a particular forest, as well as how much carbon could be saved by halting or slowing deforestation. Historical forest data, on which these predictions often are based, is often unreliable or nonexistent in tropical countries. Satellite imaging and computer modeling can give answers to these questions, but the technology can be expensive. In the first phase of funding, Norway will support political and institutional reform, along with building infrastructure and increasing capacity.

Like other donor nations, Norway is also concerned about how permanent the protections will be. What happens if they pay to protect a forest but a future administration decides to log it? Furthermore, loggers are notoriously mobile and adept at circumventing rules by bribing local authorities, if necessary. What's to prevent them from simply moving to new areas to cut trees? If you avoid deforestation in one place but then cut an equal number of trees somewhere else (sometimes known as "leakage"), carbon emissions won't have gone down at all. Similarly, there's concern that a reduction in logging in one country could lead to pressure on other countries to cut down their forests to meet demand. And there would be a financial incentive to do so if reductions in logging pushed up the price of timber.

Will this partnership protect indigenous people's rights? In theory, yes. Indonesia has more than 500 ethnic groups, and many forest communities lack secure land tenure. Large mining, logging, and palm oil operations often push local people off their traditional lands with little or no compensation. Indonesia has promised a two-year suspension on new projects to convert natural forests. They also have promised to recognize the rights of native people and local communities.

Could having such a sudden influx of money cause corruption? Yes, that's possible. But Indonesia has a good track record of managing foreign donor funds. The Aceh Reconstruction Agency, established after the 2004 tsunami, managed around (U.S.) $7 billion of donations in line with the best international standards. Indonesia has promised that the same governance principles will be used to manage REDD funds.

What do you think? Is this a viable method of protecting forests and preventing climate change? If you were a Norwegian citizen, what safeguards or guarantees would you like to see imposed on this contract?

FIGURE 12.11 Cattle ranching can increase pressure for forest destruction. However, in the proper setting, cattle also can assist forest regeneration by dispersing seeds.
© William P. Cunningham

rehabilitate the land (make an area useful to humans) but also to restore the ecosystems to naturally occurring associations. One of the best known of these projects is entomologist Dan Janzen's work in Guanacaste National Park. Like many dry tropical forests, the northwestern part of Costa Rica had been almost completely converted to ranchland. By controlling fires, however, Janzen and his coworkers are bringing back the forest. One of the keys to this success is involving local people in the project. Janzen also encourages grazing in the park. The original forest evolved, he reasons, together with ancient grazing animals that are now extinct. Ranching can be a force for forest destruction, but cattle can play a valuable role as seed dispersers (fig. 12.11).

Brazil, also, is a leader in establishing forest reserves. It now recognizes the right of traditional people—Indians, descendants of runaway slaves, traditional fishermen, peasants, and communities engaged in nondestructive extractive activities (such as rubber tapping or nut collecting)—to live in the forest. At the same time, however, Brazil is pressing ahead with building and paving a network of roads to connect the western Amazon with all-weather, high-speed roads to the Pacific. Critics warn that these projects will accelerate land invasions and will result in displacement of native people and wildlife throughout the forest.

People also are working on the grassroots level to protect and restore forests in other countries. India, for instance, has a long history of nonviolent, passive resistance movements—called *satyagrahas*—to protest unfair government policies. These protests go back to the beginning of Indian culture and often have been associated with forest preservation. Gandhi drew on this tradition in his protests of British colonial rule in the 1930s and 1940s. During the 1970s, commercial loggers began large-scale tree felling in the Garhwal region in the state of Uttar Pradesh in northern India. Landslides and floods resulted from stripping the forest cover from the hills. The firewood on which local people depended was destroyed, and the way of life of the traditional forest culture was threatened. In a remarkable display of courage and determination, the village women wrapped their arms around the trees to protect them, sparking the Chipko Andolan movement (literally, "movement to hug trees"). They prevented logging on 12,000 km² of sensitive watersheds in the Alakanada basin. Today, the Chipko Andolan movement has grown to more than 4,000 groups working to save India's forests.

Debt-for-Nature Swaps

Those of us in developed countries also can contribute toward saving tropical forests. Financing nature protection is often a problem in developing countries, where the need is greatest. One promising approach is called **debt-for-nature swaps.** Banks, governments, and lending institutions now hold nearly $1 trillion in loans to developing countries. There is little prospect of ever collecting much of this debt, and banks are often willing to sell bonds at a steep discount—perhaps as little as 10 cents on the dollar. Conservation organizations buy debt obligations on the secondary market at a discount and then offer to cancel the debt if the debtor country agrees to protect or restore an area of biological importance.

There have been many such swaps. Conservation International, for instance, bought $650,000 of Bolivia's debt for $100,000—an 85 percent discount. In exchange for canceling this debt, Bolivia agreed to protect nearly 1 million ha (2.47 million acres) around the Beni Biosphere Reserve in the Andean foothills. Ecuador and Costa Rica have had a different kind of debt-for-nature swaps. They have exchanged debt for local currency bonds that fund activities of local private conservation organizations in the country. This has the dual advantage of building and supporting indigenous environmental groups while protecting the land. Critics, however, charge that these swaps compromise national sovereignty and do little to reduce the developing world's debt or to change the situations that led to environmental destruction in the first place.

Temperate forests also are threatened

Tropical countries aren't unique in damaging and degrading their forests. Asia and the Pacific currently have had a net forest increase thanks to an ambitious reforestation effort in China (see fig. 12.8). Europe also has increased its forest area with replanting projects and forest regrowth on abandoned fields and previously harvested areas.

Although the total forest area in North America has remained nearly constant in recent years, forest management policies in the United States and Canada continue to be controversial. Large areas of the temperate rainforest of the Pacific Northwest have been set aside to protect endangered species. These forests have more standing biomass per square kilometer than any other ecosystem on earth (fig. 12.12). Because they're so wet, these forests rarely burn, and trees often live to be a thousand years old and many meters in diameter.

FIGURE 12.12 The huge old trees of the old-growth temperate rainforest accumulate more total biomass in standing vegetation per unit area than any other ecosystem on earth. They provide habitat to many rare and endangered species, but they are also coveted by loggers who can sell a single tree for thousands of dollars.
© William P. Cunningham

A unique biological community has evolved in these dense, misty forests. Dozens of species of plants and animals spend their whole lives in the forest canopy, almost never descending to ground level.

In 1994, the U.S. government adopted the Northwest Forest Plan to regulate harvesting on about 9.9 million ha (24.5 million acres) of federal lands in Oregon, Washington, and northern California. This plan was an admirable example of using good science for natural resource planning. Teams of researchers identified specific areas of ancient forest essential for sustaining viable populations of endangered species, such as the northern spotted owl and the marbled murrelet (chapter 11). The plan prohibited most clear-cut logging, especially on steep hillsides and in riparian (streamside) areas where erosion threatens water quality and salmon spawning (fig. 12.13).

Still, logging has been allowed on the "matrix" lands surrounding these islands of ancient, old-growth forests. Conservationists lament the fact that fragmentation reduces the ecological value of the remaining forest, and they claim that many of the areas now lacking old-growth status could achieve the levels of structural complexity and age required for this classification if left uncut for a few more decades.

One of the most controversial aspects of forest management in the United States in recent years has been road building in "de facto" wilderness areas. These areas are effectively wilderness—they have no roads and are not yet accessible to logging or mining—but they have no legal protection. Roads fragment forests, provide a route of entry for hunters and invasive species, often result in erosion, and destroy wilderness qualities. Shortly after passage of the Wilderness Act by the U.S. Congress in 1964, the Forest Service began a review of existing roadless (de facto wilderness) lands. Called the Roadless Area Review and Evaluation (RARE), this effort culminated in 1972 with the identification of 56 million acres (230,000 km²) suitable for wilderness protection. Some of these lands were subsequently included in state wilderness bills, but most remained vulnerable to logging, mining, and other extractive activities.

Roadless area protection gained traction in 2001, during the last days of the Clinton administration, when a national guideline called the **Roadless Rule** was established. This rule ended virtually all logging, road building, and development on virtually all the lands identified as deserving of protection in the 1972 RARE assessment. Despite repeated attempts by George W. Bush to either overturn the Roadless Rule or simply not defend it in lawsuits from timber companies, this rule continues to protect de facto wilderness in 38 states. The Obama administration maintained the rule, but actions of future administrations remain uncertain.

A much greater threat to temperate forests may be posed by climate change, insect infestations, and wildfires, all of which are interconnected. During the past few decades, the average temperature over much of North America has risen by more than 1 °F (0.5 °C). This may not sound like much, but it has caused the worst drought in 500 years. Hot, dry weather weakens trees and makes them more vulnerable to both insect attacks and fires. A research team led by ecologist Jerry Franklin has showed that tree mortality among a wide variety of species has increased dramatically across a wide area over the past few decades. Infestations by beetles in particular have killed millions of hectares of conifers throughout western North America. This includes pinyon pine forests in the southwest, lodgepole pines throughout the Rocky Mountains, and huge swaths of spruce forests in Canada

FIGURE 12.13 Clear-cuts, such as this one, threaten species dependent on old-growth forest and expose steep slopes to soil erosion.
© Gary Braasch/Stone/Getty Images

and Alaska. The billions of dead and dying trees are a huge fire danger, especially where people have built homes in remote areas.

For 70 years, the U.S. Forest Service has had a policy of aggressive fire control. The aim has been to extinguish every fire on public land before 10 a.m. Smokey Bear was adopted as the forest mascot and warned us that "only you can prevent forest fires." Recent studies, however, of fire's ecological role suggest that our attempts to suppress all fires may have been misguided. Many biological communities are fire-adapted and require periodic burning for regeneration. Eliminating fire from these ecosystems has allowed woody debris to accumulate, greatly increasing chances for very big fires (fig. 12.14).

Forests that once were characterized by 50 to 100 mature, fire-resistant trees per hectare and an open understory now have a thick tangle of up to 2,000 small, spindly, mostly dead saplings in the same area. The U.S. Forest Service estimates that 33 million ha (73 million acres), or about 40 percent of all federal forestlands, are at risk of severe fires. To make matters worse, Americans increasingly live in remote areas where wildfires are highly likely. Because there haven't been fires in many of these places in living memory, many people assume there is no danger, but by some estimates 40 million U.S. residents now live in areas with high wildfire risk.

Much of the federal and state firefighting efforts are controlled, in effect, by these homeowners who build in fire-prone areas. A government audit found that 90 percent of the Forest Service firefighting outlays go to save private property. If people who build in forested areas would take some reasonable precautions to protect themselves, we could let fires play their normal ecological role in forests in many cases. For example, you shouldn't build a log cabin with a wood shake roof surrounded by dense forest at the end of a long, narrow, winding drive that a fire truck can't safely navigate. If you're going to have a home in the forest, you should use fireproof materials, such as a metal roof and rock or brick walls, and clear all trees and brush from at least 60 m (200 ft) around any buildings.

FIGURE 12.14 By suppressing fires and allowing fuel to accumulate, we make major fires more likely. The safest and most ecologically sound management policy for some forests may be to allow natural or prescribed fires to burn periodically, as long as they don't threaten property or human life.
Source: Kari Greer/NIFC/FS/BLM

A recent prolonged drought in the western United States has heightened fire danger there, and in 2015 more than 68,000 wildfires burned 4 million ha (10 million acres) of forests and grasslands in the United States. Federal agencies spent about $2 billion to fight these fires, nearly ten times the average in the 1990s. The dilemma is how to undo years of fire suppression and fuel buildup. Fire ecologists favor small, prescribed burns to clean out debris. Loggers decry this approach as a waste of valuable timber, and local residents of fire-prone areas fear that prescribed fires will escape and threaten them. What do you think? What's the best way to restore forest health while also protecting property values and local jobs?

Ecosystem Management

In the 1990s, the U.S. Forest Service began to shift its policies from a timber production focus to **ecosystem management,** which

What Can You Do?

Lowering Your Forest Impacts

For most urban residents, forests—especially tropical forests—seem far away and disconnected from everyday life. There are things that each of us can do, however, to protect forests.

- Reuse and recycle paper. Make double-sided copies. Save office paper, and use the back for scratch paper.
- Use e-mail. Store information in digital form, rather than making hard copies of everything.
- If you build, conserve wood. Use wafer board, particle board, laminated beams, or other composites, rather than plywood and timbers made from old-growth trees.
- Buy products made from "good wood" or other certified sustainably harvested wood.
- Don't buy products made from tropical hardwoods, such as ebony, mahogany, rosewood, or teak, unless the manufacturer can guarantee that the hardwoods were harvested from agroforestry plantations or sustainable-harvest programs.
- Don't patronize fast-food restaurants that purchase beef from cattle grazing on deforested rainforest land. Don't buy coffee, bananas, pineapples, or other cash crops if their production contributes to forest destruction.
- Do buy Brazil nuts, cashews, mushrooms, rattan furniture, and other nontimber forest products harvested sustainably by local people from intact forests. Remember that tropical rainforest is not the only biome under attack. Contact the Forest Stewardship Council to learn about sustainable forestry at https://us.fsc.org/.
- If you hike or camp in forested areas, practice minimum-impact camping. Stay on existing trails, and don't build more or bigger fires than you absolutely need. Use only downed wood for fires. Don't carve on trees or drive nails into them.
- Write to your congressional representatives and ask them to support forest protection and environmentally responsible government policies. Contact the U.S. Forest Service, and voice your support for recreational and nontimber forest values.

Table 12.1	Draft Criteria for Sustainable Forestry
1.	Conservation of biological diversity
2.	Maintenance of productive capacity of forest ecosystems
3.	Maintenance of forest ecosystem health and vitality
4.	Maintenance of soil and water resources
5.	Maintenance of forest contribution to global carbon cycles
6.	Maintenance and enhancement of long-term socioeconomic benefits to meet the needs of legal, institutional, and economic framework for forest conservation and sustainable management

Source: Data from U.S. Forest Service, 2002

attempts to integrate sustainable ecological, economic, and social goals in a unified, systems approach. Some of the principles of this new philosophy include:

- Managing across whole landscapes, watersheds, or regions over ecological time scales.

- Considering human needs and promoting sustainable economic development and communities.

- Maintaining biological diversity and essential ecosystem processes.

- Utilizing cooperative institutional arrangements.

- Generating meaningful stakeholder and public involvement and facilitating collective decision making.

- Adapting management over time, based on conscious experimentation and routine monitoring.

Some critics argue that we don't understand ecosystems well enough to make practical decisions in forest management on this basis. They argue we should simply set aside large blocks of untrammeled nature to allow for chaotic, catastrophic, and unpredictable events. Others see this new approach as a threat to industry and customary ways of doing things. Still, elements of ecosystem management appear in the *National Report on Sustainable Forests* prepared by the U.S. Forest Service. Based on the Montreal Working Group criteria and indicators for forest health, this report suggests goals for sustainable forest management (table 12.1).

Section Review

1. What do we mean by *closed-canopy* forest and *primary* forest?
2. Which commodity is used most heavily in industrial economies: steel, plastic, or wood?
3. What is a *debt-for-nature swap*?

12.2 GRASSLANDS

- *Grasslands occupy more than one-quarter of the world's land surface and contribute crucially to human economy and biodiversity protection.*

- *Worldwide, the rate of grassland disturbance each year is three times that of tropical forests.*

- *Short-duration, rotational grazing can emulate the effects of wild herds and benefit grassland health.*

After forests, grasslands are among the biomes most heavily used by humans. Grasslands vary from humid to near-desert conditions and from open prairie to partially wooded savannas. Together, the various types of grasslands occupy about 27 percent of the world's land surface. Much of the U.S. Great Plains and the Prairie Provinces of Canada fall into this category (fig. 12.15). Globally, the 3.8 billion ha (12 million mi^2) of pastures and grazing lands make up about twice the area of all croplands. Livestock are also grazed on about 4 billion ha of other marginally useful land, such as forest, desert, tundra, marsh, and thorn scrub. More than 3 billion cattle, sheep, goats, camels, buffalo, and other domestic animals on these lands make a valuable contribution to human nutrition.

When grazing is not too severe, especially when grazers move from place to place, this use of grasslands can be sustainable: Historically, migratory herds of millions of buffalo were an essential part of the Great Plains ecosystem, rejuvenating grasslands as they traveled. The great migratory herds of wildebeest and zebras in Kenya and Tanzania today serve the same function. Unlike domestic grazers, they do not stay in place long enough to deplete biodiversity or expose erodible soils.

Although they may appear uniform to the untrained eye, native prairies support a rich variety of specialized plants and animals. According to the U.S. Department of Agriculture, more threatened plant species occur in grasslands than in any other major American biome. Nearly all of North America's moist grasslands, where rain is sufficient to support crops, has been converted to cropland or urban areas. Most remaining dry grasslands are managed as public range lands. Many of these areas have been grazed unsustainably for decades, resulting in lost vegetation and soil erosion.

FIGURE 12.15 Grasslands are expansive, open environments that can support surprising biodiversity.
© William P. Cunningham

Grazing can be sustainable or damaging

Historically, many societies have used grazing lands sustainably. **Pastoralists** (people who live by herding animals) have managed herds in many parts of the world, adjusting to variations in rainfall, seasonal plant conditions, and the nutritional quality of forage to keep livestock healthy and avoid overusing any particular area. Conscientious management can even improve the quality of the range, as grazing stimulates many grasses to grow rapidly, and short-term, intensive grazing creates openings that allow less common species to grow.

When grazing lands are abused by **overgrazing,** especially in arid areas, rain runs off quickly before it can soak into the soil to nourish plants or replenish groundwater. Springs and wells dry up. Seeds can't germinate in the dry, overheated soil. The barren ground reflects more of the sun's heat, changing weather patterns, driving away moisture-laden clouds, and leading to further desiccation. This process of converting once-fertile land to desert is called **desertification.**

The link between desertification and overgrazing has been recognized at least since Plato and Aristotle described land degradation in ancient Greece. More recently, range degradation has accelerated with the growth of both human and livestock populations. In many areas, as in sub-Saharan Africa, political instability also forces herders to overuse fragile lands. Overgrazing accelerates erosion, as well as reducing plant abundance and diversity (fig. 12.16). While natural grasslands often have deep, soft soils that absorb rainwater, the hard surface of bare, overgrazed pastures absorbs little rainfall. Instead, rain runs off the surface, carrying dirt particles with it. Erosion results, often leading to severe gullying

that can eventually make pastures unusable. Nearly three-quarters of all rangelands in the world show signs of either degraded vegetation or soil erosion, and climate change is only making this situation worse. But degradation can be reversed with good management. If you were a livestock herder, are there arguments that would convince you to reduce your stocks?

Overgrazing threatens many U.S. rangelands

The health of most public grazing lands, in the United States and many other countries, is not good. Political and economic pressures encourage managers to increase grazing allotments beyond the carrying capacity of the range. Lack of enforcement of existing regulations and limited funds for range improvement result in depletion of plant diversity. As livestock selectively consume palatable plants, the range comes to be dominated by unpalatable or inedible species, such as sage, mesquite, cheatgrass, and cactus. Wildlife conservation groups regard cattle grazing as the most widespread form of ecosystem degradation and the greatest threat to endangered species in the southwestern United States. Some have called for a ban on cattle and sheep grazing on all public lands, arguing that those lands support only 2 percent of all livestock producers, and the soil erosion from overused lands degrades watersheds, stream valleys, recreational fisheries, and other valuable resources. The Natural Resources Defense Council reports that only 30 percent of public rangelands are in fair condition, and 55 percent are poor or very poor (fig. 12.17).

Like federal timber management policy, grazing fees charged for use of public lands often are far below market value and are an enormous hidden subsidy to western ranchers. Holders of grazing permits generally pay the government less than 25 percent of what it would cost to lease comparable private land. The 31,000 permits on federal range bring in only $11 million in grazing fees but cost $47 million per year for administration and maintenance. The $36 million difference amounts to a massive "cow welfare" system of which few people are aware.

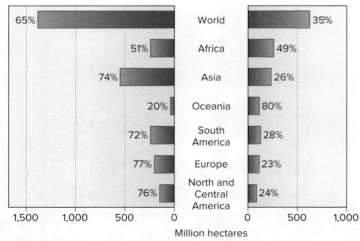

- ■ Soil degradation due to overgrazing
- ■ Soil degradation due to other causes

FIGURE 12.16 Rangeland soil degradation due to overgrazing and other causes. Notice that in Europe, Asia, and the Americas, farming, logging, mining, urbanization, and so on are responsible for about three-quarters of all soil degradation. Grazing damage is higher in Africa and Oceania, where more grazing occurs and much of the range is desert or semiarid scrub.

Source: World Resources Institute, 2004

FIGURE 12.17 Overgrazing depletes vegetation and reduces biodiversity. Exposed soils then wash away with rainfall or blow away in the wind. These cattle are grazing on dry public lands in Idaho managed by the Bureau of Land Management.

© William H. Mullins/Science Source

On the other hand, ranchers defend their way of life as an important part of western culture and history. Although few cattle go directly to market from their ranches, public lands produce almost all the beef calves subsequently shipped to feedlots. And without a viable ranch economy, they claim, even more of the western landscape would be subdivided, to the detriment of both wildlife and environmental quality. Many conservation groups agree that well-managed ranches protect wildlife habitat. What do you think? Can we protect traditional lifestyles and rural communities while also preserving natural resources?

Ranchers are experimenting with new methods

An experiment in cooperative range management is taking place in the Malpai desert (Spanish for a landscape of ancient lava fields), where Arizona, New Mexico, and Old Mexico come together (fig 12.18). In 1990, The Nature Conservancy bought the 1,300 km^2 (502 mi^2) Gray Ranch. Neighboring ranchers were worried the land would become a cattle-free wildlife refuge, further challenging an already shaky rural economy. But to many people's surprise, the Conservancy decided to turn the land into a model for scientific land use. The ranch is committed to cooperation with neighbors and to demonstrate landscape conservation.

The first area of cooperation involved fire. This desert biome needs periodic fire to suppress the growth of cactus, thorn scrub, and invasive trees, such as mesquite. But prescribed burns require coordination with neighbors. It took a lot of discussion to reach understanding and trust between people with very different backgrounds and philosophies, but eventually a Malpai Borderlands Group emerged and began to coordinate efforts.

During drought periods, western ranchers often face the choice of selling their livestock or running too many animals and degrading the land. The Nature Conservancy offered to serve as a grass bank. If the neighbors put their land into conservation easements, they could graze their cattle on the Conservancy ranch while their own lands recovered. The collaboration has worked splendidly. The health of the range has improved—not only on the Conservancy land but on neighboring ranches as well. Ranchers have cooperated in preserving endangered species. Scientific research projects have documented the role of fire in this ecosystem and its effects on wildlife. And the project now serves as a model for many other cooperative landscape conservation projects.

Rotational grazing can mimic natural regimes

Where livestock are free to roam a large pasture, they generally eat the tender, best-tasting grasses first. Tough, unpalatable species are left to flourish, and gradually these rough species dominate the pasture. To reduce pasture degradation, some farmers and ranchers are starting to practice short-term intensive grazing, which mimics wild grazers. As South African range specialist Allan Savory observed, wild grazers such as African wildebeest or American bison normally form dense herds that graze briefly but intensively in one area, then move on to another. Similarly, herd owners can practice short-duration, **rotational grazing**—confining animals to a small area for a short time (often only a day or two) before shifting them to a new location (fig. 12.19). Livestock are forced to eat everything equally, to trample the ground thoroughly, and to fertilize heavily with manure before moving. Desirable forage species are able to

FIGURE 12.18. In the Malpai borderlands, the Nature Conservancy is cooperating with neighboring ranchers to practice scientific range management on these arid lands.
© William P. Cunningham

FIGURE 12.19 Intensive, rotational grazing encloses livestock in a small area for a short time (often only one day) within a movable electric fence to force them to eat vegetation evenly and fertilize the area heavily.
© William P. Cunningham

FIGURE 12.20 Red deer (*Cervus elaphus*) are raised in New Zealand for antlers and venison.
© William P. Cunningham

compete with unpalatable weeds in this system. This approach doesn't work in all environments, but many ranchers find it a successful way to reduce damage to pastures.

Restoring fire and managing grasslands as regional units can have many benefits for both ranchers and wildlife. The Nature Conservancy has participated with private landowners in a number of innovative experiments in range restoration.

Another approach to ranching in some areas is to raise wild species, such as red deer, impala, wildebeest, or oryx (fig. 12.20). These animals forage more efficiently, resist harsh climates, often are more pest- and disease-resistant, and fend off predators better than usual domestic livestock. Native species also may have different feeding preferences and needs for water and shelter than cows, goats, or sheep. The African Sahel, for instance, can provide only enough grass to raise about 20 to 30 kg (44 to 66 lb) of beef per hectare. Ranchers can produce three times as much meat with wild native species in the same area, because these animals browse on a wider variety of plant materials.

In the United States, ranchers find that elk and bison are hardier and require less supplemental feeding than cattle or sheep. Financial returns can be high because their lean meat can bring a better market price than beef or mutton. Media mogul Ted Turner has become both the biggest private landholder in the United States and the owner of more American bison than anyone other than the government.

Section Review

1. Are pastures and rangelands always damaged by grazing animals?
2. What are some results of overgrazing?
3. What is *rotational grazing,* and how does it mimic natural processes?

12.3 PARKS AND PRESERVES

- *Many countries have set aside parks and preserves for ecological, cultural, or recreational purposes.*
- *Different levels of protection are found in nature preserves.*
- *Native people can play important roles in nature protection.*

Although most forests and grasslands serve utilitarian purposes, many nations have set aside some natural areas for ecological, cultural, or recreational purposes. Preserving land for non-extractive purposes is nothing new. Ancient Greeks and Druids, for example, protected sacred groves for religious purposes. Royal hunting grounds preserved wild areas in many countries. Although these areas were usually reserved for elite classes in society, they maintained biodiversity and natural landscapes in regions where most lands were heavily used.

The first public parks open to ordinary citizens may have been the tree-sheltered agoras in planned Greek cities. But the idea of providing natural space for recreation, or to preserve natural environments, has really developed in the past half century (fig. 12.21). According to the International Union for Conservation of Nature (IUCN), which maintains some of the best data on global protected areas, nearly 11.6 percent of the surface area of the earth is protected in some sort of park, preserve, or other natural area. This represents about 19.6 million ha (7.6 million mi^2) in 107,000 different preserves.

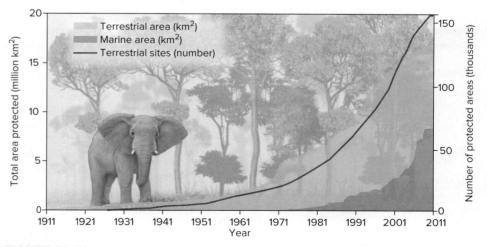

FIGURE 12.21 Growth of all protected categories worldwide, 1907–2011.
Source: UN World Commission on Protected Areas, 2012

Levels of protection vary in preserves

Preserved lands vary greatly in how much protection they receive. The IUCN identifies five categories of protection (table 12.2). In ecological reserves and wilderness areas, little or no human impact is allowed, especially where particularly sensitive wildlife or natural features are present. In some wildlife sanctuaries, visitor numbers are minimized to avoid introducing invasive species or disrupting native species. The least restrictive categories are designated for resource use and management. Most public lands in the United states, including most national forests and many areas leased for oil and gas or mining, are in this category.

Collectively, according to the World Commission on Protected Areas, Central America has 22.5 percent of its land area in some protected status (table 12.3). The Pacific region, at 1.9 percent in nature reserves, has both the lowest percentage and the second lowest total area. With land scarce on small islands, it's hard to find space to set aside for nature sanctuaries. Notice that the numbers reported in figure 12.21 and table 12.3 differ because they are based on different conservation categories.

Venezuela claims to have the highest proportion of its land area protected (70 percent) of any country in the world. About half this land is designated as preserves for indigenous people or for sustainable resource harvesting. With little formal management, there is minimal protection from poaching by hunters, loggers, and illegal gold hunters. Unfortunately, it's not uncommon in the developing world to have "paper parks" that exist only as a line drawn on a map with no budget for staff, management, or infrastructure. The United States, by contrast, has only about

Table 12.3 Area in IUCN category 1–4 protection

Region	Total Area Protected (km²)	Protected Percent	Number of Areas
North America	4,459,305	16.2	13,447
South America	1,955,420	19.3	1,456
North Eurasia	1,816,987	7.7	17,724
East Asia	1,764,648	14.0	3,265
Eastern and Southern Africa	1,696,304	14.1	4,060
Brazil	1,638,867	18.7	1,287
Australia/New Zealand	1,511,992	16.9	9,549
North Africa and Middle East	1,320,411	9.8	1,325
Western and Central Africa	1,131,153	8.7	2,604
Southeast Asia	867,186	9.6	2,689
Europe	785,012	12.4	46,194
South Asia	320,635	6.5	1,216
Central America	158,193	22.5	781
Antarctic	70,323	0.5	122
Pacific	67,502	1.9	430
Caribbean	66,210	8.2	958
Total	**19,630,148**	**11.6**	**107,107**

Source: The International Union for Conservation of Nature, 2010

Table 12.2 IUCN Categories of Protected Areas

Category	Allowed Human Impact or Intervention
1. Ecological reserves and wilderness areas	Little or none
2. National parks	Low
3. Natural monuments and archaeological sites	Low to medium
4. Habitat and wildlife management areas	Medium
5. Cultural or scenic landscapes, recreation areas	Medium to high
6. Managed resource area	High

Source: Data from World Conservation Union, 1990

15.8 percent of its land area in protected status, and less than one-third of that amount is in IUCN categories 1 or 2 (nature reserves, wilderness areas, national parks). The rest is in national forests or wildlife management zones that are designated for sustainable use. With hundreds of thousands of state and federal employees, billions of dollars in public funding, and a high level of public interest and visibility, U.S. public lands are generally well managed.

Brazil, with more than one-quarter of all the world's tropical rainforest, is especially important in biodiversity protection. Currently, Brazil has the largest total area in protected status of any country. More than 1.6 million km² or 18.7 percent of the nation's land—mostly in the Amazon basin—is in some protected status. Some of the newest of these preserves are in the northern Brazilian state of Para, where nine new protected areas were declared along the border with Suriname and Guyana. These new areas, about half of which will be strictly protected nature preserves, will link together several existing indigenous areas and nature preserves to create the largest tropical forest reserve in the world. More than 90 percent of the new 15 million ha (58,000 mi², or about the size of Illinois) Guyana Shield Corridor is in a pristine natural state. Conservation International president Russ Mittermeir says, "If any tropical rainforest on earth remains intact a century from now, it will be this portion of northern Amazonia." In contrast to this dramatic success, the Pantanal, the world's largest wetland/savanna complex, which lies in southern Brazil and is richer in some biodiversity categories than the Amazon, is almost entirely privately owned. There are efforts to set aside some of this important wetland, but so far, little is in protected status.

Some other countries with very large reserved areas include Greenland (with a 972,000 km² national park that covers most of the northern part of the island) and Saudi Arabia (with a 640,000 km² wildlife management area in its Empty Quarter). These areas are relatively easy to set aside, however, being mostly ice covered (Greenland) or desert (Saudi Arabia). Canada's Quttinirpaaq National Park on Ellesmere Island is an example of a preserve with high wilderness values but little biodiversity. Only 800 km (500 miles) from the North Pole, this remote park gets fewer than 100 human visitors per year during its brief, three-week summer season (fig. 12.22). With little evidence of human occupation, it has abundant solitude and stark beauty, but very little wildlife and almost no vegetation. By contrast, British Columbia's Great Bear Rainforest management area has a rich diversity of both marine and terrestrial life, but the valuable timber, mineral, and wildlife resources in the area make protecting it expensive and controversial.

Are protected lands representative of the world's ecosystems? Not really. Some biomes—often those that are not very useful to us—are well represented in nature preserves, while others are relatively underprotected. Figure 12.23 shows the percentage of each major biome in protected status. Not surprisingly, there's an inverse relationship between the percentage converted to human use (and where people live) and the percentage protected. Temperate grasslands and savannas (such as the American Midwest) and Mediterranean woodlands and scrub (as in Greece or southern France) are highly domesticated and therefore expensive to set aside in large areas. Temperate and boreal conifer forests (as in Siberia or Canada's vast expanse of northern forest) are relatively uninhabited, and therefore easy to put into some protected category.

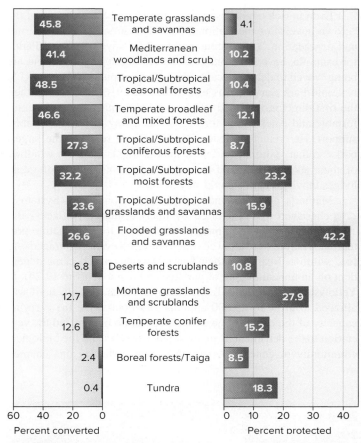

Percent converted	Biome	Percent protected
45.8	Temperate grasslands and savannas	4.1
41.4	Mediterranean woodlands and scrub	10.2
48.5	Tropical/Subtropical seasonal forests	10.4
46.6	Temperate broadleaf and mixed forests	12.1
27.3	Tropical/Subtropical coniferous forests	8.7
32.2	Tropical/Subtropical moist forests	23.2
23.6	Tropical/Subtropical grasslands and savannas	15.9
26.6	Flooded grasslands and savannas	42.2
6.8	Deserts and scrublands	10.8
12.7	Montane grasslands and scrublands	27.9
12.6	Temperate conifer forests	15.2
2.4	Boreal forests/Taiga	8.5
0.4	Tundra	18.3

FIGURE 12.23 With few exceptions, the percent of each biome converted to human use is roughly inverse to the percent protected in parks and preserves. Rock and ice, lakes, and Antarctic ecoregions are excluded. *Source: World Database on Protected Areas, 2009*

FIGURE 12.22 Canada's Quttinirpaaq National Park at the north end of Ellesmere Island offers plenty of solitude and pristine landscapes, but little biodiversity.
© William P. Cunningham

Not all preserves are preserved

Even parks and preserves designated with a high level of protection aren't always safe from exploitation or changes in political priorities. Serious problems threaten natural resources and environmental quality in many countries. In Greece, the Pindus National Park is threatened by plans to build a hydroelectric dam in the center of the park. Furthermore, excessive stock grazing and forestry exploitation in the peripheral zone are causing erosion and loss of wildlife habitat. In Colombia, dam building also threatens the Paramillo National Park. Ecuador's largest park, Yasuni National Park, which contains one of the world's most megadiverse regions of lowland Amazonian forest, has been opened to oil drilling, while miners and loggers in Peru have invaded portions of Huascaran National Park. In Palau, coral reefs identified as a potential biosphere reserve are damaged by dynamite fishing, while on some beaches in Indonesia, every egg laid by endangered sea turtles is taken by egg hunters. These are just a few of the many problems faced by parks and preserves around the world. Many of the countries with the most important biomes lack funds, trained personnel, and experience to manage the areas under their control.

Even in rich countries, such as the United States, some of the "crown jewels" of the National Park System suffer from overuse and degradation. Yellowstone and Grand Canyon National Parks, for example, have large budgets and are highly regulated, but are being "loved to death" because they are so popular. When the U.S. National Park Service was established in 1916, Stephen Mather, the first director, reasoned that he needed to make the parks comfortable and entertaining for tourists as a way of building public support. He created an extensive network of roads in the largest parks so that visitors could view famous sights from the windows of their automobiles, and he encouraged construction of grand lodges in which guests could stay in luxury.

Mather's plan was successful; the National Park System is cherished and supported by many American citizens. But sometimes entertainment seems to have trumped nature protection. Visitors were allowed—in some cases even encouraged—to feed wildlife. Bears lost their fear of humans and became dependent on an unhealthy diet of garbage and handouts (fig. 12.24). In Yellowstone and Grand Teton National Parks, the elk herd was allowed to grow to 25,000 animals, or about twice the carrying capacity of the habitat. The excess population overgrazed the vegetation to the detriment of many smaller species and the biological community in general. As we discussed earlier in this chapter,

70 years of fire suppression resulted in changes of forest composition and fuel buildup that made huge fires all but inevitable. In Yosemite, you can stay in a world-class hotel, buy a pizza, play video games, do laundry, play golf or tennis, and shop for curios, but you may find it difficult to experience the solitude or enjoy the natural beauty extolled by early conservationists as a prime reason for creating the park.

In many of the most famous parks, tremendous crowds and traffic congestion strain park resources (fig. 12.25). Some parks, such as Yosemite and Zion National Parks, have banned private automobiles from the most congested areas. Visitors must park in remote lots and ride to popular sites in clean, quiet buses that run on electricity or natural gas. Other parks are considering limits on the number of visitors admitted each day. How would you feel about a lottery system that might allow you to visit some famous parks only once in your lifetime, but to have an uncrowded, peaceful experience on your one allowed visit? Or would you prefer to be able to visit whenever you wish even if it means fighting crowds and congestion?

Originally, the great wilderness parks of Canada and the United States were distant from development and isolated from most human impacts. This has changed in many cases. Forests are clear-cut right up to some park boundaries. Mine drainage contaminates streams and groundwater. At least 13 U.S. national monuments are open to oil and gas drilling, including Texas's Padre Island, the only U.S. breeding ground for endangered Kemps Ridley sea turtles. Even in the dry desert air of the Grand Canyon, where visibility was once up to 150 km, it's often too smoggy now to see across the canyon, due to air pollution from power plants just outside the park. Snowmobiles and off-road vehicles (ORVs) create pollution and noise and cause erosion while disrupting wildlife in many parks (fig. 12.26).

FIGURE 12.24 Wild animals have always been one of the main attractions in national parks. Many people lose all common sense when interacting with big, dangerous animals. This is not a petting zoo.
© Yellowstone National Park Wildlife Photo File, American Heritage Center, University of Wyoming

FIGURE 12.25 Thousands of people wait for an eruption of Old Faithful geyser in Yellowstone National Park. Can you find the ranger who's giving a geology lecture?
© William P. Cunningham

FIGURE 12.26 Off-road vehicles cause severe, long-lasting environmental damage when driven through wetlands.
© Carl Lyttle/Getty Images RF

Chronically underfunded, the U.S. National Park System now has a maintenance backlog estimated to be at least $5 billion. Politicians from both major political parties vow during election campaigns to repair park facilities, but then find other uses for public funds once in office. Ironically, a recent study found that, on average, parks generate $4 in user fees for every $1 they receive in federal subsidies. In other words, they more than pay their own way and should have a healthy surplus if they were allowed to retain all the money they generate.

In recent years, the U.S. National Park System has begun to emphasize nature protection and environmental education over entertainment. This new agenda is being adopted by other countries as well. The IUCN has developed a **world conservation strategy** for protecting natural resources. This strategy has three objectives: (1) to maintain essential ecological processes and life-support systems (such as soil regeneration and protection, nutrient recycling, and water purification) on which human survival and development depend; (2) to preserve genetic diversity essential for breeding programs to improve cultivated plants and domestic animals; and (3) to ensure that any utilization of wild species and ecosystems is sustainable.

Marine ecosystems need greater protection

As ocean fish stocks become increasingly depleted globally (chapter 6), biologists are calling for protected areas where marine organisms are sheltered from destructive harvest methods. Research has shown that establishing no-fishing zones or marine reserves can replenish and maintain fish stocks in surrounding areas. In one study of 100 marine refuges around the world, the number of organisms inside no-fishing preserves was, on average, twice as high as in surrounding areas where fishing was allowed. The biomass of organisms in preserves was three times as great

and individual animals were, on average, 30 percent larger inside the refuge compared to outside. Closing reserves to fishing even for a few months can have beneficial results in restoring marine populations. The size necessary for a safe haven to protect flora and fauna depends on the species involved, but some marine biologists call on nations to protect at least 20 percent of their nearshore territory as marine refuges.

As we discussed in chapter 11, coral reefs are among the most threatened marine ecosystems in the world. Remote sensing surveys show that globally, living coral covers only about 285,000 km^2 (110,000 mi^2), or less than half of previous estimates. More important, 90 percent of all reefs face threats from rising sea temperatures, destructive fishing methods, coral mining, sediment runoff, and other human disturbance. Healthy coral reefs are among the richest and most biologically productive communities in existence. But when damaged, these sensitive communities can take a century or more to recover from damage. Some researchers predict that if current trends continue, in 50 years there will be no viable coral reefs anywhere in the world.

What can be done to reverse reef destruction? Marine protected areas are a critical step, and many countries are starting to establish protected or no-fishing zones (figure 12.21). Australia, for example has declared 2.3 million km^2 (890,000 mi^2) in six areas around its coastline to be marine reserves (fig. 12.27). And the Cook Islands and New Caledonia declared that 2.5 million km^2 (nearly 1 million mi^2) of their territorial waters would be managed sustainably for tourism and fishing. And an area twice as large will be managed to protect dwindling shark populations. Fourteen other small island Pacific nations, which collectively claim about

FIGURE 12.27 Australia's Great Barrier Reef is the world's largest coral reef complex. Stretching for nearly 2,000 km (1,200 mi) along Australia's northeast coast, this giant network is one of the biological wonders of the world. It is now threatened by warming oceans.
© Stephen Frink/Corbis RF

10 percent of the world's ocean area, have ambitious plans for sustainable management of their territorial waters. In 2016, President Obama expanded the Papahanaumokuakea National Monument around a chain of atolls northwest of Hawaii to 625,325 mi^2 (1.6 million km^2), making it the largest the world's largest single marine reserve.

Recently, warming oceans have become a new threat that is even more dire, global, and possibly permanent than overfishing and pollution. In Australia's vast Great Barrier Reef, heat-related coral bleaching badly damaged at least one third of the reef system in 2016. Other reefs are suffering similar This new threat makes marine preserves even more important, as safe havens for surviving reef systems.

Conservation and economic development can work together

Many of the most biologically rich communities in the world are in developing countries, especially in the tropics. These countries are the guardians of biological resources important to all of us. Unfortunately, where political and economic systems fail to provide residents with land, jobs, food, and other necessities of life, people do whatever is necessary to meet their own needs. Immediate survival takes precedence over long-term environmental goals. Clearly, the struggle to save species and ecosystems can't be divorced from the broader struggle to meet human needs.

As the opening case study for this chapter shows, residents of some developing countries are beginning to realize that their biological resources may be their most valuable assets, and that preserving those resources is vital for sustainable development. **Ecotourism** (tourism that is ecologically and socially sustainable) can be more beneficial in many places over the long term than extractive industries, such as logging and mining. The What Can You Do? box, shown later in this section, suggests some ways to ensure that your vacations are ecologically responsible.

Native people can play important roles in nature protection

The American ideal of wilderness parks untouched by humans is unrealistic in many parts of the world. As we mentioned earlier, some biological communities are so fragile that human intrusions have to be strictly limited to protect delicate natural features or particularly sensitive wildlife. In many important biomes, however, indigenous people have been present for thousands of years and have a legitimate right to pursue traditional ways of life. Furthermore, many of the approximately 5,000 indigenous or native peoples that remain today possess ecological knowledge about their ancestral homelands that can be valuable in ecosystem management. According to author Alan Durning, "encoded in indigenous languages, customs, and practices may be as much understanding of nature as is stored in the libraries of modern science."

Some countries have adopted Draconian policies to remove native people from parks (fig. 12.28). In South Africa's Kruger National Park, for example, heavily armed soldiers keep intruders

FIGURE 12.28 Some parks take Draconian measures to expel residents and prohibit trespassing. How can we reconcile the rights of local or indigenous people with the need to protect nature?
© William P. Cunningham

out with orders to shoot to kill. This is very effective in protecting wildlife. In all fairness, before this policy was instituted there was a great deal of poaching by mercenaries armed with automatic weapons. But it also means that people who were forcibly displaced from the park could be shot on sight merely for returning to their former homes to collect firewood or to hunt for rabbits and other small game. Similarly, in 2006, thousands of peasant farmers on the edge of the vast Mau Forest in Kenya's Rift Valley were forced from their homes at gunpoint by police who claimed that the land needed to be cleared to protect the country's natural resources. Critics claimed that the forced removal amounted to "ethnic cleansing" and was based on tribal politics rather than nature protection.

Other countries recognize that finding ways to integrate local human needs with the needs of nature is essential for successful conservation. This is especially important where traditional societies occupy preserve areas, whose livelihoods have long been compatible with healthy ecosystem functions. To encourage inclusion of these traditional societies, UNESCO (United Nations Educational, Scientific, and Cultural Organization) initiated in 1986 the **Man and Biosphere (MAB) program,** which encourages the designation of **biosphere reserves.** These reserves are protected areas designed to include low-impact or traditional human activities, rather than excluding them. In principle, reserves are divided into zones with different purposes. Critical ecosystem functions and endangered wildlife are protected in a central core region, where limited scientific study is the only human access allowed. Ecotourism and research facilities are located in a relatively pristine buffer zone around the core, while sustainable resource harvesting and permanent habitation are allowed in multiple-use peripheral regions (fig. 12.29). As is the case in many countries,

Being a Responsible Ecotourist

1. *Pretrip preparation.* Learn about the history, geography, ecology, and culture of the area you will visit. Understand the do's and don'ts that will keep you from violating local customs and sensibilities.

2. *Environmental impact.* Stay on designated trails and camp in established sites if available. Take only photographs and memories, and leave only goodwill wherever you go.

3. *Resource impact.* Minimize your use of scarce fuels, food, and water resources. Do you know where your wastes and garbage go?

4. *Cultural impact.* Respect the privacy and dignity of those you meet, and try to understand how you would feel in their place. Don't take photos without asking first. Be considerate of religious and cultural sites and practices. Be as aware of cultural pollution as you are of environmental pollution.

5. *Wildlife impact.* Don't harass wildlife or disturb plant life. Modern cameras make it possible to get good photos from a respectful, safe distance. Don't buy ivory, tortoise shell, animal skins, feathers, or other products taken from endangered species.

6. *Environmental benefit.* Is your trip strictly for pleasure, or will it contribute to protecting the local environment? Can you combine ecotourism with work on cleanup campaigns or delivery of educational materials or equipment to local schools or nature clubs?

7. *Advocacy and education.* Get involved in letter writing, lobbying, or educational campaigns to help protect the lands and cultures you have visited. Give talks at schools or to local clubs after you get home, to inform your friends and neighbors about what you have learned.

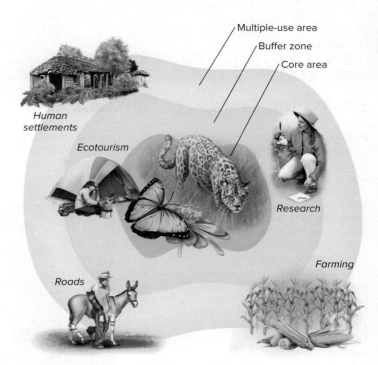

FIGURE 12.29 A model biosphere reserve. Traditional parks and wildlife refuges have well-defined boundaries to keep wildlife in and people out. Biosphere reserves, by contrast, recognize the need for people to have access to resources. Critical ecosystem is preserved in the core. Research and tourism are allowed in the buffer zone, while sustainable resource harvesting and permanent habitations are situated in the multiple-use area around the perimeter.

technology is aiding in protecting parks and involving local communities in land-use planning in Africa (see the Exploring Science box later in this section).

A well-established example of a biosphere reserve is Mexico's 545,000 ha (2,100 mi^2) Sian Ka'an Reserve on the Tulum Coast of the Yucatán. The core area includes 528,000 ha (1.3 million acres) of coral reefs, bays, wetlands, and lowland tropical forest. More than 335 bird species have been observed within the reserve, along with endangered manatees, five types of jungle cats, spider and howler monkeys, and four species of increasingly rare sea turtles. Approximately 25,000 people (about the same number who live in the Great Bear Rainforest) reside in communities and the countryside around Sian Ka'an. In addition to tourism, the economic base of the area includes lobster fishing, small-scale farming, and coconut cultivation.

The Amigos de Sian Ka'an, a local community organization, played a central role in establishing the reserve and is working to protect natural resources while also improving living standards for local people. New intensive farming techniques and sustainable harvesting of forest products enable residents to make a living without harming their ecological base. Better lobster-harvesting techniques developed at the reserve have improved the catch without depleting native stocks. Local people now see the reserve as a benefit rather than an imposition from the outside. Similar success stories from many parts of the world show how we can support local people and recognize indigenous rights while still protecting important environmental features.

Species survival can depend on preserve size and shape

Many natural parks and preserves are increasingly isolated, remnant fragments of ecosystems that once extended over large areas. As park ecosystems are shrinking, however, they are also becoming more and more important for maintaining biological diversity. Principles of landscape design and landscape structure become important in managing and restoring these shrinking islands of habitat.

For years, conservation biologists have disputed whether it is better to have a *single large* or *several small* reserves (the SLOSS debate). Ideally, a reserve should be large enough to support viable populations of endangered species, keep ecosystems intact, and isolate critical core areas from damaging external forces. For some species with small territories, several small, isolated refuges can support viable populations, and having several small reserves provides insurance against disease, habitat destruction, or other

Saving the Chimps of Gombe

How would you like to use cool technology to help save the forest habitat of the most famous population of chimps in the world? That's what Dr. Lilian Pintea does as vice president of conservation science at the Jane Goodall Institute (JGI). At JGI, Dr. Pintea integrates remote-sensing data with on-the-ground observations to study chimpanzee populations, distribution, and environment. But in addition to high-level science and international wildlife policy issues, Dr. Pintea also works with local Tanzanian villagers to develop land-use and development plans that benefit both wildlife and humans.

Lilian has an interesting history. Born in Moldova, he dreamed from an early age of working with African wildlife. A high-school research paper on snakes won him a place at Moscow State University. After postgraduate zoology studies in Bucharest, Romania, he won a Fulbright Fellowship to learn remote sensing at the University of Delaware. From there, he moved to the University of Minnesota, where he earned a PhD in conservation biology. His dissertation research used Geographic Information Systems (GIS) to map the Gombe National Park in western Tanzania.

Using the latest high-resolution satellite imagery, Dr. Pintea has created a detailed, digital map of the park to identify specific locations where Jane Goodall and her team made observations over decades of groundbreaking wildlife research. Objects as small as 0.5 m can be seen in the map, visualizing not only specific trees but streams, footpaths, and other landscape features. This digitized database is a valuable tool for analyzing geographic patterns in animal distribution, as well as determining how habitat type is related to particular chimpanzee behaviors.

In addition to carrying on Jane Goodall's pioneering research, Dr. Pintea is working with local communities on sustainable use of their natural resources. In the 50 years since Dr. Goodall started working in Gombe, population growth in villages around the park, along with an influx of refugees from war zones in nearby countries, has led to rapid deforestation and land degradation right up to park boundaries. Increasingly, the park has become an island

Dr. Lilian Pintea takes ground-based measurements in Gombe National Park.
© the Jane Goodall Institute/Lilian Pintea
www.janegoodall.org

of forest surrounded by a barren, eroding landscape. Poaching, illegal logging, and trespassing in the park are increasing problems. It isn't clear that either humans or chimps can survive over the long term if these trends continue.

About 20 years ago, JGI began working with the local communities through the Lake Tanganyika Catchment Reforestation and Education (TACARE, pronounced "take care") project to find ways to protect local natural resources on which both chimps and local people depend. The JGI staff is using its GIS expertise to help villagers develop conservation action plans. They also work with communities on health, economic development, and clean water programs to improve lives and reduce pressures on the park. For example, high-yield hybrid oil palms now produce four times as much oil from a given area as the conventional varieties. Efficient stoves and high-value crops, such as coffee and chocolate, also reduce the need for more forest clearing. Like in Brazil and Indonesia, described earlier in this chapter, this work is supported in part with funds from Norway in the REDD program.

Village land-use planning is essential for sustainability of both wildlife and human communities. GIS is used to analyze deforestation, topography, and human use patterns, and to identify forest conservation areas that can increase chimpanzee viability both inside and outside the park. These conservation areas will also help stabilize watersheds and support human livelihoods.

Recently, 13 villages within the Greater Gombe Ecosystem completed their village land-use plans, which were ratified by the Tanzanian government. Local communities have voluntarily assigned 9,690 hectares, or 26 percent, of their village voluntarily assigned lands as Village Forest Reserves. These interconnected reserves contain about two-thirds of the priority conservation area identified in the planning process. In the end, it's hoped that both wildlife and local communities will benefit from these conservation efforts. It's a dream come true for a young man from Moldova.

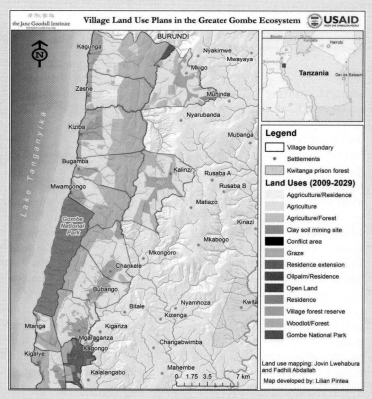

Village land-use plans for the area surrounding Gombe National Park (dark green).
© the Jane Goodall Institute/Lilian Pintea www.janegoodall.org

calamities that might wipe out a single population. But small preserves can't support species such as elephants or tigers, which need large amounts of space. Given human needs and pressures, however, big preserves aren't always possible.

One proposed solution has been to create **corridors** of natural habitat that can link multiple smaller habitat areas (fig. 12.30). Corridors could effectively create a large preserve from several small ones. Corridors could also allow populations to maintain genetic diversity or expand into new breeding territory. The effectiveness of corridors probably depends on how long and wide they are, and on how readily a species will use them.

Perhaps the most ambitious corridor project in the world today is the proposed Yellowstone to Yukon (Y2Y) corridor. Linking more than two dozen existing parks, preserves, and wilderness areas, this corridor would stretch 3,200 km (2,000 mi) from the Wind River Range in Wyoming to northern Alaska. More than half this corridor is already forested. Some 31 different First Nations and American Indian tribes occupy parts of this land, and are being consulted in ecosystem management.

One of the reasons large preserves are considered better than small preserves is that they have more **core habitat**—areas deep in the interior of a habitat area that have better conditions for specialized species than do edges. **Edge effects** is a term generally used to describe habitat edges: For example, a forest edge is usually more open, bright, and windy than a forest interior, and temperatures and humidity are more varied. For a grassland, on the other hand, edges may be wooded with more shade and perhaps more predators than in the core of the grassland area. As human disturbance fragments an ecosystem, habitat is broken into increasingly isolated islands, with less core and more edge. Small, isolated fragments of habitat often support fewer species, especially fewer rare species, than do extensive, uninterrupted ecosystems. The size and isolation of a wildlife preserve, then, may be critical to the survival of rare species.

The study of these characteristics of habitat in the landscape is known as **landscape ecology,** a science that examines the relationship between these spatial patterns and ecological processes, such as species movement or survival. Landscape ecologists explore the ways habitat size, shape, or arrangement influences the distribution—or survival—of species. Landscape ecologists also examine ecological functions of whole landscapes, including farm fields, hedgerows, and other cultivated features, rather than focusing only on pristine wilderness. By quantifying factors such as habitat extent and landscape complexity, landscape ecologists try to guide more effective design of nature preserves and parks.

A dramatic experiment in reserve size, shape, and isolation is being carried out in the Brazilian rainforest. In a project funded by the World Wildlife Fund and the Smithsonian Institution, loggers left 23 test sites when they clear-cut a forest. Test sites range from 1 ha (2.47 acres) to 10,000 ha. Clear-cuts surround some, and newly created pasture surrounds others (fig. 12.31); others remain connected to the surrounding forest. Selected species are

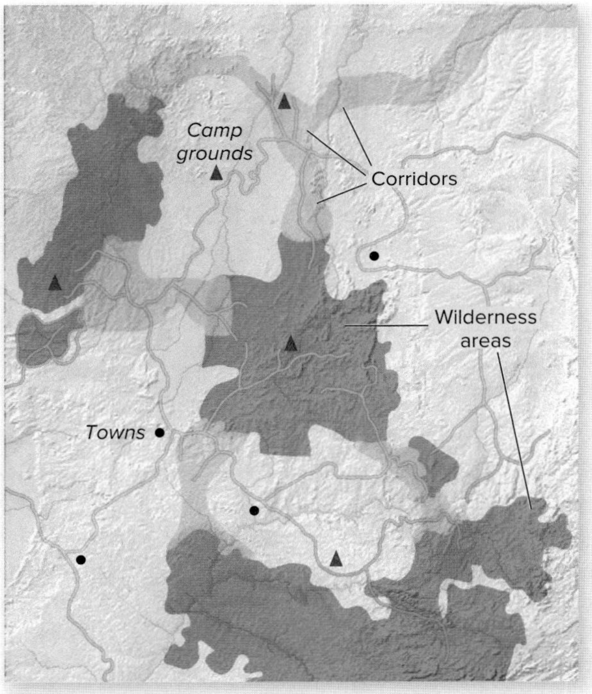

FIGURE 12.30 Corridors serve as routes of migration, linking isolated populations of plants and animals in scattered nature preserves. Although individual preserves may be too small to sustain viable populations, connecting them through river valleys and coastal corridors can facilitate interbreeding and provide an escape route if local conditions become unfavorable.

FIGURE 12.31 How small can a nature preserve be? In an ambitious research project, scientists in the Brazilian rainforest are carefully tracking wildlife in plots of various sizes, either connected to existing forests or surrounded by clear-cuts. As you might expect, the largest and most highly specialized species are the first to disappear.
Source: Courtesy of R.O. Bierregaard

regularly inventoried to monitor their survival after disturbance. As expected, some species disappear very quickly, especially from small areas. Sun-loving species flourish in the newly created forest edges, but deep-forest, shade-loving species disappear, particularly when the size or shape of a reserve reduces availability of core habitat. This experiment demonstrates the importance of maintaining core habitat in preserves.

Section Review

1. How do the size and design of nature preserves influence their effectiveness? What do landscape ecologists mean by core habitat and edge effects?
2. What is ecotourism, and why is it important?
3. What is a biosphere reserve, and how does it differ from a wilderness area or wildlife preserve?

Conclusion

Forests and grasslands cover nearly 60 percent of global land area. The vast majority of humans live in these biomes, and we obtain many valuable materials from them. And yet these biomes also are the source of much of the world's biodiversity on which we depend for life-supporting ecological services. How we can live sustainably on our natural resources while also preserving enough nature so those resources can be replenished is one of the most important questions in environmental science.

There is some good news in our search for a balance between exploitation and preservation. Although deforestation and land degradation are continuing in some developing countries, many places are more thickly forested now than they were two centuries ago. Protection of Indonesia's rainforest and Australia's Great Barrier Reef shows that we can choose to protect some biodiverse areas in spite of forces that want to exploit them. Overall, nearly 12 percent of the earth's land area is now in some sort of protected status. Although the level of protection in these preserves varies, the rapid recent increase in number and area in protected status exceeds the goals of the United Nations Millennium Project.

We haven't settled the debate between focusing on individual endangered species versus setting aside representative samples of habitat, but pursuing both strategies seems to be working. The United Nations REDD program is helping to preserve forests while also reducing global climate change. Involving indigenous people can help save landscapes while also protecting endangered cultures.

Reviewing Key Terms

Can you define the following terms in environmental science?

biosphere reserves 12.3
closed-canopy forest 12.1
core habitat 12.3
corridor 12.3
debt-for-nature-swap 12.1
desertification 12.2

ecosystem management 12.1
ecotourism 12.3
edge effects 12.3
forest 12.1
landscape ecology 12.3

Man and Biosphere (MAB) program 12.3
monoculture agroforestry 12.1
overgrazing 12.2
pastoralists 12.2

primary forest 12.1
Roadless rule 12.1
rotational grazing 12.2
savannas 12.1
world conservation strategy 12.3

Critical Thinking and Discussion Questions

1. Paper and pulp are the fastest-growing sector of the wood products market, as emerging economies of China and India catch up with the growing consumption rates of North America, Europe, and Japan. What should be done to reduce paper use?

2. Conservationists argue that watershed protection and other ecological functions of forests are more economically valuable than timber. Timber companies argue that continued production supports stable jobs and local economies. If you were a judge attempting to decide which group is right, what evidence would you need on both sides? How would you gather this evidence?

3. Divide your class into a ranching group, a conservation group, and a suburban home-builders group, and debate the protection of working ranches versus building homes or establishing nature preserves. What is the best use of the land? What landscapes are most desirable? Why? How do you propose to maintain these landscapes?

4. Calculating forest area and forest losses is complicated by the difficulty of defining exactly what constitutes a forest. Outline a definition for what counts as forest in your area, in terms of size, density, height, or other characteristics. Compare your definition to those of your colleagues. Is it easy to agree? Would your definition change if you lived in a different region?

5. Why do you suppose dry tropical forest and tundra are well represented in protected areas, while grasslands and wetlands are protected relatively rarely? Consider social, cultural, geographic, and economic reasons in your answer.

6. Oil and gas companies want to drill in a number of parks, monuments, and wildlife refuges. Do you think this should be allowed? Why or why not? Under what conditions would drilling be allowable?

7. If you were superintendent of a major national park, how would you reconcile the demand for comfort and recreation with the need to protect nature? If no one comes to your park, you will probably lose public support. But if the landscape is trashed, what's the purpose of having a park?

Data Analysis

Detecting Edge Effects

Edge effects are a fundamental consideration in nature preserves. We usually expect to find dramatic edge effects in pristine habitat with many specialized species. But you may be able to find interior/edge differences on your own college campus, or in a park or other unbuilt area near you. Here are three testable questions you can examine using your own local patch of habitat: (1) Can an edge effect be detected or not? (2) Which species will indicate the difference between edge and interior conditions? (3) At what distance can you detect a difference between edge and interior conditions? Go to Connect to find a field exercise you can do to form and test a hypothesis regarding edge effects in your own area.

TO ACCESS ADDITIONAL RESOURCES FOR THIS CHAPTER, PLEASE VISIT CONNECT AT www.connect.mheducation.com.

You will find LearnSmart, an adaptive learning system, Google Earth™ exercises, additional Case Studies, Data Analysis exercises, and an interactive ebook.

13

Restoration Ecology

▲The lower Elwha dam, before removal, blocked one of the premier salmon streams in the United States.
© Trish Drury/DanitaDelmont.com/Newscom

Learning Outcomes

After studying this chapter, you should be able to:

13.1 Illustrate ways we can help nature heal.

13.2 Explain how restoration can benefit society as well as nature.

13.3 Summarize plans to restore prairies.

13.4 Compare approaches to restoring wetlands and streams.

"When we heal the earth, we heal ourselves."

– David Orr

Restoration of the Elwha River and Its Salmon

Once an ecosystem is damaged, can we put the pieces together again? This is the question being explored on the cascading Elwha River, in Washington's Olympic National Park. For centuries, prodigious runs of six species of Pacific salmon and steelhead supported the culture and livelihood of the indigenous S'Klallam Nation. In 1910, despite tribal protests, the 33-m tall (108 ft) lower Elwha dam was built to provide electricity for local lumber mills, and 14 years later another dam, twice as high, was added at Glines Canyon, 15 km (9 miles) upstream. The two dams eliminated spawning grounds for migratory salmon and steelhead trout on most of the river. Annual runs that once included more than 400,000 fish—with some individual king salmon larger than 45 kg (100 pounds)—were reduced to just a few thousand fish clustered in the lower 8 km (5 mi) of the river. A fishery once valued at $10 million per year was largely destroyed.

Damming the river didn't just affect fish and the humans dependent on them. The entire ecosystem was diminished. Decomposing salmon carcasses provide essential nutrients for the entire aquatic and riparian systems. As much as half the forest biomass was dependent on nutrients brought in by the fish. Furthermore, the dams stopped sediment movement down the river. This starved beaches and enhanced coastal erosion. The Army Corps of Engineers has spent hundreds of millions of dollars every year to protect beaches and local harbors from erosion.

In 1968, the S'Kallam tribe and several environmental groups opposed relicensing of the dams, citing salmon losses, environmental damage, flooding of sacred tribal sites, and safety concerns about the aging structures. Furthermore, they pointed out, the dams were located within the Olympic National Park, where it's illegal to have commercial hydroelectric projects. Congressional hearings and debates dragged on for the next 24 years, but in 1992 Congress passed a law, which President George W. Bush signed, appropriating $325 million for removal of the dams and restoration of the river to suitable salmon habitat.

Still another two decades passed before deconstruction of the Elwha dam began, in 2011 (fig. 13.1). A year later demolition started at the higher Glines Canyon Dam. But how do you safely remove such large dams without releasing catastrophic mudslides from the 34 million m³ of soft sediments stored behind them? These are the largest dams ever removed in the United States, but thousands of other dams have become obsolete or are doing unacceptable ecological damage. The Elwha case is an important test of our ability to remove these dams and restore damaged river ecosystems.

The lower Elwha was structurally unsound due to age and defective construction. It wasn't safe to simply blow it up, so a diversion channel was dug around the dam to lower water levels in the reservoir behind it. Once all the water was drained away, controlled blasting destroyed the dam. The higher Glines Canyon dam was more stable, so jackhammers and diamond saws operating from a floating barge on the upstream side gradually removed large concrete chunks, which were lifted by cranes to trucks waiting above the dam. Final demolition of both dams was completed in 2013.

To control sediment and aid restoration of streamside vegetation, Park Service employees collected native seeds to revegetate and stabilize the old lakebed and stream banks. Fortunately, ground cover established quickly, and stream bank erosion was far less than had been feared. Cottonwoods grew rapidly on the newly exposed shoreline. After the first year, they were over 4 m (12 ft) tall and still growing. Invasion by non-native plants also was minimal.

Scientists are currently studying the return of small mammals, like river otters, mice, and voles in the river valley. They're also watching movement of Roosevelt elk now that these large mammals have more land to inhabit. There is also an expectation that other large creatures, such as bears, will also begin to benefit from the newly restored forest.

But will the fabulous migratory salmon runs return? Biologists are hopeful. Because the many small feeder streams that provide habitat for juvenile salmon are mostly within the National Park, they're already in good condition for reproductive success. In 2015, the best returns of fish in many years brought 4,500 adult chinook and coho salmon, as well as 1,200 steelhead trout, back to the river. In addition, several juvenile sockeye salmon were spotted in the Elwha's mouth. It's the first time this species has been reported in the Elwha's nearshore zone for nearly a century. So far, the salmon runs are only a tiny fraction of the 400,000 fish that

FIGURE 13.1 After dam demolition in 2013, the Elwha River rushes through its canyon. Restoration of the entire ecosystem remains a challenge, but salmon are already returning to the river.
© AP Photo/Peninsula Daily News-Keith Thorpe

once migrated up the river, but it's hoped that in a few years the population may be large enough for commercial fishing to resume.

Restoration ecology is a new, exciting, and experimental field that applies ecological principles to healing natural systems like the Elwha River. Full restoration of an entire ecosystem is a staggering task, but even small steps can make a remarkable difference. And we're finding that nature can be more resilient and robust than we might imagine. In this chapter, we'll examine a number of other cases in which people are working to repair damage and rehabilitate, remediate, or restore ecosystems.

13.1 HELPING NATURE HEAL

- *A strict definition of restoration is to return a biological community as nearly as possible to a pristine, predisturbance condition.*

- *A broader definition of restoration is simply to develop a self-sustaining, useful ecosystem with as many of its original elements as possible.*

- *Restoration often is fraught with philosophical questions about what we want to restore, as well as pragmatic ones about how to do it.*

Humans have disturbed nature as long as we've existed. With the availability of industrial technology, however, our impacts have increased dramatically in both scope and severity. We've chopped down forests, plowed the prairies, slaughtered wildlife, filled in wetlands, and polluted air and water. Our greatly enhanced power also makes it possible, however, to repair some of the damage we've caused. The relatively new field of **ecological restoration** attempts to do this based on both good science and pragmatic approaches. Some see this as a new era in conservation history. As we discussed in chapter 1, the earliest phases in conservation and nature protection consisted of efforts to use resources sustainably and to protect special places from degradation.

Now, thousands of projects are under way to restore or rehabilitate nature (fig. 13.2). These range from individual efforts to plant native vegetation in urban yards or parks to huge efforts to restore millions of hectares of prairie or continent-size forests. Examples we'll see in this chapter include proposals to re-create a vast Buffalo Commons, something like what Lewis and Clark saw when they crossed the North American Great Plains in 1804. Undoubtedly, the biggest reforestation project in history is the Chinese effort to create a "green wall" to hold back the encroaching Gobi and Taklamakan Deserts. More than a billion people are reported to have planted 50 billion trees in China over the past 30 years.

The success of these projects varies. In some cases, the land and biota are so degraded that restoration is impossible. In other situations, some sort of ecosystem can be re-created, but the available soil, water, nutrients, topography, or genetic diversity limit what can be done on a particular site. Figure 13.3 presents a schematic overview of restoration options. Note that restoration to an original pristine condition is rarely possible. Most often, the best choice is a compromise between ideal goals and pragmatic, achievable goals. Choosing these goals is one of the central questions in restoration ecology.

Restoration projects range from modest to ambitious

Management goals are described in a variety of ways. Table 13.1 summarizes some of the most common terms employed in ecological restoration. A strict definition of restoration would be to return a biological community as nearly as possible to a pristine,

FIGURE 13.2 People in many places are working to restore native vegetation and to protect endangered species. These scouts are replanting the banks of a trout stream.
© *William P. Cunningham*

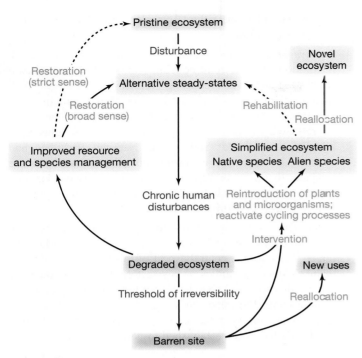

FIGURE 13.3 A model of ecosystem degradation and potential management options.

Source: Data from Walker and Moral, 2003

Table 13.1 A Restoration Glossary
Some commonly used terms in restoration ecology:
• **Restoration:** (strict sense) to return a biological community to its predisturbance structure and function.
• **Restoration:** (broad sense) to reverse degradation and reestablish some aspects of an ecosystem that previously existed on a site.
• **Rehabilitation:** to rebuild a community to a useful, functioning state but not necessarily its original condition.
• **Intervention:** to apply techniques to discourage or reduce undesired organisms and favor or promote desired species.
• **Reallocation:** to use a site (and its resources) to create a new and different kind of biological community rather than the existing one.
• **Remediation:** to clean chemical contaminants from a polluted area using relatively mild or nondestructive methods.
• **Reclamation:** to use powerful chemical or physical methods to clean and repair severely degraded or even barren sites.
• **Re-creation:** to construct an entirely new ecosystem on a severely degraded site.
• **Mitigation:** to replace a degraded site with one of more or less equal ecological value somewhere else.

predisturbance condition. Often this isn't possible. A broader definition of restoration has a more pragmatic goal simply to develop a self-sustaining, useful ecosystem with as many of its original elements as possible. *Rehabilitation* may seek only to repair ecosystem functions. This may take the form of a community generally similar to the original one on a site, or it may aim for an entirely different community that can carry out the desired functions. Sometimes it's enough to leave nature alone to heal itself, but often we need to intervene in some way to remove or discourage unwanted organisms while also promoting the growth of more desirable species. *Reintroduction* generally implies transplanting organisms from some external source (often a nursery or hatchery where native species are grown under controlled conditions) to a site where they have been reduced or eliminated.

Remediation uses chemical, physical, or biological methods to remove pollution, generally with the intention of causing as little disruption as possible. *Reclamation* employs stronger, more extreme techniques to clean up severe pollution or create a newly functioning ecosystem on a seriously degraded or barren site. *Mitigation* implies compensation for destroying a site by purchasing or creating one of more or less equal ecological value somewhere else.

In some cases, invasion by exotic species coupled with physical changes caused by human actions bring about a state that restoration ecologists call a **novel ecosystem** (see upper right corner of fig. 13.3). This new hybrid biological community may be difficult or impossible to restore to its original historical state. Weedy, invasive species may be the only ones able to survive

under new conditions. Perhaps we should learn to appreciate those invasives if they're supplying ecological services on which we or remaining populations of native species depend. A drastic intervention to try to revert to an earlier state may only doom remaining endangered species. Take the example we discussed in chapter 11 of attempts to eliminate introduced Tamarisk (or salt cedar, *Tamarix sp*) shrubs in the American Southwest. It turns out that Tamarisk plays a valuable role in stabilizing riverbanks and providing habitat for endangered species, such as the southwestern willow flycatcher.

Restore to what?

Restoration ecologists work in the real world, and although they may dream of returning a disturbed site to its untouched state, they have to deal with the constraints of a specific place, usually faced with multiple obstacles and limited budgets. Some restoration ecologists find it useful to express their goals in terms of ecosystem health or integrity, but others regard this as merely expressions of values or ideology rather than rigorous science. Often, the target of restoration is a matter of debate. In the Nature Conservancy's Hassayampa River Preserve near Phoenix, for instance, pollen grains preserved in sediments reveal that 1,000 years ago the area was a grassy marsh that was unique in the surrounding desert landscape. Some ecologists would try to rebuild a similar marsh. Corn pollen in the same sediments, however, show that 500 years ago Native Americans began farming the marsh. Which is more important, the natural or the early agricultural landscape?

It may be impossible to return to conditions of either 500 or 1,000 years ago because climate change may have made prehistoric communities incompatible with current conditions. In this case, should restoration ecologists attempt to restore areas to what they used to be, or create a community that will be more compatible with future conditions? Restoration is fraught with such philosophical questions, as well as pragmatic ones.

All restoration projects involve some common components

Different types of restoration involve separate challenges. Wetland restoration, for example, involves a variety of efforts at reestablishing hydrologic connections, plant and animal diversity, weed removal, and sometimes salinity control. Forest restoration, on the other hand, requires a different set of steps that may include controlled burning, selective cutting, weed removal, control of deer and other herbivores, and so on. Even so, there are at least five main components of restoration, and most efforts share a majority of these activities.

1. *Removing physical stressors.* The first step in most restoration efforts is to remove the cause of degradation or habitat loss. Physical stressors such as pollutants, vehicle traffic, or inadequate water supply may need to be corrected. In wetland restoration, water flow and storage usually must be corrected before other steps, such as replanting, can proceed. In forest restoration, clear-cut practices might be replaced with selective logging. In prairie restoration, cultivation might be ended so that land can be replanted with grassland plants, which then provide habitat for grassland butterflies, beetles, and birds.

2. *Controlling invasive species.* Often a few aggressive, "weedy" species suppress the growth of other plants or animals. These invasives may be considered biotic stressors. Removing invasive species can be extremely difficult, but without removal subsequent steps often fail. Some invasives can be controlled by introducing pests that eat only those species: for example, purple loosestrife (*Lythrum salicaria*) is a wetland invader that has succeeded in North America partly because it lacks predators. Loosestrife beetles (*Galerucella* sp.), introduced after careful testing to ensure that their introduction wouldn't lead to further invasions, have successfully set back loosestrife populations in many areas. Leafy spurge (*Euphorbia esula*), similarly, has invaded and blanketed vast areas of the Great Plains. Insects, including flea beetles (*Aphthona* sp.), have been a successful strategy to reduce the spread of spurge in many areas, allowing native grassland plants to compete again.

3. *Replanting.* Restoring a site or ecosystem usually involves some replanting of native plant species. Often, restorationists try to collect seeds from nearby sources (so that the plants will be genetically similar to the original plants of the area), and then they grow them in a greenhouse before transplanting to the restoration site. In some cases, restoration ecologists can encourage existing plants to grow by removing other plants that outcompete the target plants for space, nutrients, sunshine, or moisture.

4. *Captive breeding and reestablishing fauna.* In some cases, restoration involves reintroducing animals. Peregrine falcon restoration involved releasing captive-bred birds, which then managed to survive and nest on their own. In some situations, invertebrates, such as butterflies or beetles, may be released. Sometimes a top predator is reintroduced, or allowed to reinvade. Yellowstone National Park has had a 20-year experimental restoration of wolves. Evidence indicates that these predators are reducing excessive deer and elk populations, thus helping to restore vegetation, as well as reduce mid-level carnivores such as coyotes.

5. *Monitoring.* Without before-and-after monitoring, restoration ecologists cannot know if their efforts are working as hoped. Therefore, a central aspect of this science is planned, detailed, ongoing studies of key factors. Repeated counts of species diversity and abundance can tell whether biodiversity is improving. Repeated measures of water quality, salinity, temperature, or other factors can indicate whether suitable conditions have been established for target species.

Origins of restoration

As European settlers spread across North America in the eighteenth and nineteenth centuries, the woods, prairies, and wildlife populations seemed vast—much too large to be affected by anything humans could do. As we discussed in chapter 1, however, a few pioneers recognized that the rapid destruction of natural communities was unsustainable.

The most influential American forester was Gifford Pinchot. His first job after graduating from college was to manage the wealthy Vanderbilt family's Biltmore estate in North Carolina. Working on this private estate, Pinchot introduced a system of selective harvest and replanting of choice tree species that increased the value of the forest while also producing a sustainable harvest. Pinchot went on to become the first head of the U.S. Forest Service, where he became the first in U.S. history to promote resource management, including the replanting of forests, based on principles of long-term use and scientific research.

Before Pinchot, companies had practiced "cut and run" logging—devastating a region, then moving on to another without any reclamation or replanting, leaving barren expanses of stumps where once had been primeval forests. Pinchot's concepts of science-based management were a vast improvement over these practices. Subsequently, critics have complained that his policies have led to replanting of only commercially valuable trees, rather than ecologically important ones, and to a disregard for other ecological functions of a forest, such as protecting water resources and habitat. But Gifford Pinchot's ideas led to the first widespread restoration practices in American forests.

The person most often recognized as the pioneer of restoration ecology was Aldo Leopold. Born at the end of the nineteenth century, Leopold grew up hunting and fishing. He became one of the first generation of professional wildlife managers, in a time when wildlife habitat was disappearing rapidly across North America. Leopold believed that these changes should be turned around. In 1935, Leopold bought a small, worn-out farm on the banks of the Wisconsin River not far from his home in Madison (fig. 13.4). Originally intended to be merely a hunting camp, it quickly became a year-round retreat from the city, as well as a laboratory in which Leopold could test his theories about conservation, game management, and land restoration. The whole Leopold family participated in planting as many as 6,000 trees each spring. "I have read many definitions of what is a conservationist, and written not a few myself," Leopold wrote, "but I suspect that the best one is written not with a pen, but with an axe. It is a matter of what a man thinks about while chopping, or while deciding what to chop. A conservationist is one who is humbly aware that with each stroke he is writing his signature on the face of his land . . . A land ethic then, reflects the existence of an ecological conscience, and this in turn reflects a conviction of individual responsibility for the health of the land . . . Health is the capacity of the land for self-renewal. Conservation is our effort to understand and preserve this capacity."

Many modern ecologists now regard goals of restoring the health, beauty, stability, or integrity of nature as unscientific, but they generally view Aldo Leopold as a visionary pioneer and an important figure in conservation history.

Sometimes we can simply let nature heal itself

As is the case in restoring the Elwha River, the first step in conservation and ecological restoration is generally to stop whatever is causing damage. Sometimes this is all that's necessary. Nature

FIGURE 13.4 Aldo Leopold's Sand County farm in central Wisconsin served as a refuge from the city and as a laboratory to test theories about land conservation, environmental ethics, and ecologically based land management.
© William P. Cunningham

has amazing regenerative power. If the damage hasn't passed a threshold of irreversibility, natural successional processes can often rebuild a diverse, stable, interconnected biological community, given enough time.

The first official wildlife refuge established in the United States was Pelican Island, a small sand spit in the Indian River estuary not far from present-day Cape Canaveral. The island was recognized by ornithologists in the mid-1800s as being especially rich in bird life. Pelicans, terns, egrets, and other wading birds nested in huge, noisy colonies. But other people discovered the abundant birds as well. Boatloads of tourists slaughtered birds just for fun, and professional hunters shot thousands of adults during the breeding season, when the birds' plumage was most beautiful, leaving fledglings to starve to death. In 1900, the American Ornithological Union, worried about the wanton destruction of colonies, raised private funds to hire wardens to protect the birds during the breeding season. And in 1903, President Theodore Roosevelt signed an executive order establishing Pelican Island as America's first National Bird Reservation. This set an important precedent that the government could set aside public land for conservation. Pelican Island also became the first of 51 wildlife refuges created by President Roosevelt that may have saved species, such as roseate spoonbills and snowy egrets, from extinction.

Many of the forests and nature preserves described in chapter 12 suffered some degree of degradation before being granted protected status. Often, simply prohibiting logging, mining, or excessive burning is enough to allow nature to heal itself. Consider the forests of New England, for example. When the first Europeans arrived in America, New England was mostly densely forested. As settlers spread across the land, they felled the forest to create pastures and farm fields. In 1811, sheep were introduced in New England, and sheep farming expanded rapidly to provide wool to the mills in New Hampshire and Massachusetts. Just 30 years later, in 1840, Vermont had nearly 2 million sheep and 80 percent of its forests were gone. But competition soon ended this boom in wool.

In 1825, the Erie Canal opened access to western farmlands, which could raise both crops and sheep better than the cold Vermont hills. Eli Whitney invented the cotton gin, and cotton became a cheap and abundant (and more comfortable) alternative to woolen cloth. Within a few decades, most Vermont farmers had abandoned large-scale sheep farming, and abandoned pastures reverted to forest. Today, 80 percent of the land in Vermont is once again forested, and less than 20 percent is farmland (fig. 13.5).

After a century of natural succession, much of this forest has reacquired many characteristics of old-growth forest: A mixture of native species of different sizes and ages gives the forest diversity and complexity; many of the original animal species—moose, bear, bobcats, pine martins—have become reestablished. There are reports of lynx, mountain lions, and even wolves migrating south from Canada. Now, Vermont law requires that before woodlot owners log their land, they consult with a professional forester to develop a plan to sustain the biodiversity and quality of their forest.

FIGURE 13.5 A mosaic of cropland, pasture, and sugar bush clothes the hills of Vermont. Two hundred years ago, this area was 80 percent cropland and only 20 percent forest. Today, that ratio is reversed as the forests have invaded abandoned fields. Many of these forests are reaching late successional stages and are reestablishing ecological associations characteristic of old-growth forests.
© William P. Cunningham

Native species often need help to become reestablished

Sometimes rebuilding populations of native plants and animals is a simple process of restocking breeding individuals. In other cases, however, it's more difficult. Recovery of a unique indigenous seabird in Bermuda is an inspiring conservation story that gives us hope for other threatened and endangered species.

When Spanish explorers discovered Bermuda early in the fifteenth century, they were frightened by the eerie nocturnal screeching they thought came from ghosts or devils. In fact, the cries were those of extremely abundant ground-nesting seabirds now known as the hook-billed petrel, or Bermuda cahow (*Pterodroma cahow*), endemic to the island archipelago (fig. 13.6*a*). It's thought that there may originally have been half a million of these small, agile, gadfly petrels. Colonists soon found that the birds were easy to catch and good to eat. Those overlooked by humans were quickly devoured by the hogs, rats, and cats that accompanied settlers. Although Bermuda holds the distinction of having passed the first conservation laws in the New World, protecting native birds as early as 1616, the cahow was thought to be extinct by the mid-1600s.

In 1951, however, scientists found 18 nesting pairs of cahows on several small islands in Bermuda's main harbor. A protection and recovery program was begun immediately, including establishment of a sanctuary on 6-hectare (15-acre) Nonsuch Island, which has become an excellent example of ecological restoration.

Nonsuch was a near desert after centuries of abuse, neglect, and habitat destruction. All the native flora and fauna were gone, along with most of the soil in which the cahows once dug nesting burrows. This was a case of re-creating nature rather than merely protecting what was left. Sanctuary superintendent David Wingate, who devoted his entire professional career to this project, brought about a remarkable transformation of the island (fig. 13.6*b*).

The first step in restoration was to remove invasive species and reintroduce native vegetation. Wingate and many volunteers trapped and poisoned pigs and rats and other predators that threatened both wildlife and native vegetation. They uprooted millions of exotic plants and replanted native species, including mangroves and Bermuda cedars (*Juniperus bermudiana*). Initial progress was slow as trees struggled to get a foothold; once the forest knit itself into a dense thicket that deflected salt spray and ocean winds, however, the natural community began

(a) A Bermuda cahow.

(b) David Wingate examines a cahow nest.

(c) Long-tall excluder at the mouth of the burrow.

FIGURE 13.6 (a) For more than three centuries, the Bermuda cahow, or hook-billed petrel, was thought to be extinct until a few birds were discovered nesting on small islets in the Bermuda harbor. David Wingate (b) devoted his entire career to restoring cahows and their habitat. (c) A key step in this project was to build artificial burrows. A long-tail excluder device (a board with a hole just the size of the cahow) keeps other birds out of the burrow. A round cement lump at the back end of the burrow can be removed to view the nesting cahows.
© William P. Cunningham

to reestablish itself. As was the case on New Zealand's islands (chapter 11), on Nonsuch Island native plants that hadn't been seen for decades began to reappear. Once the rats and pigs that ate seedlings were removed, and competition from weedy invasives was eliminated, native seeds that had lain dormant in the soil began to germinate again.

Still, there wasn't enough soil for cahows to dig the underground burrows they need for nesting. Wingate's crews built artificial cement burrows for the birds. Each pair of cahows lays only one egg per year, and only about half the young survive under ideal conditions. It takes 8 to 10 years for fledglings to mature, giving the species a low reproductive potential. They also compete poorly against the more common long-tailed tropic birds that steal nesting sites and destroy cahow eggs and fledglings. Wingate designed wooden baffles for the burrow entrances, with holes just large enough for cahows but too small for the larger tropic birds (fig. 13.6c). The round cement cap at the back of the burrow can be removed to monitor the nesting cahows.

It takes constant surveillance to eradicate exotic plant species that continue to invade the sanctuary. Rats, cats, and toxic toads also swim from the mainland and must be removed regularly. By 2002, however, the cahow population had rebounded to about 200 individuals with 60 breeding pairs. Hurricane Fabian destroyed many nesting burrows on smaller islets in 2003. Fortunately, the restored native forest on Nonsuch Island withstood the winds and preserved the rebuilt soil. The larger island is now being repopulated with chicks, their translocation timed so they will imprint on their new home. Reestablishing a viable population of cahows (which are now Bermuda's national bird) has had the added benefit of rebuilding an entire biological community (fig. 13.7).

FIGURE 13.7 This brackish pond on Nonsuch Island is a reconstructed wetland. Note the Bermuda cedars on the shore and the mangroves planted in the center of the pond by David Wingate.
© William P. Cunningham

It's too early to know if the cahow population is large enough to be stable over the long term, but the progress to date is encouraging. Perhaps more important than rebuilding this single species is that Nonsuch has become a living museum of precolonial Bermuda that benefits many species besides its most famous resident. It is a heartening example of what can be done with vision, patience, and a great deal of hard work.

There are many other notable reintroduction programs. Once-rare peregrine falcons have been reestablished in the eastern United States, even in the "canyons" of Chicago and New York. California condors have been reintroduced and have begun to breed in the American West. Both of these species were locally extinct 30 years ago, but captive breeding programs produced enough birds to repopulate much of their former range. Similarly, Arabian Oryx have been successfully reestablished in the deserts of Saudi Arabia, and in Hawaii the endemic nene geese were raised in captivity and reintroduced to Volcano National Park, where they exist in a small but self-reproducing population (see fig. 11.19).

Section Review

1. Define restoration (strict and broad sense), rehabilitation, intervention, reallocation, remediation, reclamation, re-creation, and mitigation.
2. Describe five common components of restoration projects.
3. What species was the main restoration focus on Nonsuch Island in Bermuda?

13.2 RESTORATION IS GOOD FOR HUMAN ECONOMIES AND CULTURES

- *In most places, the largest restoration projects ever attempted have been reforestation of cut-over or degraded forest lands.*
- *Fire is an important restoration tool.*
- *You can help restore and improve the quality of your local environment.*

Restoration has become a cornerstone of managing economic resources and a source of cultural pride, not just an altruistic ecological activity. In the United States, the largest restoration projects ever attempted have been reforestation of cut-over or degraded forest lands. Building on the policies of Gifford Pinchot, lumber companies routinely replant forests they have harvested to prepare a future crop. Seedlings are grown in huge nurseries, and tractor-drawn planters allow a team of workers to plant thousands of new trees per day (fig. 13.8). Usually, this mechanical reforestation results in a monoculture of uniformly spaced trees.

These plantings are designed to produce wood quickly, but they have little resemblance to diverse, complex native forests (fig. 13.9). Still, these commercial forests supply ground cover, provide habitat for some wildlife species, and grow valuable lumber or paper pulp. As we saw in chapter 12, a recent United

FIGURE 13.8 Mechanical planters can plant thousands of trees per day.
© Still Pictures/Robert Harding

FIGURE 13.9 Monoculture forests, such as this tree plantation in Austria, often have far less biodiversity than natural forests.
© PhotoLink/Getty Images RF

Nations survey of world forests found that many countries are more thickly forested now than they were 200 years ago. Both the total biomass and the quality of the forests in most of these countries have increased as forests have been protected and replanted over the past two centuries.

Japan, for example, was almost completely deforested at the end of World War II. In the 75 years since then, Japan has carried out a massive reforestation program. Now, more than two-thirds of the land is forest-covered. Tight restrictions on logging help preserve this forest, which has great cultural value for the Japanese people. Rather than cut their own forest, Japan buys wood from its neighbors (a policy that has drawn some criticism from ecologists and human rights groups).

In 2001, at the Ninth UN Forum on Forests, the African country of Rwanda, torn by civil war and genocide in the 1990s, announced an ambitious program for country-wide restoration of its degraded forest, soil, water, and wildlife resources over the next 25 years. Poor forest management, damaging land-use practices, and war have caused the country's forests to shrink rapidly. Despite brisk economic growth in the past five years, 85 percent of the population still makes a living from subsistence farming on degraded lands. Among the new Forest Landscape Restoration Initiative priorities will be safeguarding the nation's rich wildlife, such as the critically endangered mountain gorilla, which is an important tourist attraction. For help in this project, Rwanda has turned to the International Union for Conservation of Nature (IUCN), the world's oldest and largest global environmental network. "This really is a good news story," said Stewart Maginnis, Director of Environment and Development of the IUCN. "For the first time, we're actually seeing a country recognize that part and parcel of its economic development trajectory has to be rooted in natural resources." If successful, this will be the first nationwide conservation effort in Africa, and one of very few in the world.

Tree planting can improve our quality of life

Planting trees within cities can be effective in improving air quality, providing shade, and making urban environments more pleasant. Figure 13.10 shows student volunteers planting native trees and bushes in a project called Greening the Great River. Over the

FIGURE 13.10 Student volunteers plant native trees and shrubs to create an urban forest in the Mississippi River corridor within Minneapolis and St. Paul, Minnesota. This provides wildlife habitat, as well as beautifies the urban landscape.
© William P. Cunningham

past two decades, this nonprofit organization has mobilized more than 30,000 volunteers to plant nearly half a million native trees and shrubs on vacant land within the Mississippi River corridor as it winds through Minneapolis and St. Paul, Minnesota. This project helps beautify the cities, reduces global warming, and provides habitat for wildlife. Deer, fox, raccoons, coyotes, bobcats, and even an occasional wolf or cougar are now seen in the newly revegetated area.

In 2007, the United Nations announced a "billion tree campaign" in the hope of gathering pledges to plant one billion new trees around the world. Everyone can participate in this global reforestation effort. If you look around, there's sure to be some space in your yard (or that of your friends or relatives) or an abandoned piece of land in your neighborhood that could house a tree. Most states have tree farms that will provide seedlings at little or no cost. The American Forestry Association has done an extensive remote sensing survey of the United States. They estimate that the national urban tree deficit now stands at more than 634 million trees. They suggest that everyone has a duty to plant at least one tree every year to help restore our environment (see the What Can You Do? box later in this section).

The billion-tree campaign is inspired by the work of Nobel Peace Prize laureate Wangari Maathai. As chapter 5 notes, Dr. Maathai founded the Kenyan Green Belt Movement in 1976 as a way of controlling erosion, providing fodder and food, and empowering women. This network of more than 600 local women's groups from throughout Kenya has planted more than 30 million trees while mobilizing communities for self-determination, justice, equality, poverty reduction, and environmental conservation.

Fire is often an important restoration tool

Controlled or **prescribed burning,** also known as hazard reduction burning, is setting fires under carefully controlled conditions to remove brush and flammable material from a forest or grassland. Native people used fire to open up forests and to improve grazing lands long before European colonization. Land managers now recognize that prescribed burning can reduce the threat of catastrophic conflagrations and can benefit fire-adapted biological communities.

For example, oak savannas once covered a broad band at the border between the prairies of the American Great Plains and the eastern deciduous forest (see fig 13.14). Millions of hectares of

What Can You Do?

Ecological Restoration in Your Own Neighborhood

Everyone can participate in restoring and improving the quality of their local environment.

1. *Pick up litter*
   - With your friends or fellow students, designate one day to carry a bag and pick up litter as you go. A small amount of collective effort will make your surroundings cleaner and more attractive, but also will draw attention to the local environment. Try establishing a competition to see who can pick up the most. You may be surprised how a simple act can influence others. Working together can also be fun!
   - Join a group in your community or on your campus that conducts cleanup projects. If you look around, you'll likely find groups doing stream cleanups, park cleanups, and other group projects. If you can't find such a group, you can volunteer to organize an event for a local park board, campus organization, fraternity, or other group.

2. *Remove invasive species*
   - Educate yourself on what exotic species are a problem in your area: In your environmental science class, assign an invasive species to each student and do short class presentations to educate each other on why these species are a problem and how they can be controlled.
   - Do your own invasive species removal: In your yard, your parents' yard, or on campus, you can do restoration by pulling

weeds. (Make sure you know what you're pulling!) Find a local group that organizes invasive species removal projects. Park boards, wildlife refuges, nature preserves, and organizations such as the Nature Conservancy frequently have volunteer opportunities to help eradicate invasive species.

3. *Replant native species*
   - Once the invasives are eliminated, volunteers are needed to replant native species. Many parks and clubs use volunteers to gather seeds from existing native prairie or wetlands, and to help plant seeds or seedlings. Many of the parks and natural areas you enjoy have benefited from such volunteer labor in the past.
   - Create your own native prairie, wetland, or forest restoration project in your own yard, if you have access to available space. You can learn a great deal and have fun with such a project.

4. *Plant a tree or a garden*
   - Everyone ought to plant at least one tree in their life. It's both repaying a debt to nature and a gratifying experience to see a seedling you've planted grow to a mature tree. If you don't have a yard of your own, ask your neighbors and friends if they would like a tree planted.
   - Plant a garden, or join a community garden. Gardening is good for your mind, your body, and your environment. Flowering or fruiting plants provide habitat for butterflies and pollinating insects. Especially if you live in a city, small fragments of garden can provide critical patches of green space and habitat—even a windowsill garden or potted tomatoes are a good start. You might find your neighbors appreciate the effort, too!

the American Midwest had parklike savannas, oak openings, or oak barrens. Although definitions vary, an oak **savanna** is a forest with scattered "open-grown" trees where the canopy covers 10 to 50 percent of the area and the dappled sunlight reaching the ground supports a variety of grasses and flowering plants (fig. 13.11). The most common tree species is bur oak (*Quercus macrocarpa*), but other species, including white and red oak, also occur in savannas. Due to reduced sunlight, however, some dominant prairie species, such as big and little bluestem grasses, most goldenrods, and asters, are generally missing from savannas.

Because people found them attractive places to live, most oak savannas were converted to agriculture or degraded by logging, overgrazing, and fire suppression. Wisconsin, for example, which probably had the greatest amount of savanna of any Midwestern state (at least 2 million ha) before European colonization, now has less than 200 ha (less than 0.01 percent of its original area) of high-quality savanna. Throughout North America, oak savanna rivals the tallgrass prairie as one of the rarest and most endangered plant communities.

It's difficult to restore, or even maintain, authentic oak savannas. Fire was historically important in controlling vegetation. Before settlement, periodic fires swept in from the prairies and removed shrubs and most trees. Mature oaks, however, have thick bark that allows them to survive low-intensity fires. Grazing by bison and elk may also have helped keep the savanna open. When settlers eliminated fire and grazing by native animals, savannas were invaded by a jumble of shrub and tree growth. Unfortunately, simply resuming occasional burning doesn't result in a high-quality savanna. If fires burn too frequently, they kill oak saplings—which are much more vulnerable than mature trees—and prevent forest regeneration. Excessive grazing can do the same thing. If fires are too infrequent, on the other hand, shrubs, dense tree cover, and dead wood accumulate so that when fire does occur, it will be so hot that it kills mature oaks as well as invading species. It's often necessary to clear most of the accumulated vegetation and fuel before starting fires if you want to maintain a savanna. Herbicide treatment may be necessary to prevent regrowth of invasive species until native vegetation can become established.

The Somme Prairie Grove in Cook County, Illinois, is one of the largest and oldest efforts to restore a native oak savanna. The land was purchased by the City of Chicago in the 1920s, and restoration activities began about 50 years later. A complex of wetland, prairie, and forest, much of the upland area was an oak savanna. Although the area was used as pasture for more than a century, most of it was never plowed. Limited grazing probably helped keep brush at bay and preserved the parklike character of the remaining forest. Spring burning was started in 1981 in an effort to preserve and restore the savanna. Burning alone, it was discovered, wasn't enough to eliminate aggressive invasive species, such as glossy buckthorn (*Rhamnus frangula*).

After several years of burning, the ground was mostly bare except for weedy species. In 1985, a more intensive management program was begun. Invasive trees were removed. Seeds of native species were collected from nearby areas and broadcast under the oaks. Weeds thought to be a threat to the restoration were reduced by pulling and scything; these included garlic mustard (*Alliaria officinalis*), burdock (*Arctium minus*), briers (*Rubus* sp.), and tall goldenrod (*Solidago altissima*). A research program monitored the results of this restoration effort. Four transects were established and repeatedly sampled for a wide variety of biota (trees, shrubs, herbs, cryptogams, invertebrates, birds, and small mammals) over a six-year period. Results from this survey are shown in fig. 13.12.

As you can see in this figure, the biodiversity of the forest increased significantly over the six years of sampling. Native

FIGURE 13.11 Oak savannas are parklike forests where the tree canopy covers 10 to 50 percent of the area, and the ground is carpeted with prairie grasses and flowers. This biome once covered a broad swath between the open prairies of the Great Plains and the dense deciduous forest of eastern North America.
© William P. Cunningham

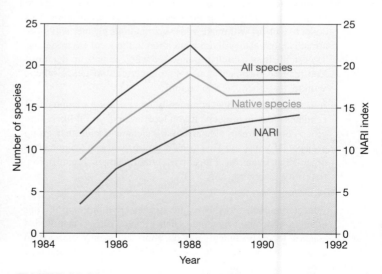

FIGURE 13.12 Biodiversity survey of the Somme Savanna restoration program. NARI, the Natural Area Rating Index, measures the frequency of native species associated with a high-quality community.
Source: Data from S. Packard and J. Balaban, 1994

species increased more than total biodiversity. NARI, the Natural Area Rating Index, is a measure of the relative abundance of native species characteristic of high-quality natural communities. Notice that this index shows the greatest increase of any of the measurements in the study. It also continues to increase after 1988, when native species began to replace invasive pioneer species. This suggests that the restoration efforts have been successful, resulting in an increasingly high-quality oak savanna.

Fire is now recognized as a key factor in preserving oak savannas and many other forest types. The Superior National Forest, which manages the Boundary Waters Canoe Area Wilderness in northern Minnesota, for example, has started an ambitious program of prescribed fires to maintain the complex mosaic of mixed conifers and hardwoods in the forest. Pioneering work in the 1960s by forest ecologist M. L. Heinselman showed that the mixture of ages and species in this forest was maintained primarily by burning. Any given area in the forest experiences a low-intensity ground fire about once per decade, on average, and an intense crown fire about once per century. As these fires burned randomly across the forest, they produced the complex forest we see today (fig. 13.13). But to maintain this forest, we need to allow natural fires to burn once again and to reintroduce prescribed fires where necessary to control fuel buildup.

Similarly, many national parks now recognize the necessity for fire to maintain forests. In Sequoia National Park, for example, rangers recognized that giant sequoias have survived for thousands of years because their thick bark is highly fire-resistant, and they shed lower branches so that fire can't get up into their crown. Seventy years of fire suppression, however, allowed a dense undergrowth to crowd around the base of the sequoias. These smaller trees provide a "ladder" for fire to climb up into the sequoia and kill them. Fuel removal and periodic prescribed fires are now a regular management tool for protecting the giant trees.

Section Review

1. What is a prescribed fire?
2. What are oak savannas, and why are they difficult to restore?
3. Why are giant sequoias now threatened by fire?

13.3 RESTORING PRAIRIES

- *Fire is also crucial for prairie restoration and maintenance.*
- *Depopulation of the American Great Plains opens up the possibility of restoring a buffalo commons.*
- *Keystone animals, such as bison, help maintain prairies.*

Before European settlement, prairies covered most of the middle third of what is now the United States (fig. 13.14). The eastern edge of the Great Plains was covered by tallgrass prairies where big bluestem (*Andropogon gerardii*) reached heights of 2 m (6 ft). Their roots could extend more than 4 m into the soil and formed a dense, carbon-rich sod. This prairie has almost entirely disappeared, having been plowed and converted to corn and soybean fields. Less than 2 percent of the original 1 million km^2 (400,000 mi^2) of tallgrass prairie remains in its original condition. In Iowa, for example, which once was almost entirely covered by this biome, the largest

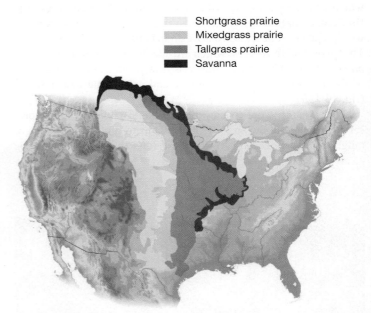

- Shortgrass prairie
- Mixedgrass prairie
- Tallgrass prairie
- Savanna

FIGURE 13.14 The eastern edge of the Great Plains was covered by tallgrass prairie, where some grasses reached heights of 2 m (6 ft) and had roots more than 4 m long that formed a dense, carbon-rich sod. Less than 2 percent of the original 10 million km^2 (400,000 mi^2) of tallgrass prairie remains in its original condition. The middle of the Great Plains contained a mixed prairie with both bunch and sod-forming grasses. Few grasses in this region grew to heights of more than 1 m. The westernmost region of the Great Plains, in the rain shadow of the Rocky Mountains, had a shortgrass prairie where sparse bunch grasses rarely grew to more than 30 or 40 cm tall.

FIGURE 13.13 The conifer-hardwood mixture and complex matrix of ages and species in the Great Lakes Forest is dependent on regular but random fire. Maintaining this forest requires prescribed burning.
© William P. Cunningham

sample of unplowed tallgrass prairie is only about 80 ha (200 acres), about the extent of many college campuses. The middle of the Great Plains contained a mixed prairie with both bunch and sod-forming grasses. The drier climate west of the 100th meridian meant that few grasses grew to heights of more than 1 m. The westernmost band of the Great Plains, in the rain shadow of the Rocky Mountains, received only 10 to 12 inches (25 to 30 cm) of rain per year. In this shortgrass region, the sparse bunch grasses rarely grew to more than 30 or 40 cm tall.

Like the oak savanna immediately to the east, these prairies were maintained by frequent fires and grazing by bison, elk, and other native wildlife. Native American people understood the role of fire in regenerating the prairie. They frequently set fires to provide good hunting grounds and to ease travel.

Fire is also crucial for prairie restoration

One of the earliest attempts to restore native prairie occurred at the University of Wisconsin. Starting in 1934, Aldo Leopold and others worked to re-create a tallgrass prairie on an abandoned farm field at the University Arboretum in Madison. Student volunteers and workers from the Civilian Conservation Corps gathered seed from remnant prairies along railroad rights-of-way and in pioneer cemeteries, and then hand-planted and cultivated them (fig. 13.15a). Prairie plants initially had difficulty getting established and competing against exotics, until it was recognized that fire is an essential part of this ecosystem. Fire not only kills many weedy species, but it also removes nutrients (especially nitrogen). This gives native species, which are adapted to low-nitrogen soils, an advantage. The Curtis Prairie is now an outstanding example of the tallgrass prairie community and a valuable research site (fig. 13.15b).

The Nature Conservancy (TNC) has established many preserves throughout the eastern Great Plains to protect fragments of tallgrass prairie. The biggest of these (and the largest remaining fragment of this once widespread biome) is in Oklahoma, where TNC has purchased or obtained conservation easements on about 18,000 ha (45,000 acres) of land just northwest of Tulsa that was grazed but never plowed. In cooperation with the University of Tulsa, TNC has established the Tallgrass Prairie Ecological Research Station, which is carrying out a number of experimental ecological restoration projects on both public and private land. Approximately three dozen prescribed burns are conducted each year, totaling 15,000 to 20,000 acres (fig. 13.16). Since 1991, over 350 randomly selected prescribed burns have been conducted, totaling 210,000 acres. About one-third of each pasture is burned each year. About half of the burns are done in spring and the rest in late summer.

In addition, 300 bison were reintroduced to the preserve in the early 1990s. In 2016, the herd numbered about 2,700 animals on about 25,000 acres (10,900 ha). Patch burning and bison grazing create a habitat that supports the diverse group of plants and animals that make up the tallgrass prairie ecosystem. So far, more than three dozen research projects are active on the preserve, and 78 reports from these studies have been published in scientific journals.

The second largest example of this biome in the United States is the Tallgrass Prairie National Preserve in the Flint Hills region of Kansas. Because the land in this area is too rocky for agriculture, the land was grazed but never plowed. Most of this preserve was purchased originally by the Nature Conservancy, but it is now being managed by the National Park Service. Like its neighbor to the south, a part of this preserve—known as the Konza Prairie—is reserved for scientific research. A long-term ecological research (LTER) program, funded by the National Science Foundation and managed by Kansas State University, sponsors a wide variety of basic scientific research and applied ecological restoration experiments. Bison reintroduction and varied fire regimes are being studied at this prairie. The role of grasslands in carbon sequestration and responses of this biological community to climate change are of special interest.

(a)

(b)

FIGURE 13.15 In 1934, workers from the Civilian Conservation Corps dug up old farm fields and planted native prairie seeds for the University of Wisconsin's Curtis Prairie (a). The restored prairie (b) has taught ecologists much about ecological succession and restoration.
Source: a) Courtesy of University Wisconsin-Madison Archives and, b) Source: Courtesy University of Wisconsin Arboretum

FIGURE 13.16 A burn-crew technician sets a back fire to control the prescribed fire that will restore a native prairie.

Source: Courtesy Minnesota Department of Natural Resources

Huge areas of shortgrass prairie are being preserved

Much of the middle, or mixed grass, section of the Great Plains has been converted to crop fields irrigated by water from the Ogallala Aquifer (chapter 17). As this fossil water is being used up, it may become impossible to continue farming over much of the area using current techniques. Many areas have declining populations (fig. 13.17). As farms and ranches fade away, the small towns that once supplied them also dry up. A great deal of worry and debate exist about the future of this region of the country. Some people argue that the best use for the land may be to restore it to the original prairie, as much as possible.

There are two competing approaches to saving shortgrass prairie: with people or without them. Working for conservation that includes people, the Nature Conservancy (TNC) is cooperating with ranchers on joint conservation and restoration programs. On the 24,000 ha (60,000 acre) Matador ranch in Montana, which TNC bought in 2000, 13 neighboring ranchers graze cattle in exchange for specific conservation measures on their home range. The ranchers have agreed, for instance, to protect about 900 ha of prairie dog colonies and sage grouse leks (dancing grounds). All the ranchers have also agreed to control weeds, resulting in almost 120,000 ha of weed-free range (fig. 13.18). Together with the Conservancy, ranchers are experimenting with fire to improve the prairie, and they plan and manage a grass-banking arrangement in which ranchers access more Conservancy land in drought emergencies. In one case, TNC deeded one of its ranches to a young couple who promised to manage it sustainably and to protect some rare wetlands on the property. The Conservancy believes that keeping ranch families on the land is the best way to preserve both the social fabric and the biological resources of the Plains.

Near the Matador land, another group is pursuing a very different strategy for preserving the buffalo commons. The American Prairie Foundation (APF), which is closely linked to the World Wildlife Fund, has also bought about 24,000 ha of former ranchland. Rather than keep it in cattle production, however, this group intends to pull out fences, eliminate all the ranch buildings, and turn the land back into wilderness. Ultimately, the APF hopes to create a reserve of at least 1.5 million ha in the Missouri Breaks region between the Charles M. Russell National Wildlife Refuge and the Fort Belknap Indian Reservation (fig. 13.19). The APF plans to reintroduce native wildlife, including elk, bison, wolves, and grizzly bears, to its lands. Neighboring ranchers don't mind

FIGURE 13.17 Much of the Great Plains is being depopulated as farms and ranches are abandoned and small towns disappear. This provides an opportunity to restore a buffalo commons much like that discovered two centuries ago by Lewis and Clark.

© *William P. Cunningham*

FIGURE 13.18 Millions of hectares of shortgrass prairie are being converted to nature preserves. Some may be populated by bison and other wildlife rather than cattle, as envisioned by artist George Catlin in 1832.

© *William P. Cunningham*

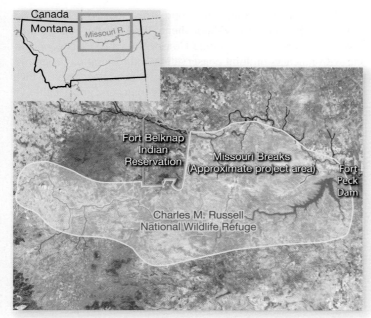

FIGURE 13.19 Both the Nature Conservancy and the American Prairie Foundation have bought large tracts of land in the Missouri Breaks between the Charles M. Russell National Wildlife Refuge and the Fort Belknap Indian Reservation. Ultimately, conservationists hope to protect and restore as much as 2 million ha of shortgrass prairie.

having elk or bison nearby, but they object to reintroducing wolves and grizzlies, predators their parents and grandparents exterminated a century ago. Ranchers also bristle at the funding for the APF project. Many of the project's large donations come from Wall Street or California's Silicon Valley. Wealthy individuals, such as media mogul Ted Turner, are using their money to make striking changes in how western land is used. Often locals, who struggle just to stay on the land, resent outsiders coming in and competing to buy the range.

With donations of more than $12 million so far, and fundraising goals of at least $100 million, the APF points out that it isn't forcing anyone from the land. The group is only buying from ranchers who want to sell. Locals worry, nonetheless, that the land will be restricted for hunting and other uses. The APF says it will allow tourism, bird-watching, and hunting on nearly all its land. Small towns also are anxious about how it will affect the local economy to take land out of production. The APF points out that without federal subsidies the average return on ranchland in the Missouri Breaks is less than $5 per acre. Tourism and hunting already bring in more income per acre than does raising cows.

Whether conservation should include human residents and economic activities has often been a deep divide among conservationists, who otherwise agree on the need to preserve biodiversity and ecosystems. Should people expect to maintain an economy in a preserve? How much should those economies be subsidized? Should the ecosystem be restored, predators and all? Or is it complete enough without predators? How would you decide these questions if you were in charge?

Bison help maintain prairies

American prairies coevolved with grazing animals. In particular, a keystone species for the Great Plains was the American buffalo (*Bison bison*). Perhaps 60 million of these huge, shaggy animals once roamed the plains from the Rocky Mountains to the edge of the eastern deciduous forest and from Manitoba to Texas. By 1900, there were probably fewer than 150 wild bison left in the United States, mostly in Yellowstone National Park. Wildlife protection and breeding programs have rebuilt the population to about 500,000 animals, but probably less than 4 percent of them are genetically pure.

Like fire, the bison's intensive grazing helped maintain native plant species. When put on open range, domestic cattle graze selectively on the species they prefer, giving noxious weeds a selective advantage. Bison, on the other hand, tend to move in dense herds, eating almost everything in their path and heavily fertilizing the area over which they pass (fig. 13.20). Their trampling and intense grazing disturb the ground and provide habitat for pioneer species, many of which disappear when bison are removed. Bison also create areas for primary succession by digging out wallows in which they take dust baths.

Having grazed an area heavily, bison will tend to move on, and if they have enough space to roam in, they won't come back for several years. This pattern of intensive, short-duration grazing and localized fertilization creates a mosaic of different successional stages that enhances biodiversity. This is one reason supporting the idea of rotational grazing in sustainable livestock management. Bison increase plant productivity by increasing the availability of light and reducing water stress, both of which increase photosynthesis rates.

Grazing also affects the nutrient cycling in prairie ecosystems. Nitrogen and phosphorus are essential for plant productivity.

FIGURE 13.20 Bison can be an important tool in prairie restoration. Their trampling and intense grazing disturb the ground and provide an opening for pioneer species. The buffalo chips they leave behind fertilize the soil and help the successional process. Where there were only about two dozen wild bison a century ago, there are now more than 400,000 on ranches and preserves across the Great Plains.
© Fuse/Getty Images RF

By consuming plant biomass, bison return these nutrients to the soils in urine and buffalo chips. Bison are more efficient nutrient recyclers than the slow release from plant litter decay. Fire releases nitrogen by burning plant material. Bison, on the other hand, limit nitrogen loss by reducing the aboveground plant biomass and increasing the patchiness of the fire. These changes in nutrient cycling and availability in prairie ecosystems lead to increased plant productivity and species composition.

Many ranchers are coming to recognize the benefits of raising native animals. Where cows require shelter from harsh weather and lots of water to drink, bison are well adapted to the harsh conditions of the prairie. Bison meat is lean and flavorful. It brings a higher price than beef in the marketplace. It's estimated that there are now about 400,000 bison on ranches and farms in North America. This is probably less than 1 percent of the original population, but it's 10,000 times more than the tiny remnant left after the wanton slaughter in the nineteenth century.

Section Review

1. Why are fires essential for prairies?
2. Where did prairies exist in North America before European settlement?
3. Why are bison beneficial for prairies?

13.4 RESTORING WETLANDS AND STREAMS

- *Wetlands and streams provide essential ecological services, but they've been abused and degraded by drainage and filling.*
- *Some major wetlands such as the Iraq marshes, Gulf of Mexico wetlands, and the American Everglades are being restored.*
- *Many streams need rebuilding.*

Wetlands, rivers, and streams provide important ecological services. They play irreplaceable roles in the hydrologic cycle. They also are often highly productive, and provide food and habitat for a wide variety of species. Louisiana's freshwater swamps and coastal marshes (fig. 13.21) play important roles in both biological and human communities. Although wetlands currently occupy less than 5 percent of the land in the United States, the Fish and Wildlife Service estimates that one-third of all endangered species spend at least part of their lives in wetlands. Coastal wetlands are vital for absorbing storm surges. Storage of floodwaters in wetlands is worth an estimated $3 billion to $4 billion per year. Wetlands also improve water quality by acting as natural water purification systems, removing silt and absorbing nutrients and toxins.

Recognition of these services has come only recently, though. For many years, wetlands were considered disagreeable, dangerous, and useless. This attitude was reflected in public policies, such as the U.S. Swamp Lands Act of 1850, which allowed individuals to buy swamps and marshes for as little as 10 cents per acre. Until recently, federal, state, and local governments encouraged wetland drainage and filling to create land for development. In an effort to boost crop production, the government paid farmers to ditch and tile millions of acres of wet meadows and potholes.

FIGURE 13.21 A Louisiana cypress swamp of the sort being threatened by coastal erosion and salt infiltration. Wetlands provide habitat for a wide variety of species, and play irreplaceable ecological roles.
Source: U.S. Army Corps of Engineers

Building levees and channelizing the Mississippi River sped flood water and the sediment it carried out into the Gulf of Mexico. Soil that would have once built up the delta and nourished coastal wetlands was lost into the depths of the Gulf. Land subsidence and erosion from wave action currently result in the loss of about 75 km^2(18,500 acres) per year in Louisiana alone. Furthermore, cutting channels for oil and gas exploration and shortcuts for navigation through Gulf Coast marshes (fig. 13.22) has caused additional destruction of wetlands that once protected New Orleans from hurricanes.

FIGURE 13.22 Louisiana has nearly 40 percent of the remaining coastal wetlands in the United States, but diversion of river sediments that once replenished these marshes and swamps, together with channels and boat wakes, are causing the Gulf shoreline to retreat about 4 m (13.8 ft) per year.
© Chris Harris

Altogether, the draining and filling of wetlands has destroyed millions of hectares of wetlands in the United States. In Iowa, for example, 99 percent of the original wetlands have disappeared since European settlement. The 1972 Clean Water Act began protecting streams and wetlands by requiring discharge permits for dumping waste into surface waters. In 1977, federal courts interpreted this rule to prohibit both pollution and filling of wetlands (but not drainage). The 1985 Farm Bill went farther with a "swamp buster" provision that blocked agricultural subsidies to farmers who drain, fill, or damage wetlands. Many states now have "no net loss" wetlands policies. Between 1998 and 2004, the United States had a net gain of about 80,000 ha (197,000 acres) of wetlands. This total area concealed an imbalance, however. Continued losses of 210,000 ha of swamp and marsh wetlands were offset by a net gain of 290,000 ha of small ponds and shallow-water wetlands (which are easy to construct and good for duck production). And since 2004, another 700,000 ha have been added to the national total. This is a good start, but the United States needs much more wetland restoration to undo decades of damage. In this section, we'll look at some efforts to accomplish this goal.

Restoring water and sediment flows help wetlands heal

As is the case with other biological communities, sometimes all that's needed is to stop the destructive forces. For wetlands, this often means simply to restore water and sediment supplies that have been diverted elsewhere.

To restore Gulf Coast wetlands, gaps are being cut in the levees below New Orleans, allowing fresh, muddy Mississippi River water to reenter the wetlands. Sediment deposited as the water slows is rebuilding eroded marshes, and fresh water encourages revegetation. In a few experimental restoration areas, such as the Caernarvon Diversion east of New Orleans, simply replenishing freshwater flow has reduced saltwater intrusion and already helped expand the area of healthy wetland.

Much of the restoration of the Louisiana wetlands depends on an assumption that controlling water flow and sediment deposition will result in the restoration of a healthy biological community. This may turn out to be true, but monitoring is needed to evaluate results. How do we know whether restoration is working or not? (See the Exploring Science box further on in this section.)

One of the benefits of removing dams, such as those on the Elwha in the opening case study for this chapter, is replenishment of beaches. In many coastal areas of the United States, as well as elsewhere in the world, damming rivers has trapped sediment upstream that would once have been carried down to the sea to form deltas, beaches, and barrier islands. Coastal currents, wave action, and rising sea levels all are causing erosion that depletes beaches. Some communities pay millions of dollars annually to dredge sediment from the ocean, or to truck in sand from elsewhere. A more natural solution is to restore stream sediment flow.

In the first two years after the Elwha dams were removed, about 4 million cubic yards (nearly 10 million metric tons) of sediment was carried to the river mouth where it has formed about 70 acres (300,000 m2) of new delta, estuary, and beach. This additional habitat has benefited crabs, clams, sea birds, fish, and other coastal creatures. It also provides new recreational opportunities for human neighbors.

Another simple physical solution to wetland degradation can be seen in the American Midwest. When the U.S. Army Corps of Engineers built 26 locks and dams on the upper Mississippi River to facilitate barge traffic in the 1950s, it created a series of large impoundments. Sediment from the surrounding farm fields began to fill these pools. This might have created a valuable network of wetlands, but waves and currents created by wind, floods, and river traffic kept the sediment constantly roiled up so that wetland plants couldn't take root. The result has been a series of wide, shallow, semisolid mud puddles that are too thick for fish but too liquid for vegetation. To remedy this situation, the Army Corps is experimenting with wing dams (see the thin white lines connecting islands in fig. 13.23) that separate the backwaters from the main river channel. It's hoped that these dams will allow the mud to solidify and turn into marshlands.

Replumbing the Everglades is one of the costliest restoration efforts ever

A huge, expensive restoration is now under way in the Florida Everglades (fig. 13.24). Famously described as a "river of grass," the Everglades covers nearly four million acres (1.6 million ha)

FIGURE 13.23 Dams on the upper Mississippi River have created a series of large lakes. Currents created by wind, floods, and river traffic keep the soft sediment stirred up so that wetland plants can't take root. To restore these backwaters, the Army Corps is experimenting with wing dams (see the thin white lines between islands) that will allow the mud to solidify and turn into marshlands.
Source: U.S. Army Corps of Engineers

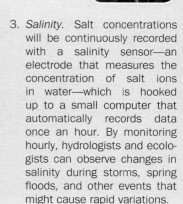

Measuring Restoration Success

The science of restoration is complex, with few simple answers. Restoration is also highly optimistic, because it often addresses environmental problems that are huge and persistent. But increasingly we recognize that ecological restoration is a necessity if we are to preserve economies, cultures, and ways of life.

Restoring Louisiana's coastal wetlands has been discussed for half a century, but the projects have gained a new urgency since 2005, when Hurricane Katrina flooded New Orleans. Restoring this vast system is almost an inconceivably large project, but without restoration Louisiana will continue to lose communities, roads, and economic activity. Hundreds of small projects have been planned, and some are in progress. A prominent project is the Caernarvon Diversion, a series of structures built on the south bank of the Mississippi east of New Orleans. At a cost of $4.5 million, the project is using culverts (1.25-m diameter pipes) to divert water from the Mississippi, together with fill to plug or block abandoned gas-field canals. These plugs slow the escape of river water.

Student volunteers sample plant biomass and species composition to evaluate the health of a coastal wetland.
© Mary Ann Cunningham

Engineering this project involved a decade of planning, but monitoring is the most extensive part of the project. In fact, monitoring is a key aspect of all restoration. Otherwise, how would we know if the project has been successful—and if it's worth doing again?

In addition to the project area, two "reference areas" are also being monitored. Ideally, the project area will improve considerably over the reference areas, and over the baseline conditions before the project started. The project plan outlines a 20-year monitoring, with three central concerns:

1. *Ratio of land to open water.* Land and water are being mapped using aerial photos. Using a GIS, the monitoring team can calculate the amount of land and water in 2000, and then again in 2006 and in 2018. In theory, the amount of land will increase from one time period to the next. With a GIS, they can also overlay one year's map on top of another year's map. By subtracting one layer from the other, analysts identify not only the amount of change but which areas have changed.

2. *Plant species composition and relative abundance.* Before water diversions began, ecologists designated a series of square plots of wetland, 2 m on a side. These 4 m² plots were placed in reference areas, as well as in the project area. Plant ecologists then visited each plot to list all species in each one. They also estimated the relative abundance of the different species. By leaving permanent markers at the plots, they can revisit the same location years later to monitor change. By sampling a number of replicate plots, they can get a sense of aggregate change in the area.

3. *Salinity.* Salt concentrations will be continuously recorded with a salinity sensor—an electrode that measures the concentration of salt ions in water—which is hooked up to a small computer that automatically records data once an hour. By monitoring hourly, hydrologists and ecologists can observe changes in salinity during storms, spring floods, and other events that might cause rapid variations.

To more clearly organize and structure the data collection and interpretation, scientists framed specific hypotheses (testable statements). For example, one goal is to increase the abundance and diversity of plants. A testable hypothesis is: "After the project implementation, diversity will be significantly greater than before project implementation." By gathering data before and after, the restoration team can test this hypothesis by comparing the average number of plant species, or the average abundance of each species, in the different plots. A simple yes or no answer will diagnose whether the project is working as planned.

Similarly, reducing salinity is a goal. A testable hypothesis is: "After project implementation, salinity will be significantly less than before project implementation." Again, this hypothesis can be tested by comparing before-project samples and after-project samples for mean salinity value and variation from the average.

Restoration often relies on inputs from a variety of sciences, including hydrology, ecology, geology, and other fields. Restoration is also experimental, partly because it is a new science, and much remains uncertain in restoration projects. But restoration is an extremely important—and exciting—field within environmental science.

FIGURE 13.24 The Florida Everglades, often described as a "river of grass," is threatened by water pollution and diversion projects.
© Jeff Greenberg/Science Source

across southern Florida. This vast marshland is created by a slow-moving sheet of water, which can be 80 km (50 mi) wide, but only a few centimeters deep. This river starts in springs near Orlando in the center of the state, and then moves through Lake Okeechobee and flows southward to the Gulf (fig. 13.25). Spreading over most of the southern tip of the peninsula, this landscape supports myriad fish, invertebrates, birds, alligators, and the rare Florida panther.

Farmers found the rich, black muckland of the Everglades could grow fantastic crops if it was drained. Ditching and diverting the water started more than a century ago. A series of floods that threatened the wealthy coastal cities also triggered a demand for more water management. Once-meandering rivers were straightened to shunt surplus water out to sea. Altogether, the Army Corps of Engineers built more than 1,600 km of canals,

1,000 km of levees, and 200 water-control structures to intercept normal water flow, drain farmlands, and divert floodwater (fig. 13.26). Ironically, many of the cities that demanded the water be diverted are now experiencing water shortages during the dry season. Water that might have been stored in the natural wetlands is no longer available. Nature, also, is suffering from water shortages. The Everglades National Park has lost 90 percent of its wading birds, and there are worries that the entire aquatic ecosystem may be collapsing.

After decades of debate and acrimony, the various stakeholders finally agreed in 2000 on a massive reengineering of the south Florida water system. The plan aimed to return some water to the Everglades, yet retain control to prevent flooding. More than 400 km of levees and canals would be removed. New reservoirs would store water currently lost to the ocean, and 500 million liters of water per day would be pumped into underground aquifers for later release. Rivers already are being dechannelized to restore natural meanders that store storm water and provide wildlife habitat (fig. 13.26). It's hoped that simply restoring some of the former flow to the Everglades will allow the biological community to recover, although whether it will be that easy remains to be seen. Altogether, this project is expected to cost at least $8 billion, making it one of the most expensive restoration projects ever undertaken.

Although announced with great fanfare, the restoration project was almost immediately derailed by an extraordinary lobbying blitz from the sugar industry—the largest polluter in the Everglades and one of the largest political donors in the state. Specifically, the industry opposed requirements to pay for cleaning up pollution. The original plan was to reduce phosphate

FIGURE 13.26 The naturally meandering Kissimmee River (*right channel*) was straightened by the Army Corps of Engineers (*left*) for flood control 30 years ago. Now the Corps is attempting to reverse its actions and restore the Kissimmee and its associated wetlands to their original state.
Source: South Florida Water Management District

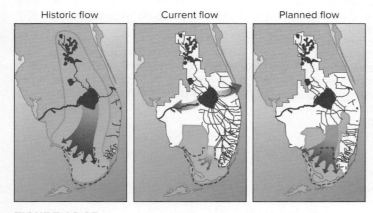

Historic flow   Current flow   Planned flow

FIGURE 13.25 Planned results of the Comprehensive Everglades Restoration Plan. The red dashed outline shows the national park boundary.

pollution, and for major polluters to pay for the effort. However, a law passed by the Florida legislature, and signed in 2004 by Governor Jeb Bush, pushed back the phosphorus deadlines to 2025 and required the public to pay for it. The Federal Government warned that the bill was so egregious that it would derail the whole project, but Governor Bush and the legislature refused to budge.

Over the past decade, a number of small projects have been undertaken, but the results have been modest. Delays, corruption, lack of political will, and inflated costs have held back progress. In 2012, however, federal and state regulators, after long negotiations, settled on a revamped plan to clean up the polluted water. Public officials say that they are confident that the Everglades are now back on track, but environmentalists are doubtful. Whether restoration will ever be successful remains to be seen. This case shows how large-scale restoration projects involving big financial interests often are complicated by politics.

Restoration of the Chesapeake Bay is another long, expensive, and contentious project (see chapter 3). Volunteers are replanting 200,000 ha of coastline, salt marsh, and shallow water areas (fig. 13.27).

Wetland mitigation is challenging

Working in large, complex, highly political ecosystems that have been damaged by a large variety of human actions is especially challenging. Smaller ecosystems can be much easier to restore or replace. An encouraging example is the re-creation of **prairie potholes** in the Great Plains. Before European settlement, millions of these shallow ponds and grassy wetlands once spread across the prairies from Alberta, Saskatchewan, and Manitoba, to northeastern Kansas and western Iowa (fig. 13.28). They produced more than half of all North American migratory waterfowl and mediated flooding by storing enormous amounts of water.

FIGURE 13.27 Volunteers plant dune grasses as part of Chesapeake Bay restoration. The large cylinder is a "bio log" made of biodegradable material that will help stabilize the shore.
Source: NOAA

FIGURE 13.28 Millions of prairie potholes once covered the Great Plains from Iowa and South Dakota, north to Canada's Prairie Provinces. More than half of these shallow ponds have been drained for agricultural crop production, but now many are being replaced.
Source: Photo by Don Poggensee, USDA Natural Resources Conservation Service

As agriculture expanded across the Plains, however, at least half of all prairie wetlands were drained and filled. In the 1900s, Canada's Prairie Provinces had about 10 million potholes; by 1964, an estimated 7 million were gone. The United States had similar losses.

Passage of the Migratory Bird Hunting Stamp Act (popularly known as the Duck Stamp Act) in 1934 marked a turning point in wetland conservation. Over the past 70 years, this program has collected $700 million that has been used to acquire more than 5.2 million acres of habitat for the National Wildlife Refuge System. In 1934, when the Duck Stamp Act was passed, drought, predators, and habitat destruction had reduced migratory duck populations in the United States to about 26 million birds. John Phillips and Frederick Lincoln, in a 1930 report on the state of American waterfowl, wrote, "We believe it soon will be too late to save [wild-fowl] in numbers sufficient to be of any real importance for recreation in the future." By 2006, however, conservation efforts, a wetter climate, and habitat restoration had increased duck production to about 44 million birds. Duck stamps now provide about $25 million annually for wildlife conservation.

While it's relatively easy to dig new ponds, it's much more difficult to replace other wetland types. As we mentioned earlier in this chapter, between 1998 and 2004 the United States lost 210,000 ha of swamp and marshes. This was offset by repair or construction of 290,000 ha of small ponds and shallow-water wetlands, such as restored prairie potholes. Replacing a damaged wetland with a substitute is called **wetland mitigation.** It's required whenever development destroys a natural wetland. In

FIGURE 13.29 In the past, developers weren't required to plant native vegetation in wetland mitigation. They were allowed to simply dig a hole and wait for it to fill with rainwater and invasive species. New rules now require a more ecological approach.
© William P. Cunningham

the past, this process often didn't necessarily replace the native species and ecological functions represented by the original biological community. Figure 13.29, for example, shows a replacement wetland created by a housing developer in Minnesota. In building a housing project, the developer destroyed about 10 ha of a complex, native wetland that contained rare native orchids and several scarce sedge species. To compensate for this loss, the developer simply dug a hole and waited for it to fill with rainwater. He wasn't required to replant wetland species. The law assumes that natural succession will revegetate the disturbed area. In fact, it was soon revegetated, but entirely with exotic invasive species.

Constructed wetlands can filter water

Many cities are finding that artificial wetlands provide a low-cost way to filter and treat sewage effluent. Arcata, California, for instance, needed an expensive sewer plant upgrade. Instead, the city transformed a 65-ha garbage dump into a series of ponds and marshes that serve as a simple, low-cost waste treatment facility. Arcata saved millions of dollars and improved its environment simultaneously. The marsh is a haven for wildlife and has become a prized recreation area for the city. Eventually, the purified water flows into Humbolt Bay, where marine life flourishes. Similarly, small pools and rain gardens—shallow pits lined with porous surface material and planted with water-tolerant vegetation—are used to collect storm runoff and allow it to seep into the ground rather than run into rivers or lakes. And constructed marshes allow industrial cooling water to equilibrate before entering streams or other surface water bodies. All these created wetlands can be useful to both humans and wildlife. For more on this topic, see chapter 18.

Many streams need rebuilding

Pollution, pathogens and diseases, industrial toxins, invasive organisms, erosion, and a host of other factors degrade streams and rivers. The United States has more than 5.6 million km of rivers and streams. In a 1994 EPA survey of nearly 1 million km of rivers and streams, only 56 percent fully supported multiple uses, including drinking-water supply, fish and wildlife habitat, recreation, and agriculture, as well as flood prevention and erosion control. Sedimentation and excess nutrients were the most significant causes of degradation in the remaining 44 percent. Presumably, these results could be extrapolated to the rest of the nation's waterways. Given these statistics, the need for stream restoration is obvious.

One response to erosion and flooding in urban streams has been to turn them into cement channels that rush rainwater off into some larger body of water (fig. 13.30) or to bury them in underground culverts. The result is an artificial system with little resemblance to the living biological community that once made up the stream. Some cities have come to recognize, however, that natural streams can increase property values and improve the livability of the urban environment. Buried streams are being "daylighted," and channelized ditches are being turned back into living biological communities.

A variety of restoration techniques have been developed for streams. This field has become an important source of jobs for environmental science majors. A simple approach in which everyone can participate is to reduce sediment influx by planting ground cover on uplands and filling gullies with rocks or brush. For small streams, sometimes the quickest way to rebuild a channel is to use heavy earthmoving equipment to simply dig a new

FIGURE 13.30 Many former streams have been turned into concrete-lined ditches to control erosion and speed runoff. The result is an artificial system with little resemblance to the living biological community that once made up the stream.
Source: Federal Interagency Stream Corridor Restoration Handbook, NRCS, USDA

one. This can be very disruptive, however, stirring up sediment that can be harmful for fish and other aquatic organisms. Alternative, less intrusive stream improvement methods are available. Most of these methods involve placing barriers (weirs, vanes, dams, log barriers, brush bundles, root wads, or other obstructions) in streams to deflect current away from the banks or trap sediment. Often, these barriers will cause currents to scour out deep pools in the stream bottom that provide places for fish to hide and rest.

Other techniques also create fish habitat with **coarse woody debris.** Logs, root wads, brush bundles, and boulders can shelter fish. An expensive but effective way to create fish hiding places is the so-called "lunker" structure. This is a wood framework that rests on the stream bottom and is anchored securely to the shore. The top of the box is covered with rock, soil, and vegetation. Openings in the structure provide hiding places for fish (fig. 13.31).

Sometimes what's needed is to speed up the current rather than slow it down. Figure 13.32a shows a spring-fed Minnesota trout stream that was degraded by crop production in its uplands, and grazing that broke down the banks and filled the stream with sediment. The stream, which had been about 1 m wide and 1 m deep, spread out to be 5 m wide and only about 10 cm deep. The formerly cold, swiftly moving water was warmed by the sun as it passed through these shallows so that it became uninhabitable for trout. Furthermore, there was no place to hide from predators. Because this was the last remaining trout stream in the Minneapolis/St. Paul metropolitan area, a decision was made to restore it. Two very different restoration approaches are shown in figures 13.32b and c. Trout Unlimited, an angler's group, offered to bring in a backhoe and completely rebuild the stream. A demonstration section they reconstructed is shown in figure 13.32b. They narrowed the channel with large stone blocks, which also provide a good surface for anglers to walk along the stream. There

FIGURE 13.31 A lunker structure is a multilevel wooden framework that rests on the stream bottom and can be anchored to the shore. The top of the box is covered with rock, soil, and vegetation. Openings in the structure provide hiding places for fish.

isn't much biological production in this stream model, but that isn't important to many anglers. They expected the Department of Natural Resources to stock the stream with hatchery-raised fish, which usually are caught before they have time to learn to forage for natural food.

Other groups involved in stream restoration objected to the artificial nature and lack of a native biological community in this design. Figure 13.32c shows an alternative approach that was

(a)

(b)

(c)

FIGURE 13.32 Different visions for restoring a trout stream. (a) Degraded by a century of agriculture and grazing, the stream had become too wide, shallow, and warm for native trout. (b) Trout Unlimited rebuilt a section of the stream to show their preferred option, which featured banks made of large stone blocks. (c) Other environmental organizations preferred a more organic approach. They used straw bales to narrow the stream and increase current flow. This washed away the soft sediment and re-created the deep, narrow, cool, deeply shaded channel that favored native trout.
© William P. Cunningham

ultimately adopted for most of the stream. This is the same section shown in figure 13.32a. Straw bales were placed in the deepest part of the stream. This narrowed the stream and increased the speed of the current, which then cut down into the soft sediment. Within three months, the stream had deepened about 50 cm. The shallow area between the straw bales and the original shore quickly filled with sediment and became a cattail marsh. The bales were anchored in place with green willow stakes, which rooted in the moist soil and sprouted to make saplings that arched over the stream and shaded it from the sun.

Seeds of native wetland vegetation were scattered on top of the straw bales. They sprouted, and after the first summer the stream was surrounded by a dense growth of vegetation that keeps the water cool, provides fish shelter, and supports a rich community of invertebrates that feed the native trout. An aquatic invertebrate survey found that while the stream was wide and shallow, 58 percent of the aquatic species were snails and copepods, whereas stone flies, caddis flies, and other preferred trout food made up only 20 percent of all invertebrates. After the straw bale restoration, snails and copepods made up only 15 percent of the invertebrate population, and stone flies and caddis flies had increased to 47 percent.

Stabilizing banks is an important step in stream restoration. Where banks have been undercut by erosion and stream action, they will continue to be unstable and cave into the stream. Ideally, the bank should be recontoured to a slope of no more than 45 degrees (fig. 13.33). Soil can then be held in place by rocks, planted vegetation, or other ground cover. It may be necessary to install erosion control fabric or mulch to hold soil until vegetation is established.

If space isn't available for recontouring, steep banks may have to be supported by rock walls, riprap, or embedded tree trunks.

Severely degraded or polluted sites can be repaired or reconstructed

In a relatively small area—say an old industrial site—it may be economical to simply excavate and replace contaminated soil. If the pollutants are organic, it may be possible to pass contaminated soil through an incinerator to eliminate toxins. After this treatment, the soil won't be worth much for growing vegetation, however.

For polluted surface or groundwater, bacteria can remove organic compounds such as oil and other contaminants. Naturally occurring bacteria in groundwater, when provided with oxygen and nutrients, can decontaminate many kinds of toxins. Experiments have shown that pumping air into aquifers can be more effective than pumping water out for treatment. For hostile environments or exotic, human-made chemicals that can't be metabolized by normal organisms, it's sometimes possible to genetically engineer new varieties of bacteria that can survive in extreme conditions and consume materials that would kill ordinary species.

Bioremediation is a growing strategy that uses living things, especially plants or bacteria, to selectively eliminate toxins from the soil. A number of plant species can selectively eliminate toxins from the soil, air, or water. Some types of mustard, for example, can extract lead, arsenic, zinc, and other metals from contaminated soil. Radioactive strontium and cesium have been removed from soil near the Chernobyl nuclear power plant using common sunflowers. And sunflowers and other plants can capture toxins, metals, and carcinogenic compounds from other industrial sites (fig. 13.34).

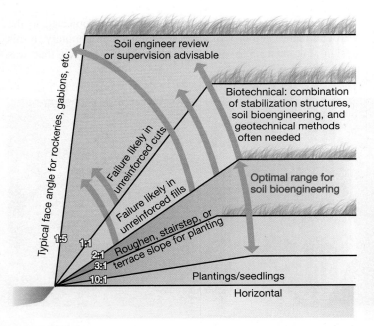

FIGURE 13.33 Steep streambanks need to be reinforced or recontoured to avoid erosion. Shallow slopes can be stabilized with vegetation or mulch. Steeper slopes need stabilization structures or reinforcement to hold the soil.

FIGURE 13.34 Sunflowers capture heavy metals and unburned hydrocarbons from a natural gas pumping station.
© ProSha/Shutterstock.com

In some cases, bioremediation could have multiple benefits. A weedy species called field pennycress or stinkweed (*Thlaspi arvense*), for example, grows well on degraded, polluted soil, absorbing metals as it grows. Its seeds contain high concentrations of oils that make it competitive with canola as an oil seed crop. The oils aren't edible but can be easily converted into biodiesel fuel or can be used for bioplastic production. Furthermore, glucosinolates in the oil can be converted into a soil fumigant that could be an eco-friendly alternative to methyl bromide. Combining all these features, pennycress could be grown on contaminated inner-city lands, producing either fuel or chemical feedstocks while also decontaminating the soil.

Many cities are finding that decontaminating urban "brown fields" (abandoned, contaminated industrial sites) can turn unusable inner-city property into valuable assets. This is a good way to control urban sprawl and make use of existing infrastructure. Cleaning up hazardous and toxic wastes is now a big business in America, and probably will continue to be so for a long time in the future. This is a growth industry in places where most other industry is disappearing.

Reclamation implies using intense physical or chemical methods to clean and repair severely degraded or even totally barren sites. Historically, reclamation meant irrigation projects that brought wetlands and deserts (considered useless wastelands) into agricultural production. In the early part of this century, the Bureau of Reclamation and the Army Corps of Engineers dredged, diked, drained, and provided irrigation water to convert millions of acres of wild lands into farm fields. Many of those projects were highly destructive to natural ecosystems. Ironically, we are now using ecological restoration to restore some of these "reclaimed" lands to a more natural state.

Today, reclamation means the repairing of human-damaged land. The Surface Mining Control and Reclamation Act (SMCRA), for example, requires mine operators to restore the shape of the land to its original contour and revegetate it to minimize impacts on local surface water and groundwater. According to the U.S. Office of Surface Mining, more than 8,000 km^2 (3,000 mi^2) of former strip mines have been reclaimed and returned to beneficial uses, such as recreation areas, farming and rangeland, wetlands, wildlife refuges, and sites for facilities such as hospitals, shopping centers, schools, and office and industrial parks.

Ideally, if topsoil is set aside during surface mining, overburden and tailings (waste rock discarded during mining and ore enrichment) could be returned to the pit, smoothed out, and covered with good soil that will support healthy vegetation. Unfortunately, topsoil is often buried deeply during mining, and what ends up on the surface is crushed rock that won't revegetate very well without a great deal of fertilizer and water.

The largest mine pits will never be returned to their original contour. Figure 13.35 shows a view of the Berkely mine pit in Butte, Montana. From the 1860s to the 1980s, this area was one of the world's richest sources of metals, including copper, silver, lead, zinc, manganese, and gold. In the early days, all mining was

FIGURE 13.35 The Berkely mine pit in Butte, Montana, may be the most toxic water body in the United States. Water entering the pit is now being treated, and eventually there may be an effort to pump water out of the pit and decontaminate it, but the pit itself will probably never be filled in.
© *William P. Cunningham*

in deep shafts. In 1955, the Anaconda Mining Company switched to open-pit mining and dug the hole you see now. After mining ended in 1981, groundwater, previously controlled with pumps, began to fill the pit. It has now accumulated to make a lake 1.6 km wide and 300 m deep.

The water has a pH of 2.5 (about the same as vinegar) and is laden with heavy metals and toxic chemicals, such as arsenic, cadmium, zinc, and sulfuric acid. In 1995, a whole flock of migrating snow geese was found dead after landing by mistake in the pit. The pit is now a Superfund site. A water treatment plant has been built on the far shore of the pit (you can see a thin, white stream of treated water from the plant cascading into the lake). By 2018, or whenever the water level hits the critical elevation of 1,649 m above sea level (the height at which it will threaten groundwater used by the city of Butte), the plant will begin to treat the contents of the pit. This is considered a reclamation project, but it's highly unlikely that the pit will ever be cleaned completely.

Section Review

1. Why is wetland restoration important?
2. What is wetland mitigation?
3. What is bioremediation?

Conclusion

Humans have caused massive damage and degradation to a wide variety of biological communities, but there are many ways to repair this damage and to restore or rehabilitate nature. Ideally, we might prefer to return a site to its pristine, predisturbance condition, but that often isn't possible. A more pragmatic goal is simply to develop a useful, stable, self-sustaining ecosystem with as many of its original ecological elements as possible. Sometimes it's enough to leave nature alone to heal itself, but often we need to intervene in some way to remove or discourage unwanted organisms while also promoting the growth of more desirable species.

Restoration pioneer Aldo Leopold wrote, "A thing is right when it tends to preserve the integrity, stability, and beauty of the biotic community. It is wrong when it tends otherwise." Some modern ecologists object that ecosystems are highly dynamic and species appear and disappear stochastically and individually. Characteristics such as integrity, health, stability, beauty, and moral responsibility tend to be human interpretations rather than scientific facts. Still, we need to have goals for restoration that the public can understand and accept.

Many ecological restoration projects are now underway. Some are huge efforts, such as restoring the Elwha River, rehabilitating the Florida Everglades and Chesapeake Bay, or returning a huge swath of shortgrass prairie to its primeval state. Others are much more modest: building a rain garden to trap polluted storm runoff, planting trees on empty urban land, or turning a part of your yard into a native prairie or woodland. Within this wide range, there are lots of opportunities for all of us to get involved.

Reviewing Key Terms

Can you define the following terms in environmental science?

buffalo commons 13.1
bioremediation 13.4
coarse woody debris 13.4
ecological restoration 13.1
intervention 13.1

mitigation 13.1
novel ecosystem 13.1
oak savanna 13.2
prairie potholes 13.4

prescribed burning 13.2
reallocation 13.1
reclamation 13.4
re-creation 13.1

rehabilitate 13.1
remediation 13.1
restoration 13.1
wetland mitigation 13.4

Critical Thinking and Discussion Questions

1. Should we be trying to restore biological communities to what they were in the past, or modify them to be more compatible with anticipated future conditions? What future conditions would you consider most likely to be problematic?

2. How would you balance human preferences (aesthetics, utility, cost) with biological considerations (biodiversity, ecological authenticity, evolutionary potential)?

3. The Nature Conservancy's Hassayampa River Preserve near Phoenix illustrates a situation in which there may be more than one historic condition to which we may wish to restore a landscape. How would you reconcile these different values and goals?

4. Restoring savannas often requires the use of herbicides to remove invasive species. Some people regard this as dangerous and unnatural. How would you respond?

5. The Nature Conservancy believes it is essential to keep productive ranches on the land, both to sustain rural society and for effective protection of the range. The American Prairie Foundation (APF) is buying up ranches and converting them to wilderness where wild animals can roam freely. Which of these approaches would you favor?

6. Sometimes the quickest and easiest way to restore a stream is simply to reconstruct it with heavy equipment. You might even create something more interesting and useful (at least from a human perspective) than the original. Is it okay to replace real nature with something synthetic?

Data Analysis

Concept Maps

The figure below is a form of graphic representation we haven't used very often in this book. It's a concept map, or a two-dimensional representation of the relationship between key ideas. It could also be considered a decision flowchart because it is an organized presentation of different policy options. This kind of chart shows how we might think about a situation, and suggests affinities and associations that might not otherwise be obvious.

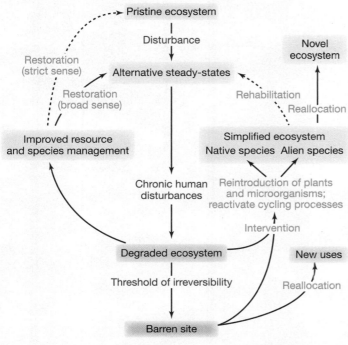

A model of ecosystem degradation and potential management options.
Source: Data from Walker and Moral, 2003

1. Using one of the examples presented in this chapter—or another familiar example—replace the descriptions in the colored boxes with brief descriptions of an actual ecosystem. Does this help you see the relationship between different states for the community you've chosen?

2. Now replace the terms associated with the arrows between boxes with actions that cause changes in your particular biological community, as well as restoration treatments that could accomplish the proposed restoration outcomes.

3. What do you suppose the authors meant by a "threshold of irreversibility"? If the system is irreversible, why are there arrows for reallocation or intervention?

4. The box for simplified ecosystem has two subcategories, native species and alien species. What does this mean? What would be some examples for the ecological community you've chosen?

5. There are two arrows labeled "reallocation" in this diagram. One leads to "new uses," while the other leads to "novel ecosystem." What's the difference? Why are the two boxes separated in the diagram?

6. The arrow labeled "restoration (strict sense)" is a dotted line. What do you think the authors meant by this detail?

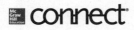

TO ACCESS ADDITIONAL RESOURCES FOR THIS CHAPTER, PLEASE VISIT CONNECT AT www.connect.mheducation.com.

You will find LearnSmart, an adaptive learning system, Google Earth™ exercises, additional Case Studies, Data Analysis exercises, and an interactive ebook.

14

Geology and Earth Resources

▲ The tops of at least 500 mountains in the southern Appalachians have been sheared off to access buried coal seams. More than a million acres (0.5 million ha) of forest have been destroyed and about 2,000 miles (3,200 km) of streams have been polluted or buried by waste rock pushed down into "valley fills" by mountaintcp removal mining.

© Vivian Stockman/www.ohvec.org

Learning Outcomes

After studying this chapter, you should be able to:

14.1 Summarize the processes that shape the earth and its resources.

14.2 Describe several earth resources of economic value.

14.3 Outline the environmental effects of resource extraction.

14.4 Identify several geological hazards, and explain how they occur.

"We live in a society exquisitely dependent on science and technology, in which hardly anyone knows anything about science and technology."

– Carl Sagan

Moving Mountains for Coal

Of the many geologic resources we depend on, coal has long been one of our most important. Coal powered the Industrial Revolution and the steam trains that crossed continents. The places and people of coal-mining regions, though, often have been devastated in our constant push to extract coal faster and cheaper.

Coal is the rock-hard remains of ancient swamp muck—leaves and other organic matter that accumulated in expansive coastal wetlands. Most coal accumulated more than 300 million years ago, during a 60-million-year period called the Carboniferous, named for the coal beds common in rocks of that age. Coal layers are sandwiched between other sedimentary rocks, mainly sandstone, shale (compressed mud and clay), and limestone, each layer deposited at different stages as ancient sea levels rose or fell over time.

Traditionally, coal was mined by digging into the coal seams between the rock layers. Usually, miners dug deep mine shafts to access the coal. The development of colossal earth-moving equipment in recent decades has made coal mining far faster and cheaper, in part because a few giant machines replaced legions of miners. It is now easiest simply to blast and scrape away layers of rock (overburden) and then scoop up the coal with giant shovels. This method has made mining highly profitable to mining companies, although it provides little income to mining communities.

Mountaintop removal mining (MTR) is the term used for this approach in the mountains of Appalachia. As the term suggests, entire ridge tops are removed, and the overburden often is pushed downhill, where it fills valleys and buries streams. The first step in MTR is to clear-cut forests (fig. 14.1). Wildlife habitat is destroyed, and the steep, newly bare slopes become prone to erosion, floods, and landslides. Thousands of tons of high-powered explosives are then used to tear off the top of the mountain. Every week, the explosive equivalent of a Hiroshima-sized atomic bomb (15,000 tons of TNT) is detonated in MTR mines in the four states (Kentucky, Tennessee, Virginia, and West Virginia) where this technique is most common. Nearby communities complain of explosions and dust; flying boulders have destroyed property and even killed people.

As much as 240 m (800 ft) of rock may be removed from the mountaintop to reach a coal seam. Sometimes the overburden can be piled back on top of the mountain or ridge. Often, it's simply shoved into the valley below. Nearly 500,000 ha (1.2 million acres) of land in Appalachia on more than 500 mountain peaks and ridge lines have been turned into mining wastelands. About 3,200 km (2,000 mi) of headwater streams have been buried or severely polluted by the "valley fill" created by MTR. The broken debris contains sulfur, arsenic, mercury, and other toxic metals, which derive from the ancient swamp plants and soils. Rainwater percolating through the rubble dissolves the sulfur to produce sulfuric acid.

Coal must be crushed and washed to remove these impurities. The wastewater, a slurry of coal dust laced with sulfuric acid and heavy metals, can't be dumped into local rivers. So millions of gallons of contaminated wastewater are stored in huge ponds behind earthen dams. A number of these dams have failed, flooding communities down-valley with toxic sludge.

In theory, mining companies must restore land to its original condition when mining is finished, but when a mountain has been flattened and the rock has been dumped into the valley below, it's impossible to re-create former conditions. The denuded mountaintop can be graded, but with topsoil gone, vegetation recovers poorly.

The demise of MTR mining may not come from public protests or environmental regulations, but from economics. Although there are many drawbacks from fracking, it has produced an abundance of cheap, relatively clean-burning natural gas. In the past six years, gas prices have fallen 80 percent, and the costs for wind and solar power have fallen by about half in that same period. Hundreds of coal-burning power plants have either been torn down or converted to gas. Twenty-six coal companies have declared bankruptcy and 264 mines have closed since 2008. We'll discuss

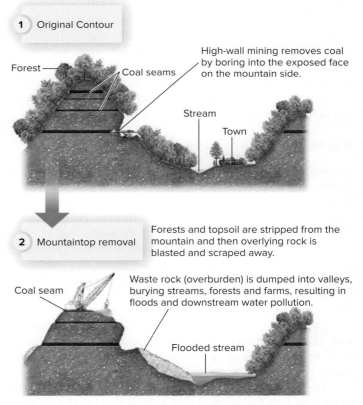

FIGURE 14.1 Mountaintop removal mining removes ridgetops to access coal. Restoration and revegetation should follow mining, although streams and ridgetop forests cannot readily be replaced.

energy policy and sources in more detail in chapters 19 and 20, but it looks as if the era of mountain destruction may be coming to an end, along with much other coal mining.

Coal is just one of many earth resources on which our industrial society depends. What are the other resources? Where are they? What are the consequences of extracting them? Are there better ways to use, or to conserve, these resources? In this chapter, we'll review some basic ideas in geology and ways in which we humans have become geologic agents.

14.1 EARTH PROCESSES AND MINERALS

- *Convection in the earth's mantle causes tectonic plate movement, earthquakes, and volcanoes.*
- *Oceanic crust is thinner, denser, and younger than continental crust.*
- *The rock cycle involves formation of three general rock types.*

All of us benefit from geological resources. Right now you are probably wearing several geological products: Plastics, including glasses and synthetic fabric, are made from oil; iron, copper, and aluminum mines produced your snaps and zippers and perhaps the chair you're sitting on; silver, gold, and diamond mines may have produced your jewelry. All of us also share responsibility for the environmental and social impacts of mining and drilling. Fortunately, there are many ways to reduce these costs, including recycling and using alternative materials.

Where do these resources come from? To understand this question, we must first examine the earth's structure and the processes that shape it.

Earth is a dynamic planet

Although we think of the ground under our feet as solid and stable, our planet is a dynamic and constantly changing structure. Titanic forces inside the earth cause continents to split, move, and crush into each other in slow, inexorable collisions. All this motion results from convection currents deep in the earth.

The earth is a layered sphere. The **core,** or interior, is composed of a dense, intensely hot mass of metal—mostly iron—thousands of kilometers in diameter (fig. 14.2). Solid in the center but more fluid in the outer core, this immense mass generates the magnetic field that envelops the earth.

Surrounding the molten outer core is a hot, pliable layer of rock called the **mantle.** The mantle is much less dense than the core, because it contains a higher concentration of lighter elements, such as oxygen, silicon, and magnesium.

The outermost layer of the earth is the lightweight, brittle **crust.** The crust below oceans is relatively thin (8–15 km), dense, and young (less than 200 million years old) because it is constantly recycled. This oceanic crust consists of rocks rich in iron, silicon, and magnesium, such as basalt. Continental crust is thicker (25–75 km), less dense, and older, some of it 3.8 billion years old. Rocks in the continental crust, such as granite or sandstone, generally are composed of lighter elements, such as oxygen, silicon, and aluminum, compared to the rest of the earth (table 14.1).

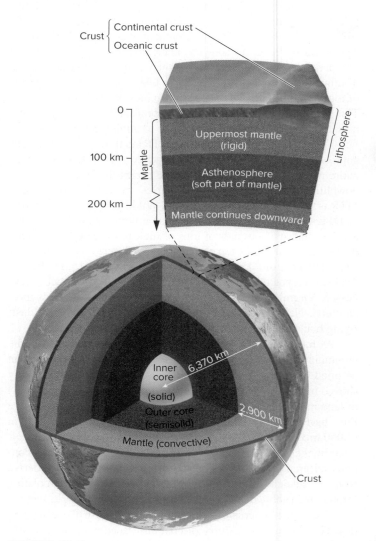

FIGURE 14.2 Earth's cross-section. Slow convection in the mantle causes the thin, brittle crust to move.

Table 14.1	Eight Most Common Chemical Elements (Percent)		
Whole Earth		**Crust**	
Iron	33.3	Oxygen	45.2
Oxygen	29.8	Silicon	27.2
Silicon	15.6	Aluminum	8.2
Magnesium	13.9	Iron	5.8
Nickel	2.0	Calcium	5.1
Calcium	1.8	Magnesium	2.8
Aluminum	1.5	Sodium	2.3
Sodium	0.2	Potassium	1.7

Tectonic processes move continents

The convection currents in the mantle provide the force to move the overlying crust, a mosaic of huge blocks called **tectonic plates** (fig. 14.3). These plates slide slowly across the earth's surface like ice sheets on water, in some places breaking up into smaller pieces, and in other places crashing ponderously into each other to create new, larger landmasses. Ocean basins form where continents crack and pull apart. The Atlantic Ocean, for example, is growing slowly as Europe and Africa move away from the Americas. **Magma** (molten rock) forced up through the cracks forms new oceanic crust that piles up underwater in **mid-ocean ridges.** Creating the largest mountain range in the world, these ridges wind around the earth for 74,000 km (46,000 mi) (fig. 14.3). Although concealed from our view, this jagged range boasts higher peaks, deeper canyons, and sheerer cliffs than any continental mountains.

At the margins of converging plates, mountain ranges are pushed up, like those on the West Coast of North America and in Japan. The Himalayas are still rising as the Indian subcontinent grinds slowly into Asia. (These mountains are unusually tall because they occur at the convergence of two continental plates.) At many plate boundaries, two plates slide past each other. Southern California is slowly sliding north toward Alaska. In about 30 million years, Los Angeles will pass San Francisco. At these boundaries, earthquakes are common.

When an oceanic plate converges with a continental landmass, the continental plate usually rides up over the seafloor, while the oceanic plate is **subducted,** or pushed down into the mantle. As the subducted plate is forced downward, heat and pressure melt and recrystallize minerals. At great depth, melting material becomes fluid magma, which can rise back toward the surface (fig. 14.4). Deep ocean trenches mark these subduction zones (fig. 14.3). Volcanoes form where the magma erupts through vents and fissures in the overlying crust. Trenches and volcanic mountains ring the Pacific Ocean rim from Indonesia to Japan to Alaska and down the west coast of the Americas, forming a "ring of fire" where oceanic plates are being subducted under the continental plates. This ring is the source of more earthquakes and volcanic activity than any other region on the earth.

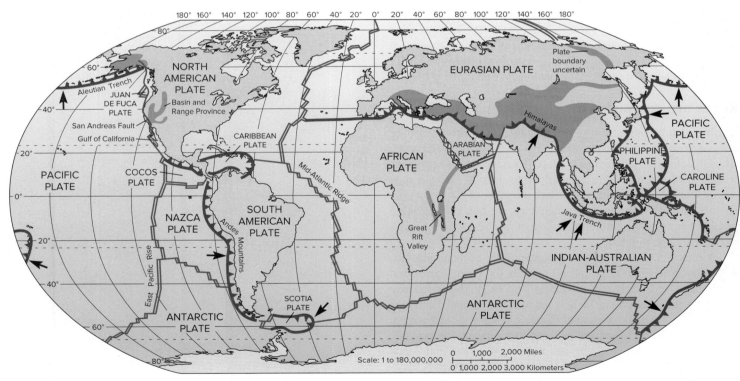

FIGURE 14.3 Map of tectonic plates. Plate boundaries are dynamic zones, characterized by earthquakes and volcanism and the formation of great rifts and mountain ranges. Arrows indicate direction of subduction where one plate is diving beneath another. These zones are sites of deep trenches in the ocean floor and high levels of seismic and volcanic activity.

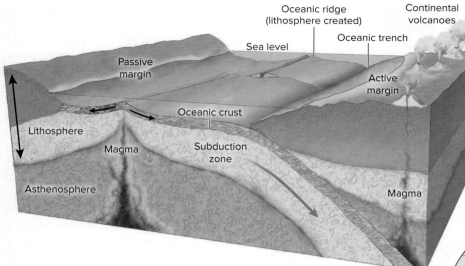

FIGURE 14.4 Tectonic plate movement. Where thin, oceanic plates diverge, upwelling magma forms mid-ocean ridges. A chain of volcanoes, like the Hawaiian Islands, may form as plates pass over a "hot spot." Where plates converge, melting can cause volcanoes, such as the Cascades.

Over millions of years, continents can drift long distances. North America spent millions of years near the equator. Antarctica and Australia once were connected to Africa, also near the equator, and supported luxuriant forests. Geologists suggest that several times in the earth's history, most or all of the continents have gathered to form supercontinents, which have ruptured and re-formed over hundreds of millions of years (fig. 14.5). The redistribution of continents has profound effects on the earth's climate and may help explain the periodic mass extinctions of organisms that mark the divisions between many major geological periods (fig. 14.6).

Rocks are composed of minerals

A **mineral** is a naturally occurring, inorganic, solid element or compound with a specific chemical composition and crystal structure. Organic materials such as coal, produced by living organisms or biological processes, are generally not minerals, because they vary in chemical composition and lack a regular crystal structure. The ores from which metals are extracted are minerals. Purified metals such as iron, aluminum, or copper lack a crystal structure and thus are not minerals.

A **rock** is a solid, cohesive, aggregate of one or more minerals. Within the rock, individual mineral crystals (or grains) are mixed together and held firmly in a solid mass (fig. 14.7). The grains may be large or small, depending on how the rock was formed. Each rock type has a characteristic mixture of minerals, grain sizes, and ways in which the grains are mixed and held together. Granite, for example, is a mixture of quartz, feldspar, and mica crystals. Different kinds of rocks have distinct percentages of these or other minerals. Fine-grained crystalline rocks indicate a relatively rapid cooling environment near the surface. Large-grained rocks crystallized and solidified slowly, deep underground.

Rocks and minerals are recycled constantly

Like continents, rocks are endlessly reshaped and re-formed. They are crushed, folded, melted, and recrystallized. Heat, pressure, erosion, deposition, and other processes create new minerals and rocks. We call this cycle of creation, destruction, and metamorphosis the **rock cycle** (fig. 14.8).

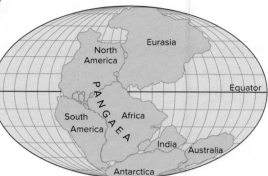

Pangaea, 250 MYA

Laurasia and Gondwana, 210 MYA

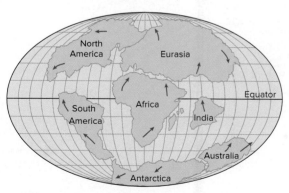

Most modern continents had formed by 65 MYA

FIGURE 14.5 Pangaea, an ancient supercontinent of 250 million years ago (MYA), combined all the world's continents in a single landmass. Continents have combined and separated repeatedly.

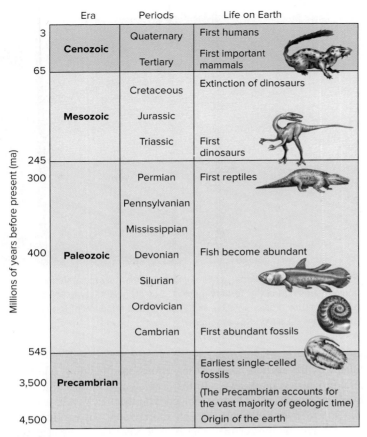

Era	Periods	Life on Earth
Cenozoic	Quaternary	First humans
	Tertiary	First important mammals
Mesozoic	Cretaceous	Extinction of dinosaurs
	Jurassic	
	Triassic	First dinosaurs
Paleozoic	Permian	First reptiles
	Pennsylvanian	
	Mississippian	
	Devonian	Fish become abundant
	Silurian	
	Ordovician	
	Cambrian	First abundant fossils
Precambrian		Earliest single-celled fossils
		(The Precambrian accounts for the vast majority of geologic time)
		Origin of the earth

Millions of years before present (ma): 3, 65, 245, 300, 400, 545, 3,500, 4,500

FIGURE 14.6 Periods and eras in geological time, and major life-forms that mark some periods.

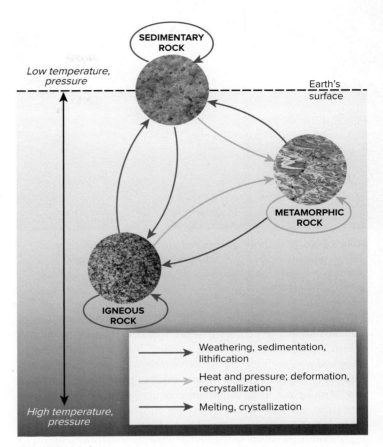

FIGURE 14.8 The rock cycle includes a variety of geological processes that can transform any rock.

FIGURE 14.7 Crystals of different minerals create beautifully colored patterns in a rock sample seen in a polarizing microscope.
© Corbis RF

There are three major rock classifications: igneous, metamorphic, and sedimentary. The most common rock-type in the earth's crust is solidified from magma, welling up from the earth's interior. These rocks are classed as **igneous rocks** (from *igni,* Latin for fire). Magma extruded to the surface from volcanic vents cools quickly to make basalt, rhyolite, andesite, and other fine-grained rocks. Magma that cools slowly in chambers deep below the surface or that intrudes into overlying layers makes granite, gabbro, or other coarse-grained crystalline rocks.

Extreme heat and pressure can transform mineral structures to create new forms called **metamorphic rock.** Metamorphism generally happens as buried layers of rocks are squeezed, folded, and heated by tectonic processes. Metamorphic rocks often show patterns of banding and folding, which make them desirable for decorative building facades. Some common metamorphic rocks are marble (recrystallized from limestone), quartzite (from sandstone), and slate (from mudstone and shale). Metamorphic rocks produce a variety of economically important minerals such as talc, graphite, and gemstones.

Weathering breaks down rocks

Sedimentary rocks, the third major type, are accumulations of sand, mud, or other material deposited over time from another source. Although we consider rocks hard and durable, they erode steadily when exposed to air, water, changing temperatures, and reactive chemical agents. The gradual breaking down of exposed rocks is called **weathering** (fig. 14.9). *Mechanical weathering* is the physical breakup of rocks into smaller particles without a change in chemical composition of the constituent minerals.

FIGURE 14.9 Weathering slowly reduces an igneous rock to loose sediment. Here, exposure to moisture expands minerals in the rock, and frost may also force the rock apart.
© Photo by David McGeary

FIGURE 14.10 Different colors of soft sedimentary rocks deposited in ancient seas during the Tertiary period 63 to 40 million years ago have been carved by erosion into the fluted spires and hoodoos of the Pink Cliffs of Bryce Canyon National Park.
© Digital Vision/Getty Images RF

You have probably seen mountain valleys scraped by glaciers or river and shoreline pebbles that are rounded from being rubbed against one another as they are tumbled by waves and currents. *Chemical weathering* is the selective removal or alteration of specific components that leads to the weakening and disintegration of rock. Among the more important chemical weathering processes are oxidation (combination of oxygen with an element to form an oxide or hydroxide mineral) and hydrolysis (hydrogen atoms from water molecules combine with other chemicals to form acids). The products of these reactions are more susceptible to both mechanical weathering and dissolution in water. For instance, when acidic rainfall (formed naturally when CO_2 combines with rain) percolates through porous limestone layers in the ground, it dissolves the calcium carbonate (limestone) and creates caves and sink holes.

Particles of rock are transported by wind, water, ice, and gravity until they come to rest again in a new location. The deposition of these materials is called **sedimentation.** Most sedimentary rocks were accumulated in broad, shallow coastal seas or river deltas. These deposits can be hundreds of meters thick. As layers accumulate over tens or hundreds of millions of years, the mass compacts and hardens deeper layers, so that loose sand becomes sandstone, clay and mud become shale, or organic shell fragments become limestone. Color and texture of rocks help identify environmental conditions in which they were deposited—near shore or in deep water, in arid and evaporating conditions, in shallow water, or exposed to the air. Some sedimentary rocks, such as sandstone and limestone, are relatively easily eroded and can be shaped into striking features (fig. 14.10).

Sedimentary "evaporite" minerals also form as salt layers accumulate in shallow, evaporating water bodies. The mineral halite, or rock salt, is deposited this way. Halite is the same as ordinary table salt (sodium chloride). Gypsum, the material used to make drywall in houses, is calcium sulfate, also deposited in shallow, evaporating water bodies.

Humans have become a major force in shaping landscapes. Geomorphologist Roger Hooke, of the University of Maine, estimates that activities such as housing excavations, road building, and mineral production move about 30 to 35 gigatons (Gt, or billion tons) per year worldwide. When combined with the 10 Gt each year that we add to river sediments through soil erosion, our earth-shaping impact may be greater than that of any other single agent except plate tectonics.

Section Review

1. Describe the layered structure of the earth.
2. Why are there many volcanoes and earthquakes along the "ring of fire" that rims the Pacific Ocean?
3. Describe the processes that produce the three rock types in the rock cycle.

14.2 EARTH RESOURCES

- *Metals are valuable because they are strong and easily shaped.*
- *Metals derive from ores, which often include sulfur.*
- *Nonmetal resources are diverse and valuable.*
- *Conservation reduces energy consumption as well as pollution.*

The earth is unusually rich in mineral variety. Mineralogists have identified some 4,400 different mineral species—far more, we believe, than on any of our neighboring planets. What makes the difference? The processes of plate tectonics and the rock cycle on this planet have gradually concentrated uncommon elements and allowed them to crystallize into new minerals. But this accounts for only about one-third of our geological legacy. The biggest distinction is life. Most of our minerals are oxides, but there was little free oxygen in the atmosphere until it was released by photosynthetic organisms, thus triggering evolution of our great variety of minerals.

We use a tremendous variety of nonmetal minerals, from high-value gemstones, to industrial resources such as mica, talc, graphite, and sulfur (used in chemical industries), to everyday building materials such as sand, gravel, salts, and limestone. Of these, sand and gravel production, needed for road construction and fill, comprise by far the greatest volume and dollar value of all nonmetal mineral resources and a far greater volume than all metal ores. Sand and gravel are used mainly in brick and concrete construction and paving, as road fill, and for sandblasting. High-purity silica sand is our source of glass. These materials usually are retrieved from surface pit mines and quarries, where they were deposited by glaciers, winds, or ancient oceans.

Metals are especially valuable resources

Most attention to economic minerals is concerned with metal ores, which are valuable, versatile, and used in great quantities. The metals consumed in greatest quantity by world industry are iron (740 million metric tons annually), aluminum (40 million metric tons), manganese (22.4 million metric tons), copper and chromium (8 million metric tons each), and nickel (0.7 million metric tons). Most of these metals are consumed in the United States, Japan, and Europe, in that order. The largest sources are China, Australia, Russia, Canada, and the United States.

Metal-bearing ores are minerals with unusually high concentrations of metallic elements. Lead, for example, often comes from the mineral galena (PbS), and copper comes from sulfide ores, such as bornite (Cu_5FeS_4). Iron occurs in a variety of minerals, such as hematite (Fe_2O_3) and goethite (Fe_2O_3s H_2O).

Metals have been so important in human affairs that major epochs of human history are commonly known by their dominant materials and the technology involved in using them: the Stone Age was followed by the Bronze Age and the Iron Age. The mining, processing, and distribution of these materials have broad implications for both our culture and our environment (fig. 14.11). Metals are useful because they are light, strong, and malleable (easily shaped). Unlike stone, metals can hold a sharp edge. Most metals also conduct electricity well (table 14.2).

A new class of metals, known as rare earth metals, has become key to the production of small electronics, lightweight motors, as in wind turbines, and batteries for electric or hybrid cars. Because they are sparsely distributed, extracting them causes considerable environmental disturbance. Because they often occur, like other metals, in sulfur-rich ores, mining and processing these metals can produce wastewater rich in sulfuric acid. International competition for these metals has made them a new subject for trade tensions (see the Exploring Science box later in this section).

Fossil fuels originated as peat and plankton

Coal, oil, and natural gas are another class of valuable and widely traded earth resources. As noted in the opening case study, these fossil fuels are deposits from living organisms. They are not

FIGURE 14.11 The availability of metals and the ways we extract and use them have profound effects on our society and environment.
© Digital Vision/PunchStock RF

minerals because they are not crystalline, and they are "organic" in that they are carbon-based compounds. Although we worry about supplies of these fossil fuels, they are globally distributed, and our estimates of their abundance has risen recently, with new exploration technology (see chapter 19).

Table 14.2	Primary Uses of Some Major Metals Consumed in the United States
Metal	**Use**
Aluminum	Packaging foods and beverages (38%), transportation, electronics
Chromium	High-strength steel alloys
Copper	Building construction, electric and electronic industries
Iron	Heavy machinery, steel production
Lead	Leaded gasoline, car batteries, paints, ammunition
Manganese	High-strength, heat-resistant steel alloys
Nickel	Chemical industry, steel alloys
Platinum-group	Automobile catalytic converters, electronics, medical uses
Gold	Medical, aerospace, electronic uses; accumulation as monetary standard
Silver	Electronics, jewelry

Rare Earth Minerals

What earth materials allow your cell phone and iPad to be so small? A group of elements known as "rare earth" metals are largely responsible. This group of metallic elements, including scandium, yttrium, and 15 lanthanides, such as neodymium, dysprosium, and gadolinium, have unusual properties. They conduct electricity efficiently. Small amounts of these metals can make motors 90 percent lighter and lights 80 percent more efficient. These make possible our small cell phones, high-efficiency lightbulbs, hybrid cars, superconductors, high-strength magnets, lightweight batteries, lasers, energy-conserving lamps, and a variety of medical devices.

Lightweight metals and batteries make it possible to build high-capacity wind turbines, whose motors must be extremely efficient and lightweight. The batteries in a hybrid car are heavy, but vastly less so because they contain rare earth metals. A Toyota Prius uses about a kilogram of neodymium and dysprosium for its electric motor and as much as 15 kg of lanthanum for its battery pack.

Despite their name, these elements are not rare. They are widely distributed globally, although commercially viable concentrations are relatively uncommon. China, Russia, and the United States possess more than 68 percent of the world's known reserves of these materials.

China currently produces about 95 percent of all rare earth metals (fig. 1). China dominates production partly because these metals are essential for manufacturing electronics, and partly because the Chinese government has been willing to overlook the environmental damage caused by extracting and processing these metals. About half of all Chinese production of rare earth metals occurs in a single mine in Baotou in Inner Mongolia; most of the rest come from small, often unlicensed mines in southern China.

Unregulated mineral extraction can poison water resources and soils. Like gold, silver, and other precious metals, rare earth elements are often separated from ore by

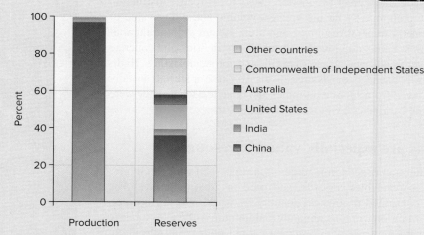

FIGURE 1 China controls a little more than one-third of known rare earth metals, but currently produces 97 percent of these important materials.

Source: USGS, 2010

crushing ore-bearing rocks and washing the ore in strong acids. These acids extract metals from the ore, but when the metals are later separated from the acid slurry, tremendous amounts of toxic wastewater are produced. Often, metals are extracted by pumping acids straight into holes drilled in the ground. Acids dissolve metal from ores in place, and a toxic, metal-bearing slurry is then pumped to the surface for processing. Acidic wastewater is frequently stored behind earthen dams, which can leak into surface and groundwater. Processing also releases the sulfur and radioactive uranium and thorium that frequently occur with rare earth elements.

These metals have become strategic resources globally. China has restricted exports of rare earth metals in recent years to protect its electronics industry. This move has made other nations worry about supplies for tactical needs (such as military guidance systems), as well as consumer products. Many firms are simply moving to China to ensure access to the resource. The General Motors division that

deals with miniaturized magnet research, for example, shut down its U.S. office and moved its entire staff to China about a decade ago. The Danish wind turbine company Vestas also has moved much of its production to China. For several years, automobile companies have said they're developing electric engines that don't need rare earths, but these engines haven't made it to the market yet.

In response to threatened shortages and rising prices, other countries are opening domestic mines. India is currently producing 5,000 tons of rare earths (5 percent of the world total). In California, Molycorp Minerals is reopening a closed mine that they hope will meet 10 percent of global demand. Mining may soon open in Australia and Canada's Northwest Territories. Greenland hopes to produce up to 25 percent of rare earth metals from recently discovered ore bodies. However, it remains to be seen whether there will be effective environmental controls for this coming expansion, or if they will break the Chinese monopoly on this strategic resource.

How did these resources form? All result from accumulation of organic material, mainly in shallow, warm swamps or seashore environments. Over hundreds of millions of years, there have been long periods, such as the Carboniferous period (about 360–300 million years ago), when these environments were widespread.

Coal derives from leaves and other plant material that accumulated in the bed of ancient swamps. During the Carboniferous era, when most coal was deposited, tree ferns and other large plants were abundant, as were giant dragonflies with wing spans up to 75 cm (30 in). (Dinosaurs were not yet around; they evolved about

50 million years later.) Modern swamps can accumulate tens of meters of undecayed plant material, or peat. Ancient swamps similarly accumulated peat to great depths but in vast expanses. Coal beds are generally interleaved with layers of sandstone, shale, and limestone. These occurred as rising sea levels would flood coastal swamps, allowing clay deposits, then limestone, to form. Eventually the seas receded, and nearshore sands were deposited. Rising sea levels then created coastal wetlands, and gradually the cycle was repeated. In some places, up to a hundred of these interleaved cycles are evident from the Carboniferous period. Pressure, heat from the earth's interior, and time have converted these swampy deposits into rock-hard coal seams (fig. 14.12).

Oil, in contrast, derives from the remains of tiny plankton and algae (fig. 14.13). Also growing in generally warm, coastal environments, these microscopic organisms accumulated on the sea floor over millions of years. Deposits of dead algae and plankton were especially rich in delta environments, where rivers discharged nutrient-rich waters and sediment from onshore. Many of our richest oil deposits today occur in areas that have been deltas for hundreds of millions of years—the Mississippi Delta, the Niger Delta, and the McKenzie Delta off the coast of Alaska and northern Canada, for example.

Sand, mud, or clay intermixed with accumulating organic material so that oil compounds generally occur within shale or sandstone formations. Over millions of years, as with coal, pressure and heat and time have transformed these materials first into kerogen, a thick precursor to oil, then to simpler, more fluid molecules of oil. Less-developed oil formations, such as Alberta's tar sands, contain mostly thick, hard-to-extract kerogen (chapter 19). More developed oil bodies, as in Texas or Saudi Arabia, have more fluid forms of oil. Fluid molecules seep gradually upward through porous sandstone because they are less dense than the surrounding rocks. Folds or faults often trap accumulations of oil. The smallest, volatile molecules, mainly methane (CH_4, the main component of natural gas), collect in these folds above the oil. Methane can also collect above coal formations. Coal-bed methane has been widely exploited in the American West.

Conserving resources saves energy and materials

Conservation offers great potential for extending our supplies of all these earth resources. Conservation can also help reduce the effects of mining and processing, including water use and contamination, air pollution, and energy use.

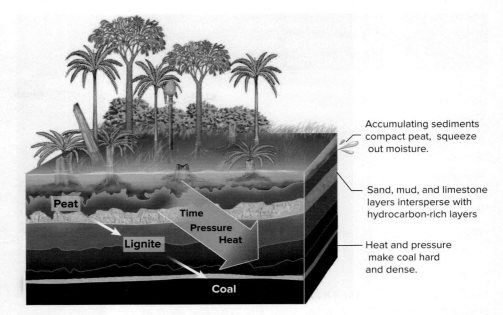

FIGURE 14.12 Coal was formed from vegetation in vast, ancient swamps. Pressure and heat gradually eliminated water and solidified deposits from soft peat to brownish lignite to rock-like coal.

Some waste products already are being exploited, especially for scarce or valuable metals. Aluminum, for instance, is extracted from bauxite ore by electrolysis, an expensive, energy-intensive process. Recycling waste aluminum such as beverage cans, on the other hand, consumes one-twentieth of the energy of extracting new aluminum. Today, nearly two-thirds of all aluminum beverage cans in the United States are recycled, up from only 15 percent 20 years ago. The high value of aluminum scrap ($650 a ton versus $60 for steel, $200 for plastic, $50 for glass, and $30 for paperboard) gives consumers plenty of incentive to deliver their cans for collection. Recycling is so rapid and effective that half of all the aluminum cans now on a grocer's shelf will be made into another can within two months. The energy cost of extracting other materials is shown in table 14.3.

Platinum, the catalyst in automobile catalytic exhaust converters, is valuable enough to be regularly retrieved and recycled

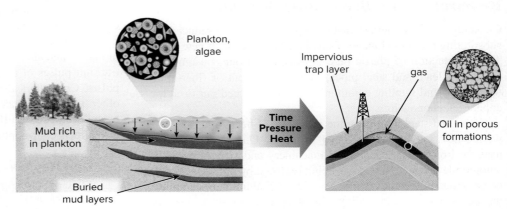

FIGURE 14.13 Oil and gas derive from buried deposits of plankton, algae, mud, and sand in ancient coastal seas. Over millions of years these low-density fluids can seep upward until they reach an impervious trap layer, which blocks and concentrates oil and gas.

Table 14.3	Energy Requirements in Producing Various Materials from Ore and Raw Source Materials		
	Energy Requirement (Mj/Kg)		
Product	New	From Scrap
Glass	25	25
Steel	50	26
Plastics	162	N/A
Aluminum	250	8
Titanium	400	N/A
Copper	60	7
Paper	24	15

Source: E. T. Hayes, Implications of Materials Processing, 1997.

FIGURE 14.14 The richest metal source we have—our mountains of scrapped cars—offers a rich, inexpensive, and ecologically beneficial resource that can be "mined" for a number of metals.
© Joseph Nettis/Science Source

from used cars (fig. 14.14). Other metals commonly recycled are gold, silver, copper, lead, iron, and steel. The latter four are readily available in a pure and massive form, including copper pipes, lead batteries, and steel and iron auto parts. Gold and silver are valuable enough to warrant recovery, even through more difficult means. See chapter 21 for further discussion of recycling.

Although total U.S. steel production has fallen in recent decades—largely because of inexpensive supplies from new and efficient steel mills overseas—a new type of mill subsisting entirely on a readily available supply of scrap steel and iron is a growing industry. Minimills, which remelt and reshape scrap iron and steel, are smaller and cheaper to operate than traditional integrated mills that perform every process from preparing raw ore to finishing iron and steel products. Minimills produce steel at $225 to $480 per metric ton, whereas steel from integrated mills averages $1,425 to $2,250 per metric ton. The energy cost is likewise lower in minimills: 5.3 million Btu/ton of steel compared to 16.08 million Btu/ton in integrated mill furnaces. Recycling is slowly increasing as raw materials become more scarce and markets for waste material grow.

Resource substitution reduces demand

Consumption of minerals, metals, and fuels can be reduced by substituting traditional materials and technologies with new ones. The introduction of plastic pipe has decreased our consumption of copper, lead, and steel pipes. The development of fiber-optic technology and satellite communication has reduced our need for copper telephone wires.

Iron and steel have been the backbone of heavy industry, but we are now moving toward other materials. One of our primary uses for iron and steel has been machinery and vehicle parts. In automobile production, steel is being replaced by polymers (long-chain organic molecules similar to plastics), aluminum, ceramics, and new, high-technology alloys. All of these reduce vehicle weight and cost while increasing fuel efficiency. Some of the newer alloys that combine steel with titanium, vanadium, or other metals wear

much better than traditional steel. Ceramic engine parts provide heat insulation around pistons, bearings, and cylinders, keeping the rest of the engine cool and operating efficiently. Plastics and polymers reinforced with glass fiber are used in body parts and some engine components.

Electronics and telephone technology, once major consumers of copper and aluminum, now use ultra-high-purity glass cables to transmit pulses of light, instead of metal wires carrying electron pulses. Computers also contain a fraction of the copper, lead, and other metals they once contained. Once again, this technology has been developed for efficiency, convenience, and lower cost, but it also affects consumption of our most basic metals.

Section Review

1. Why are rare earth metals important, and why are we concerned about supplies?
2. What are the origins of coal and oil?
3. Identify several reasons for recycling metals.

14.3 ENVIRONMENTAL EFFECTS OF RESOURCE EXTRACTION

- *Mining produces dust, acids, and toxic metal pollutants.*
- *Metal is extracted by smelting or chemical extraction.*
- *The value of minerals and the cost of cleanup lead to conflict over these resources.*

Each of us depends daily on geological resources mined from all around the world. We use scores of metals and minerals, many of which we've never even heard of, in our lights, computers, watches, fertilizers, and cars. Mining and purifying all these resources can have severe environmental and social consequences. The most obvious effect of mining is often the disturbance or removal of the land surface. Farther-reaching effects, though, include air and water pollution. The EPA lists more than 100 toxic air pollutants, from acetone to xylene, released from U.S. mines every year. Nearly 80,000 metric tons of particulate matter (dust) and 11,000 tons of sulfur dioxide are released from nonmetal mining alone. Pollution from chemical and sediment runoff is a major problem in many local watersheds.

Mining can affect water quality in several ways. Gold, copper, and other metals are often found in sulfide ores. When these minerals are exposed to air and water, they produce sulfuric acid, which is highly mobile and strongly acidic. Ores contain very low concentrations of gold, platinum, or other valuable metals, just 10 to 20 parts per billion, that may be economically extractable. Consequently, vast quantities of ore must be crushed and washed to extract metals. Toxic substances such as cyanide are often used to chemically separate metals from ore.

Different mining techniques pose different risks to water and air

There are many techniques for extracting geological materials. The most common methods are open-pit mining, strip mining, and underground mining. Less widespread but possibly more ancient is placer mining, in which nuggets of gold, diamonds, or coal are washed from stream sediments. Since the California gold rush of 1849, placer miners have used water cannons to blast away gravel hillsides. Mobilized sand and gravel clog streams and destroy fisheries, but this inexpensive method is still used in Alaska, Canada, and many other regions.

Underground mining is another ancient and much more dangerous method. Mine tunnels occasionally collapse. In underground coal mines, coal dust, as well as natural gas associated with coal beds, can cause explosions and uncontrollable fires. One mine fire in Centralia, Pennsylvania, has been burning since 1962; control efforts have cost at least $40 million, but the fire continues to expand. China, which depends on coal for much of its heating and electricity, has hundreds of smoldering mine fires; one has been burning for 400 years. Hundreds of coal mines smolder in the United States, China, Russia, India, South Africa, and Europe. The inaccessibility and size of these fires can make them impossible to extinguish or control. According to a recent study from the International Institute for Aerospace Survey in the Netherlands, these fires consume up to 200 million tons of coal every year and emit as much carbon dioxide as all the cars in the United States.

Open-pit mines are used to extract massive beds of metal ores, coal, and other minerals. The size of modern open pits can be hard to comprehend. The Bingham Canyon copper mine, near Salt Lake City, Utah, is 800 m (2,640 ft) deep and nearly 4 km (2.5 mi) wide at the top. More than 5 billion tons of copper ore and waste material have been removed from the hole since 1906. A chief environmental challenge of open-pit mining is that groundwater accumulates in the pit. In metal mines, a toxic soup results. No one yet knows how to detoxify these lakes, which can endanger wildlife.

Surface or strip mines produce more than half the coal used in the United States (fig. 14.15). Because coal is often found in expansive, horizontal beds, the entire land surface can be stripped away to cheaply and quickly expose the coal. The overburden, or surface material, is placed back into the mine. Historically, this debris, or spoil, was dumped in long ridges called spoil banks, which eroded easily and allowed sulfur, mercury, and other contaminants to enter surface and groundwater. Lacking topsoil, spoil banks revegetate slowly, if at all.

The 1977 federal Surface Mining Control and Reclamation Act (SMCRA) has required better restoration of strip-mined

FIGURE 14.15 Some giant mining machines stand as tall as a 20-story building and can scoop up thousands of cubic meters of rock per hour.

© James P. Blair/National Geographic Image Collection

lands, especially where mines replaced prime farmland. Since then, the record of strip-mine reclamation has improved substantially. Complete mine restoration is expensive, often more than $10,000 per hectare. Compacted, acidic soil makes restoration difficult.

Mountaintop removal (opening case study) also removes overburden to expose coal deposits. Restoration is even more difficult, however, because it occurs on long, sinuous ridge-tops (fig. 14.16), with pulverized overburden pushed into adjacent river valleys. The debris can be laden with selenium, arsenic, sulfur, and other toxic substances. At least 900 km (560 mi) of streams have been buried in West Virginia alone. In 2011, the EPA stopped one of the largest proposed mountaintop mines in U.S. history, the Spruce No. 1 in West Virginia, because the company's plan to dispose of mining waste in valley streambeds violated the Clean Water Act.

Processing also produces acids and metals

Metals are extracted from ores by heating or with chemical solvents. Both processes release large quantities of toxic materials that can be even more environmentally hazardous than mining. Smelting—roasting ore to release metals—is a major source of air pollution. One of the most notorious cases of ecological devastation from smelting was a wasteland near Ducktown, Tennessee. In the mid-1800s, mining companies began excavating the rich

FIGURE 14.16 Mountaintop removal mining is a highly destructive, and deeply controversial, method of extracting Appalachian coal.
© AP Photo/Bob Bird

copper deposits in the area. To extract copper from the ore, the company built huge, open-air wood fires, using timber from the surrounding forest. Dense clouds of sulfur dioxide released from sulfide ores poisoned the vegetation and acidified the soil over a 130 km^2 (50 mi^2) area. Rains washed the soil off the denuded land, creating a barren moonscape.

Sulfur emissions from Ducktown smelters were reduced in 1907 after Georgia sued Tennessee over air pollution. In the 1930s, the Tennessee Valley Authority (TVA) began treating the soil and replanting trees to cut down on erosion. Costly restoration efforts are still ongoing, a century after initial efforts to control sulfur emissions. While cases like this are unusual today in North America, they continue to occur elsewhere. The smelters of Norilsk, in Siberia, continue to make that city one of the world's most devastated places.

Heap-leach extraction is often used to dissolve and remove high-value metals, especially gold, from ore. A cyanide solution is sprayed on a large heap of crushed rock to dissolve gold. The solution is captured to recover the gold, but cyanide can leak into surface or groundwater. At the Beartrack Mine in east-central Idaho, for example, $220 million in gold was extracted with cyanide. When the mine closed in 2000, the company left 15 million metric tons of mine waste and huge ponds laced with cyanide. Meridian Gold, the mine owner, posted a $2 million bond to cover the costs of reclamation, but it may well cost 30 times that much to restore the area. Meanwhile, the acids, metals, and cyanide in the heap threaten the nearby Salmon River and its $2 billion annual tourism and fishing industry.

Ore washing and leaching consume a great deal of water. This is a concern in arid mining regions, such as the western United States. The USGS estimates that in Nevada, mining consumes about 230,000 m^3 (60 million gal) of fresh water per day. After use in ore processing, much of this water contains sulfuric acid, arsenic, and heavy metals, which can contaminate lakes and streams.

The Mineral Policy Center in Washington, D.C., estimates that mine drainage contaminates 19,000 km (12,000 mi) of rivers and streams in the United States. The EPA estimates that cleaning up impaired streams, along with 550,000 abandoned mines, in the United States could cost $70 billion. Worldwide, mine closing and rehabilitation costs are estimated in the trillions of dollars. Because of the volatile prices of metals and coal, many mining companies have gone bankrupt before restoring mine sites, leaving the public responsible for cleanup (fig. 14.17).

In 2015, a contractor working for the EPA was trying to drain toxic water from the abandoned Gold King mine outside the historic town of Silverton, Colorado. The mine, which closed in 1923 had been leaking into a nearby creek for decades and had killed almost all aquatic life there. The workers misjudged the volume and pressure of water in the mine. A blowout occurred, and at least 11 million liters (3 million gal) of highly toxic sludge gushed out into the Animas River, which flows into the Colorado River, a body that provides drinking water to 40 million people. The EPA apologized profusely for the error and tried to clean up the mess.

FIGURE 14.17 Thousands of "zombie" mines on public lands poison streams and groundwater with acid, metal-laced drainage.
© Bryan F. Peterson

Agency opponents howled that the EPA was deliberately trying to poison them; they never questioned why the mine was filled with toxic waste in the first place.

High-value minerals can support corruption

Political and human costs also result from mineral extraction and trade. Durable, highly valuable, and easily portable, gemstones and precious metals have long been a way to store and transport wealth. These properties have also led to illicit and criminal trade. Often, these valuable materials have bankrolled political dictators, criminal gangs, and terrorism around the world. In recent years, brutal civil wars in Africa have been financed—and often motivated—by gold, diamonds, tantalum ore (used in consumer electronics such as cell phones), and other high-priced commodities.

Although international agreements often ban trade that finances war and genocide, illegal trade continues. Much of the "blood diamond" trade ends up in the $100-billion-per-year global jewelry trade, two-thirds of which sells in the United States. Many people who treasure a diamond ring or a gold wedding band as a symbol of love and devotion are unaware that it may have been obtained through inhumane labor conditions and environmentally destructive mining and processing methods. Civil rights organizations are campaigning to require better documentation of the origins of gems and precious metals to prevent their use as financing for crimes against humanity. Conflict over oil fields, as in Iraq and Kuwait, is another form of violence over mineral resources. These conflicts are one argument that is often made for reducing our dependence on oil. Even where outright war is absent, violence and human rights abuses are widespread in oil-producing regions, as in Nigeria and Ecuador.

In 2004, a group of Nobel Peace Prize laureates asked the World Bank to overhaul its policies on lending for resource extractive industries. "War, poverty, climate change, and ongoing violations of human rights—all of these scourges are all too often linked to the oil and mining industries," wrote Archbishop Desmond Tutu, winner of the 1984 Nobel Peace Prize for helping to eliminate apartheid in South Africa. In response, the World Bank appointed an Extractive Industries Review committee. In its final report, the committee recommended that some areas of exceptionally high biodiversity value should be "no-go" zones for extractive industries, and it recommended that the rights of those affected by extractive projects need better protection. Such recommendations remain hard to follow, however, when so much money is at stake.

Every other year the Blacksmith Institute (www.blacksmithinstitute.org) compiles a list of the world's worst pollution problems. Among the top threats to human health are lead, mercury, chromium, and arsenic. Lead smelting and small-scale gold mining, which uses abundant mercury and has no environmental controls, are among the worst causes. These problems are especially disastrous in the developing world and in states of the former Soviet Union, where little money or political will is available to deal with environmental or health effects of pollution.

In the United States, where legal controls often are in place for environmental and public protection, mining also remains controversial. Often, the public bears the cost of cleanup, and must

Quantitative Reasoning

If the United States has 550,000 abandoned mines, and the estimated cleanup cost is $70 billion, what is the average cleanup cost per mine? How much is that per person, if there are just over 300 million Americans?

Should We Revise Mining Laws?

In 1872, the U.S. Congress passed the General Mining Law, intended to encourage prospectors to open up the public domain and promote commerce. This law, which has been in effect for more than a century, says, "All valuable mineral deposits in lands belonging to the United States are hereby declared to be free and open to exploration and purchase . . . by citizens of the United States." Claim holders can "patent" (buy) the land for $2.50 to $5 per acre (0.4 hectares), depending on the type of claim. Once the patent fee is paid, the owners can do anything they want with the land, just like any other private property. Although $2.50 per acre may have been a fair market value in 1872, many people regard it as ridiculously low today, amounting to a giveaway of public property.

In Nevada, for example, a Canadian mining company paid $9,000 for federal land that contains an estimated $20 billion worth of precious metals. Similarly, Colorado investors bought about 7,000 ha (17,000 acres) of rich oil-shale land in 1986 for $42,000 and sold it a month later for $37 million. You don't actually have to find any minerals to patent a claim. A Colorado company paid a total of $400 for 65 ha (160 acres) it claimed would be a gold mine. Almost 20 years later, no mining has been done, but the property—which just happens to border the Keystone Ski Area—is being subdivided for condos and vacation homes.

According to the Bureau of Land Management (BLM), some $4 billion in minerals are mined each year on U.S. public lands. Under the 1872 law, mining companies pay nothing to the government for the ores they take. Furthermore, they can deduct a depletion allowance from taxes on mineral profits. Former Senator Dale Bumpers of Arkansas, who calls the 1872 mining law "a license to steal," has estimated that the government could derive $320 million per year by charging an 8 percent royalty on all minerals and probably could save an equal amount by requiring larger bonds to be posted to clean up after mining is finished.

On the other hand, mining companies argue they would be forced to close down if they had to pay royalties. Jobs would be lost, and the economies of western mining towns would collapse if mining becomes uneconomic. We provide subsidies and economic incentives to many industries to stimulate economic growth. Why not support mining for metals essential for our industrial economy? Mining is a risky and expensive business. Without subsidies, mines would close down and we would be completely dependent on unstable foreign supplies.

Mining critics respond that other resource-based industries have been forced to pay royalties on materials they extract from public lands. Coal, oil, and gas companies pay 12.5 percent royalties on fossil fuels obtained from public lands. Timber companies—although they don't pay the full costs of the trees they take—have to bid on logging sales and clean up when they are finished. Even gravel companies pay for digging up the public domain. Ironically, we charge for digging up gravel, but give gold away for free.

Numerous bills have been introduced in Congress to revise the mining law. Bills supported by public interest groups would require companies mining on federal lands to pay a higher royalty on their production. They also would eliminate the patenting process, impose stricter reclamation requirements, and give federal managers authority to deny inappropriate permits. In contrast, bills offered by western legislators and backed by mining supporters would leave most provisions of the 1872 bill in place.

A current controversy involves a subsidiary of Antofagasta, a Chilean company that wants to dig a 300 m (1,000 ft) deep open pit copper–nickel mine next to (and in one section under) a river flowing into the Boundary Canoe Area Wilderness in northern Minnesota. The million acre (400,000 ha) wilderness is the most heavily visited wilderness in the United States, and the only one devoted to canoeing on lakes and rivers currently clean enough to drink. The ore is less than 1 percent copper, so more than 99 tons of rock have to be dug up to get one ton of ore. The ore is copper sulfide. When crushed and exposed to air, the sulfide oxidizes to form sulfuric acid. Opponents argue that the acids, heavy metals (including mercury, arsenic, lead, and selenium) leached out of waste rock will poison water in the wilderness. The company owns the mineral rights, but wants to pile up the 400 million tons of waste they'll produce on state and federal land. The mine would only be in operation for about 20 years, but waste water would need to be treated forever. Is this a worthwhile trade off?

What do you think we should do about our mining laws? Do they remain necessary and justifiable, or should they be revoked or revised? How would you write the law if you had the chance?

A Chilean company with a poor environmental record wants to dig an enormous open-pit copper–nickel mine immediately adjacent to Minnesota's Boundary Waters Canoe Area Wilderness. Does 20 years of a few dozen local jobs justify risking a million acres of wilderness?

© William P. Cunningham

---

also provide the resources nearly free of charge. On the other hand, mining on public lands is a way of boosting the economy, as long as costs don't outweigh benefits (see What Do You Think? reading, "Should We Revise Mining Laws?").

Section Review

1. Why is sulfuric acid common in mine runoff?
2. Describe smelting and some of its environmental effects.
3. Why might restoration be hard after mountaintop removal?

14.4 Geological Hazards

- *Earthquakes result from sudden shifts of tectonic plates.*
- *Earthquakes and floods cause the greatest damage and mortality of all geologic hazards.*
- *Erosion and landslides frequently cause property damage.*

Earthquakes, volcanic eruptions, floods, and landslides are among the geological forces that have shaped the world around us (fig. 14.18). Catastrophic events have triggered mass extinctions and reset the course of evolution: The impact of a giant asteroid off the coast of Yucatán 65 million years ago is believed to have ended the age of dinosaurs. It created a tsunami hundreds of meters high that likely swept around the world several times before subsiding. This impact also ejected so much dust into the air that sunlight was blocked for years, and a global winter decimated much of the life on the earth. Similarly, volcanic eruptions 250 million years ago, which covered 2 million km² of Siberia with basalt up to 2 km deep, may have caused the loss of 90 percent of species at the end of the Permian era.

Fortunately, such massive events are rare. Still, geological hazards are a huge threat. Among them, floods take the largest number of human lives, followed by earthquakes.

Earthquakes usually occur on plate margins

Earthquakes are sudden movements in the earth's crust that occur along faults (planes of weakness) where one plate slides past another. Only floods have caused more widely devastating events. A 2010 earthquake in Haiti killed perhaps a quarter million people; the impoverished island nation is still reeling from its effects. Still worse was a 1976 earthquake in Tangshan, China. Government officials reported 655,000 deaths (some geologists doubt the total was that high).

The San Andreas fault in California is one of the most notorious and highly visible faults in the world (fig. 14.19). It runs about 1,300 km (810 mi) from the Salton Sea in southern California to Point Reyes, just north of San Francisco, where it veers offshore to follow the coast north nearly to Oregon. The fault occurs where the Pacific plate is moving to the northwest, past the North American plate, which is moving south. A number of major quakes have originated along this fault, including those that struck San Francisco in 1906 and 1989.

Earthquakes vary dramatically in the amount of energy and movement released (table 14.4). When movement along faults occurs gradually and relatively smoothly, it is called creep or seismic slip and may be undetectable to the casual observer. When friction prevents rocks from slipping easily, stress builds up until it is finally released with a sudden jerk, which can cause the ground to shake like jelly.

FIGURE 14.19 The Elkhorn Escarpment shows where the San Andreas fault runs across the Carrizo Plain northwest of Los Angeles.
© StockTrek/Getty Images RF

Most earthquake deaths result from the collapse of weakly constructed buildings (fig. 14.20). Cities built on soft landfill or poorly consolidated soil, like much of San Francisco or Port-au-Prince, Haiti, usually suffer the greatest damage from earthquakes. Water-saturated soil can liquefy when shaken. Under these conditions, buildings sometimes sink out of sight or fall like rows of dominoes.

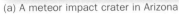

(a) A meteor impact crater in Arizona

(b) Volcanic eruption

(c) Glacier in Alaska

FIGURE 14.18 Geological events such as meteor or asteroid impacts (a), massive volcanic eruptions (b), or climate change (c) are thought to trigger mass extinctions that mark major eras in the earth's history.
(a) © StockTrek/Getty Images RF; (b–c) © Corbis RF

Table 14.4	Worldwide Frequency of Earthquakes	
Richter Scale Magnitude*	Description	Average Number per Year
2–2.9	Unnoticeable	300,000
3–3.9	Smallest	felt 49,000
4–4.9	Minor earthquake	6,200
5–5.9	Damaging earthquake	800
6–6.9	Destructive earthquake	120
7–7.9	Major earthquake	18
>8	Great earthquake	1 or 2

* For every unit increase in the Richter scale, ground displacement increases by a factor of 10, while energy release increases by a factor of 30. There is no upper limit to the scale, but the largest earthquake recorded was 9.5, in Chile in 1960.

Source: B. Gutenberg in Earth by F. Press and R. Siever, 1978, W. H. Freeman & Company.

Increasingly, builders in earthquake zones are designing buildings to withstand tremors. The primary methods used are heavily reinforced structures, strategically placed weak spots in the building that can absorb vibration from the rest of the building, and pads or floats beneath the building on which it can shift harmlessly with ground motion.

Human-induced earthquakes are becoming more common

We used to think that humans couldn't cause earthquakes, but surges in seismic activity associated with well drilling and disposal of waste water are changing our assumptions. Between 2000 and 2007, Oklahoma, for example, averaged about 20 measurable earthquakes

FIGURE 14.20 Earthquakes are most devastating where building methods can't withstand shaking. The 2010 earthquake in Haiti is thought to have killed 230,000 people. Collapsed buildings, food and water shortages, infectious diseases, and exposure to the elements all contributed to the death toll.
Source: USGS photo by Walter D. Mooney

per year. Starting about 2008, that number began shooting up to more than 6,000 tremors in 2015. This was more than all the other 48 continental states combined. Oklahoma has become the earthquake capital of the U.S. Some towns that had never experienced a quake had as many as 50 in a single day in 2015. This swarm of quakes correlates with a rush of oil and gas well drilling, and in particular with a proliferation of waste water wells. Between 2008 and 2014, oil and gas production in Oklahoma more than doubled.

Oklahoma has a waste water disposal problem. Some oil wells there produce as much as 65 barrels of salty waste water for every barrel of oil. The waste is much too toxic to dump into surface water, and it would cost more than the oil is worth to treat it, so most of it is pumped back down into deep geological formations. In north central Oklahoma alone, about 1,000 wells injected some 17 billion liters (850 million barrels) of wastewater into deep strata in 2014. The result, many seismologists believe, was a flood of earthquakes.

Arkansas, Kansas, and Colorado also have experienced dramatic upticks in seismic activity associated with disposal wells. Geologists now think injection water lubricates fault lines causing slippage that releases energy and results in quakes. Not all injection wells are equally guilty. In Arkansas, it was only necessary to plug four wells to prevent earthquakes. Oklahoma is considering limits on deep well disposal, but the economy may do that for them. In 2016, with oil at less than $30 per barrel, thousands of wells were capped, and earthquakes decreased by at least 20 percent. But will seismic activity go back to the low level of a decade ago, or have we unleashed a permanent geological hazard?

Tsunamis can be more damaging than the earthquakes that trigger them

Tsunamis are powerful waves triggered by earthquakes or landslides. The name is derived from the Japanese word for "harbor wave," because the waves often are noticed only when they approach shore. Tsunamis can be more damaging than the earthquakes that create them. The tsunami that struck the coast of Japan on March 11, 2011, for example, was triggered by a magnitude 9.0 underwater earthquake about 72 km (45 mi) out to sea. Although the earthquake shook buildings, it did relatively little damage because Japan has high construction standards. The waves that followed, however, were up to 38 m (124 ft) high. In some low-lying areas, the waves traveled up to 10 km (6 mi) inland, washing away buildings, boats, cars, and even whole trains. As the wall of water smashed through towns and villages, it carried a thick slurry of vehicles, building materials, mud, and debris that was much more destructive than the water alone. At least 25,000 people were listed as dead or missing, hundreds of thousands of buildings were damaged or destroyed (fig. 14.21).

The most long-lasting effect of that tsunami will be the damage it did to the Fukushima Daiichi nuclear power plant on Japan's northeast coast. The reactors shut down after the earthquake, as they were designed to do, but the tsunami destroyed the cooling systems (as well as the backup generators) needed to keep the reactors under control. Three of the six reactors at the Fukushima Daiichi complex were damaged by fires and hydrogen gas explosions after partial meltdown of the reactor cores, and a fourth was

FIGURE 14.21 In 2011, a magnitude 9.0 earthquake just off the coast of Japan created a massive tsunami that destroyed homes, killed at least 25,000 people, and destroyed four nuclear power plants.
© JIJI Press/AFP/Getty

damaged when water levels dropped in the waste fuel storage pool, exposing fuel rods to the air.

Some 230,000 residents were evacuated from around the damaged reactors, and many will never move back. It's estimated that five million tons of debris (some of it radioactive) washed into the sea. Within a year, some of that waste reached North America. Japan estimated economic losses of as much as $300 billion, not including costs such as the cleanup of the nuclear reactors, several of which will probably have to be entombed in concrete forever.

Arguably, the highest tsunami risk in North America is the Pacific coast from British Columbia to northern California. A complex of subduction zones called the Cascadia fault stretches for 1,000 km (620 mi) along this coast. The last recorded major earthquake along this fault was in 1700. Records of an "orphan" tsunami in Japan match Native American memories of whole villages being swallowed by the sea, and physical evidence of up to 2 meters of subsidence along the coast. The fault has been quiet for 300 years, but similar subduction zones tend to average earthquakes about every 100 to 200 years. A mega earthquake and tsunami along the fault today could cause huge damage because most cities, including Seattle, Portland, and Vancouver, are poorly prepared. Some geologists predict a 10 to 15 percent chance of a magnitude 9 earthquake and a tsunami 30 meters or more high in the next 50 years. Millions of people could be killed or displaced. Volcanic activity along this section of the "ring of fire" also presents a substantial risk.

Volcanoes eject gas and ash, as well as lava

Volcanoes and undersea magma vents produce much of the earth's crust. Over hundreds of millions of years, gaseous emissions from these sources formed the earth's earliest oceans and atmosphere.

Many of the world's fertile soils are weathered volcanic materials. Volcanoes have also been an ever-present threat to human populations. One of the most famous historic volcanic eruptions was that of Mount Vesuvius in southern Italy, which buried the cities of Herculaneum and Pompeii in 79 C.E. The mountain had been giving signs of activity before it erupted, but many citizens chose to stay and take a chance on survival. On August 24, the mountain buried the two towns in ash. Thousands were killed by the dense, hot, toxic gases that accompanied the ash flowing down from the volcano.

Nuées ardentes (French for "glowing clouds") are deadly, denser-than-air mixtures of hot gases and ash like those that inundated Pompeii and Herculaneum. These pyroclastic clouds can have temperatures exceeding 1,000°C, and they move at more than 100 km/hour (60 mph). In November 2010, pyroclastic clouds rolled down the slopes of Mt. Merapi just outside Yogyakarta, Indonesia. At least 325 people were killed, and more than 300,000 were forced from their homes (fig. 14.22).

It is not just a volcano's dust that blocks sunlight. Sulfur emissions from volcanic eruptions combine with rain and atmospheric moisture to produce sulfuric acid (H_2SO_4). The resulting droplets of H_2SO_4 interfere with solar radiation and can significantly cool the world climate. In 1991, Mt. Pinatubo in the Philippines emitted 20 million tons of sulfur dioxide, which combined with water to form tiny droplets of sulfuric acid. This acid aerosol reached the stratosphere, where it circled the globe for two years. The high, thin haze cooled the entire earth by 1°C and postponed global warming for several years. It also caused a 10 to 15 percent reduction in stratospheric ozone, allowing increased ultraviolet light to reach the earth's surface.

FIGURE 14.22 Volcanic ash covers damaged houses and dead vegetation after the November 2010 eruption of Mt. Merapi *(background)* in central Java. More than 300,000 residents were displaced and at least 325 were killed by this multiday eruption.
© Bay Ismoyo/AFP/Getty Images

Landslides and mass wasting can bury villages

Gravity constantly pulls downward on every material everywhere on earth, causing a variety of phenomena collectively termed **mass wasting** or mass movement, in which geological materials are moved downslope from one place to another. The resulting movement is often slow and subtle, but some slope processes such as rockslides, avalanches, and land slumping can be swift, dangerous, and very obvious. *Landslide* is a general term for rapid downslope movement of soil or rock. In the United States alone, over $1 billion in property damage is done every year by landslides and related mass wasting.

Activities such as road construction, forest clearing, agricultural cultivation, and building houses on steep, unstable slopes increase both the frequency and the damage done by landslides. In some cases, people are unaware of the risks they face by locating on or under unstable hillsides. In other cases, they simply ignore clear and obvious danger. Southern California, where people build expensive houses on steep hills of relatively unconsolidated soil, is often the site of large economic losses from landslides. Especially after chaparral fires remove plant cover, winter rains can saturate and loosen the exposed soil. These conditions frequently cause slope failure (fig. 14.23). In narrow canyons, mudslides can carry away whole neighborhoods and bury downslope areas in debris flows.

Floods are the greatest geological hazard

Like earthquakes and volcanoes, floods are normal events that cause damage when people get in the way. As rivers carve and shape the landscape, they build broad floodplains and level expanses that are periodically inundated. Many cities have been built on these flat, fertile plains, which are both good for agriculture and convenient to the river. When floods occur irregularly, people develop a false sense of security. But eventually most floodplains do flood.

Among direct natural disasters, floods take the largest number of human lives and cause the most property damage. A flood on the Yangtze River in China in 1931 killed 3.7 million people, making it the most deadly natural disaster in recorded history. In another flood, on China's Yellow River in 1959, about 2 million people died mostly due to resultant famine and disease.

Some more recent flooding disasters include Bangladesh in 2004, where 25 million people were displaced when torrential monsoon rains flooded about one-fourth of the low-lying country (fig. 14.24). Similarly, in 2008 the cyclone Nargis hit the Irrawaddy Delta of Myanmar. The government made the situation worse by refusing international assistance. It took weeks for aid to reach some areas. An estimated 140,000 people died, either

FIGURE 14.23 Landslides tend to occur on steep, unconsolidated hillsides, like this one in Laguna Beach, California.
© Nick UT/AP Wide World Photos

FIGURE 14.24 In 2004, unusually strong monsoon rains caused rivers to overflow their banks, and about one-quarter of Bangladesh was under water. About 25 million people were displaced.
© Digital Vision/PunchStock RF

directly from the flood or from starvation and disease that followed. Pakistan's floods of 2010 killed 1,800 people, and 21 million people were displaced.

Are these recent disasters related to global climate change? Many climate scientists argue that increasing heat storage in the atmosphere is beginning to cause more extreme weather events, including severe droughts in some places and more intense rainfall in others. Other human activities also increase the severity and frequency of floods. Covering the land with hardened surfaces, such as roads, parking lots, and building roofs, reduces water infiltration into the soil and speeds the rate of runoff into streams and lakes. Clearing forests for agriculture and destroying natural wetlands also increases both the volume and the rate of water discharge after a storm.

In an effort to control floods, many communities build levees and floodwalls to keep water within riverbanks, and river channels are dredged and deepened to allow water to recede faster. Every flood-control structure simply transfers the problem downstream, however, and the water has to go somewhere. If it doesn't soak into the ground upstream, it will simply exacerbate floods somewhere downstream.

Rather than spend money on levees and floodwalls, many people think it would be better to restore wetlands, replace groundcover on water courses, build check dams on small streams, move buildings off the floodplain, and undertake other nonstructural ways of reducing flood danger. Floodplains could be used for wildlife habitat, parks, recreation areas, and other uses not susceptible to flood damage.

The National Flood Insurance Program administered by the Federal Emergency Management Agency (FEMA) was intended to aid people who cannot buy insurance at reasonable rates, but its effects have been to encourage building on the floodplains by making people feel that, whatever happens, the government will take care of them. Many people would like to relocate homes and businesses out of harm's way after the recent floods, or improve them so they will be less susceptible to flooding, but owners of damaged property can collect only if they rebuild in the same place and in the same way as before. This perpetuates problems rather than solves them.

Beaches erode easily, especially in storms

Beach erosion occurs on all sandy shorelines because the motion of the waves is constantly redistributing sand and other sediments. One of the world's longest and most spectacular sand beaches runs down the Atlantic Coast of North America from New England to Florida and around the Gulf of Mexico. Much of this beach lies on some 350 long, narrow barrier islands that stand between the mainland and the open sea.

Early inhabitants recognized that exposed, sandy shores were hazardous places to live, and they settled on the bay side of barrier islands or as far upstream on coastal rivers as was practical. Modern residents, however, place a high value on living where they have an ocean view and ready access to the beach. Construction directly on beaches and barrier islands can cause irreparable

damage to the whole ecosystem. Under normal circumstances, fragile vegetative cover holds the shifting sand in place. Damaging this vegetation and breaching dunes with roads can destabilize barrier islands. Storms then wash away beaches or even whole islands. Hurricane Katrina in 2005 caused $100 billion in property damage along the Gulf Coast of the United States, mostly from the storm washing over barrier islands and coastlines (fig. 14.25). In the future, intensified storms and rising sea levels caused by global warming will make barrier islands and low-lying coastal areas even riskier places to live.

Cities and individual property owners often spend millions of dollars to protect beaches from erosion and repair damage after storms. Sand is dredged from the ocean floor or hauled in by the truckload, only to wash away again in the next storm. Building artificial barriers, such as groins or jetties, can trap migrating sand and build beaches in one area, but they often starve downstream beaches and make erosion there even worse.

As is the case for inland floodplains, many government policies encourage people to build where they probably shouldn't. Subsidies for road building and bridges, support for water and sewer projects, tax exemptions for second homes, flood insurance, and disaster relief are all good for the real estate and construction businesses but invite people to build in risky places. In some areas, beach houses have been rebuilt, mostly at public expense, four or five times in the past 20 years. Does it make sense to keep rebuilding in these places?

The Coastal Barrier Resources Act of 1982 prohibited federal support, including flood insurance, for development on sensitive

FIGURE 14.25 The aftermath of Hurricane Katrina on Dauphin Island, Alabama. Since 1970, this barrier island at the mouth of Mobile Bay has been overwashed at least five times. More than 20 million yd^3 (15 million m^3) of sand has been dredged or trucked in to restore the island. Some beach houses have been rebuilt, mostly at public expense, five times. Does it make sense to keep rebuilding in such an exposed place?
Source: USGS/NASA

islands and beaches. In 1992, however, the U.S. Supreme Court ruled that ordinances forbidding floodplain development amount to an unconstitutional "taking," or confiscation, of private property. The debate over property rights and sound environmental policy is likely to continue.

Section Review

1. What are the main reasons for mortality with earthquakes?
2. Describe different types of volcanic emissions and their effects.
3. Is it a good idea to build on barrier islands? Why or why not?

Conclusion

We need materials from the earth to sustain our modern lifestyle, but many of the methods we use to get those materials have severe environmental consequences. We can extend resources through recycling and by substituting old methods with new materials and more efficient ways of using them. We also can do much to repair the damage caused by resource extraction, although open-pit mines and mountains with their tops removed will never be returned to their original condition. Water contamination is one of the main environmental costs of mining. Pollutants include acids, cyanide, mercury, heavy metals, and sediment. Air pollution from smelting also can cause widespread damage.

Geological hazards, including floods, earthquakes, volcanoes, landslides, and coastal erosion, can cause tremendous damage and loss of life. Usually, poor building practices or land use decisions contribute to these disasters. Many of these events also occur rarely, so the risks are easy to ignore until catastrophe strikes. Although the earth seems stable most of the time, floods, earthquakes, and other hazards remind us that the same forces that have modified the earth in the past are still at work, continuing to reshape the land we live on.

Reviewing Key Terms

Can you define the following terms in environmental science?

core 14.1
crust 14.1
earthquakes 14.4
igneous rocks 14.1
magma 14.1

mantle 14.1
mass wasting 14.4
metamorphic rock 14.1
mid-ocean ridges 14.1
mineral 14.1

rock 14.1
rock cycle 14.1
sedimentary rock 14.1
sedimentation 14.1
subduction 14.1

tectonic plates 14.1
tsunami 14.4
volcano 14.4
weathering 14.1

Critical Thinking and Discussion Questions

1. Look at the walls, floors, appliances, interior, and exterior of the building around you. How many earth materials were used in their construction?

2. Is your local bedrock igneous, metamorphic, or sedimentary? How do you know? If you don't know, who might be able to tell you?

3. Suppose you live in a small, mineral-rich country, and a large, foreign mining company proposes to mine and market your minerals. How should revenue be divided between the company, your government, and citizens of the country? Who bears the greatest costs? How should displaced people be compensated? Who will make sure that compensation is fairly distributed?

4. Geological hazards affect a minority of the population that builds houses on unstable hillsides, in flood-prone areas, or on faults. What should society do to ensure safety for these people and their property?

5. A persistent question in this chapter is how to reconcile our responsibility as consumers with the damage from mineral extraction and processing. How responsible are you? What are some steps you could take to reduce this damage?

6. If gold jewelry is responsible for environmental and social devastation, should we stop wearing it? Should we worry about the economy in producing areas if we stop buying gold and diamonds? What further information would you need to answer these questions?

Data Analysis

Mapping Geological Hazards

Find an animated map of volcanoes in the animated National Atlas at www.nationalatlas.gov/dynamic.html. Where are most of the volcanoes in the lower 48 states? Roll over the red dots to see photos and elevations of the volcanoes.

You can explore recent earthquakes, where they were, and how big they were, by looking at the USGS earthquake information page at www.earthquake.usgs.gov/. Click on the map to investigate recent events, and go to Connect to find study questions for both volcanoes and earthquakes.

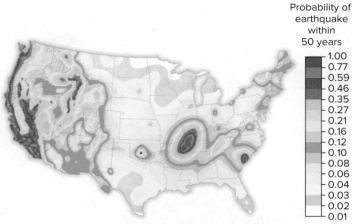

Probability of earthquake within 50 years

1.00
0.77
0.59
0.46
0.35
0.27
0.21
0.16
0.12
0.10
0.08
0.06
0.04
0.03
0.02
0.01

Go to Connect to explore earthquake and volcano occurrences.

TO ACCESS ADDITIONAL RESOURCES FOR THIS CHAPTER, PLEASE VISIT CONNECT AT www.connect.mheducation.com.

You will find LearnSmart, an adaptive learning system, Google Earth™ exercises, additional Case Studies, Data Analysis exercises, and an interactive ebcok.

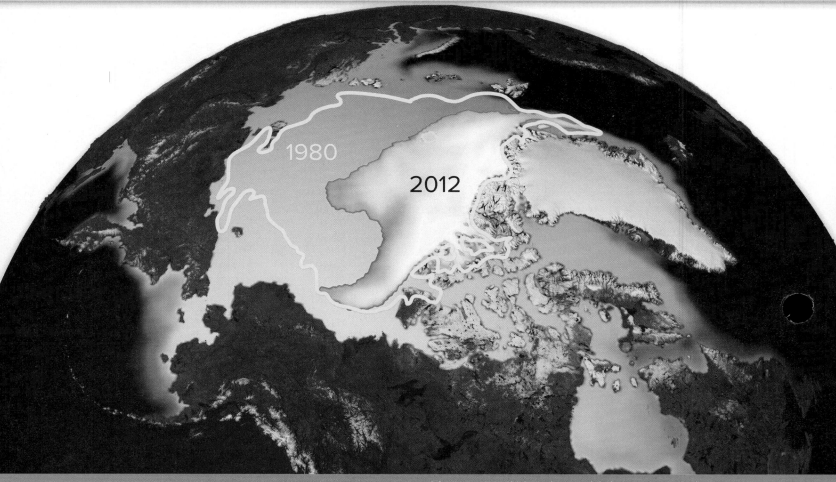

1980

2012

▲ Arctic sea ice, which moderates global temperatures, is disappearing at an accelerating rate. Summer 2012 had the lowest ice extent and thickness on record.
Source: NASA's Scientific Visualization Studio

Learning Outcomes

After studying this chapter, you should be able to:

15.1 Describe the general composition and structure of the atmosphere.

15.2 Explain processes that shape regional and seasonal weather patterns.

15.3 Outline some factors in natural climate variability.

15.4 Explain how we know that recent climate change is human-caused.

15.5 List some effects of climate change.

15.6 Identify some solutions being developed to slow climate change.

"I was born in 1992. You have been negotiating all my life. You cannot tell me you need more time."

– Christina Ora, youth delegate from the Solomon Islands addressing the plenary at COP15, 2009

When Wedges Do More than Silver Bullets

Each year, we hear new reports of record heat waves, droughts, declining water supplies, unpredictable weather, and volatile crop prices. Arctic sea ice, which reflects solar energy and helps stabilize global climate, has declined by half in summer and continues to set new low records in extent and thickness. All these changes are consistent with projections of climate models, as increasing concentrations of greenhouse gases retain more and more energy in our atmosphere (fig. 15.1).

Climate data show that the past 30 years were the warmest period in 1,300 years. Even slight warming allows new crop pests and weeds to survive winters, or it dries the soil enough to force farmers to irrigate crops more where irrigation is possible, or to abandon farms where it's not. In California and other parts of the western United States, cities rely on mountain snowmelt for water. Declining snow cover has become a severe problem, but still we have a hard time agreeing to policies to reduce greenhouse gas emissions.

Meanwhile, data indicate that if we don't reduce our carbon output in the next few years, we will permanently lose ice caps and permafrost, which will accelerate change. Soon, we will be on a path for irreversible and unavoidable increases of 5–7°C within the coming century, with sea-level rises of 1 m or more by 2100.

Among climate scientists, it is agreed that our carbon emissions are triggering climate change, and that change will be extraordinarily costly, in both human and economic terms. Remaining debates are only about details: how fast sea levels are likely to rise, where drought will be worst, or how to fine-tune the climate models.

Among policymakers, it's another matter. Politicians are responsible for establishing new rules to reduce our carbon output, but many still have a hard time understanding the connection between climate change and the increasing incidence of forest fires, drought, water shortages, heat waves, and pest outbreaks. The energy industry also generously funds campaigns to sow doubt in the minds of policymakers and the public.

Many politicians have hoped for a silver bullet—a technology that will fix the problem all at once—such as nuclear fusion, or space-based solar energy, or giant mirrors that would reflect solar energy away from the earth's surface. These are intriguing ideas, but none are workable *now,* and climate scientists are warning us that immediate action is critical to avoid tipping points such as the loss of polar ice, ancient arctic permafrost, and glaciers.

Wedges Can Work Now

To help us out of this quagmire of indecision, a Princeton ecologist and an engineer proposed a completely different approach to imagining alternatives: **wedge analysis,** or breaking down a large problem into smaller, bite-size pieces. By calculating the contribution of each wedge, we can add them up, see the magnitude of their collective effect, and decide that it's worth trying to move forward.

Stephen Pacala and Robert Socolow, of Princeton University's Climate Mitigation Initiative, introduced the wedge idea in a 2004 article in the journal *Science.* They showed that currently available technologies—efficient vehicles, buildings, power plants, alternative fuels—could solve our problems today, if we just take them seriously. Pacala and Socolow have further refined their ideas in subsequent papers, and others have picked up the wedge idea to envision strategies for problems such as reducing transportation energy use or water consumption.

Pacala and Socolow described three possible trajectories in our carbon emissions. The "business as usual" scenario follows the current pattern of constantly increasing CO_2 output. This trajectory heads toward a tripling of CO_2 by 2100, accompanied by temperature increases of around 5°C (9°F) and a sea-level rise of 0.5–1 m (fig. 15.2).

A second trajectory is a "stabilization scenario." In this scenario, we prevent further increases in CO_2 emissions, and we nearly double CO_2 in the atmosphere by 2100. Temperatures increase by about 2–3°C, and sea level rises by about 29–50 cm.

A third trajectory, declining CO_2 emissions, could result from new energy sources and better land management.

Change in temperature, sea level, and Northern Hemisphere snow cover

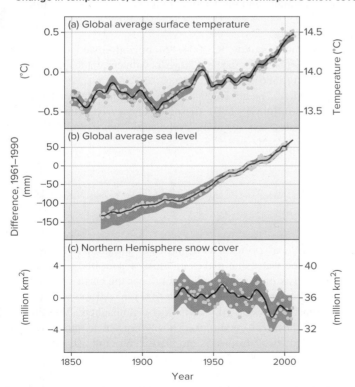

FIGURE 15.1 Observed changes in temperature, sea level, and snow cover. Blue shading shows uncertainty (range of possible values) for global averages (black lines), which are derived from observed data (gray dots).
Source: IPCC Fifth Assessment Report

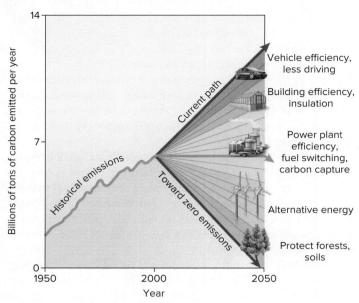

FIGURE 15.2 We could stabilize or even reduce carbon emissions now, if we focus on multiple modest strategies.

To achieve stabilization, we need to reduce our annual carbon emissions by 7 billion tons (or 7 gigatons, Gt) per year within 50 years. This 7 Gt is easier to consider as seven 1-Gt slices.

Cutting one of those gigatons could be accomplished by increasing fuel economy in our cars from 30 to 60 mpg. Another gigaton could be eliminated if we reduced reliance on cars (with more public transit or less suburban sprawl, for example) and cut driving from an average 10,000 miles to 5,000 miles per year. Better insulation and efficient appliances in our houses and office buildings would equal another wedge. Increased efficiency in our coal power plants would equal another wedge.

These steps add up to four-sevenths of the stabilization triangle, using currently available technologies. The remaining three-sevenths can be accomplished by capturing and storing carbon at power plants, improving plant operation, and slashing reliance on coal. Another set of seven wedges, including alternative energy, preventing deforestation, and reducing soil loss, could put us on a trajectory to reduce our CO_2 emissions and prevent disastrous rates of climate change.

These strategies also offer economic advantages. Improved efficiency and reduced energy consumption mean long-term cost savings. Efficient cars cut household expenses. Sustainable energy provides long-term employment stability, rather than the boom-and-bust instability of conventional fuels. We continually replace buildings, roads, and vehicles; if we start now to build them better, we could drastically cut our costs in the near future. Cleaner power sources will also reduce asthma and other respiratory illnesses, saving health care costs and improving quality of life.

Perhaps it's not surprising then that thousands of local communities are stepping up to lead the way on these initiatives, even while national governments dither. You don't have to care about climate change to agree about saving money and reducing smog. If you do care about climate change, it feels good to stop worrying and start acting.

In this chapter, we'll examine the composition and behavior of our atmosphere and the factors that make it change over time.

Further Reading:

Pacala, S., and Socolow, R. 2004. "Stabilization wedges: Solving the climate problem for the next 50 years with current technologies," *Science*, 305 (5686): 968–972.

Jacobson, M. Z., et al. 2015. "100% clean and renewable wind, water, and sunlight (WWS) all-sector energy roadmaps for the 50 United States." *Energy & Environmental Science*, 8: 2093–2117. See also: http://cmi.princeton.edu/wedges/.

15.1 WHAT IS THE ATMOSPHERE?

- *The atmosphere is layered, with more massive molecules near the ground surface.*
- *The troposphere is heated by the warmed earth surface; convection cells result as warmed air rises.*
- *The composition of the atmosphere influences how much heat energy is stored.*
- *Atmospheric circulation redistributes heat and moisture.*

Of all the planets in our solar system, only the Earth has an atmosphere that makes life possible, as far as we know. The atmosphere retains solar heat, protects us from deadly radiation in space, and distributes the water that makes up most of your body. The atmosphere consists of gas molecules, held near the earth's surface by gravity and extending upward about 500 km. All the weather we see is in just the lowest 10–12 km of the atmosphere. **Weather** is a term for the short-lived and local patterns of temperature and moisture that result from this circulation. In contrast, **climate** is long-term patterns of temperature and precipitation. Understanding the difference between short-term variations and long-term patterns is important in understanding our climate.

The earth's earliest atmosphere probably consisted mainly of lightweight hydrogen and helium. Over billions of years, most of that hydrogen and helium diffused into space. Volcanic emissions added carbon, nitrogen, oxygen, sulfur, and other elements to the atmosphere. Virtually all the molecular oxygen (O_2) that we breathe was probably produced by photosynthesis in blue-green bacteria, algae, and green plants.

Clean, dry air is mostly nitrogen and oxygen (table 15.1). Water vapor concentrations vary from near zero to 4 percent, depending on air temperature and available moisture. Minute particles and liquid droplets—collectively called **aerosols**—also are suspended in the air. Atmospheric aerosols are important in capturing, distributing, or reflecting energy.

The atmosphere has four distinct zones of contrasting temperature, which result from differences in absorption of solar energy (fig. 15.3). The layer of air immediately adjacent to the earth's surface is called the **troposphere** (*tropein* means to "turn" or "change" in Greek). Air in this layer is in constant motion, redistributing heat and moisture around the globe. Within the troposphere, air absorbs energy from the sun-warmed earth's surface, and from moisture evaporating from oceans. Warmed air circulates in great vertical and horizontal **convection currents,** which occur when warm, low-density air rises above a cooler, denser layer. (You can observe a similar process in a pot of simmering water on the stove: Water heated at the hot bottom of the pot rises up above the cooler layers at the top, creating convective circulation patterns.) Convection constantly redistributes heat and moisture around the globe. The depth of the troposphere ranges from about 18 km (11 mi) over the equator, where heating and convection are intense, to about 8 km (5 mi) over the poles, where air is cold and dense. The troposphere contains about 75 percent of the total mass of the atmosphere. Within the troposphere, temperatures drop rapidly with increasing distance from the earth, reaching about −60°C (−76°F). At this point, air is no longer warmer than its surroundings, and it ceases to rise. We call this boundary, where mixing ends, the tropopause.

The **stratosphere** extends from the tropopause up to about 50 km (31 mi). This layer is vastly more dilute than the troposphere,

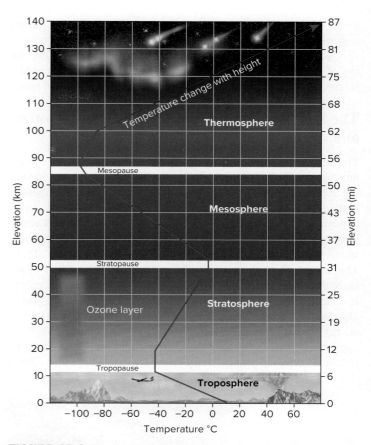

FIGURE 15.3 Layers of the atmosphere vary in temperature and composition. Our weather happens in the troposphere. Stratospheric ozone is important for blocking ultraviolet solar energy.

but it has similar composition—except that it has almost no water vapor and nearly 1,000 times more **ozone** (O_3). Ozone is a pollutant near the earth's surface, but in the stratosphere it serves a very important function: Ozone absorbs certain wavelengths of ultraviolet solar radiation, known as UV-B (290–330 nm; see fig. 3.11). This absorbed energy warms the upper stratosphere, so temperature increases with elevation. Stratospheric UV absorption also protects life on the earth's surface, because UV radiation damages living tissues, causing skin cancer, genetic mutations, and crop failures. A number of air pollutants, including Freon, once used in refrigerators, and bromine compounds, used as pesticides, deplete stratospheric ozone, especially over Antarctica. This has allowed increased amounts of UV radiation to reach the earth's surface (see fig. 16.19).

Unlike the troposphere, the stratosphere is relatively calm, because warm layers lie above colder layers. There is so little mixing that when volcanic ash or human-caused contaminants reach the stratosphere, they can remain in suspension there for years.

Above the stratosphere, the temperature diminishes again in the mesosphere, or middle layer. The thermosphere (heated layer) begins at about 80 km. This is a region of highly ionized (electrically charged) gases, heated by a steady flow of high-energy solar and cosmic radiation. In the thermosphere, intense pulses of high-energy radiation cause electrically charged particles (ions) to

Table 15.1	Present Composition of the Lower Atmosphere*	
Gas	Symbol or Formula	Percent by Volume
Nitrogen	N_2	78.08
Oxygen	O_2	20.94
Argon	Ar	0.934
Carbon dioxide	CO_2	0.035
Neon	Ne	0.00182
Helium	He	0.00052
Methane	CH_4	0.00015
Krypton	Kr	0.00011
Hydrogen	H_2	0.00005
Nitrous oxide	N_2O	0.00005
Xenon	Xe	0.000009

*Average composition of dry, clean air.

glow. We call this phenomenon the *aurora borealis* and *aurora australis*, or northern and southern lights.

No sharp boundary marks the end of the atmosphere. The density of gas molecules decreases with distance from the earth until it becomes indistinguishable from the near-vacuum of interstellar space.

The land surface absorbs solar energy to warm our world

The sun supplies the earth with an enormous amount of energy, but that energy is not evenly distributed over the globe. Incoming solar radiation (insolation) is much stronger near the equator than at high latitudes. Of the solar energy that reaches the outer atmosphere, about one-quarter is reflected by clouds and atmospheric gases, and another quarter is absorbed by carbon dioxide, water vapor, ozone, methane, and a few other gases (fig. 15.4). This energy absorption warms the atmosphere. About half of the incoming energy reaches the earth's surface. Some of this energy is reflected by bright surfaces, such as snow and ice. The rest is absorbed by the earth's surface and by water. Surfaces that *reflect* energy have a high **albedo** (reflectivity). Most of these surfaces appear bright to us because they reflect light and other forms of radiative energy. Surfaces that *absorb* energy have a low albedo

and generally appear dark. Black soil, pavement, and open water, for example, have low albedos (table 15.2).

Absorbed energy heats the absorbing surface (such as an asphalt parking lot that becomes hot in the summer sun), evaporates water, or provides the energy for photosynthesis in plants. Following the second law of thermodynamics, absorbed energy is gradually reemitted as lower-quality heat energy. A brick building, for example, absorbs energy in the form of light and reemits that energy in the form of heat.

Table 15.2 Albedo (Reflectivity) of Surfaces	
Surface	**Albedo (%)**
Fresh snow	80–85
Dense clouds	70–90
Water (low sun)	50–80
Sand	20–30
Water (sun overhead)	5
Forest	5–10
Black soil	3
Earth/atmosphere average	30

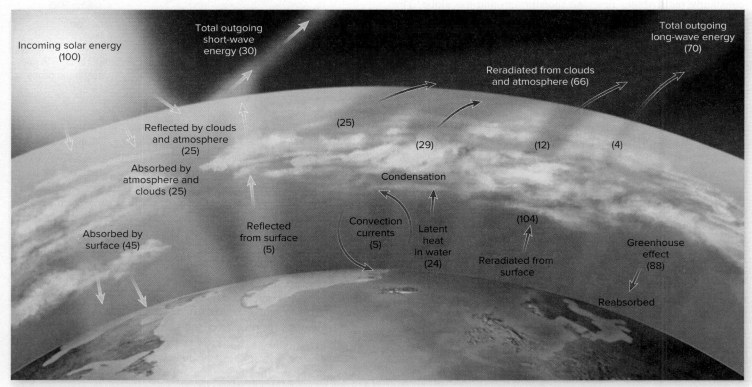

FIGURE 15.4 Energy balance in our atmosphere. Of the total incoming solar energy, atmospheric gases and clouds reflect about 25 percent, and the surface reflects roughly 5 percent. Gases and clouds also absorb about 25 percent, which re-radiates back to space as long-wave energy (red shading). The earth's surface absorbs the remaining 45 percent. Before escaping to space, this absorbed energy warms the air and contributes to convection (5 percent), evaporates water (24 percent), or re-radiates as long-wave energy, which has a greater warming effect than short-wave solar radiation. The equivalent of 104 percent of the incoming energy is radiated from the surface, with most of that absorbed and returned by clouds and gases (88 percent). This radiation warms the atmosphere. These values largely cancel each other out in the energy balance (104–88), with a net 12 percent released to space from the surface and 4 percent from clouds and atmospheric gases.

Water has low albedo and is extremely efficient at absorbing energy and storing it as heat. This is why increasing open water at the poles (shown in this chapter's opening photo) worries climatologists. For hundreds of thousands of years, the Arctic has been mostly white, reflecting most energy that reached the icy surface. Now the open water absorbs and stores that energy as heat, further accelerating ice melting and atmospheric warming. This is a good example of a **positive feedback loop,** in which melting leads to further melting, with probably dramatic consequences.

Gases in the atmosphere capture heat

Most solar energy is intense, high-energy light with short wavelengths near the ultraviolet and visible spectrum (see fig. 3.11). These short wavelengths pass relatively easily through the atmosphere to the earth's surface. Some of this energy is reflected back into space by the albedo effect. The rest of this energy is absorbed by the earth's surface and converted into heat, which is lower-intensity, longer-wavelength radiation in the far-infrared part of the spectrum. Certain gases in the atmosphere, especially carbon dioxide and water vapor, absorb much of this long-wavelength energy and re-release it in the lower atmosphere. This long-wave energy provides most of the heat in the lower atmosphere (see the red shading in fig. 15.4). If these gases were not in the atmosphere, the earth's average surface temperature would be about 18°C (33°F) colder than it is now, and there would be no liquid water on earth.

The retention of long-wave terrestrial energy in the atmosphere is called the **greenhouse effect** because the atmosphere, loosely comparable to the glass of a greenhouse, allows sunlight to travel through while trapping heat inside. The greenhouse effect is a natural atmospheric process that is necessary for life on earth. **Greenhouse gases** is a general term for trace gases that are especially effective at capturing the long-wavelength heat energy from the earth's surface. Water vapor (H_2O) is the most abundant greenhouse gas, and it is always present in the atmosphere at about the same amount (when levels of H_2O get high, it rains, removing that H_2O). Carbon dioxide (CO_2), methane (CH_4), nitrous oxide (N_2O), and a variety of fluorine gases are among the most effective greenhouse gasses.

Energy is redistributed around the globe

Incoming solar energy that passes all the way through our atmosphere is absorbed either by the ground surface or by water. This incoming solar energy is unevenly distributed around the globe. Near the equator, the sun shines straight overhead, and solar heating (and water evaporation) is intense. At mid-latitudes, around 30° to 60° north or south, the angle of the sun is lower, and heating is less intense. Near the poles, the sun shines low on the horizon much of the year and little heat is available for warming the ground or evaporating water.

Solar energy absorbed by the ground eventually re-radiates as long-wave infrared energy or heat, and this energy warms the lowest layers of air. Warmed air expands, becomes less dense than the cooler air above it, and begins to rise. Eventually, this warm air cools and then sinks, forming a circulation pattern known as a convection cell.

Circulation is more vigorous near the equator than at higher latitudes because of the more intense heating from the sun. We call the equatorial convection cells (one north and one south of the equator) Hadley cells (fig. 15.5). We call the mid-latitude and polar cells Ferrell cells and polar cells, respectively. These have less intense heating and less vigorous circulation than the equatorial Hadley cells.

Where air rises in convection cell currents, air pressure at the surface is low. Where air is sinking, or subsiding, air pressure is high (fig. 15.6). On a weather map, you can see these high- and low-pressure centers—spiraling currents of rising and sinking air—move across continents. In the Northern Hemisphere, these pressure centers generally move from west to east.

Incoming solar energy is also used to evaporate water. Each gram of evaporating water absorbs 580 calories of energy as it transforms from liquid to gas. This is known as **latent heat.** This rising air cools with altitude, and water vapor condenses to liquid droplets (which you can see as clouds). As it condenses, each gram of water releases 580 calories of heat. Thus, by moving water vapor, latent heat can warm and accelerate rising air currents in different locations.

Imagine the sun shining on the Gulf of Mexico in early spring. Warm sunshine and plenty of water allow continuous evaporation that converts an immense amount of solar energy into latent heat. Convection cells carry that warm humid air northward. Somewhere over the Midwest, this Gulf air is likely to encounter a cold, dry air mass moving south from Canada. The collision of these two air masses can cause the warm air to rise vigorously. Rain, snow, thunderstorms, or even tornadoes may result.

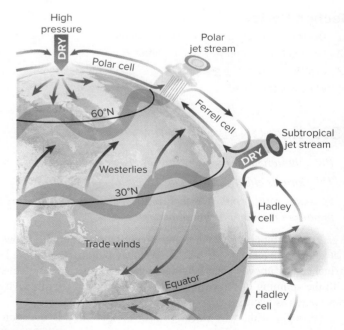

FIGURE 15.5 Convection cells circulate air, moisture, and heat around the globe. Jet streams develop where cells meet, and surface winds result from convection. Convection cells expand and shift seasonally.

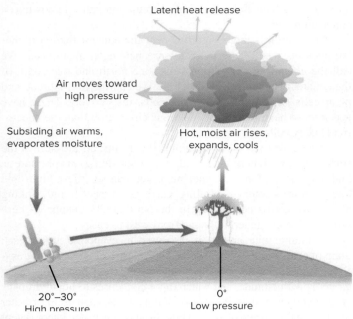

Latent heat release

Air moves toward
high pressure

Subsiding air warms,
evaporates moisture

Hot, moist air rises,
expands, cools

20°–30°
High pressure

0°
Low pressure

FIGURE 15.6 Convection currents distribute latent energy (heat in evaporated water) around the globe.

Atmospheric circulation carries a great deal of heat and moisture from warm humid places to colder, drier places. The redistribution of heat and water around the globe is essential to life on earth.

Pressure differences are an important cause of wind. There is always someplace with high-pressure (sinking) air and someplace with low-pressure (rising) air. Air moves from high-pressure centers toward low-pressure areas, and we call this movement wind.

Section Review

1. Describe the troposphere and stratosphere.
2. What is albedo, and why is it important?
3. Explain the idea of a greenhouse gas, and list four of them.
4. What is latent heat?

15.2 REGIONAL PATTERNS OF WEATHER

- *Precipitation happens when moisture cools and condenses.*
- *The Coriolis effect describes the apparent curving of winds.*
- *Seasonal rain, frontal weather, and cyclonic storms are three important types of precipitations.*

Weather is temperature, wind, and precipitation (rain and snow) that changes over days or weeks. Weather patterns are strongly influenced by uneven solar heating on the earth's surface, together with the spinning of the earth. In this section, we'll examine why those patterns occur.

To begin with, why does it rain? To answer this question, remember two things: Water condenses as air cools, and air cools as it rises. Any time air is rising, clouds, rain, or snow might form. Cooling occurs because of changes in pressure with altitude: Air

cools as it rises (as pressure decreases); air warms as it sinks (as pressure increases). Air rises in convection currents where solar heating is intense, such as over the equator. Air rises when two air masses collide and one must rise over the other. Air also rises when it encounters mountains. If the air is moist (if it has recently come from over an ocean or an evaporating forest region, for example), condensation and rainfall are likely as the air is lifted (fig. 15.6). Regions with intense solar heating, frequent colliding air masses, or mountains tend to receive a great deal of precipitation.

Where air is sinking, on the other hand, it tends to warm because of increasing pressure. As it warms, available moisture evaporates. Rainfall occurs relatively rarely in areas of high pressure. High pressure and clear, dry conditions occur where convection currents are sinking. High pressure also occurs where air sinks after flowing over mountains. Figure 15.6 shows sinking, dry air at about 30° north and south latitudes. If you look at a world map, you will see a band of deserts at approximately these latitudes.

Another ingredient is usually necessary to initiate condensation of water vapor: condensation nuclei. Tiny particles of smoke, dust, sea salts, spores, and volcanic ash all act as condensation nuclei. These particles form a surface on which water molecules can begin to coalesce. Without them, even supercooled vapor can remain in gaseous form.

The Coriolis effect explains why winds seem to curve

In the Northern Hemisphere, winds generally appear to bend clockwise (right), and in the Southern Hemisphere they appear to bend counterclockwise (left). Examples include the trade winds that brought Columbus to the Americas and the mid-latitude Westerlies that bring hurricanes north from Florida to North Carolina (see fig. 15.5). Ocean currents similarly curve clockwise in the Northern Hemisphere (the Gulf Stream) and counterclockwise in the Southern Hemisphere (the Humboldt Current near Peru). This curving pattern results from the fact that the earth rotates in an eastward direction as the winds move above the surface.

The apparent curvature of the winds is known as the **Coriolis effect.** How does this effect work? Imagine you are looking down on the North Pole of the rotating earth. The earth is spinning like a merry-go-round in a playground, with the North Pole at its center and the equator around the edge. As it spins counterclockwise (eastward), the spinning edge moves very fast (a full rotation, 39,800 km, every 24 hours for the real earth, or more than 1,600 km/hour). Near the center, though, there is very little eastward velocity, because the distance around a circle near the pole is relatively short.

Suppose you were standing at the center of an eastward spinning merry-go-round, looking at a person standing at the edge. You are looking straight at the other person, but she is moving much faster (in an eastward direction) than you are. If you threw a ball toward the person standing on the edge, she would have moved eastward by the time the ball reached the edge. To you, it would appear that the ball had swerved to the right, even though you actually threw it in a straight line. If you stood at the outer edge of a spinning merry-go-round and threw toward the center, the ball would leave your hand with your fast-moving eastward velocity, and it would maintain this eastward movement as it traveled toward the center. Again it would appear to you to bend to the right.

Winds move freely above the earth's surface much as the ball does. That is why on a weather map, weather patterns are shown curving to the right in the Northern Hemisphere. Ocean currents also curve to the right in the North. In the Southern Hemisphere, winds and currents curve to the left (It's a myth that bathtubs and sinks spiral in opposite directions in the Northern and Southern Hemispheres. Those movements are far too small to be affected by the spinning of the earth.)

On a global scale, this effect produces predictable weather patterns and currents. On a regional scale, the Coriolis effect produces "cyclonic" winds, or wind movements controlled by the earth's spin. Cyclonic winds spiral clockwise out of an area of high pressure (sinking air) in the Northern Hemisphere. They spiral counterclockwise into a low-pressure zone, because the air is moving upward: Imagine you were looking up from the ground into a rising column of air that is spinning to the right as you look up. On a weather map, looking down from the sky on that same rising air, you see it spinning counter-clockwise. This is why hurricanes (extremely strong cyclonic storms) in the North spin counterclockwise.

At the top of the troposphere are **jet streams,** hurricane-force winds that circle the earth. These powerful winds follow an undulating path approximately where the vertical convection currents known as the Hadley and Ferrell cells meet. The approximate path of one jet stream over the Northern Hemisphere is shown in figure 15.7. Although we can't perceive jet streams on the ground, they are important to us because they greatly affect weather patterns. Deeply wavy curvy jet-streams can bring cold polar temperatures far south, leading to periods that are colder than normal. There is some evidence that a warming climate may lead to wavier jet streams.

Ocean currents modify our weather

Ocean currents strongly influence climate conditions on land, depending upon whether the ocean water is warm or cold. Surface ocean currents result from wind pushing on the ocean surface, as well as from the Coriolis effect. As surface water moves, deep water wells up to replace it, creating deeper ocean currents. Differences in water density—depending on the temperature and saltiness of the water—also drive ocean circulation. Huge cycling currents called gyres carry water north and south, redistributing heat from low latitudes to high latitudes (see fig. 17.4). The Alaska current, flowing from Alaska southward to California, keeps San Francisco cool and foggy during the summer.

The Gulf Stream, one of the best-known currents, carries warm salty Caribbean water north past Canada to northern Europe. This current is immense, some 800 times the volume of the Amazon, the world's largest river. The heat transported from the Gulf Stream keeps Europe much warmer than it should be for its latitude. As the warm Gulf Stream passes Scandinavia and swirls around Iceland, the water cools and evaporates, becomes more dense, and sinks downward, creating a strong, deep, southward current. Oceanographer Wallace Broecker calls this the ocean conveyor system.

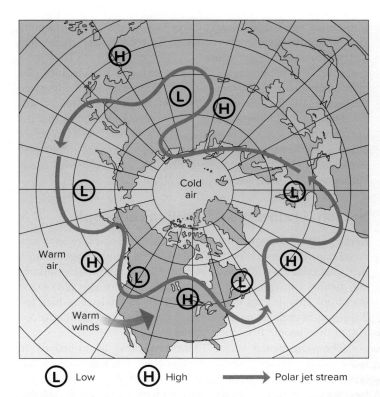

L Low   H High   ➝ Polar jet stream

FIGURE 15.7 A typical pattern of the northern circumpolar jet stream. This constantly shifting current of high altitude winds draws cold air southward along its snaking path and warm air northward. Low-pressure (stormy) and high-pressure (clear) weather patterns develop along these undulations.

Quantitative Reasoning

Find London and Copenhagen on a globe or a map. Then, in Canada find Churchill, Manitoba, and Edmonton, Alberta, at roughly the same latitudes. Temperatures in London and Copenhagen rarely fall much below freezing, but Churchill and Edmonton have freezing weather seven to eight months each year. What factors described on these pages could explain the difference?

Ocean circulation patterns were long thought to be unchanging, but now oceanographers believe that currents can change over time. About 11,000 years ago, for example, as the earth was gradually warming at the end of the Pleistocene ice ages, a huge body of meltwater, called Lake Agassiz, collected along the south margin of the North American ice sheet. At its peak it contained more water than all the current freshwater lakes in the world. Drainage of this lake to the east was blocked by ice covering what is now the Great Lakes. When that ice dam suddenly gave way, it's estimated that some 163,000 km^3 of fresh water roared down the St. Lawrence Seaway and out into the North Atlantic, where it layered on top of the ocean and prevented the sinking of deep, cold, dense seawater. This, in turn, apparently stopped the oceanic conveyor and plunged the whole planet back into an ice age (called the Younger Dryas, after a small tundra flower that became more common in colder conditions) that lasted for another 1,300 years.

Could this happen again? Freshwater inputs from melting glaciers and higher river flow around the Arctic are now flooding into the North Atlantic just where the Gulf Stream sinks and creates the deep south-flowing current. Already evidence shows that the deep return flow has weakened by about 30 percent. Even minor changes in the strength or path of the Gulf Stream might give northern Europe a climate more like that of Siberia—an ironic consequence of polar warming.

Seasonal rain supports billions of people

Large parts of the world, especially near the tropics, receive seasonal rains that sustain both ecosystems and human communities. Seasonal rains give life, but when they fail to arrive, crop failures and famine can result (fig. 15.8). Seasonal rains can also cause disastrous flooding, as in the 2010 floods in Pakistan, which left 2 million homeless, or the 2003 floods in China, which forced 100 million people from their homes.

The most regular seasonal rains are known as **monsoons.** In India and Bangladesh, monsoon rains come when seasonal winds blow hot, humid air from the Indian Ocean (fig. 15.9). The hot land surface produces strong convection currents that lift this air, causing heavy rain across the subcontinent. When the rising air reaches the Himalayas, it rises even further, creating some of the heaviest rainfall in the world. During the five-month rainy season of 1970, one weather station in the foothills of the Himalayas recorded 25 m (82 ft) of rain!

Tropical and subtropical regions around the world have seasonal rainy and dry seasons (see Chapter 5). The main reason for this variable climate is that the region of most intense solar heating and evaporation shifts through the year. Remember that the earth's axis of rotation is at an angle. In December and January, the sun is most intense just south of the equator; in June and July, the sun is most intense just north of the equator. Wherever the sun shines most directly, evaporation and convection currents—and rainfall and thunderstorms—are very strong.

As the earth orbits the sun, the tilt of its axis creates seasons with varying amounts of wind, rain, and heat or cold. Seasonal rains support seasonal tropical forests, and they fill some of the world's greatest rivers, including the Ganges and the Amazon. As the year shifts from summer to winter, solar heating weakens, the rainy season ends, and little rain may fall for months.

Frontal systems occur where warm and cold air meet

The boundary between two air masses of different temperature and density is called a front. When cooler air pushes away warmer air, we call the moving boundary a **cold front.** Cold, dense air of a cold front tends to hug the ground and push under the lighter, warmer air as it advances. As warm air is forced upward, it cools, and its

FIGURE 15.8 A shortage of monsoon rains brings drought, starvation, and death to both livestock and people in the Sahel desert margin of Africa. Although drought is a fact of life in Africa, many governments fail to plan for it, and human suffering is much worse than it needs to be.
© Mamadou Toure Behan/AFP/Getty Images

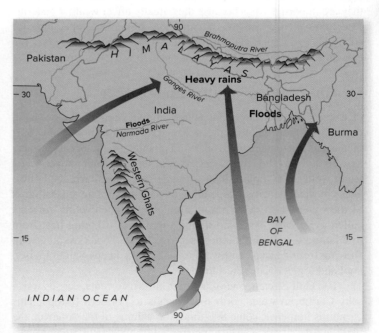

FIGURE 15.9 Summer monsoon air flows over the Indian subcontinent. Warming air rises over the plains of central India in the summer, creating a low-pressure cell that draws in warm, wet oceanic air. As this moist air rises over the Western Ghats or the Himalayas, it cools and heavy rains result. These monsoon rains flood the great rivers, bringing water for agriculture but also causing much suffering.

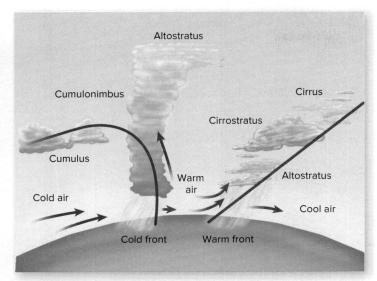

FIGURE 15.10 An advancing cold front has a steep profile, often producing vigorous storms as it forces warmer air upward. An advancing warm front, in contrast, rises over cooler air masses in a long, wedge-shaped zone. Warm fronts often produce light rain and cloudy days.

cargo of water vapor condenses to water droplets or ice crystals. Air masses near the ground move slowly because of friction and turbulence near the ground surface, so upper layers of a moving air mass often move ahead of the lower layers (fig. 15.10). Notice that the region of cloud formation and precipitation is relatively narrow. Cold fronts can generate strong convective currents as they push warmer air rapidly upward. Violent winds and thunderstorms can result, with towering thunderheads. The weather after the cold front passes is usually clear, dry, and invigorating.

In a **warm front,** the advancing air mass is warmer than surrounding air. Because warm air is less dense than cool air, an advancing warm front will slide up over cooler air masses, creating a long, wedge-shaped profile with a broad band of clouds and

precipitation (fig. 15.10, *right*). Gradual lifting and cooling in a warm front lacks the violent updrafts and strong convection currents that accompany a cold front. A warm front can have many layers of clouds at different heights. The highest layers are often wispy cirrus ("mare's tail") clouds, composed mainly of ice crystals, which can extend 1,000 km (600 mi) ahead of the front we detect at the ground level. Often, we see these wispy cirrus clouds a day or two before the warm front arrives on the ground. A moist warm front can bring days of drizzle and cloudy skies.

Cyclonic storms can cause extensive damage

Vigorously rising air can produce especially strong storm systems. If water vapor is abundant, as it is over a warm ocean, rising and cooling air releases latent heat. This released heat intensifies circulation, producing swirling winds that capture still more moisture and latent heat energy. These storms swirl in a direction dictated by the Coriolis effect (clockwise in the Northern Hemisphere, counterclockwise in the South) and are called **cyclonic storms.**

Over the warm low-latitude Atlantic, these cyclonic storms are called **hurricanes,** storms that can be hundreds of kilometers across with winds up to 320 km/hr (200 mph). Powerful winds drive storm surges (high water) far inland (fig. 15.11a). Hurricane winds can tear apart buildings, but most death and damages result from storm surges. Flooding after Hurricane Katrina, which devastated coastal Louisiana and the U.S. Gulf Coast in 2005, was as much as 9 m (29 ft) deep in some areas. Aided by canals for shipping and oil drilling, the surge destroyed large parts of New Orleans and several other cities, many of which still have not recovered (fig. 15.11b). Flooding in Superstorm Sandy in 2012, which traveled north as far as New York and Massachusetts, caused some $50 billion in damage. This was one of the most expensive storm events ever.

Tornadoes, swirling funnel clouds that form over land, also are considered cyclonic storms. Though never as large or powerful as hurricanes, tornadoes can be just as destructive in the limited areas where they touch down (fig. 15.11c). Tornadoes are

(a) Hurricane Katrina, 2005

(b) Gulf Shores, Alabama, 2005

(c) A tornado touches down

FIGURE 15.11 (a) Hurricane Katrina was hundreds of kilometers wide as it approached Louisiana in 2005. Note the hole, or eye, in the center of the storm. (b) Destruction caused by Hurricane Katrina in 2005. More than 230,000 km^2 (90,000 mi^2) of coastal areas were devastated by this massive storm, and many cities were almost completely demolished. (c) Tornadoes are much smaller than hurricanes, but can have stronger local winds.
(a) © StockTrek/Getty Images RF; (b) © Marc Serota/Corbis; (c) © 2010 Willoughby Owen/Getty Images RF

generated on the American Great Plains by giant "supercell" frontal systems where strong, dry-air cold fronts from Canada collide with warm, humid air moving north from the Gulf of Mexico.

Greater air temperature differences cause more powerful storms. This is why most tornadoes occur in the spring, when arctic cold fronts reach far south over the warming plains. As warm air rises rapidly over dense, cold air, intense vertical convection currents generate towering thunderheads with anvil-shaped leading edges and domed tops up to 20,000 m (65,000 ft) high. Water vapor cools and condenses as it rises, releasing latent heat and accelerating updrafts within the supercell. Sometimes penetrating into the stratosphere, the tops of these clouds can encounter jet streams, which help create even stronger convection currents.

Section Review

1. Why does it rain?
2. What is the Coriolis effect, and how might it cause trade winds?
3. Describe a monsoon, a cold front, and a warm front.
4. What is a cyclonic storm?

15.3 NATURAL CLIMATE VARIABILITY

- *Milankovitch cycles drive long-term climate changes.*
- *Ice cores contain CO_2 and oxygen isotopes used to reconstruct past temperatures and atmospheric composition.*
- *El Niño is one example of an ocean–atmosphere oscillation.*

Climatologists have studied many different sources of evidence to understand long-term climate changes. The main drivers in these long-term changes are known as **Milankovitch cycles,** after Serbian scientist Milutin Milankovitch, who first described them in the 1920s. These cycles influence how close the earth is to the sun and control how much of the sun's energy we receive, which in turn affects earth's climate. Long-term climate shifts can be seen in sedimentary rocks, as temperatures and rainfall shifted over millions of years (fig. 15.12).

These cycles are regular shifts in the earth's orbit and tilt (fig. 15.13), which change the distribution and intensity of sunlight reaching the earth's surface. The earth's elliptical orbit stretches and shortens in a 100,000-year cycle, while the axis of rotation changes its angle of tilt in a 40,000-year cycle. Over a 26,000-year period, the axis wobbles like an out-of-balance spinning top. Bands of sedimentary rock laid in the oceans seem to match these Milankovitch cycles. Periodic cold spells associated with worldwide expansion of glaciers every 100,000 years or so also correspond to these cycles.

Ice cores tell us about climate history

On shorter time frames of millennia and centuries, geologists and climatologists follow other lines of evidence. One is examining the widths of tree rings, which let them count back to find dry, drought-stressed years, or humid years of fast growth. Another is remains of plants, pollen, or microscopic organisms in lake beds and marine mud. Over really long time periods of millions of years, geologists use fossils and rock types to infer what the climate must have been like.

FIGURE 15.12 Banded rocks in these cliffs on the coast of France show a record of long-term climate cycles.
© Muriel Hazan/Science Source

Ice cores are among our key sources of data, because they provide reasonably long and detailed climate records. Every time it snows, small amounts of air are trapped in the snow layers. In Greenland, Antarctica, and other places where cold is persistent, yearly snows slowly accumulate over the centuries. New layers compress lower layers into ice, but still tiny air bubbles remain, even thousands of meters deep into glacial ice (fig. 15.13). Each bubble is a tiny sample of the atmosphere at the time the bubble was sealed off.

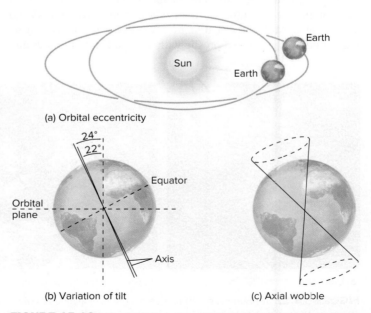

(a) Orbital eccentricity

(b) Variation of tilt

(c) Axial wobble

FIGURE 15.13 Milankovitch cycles, which may affect long-term climate conditions: (a) changes in the elliptical shape of the earth's orbit, (b) shifting tilt of the axis, and (c) wobble of the earth.

Climatologists have discovered that by drilling deep into an ice sheet, they can extract ice cores, from which they can collect and analyze air from these bubbles. Samples taken every few centimeters show how the atmosphere has changed over time. Ice core records have revolutionized our understanding of climate history (fig. 15.14). We can now see how concentrations of atmospheric CO_2 have changed over time. We can detect ash layers and spikes in sulfate concentrations that record volcanic eruptions. Most important, we can look at changes in air temperature using isotopes of oxygen in the water. In cold years, water molecules with slightly lighter oxygen atoms evaporate more easily than water with slightly heavier isotopes. Consequently, by looking at the proportions of heavier and lighter oxygen atoms (isotopes), climatologists can reconstruct temperatures over time and plot temperature changes against concentrations of CO_2 and other atmospheric gases.

The first very long record was from the Vostok ice core, which reached 3,100 m into the Antarctic ice and gives us a record of temperatures and atmospheric CO_2 over the past 420,000 years. A team of Russian scientists worked for 37 years at the Vostok site, about 1,000 km from the South Pole, to extract this ice core. A similar core has been drilled from the Greenland ice sheet. More recently, the European Project for Ice Coring in Antarctica (EPICA) has produced a record reaching back over 800,000 years (fig. 15.15). All these cores show that climate has varied dramatically over time but that there is a close correlation between atmospheric temperatures and CO_2 concentrations. From these ice cores, we know that CO_2 concentrations have varied between 180 to 300 ppm (parts per million) in the past 800,000 years. Therefore, we know that today's concentrations, 400 ppm, are about one-third higher than the earth has seen in nearly a million years. We also know that present temperatures are as warm as any in the ice core record. Further warming in the coming decades is likely to exceed anything in the ice core records, as the CO_2 gradually traps heats in the atmosphere.

The climate is warmer now than it has been since the development of civilization, agriculture, and urbanization as we know them. We know from historical accounts that slight climate shifts can be destabilizing for human communities. During the "little ice age" that began in the 1400s, a cooling climate caused crops to fail repeatedly in agricultural regions of northern Europe. Scandinavian settlements in Greenland founded during the warmer period around 1000 C.E. lost contact with Iceland and Europe as ice blocked shipping lanes. It became too cold to grow crops, and fish that once migrated along the coast stayed farther south. The Greenland settlers died out, perhaps in battles with Inuit people who were driven south from the high Arctic by colder weather.

Evidence from ice cores drilled in the Greenland ice cap suggests that world climate can change abruptly. It appears that during the last major interglacial period, 135,000 to 115,000 years ago, temperatures flipped suddenly from warm to cold or vice versa over a period of decades rather than centuries.

El Niño is an ocean–atmosphere cycle

On a shorter time scale, there are oscillations, or periodic shifts, in ocean currents and atmospheric circulation over years or decades. One important example is known as El Niño/Southern Oscillation.

Peruvian fishermen were the first to name these irregular cycles, because the weakened upwelling currents and warming water lead to the disappearance of the anchovy fishery. They named these events **El Niño** (Spanish for "the Christ child") because they were observed around Christmas time. Increased attention to these patterns has shown that sometimes, between El Niño events, coastal waters become extremely cool, and these extremes have come to be called **La Niña** (or "the little girl"). Together, this cycle is called the El Niño Southern Oscillation (ENSO).

(a)

(b)

FIGURE 15.14 (a) Ice cores contain air bubbles that give samples of the atmosphere thousands of years ago, as in this sample of 45,000-year-old ice from Antarctica (b). Dr. Mark Twickler, of the University of New Hampshire, holds a section of the 3,000 m Greenland ice sheet core, which records 250,000 years of climate history.

Courtesy. (a) © *Karin Kirk*; (b) *Source Candace Kohl, University of California, San Diego,*

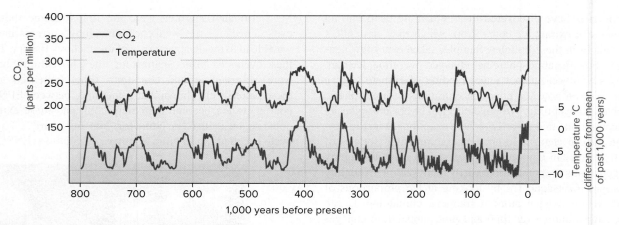

FIGURE 15.15 Atmospheric CO_2 concentrations (red line) correspond with temperatures (blue, derived from oxygen isotopes) in the Antarctic EPICA ice core. This 800,000-year record has no evidence of temperatures or CO_2 higher than that anticipated within the coming century.

Sources: UN Environment Programme; J. Bouzel et al., 2007; EPICA Dome C Ice Core 800KYr Deuterium Data and Temperature Estimates.

The core of the ENSO system is a huge pool of warm surface water in the Pacific Ocean that sloshes slowly back and forth between Indonesia and South America. In most years, steady equatorial trade winds push ocean surface currents westward, holding this warm pool in the western Pacific, roughly between Tahiti and Indonesia (fig. 15.16). Air above this warm region of the ocean is heated and moist, so it rises, producing strong convection cells in the atmosphere. Towering thunderheads created by rising air bring torrential summer rains to the tropical rainforests of northern Australia and Southeast Asia. Surface winds are drawn in from across the Pacific to replace the rising air. In the eastern Pacific, dry subsiding air currents contribute to arid conditions from Chile to southern California.

Trade winds also help drive ocean currents westward, setting up currents in the Pacific. Upwelling currents along the coast of South America draw cold, nutrient-rich water up from deep in the ocean. These nutrients support the dense schools of anchovies and other fish that traditionally support many fishermen in Peru.

Every three to five years, for reasons we don't fully understand, these convection currents weaken. The pool of warm surface water surges back east across the Pacific, toward Peru and California. The shift in position of the low-pressure area (the rising convection currents) has repercussions in weather systems across North and South America and perhaps around the world.

How does the ENSO cycle affect you? During an El Niño year, the northern jet stream—which normally is over Canada—splits and is drawn south over the United States. This pulls moist air inland from the Pacific and Gulf of Mexico, bringing intense storms and heavy rains from California across the Midwestern states. La Niña years bring extreme hot, dry weather to these same areas. El Niño events have brought historic floods to the Mississippi River basin, but Oregon, Washington, and British Columbia tend to be warm and dry in El Niño years. Droughts in Australia and Indonesia during El Niño episodes cause disastrous crop failures and forest fires.

ENSO-related droughts and floods are expected to intensify and become more irregular with global climate change, in part because the pool of warm water is warming and expanding. Extreme El Niños, such as the one in 1997/1998, may occur more frequently than they

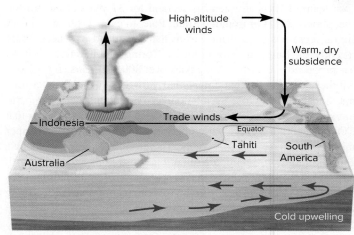

(a) Normal

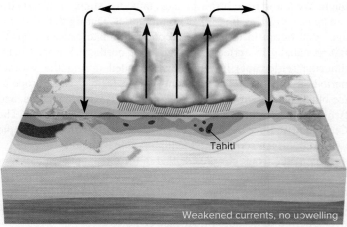

(b) El Niño

FIGURE 15.16 Normally, surface trade winds drive currents from South America toward Indonesia, and cold, deep water wells up near Peru. During El Niño years, winds and currents weaken, and warm, low-pressure conditions shift eastward, bringing storms to the Americas.

have in the past. High sea surface temperatures spawn larger and more violent storms such as hurricanes. On the other hand, increased cloud cover would raise the albedo while upwelling convection currents generated by these storms could pump heat into the stratosphere. This might have an overall cooling effect—a negative feedback in the warming climate system.

Climatologists have observed many decade-scale oscillations. The Pacific Decadal Oscillation (PDO), for example, involves a vast pool of warm water that moves back and forth across the North Pacific every 30 years or so. From about 1977 to 1997, surface water temperatures in the middle and western parts of the North Pacific Ocean were cooler than average, while waters off the western United States were warmer. During this time, salmon runs in Alaska were bountiful, while those in Washington and Oregon were greatly diminished. In 1997, however, ocean surface temperatures along the coast of western North America turned significantly cooler, perhaps marking a return to conditions that prevailed between 1947 and 1977. Under this cooler regime, Alaskan salmon runs declined while those in Washington and Oregon improved somewhat. A similar North Atlantic Oscillation (NAO) occurs between Canada and Europe.

Section Review

1. What are three Milankovitch cycles?
2. How might past climate be reconstructed from ice cores?
3. How do ocean and atmosphere interact in El Niño cycles?

15.4 ANTHROPOGENIC CLIMATE CHANGE

- *The Keeling curve documents changing CO_2 concentrations.*
- *Human activities alter greenhouse gas concentrations.*
- *There is clear evidence for a human role in current climate change.*

Many scientists regard anthropogenic (human-caused) global climate change to be the most important environmental issue of our times. The idea that humans might alter world climate is not new. In 1895, Svante Arrhenius, who subsequently received a Nobel Prize in Chemistry, predicted that CO_2 released by coal burning could cause global warming. At the time, this idea was mostly theoretical, though, and real impacts seemed unlikely.

The first data showing human impacts on atmospheric CO_2 came from an observatory on top of the Mauna Loa volcano in Hawaii. The observatory, established in 1957 as part of an International Geophysical Year, was built in a remote location to capture data on air chemistry in an undisturbed, pristine environment. Surprisingly, measurements showed CO_2 concentrations increasing about 0.5 percent per year. Levels have risen from 315 ppm in 1958 to over 400 ppm in 2016 (fig. 15.17). This graph, first produced by Charles David Keeling at the Mauna Loa observatory, is one of the first and most important pieces of evidence that demonstrates Svante Arrhenius's prediction.

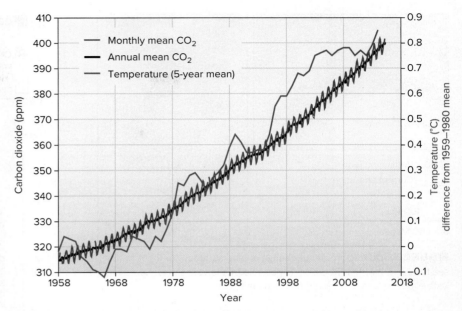

FIGURE 15.17 Measurements of atmospheric CO_2 taken at the top of Mauna Loa, Hawaii, show an increase of 1.5 to 2.5 percent each year in recent years. For carbon dioxide, monthly mean (red) and annual mean (black) carbon dioxide are shown. Temperature represents a five-year mean variation from the 1950–1980 average.
Source: Data from NOAA Earth System Research Laboratory, 2015.

Keeling's graph has some distinctive patterns. One is the annual variation in CO_2 concentrations: Every May, CO_2 levels drop slightly as plant growth on the vast northern continents capture CO_2 in photosynthesis; during the northern winter, levels rise again as respiration releases CO_2. Another pattern is that CO_2 levels are rising at an accelerating rate, currently more than 2 ppm each year. We are on track to double the pre-industrial concentration of CO_2, which was 280 ppm, in about a century's time.

The IPCC assesses climate data for policymakers

The climate system is extraordinarily complex, so a great deal of effort has been invested in analyzing observations like those from Mauna Loa. Since 1988, the **Intergovernmental Panel on Climate Change (IPCC)** has brought together scientists and government representatives from 130 countries to review scientific evidence on the causes and likely effects of climate change. The fifth Assessment Report (known as AR5) was published in 2013-2014, representing six years of work by 2,500 scientists. The IPCC reports include the findings that there is a 90 percent probability (it is "very likely") that recently observed climate changes result from human activities; for some changes, the report stated that it was "virtually certain" (a 99 percent probability) that they were anthropogenic (human-caused).

Among climate scientists who work with the data, there is no disagreement about whether human activities are causing current rapid climate changes. The IPCC projects warming of about 1–6°C by 2100, depending on what policies we follow to curb climate change. The IPCC's "best estimate" for the most likely scenario is 2–4°C (3–8°F). To put that in perspective, the average global temperature change between now and the middle of the last glacial period is about 5°C.

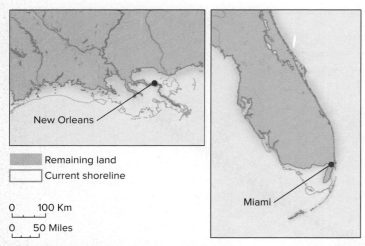

FIGURE 15.18 Approximate change in land surface with the 1-m (3-ft) sea-level rise that the IPCC says is possible by the year 2100. Some analysts expect a 2-m (6-ft) rise if no action is taken.

Droughts, heat stress, and increasing hurricane frequency (caused by warming oceans and atmosphere) could have disastrous human and economic costs. Melting ice on the Arctic Ocean, Greenland, and Antarctica is expected to contribute up to 1–2 m (about 3–6 ft) of sea-level rise. This increase would flood populous coastal regions, including low-lying cities such as Miami, New Orleans (fig. 15.18), Boston, New York, London, and Mumbai.

Human activities increase greenhouse gases

Earlier in this chapter, we described the greenhouse effect: gases in our atmosphere prevent long-wavelength (terrestrial) energy leaving the earth's surface from escaping to space (see fig. 15.5). The energy retained in our atmosphere keeps our earth warm enough for life as we know it. Certain gases such as water (H_2O) are especially effective at blocking or absorbing this long-wavelength energy. Human activity is not altering the overall concentrations of water in the atmosphere, but fossil fuel combustion, forest clearing, and agriculture have multiplied concentrations of several other greenhouse gases (fig. 15.19). Concentrations of these gases are low, less than 0.1 percent of the atmosphere, but multiplying even these low concentrations has considerably increased energy storage, raising both temperatures and storm activity (see section 15.1) in the atmosphere.

Carbon dioxide is the most important greenhouse gas because it is produced abundantly, it lasts decades to centuries in the atmosphere, and it is very effective at capturing long-wave energy. Emissions of CO_2 more than doubled in the 45 years from 1970 to 2015, from about 14 Gt/yr to more than 35 Gt/yr (fig. 15.19). Carbon dioxide contributes to about three-quarters (75 percent) of human-caused climate impacts. Burning fossil fuels is by far the greatest source of CO_2. Deforestation and other land-use changes are the second biggest

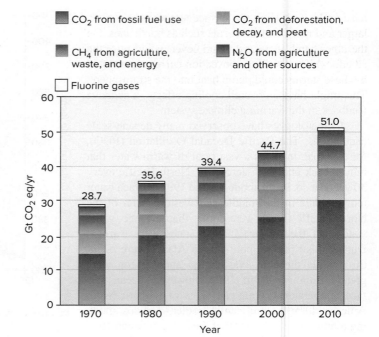

FIGURE 15.19 Contributions to climate change by different gases.
Source: IPCC, 2007

factor. Deforestation releases carbon stored in standing trees, and organic material in the exposed soil oxidizes and decays, producing still more CO_2 and CH_4. Cement production is also an important contributor, and cement for construction has recently helped push China into the lead for global CO_2 emissions (fig. 15.20).

Methane (CH_4) from agriculture and other sources is the second most important greenhouse gas, accounting for 14 percent of our greenhouse output. Methane absorbs 23 times as much energy per gram as CO_2 does, and it is accumulating at a faster rate than CO_2. Methane is produced when plant matter decays in oxygen-free conditions, as in the bottom of a wetland. (Where oxygen is abundant, decay produces mainly CO_2.) Methane is also released from natural gas wells. Rice paddies are a rich source of CH_4, as are ruminant animals, such as cattle. In a cow's stomach, which

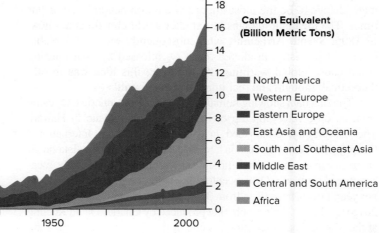

FIGURE 15.20 Carbon emissions by region since 1850. The two largest emitters, China (24%) and the United States (21%), produce nearly half of all emissions.
Source: Data from Boden, T. A., G. Marland, and R.J. Andres. 2010. Carbon Dioxide Information Analysis Center, Oak Ridge National Laboratory, U.S. Department of Energy

has little oxygen, digestion produces CH_4, which cows then burp into the atmosphere. A single cow can't produce much CH_4, but the global population of nearly 1 billion cattle produces enough methane to double the concentration naturally present in the atmosphere.

Nitrous oxide (N_2O), our third most important greenhouse gas, accounts for 8 percent of greenhouse gases. This gas is also released from agricultural processes, plant decay, vehicle engines, denitrification of soils, and other sources. Even though we don't produce as much N_2O as we do other greenhouse gases, this is an important gas because it is especially effective at capturing heat. Many other gases, including chlorofluorocarbons, sulfur hexafluoride, and other fluorine gases, make smaller contributions. Like N_2O and CH_4, these are emitted in relatively small amounts, but their ability to absorb specific energy wavelengths gives them a disproportionate effect.

One way to compare the importance of these various sources is to convert them all to equivalents of our most important greenhouse gas, CO_2. The units used on the Y-axis in fig. 15.19, gigatons of CO_2-equivalent per year (Gt CO_2 eq/yr), let us compare the effects of these sources. All four have increased, but fossil fuel burning rose the most between 1970 and 2010. This is a reason transportation and coal-burning power plants are two of the key sectors addressed in efforts to slow climate change.

The large orange bars in fig. 15.19 show that burning fossil fuels is also our most abundant source of greenhouse gases. Electricity production, transportation, heating, and industrial activities that depend on fossil fuels together produce 50 percent of our greenhouse gases. Deforestation and agriculture account for another 30 percent. The remaining 20 percent is produced by industry.

Positive feedbacks accelerate change

As noted earlier in this chapter, the melting of polar ice is a concern because it will increase energy absorption (water has a lower albedo than ice) and enhance warming globally. These and other feedbacks, and some tipping points at which sudden change occurs, are critical factors in climate change.

Another important feedback is the CO_2 release from warming and drying peat. Peat is soggy, semidecayed plant matter accumulated over thousands of years across the vast expanses of tundra in Canada and Siberia. As this peat thaws and dries, it oxidizes and decays, releasing more CO_2 and CH_4. A more ominous consequence of melting in the expanses of frozen arctic lands may be the release of vast stores of frozen, compressed CH_4 (methane hydrate) now locked in permafrost and ocean sediments. Release of these two carbon stores could add as much CO_2 to the atmosphere as all the fossil fuels ever burned.

Negative feedbacks are also possible: Increased ocean evaporation could intensify snowfall at high latitudes, restoring some of the high-albedo snow surfaces.

How do we know that recent change is caused by humans?

There are three main arguments that show how current climate change must be related to human activities. The first is that all the climate changes that we see happening correspond well with changes in human activities. For example, as our fossil fuel consumption has increased, the global average air temperature has increased. The increase in the world's GDP matches an increase in CO_2 emissions. Changes in atmospheric greenhouse gases occurred simultaneously with increases in human population and in the burning of coal and oil since the eighteenth century (fig. 15.21).

But just because changes are happening simultaneously does not mean that the change in one thing is *causing* the change in another. In an uncontrolled experiment, a model is usually the best way to demonstrate cause and effect. A computer model is a set of equations that includes variables for all the known natural fluctuations (such as the Milankovitch cycles). You also include variables for all the known human-caused inputs (CO_2, methane, aerosols, soot, and so on). Then, you run the model and see if it can re-create past changes in temperatures.

If it does a good job of replicating what happened in the past, then your model is a good description of how the system works—how the atmosphere responds to more CO_2, how oceans absorb heat, how reduced snow cover contributes feedback, and so on.

If you can create a model that represents the system quite well, then you can re-run the model, but this time you leave out the extra CO_2 and other factors we know that humans have contributed. If the model *without* human inputs is *inconsistent* with observed changes in temperature, and if the model *with* human inputs is *consistent* with observations, then you can be extremely confident, beyond the shadow of a reasonable doubt, that the human inputs have made the difference. Models with human activities incorporated into them (pink) are the only way to explain recently observed increases in air temperatures (black line, fig. 15.22).

We also know that current climate change must be linked to human activity because the current rates of change are much higher than anything we see in the geologic record. If we look back at data in the ice cores, we can see how rapidly air temperature and atmosphere CO_2 change when the earth switches from being in a glacial to an interglacial period (fig. 15.15). Even though the change between these periods appears very rapid, appearing almost as a straight

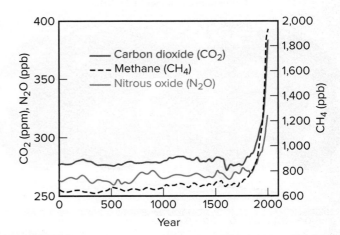

FIGURE 15.21 Changes in atmospheric concentration of the major greenhouse gases over time. Rapid increases began with the industrialization after 1800.
Source: USGS 2009,

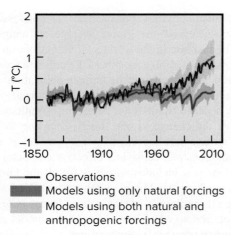

Observations
Models using only natural forcings
Models using both natural and anthropogenic forcings

FIGURE 15.22 Comparison of observed and modeled changes in global air temperature anomalies with results simulated by climate models using either natural or both natural and anthropogenic forcings. Decadal averages of observations are shown for the period 1906–2005 (black line). The blue line shows the average output using only the natural forcings due to solar activity and volcanoes and the blue shaded band is the 5 to 95 percent confidence interval. The red line shows the average output using both natural and human factors, and the pink shaded band shows the 5 to 95 percent confidence interval.
Source: IPCC 2013

vertical line on the graphs, it is still much slower than current rates of change. The earth's current air temperature is rising ten times faster than during the transition from a glacial to interglacial period, and atmospheric CO_2 concentrations are increasing one thousand times faster. This current rapid rate of change is one of the most disconcerting aspects of current climate change, as nothing on earth has ever before experienced change this fast.

Section Review

1. Approximately how much has atmospheric CO_2 changed since 1959?
2. List four important greenhouse gases, and identify major sources.
3. How do we know recent climate change is human-caused?

15.5 WHAT EFFECTS ARE WE SEEING?

- *Polar regions and continental interiors are warming most rapidly.*
- *Effects of climate change include loss of glaciers and intensified drought.*
- *Public confusion about climate change persists for various reasons.*

The American Geophysical Union, one of the nation's largest and most respected scientific organizations, says, "As best as can be determined, the world is now warmer than it has been at any point in the last two millennia, and, if current trends continue, by the end of the century it will likely be hotter than at any point in the last two million years."

Fortunately, as shown by Socolow and Pacala (opening case study) and others, we do have options, if we choose to use them. Mitigating climate change doesn't mean reverting to the Stone Age; it mostly means investing our resources in different kinds of energy. In this section, we'll examine some of the consequences of recent climate changes and some of the reasons so many scientists urge us to take action soon. We'll then consider some of the many steps we can take as individuals and as a society to work for a better future.

There are many effects of current climate change

Over the last century, the average global temperature climbed about 0.9°C (1.6°F) (fig. 15.23). Sixteen of the warmest years in the past 150 have occurred since 1998. New records for hot years are observed with increasing frequency. Here are some effects that have been observed:

- Summer heat waves are increasing in severity and frequency, with more dangerously hot days and increasing risks of forest fire. Precipitation patterns are also changing, more storm intensity and flooding in some areas and less rainfall in others (fig. 15.23).
- Drought has expanded and deepened in many agricultural regions, including much of the United States. In North America, recent wet winters and hot, dry summers are consistent with

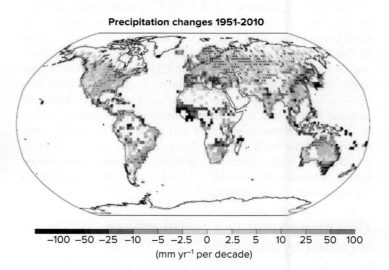

FIGURE 15.23 Changes in temperature and precipitation.
Source: IPCC, 2013

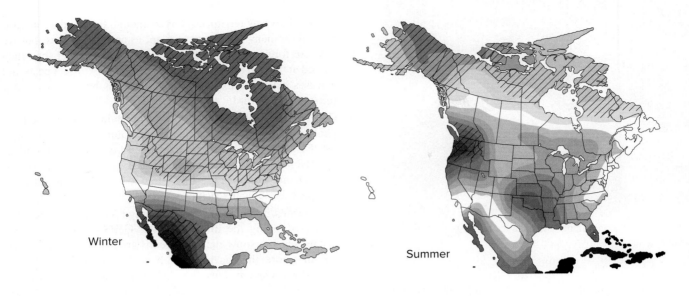

Winter

Summer

Percent change

<240 −35 −30 −25 −20 −15 −10 −5 0 5 10 15 20 25 30 35 >40

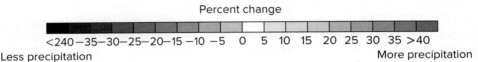

(a)

Less precipitation

More precipitation

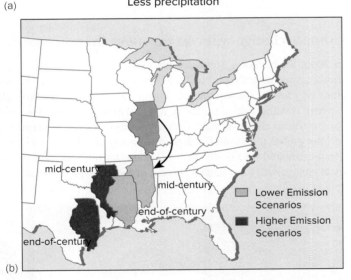

(b)

FIGURE 15.24 Models predict warmer, wetter winters and drier summers by 2100, compared to recent averages (a). Hatching marks areas with highest confidence in model projections. Midwestern farm states, the core of our food economy, will have a summer climate similar to current summers in Louisiana or Texas (b).
Source: U.S. Global Change Research Program, 2009: www.globalchange.gov/usimpacts

climate models (fig. 15.24). In Africa, droughts have increased about 30 percent since 1970. Extreme droughts in the Amazon rainforest, associated with high temperatures, have killed billions of trees, releasing an estimated 8 billion metric tons of CO_2 (more than China produced in 2009). A 2°C temperature rise (the best-case scenario) will destroy 20–40 percent of the Amazon forest, turning it from a carbon sink to a carbon source.

- Regions near the poles have warmed much faster than the rest of the world. In Alaska, western Canada, and eastern Russia, average temperatures have increased as much as 4°C (7°F) over the past 50 years. Permafrost is melting; houses, roads, pipelines, sewage systems, and transmission lines are being damaged as the ground sinks beneath them. Beetle infestations, made possible by warmer winters, are killing millions of hectares of pine and spruce across western North America.

- Arctic sea ice is only half as thick now as it was 30 years ago, and the area covered by sea ice has decreased by more than 1 million km^2 (an area larger than Texas and Oklahoma combined) in just three decades. By 2040, the Arctic Ocean could be totally ice-free in the summer. This is bad news for polar bears, which depend on the ice to hunt seals. An aerial survey in 2005 found bears swimming across as much as 260 km (160 mi) of open water to reach the pack ice. The United States has put polar bears on the endangered species list because of loss of Arctic sea ice (fig. 15.25). Loss of sea ice is also devastating for Inuit people, whose traditional lifestyle depends on ice for travel and hunting.

- Ice shelves on the Antarctic Peninsula are breaking up and disappearing rapidly, and Emperor and Adélie penguin populations have declined by half over the past 50 years as the ice shelves on which they depend for feeding and breeding disappear. Ninety percent of the glaciers on the Antarctic Peninsula are now retreating an average of 50 m per year. The Greenland ice cap also is melting twice as fast as it did a few years ago. Because ice shelves are floating, they don't affect sea level when they melt. Greenland's massive ice cap, however, holds enough water to raise sea level by about 7 m (about 23 ft) if it all melts. Melting glaciers and ice caps are contributing about 1 mm per year to sea-level rise.

FIGURE 15.25 Diminishing Arctic sea ice prevents polar bears from hunting seals, their main food source.

© Corbis RF

- Glaciers are disappearing at an accelerating rate. Glacier National Park, which had 150 glaciers in 1910, now has fewer than 30 glacier remnants (fig. 15.26). Half of the world's small glaciers will disappear by 2100, according to a study of 120,000 such glaciers. The loss of glaciers is especially worrisome in Bolivia, Chile, Nepal, and other regions that depend on meltwater for drinking and irrigation.

- Sea level has risen worldwide approximately 15–20 cm (6–8 in.) in the past century, with accelerating sea-level rise in the past 20 years. Thermal expansion of seawater is the main cause, followed by melting glaciers.

- Biologists report that many animals are breeding earlier or extending their ranges into new territory as the climate changes. In Europe and North America, for example, 57 butterfly species have either died out at the southern end of their range, or extended the northern limits, or both. Plants also are

moving into new territories. Given enough time and a route for migration, many species may adapt to new conditions, but we now are forcing many of them to move much faster than they moved at the end of the last ice age (fig. 15.27). Insect pests and diseases have also expanded their range as hard winters have retreated northward.

- Coral reefs worldwide are "**bleaching**" (dying as they lose their photosynthetic algae), as water temperatures rise above 30°C (85°F). With reefs nearly everywhere threatened by pollution, overfishing, and other stressors, biologists worry that rapid climate change could be the final blow for many species in these complex, biologically rich ecosystems.

- The oceans have been both absorbing CO_2 and storing heat, thus postponing the impacts of our GHG emissions. Deep-diving sensors show that the oceans are absorbing 0.85 watts per m^2 more than is radiated back to space. This absorption slows current warming, but it also means that even if we reduce our greenhouse gas emissions today, it will take centuries to dissipate that stored heat. Absorbed CO_2 is also acidifying the oceans. Because shells of mollusks and corals dissolve at low pH, ocean acidity is likely to alter marine communities.

Climate change will cost far more than prevention

In 2006, Sir Nicholas Stern, former chief economist of the World Bank, issued a study on behalf of the British government on the costs of global climate change. It was one of the most strongly worded warnings to date from a government report. He said, "Scientific evidence is now overwhelming: Climate change is a serious global threat, and it demands an urgent global response." Stern estimated that if we don't act soon, immediate costs of climate change will be at least 5 percent of the global GDP each year.

FIGURE 15.26 Alpine glaciers everywhere are retreating rapidly. These images show the Grinnell Glacier in 1914 and 1998. By 2030 there will probably be no glaciers in Glacier National Park.

Source: (a) Fred Kiser Photo 1910 courtesy of GNP Archives, USGS; Source: (b) Lisa McKeon, USGS

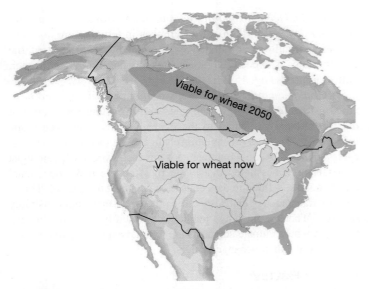

FIGURE 15.27 Most of the central United States is suitable for growing wheat now, but if current trends continue, the climatic conditions for wheat could be in central Canada by 2050.

If a wider range of risks is taken into account, the damage could equal 20 percent of the annual global economy. That would disrupt our economy and society on a scale similar to the great wars and economic depression of the first half of the twentieth century. The fourth IPCC report, meanwhile, estimated that preventing CO_2 doubling and stabilizing the world climate would cost only 0.12 percent of annual global GDP per year.

The Stern report, updated in 2009, estimates that reducing greenhouse gas emissions now to avoid the worst impacts of climate change would cost only about 1 percent of the annual global GDP. That means that $1 invested now could save us $20 later in this century. The actions we take—or fail to take—in the next 10 to 20 years will have a profound effect on those living in the second half of this century and in the next. Energy production, Stern suggests, will have to be at least 80 percent decarbonized by 2050 to stabilize our global climate.

Those of us in the richer countries will likely have resources to blunt problems caused by climate change, but residents of poorer countries will have fewer options. The Stern report says that without action, at least 200 million people could become refugees as their homes are hit by drought or floods. Furthermore, there's a question of intergenerational equity. What kind of world are we leaving to our children and grandchildren? What price will they pay if we fail to act?

The Stern review recommends four key elements for combating climate change: (1) *emissions trading* to promote cost-effective emissions reductions; (2) *technology sharing* that would double research investment in clean-energy technology and accelerate the spread of that technology to developing countries; (3) *reduction of deforestation,* which is a quick and highly cost-effective way to reduce emissions; and (4) *helping poorer countries* by honoring pledges for development assistance to adapt to climate change.

Rising sea levels will flood many cities

About one-third of the world's population now live in areas that would be flooded if all of Greenland's ice were to melt. Even the 75 cm (30 in.) sea-level rise expected by 2050 will flood much of south Florida, Bangladesh, Pakistan, and many other low-lying coastal areas. Most of the world's largest urban areas are on coastlines. Wealthy cities such as New York or London can probably afford to build dikes to keep out rising seas, but poorer cities such as Jakarta, Kolkata, or Manila might simply be abandoned as residents flee to higher ground. Small island countries such as the Maldives, the Bahamas, Kiribati, and the Marshall Islands could become uninhabitable if sea levels rise a meter or more. The South Pacific nation of Tuvalu has already announced that it is abandoning its island homeland. All 11,000 residents will move to New Zealand, perhaps the first of many climate-change refugees.

Insurance companies worry that the $2 trillion in insured property along U.S. coastlines is at increased risk from a combination of high seas and catastrophic storms. At least 87,000 homes in the United States within 150 m (500 ft) of a shoreline are in danger of coastal erosion or flooding in the next 50 years. Accountants warn that loss of land and structures to flooding and coastal erosion together with damage to fishing stocks, agriculture, and water supplies could rapidly multiply the financial costs of climate change.

Why do we still debate climate evidence?

Scientific studies have long been unanimous about the direction of climate trends, but commentators on television, newspapers, and radio continue to fiercely dispute the evidence. Why is this? One reason may be that change is threatening, and many of us would rather ignore it or dispute it than acknowledge it. Another reason may be a lack of information. Another is that while scientists tend to look at trends in data, the public might be more impressed by one or two recent events, such as an especially snowy winter in their local area. On radio and TV, colorful opinions capture more attention than data and graphs. And there is an established industry dedicated to disputing scientific evidence (fig. 15.28), including the Heartland Institute. Climate scientists offer the following responses to some of the following claims in the popular media:

"Reducing climate change requires abandoning our current way of life." Reducing climate change requires that we use different energy. By replacing coal-powered electricity with wind, solar, natural gas, and improved efficiency, we can drastically cut our emissions but keep our computers, TVs, cars, and other conveniences. Reducing coal dependence will also reduce financial costs of pollution damage to health and vegetation.

"There is no alternative to current energy systems." Alternative energy and conservation could easily reduce our fossil fuel consumption to safe levels if we chose to invest in them. Already, new wind and even solar electricity are cheaper than new oil or gas-burning power plants. Alternative energy and improved efficiency have also created new markets and jobs.

"Alternative energy requires subsidies." Fossil fuels receive far more subsidies, in tax incentives and other forms, than

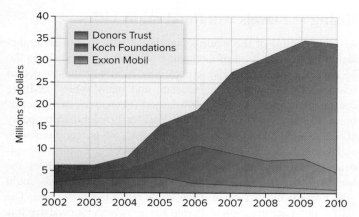

FIGURE 15.28 Funding given to groups promoting climate denial in popular media, blogs, and political lobbying. Sources include oil and gas industries and anonymous contributions to the Donor's Trust.
Source: Greenpeace

alternatives do. Global direct subsidies reached $550 billion in 2014, compared to about $120 billion for renewables, according to the International Energy Agency. In early stages of energy development, according to a Yale study, oil and gas received five times as much in direct subsidies as alternative energy, and that nuclear received ten times as much, after adjustment for inflation and economic growth. Energy subsidies are important in all societies, but alternative energies have received far less investment than other forms. Shifting more subsidies toward wind, solar, and conservation would make these alternatives economical, and more profitable than the energy and transportation technologies of the 1940s.

"A comfortable lifestyle requires high CO_2 output." Most northern Europeans produce less than half the CO_2 of North Americans, per capita. Yet they have higher standards of living (in terms of education, health care, life span, vacation time, financial security) than North Americans. Residents of San Francisco consume about one-sixth as much energy as residents of Kansas City, yet quality of life is not necessarily six times greater in Kansas City than in San Francisco.

"Natural changes such as solar variation can explain observed warming." Solar input fluctuates, but the changes are slight; these and other natural changes do not coincide with the direction of changes in temperatures or with the amount of warming (see fig. 15.17).

"The climate has changed before, so this is nothing new." Today's CO_2 level of roughly 400 ppm exceeds by at least 30 percent anything the earth has seen for nearly a million years, and perhaps as long as 15 million years. Antarctic ice cores indicate that CO_2 concentrations for the past 800,000 years have varied from 180 to 300 ppm (see fig. 15.15). This natural variation in CO_2 appears to be a feedback in glacial cycles, resulting from changes in biotic activity in warm periods. Because temperature closely tracks CO_2, temperatures by 2100 are likely to exceed anything in the past million years. The rate of change is probably also unprecedented. Changes that took 1,000 to 5,000 years at the end of ice ages are now occurring on the scale of a human lifetime.

"Temperature changes are leveling off." Over short time frames, temperature trends vary (fig. 15.1), but over decades the trends in surface air temperatures and in sea level continue to rise.

"We had cool temperatures and snowstorms last year, not heat and drought." Regional differences in temperature and precipitation trends are predicted by climate models. Most of the United States is expected to see wetter, warmer winters and drier, hotter summers (fig. 15.26).

"Climate scientists have gaps in their knowledge and have made errors in the past." The gaps and uncertainties in climate data are minute compared to the evident trends. There are many unknowns—details of precipitation change or interaction of long-term cycles such as El Niño—but the trends are unequivocal. Climatologist James Hansen has noted that while most people make occasional honest mistakes, fraud in data collection is almost unheard of. This is because transparency in the scientific process ensures public visibility of errors. There is much less public accountability in popular media, however, where climate scientists are regularly subjected to personal attacks from climate-change deniers.

Section Review

1. Why might drought be associated with climate warming?
2. List five to ten effects of changing climate.
3. List several reasons for public disputes over climate evidence.

15.6 ENVISIONING SOLUTIONS

- *The Paris Agreement seeks to keep warming below 2° C.*
- *About seven 1-Gt "wedges" are needed to stabilize climate.*
- *Regional initiatives show great promise.*

Dire warnings of climate change are intimidating, but in response individuals and communities around the world have been working on countless promising and exciting strategies to reduce these effects. All of these efforts, at all scales, are valuable. In this section, we'll look at some of these strategies. Curbing climate change is a daunting task, but it is also full of opportunity.

The most celebrated global agreement on controlling climate change has been the **Kyoto Protocol,** which followed a 1997 meeting in Kyoto, Japan. This agreement called for countries to voluntarily set their own targets for reducing emissions of CO_2, CH_4, and N_2O, and fluorine gases. Although the Kyoto Protocol was not successful in terms of reducing overall global emissions, many countries, mostly in Europe, met or exceeded their target reduction of 5 to 10 percent below 1990 emissions by 2012. Poorer nations, such as India and China, were exempt, allowing them to expand their economies and improve standards of living. The United States, until recently the world's greatest producer of greenhouse gases, was almost unique among wealthy countries in declining to sign on to the agreement.

The Paris Climate Agreement establishes new goals

The most important international agreement since the Kyoto Protocol was the 2015 Paris Climate Agreement. The 195 countries in attendance at the Paris meeting agreed on a number of major points, including these:

- Holding the global average temperature increase to well below 2°C (above pre-industrial levels) is necessary. Below 1.5°C should be the target.

- Zero carbon emissions is a global goal, to avoid compounding climate effects already committed by past emissions.

- Each participating country establishes voluntary emission reductions goals, and those goals, and progress toward them, must be publicly visible.

- Reduction plans submitted thus far are not sufficient to keep warming below 2°C, so plans must be revised every five years.

- Climate finance is necessary: advanced economies agreed to strive toward donations of $100 billion per year to a "green carbon fund" to support low-carbon development in emerging economies.

This agreement calls for all countries to invest a great deal of effort and resources in emissions reductions. Many people are skeptical that countries will follow through on pledges. But others point to ongoing innovations in renewable energy, in greenhouse gas regulation, and in legislative strategies such as the Clean Power Plan that can make goals achievable. And many developing regions are not waiting for the green carbon fund to start investing in clean energy. In Bangladesh and India, for example, both central banks and private funds are investing in "green bonds" that support renewable energy and that are producing hundreds of thousands of new jobs. Perhaps most important, agreement on the need for better policies is almost universally accepted now. And there are increasing numbers of local and regional examples. These are showing that efficiency and renewable energy can be good for financial stability, jobs, energy security, and public health, as in the case of Germany's renewable energy transition (see chapter 20 and What Do You Think? below).

The Paris accord required ratification by at least 55 countries, accounting for 55 percent of global emissions, to come into force. Remarkably, this goal was achieved within a year. This achievement indicates that countries wish to appear part of the solution. Whether they follow through on pledges remains to be seen, however.

Financial strategies are also important. Among these are carbon taxes and markets for trading carbon emission credits. Carbon taxes or fees can be assessed, as a set cost per ton of carbon emitted, for purchases of fossil fuels or other activities. These taxes have helped reduce consumption and fund renewable

What Do You Think?

States Take the Lead on Climate Change

In 2006, California passed a groundbreaking law that places a cap on emissions of carbon dioxide and other global warming gases from utilities, refineries, and manufacturing plants. The law aims to roll back the state's greenhouse gas releases to 1990 levels (a reduction of 15 percent) by 2020, and to 80 percent below 1990 levels by 2050. Reductions involve enforceable caps on emissions, monitored through regular industry emissions reports. Companies that cut emissions below their maximum allowance can profit by selling credits to other companies that have not met their caps. Putting a price on carbon emissions is creating incentives for innovation, which can now be cheaper than polluting. At the same time the cost of implementing the plan is low, and industries can meet standards in any way they choose. The legislation addresses a wide range of carbon sources, including agriculture, cement production, electricity generation, and suburban sprawl. Utilities and corporations are also prohibited from buying power from out-of-state suppliers whose sources don't meet California's emission standards. All these can be seen online at the "California Climate Change Portal."

This rule is the most aggressive climate-change effort of any state, but California voters strongly support it. When the energy industry challenged it in a ballot initiative in 2010, claiming it would cost the state jobs, 62 percent of voters still voted to keep the law. California has often led the way in improving air quality. In 2004, the state passed revolutionary legislation that required automakers to cut tailpipe emissions of carbon dioxide from cars and trucks, which has since been picked up by New York and other states. When car manufacturers failed to comply, California sued the six largest automakers in 2006, charging that they were costing the state billions of dollars in health and environmental damages.

What inspires such revolutionary steps? One factor is that California's economy relies almost entirely on declining winter snowpack for both urban water use and farm irrigation. Recent years of severe droughts have affected much of the state and worried cities and counties. Californians also have gotten tired of waiting for action in Washington, where the dominant view has been that climate controls will cost jobs. Contrary to this argument, California has seen rapid job growth in clean energy. Between 1995 and 2008, clean-energy businesses grew by 45 percent, ten times the state's average growth rate. Clean energy employs over 500,000 people and has brought in over $9 billion in venture capital, or 60 percent of all clean-energy investments nationwide.

Following this lead, most U.S. states and more than 500 cities have taken steps to promote renewable energy and reduce greenhouse gas emissions. Massachusetts announced in 2010 that, like California, it will cut greenhouse gases by 25 percent by 2020. Strategies the states are taking include efficient building standards, support for alternative energy, more-efficient distribution grids, land-use planning standards, support for retrofitting old houses, and auto insurance incentives for efficient vehicles.

Carbon trading has also caught on, with 27 states and four Canadian provinces participating in three regional carbon-trading compacts—the Midwestern Greenhouse Gas Reduction Accord, the Western Climate Initiative, and the northeastern Regional Greenhouse Gas Initiative (RGGI). The northeastern compact (RGGI) began trading carbon credits for 233 plants in 2008. By 2010, carbon credit auctions produced more than $700 million in revenue to support conservation and alternative energy initiatives in participating states.

Carbon trading is not perfect: Carbon prices are often too low to provide real incentives for some industries; many question whether a "right to pollute" is the best strategy; and carbon revenues risk being diverted to states' general funds. However, these compacts are widely considered successful—and palatable—approaches to reducing emissions.

New rules are a challenge to industry, but they can also lead to greater efficiency in operations, and changes are generally manageable if they are predictable and evenly applied. Still, these rules have been difficult to establish in Washington. If you were in Congress, what evidence would you want to see in order to buy into some of these state-led innovations?

energy in Sweden, Denmark, Canada, and elsewhere. In carbon markets, on the other hand, a set number of permits are issued for emitting carbon. Industries that reduce emissions can then sell their permits for a profit. This approach uses the market to set a price for carbon. Markets have grown rapidly: By 2016, trade had grown to 140 billion tons worth $53 billion, with markets in the European Union, California, the northeastern United States, and other regions. But markets have failed, thus far, to set a price high enough to encourage much change. As of 2016, the price was only about $10 per metric ton of CO_2 equivalent.

Stabilization wedges could work now

As discussed in the opening case study, the idea of stabilization wedges is that they can work just by expanding currently available technologies. To stabilize carbon emissions, we would need to cut about 7 Gt in 50 years; to reduce CO_2, as called for in the Kyoto Protocol, we could add another seven wedges (see fig. 15.2).

Because most of our CO_2 emissions come from fossil fuel combustion, energy conservation and a switch to renewable fuels are important. Doubling vehicle efficiency and halving the miles we drive would add up to 1.5 of the 1-Gt wedges. Installing efficient lighting and appliances, and insulating buildings, could add up to another 2 Gt. Capturing and storing carbon released by power plants, gas wells, and other sources could save another gigaton.

Pacala and Socolow's original 14 wedges are paraphrased in table 15.3. As the authors note, nobody will agree that all the wedges are a good idea, and all have some technological limitations, but none are as far off as revolutionary technologies such as nuclear fusion. Some analysts have subsequently proposed still additional wedges, and technologies that make these wedges possible, or that point to new ones, are changing rapidly.

Greenhouse gases can be captured and stored

Carbon capture and storage, one of the important stabilization wedges, is beginning to be widely practiced. Norway's state oil company, Statoil, which extracts oil and gas from beneath the North Sea, has been pumping more than 1 million metric tons of CO_2 per year into an aquifer 1,000 m below the seafloor at one of its North Sea gas wells. Injecting CO_2 increases pressure on oil reservoirs and enhances oil recovery. It also saves money because the company would have to pay a $50 per ton carbon tax on its emissions. Around the world, deep saltwater aquifers could store a century's worth of CO_2 at current fossil fuel consumption rates.

Carbon capture and injection is widely practiced for improving oil and gas recovery, so the technology is available (fig. 15.29). There are concerns about leaking from deep storage, but the main concern is that this is an expensive technology that, even under optimistic scenarios, can address only a modest portion of our carbon emissions.

Most attention is focused on CO_2 because it is our most abundant greenhouse gas, but methane is also important because, although we produce less of it, methane is a much more powerful absorber of infrared energy. Some atmospheric scientists think the best short-term strategy might be to focus on methane.

Table 15.3 Actions to Reduce Global CO_2 Emissions by 1 Billion Tons over 50 Years
1. Double the fuel economy for 2 billion cars from 30 to 60 mpg.
2. Cut average annual travel per car from 10,000 to 5,000 miles.
3. Improve efficiency in heating, cooling, lighting, and appliances by 25 percent.
4. Update all building insulation, windows, and weather stripping to modern standards.
5. Boost efficiency of all coal-fired power plants from 32 percent today to 60 percent (through co-generation of steam and electricity).
6. Replace 800 large coal-fired power plants with an equal amount of gas-fired power (four times current capacity).
7. Capture CO_2 from 800 large coal-fired or 1,600 gas-fired, power plants and store it securely.
8. Replace 800 large coal-fired power plants with an equal amount of nuclear power (twice the current level).
9. Add 2 million 1 MW windmills (50 times current capacity).
10. Generate enough hydrogen from wind to fuel a billion cars (4 million 1 MW windmills).
11. Install 2,000 GW of photovoltaic energy (700 times current capacity).
12. Expand ethanol production to 2 trillion liters per year (50 times current levels).
13. Stop all tropical deforestation and replant 300 million ha of forest.
14. Apply conservation tillage to all cropland (10 times current levels).

Source: Data from Pacala and Socolow, 2004

Methane from landfills, oil wells, and coal mines is now being collected in some places for fuel. Rice paddies are another major methane source. Changing flooding schedules and fertilization techniques can reduce some of these emissions. Reducing gas pipeline leaks would conserve this resource, as well as reduce warming. Finally, ruminant animals (such as cows, camels, sheep) are a major source of methane. Modifying human diets, including less beef consumption, would reduce methane significantly.

Regional initiatives show commitment to slowing climate change

Many countries are working to reduce greenhouse emissions. The United Kingdom, for example, by 2000 had already rolled back CO_2 emissions to 1990 levels and vowed to reduce them by 60 percent by 2050. Britain already has started to substitute natural gas for coal, promote energy efficiency in homes and industry, and raise its already high gasoline tax. Plans are to "decarbonize" British society and to decouple GNP growth from CO_2 emissions. A revenue-neutral carbon levy is expected to lower CO_2 releases and trigger a transition to renewable energy over the next five decades. In 2007, New Zealand's prime minister, Helen Clark, pledged that her country would be **carbon neutral** by 2025, through a combination of wind and geothermal energy, carbon capture on farms, and other strategies.

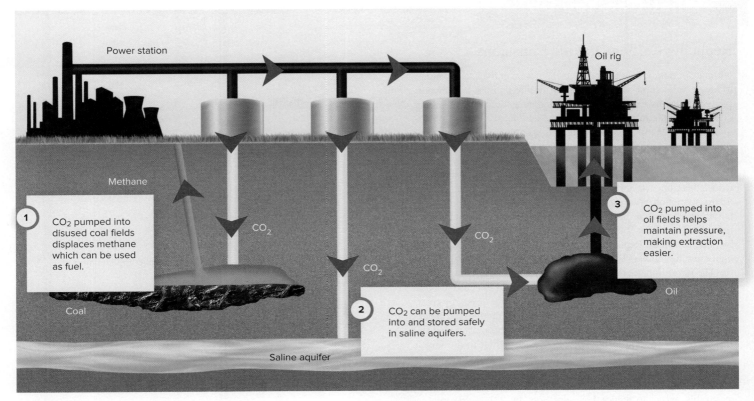

Power station

Oil rig

Methane

1 CO_2 pumped into disused coal fields displaces methane which can be used as fuel.

Coal

CO_2

CO_2

CO_2

2 CO_2 can be pumped into and stored safely in saline aquifers.

3 CO_2 pumped into oil fields helps maintain pressure, making extraction easier.

Oil

Saline aquifer

FIGURE 15.29 Carbon capture and storage involves pumping CO_2 into permanent storage, usually a salty aquifer or other geologic formation. Captured CO_2 can also be used to increase pressure on oil and gas wells, increasing recovery.

Germany has raised its targets, after years of growth in renewable energy development, to 80 percent renewable electricity by 2020. Germany has steadily focused on switching from coal to wind, solar, and gas and encouraging energy efficiency throughout society. Denmark, a world leader in wind power, now gets over 40 percent of its electricity from windmills. Plans are to generate half of the nation's electricity from offshore wind farms by 2030. Atmospheric scientist Steve Schneider called this a "no regrets" policy—even if we didn't need to stabilize our climate, many of these steps save money, conserve resources, and have other environmental benefits. Nuclear power also is being promoted as an energy alternative that produces no greenhouse gases directly (during energy production) and that provides high-volume, centralized power production. Nuclear remains an imperfect option because greenhouse gases and other pollutants are produced in mining, processing, and transporting nuclear fuel. There are also security worries and unresolved problems of how to store wastes safely. Still, this is an option favored by many states and utility companies.

Countless individual cities and states have announced their own plans to combat global warming. Among the first of these were Toronto, Copenhagen, and Helsinki, which pledged to reduce CO_2 emissions 20 percent from 1990 levels by 2010. Many of the cities making these pledges have met or exceeded their targets. The European Union has cut GHG production by 20 percent since 1990, while the region's GDP have increased by nearly 50 percent because of investments in conservation and alternative energy. Current U.S. emissions are about 9 percent higher than in 1990, but have declined slightly since 2008, largely because of a decline in coal use. Each of us can make a contribution in this effort. Simply driving less and buying efficient vehicles could save about 1.5 billion tons of carbon emissions by 2054 (fig. 15.30).

In the midst of all the debate about how serious the consequences of global climate change may or may not be, we need

FIGURE 15.30 Burning fossil fuels produces about half our greenhouse gas emission, and transportation accounts for about half of our fossil fuel consumption. Driving less, choosing efficient vehicles, carpooling, and other conservation measures are among our most important personal choices in the effort to control global warming.
© PhotoLink/Getty Images RF

Reducing Carbon Dioxide Emissions

Ultimately, policymakers need to establish incentives and policies to lower carbon emissions, so participating in politics, and encouraging politicians to support sound climate policies is something everyone can do. In addition, individual actions also impact on climate change. Many of the existing options save money in the long run, as well as reducing pollution and resource consumption.

Household energy consumption is responsible for nearly 40 percent of U.S. greenhouse gas production. The most obvious ways to reduce these impacts are in the areas of heating, lighting , and domestic transportation. You can drive less, walk, bike, take public transportation, carpool, or buy an efficient vehicle. Average annual CO_2 reductions are about 9 kg (20 lb) for each gallon of gasoline saved by these measures. Replacing standard incandescent light bulbs with light-emitting diode (LED) bulbs is another easy and money-saving fix. The EPA estimates that we could save 4 million metric tons (9 billion pounds) of CO_2 per year if each U.S. household switched just one bulb to a more efficient variety. These steps also save on energy costs.

Some changes are easier to make than others, though. It's generally easy to replace a light bulb or turn down a thermostat slightly. Buying a new car and insulating a house are harder to do. Given behavioral differences in our willingness to adopt energy-saving strategies, what steps are both effective and likely to be taken?

A study of behavior and household options found that we could reasonably expect to reduce U.S. emissions by 233 metric tons of carbon with a number of simple, inexpensive, and widely acceptable changes. These strategies—another example of wedge analysis at the household level—could reduce total emissions by 7.4 percent in 10 years without any new regulations, technology, or reductions in well-being. Transportation efficiency would make the most rapid difference (fig. 1). Weatherization also saves energy and money. Smaller steps such as unplugging appliances and adjusting thermostats have less dramatic impacts but also have an important impact because they add up to considerable savings. Which of these would be easiest for you to adopt?

To read more, see T. Dietz et al., 2009. Household actions can provide a behavioral wedge to rapidly reduce U.S. carbon emissions. *Proceedings of the National Academy of Sciences*, 106(44): 18452–56.

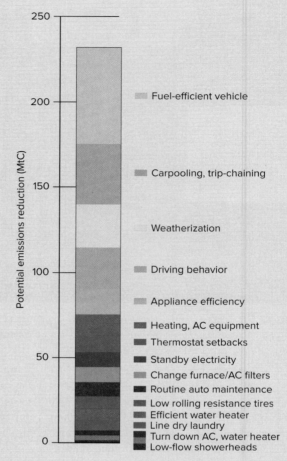

FIGURE 1 Potential impact on emissions in 10 years if available strategies were widely adopted.
Source: Adapted from Dietz et al., 2009

to remember that many of the proposed solutions are advantageous in their own right. Moving from fossil fuels to renewable energy sources such as solar or wind power, for example, would free us from dependence on foreign oil, with its volatile prices, and improve air quality. Planting trees makes cities pleasant places to live and provides habitat for wildlife. Making buildings more energy efficient and buying efficient vehicles saves money now and in the long run. Walking, biking, and climbing stairs are good for your health, and they help reduce traffic congestion and energy consumption. Reducing waste, recycling, and other green practices improve our environment, as well as help fight climate change.

We have countless options for avoiding a climate catastrophe. As the Irish statesman and philosopher Edmund Burke said, "Nobody made a greater mistake than he who did nothing because he could do only a little."

Section Review

1. What are some points in the Paris Climate Agreement?
2. List seven climate "wedges."
3. Which is expected to cost more, controlling climate change or adapting to it?

Conclusion

Climate change may be the most far-reaching issue in environmental science today. Although the challenge is almost inconceivably large, solutions are possible if we choose to act as both individuals and as a society. Temperatures are now higher than they have been in thousands of years, and climate scientists say that if we don't reduce greenhouse gas emissions soon, drought, floods, and conflict may be inevitable. The "stabilization wedge" proposal is a list of immediate and relatively modest steps that could be taken to accomplish needed reductions in greenhouse gases.

Understanding the climate system is essential to understanding the ways in which the changing composition of the atmosphere (more carbon dioxide, methane, and nitrous oxide, in particular) matters to us. Basic concepts to remember about the climate system include how the earth's surfaces absorb solar heat, how atmospheric convection transfers heat, and that different gases in the atmosphere absorb and store heat that is reemitted from the earth. Increasing heat storage in the lower atmosphere can cause increasingly vigorous convection, more extreme storms and droughts, melting ice caps, and rising sea levels. Changing patterns of monsoons, cyclonic storms, frontal weather, and other precipitation patterns could have extreme consequences for humans and ecosystems.

Despite the importance of natural climate variation, observed trends in temperature and sea level are more rapid and extreme than other changes in the climate record. Exhaustive modeling and data analysis by climate scientists show that these changes can be explained only by human activity. Increasing use of fossil fuels is our most important effect, but forest clearing, decomposition of agricultural soils, and increased methane production are also extremely important.

International organizations, national governments, and local communities have all begun trying to reverse these changes. Individual actions and commitment are also essential if we are to avoid dramatic and costly changes in our own lifetimes.

Reviewing Key Terms

Can you define the following terms in environmental science?

aerosols 15.1	El Niño 15.3	jet streams 15.2	positive feedback loop 15.1
albedo 15.1	greenhouse effect 15.1	Kyoto Protocol 15.6	stratosphere 15.1
climate 15.1	greenhouse gases 15.1	La Niña 15.3	tornado 15.2
cold front 15.2	hurricanes 15.2	latent heat 15.1	troposphere 15.1
convection currents 15.1	Intergovernmental Panel	Milankovitch cycles 15.3	warm front 15.2
Coriolis effect 15.2	on Climate Change	monsoon 15.2	weather 15.1
cyclonic storms 15.2	(IPCC) 15.4	ozone 15.1	wedge analysis 15

Critical Thinking and Discussion Questions

1. Weather patterns change constantly over time. From your own memory, what weather events can you recall? Can you find evidence in your own experience of climate change? What does your ability to recall climate changes tell you about the importance of data collection?

2. Many people don't believe that climate change is going on, even though climate scientists have amassed a great deal of data to demonstrate it. What factors do you think influence the degree to which a person believes or doesn't believe climatologists' reports?

3. How does the decades-long, global-scale nature of climate change make it hard for new policies to be enacted? What factors might be influential in people's perception of the severity of the problem?

4. What forces influence climate most in your region? In neighboring regions? Why?

5. Of the climate wedges shown in table 15.3, which would you find most palatable? Least tolerable? Why? Can you think of any additional wedges that should be included?

6. Would you favor building more nuclear power plants to reduce CO_2 emissions? Why or why not?

Data Analysis

Examining the IPCC Assessment Reports

The Intergovernmental Panel on Climate Change (IPCC) has a rich repository of figures and data, and because these data are likely to influence some policy actions in your future, it's worthwhile taking a few minutes to look at the IPCC reports.

The most brief and to the point is the Summary for Policy Makers (SPM) that accompany the Assessment Reports. You can find the summary at http://ipcc.ch/. Go to Connect to examine these documents closely and to show your understanding of what they say.

Change in temperature, sea level, and Northern Hemisphere snow cover

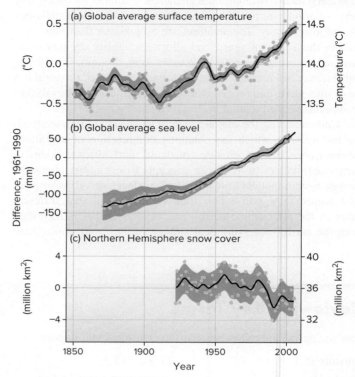

See the evidence. View the IPCC report at http://www.ipcc.ch/publications_and_data/ar4/syr/en/spm.html

---

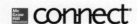 **TO ACCESS ADDITIONAL RESOURCES FOR THIS CHAPTER, PLEASE VISIT CONNECT AT www.connect.mheducation.com.**

You will find LearnSmart, an adaptive learning system, Google Earth™ exercises, additional Case Studies, Data Analysis exercises, and an interactive ebook.

16

Air Pollution

▲ Like many fast-growing cities, Beijing is struggling to control its air pollution.
© Mark Ralston/AFP/Getty Images

Learning Outcomes

After studying this chapter, you should be able to:

16.1 Identify natural and human-caused sources of air pollution.
16.2 Explain how atmospheric circulation affects air quality.
16.3 Compare the effects of air pollution.
16.4 Evaluate air pollution control efforts and progress.

"You shouldn't have to leave your neighborhood to live in a better one."

– Majora Carter, Activist and Community Advocate

Beijing Looks for Answers to Air Pollution

As the capital of China, Beijing is famous for its history, culture, cuisine, and architecture. Increasingly, the city also has a global reputation for its air pollution. Visibility is often limited to a few hundred meters, and air pollution index values rank between "unhealthy" and "hazardous" over half of the time (200 days/year in 2014). In December 2015, schools closed because of air pollution. Causes of pollution are well known: coal-burning power plants are the primary source, followed by industry, vehicles, and home oil and coal burners. Epidemiologists—and the city's residents—have long been interested in understanding the effects of these conditions for the 25 million people in greater Beijing, and for hundreds of millions more in other Chinese cities. The Chinese government calculated that the economic cost of air pollution amounts to around 3 percent of China's GDP.

Understanding the overall impacts of this pollution requires a lot of data. The Chinese government has developed a network of automated sampling stations, 945 in 190 cities across the country as of 2016, with hundreds more not yet in the network. Each station evaluates ambient (surrounding) air every hour and automatically submits hourly reports on six of the most important air pollutants.

For human health impacts, the most important pollutant in most cities is PM2.5—particulate matter (tiny airborne particles) less than 2.5 micrometers (0.0025 mm) in diameter. These tiny particulates can penetrate deep into lungs, obstructing breathing and exacerbating heart disease, asthma, and other conditions. Also important is slightly larger but still breathable particulate matter 10 micrometers (0.01 mm) or less in diameter (PM10). Together, PM2.5 and PM10 are the main health concerns in most Chinese cities, and in most other world cities, increasing illness and contributing to deaths from stroke, heart disease, lung cancer, and obstructive pulmonary disease.

Among the other major considerations for both human health environmental quality are ozone (O_3), nitrogen dioxide (NO_2), sulfur dioxide (SO_2), and carbon monoxide (CO). Ozone, less stable than normal oxygen (O_2) molecules, is highly reactive. It irritates sensitive tissues such as eyes, throat, and lungs. Nitrogen dioxide and sulfur dioxide, also highly reactive, contribute to the formation of acid rain and damage plant tissues and buildings, as well as health. Carbon monoxide bonds to blood cells, restricting the flow of oxygen in the blood stream.

How many deaths result from these pollutants? A 2015 study by University of California, Berkeley scientists Robert Rhode and Richard Muller examined this question. The study drew on data from over 1500 air monitoring sites, including most of the 945 nationally networked sites and hundreds of additional sites in China and neighboring countries. Hourly data from these sites were collected for four straight months in 2014, and then combined with weather data to map the movement of airborne pollutants. Rhode and Muller could then model hourly pollution exposures by population for all of eastern China, where 97 percent of the population lives. After evaluating population size and exposure to the major pollutants, they used World Health Organization models to project the likely number of deaths per year.

For the whole of China, this study calculated that air pollution contributes to 1.6 million premature deaths per year, about 17 percent of all deaths. Numbers like these help strengthen incentives for Beijing and other cities to push harder for air pollution controls. Recent policy shifts have included plans to eliminate coal-burning power plants (replacing them with renewable energy and also with nuclear power) and establish policies promoting electric cars and energy efficiency. The central government also promises to improve the enforcement of rules, so as to ensure that polluters comply with established health standards.

Beijing is one of hundreds of cities in the rapidly industrializing world to suffer from these conditions. In China, India, Southeast Asia, and elsewhere, rapidly growing cities are powered by coal plants or by poorly regulated oil and gas burners. Coal is often used for household heating and for industrial furnaces. Oil and gas production are also major pollutant sources, and regions that produce or process oil and gas often have especially poor air quality.

You can see real-time air quality index reports for cities in China and others worldwide by looking up the Beijing-based nongovernmental organization AQICN.org online (fig. 16.1). This group

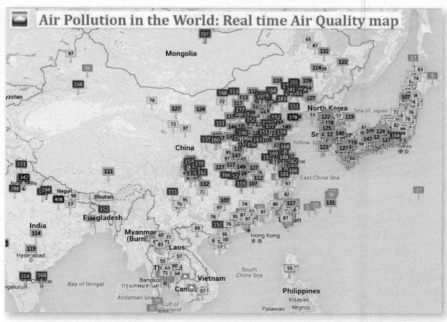

FIGURE 16.1 Improved data monitoring makes it possible to evaluate air quality in Beijing and elsewhere. The Beijing-based organization AQICN.org created this map to deliver real-time air quality data for China and much of the rest of the world. Colors range from purple and red (unsafe air quality) to green (good air quality).
Source: AQICN.org, U.S. Embassy Beijing, China Ministry of Environmental Protection

began in 2007 by reporting data from the U.S. embassy in Beijing but continues by reporting real-time air monitoring data provided by the Chinese Ministry of Environmental Protection and other agencies worldwide. Like the Rode and Muller study, this group draws on increasingly available monitoring data. By delivering it in standard units, the site allows easy comparison of world cities. The same group provides world air quality index data at WAQI.info.

Further Reading:

Rohde, R. A., and Muller, R. A. 2015. *Air Pollution in China: Mapping of Concentrations and Sources,* PLoS ONE 10(8): e0135749. doi:10.1371/journal.pone.0135749.

16.1 MAJOR POLLUTANTS IN OUR AIR

- *Air pollutants affect health and environmental quality.*
- *The EPA regulates six major criteria pollutants.*
- *Mercury and organic compounds are examples of hazardous air pollutants.*

Beijing, Delhi, Bangkok, and other rapidly growing cities face serious air pollution challenges today, but only a few decades ago, many American and European cities also endured similar conditions. Chronic bad air, and occasional severe smog events (see section 16.2) gradually led to the adoption of pollution controls. Legal enforcement has improved, and many students today don't realize how bad air quality once was in their home towns. Many American and European cities still have bad air: major port cities, oil and gas extraction areas, and industrial cities are particularly bad (fig. 16.2). But in the past 40 years, air quality protections have increased in number and in effectiveness, greatly improving public health. In most developed economies, there are established, legally enforceable rules to protect the air we all breathe. Industry has also relocated to hungry regions of the developing world, where environmental and health protections are poorly enforced.

In this chapter, we examine what major pollutants are, what their sources are, and how we can implement policies to control them. Air pollution impairs human health, damages crops and ecosystems, and corrodes buildings and infrastructure. Greenhouse gases pollutants are altering our climate. **Aesthetic degradation,** such as odors and lost visibility are also important consequences of air pollution. These factors rarely threaten life or health directly, but they can strongly impact our quality of life. They also increase stress, which affects health.

Many natural factors degrade air quality degradation, including ash and gases from volcanoes (fig. 16.3) and desert dust. Trees emit volatile organic compounds (terpenes and isoprenes), and decaying vegetation in swamps produces methane, as do termites and ruminant animals. Forest fires produce particulate matter, nitrogen oxides, and carbon monoxide. For the purposes of this discussion, however, these sources are treated as background levels. They are normally too diffuse to cause severe damage to living systems, and most of the contaminants discussed in this chapter

are produced in greater abundance by human activities. When we refer to pollutants, we generally mean human-caused emissions.

In many Chinese cities, airborne dust, smoke, and soot often are ten times higher than levels considered safe for human health. Of the 20 smoggiest cities in the world, 16 are in China. China's city dwellers are four to six times more likely than rural people to die of lung cancer. Poorly regulated industrial cities of India,

FIGURE 16.2 Poor air quality and photochemical smog are less common in U.S. cities than they were 40 years ago, although they persist in many major cities and industrial or port areas.
© BananaStock/PunchStock RF

FIGURE 16.3 Natural pollution sources, such as volcanoes, can be important health hazards.
Source: U.S. Geological Survey

Russia, Pakistan, and many other countries cause similar hazards. While pollution contributes to about China 1.6 million deaths in China each year, it contributes to an estimated 1.4 million deaths in India. In India, the primary risk has historically been particulate matter from household fires and other biomass burning, but rapid industrialization is expanding the range of pollutants.

The Clean Air Act designates standard limits

Air pollution is one of our most diffuse and hard to control environmental issues. Industry continues to challenge clean air rules in every country, and public attention is always needed to protect the safeguards we now rely on. However, epidemiologists and economists are increasingly making it clear that pollution control rapidly pays for itself in social costs of health care and infrastructure damage. Cost-benefit analysis of air regulation is discussed in section 16.4.

There have been countless efforts throughout history to control objectionable smoke, odors, and noise, but nearly all of these efforts were local. For example, the 1963 Clean Air Act, the first national air quality legislation in the United States, was careful to preserve states' rights to set and enforce air quality regulations. But polluting industries can easily move across state boundaries

to avoid local regulations. It soon became obvious that piecemeal, local standards did not resolve the problem, because neither pollutants nor the markets for energy and industrial products are contained within state boundaries.

Amendments to the U.S. Clean Air Act in 1970 essentially rewrote the law. Congress designated new standards, to be applied evenly across the country, for six major pollutants: sulfur dioxide, nitrogen oxides, carbon monoxide, ozone (and its precursor volatile organic compounds), lead, and particulate matter. These standards were set according to health criteria and environmental quality. Transportation and power plants are the dominant sources of criteria pollutants (fig. 16.4). We'll examine each of these, and then we'll look at additional pollutants that are also monitored under the Clean Air Act.

Conventional pollutants are most abundant

National ambient air quality standards (NAAQS) identify maximum allowable limits for these (**ambient air** is the air around us). These six **conventional** or **criteria pollutants.** The Clean Air Act addressed these first because they contributed the largest volume of air quality degradation and also are considered the most serious threats to human health and welfare. Primary standards (table 16.1) are intended to protect human health. Secondary standards are also set to protect crops, materials, climate, visibility, and personal comfort.

In addition to the six conventional pollutants, the Clean Air Act regulates an array of **unconventional pollutants,** compounds that are produced in less volume than conventional pollutants but that are especially toxic or hazardous, such as asbestos, benzene, mercury, polychlorinated biphenyls (PCBs), and vinyl chloride. Most of these are uncommon in nature or have no natural sources.

We also distinguish pollutants according to how they are produced. **Primary pollutants** are those released directly from

Table 16.1	National Ambient Air Quality Standards (NAAQS)	
Pollutant	Primary (Health-Based) Averaging Time	Standards (Allowable Concentrations)
PM^a	Annual geometric mean[b]	50 µg/m³
	24 hours	150 µg/m³
SO_2	Annual arithmetic mean[c]	80 µg/m³ (0.03 ppm)
	24 hours	120 µg/m³ (0.14 ppm)
CO	8 hours	10 mg/m³ (9 ppm)
	1 hour	40 mg/m³ (35 ppm)
NO_2	Annual arithmetic mean	80 µg/m³ (0.05 ppm)
O_3	Daily max 8 hour avg.	157 µg/m³ (0.08 ppm)
Lead	Maximum quarterly avg.	1.5 µg/m³

[a] Total suspended particulate material, PM2.5 and PM10.

[b] The geometric mean is obtained by taking the *n*th root of the product of *n* numbers. This tends to reduce the impact of a few very large numbers in a set.

[c] An arithmetic mean is the average determined by dividing the sum of a group of data points by the number of points.

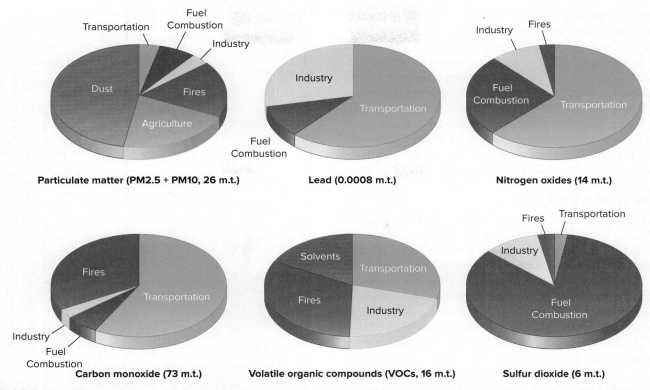

Particulate matter (PM2.5 + PM10, 26 m.t.)

Lead (0.0008 m.t.)

Nitrogen oxides (14 m.t.)

Carbon monoxide (73 m.t.)

Volatile organic compounds (VOCs, 16 m.t.)

Sulfur dioxide (6 m.t.)

FIGURE 16.4 Major sources of primary "criteria" pollutants in the United States, and amount produced in metric tons (m.t.). Volatile organic compounds and nitrogen oxides are important precursors of ozone, one of the six criteria pollutants.
Source: EPA Air Emission Sources, 2016

the source into the air in a harmful form (fig. 16.5). **Secondary pollutants** are converted to a hazardous form after they enter the air or are formed by chemical reactions as components of the air mix and interact. Solar radiation often provides the energy for these reactions. Photochemical oxidants and atmospheric acids formed by these mechanisms (see fig. 16.2) are among our most important pollutants in terms of health and ecosystem damage.

Many pollutants come from a point source, such as a smokestack. **Fugitive emissions** are those that do not go through a smokestack. By far the most massive example of this category is dust from soil erosion, strip mining, rock crushing, and building construction (and destruction). Fugitive industrial emissions are hard to monitor, but they are extremely important sources of air pollution. Leaks around valves and pipe joints, and evaporation of volatile compounds from oil-processing facilities, contribute as much as 90 percent of the hydrocarbons and volatile organic chemicals emitted from oil refineries and chemical plants.

Sulfur Dioxide (SO₂)

Natural sources of sulfur in the atmosphere include evaporation of sea spray, erosion of sulfate-containing dust from arid soils, fumes from volcanoes and hot springs, and biogenic emissions of hydrogen sulfide (H_2S) and organic sulfur-containing compounds. Total yearly emissions of sulfur from all sources amount to some 114 million metric tons. Worldwide, anthropogenic sources represent about two-thirds of all the airborne sulfur, but in most urban areas they contribute as

much as 90 percent of the sulfur in the air. The predominant form of anthropogenic sulfur is **sulfur dioxide** (SO_2) from combustion of sulfur-containing fuel (coal and oil), purification of sour (sulfur-containing) natural gas or oil, and industrial processes, such as smelting of sulfide ores. China and the United States are the largest sources of anthropogenic sulfur, primarily from coal burning and smelting.

FIGURE 16.5 Primary pollutants are released directly from a source into the air. Coal-burning power plants like this one produce about two-thirds of the sulfur oxides, one-third of the nitrogen oxides, and one-half of the mercury emitted in the United States each year.
© *Bildagentur Zoonar GmbH/Shutterstock.com*

Sulfur dioxide is a colorless, corrosive gas that is directly damaging to both plants and animals (fig. 16.6). Once in the atmosphere, it can react with atmospheric oxygen and water vapor to form sulfuric acid (H_2SO_4), a major component of acid rain. Very small solid particles or liquid droplets can transport the acidic sulfate ion (SO_4^{-2}) long distances through the air or deep into the lungs, where it is very damaging. Sulfate particles and droplets reduce visibility, sometimes dramatically. Some of the smelliest and most noxious air pollutants are sulfur compounds, such as hydrogen sulfide from pig manure lagoons or mercaptans (organosulfur thiols) from paper mills (fig. 16.7).

Nitrogen Oxides (NO$_x$)

Nitrogen oxides are highly reactive gases formed when nitrogen in fuel or in air is heated (during combustion) to temperatures above 650°C (1,200°F) in the presence of oxygen. The initial product, **nitric oxide** (NO), oxidizes further in the atmosphere to **nitrogen dioxide** (NO_2), a reddish-brown gas that gives photochemical smog its distinctive color. In addition, **nitrous oxide** (N_2O) is an intermediate form that results from soil denitrification. Nitrous oxide absorbs ultraviolet light and is an important greenhouse gas (chapter 15). Because nitrogen readily changes from one of these forms to another by gaining or losing O atoms, the general term NO_x is used to describe the gases NO and NO_2. Nitrogen oxides react with water to make nitric acid (HNO_3), a major component of acid rain.

Anthropogenic sources account for 60 percent of the global emissions of about 230 million metric tons of reactive nitrogen compounds each year (see table 16.1). About 95 percent of all human-caused NO_x in the United States is produced by fuel combustion in

FIGURE 16.7 The most noxious pollutants from this paper mill are pungent organosulfur thiols and sulfides. Chlorine bleaching can also produce extremely dangerous organochlorines, such as dioxins.
© William P. Cunningham

transportation and electric power generation. Because we continue to drive more miles every year, and to consume abundant electricity, we have had less success in controlling NO_x than other pollutants.

Excess nitrogen from agricultural fertilizer use and production is also an important, but little understood, contributor to airborne NO_x. Fertilizers washing from farmlands also cause excess fertilization and eutrophication of inland waters and coastal seas. Environmental dispersal of nitrogen from fertilizers also may be adversely affecting terrestrial plants by fertilizing weedy and invasive plants.

Carbon Monoxide (CO)

Carbon monoxide (CO) is a colorless, odorless, nonirritating, but highly toxic gas. CO is produced mainly by incomplete combustion of fuel (coal, oil, charcoal, or gas), as in furnaces, incinerators, engines, or fires, as well as in the decomposition of organic matter. CO blocks oxygen uptake in blood by binding irreversibly to hemoglobin (the protein that carries oxygen in our blood), making hemoglobin unable to hold oxygen and deliver it to cells. Human activities produce about half of the 1 billion metric tons of CO released to the atmosphere each year. In the United States, two-thirds of the CO emissions are created by internal combustion

FIGURE 16.6 High concentrations of sulfur dioxide damage plants directly. This soybean plant was exposed to 2.1 mg/m³ sulfur dioxide for 24 hours. White patches show where chlorophyll has been destroyed.
© William P. Cunningham

engines in transportation. Land-clearing fires and cooking fires also are major sources. About 90 percent of the CO in the air is converted to CO_2 in photochemical reactions that produce ozone. Catalytic converters on vehicles are one of the important methods to reduce CO production by ensuring complete oxidation of carbon to carbon dioxide (CO_2).

Carbon dioxide is the predominant form of carbon in the air. Growing recognition of the health and environmental risks associated with climate change (chapter 15) have led to recent regulations on CO_2, which are discussed below.

Ozone (O_3) and Photochemical Oxidants

Ozone (O_3) high in the stratosphere provides a valuable shield for the biosphere by absorbing incoming ultraviolet radiation. But at ground level, O_3 is a strong oxidizing reagent that damages vegetation, building materials (such as paint, rubber, and plastics), and sensitive tissues (such as eyes and lungs). Ozone has an acrid, biting odor that is a distinctive characteristic of photochemical smog. Ground-level O_3 is a product of photochemical reactions (reactions initiated by sunlight) between other pollutants, such as NO_x or volatile organic compounds. A general term for products of these reactions is **photochemical oxidants** (fig. 16.8). One of the most important of these reactions involves splitting nitrogen dioxide (NO_2) into nitrous oxide (NO) and oxygen (O). This single O atom is then available to combine with a molecule of O_2 to make ozone (O_3).

Hydrocarbons in the air contribute to the accumulation of ozone by combining with NO to form new compounds, leaving single O atoms free to form O_3 (see fig. 16.7). Many of the NO compounds are damaging photochemical oxidants. A general term for organic chemicals that evaporate easily or exist as gases in the air is **volatile organic compounds (VOCs).** Plants are the largest source of VOCs, releasing an estimated 350 million tons of isoprene (C_5H_8) and 450 million tons of terpenes ($C_{10}H_{15}$) each year, but normally concentrations are low.

The greater health threats from VOCs involve a large number of other synthetic organic chemicals, such as benzene, toluene, formaldehyde, vinyl chloride, phenols, chloroform, and trichloroethylene, which are released into the air by human activities. About 28 million tons of these compounds are emitted each year in the United States, mainly unburned or partially burned hydrocarbons from transportation, power plants, chemical plants, and petroleum refineries (fig. 16.9). These chemicals play an important role in the formation of photochemical oxidants.

Lead

Our most abundantly produced metal air pollutant, lead, is toxic to our nervous systems and other critical functions. Lead binds to enzymes and to components of our cells, such as brain cells, which then cannot function normally. Airborne lead is produced by a wide range of industrial and mining processes. The main sources are

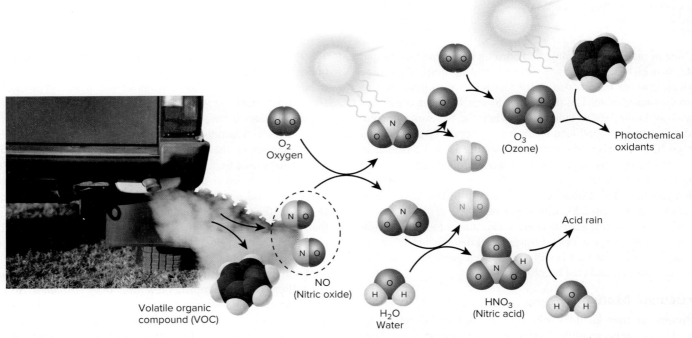

FIGURE 16.8 The heat of fuel combustion causes nitrogen oxides to form from atmospheric N_2 and O_2. Nitrogen dioxide (NO_2) interacts with water (H_2O) to form nitric acid (HNO_3), a component of acid rain. In addition, solar radiation can force NO_2 to release a free oxygen atom, which joins to atmospheric O_2, creating ozone (O_3). Fuel combustion also produces incompletely burned hydrocarbons (including volatile organic compounds, VOCs). Both O_3 and VOCs contribute to photochemical oxidants, in reactions activated by sunlight. The VOC shown here is benzene, a ring of six carbon atoms with a hydrogen atom attached to each carbon.
© Hisham F. Ibrahim/Getty Images RF

FIGURE 16.9 Many of our most serious pollutants are hazardous organic compounds that escape as fugitive emissions from petroleum facilities, such as this one in Baton Rouge, Louisiana.
© Mary Ann Cunningham

smelting of metal ores, mining, and burning of coal and municipal waste, in which lead is a trace element, and burning of gasoline to which lead has been added. Until recently, leaded gasoline was the main source of lead in the United States, but leaded gas was phased out in the 1980s. Since 1986, when the ban was enforced, children's average blood lead levels have dropped 90 percent and average IQs have risen three points. Banning leaded gasoline in the United States was one of the most successful pollution-control measures in American history. Now, 50 nations have renounced leaded gasoline. The global economic benefit of this step is estimated to be more than $200 billion per year.

Worldwide atmospheric lead emissions amount to about 2 million metric tons per year, or two-thirds of all metallic air pollution. Globally, most of this lead is still from leaded gasoline, as well as metal ore smelting and coal burning.

Particulate Matter

Particulate matter includes solid particles or liquid droplets suspended in a gaseous medium. Very fine solid or liquid particulates suspended in the atmosphere are **aerosols.** This includes dust, ash, soot, lint, smoke, pollen, spores, algal cells, and many other suspended materials. Particulates often are the most obvious form of air pollution, because they reduce visibility and leave dirty deposits on windows, painted surfaces, and textiles.

Particulates small enough to be breathed in are monitored under the Clean Air Act. Particles smaller than 2.5 micrometers in diameter, such as those found in smoke and haze, and produced by fires, power plants, or vehicle exhaust, are among the most dangerous particulates because they can be drawn into the lungs, where they damage respiratory tissues. Asbestos fibers and cigarette smoke are among these dangerous fine particles. Reducing sulfur in coal and diesel fuel, which produces aerosol droplets of sulfuric acid, is one important strategy for controlling PM2.5 particulates.

Coarse inhalable particles larger than 2.5 micrometers but less than 10 micrometers in diameter are known as PM10. Heavier than PM2.5, they do not travel as far, and they are typically found near roads or other dust sources. The American Dust Bowl of the 1930s involved mainly this kind of particulates. At that time, farmland soils were often left bare, especially during severe drought, and billions of tons of topsoil blew away from farmlands. Soil conservation on farmlands is one strategy for reducing PM10; another strategy is better management of dust at construction sites.

Dust storms can travel remarkable distances. Dust from Africa's Sahara desert regularly crosses the Atlantic and raises particulate levels above federal health standards in Miami and San Juan, Puerto Rico (fig. 16.10). Amazon rainforests receive mineral nutrients carried in dust from Africa; more than half the 50 million tons of dust transported to South America each year has been traced to

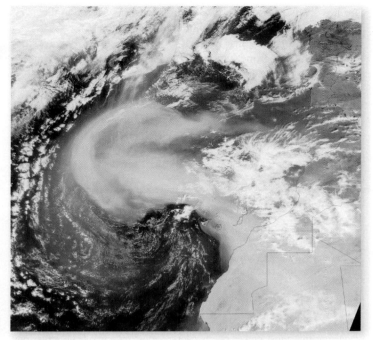

FIGURE 16.10 A massive dust storm extends more than 1,600 km (1,000 mi) from the coast of western Sahara and Morocco (the land mass on the right side of the image). Storms such as this regularly reach the Americas, and they have been linked to both the decline of coral reefs in the Caribbean and the frequency and intensity of hurricanes formed in the eastern Atlantic Ocean.
Source: Image courtesy of Norman Kuring, SeaWIFS Project/NASA

the bed of the former Lake Chad in Africa. In China, vast dust storms blow out of the Gobi desert every spring, choking Beijing and closing airports and schools in Japan and Korea. The dust plume follows the jet stream across the Pacific to Hawaii and then to the west coast of North America, where it sometimes makes up as much as half the particulate air pollution in Seattle, Washington. Some Asian dust storms have polluted the U.S. skies as far east as Georgia and Maine.

Mercury, from coal, is particularly dangerous

In addition to criteria pollutants or conventional pollutants, many other substances are regulated to protect public health and our environment. Standards for these pollutants continue to evolve, as do definitions of which pollutants require regulation. These changes reflect increases in certain pollutants, such as airborne mercury; the introduction of new pollutants, such as newly developed organic compounds; and increasing recognition of risks, as in the case of carbon dioxide.

Many toxic metals are released into the air by burning coal and oil, mining, smelting of metal ores, or manufacturing. Lead, mercury, cadmium, nickel, arsenic (a highly toxic metalloid), and others are released in the form of metal fumes or suspended particulates by fuel combustion, ore smelting, and burning garbage, coal, or other metal-laden materials. Among these, lead and mercury are the most abundantly produced toxic metals.

Mercury is a powerful neurotoxin that damages the brain and central nervous system at high doses. Minute amounts can cause nerve damage and developmental defects in children. About 75 percent of human exposure to mercury comes from eating fish. This is because aquatic bacteria are mainly responsible for converting airborne mercury (which falls or washes into water bodies) into methyl mercury, a form that is taken up and stored by living organisms. Methyl mercury accumulates in the tissues of fish, becoming especially concentrated in top predators. Large, long-lived, predatory fish contain especially high levels of mercury in their tissues, and these are most dangerous to eat. In a survey of freshwater fish from 260 lakes across the United States, the EPA found that every fish sampled contained some level of mercury.

Tuna fish alone is responsible for about 40 percent of all U.S. exposure to mercury (fig. 16.11). Swordfish, shrimp, and other seafood are also significant sources of mercury in our diet, and these species should be avoided. They also should be avoided because their populations are dwindling in many areas.

Global air circulation also deposits airborne mercury on land. Half or more of the mercury that falls on North America probably comes from Asian coal-burning power plants. Similarly, North American mercury travels to Europe. A 2009 report by the U.S. Geological Survey found that mercury levels in Pacific Ocean tuna have risen 30 percent in 20 years, with another 50 percent increase projected by 2050. Increased coal burning in China, which for years built new coal-burning power plants at the rate of one or two per week, is understood to be the main cause of growing mercury emissions in the Pacific.

The U.S. National Institutes of Health (NIH) estimates that one in 12 American women has more mercury in her blood than the 5.8 µg/l considered safe by the EPA. Between 300,000 and 600,000 of the 4 million children born each year in the United States are exposed in the womb to mercury levels that could cause diminished intelligence or developmental impairments. According to the NIH, elevated mercury levels cost the U.S. economy $8.7 billion each year in higher medical and educational costs and in lost workforce productivity. The EPA has been slow to regulate mercury, however (see the What Do You Think? box).

FIGURE 16.11 Airborne mercury accumulates in seafood, especially in top predators such as tuna.
© McGraw-Hill Education

Carbon dioxide, methane, and halogens are key greenhouse gases

Some 370 billion tons of CO_2 are emitted each year from respiration (oxidation of organic compounds by plant and animal cells; see table 16.1). These releases are usually balanced by an equal uptake by photosynthesis in green plants. At normal concentrations, CO_2 is nontoxic and innocuous, but atmospheric levels are steadily increasing (about 0.5 percent per year) due to human activities and are now causing global climate change, with serious implications for both human and natural communities (chapter 15).

Regulating CO_2 has been a subject of intense debate since the 1990s. On the one hand, policymakers have widely acknowledged that climate change is likely to have disastrous effects. On the other hand, CO_2 is difficult to consider limiting because we produce abundant quantities, reductions involve changes to both technology and behavior, and CO_2 production historically has been closely tied to our economic productivity. Although future

What Do You Think?

Politics, Public Health, and the Minamata Convention

The dangers of mercury have been understood for decades, but action has been slow. In the United States, it wasn't until 1994 that the EPA declared mercury a hazardous pollutant, to be regulated under the Clean Air Act. Municipal and medical incinerators were required to reduce their mercury emissions by 90 percent within 5 years. Industrial and mining operations also agreed to cut emissions. However, the law did not address the 1,032 coal-burning power plants, which produced nearly half of total annual U.S. emissions.

In 2000, the EPA finally declared mercury from coal plants, like that from other sources, a public health risk. But the agency opted for a "cap and trade" market approach, expected to reduce mercury 70 percent in about 30 years, unlike the 90 percent reductions in 5 years required of other sources. Cap and trade approaches set limits (caps) and allow utilities to buy and sell unused pollution credits. This strategy uses a profit motive rather than rules, and it allows industries to find the cheapest reduction methods. It also allows continued emissions if credits are cheaper than emission controls. Public health advocates argue that this slow-going approach is inappropriate for a substance like mercury, which is toxic at very low levels, and which should be eliminated as quickly as possible.

Meanwhile, on the global stage, United Nations negotiators began working in earnest to control mercury by 2003, but negotiations stalled until 2009, with U.S. delegates insisting on voluntary emissions controls. Voluntary regulations are much like students writing their own tests and assigning their own grades. It works well for individual students, but it doesn't ensure that progress is being made.

In 2009, with a change in U.S. presidential administrations, progress started up again, and finally in 2013–14, delegates from 140 countries signed the Minamata Convention on Mercury. The agreement is named for the city of Minamata, Japan, which experienced tragic cases of severe mercury poisoning in the 1950s. Like many industries at the time, a chemical factory in Minamata regularly discharged mercury-laden waste into the bay. In Japan, where people consume a great deal of seafood, this meant that the local population was directly exposed to unusually concentrated levels of mercury. Babies whose mothers ate mercury-contaminated fish suffered profound neurological disabilities, including deafness, blindness, mental retardation, and cerebral palsy (fig. 1). In adults, mercury poisoning caused numbness, loss of muscle control, and dementia. The connection between "Minamata disease" and mercury was established in the 1950s, but waste dumping didn't end for another ten years.

The Minamata Convention establishes global rules for monitoring and reporting on mercury emissions. It calls for elimination of mercury in many uses where alternatives exist, such as household batteries and thermometers. It also requires control of mercury released in gold mining, where it is used to separate gold from ores and is often released into rivers with little or no regulation in developing areas. The convention also requires that countries monitor and limit pollution from coal-burning power plants.

The U.S. now falls under the international agreement, and standards will be monitored by the international community, which after all is affected by U.S. emissions. Subsequently, in 2016, the EPA issued rules requiring that most power plants capture most mercury from emissions, along with fine particulate matter.

The Minamata Convention would seem an easy agreement to reach because the dangers of mercury are well understood. But broad agreements to control diffuse and widespread pollutants are often difficult. Regulating air pollution often pits a diffuse public interest (improving general health levels or child development) against specific private interests (utilities that must pay millions of dollars per year to control pollutants).

If you were in charge, how would you set rules for controlling mercury emissions? Would you impose rules or allow for trading of mercury emission permits, or would you allow voluntary self-regulation? How would you negotiate the responsibility for controlling pollutants?

FIGURE 1 Minamata disease is a severe form of mercury poisoning, which causes nerve damage and other conditions. The international treaty to control mercury emissions was named after the city of Minamata, Japan.
© AP Photo

economic growth is likely to depend on efficiencies and new technologies, these concerns remain an important part of the debate.

Methane (CH_4) is an important greenhouse gas that has 25 times the global warming effect per unit mass than CO_2 has (table 16.2). Like CO_2, methane was not a primary consideration when the Clean Air Act was passed in 1970, but now these gases are dominant factors in climate change. Methane occurs naturally as organic matter, such as plant material, decays in the absence of oxygen. Wetlands and rice paddies produce methane, as does anaerobic (oxygen-free) digestion by termites and ruminant animals such as cattle. Dramatic growth in livestock populations in recent decades and increasing rice production have sharply increased CH_4 concentrations in the atmosphere. Leaking natural gas wells and pipelines are also an important new source of atmospheric CH_4. Since the development of hydrofracturing (fracking) for natural gas and oil wells after about 2008, oil and gas producing regions around the world have introduced a great, but not well measured, amount of methane to the atmosphere (see chapter 15).

Since the midterm elections of 2010, many members of Congress have been intent on eliminating this and other pollution regulation, arguing that it is too costly for industry and the economy (see further discussion in section 16.4). Energy companies and their representatives, in particular, have lobbied to prevent legal limits on greenhouse gases. The 2011 congressional budget proposed to slash EPA funding by one-third, in part to reduce pollution monitoring and regulation.

The question of whether the EPA should regulate greenhouse gases was so contentious that it went to the Supreme Court in 2007. The Court ruled that it was the EPA's responsibility to limit these gases, on the grounds that greenhouse gases endanger public health and welfare within the meaning of the Clean Air Act. The Court, and subsequent EPA documents, noted that these risks include increased drought, more frequent and intense heat waves and wildfires, sea-level rise, and harm to water resources, agriculture, wildlife, and ecosystems. In addition to these risks, the U.S. military has cited climate change as a security threat. A coalition of generals and admirals signed a report from the Center for Naval Analyses stating that climate change "presents significant national security challenges" including violence resulting from scarcity of water, and migration due to sea-level rise and crop failure.

Since the Supreme Court ruling, the EPA is charged with regulating six greenhouse gases: carbon dioxide, methane, nitrous oxide, hydrofluorocarbons, perfluorocarbons, and sulfur hexafluoride (table 16.2). These are gases whose emissions have grown dramatically in recent decades.

Three of these six greenhouse gases contain halogens, a group of lightweight, highly reactive elements (fluorine, chlorine, bromine, and iodine). Because they are generally toxic in their elemental form, they are commonly used as fumigants and disinfectants, but they also have hundreds of uses in industrial and commercial products. Chlorofluorocarbons (CFCs) have been banned for most uses in industrialized countries, but about 600 million tons of these compounds are used annually worldwide in spray propellants and refrigeration compressors and for foam blowing. They diffuse into the stratosphere, where they release chlorine and fluorine atoms that destroy ozone molecules that protect the earth from ultraviolet radiation (see section 16.2).

Halogen compounds are also powerful greenhouse gases: They trap more energy per molecule than does CO_2, and they persist in the atmosphere for decades to centuries. Perfluorocarbons will persist in the atmosphere for thousands of years. The global warming potential (per molecule, over time) of some CFCs is thousands of times greater than that of CO_2 (fig. 16.12).

Developing rules and standards for greenhouse gases will take time and considerable debate. Many strategies have been proposed, including subsidies for alternative energy, reducing tax breaks and other subsidies for fossil fuels, imposing a tax on coal, oil, and gas, and cap-and-trade systems, including carbon-trading markets. The last of these options has been the most acceptable,

Table 16.2 Global Warming Potential (GWP) of Several Greenhouse Gases		
GAS	Global warming potential[1]	Atmospheric lifetime (years)[2]
Carbon dioxide (CO_2)	1	100
Methane (CH_4)	25	124
Nitrous oxide (N_2O)	298	1144
CFC-12 (CCl_2F_2)	10,900	100
HCFC-142b (CH_3CClF_2)	2,310	18
Sulfur hexafluoride (SF_6)	22,800	3200

[1] A measure of radiative effects, integrated over a 100-yr time horizon, relative to an equal mass of CO_2 emissions. CO_2 is set as 1 for comparison.

[2] Average residence times shown; actual range for CO_2 is decades to centuries.

Source: Carbon Dioxide Information Analysis Center, 2011

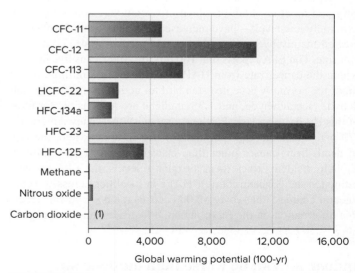

FIGURE 16.12 Global warming potential of several CFCs, compared to major greenhouse gases. Effects are integrated over a 100-year time frame, in comparison to CO_2.

Source: Carbon Dioxide Information Analysis Center, 2011

and carbon trading is now worth billions of dollars every year. Data remain inconclusive regarding whether this has produced an overall decline in emissions.

Hazardous air pollutants (HAPs) can cause cancer and nerve damage

Although most air contaminants are regulated because of their potential adverse effects on human health or environmental quality, a special category of toxins is monitored by the U.S. EPA because they are particularly dangerous. Called **hazardous air pollutants (HAPs),** these chemicals include carcinogens, neurotoxins, mutagens, teratogens, endocrine system disrupters, and other highly toxic compounds (chapter 8). The most persistent compounds require special reporting and management because they remain in ecosystems for long periods of time and accumulate in animal and human tissues. The tendency to bioaccumulate makes many of these hazardous air pollutants especially dangerous. Most of these chemicals are either metal compounds, chlorinated hydrocarbons, or volatile organic compounds. Gasoline vapors, solvents, and components of plastics are all HAPs that you may encounter on a daily basis.

Only about 50 locations in the United States regularly measure concentrations of HAPs in ambient air. Often, the best source of information about these chemicals is the **Toxic Release Inventory (TRI)** collected by the EPA as part of the community right-to-know program. Established by Congress in 1986, the TRI requires 23,000 factories, refineries, hard rock mines, power plants, and chemical manufacturers to report on toxin releases (above certain minimum amounts) and waste management methods for 667 toxic chemicals. Although this total is less than 1 percent of all chemicals registered for use, and represents a limited range of sources, the TRI is widely considered the most comprehensive source of information about toxic pollution in the United States (fig. 16.13).

Most HAP releases are decreasing, but discharges of mercury and dioxins—both of which bioaccumulate and are toxic at extremely low levels—have increased in recent years. Dioxins are created mainly by burning plastics and medical waste containing chlorine. The EPA reports that 100 million Americans live in areas where the cancer rate from HAPs exceeds 10 in 1 million, or ten times the normally accepted standard for action. Benzene, formaldehyde, acetaldehyde, and 1,3 butadiene are responsible for most of this HAP cancer risk. Furthermore, twice that many Americans (70 percent of the U.S. population) live in areas where the risk of death from causes other than cancer exceeds 1 in 1 million. To help residents track local air quality levels, the EPA recently estimated the concentration of HAPs in localities across the continental United States (over 60,000 census tracts). You can access this information on the Environmental Defense Fund web page at http://scorecard.goodguide.com/env-releases/hap/us.tcl.

Indoor air can be worse than outdoor air

We have spent a considerable amount of effort and money to control the major outdoor air pollutants, but we have only recently begun to address indoor air pollutants. The EPA has found that

FIGURE 16.13 Harmful air toxics from large industrial sources, such as chemical plants, petroleum refineries, and paper mills, have been reduced by nearly 70 percent since the EPA began regulating them. Many smaller sources remain unregulated.
© Royalty-Free/Corbis

concentrations of toxic air pollutants are often higher indoors than outdoors. Furthermore, people generally spend more time inside than out, so they are exposed to higher doses of these pollutants.

In some cases, indoor air in homes has concentrations of chemicals that would be illegal outside or in the workplace. The EPA has found that concentrations of such compounds as chloroform, benzene, carbon tetrachloride, formaldehyde, and styrene can be 70 times higher in indoor air than in outdoor air, as plastics, carpets, paints, and other common materials off-gas these compounds. Finding less-toxic paints and fabrics can make indoor spaces both healthier and more pleasant.

In the less-developed countries of Africa, Asia, and Latin America, where such organic fuels as firewood, charcoal, dried dung, and agricultural wastes provide the majority of household energy, smoky and poorly ventilated heating and cooking fires are the greatest source of indoor air pollution (fig. 16.14). The World Health Organization (WHO) estimates that 2.5 billion people—over a third of the world's population—are adversely affected by pollution from this source. Women and small children spend long hours each day around open fires or unventilated stoves in enclosed spaces. Levels of carbon monoxide, particulates, aldehydes, and other toxic

FIGURE 16.14 Smoky cooking and heating fires may cause more ill health effects than any other source of indoor air pollution except tobacco smoking. Some 2.5 billion people, mainly women and children, spend hours each day in poorly ventilated kitchens and living spaces where carbon monoxide, particulates, and cancer-causing hydrocarbons often reach dangerous levels.
© Chinese Tourism Press/Stone/Getty Images

chemicals can be 100 times higher than would be legal for outdoor ambient concentrations in the United States. Designing and building cheap, efficient, nonpolluting energy sources for the developing countries would not only save shrinking forests but would make a major impact on health as well.

Section Review

1. Define *primary air pollutants* and *secondary air pollutants*.
2. What are the six criteria pollutants in the original Clean Air Act? Why were they chosen?
3. List several additional hazardous air toxins that are regulated.

16.2 ATMOSPHERIC PROCESSES

- *Inversions trap still, contaminated air.*
- *Long-distance transport distributes pollutants.*
- *Reactive chlorine destroys stratospheric ozone.*

Topography, climate, and physical processes in the atmosphere play an important role in the transport, concentration, dispersal, and removal of many air pollutants. Cities concentrate dust and pollutants in urban

"dust domes"; winds cause mixing between air layers, precipitation, and atmospheric chemistry. All these factors determine whether pollutants will remain in the locality where they are produced or go elsewhere. Often, a change in weather produces dramatic changes in air quality (fig. 16.15). In this section, we examine how atmospheric processes influence the concentration and movement of pollutants.

FIGURE 16.15 Air quality can change dramatically when a shift in the weather brings fresh air to a city. Here, Shanghai is shown before and after a cold front brought in clean air, pushing smog out of the city.
© William P. Cunningham

Temperature inversions trap pollutants

Normally, air near the ground is warmed in the daytime, as the ground absorbs solar energy and radiates it into the lower layers of air. As air near the ground warms, it expands, becomes less dense, and rises, causing turbulence or winds that circulate air. Circulation also removes pollutants generated near the ground in a city, so it is important to keeping a city livable. Sometimes, however, **temperature inversions** occur. At night or on cold days, air near the ground cools, or cold air may sink down into a valley from surrounding hills. A stable layer of cold air settles near the ground, while warmer air sits above. In these stable conditions, pollutants can accumulate to very high concentrations. Inversions might last from a few hours to a few days, until the weather warms or a warm front moves in.

The most stable inversion conditions are usually created by rapid nighttime cooling in a valley or basin where air movement is restricted. Los Angeles is a classic example, with conditions that create both temperature inversions and photochemical smog (fig. 16.16). The city is surrounded by mountains on three sides and the climate is dry, with abundant sunshine for photochemical oxidation and ozone production. Millions of automobiles and trucks create high pollution levels. Skies are generally clear at night, allowing heat to radiate from the ground. The ground and the lower layers of air cool quickly at night, while upper air layers remain relatively warm. During the night, cool, humid, onshore breezes also slide in under the contaminated air, which is trapped by a wall of mountains to the east and by the cap of warmer air above.

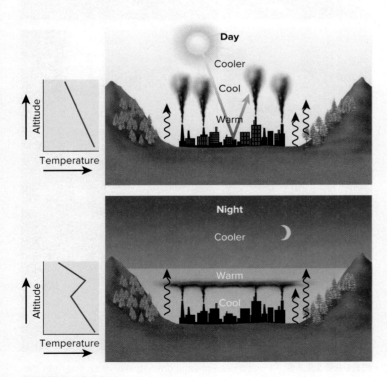

FIGURE 16.16 Atmospheric temperature inversions occur where ground-level air cools more quickly than air at upper levels. This temperature differential prevents mixing and traps pollutants close to the ground.

Morning sunlight is absorbed by the concentrated aerosols and gaseous chemicals caught near the ground by the inversion. This complex mixture quickly cooks up a toxic brew of hazardous compounds. As the ground warms later in the day, convection currents break up the temperature gradient and pollutants are carried back down to the surface, where more contaminants are added. Nitric oxide (NO) from automobile exhaust is oxidized to a brownish haze of nitrogen dioxide (NO_2). As nitrogen oxides are used up in reactions with unburned hydrocarbons, the ozone level begins to rise. By early afternoon an acrid brown haze fills the air, making eyes water and throats burn. In the 1970s, before pollution controls were enforced, afternoon concentrations of NO_x and ozone in the Los Angeles basin often would reach levels hazardous to health (see fig. 16.2). Usually, inversions are minor nuisances for healthy people. Sometimes, however, they can have dramatic effects (see Exploring Science: The Great London Smog, later in this section).

Wind currents carry pollutants worldwide

Dust and contaminants can be carried great distances by the wind. Areas downwind from industrial complexes often suffer serious contamination, even if they have no pollution sources of their own (fig. 16.17). Pollution from the industrial belt between the Great Lakes and the Ohio River Valley, for example, regularly contaminates the Canadian Maritime Provinces, and sometimes can be traced as far as Ireland. As noted earlier, long-range transport is a major source of Asian mercury in North America.

Studies of air pollutants over southern Asia reveal a 3-km-thick toxic cloud of ash, acids, aerosols, dust, and photochemical reactants that regularly covers the entire Indian subcontinent and can last for much of the year. Nobel laureate Paul Crutzen estimates that up to 2 million people in India alone die each year from atmospheric pollution. Produced by forest fires, the burning of agricultural wastes, and dramatic increases in the use of fossil fuels, the Asian smog layer cuts by up to 15 percent the amount of solar energy reaching the earth's surface beneath it. Meteorologists suggest that the cloud—80 percent of which is human-made—could disrupt monsoon weather patterns and may be disturbing rainfall and reducing rice harvests over much of South Asia. As UN Environment Programme executive director Klaus Töpfer said, "There are global implications because a pollution parcel like this, which stretches 3 km high, can travel half way round the globe in a week."

An increase in monitoring activity has revealed industrial contaminants in places usually considered among the cleanest in the world. Samoa, Greenland, Antarctica, and the North Pole all have heavy metals, pesticides, and radioactive elements in their air. Since the 1950s, pilots flying in the high Arctic have reported dense layers of reddish-brown haze clouding the arctic atmosphere. Aerosols of sulfates, soot, dust, and toxic heavy metals, such as vanadium, manganese, and lead, travel to the pole from the industrialized parts of Europe and Russia.

A process called "grasshopper" transport, or atmosphere distillation, helps deliver contaminants to the poles. Volatile compounds evaporate from warm areas, travel through the atmosphere, and

The Great London Smog and Pollution Monitoring

London was once legendary for its pea-soup fogs. In the days of Charles Dickens and Sherlock Holmes, darkened skies and blackened buildings, saturated with soot from hundreds of thousands of coal-burning fireplaces, were a fact of life. Londoners had been accustomed to filthy air since the Industrial Revolution, but over a period of four days in 1952 they experienced the worst air pollution disaster on record, which helped dramatically change approaches to air pollution.

On a cold afternoon in early December 1952, a dense blanket of coal smoke and fog settled on the city. Because it was a cold day, home coal burners and industrial furnaces began pumping out smoke at full force. During the afternoon, visibility plummeted and traffic came to a halt as drivers were blinded by the smoke and fog. Hundreds of cattle at a cattle market were the first to go. With lungs blackened by soot, they suffocated while standing in their pens. Concerts were canceled because of blackened air in the halls, and books in the British Museum were tainted with soot. Visibility fell to one foot in some places by the third day of the inversion (fig. 1a).

Thousands died, especially the ill and elderly. Hospitals filled with victims of bronchitis, pneumonia, lung inflammations, and heart failure. Patients' lungs were clogged by smoke and microscopic soot particles, their lips turned blue, and many asphyxiated due to lack of oxygen.

A temperature inversion initiated the crisis. Normally, atmospheric circulation moves polluted air away from the city and out over the countryside. Stable inversions, with cold air resting near the ground, can be unpleasantly chilly and damp. In this case, the inversion trapped coal dust particulates and tiny droplets of sulfuric acid (from sulfur in coal) in the city.

After four days, a change in the weather brought fresh winds into London, and the inversion dissipated. At least 4,700 deaths were attributable to air pollution during and immediately after the inversion. More-recent studies have found that lingering ailments killed perhaps another 8,000 in the months that followed, bringing the total death toll to over 12,000.

London had been debating what to do about air pollution for centuries, since at least 1300, when controls on smoke were

attempted (and failed). Pollution was normal, too pervasive to change. The crisis of 1952 pushed the city to come up with new policies. Coal fireplaces were phased out, replaced with oil burners. In 1956, Parliament passed a Clean Air Act, and in 1968 the act was strengthened and expanded to address additional industrial emissions. The United States, watching closely, adopted a Clean Air Act in 1963, with major amendments in 1970.

These laws would not be possible without good quality air monitoring data. As with any environmental policy, regular and reliable collection of data is a first requirement for setting emissions limits. Atmospheric chemists can then compare observed sulfur dioxide levels, for example, to acceptable health limits (fig. 1b). Epidemiologists can tie rates of hospitalization and deaths to high levels of sulfur dioxide in the air. Monitoring data make it possible to justify and legally defend emissions limits. Legal enforcement of rules is the main reason the events of London in 1952 are now, for many places, fading into history

(a)

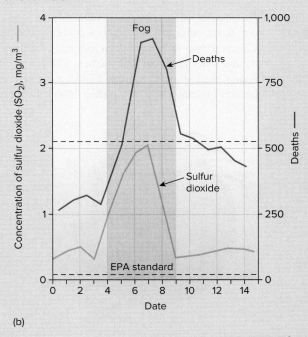

(b)

FIGURE 1 Sulfur dioxide was a major cause of deaths during the London smog of December 1952 (a). During this event, sulfur dioxide levels reached 2 mg/m³ of air, well above the EPA standard limit of 0.08 mg/m³ (dashed line, (b))

a. © Central Press/Hulton Archive/Getty Images

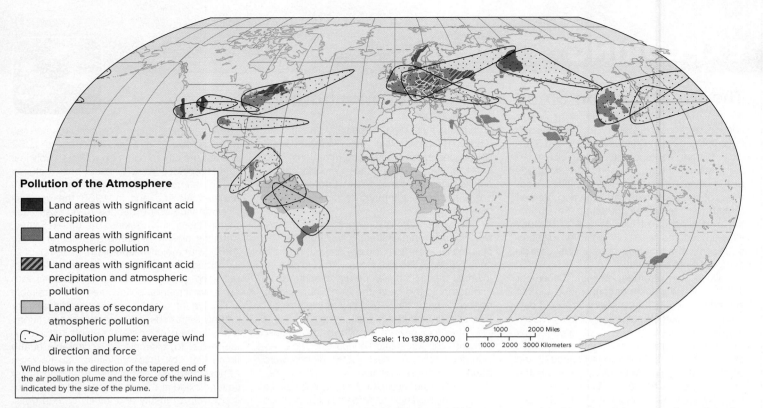

Pollution of the Atmosphere

Land areas with significant acid precipitation

Land areas with significant atmospheric pollution

Land areas with significant acid precipitation and atmospheric pollution

Land areas of secondary atmospheric pollution

Air pollution plume: average wind direction and force

Wind blows in the direction of the tapered end of the air pollution plume and the force of the wind is indicated by the size of the plume.

Scale: 1 to 138,870,000

0   1000   2000 Miles

0   1000   2000   3000 Kilometers

FIGURE 16.17 Long-range transport carries air pollution from source regions thousands of kilometers away into formerly pristine areas. Secondary air pollutants can be formed by photochemical reactions far from primary emissions sources.

then condense and precipitate in cooler regions (fig. 16.18). Over several years, contaminants accumulate in the coldest places, generally at high latitudes where they bioaccumulate in food chains. Whales, polar bears, sharks, and other top carnivores in polar regions have been shown to have dangerously high levels of pesticides, metals, and other HAPs in their bodies. The Inuit people of Broughton Island, well above the Arctic Circle, have higher levels of polychlorinated biphenyls (PCBs) in their blood than any other known population except victims of industrial accidents. Far from any source of this industrial by-product, these people accumulate PCBs from the flesh of fish, caribou, and other animals they eat.

Chlorine destroys ozone in the stratosphere

Ozone near the ground is a pollutant because it irritates living tissues and contributes to smog. Far out in the stratosphere, however, ozone absorbs harmful ultraviolet (UV) solar energy, high-energy wavelengths that can damage plant and animal tissues, including the eyes and the skin. **Stratospheric ozone** provides a critical protective shield against UV radiation.

In 1985, the British Antarctic Atmospheric Survey announced a startling and disturbing discovery: Stratospheric ozone concentrations over the South Pole were dropping precipitously during September and October every year as the sun reappeared at the end of the long polar winter (fig. 16.19). This ozone depletion has been occurring at least since the 1960s but was not recognized because earlier researchers programmed their instruments to ignore changes in ozone levels that were presumed to be erroneous.

Chlorine-based aerosols, especially **chlorofluorocarbons (CFCs)** and other halon gases, are the principal agents of ozone depletion. Nontoxic, nonflammable, chemically inert, and cheaply produced, CFCs were extremely useful as industrial gases and in refrigerators, air conditioners, Styrofoam inflation, and aerosol spray cans for many years. From the 1930s until the 1980s, CFCs were used all over the world and widely dispersed through the atmosphere.

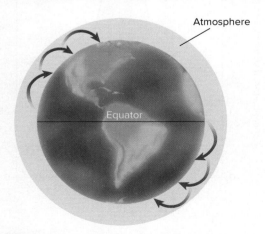

Atmosphere

Equator

FIGURE 16.18 Air pollutants evaporate from warmer areas and then condense and precipitate in cooler regions. Eventually, this "grasshopper" redistribution leads to accumulation in the Arctic and Antarctic.
Source: NASA

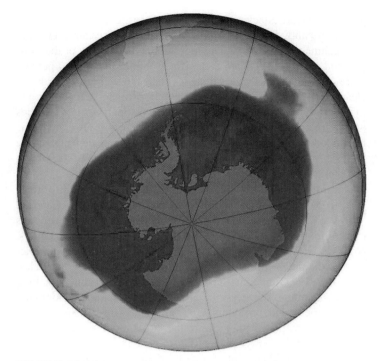

FIGURE 16.19 The region of stratospheric ozone depletion grew steadily to an area of nearly 30 million km² in 2006 (shown here). This ozone "hole" has shown signs of decline since the Montreal Protocol went into effect. *Source: NASA*

What we often call an ozone "hole" is really a vast area of reduced concentrations of ozone in the stratosphere. The thinning happens in the early Antarctic spring: Antarctica's exceptionally cold winter temperatures (–85 to –90°C) help break down ozone. During the long, dark winter months, strong winds circle the pole, isolating Antarctic air and allowing stratospheric temperatures to drop low enough to create ice crystals at high altitudes—something that rarely happens elsewhere in the world. Ozone and chlorine-containing molecules are absorbed on the surfaces of these ice particles. When the sun returns in the spring, it provides energy to liberate chlorine ions, which readily bond with ozone, breaking it down to molecular oxygen (table 16.3). It is only during the Antarctic spring (September through December) that conditions are ideal for rapid ozone destruction. During that season, temperatures are still cold enough for high-altitude ice crystals, but the sun gradually becomes strong enough to drive photochemical reactions.

As the Antarctic summer arrives, temperatures moderate, the circumpolar vortex breaks down, and air from warmer latitudes mixes with Antarctic air, replenishing ozone concentrations in the ozone hole. Slight decreases worldwide result from this mixing, however. Ozone re-forms naturally, but not nearly as fast as it is destroyed. Because the chlorine atoms are not themselves consumed in reactions with ozone, they continue to destroy ozone for years. Eventually, they can precipitate out, but this process happens very slowly in the stable stratosphere.

About 10 percent of all stratospheric ozone worldwide has been destroyed in recent years, and levels over the Arctic have

Table 16.3	Stratospheric Ozone Destruction by Chlorine Atoms and UV Radiation
Step	**Products**
1. CFCl$_3$ (chlorofluorocarbon) + UV energy	CFCl$_2$ + Cl
2. Cl + O$_3$	ClO + O$_2$
3. O$_2$ + UV energy	2O
4. ClO + 2O	O$_2$ + Cl
5. Return to step 2	

averaged 40 percent below normal. Ozone depletion has been observed over the North Pole as well, although it is not as concentrated as that in the south.

The Montreal Protocol was a resounding success

The discovery of stratospheric ozone losses brought about a remarkably quick international response. In 1987, an international meeting in Montreal, Canada, produced the Montreal Protocol, the first of several major international agreements on phasing out most use of CFCs by 2000. As evidence accumulated, showing that losses were larger and more widespread than previously thought, the deadline for the elimination of all CFCs (halons, carbon tetrachloride, and methyl chloroform) was moved up to 1996, and a $500 million fund was established to assist poorer countries in switching to non-CFC technologies. Fortunately, alternatives to CFCs for most uses already exist. The first substitutes are hydrochlorofluorocarbons (HCFCs), which release much

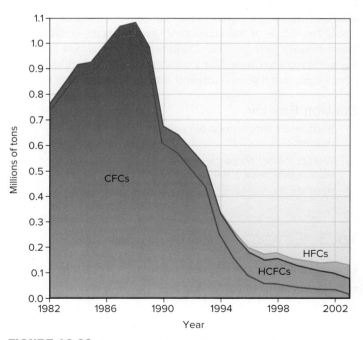

FIGURE 16.20 The Montreal Protocol has been remarkably successful in eliminating CFC production. The remaining HFC and HCFC use is primarily in developing countries, such as China and India.

less chlorine per molecule. These HCFCs are also being phased out, as newer halogen-free alternatives are developed.

The Montreal Protocol is often cited as the most effective international environmental agreement ever established. Global CFC production has been cut by more than 95 percent since 1988 (fig. 16.20). Some of that has been replaced by HCFCs, which release chlorine, but not as much as CFCs. The amount of chlorine entering the atmosphere already has begun to decrease.

The size of the ozone "hole" increased steadily from its discovery until the mid-1990s, when the Montreal Protocol began having an effect. Since then, it has varied from year to year, but the trend has been to stabilize or decrease in recent years. In one of the world's most remarkable success stories, stratospheric O_3 levels should be back to normal by about 2049. There is variation in this trend, however. The 2006 ozone hole was the largest ever. Ironically, climate warming (heat retention) in the lower atmosphere has contributed to cooling in the stratosphere. This cooling increases ice crystal formation over the Antarctic and results in more ozone depletion.

The Montreal Protocol had an added benefit in slowing climate change, because CFCs and other ozone-destroying gases are also powerful greenhouse gases (see fig. 16.12). Reductions in emissions of these gases under the Montreal Protocol amount to one-quarter of all greenhouse gas emissions worldwide. This reduction has had a greater impact on climate-changing gases than the Kyoto Protocol.

There's another interesting connection to climate change. Under the Montreal Protocol, China, India, Korea, and Argentina were allowed to continue to produce 72,000 tons (combined) of CFCs per year until 2010. Most of the funds appropriated through the Montreal Protocol are going to these countries to help them phase out CFC production and destroy their existing stocks. Because CFCs are potent greenhouse gases, this phase-out also makes these countries eligible for credits in the climate trading market. In 2006, nearly two-thirds of the greenhouse gas emissions credits traded internationally were for HFC-23 elimination, and almost half of all payments went to China. Some critics think this is double-dipping; others argue that it doesn't matter if serious risks are being reduced.

Section Review

1. What is an atmospheric temperature inversion, and why is it a problem?
2. What is the difference between ambient and stratospheric ozone? What is destroying stratospheric ozone?
3. What did the Montreal Protocol aim to accomplish?

16.3 Effects of Air Pollution

- *Lung tissues are vulnerable to acids and particulates.*
- *Sulfur and nitrogen produce acidic deposition.*

Air pollution is a problem of widespread interest because it affects so many parts of our lives. The most obvious effects are on our health. Damage to infrastructure, vegetation, and aesthetic quality—especially visibility—are also important considerations.

The World Health Organization estimates that some 5 to 6 million people die prematurely every year from illnesses related to air pollution. Heart attacks, respiratory diseases, and lung cancer all are significantly higher in people who breathe dirty air, compared to matching groups in cleaner environments. Residents of the most polluted cities in the United States, for example, are 15 to 17 percent more likely to die of these illnesses than those in cities with the cleanest air. This can mean as much as a five- to ten-year decrease in life expectancy for those who live in the worst parts of Los Angeles or Baltimore, compared to a place with clean air. Of course, the likelihood of suffering ill health from air pollutants depends on the intensity and duration of exposure, as well as the age of the person and their prior health status.

In industrialized countries, one of the biggest health threats from air pollution is from soot or fine particulate material. We once thought that particles smaller than 10 micrometers (10 millionths of a meter) were too small to be trapped in the lungs. Now we know that fine PM2.5 particles (less than 2.5 micrometers in diameter) pose even greater risks than coarse particles. They have been linked with heart attacks, asthma, bronchitis, lung cancer, immune suppression, and abnormal fetal development, among other health problems. Fine particulates have many sources. Until recently, power plants were the largest source, but clean air rules will require power plants to install filters and precipitators to remove at least 70 percent of their particulate emissions.

The U.S. EPA estimates that at least 160 million Americans—more than half the population—live in areas with unhealthy concentrations of fine particulate matter. PM2.5 levels have decreased about 30 percent over the past 25 years, but health conditions will improve if we can make further reductions.

Diesel engines have long been a major source of both soot and SO_2 in the United States (fig. 16.21). Until 2006, diesel-powered engines were allowed to use fuel that was up to 3,400 parts per

FIGURE 16.21 Soot and fine particulate material from diesel engines, wood stoves, power plants, and other combustion sources have been linked to asthma, heart attacks, and a variety of other diseases.
© *McGraw-Hill Education/John Thoeming, photographer*

million sulfur. In 2006, the allowable limit was cut to 500 ppm sulfur, and by 2010, the EPA phased in rules requiring that trucks, buses, and cars use ultra-low-sulfur fuel, with sulfur concentrations of less than 15 ppm. This dramatic reduction in sulfur also allows the use of advanced emission control technology in engines, further reducing health risks and odors of diesel emissions. These standards now also apply to off-road vehicles, such as tractors, bulldozers, locomotives, and barges, whose engines previously emitted more soot than all the nation's cars, trucks, and buses together.

How does pollution make us sick?

The most common route of exposure to air pollutants is by inhalation, but direct absorption through the skin or contamination of food and water also are important pathways. Because they are strong oxidizing agents, sulfates, SO_2, NO_x, and O_3 act as irritants that damage delicate tissues in the eyes and respiratory passages. Fine particulates, irritants in their own right, penetrate deep into the lungs and carry metals and other HAPs on their surfaces. Inflammatory responses set in motion by these irritants impair lung function and trigger cardiovascular problems as the heart tries to compensate for lack of oxygen by pumping faster and harder. If the irritation is really severe, so much fluid seeps into the lungs through damaged tissues that the victim actually drowns.

Carbon monoxide binds to hemoglobin and decreases the ability of red blood cells to carry oxygen. Asphyxiants such as this cause headaches, dizziness, and heart stress, and can be lethal if concentrations are high enough. Lead also binds to hemoglobin, reducing its oxygen-carrying capacity at high levels. At lower levels, lead causes long-term damage to critical neurons in the brain that results in mental and physical impairment and developmental retardation.

Some important chronic health effects of air pollutants include bronchitis and emphysema. **Bronchitis** is a persistent inflammation of bronchi and bronchioles (large and small airways in the lung) that causes mucus buildup, a painful cough, and involuntary muscle spasms that constrict airways. Severe bronchitis can lead to emphysema, an irreversible **chronic obstructive lung disease** in which airways become permanently constricted and alveoli are damaged or even destroyed. Stagnant air trapped in blocked airways swells the tiny air sacs in the lung (alveoli), blocking blood circulation. As cells die from lack of oxygen and nutrients, the walls of the alveoli break down, creating large empty spaces incapable of gas exchange (fig. 16.22). Thickened walls of the bronchioles lose elasticity, and breathing becomes more difficult. Victims of emphysema make a characteristic whistling sound when they breathe. Often, they need supplementary oxygen to make up for reduced respiratory capacity.

Plants suffer cell damage and lost productivity

Uncontrolled industrial fumes from furnaces, smelters, refineries, and chemical plants destroy vegetation and have created desolate, barren landscapes around mining and manufacturing centers. The copper–nickel smelter at Sudbury, Ontario, is a spectacular and notorious example of air pollution effects on vegetation and ecosystems. In 1886, the corporate ancestor of the International Nickel Company (INCO) began open-bed roasting of sulfide ores

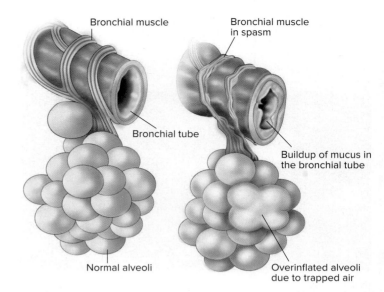

FIGURE 16.22 Bronchitis and emphysema can result in constriction of airways and permanent damage to tiny, sensitive air sacs called alveoli, where oxygen diffuses into blood vessels.

at Sudbury. Sulfur dioxide and sulfuric acid released by this process caused massive destruction of the plant community within about 30 km of the smelter. Rains washed away the exposed soil, leaving a barren moonscape of blackened bedrock (fig. 16.20a). Super-tall, 400-m smokestacks were installed in the 1950s, and sulfur scrubbers were added 20 years later. Emissions were reduced by 90 percent and the surrounding ecosystem is beginning to recover (fig. 16.23b).

There are two probable ways that air pollutants damage plants. They can be directly toxic, damaging sensitive cell membranes much as irritants do in human lungs. Within a few days of exposure to toxic levels of oxidants, mottling (discoloration) occurs in leaves due to chlorosis (bleaching of chlorophyll), and then necrotic (dead) spots develop (see fig. 16.5). If injury is severe, the whole plant may be killed. Sometimes these symptoms are so distinctive that positive identification of the source of damage is possible. Often, however, the symptoms are vague and difficult to separate from diseases or insect damage.

Certain combinations of environmental factors have **synergistic effects** in which the injury caused by exposure to two factors together is more than the sum of exposure to each factor individually. For instance, when white pine seedlings are exposed to subthreshold concentrations of ozone and sulfur dioxide individually, no visible injury occurs. If the same concentrations of pollutants are given together, however, visible damage occurs. In alfalfa, however, SO_2 and O_3 together cause less damage than either one alone. These complex interactions point out the unpredictability of future effects of pollutants.

Acid deposition damages ecosystems

Most people in the United States became aware of problems associated with **acid deposition** (the deposition of wet acidic solutions or dry acidic particles from the air) within the last decade or so, but

(a) 1975

(b) 2005

FIGURE 16.23 In 1975, acid precipitation from the copper–nickel smelters (tall stacks in background) had killed all the vegetation and charred the pink granite bedrock black for a large area around Sudbury, Ontario (a). By 2005, forest cover was growing again, although the rock surfaces remain burned black (b).
© William P. Cunningham

English scientist Robert Angus Smith coined the term *acid rain* in his studies of air chemistry in Manchester, England, in the 1850s. By the 1940s, it was known that pollutants, including atmospheric acids, could be transported long distances by wind currents. This was thought to be only an academic curiosity until it was shown that precipitation of these acids can have far-reaching ecological effects.

We describe acidity in terms of pH (see chapter 3). Values below 7 are acidic, while those above 7 are basic. Normal, unpolluted rain generally has a pH of about 5.6 due to carbonic acid created by CO_2 in air. Sulfur, chlorine, and other elements also form acidic compounds as they are released in sea spray, volcanic emissions, and biological decomposition. These sources can lower the pH of rain well below 5.6. Other factors, such as alkaline dust can raise it above 7. In industrialized areas, anthropogenic acids in the air usually far outweigh those from natural sources.

Aquatic Effects Lakes and streams can be especially sensitive to acid deposition, especially where vegetation or bedrock makes them naturally acidic to start with. This problem was first publicized in Scandinavia, which receives industrial and automobile emissions—principally H_2SO_4 and HNO_3—generated in northwestern Europe. The thin, acidic soils and oligotrophic lakes and streams in the mountains of southern Norway and Sweden have been severely affected by this acid deposition. Some 18,000 lakes in Sweden are now so acidic that they will no longer support game fish or other sensitive aquatic organisms.

Generally, reproduction is the most sensitive stage in fish life cycles. The eggs and young of many species are killed when the pH drops to about 5.0. This level of acidification also can disrupt the food chain by killing aquatic plants, insects, and invertebrates on which fish depend for food. At pH levels below 5.0, adult fish die as well. Trout, salmon, and other game fish are usually the most sensitive. Carp, gar, suckers, and other less desirable fish are more resistant.

In the early 1970s, evidence began to accumulate suggesting that air pollutants were acidifying many lakes in North America. Studies in the Adirondack Mountains of New York revealed that about half of the high-altitude lakes (above 1,000 m or 3,300 ft) were acidified and had no fish. Areas showing lake damage correlated closely with average pH levels in precipitation (fig. 16.24).

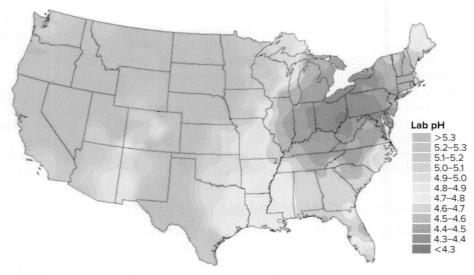

Lab pH
>5.3
5.2–5.3
5.1–5.2
5.0–5.1
4.9–5.0
4.8–4.9
4.7–4.8
4.6–4.7
4.5–4.6
4.4–4.5
4.3–4.4
<4.3

FIGURE 16.24 Acid precipitation over the United States in 2000. Many areas have improved markedly since then, because of the new rules on sulfur dioxide emissions and other pollutants.
Source: National Acid Depositions Program

Some 48,000 lakes in Ontario are currently endangered, and nearly all of Quebec's surface waters, including about 1 million lakes, are believed to be highly sensitive to acid deposition.

Forest Damage   In the early 1980s, disturbing reports appeared of rapid forest declines in both Europe and North America. One of the earliest was a detailed ecosystem inventory on Camel's Hump Mountain in Vermont. A 1980 survey showed that seedling production, tree density, and the viability of spruce-fir forests at high elevations had declined about 50 percent in 15 years. A similar situation was found on Mount Mitchell in North Carolina, where almost all red spruce and Fraser fir above 2,000 m (6,000 ft) are in a severe decline. Nearly all the trees are losing needles and about half of them are dead (fig. 16.25). The stress of acid rain and fog, other air pollutants, and attacks by an invasive insect called the woody aldegid are killing the trees.

Many European countries reported catastrophic forest destruction in the 1980s. It still isn't clear what caused this injury. In the longest-running forest-ecosystem monitoring record in North America, researchers at the Hubbard Brook Experimental Forest in New Hampshire have shown that forest soils have become depleted of natural buffering reserves of basic cations such as calcium and magnesium through years of exposure to acid rain. Replacement of these cations by hydrogen and aluminum ions seems to be one of the main causes of plant mortality.

Buildings and Monuments   In cities throughout the world, some of the oldest and most glorious buildings and works of art are being destroyed by air pollution. Smoke and soot coat buildings, paintings, and textiles. Limestone and marble are destroyed by atmospheric acids at an alarming rate. The Parthenon in Athens, the Taj Mahal in Agra, the Colosseum in Rome, frescoes and statues in Florence,

FIGURE 16.26 Atmospheric acids, especially sulfuric and nitric acids, have almost completely eaten away the face of this medieval statue. Each year, the total loss from air pollution damage to buildings and materials amounts to billions of dollars.
© Ryan McGinnis/Alamy Stock Photo

medieval cathedrals in Europe (fig. 16.26), and the Lincoln Memorial and Washington Monument in Washington, D.C., are slowly dissolving and flaking away because of acidic fumes in the air. Medieval stained glass windows in Cologne's gothic cathedral are so porous from etching by atmospheric acids that pigments disappear and the glass literally crumbles away. Restoration costs for this one building alone are estimated at up to 3 billion euros (U.S. $4 billion).

Air pollution also damages ordinary buildings and structures. Corroding steel in reinforced concrete weakens buildings, roads, and bridges. Paint and rubber deteriorate due to oxidation. Limestone, marble, and some kinds of sandstone flake and crumble. The Council on Environmental Quality estimates that U.S. economic losses from architectural damage caused by air pollution amount to about $4.8 billion in direct costs and $5.2 billion in property value losses each year.

Section Review

1. List several illnesses that are made worse by dirty air.
2. Explain the idea of "synergistic effects."
3. What is acid deposition? Identify two of the pollutants that cause it.

16.4 POLLUTION CONTROL

- *Pollutants can be captured after production or avoided.*
- *Clean air legislation has reduced CO and SO_4 but not NO_x.*
- *Data show that the CAA has saved money and has not diminished economic growth.*

"Dilution is the solution to pollution" was our main approach to air pollution control for most of history. Tall smokestacks were built to send emissions far from the source, where they became

FIGURE 16.25 A Fraser fir forest on Mount Mitchell, North Carolina, killed by acid rain, insect pests, and other stressors.
© William P. Cunningham

unidentifiable and largely untraceable. But dispersed and diluted pollutants are now the source of some of our most serious pollution problems. We are finding that there is no "away" to which we can throw our waste products. Although most of the discussion in this section focuses on industrial solutions, each of us can make important personal contributions to this effort (see the What Can You Do? box below).

Because most air pollution in the developed world is associated with transportation and energy production, the most effective strategies involve conservation and renewable energy. Reducing electricity consumption, insulating homes and offices, and developing better public transportation could all greatly reduce air pollution. Alternative energy sources, such as wind and solar power, produce energy with little or no pollution, and these and other technologies are becoming increasingly available (see chapter 20). In addition to conservation, pollution control technology can capture pollutants at the source.

Pollutants can be captured after combustion

Particulate removal involves filtering air emissions. Filters trap particulates in a mesh of cotton cloth, spun glass fibers, or asbestos-cellulose. Industrial air filters are generally giant bags 10 to 15 m long and 2 to 3 m wide. Effluent gas is blown through the bag, much like the bag on a vacuum cleaner. Every few days or weeks, the bags are opened to remove the dust cake. Electrostatic precipitators are the most common particulate controls in power plants. Ash particles pick up an electrostatic surface charge as they

pass between large electrodes in the effluent stream (fig. 16.27). Charged particles then collect on an oppositely charged collecting plate. These precipitators consume a large amount of electricity, but maintenance is relatively simple, and collection efficiency can be as high as 99 percent. The ash collected by both of these techniques is a solid waste (often hazardous due to the heavy metals and other trace components of coal or other ash source) and must be buried in landfills or other solid-waste disposal sites.

Sulfur removal is important because sulfur oxides are among the most damaging of all air pollutants in terms of human health and ecosystem impacts. As with particulate matter, sulfur dioxide can be captured after combustion, or fuel (especially coal) can be cleaned before burning. Coal can be crushed, washed, and gasified to remove sulfur and metals before combustion. These processes cost money, but when sulfur is captured from fuels, it can also be sold as a useful industrial product. Elemental sulfur, sulfuric acid, and ammonium sulfate can all be produced using catalytic converters to oxidize or reduce sulfur.

Switching to low-sulfur fuels is the surest way to reduce sulfur emissions: from soft coal with a high sulfur content to low-sulfur coal, or from coal to oil or gas or renewable energy. We often use high-sulfur coal for political or economical reasons. In the United States, most high-sulfur coal is produced in Appalachia, a region with chronic poverty but substantial political influence. China, similarly, has long relied on sulfur-rich domestic coal. Both the U.S. and China have begun switching to lower-sulfur coal, and to cleaner oil or gas, but the change takes time. Sulfur emissions from transportation have been reduced by new rules requiring low-sulfur diesel fuel.

Nitrogen oxides (NO_x) can be reduced in both internal combustion engines and industrial boilers by as much as 50 percent by carefully controlling the flow of air and fuel. Staged burners, for

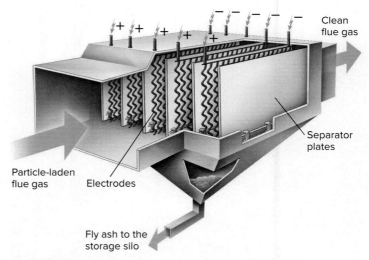

FIGURE 16.27 An electrostatic precipitator can remove 99 percent of unburned particulates in the effluent (smoke) from power plants. Electrodes transfer a static electric charge to dust and smoke particles, which then adhere to collector plates. Particles are then shaken off of the plates and collected for reuse or disposal.

example, control burning temperatures and oxygen flow to prevent formation of NO_x. The catalytic converter on your car uses platinum-palladium and rhodium metals to catalyze reactions that convert NO_x, unburned hydrocarbons, and carbon monoxide, to water, nitrogen gas, and carbon dioxide. Catalytic converters became required on new car engines in the United States in the 1980s, and EPA data shows that new vehicles today emit 96 percent less carbon monoxide, 98 percent less hydrocarbons, and 90 percent less NO_x than new vehicles in the 1970s.

Hydrocarbon controls mainly involve complete combustion or controlling evaporation. Hydrocarbons and volatile organic compounds are produced by incomplete combustion of fuels or by solvent evaporation from chemical factories, paints, dry cleaning, plastic manufacturing, printing, and other industrial processes. Closed systems that prevent escape of fugitive gases can reduce many of these emissions. In automobiles, for instance, positive crankcase ventilation (PCV) systems collect oil that escapes from around the pistons and unburned fuel and channels them back to the engine for combustion. Controls on fugitive losses from industrial valves, pipes, and storage tanks can have a significant impact on air quality. Afterburners are often the best method for destroying volatile organic chemicals in industrial exhaust stacks.

Clean air legislation is controversial but effective

Since 1970, the Clean Air Act has been modified, updated, and amended many times. Amendments have involved acrimonious debate. As in the case of CO_2 restrictions, discussed earlier, victims of air pollution demand more protection, while industry and energy groups insist that controls are too expensive. Bills have sometimes languished in Congress for years because of disputes over burdens of responsibility, cost, and definitions of risk. The EPA reports that simply by enforcing existing clean air legislation, the United States could prevent at least 6,000 deaths and 140,000 asthma attacks every year.

The most significant amendments were in the 1990 update, which addressed a variety of issues, including acid rain, urban air pollution, and toxic air emissions. These amendments also restricted ozone-depleting chemicals in accordance with the Montreal Protocol.

One of the most contested aspects of the act has been the "new source review," which was established in 1977. This provision was adopted because industry argued that it would be intolerably expensive to install new pollution-control equipment on old power plants and factories that were about to close down anyway. Congress agreed to "grandfather" existing equipment, or exempt it from new pollution limits, with the stipulation that when they were upgraded or replaced, more stringent rules would apply. The result was that owners have kept old facilities operating precisely because they were exempted from pollution control. In fact, corporations poured millions into aging power plants and factories, expanding their capacity, to avoid having to build new ones. Forty years later, many of those grandfathered plants are still going strong and continue to be among the biggest contributors to smog and acid deposition.

Clean air protections help the economy and public health

Despite these disputes, the Clean Air Act has been extremely successful in saving money and lives. The EPA estimates that between 1970 and 2014, lead emissions fell 99 percent, SO_2 declined 84 percent, and CO shrank 72 percent (fig. 16.28). Filters, scrubbers, and electrostatic precipitators on power plants and other large stationary sources are responsible for most of the particulate and SO_2 reductions. Catalytic converters on cars are responsible for most of the CO and O_3 reductions. For 23 of the largest U.S. cities, air quality now reaches hazardous levels 93 percent less frequently than a decade ago. Forty of the 97 metropolitan areas that failed to meet clean air standards in the 1980s are now in compliance, many for the first time in a generation.

In a 2011 study of the economic costs and benefits of the 1990 Clean Air Act, the EPA found that the direct benefits of air quality protection by 2020 will be $2 trillion, while the direct costs of implementing those protections was about one-thirtieth of that, or $65 billion (fig. 16.29). The direct benefits were mainly in prevented costs of premature illness, death, and work losses (table 16.4). About half of the direct costs were improvements in cars and trucks, which now burn more cleanly and efficiently than they did in the past. This cost has been distributed to vehicle owners, who also benefit from lower expenditures on fuel. A quarter of costs involved cleaner furnaces and pollutant capture at electricity-generating power plants and other industrial facilities. The remaining costs involved pollution reductions at smaller businesses, municipal facilities, construction sites, and other sources. Overall, emission controls have not dampened economic productivity, despite widespread fears to the contrary. Emissions of criteria pollutants have declined in recent decades, whereas economic indicators have grown (fig. 16.30).

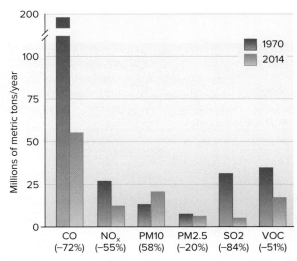

FIGURE 16.28 Air pollution trends in the United States, 1970 to 2014. Since Congress passed the Clean Air Act , most criteria air pollutants decreased significantly, even while population has increased. Pollution protections also aid the economy by reducing illness and creating jobs in pollution prevention. *Source: Environmental Protection Agency, 2016.*

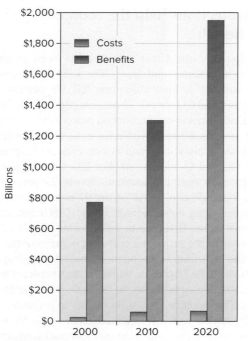

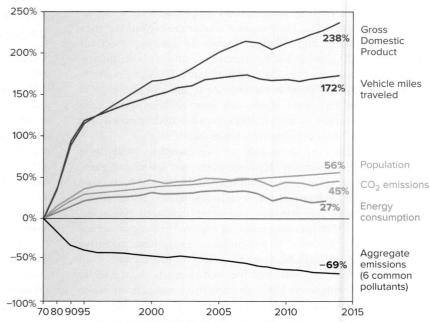

FIGURE 16.29 Direct costs and benefits of the Clean Air Act provisions by 2000, 2010, and 2020, in billions of 2006 U.S. dollars.
Source: EPA 2011, Clean Air Impacts Summary Report

FIGURE 16.30 Comparison of growth measures and emissions of criteria air pollutants, 1970–2014.
Source: EPA, 2016

In addition to these savings, the Clean Air Act has created thousands of jobs in developing, installing, and maintaining technology and in monitoring. At a time when many industries are providing fewer jobs, owing to greater mechanization, jobs have been growing in clean technologies and pollution control and monitoring. At the same time, reductions in acid rain have decreased losses to forest resources and building infrastructure.

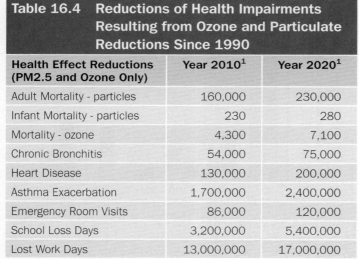

Table 16.4	Reductions of Health Impairments Resulting from Ozone and Particulate Reductions Since 1990	
Health Effect Reductions (PM2.5 and Ozone Only)	**Year 2010[1]**	**Year 2020[1]**
Adult Mortality - particles	160,000	230,000
Infant Mortality - particles	230	280
Mortality - ozone	4,300	7,100
Chronic Bronchitis	54,000	75,000
Heart Disease	130,000	200,000
Asthma Exacerbation	1,700,000	2,400,000
Emergency Room Visits	86,000	120,000
School Loss Days	3,200,000	5,400,000
Lost Work Days	13,000,000	17,000,000

[1] Number of cases reduced

Source: EPA, 2011 Clean Air Impacts Summary Report

Market mechanisms have been part of the solution, especially for sulfur dioxide, which is widely considered to have benefited from a cap-and-trade approach. This strategy sets maximum limits for each facility and then lets facilities sell pollution credits if they can cut emissions, or facilities can buy credits if they are cheaper than installing pollution-control equipment. When trading began in 1990, economists estimated that eliminating 10 million tons of sulfur dioxide would cost $15 billion per year. Left to find the most economical ways to reduce emissions, however, utilities have been able to reach clean air goals for one-tenth that price. A serious shortcoming of this approach is that while trading has resulted in overall pollution reduction, some local "hot spots" remain where owners have found it cheaper to pay someone else to reduce pollution than to do it themselves.

Particulate matter (mostly dust and soot) is produced by agriculture, fuel combustion, metal smelting, concrete manufacturing, and other activities. Industrial cities, such as Baltimore, Maryland, and Baton Rouge, Louisiana, also have continuing problems. Eighty-five other urban areas are still considered nonattainment regions. In spite of these local failures, however, 80 percent of the United States now meets the National Ambient Air Quality Standards (fig. 16.31). This improvement in air quality is perhaps the greatest environmental success story in our history.

In developing areas, rapid growth can outpace pollution controls

The outlook is not so encouraging in many parts of the world. The major metropolitan areas of many developing countries are growing at explosive rates to incredible sizes (chapter 22), and environmental

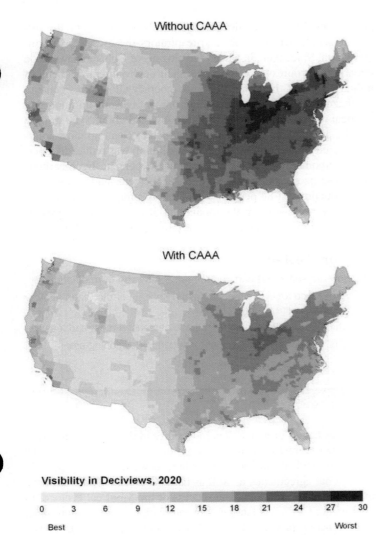

Without CAAA

With CAAA

Visibility in Deciviews, 2020

| 0 | 3 | 6 | 9 | 12 | 15 | 18 | 21 | 24 | 27 | 30 |

Best   Worst

FIGURE 16.31 Projected visibility impairments, shown with dark colors, would be considerably worse in 2020 without the 1990 Clean Air Act amendments (CAAA, *top*) than they will be with the amendments (*bottom*). Units are deciviews, a measure of perceptible change in visibility.
Source: EPA 2011, Clean Air Impacts Summary Report

quality can be abysmal. In Mexico City, notorious for bad air, pollution levels exceed WHO health standards 350 days per year, and more than half of all city children have lead levels in their blood high enough to lower intelligence and retard development. Mexico City's 131,000 industries and 2.5 million vehicles spew out more than 5,500 tons of air pollutants daily. In Santiago, Chile, suspended particulates exceed WHO standards of 90 mg/m^3 about 299 days per year. Rapid growth and industrialization in China, India, and many other parts of the developing world are producing emissions much faster than pollution-control agencies can manage.

Even when laws on the books are strong, enforcement is often weak. In China, despite legal restrictions, many of China's 400,000 factories have no air pollution controls. Experts estimate that home coal burners and factories emit 10 million tons of soot and 15 million tons of sulfur dioxide annually and that emissions

have increased rapidly over the past 20 years. Sixteen of the 20 cities in the world with the worst air quality are in China. Shenyang, an industrial city in northern China, is thought to have the world's worst continuing particulate problem, with peak winter concentrations over 700 mg/m^3 (nine times U.S. maximum standards). Airborne particulates in Shenyang exceed WHO standards on 347 days per year. In many Chinese cities, concentrated air pollution is linked to high incidence of cancer.

Every year the organization Pure Earth (formerly the Blacksmith Institute) compiles a list of the world's worst-polluted places. Globally, smelters, mining operations, petrochemical industries—which release hazardous organic compounds to the air and water—and chemical manufacturing are frequently the worst sources of pollutants. Often, these are in impoverished and developing areas of Africa, Asia, or the Americas, where government intervention is weak and regulations are nonexistent or poorly enforced. Funds and political will are usually unavailable to deal with pollution, much of which is involved with materials going to wealthier countries or waste that is received from developed countries. You can learn more about these places at www.pureearth.org.

Norilsk, Russia (one site highlighted on Pure Earth's list of worst places), is a notorious example of toxic air pollution. Founded in 1935 as a prison labor camp, this Siberian city is considered one of the most polluted places on earth. Norilsk houses the world's largest nickel mine and heavy metals smelting complex, which discharge over 4 million tons of cadmium, copper, lead, nickel, arsenic, selenium, and zinc into the air every year. The snow turns black as quickly as it falls, the air tastes of sulfur, and the average life expectancy for factory workers is ten years below the Russian average (which already is the lowest of any industrialized country). Difficult pregnancies and premature births are far more common in Norilsk than elsewhere in Russia. Children living near the nickel plant are ill twice as much as Russia's average, and birth defects are reported to affect up to 10 percent of the population. Why do people stay in such a place? Many were attracted by high wages and hardship pay, and now that they're sick, they can't afford to move.

Air quality improves where controls are implemented

Despite global expansion of chemical industries and other sources of air pollution, there have been some spectacular successes in air pollution control. Sweden and West Germany (countries affected by forest losses due to acid precipitation) cut their sulfur emissions by two-thirds between 1970 and 1985. Austria and Switzerland have gone even farther, regulating even motorcycle emissions. The Global Environmental Monitoring System (GEMS) reports declines in particulate levels in 26 of 37 cities worldwide. Sulfur dioxide and sulfate particles, which cause acid rain and respiratory disease, have declined in 20 of these cities.

Twenty years ago, Cubatao, Brazil, was described as the "Valley of Death," one of the most dangerously polluted places in the world. Every year, a steel plant, a huge oil refinery, and fertilizer and chemical factories churned out thousands of tons of air pollutants that were trapped between onshore winds and the uplifted

FIGURE 16.32 Cubatao, Brazil, was once considered one of the most polluted cities in the world. Better environmental regulations and enforcement, along with massive investments in pollution-control equipment, have improved air quality significantly.
© William P. Cunningham

plateau on which São Paulo sits (fig. 16.32). Trees died on the surrounding hills. Birth defects and respiratory diseases were alarmingly high. Since then, however, the citizens of Cubatao have made remarkable progress in cleaning up their environment. The end of military rule and the restoration of democracy allowed residents to publicize their complaints. The environment became an important political issue. The state of São Paulo invested about $100 million and the private sector spent twice as much to clean up most pollution sources in the valley. Particulate pollution was reduced 75 percent, ammonia emissions were reduced 97 percent, hydrocarbons that cause ozone and smog were cut 86 percent, and sulfur dioxide production fell 84 percent. Fish are returning to the rivers, and forests are regrowing on the mountains. It proves that progress is possible. Similar successes could certainly be obtained elsewhere.

Section Review

1. How are sulfur and particulate matter removed from effluent?
2. What is the ratio of direct costs and benefits of the Clean Air Act? What costs are mainly saved?
3. Which conventional pollutants have decreased most and least?
4. What are some sources of air pollution in developing areas?

Conclusion

Air pollution is often the most obvious and widespread type of pollution. Everywhere on earth, from the most remote island in the Pacific, to the highest peak in the Himalayas, to the frigid ice cap over the North Pole, there are traces of human-made contaminants, remnants of the 2 billion metric tons of pollutants released into the air worldwide every year by human activities.

The adverse effects of air pollution include respiratory diseases, birth defects, heart attacks, developmental disabilities in children, and cancer. Environmental impacts include the destruction of stratospheric ozone, the poisoning of forests and waters by acid rain, and corrosion of building materials.

We have made encouraging progress in controlling air pollution, however, progress that has offered economic benefits as well as health benefits. Many students aren't aware of how much worse air quality was in the industrial centers of North America and Europe a century or two ago compared to today. Cities such as London, Pittsburgh, Chicago, Baltimore, and New York had air

quality as bad as or worse than most cities in the developing world now. The progress in reducing air pollution in these cities gives us hope that residents can do so elsewhere as well.

Global agreements, as well as national clean air policies, have made important progress. The success of the Montreal Protocol in eliminating CFCs is a landmark in international cooperation on an environmental problem. Stratospheric ozone depletion has slowed and should end in about 50 years. The Minamata Convention to control mercury is another example of important global cooperation.

Developing areas face severe challenges in air quality. Most of the worst air pollution in the world occurs in large cities of developing countries. However, there are dramatic cases of pollution in developing countries. Problems that once seemed overwhelming can be overcome. In some cases, this requires lifestyle changes or different ways of doing things to bring about progress, but as the Chinese philosopher Lao Tsu wrote, "A journey of a thousand miles must begin with a single step."

Reviewing Key Terms

Can you define the following terms in environmental science?

acid deposition 16.3
aerosols 16.1
aesthetic degradation 16.1
ambient air 16.1

bronchitis 16.3
carbon monoxide 16.1
chlorofluorocarbons (CFCs) 16.2

chronic obstructive lung disease 16.3
conventional or criteria pollutants 16.1

fugitive emissions 16.1
hazardous air pollutants (HAPs) 16.1
nitric oxide 16.1

Critical Thinking and Discussion Questions

1. What might be done to improve indoor air quality? Should governments mandate such changes? What values or world-views are represented by different sides of this debate?

2. Debate the following proposition: Our air pollution blows onto someone else; therefore, installing pollution controls will not bring any direct economic benefit to those of us who have to pay for them.

3. Utility managers once claimed that it would cost $1,000 per fish to control acid precipitation in the Adirondack lakes and that it would be cheaper to buy fish for anglers than to put scrubbers on power plants. Suppose that is true. Does it justify continuing pollution?

4. Developing nations claim that richer countries created global warming and stratospheric ozone depletion, and therefore they should bear the responsibility for fixing these problems. How would you respond?

5. If there are thresholds for pollution effects, is it reasonable or wise to depend on environmental processes to disperse, assimilate, or inactivate waste products?

6. How would you choose between government "command and control" regulations versus market-based trading programs for air pollution control? Are there situations where one approach would work better than the other?

Data Analysis

How Is the Air Quality in Your Town?

How does air quality in your area compare to that in other places? You can examine trends in major air pollutants—both national and local trends in your area—on the EPA's website. The EPA is the principal agency in charge of protecting air quality and informing the public about the air we breathe and how healthy it is.

Go to Connect to find a link to data and maps showing trends in SO_2 emissions since 1980. At the same site, you can see trends in NO_x, CO, lead, and other criteria pollutants. Examine national trends, and then look at your local area on the map on the same page to answer questions about trends in your area and to also compare your area to others.

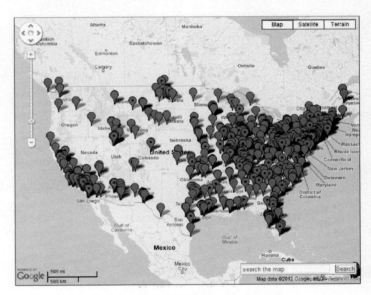

Examine pollutant trends in your area on the EPA website.

TO ACCESS ADDITIONAL RESOURCES FOR THIS CHAPTER, PLEASE VISIT CONNECT AT
www.connect.mheducation.com.

You will find LearnSmart, an adaptive learning system, Google Earth™ exercises, additional Case Studies, Data Analysis exercises, and an interactive ebook.

▲Between 2000 and 2011, the surface level of lake Mead, the largest reservoir on the Colorado River, fell more than 100 ft (30.5 m) during the worst drought in recorded history. If the water falls another 100 ft, the reservoir will reach "dead pool" levels at which it can provide neither the water nor the electrical power on which millions of people depend.
© David McNew/Getty Images

Learning Outcomes

After studying this chapter, you should be able to:

17.1 Summarize why water is a precious resource and why shortages occur.

17.2 Summarize water availability and use.

17.3 Explain causes of freshwater shortages.

17.4 Describe how we might get by with less water.

"I tell you gentlemen; you are piling up a heritage of conflict and litigation of water rights, for there is not sufficient water to supply the land."

– John Wesley Powell

When Will Lake Mead Go Dry?

The Colorado River is the lifeblood of the American Southwest. More than 30 million people and a $1.2 trillion regional economy in cities such as Los Angeles, Phoenix, Las Vegas, and Denver depend on its water (fig. 17.1). Hydroelectric dams, which created the two large reservoirs Lake Powell and Lake Mead, provide electricity to the region. But the reliability of this essential resource is in doubt. Drought, climate change, and rapid urban growth are creating worries about the sustainability of the water supply for the entire watershed.

In 2008, Tim Barnett and David Pierce from the Scripps Institute in California published a provocative article projecting that by 2018 or so, both Lake Mead and Lake Powell could fall so low that neither would be able to produce power or provide water for urban or agricultural use. These huge lakes constitute more than 85 percent of the water storage for the entire Colorado system. If they reach "dead pool" levels (at which water could not drain by gravity), it would be a catastrophe for the whole region. This warning was based on historical records and on climate models that suggest a 10 to 30 percent runoff reduction in the area over the next 50 years.

The roots of this problem can be traced to the Colorado Compact of 1922, which allocated a portion of the river to each state in the watershed (outlined in fig. 17.1). Unfortunately, the agreement dramatically overestimated the amount of water available to share. It was written at the end of the wettest year in more than a millennia, and negotiators estimated the annual river flow at 18 million acre-feet (22 billion m^3), about 20 percent more than the twentieth-century average.

The error didn't matter much at the time, because none of the states had the technology or need to withdraw its entire share. As cities have grown and agriculture has expanded since 1922, competing claims for water have grown, and tensions have risen. Massive water diversion projects now drain the river nearly dry. The Colorado River Aqueduct delivers water to Los Angeles; the All-American Canal irrigates California's Imperial Valley; and the Central Arizona Project transports water over the mountains and across the desert to Phoenix and Tucson. Back in 1944, hardly any water was left in the river by the time it reached Mexico. So, the United States agreed to allocate 1.5 million acre-feet to Mexico, although it might have been of dubious quality.

Now, climate change is expected to decrease western river flows drastically. The Southwest has seen drought conditions in most of the last 20 years. The maximum water level in Lake Mead (an elevation of 1,220 feet or 372 m) was last reached in 2000. Since then, the lake level has been dropping about 12 feet (3.6 m) per year, and it fell to 1,071 feet in July 2016, the lowest level since the lake was filled in the 1930s. The minimum power level (the height at which electricity can be produced) is 1,050 feet (320 m). The minimum level at which water can be drawn off by gravity is 900 feet (274 m). Barnett and Pierce estimated that without changes in current management plans, there was a 50 percent chance that minimum power pool levels in both Lakes Mead and Powell would be reached by 2017, and there is an equal chance that live storage in both lakes will be gone by about 2021. By 2016, their projections were proving accurate and on schedule, and the region was in "exceptional drought." Models suggest an 80 percent likelihood of a "megadrought," one that lasts at least 35 years, in the coming century.

Already, we're at or beyond the sustainable limits of the river. Currently, Lake Powell holds less than half its maximum volume, and Lake Mead is less than 40 percent full. The shores of both lakes now display a wide "bathtub ring" of deposited minerals left by the receding water (see the chapter-opening photo). One suggestion has been to drain Lake Powell to ensure a water supply for Lake Mead. This solution would keep the lights on and faucets running in Los Angeles, Phoenix, Las Vegas, and other cities. It is strenuously opposed by many of the 3 million recreational users of Lake Powell, however.

The American Southwest isn't alone in facing this problem. The United Nations warns that water supplies are likely to become one of the most pressing environmental issues of the twenty-first

FIGURE 17.1 The Colorado River flows 2,330 km (1,450 mi) through seven western states. Its water supports 30 million people and a $1.2 trillion regional economy, but drought, climate change, and rapid urban growth threaten the sustainability of this resource.

century. By 2025, two-thirds of all humans could be living in places where water resources are inadequate. In this chapter, we'll look at the sources of our fresh water, what we do with it, and how we might protect its quality and extend its usefulness.

For further reading, see:

Barnett, T. P., and D. W. Pierce. 2008. When Will Lake Mead Go Dry? *Journal of Water Resources Research,* vol. 44, W03201.

17.1 WATER RESOURCES

- *Less than 3 percent of all the water on the earth is fresh, and 87 percent of that water is locked up in ice and snow.*
- *The hydrologic cycle constantly purifies and redistributes fresh water.*
- *Some places get too much rain, while others get almost none at all.*

Water is a marvelous substance—flowing, rippling, swirling around obstacles in its path, seeping, dripping, trickling, constantly moving from sea to land and back again. Water can be clear, crystalline, icy green in a mountain stream, or black and opaque in a cypress swamp. Water bugs skitter across the surface of a quiet lake; a stream cascades down a stairstep ledge of rock; waves roll endlessly up a sand beach, crash in a welter of foam, and then recede. Rain falls in a gentle mist, refreshing plants and animals. A violent thunderstorm floods a meadow, washing away streambanks. Water is a most beautiful and precious resource.

Water is also a great source of conflict. Some 2 billion people now live in countries with insufficient fresh water. Some experts estimate this number could double in 25 years. To understand this resource, let's first ask, "Where does our water come from, and why is it so unevenly distributed?"

The hydrologic cycle constantly redistributes water

The water we use cycles endlessly through the environment. The total amount of water on our planet is immense—more than 1.4 billion km³ (370 quintillion gal) (table 17.1). This water evaporates from moist surfaces, falls as rain or snow, passes through living organisms,

Table 17.1   Some Units of Water Measurement
One cubic kilometer (km³) equals 1 billion cubic meters (m³), 1 trillion liters, or 264 billion gallons.
One acre-foot is the amount of water required to cover an acre of ground 1 foot deep. This is equivalent to 325,851 gallons, or 1.2 million liters, or 1,234 m³, about the amount consumed annually by a family of four in the United States.
One cubic foot per second of river flow equals 28.3 liters per second or 449 gallons per minute.

Note: See the table at the end of the book for conversion factors.

and returns to the ocean in a process known as the **hydrologic cycle** (see fig. 3.19). Every year, about 500,000 km³, or a layer 1.4 m thick, evaporates from the oceans. More than 90 percent of that moisture falls back on the ocean. The 47,000 km³ carried onshore joins some 72,000 km³ that evaporate from lakes, rivers, soil, and plants to become our annual, renewable freshwater supply. Plants play a major role in the hydrologic cycle, absorbing groundwater and pumping it into the atmosphere by transpiration (transport plus evaporation). In tropical forests, as much as 75 percent of annual precipitation is returned to the atmosphere by plants.

Solar energy drives the hydrologic cycle by evaporating surface water, which becomes rain and snow. Because water and sunlight are unevenly distributed around the globe, water resources are very uneven. At Iquique in the Chilean desert, for instance, no rain has fallen in recorded history. At the other end of the scale, 27 m (1,041 in) of rain were recorded in a single year at Cherrapunji in India. Figure 17.2 shows broad patterns of precipitation around the world. Most of the world's rainiest regions are tropical, where heavy rainy seasons occur, or in coastal mountain regions. Deserts occur on every continent just outside the tropics (the Sahara, the Namib, the Gobi, the Sonoran, and many others). Rainfall is also slight at very high latitudes, another high-pressure region.

Water supplies are unevenly distributed

Rain falls unevenly over the planet (fig. 17.2). As mentioned earlier, some places get almost no precipitation, while others receive heavy rain almost daily. Three principal factors control these global water deficits and surpluses. First, global atmospheric circulation creates regions of persistent high air pressure and low rainfall about 20° to 40° north and south of the equator (chapter 15). These same circulation patterns produce frequent rainfall near the equator and between about 40° and 60° north and south latitudes. Second, proximity to water sources influences precipitation. Where prevailing winds come over oceans or large lakes, they bring moisture to land. Areas far inland—in a windward direction—are usually relatively dry.

A third factor in water distribution is topography. Mountains act as both cloud formers and rain catchers. As air sweeps up the windward side of a mountain, air pressure decreases and air cools. As the air cools, it reaches the saturation point, and moisture condenses as either rain or snow. Thus, the windward side of a mountain range, as in the Pacific Northwest, is usually wet much of the year. Precipitation leaves the air drier than it was on its way

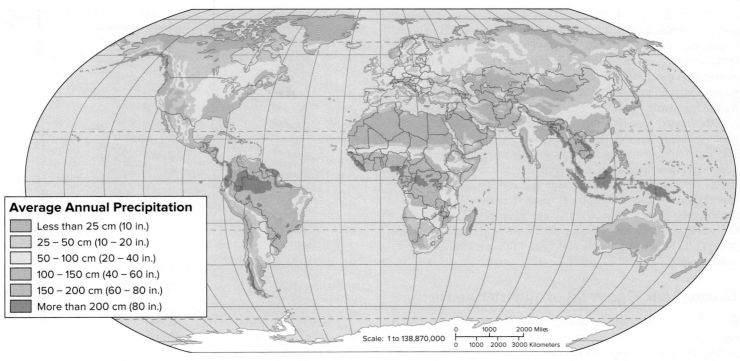

FIGURE 17.2 Average annual precipitation. Note wet areas that support tropical rainforests occur along the equator, while the major world deserts occur in zones of dry, descending air between 20° and 40° north and south.

Average Annual Precipitation

- Less than 25 cm (10 in.)
- 25 – 50 cm (10 – 20 in.)
- 50 – 100 cm (20 – 40 in.)
- 100 – 150 cm (40 – 60 in.)
- 150 – 200 cm (60 – 80 in.)
- More than 200 cm (80 in.)

Scale: 1 to 138,870,000

up the mountain. As the air passes the mountaintop and descends the other side, air pressure rises, and the already-dry air warms, increasing its ability to hold moisture. Descending, warming air rarely produces any rain or snow. Places in the **rain shadow,** the dry, leeward side of a mountain range, receive little precipitation. A striking example of the rain shadow effect is found on Mount Waialeale, on the island of Kauai, Hawaii (fig. 17.3). The windward side of the island receives nearly 12 m of rain per year, while the leeward side, just a few kilometers away, receives just 46 cm.

Usually, a combination of factors affects precipitation. In Cherrapunji, India, monsoon winds carrying moisture from the warm Indian Ocean crash against the high ridges of the Himalayas to release a seasonal deluge. Iquique, Chile, lies in the rain shadow of the Andes and in a high-pressure desert zone. Prevailing winds are from the east, so even though Iquique lies near the ocean, it is far from the winds' moisture source—the Atlantic. In the American Southwest, Australia, and the Sahara, high-pressure atmospheric conditions tend to keep the air and land dry. The global map of precipitation represents a complex combination of these forces of atmospheric circulation, prevailing winds, and topography.

Human activity also explains some regions of water deficit. As noted earlier, plant transpiration recycles moisture and produces rain. When forests are cleared, falling rain quickly enters streams and returns to the ocean. In Greece, Lebanon, parts of Africa, the Caribbean, South Asia, and elsewhere, desert-like conditions have developed since the original forests were destroyed.

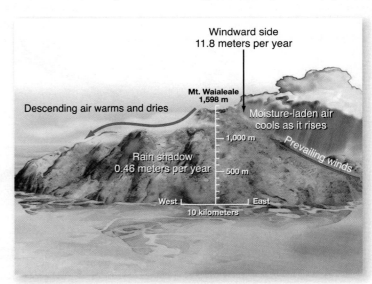

FIGURE 17.3 Rainfall on the east side of Mount Waialeale in Hawaii is more than 20 times as much as on the west side. Prevailing trade winds bring moisture-laden sea air onshore. The air cools as it rises up the flanks of the mountain, and the water it carries precipitates as rain, as much as 11.8 m (38 ft) per year!

Quantitative Reasoning

How do the 27 meters of rain that fall in a normal year in Cherrapunji, India, compare to the rainfall where you live? If you don't know your average annual precipitation, you can approximate it from the map in figure 17.2. How might your life change if you lived in a place with 27 meters of rain per year?

Water distribution often is described in terms of interacting compartments in which water resides, sometimes briefly and sometimes for eons (table 17.2). The length of time water typically stays in a compartment is its **residence time.** On average, a water molecule stays in the ocean for about 3,000 years, for example, before it evaporates and starts through the hydrologic cycle again.

Residence time is short in the atmosphere: An individual water molecule resides in the atmosphere for about ten days, on average. While water vapor makes up only a small amount (4 percent maximum at normal temperatures) of the total volume of the air, movement of water through the atmosphere provides the mechanism for distributing fresh water over the landmasses and replenishing terrestrial reservoirs. The atmosphere is also one of the smallest of the major water reservoirs of the earth in terms of water volume, containing less than 0.001 percent of the total water supply.

Oceans hold 97 percent of all water on earth

If explorers from another galaxy were to discover our planet, they'd probably call it Aqua rather than Earth, because water is its dominant characteristic. Oceans cover 71 percent of our planet's surface and contain more than 97 percent of all the *liquid* water in the world. (The water of crystallization in rocks is far larger than the amount of liquid water.) Seawater is too salty for most human uses, but contains 90 percent of the world's living biomass. Although the ocean basins really form a continuous reservoir, shallows and narrows between them reduce water exchange, so they have different compositions, different climatic effects, and even different surface elevations.

Table 17.2 Earth's Water Compartments

Compartment	Volume (1,000 km³)	Percent of Total Water	Average Residence Time
Total	1,386,000	100	2,800 years
Oceans	1,338,000	96.5	3,000 to 30,000 years*
Ice and snow	24,364	1.76	1 to 100,000 years*
Saline groundwater	12,870	0.93	Days to thousands of years*
Fresh groundwater	10,530	0.76	Days to thousands of years*
Fresh lakes	91	0.007	1 to 500 years*
Saline lakes	85	0.006	1 to 1,000 years*
Soil moisture	16.5	0.001	2 weeks to 1 year*
Atmosphere	12.9	0.001	1 week
Marshes, wetlands	11.5	0.001	Months to years
Rivers, streams	2.12	0.0002	1 week to 1 month
Living organisms	1.12	0.0001	1 week

*Depending on depth and other factors.

Source: Data from UNEP, 2002

The ocean plays a crucial role in moderating the earth's temperature (fig. 17.4). Vast riverlike currents transport warm water from the equator to higher latitudes, and cold water flows from the poles to the tropics. The Gulf Stream, which flows northeast from the coast of North America toward northern Europe (fig. 17.5),

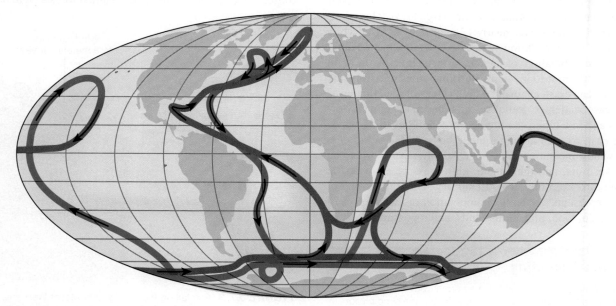

FIGURE 17.4 Ocean currents act as a global conveyor system, redistributing warm (red) and cold (blue) currents around the globe. These currents moderate our climate. For example, the Gulf Stream keeps northern Europe much warmer than northern Canada. Ocean colors show salinity variation from low (*blue*) to high (*yellow*).
Source: NASA

FIGURE 17.5 Ocean currents, such as the warm Gulf Stream, redistribute heat as they flow around the globe. Here, orange and yellow indicate warm water temperatures (25–30°C); blue and green are cold (0–5°C). As this warm salty water flows north, it cools and becomes very dense relative to the surrounding water, causing it to sink, which helps drive deep ocean circulation patterns.
Source: The sea surface temperature image was created at the University of Miami using the 11- and 12-micron bands of MODIS, by Bob Evans, Peter Minnett, and their coworkers.

flows at a steady rate of 10–12 km per hour (6–7.5 mph) and carries over 100 times more water than all the rivers on earth put together.

In tropical seas, surface waters are warmed by the sun, diluted by rainwater and runoff from the land, and aerated by wave action. In higher latitudes, surface waters are cold, saltier, and much more dense. This dense water subsides or sinks to the bottom of deep ocean basins and flows toward the equator. Warm surface water of the tropics stratifies or floats on top of this cold, dense water as currents carry warm water to high latitudes. Upwelling zones, where cold deep water comes back to the surface to replace water carried toward the poles by the ocean conveyor system, also bring up nutrients and support prolific sea life. Sharp boundaries form between different water densities, different salinities, and different temperatures, retarding mixing between these layers.

Glaciers, ice, and snow contain most surface fresh water

Of the 2.4 percent of all water that is fresh, 87 percent is tied up in glaciers, ice caps, and snowfields (fig. 17.6). Although most of this ice is located in Antarctica, Greenland, and the floating ice cap in the Arctic, alpine glaciers and snowfields supply water to billions of people. The winter snowpack on the western slope of the Rocky Mountains, for example, provides 75 percent of the flow in the Colorado River described in the opening case study of this chapter. Drought conditions already have reduced snowfall (and runoff) in the western United States, and global warming is projected to cause even further declines.

As Chapter 15 discusses, climate change is shrinking glaciers and snowfields nearly everywhere (fig. 17.7). In Asia, the Tibetan glaciers, which are the source of six of the world's largest rivers and which supply drinking water for three billion people, are shrinking rapidly. There are warnings that these glaciers could vanish in a few decades, which would bring enormous suffering and economic loss in many places.

Groundwater stores large resources

After glaciers, the next largest reservoir of fresh water is held in the ground as **groundwater.** Precipitation that does not evaporate back into the air or run off over the surface percolates through the soil and into fractures and spaces of permeable rocks in a process called **infiltration** (fig. 17.8). Upper soil layers that hold both air and water make up the **zone of aeration.** Moisture for plant growth comes primarily from these layers. Depending on rainfall amount, soil type, and surface topography, the zone of aeration may be very shallow or quite deep. Lower soil layers where all spaces are filled with water make up the **zone of saturation.** The top of this zone is the **water table.** The water table is not flat, but undulates according to the surface topography and subsurface structure. Water tables also rise and fall seasonally, depending on precipitation, infiltration, and withdrawal rates.

Porous layers of sand, gravel, or rock lying below the water table are called **aquifers.** Aquifers are always underlain by relatively impermeable layers of rock or clay that keep water from seeping out at the bottom (fig. 17.9).

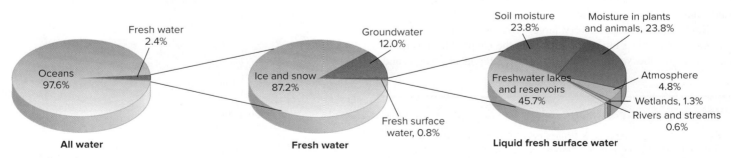

FIGURE 17.6 Less than 1 percent of fresh water, and less than 0.02 percent of all water, is the fresh, liquid surface water on which terrestrial life depends.
Source: U.S. Geological Survey

FIGURE 17.7 Glaciers and snowfields provide much of the water on which billions of people rely. The snowpack in the western Rocky Mountains, for example, supplies about 75 percent of the annual flow of the Colorado River. Global climate change is shrinking glaciers and causing snowmelt to come earlier in the year, disrupting this vital water source.
© William P. Cunningham

Transpiration from plant surfaces

Precipitation

Evaporation from land and water surfaces

Runoff

Infiltration

Zone of aeration

Air pocket

Water table
Groundwater

Zone of saturation

FIGURE 17.8 Precipitation that does not evaporate or run off over the surface percolates through the soil in a process called infiltration. The upper layers of soil hold droplets of moisture between air-filled spaces. Lower layers, where all spaces are filled with water, make up the zone of saturation, or groundwater.

Folding and tilting of the earth's crust by geological processes can create shapes that generate water pressure in confined aquifers (those trapped between two impervious, confining rock layers). When a pressurized aquifer intersects the surface, or if it is penetrated by a pipe or conduit, the result is an **artesian well** or spring, from which water gushes without being pumped.

Areas where water infiltrates into an aquifer are called **recharge zones.** The rate at which most aquifers are refilled is very slow, however, and groundwater presently is being removed faster than it can be replenished in many areas. Urbanization, road building, and other development often block recharge zones and prevent replenishment of important aquifers. Contamination of surface water in recharge zones and seepage of pollutants into abandoned wells have polluted aquifers in many places, making them unfit for most uses (chapter 18). Many cities protect aquifer recharge zones from pollution or development, both as a way to drain off rainwater and as a way to replenish the aquifer with pure water.

Some aquifers contain very large volumes of water. The groundwater within 1 km of the surface in the United States is more than 30 times the volume of all the freshwater lakes, rivers, and

reservoirs on the surface. Although water can flow through limestone caverns in underground rivers, most movement in aquifers is a dispersed and almost imperceptible trickle through tiny fractures and spaces. Depending on geology, it can take anywhere from a few hours to several years for contaminants to move a few hundred meters through an aquifer.

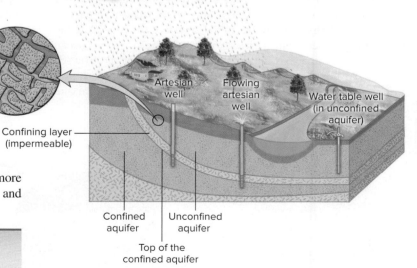

Artesian well

Flowing artesian well

Water table well (in unconfined aquifer)

Confining layer (impermeable)

Confined aquifer

Unconfined aquifer

Top of the confined aquifer

FIGURE 17.9 An aquifer is a porous or cracked layer of rock. Impervious rock layers (aquicludes) keep water within a confined aquifer. Pressure from uphill makes an artesian well flow freely. Pumping can create a cone of depression, which leaves shallower wells dry.

Quantitative Reasoning

Figure 17.6 shows that only 2.4 percent of all water on the earth is fresh (nonsalty). How much of that water is liquid in rivers and lakes? Assuming this is the main source of drinking water for most animals, what percentage is this of the total world water supply?

Rivers, lakes, and wetlands cycle quickly

Precipitation that does not evaporate or infiltrate into the ground runs off over the surface, drawn back toward the sea by the force of gravity. Rivulets accumulate to form streams, and streams join to form rivers. Although the total amount of water contained at any one time in rivers and streams is small compared to that in the other water reservoirs of the world (table 17.2), these surface waters are vitally important to humans and most other organisms. Most rivers, if they were not constantly replenished by precipitation, meltwater from snow and ice, or seepage from groundwater, would begin to diminish in a few weeks.

We measure the size of a river in terms of its **discharge,** the amount of water that passes a fixed point in a given amount of time. This is usually expressed as liters or cubic feet of water per second. The 16 largest rivers in the world carry nearly half of all surface runoff on earth. The Amazon is by far the largest river in the world (table 17.3), carrying roughly ten times the volume of the Mississippi. Several Amazonian tributaries, such as the Madeira, Rio Negro, and Ucayali, would be among the world's top rivers in their own right.

Ponds are generally considered to be small temporary or permanent bodies of water shallow enough for rooted plants to grow over most of the bottom. Lakes are inland depressions that hold standing fresh water year-round. Maximum lake depths range from a few meters to over 1,600 m (1 mi) in Lake Baikal in Siberia. Surface areas vary in size from less than one-half hectare (one acre) to large inland seas, such as Lake Superior or the Caspian Sea, covering hundreds of thousands of square kilometers. Both ponds and lakes are relatively temporary features on the landscape because they eventually fill with silt or are emptied by cutting an outlet stream through the barrier that creates them.

While lakes contain nearly 100 times as much water as all rivers and streams combined, they are still a minor component of the total world water supply. Their water is much more accessible than groundwater or the water in glaciers, however, and they are important in many ways for humans and other organisms.

Wetlands play a vital and often unappreciated role in the hydrologic cycle. Their lush plant growth stabilizes soil and holds back surface runoff, allowing time for infiltration into aquifers and producing even, year-long stream flows. In the United States, about 20 percent of the 1 billion ha of land area was once wetland. In the past 200 years, more than one-half of those wetlands have been drained, filled, or degraded. Agricultural drainage accounts for the bulk of the losses.

When wetlands are disturbed, their natural water-absorbing capacity is reduced and surface waters run off quickly, resulting in floods and erosion during the rainy season, and creating dry, or nearly dry, streambeds the rest of the year. This has a disastrous effect on biological diversity and productivity, as well as on human affairs.

Section Review

1. What is the hydrologic cycle, and how does it redistribute water around the globe?
2. What percentage of water on the planet is fresh (nonsalty)?
3. What is an aquifer, and how is it recharged?

17.2 WATER AVAILABILITY AND USE

- *Many countries suffer water scarcity and stress.*
- *Water use is increasing.*
- *Agriculture is the greatest water user worldwide.*

Clean, fresh water is essential for nearly every human endeavor. Perhaps more than any other environmental factor, the availability of water determines the location and activities of humans on earth (fig. 17.10). **Renewable water supplies** are made up, in general, of surface runoff plus the infiltration into accessible freshwater aquifers. About two-thirds of the water carried in rivers and streams every year occurs in seasonal floods that are too large or violent to be stored or trapped effectively for human uses. Stable runoff is the dependable, renewable, year-round supply of surface water. Much of this occurs, however, in sparsely inhabited regions or where the lack of technology, finances, or other factors make it difficult to use it productively. Still, the readily accessible, renewable water supplies are very large, amounting to some 1,500 km^3 (about 400,000 gal) per person per year worldwide.

Many countries suffer water scarcity or water stress

The United Nations considers 1,000 m^3 (264,000 gal) of water per person per year to be the minimum necessary to meet basic human needs. **Water scarcity** occurs when the demand for water exceeds the available amount or when poor quality restricts its use. **Water stress** occurs when renewable water supplies are inadequate to satisfy essential human or ecosystem needs, bringing about increased competition among potential demands. Water stress is most likely

Table 17.3	Major Rivers of the World	
River	**Countries in River Basin**	**Average Annual Discharge at (m³/sec)**
Amazon	Brazil, Peru	175,000
Orinoco	Venezuela, Colombia	45,300
Congo	Congo	39,200
Yangzi	Tibet, China	28,000
Bramaputra	Tibet, India, Bangladesh	19,000
Mississippi	United States	18,400
Mekong	China, Laos, Burma, Thailand, Cambodia, Vietnam	18,300
Parana	Paraguay, Argentina	18,000
Yenisey	Russia	17,200
Lena	Russia	16,000

1 m^3 = 264 gallons

Source: World Resources Institute

FIGURE 17.10 Water has always been the key to survival. Who has access to this precious resource and who doesn't has long been a source of tension and conflict.
© Ray Ellis/Science Source

to occur in poor countries where the per capita renewable water supply is low.

As you can see in figure 17.2, South America, West Central Africa, and South and Southeast Asia all have areas of very high rainfall. The highest per capita water supplies generally occur in countries with wet climates and low population densities. Iceland, for example, has about 600 million liters per person per year. In contrast, Bahrain, where temperatures are extremely high and rain almost never falls, has essentially no natural fresh water. Almost all of Bahrain's water comes from imports and desalinized seawater. Egypt, in spite of the fact that the Nile River flows through it, has only about 42,000 liters of water annually per capita, or nearly 15,000 times less than Iceland.

Ironically, Cherrapunji, India, described at the beginning of this chapter as the wettest place on earth, is now experiencing water shortages during the dry season. Growing population, increased per capita water use, deforestation of surrounding hillsides, higher industrial demand, and changing monsoon patterns caused by global warming all contribute to this problem.

There is no simple definition of drought. What would be an amazingly wet year in a desert might be considered a catastrophic drought elsewhere. In general, a drought is an extended period of consistently below-average precipitation that has a substantial impact on ecosystems, agriculture, and economies.

As the opening case study shows, much of the western United States has been exceptionally dry over the past decade. Many places are experiencing water crises. Is this just a temporary cycle or a long-term climatic change? In 2012, the United States

experienced the most severe drought in at least 50 years, which according to the U.S. Department of Agriculture (USDA) reduced yields on 80 percent of agricultural land in the country. Increased temperatures, unusual weather patterns, population growth, urban sprawl, and wasteful uses all are contributing to shortages in many areas. The effects on water supplies may well be the most serious consequences of global climate change.

Current drought conditions in the United States are now approaching those of the "dust bowl" days of the 1930s, when wind stripped topsoil from millions of hectares of land, and billowing dust clouds turned day into night. Thousands of families were forced to leave farms and migrate to other areas.

Droughts in the American West aren't a new phenomenon. In fact, the dry spells in recent years have been of relatively short duration compared to historic events. The dust bowl of the 1930s, for example, lasted only about six years. By contrast, the megadroughts that destroyed the Anasazi (or Ancient Pueblo) cultures in the twelfth and thirteenth centuries (fig. 17.11) lasted 25 to 50 years.

If the government had listened to Major John Wesley Powell, settlement patterns in the western United States would be very different from what we see there today. Powell, who led the first expedition down the Colorado River, went on to be the first head of the U.S. Geological Survey. In that capacity, he did a survey of the agricultural and settlement potential of the western desert. His conclusion, quoted at the beginning of this chapter, was that there isn't enough water there to support a large human population.

Powell recommended that the political organization of the West be based on watersheds, so that everyone in a given jurisdiction would be bound together by the available water. He thought that farms should be limited to local surface water supplies, and that cities should be small oasis settlements. Instead, we've built huge metropolitan areas, such as Los Angeles, Phoenix, Las Vegas, and

FIGURE 17.11 The stunning cliff dwellings of Mesa Verde National Park were abandoned by the Anasazi, or ancient Puebloan people, during a severe megadrought between 1275 and 1299. Will some of our modern cities face the same fate?
© William P. Cunningham

Denver, in places where there is little or no natural water supply. Will those cities survive predicted shortages?

Water use is increasing

Human water use has been increasing about twice as fast as population growth over the past century. Water use is stabilizing in industrialized countries, but demand will increase in developing countries where supplies are available. The average amount of water withdrawn worldwide is about 646 m^3 (170,544 gal) per person per year. This overall average hides great discrepancies in the proportion of annual runoff withdrawn in different areas. Some countries with a plentiful water supply withdraw a very small percentage of the water available to them. Canada, Brazil, and the Congo, for instance, withdraw less than 1 percent of their annual renewable supply.

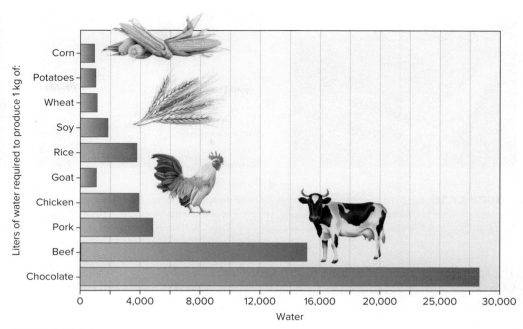

FIGURE 17.12 Water required to produce 1 kg of some important foods.

By contrast, in desert countries such as Libya, Yemen, and Israel, where water is one of the most crucial environmental resources, groundwater and surface water withdrawal together amount to more than 100 percent of their renewable supply. They are essentially "mining" water—extracting groundwater faster than it is being replenished. Obviously, this isn't sustainable in the long run.

The total annual renewable water supply in the United States amounts to an average of about 9,000 m^3 (nearly 2.4 million gal) per person per year. We now withdraw about one-fifth of that amount, or some 5,000 liters (1,300 gal) per person per day, including industrial and agricultural water. By comparison, the average water use in Haiti is less than 30 liters (8 gal) per person per day.

In contrast to energy resources, which usually are consumed when used, water can be used over and over if it is not too badly contaminated. Water **withdrawal** is the total amount of water taken from a water body. Much of this water could be returned to circulation in a reusable form. Water **consumption,** on the other hand, is loss of water due to evaporation, absorption, or contamination. Essentially, consumption is when the water gets moved into a different part of the water cycle.

Agriculture is a dominant water use

Clean, fresh water is essential for nearly every human endeavor. Worldwide, agriculture claims about 70 percent of total water withdrawal, and that use is increasing rapidly. Agricultural use ranges from 93 percent of all water withdrawn in India to only 4 percent in Kuwait, which cannot afford to spend its limited water on crops.

Most of us never think about the water required to grow and prepare the food we eat (fig. 17.12). Raising a kilogram of beef in a concentrated feeding operation, for example, takes more than 15,000 liters of water, because cattle in these facilities are fed

grain, which they process very inefficiently, that often was grown with water-intensive irrigation systems. The result is that the average hamburger takes 2,400 liters (630 gal) to produce. Raising goats requires at least 90 percent less water from all sources than cattle because they eat a much wider diet and more effectively metabolize what they do eat.

Rice cultivation (which generally occurs in wet paddies) takes about three times as much water as raising potatoes or wheat. For some foods, the greatest water use isn't in growing the crop, but in preparing it for our consumption. Chocolate, for example, is highly processed before we eat it.

Irrigation can be very inefficient. Traditionally, the main method has been flood or furrow irrigation, in which water runs across the land surface (fig. 17.13a). Up to half of this water can be lost through evaporation. Much of the rest runs off before it is used by plants. In arid lands, flood irrigation helps remove toxic salts from soil, but these salts contaminate streams, lakes, and wetlands downstream. Repeated flooding also waterlogs the soil, eventually reducing crop growth. Sprinkler systems can also be wasteful (fig. 17.13b). Water spraying high in the air quickly evaporates, rather than watering crops. In recent years, growing pressure on water resources has led to more-efficient sprinkler systems that hang low over crops to reduce evaporation.

Drip irrigation (fig. 17.13c) is a promising technology for reducing irrigation water use. These systems release carefully regulated amounts of water just above plant roots, so that nearly all water is used by plants. Only about 1 percent of the world's croplands currently use these expensive systems, however.

Irrigation infrastructure, such as dams, canals, pumps, and reservoirs, is expensive. Irrigation is also the economic foundation of many regions. In the United States, for nearly a century, the federal government has taken responsibility for providing irrigation.

(a) Flood irrigation

(b) Rolling sprinklers

(c) Drip irrigation

FIGURE 17.13 Agricultural irrigation consumes more water than any other use. Methods vary from flood and furrow (a), which uses extravagant amounts of water but also flushes salts from soils, to sprinklers (b), to highly efficient drip systems (c).
Source: (a) Photo by Jeff Vanuga, USDA Natural Resources Conservation Service; (b) © Royalty-Free/Corbis; (c) © William P. Cunningham

The argument for doing so is that irrigated agriculture is a public good that cannot be provided by individual farmers. A consequence of this policy has frequently been heavily subsidized crops whose costs, in water and in dollars, far outweigh their value.

Domestic and industrial water uses tend to be far less than agricultural use

Worldwide, domestic water use accounts for only about 6 percent of water withdrawals. Because most of this water is returned to a river or lake after use, actual consumption is low, about 10 percent of the world total, on average. Where sewage treatment is unavailable, however, water can be badly degraded by urban uses. In wealthy countries, each person uses about 500 to 800 liters per day (180,000 to 280,000 liters per year), far more than in developing countries (30 to 150 liters per day). In North America, the largest single use of domestic water is toilet flushing (fig. 17.14). On average, each person in the United States uses about 50,000 liters (13,000 gal) of drinking-quality water annually to flush toilets. Showering and washing accounts for nearly a third of water use, followed by laundry. In western cities such as Palm Desert and Phoenix, landscape irrigation is also a major water user.

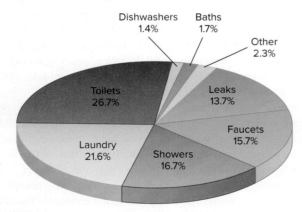

Dishwashers 1.4%
Baths 1.7%
Other 2.3%
Toilets 26.7%
Leaks 13.7%
Laundry 21.6%
Showers 16.7%
Faucets 15.7%

FIGURE 17.14 Typical household water use in the United States.
Source: Data from the American Water Works Association, 2010

Urban and domestic water use have grown approximately in proportion with urban populations, about 50 percent between 1960 and 2010. Although individual water use seems slight on the scale of world water withdrawals, the cumulative effect of inefficient appliances, long showers, lavish lawn-watering, and other uses is enormous. California has established increasingly stringent standards for washing machines, toilets, and other appliances, in order to reduce urban water demands. Many other cities and states are following this lead to reduce domestic water use. Some people are replumbing houses so that gray water (that used for showers, dishwashing, etc.) can be reused for irrigating plants or flushing toilets rather than simply dumped directly in a sewer. We'll discuss more water conservation practices later in this chapter.

Industry accounts for 20 percent of global freshwater withdrawals. Industrial use rates range from 70 percent in industrialized parts of Europe to less than 5 percent in countries with little industry. Power production, including hydroelectric, nuclear, and thermoelectric power, makes up 50 to 70 percent of industrial uses, and industrial processes make up the remainder. As with domestic water, little of this water is made unavailable after use, but it is often degraded by defouling agents, chlorine, or heat when it is released to the environment.

The greatest industrial producer of degraded water is mining and energy production. Ores must be washed and treated with chemicals such as mercury and cyanide (chapter 14). As much as 80 percent of water used in mining and processing is released with only minimal treatment. Fracturing (or fracking) oil and natural gas wells takes as much as 17 million liters (4.5 million gal) for a single well (chapter 19). Up to 40 percent of this water comes back to the surface contaminated with petrochemicals, arsenic, radionuclides, and other toxic materials. Providing this water is difficult in dry regions, and disposal of contaminated water is a problem everywhere.

Section Review

1. Define water scarcity and water stress.
2. What is drip irrigation, and why is it beneficial?
3. What percentage of all water withdrawals are for agricultural uses?

17.3 FRESHWATER SHORTAGES

- *Billions of people lack access to clean water.*
- *Groundwater is being depleted in many places.*
- *Diversion projects redistribute water, but often with social and environmental costs.*

Clean drinking water and basic sanitation are necessary to prevent communicable diseases and to maintain a healthy life. For many of the world's poorest people, one of the greatest environmental threats to health is polluted water. In Mali, for example, 88 percent of the population lacks safe water; in Ethiopia, 94 percent do. Rural people often have less access to good water than do city dwellers. This often results in diarrhea and a variety of other diseases linked to contaminated water and lack of sanitation. Every year, about 1.6 million people, 90 percent of them children under 5 years old, die from these diseases.

Shortage of water resources compounds these problems. The United Nations estimates that about a third of the world's population lives in countries where water supplies don't meet essential needs (fig. 17.15). By 2025, two-thirds of the world's people may be living in countries that are water-stressed—defined by the United Nations as consuming more than 10 percent of their renewable freshwater resources. As noted earlier, the World Health Organization considers an average of 1,000 m^3 (264,000 gal) per person per year to be a necessary amount of water for modern domestic, industrial, and agricultural uses. Some 45 countries, most of them in Africa or the Middle East, cannot meet the minimum essential water needs of all their citizens.

Severe conditions of water stress and scarcity also occur in north Africa and in central and southern Asia, where rainfall is low and poor countries can't afford to build expensive infrastructure. Growing populations, industrial demands, and agricultural irrigation only compound these stresses.

More than two-thirds of the world's households have to fetch water from outside the home (fig. 17.16). This is heavy work, done mainly by women and children, often taking hours every day. Improved public water systems bring many benefits to these poor families, but availability doesn't always mean affordability. A typical poor family in Lima, Peru, for instance, uses only one-sixth as much water as a middle-class American family but pays three times as much for it. If they followed government recommendations to boil all water to prevent cholera, up to one-third of the poor family's income could be used just in acquiring and purifying water. These challenges are widespread in developing areas (Chapter 18).

Groundwater is an essential but declining resource

Groundwater provides nearly 40 percent of the fresh water for agricultural and domestic use in the United States. Nearly half of all Americans and about 95 percent of the rural U.S. population

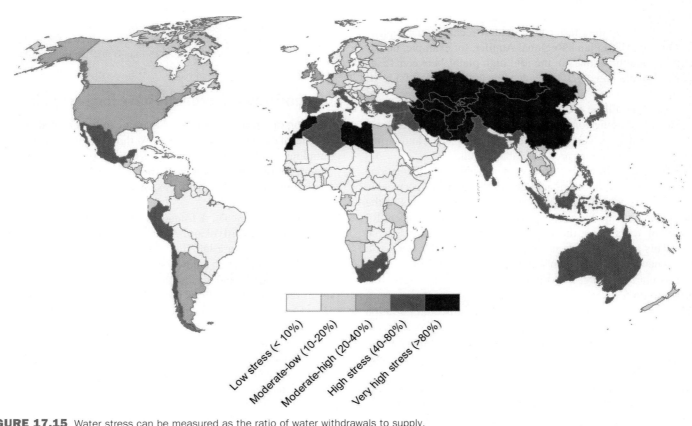

FIGURE 17.15 Water stress can be measured as the ratio of water withdrawals to supply.
Source: Data World Resources Institute

Low stress (< 10%) Moderate-low (10-20%) Moderate-high (20-40%) High stress (40-80%) Very high stress (>80%)

FIGURE 17.16 Women and children often spend hours every day collecting water—which often is unsafe for drinking—from local water sources.
© Layton Thompson

depend on groundwater for drinking and other domestic purposes. Overuse of these supplies causes several kinds of problems, including drying of wells and natural springs, disappearance of surface water features such as wetlands, and loss of aquifer storage capacity.

In many areas of the United States, groundwater is being withdrawn from aquifers faster than natural recharge can replace it. The Ogallala High Plains Regional Aquifer, for example, underlies eight agricultural states in the arid high plains between Texas and North Dakota (fig. 17.17). As deep as 400 m (1,200 ft) in its center, this porous bed of sand, gravel, and sandstone once held more water than all the freshwater lakes, streams, and rivers on earth. Steady pumping for irrigation and other uses has removed so much water that wells have dried up in many places, and farms, ranches, even whole towns are being abandoned.

Aquifers in shallow, fractured, or permeable rock formations can recharge rapidly, but deep aquifers, or those in less permeable formations, may take thousands of years to refill. Portions of the southern Ogallala aquifer contains water that has been present for thousands of years, since the wetter climate conditions of glacial periods. This ancient resource is often called "fossil water," and is not a renewable resource on a human time scale. Where aquifers are regularly recharged by rainfall on the ground surface, it is important to protect recharge areas.

Many cities and states try to prevent development or pollution in aquifer recharge zones. Arid Los Angeles, for example, spent its first century or so building drains and canals to remove rainwater, draining it efficiently as possible to the ocean. Now that water shortages are clearly the city's future, it is working to establish unpaved areas where rainwater can infiltrate groundwater. Beijing, another chronically thirsty and fast-growing city, is striving to find ways to become a "sponge city" that captures and stores water in the ground, rather than rushing stormwater out of the city.

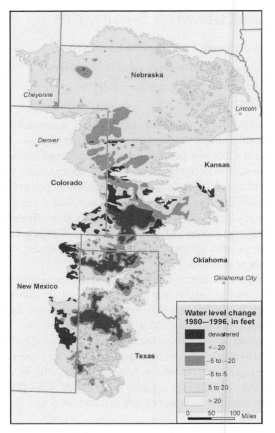

FIGURE 17.17 The Ogallala/High Plains regional aquifer supports a multimillion-dollar agricultural economy, but withdrawal far exceeds recharge. Some areas are down to less than 3 m of saturated thickness.

Groundwater overdrafts have long-term impacts

Drawing out groundwater faster than it can be replaced can cause long-term damage to the resource. At the scale of individual wells, pumping lowers the water table, forming a cone of depression (figure 17.18). Often, new wells must be drilled deeper, to reach the falling water table, and this makes nearby older, shallower wells run dry.

On a larger scale, a frequent consequence of aquifer depletion is **saltwater intrusion.** Along coastlines and in areas where saltwater deposits are left from ancient oceans, overuse of freshwater reservoirs often allows saltwater to intrude into aquifers used for domestic and agricultural purposes. Miami and other cities in Florida now face this problem. As freshwater is withdrawn from Florida's porous limestone aquifers, seawater is moving in to replace it.

Withdrawing large amounts of groundwater causes porous formations to collapse, resulting in **subsidence,** or settling of the surface. The cause is loss of water pressure in the pore spaces of deep formations. Once these pore spaces have collapsed, they can never be refilled, and the resource is lost. The U.S. Geological Survey estimates that the San Joaquin Valley in California has sunk more than 10 m in the last 50 years because of excessive

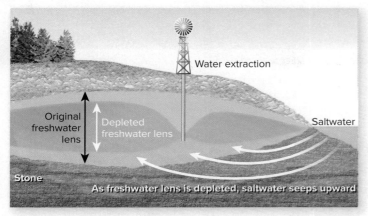

FIGURE 17.18 Saltwater intrusion into a coastal aquifer as the result of groundwater depletion. Many coastal regions of the United States are losing freshwater sources due to saltwater intrusion.

groundwater pumping (figure 17.19). Since this photo was taken, continuing groundwater extraction has lowered the ground surface by several more meters in some parts of the valley.

Around the world, many cities are experiencing subsidence. Most were built on former lake beds, river deltas or other unconsolidated sediments. Flooding is often a problem where coastal areas sink below sea level. Mexico City is one of the most notorious examples. Built on an old lake bed, it has probably been sinking since Aztec times. In recent years, however, rapid population growth and urbanization have caused groundwater overdrafts. Some areas of the city have sunk as much as 8.5 m (25.5 ft). The Shrine of Guadalupe, the cathedral, and many other historic monuments are sinking at odd and perilous angles.

Diversion projects redistribute water

Regional conditions within a country can be much worse than the countrywide averages shown in figure 17.15. For example, conditions in the American Southwest (described in the opening case study) are especially severe. Similarly, China is vulnerable to water shortages in dry northern regions, including the capital Beijing. Two-thirds of the country's population lives in the north, but two-thirds of its rainfall is in the south. China has undertaken vast water transfer projects, redirecting water from the Yangtze River region hundreds of km northward to Beijing. By most accounts, this project, known as the South to North Water Transfer Project, is the largest, and most expensive, infrastructure project ever undertaken.

Dams and canals are a foundation of civilization because they store and redistribute water for farms and cities. Many great civilizations, including the ancient empires of Sumeria, Egypt, and India, have been organized around large-scale water control systems. As modern dams and water diversion projects have grown in scale and number, though, their environmental costs have raised serious questions about efficiency, costs, and the loss of river ecosystems.

One of the most disastrous diversions in world history is that of the Aral Sea. Situated in arid Central Asia, on the border of Kazakhstan and Uzbekistan, the Aral Sea is a huge saline lake fed

FIGURE 17.19 A photo taken in 1977 showing subsidence caused by groundwater pumping in California's San Joaquin Valley. The 1925 sign indicates the ground level in that year.
Source: Richard Ireland, U.S. Geological Survey

by rivers from distant mountains. Starting in the 1950s, the Soviet Union began diverting these rivers to water cotton and rice fields. Gradually, the Aral Sea has shrunk, leaving vast salt flats contaminated by toxic chemical runoff from the farm fields. (fig. 17.20). The economic value of the cotton and rice probably never equaled the cost of lost fisheries and villages, and the health of local residents.

Recently some river flow has been restored to the "Small Aral" or northern lobe of the once-great sea. Water levels have risen 8 m and native fish are being reintroduced. It's hoped that one day commercial fishing may be resumed. The fate of the larger, southern remnant is more uncertain. There may never be enough water to refill it, and if there were, the toxins left in the lake bed could make it unusable anyway.

As the case study at the beginning of this chapter shows, many cities in the U.S. Southwest are facing a crisis with the drying of Lake Mead. Las Vegas, Nevada, which gets 40 percent of its water from the lake, has started a $3.5 billion, 525 km (326 m) pipeline to tap aquifers in the northeastern part of the state. Local ranchers fear that groundwater pumping will decimate the rangeland,

Measuring Invisible Water

Farms and cities around the world depend on groundwater. Powerful pumps extract water, sometimes from hundreds of meters below the ground, to irrigate rice fields in California's Central Valley and to flood alfalfa fields in the baking Imperial valley, to water cotton fields in north Texas, and to fill urban water lines in Miami. In California and many other places, surface water is tightly monitored and regulated, but groundwater is unregulated. If you can get it out, it's yours. This is a classic tragedy of the commons, with predictable results: Groundwater is declining steadily. Some of it is irreplaceable. Much of it is unlikely to recover for centuries—especially as the climate warms and rain becomes less reliable.

Farmers and water districts know their groundwater is declining, but it's hard to know how much, or how fast. The only way to watch groundwater resources has been to monitor wells, to see when they started to run dry. It is expensive and difficult to collect records from tens of thousands of scattered well logs, and to assess regional changes from these records. But as the resource dwindles, it becomes essential to understand it better.

A new approach is available in the Gravity Recovery and Climate Experiment (GRACE) satellites. The system is a pair of satellites, which travel in a polar orbit (passing over the poles as the earth rotates below them) 500 km (310 mi) above the earth. The two satellites are separated by about 220 km (137 mi). Subtle differences in gravitational pull cause acceleration or deceleration in the satellites, which they record as slight changes in the distance between the pair. The GRACE satellites have been monitoring variations in gravity since 2002, and recent years of drought have turned the world's spotlight on the data they are producing.

What causes variation in gravity on the earth's surface? The main factor is geological formations. Dense rock types, such as those rich in iron and other metals, have more gravitational pull than less massive ones, such as limestone formations. But changes in ice and water storage, over months or years, are also detectable as gravitational changes. Greenland's mass has declined as its ice melts and runs off into the Arctic Ocean. Major lakes lose mass as water is

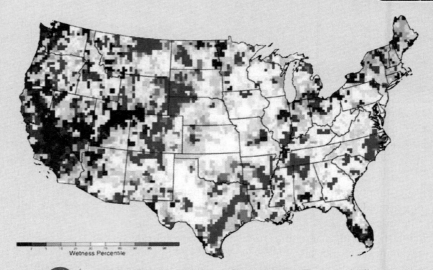

GRACE-Based Shallow Groundwater Drought Indicator

FIGURE 1 Groundwater and soil moisture, in much of the American West, is low compared to the range of normal conditions. Dark brown (2nd percentile) indicates areas in which 98 percent of historical conditions are wetter.

Source: University of Nebraska GRACE Data Assimilation Project

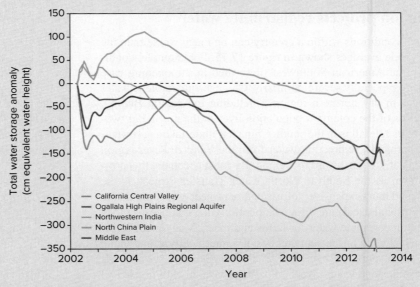

FIGURE 2 Withdrawals are reducing water storage in many of the world's most important aquifers, including those in politically tense regions. Groundwater change calculated from NASA's Gravity Recovery and Climate Experiment (GRACE) satellites.

Source: Adapted from J. S. Famiglietti, 2014. "The Global Groundwater Crisis," Nature Climate Change, 4: 945–948.

diverted or evaporates. Even changes in the density of ocean currents can be detected by the GRACE satellites. And now it is possible to monitor changes in groundwater storage in major aquifers from Los Angeles, to Beijing, to Delhi.

In the U.S., GRACE data show that droughts extend well beyond California, covering the Colorado River basin, and populous regions of the eastern and southern states (fig. 1). GRACE data also show that heavy rains in early 2016 did little to alter the deep drawdowns of previous years.

Global analysis shows that most of the world's major aquifers are becoming seriously overdrawn. Northeastern India is among the worst in drawdown (fig. 2). This information can be critical for policymakers who need to decide whether to subsidize irrigation or promote water conservation measures.

Satellites are providing more detailed and more global perspectives on our environment. You can learn more by finding NASA and GRACE websites online. You can also learn more by studying earth science, geography, physics, or any of the many fields that make use of satellite data.

For further reading: J. S. Famiglietti, 2014. "The global groundwater crisis," *Nature Climate Change* 4: 945–948.

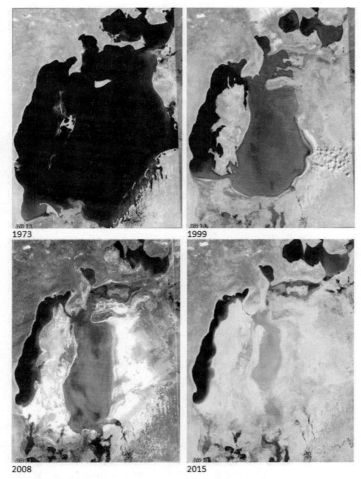

FIGURE 17.20 For 30 years rivers feeding the Aral Sea were diverted to irrigate cotton and rice fields. The Sea has lost more than 90 percent of its water. The "Small Aral" (upper right lobe) has separated from the main lake, and is now being refilled.

Source: USGS Earth Resources Observation and Science (EROS) Center

destroy native vegetation, and cause massive dust storms. They point to Owens Valley in California, where a similar water grab by Los Angeles in 1913 dried up the river and destroyed both natural vegetation and the ranching economy.

Las Vegas is also digging a $3.5 billion tunnel that will burrow into Lake Mead, 100 m (300 ft) below the normal outlet. Even if the lake reaches the "dead pool" level as warned at the beginning of this chapter, the city will still be able to draw off water. Of course, this might prevent refilling the reservoir to provide water and power to downstream users. If you lived downstream, what would you say about this project?

Dams have diverse environmental and social impacts

More than half of the world's 227 largest rivers have been dammed or diverted (fig. 17.21). Of the 50,000 large dams in the world, 90 percent were built in the twentieth century. Half of those are in China, and China continues to build and plan dams on its remaining rivers. Dams and diversion projects are claimed as essential for flood control, water storage, and electricity production. However, the costs of relocating villages, lost fishing and farming, and water losses to evaporation are enormous. Economically speaking, at least one-third of the world's large dams probably should never have been built.

Though dams provide hydroelectric power and water to distant cities, they also can have unintended consequences. Reservoirs in hot, dry climates lose tremendous amounts of water to evaporation. Lake Mead, on the Colorado River, loses more than 1 billion m^3 of water to evaporation and seepage every year. Lake Powell loses nearly as much through evaporation and through seepage into the sandstone bedrock that lines the reservoir. Some studies estimate that 6 percent of the river's water is lost to evaporation and seepage in reservoirs.

In Asia and Africa, where there are still undammed rivers and governments hungry for industrial growth, dams are planned

FIGURE 17.21 Hoover Dam powers Las Vegas, Nevada. Lake Mead, behind the dam, loses about 1.3 billion m³ per year to evaporation.
Source: Photo by Lynn Betts, USDA Natural Resources Conservation Service

the ocean in two or three weeks. Reservoirs slow this journey to as much as three months, throwing off the time-sensitive physiological changes that allow the fish to survive in salt water when they reach the ocean. Reservoirs expose young salmon to predators, and warm water in reservoirs increases disease in both young and older fish. About half of the roughly 400 stocks of salmon, steelhead, and sea-run cutthroat trout in the United States are endangered, and more than 100 are already extinct due to dam construction.

Some dams have fish ladders—a cascading series of pools and troughs—that allow fish to bypass the dam. Another option is to move both adults and juveniles by barge or truck. This can result in the strange prospect of barges of wheat moving downstream while passing barges of fish move in the opposite direction. These options are expensive and only partially effective in restoring blocked salmon runs.

Dams may have a limited lifespan

As water shortages on the Colorado River become more severe, there have been increasing calls to save Lake Mead and the Hoover Dam by eliminating Lake Powell, the reservoir behind the Glenn Canyon Dam (see opening case study). Already low water has cut power production in half at the Glenn Canyon dam. This may be the most prominent of many U.S. dams whose purpose has been questioned. In 1998, the Army Corps of Engineers announced that it would no longer build large dams and diversion projects, in part because few available sites remain but also because political opposition is greater than it was a generation ago. Now we are starting to remove dams and restore rivers and habitat. Former interior secretary Bruce Babbitt said, "Of the 75,000 large dams in the United States, most were built a long time ago and are now obsolete, expensive, and unsafe. They were built with no consideration of the environmental costs. As operating licenses come up for renewal, removal and restoration to original stream flows will be one of the options." The tallest dam demolished in the United States, so far, is the 64 m (210 ft) Glines Canyon Dam on the Elwha River in Washington State (see chapter 13, Case Study). For the first time in a century, the Elwha flows freely to the sea, and, it's hoped, some of the 400,000 salmon per year that once migrated up this river may be restored.

In addition to low water levels and declining power production, sediment is a common threat to dams. Rivers with high sediment loads can fill reservoirs quickly (fig. 17.22). In 1957, the Chinese government began building the Sanmenxia Dam on China's Huang He (Yellow River). From the beginning, engineers warned that the river carried so much sediment that the reservoir would have a very limited useful life. Dissent was crushed, however, and by 1960 the dam began filling the river valley and inundating fertile riparian fields that once had been part of China's traditional granaries.

Within two years, sediment accumulation behind the dam had become a serious problem. It blocked the confluence of the Wei and Yellow Rivers and backed up the Wei so much that it threatened to flood the historic city of Xi'an. By 1962, the reservoir was almost completely filled with sediment, and hydropower production dropped by 80 percent. The increased elevation of the riverbed raised the underground water table and caused salinization of

on many major rivers such as the Mekong and tributaries of the Amazon, the Congo, and others. Evaporation losses are a problem in many of these areas, as are questions of water pollution. In tropical areas, where vegetation is abundant, decaying plant matter in reservoirs produces methane. Recent studies have calculated that methane from tropical dams has greater global warming potential than equivalent coal or oil power plants would have. Local governments and dam-building companies see a great deal of money to be made in these projects, however.

International Rivers, an environmental and human rights organization, reports that dam projects have forced more than 23 million people from their homes and land and that years later many are still suffering the impacts of dislocation. Many of the people being displaced are ethnic minorities. On India's Narmada River, for example, a proposed series of 30 dams have displaced about 1 million villagers and tribal people. Many who have been relocated have never successfully integrated either socially or economically. Protests around this project have raged for 20 years or more.

There is also increasing concern that big dams in seismically active areas can trigger earthquakes. In more than 70 cases worldwide, large dams have been linked with increased seismic activity. Geologists suggest that filling the reservoir behind the nearby Zipingpu Dam on the Min River caused the devastating 7.9-magnitude Sichuan earthquake that killed an estimated 90,000 people in 2008. If true, it would be the world's deadliest dam-induced earthquake.

Dams are also lethal for migratory fish, such as salmon. Adult fish are blocked from migrating to upstream spawning areas. And juvenile fish die if they go through hydroelectric turbines. The slack water in reservoirs behind dams is also a serious problem. Juvenile salmon evolved to ride the surge of spring runoff downstream to

FIGURE 17.22 This dam is now useless because its reservoir has filled with silt and sediment.
Source: Photo by Tim McCabe, USDA Natural Resources Conservation Service

Climate change threatens water supplies

The Intergovernmental Panel on Climate Change (IPCC) warns us that climate change threatens to exacerbate water shortages caused by population growth, urban sprawl, wasteful practices, and pollution. The IPCC Fifth Assessment Report predicted with "very high confidence" that reduced precipitation and higher evaporation rates caused by higher temperatures will result in a 10 to 30 percent runoff reduction over the next 50 years in some dry regions at midlatitudes (see Chapter 15). IPCC models generally show dry areas becoming drier and wet areas becoming wetter. Areas with declining precipitation include the southwestern United States, northern Africa, South Africa, central Asia, and southern Europe.

Change in precipitation is just part of the story, however. Rising temperatures will increase evaporation, as well as consumption of water by plants (transpiration). Evaporation and transpiration will reduce available water resources even in many areas where precipitation increases. Figure 17.23 shows expected changes in stream flow in 2100, compared to the average for 1980–2010, with a 2°C increase in mean temperatures. The figure indicates agreement among 55 different models of climate and hydrology, which represent different assumptions about how atmospheric circulation and rainfall may change.

wells and farm fields. By 1991, the riverbed was 4.6 m above the surrounding landscape. The river is kept in check only by earthen dams that frequently fail and flood the surrounding countryside. By the time the project was complete, more than 400,000 people had been relocated, far more than planners expected.

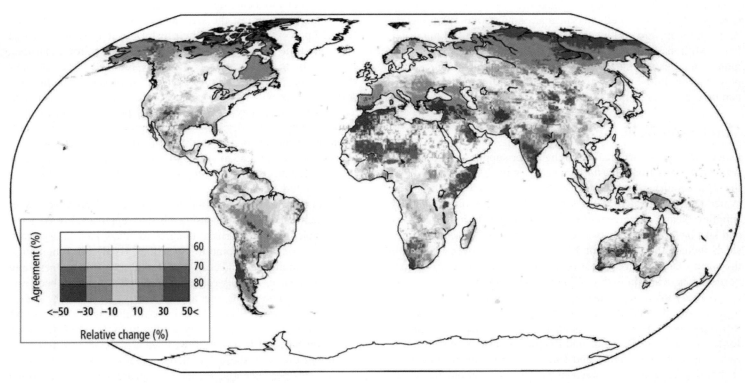

FIGURE 17.23 Changing streamflow with climate change. Colors show percentage change in mean annual streamflow for a global mean temperature rise of 2°C. Color intensity shows agreement among 55 combined climate-hydrologic models.
Source: Figure 3-4 from Jiménez Cisneros, B.E., T. Oki, N.W. Arnell, G. Benito, J.G. Cogley, P. Döll, T. Jiang, and S.S. Mwakalila, 2014: Freshwater resources. In: Climate Change 2014: Impacts, Adaptation, and Vulnerability. Part A: Global and Sectoral Aspects. Contribution of Working Group II to the Fifth Assessment Report of the Intergovernmental Panel on Climate Change [Field, C.B., V.R. Barros, D.J. Dokken, K.J. Mach, M.D. Mastrandrea, T.E. Bilir, M. Chatterjee, K.L. Ebi, Y.O. Estrada, R.C. Genova, B. Girma, E.S. Kissel, A.N. Levy, S. MacCracken, P.R. Mastrandrea, and L.L. White (eds.)]. Cambridge University Press, Cambridge, United Kingdom and New York, NY, USA.

In many parts of the world, severe droughts are already resulting in depleted rivers, empty reservoirs, and severe water shortages for millions of people. Australia is suffering from extreme heat waves, dying vegetation, massive wildland fires, and increasing water deficits. The Australian government has acknowledged that recent droughts and devastating wildfires have resulted from climate change. Wealthy countries have a higher adaptive capacity—an ability to provide infrastructure and aid to their populations, but they cannot stop widespread wildfires or drought.

Declining water supplies lead to conflict and humanitarian crises. The Syrian civil war and refugee crisis followed years of the most extreme drought on record in the region. Agricultural failure resulted as irrigation water dried up, and farmers were forced to migrate to cities that could not provide them with jobs. The Syrian situation was made worse by a generation of subsidies for water-wasteful cotton production, which collapsed especially rapidly in a drying climate. But Syria is just one of many countries that subsidize extravagant agricultural water consumption. Many countries continue to do so despite an awareness of dwindling supplies—including parts of Central Asia, southern Europe, and also the United States, which raises cotton in the arid Plains of north Texas and Oklahoma and raises rice and other water-intensive crops in the intensely arid Central Valley of California.

Changing precipitation and surface water resources also affect groundwater recharge. Yemen, for example, depends entirely on groundwater, which is being depleted much more rapidly than it is replenished. Some wells in Sanaa are now 800 to 1,000 meters deep, and many are no longer usable because of the sinking water table. Millions of people may have to abandon Sanaa and other mountain cities for the coastal plain. The shrinking resource base played a role in the civil wars and the rise in international terrorism in Yemen in recent years.

Would you fight for water?

Defense analysts warn that declining water supplies could lead to wars between nations. *Fortune* magazine wrote, "Water will be to the 21st century what oil was to the 20th." For its 2009 World Water Day, the United Nations focused on transboundary water supplies. Nearly 40 percent of the world's population lives in river and lake basins shared by two or more countries. These 263 watersheds include the territory of 145 countries and cover nearly half the earth's land surface. Great reservoirs of fresh water also cross borders. There are more than 270 known transboundary aquifers.

Already we've seen skirmishes—if not outright warfare—over water access. An underlying factor in hostilities between Israel and its neighbors has been control of aquifers and withdrawals from the Jordan River. India, Pakistan, and Bangladesh also have confronted each other over water rights; and Turkey and Iraq threatened to send armies to protect access to water in the Tigris and Euphrates rivers.

Water scarcity is an underlying cause of the conflict across sub-Saharan Africa, including ongoing genocide in the Darfur region of Sudan. When rain was plentiful, Arab pastoralists and African farmers coexisted peacefully. Drought upsets that truce. The hundreds of thousands who have fled to Chad could be considered climate refugees as well as war victims. How many more tragedies such as these might we see in the future as people struggle for declining water resources?

Section Review

1. How does agriculture affect water resources in the Aral Sea?
2. Describe some effects of dams and diversion projects.
3. How might climate change affect our water supplies?

17.4 Water Conservation

- *Conservation can save water.*
- *Recycled water is useful for many purposes.*
- *Water prices and policies can encourage efficient water use.*

In many cases, we may simply have to adapt to less water. An example is a breakthrough agreement for the Klamath River in California. To keep enough water in the river to rebuild fish populations sufficient for sustainable tribal, recreational, and commercial fisheries, farmers have to reduce their withdrawals. A key provision for farmers is to have a reliable and certain water allocation. When the irrigation gates were closed in 2001, most farmers had already planted their crops. Most of their expenses for the year were already invested. To cut off water to their crops at that point meant financial ruin.

A major feature of the settlement is that farmers agree to a 10 to 25 percent reduction in their historic water use in exchange for a one-time payment to help finance conservation measures. The benefit to the farmers is a greatly reduced threat of their water being completely shut off again to protect fish. In most years, farmers will have to get by on less water than usual. In really dry years, they'll either have to pump groundwater or fallow—temporarily dry up—some cropland. With clearer rules in place, farmers can shift in dry years from planting low-value crops, such as alfalfa, to using just part of their land to grow higher-value crops, thus keeping their income up while still using less water and fertilizer.

A description often used for such a plan is a "land bank." Other Californian water districts are using this same approach. Los Angeles, for example, is paying farmers to agree to fallow land in dry years. Farmers get enough income to cover their fixed costs—buildings, equipment, mortgages, and taxes—while still staying in business until better years come. The city can ensure a supply of water in bad years at a much lower cost than other alternatives.

Farmers in the Klamath basin have also agreed to a similar approach for wetlands. They'll take turns flooding fields on a rotating basis so that waterfowl have a place to rest and feed. Some call this a "walking" wetlands program. No one loses his or her fields permanently, but there's a guarantee of more habitat for birds (fig. 17.24). Money to pay for both wetland mitigation and crop reductions will come from a $1 billion budget provided mainly for endangered species protection. This plan also contains guarantees for stabilized power costs for family farms, ranches, and the two Klamath wildlife refuges.

How Does Desalination Work?

As you've learned in this chapter, the ocean has a vast amount of water. Unfortunately, it's too salty for most human uses. There are ways to remove salt and other minerals from water, but they tend to be energy-intensive and expensive. The most common form of desalination is distillation: Water is boiled, and the steam is collected and condensed to make fresh water. This works well. Glass-distilled water remains the standard for purity in most science labs. The world's largest desalination plant is the Jebel Ali Plant in the United Arab Emirates. It uses multistage flash distillation and is capable of producing 300 million cubic meters of water per year. Distillation produces over 85 percent of all desalinated water in the world.

In a multistage distillation facility, water evaporates in a series of spaces called stages containing heat exchangers and condensate collectors. Each compartment has a partial vacuum, which causes water to boil at a much lower temperature than 100°C (212°F). As water passes from one stage to another, temperatures and pressures are adjusted to match the boiling point as salt concentration increases. The total evaporation in all the stages is about 15 percent of the input water. The main limitations of this design are the corrosion caused by the warm brine as well as

the energy required to heat and pump water and to create a vacuum. In oil-rich countries, such as Saudi Arabia, where water costs more than pumping oil out of the ground, energy costs aren't important, but in other places it becomes a concern. However, coupling a desalination facility with a power plant can cut costs by as much as two-thirds. Waste heat from the power plant is used to preheat seawater, which simultaneously provides cooling for the power plant. Disposal of warm, salty brine can have serious adverse impacts, however, on local coastal areas.

The other principal method of desalination is reverse osmosis. This filtration process removes large molecules from solutions by applying pressure to the solution on one side of a selectively permeable membrane. Every cell in your body is enclosed by a selectively permeable membrane. It's one of the things that make life possible. The plasma membrane around each cell has tiny pores that allow small molecules, such as water, to pass through but that exclude large molecules, such as salt. Ordinarily, osmosis causes small molecules to move from an area of high concentration (pure water) to areas of lower concentration (a salt solution, which has fewer water molecules per unit area because some

space is occupied by the salt). You may have observed osmosis in a biology lab. If you put an amoeba in a high-salt solution, it shrivels (water is drawn out). If you put it in pure water, it bloats and explodes (the interior is saltier than the water).

Reverse osmosis drives this process backward by applying a pressure to the high-salt side of a semipermeable membrane to filter the water. In practice, the membranes are packed in concentric coils inside a tube. A large facility may have tens of thousands of these tubes. Pore sizes can vary from 0.1 nanometers (3.9×10^{-9} in) to 5,000 nanometers (0.0002 in) depending on the solution to be filtered. Reverse osmosis systems can range from industrial-size facilities capable of purifying hundreds of thousands of gallons per day, to a pen-size straw that you can carry in your pocket to sip water from a contaminated source.

Although there are many more reverse osmosis facilities than thermal desalination plants, they produce a relatively small percentage of all desalinated water. Still, they can be more mobile and easier to operate than a distillation plant. And if the pores are less than 1 nanometer, the water produced can be cleaner than distillation (although the production rate is very low with such small pore sizes).

FIGURE 17.24 A flock of snow geese rises from the Lower Klamath National Wildlife Refuge. Millions of migrating birds use these wetlands for feeding and resting.
© Alan and Sandy Carey/Getty Images RF

Increasing water supplies

There have been many attempts to enhance local supplies and redistribute water. As the opening case study for this chapter shows, building dams and enormous aqueducts has allowed us to establish farms and major cities (at least temporarily) in arid regions. Creating rain in dry regions has been tried, with mixed success, by cloud seeding—distributing condensation nuclei in humid air to help form raindrops. There are concerns, however, about diverting rain from areas where it might otherwise have fallen, as well as about contamination from cloud-seeding materials.

A technology that might have great potential for increasing freshwater supplies is desalination of ocean water or brackish saline lakes and lagoons. Worldwide, over 18,000 desalination plants produce over 50 billion liters (23 billion gallons of water a day (see the Exploring Science box later in this section). Middle Eastern oil-rich states produce about 60 percent of desalinated water. Saudi Arabia is the largest single producer, at about one-third of the world total. The United States is second, at 20 percent of the world supply. Some American cities, such as Tampa, Florida, and San Diego, California, now depend on desalination for a significant part of their water resources. Although desalination is still three to four times more expensive than most other sources of fresh water, it provides a welcome water source in such places as Oman and Bahrain, where there is no other access to fresh water. If cheap, clean, inexhaustible energy were available, however, the oceans could provide all the water we would ever need. Interestingly, clean, fresh water could be a by-product of solar-thermal generating plants (see Chapter 19).

Domestic conservation has important impacts

We could probably save as much as half of the water we now use for domestic purposes without great sacrifice or serious changes in our lifestyles. Simple steps, such as taking shorter showers, stopping leaks, and washing cars, dishes, and clothes as efficiently as possible, can go a long way toward forestalling the water shortages that many authorities predict. Isn't it better to adapt to more conservative uses now, when we have a choice, than to be forced to do it by scarcity in the future?

The use of resource-conserving appliances, such as low-volume shower heads and efficient dishwashers and washing machines, can reduce water consumption greatly (see the What Can You Do? box later in this section). If you live in an arid region, you might consider whether you really need a lush green lawn that requires constant watering, feeding, and care. Planting native ground cover in a "natural lawn" or developing a rock garden or landscape in harmony with the surrounding ecosystem can be both ecologically sound and aesthetically pleasing (fig. 17.25). There are about 30 million ha (75 million acres) of cultivated lawns, golf courses, and parks in the United States. They receive more water, fertilizer, and pesticides per hectare than any other kind of land.

FIGURE 17.25 By using native plants in a natural setting, residents of Phoenix save water and fit into the surrounding landscape.
© William P. Cunningham

The largest U.S. domestic water use, after watering yards, is toilet flushing (fig. 17.14). There are now several types of waterless or low-volume toilets. Waterless composting systems can digest both human and kitchen wastes by aerobic bacterial action, producing a rich, non-offensive compost that can be used as garden fertilizer. There are also low-volume toilets that use recirculating oil or aqueous chemicals to carry wastes to a holding tank, from which they are periodically taken to a treatment plant. Anaerobic digesters use bacterial or chemical processes to produce usable methane gas from domestic wastes. These systems provide valuable energy and save water but are more difficult to operate than conventional toilets. Few cities are ready to mandate waterless toilets, but a number of cities (including Los Angeles, Orlando, Austin, and Phoenix) have ordered that water-saving toilets, showers, and faucets be installed in all new buildings. The motivation was twofold: to relieve overburdened sewer systems and to conserve water.

Recycling can reduce consumption

Significant amounts of water can be reclaimed and recycled. In California, water recovered from treated sewage constitutes the fastest-growing water supply, growing about 30 percent per year. Despite public squeamishness, purified sewage effluent is being

What Can You Do?

Saving Water and Preventing Pollution

Each of us can conserve much of the water we use and avoid water pollution in many simple ways.

- Take short showers; don't flush every time you use the toilet; wash your car infrequently.
- Don't let the faucet run while washing hands, dishes, food, or brushing your teeth. Draw a basin of water for washing and another for rinsing dishes. Don't run the dishwasher when half full.
- Dispose of used motor oil and household hazardous waste responsibly. Don't dump anything down a storm sewer that you wouldn't want to drink.
- Avoid using toxic or hazardous chemicals for simple cleaning or plumbing jobs. A plunger or plumber's snake will often unclog a drain just as well as caustic acids or lye.
- If you have a lawn, use water sparingly. Water your grass and garden at night, not in the middle of the day. Consider planting native plants, low-maintenance ground cover, a rock garden, or some other xeriphytic landscaping.
- Use water-conserving appliances: low-flow showerheads, low-flush toilets, and aerated faucets.
- Use recycled (gray) water for lawns, house plants, and car washing.
- Check your toilet for leaks. A leaky toilet can waste 50 gallons per day. Add a few drops of dark food coloring to the tank and wait 15 minutes. If the tank is leaking, the water in the bowl will have changed color.

FIGURE 17.26 Recycled water is being used in California and Arizona for everything from agriculture, to landscaping, to industry. Some cities even use treated sewage effluent for human drinking-water supplies.
© William P. Cunningham

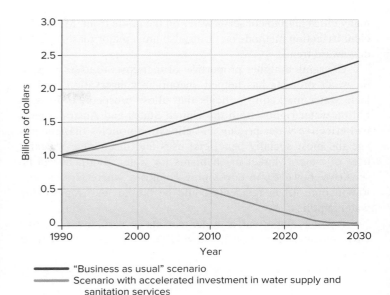

"Business as usual" scenario
Scenario with accelerated investment in water supply and
        sanitation services
Scenario with accelerated investment and efficiency reforms

FIGURE 17.27 Three scenarios for government investments on clean water and sanitation services, 1990 to 2030.
Source: World Bank estimates based on the research paper by Dennis Anderson and William Cavendish, "Efficiency and Substitution in Pollution Abatement: Simulation Studies in Three Sectors."

used for everything from agricultural irrigation to flushing toilets (fig. 17.26). In a statewide first, San Diego is currently piping water from the local sewage plant directly into a drinking-water reservoir. Residents of Singapore and Queensland, Australia, also are now drinking purified sewage effluent.

In many developing countries, as much as 70 percent of all the agricultural water used is lost to leaks in irrigation canals, application to areas where plants don't grow, runoff, and evaporation. Better farming techniques, such as leaving crop residue on fields and ground cover on drainage ways, intercropping, use of mulches, and low-volume irrigation, could reduce these water losses dramatically.

Nearly half of all industrial water use is for cooling of electric power plants and other industrial facilities. Some of this water use could be avoided by installing dry cooling systems similar to a car's radiator. In many cases, cooling water could be reused for irrigation or other purposes in which water does not have to be drinking quality. The waste heat carried by this water could be a valuable resource if techniques were developed for using it.

Prices and policies have often discouraged conservation

Through most of U.S. history, water policies have generally worked against conservation. In the well-watered eastern United States, water policy was based on riparian usufructuary (use) rights—those who lived along a river bank had the right to use as much water as they liked as long as they didn't interfere with its quality or availability to neighbors downstream. It was assumed that the supply would always be endless and that water had no value until it was used. In the drier western regions where water

often is a limiting resource, water law is based primarily on the Spanish system of prior appropriation rights, or "first in time are first in right." Even if the prior appropriators are downstream, they can legally block upstream users from taking or using water flowing over their property. But the appropriated water has to be put to "beneficial" use by being consumed. This creates a policy of "use it or lose it." Water left in a stream, even if essential for recreation, aesthetic enjoyment, or to sustain ecological communities, is not being appropriated or put to "beneficial" (that is, economic) use. Under this system, water rights can be bought and sold, but water owners frequently are reluctant to conserve water for fear of losing their rights.

In most federal "reclamation" projects, customers are charged only for the immediate costs of water delivery. The costs of building dams and distribution systems is subsidized, and the potential value of competing uses is routinely ignored. Farmers in California's Central Valley, for instance, for many years paid only about one-tenth of what it cost the government to supply water to them. This didn't encourage conservation. Subsidies created by underpriced water amounted to as much as $500,000 per farm per year in some areas.

Growing recognition that water is a precious and finite resource has changed policies and encouraged conservation across the United States. Despite a growing population, the United States is now saving some 144 million liters (38 million gal) per day—or enough water to fill Lake Erie in a decade—compared to per capita consumption rates of 20 years ago. With 90 million more people in the United States now than there were in 1980, we get by with 10 percent less water. New requirements for water-efficient fixtures and low-flush toilets in many cities help to conserve water on the home front. More efficient irrigation methods on farms also are a major reason for the downward trend.

Charging a higher proportion of real costs to users of public water projects has helped encourage conservation, and so have water marketing policies that allow prospective users to bid on water rights. Both the United States and Australia have had effective water pricing and allocation policies that encourage the most socially beneficial uses and discourage wasteful uses of water. Market mechanisms for water allotment can be sensitive, however, in developing countries where farmers and low-income urban residents could be outbid for irreplaceable water supplies.

It will be important, as water markets develop, to be sure that environmental, recreational, and wildlife values are not sacrificed to the lure of high-bidding industrial and domestic uses. Given prices based on real costs of using water and reasonable investments in public water supplies, pollution control, and sanitation, the World Bank estimates that everyone in the world could have an adequate supply of clean water by the year 2030 (fig. 17.27). We will discuss the causes, effects, and solutions for water pollution in Chapter 18.

Section Review

1. How have farmers in California's Klamath River basin reduced their water use?
2. Describe some examples of domestic water conservation.
3. Give an example in which water policies and prices have encouraged conservation.

Conclusion

Water is a precious resource. As human populations grow and climate change affects rainfall patterns, water is likely to become even scarcer in the future. Already about 2 billion people live in water-stressed countries (where water supplies are inadequate to meet all demands), and at least half those people don't have access to clean drinking water. Depending on population growth rates and climate change, by 2050 there could be 7 billion people (about 60 percent of the world population) living in areas with water stress or scarcity. Conflicts over water rights are becoming more common between groups within countries and between neighboring countries that share water resources. This is made more likely by the fact that most major rivers cross two or more countries before reaching the sea. Many experts agree with *Fortune* magazine that "water will be to the 21st century what oil was to the 20th."

There are many ways to make more water available. Huge projects, such as the giant dams and diversion projects on America's Colorado River, have allowed us to create farms and cities in arid areas. However, impounding water and shipping it between watersheds can have severe ecological and social effects. Desalination could also provide an endless supply of fresh water if we had a clean, inexpensive, renewable energy source. Perhaps a better way is to practice conservation and water recycling. These efforts, also, are underway in many places and show great promise for meeting our needs for this irreplaceable resource. There are things you can do as an individual to save water and prevent pollution. Even if you don't have water shortages now where you live, it may be wise to learn how to live in a water-limited world.

Reviewing Key Terms

Can you define the following terms in environmental science?

aquifers 17.1	hydrologic cycle 17.1	residence time 17.1	water table 17.1
artesian well 17.1	infiltration 17.1	saltwater intrusion 17.3	withdrawal 17.2
consumption 17.2	rain shadow 17.1	subsidence 17.3	zone of aeration 17.1
discharge 17.1	recharge zone 17.1	water scarcity 17.2	zone of saturation 17.1
groundwater 17.1	renewable water supplies 17.2	water stress 17.2	

Critical Thinking and Discussion Questions

1. What changes might occur in the hydrologic cycle if our climate were to warm or cool significantly?

2. Why does it take so long for the deep ocean waters to circulate through the hydrologic cycle? What happens to substances that contaminate deep ocean water or deep aquifers in the ground?

3. Are there ways you could use less water in your own personal life? What obstacles prevent you from taking these steps?

4. Should we use up underground water supplies now or save them for some future time?

5. How should we compare the values of free-flowing rivers and natural ecosystems with the benefits of flood control, water diversion projects, hydroelectric power, and dammed reservoirs?

6. Would it be feasible to change from flush toilets and using water as a medium for waste disposal to some other system? What might be the best way to accomplish this?

Data Analysis

Graphing Global Water Stress and Scarcity

According to the United Nations, water stress is when annual water supplies drop below 1,700 m^3 per person. Water scarcity is defined as annual water supplies below 1,000 m^3 per person. More than 2.8 billion people in 48 countries will face either water stress or scarcity conditions by 2025. Of these 48 countries, 40 are expected to be in West Asia or Africa. By 2050, far more people could be facing water shortages, depending on both population projections and scenarios for water supplies based on global warming and consumption patterns.

To explore some of the issues and questions about future water scarcity, go to Connect, and answer questions about this figure and others from this chapter.

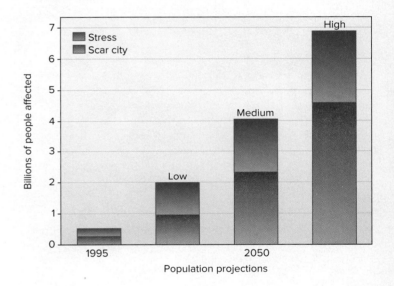

18

Water Pollution

▲ Daily bathing in the Ganges River confers ritual purity, but pollution makes this practice physically risky.
© dbimages/Alamy Stock Photo

Learning Outcomes

After studying this chapter, you should be able to:

18.1 Describe the types and effects of water pollutants.
18.2 Discuss water quality issues today.
18.3 Explain water pollution control.
18.4 Summarize some water legislation.

"If there is magic on this planet, it is contained in water."

– Loren Eiseley

India's Holy River

Early each morning, residents of Varanasi, India, go to the river to purify themselves in the water of the holy Ganges. Mother Ganga, as the river is known, is considered a goddess on earth, and the river confers purity on bathers making their daily ablutions. Pilgrims travel to the Ganges from across India, bringing home bottles of the river's sacred water. After death, many hope to be cremated on the banks of the river, especially in the ancient and holy city of Varanasi, and then have Mother Ganga carry away their ashes.

The Ganges is also one of the world's most important rivers. Starting as an icy stream rushing from the mouth of the Gangotri Glacier high in the Himalayas, the river falls over 3,500 m (over 11,300 ft) to the plains at the foot of the mountains. Flowing through India's most populous region (fig. 1), the river provides water to more than 500 million people—nearly 40 percent of all Indians and 1 in 15 of the world's entire human population. The Ganges river plain holds some of India's most fertile farmland, much of it irrigated by the river. The river also supplies water to industries and power plants, and its hydroelectric dams provide electricity for India's cities. Far downstream, 2,500 km from its glacial headwaters, the river deposits Himalayan sediment in the Ganges delta, home to rich mangrove forests and millions of farmers and fishers.

The river also carries away the waste of hundreds of polluting industrial sites—chemical factories, fabric dying industries, tanneries, mines, and other sources—as well as plastic and other trash, animal carcasses, agricultural runoff (including fertilizers and pesticides), and the ashes of cremated bodies. And it carries sewage, most of it untreated, from the half a billion people living in the watershed. Even while it is a holy river, the Ganges is one of the world's most severely polluted rivers. Sewage effluent accounts for some 80 percent of the river's pollution, and industrial effluent makes up the other 20 percent.

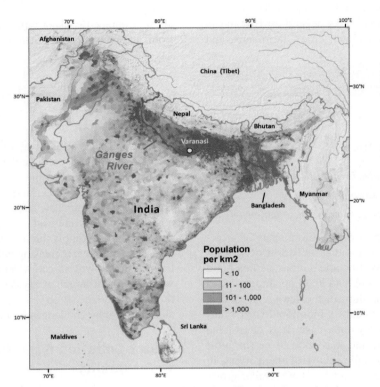

FIGURE 18.1 Population density in South Asia. The densely populated Ganges River Plain follows the southern edge of the Himalayas, which are the source of most of Asia's most important rivers.
Source: Center for International Earth Science Information Network (CIESIN) Columbia University. 2015. Gridded Population of the World, Version 4.

The effects of water pollution, and the strategies for controlling it, depend on what the pollutants are. Organic waste and nutrients support bacteria, which can cause illnesses ranging from diarrhea to cholera. Coliform bacteria, which grow in nutrient-rich water, occur at 100 times the legal standards in many areas of the Ganges. Decomposition of organic waste also depletes oxygen in the water, asphyxiating fish and other aquatic life and leading to "dead zones" with little oxygen and few living organisms. Industrial waste, especially chemical effluent or metals, such as chromium or arsenic, is toxic in low concentrations, especially with frequent exposure. These pollutants, and agricultural pesticides, cause cancer or damage nerves or organs.

Cleaning the Ganges is a leading goal of many Indians, including Prime Minister Narendra Modi. The river has extraordinary cultural and historical importance—the bathing steps in Varanasi are just the most famous of the countless cultural and religious places. While the Ganges is undeniably filthy, it also deserves the respectful treatment owed to a sacred river. In 2014, Mr. Modi's newly elected government initiated a plan to protect the river, its cultural heritage, and public health, starting with new efforts to monitor pollution sources, and then to control them in the dozens of major cities, hundreds of factories and thousands of sewage outfalls that line the river and its tributaries.

This was far from the first such effort. Studies in 1981 led to the Ganga Action Plan, an effort to identify and address dominant pollution sources. Over the decades, though, ineffective enforcement has led to a lack of action at all levels of government and industry. Another cleanup campaign was announced in 2010 to treat municipal sewage waste and regulate industrial effluent. A year later, the World Bank gave $1 billion to support cleanup efforts. Prime Minister Modi's 2014 initiative included higher budgets, an army battalion designated as an Eco Task

Force, new legislative efforts, and better coordination among ministers. With international funding, the National Ganga River Basin Authority intends to stop all untreated sewage and industrial inputs by 2020. Industrial cooperation has been slow thus far, however.

Like many rivers of the world's rivers, the Ganges also faces serious threats from climate change. With warming conditions, Himalayan glaciers have been retreating for decades. Many of the world's major agricultural areas and cities depend on Himalayan ice melt, and declining supplies threaten cities, farmlands, and

hydropower. Lower water levels also reduce a river's ability to absorb and dilute pollutants.

Clean water is one of the most important resources we have. Protecting clean water, or restoring it, is one of our most difficult public challenges. On the other hand, water is constantly recycled and replenished. If we can reduce pollution inputs, rivers become clean again. This has been observed in many countries that have managed to institute and enforce policies for protecting water quality. In this chapter, we examine the major types of water pollution, where they come from, and how to control them to protect clean water for everyone.

18.1 WATER POLLUTION

- *Waterborne pathogens kill hundreds of millions of people.*
- *Nutrient enrichment leads to eutrophication.*
- *Organic pollutants include drugs, pesticides, and other chemicals.*

The pollutants plaguing the Ganges reflect the impacts of growing, urbanizing, industrializing populations the world over. Rapidly growing cities, especially in low-income regions, cannot build sewage treatment facilities fast enough to keep up. Indian cities have the capacity to treat only one third of the sewage produced, and more than two-thirds of the country's monitored river miles are severely polluted. India is an extreme case: It is one of the world's fastest-growing countries, and it is soon to be the most populous, with over 1.3 billion people, many of them living in deep poverty. But many developing countries share similar growing pains. How can these conditions be improved?

Most students today are too young to appreciate that water in most industrialized countries was once far more polluted and dangerous than it is now. Until the 1970s, it was common to dump toxic solvents, organic chemicals, mine waste, and raw sewage into U.S. surface waters, or onto the ground, where it seeps into groundwater. We are still paying billions of dollars every year to clean up those contaminants of the past. These practices began to change in the United States when President Nixon signed the Clean Water Act in 1972. This law has been called the United States' most successful and popular environmental legislation. It established a goal that all the nation's waters should be "fishable and swimmable." While this goal is far from being achieved, the Clean Water Act remains popular because it protects public health (thus saving taxpayer dollars) and reduces environmental damage. Clean water is also valued for its aesthetic appeal. The view of a clean lake, river, or seashore makes people happy, and water provides for recreation, so many people feel their quality of life has improved as water quality has been restored. It is clear that strong policies with reliable enforcement can make a difference.

We still have a long way to go in improving water quality. In the U.S. and other wealthy countries, pollution from factory pipes has been vastly reduced in recent decades, but erosion from farm fields, construction sites, and streets has gotten worse in many areas. Airborne mercury, sulfur, and other substances continue to contaminate lakes and wetlands. Concentrated livestock production and agricultural runoff threaten underground water as well as surface water systems. Increasing industrialization in developing countries has led to widespread water pollution in impoverished regions with little environmental regulation. Much of this chapter outlines the U.S. experience with water quality, but the same types of pollutants, and similar challenges for pollution control, exist worldwide.

Water pollution is anything that degrades water quality

Any physical, biological, or chemical change in water quality that adversely affects living organisms or makes water unsuitable for desired uses might be considered **water pollution.** There are natural sources of water contamination, such as poison springs, oil seeps, and sediment from erosion, but in this chapter we focus primarily on human-caused factors that affect water quality.

Pollution-control standards and regulations usually distinguish between point and nonpoint pollution sources. Factories, power plants, sewage treatment plants, underground coal mines, and oil wells are classified as **point sources** because they discharge pollution from specific locations, such as drain pipes, ditches, or sewer outfalls (fig. 18.2). These sources are discrete and identifiable, so they are relatively easy to monitor and regulate. It is generally possible to divert effluent from the waste streams of these sources and treat it before it enters the environment.

In contrast, **nonpoint sources** of water pollution are scattered or diffuse, having no specific location where they discharge into a particular body of water. Nonpoint sources include runoff from farm fields and livestock feedlots and pastures (fig. 18.3), golf courses, lawns and gardens, construction sites, logging operations,

FIGURE 18.2 Sewer outfalls, industrial effluent pipes, acid draining out of abandoned mines, and other point sources of pollution are generally easy to recognize.
© Simon Fraser/SPL/Science Source

FIGURE 18.3 This scene looks peaceful and idyllic, but allowing cows free access to streams and ponds is a major cause of bank erosion and water contamination. Nonpoint sources such as this have become a leading unresolved cause of water pollution.
© Image Source RF

roads, streets, and parking lots. Whereas point sources may be fairly uniform and predictable throughout the year, nonpoint sources are often highly episodic. The first heavy rainfall after a dry period may flush high concentrations of gasoline, lead, oil, and rubber residues off city streets, for instance, while subsequent runoff may have lower levels of these pollutants. In some areas, spring snowmelt carries high levels of atmospheric acid deposition into streams and lakes. The irregular timing of these events, their multiple sources, and their scattered locations make them much more difficult to monitor, regulate, and treat than point sources.

A particularly diffuse type of nonpoint pollution is **atmospheric deposition** of contaminants carried by air currents and precipitated into watersheds or directly onto surface waters as rain, snow, or dry particles. The Great Lakes, for example, have been found to be accumulating industrial chemicals such as PCBs and dioxins, as well as agricultural toxins such as the insecticide toxaphene, that cannot be accounted for by local sources alone. The nearest sources for many of these chemicals are sometimes thousands of kilometers away (chapter 16).

Amounts of these pollutants can be quite large. It is estimated that there are 600,000 kg of the herbicide atrazine in the Great Lakes, most of which is thought to have been deposited from the atmosphere. Several studies have indicated health problems among people who regularly eat fish, which accumulate toxic compounds in the food chain, from the Great Lakes.

Ironically, lakes can be pollution sources as well as recipients. In the past 12 years, about 26,000 metric tons of PCBs have "disappeared" from Lake Superior. Apparently these compounds evaporate from the lake surface and are carried by air currents to other areas where they are redeposited.

Although the types, sources, and effects of water pollutants are often interrelated, it is convenient to divide them into major categories for discussion (table 18.1). Examine this table closely as you consider important sources and effects of each type of pollutant in the sections that follow.

Table 18.1   Major Categories of Water Pollutants

Category	Examples	Sources
A. Causes Health Problems		
1. Infectious agents	Bacteria, viruses, parasites	Human and animal excreta
2. Organic chemicals	Pesticides, plastics, detergents, oil, and gasoline	Industrial, household, and farm use
3. Inorganic chemicals	Acids, caustics, salts, metals	Industrial effluents, household cleansers, surface runoff
4. Radioactive materials production, natural sources	Uranium, thorium, cesium, iodine, radon	Mining and processing of ores, power plants, weapons
B. Causes Ecosystem Disruption		
1. Sediment	Soil, silt	Land erosion
2. Plant nutrients	Nitrates, phosphates, ammonium	Agricultural and urban fertilizers, sewage, manure
3. Oxygen-demanding wastes	Animal manure and plant residues	Sewage, agricultural runoff, paper mills, food processing
4. Thermal	Heat	Power plants, industrial cooling

Infectious agents, or pathogens, cause diseases

The most serious water pollutants in terms of human health worldwide are pathogenic organisms (chapter 8). Among the most important waterborne diseases are typhoid, cholera, bacterial and amoebic dysentery, enteritis, polio, infectious hepatitis, and schistosomiasis. Malaria, yellow fever, and filariasis are transmitted by insects that have aquatic larvae. Altogether, at least 25 million deaths each year are blamed on these water-related diseases. Nearly two-thirds of the mortalities of children under 5 years old are associated with waterborne diseases.

The main source of these pathogens is untreated or improperly treated human wastes. Animal wastes from feedlots or fields near waterways and food-processing factories with inadequate waste treatment facilities also are sources of disease-causing organisms.

In wealthier countries, sewage treatment plants and other pollution-control techniques have reduced or eliminated most of the worst sources of pathogens in inland surface waters, and drinking water is generally disinfected by chlorination, so epidemics of waterborne diseases are rare in these countries. The United Nations estimates that 90 percent of the people in developed countries have adequate (safe) sewage disposal, and 95 percent have clean drinking water.

The situation is quite different in poor countries, as noted in the opening Case Study for this chapter. The United Nations estimates that at least 2.5 billion people in these countries lack adequate sanitation, and 780 million lack access to clean drinking water. Conditions are especially bad in remote rural areas where sewage treatment is usually primitive or nonexistent and purified water is either unavailable or too expensive. The World Health Organization estimates that 80 percent of all sickness and disease in less-developed countries can be attributed to waterborne infectious agents and inadequate sanitation.

The World Bank calculates that if everyone had pure water and satisfactory sanitation, 200 million fewer episodes of diarrheal illness would occur each year, and 2 million childhood deaths would be avoided. Furthermore, 450 million people would be spared debilitating roundworm or fluke infections. Access to clean water is a major focus of the UN's Sustainable Development Goals (chapter 1).

Detecting specific pathogens in water is difficult, time-consuming, and costly; thus, water quality control personnel usually analyze water for the presence of **coliform bacteria,** any of the many types that live in the colon or intestines of humans and other animals. The most common of these is *Escherichia coli* (or *E. coli*). Many strains of bacteria live symbiotically, and beneficially, in mammals, but some, such as *Shigella, Salmonella,* or *Lysteria,* can cause fatal diseases. It is usually assumed that if any coliform bacteria are present in a water sample, infectious pathogens are present also.

How do we test for coliform bacteria? Water is filtered through a paper filter, which is then placed in a dish containing a nutrient medium that supports bacterial growth. After 24 hours in an incubator, living cells will have produced small colonies. If *any* colonies are found in drinking water samples, the U.S. Environmental Protection Agency considers the water unsafe and requiring disinfection. The EPA-recommended maximum coliform count for swimming water is 200 colonies per 100 ml of water, but some cities and states allow higher levels. If the limit is exceeded, the contaminated pool, river, or lake usually is closed to swimming (fig. 18.4).

Low oxygen levels indicate nutrient contamination

The amount of oxygen dissolved in water is a good indicator of water quality and of the kinds of life it will support. Water with an oxygen content above 6 parts per million (ppm) will support game fish and other desirable forms of aquatic life. Water with less than 2 ppm oxygen will support mainly worms, bacteria, fungi, and other detritus feeders and decomposers. Oxygen is added to water by diffusion from the air, especially when turbulence and mixing rates are high, and by photosynthesis of green plants, algae, and cyanobacteria. Oxygen is removed from water by respiration and chemical processes that consume oxygen.

Organic waste, such as sewage, paper pulp, or food waste, is rich in nutrients, especially nitrogen and phosphorus. These nutrients stimulate the growth of oxygen-demanding decomposing bacteria. **Biochemical oxygen demand (BOD)** is thus a useful test for the presence of organic waste in water. Most BOD tests involve incubating a water sample for five days, and then comparing oxygen levels in the water before and after incubation. An alternative method, called the chemical oxygen demand (COD), uses a strong oxidizing agent (dichromate ion in 50 percent sulfuric acid) to completely break down all organic matter in a water sample. This method is much faster than the BOD test, but it records inactive organic matter as well as bacteria, so it is less useful.

FIGURE 18.4 The national goal of making all surface waters in the United States "fishable and swimmable" has not been fully met, but scenes like this have been reduced by pollution-control efforts.
© Roger A. Clark/Science Source

A third method of assaying pollution levels is to measure **dissolved oxygen (DO) content** directly, using an oxygen electrode. The DO content of water depends on factors other than pollution (for example, temperature and aeration), so it indicates general conditions in the water body, rather than nutrients and bacteria in particular.

The effects of oxygen-demanding wastes on rivers depends to a great extent on the volume, flow, and temperature of the river water. Aeration occurs readily in turbulent, rapidly flowing rivers, which are therefore often able to recover quickly from oxygen-depleting processes. Downstream from a point source, such as a municipal sewage plant discharge, a characteristic decline and restoration of water quality can be detected either by measuring dissolved oxygen content or by observing the flora and fauna that live in successive sections of the river.

The oxygen decline downstream of a pollutant source is called the **oxygen sag** (fig. 18.5). Upstream from the pollution source, oxygen levels support normal populations of clean-water organisms. Immediately below the source of pollution, oxygen levels begin to fall as decomposers consume waste materials. Rough fish, such as carp, bullheads, and gar, are able to survive in this oxygen-poor environment where they eat both decomposer organisms and the waste itself. Farther downstream, the water may become so oxygen-depleted that almost nothing but the most resistant microorganisms and invertebrates can survive (below 2 ppm). We often call this low-oxygen environment a "dead zone." Eventually most of the nutrients are used up, decomposer populations decline, and oxygen levels recover as oxygen diffuses into water from the air, or is released by aquatic plants. Depending on the volumes and flow rates of the effluent plume and the river receiving it, normal communities may not appear for several miles downstream.

Nutrient enrichment leads to cultural eutrophication

Water clarity (transparency) is affected by sediments, chemicals, and the abundance of plankton organisms, and is a useful measure of water quality and water pollution. Rivers and lakes that have clear water and low biological productivity are said to be **oligotrophic** (*oligo* = little + *trophic* = nutrition). By contrast, **eutrophic** (*eu* + *trophic* = truly nourished) waters are rich in organisms and organic materials. Eutrophication is an increase in nutrient levels and biological productivity. Some amount of eutrophication is a normal part of successional changes in most lakes. Tributary streams bring in sediments and nutrients that stimulate plant growth. Over time, ponds or lakes may fill in, eventually becoming marshes. The natural rate of eutrophication and succession depends on water chemistry and depth, volume of inflow, mineral content of the surrounding watershed, and the biota of the lake itself.

As with BOD, nutrient enrichment sewage, fertilizer runoff, even decomposing leaves in street gutters can produce a human-caused increase in biological productivity called **cultural eutrophication.** Cultural eutrophication can also result from higher temperatures, more sunlight reaching the water surface, or a number of other changes. Increased productivity in an aquatic system sometimes can be beneficial. Fish and other desirable species may grow faster, providing a welcome food source.

Often, however, eutrophication has undesirable results. Elevated phosphorus and nitrogen levels stimulate "blooms" of algae or thick growths of aquatic plants (fig. 18.6). Bacterial populations also increase, fed by larger amounts of organic matter. The water

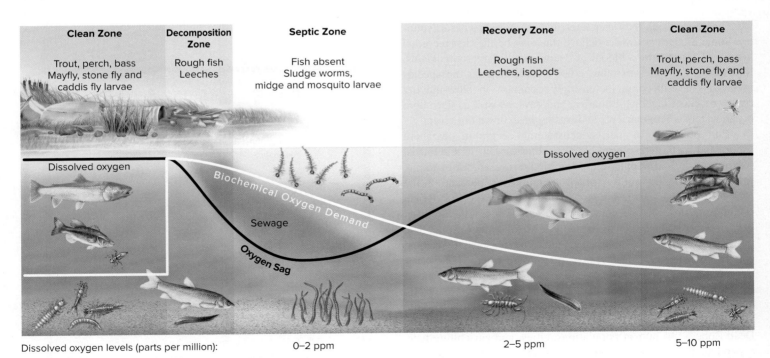

Clean Zone	Decomposition Zone	Septic Zone	Recovery Zone	Clean Zone
Trout, perch, bass Mayfly, stone fly and caddis fly larvae	Rough fish Leeches	Fish absent Sludge worms, midge and mosquito larvae	Rough fish Leeches, isopods	Trout, perch, bass Mayfly, stone fly and caddis fly larvae

Dissolved oxygen

Biochemical Oxygen Demand

Dissolved oxygen

Sewage

Oxygen Sag

Dissolved oxygen levels (parts per million):   0–2 ppm   2–5 ppm   5–10 ppm

FIGURE 18.5 Oxygen sag downstream of an organic source. A great deal of time and distance may be required for the stream and its inhabitants to recover.

FIGURE 18.6 Eutrophic lake. Nutrients from agriculture and domestic sources have stimulated growth of algae and aquatic plants. This reduces water quality, alters species composition, and lowers the lake's recreational and aesthetic values.
© William P. Cunningham

often becomes cloudy or turbid and has unpleasant tastes and odors. In extreme cases, plants and algae die and decomposers deplete oxygen in the water. Collapse of the aquatic ecosystem can result.

Eutrophication can cause toxic tides and "dead zones"

According to the Bible, the first plague to afflict the Egyptians when they wouldn't free Moses and the Israelites was that the water in the Nile turned into blood. All the fish died and the people were unable to drink the water, a terrible calamity in a desert country. Some modern scientists believe this may be the first recorded history of a **red tide,** a bloom of deadly aquatic microorganisms. Red tides—and tides of other colors, depending on the species involved—have become increasingly common in slow-moving rivers, brackish lagoons, estuaries, and bays, as well as nearshore ocean waters where nutrients and wastes wash down our rivers.

Eutrophication in marine ecosystems occurs in nearshore waters and partially enclosed bays or estuaries. During the tourist season, the coastal population of the Mediterranean, for example, swells to 200 million people. Eighty-five percent of the effluent from large cities go untreated into the sea. Beach pollution, fish kills, and contaminated shellfish result. Enclosed bays or water bodies, such as the Gulf of Mexico, the Caspian Sea, the Baltic, and China's Bohai Bay, tend to be in especially critical condition. Extensive "dead zones" often form where rivers carry oxygen-depleting nutrients into estuaries and shallow seas. One of the largest hypoxic (oxygen-depleted) zones in the world occurs during summer months in the Gulf of Mexico at the mouth of the Mississippi River (see Exploring Science below, "Studying the Dead Zone"). As human populations, cities, and agriculture have expanded, these hypoxic zones have increased in coastal areas around the world.

Metals are important inorganic pollutants

Some toxic inorganic chemicals are released from rocks by weathering, are carried by runoff into lakes or rivers, or percolate into groundwater aquifers. This pattern is part of natural mineral cycles (chapter 3). Humans often accelerate the transfer rates in these cycles thousands of times above natural background levels through the mining, processing, using, and discarding of minerals. These processes introduce a variety of heavy metals to waterways, such as mercury, lead, tin, and cadmium. Other toxic elements, such as selenium and arsenic, also have reached hazardous levels in some waters.

The primary metal contaminants of health concern include mercury, lead, cadmium, tin, and nickel, all highly toxic in minute concentrations. These are widely used in industrial processes, such as battery production or metal processing and plating, and they are released in coal combustion. Metals persist in the environment, so they accumulate in food webs and have stronger effects upon top predators—including humans.

Mercury, released from coal-burning power plants, is the most widespread toxic metal contamination problem in North America and many other industrialized regions. As noted in chapter 16, a 2007 study tested more than 2,700 fish from 636 rivers and streams in 12 western states, and mercury was found in every one of them. More than half the fish contained mercury levels unsafe for women of childbearing age, and three-quarters exceed the safe limit for young children.

Consuming fish is our main source of human mercury exposure. Fifty states have issued advisories about eating freshwater or ocean fish, most often because of mercury contamination (fig. 18.7). Top marine predators, such as shark, swordfish, marlin, king mackerel, and blue-fin tuna, should be avoided completely. Public health officials estimate that 600,000 American children now have mercury levels in their bodies high enough to cause mental and

FIGURE 18.7 Health departments across the United States have issued warnings about eating locally caught freshwater fish. Mercury contamination is often the reason.

Studying the Dead Zone

In the 1980s, shrimp boat crews noticed that certain locations off the Gulf Coast of Louisiana were emptied of all aquatic life. Because the region supports shrimp, fish, and oyster fisheries worth $250 to $450 million per year, these "dead zones" were important to the economy, as well as to the Gulf's ecological systems. In 1985, marine scientist Nancy Rabelais began mapping areas of low oxygen concentrations in the Gulf waters. Her results, published in 1991, showed that vast areas just above the floor of the Gulf have a summer oxygen concentration of less than 2 parts per million (ppm), a level that eliminated all animal life except microorganisms and primitive worms. Healthy aquatic systems usually have about 10 ppm dissolved oxygen. What caused this hypoxic (oxygen-starved) area to develop?

Rabelais and her team tracked the phenomenon for several years, and it became clear that the dead zone was growing larger over time, that poor shrimp harvests coincided with years when the zone was large, and that the size of the dead zone, which ranges from 5,000 to 20,000 km^2 (about the size of New Jersey), depended on rainfall and runoff rates from the Mississippi River. Excessive nutrients, mainly nitrogen, from farms and cities far upstream were the suspected culprit.

How did Rabelais and her team know that nutrients were the problem? They observed that each year, seven to ten days after large spring rains in the agricultural parts of the upper Mississippi watershed, oxygen concentrations in the Gulf drop from 5 ppm to below 2 ppm. These rains are known to wash soil, organic debris, and last year's nitrogen-rich fertilizers from farm fields. Scientists also knew that saltwater ecosystems normally have little available nitrogen, a key nutrient for algae and plant growth. Pulses of agricultural runoff were followed by a profuse growth of algae and phytoplankton (tiny floating plants). This burst of biological

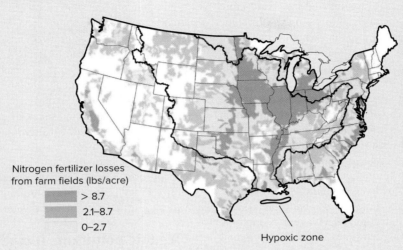

Nitrogen fertilizer losses from farm fields (lbs/acre)

- > 8.7
- 2.1–8.7
- 0–2.7

Hypoxic zone

The Mississippi River drains 40 percent of the conterminous United States, including the most heavily farmed states. Nitrogen fertilizer produces a summer "dead zone" in the Gulf of Mexico.

Source: U.S. EPA Water Atlas

activity produces an excess of dead plant cells and fecal matter that drifts to the seafloor. Shrimp, clams, oysters, and other filter feeders normally consume this debris but they can't keep up with the sudden flood of material. Instead, decomposing bacteria in the sediment break down the debris and consume most of the available dissolved oxygen as well. Putrefying sediments also produce hydrogen sulfide, which further poisons the water near the seafloor.

Human activities have increased the flow of nitrogen reaching U.S. coastal waters by four to eight times since the 1950s. Phosphorus, another key nutrient, has tripled. This case study shows how water pollution can connect far-distant places, such as Midwestern farmers and Louisiana shrimpers.

In well-mixed water bodies, such as the open ocean, oxygen from upper water layers is frequently mixed into lower layers. Warm protected water bodies are often stratified, however, as abundant sunlight keeps the upper layers warmer and less dense than lower layers. Denser lower layers can't mix with upper layers unless strong currents or winds stir the water.

Many enclosed coastal waters, including Chesapeake Bay, Long Island Sound, the

Mediterranean Sea, and the Black Sea, tend to be stratified and suffer hypoxic conditions that destroy bottom and near-bottom communities. There are about 200 dead zones around the world, and the number has doubled each decade since dead zones were first observed in the 1970s. The Gulf of Mexico is second in size behind a 100,000 km^2 dead zone in the Baltic Sea.

Can dead zones recover? Yes. Water is a forgiving medium, and organisms use nitrogen quickly. In 1996 in the Black Sea region, farmers in collapsing communist economies cut their nitrogen applications by half out of economic necessity; the Black Sea dead zone disappeared, while farmers saw no drop in their crop yields. In the Mississippi watershed, farmers can afford abundant fertilizer, and fear they can't risk under-fertilizing. Because of the great geographic distance between the farm states and the Gulf, Midwestern states have been slow to develop an interest in the dead zone. At the same time, concentrated feedlot production of beef and pork is rapidly increasing, and feedlot runoff is the fastest growing, and least regulated, source of nutrient enrichment in rivers.

In 2001, federal, state, and tribal governments forged an agreement to cut nitrogen inputs by 30 percent and reduce the size of the dead zone to 5,000 km^2. This agreement represented a remarkably quick political response to scientific results, but it doesn't appear to be enough. Models suggest that it would take a 40 to 45 percent reduction in nitrogen to achieve the 5,000 km^2 goal. Meanwhile, the size of the dead zone has failed to decline. The long-term average is 13,000 km^2, and in spring of 2015, it covered over 17,000 km^2, three times the agreed-upon target. Translating science to policy is a slow process. But without the science, it's impossible to know if conditions are improving or not.

developmental problems, and that one in six U.S. women have blood-mercury concentrations that would endanger a fetus.

Mine drainage and leaching of mining wastes are serious sources of metal pollution in water. A survey of water quality in eastern Tennessee—where there has been a great deal of surface mining—found contamination by acids and metals from mine drainage in 43 percent of all surface streams and lakes and more than half of all groundwater used for drinking supplies. In some cases, metal levels were 200 times higher than what is considered safe for drinking water.

Perhaps the largest human population threatened by naturally occurring arsenic in groundwater is in West Bengal, India, and eastern Bangladesh (fig. 18.8). Arsenic occurs naturally in the sediments that make up the Ganges River delta. Rapid population growth, industrialization, and intensification of irrigated agriculture have depleted or polluted limited surface water supplies. In an effort to provide clean drinking water for local residents, thousands of tube wells were sunk in the 1960s throughout the area. Much of this humanitarian effort was financed by loans from the World Bank.

By the 1980s, health workers became aware of widespread signs of chronic arsenic poisoning among Bengali villagers. Symptoms include watery and inflamed eyes, gastrointestinal cramps, gradual loss of strength, dry skin and skin tumors, anemia, confusion, and eventually death. Health workers estimate that the total number of potential victims in India and Bangladesh may exceed 100 million people. Fortunately, arsenic can be removed from water supplies once it is discovered (see the Exploring Science box titled "Inexpensive Water Purification" in section 18.3).

Acidic runoff can destroy aquatic ecosystems

Acids are released as by-products of industrial processes, such as leather tanning, metal smelting and plating, petroleum distillation, and organic chemical synthesis. Coal mining is an especially important source of acid water pollution. Sulfur compounds in coal react with oxygen and water to make sulfuric acid. Thousands of kilometers of streams in the United States have been acidified by acid mine drainage, some so severely that they are essentially lifeless.

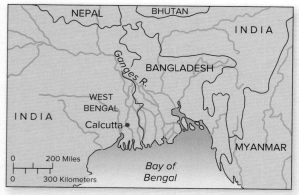

FIGURE 18.8 West Bengal and adjoining areas of Bangladesh have hundreds of millions of people who may be exposed to dangerous arsenic levels in well water.

Coal and oil combustion also leads to formation of atmospheric sulfuric and nitric acids (chapter 16), which are disseminated by long-range transport processes and deposited via precipitation (acidic rain, snow, fog, or dry deposition) in surface waters. Where soils are rich in such alkaline material as limestone, these atmospheric acids have little effect because they are neutralized. In high mountain areas or recently glaciated regions where bedrock is granitic and lakes are oligotrophic, however, there is little buffering capacity (ability to neutralize acids), and aquatic ecosystems can be severely disrupted. These effects were first recognized in the mountains of northern England and Scandinavia about 50 years ago.

Aquatic damage due to acid precipitation has been reported in about 200 lakes in the Adirondack Mountains of New York State and in several thousand lakes in eastern Quebec, Canada. Game fish, amphibians, and sensitive aquatic insects are generally the first to be killed by increased acid levels in the water. If acidification is severe enough, aquatic life is limited to a few resistant species of mosses and fungi. Increased acidity may result in leaching of toxic metals, especially aluminum, from soil and rocks, making water unfit for drinking or irrigation as well.

Organic pollutants include drugs, pesticides, and industrial products

Thousands of different natural and synthetic organic chemicals are used in the chemical industry to make pesticides, plastics, pharmaceuticals, pigments, and other products that we use in everyday life. Many of these chemicals are highly toxic (chapter 8). Exposure to very low concentrations (perhaps even parts per quadrillion, in the case of dioxins) can cause birth defects, genetic disorders, and cancer. Some can persist in the environment because they are resistant to degradation and toxic to organisms that ingest them. Contamination of surface waters and groundwater by these chemicals is a serious threat to human health.

The two most important sources of toxic organic chemicals in water are improper disposal of industrial and household wastes and runoff of pesticides from farm fields, forests, roadsides, golf courses, and other places where they are used in large quantities. The U.S. EPA estimates that about 500,000 metric tons of pesticides are used in the United States each year. Much of this material washes into the nearest waterway, where it passes through ecosystems and may accumulate in high levels in nontarget organisms. The bioaccumulation of DDT in aquatic ecosystems, which decimated populations of fish-eating birds such as Bald Eagles in the 1960s, was one of the first of these pathways to be understood. Dioxins and other chlorinated hydrocarbons (hydrocarbon molecules that contain chlorine atoms) have been shown to accumulate to dangerous levels in the fat of salmon, fish-eating birds, and humans. Many of these compounds disrupt endocrine hormone pathways that have widespread effects on growth and development.

Atrazine, the second most widely used herbicide in America, has been shown to disrupt normal development in frogs at concentrations as low as 0.1 ppb (chapter 8). This level is very common in farming regions. Could this be a problem for humans as well as frogs? Another of the most widely distributed water contaminants is triclosan, an antibiotic used in thousands of products including toothpaste, hand soap, face wash, body lotion, and cosmetics. Triclosan survives wastewater treatment and the final chlorination step converts it to dioxin, which is both a carcinogen and an endocrine hormone disrupter. According to the EPA, at least 75 percent of all Americans have trace amounts of this compound in their bodies. Washing with ordinary soap is just as effective as using antibiotic compounds. We could easily reduce this ubiquitous pollutant.

Countless other organic compounds also enter our water. How do they get there? In some cases, people simply dump unwanted food, medicines, and health supplements down the toilet or sink. More often, we consume more than our bodies can absorb, and we excrete the excess, which passes through sewage treatment facilities relatively unchanged. Numerous studies have found quite high levels of caffeine, birth-control hormones, antibiotics, recreational drugs, and other compounds downstream from major cities. This often results in developmental and behavioral changes in fish and other aquatic organisms.

In a nationwide study of pharmaceuticals and hormones in 130 streams, the U.S. Geological Service Scientists found detectable levels of 95 contaminants, including antibiotics, natural and synthetic hormones, detergents, plasticizers, insecticides, and fire retardants (fig. 18.9). One stream had 38 of the compounds tested. Drinking water standards exist for only 14 of the 95 substances. A similar study found the same substances in groundwater, which

is much harder to clean than surface waters. What are the effects of these widely used chemicals on our environment or on people consuming the water? Nobody knows.

Oil spills are common and often intentional

Oil spills are the dominant source of toxic organic compounds into water bodies. Although they are invisible to most of us, oil spills occur constantly around the world, as ships carrying oil leak or spill their cargo, or as ships leak engine oils and fuel oil into the ocean. Onshore, oil pipelines can rupture, spilling oil into surface waters. Likely ruptures in oil pipelines were the chief obstacle in long debates over the proposed Keystone XL pipeline, which was planned to carry Canadian oil across sensitive Midwestern aquifer regions to Texas refineries.

Oil spills can be dramatic, large-scale disasters. Oil well blowouts, such as the 2010 explosion of the *Deepwater Horizon* in the Gulf of Mexico, devastate ecosystems and local fishing economies (fig. 18.10). According to the Smithsonian Institution, however, by far the greatest amount of oil leaked into the ocean every year comes from routine, intentional oil dumping. By their estimates, vehicle maintenance, urban runoff, bilge pumping, and other human activities release about 2.4 billion liters (644 million gal) of oil annually, most of which eventually ends up in the sea. This is ten times as much as from natural oil seeps and about 20 times as much as shipwrecks and well spills. However, as we move into more remote and more dangerous areas for drilling and shipping oil, accidents may become a larger and more destructive cause of ocean pollution (see chapter 19).

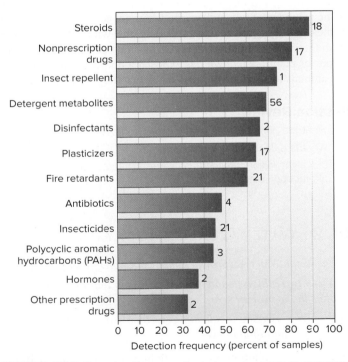

FIGURE 18.9 Detection frequency of organic contaminants in wastewater in a USGS survey. Numbers indicate maximum concentrations, in micrograms/ liter. Dominant substances included caffeine (a steroid), drugs such as ibuprofen and acetaminophen, DEET (an insect repellent), and triclosan, from antibacterial soaps.
Source: USGS

FIGURE 18.10 Oil spills kill thousands of birds, fish, and reptiles, and as we hunt for oil in increasingly remote and extreme places, accidents are likely to increase.
Source: NOAA and Georgia Department of Natural Resources

Sediment also degrades water quality

Rivers have always carried sediment to the oceans, but erosion rates in many areas have been greatly accelerated by human activities. Some rivers carry astounding loads of sediment. Erosion and run-off from croplands contribute about 25 billion metric tons of soil, sediment, and suspended solids to world surface waters each year. Forests, grazing lands, urban construction sites, and other sources of erosion and runoff add at least 50 billion additional tons. This sediment fills lakes and reservoirs, obstructs shipping channels, clogs hydroelectric turbines, and makes purification of drinking water more costly. Sediments smother gravel beds in which insects take refuge and fish lay their eggs. Sunlight is blocked so that plants cannot carry out photosynthesis, and oxygen levels decline. Murky, cloudy water also is less attractive for swimming, boating, fishing, and other recreational uses (fig. 18.11).

In many river systems, though, sediment is also beneficial. Mud carried by rivers nourishes floodplain farm fields. Sediment deposited in the ocean at river mouths creates valuable deltas and islands. The Ganges River, for instance, builds up islands in the Bay of Bengal that are eagerly colonized by land-hungry people of Bangladesh. Louisiana's coastal wetlands require constant additions of sediment from the muddy Mississippi to counteract coastal erosion and subsidence. In recent decades, this sediment source has been lost, as levees direct river sediment out to the Gulf, in order to keep a deep shipping channel open. As a consequence, Louisiana's wetlands are disappearing at a disastrous rate.

Thermal pollution threatens sensitive organisms

Thermal pollution (heated water entering water bodies) results from a variety of sources, such as warm water discharged from power plants, or runoff from warm, paved urban land surfaces. Normally, water temperatures are much more stable than air temperatures, so aquatic organisms tend to be poorly adapted to rapid temperature changes. Raising or lowering the temperature of tropical oceans by even one degree can be lethal to some corals and other reef species. For many organisms, warming water is a problem because it reduces oxygen levels. Cold water holds more dissolved oxygen than warm water. As temperatures increase, species requiring high oxygen levels, such as trout, may not survive.

Industry and power plants often create thermal pollution: The cheapest way to remove heat from an industrial facility is to draw cool water from an ocean, river, lake, or aquifer, run it through a heat exchanger to extract excess heat, and then return the heated water back to the original source. A **thermal plume** of heated water is often discharged into rivers and lakes, where raised temperatures can disrupt natural ecosystems. Warm-water species proliferate, and cold-water species are driven away. Nearly half the water we withdraw is used for industrial cooling. Electric power plants, metal smelters, petroleum refineries, paper mills, food-processing factories, and chemical manufacturing plants all use and release large amounts of cooling water.

To minimize thermal pollution, power plants frequently are required to construct artificial cooling ponds or cooling towers in which heat is released into the atmosphere and water is cooled before being released into natural water bodies.

Some species find thermal pollution attractive. Warm water plumes from power plants often attract fish, birds, and marine mammals that find food and refuge there, especially in cold weather. This artificial environment can be a fatal trap, however. Organisms dependent on the warmth may die if they leave the plume or if the flow of warm water is interrupted by a plant shutdown. Endangered manatees in Florida, for example, are attracted to the abundant food and warm water in power plant thermal plumes and are enticed into spending the winter much farther north than they normally would. On several occasions, a midwinter power plant breakdown has exposed a dozen or more of these rare animals to a sudden thermal shock that they could not survive.

Section Review

1. What are point sources and nonpoint sources?
2. What is a dead zone? What causes it?
3. List eight major categories of water pollutants and give an example for each category.

18.2 WATER QUALITY TODAY

- *The Clean Water Act provides important water quality protection.*
- *About two billion people gained access to improved water supplies in the past 20 years.*
- *Problems remain in cleaning surface waters and providing modern sanitation to everyone.*

Surface-water pollution is often both highly visible and one of the most common threats to environmental quality. In more developed countries, reducing water pollution has been a high priority over the past few decades. Billions of dollars have been spent on control programs, and considerable progress has been made. But much remains to be done; poor water quality often remains a

FIGURE 18.11 A plume of sediment and industrial waste flows from this drainage canal into Lake Erie.
© Lawrence Lowry/Science Source

serious problem. In this section, we will look at progress as well as continuing obstacles in this important area.

The Clean Water Act protects our water

In wealthy countries with strong enforcement of environmental policies, there has been encouraging progress in protecting and restoring water quality in rivers and lakes over the past 40 years. In 1948, only about one-third of Americans were served by municipal sewage systems, and most of those systems discharged sewage without any treatment or with only primary treatment (the bigger lumps of waste are removed). Most people depended on cesspools and septic systems to dispose of domestic wastes.

In 1972, the U.S. Clean Water Act established a National Pollution Discharge Elimination System (NPDES), which requires an easily revoked permit for any industry, municipality, or other entity discharging wastes in surface waters. The permit requires disclosure of discharges, and it gives regulators valuable data and evidence for litigation. As a consequence, only about 10 percent of our water pollution now comes from industrial or municipal point sources. Sewage treatment has been one of the biggest improvements.

Since the Clean Water Act was passed in 1972, the United States has spent more than $180 billion in public funds and perhaps ten times as much in private investments on water pollution control. Most of that effort has been aimed at point sources, especially to build or upgrade thousands of municipal sewage treatment plants. As a result, nearly everyone in urban areas is now served by municipal sewage systems and no major city discharges raw sewage into a river or lake except as overflow during heavy rainstorms.

This campaign has led to significant improvements in surface-water quality in many places. Fish and aquatic insects have returned to waters that formerly were depleted of life-giving oxygen. Swimming and other water-contact sports are again permitted in rivers, lakes, and at ocean beaches that once were closed by health officials.

The Clean Water Act goal of making all U.S. surface waters "fishable and swimmable" hasn't been fully met, but in 1999 the EPA reported that 91.4 percent of all monitored river miles and 87.5 percent of all assessed lake acres are suitable for their designated uses (fig. 18.12). This sounds good, but remember that not all water bodies are monitored. Furthermore, the designated goal for some rivers and lakes is merely to be "boatable." Water quality doesn't have to be very high to be able to put a boat in it. Even in "fishable" rivers and lakes there isn't a guarantee that you can catch anything other than rough fish like carp or bullheads, nor can you be sure that what you catch is safe to eat. Even with billions of dollars of investment in sewage treatment plants, elimination of much of the industrial dumping and other gross sources of pollutants, and a general improvement in water quality, the EPA reports that 21,000 water bodies still do not meet their designated uses.

In 1998, a new regulatory approach to water quality assurance was instituted by the EPA. Rather than issue standards on a river-by-river approach or factory-by-factory permit discharge, the focus is being changed to watershed-level monitoring and protection.

FIGURE 18.12 Not all rivers and lakes are "fishable or swimmable," but we've made substantial progress since the Clean Water Act was passed in 1972.
© Tom Finkle

States are required to identify waters not meeting water quality goals and to specify **total maximum daily loads (TMDLs)** for each pollutant and each listed water body. A TMDL is the amount of a particular pollutant that a water body can receive from both point and nonpoint sources. It considers seasonal variation and includes a margin of safety. Some 4,000 watersheds are monitored for water quality. You can find information about your watershed from the EPA website. The intention of this program is to give the public more and better information about the health of their watersheds. In addition, states will have greater flexibility as they identify impaired water bodies and set priorities, and new tools will be used to achieve goals.

By 1999, all 56 states and territories had submitted TMDL lists, and the EPA had approved most of them. Of the 3.5 million mi (5.6 million km) of rivers monitored, only 300,000 mi (480,000 km) currently fail to meet their clean-water goals. Similarly, of 40 million lake acres (99 million ha), only 12.5 percent (in about 20,000 lakes) failed to meet their goal. To give states more flexibility in planning, the EPA has proposed new rules that include allowances for reasonably foreseeable increases in pollutant loadings to encourage "Smart Growth." In the future, TMDLs also will include load allocations from all nonpoint sources, including air deposition and natural background levels.

Nonpoint sources are difficult to control

The greatest impediments to achieving national goals in water quality in both the United States and Canada are sediment, nutrients, and pathogens (fig. 18.13), especially from nonpoint discharges of pollutants. These sources are harder to identify and to reduce or treat than are specific point sources. About three-fourths of the water pollution in the United States comes from soil erosion, fallout of air pollutants, and surface runoff from urban areas, farm fields, and feedlots. In the United States, as much as 25 percent of

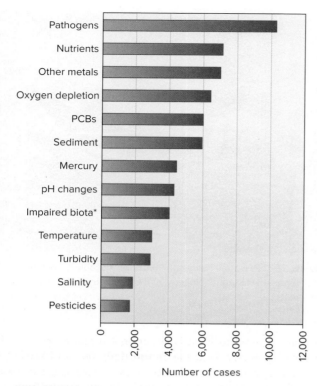

FIGURE 18.13 Leading causes of surface-water impairment in the United States. The * represents undetermined causes.
Source: U.S. EPA, 2015

the 46,800,000 metric tons (52 million tons) of fertilizer spread on farmland each year is carried away by runoff.

Lake Erie provides an example of how complex water quality protection can be. When urban industrial pollution was cleaned up in the 1970s, it was thought the lake's problems were solved. Bacterial counts and algal blooms decreased more than 90 percent. Water that once was murky brown cleared up.

Decades later, runoff from farms is now stimulating algal blooms once again. Dramatic summer air photos of green swirling algae on Lake Erie show the effects of agricultural runoff. It was much easier to stop highly concentrated "end of the pipe" effluents from specific industrial sources than low-level, nonpoint runoff from many individual farms.

Cattle in feedlots produce some 129,600,000 metric tons (144 million tons) of manure each year, and the runoff from these sites is rich in viruses, bacteria, nitrates, phosphates, and other contaminants. A single cow produces about 30 kg (66 lbs) of manure per day, or about as much as that produced by ten humans. Some feedlots have 100,000 animals with little provision for capturing or treating runoff water. Imagine drawing your drinking water downstream from such a facility. Pets also can be a problem. It is estimated that the wastes from about a half million dogs in New York City are disposed of primarily through storm sewers, and therefore do not go through sewage treatment.

Loading of both nitrates and phosphates in surface water has decreased from point sources but has increased about fourfold

since 1972 from nonpoint sources. Fossil fuel combustion has become a major source of nitrates, sulfates, arsenic, cadmium, mercury, and other toxic pollutants that find their way into water. Carried to remote areas by atmospheric transport, these combustion products now are found nearly everywhere in the world. Toxic organic compounds, such as DDT, PCBs, and dioxins, also are transported long distances by wind currents.

Water pollution is especially serious in developing countries

Japan, Australia, and most of western Europe also have improved surface-water quality in recent years. Sewage treatment in the wealthier countries of Europe generally equals or surpasses that in the United States. Sweden, for instance, serves 98 percent of its population with at least secondary sewage treatment (compared with 70 percent in the United States), and the other 2 percent have primary treatment. Poorer countries, unfortunately, have much less to spend on sanitation (fig. 18.14).

As we saw in Chapter 17, for example, more than 100 million Chinese live in areas without sufficient fresh water. Pollution makes much of the limited water unusable. An estimated 70 percent of China's surface water is unsafe for human consumption, and that the water in half the country's major rivers is so contaminated that it's unsuited for any use, even agriculture. Economic growth has been a much bigger priority than environmental quality (fig. 18.15).

According to the Chinese Environmental Protection Agency, the country's ten worst polluted cities are all in Shanxi Province. Factories have been allowed to exceed pollution discharges with impunity. For example, 3 million tons of wastewater are produced every day in the province, with two-thirds of it discharged directly into local rivers without any treatment. Locals complain that the rivers, which once were clean and fresh, are now visibly polluted with industrial waste. Among the 26 rivers in the province, 80 percent are rated Grade V (unfit for any human use) or higher. More than

FIGURE 18.14 Ditches in this Haitian slum serve as open sewers into which all manner of refuse and waste are dumped. The health risks of living under these conditions are severe.
© *Robert Nickelesberg/Getty Images*

FIGURE 18.15 As high-rise apartment blocks move in to China's agricultural areas, development takes priority over environmental protection. Half of the water in China's major rivers is too polluted to be suitable for any human use.
© Mary Ann Cunningham

half the wells in Shanxi are reported to have dangerously high arsenic levels. Many of the 180,000 reported public protests in China in 2012 involved complaints about air and water pollution. In 2013, more than 16,000 dead pigs were pulled from the Hangpu River just upstream from Shanghai's drinking water intake. They apparently were dumped by farmers to avoid fines for having diseased pigs.

As discussed in the opening case study for this chapter, water quality remains a severe problem in India. Two-thirds of India's surface water is so contaminated that even coming into contact with it is considered dangerous to human health. Yet millions of people drink and bathe in this water. Although the Ganges receives most attention, its tributaries are often much worse. Consider the Yamuna River, which flows through New Delhi and past the magnificent Taj Mahal in Agra. About 57 million people depend on the Yamuna for agricultural, domestic, and industrial use. Much of the runoff from these activities goes back into the river either untreated or only partially cleaned. Coliform bacterial counts can be millions of times the level considered safe for drinking or bathing. Although the Indian government has spent more than $500 million in recent years to upgrade the sewage system, the Yamuna, and Ganges into which it flows, remain badly polluted.

Water treatment improves safety

Progress is not impossible, despite the challenges. China has spent at least $125 billion over the past five years to reduce water pollution and provide clean water to its cities and villages. Already there are indications of success. In 1990, the United Nations declared, as one of its Millennium Development Project targets, the goal of reducing by half the proportion of people without sustainable access to safe drinking

water by 2015. That goal was met in 2010. Over those 20 years, more than 2 billion people gained access to "improved" water. Half of that progress occurred in India and China.

"Improved" water sources are not necessarily free of all pathogens and contaminants. Health agencies classify surface water and hand-dug, shallow wells as "unimproved" and especially likely to be polluted. "Improved" water supplies include tube (drilled and sealed) wells, water piped from springs and other clean sources, and treated water supplies. These sources may not be completely safe to drink, but they're usually better than lakes, rivers, or ponds. Improved water also doesn't necessarily mean it's available in the home. You may still have to carry it from a central spigot or well. Still, that's better than having to carry it from a source hours away. Having water piped into the home saves hours of labor and frees women and children for studies or income-producing work.

Overall, about 96 percent of all urban areas in the world enjoy improved water. The greatest remaining problems are in rural areas, especially in sub-Saharan Africa (fig. 18.16). Although we've made progress in supplying this vital resource, about 780 million people still lack access to improved water. Two-thirds of that number are in just ten countries (fig. 18.17).

Water treatment is not always adequate even in wealthier countries. Data collected by the EPA in 2008 show that about 30 million people get water from community systems that don't meet all health-based drinking water standards. Most of these systems are small, under-funded systems. Every year epidemiologists estimate that around 1.5 million Americans fall ill from infections caused by fecal contamination. Preventive measures such as protecting water sources and aquifer recharge zones, providing basic treatment for all systems, installing modern technology and distribution networks, consolidating small systems, and strengthening the Clean Water Act and the Safe Drinking Water Act would cost far less. Unfortunately, in the present climate of budget cutting and antiregulation, these steps seem unlikely.

In 2016, water contamination put the city of Flint, Michigan, in the news. Two years earlier, as part of a cost-saving measure,

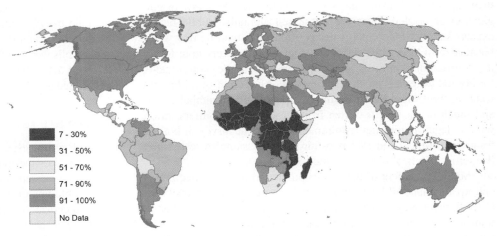

■	7 - 30%
■	31 - 50%
□	51 - 70%
▨	71 - 90%
▨	91 - 100%
□	No Data

FIGURE 18.16 Improved drinking water coverage in rural areas. Hundreds of millions of rural residents, especially in sub-Saharan Africa, lack access to safe, clean drinking water.
Sources: UNESCO and WHO, 2012

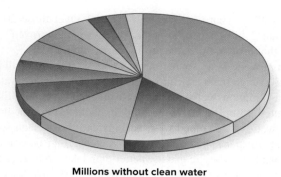

Millions without clean water

▮ Rest of World 292	▮ Democratic Republic of the Congo 36
▮ China 119	▮ Bangladesh 28
▮ India 97	▮ United Republic of Tanzania 21
▮ Nigeria 66	▮ Sudan 18
▮ Ethiopia 46	▮ Kenya 17
▮ Indonesia 43	

FIGURE 18.17 Just ten countries represent two-thirds of the 780 million people who still lack access to improved drinking water sources. *Sources: UNESCO and WHO, 2012*

the city had begun drawing water from the chlorine-contaminated Flint River, rather than its cleaner Lake Huron source. Because chlorine leads to corrosion of aging lead pipes, most cities are required to add anti-corrosion agents (mainly orthophosphate) to the water. Flint overlooked this step, and household taps began to deliver water contaminated with lead from city water lines. Two years later, as Flint residents reported widespread illness and rashes, and untold rates of impaired child development, the crisis erupted, rapidly spreading across the state and prompting new investigations of water quality in cities across the U.S.

Is bottled water safer?

Every year, Americans buy about 28 billion bottles of water at a cost of about $15 billion with the mistaken belief that it's safer than tap water. Worldwide, some 160 billion liters (42 billion gallons) of bottled water are consumed annually. Public health experts say that municipal water is often safer than bottled water because most large cities test their water supplies every hour for up to 25 different chemicals and pathogens, while the requirements for bottled water are much less rigorous. About half of all bottled water in the United States is simply reprocessed municipal water, and much of the rest is drawn from groundwater aquifers, often with no legal monitoring requirements. A recent survey of bottled water in China found that two-thirds of the samples tested had dangerous levels of pathogens and toxins.

Some 80 percent of the bottles purchased in the United States end up in a landfill, even though the plastic is easily recyclable. Overall, the average energy cost to make the plastic, fill the bottle, transport it to market, and then deal with the waste would be "like filling up a quarter of every bottle with oil," says water expert Peter Gleick. Furthermore, it takes three to five times as much water to make the bottles as they hold. In blind tasting tests, most adults

either can't tell the difference between municipal and bottled water, or they actually prefer municipal water. Furthermore, if water is held in a plastic bottle for weeks or months, plasticizers and other toxic chemicals can leach from the bottle into the water (see chapter 8).

Whereas municipal water is monitored and treated, bottled water mainly benefits the corporations that sell it, and its safety is less sure than city water. Generally, drinking tap water will do a favor for your environment, your budget, and, possibly, your health.

Groundwater is hard to monitor and clean

About half the people in the United States, including 95 percent of those in rural areas, depend on underground aquifers for their drinking water. This vital resource is threatened in many areas by overuse and pollution and by a wide variety of industrial, agricultural, and domestic contaminants. For decades it was widely assumed that groundwater was impervious to pollution because soil would bind chemicals and cleanse water as it percolated through. Springwater or artesian well water was considered to be the definitive standard of water purity. But that is no longer true in many areas.

The U.S. EPA estimates that every day some 4.5 trillion liters (1.2 trillion gal) of contaminated water seep into the ground in the United States from septic tanks, cesspools, municipal and industrial landfills and waste disposal sites, surface impoundments, agricultural fields, forests, and abandoned wells (fig. 18.18). The most toxic of these are probably waste disposal sites. Agricultural chemicals and wastes are responsible for the largest total volume of pollutants and area affected. Because deep underground aquifers often have residence times of thousands of years, many contaminants are extremely stable once underground. It is possible, but expensive, to pump water out of aquifers, clean it, and then pump it back.

In farm country, especially in the Midwest's Corn Belt, fertilizers and pesticides commonly contaminate aquifers and wells. Herbicides such as atrazine and alachlor are widely used on corn and soybeans and show up in about half of all wells in Iowa, for example. Nitrates from fertilizers often exceed safety standards in rural drinking water. These high nitrate levels are particularly dangerous to infants (nitrate combines with hemoglobin in the blood and results in "blue baby" syndrome). They also are transformed into cancer-causing nitrosamines in the human gut. In Florida, 1,000 drinking water wells were shut down by state authorities because of excessive levels of toxic chemicals, mostly ethylene dibromide (EDB), a pesticide used to kill nematodes (roundworms) that damage plant roots.

There are few controls on ocean pollution

Coastal zones, especially bays, estuaries, shoals, and reefs near large cities or the mouths of major rivers, often are overwhelmed by human-caused contamination. Suffocating and sometimes poisonous blooms of algae regularly deplete ocean waters of oxygen and kill enormous numbers of fish and other marine life. High levels of toxic chemicals, heavy metals, disease-causing organisms, oil, sediment, and plastic refuse are adversely affecting some of the most attractive and productive ocean regions. The potential losses caused by this pollution amount to billions of dollars each year.

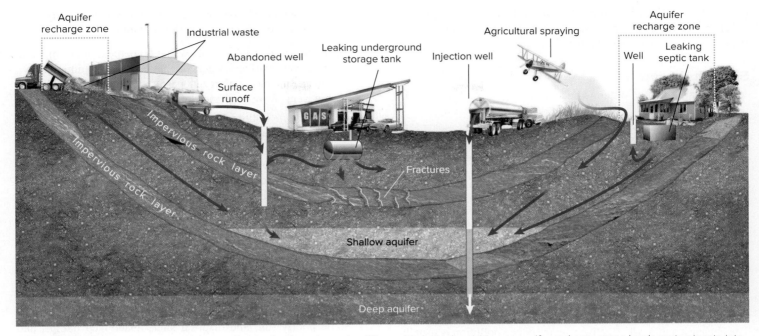

FIGURE 18.18 Sources of groundwater pollution. Septic systems, landfills, and industrial activities on aquifer recharge zones leach contaminants into aquifers. Wells provide a direct route for injection of pollutants into aquifers.

Discarded bits of plastic debris are lightweight and non-biodegradable. They are carried thousands of miles on ocean currents and last for years (fig. 18.19). Even the most remote beaches of distant islands are likely to have bits of polystyrene foam containers or polyethylene packing material that were discarded half a world away. It has been estimated that some 6 million metric tons of plastic bottles, packaging material, and other litter are tossed from ships every year into the ocean, where they ensnare and choke seabirds, mammals (fig. 18.20), and even fish. In one day, volunteers in Texas

FIGURE 18.19 Beach pollution, including garbage, sewage, and contaminated runoff, is a growing problem associated with ocean pollution.
© John Lund/The Image Bank/Getty Images

FIGURE 18.20 A deadly necklace. Marine biologists estimate that castoff nets, plastic beverage yokes, and other packing residue kill hundreds of thousands of birds, mammals, and fish each year.
Courtesy of Joe Lucas/Marine Entanglement Research Program/National Marine Fisheries Service NOAA

gathered more than 300 tons of plastic refuse from Gulf Coast beaches. For further discussion of ocean pollution and the Great Pacific garbage patch, see chapter 21.

Section Review

1. How has the Clean Water Act reduced sewage discharge into public waterways?
2. Which countries have had the greatest success in providing safe water to their citizens?
3. What geographic region lacks safe water for the greatest remaining proportion of its population?

18.3 WATER POLLUTION CONTROL

- *Municipal sewage treatment involves primary, secondary, and tertiary treatment.*
- *Natural or artificial wetlands can treat wastes economically.*
- *Bioremediation can be effective in urban settings.*

The cheapest and most effective way to reduce pollution is usually to avoid producing it or releasing it in the first place. Elimination of lead from gasoline has dramatically cut the amount of lead in surface waters in the United States. Banning DDT and PCBs in the 1970s has resulted in significant reductions in levels in wildlife. Mercury in the U.S. waterways has declined along with emissions from coal power plants, the source of over half of U.S. mercury emissions. Better land management has succeeded in reducing nutrient loads and sediment in critical watersheds, especially Chesapeake Bay and New York's Catskills mountains, which provides drinking water to New York City. In the Catskills, economic incentives to help farmers control runoff have allowed New York City to avoid building filtration plants that would have cost billions of dollars.

Industry can reduce pollution by recycling or reclaiming materials that otherwise might be discarded in the waste stream. Both of these approaches usually have economic as well as environmental benefits. It turns out that a variety of valuable metals can be recovered from industrial wastes and reused or sold for other purposes. The company benefits by having a product to sell, and the municipal sewage treatment plant benefits by not having to deal with highly toxic materials mixed in with millions of gallons of other types of wastes.

Controlling nonpoint sources requires land management

Among the greatest remaining challenges in water pollution control are diffuse, nonpoint pollution sources. Unlike point sources, such as sewer outfalls or industrial discharge pipes, which represent both specific locations and relatively continuous emissions, nonpoint sources have many origins and numerous routes by which contaminants enter ground and surface waters. It is difficult to identify—let alone monitor and control—all these sources and routes. Some main causes of nonpoint pollution are:

- *Agriculture:* The EPA estimates that 60 percent of all impaired or threatened surface waters are affected by sediment from eroded fields and overgrazed pastures; fertilizers, pesticides, and nutrients from croplands; and animal wastes from feedlots.
- *Urban runoff:* Pollutants carried by runoff from streets, parking lots, and industrial sites contain salts, oily residues, rubber, metals, and many industrial toxins (fig. 18.21). Yards, golf courses, park lands, and urban gardens often are treated with far more fertilizers and pesticides per unit area than farmlands. Excess chemicals are carried by storm runoff into waterways.
- *Construction sites:* New buildings and land development projects such as highway construction affect relatively small areas but produce vast amounts of sediment, typically 10 to 20 times as much per unit area as farming.
- *Land disposal:* When done carefully, land disposal of certain kinds of industrial waste, sewage sludge, and biodegradable garbage can be a good way to dispose of unwanted materials. Some poorly run land disposal sites, abandoned dumps, and leaking septic systems, however, contaminate local waters.

Generally, agricultural soil conservation methods (chapter 9) also help protect water quality. Applying precisely determined amounts of fertilizer, irrigation water, and pesticides saves money and reduces contaminants entering the water. Preserving wetlands, which act as

FIGURE 18.21 People often dump waste oil and other pollutants into street drains without thinking about where their wastes go. Painting reminders, such as this one, is a good project for students and youth groups.
© William P. Cunningham

natural processing facilities for removing sediment and contaminants, also helps protect surface water and groundwater.

In many American cities, storm sewers (which carry runoff from streets and parking lots) are connected to sanitary sewers, which drain bathrooms and kitchens. Storm sewers are routed to water treatment plants because runoff from streets, yards, and parking lots contains a tremendous amount of contaminants, such as litter, fertilizers, pesticides, oils, rubber, tars, gasoline residue, and pet waste. In moderate rainfall, combined sewers protect surface waters. Heavy storms often overload the system, however. Treatment plants overwhelmed by storm runoff then release everything untreated—raw sewage and toxic surface runoff—directly into lakes and rivers. These events are called combined sewer overflows (or CSOs), and in many cities, CSOs are the principal cause of surface water contamination. To prevent CSOs, cities are spending hundreds of millions of dollars to separate storm and sanitary sewers. These are huge, disruptive projects. When they are finished, surface runoff is diverted into surface waters, where other pollution problems may arise, but direct release of raw sewage is less frequent.

Reducing materials carried by storm runoff is critical for protecting water quality. All of us can try to reduce the debris, litter, and waste that enters storm drains—that includes everything from candy wrappers to piles of autumn leaves, which contribute nutrients to lakes and rivers. We should also minimize the use of fertilizers and pesticides in yards, which become water contaminants, and never pour into street drains any waste oil, paint, or other substances that you wouldn't want to see fish swimming in. Regular street sweeping greatly reduces water contamination.

A good example of watershed management is seen in Chesapeake Bay, the United States' largest estuary. Once fabled for its abundant oysters, crabs, shad, striped bass, and other valuable fisheries, the Bay had deteriorated seriously by the early 1970s. Citizens' groups, local communities, state legislatures, and the federal government together established an innovative pollution-control program that made the bay the first estuary in America targeted for protection and restoration (see chapter 3).

Controlling runoff from agricultural and urban lands is central to the Chesapeake Bay restoration plan. Pollution prevention measures such as banning phosphate detergents also are important, as are upgrading wastewater treatment plants and making sure plants are managed well. Since the 1980s, annual phosphorous discharges into the bay dropped 40 percent. Nitrogen levels, however, have remained constant or have even risen in some tributaries. The bay is still decades away from meeting its goals, which include reducing nitrogen and phosphate levels and restoring fisheries, but as former EPA administrator Carol Browner has said, the project demonstrates the "power of cooperation" in environmental protection.

Human waste disposal occurs naturally when concentrations are low

As we have already seen, human and animal wastes usually create the most serious health-related water pollution problems. More than 500 types of disease-causing (pathogenic) bacteria, viruses, and parasites can travel from human or animal excrement through water. In this section, we will look at how to prevent the spread of these diseases.

Natural Processes

In the poorer countries of the world, most rural people simply go out into the fields and forests to relieve themselves as they have always done. Where population densities are low, natural processes eliminate wastes quickly, making this a feasible method of sanitation. The high population densities of cities make this practice unworkable, however. Even major cities of many less-developed countries are often littered with human waste that has been left for rains to wash away or for pigs, dogs, flies, beetles, or other scavengers to consume. This is a major cause of disease, as well as being extremely unpleasant. Studies have shown that a significant portion of the airborne dust in Mexico City is actually dried, pulverized human feces.

Where intensive agriculture is practiced—especially in wet rice paddy farming in Asia—it has long been customary to collect "night soil" (human and animal waste) to be spread on the fields as fertilizer. This waste is a valuable source of plant nutrients, but it is also a source of disease-causing pathogens in the food supply. It is the main reason that travelers in less-developed countries must be careful to surface sterilize or cook any fruits and vegetables they eat. Collecting night soil for use on farm fields was common in Europe and America until about 100 years ago, when the association between pathogens and disease was recognized.

Until about 70 years ago, most rural American families and many residents of towns and small cities depended on a pit toilet or "outhouse" for waste disposal. Untreated wastes tended to seep into the ground, however, and pathogens sometimes contaminated drinking water supplies. The development of septic tanks and properly constructed drain fields considerably improved public health (fig. 18.22). In a typical septic system, wastewater is first drained into a septic tank. Grease and oils rise to the top and solids settle to the bottom, where they are subject to bacterial decomposition. The clarified effluent from the septic tank is channeled out through a drainfield of small perforated pipes embedded in gravel just below the surface of the soil. The rate of aeration is high in this drainfield so that pathogens (most of which are anaerobic) will be killed, and soil microorganisms can metabolize any nutrients carried by the water. Excess water percolates up through the gravel and evaporates. Periodically, the solids in the septic tank are pumped out into a tank truck and taken to a treatment plant for disposal.

Where land is available and population densities are not too high, this can be an effective method of waste disposal. It is widely used in rural areas, but aging, leaky septic systems can be a huge cumulative problem. The Chesapeake Bay watershed has 420,000 individual septic systems, which constitute a major source of nutrients that create a dead zone over much of the Bay. Maryland alone plans to spend $7.5 million annually to upgrade failing septic systems.

Municipal Sewage Treatment

Over the past 100 years, sanitary engineers have developed ingenious and effective municipal wastewater treatment systems to protect human health, ecosystem stability, and water quality. This

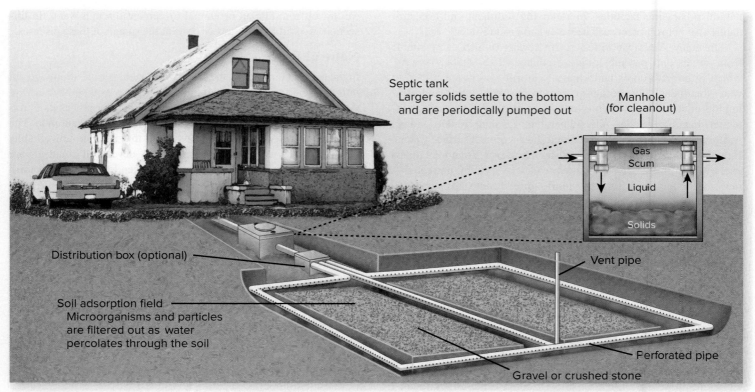

FIGURE 18.22 A domestic septic tank and drain field system for sewage and wastewater disposal. To work properly, a septic tank must have healthy microorganisms, which digest toilet paper and feces. For this reason, antimicrobial cleaners and chlorine bleach should never be allowed down the drain.

Labels in figure:
- Septic tank: Larger solids settle to the bottom and are periodically pumped out
- Manhole (for cleanout)
- Gas
- Scum
- Liquid
- Solids
- Distribution box (optional)
- Vent pipe
- Soil adsorption field: Microorganisms and particles are filtered out as water percolates through the soil
- Perforated pipe
- Gravel or crushed stone

topic is an important part of pollution control, and is a central focus of every municipal government; therefore, let's look more closely at how a typical municipal sewage treatment facility works.

Primary treatment is the first step in municipal waste treatment. It physically separates large solids from the waste stream. As raw sewage enters the treatment plant, it passes through a metal grating that removes large debris (fig. 18.23). A moving screen then filters out smaller items. Brief residence in a grit tank allows sand and gravel to settle. The waste stream then moves to the primary sedimentation tank, where about half the suspended organic solids settle to the bottom as sludge. Many pathogens remain in the effluent, which is not yet safe to discharge into waterways or onto the ground.

Secondary treatment consists of biological degradation of the dissolved organic compounds. The effluent from primary treatment flows into a trickling filter bed, an aeration tank, or a sewage lagoon. The trickling filter is simply a bed of stones or corrugated plastic sheets through which water drips from a system of perforated pipes or a sweeping overhead sprayer. Bacteria and other microorganisms in the bed catch organic material as it trickles past and aerobically decompose it.

Aeration tank digestion is also called the activated sludge process. Effluent from primary treatment is pumped into the tank and mixed with a bacteria-rich slurry. Air pumped through the mixture encourages bacterial growth and decomposition of the organic material (fig. 18.24a and b). Water flows from the top of

the tank, and sludge is removed from the bottom. Some of the sludge is added to incoming primary effluent, to inoculate it with useful bacteria. The remainder would be valuable fertilizer if it were not contaminated by metals, toxic chemicals, and pathogenic organisms.

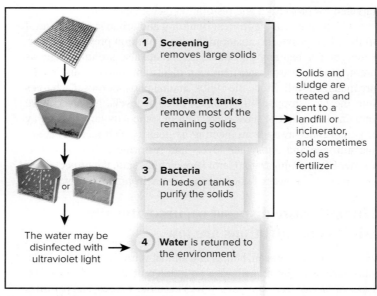

1. **Screening** removes large solids
2. **Settlement tanks** remove most of the remaining solids
3. **Bacteria** in beds or tanks purify the solids
4. **Water** is returned to the environment

Solids and sludge are treated and sent to a landfill or incinerator, and sometimes sold as fertilizer

The water may be disinfected with ultraviolet light

FIGURE 18.23 The process of conventional sewage treatment.

(a)

(b)

FIGURE 18.24 In conventional sewage treatment, aerobic bacteria digest organic materials in trickling filter beds in high-pressure aeration tanks. This is described as secondary treatment.
(a) © ThinkStock/Corbis RF, (b) © Steve Allen/Brand X Pictures RF

The toxic content of most sewer sludge necessitates disposal by burial in a landfill or incineration. Sludge disposal is a major cost in most municipal sewer budgets. In some communities, this is accomplished by land farming, composting, or anaerobic digestion, but these methods don't inactivate metals and some other toxic materials.

Where space is available for sewage lagoons, the exposure to sunlight, algae, aquatic organisms, and air does the same job more slowly but with less energy cost. Effluent from secondary treatment processes is usually disinfected with chlorine, UV light, or ozone to kill harmful bacteria before it is released to a nearby waterway.

Tertiary treatment removes plant nutrients, especially nitrates and phosphates, from the secondary effluent. Although wastewater is usually free of pathogens and organic material after secondary treatment, it still contains high levels of inorganic nutrients, such as nitrates and phosphates. When discharged into surface waters, these nutrients stimulate algal blooms and eutrophication. To preserve water quality, these nutrients also must be removed. Passage through a wetland or lagoon can accomplish this. Alternatively, chemicals often are used to bind and precipitate nutrients.

Low-Cost Waste Treatment

The municipal sewage systems used in developed countries are often too expensive to build and operate in the developing world where low-cost, low-tech alternatives for treating wastes are needed (see the Exploring Science box later in this section). One option is **effluent sewerage,** a hybrid between a traditional septic tank and a full sewer system. A tank near each dwelling collects and digests solid waste just like a septic system. Rather than using a drainfield, however, to dispose of liquids—an impossibility in crowded urban areas—effluents are pumped to a central treatment plant. The tank must be emptied once a year or so, but because only liquids are treated by the central facility, pipes, pumps, and treatment beds can be downsized and the whole system is much cheaper to build and run than a conventional operation.

Another alternative is to use natural or artificial wetlands to dispose of wastes. **Constructed wetlands** can cut secondary treatment costs to one-third of mechanical treatment costs, or less. Variations on this design are now operating in many places (fig. 18.25). Effluent from these operations can be used

FIGURE 18.25 This constructed wetland purifies water and provides an attractive landscape at the El Monte Sagrado Resort in Taos, New Mexico.
© William P. Cunningham

Inexpensive Water Purification

When Ashok Gadgil was a child in Bombay, India, five of his cousins died in infancy from diarrhea spread by contaminated water. Although he didn't understand the implications of those deaths at the time, as an adult he realized how heartbreaking and preventable those deaths were. After earning a degree in physics from the University of Bombay, Gadgil moved to the University of California at Berkeley, where he was awarded a PhD in 1979 and is now a professor of civil and environmental engineering. His research focuses on solar energy and pollution issues.

But Dr. Gadgil wanted to do something about the problem of waterborne diseases and pollutants in India and other developing countries. Although progress has been made in bringing clean water to poor people in many countries, at least 780 million people worldwide still lack access to safe drinking water. After studying ways to purify water, he decided that UV light treatment had the greatest potential for removing infectious agents. It requires far less energy than boiling, and it takes less sophisticated chemical monitoring than chlorination.

There are many existing UV water treatment systems, but they generally involve water flowing around an unshielded fluorescent lamp. Minerals in the water collect on the glass lamp and must be removed regularly to maintain effectiveness. Regular disassembly, cleaning, and reassembly of the apparatus are difficult in primitive conditions. The solution, Gadgil realized, was to mount the UV source above the water where it couldn't develop mineral deposits. He designed a

Women carry water from the village WaterHealth kiosk. More than 6 million people's lives have been improved by this innovative system of water purification.
© WaterHealth India

system in which water flows through a shallow, stainless steel trough. The apparatus can be gravity fed and requires only a car battery as an energy source.

Another crisis in South Asia is arsenic contamination in well water. About 100 million people in Bangladesh and West Bengal are exposed to dangerous levels of arsenic in their drinking water (see section 18.1 for more details on this disaster). Arsenic can be removed from water by being absorbed by iron or aluminum oxides. Professor Gadgil and his team have devised an inexpensive, effective way to do this. They pass a weak electric current (less than 3 volts) through iron plates as the water passes over them. This releases iron oxides (rust) that bind to arsenic. The heavy iron particles can then be filtered cut easily. A combined UV treatment and electrochemical arsenic precipitation system can purify 15 liters (4 gallons) of water per minute, killing more than 99.9 percent of all bacteria and viruses and removing an equal percentage of arsenic. This produces enough clean water for a village of 1,000 people at a cost of only about 5 cents per ton (950 liters).

WaterHealth International, the company founded to bring this technology to market, now makes several versions of Gadgil's treatment system for different applications. A popular version provides a complete water purification system, including a small kiosk, jugs for water distribution, and training on how to operate everything. A village-size system costs about $5,000. Grants and loans are available for construction, but villagers own and run the facility to ensure there is local responsibility. Each family in the cooperative pays about $1 per month for pure water. These systems have been installed in thousands of villages in India, Bangladesh, Africa, and the Philippines. Currently, about 6.6 million people are getting clean, healthy water at an easily affordable price from the simple systems Dr. Gadgil invented.

to irrigate crops or even raise fish for human consumption if care is taken to first destroy pathogens. Usually 20 to 30 days of exposure to sun, air, and aquatic plants is enough to make the water safe. These systems also can make an important contribution to human food supplies. A 2,500-ha (6,000-acre) waste-fed aquaculture facility in Calcutta, for example, supplies about 7,000 metric tons of fish annually to local markets. The World Bank estimates that some 3 billion people will be without sanitation services by the middle of the next century under a business-as-usual scenario (fig. 18.26). With investments in innovative programs, however, sanitation could be provided to about half those people and thus a great deal of misery and suffering could be avoided.

Water remediation may involve containment, extraction, or phytoremediation

Remediation means finding remedies for problems. Just as there are many sources for water contamination, there are many ways to clean it up. New developments in environmental engineering are providing promising solutions to many water pollution problems.

Containment methods confine or restrain dirty water or liquid wastes *in situ* (in place) or cap the surface with an impermeable layer to divert surface water or groundwater away from the site and to prevent further pollution. Where pollutants are buried too deeply to be contained mechanically, materials sometimes can be injected to precipitate, immobilize, chelate, or solidify them.

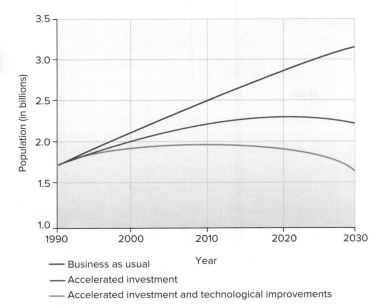

Population (in billions)

— Business as usual
— Accelerated investment
— Accelerated investment and technological improvements

FIGURE 18.26 World population without adequate sanitation—three scenarios in the year 2030. If business as usual continues, more than 3 billion people will lack safe sanitation. Accelerated investment in sanitation services could lower this number. Higher investment, coupled with technological development, could keep the number of people without adequate sanitation from growing even though the total population increases.
Source: World Bank estimates based on research paper by Dennis Anderson and William Cavendish, "Efficiency and Substitution in Pollution Abatement: Simulation Studies in Three Sectors."

Bentonite slurries, for instance, can effectively stabilize liquids in porous substrates. Similarly, straw or other absorbent material is spread on surface spills to soak up contaminants.

Extraction techniques pump out polluted water so it can be treated. Many pollutants can be destroyed or detoxified by chemical reactions that oxidize, reduce, neutralize, hydrolyze, precipitate, or otherwise change their chemical composition. Where chemical techniques are ineffective, physical methods may work. Solvents and other volatile organic compounds, for instance, can be stripped from solution by aeration and then burned in an incinerator. Some contaminants can be removed by semipermeable membranes or resin filter beds that bind selectively to specific materials. Some of the same techniques used to stabilize liquids *in situ* can also be used *in vitro* (in a reaction vessel). Metals, for instance, can be chelated or precipitated in insoluble, inactive forms.

Often, living organisms can be used effectively and inexpensively to clean contaminated water. We call this bioremediation (chapter 21). For instance, restored wetlands along streambanks or lake margins can be very effective in filtering out sediment and removing pollutants. They generally cost far less than mechanical water treatment facilities, and they provide wildlife habitat as well.

Lowly duckweed (*Lemna* sp.), the green scum you often see covering the surface of eutrophic ponds, grows fast and can remove large amounts of organic nutrients from water. Under optimal conditions, a few square centimeters of these tiny plants can grow to cover nearly a hectare (about 2.5 acres) in four months. Large duckweed lagoons are being used as inexpensive, low-tech

sewage treatment plants in developing countries. Where conventional wastewater purification typically costs $300 to $600 per person served, a duckweed system can cost one-tenth as much. The duckweed can be harvested and used as feed, fuel, or fertilizer. Up to 35 percent of its dry mass is protein—about twice as much as alfalfa, a popular animal feed.

Where space for open lagoons is unavailable, bioremediation can be carried out in tanks or troughs. This has the advantage of controlling conditions more precisely and doesn't release organisms into the environment. Some of the most complex, holistic systems for water purification are designed by Ocean Arks International (OAI) in Falmouth, Massachusetts. Their "living machines" combine living organisms—chosen to perform specific functions—in contained environments. In a typical living machine, water flows through a series of containers, each with a distinct ecological community designed for a particular function. Wastes generated by the inhabitants of one vessel become the food for inhabitants of another. Sunlight provides the primary source of energy.

OAI has created or is in the process of building water treatment plants in a dozen states and foreign countries. Designs range from remediating toxic wastes from Superfund sites to simply treating domestic wastes. Starting with microorganisms in aerobic and anaerobic environments where different kinds of wastes are metabolized or broken down, water moves through a series of containers containing hundreds of different kinds of plants and animals, including algae, rooted aquatic plants, clams, snails, and fish, each chosen to provide a particular service. Technically, the finished water is drinkable, although few people feel comfortable drinking it. More often, the final effluent is used to flush toilets or for irrigation. Called ecological engineering, this novel approach can save resources and money, as well as help clean up our environment (fig. 18.27).

FIGURE 18.27 Bioreactors, such as these "living machines" from Ocean Arks International use living communities that mimic natural ecosystems to treat water. Polluted water (right flask) can be purified to drinking water quality (left flask).
© *Peter Essick/Aurora/Getty Images*

Steps You Can Take to Improve Water Quality

Individual actions have important effects on water quality. Here are some steps you can take to make a difference.

- Compost your yard waste and pet waste. Nutrients from decayed leaves, grass, and waste are a major urban water pollutant. Many communities have public compost sites available.

- Don't fertilize your lawn or apply lawn chemicals. Untreated grass can be just as healthy, and it won't poison your pets or children.

- Make sure your car doesn't leak fluids, oil, or solvents on streets and parking lots, from which contaminants wash straight into rivers and lakes. Recycle motor oil at a gas station or oil-change shop.

- Create a "rain garden" to capture and filter surface runoff. This helps recharge groundwater aquifers and keeps nutrients and toxins out of rivers and lakes (fig. 18.28).

- Don't buy lawn mowers, personal watercraft, or other vehicles with two-cycle engines, which release abundant fuel and oil into air and water. Instead, buy more efficient four-stroke engines.

- Visit your local sewage treatment plant. Often, public tours are available or group tours can be arranged, and these sites can be fascinating.

- Keep informed about water policy debates at local and federal levels. Policies change often, and public input is important.

FIGURE 18.28 A rain garden is a shallow depression situated to collect runoff from streets or parking lots. It's planted with species that can survive in saturated soils. This vegetation helps evaporate and cleanse runoff, while temporary storage in the basin allows groundwater recharge. You might build a rain garden in your yard, or on your campus, or elsewhere in your city.
© William P. Cunningham

Section Review

1. Describe the process of municipal sewage treatment.
2. What is a constructed wetland?
3. How does bioremediation clean water?

18.4 WATER LEGISLATION

- *The Clean Water Act has been successful in many ways.*
- *The Clean Water Act is disputed and needs updating.*
- *Other laws also help protect water quality.*

As the opening case study for this chapter shows, water pollution control has been among the most broadly popular and effective of all environmental legislation in the United States. It has not been without controversy, however. In this section, we will look at some of the major issues concerning water quality laws and their provisions (table 18.2).

The Clean Water Act was ambitious, bipartisan, and largely successful

Passage of the U.S. Clean Water Act of 1972 was a bold, bipartisan step determined to "restore and maintain the chemical, physical, and biological integrity of the Nation's waters" that made clean water a national priority. Along with the Endangered Species Act and the Clean Air Act, this is one of the most significant and effective pieces of environmental legislation ever passed by the U.S. Congress. It also is an immense and complex law, with more than 500 sections, regulating everything from urban runoff, industrial discharges, and municipal sewage treatment to land-use practices and wetland drainage.

The ambitious goal of the Clean Water Act was to return all U.S. surface waters to "fishable and swimmable" conditions. For specific "point" sources of pollution, such as industrial discharge pipes or sewage outfalls, the act requires discharge permits and **best practicable control technology (BPT).** It sets national goals of **best available, economically achievable technology (BAT)** for toxic substances and zero discharge for 126 priority toxic pollutants. As we discussed earlier in this chapter, these regulations have had a positive effect on water quality. Although surface waters are not yet swimmable or fishable everywhere, surface-water quality in the United States has significantly improved, on average, over the past quarter century. Perhaps the most important result of the act has been the investment of $54 billion in federal funds and more than $128 billion in state and local funds for municipal sewage treatment facilities.

Not everyone is happy with the Clean Water Act. Industries, state and local governments, farmers, land developers, and others who have been forced to change their operations or spend money on water protection often object to the costs of controlling pollution. One of the most controversial provisions of the act has been Section 404, which regulates the draining or filling of wetlands. Although the original bill mentions wetlands only briefly, this section has evolved through judicial interpretation and regulatory policy to become one of the principal federal tools for wetland protection. Many people applaud the protection granted to these

Table 18.2 Some Important U.S. and International Water Quality Legislation

1. *Federal Water Pollution Control Act* (1972). Established uniform nationwide controls for each category of major polluting industries.

2. *Marine Protection Research and Sanctuaries Act* (1972). Regulates ocean dumping and established sanctuaries for the protection of endangered marine species.

3. *Ports and Waterways Safety Act* (1972). Regulates oil transport and the operation of oil handling facilities.

4. *Safe Drinking Water Act* (1974). Requires minimum safety standards for every community water supply. Among the contaminants regulated are bacteria, nitrates, arsenic, barium, cadmium, chromium, fluoride, lead, mercury, silver, pesticides; radioactivity and turbidity also are regulated. This act also contains provisions to protect groundwater aquifers.

5. *Resource Conservation and Recovery Act (RCRA)* (1976). Regulates the storage, shipping, processing, and disposal of hazardous wastes and sets limits on the sewering of toxic chemicals.

6. *Toxic Substances Control Act (TOSCA)* (1976). Categorizes toxic and hazardous substances, establishes a research program, and regulates the use and disposal of poisonous chemicals.

7. *Comprehensive Environmental Response, Compensation, and Liability Act (CERCLA)* (1980) and Superfund Amendments and Reauthorization Act (SARA) (1984). Provide for sealing, excavation, or remediation of toxic and hazardous waste dumps.

8. *Clean Water Act* (1985) (amending the 1972 Water Pollution Control Act). Sets as a national goal the attainment of "fishable and swimmable" quality for all surface waters in the United States.

9. *London Dumping Convention* (1990). Calls for an end to all ocean dumping of industrial wastes, tank washing effluents, and plastic trash. The United States is a signatory to this international convention.

---

ecologically important areas that were being filled in or drained at a rate of about half a million hectares per year before the passage of the Clean Water Act. Farmers, land developers, and others who are prevented from converting wetlands to other uses often are outraged by what they consider the "taking" of private lands.

Another sore point for opponents of the Clean Water Act is what are called "unfunded mandates," or requirements for state or local governments to spend money that is not repaid by Congress. The $128 billion already spent by cities to install sewage treatment and to improve sewer systems to meet federal standards far exceeds the $54 billion in congressional assistance for these projects. Estimates are that local units of government could be required to spend another $130 billion to finish the job without any further federal funding. Small cities that couldn't afford or chose not to participate in earlier water quality programs, in which the federal government paid up to 90 percent of the costs, are especially hard hit by requirements that they upgrade municipal sewer and water systems. They now are faced with carrying out those same projects entirely with their own funds.

Clean water reauthorization remains contentious

In the decades since the Clean Water Act was passed in 1972, new challenges have emerged, and funding provisions have expired. Between 1989 and 1994, the CWA provided $7.2 billion to help communities build and maintain wastewater treatment facilities. Since 1994, that funding has not been reauthorized, leaving local governments to cover most costs. Available federal funding has depended on annual extensions, which are variable and unreliable, depending on the budget, or mood, of Congress. The loss of this funding is a problem, as many of the nation's water treatment plants, now over 40 years old, need updating, repairs, or expansion to serve the cities growing around them.

At the same time, new contaminants have emerged, such as organic compounds discussed earlier in this chapter. We still have no standard policies for controlling nonpoint runoff from streets and farm fields, which have become our most important water pollutants in many areas. There is need for better use of new stormwater management systems, such as rain gardens or permeable paving to inexpensively reduce runoff. In addition, climate change is altering water resources. In some areas, storms are becoming more intense, increasing stress on water systems. In other areas, water supplies are dwindling, making pollutants more concentrated than in the past.

The Clean Water Act was updated in 1977, 1981, and 1987, and there have been calls since then to update and reauthorize the law. Persistent, deep disputes over polluters' responsibility, and the powers of regulators, have prevented new updates. Groups such as the Natural Resources Defense Council have argued that enforcement of current laws is weak, that required reporting is often incomplete. In recent years, political impasses have made effective compromises difficult to achieve.

Other important legislation protects water quality

In addition to the Clean Water Act, several other laws help to regulate water quality in the United States and abroad. Among these is the Safe Drinking Water Act, which regulates water quality in commercial and municipal systems. Critics complain that standards and enforcement policies are too lax, especially for rural water districts and small towns. Some researchers report pesticides, herbicides, and lead in drinking water at levels they say should be of concern. For instance, in one study of 374 communities across 12 states, the herbicide Atrazine was detected (though not necessarily at seriously threatening levels) in 96 percent of all surface-water samples.

The Superfund program for remediation of toxic waste sites was created in 1980 by the Comprehensive Environmental Response, Compensation, and Liability Act (CERCLA) and was amended by the Superfund Amendments and Reauthorization Act

(SARA) of 1984. This program is designed to provide immediate response to emergency situations and to provide permanent remedies for abandoned or inactive sites. These programs provide many jobs for environmental science majors in the monitoring and removal of toxic wastes and landscape restoration. A variety of methods have been developed for remediation of problem sites.

Section Review

1. Identify five important examples of water quality legislation.
2. What is best practicable control technology?
3. Why does the Clean Water Act need reauthorization to be effective?

Conclusion

Half a century ago, rivers in the United States were badly polluted with untreated sewage and with toxic industrial wastes. Many cities still dumped raw sewage into local rivers and lakes, so that warnings had to be posted to avoid any bodily contact. Today these conditions persist in rapidly developing countries, such as China and India, where water pollution remains a serious threat to the health of both humans and ecosystems. Infectious pathogens, often found in sewage, are the most common health risk in water. Nutrients are also key contaminants, because nutrients support algae and bacteria, many of them disease causing. Eutrophication and hypoxic "dead zones" also result from excessive nutrients. Other important contaminants include organic compounds, such as pesticides, metals, salts, and acids. All these different types of contaminants have different sources, effects, and remedies.

Billions of people still lack access to clean drinking water or adequate sanitation, but in many places conditions have improved. Even in the U.S. and other wealthy countries, not all rivers and lakes are "fishable or swimmable," but federal, state, and local pollution controls have greatly improved our water quality in most of the U.S. over what we had a generation ago. Progress since the 1972 Clean Water Act, and improvements in wastewater treatment, show that water quality can improve dramatically if we develop good policies and sufficient funding. The challenge now is to extend this progress and improving human health and environmental quality around the world.

Reviewing Key Terms

Can you define the following terms in environmental science?

atmospheric deposition 18.1
best available, economically achievable technology (BAT) 18.4
best practicable control technology (BPT) 18.4

biochemical oxygen demand (BOD) 18.1
coliform bacteria 18.1
constructed wetlands 18.3
cultural eutrophication 18.1
dissolved oxygen (DO) content 18.1

effluent sewerage 18.3
eutrophic 18.1
nonpoint sources 18.1
oligotrophic 18.1
oxygen sag 18.1
point sources 18.1
primary treatment 18.3

red tide 18.1
secondary treatment 18.3
tertiary treatment 18.3
thermal plume 18.1
total maximum daily load (TMDL) 18.2
water pollution 18.1

Critical Thinking and Discussion Questions

1. Cost is the greatest obstacle to improving water quality. How would you decide how much of the cost of pollution control should go to private companies, government, or individuals?

2. How would you define *adequate sanitation*? Think of some situations in which people might have different definitions for this term.

3. What sorts of information would you need to make a judgment about whether water quality in your area is getting better or worse? How would you weigh the different sources, types, and effects of water pollution?

4. Imagine yourself in a developing country with a severe shortage of clean water. What would you miss most if your water supply were suddenly cut by 90 percent?

5. Proponents of deep well injection of hazardous wastes argue that it will probably never be economically feasible to pump water out of aquifers more than 1 kilometer below the surface. Therefore, they say, we might as well use those aquifers for hazardous waste storage. Do you agree? Why or why not?

6. Arsenic contamination in Bangladesh results from geological conditions, World Bank and U.S. aid, poverty, government failures, and other causes. Who do you think is responsible for finding a solution? Why? Would you answer differently if you were a poor villager in Bangladesh?

Data Analysis

Examining Pollution Sources

Water pollution has many sources and there are numerous approaches to reduce or remediate various contaminants. We have reviewed lots of these sources and approaches in this chapter. To understand these issues in greater depth, it's helpful to study and compare the graphs in this chapter that list the causes and consequences of water pollution. Visit Connect to find a set of questions about specific figures that will test your comprehension of this important topic.

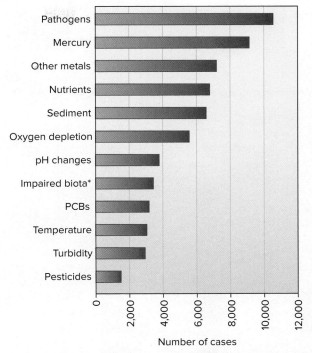

Leading causes of surface-water impairment in the United States. *Note: The * represents undetermined causes.*
Source: Data EPA, 2015

19 Conventional Energy

▲Fossil fuels have provided a large majority of our energy needs for two centuries, but now worries about air pollution, toxic emissions, environmental damage, and climate change are forcing us to look for alternative energy sources.
© Clynt Garnham Energy/Alamy Stock Photo

Learning Outcomes

After studying this chapter, you should be able to:

19.1 Define *energy* and *work*, and how our energy use has varied over time.

19.2 Describe the benefits and disadvantages of using coal.

19.3 Explain the consequences and rewards of exploiting oil.

19.4 Identify the advantages and disadvantages of natural gas.

19.5 Summarize the potential and risk of nuclear power.

"The pessimist complains about the wind; the optimist expects it to change; the realist adjusts the sails."

– William Arthur Ward

The End of Coal?

In 2010, with great fanfare, Arch Coal, the U.S.'s second-largest coal company, announced a land lease in Montana's Powder River Basin for what was predicted to be the nation's largest open-pit coal mine. The Otter Creek property was thought to hold 1.4 billion tons of coal, which was expected to bring $250 million annually to the state of Montana, or $5 billion over the 20-year life of the project. Most of the coal would be exported west by way of a railroad line, built specifically for this purpose, to shipping terminals in Washington or Oregon. From there, the coal would be sent to Asia.

But in early 2016 Arch filed for bankruptcy and announced it was abandoning both the Otter Creek mine and its accompanying rail line. Environmental activists celebrated the decision. For several years, Montana ranchers, farmers, and environmentalists, together with tribal leaders from the adjacent Northern Cheyenne Reservation, had worked to stop the mine. Furthermore, neighbors along the proposed rail line, as well as citizens in its West Coast terminals, had objected to the traffic congestion, noise, and dust from shipping as well as the climate disaster represented by burning an additional 1.4 billion tons of coal that would result.

Although public opposition played a major role in the demise of the Otter Creek mine, economics was its ultimate death blow. In the previous three years, Arch had lost nearly $2 billion, and in the next five years another $5 billion in debt would be due, something the major banks were refusing to refinance.

This tale raises the question "Are we in the midst of a major shift away from coal (and perhaps other fossil fuels) toward renewable energy?" Perhaps the "Beyond Coal" campaign, waged by environmental groups, is making progress.

The Arch story isn't an isolated incident. Over the past few years, 26 major coal companies have gone bankrupt, and 264 active mines have closed. Among the woes plaguing the industry are competition from cheap natural gas, falling coal prices, tougher regulations, and uncertainty about export markets. Not only is coal demand falling in Asia (mainly due to air pollution concerns), but all six proposed West Coast coal terminals have either been rejected by courts, following objections by local communities, or are mired in appeals and legal wrangling.

The end of the age of coal would bring a huge change to society. For 200 years, coal fueled the Industrial Revolution, kept lights on, and warmed houses. It employed millions of people worldwide, but also killed thousands of them every year due to mining disasters and disease. Disputes about wages and working conditions in the mines also played a central—and bloody—role in the rise of the labor movement in the U.S. Until a few years ago, coal supplied more than half of all electricity in the United States. Already, people in areas such as Appalachia and the Great Plains, where coal mining has been a major economic engine, feel they're caught in the crossfire of a "war on coal."

However, climate scientists warn that we need to leave 80 percent of our remaining coal, oil, and gas reserves in the ground if we're to limit climate change to less than 2°C—the threshold they calculate is needed to maintain a habitable Earth. How we'll persuade energy companies to forego the trillions of dollars in profits and stranded assets connected to these untouchable deposits, as well as the infrastructure to mine, process, and burn them, is a huge question.

In this chapter, we'll look at how fossil fuels and other conventional sources provide energy, and likewise focus on their benefits and limitations. Then, in Chapter 20, we'll discuss renewable energy solutions.

FIGURE 19.1 Over half the coal produced in the U.S. comes from the northern Great Plains, where it's excavated almost entirely from open-pit surface mines such as this.
© Moritz Werthschulte/Moment/Getty Images

19.1 ENERGY RESOURCES AND USES

- *Fossil fuels provide most of the world's energy.*
- *Transportation uses the greatest amount of energy in the U.S.*
- *Energy consumption is rising rapidly in the developing world.*

Energy drives our economy today, and many of our most important questions in environmental science have some link to energy resources—from air pollution, climate change, and mining impacts, to technological innovations in alternative energy sources.

We have undergone dramatic transitions in energy use. Fire was probably the first external energy source used by humans. Charcoal from fires has been found at sites occupied by our early ancestors 1 million years ago. Muscle power provided by domestic animals has been important at least since the dawn of the Neolithic Age 10,000 years ago. Wind and water power have been used nearly as long. Firewood was by far the largest source of energy for cooking and heating in the United States from colonial days until the mid-nineteenth century. The invention of the steam engine, together with diminishing supplies of wood, caused a switch to coal as the major energy source during the Industrial Revolution. Coal, in turn, was largely replaced by oil in the twentieth century due to the ease of shipping and burning liquid fuels. Now, as the opening case study for this chapter shows, we may be coming to the end of coal as a major energy source. And as easily accessible petroleum supplies have been depleted, we look increasingly to remote, dangerous, or politically unstable places for the oil on which we have become dependent.

Our dependence on—some would say addiction to—fossil fuels creates serious critical geopolitical and economic problems. The costs of propping up regimes—often totalitarian ones—that supplied our oil have been high. And the extraction, shipping, and burning of carbon-based fuels to support our lifestyles are now causing unsustainable environmental impacts, ranging from oceanic oil spills, such as the *Deepwater Horizon* disaster in 2010 (see section 19.3), to mountaintop removal for coal extraction, water pollution from extracting tar sands, or air pollution and global climate change from burning all those fuels. Although we've worried in the past about running out of fossil fuels, analysts now warn us that we have far more coal, oil, and natural gas than we can safely use. We need to switch to non-carbon energy sources as quickly as possible. And, as mentioned earlier, we will probably need to leave 80 percent of the remaining fossil fuels in the ground. Chapter 20 looks at some options for conservation and sustainable energy.

How do we measure energy?

To understand the magnitude of energy use, it is helpful to know the units used to measure it. **Work** is the application of force over distance, and we measure work in **joules** (table 19.1). **Energy** is the capacity to do work. **Power** is the rate of energy flow or the rate of work done: for example, one **watt** (W) is one joule per second. If you use a 100-watt lightbulb for 10 hours, you have used 1,000 watt-hours, or one kilowatt-hour (kWh). Most American households use about 11,000 kWh per year (table 19.2).

Table 19.1 Some Energy Units

1 joule (J) = the force exerted by a current of 1 amp per second flowing through a resistance of 1 ohm
1 watt (W) = 1 joule (J) per second
1 kilowatt-hour (kWh) = 1 thousand (10^3) watts exerted for 1 hour
1 megawatt (MW) = 1 million (10^6) watts
1 gigawatt (GW) = 1 billion (10^9) watts
1 petajoule (PJ) = 1 quadrillion (10^{15}) joules
1 PJ = 947 billion Btu, or 0.278 billion kWh
1 British thermal unit (Btu) = energy to heat 1 lb of water 1°F
1 standard barrel (bbl) of oil = 42 gal (160 liter) or 5.8 million Btu
1 metric ton of standard coal = 27.8 million Btu or 4.8 bbl oil

A typical power plant might supply about 1,000 megawatts (MW), or 1 gigawatt (GW) of electricity. This is enough for about 640,000 U.S. households, or about 1.3 million households in Europe. Some larger power plants are designed to supply 2,000 MW or more. Wind turbines produce around 1–5 MW, so it takes a lot of wind turbines to add up to one conventional power plant. But it takes a lot of coal mines and oil fields to supply fuel to a conventional power plant. Commercial-scale solar plants typically produce around 150–500 MW of electricity.

The energy industry in the United States also uses traditional measures such as the British thermal unit (Btu), the amount of energy it takes to heat one pound of water one degree F. Oil is measured in barrels (bbl). One barrel equals 42 gallons (160 l). Natural gas, which is a gas like air, not liquid like gasoline, is measured in cubic meters or cubic feet (m^3 or ft^3). Coal is measured in tons (2,000 lb) or metric tons (1,000 kg, or 2,200 lb).

Fossil fuels still supply most of the world's energy

Fossil fuels (petroleum, natural gas, and coal) still supply about 80 percent of world commercial energy but that percentage is expected to decline in the next few decades (fig. 19.2). Oil supplies

Table 19.2 Some Energy Uses

Uses	kWh/year*
Computer	100
Television	125
100 W light bulb	250
15 W fluorescent bulb	40
Dehumidifier	400
Dishwasher	600
Electric stove/oven	650
Clothes dryer	900
Refrigerator	1100

* Averages shown; actual rates vary greatly.
Source: U.S. Department of Energy.

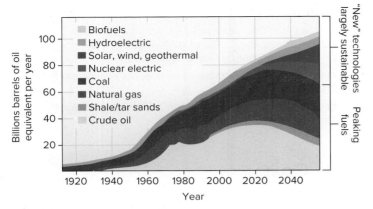

FIGURE 19.2 Fossil fuels, which now supply about 88 percent of all commercial energy in the world, are likely to decline as their costs increase and renewable energy gets cheaper.

about 35 percent of our current supply, followed by natural gas (25 percent) and coal (20 percent). Renewable sources—solar, wind, geothermal, biomass, and hydro—now make up about 10 percent of our power, while nuclear reactors provide a little over 8 percent. Renewables are the most rapidly growing energy sector,

but they have a long way to go to replace fossil fuels entirely. One reason for this disparity is the subsidies we provide for conventional fuels. No one knows the true amount of this support, but in 2016 the International Energy Agency estimated that global fossil fuel subsidies amounted to about $550 billion per year, or at least four times the support for wind, solar, and all other renewable energy sources.

For many years, the richer countries, with about 20 percent of the world population, consumed roughly 80 percent of all commercial energy, while the other 80 percent of the world had only 20 percent of the total supply. That situation is changing now. By 2035, energy experts predict, emerging economies such as China and India will be consuming about 60 percent of all commercial energy.

Transportation consumes the largest share of primary energy in the United States (fig. 19.3). Almost three-quarters of transportation energy is used by motor vehicles, almost all of it from petroleum. Nearly 3 trillion passenger miles and 600 billion ton-miles of freight are carried annually by motor vehicles in the United States. About 75 percent of all freight traffic in the United States is carried by trains, barges, ships, and pipelines, but because they are very efficient, they use only 12 percent of all transportation fuel.

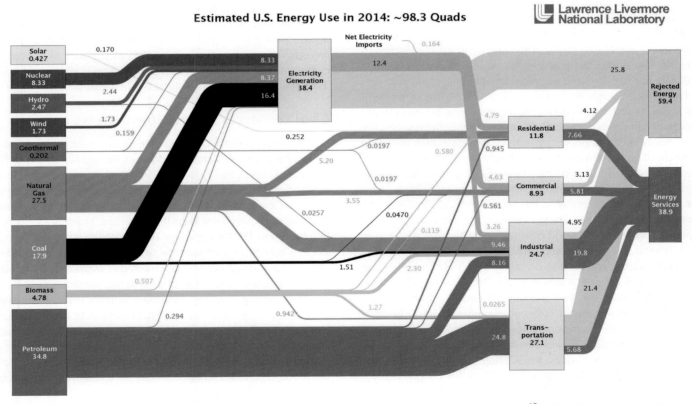

FIGURE 19.3 Sources and uses of energy in the United States in 2014. Units are quads, or quadrillion (10^{15}) British thermal units (BTUs). Line widths are proportional to the amount of energy. "Rejected energy," lost through leakage and inefficiency, accounts for 55.6 percent of energy production. Most coal-powered plants, for example, waste about 65 percent of the energy in coal fuel.
Source: Lawrence Livermore National Laboratory and U.S. Department of Energy, 2015

Although declining in recent years, industry is still the second-largest major category of energy use. Mining, milling, smelting, and forging of primary metals consume about one-quarter of that industrial energy share. The chemical industry uses a large amount of fossil fuels, but only half of that is for energy generation. The remainder is raw material for plastics, fertilizers, solvents, lubricants, and hundreds of thousands of organic chemicals in commercial use. The manufacture of cement, glass, bricks, tile, paper, and processed foods also consumes large amounts of energy. Although coal provides about one-quarter of our total energy in the United States, it supplies about half our electricity.

Residential and commercial customers use roughly 20 percent of the primary energy consumed in the United States, mostly for space heating, air conditioning, lighting, and water heating. Transportation requires about 27 percent of all energy used in the United States each year. About 98 percent of that energy comes from petroleum products refined into gasoline and diesel fuel, and the remaining 2 percent is provided by natural gas and electricity.

Producing and transporting energy also consumes energy. More than half the energy in primary fuels is lost during conversion to more useful forms, while being shipped to the site of end use, or during use. Electricity is generally promoted as a clean, efficient source of energy because, when it is used to run a resistance heater or an electrical appliance, almost 100 percent of its energy is converted to useful work and no pollution is given off. But what happens before electricity reaches us? Coal-fired power plants supply about half our electrical energy, and large amounts of pollution are released during mining and burning of that coal. Furthermore, nearly two-thirds of the energy in the coal is lost in thermal conversion in the power plant. About 10 percent more is lost during transmission and stepping down to household voltages.

How much energy do you use every year? Most of us don't think about it much, but maintaining the luxuries we enjoy usually involves an enormous energy input. On average, each person in the United States or Canada uses more than 300 gigajoules (GJ) (the equivalent of about 60 standard barrels or 8 metric tons of oil) per year. By contrast, in the poorest countries of the world, such as Bangladesh, Yemen, and Ethiopia, each person, on average, consumes less than one GJ per year. Put another way, each of us in the richer countries consumes nearly as much energy in a single day as the poorest people in the world consume in a year.

In general, income and standards of living rise with increasing energy availability, but the correlation isn't absolute (see the Data Analysis at the end of this chapter). Some energy-rich countries, such as Qatar, use vast amounts of energy, although their level of human development isn't correspondingly high. Perhaps more important is that some countries, such as Norway, Denmark, and Japan, have a much higher standard of living by almost any measure than the United States, while using less than half as much energy. This suggests abundant opportunities for energy conservation without great sacrifices. In 2016, Sweden announced it intends to become entirely oil free, with an interim goal of carbon neutrality by 2045. Combined strategies such as wind, solar, and hydroelectricity, biogas (methane) for transportation, biomass, and efficiency make this goal possible.

Section Review

1. Why has Arch Coal abandoned its plans for the Cabin Creek mine in Montana (opening case study)? What might our energy future be if other companies follow a similar path?
2. What are the major sources of commercial energy worldwide and in the United States?
3. How does energy use in the United States compare with that in other countries?

19.2 COAL

- *We have a huge amount of coal.*
- *Burning coal releases many pollutants.*
- *Clean coal technology remains uneconomical.*

Coal is fossilized plant material preserved by burial in sediments and altered by geological forces that compacted and condensed it into a carbon-rich fuel. Because it derives from fibrous plant residue, coal contains larger and more complex carbon compounds than oil. Coal is found in every geologic system since the Silurian Age 400 million years ago, but graphite deposits in very old rocks suggest that coal formation may date back to Precambrian times. Most coal was laid down during the Carboniferous period (286 million to 360 million years ago) when the earth's climate was warmer and wetter than it is now. Because coal takes so long to form, it is essentially a nonrenewable resource. Because they originate in anoxic swamp beds, coal deposits contain sulfur (common in low-oxygen biological systems), as well as metals such as mercury and cadmium, which plants accumulate from soils (see section 14.2). These become important pollutants when coal is burned.

Coal resources are greater than we can use

World coal deposits are enormous, ten times greater than conventional oil and gas resources combined. Most of the world's coal is in North America, Europe, and Asia (fig. 19.4), and just three countries, the United States, Russia, and China, account for two-thirds of all proven reserves. Coal seams can be 100 m thick and can extend across tens of thousands of square kilometers that were vast, swampy forests in prehistoric times. The total resource is estimated to be 10 trillion metric tons. At current rates of use, this would amount to several thousand years' supply. But it's clear that we couldn't survive if all that carbon (and associated pollutants) were emitted into the atmosphere.

Coal mining is a dirty, dangerous activity. Underground mines are notorious for cave-ins, explosions, and lung diseases, such as black lung suffered by miners. Surface mines (called strip mines, where large machines scrape off overlying sediment to expose coal seams) are cheaper and generally safer for workers than tunneling, but leave huge holes where coal has been removed and vast piles of discarded rock and soil.

An especially damaging technique employed in Appalachia is called mountaintop removal. Typically, the whole top of a mountain ridge is scraped off to access buried coal (fig. 19.5). In 2010 the EPA announced it would ban "valley fill," in which waste rock

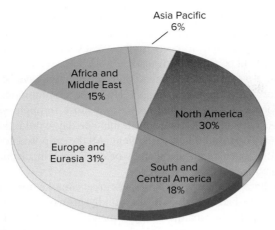

FIGURE 19.4 Proven-in-place coal reserves by region, 2014.
Source: British Petroleum, 2016

FIGURE 19.5 One of the most environmentally destructive methods of coal mining is mountaintop removal. Up to 100 m of the mountain is scraped off and pushed into the valley below, burying forests, streams, farms, cemeteries, and sometimes houses.
© Melissa Farlow/National Geographic/Getty Images

is pushed into nearby valleys, but existing operations are "grand-fathered in" (see Chapter 14 for further discussion). Mine recla-mation is now mandated in the United States, but success varies depending on how strongly rules are enforced.

A little more than one-third of all electricity in the United States is now generated by coal-fired power plants (fig. 19.6). Burning coal releases huge amounts of air pollution. Every year the roughly 1 billion tons of coal burned in the United States (83 percent for electric power generation) releases close to a trillion metric tons of CO_2. This is about half of the industrial CO_2 released by the United States each year.

Coal also contains toxic impurities, such as mercury, arsenic, chromium, lead, and uranium, which are released into the air

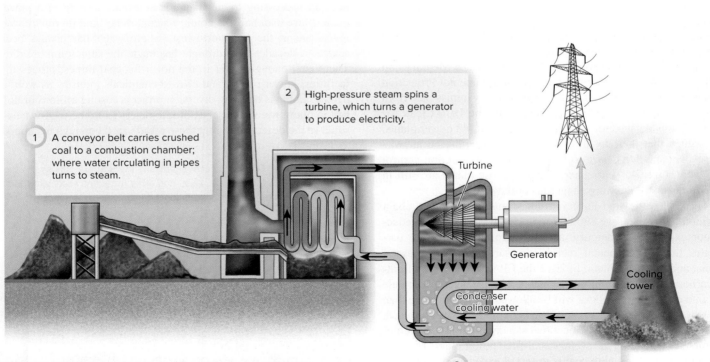

1 A conveyor belt carries crushed coal to a combustion chamber; where water circulating in pipes turns to steam.

2 High-pressure steam spins a turbine, which turns a generator to produce electricity.

Turbine

Generator

3 Water circulated from a lake, river, or cooling tower cools and condenses steam.

Condenser cooling water

Cooling tower

FIGURE 19.6 A coal-fired power plant works by generating steam, which turns an electrical generator. Nearly all electric power, apart from solar photovoltaic energy, is produced by spinning a turbine that turns a generator. Until recently, coal plants produced most of our electricity.

during combustion. The coal burned every year in the United States releases 18 million metric tons of sulfur dioxide (SO_2), 5 million metric tons of nitrogen oxides (NO_x), 4 million metric tons of airborne particulates, 600,000 metric tons of hydrocarbons and carbon monoxide, and 40 tons of mercury. This is about three-quarters of the SO_2 and one-third of the NO_x released by the United States each year. Sulfur and nitrogen oxides combine with water in the air to form sulfuric and nitric acids, making coal burning the largest single source of acid rain in many areas (Chapter 16).

Most people aren't aware of it, but coal-burning plants emit radioactivity from uranium and thorium. You'd get more radioactivity living for a lifetime next to a coal power plant than next to a nuclear plant—assuming no accidents at the nuclear plant. It's possible to make either gas or liquid fuels out of coal, but these processes are even dirtier and more expensive than burning the coal directly. Both coal-to-liquid and coal-to-gas would be environmental disasters.

Another concern with coal combustion is that it produces vast amounts of toxic ash, which is expensive to manage safely. In 2009 an earthen dam, which held a pond for storing coal ash, broke in eastern Tennessee and released a billion gallons (3.8 billion liters) of coal ash sludge into a tributary of the Tennessee River. The ash contained dangerous levels of arsenic, mercury, and toxic hydrocarbons. After the spill, the U.S. EPA revealed that this impoundment was only one of hundreds of equally risky coal ash dumps across the country.

Coal use is declining in the U.S.

Recent improvements in technology have made other choices, from gas and oil to wind and solar, cheaper than coal. At the same time, most of our coal-fired power plants are aging. It's expensive to retrofit them to meet new public health standards. The Mercury and Air Toxics Standards announced by the Environmental Protection Agency in 2012 will slash allowable mercury emissions from coal-fired power plants. This was required by the 1970 Clean Air Act, but it was delayed for decades by owners of old power plants, who argued that their facilities were about to be closed anyway and so they shouldn't have to install expensive pollution control equipment. Over 45 years later, many of those plants are still in operation and still emitting dangerous pollutants.

The EPA estimated its new rules would cost utilities about $9 billion but would save $90 billion in health care costs by reducing our exposure to mercury, arsenic, chromium, and fine particulates that cause mental retardation, cardiovascular diseases, asthma, and other disorders. In 2012 the EPA also proposed limiting carbon emissions from power plants. If this Clean Power rule goes into effect, new facilities will be allowed to emit no more than 1,000 lbs (454 kg) of CO_2 per megawatt hour of electricity produced. Natural gas plants can easily meet that standard, but it's about half the amount released by the average coal-fired power plant. The only way to meet this limit with coal is to install expensive carbon capture and storage equipment, which we'll discuss later in this section.

With coal consumption falling in North America and Europe, coal companies are looking desperately for new markets overseas.

As discussed in the opening case study for this chapter, Arch Coal and other western mining companies hoped to ship 20 million tons of coal per year train to ports on the West Coast (fig. 19.7) from which it could be shipped to Asia.

Half a dozen sites in Oregon and Washington have been proposed for coal-shipping terminals. However, communities where these ports could be located as well as those through which as many as 50 or 60 mile-long unit trains per day would pass have strenuously opposed this development. All the proposed coal ports have either been rejected outright or are mired in disputes that could last for years.

Furthermore, China may not be as great a market for American coal as some had hoped. With its rapidly growing economy and increasing personal wealth, energy demand rose rapidly in China at the beginning of this century. Coal consumption in China grew by about 50 percent between 2000 and 2010. The country built about one new coal-fired power plant every week through that decade. At the end of 2013, according to Greenpeace, China had 45 percent of all the world's coal-burning power plants. As a result, China became the world's largest consumer of coal and also the largest source of anthropogenic CO_2. But catastrophic air pollution is making China's leaders rethink their energy policies. At least 1.2 million people die every year in China as a result of air pollution. And climatologists warn that millions of people in the important commercial centers around Shanghai and Hong Kong are at risk from storms and flooding triggered by climate change.

Water shortages are another serious limitation for China's coal use. Water is needed for every aspect of coal use, from mining and processing to steam production and cooling in a power plant. More than half the country's coal mines and thermal power plants are in the arid northwest, where water has always been scarce. A decades-long drought has made the situation more dire. There isn't enough water in the north for coal mines, processing facilities, power plants, and coal-to-chemicals industry as well as people. China is building diversion projects costing billions of dollars to pump water thousands of kilometers from the well-watered south to the desert north, but these efforts aren't enough to support

FIGURE 19.7 Many communities in the U.S. are opposed to the traffic disruptions, noise, dust, and other irritations associated with coal shipments. Exports from coastal coal-loading terminals would only exacerbate these conflicts.
© Kevin Burke/Digital Stock/Corbis RF

continued coal use. In 2016, China reported that its coal use had declined by nearly 7 percent in the prior two years, and that it was closing more than 1,000 coal mines and stopping approval of any new mines in an effort to control air pollution.

Is clean coal technology the answer?

Early in his administration, President Obama promised an "all of the above" policy for energy. Billions of dollars were appropriated for "Clean Coal" technology in an effort to preserve jobs in the fossil fuel industry. Most of those projects called for **Carbon capture and storage (CCS),** in which CO_2 is captured and pumped into deep geologic formations. While this technology has been used for many years for enhanced oil recovery from depleted oil wells, it is too expensive to capture and transport the huge volumes of CO_2 produced by power plants to oil wells more than a few kilometers away.

There have been many attempts to develop clean coal technology. President George W. Bush described his FutureGen project as the world's first zero-emission coal plant, but abandoned it when it became too expensive. Obama pinned his hopes on a huge project in Kemper County, Mississippi. Originally estimated to cost $2.4 billion when construction started in 2010, the Kemper project was 2 years behind schedule and $4 billion over budget in 2016. Poor management and shoddy construction make it doubtful that the plant will ever capture carbon profitably. Rather than bring economic development and jobs to this impoverished county, rising electric rates have driven away business and burdened residents with one of the most expensive power plants ever built. China claims to have about a dozen carbon-capture coal gasification plants, but observers report that although these facilities are capturing CO_2, none are actually storing it because they aren't required to by the central government. Instead they simply vent it to the air. Some people describe this program as "catch and release."

Another proposed approach is integrated gasification combined cycle (IGCC) combustion, in which ultra-supercritical boilers can make carbon capture easier. In theory, this system sounds good. Gasification involves heating a coal slurry at high pressure in the presence of almost pure oxygen. The coal doesn't burn; instead it reacts with the oxygen and breaks down into a variety of gases, mostly hydrogen and carbon dioxide. Hydrogen burns very cleanly, but is highly explosive and hard to handle. Carbon dioxide can be captured for storage, and the remaining stack gas is cooled and purified of contaminants, such as sulfur, mercury, and arsenic. In theory, then, clean coal technology could be useful, but it continues to be uneconomical and difficult. Many believe the best plan is to simply focus on cleaner alternatives.

Because of the serious concerns about climate change, air pollution, and destructive mining techniques that arise from fossil fuel use, a powerful social movement in the United States is calling for an end of fossil fuel use for energy (fig. 19.8). Many colleges and other public institutions have promised to divest (sell stocks and other investments) in fossil fuels. This may make sense for strictly economic reasons aside from the environmental and social arguments for doing do. Is your school part of this divestment movement?

FIGURE 19.8 At the 2014 People's Climate March, more than 400,000 people gathered to call for alternatives to fossil fuels. The same weekend, New York City announced a goal of 80 percent CO_2 reduction by 2050.
© Tom Finkle

Section Review

1. Where are the largest coal deposits located?
2. What are the greatest disadvantages of burning coal?
3. What is carbon capture and how might it work?

19.3 OIL

- *Tar sands, deep ocean deposits, and tight shale formations that once were thought impossible to access have expanded our oil supplies.*
- *The Arctic Ocean could be the second-largest oil source in the world.*
- *Oil refineries supply many useful products but also often have adverse environmental effects.*

Like coal, petroleum is derived from organic molecules created by living organisms millions of years ago and buried in sediments, where high pressures and temperatures concentrated and transformed them into energy-rich compounds. While coal originates as vegetation in peat swamps, though, oil derives mainly from marine algae and plankton in coastal seas. Oil therefore tends to be associated with sedimentary sand or shale (mud) deposits. Like coal, oil often contains sulfur because of the anoxic environment in which it was deposited. Depending on its age and history, a petroleum deposit will have varying mixtures of oil, gas, and solid tarlike materials. Extracting and processing these deposits can be expensive, complex, and often dangerous (fig. 19.9). But the different carbon molecules in oil are readily used and transformed, once extracted. This makes oil useful for countless purposes, from gasoline and other fuels to the raw material for a host of petrochemicals, including plastics and pesticides.

In an oil deposit, liquid and gaseous hydrocarbons can migrate up through cracks and pores out of the sediments in which they formed and into surrounding rock layers. Oil and gas move upward (over millions of years) because they are less dense than the surrounding rock.

FIGURE 19.9 Oil still provides more than half our energy, but extraction is dangerous work. Here, workers thread pipe on Brazil's Enchova platform, on which 84 workers have died in explosions and fires. Safety concerns remain widespread in the industry.
© Stephanie Maze/Corbis Documentary/Getty Images

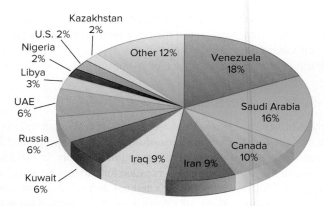

FIGURE 19.10 Proven oil reserves. The Middle East now accounts for a much smaller percent of the known, economically recoverable oil than in previous years. The numbers here add to more than 100 percent due to rounding.
Source: CIA Factbook, 2016

Oil and gas deposits generally accumulate under layers of shale or other impermeable sediments, especially where folding and deformation of systems create pockets that trap upward-moving, low-density hydrocarbons. In these folds, oil fills spaces in a porous sandstone or limestone, much like water saturating a sponge. Many major oil fields include thousands of wells that dominate the landscape over very large areas.

In the 1940s, Dr. M. King Hubbert, a Shell Oil geophysicist, predicted that oil production in the United States would peak in the 1970s, based on estimates of U.S. reserves at the time. Until recently, Hubbert's prediction of **peak oil** seemed correct. Discovery of new oil declined, the United States went from being the world's greatest oil exporter to being the largest importer. In other regions, oil production may have already peaked, and global supplies are likely to peak at some point. The larger concern now, though, is how to wean ourselves off of climate-disrupting oil long before it is all burned.

Estimates of our recoverable oil supplies have expanded dramatically as we've developed techniques for obtaining oil from ever more remote and extreme places. Not long ago, countries in the Middle East accounted for nearly 90 percent of all known oil deposits. Huge new discoveries such as Canada's Athabascan **tar sands,** Brazil's Santos Basin, Venezuela's Orinoco Belt, Angola's Kwanza Basin, the United States' Bakken formation, and the Arctic sea bed, for example, suggest we may have far more oil than Dr. Hubbert thought. Between 1993 to 2003, proven (commercially extractable) reserves grew by nearly two-thirds, from 1.04 trillion barrels to 1.69 trillion barrels. Places that we never thought would have oil have become major suppliers (fig. 19.10). Venezuela and Canada may have nearly as much petroleum in their tar sands and heavy oils as the previously estimated world supply. Some of those announced discoveries could be simply industry hyperbole, but rather than facing an end of oil, we could be looking at an impending glut. Even the United States, which has been an oil importer for decades, is once again an oil exporter.

Extreme oil and tar sands have extended our supplies

Drilling technology has made tremendous strides in recent years. It's now possible to drill into the extreme temperatures and pressures encountered at a depth of 12,000 m (40,000 ft). Drilling can extend horizontally up to 32 km (20 mi) from the well head, and as many as 24 individual wells can be supported from a single drilling pad. **Hydraulic fracturing ("fracking")** can release oil from "tight" formations through which passage of liquid would otherwise be obstructed. Fracking is also now used extensively in the quest for natural gas. We'll discuss some of the negative impacts of this technology in section 19.4.

Most of us hadn't thought much about the dangers of deep ocean oil wells in remote places until the 2010 explosion and sinking of the *Deepwater Horizon* in the Gulf of Mexico (fig. 19.11). The rig was drilling the Macondo well for BP in 1-mi (1.6-km) deep water. At least 5 billion barrels (800 million liters) of oil were spilled during the 4 months it took to plug the leak. A massive effort was made to scoop up the oil, burn, or disperse it with chemicals, At least 6.8 liters (1.8 million gallons) of toxic dispersant was sprayed on the ocean. Eleven workers on the drill rig were killed, and the fishing and tourism industry lost billions of dollars. The damage to marine and wildlife habitats is incalculable. Altogether, criminal and civil settlements have cost BP and its contractors about $70 billion. It was the worst oil spill in American history. But much deeper wells in much more remote areas are being drilled today.

Melting of Arctic sea ice has triggered a scramble for territorial claims in the Arctic ocean. All the surrounding countries have asserted ownership of the seafloor out to (or beyond) their continental shelf. It's thought that the Arctic may have 15 percent of the world's undiscovered oil and 30 percent of all natural gas. Exploratory drilling has already started. In 2012, Shell Oil sent two floating drill rigs into the Arctic ocean. They were so badly damaged by storms and ice, however, that one had to be towed to Korea and the other all the way to Singapore for repairs. All drilling has been canceled for the time being.

FIGURE 19.11 In 2010, the oil drilling rig *Deepwater Horizon* exploded and sank, spilling at least 5 million barrels (800 million liters) of crude oil into the Gulf of Mexico. It was drilling in water 1 mile (1.6 km) deep, but other wells are now more than twice as deep.
© U.S. Coast Guard Handout/Getty Images

FIGURE 19.12 Alberta tar sands are now the largest single source of oil for the United States, but there are severe environmental and social costs of extracting this oil.
© Larry MacDougal/Canadian Press via AP Images

This demonstrates the difficulties of working in the Arctic. Although most of the pack ice may be melted in the summer, it's still a very remote and dangerous place. A leak similar to the BP spill in the Gulf of Mexico would be very difficult to contain in the frigid and often stormy Arctic thousands of kilometers from the nearest supply base.

Risks and costs of oil extraction

Canada's Athabasca oil sands are another example of the dangers and costs of our search for oil. These deposits are estimated to contain 1.7 trillion barrels of oil lying beneath some 141,000 km^2 (54,000 mi^2, or about the size of Greece) of boreal forest in northern Alberta. There are severe environmental impacts in producing this oil. A typical facility producing 125,000 bbl of oil per day will consume several square miles of forest for mine pits and tailings ponds over its lifetime (fig. 19.12). It also creates about 15 million m^3 of toxic sludge, releases 5,000 tons of greenhouse gases, and consumes or contaminates at least 75 million liters (20 million gal) of water each year. Native Cree, Chipewyan, and Metis people are protesting the cancer-causing water pollution and destruction of forests and wildlife, which are destroying their traditional ways of life.

Oil prices are often extremely unstable. In July, 2008, for example, the price for Brent crude (the benchmark for world markets) reached U.S. $147 per barrel, and Americans saw gasoline prices over $4 per gallon. But then the housing market imploded, the whole world banking system teetered on the brink of collapse, and by December of that year, the price of oil had plummeted to $30 per barrel. The price bounced up and down for several years as the world economy lurched and staggered, surging above $100 per bbl in mid 2014, only to plunge again to just over $30 per barrel in early 2016.

Some of this volatility is driven by market forces, but much is geopolitical. Saudi Arabia, for example, flooded the market with cheap oil to win market share and to drive out competitors producing oil from expensive deposits. Events in North Dakota have also disrupted oil markets. The Bakken formation underlying parts of North Dakota, Montana, and Saskatchewan is composed of "tight shale" deposits that are unsuited to conventional wells. Horizontal drilling and fracking, however, unleashed a major oil boom in this formation. North Dakota, which ranked 38th in the country in per capita income in 2002, saw its economy double in a decade. By 2012, North Dakota had overtaken Alaska to become the number two oil-producing state in the country (after only Texas). At its peak, the state was shipping more than a billion barrels of oil per day and had a budget surplus of $1 billion.

By early 2016, however, with oil at only $30 per barrel, production in the Bakken crashed. Only 30 drill rigs were still in operation in 2016 compared to five times as many a few years earlier. Thousands of workers lost their jobs, and many left the state. Williston, the epicenter of the Bakken, which had tripled its population in the previous four years, had invested millions in new schools, housing, parks, roads, and services. Suddenly, its income dried up and the city and state are having difficulty meeting payments.

The drag on oil markets isn't limited to North Dakota. A 2016 analysis of major U.S. oil companies warned that one-third of them could go bankrupt. Collectively, they owed more than $150 billion to creditors. And, as is the case with coal companies, most banks are refusing to lend money or restructure debt for oil development. In 2016, President Obama reversed a previous policy and banned offshore oil drilling along the U.S. Atlantic Coast. A safer and more economical energy source, he said, would be solar, wind, and biomass.

Even Saudi Arabia is planning for a life without oil. With prices so low, the desert kingdom is running low on cash to support its lavish lifestyle and generous welfare system. In 2016, the Saudis offered to sell part of its US$10 trillion national oil company (the most valuable corporation in the world). They realize it's wise to diversify and not depend on a single commodity.

Low oil prices can be bad for the environment because they encourage people to buy inefficient vehicles and drive more. But they can also be an opportunity to avoid dangerous and long-lasting projects, such as pipelines and power plants. It's easier to persuade investors to leave **unburnable carbon** in the ground when it isn't worth much, than when prices are high and the resources are worth trillions.

Shipping oil can be dangerous and disruptive

As is the case with coal, access to markets is a key challenge for oil companies. Although Canadian First Nations' people have complained about the environmental and social effects of tar sands extraction for decades, the battle over the Keystone XL pipeline, really brought this fuel source to public attention. The pipeline was proposed to carry tar sands oil from mines in Alberta to Houston, Texas. Pipeline supporters claimed it would bring energy security to the United States and provide 20,000 jobs. Opponents countered that the pipeline wouldn't help the United States energy supply because the oil was being shipped to Texas for sale abroad. And critics claimed that only about 50 permanent jobs would be created by this project.

Blowouts and leaks on the 1,400-km (875-mi) Keystone pipeline were a serious concern. Pipelines carrying tar sands oil have a much higher rupture rate than those for conventional oil. The residual sand is more abrasive, the oil is more acidic and corrosive, and heavy oil must be heated to higher temperatures and pressures to be shipped, all making tar sands pipelines more accident prone than other lines. The proposed route would have crossed the Ogallala Aquifer, which provides drinking water and irrigation to most of the central plains. A major oil spill above the largest aquifer in the U.S. would be an unmitigated disaster. In 2016, President Obama announced that he would not approve the Keystone pipeline.

TransCanada, the company behind Keystone, also has been pursuing a northern "Gateway" route that would cross the Canadian Rockies on its way to a terminal in the fjords of British Columbia's Great Bear Rainforest. In 2016, Canadian courts rejected the Gateway pipeline because it ignored First Nation's concerns. The only remaining Canadian pipeline proposals are Kinder Morgan's TransMountain extension through British Columbia and TransCanada's pipeline to New Brunswick.

The United States has large supplies of unconventional oil

In addition to the oil deposits in North Dakota we've just discussed, large deposits of solid organic material called **kerogen** (which is similar to **bitumen**), are found in sedimentary rocks in many places in the U.S. When heated to about 480°C (900°F), the kerogen liquefies and can be extracted from the stone. Kerogen-containing **oil shale** beds up to 600 m (1,800 ft) thick occur in the Green River Formation in Colorado, Utah, and Wyoming, and lower-grade deposits are found over large areas of the eastern United States. These deposits

might yield the equivalent of several trillion barrels of oil, but whether we can or should mine it is questionable.

Mining and extracting kerogen creates even more environmental problems than tar sands. It is expensive, it uses vast quantities of water (a scarce resource in the arid West), and it produces about twice as much CO_2 per unit of energy as conventional oil. Because the shale doubles in size when heated, there have been serious discussions about filling whole canyons, rim to rim, with oil shale waste. So far it hasn't been economically feasible to mine these oil shales. Nevertheless, the Interior Department has discussed opening leases for oil shale research and development in Colorado and Utah. What do you think? How much of this oil can, or should we, extract?

Refineries are major sources of air pollution

Oil produces a host of useful products. Crude oil is "cracked" or broken into smaller molecules by heating. Distilling operates on the principle that progressive cooling separates larger molecules from smaller ones (fig. 19.13).

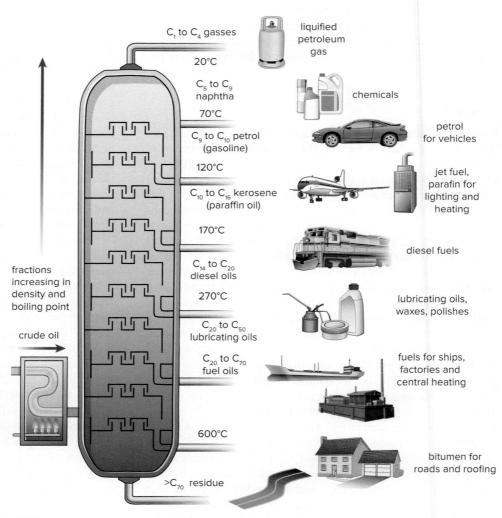

FIGURE 19.13 Distillation operates on the principle that longer-chain hydrocarbons condense at a higher temperature than shorter-chain hydrocarbons. Fuel oils, for example, have very long carbon chains (20–70 C atoms in a molecule, shown as C_{20} to C_{70}), while natural gas has only 4 (C_4).

Some people argue that rather than burn petroleum for energy, we should save it for useful products that can be created from it, such as plastics, lubricating oils, paints, and other chemicals. It is also important to recognize that refining—at least as it's currently done in the United States and most other regions—releases a great deal of volatile organic compounds into the air. Some of the worst air quality in America is found near the heavy concentrations of petrochemical industries in Texas, Louisiana, and New Jersey. High levels of cancer, asthma, and other diseases often occur in these areas. BP, which is blamed for most of the bad decisions that led to the 2010 Gulf oil spill, has a terrible record of refinery operations. In the past three years, according to the Center for Public Integrity, BP accounted for 97 percent of all "flagrant violations" found in the refining industry. In 2005, for example, an explosion at a BP Texas City refinery killed 15 workers and injured 170 others. This is just one instance where BP's disregard for safety regulations proved disastrous, and deadly.

Section Review

1. How have our views about the size of world oil resources changed in the past few decades?

2. What country is considered to have the world's largest oil reserves according to the CIA Factbook?

3. What are tar sands, and what problems are associated with their mining, processing, and shipping?

19.4 Natural Gas

- *We're finding large new supplies of natural gas in many places.*

- *Rapidly falling gas prices make coal and nuclear power less attractive but also hold back development of renewable energy.*

- *Fracking (hydraulic fracturing) has greatly expanded our gas supplies but has serious drawbacks.*

Natural gas is the world's third-largest commercial fuel, currently making up about one-quarter of global energy consumption. Natural gas consists of small, volatile organic compounds, mainly methane (CH_4), that accumulate at the top of oil or coal deposits. Because it consists of simpler, purer compounds than coal or oil, gas burns more cleanly than either coal or oil, and it produces only half as much CO_2 as an equivalent amount of coal during combustion. Natural gas's rapidly falling prices are also a large part of the recent decline in coal use. Many people hope this shift from coal to gas will help slow climate warming. Others are not so sure it will have that effect.

Most of the world's currently known natural gas is in a few countries

More than half of all the world's proven natural gas reserves are in the Middle East and the former Soviet Union (fig. 19.14). Both eastern and western Europe are highly dependent on imported gas. The total ultimately recoverable natural gas resources are thought to be 10,000 trillion ft³, corresponding to about 80 percent as much energy as the estimated recoverable reserves of crude oil. The proven world reserves of natural gas are 6,200 trillion ft³

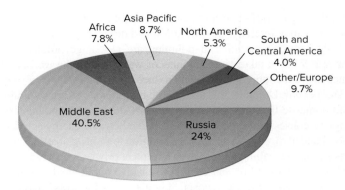

FIGURE 19.14 Proven natural gas reserves by region, 2012.
Source: Data from British Petroleum, 2014

(176 million metric tons). Because gas consumption rates are only about half of those for oil, current proven gas reserves represent roughly a 60-year supply at present usage rates.

As it breaks down, coal is slowly transformed into methane. Accumulation of this explosive gas is one of the things that makes coal mining so dangerous. In many places where mining coal seams isn't economically feasible, it is relatively cheap and easy to produce from these "coal-bed" methane deposits. In Wyoming's Powder River basin, for example, 140,000 wells have been proposed for methane extraction. Together with the vast network of roads, pipelines, pumping stations, and service facilities, this industry is having a serious impact on ranching, wildlife, and recreation in formerly remote areas. In Wyoming's Upper Green River Basin, for example, 50,000 pronghorn antelope and 10,000 elk migrate through a narrow corridor every year on their way between summer and winter ranges. The Jonah Gas Field (fig. 19.15) lies across this migration route, and biologists worry

FIGURE 19.15 An aerial view of the Jonah Field in the Upper Green River Basin.
© Peter Aengst/The Wilderness Society/Lighthawk

that the noise, traffic, polluted wastewater pits, and activity around the wells may interrupt the migration and doom the herd.

Water consumption and pollution are also huge problems in the arid West. It takes large amounts of water to drill the wells, and once in production, each well can produce up to 75,000 liters of salty water per day. Dumping this toxic waste into streams poisons wildlife and domestic livestock. In several western states, ranchers, hunters, anglers, conservationists, water users, and renewable energy activists have banded together in an unlikely coalition to fight against coal-bed gas extraction, calling on Congress to protect private property rights, preserve water quality, and conserve sensitive public lands. "It may be a clean fuel," says one rancher, "but it's a dirty business."

Deep gas-bearing shale formations underlie vast areas of Appalachia, Texas, and other regions (fig. 19.16), and other regions. The Marcellus and Devonian Shales, underlying Pennsylvania and New York, have been the focus of considerable debate. The U.S. Geological Survey estimates that the Marcellus/Devonian formation may contain a 100-year supply for the United States at current consumption rates. With such large amounts of gas now available, prices have fallen sharply in recent years, making the resource much more attractive.

These shale deposits are generally "tight" formations through which gas doesn't flow easily. But pumping a mixture of water, sand, and various chemicals into rock formations at extremely high pressure has produced a flood of new, cheap gas. This has been a boon to consumers, but the multitude of wells, water pollution, and threats to water supplies on which millions of people depend, raise

thorny problems (see What Do You Think?, "Fracking" below). Many utilities are converting old, dirty coal-burning power plants to cleaner natural gas. Cheap gas also makes nuclear power much less attractive, but it also impedes the development of renewable sources, such as solar, wind, and geothermal energy.

An extreme example of the risks of gas leaks occurred in California in October, 2015. A well pipe ruptured in a depleted oil well being used for gas storage in Porter Ranch, a Los Angeles suburb. In the four months it took to find and plug the leak, more than 97,000 tons (5 billion ft^3) of methane and ethane were emitted from the well. In terms of greenhouse effects, this was equal to the annual emissions from about 600,000 cars.

Other countries are now considering or implementing fracking. Poland has recently started fracking extensive shale deposits, and Russia, China, and Australia are beginning similar developments. On the other hand, several European countries, including England, Germany, and Spain have banned fracking. Spread of this technology could have very large impacts on both energy prices and global climate.

Fracking and other gas-drilling operations leak methane to the atmosphere, from fractured rock formations, cracks in well casings, and leaking valves and joints in pipelines. Methane is more than 20 times more effective, per molecule, in retaining heat in the atmosphere, so the climate implications are substantial, but estimates vary on the amount of methane and CO_2 released by gas well drilling and fracking (fig. 19.17). Even under the most optimistic estimates, however, shale gas has a larger climate impact than either coal or oil.

FIGURE 19.16 Shale gas plays in the United States. The cost of extracting these resources varies with depth.
Source: Energy Information Association

The Fracking Debate

Vast amounts of methane may lie in relatively shallow sediments under large areas of North America. This gas burns more cleanly than coal or oil, is easier to ship, and produces less CO_2 during combustion. But much of the resource is held in "tight" formations through which gas doesn't flow easily. To boost well output, mining companies rely on hydraulic fracturing ("fracking"). A mixture of water, sand, and potentially toxic chemicals is pumped into the ground and rock formations at extremely high pressure. The pressurized fluid cracks sediments and releases the gas. Fracturing rock formations often disrupts aquifers, however, and contaminates water wells.

For years, methane extraction from coal beds and other formations was a problem only in western states, but this controversial technology is now moving to the East Coast as well. Questions raised in the West concerning a multitude of wells, water pollution, and threats to both drinking water supplies and public health are now being asked in eastern states. While well drilling in the West is declining, it's being replaced by intense activity in the East. In the past few years, tens of thousands of wells have been drilled into the Marcellus Shale in Pennsylvania, West Virginia, New York, and Ohio. Much of this area lies in the environmentally sensitive Chesapeake Bay watershed or the Delaware River basin. Contamination of groundwater or surface runoff in these watersheds is of special concern (see chapter 3).

Drilling companies have generally refused to reveal the chemical composition of the fluids they use in fracking. They claim it's a proprietary secret, but it's well known that a number of petroleum distillates, such as diesel fuel, benzene, toluene, xylene, polycyclic aromatic hydrocarbons, glycol ethers, as well as hydrochloric acid or sodium hydroxide, may be used. Many of these chemicals are known to be toxic to humans and wildlife. The U.S. EPA recently forced mining companies to reveal the contents, but not the specific fractional composition, of their fracking fluids. Because hydraulic fracturing has a special exemption from the federal Safe Drinking Water Act (in an amendment known as the Halliburton clause, after the oil services company that is thought to have written it), it's up to states and local units of government to protect public health. Several states and many towns and cities have demanded details on the chemicals being pumped into the ground.

A study released in 2011 by the National Academy of Sciences reported that drinking water samples from shallow wells near methane drilling sites in Pennsylvania and New York had 17 times as much methane as those from sites far from drilling. And a study by researchers at Dartmouth concluded that 3 to 8 percent of the methane from shale gas wells escapes to the atmosphere in leaks and venting over the life of the well. These methane emissions are up to twice those from conventional gas wells.

Homeowners near fracking operations sometimes report that so much methane has seeped into their wells that they can ignite water from their faucets. Furthermore, a study published in 2013 reported that pumping fracking wastewater into a deep well near Youngstown, Ohio, triggered nearly 12 earthquakes per month. After water injection ended, the earthquakes stopped. And fracking in the Marcellus Shale often releases radioactivity from uranium in the deposits. A number of cities have passed ordinances prohibiting drilling within city limits or near schools or hospitals.

What do you think? Does having access to natural gas justify the social and environmental costs of its extraction? If you were voting on this issue, what restrictions would you impose on the companies drilling wells in your hometown?

There are huge deposits of natural gas in North America, but the costs of extracting this gas could be unacceptable in many instances.
Source: Courtesy Mike Williams, Ohio Department of Natural Resources, Division of Mineral Resources Management

Gas can be shipped to market

In many places, gas and oil are found together in sediments, and both can be recovered at the same time. In remote areas, however, where there are no gas-shipping facilities, the gas often is simply flared (burned) off—a terrible waste of a valuable resource.

Quantitative Reasoning

As you can estimate from figure 19.17, shale gas mining releases only half as much CO_2 as surface-mined coal, but about 20 times as much methane. If methane has an infrared absorbing capacity 20 times as high as CO_2, what is the relative climate impact of these two fuels?

The World Bank estimates that 100 billion m^3 of gas, or 1.5 times the amount used annually in Africa, are flared every year. Increasingly, however, these "stranded" gas deposits are being captured and shipped to market.

World consumption of natural gas is growing by about 2.2 percent per year, considerably faster than either coal or oil. Much of this increase is in the developing world, where concerns about urban air pollution encourage the switch to cleaner fuel. Gas can be shipped easily and economically through buried pipelines. The United States has been fortunate to have abundant gas resources accessible by an extensive pipeline system. Until 2001, Canada was the primary source of natural gas for the United States, providing about 105 billion m^3 per year. **Liquefied natural gas (LNG)** imports were once expected to supply a significant

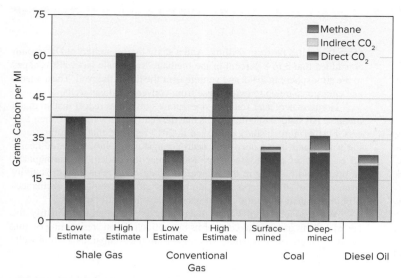

FIGURE 19.17 Comparison of greenhouse gas emissions from different fuel sources over a 20-year period. High and low estimates are provided for shale and conventional gas due to variations in field measurements. Under either scenario, shale gas has far higher climate-changing potential than any other fuel source.

Source: Data from Howarth, et al. 2011, https://www.fe.doe.gov/programs/gasregulation/authorizations/Orders_Issued_2012/75._Howarth-EtAl-2011.pdf.

portion of the United States supply. Since fracking has created a U.S. surplus, these ports are now being used as export, rather than import facilities.

In other places, gas lines have been subject to political or economic pressures. Russia, for example, has cut off gas supplies to Ukraine and Belarus in a dispute over prices and has threatened shipments to northern Europe over policy differences. Recent political unrest in North Africa made southern Europe nervous about its access to gas.

Intercontinental gas shipping can be difficult and dangerous because gas is volatile and easily ignited, and because it is transported under pressure, in liquid form, in order to make shipping economical. At −160°C (−260°F), the liquid's volume is about 1/600 that of gas. Special refrigerated ships transport LNG (fig. 19.18), but finding sites for terminals to load and unload these ships is difficult. Many cities are unwilling to accept the risk of an explosion of the volatile cargo. A fully loaded LNG ship contains about as much energy as a medium-size atomic bomb. Furthermore, huge amounts of seawater are used to warm and re-gasify the LNG. This can have deleterious effects on coastal ecology. In the United States, to override local objections, the federal government has assumed jurisdiction over LNG terminal siting.

Other unconventional gas sources

Natural gas resources have been less extensively investigated than petroleum reserves. There may be extensive "unconventional" sources of gas in unexpected places. Prime examples are recently discovered methane hydrate deposits in arctic permafrost and beneath deep ocean sediments. **Methane hydrates** are composed of small bubbles or individual molecules of natural gas trapped in a crystalline matrix of frozen water. At least 50 oceanic deposits and hundreds of land deposits are known. Altogether, they are thought to hold some 10,000 gigatons (10^{13} tons) of carbon, or twice as much as the combined amount of all coal, oil, and conventional natural gas. There are already reports of methane bubbling up over wide areas of the melting Arctic Ocean and adjacent permafrost regions. This could trigger a catastrophic feedback spiral of global warming, because methane has much greater heat-absorbing capacity than CO_2. And as the climate warms, even more methane will be emitted.

In 2013, Japan reported successful gas production from methane hydrate deposits located about 300 m (nearly 1,000 ft) below the seafloor in the Eastern Nankai Trough off its Pacific Coast. This could represent a 100-year supply for the energy-poor island nation. But there are worries about whether this volatile gas can be safely extracted and shipped to markets. Furthermore, mining the vast supplies of frozen methane worldwide could have potentially explosive consequences for both global energy markets and efforts to control climate change.

Methane also can be produced by digesting garbage or manure. Some U.S. cities collect methane from landfills and sewage sludge digestion. Burning the gas from these sources is better than simply letting it diffuse into the atmosphere. And in developing countries, small-scale manure digesters provide a valuable, renewable source of gas for heating, lighting, and cooking (chapter 20).

FIGURE 19.18 Global trade in natural gas involves shipments of pressurized, liquefied gas in specialized ships, such as this one at an Australian terminal. The potential hazards of handling volatile gas under pressure makes it difficult to find sites for these ports.
© Regis Martin/The New York Times/Redux Pictures

Section Review

1. What is fracking, and what problems does it cause?
2. Where is the Marcellus Shale, and why is it important?
3. What are methane hydrates, and why are they of concern?

19.5 NUCLEAR POWER

- *Nuclear power was expected to be too cheap to meter, but now appears to be too expensive to matter.*
- *Pressurized water reactors are by far the most common design.*
- *We haven't yet agreed on the best way to safely, and permanently, store nuclear waste.*

In 1953, President Dwight Eisenhower presented his "Atoms for Peace" speech to the United Nations. He announced that the United States would build nuclear-powered electrical generators to provide clean, abundant energy. He predicted that nuclear energy would fill the deficit caused by predicted shortages of oil and natural gas. It would provide power "too cheap to meter" for continued industrial expansion of both the developed and developing world, and it would be a supreme example of "beating swords into plowshares." Technology and engineering would tame the evil genie of atomic energy and use its enormous power to do useful work.

Glowing predictions about the future of nuclear energy continued into the early 1970s. Between 1970 and 1974, American utilities ordered 140 new reactors for power plants (fig. 19.19). Some advocates predicted that by the end of the century there would be 1,500 reactors in the United States alone. In 1970, the International Atomic Energy Agency (IAEA) projected worldwide nuclear power generation of at least 4.5 million megawatts (MW) by the year 2000, 18 times more than our current nuclear capacity and twice as much as present world electrical capacity from all sources.

Rapidly increasing construction costs, safety concerns, and the difficulty of finding permanent storage sites for radioactive waste

have made nuclear energy less attractive than promoters expected in the 1950s. Of the 140 reactors on order in 1975, 100 were never built (fig. 19.20). Even closing down a plant is expensive: The costs of demolishing a retired plant may be 10 times as much as building it in the first place. Ten nuclear reactors have been shut down in the United States, and deconstruction of most of them is now underway. Although these plants were generally small, costs have averaged several hundred million dollars each. To some people, it looks as if the much-acclaimed nuclear power industry might have been a very expensive wild-goose chase that will never produce enough energy to compensate for the money invested in research, development, mining, fuel preparation, and waste storage. New power sources, such as wind, solar, and biomass energy, are now considerably cheaper than new nuclear (chapter 20).

The nuclear power industry has been campaigning for greater acceptance, arguing that reactors don't release greenhouse gases during ordinary operation of the reactor. On the other hand, the mining, processing, and shipping of nuclear fuel, together with decommissioning of old reactors and perpetual storage of wastes, produces substantial amounts of CO_2 emissions and other pollutants.

Nevertheless, a number of prominent environmentalists have endorsed nuclear power as a solution to global climate change. In 2012, the Nuclear Regulatory Commission approved permits for two new nuclear reactors to be built in Georgia by the Southern Company. These reactors will be supported by $8 billion in loan guarantees from the federal government in addition to insurance caps on catastrophic accidents. If ever completed, these will be the first new nuclear power plants built in three decades in the United States.

How do nuclear reactors work?

The most commonly used fuel in nuclear power plants is Uranium[235], a naturally occurring radioactive isotope of uranium. Ordinarily, U^{235} makes up only about 0.7 percent of uranium ore, too little to sustain a chain reaction in most reactors. It must be

FIGURE 19.19 New York's Indian Point nuclear power plant is ranked the riskiest in the country by the U.S. Nuclear Regulatory Commission due to its age and location on the Hudson River just 24 miles (38 km) north of Manhattan Island. What would it cost to evacuate New York City if these reactors melt down?
© William P. Cunningham

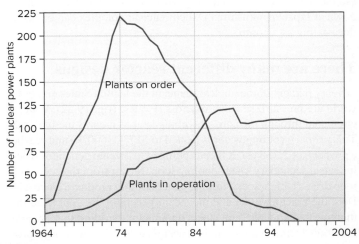

FIGURE 19.20 The changing fortunes of nuclear power in the United States are evident in this graph showing the number of nuclear plants on order and in operation.

purified and concentrated by mechanical or chemical procedures. Mining and processing uranium to create nuclear fuel is even more dirty and dangerous than coal mining. In some uranium mines, 70 percent of the workers—most of whom were Native Americans—died from lung cancer caused by high radon and dust levels. In addition, mountains of radioactive tailings and debris have been left unprotected at fuel preparation plants.

When the U^{235} concentration reaches about 3 percent, the uranium is formed into cylindrical pellets slightly thicker than a pencil and about 1.5 cm long. These small pellets pack an amazing amount of energy. Each 8.5-gram pellet is equivalent to a ton of coal or four barrels of crude oil.

The pellets are stacked in hollow metal rods approximately 4 m long. About 100 of these rods are bundled together to make a **fuel assembly.** Thousands of fuel assemblies containing 100 tons of uranium are bundled together in a heavy steel vessel called the reactor core. Radioactive uranium atoms are unstable—that is, when struck by a high-energy subatomic particle called a neutron, they undergo **nuclear fission** (splitting), releasing energy and more neutrons. When uranium is packed tightly in the reactor core, the neutrons released by one atom will trigger the fission of another uranium atom and the release of still more neutrons (fig. 19.21). Thus, a self-sustaining **chain reaction** is set in motion and vast amounts of energy are released.

The chain reaction is moderated (slowed) in a power plant by a neutron-absorbing cooling solution that circulates between the fuel rods. In addition, **control rods** of neutron-absorbing material, such as cadmium or boron, are inserted into spaces between fuel assemblies to shut down the fission reaction or are withdrawn to allow it to proceed. Water or some other coolant is circulated between the fuel rods to remove excess heat.

The greatest danger in one of these complex machines is a cooling system failure. If the pumps fail or pipes break during operation, the nuclear fuel quickly overheats and a "meltdown" can result that releases deadly radioactive material. Although nuclear power plants cannot explode like a nuclear bomb, the radioactive releases from a worst-case disaster like the 2011 meltdown at Japan's Fukushima Daiichi nuclear complex can be just as devastating as a bomb.

There are many different reactor designs

Seventy percent of the nuclear plants in the United States and in the world are pressurized water reactors (PWRs) (fig. 19.22). Water is circulated through the core, absorbing heat as it cools the fuel rods. This primary cooling water is heated to 317°C (600°F) and reaches a pressure of 2,235 psi. It then is pumped to a steam generator where it heats a secondary water-cooling loop. Steam from the secondary loop drives a high-speed turbine generator that produces electricity. Both the reactor vessel and the steam generator are contained in a thick-walled concrete and steel containment building that prevents radiation from escaping and is designed to withstand high pressures and temperatures in case of accidents. Engineers operate the plant from a complex, sophisticated control room containing many gauges and meters to tell them how the plant is running.

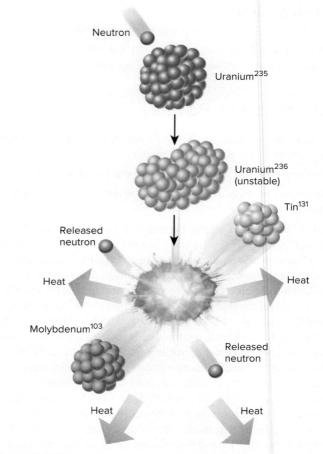

FIGURE 19.21 The process of nuclear fission is carried out in the core of a nuclear reactor. In the sequence shown here, the unstable isotope, uranium[235], absorbs a neutron and splits to form tin[131] and molybdenum[103]. Two or three neutrons are released per fission event and continue the chain reaction. The total mass of the reaction product is slightly less than that of the starting material. The residual mass is converted to energy (mostly heat).

Overlapping layers of safety mechanisms are designed to prevent accidents, but these fail-safe controls make reactors very expensive and very complex. A typical nuclear power plant has 40,000 valves, compared to one-tenth as many in a fossil-fuel-fired plant of similar size. In some cases, the controls are so complex that they confuse operators and cause accidents rather than prevent them. Under normal operating conditions, a PWR releases very little radioactivity and is probably less dangerous for nearby residents than a coal-fired power plant.

A simpler but dirtier and more dangerous reactor design is the boiling water reactor (BWR). In this model, water from the reactor core boils to make steam, which directly drives the turbine generators. This means that highly radioactive water and steam leave the containment structure. Controlling leaks is difficult, and the chances of releasing radiation in an accident are very high, as was the case in the BWR in Fukushima Daiichi, Japan.

In Britain, France, and the former Soviet Union, a common reactor design uses graphite, both as a moderator and as the

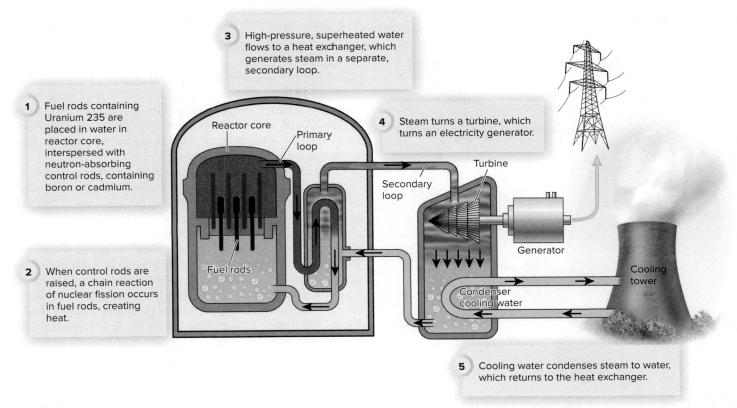

1 Fuel rods containing Uranium 235 are placed in water in reactor core, interspersed with neutron-absorbing control rods, containing boron or cadmium.

2 When control rods are raised, a chain reaction of nuclear fission occurs in fuel rods, creating heat.

3 High-pressure, superheated water flows to a heat exchanger, which generates steam in a separate, secondary loop.

4 Steam turns a turbine, which turns an electricity generator.

5 Cooling water condenses steam to water, which returns to the heat exchanger.

Reactor core

Primary loop

Turbine

Secondary loop

Fuel rods

Generator

Cooling tower

Condenser cooling water

FIGURE 19.22 Pressurized water nuclear reactor. Water is superheated and pressurized as it flows through the reactor core. Heat is transferred to nonpressurized water in the steam generator. The steam drives the turbogenerator to produce electricity.

structural material for the reactor core. In the British MAGNOX design (named after the magnesium alloy used for its fuel rods), gaseous carbon dioxide is blown through the core to cool the fuel assemblies and carry heat to the steam generators. In the Soviet design, called RBMK (the Russian initials for a graphite-moderated, water-cooled reactor), low-pressure cooling water circulates through the core in thousands of small metal tubes.

These designs were originally thought to be very safe, because graphite has high capacity for both capturing neutrons and dissipating heat. Designers claimed that these reactors could not possibly run out of control; unfortunately, they were proven wrong. The small cooling tubes are quickly blocked by steam if the cooling system fails, and the graphite core burns when exposed to air. Two of the most disastrous reactor accidents in the world involved fires in graphite cores that allowed the nuclear fuel to melt and escape into the environment. A 1956 fire at the Windscale Plutonium Reactor in England contaminated hundreds of square kilometers of countryside. Similarly, burning graphite in the Chernobyl nuclear plant in Ukraine made the fire much more difficult to control than it might have been in another reactor design.

The most serious accident at a North American commercial reactor occurred in 1979 when the Three Mile Island nuclear plant near Harrisburg, Pennsylvania, suffered a partial meltdown of the reactor core. The containment vessel held in most of the radioactive material. No deaths or serious injuries were verified, but the accident was a serious blow to future nuclear development.

Another unsettling revelation from Europe was that terrorists in Belgium had targeted nuclear power stations for attacks. These plans were foiled, but the potential for other similar attacks make many people worried.

Breeder reactors might extend the life of our nuclear fuel

For more than 30 years, nuclear engineers have been proposing high-density, high-pressure **breeder reactors** that produce fuel rather than consume it. These reactors create fissionable plutonium and thorium isotopes from the abundant, but stable, forms of uranium (fig. 19.23). The starting material for this reaction is plutonium reclaimed from spent fuel from conventional fission reactors. After about ten years of operation, a breeder reactor would produce enough plutonium to start another reactor. Sufficient uranium currently is stockpiled in the United States to produce electricity for 100 years at present rates of consumption, if breeder reactors can be made to work safely and dependably.

Several problems have held back the breeder reactor program in the United States. One problem is the concern about safety. The reactor core of the breeder must be at a very high density for the breeding reaction to occur. Water does not have enough heat capacity to carry away the high heat flux in the core, so liquid sodium generally is used as a coolant. Liquid sodium is very corrosive and difficult to handle. It burns with an intense flame if

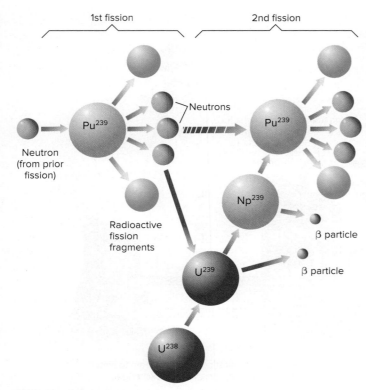

1st fission   2nd fission

Neutron (from prior fission)

Pu^{239}

Neutrons

Pu^{239}

Radioactive fission fragments

Np^{239}

β particle

U^{239}

β particle

U^{238}

FIGURE 19.23 Reactions in a "breeder" fission process. Neutrons from a plutonium fission change U^{238} to U^{239} and then to Pu^{239} so that the reactor creates more fuel than it uses.

exposed to oxygen, and it explodes if it comes into contact with water. Because of its intense heat, a breeder reactor will melt down and self-destruct within a few seconds if the primary coolant is lost, as opposed to a few minutes for a normal fission reactor.

Another very serious concern about breeder reactors is that they produce excess plutonium that can be used for bombs. It is essential to have a spent-fuel reprocessing industry if breeders are used, but the existence of large amounts of weapons-grade plutonium in the world would surely be a dangerous and destabilizing development. The chances of some of that material falling into the hands of terrorists or other troublemakers are very high. Japan planned to purchase 30 tons of this dangerous material from France and ship it halfway around the world through some of the most dangerous and congested shipping lanes on the planet to fuel a breeder program. In 1995, a serious accident at Japan's Moju breeder reactor caused reevaluation of the whole program.

A proposed $1.7 billion breeder-demonstration project in Clinch River, Tennessee, was on and off for 15 years. At last estimate, it would cost up to five times the original price if it is ever completed. In 1986, France put into operation a full-sized commercial breeder reactor, the SuperPhénix, near Lyons. It cost three times the original estimate to build and produces electricity at twice the cost per kilowatt of conventional nuclear power. After only a year of operation, a large crack was discovered in the inner containment vessel of the SuperPhénix, and in 1997 it was shut down permanently.

We lack safe storage for radioactive wastes

The first disposal problem in the nuclear fuel cycle is the enormous piles of radioactive mine wastes and abandoned mill tailings at uranium mines. Producing 1 ton of uranium fuel typically generates 100 tons of tailings and 3.5 million liters of liquid waste. There now are approximately 200 million tons of radioactive waste in piles around mines and processing plants in the United States. This material is carried by the wind or washed into streams, contaminating areas far from its original source.

In addition to the leftovers from fuel production, the United States has about 100,000 tons of low-level waste (contaminated tools, clothing, building materials, and so on) and about 77,000 tons of high-level (very radioactive) wastes. The high-level wastes consist mainly of spent fuel rods from commercial nuclear power plants and assorted wastes from nuclear weapons production. While they're still intensely radioactive, spent fuel assemblies are stored in deep, water-filled pools at the power plants. These pools were originally intended only as temporary storage until the wastes were shipped to reprocessing centers or permanent disposal sites.

In 1987, after a years-long search the U.S. Department of Energy announced plans to build the first high-level waste repository under a barren desert ridge called Yucca Mountain in Nevada. Waste would be buried deep in the ground, where it was hoped it would remain unexposed to groundwater and earthquakes for the thousands of years required for the radioactive materials to decay to a safe level. But objections from Nevadans, who objected to receiving waste from far-away states, and continuing worries about the stability of the site led the Obama administration to cut off funding for the project in 2009 after 20 years of research and $100 billion in exploratory drilling and development. For the foreseeable future, the high-level wastes are being held in storage pools and dry casks located at 131 sites in 39 states (fig. 19.24), a long-term temporary, and not very secure strategy.

FIGURE 19.24 Spent fuel is being stored temporarily in large, aboveground "dry casks" at many nuclear power plants.
Source: Office of Civilian Radioactive Waste Management/U.S. Dept. of Energy

Russia has offered to store nuclear waste from other countries. Plans are to transport wastes to the Mayak in the Ural Mountains. The storage site is near Chelyabinsk, where an explosion at a waste facility in 1957 contaminated about 24,000 km² (9,200 mi²) of countryside. The region is now considered the most radioactive place on earth, so the Russians think it can't get much worse. They expect that storing 20,000 tons of nuclear waste should pay about US$20 billion.

Some nuclear experts believe that monitored, retrievable storage would be a much better way to handle wastes. This method involves holding wastes in underground mines or secure surface facilities where they can be watched. If canisters begin to leak, they could be removed for repacking. But safeguarding the wastes would be expensive, and the sites might be susceptible to wars or terrorist attacks. If the owners of nuclear facilities had to pay the full cost for fuel, waste storage, and insurance against catastrophic accidents, no one would be interested in this energy source. Rather than being too cheap to meter, it would be too expensive to matter.

Decommissioning nuclear plants is costly

Old power plants themselves eventually become waste when they have outlived their useful lives. Most plants are designed for a life of only 30 years. After that, pipes become brittle and untrustworthy because of the corrosive materials and high radioactivity to which they are subjected. Plants built in the 1950s and early 1960s already are reaching the ends of their lives. You don't just lock the door and walk away from a nuclear power plant; it is much too dangerous. It must be taken apart, and the most radioactive pieces have to be stored just like other wastes. This includes not only the reactor and pipes but also the meter-thick, steel-reinforced concrete containment building. The pieces are cut apart by remote-control robots because it's too dangerous to work on them directly.

Altogether, the U.S. reactors now in operation might cost somewhere between $200 billion and $1 trillion to decommission. No one knows how much it will cost to store the debris for thousands of years or how it will be done. These aren't the only reactors needing decommissioning: Plutonium weapons production plants and nuclear submarines also have to be decommissioned. Originally, the Navy proposed to just tow old submarines out to sea and sink them. The risk that the nuclear reactors would corrode away, however, and leak radioactive isotopes into the ocean makes this method of disposal unacceptable.

The changing fortunes of nuclear power

Decommissioning nuclear plants is expensive, but it has now become cheaper than keeping them running. In 2016 Pacific Gas and Electric announced that it would close California's last nuclear power plant, Diablo Canyon, when its license expires in 2025. The plant has been controversial because of its vulnerability to earthquakes and tsunamis, but the reason for decommissioning it is economic. The cost of decommissioning the plant is expected to be at least $3.8 billion, but PG&E concludes that it is cheaper to build and operate renewables than it is to upgrade and maintain Diablo Canyon.

The cost of producing one MWh of electricity at the plant is $70, far more than the $50/MWh now needed to build and operate commercial-scale solar in California. To make a profit, PG&E needs to keep its production costs low, so it makes sense to close the nuclear plant. By planning ahead, the company can work with unions and employees to build new solar or wind capacity and efficiency measures while it prepares to shut down and deconstruct the nuclear plant.

A longer-running debate has been about safety. Before the 1978 Three Mile Island accident, two-thirds of Americans supported nuclear power. After Chernobyl exploded in 1986, less than one-third of Americans favored nuclear power. In Japan, where nuclear reactors provided about one-third of the country's electricity, there were plans to expand nuclear energy to about half the nation's power supply. That changed with the magnitude-9 earthquake and huge tsunami that hit the northeast coast of Japan on March 11, 2011 (see What Do You Think?, "Twilight for Nuclear Power?").

Nuclear advocates—and even some conservationists—have been promoting nuclear reactors as clean and environmentally friendly because they don't emit greenhouse gases. There's been talk about a "nuclear renaissance," not only in the United States but in other countries as well. In 2010, the Obama administration pledged an $8.3 billion loan to support construction of two nuclear reactors in Georgia. And in 2016, the Tennessee Valley Authority brought online its first new power plant in three decades. Supporters of nuclear energy hailed this as the start of a new era, but critics pointed out that the TVA's Watt's Bar reactor had a history of enormous cost overruns, antiquated design, and interminable construction delays, having been listed as "under construction" for 43 years. When it was launched in 1972, it was expected to cost $400 million. When it opened in 2016, it had cost $6.1 billion. Analysts increasingly find that nuclear power remains costly, in comparison to conservation and renewable energy, if we account for the subsidies involved in its operation.

Fusion energy has been proposed as an alternative to nuclear fission. **Nuclear fusion** energy is released when two smaller atomic nuclei fuse into one larger nucleus (in contrast, nuclear *fission* involves breaking the nucleus apart). Nuclear fusion reactions are the energy source for the sun and for hydrogen bombs. The fuels for these reactions are deuterium and tritium, two heavy isotopes of hydrogen. At temperatures of 100 million °C and pressures of several billion atmospheres, fusion of deuterium and tritium will occur. Under these conditions, the electrons are stripped away from atoms and the forces that normally keep nuclei apart are overcome. As nuclei fuse, some of their mass is converted into energy, some of which is in the form of heat. In theory, the advantages of fusion reactions would include production of fewer radioactive wastes, the elimination of products that could be made into bombs, and a fuel supply (deuterium and tritium) that is much more abundant and safer than uranium.

There are two main schemes for fusion: magnetic confinement and inertial confinement. Inertial confinement involves a small pellet (or a series of small pellets) bombarded from all sides at once with extremely high-intensity laser light (fig. 19.25a). The

Twilight for Nuclear Power?

On Friday, March 11, 2011, at 2:46 p.m. Tokyo time, a magnitude-9.0 earthquake hit northern Japan. The largest earthquake in Japan's recorded history damaged buildings and roads, but even worse, it generated tsunami waves up to 30 m (98 ft) high that crushed buildings, toppled power lines, and washed away cars, boats, and millions of tons of debris. Authorities reported 15,846 deaths, 6,011 injuries, 3,320 people missing, and 125,000 buildings damaged or destroyed by the giant waves.

One of the many disastrous results of this catastrophe was the destruction of four of the six nuclear reactors at the Fukushima Daiichi power station 170 mi (273 km) north of Tokyo. The reactors shut down, as they were designed to do, when the earthquake hit, but that eliminated the electricity needed to pump cooling water through the intensely hot reactor core. Backup generators and connections to the regional power grid that would have provided emergency power were destroyed by the tsunami. The reactors quickly overheated, and the fuel rods began to melt in three of the six reactors cores. Hydrogen explosions in the reactor buildings at the complex destroyed roofs and walls and scattered radioactive debris around the area. In addition, spent fuel rods in storage pools of two units also overheated and caught fire, releasing even more radiation.

Plant operators sprayed seawater onto reactors to cool the reactors and put out fires, but that washed radioactive pollution into the ocean and contaminated seafood on which Japan depends. High radioactivity caused authorities to order the evacuation of 140,000 people living within a 12-mile (20 km) radius around the facility. But the toll could have been worse. If the melting fuel and fires hadn't been contained, the radiation release could have been ten times greater than the 1986 disaster at Chernobyl in Ukraine.

At one point, government officials seriously considered evacuating the Tokyo metropolitan area. That might have meant moving up to 40 million people, which would have been the largest single mass relocation in world history. However, westerly winds blew most of the radiation out to sea, and abandoning Tokyo wasn't necessary.

Still, cleanup will take decades, and the zone near the reactors may never be habitable again. Years after the disaster, the cooling pools are still leaking and millions of liters of highly radioactive water have accumulated in leaky, above-ground storage tanks. Millions of tons of contaminated debris swept out to sea by the tsunami has traveled around the world, some of it showing up on North American shores. The Japanese government is considering taking over remediation efforts, which are likely to take 30–40 years and cost over $100 billion. Commercial fishing is still not allowed in a wide area near the Fukushima plant.

Altogether, Japanese officials estimate that losses may reach $300 billion. This disaster is causing people in many countries to reconsider nuclear power. In Japan, which once got about one-third of its electricity from nuclear plants and had plans to expand that share to more than half the nation's power supply, more than 80 percent of the population now say they oppose nuclear power. After the disaster, all of the nation's 54 reactors were shut down. An intense debate continues about whether to ever restart them.

After Fukushima, Germany immediately shut down eight reactors and promised to close all the rest of its nuclear plants by 2022. China has suspended approvals for new reactors. Italy, Switzerland, and Spain voted to keep their countries nonnuclear. And in France, which gets three-quarters of its electricity from nuclear power, 62 percent of the population favored a phase-out of this energy source.

Could this be the death knell for nuclear power? Although public opinion swung strongly against this technology after Chernobyl, some people have been arguing that we need nuclear power at least as a temporary stopgap to replace fossil fuels in an effort to stop global climate change. Others argue that nuclear power consumes funds that should be spent on more sustainable alternatives. What do you think? How would you weigh the risks and benefits of nuclear power? What further information would you need to evaluate options? What safeguards should we employ to reduce our risks?

Three of the four nuclear reactors at Japan's Fukushima Diiachi that were destroyed by fuel melting and hydrogen explosions after the 2011 tsunami knocked out emergency cooling systems.
© DigitalGlobe via Getty Images

---

sudden absorption of energy should cause an implosion (an inward collapse of the material) that will increase densities by 1,000 to 2,000 times and raise temperatures above the critical minimum. So far, however, no lasers powerful enough to create fusion conditions have been built.

Magnetic confinement involves the containment and compression of plasma in a powerful magnetic field inside a vacuum chamber. Compression by the magnetic field should raise temperatures and pressures enough for fusion to occur. The most promising example of this approach, so far, has been a Russian design called *tokomak* (after the Russian initials for "torodial magnetic chamber"), in which the vacuum chamber is shaped like a large donut (fig. 19.26b). In 2016, Chinese researchers announced they were able to produce a plasma three times hotter than the core of

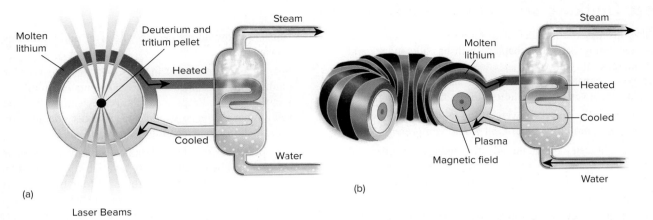

FIGURE 19.25 Nuclear fusion designs. (a) Inertial confinement is created by laser beams that bombard and ignite fuel pellets. Molten lithium absorbs heat and transfers it to a steam generator. (b) In the tokomak design, a powerful magnetic field confines and compresses the plasma to reach critical temperatures and pressures.

the sun—approximately 50 million °C (90 million °F)—and were able to maintain this temperature for a record-breaking 102 seconds. Despite 50 years of research and a $25 billion investment, however, fusion reactors have not yet been able to produce more energy than they consume.

Section Review

1. Draw and label a diagram of a typical PWR nuclear plant.
2. How is energy released from uranium atoms?
3. What are some major obstacles to expansion of nuclear power?

Conclusion

Our energy future is far from certain. We have huge amounts of coal, petroleum, and natural gas. This provided a lifestyle of luxury and convenience for those of us lucky enough to live in the industrialized countries of the world, but it has created titanic environmental problems—including acid rain, strip-mined landscapes, massive oil spills, huge payments to unstable countries, and, perhaps most importantly, global climate change. There are still very large supplies of unconventional fossil fuels, including tar sands, oil shale, coal-bed methane, and methane hydrates, but the environmental costs of extracting those resources may preclude their use.

What, then, should we do? Some people hold out the promise of technological solutions to this dilemma. They point to coal gassification with carbon capture and storage, nuclear power, and possibly fusion reactors as answers to our energy problem. Others, however, point to nuclear disasters and unacceptable waste storage options. Many argue that we ought to move immediately toward conservation and renewable energy, such as solar, wind, biofuels, small-scale hydro, and geothermal power. Even if we do this, however, it will probably take decades to end our dependence on fossil fuels. Therefore, it's important to understand the relative benefits and disadvantages of each of our conventional energy sources. As consumers, each of us needs to examine our energy use and its environmental impacts. In chapter 20, we'll investigate conservation and renewable energy options.

Reviewing Key Terms

Can you define the following terms in environmental science?

bitumen 19.3
breeder reactor 19.5
carbon capture and storage (CCS) 19.2
chain reaction 19.5
control rods 19.5

energy 19.1
fossil fuels 19.1
fuel assembly 19.5
hydraulic fracturing ("fracking") 19.3
joule 19.1

kerogen 19.3
liquefied natural gas (LNG) 19.4
methane hydrates 19.4
nuclear fission 19.5
nuclear fusion 19.5

oil shale 19.3
power 19.1
tar sands 19.3
unburnable carbon 19.3
watt 19.1
work 19.1

Critical Thinking and Discussion Questions

1. We have discussed a number of different energy sources and energy technologies in this chapter. Each has advantages and disadvantages. If you were an energy policy analyst, how would you compare such different problems as the risk of a nuclear accident versus air pollution effects from burning coal?

2. If your local utility company were going to build a new power plant in your community, what kind would you prefer? Why?

3. The nuclear industry is placing ads in popular magazines and newspapers claiming that nuclear power is environmentally friendly because it doesn't contribute to the greenhouse effect. How do you respond to that claim?

4. Our energy policy effectively treats some strip-mine and well-drilling areas as national sacrifice areas, knowing they will never be restored to their original state when extraction is finished. How do we decide who wins and who loses in this transaction?

5. Storing nuclear wastes in dry casks outside nuclear power plants is highly controversial. Opponents claim the casks will inevitably leak. Proponents claim they can be designed to be safe. What evidence would you consider adequate or necessary to choose between these two positions?

6. The policy of the United States has always been to make energy as cheap and freely available as possible. Most European countries charge three to four times as much for gasoline as we do. Who benefits and who or what loses in these different approaches? How have our policies shaped our lives? What does existing policy tell you about how governments work?

Data Analysis

Comparing Energy Use and Standards of Living

In general, income and standard of living increase with energy availability. This makes sense because cheap energy makes it possible to obtain goods and services that enrich our lives, and to use machines to extend our productivity. However, energy use per capita isn't strictly tied to quality of life. Some countries use energy extravagantly without corresponding increases in income or standard of living. Go to Connect to examine a graph comparing per capita income and energy consumption and to answer questions about the relationship between these factors.

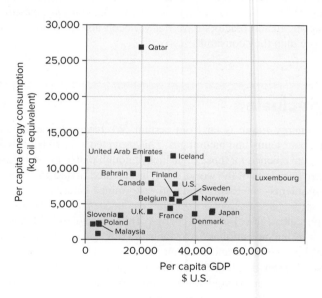

20

Sustainable Energy

▲Germany currently has the largest amount of photovoltaic solar power in the world. This renewable energy source provides nearly 60 percent of peak energy demand for the country.
© peart/E+/Getty Images

Learning Outcomes

After studying this chapter, you should be able to:

20.1 Describe strategies for energy efficiency.
20.2 Explain how we could tap solar energy.
20.3 Identify applications for wind power.
20.4 Discuss the potential of biomass and fuel cells.
20.5 Summarize the costs and benefits of hydropower.

"We know the country that harnesses the power of clean, renewable energy will lead the twenty-first century."

– Barack Obama

A Renewable Energy Transition

In an effort to combat climate change, Germany has made an ambitious commitment to a renewable energy transition (Energiewende). The aim is to cut both fossil fuel use and carbon emissions 80 percent by 2050, and to replace most current energy use with conservation and sustainable energy. How will they accomplish this formidable task?

Although much of Germany has a cloudy, cool climate, the country has sufficient solar energy and wind power to fuel its economy and lifestyle. This transition is made easier because prices for wind and solar have fallen to about half that of fossil fuels and nuclear on which the country has depended in the past. Germany has an abundant supply of coal, and this carbon-intensive fuel once made up 40 percent of all electric generation. But most of that coal will have to stay in the ground. By 2050, coal is expected to fall to 8 percent of the total energy supply. Furthermore, Germany's heavy dependence on imported natural gas from Russia has made it vulnerable to political pressure. In 2013, Germany spent some 90 billion euros on energy imports. Reducing this reliance on imported fuel and nurturing local jobs in renewables makes a lot of sense.

Germany has been an early adopter and leader in both solar and wind for many years. In 2014, Germany had nearly as much solar power as all the rest of the E.U. put together, and it had more solar than all the rest of the world outside the E.U. combined. Contrary to what some critics predicted, the introduction of this new technology hasn't bankrupted the country. The renewable share of electricity generation in Germany rose seven-fold between 1991 and 2014, while the economy grew 180 percent (fig. 20.1). Currently, nearly 400,000 German workers are employed in renewable energy, compared to less than half that many in the fossil fuel industry. By 2050, Germany expects to get 100 percent of its electricity from renewable sources. This transition is aided by the fact that the price of German solar panels has decreased by 74 percent since 2006.

Although some environmentalists in the U.S. argue that nuclear power is the only way to meet reduced carbon goals, Germany rejects an atomic path because of the dangers it poses to this densely packed country. Germans were heavily affected by the 1986 meltdown of the Chernobyl reactor in the Ukraine. They don't want to repeat that experience. Furthermore, they have no uranium deposits of their own, and no good place to store radioactive waste.

Much of the emphasis on renewable energy in Germany has been on individual and community ownership rather than in large utilities. From fewer than 100 citizen-owned energy cooperatives in 2006, this movement has grown ten-fold in just a decade. This not only fosters community involvement, but also serves to encourage social justice. Much of the growth in solar energy has been focused on individual homeowners, partly because the shift to renewables has been opposed by large utilities, but has been welcomed by ordinary citizens. Rooftop PV panels are encouraged by generous feed-in tariffs that pay homeowners near-retail prices for the power they generate. Most U.S. utilities are actively hostile to both feed-in tariffs and cooperative or home ownership of power generation.

Conservation is another key to renewable energy in Germany, and the country has been a global leader in this area. As we'll discuss in this chapter, **passive houses** consume less energy while providing more comfort than conventional buildings. Total energy consumption in Germany is forecast to fall by half over the next 40 years, while the renewable contribution is projected to increase 10-fold from 6 percent to more than 60 percent of the total mix.

Biomass is the most versatile of all types of renewable energy because it can provide heat, electricity, and motor fuel.

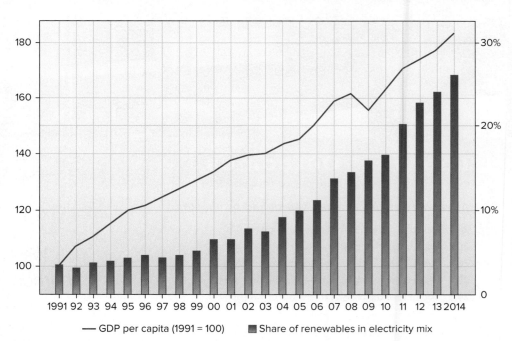

— GDP per capita (1991 = 100)   ■ Share of renewables in electricity mix

FIGURE 20.1 Over the past 15 years, the renewable contribution to the German energy supply has increased seven-fold, while the economy has grown by 180 percent.
Source: Heinrich Böll Foundation 2015

Not surprisingly, biomass is expected to make up nearly two-thirds of Germany's renewable energy consumption by 2020. Within the E.U., Germany is the greatest producer of wood, and wood is by far the greatest source of bioenergy in the country. Roughly 40 percent of German timber production is used as a source of energy. Germany is also the leading biogas market. In 2013, more than

50 percent of Europe's electricity from biogas was produced there, with further dynamic growth to come.

If other countries were to follow Germany's example, we might have a chance to hold climate warming to no more than 2°C, a level that would avoid the most disastrous consequences of climate change.

20.1 ENERGY EFFICIENCY

- *There are many ways to save energy.*
- *Green buildings can cut energy costs dramatically.*
- *Transportation is important factor in energy efficiency.*

As chapter 19 demonstrates, the world has vast supplies of fossil fuels, but digging up, transporting, and burning all that carbon can have catastrophic environmental and social effects, chief of which is global climate change. To maintain a livable world, we need to leave 80 percent of all remaining fossil fuels in the ground. Fortunately, there are abundant, readily available renewable energy sources that we can access immediately. Using currently available technology in sites where energy facilities are socially, economically, and politically acceptable, there's more than enough power from the sun, wind, geothermal, biomass, and other sources to meet all our present energy needs (fig. 20.2). We'll look at each of those sources in this chapter.

Often the easiest and cheapest ways to avoid energy shortages and to relieve the environmental and health effects of our current energy technologies is simply to use less. As the opening case study for this chapter shows, Germany has increased its energy efficiency by more than 50 percent over the past 25 years, while decreasing its primary energy consumption (fig. 20.3). These improvements reduce Germany's reliance on imported energy from distant and dangerous places. Even those who aren't interested in environmental or social issues may be interested in reducing energy consumption for economic reasons.

There are many ways to save energy

Conservation is partly about behavior—turning off lights and adjusting thermostats—but it's also a technological challenge. Often our use of energy is extremely inefficient: Most energy used to operate internal combustion engines and conventional incandescent light bulbs, for example, is lost as heat, rather than producing motion or light. Better appliances, buildings, and vehicles are critical to efficiency improvements—these help explain why a number of European countries have higher standards of living than the United States despite using about half as much energy per person.

Light-emitting diodes (LEDs), for example, have transformed lighting efficiency. They cost more than a compact fluorescent bulb, but they use about half as much electricity, and last about five times as long. Compared to a standard incandescent bulb, LEDs use one-tenth as much energy and last up to 50 times longer. LEDs can produce millions of colors and be adjusted in brightness to suit ambient conditions. They are being used now in everything from flashlights and holiday lights to advertising signs, brake lights, exit signs, and street lights. Many cities have replaced traffic lights and street lights with LEDs. These retrofits typically pay for themselves (in dramatically lowered energy costs) in about two years.

Few of us think about how much electricity is used by appliances in standby mode. You may think you've turned off your TV, DVD player, cable box, or printer, but they're really continuing to draw power in an "instant-on" mode (fig. 20.4). For the average home, these "vampire currents" can amount to a quarter of the monthly electric bill. Putting your computer to sleep saves about 90 percent of the energy it uses when fully on, but

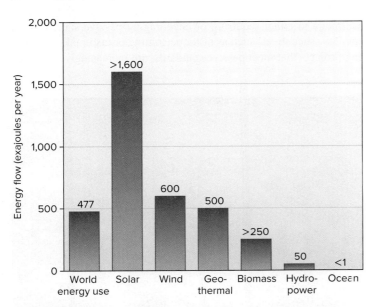

FIGURE 20.2 Potential energy available from renewable resources using currently available technology in presently accessible sites. Together, these sources could supply more than six times the current world energy use.
Source: Adapted from UNDP and International Energy Agency

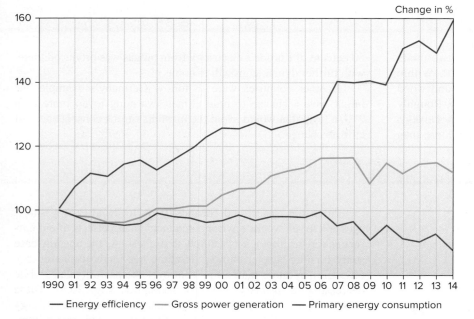

Change in %

— Energy efficiency　— Gross power generation　— Primary energy consumption

FIGURE 20.3 Changes in energy Germany's efficiency, power generation, and consumption, relative to 1990. Germany is getting more value from less energy because of efficiency improvements.
Source: Heinrich Böll Foundation 2015

For decades, Rocky Mountain Institute founder, Amory Lovins, has suggested that accounting for **negawatts** (or the amount of energy saved through conservation) should be a priority in energy policy. He argued that it's much cheaper to conserve energy than to build new generating stations and transmission lines to meet demand. The Federal Energy Regulatory Commission (FERC) has ruled that negawatts can be traded as a commodity. This means that a company that avoids using energy can market its savings just as a power company would do with real megawatts.

Green buildings dramatically reduce energy costs

Innovations in green building have been stirring interest in both commercial and household construction. Much of the innovation has occurred in large commercial structures, which have larger budgets—and more to save through efficiency—than most homeowners have. Energy audits can help show you where energy losses are occurring (fig. 20.5). Sealing leaks with caulk or weather stripping is one of the quickest and most cost-effective things you can do to save energy, and to make living spaces more comfortable.

turning it completely off is even better. Plugging appliances into an inexpensive power strip allows you to switch them off when not in use.

Industrial energy savings are another important part of our national energy budget. More efficient electric motors and pumps, new sensors and control devices, advanced heat-recovery systems, and materials recycling have reduced industrial energy requirements significantly. In the early 1980s, U.S. businesses saved $160 billion per year through conservation. When oil prices collapsed, however, many businesses returned to wasteful ways.

High-efficiency buildings are becoming the norm in Europe. Many regions in Germany require that new buildings conform to **passive house standards.** These standards are strict energy use limits, along with guidelines for building practices that can reduce energy consumption to just 10 percent of what normal buildings use. Many cities and states across Europe have begun to follow Germany in either requiring or encouraging passive standards. In the U.S., incentives, such as faster permitting or tax breaks to building projects that meet passive standards, are increasingly common.

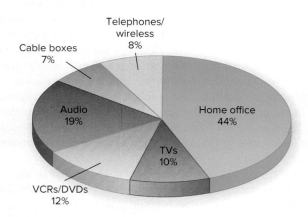

FIGURE 20.4 Typical standby energy consumption by household electrical appliances.
Source: U.S. Department of Energy

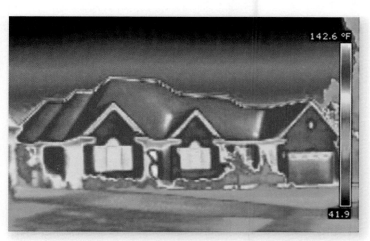

FIGURE 20.5 Infrared photography shows heat loss in a building.
Source: Courtesy John Cunningham and Meadowlark Energy

Relatively rare a few years ago, passive standard houses and office buildings are becoming more common in Europe, North America, and Asia. In a place where building standards are already high—as they are in Germany—a passive house can cost only 10 percent more than a standard house, and that extra cost pays for itself in just a few years of reduced energy use. Where building standards are more lax and housing is inexpensive, the cost difference can be much greater, but so can the savings. Residents also say passive houses are more comfortable, with even heating, better natural light, and less noise than conventional housing (fig. 20.6).

What does passive housing involve? The main principles are these:

1. No thermal bridges (wood or metal components conduct heat through walls, roof, or foundation)
2. Roof and wall insulation 24–32 cm (10–14 in) thick
3. Well-sealed windows and doors to prevent heat loss or gain
4. Heat exchangers, to warm or cool fresh air as it is exchanged with stale air leaving the building
5. Triple-pane windows that reduce radiative heat loss or gain
6. Windows well positioned to let in daylight

When a passive building produces electricity—for example, with solar panels on the roof—it can become a "net-zero" building that uses no more energy than it produces. Some of these are even "energy-plus" buildings, producing more energy than they consume on an annual basis. Additional measures are becoming common in many areas: Appliances often have timers so they can run overnight, when energy use is low. Thermostats can automatically adjust the temperature when you are not home or when you are sleeping. Smart metering gives you information on how much energy is being used and how much it costs. Using one of these systems, you might program your water heater to operate only after midnight, when surplus wind power is available (see the **Exploring Science** box below, "Greening Gotham").

Transportation could be far more efficient

One of the areas in which most of us can accomplish the greatest energy conservation is in our transportation choices. You may not be able to build an energy-efficient house or persuade your utility company to switch from coal or nuclear to solar energy, but you can decide every day how you travel to school, to work, or for shopping or entertainment. Automobiles and light trucks account for 40 percent of the U.S. oil consumption and produce one-fifth of its carbon dioxide emissions.

The Bureau of Transportation Statistics reports that there are now more vehicles in the United States (254 million) than licensed drivers (190 million). More importantly, those vehicles are used for an average of 1 billion trips per day. Many of us drive now for errands or short shopping trips that previously might have been made on foot. Some of that is due to the design of our cities (chapter 22). Suburban subdivisions have replaced compact downtown centers in most cities. Shopping areas are surrounded by busy streets and vast parking lots that are highly pedestrian unfriendly. But sometimes we use fuel inefficiently simply because we haven't thought about alternatives. The Census Bureau reports that three-quarters of all workers commute alone in private vehicles. About 20 percent carpool, while less than 5 percent use public transportation, and a mere 0.38 percent walk or travel by bicycle.

What can you do if you want to be environmentally responsible? The cheapest, least environmentally damaging, and healthiest alternative for short trips is walking or bicycling. You need to get some exercise every day, so why not make walking part of it? For trips less than 2 km, it's often quicker to go by bicycle than to find

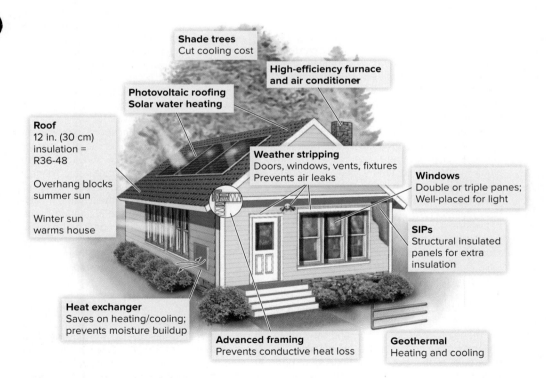

Shade trees
Cut cooling cost

High-efficiency furnace and air conditioner

Photovoltaic roofing
Solar water heating

Roof
12 in. (30 cm)
insulation =
R36-48

Overhang blocks
summer sun

Winter sun
warms house

Weather stripping
Doors, windows, vents, fixtures
Prevents air leaks

Windows
Double or triple panes;
Well-placed for light

SIPs
Structural insulated
panels for extra
insulation

Heat exchanger
Saves on heating/cooling;
prevents moisture buildup

Advanced framing
Prevents conductive heat loss

Geothermal
Heating and cooling

FIGURE 20.6 Energy-efficient buildings can lower energy costs dramatically. Many features can be added to older structures. New buildings that start with energy-saving features (such as SIPs or advanced framing) can save even more money.

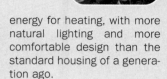

Greening Gotham: Can New York Reach Its 80 by 50 Goal?

New York City has a gritty reputation, and it's not known as a world leader in energy conservation. As the biggest city in the United States, New York uses more energy than almost any other, almost all of it derived from oil, natural gas, and nuclear power. As a northern city, New York needs about six months of heating and four months of air conditioning each year for its aging, often leaky building stock, and it has limited solar potential.

At the same time, New York City planners know that climate change threatens the city with rising sea levels, increased hurricane frequency, intense heat waves, and economic instability as a warming climate undermines agriculture, transportation, health, and other underpinnings of the city's economic foundations. A serious heat wave can be especially deadly for low-income or elderly urban residents.

If these threats are to be reduced or averted, New York and other cities have to do their share to cut greenhouse gas emissions, and they need to do it soon. The Intergovernmental Panel on Climate Change (IPCC) reports that to avoid the worst impacts of climate change, the world needs to reduce its greenhouse gas emissions 80 percent by the year 2050. This is a daunting goal for any city. New York's energy infrastructure, like most of its building stock, was established half a century ago, or more. Can the city replace or upgrade its aging infrastructure on a tight budget and a tight timeline?

In 2014, the city's Office of Sustainability published a new policy stating that New York would aim for the IPCC's goal of 80 percent reductions by 2050, often called an "80 by 50" goal. The goal is ambitious, and the path to get there is not entirely clear, but Mayor Bill de Blasio and his administration agreed that it was necessary to protect the city's long-term interests. They also knew that if New York could lead the way, other cities would take the idea seriously and follow.

Knickerbocker Commons, in Brooklyn, is the first mid-sized apartment building designed to passive house standards in the United States.

© Architect Chris Benedict, R.A. 803 Knickerboker Avenue

New York already uses less energy per person than any other major American city. Densely built apartments and offices share walls and conserve heat. Public transportation uses a minute fraction of the energy private cars do, per rider. Bicycle infrastructure has transformed the city in recent years, providing cheap and efficient transportation options. Many people walk to work, school, and stores because distances are short. People with small kitchens also tend to buy and waste less food. In New York, the biggest challenges are leaky, thin-walled buildings and a lack of easy places to plan wind, solar, or other alternative energy sources.

New York is starting with its building stock. Financial incentives are helping landlords upgrade heating and cooling systems. The public housing authority has begun installing geothermal systems, better windows and building sealing, and more efficient heating systems. Upgrades save new Yorkers money as well as energy, and they make buildings more comfortable. City housing has even built its first apartment building to passive house standards, which prescribe buildings so tightly made that they use almost no

energy for heating, with more natural lighting and more comfortable design than the standard housing of a generation ago.

The city also works closely with large institutions in finding ways to reduce electricity consumption at peak hours, help them finance improved lighting, windows, and thermostats, and employ measures to systematically turn off lights and turn down thermostats. Landlords are also investing in co-generation—generating electricity with waste heat from heating and cooling systems—which reduces energy demand on public utilities.

New policies are as important as new technology and tighter buildings. Reducing paperwork and bureaucratic obstacles to upgrades saves time and money. New rules at the state level can give utilities new incentives to integrate solar, wind, geothermal, and other renewable energy sources into their energy production system. The city also collaborates with public advocacy organizations to help define new incentives for the public to invest in solar and wind energy, as well as energy conservation. The city is also negotiating with the state of New York to develop new rules that promote shared solar energy investments, and the easier permitting of offshore wind. Because, without some investment in alternatives, all the insulation in the world won't do the job that needs to be done.

In many ways, the story of energy in the future is going to be different from the story in the past. Just as oil and automobiles transformed life a century ago, alternative energy sources, consumption practices, and ownership policies are likely to transform how we live and work. Mayor de Blasio and his team are betting that the future scenario will have exciting co-benefits, including safer, more reliable, and more affordable energy, if we can just commit to doing it right.

FIGURE 20.7 These rental bikes provide a convenient and efficient way to get around in a bicycle-friendly city.
© William P. Cunningham

a parking space for your car (fig. 20.7). Public transit is also an excellent environmental option. The American Public Transportation Association reports that taking mass transit instead of driving a private car to work or school can reduce your energy consumption and greenhouse gas emissions by 30 percent per year.

Auto manufacturers play a key role in energy efficiency. In response to the 1970s oil price shocks, automobile gas-mileage averages in the United States more than doubled from 13.3 mpg in 1973 to 25.9 mpg in 1988. Then falling fuel prices in the 1990s discouraged further conservation, and mileage hardly improved for nearly 20 years. In 2012, however, the Department of Transportation and the Environmental Protection Agency issued regulations requiring cars and light trucks to achieve an average of 54.5 mpg (23 km/liter) by 2025. The agencies estimate this efficiency standard will save consumers $1.7 trillion in fuel costs, reduce oil consumption by 2 million barrels per day, and eliminate 6 billion metric tons of CO_2 emissions by 2025. The agencies say the technology to achieve this standard already exists, it's simply a matter of implementing it. The federal government will provide assistance and incentives to achieve these goals.

Diesels make up about half the vehicles sold in Europe because of their superior efficiency. Many automotive companies claim that recent advances have made diesels much cleaner and quieter than they were a generation ago. However, it was revealed in 2015 that some car companies had rigged their engines to run more efficiently during tests than they actually do on the street. This has hurt consumer confidence in diesel technology, but greater efficiency still makes diesels advantageous.

Federal and state rules and incentives have led to a growing variety of efficient automobile choices. You probably know that hybrid gasoline–electric engines offer excellent fuel economy and low emissions. The Toyota Prius C is rated by the EPA at 53 mpg (23 km/liter) in the city and 46 mpg (19 km/liter) on the highway. Nearly every major auto manufacturer has hybrid models available, although some of these vehicles get only slightly better mileage than their regular gasoline engine varieties.

Plug-in hybrids reduce gasoline dependency still more. These vehicles have larger battery arrays than an ordinary hybrid. You can recharge the battery at a charging station rather than depend only on the internal engine to do the charging., and in most city driving, the battery is enough for all daily needs. The EPA rates the plug-in Prius as having the equivalent of 95 mpg. Recharging the batteries from ordinary household current at night can allow these vehicles to travel up to 100 km (60 mi) on the electric motor alone. Because most Americans drive only about 30 miles per day, they should rarely have to buy any gasoline. Electricity is also cheaper: In most places, electricity costs the equivalent of about 50¢ per gallon of gasoline. Operating a large fleet of electric vehicles could mean higher electricity demand, but producing electricity at a central power plant is far cleaner than burning gas in thousands of vehicle engines. And if the electricity comes from renewable sources, such as solar or wind, most of your transportation could be almost entirely pollution-free.

All-electric vehicles are even more efficient than hybrids. Some are rated by the EPA as getting the equivalent of 100 to 120 mpg in combined city and highway driving. Most have a limited driving range, but some, such as the Tesla Model S, claim as much as 265 mi (426 km) on a single charge. Some countries, such as the Netherlands, Denmark, Norway, and China have encouraged use of electric vehicles by creating networks of charging stations in cities and along highways, and by offering owners of electric vehicles free street parking and charging. With hefty tax breaks, promotional leases, and cheaper operating costs—they never need an oil change—these vehicles might offer driving costs significantly lower than conventional cars.

Quantitative Reasoning

In 2016, the combined corporate average fuel economy for cars and light trucks in the United States was 34.1 mpg. If the 190 million drivers traveled an average of 10,000 miles that year, how much gasoline did they use? If burning a gallon of gasoline creates 14 pounds of CO_2, how much CO_2 did they produce in their automobile travel?

Cogeneration produces both electricity and heat

One of the fastest-growing sources of new energy is **cogeneration,** the production of electricity from waste heat in a steam heating plant. These systems are also called combined heat and power (CHP) plants. Many colleges, universities, cities, and industrial facilities have a central heating plant that distributes steam or hot water.

By producing useful electricity from waste heat in these plants, the net energy yield from the primary fuel is increased from 30–35 percent to 80–90 percent. In 1900, half the electricity generated in the United States came from plants that also provided industrial steam or district heating. Cogeneration became less common with falling fuel prices and the growth of electric utilities.

In recent decades, interest in this technology is being renewed, as industries and cities search for efficiency improvements. District heating systems are being rejuvenated, and plants that burn municipal wastes are being studied. Small power-generating units to service neighborhoods or apartment buildings are being built that burn methane (from biomass digestion), natural gas, diesel fuel, or other fuel (fig. 20.8). The capacity for cogeneration is now over 70 GW, or about 7 percent of U.S. electricity supplies.

Smart metering can save money and energy

Smart meters are electricity meters that record data rapidly, for example, every hour, and that communicate with the electric utility grid. A smart meter rewards customers who use most of their energy when there is excess supply, and it is cheap. For example, in the middle of the night, when consumption is low, utility companies offer customers low rates to take unwanted power off their hands. With a smart house system, you could time your appliances (dishwasher, water heater, and so on) to run only when a certain price is available, or only when the electricity comes from wind or solar sources.

Better meters and smart electric systems would also enable us to take advantage of the potential of plug-in hybrids and electric vehicles, which could serve as an enormous, distributed battery array. Automobiles in particular could be programmed to charge only at night when electric prices are at their lowest. If we had millions of vehicles connected to a smart grid, they would collectively have enough storage capacity to smooth out peaks and valleys in the energy supply. Utilities report that they could provide power for 4 million plug-in vehicles just with their existing surplus power supplies. And if the batteries were recharged from solar or wind facilities, our fossil fuel consumption would be sharply reduced.

Smart meters are becoming more common because they save money for utilities as well as for consumers. The Pacific Gas and Electric utility in California is installing 9 million smart meters for its customers, to encourage conservation during peak hours. Europe expects to have 200 million smart electric meters and 45 million smart gas meters installed by 2020. Many customers love saving money and knowing exactly how much power costs for each appliance.

Whether or not you are buying large energy consumers such as a house or car, and even if you don't have much influence over utility operations, there are things that all of us can do to save energy every day (see the What Can You Do? box below).

FIGURE 20.8 A technician adjusts a gas microturbine that produces on-site heat and electricity for businesses, industry, or multiple housing units.
© Copyright 2016 Capstone Turbine Corporation All rights reserved.

What Can You Do?

Steps You Can Take to Save Energy

1. Drive less: make fewer trips, use telecommunications and mail instead of going places in person.
2. Walk, ride a bicycle, or use public transportation.
3. Use stairs instead of elevators.
4. Join a car pool or drive a smaller, more efficient car; reduce speeds.
5. Insulate your house or add more insulation to the existing amount.
6. Turn thermostats down in the winter and up in the summer, and dress appropriately for the season.
7. Weather-strip and caulk around windows and doors; add storm windows or plastic sheets over windows.
8. Create a windbreak on the north side of your house; plant deciduous trees or vines on the south side.
9. During the winter, close windows and drapes at night; during summer days, close windows and drapes if using air conditioning.
10. Turn off lights, television sets, and computers when not in use.
11. Stop faucet leaks, especially hot water.
12. Take shorter, cooler showers; install water-saving faucets and showerheads.
13. Recycle glass, metals, and paper; compost organic wastes.
14. Eat locally grown food in season.
15. Buy locally made, long-lasting materials.

Section Review

1. List five ways we could conserve energy individually or collectively.
2. What are some ways building design can reduce energy loss?
3. Why is transportation efficiency important in energy conservation?
4. What is cogeneration?

20.2 SOLAR ENERGY

- *Solar energy can be collected passively or actively.*
- *Solar thermal systems create steam to generate electricity.*
- *Photovoltaics generate electricity directly.*

The sun serves as a giant nuclear furnace in space, constantly bathing our planet with a free energy supply. Solar heat drives winds and the hydrologic cycle. All biomass, as well as fossil fuels and our food (both of which are derived from biomass), results from conversion of light energy (photons) into chemical bond energy by photosynthetic bacteria, algae, and plants. The average amount of solar energy arriving at the top of the atmosphere is 1,330 watts per square meter. About half of this energy is absorbed or reflected by the atmosphere, but over the course of a year, in most populated regions, we receive around 2,000 kilowatt-hours per year per square meter (Figure 20.9). This is nearly 1/5 the amount of electricity used by a U.S. household every year (about 11,000 kWh per year), or 1/2 the electricity use of a household in Europe (about 4,000 kWh per year). The amount reaching the earth's surface is some 10,000 times all the commercial energy used each year. Because this tremendous infusion of energy comes in a diffuse, low-intensity form, we have only recently learned to convert it to electricity with economical efficiency. But as the opening case study for this chapter shows, there may be ways to use this vast power source, so we might never again have to rely on fossil fuels.

Direct Normal Irradiation (DNI)

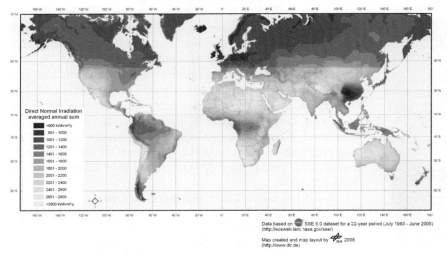

FIGURE 20.9 Cumulative average annual solar radiation. Within six hours, deserts receive more energy from the sun than humans consume in a year.
Source: German Aerospace Center, 2012

Solar heat collectors can be passive or active

Our simplest and oldest use of solar energy is **passive heat absorption,** using natural materials or absorptive structures with no moving parts to simply gather and hold heat. For thousands of years people have built thick-walled stone and adobe dwellings that slowly collect heat during the day and gradually release that heat at night (fig. 20.10). After cooling at night, these massive building materials maintain a comfortable daytime temperature inside, even as they absorb external warmth.

A modern adaptation of this principle is a glass-walled "sun-space" or greenhouse on the south side of a building. Massive energy-storing materials, such as brick walls, stone floors, or barrels of heat-absorbing water, collect heat to be released slowly at night. An interior, heat-absorbing wall called a Trombe wall is an effective passive heat collector. Some Trombe walls are built of glass blocks enclosing a water-filled space or water-filled circulation tubes, so heat from solar rays can be absorbed and stored while light passes through to inside rooms.

Active solar systems generally pump a heat-absorbing fluid (air, water, or an antifreeze solution) through a collector, such as a flat, glass-covered black surface, or glass vacuum tubes. Water (or another fluid) pumped through the collector picks up heat for space heating or to provide hot water. Active solar thermal (heating) systems are often located adjacent to or on top of buildings, and they can be readily retrofitted onto existing buildings. A collector with about 5 m^2 of surface can reach 95°C (200°F) and can provide enough hot water for an average family. China produces about 80 percent of the world's solar water heaters, which cost about $160 each and pay for themselves in only two or three years. At least 30 million Chinese homes get hot water and/or space heat from solar energy, and China now has more than 80 percent of the solar hot water market. In Europe, municipal solar systems provide district heating and hot water for whole cities (fig. 20.11).

In a symbolic act to illustrate his commitment to solar energy, President Obama restored the solar electric panels and a solar water heater to the White House roof that were removed 30 years earlier by the Reagan administration.

High-temperature solar produces electricity

High-temperature solar thermal plants are suitable for industrial-size facilities. One of the most common designs for **concentrating solar power (CSP) systems** uses long trough-shaped parabolic mirrors to reflect and concentrate sunlight on a central tube containing a heat-absorbing fluid (fig. 20.12). The fluid reaches extremely high temperatures, then passes through a heat exchanger, where it heats water and generates steam to turn a turbine to produce electricity. Because these systems are large and expensive to build, they are sited in desert areas with abundant sunshine. The German Aerospace Center

FIGURE 20.10 Taos Pueblo in northern New Mexico uses adobe construction to keep warm at night and cool during the day.
© William P. Cunningham

FIGURE 20.12 One form of concentrated solar power uses long trough-shaped parabolic mirrors to heat an absorbing fluid, which is piped to a central steam generator to turn turbines that generate electricity.
Source: Warren Gretz/NREL/U.S. Dept. of Energy

points out that in just six hours, the world's deserts receive more solar energy than all humans consume in an entire year. In the 1980s, a grand plan was proposed to build about 20 large CSP plants in North Africa and the Middle East linked together by high voltage, direct current transmission lines with offshore wind farms, geothermal installations, and hydroelectric plants to supply Europe with renewable energy. Unfortunately, financing difficulties and political instability in the region doomed this project.

CSP plants have important advantages. Heat from the transfer fluid can be stored in a medium, such as molten salt, for later use. This allows the system to continue to generate electricity on cloudy days or at night. Some CSP plants can produce power nearly around the clock. In addition, if plants are located near coastlines, waste heat from the turbines can be used to evaporate seawater to create pure drinking water—something sorely lacking in arid regions. A 250 MW collector field could provide 200 MW of electricity plus 100,000 m^3 (about 26 million gal) of distilled water per day.

FIGURE 20.11 Solar water heaters can be scaled up to provide hot water and space heating for whole cities.
© Tom Finkle

Some of the oil-rich Persian Gulf states are building CSP plants as they prepare for a fossil fuel–free future. The United Arab Emirates, for example, built a 100 MW CSP plant in Abu Dhabi in 2014. The 258,000 mirrors in the Shams (Arabic for Sun) plant can provide enough electricity for about 20,000 homes. Shams I is expected to be followed by units II and III for an eventual total of 300 MW. And Saudi Arabia has pledged US$109 billion for solar power. If carried through, this investment could produce 41 GW of solar energy (as much as Germany has now) and an additional 21 GW of geothermal and wind power. The Saudi decision was triggered by a 2012 estimate that the kingdom could become a net oil importer by 2038 if it didn't diversify its energy portfolio.

The parabolic collectors in the CSP plant follow the sun to maximize solar energy absorption. But, you may ask, aren't highly polished mirrors vulnerable to wind and sandstorms? On days when storms are forecast, the mirrors can be rotated into a protective position. Solar-thermal power plants in California's Mojave Desert have been operating for over 20 years and have withstood hailstorms, sandstorms, and gale-force winds. And wouldn't it take huge areas of land to capture solar energy? According to the German Aerospace Center, supplying 17 percent of Europe's energy requirements would take 2,500 km^2, or less than 0.3 percent of the Sahara Desert.

A disadvantage of CSP is that desert sites are often biologically rich and vulnerable, and they recover very slowly from disturbance. In 2010, state regulators approved 13 large solar thermal facilities and wind farms for California's Mojave Desert. Subsequently, however, most of these projects were canceled or delayed by protests over land use. Biologists argued that the installations would harm rare or endangered species, such as the desert tortoise or the fringe-toed lizard. Native American groups protested that some areas were sacred cultural sites. Wilderness advocates, who loved the solitude and mystery of the desert, opposed the intrusion of large industrial facilities. In response to these challenges, millions of hectares of desert have been added to new or existing protected areas to forestall energy development.

Another high-temperature system uses thousands of smaller mirrors arranged in concentric rings around a tall central **power tower** (fig. 20.13). The mirrors track the sun and focus light on a heat absorber at the top of the "power tower," where a transfer medium is heated to temperatures as high as 500°C (1,000°F), which then drives a steam-turbine electric generator. Under optimal conditions, a 50 ha (130 acre) mirror array should be able to generate 100 MW of clean, renewable power. The Ivanpah Solar Station, built by Bright Source in California's Mojave Desert 60 km (40 mi) southwest of Las Vegas, with a total capacity of 392 MW, is currently the world's largest thermal solar installation. Covering 4,000 acres (1,600 ha), the plant has 173,500 heliostats focusing energy on three towers, and produced 209 GW hours of electricity in 2015.

Because all the mirrors are focused on a single point, the heat transfer medium has to be capable of absorbing much higher energy levels than in solar troughs. So far, most of these plants use liquid sodium or molten nitrate salt for heat absorption. These molten salts are much more corrosive and difficult to handle, however, than the lower-temperature fluids suitable for a solar trough.

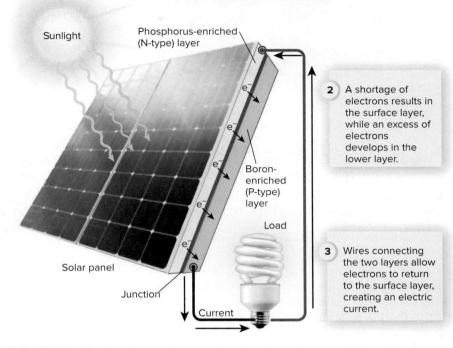

1 Photons striking the panel surface excite electrons (e⁻), which move to the lower layer of panel.

Sunlight

Phosphorus-enriched (N-type) layer

2 A shortage of electrons results in the surface layer, while an excess of electrons develops in the lower layer.

Boron-enriched (P-type) layer

Load

Solar panel

Junction

Current

3 Wires connecting the two layers allow electrons to return to the surface layer, creating an electric current.

FIGURE 20.14 When solar energy strikes a photovoltaic (PV) cell, an electron is dislodged from atoms in the p-layer in the silicon crystal. These electrons cross an electrostatic junction between different semiconductor materials. This creates a surplus of electrons in the n-layer and a shortage of electrons (or a positive charge) in the p-layer.

FIGURE 20.13 A power tower is a form of concentrated solar thermal electrical generation. Thousands of movable mirrors focus sunlight on the central tower, where a liquid is heated to drive a steam turbine.
© Kevin Burke/Corbis RF

Photovoltaic cells generate electricity directly

Photovoltaic (PV) cells capture solar energy and convert it directly to electrical current by separating electrons from their parent atoms and accelerating them across a one-way electrostatic barrier formed by the junction between two different types of semiconductor material (fig. 20.14). The first PV cells were made by slicing thin wafers from giant crystals of extremely pure silicon.

Over the past 25 years the efficiency of energy captured by PV cells has increased from less than 1 percent of incident light to more than 15 percent under field conditions and over 75 percent in the laboratory. Promising experiments are underway using exotic metal alloys, such as gallium arsenide, and semiconducting polymers of polyvinyl alcohol, which are more efficient in energy conversion than silicon crystals. Increasing efficiency coupled with falling prices has led to rapid growth in the solar industry (fig. 20.15). Some energy experts suggest that solar collectors could provide 25 percent of global electricity in a decade.

In 2016, the price of PV cells fell to only 55¢ per watt, and some companies were offering installed panels for less than $1 per watt, a price that makes them competitive with fossil fuels and nuclear power in many situations. As further research

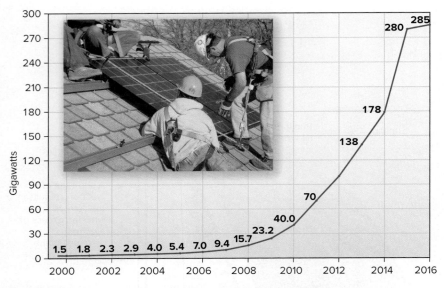

The chart shows data plotted with the following labeled points: 1.5, 1.8, 2.3, 2.9, 4.0, 5.4, 7.0, 9.4, 15.7, 23.2, 40.0, 70, 138, 178, 280, 285

(Y-axis: Gigawatts, labeled 0, 30, 60, 90, 120, 150, 180, 210, 240, 270, 300)
(X-axis: 2000, 2002, 2004, 2006, 2008, 2010, 2012, 2014, 2016)

FIGURE 20.15 Total world solar PV capacity, 2000–2012. Solar power is growing exponentially and has by far the greatest potential of any renewable energy.
Source: MREA/NREL/U.S. Dept. of Energy

improves their efficiency and life span, industry experts believe they could produce electricity for less than 5¢ per kilowatt-hour by 2020. This low price makes solar suitable for utility-scale baseload power arrays (fig. 20.16a).

As the opening case study for this chapter notes, Germany led the world in solar power for most of the past decade. In 2015, however, China vaulted ahead of Germany to a total of 43 GW of PV power (compared to 40.1 in Germany and 26.8 GW in the U.S.). Japan and India also have moved aggressively into solar energy. The 1.5 million PV systems in Germany produced about 31 percent of the country's electricity in 2016. By 2050, Germany expects to get 100 percent of its electricity from renewables. By some estimates, half of all jobs in Germany are now in the solar industry. The price of PV panels has fallen faster than that for CSP facilities so that far more PV installations are currently under construction than thermal plants.

One of the most promising developments in PV technology in recent years is the invention of silicon collectors. First described in 1968 by Stanford Ovshinky, amorphous, noncrystalline, thin-film silicon semiconductors can be made into lightweight, paper-thin films that require much less material than conventional crystalline silicon cells. They also are much cheaper to manufacture and can be made in a variety of shapes and sizes, permitting ingenious applications. The largest PV installations in the United States currently are the Desert Sunlight and Topaz Solar Farms in southern California. Each has about 9 million stationary, cadmium telluride thin-film PV panels and produces 550 MW of power. They're among the largest such facilities in the world. Flexible roof tiles with amorphous silicon collectors layered on their surface can be installed on residential rooftops (fig. 20.16b). In developing countries, solar power could make electricity available to

some of the nearly 2 billion people who aren't served by a power grid.

Rooftop solar energy has huge potential. It's estimated that in the United States more than 1,000 mi^2 (2,590 km^2) of roofs suitable for photovoltaic systems could generate about three-quarters of our present electrical consumption. Just the roofs of warehouses, shopping malls, office buildings, and big-box retail stores (fig. 20.16c) could produce an estimated 62 GW of electricity. Already, Walmart, Target, IKEA, Apple, Costco, and Kohl's have installed solar arrays on their buildings.

For individual households, a photovoltaic array of about 30–40 m^2 (3 to 4 kw) will generate enough electricity for an efficient house. The initial cost of this technology has been a barrier for most homeowners. When solar was $5 per watt installed, a home-sized system cost $15,000 to $20,000. But these prices are falling rapidly. At $1 per w, a rooftop array might pay for itself in five to seven years. Innovative financing programs are helping make energy independence a reality. First introduced in Berkeley, California, **property assessed clean energy (PACE)** uses city bonds to pay for renewable energy and conservation expenses. The bonds are paid off through a 20-year assessment on property taxes. Lower utility bills can offset tax increases, so that switching to renewable energy is relatively painless for the property owner.

Solar City, a company founded by entrepreneur Elon Musk (who also co-created PayPal, Tesla Motors, and SpaceX) is the largest residential solar company in the U.S. It offers three options: (1) they install solar panels on your roof and you buy electricity by the month; (2) you lease your rooftop system from them, and get to use or sell any energy it captures; or (3) you pay up front and own the system and its output. In the first two options, you don't pay anything initially, but you also don't have any equity. In the third option, you own the array and can sell it with the house if you move. Solar City operates in 20 states in the western U.S. and in New England. By 2015, it had installed enough solar panels to produce 6.2 GW of power, or about the capacity of 6-8 nuclear plants.

Economical battery storage is an important goal of solar producers. Along with his solar and electric vehicle systems, Elon Musk has also built a giant battery factory in Nevada. He hopes to produce millions of "power walls," or battery arrays large enough to supply a house with power for about 24 hours. If he can produce these batteries at a reasonable price and reliability, it could erase the intermittency of solar and wind energy, and make it possible to unhook from the utility grid. Another form of storage being explored in Germany is called Gas-2-power. Excess electricity can be used to hydrolyze water to hydrogen and oxygen gas. The hydrogen can be stored and burned for heat, or used in fuel cells to generate more electricity when the need arises.

What can you do if you don't own a suitable roof for solar energy? This is a common question. Renters don't own their rooftops, and many houses don't have good sun exposure. An emerging

(a) Active photovoltaic array

(c) Rooftop solar

FIGURE 20.16 (a) Active PV arrays follow the sun to increase efficiency. (b) Thin-film PV collectors can be printed on flexible backing and used like ordinary roof tiles. (c) Millions of square meters of roof tops on schools and commercial buildings could be fitted with solar panels. *Sources: (a) © Fotosearch/Photolibrary RF; (b) Stellar Sun Shop/NREL/U.S. Dept. of Energy; (c) Craig Miller Productions/NREL/U.S. Dept. of Energy.*

Public policy can promote renewable energy

Energy policies in some states include measures to promote renewable energy. The most important has been establishment of **feed-in tariffs,** which require utilities to buy surplus power from small producers at a fair price. These tariffs (payments) are often essential to make small solar installations economical. Rather than pay for an expensive energy storage system, your meter simply spins backward during the day when you sell surplus electricity to your local utility. At night, when your solar panels aren't making any electricity, you buy some back from your power company. Depending on how big your system is, and how frugal you are with electrical use, you might end up making money. Utilities usually resist feed-in tariffs, which divert some profits to homeowners and also complicate management of the electric grid, since solar power can be variable. Policies are rapidly evolving as systems change.

There are countless other policy approaches. Among these are (1) "distributional surcharges" in which a small charge is levied on all utility customers to help finance renewable energy research and development, (2) **renewable portfolio standards** that require power suppliers to obtain a minimum percentage of their energy from sustainable sources, and (3) **green pricing,** which allows utilities to profit from conservation programs by charging higher prices for energy from renewable sources. Some or all of these may exist where you live.

(b) Flexible photovoltaic collectors

alternative for these situations is a **solar garden,** or solar farm. These are community-shared solar arrays connected to the utility grid (fig. 20.17). Members buy power from the utility as usual, but they get a rebate from the garden based on their share of ownership. As with power on a private roof, members can either pay up front for a share, or pay on a monthly basis. There are many different business models for solar gardens.

Solar gardens can also foster social justice by including people who don't have the cash or credit rating to own a solar system of their own. And a benefit over individual rooftop ownership is that the association maintains the facility—washing off dust in the summer, or cleaning off snow in the winter, for example.

FIGURE 20.17 A solar garden is a community-based solar array connected to the utility grid. Members share the proceeds from net metering. © William P. Cunningham

Iowa, for example, has a Revolving Loan Fund supported by a surcharge on investor-owned gas and electric utilities. This fund provides low-interest loans for renewable energy and conservation. Many utilities now offer green energy options. Customers agree to pay a couple of dollars extra on monthly bills, and utilities promise to use the money to build or buy renewable energy. Buying a 100-kW "block" of wind power provides the same environmental benefits as planting a half acre of trees or not driving an automobile 4,000 km (2,500 mi) per year.

Distributed power generation is decentralized

In Germany, a world leader in solar power, more than half of all solar power is owned by individual homeowners or community cooperatives, much of it for local use (fig. 20.18). Distributed solar

FIGURE 20.18 Household solar panels provide distributed energy that could free us from energy monopolies.
Source: groSolarA/NREL/U.S. Dept. of Energy

power generation, produced at the site where it will be used, has tremendous potential in efficiency and cost savings. Distributed systems can reduce the need for power lines, which are expensive and inefficient. They can eliminate disruptive and dangerous infrastructure like coal trains, oil and gas pipelines, and oil and gas terminals. Distributed production is also democratic: Rather than have a huge monopoly controlling power production and transmission, you could own your own system or join with your neighbors in a local cooperative.

Distributed production can also be threatening to power utilities accustomed to monopoly control. Large utilities have enormous sunk costs in power plants and transmission grids, and they resist competition from distributed and renewable power producers. Many utilities charge for use of the utility network, and some charge private wind or solar systems a fee to hook up to the utility grid even if they only use the connection for emergencies. These fees can be a serious obstacle to renewable energy adoption. In Germany, some cities are buying back their transmission lines to reduce the power of big utilities to control who produces energy, and what kind of energy is produced.

Quantitative Reasoning

According to the International Energy Agency, total world subsidies for oil, coal, and natural gas amounted to US$493 billion in 2014. This was at least four times as much as for renewable energy. Where might we be if that ratio had been reversed?

Section Review

1. What is the difference between active and passive solar energy?
2. How do photovoltaic cells generate electricity?
3. What are some policy strategies to support solar energy?

20.3 WIND

- *Wind power could meet all our energy needs.*
- *China is now the world's largest producer of wind turbines.*
- *Turbines pose risks to birds and bats, but careful placement and operation can reduce collisions.*

Wind power currently provides the world's largest amount of renewable energy, nearly 2.5 times as much as solar. Over the past decade, China has become the world's leading producer of wind turbines, and clean technology provides more than 1 million Chinese jobs. In 2015, China added an astonishing 30.5 GW of new wind power (fig. 20.19), giving the country 145 GW of installed wind capacity, or more than all of Europe put together. The United states is second in the world with 74 GW of wind power, while Germany is third with 45 GW. Spain and India fill out the top five wind power leaders.

Modern **wind turbines** are far different from those employed a generation ago. The largest wind turbines are 40 times more

FIGURE 20.19 China is the world's leading producer of wind turbines, and now has about one-third of all installed wind power. This clean technology provides more than 1 million jobs in China.
© Liu Jin/AFP/Getty images

FIGURE 20.20 Vertical axis wind turbines are lighter and cheaper than similarly sized solar arrays. They can operate in slower wind speeds, and can be sized to be more appropriate for a single home or small business.
© William P. Cunningham

powerful today than 20 years ago. Some now have towers up to 150 m tall with 82-m-long blades that reach as high as a 50-story building. Each can generate 8 MW of electricity, or enough for 3,000 typical American homes. These giant machines are mostly used offshore. Out of commission for maintenance only about three days per year, many can produce power 90 percent of the time. Theoretically up to 60 percent efficient, modern windmills typically produce about 35 percent of peak capacity under field conditions.

Germany currently gets roughly 8.6 percent of its electricity from wind turbines, almost all of which is onshore where breezes are strong and reliable. By 2020, Germany plans to roughly triple the share of wind power (both onshore and offshore). But the fledgling offshore sector differs greatly from traditional onshore wind. While the latter mostly consists of midsize firms and distributed wind projects owned largely by communities and small investors, the former is almost entirely in the hands of large corporations, many of which have opposed the switch to renewables up to now.

Currently, land-based wind farms are the cheapest source of *new* power generation, costing as little as 3¢/kWh compared to 4–5¢/kWh for coal, and five times that much for nuclear fuel. If we had a carbon "cap and trade" program as many European countries do, wind energy could be cheaper than fossil fuels nearly everywhere.

While giant modern turbines are most efficient, they don't fit everywhere. Vertical axis wind turbines (VAWTs) are lighter and cheaper than horizontal axis wind turbines (HAWTs). Because the main components are located at the base of the machine, VAWTs are lighter and easier to service and repair. They can operate in slower wind speeds, and can be sized to be more appropriate for a single home or small business (fig. 20.20). The price for a 3-kW

turbine can be less than for an equal-sized solar array. Research at Caltech has shown that a carefully designed wind farm using VAWTs can have an output power ten times that of conventional HAWT machines. In the past, the blades of VAWTs had a tendency to twist and break, but better designs and modern composite materials have largely solved that problem.

Wind could meet all our energy needs

Wind power is the world's fastest-growing energy source and could replace all the commercial energy we now use. The Global Wind Energy Council calculates that wind could supply 40 times the current world electricity supply and five times all global energy consumption if all the potential were tapped. With 432 GW of total installed capacity in 2016, wind power is producing nearly 1,000 terawatt-hours (1 TWh = 10^{12} Wh) of electricity annually. That total could increase to 1,750 TWh by 2020. Wind has a number of advantages over most other power sources. Wind farms have much shorter planning and construction times than fossil fuel or nuclear power plants. Wind farms are modular (more turbines can be added if loads grow) and they have no fuel costs or air emissions.

Wind does have limitations, however. Like solar energy, it is an intermittent source. Furthermore, not every place has strong enough or steady enough wind to make this an economical resource. Although modern windmills are more efficient than those of a few years ago, it takes a wind velocity between 7 m per second (16 mph) and 27 m per second (60 mph) to generate useful amounts of electricity.

Large areas of the Great Plains and mountain states have persistent winds suitable for commercial development (fig. 20.21). Thirty-seven states now have utility-scale wind farms. Texas, with 15.6 MW, leads the nation, followed by California, Iowa, Illinois, and Oregon, but a huge potential remains to be tapped. Wind power is now so plentiful in Texas that on some occasions, the price has dropped into negative territory. That is, wind farms have to pay consumers to use their excess power.

Although the United States does have good wind potential on the mid-Atlantic continental shelf, the bulk of North America's wind potential is situated on land. Offshore installations, which are costly because of the need to operate in deep water and withstand storms and waves. Wind turbine construction on land, on the other hand, is relatively simple and cheap. Still, the first offshore wind farm in North America, in Nantucket Sound, received approval from federal regulators and financial backing from investors in 2013. The U.S. Department of Energy has set a goal of 54 GW of offshore energy by 2030.

There is growing demand for wind projects from farmers, ranchers, and rural communities because of the economic benefits that wind energy brings. One thousand megawatts of wind power (equivalent to one large nuclear or fossil fuel plant) can create more than 3,000 permanent jobs, while paying about $4 million in rent to landowners and $3.6 million in tax payments to local governments.

As figure 20.22 shows, wind power takes about one-third as much area and creates about five times as many jobs to create the same amount of electrical energy as coal, when the land consumed by mining is taken into account. Furthermore, with each tower taking only about 0.1 ha (0.25 acre) of cropland, farmers can continue to cultivate 90 percent of their land while getting $2,000 or more in annual rent for each wind machine. An even better return results if the landowner builds and operates the wind generator, selling the electricity to the local utility. Annual profits can be as much as $100,000 per turbine, a wonderful bonus for the use of 10 percent of your land.

Cooperatives are springing up to help landowners finance, build, and operate their own wind generators. About 20 Native American tribes, for example, have formed a coalition to study wind power. Together their reservations (which were sited in the windiest, least productive parts of the Great Plains) could generate at least 350,000 MW of electrical power, equivalent to about half of the current total U.S. installed capacity.

In some places, high bird and bat mortality has been reported around wind farms. This seems to be particularly true in California, where rows of generators were placed at the summits of mountain passes, areas where wind velocities are high but where migrating birds and bats are likely to fly into rotating blades. New generator designs and more careful tower placement seems to have reduced this problem in most areas. Shutting down turbines during

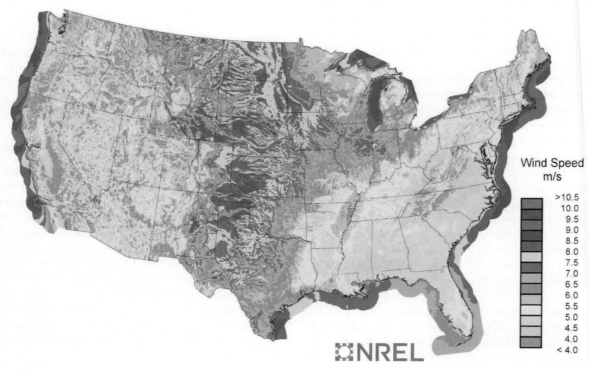

FIGURE 20.21 U.S. wind resource map. Mountain ranges and areas of the High Plains have the highest wind potential, but much of the country has fair to good wind supply.
Source: U.S. Department of Energy

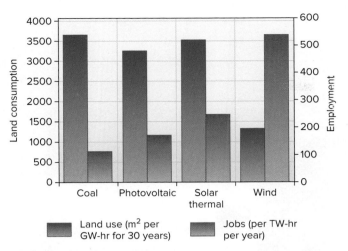

FIGURE 20.22 If you include the land required for mining, wind power takes about one-third as much area and creates about five times as many jobs to create the same amount of electrical energy as coal.

migrations can help, too. Although national polls in the United States show that 82 percent of the public supports additional wind power, the rate of support is often considerably less among people who live close to the towers.

It is also important to note that oil, gas, coal, and nuclear power cause more wildlife mortality than renewables do. Oil development is the leading cause of habitat destruction in much of western South America and West Africa. In North America, the vulnerable Sage Grouse and dwindling pronghorn antelope populations are threatened by development of vast oil and gas fields. Oil spills such as that in the Gulf of Mexico in 2010 (see chapter 19) cause uncounted deaths of birds, fish, and marine ecosystems worldwide every year. Ethanol and biofuels, meanwhile, are responsible for destruction of millions of hectares of wildlife habitat across North America, Malaysia, Ecuador, Congo, and other regions (see section 20.4).

Some people object to the sight of large machines looming over the landscape, and there's controversy about how close to houses and schools wind turbines should be allowed. There are claims that the low-frequency sound waves from moving blades cause headaches, insomnia, digestive problems, panic attacks, and other health issues. Others dismiss these symptoms as too vague and general to assign to a specific source. It's difficult to study the issue scientifically because the low-frequency sounds attributed to wind towers are indistinguishable from wind, highway noise, and other ambient sounds. To some observers, on the other hand, windmills offer a welcome alternative to nuclear or fossil fuel–burning plants.

Section Review

1. How much energy could we obtain from wind?
2. What regions of the United States have the strongest and most reliable wind?
3. What state currently has the greatest installed wind capacity?

20.4 FUEL CELLS AND BIOMASS

- *Fuel cells use a semipermeable membrane to separate electrons from positively charged ions.*
- *We can burn biomass or turn it into biofuels.*
- *Methane digestion is an important way to use waste material.*

Fuel cells are devices that use ongoing electrochemical reactions to produce an electric current. They are similar to batteries except that batteries are recharged by adding electrical current, whereas fuel cells are recharged by adding fuel to power a chemical reaction. Fuel cells are not new; the basic concept was recognized in 1839 by William Grove, who was studying the electrolysis of water. He suggested that rather than use electricity to break apart water and produce hydrogen and oxygen gases, it should be possible to reverse the process by joining oxygen and hydrogen to produce water and electricity. The term *fuel cell* was coined in 1889 by Ludwig Mond and Charles Langer, who built the first practical device using a platinum catalyst to produce electricity from air and coal gas. The concept languished in obscurity until the 1950s when the U.S. National Aeronautics and Space Administration (NASA) was searching for a power source for spacecraft. Research funded by NASA eventually led to development of fuel cells that now provide both electricity and drinkable water on every space shuttle flight. The characteristics that make fuel cells ideal for space exploration—small size, high efficiency, low emissions, net water production, no moving parts, and high reliability—also make them attractive for a number of other applications.

Fuel cells produce electricity chemically

All fuel cells consist of a positive electrode (the cathode) and a negative electrode (the anode) separated by an electrolyte, a material that allows the passage of charged atoms, called ions, but is impermeable to electrons (fig. 20.23). In the most common systems, hydrogen or a hydrogen-containing fuel, such as natural gas (CH_4) is passed over the anode while oxygen is passed over the cathode. At the anode, a reactive catalyst, such as platinum, strips an electron from each hydrogen atom, creating a positively charged hydrogen ion (a proton). The hydrogen ion can migrate through the electrolyte to the cathode, but the electron is excluded. Electrons pass through an external circuit, and the electrical current generated by their passage can be used to do useful work. At the cathode, the electrons and protons are reunited and combined with oxygen to make water.

The fuel cell provides direct-current electricity as long as it is supplied with hydrogen and oxygen. For most uses, oxygen is provided by ambient air. Hydrogen can be supplied as a pure gas, but storing hydrogen gas is difficult and dangerous because of its volume and explosive nature. Liquid hydrogen takes far less space than the gas, but must be kept below −250°C (−400°F), not a trivial task for most mobile applications. The alternative is a device called a **reformer** or converter that strips hydrogen from fuels such as natural gas, methanol, ammonia, gasoline, ethanol, or even vegetable oil. Many of these fuels can be derived from biomass crops.

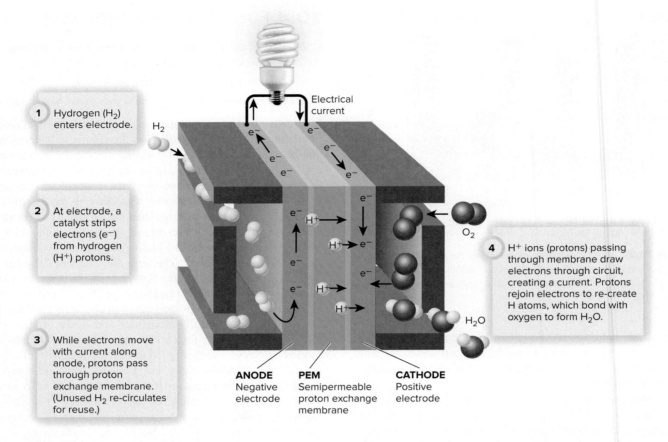

1. Hydrogen (H_2) enters electrode.

2. At electrode, a catalyst strips electrons (e^-) from hydrogen (H^+) protons.

3. While electrons move with current along anode, protons pass through proton exchange membrane. (Unused H_2 re-circulates for reuse.)

4. H^+ ions (protons) passing through membrane draw electrons through circuit, creating a current. Protons rejoin electrons to re-create H atoms, which bond with oxygen to form H_2O.

Electrical current

H_2

e^-

H^+

O_2

H_2O

ANODE
Negative electrode

PEM
Semipermeable proton exchange membrane

CATHODE
Positive electrode

FIGURE 20.23 Fuel cell operation. Electrons are removed from hydrogen atoms at the anode to produce hydrogen ions (protons) that migrate through a semipermeable electrolyte medium to the cathode, where they reunite with electrons from an external circuit and oxygen atoms to make water. Electrons flowing through the circuit connecting the electrodes create useful electrical current.

Methane captured from landfills and wastewater treatment plants could be used as a hydrogen source. Hydrogen could be also provided by hydrolyzing water (separating hydrogen and oxygen) using solar or wind-powered electricity.

A fuel cell run on pure oxygen and hydrogen produces no waste products except drinkable water and radiant heat. When a reformer is coupled to the fuel cell, some pollutants are released (most commonly carbon dioxide), but the levels are typically far less than conventional fossil fuel combustion in a power plant or automobile engine. Although the theoretical efficiency of electrical generation of a fuel cell can be as high as 70 percent, the actual yield is closer to 40 or 45 percent. On the other hand, the quiet, clean operation and variable size of fuel cells make them useful in buildings where waste heat can be captured for water heating or space heating. There have also been efforts to develop fuel cells for vehicles, although it remains more cost-effective to use batteries for electric or plug-in hybrid vehicles.

The current from a fuel cell is proportional to the size (area) of the electrodes, while the voltage is limited to about 1.23 volts per cell. A number of cells can be stacked together until the desired power level is reached. A fuel cell stack that provides almost all of the electricity needed by a typical home (along with hot water

and space heating) would be about the size of a refrigerator. A 200-kilowatt unit fills a medium-size room and provides enough energy for 20 houses or a small factory (fig. 20.24).

Biomass is an ancient and modern energy source

Photosynthetic organisms have been collecting and storing the sun's energy for more than 2 billion years as **biomass** (any organic material). Plants capture about 0.1 percent of all solar energy that reaches the earth's surface, transforming solar energy, via photosynthesis, into chemical bonds in organic molecules of plant tissues, or biomass. The magnitude of this resource is difficult to measure. Most experts estimate useful biomass production at 15 to 20 times the amount we currently get from all commercial energy sources. It would be ridiculous to consider consuming all green plants as fuel, but biomass has the potential to become a prime source of energy. Biomass has many advantages over nuclear and fossil fuels because of its renewability and easy accessibility, as long as we don't over-exploit the living systems that produce it.

Wood fires were our primary source of heating and cooking for thousands of years. As recently as 1850, wood supplied

FIGURE 20.24 The Long Island Power Authority has installed 75 stationary fuel cells to provide reliable backup power.
© Long Island Power Authority

90 percent of the fuel used in the United States. Wood now provides less than 1 percent of the energy in advanced economies, but 2 billion people—nearly 30 percent of the world population—depend on firewood and charcoal (partially burned, carbonized wood) as their primary energy source.

In many countries, firewood gathering and charcoal production is a dominant cause of forest destruction and habitat degradation. Firewood gathering is also a socially unequal burden. Poor people often spend a high proportion of their income on cooking and heating fuel. In rural families, women and children may spend hours every day searching for fuel. Development agencies are working to design and distribute highly efficient stoves, both as a way to improve the lives of poor people and to reduce forest losses.

In wealthy countries, biomass also makes a significant contribution to renewable energy supplies. In Denmark, the energy-independent islands of Samsø and Ærø get about half their space heating from biomass, both from agricultural wastes (such as straw) and biomass crops, such as reeds and elephant grass growing on land unsuitable for crops (all the rest comes from solar and wind). Burning these crops in an industrial boiler for district heating makes it easier to install and maintain pollution-control equipment than in individual stoves. Most plant material contains little sulfur, so it doesn't contribute to acid rain. And because it burns at a lower temperature than coal, it doesn't create as much nitrogen oxides. These crops also are carbon neutral—that is, they absorb as much CO_2 in growing as they emit when burned.

Some utilities are installing **flex-fuel boilers** in their power plants that can burn wood chips, agricultural waste, or other biomass fuels (fig. 20.25). As chapter 19 points out, co-combustion of coal together with biomass can have benefits over burning either alone. Including biomass in the mix reduces greenhouse gas emissions while also improving combustion properties. Even higher efficiencies can be achieved by capturing waste heat for beneficial use. A district heating plant in St. Paul, Minnesota, for example, uses 275,000 tons of wood per year (mostly from urban trees killed by storms and disease) to provide heating, air conditioning, and electricity to 25 million square feet of offices and living space in 75 percent of all downtown buildings. Although the efficiency of electrical generation in this plant is less than 40 percent (as it is in most power plants), the net yield is about 80 percent because waste heat is used rather than discarded.

Methane from biomass can be clean and efficient

Where wood and other fuels are in short supply, people often dry and burn animal manure. This is an efficient use of waste biomass, but it can intensify food shortages if manure is needed to fertilize crops. In India, for example, where fuelwood supplies have been chronically short for generations, a limited manure supply must fertilize crops and provide household fuel. Cows in India produce more than 800 million tons of dung per year, more than half of which is dried and burned in cooking fires. While dry dung makes an excellent, even-burning fuel, direct burning in an open fire may

FIGURE 20.25 This Michigan power plant uses wood chips to fuel its boilers. Where wood supplies are nearby, this is a good choice both economically and environmentally.
© William P. Cunningham

not be the most efficient means of using the energy and nutrients in dung. By some estimates, applying that dung to fields as fertilizer could boost crop production of edible grains by 20 million tons per year, enough to feed about 40 million people.

In China, animal waste is widely used to produce methane gas in simple, inexpensive methane digesters designed for villages and homes. **Methane** (CH_4) is the main component of natural gas. It is produced by anaerobic decomposition (digestion by anaerobic bacteria) of any moist organic material. Many people are familiar with the fact that swamp gas is explosive. Swamps are simply large methane digesters, basins of wet plant and animal wastes sealed from the air by a layer of water. Under these conditions, organic materials are decomposed by anaerobic (oxygen-free) rather than aerobic (oxygen-using) bacteria, producing methane instead of carbon dioxide. This same process may be reproduced artificially by placing organic wastes in a container and providing warmth and water (fig. 20.26). After gas is removed, the remaining solid material can then be used as clean fertilizer. Some 35 million Chinese households use biogas for cooking and lighting, and that number is expected to double in 20 years. These systems can be even more efficient at large scales. Two large municipal facilities in Nanyang will soon provide fuel for more than 20,000 families.

Methane digesters are increasingly widespread in developed countries as well, especially in Switzerland, Germany, and other parts of Western Europe, where organic household waste is collected separately from recycling and other garbage, and where agricultural waste is abundant (see chapter 21, fig. 21.14). Any kind of organic waste material can be used to generate biogas—livestock manure, kitchen and garden scraps, and even municipal garbage and sewage. In the United States, municipal biogas production is rare, but dairies and hog farms increasingly produce methane from their abundant supplies of animal manure. The Haubenschild farm in central Minnesota, for instance, uses manure from 850 Holsteins to generate all the power needed for their dairy operation and still have enough excess electricity for an additional 80 homes. In a typical January, the farm saves about 35 tons of coal and 1,200 gallons of propane, while making more than $4,000 from electric sales. Cattle feedlots and chicken farms in the United States also offer tremendous potential fuel sources. Feedlot wastes and collectible crop residues contain 4.8 billion gigajoules (4.6 quadrillion BTUs) of energy every year, more than all the energy U.S. farmers use.

Municipal landfills, which contain vast amounts of buried (anaerobic) organic waste, are also active sites of methane production, contributing as much as 20 percent of the annual output of methane to the atmosphere. This is a waste of a valuable resource and a threat to the environment because methane is an important greenhouse gas (chapter 15). About 300 landfills in the United States currently burn methane and generate enough electricity together for a million homes. Another 600 landfills have been identified as potential sources for methane development.

A number of colleges around the United States are using biomass and biogas to wean themselves off fossil fuels. In Vermont, Middlebury College feeds locally harvested wood chips into a gasification plant (which heats chips to produce burnable gas) that provides both heat and electricity to the campus, and the college is buying methane from a local dairy as well. These steps have led the way in helping Middlebury become carbon neutral. And the University of New Hampshire in Durham plans to provide 80 percent of its heating and electrical needs by buying and burning methane gas given off by a landfill a few miles away. At the University of Minnesota's Morris campus, a gasification plant uses about 1,700 tons of corn stover (stalks and cobs) and other local agricultural material every year to provide as much as 80 percent of the school's heating and cooling needs (fig 20.27), while a wind turbine provides most of its electricity. Together these alternative energy systems replace at least $1 million per year in fossil fuels (mostly natural gas) and

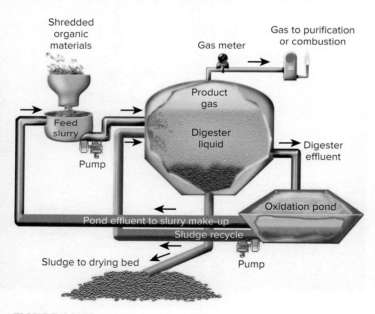

FIGURE 20.26 Continuous unit for converting organic material to methane by anaerobic fermentation. One kilogram of dry organic matter will produce 1–1.5 m^3 of methane, or 2,500–3,600 million calories per metric ton.

FIGURE 20.27 Biomass crops can provide heat and electricity as well as liquid fuels. Cellulosic ethanol is much more efficient than grain ethanol and avoids using food crops for fuel.
© William P. Cunningham

could make the campus not only carbon neutral but even carbon negative in a few years. These are just a few of the efforts across the country of campuses to "walk the walk" by not only teaching about environmental issues but also changing the way they operate. Altogether, 614 colleges and universities representing about one-third of the students in the United States have made a commitment to reduce their carbon footprint. Could you convince the administration at your school to join this movement?

Ethanol and biodiesel can contribute to fuel supplies

Biofuels, such as ethanol and biodiesel, are by far the biggest recent news in biomass energy. Globally, production of these two fuels is booming, from Brazil, which gets about 40 percent of its transportation energy from ethanol generated from sugarcane, to Southeast Asia, where millions of hectares of tropical forest have been cleared for palm oil plantations, to the United States, where about one-fifth of the corn (maize) crop currently is used to make ethanol. Crops with a high oil content, such as soybeans, sunflower seed, rapeseed (usually called canola in America), and palm oil fruits are relatively easy to make into biodiesel. In some cases, the oil needs only minimal cleaning to be used in a standard diesel engine. Yields per hectare for many of these crops are low, however. It would take a very large land area to meet our transportation needs with soy or sunflowers, for example. Furthermore, diversion of these oils for vehicles deprives humans of an important source of edible oils.

Oil palms are eight to ten times more productive per unit area than soy or sunflower (although palm fruit is more expensive to harvest and transport). Currently, millions of hectares of species-rich forests in Southeast Asia are being destroyed to create palm oil plantations. Indonesia already has 6 million ha of palm oil plantations and Malaysia has nearly as much. Together these two countries produce nearly 90 percent of the world's palm oil. The burning of Indonesian forests and the peat lands on which they stand currently releases some 1.8 billion tons of carbon dioxide every year. Indonesia is currently third in the world—behind the United States and China—in human-made greenhouse gas emissions. At least 100 species, including orangutans, Sumatran tigers, and the Asian rhinoceros, are threatened by habitat loss linked to palm oil expansion. In 2011, however, Indonesia signed an agreement with Norway to protect its forests in exchange for $1 billion in development aid (see chapter 12).

A shrubby tree called *Jatropha curcas* has recently been promoted as a good alternative to both palm oil and soybean oil. This native of Mexico and the Caribbean has nuts with a high (but toxic) oil content that can be easily converted to diesel fuel. Indian scientists have bred prolific new varieties they claim grow well on marginal soil with few inputs. Field tests suggested that *Jatropha* might produce as much as 15,000 liters of oil per hectare (1,600 gal per acre, or about three times as much as palm oil). India has set aside 50 million ha (123 million acres) of land for *Jatropha* that it hopes will provide 20 percent of its diesel fuel. In 2008, Air New Zealand flew a Boeing 747 using a 50–50 blend of *Jatropha* oil and aviation fuel. But test plantings in other areas suggest that *Jatropha* needs more water and fertilizer, is more sensitive to pests and diseases, and has lower yield than promised. As so often happens, what's touted as a miracle solution may not be so great when we look closer.

Cellulosic ethanol could be better than using food crops for fuel

Crops such as sugarcane and sugar beets have a high sugar content that can be fermented into ethanol, but sugar is expensive and the yields from these crops are generally low, especially in temperate climates. Starches in grains, such as corn, have higher yields and can be converted into sugars that can be turned by yeast into ethanol (this is the same process used to make drinking alcohol), butanol (which burns in engines much like gasoline), or methanol (fig. 20.28). The idea of burning ethanol in vehicles isn't new.

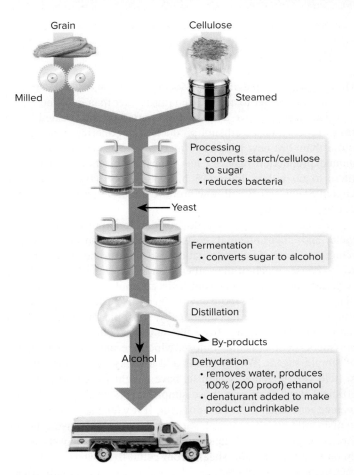

FIGURE 20.28 Ethanol (or ethyl alcohol) can be produced from a wide variety of sources. Maize (corn) and other starchy grains are milled (ground) and then processed to convert starch to sugar, which can be fermented by yeast into alcohol. Distillation removes contaminants and yields pure alcohol. Cellulosic crops, such as wood or grasses, can also be converted into sugars, but the process is more difficult. Steam blasting, alkaline hydrolysis, enzymatic conditioning, and acid pretreatment are a few of the methods for breaking up woody material. Once sugars are released, the processes are similar.

FIGURE 20.29 More than 200 ethanol plants distill biofuels from corn and other crops in the United States.
© William P. Cunningham

Henry Ford designed his 1908 Model T to run on ethanol, which at the time was easier to acquire and much safer than gasoline.

The need to move away from imported oil has created boom times for corn-based ethanol production in America. Since 1980, more than 200 new ethanol refineries have been built, and U.S. annual production capacity has grown from about 500 million liters to 60 billion liters (15 billion gallons) (fig. 20.29). The United States and Brazil now produce about 95 percent of all the ethanol in the world.

There has been a great deal of debate about the net energy yield and environmental costs of ethanol from corn, which requires a great deal of irrigation, pesticides, and fertilizer, and which causes much of the soil erosion in the United States. There is greater agreement that **cellulosic ethanol,** made from woody plant tissue, would be more environmentally sustainable, if we can find ways to produce it economically. Most plants put the bulk of the energy they capture from the sun into cellulose and related polymers, such as hemicellulose, which are made of long chains of simple sugars. Woody plants add a sticky glue, called lignin, to hold cells together. Chemists have sought for decades to find efficient ways to release those simple sugars so they can be fermented into ethanol. But it's difficult: If it were easy for microbes to dismantle woody material, we probably wouldn't have standing forests or prairies—the landscape would simply be covered with green slime. Extracting sugars from cellulosic materials involves chopping or shredding biomass, followed by treatment with bacteria or fungi (or enzymes derived from them).

Ethanol production is possible but expensive, and in Denmark, where fuel costs are high and government subsidies are abundant, cellulosic ethanol is produced. So far, there are no commercial-scale cellulosic ethanol factories operating in North America, but the Department of Energy has provided $385 million in grants for six cellulosic biorefinery plants. These pilot projects will test a wide variety of feedstocks, including rice and wheat

straw, milo stubble, switchgrass hay, almond hulls, corn stover, and wood chips. A favorite potential fuel source is switchgrass (*Panicum virgatum*), a tall grass native to the Great Plains that is easy to plant mechanically. Switchgrass is a perennial, with deep roots that store carbon (and thus capture atmospheric greenhouse gases). Long-lasting perennial roots hold soil in place, unlike annual corn crops that require tillage for planting and weed control. Making fewer trips through the field with a tractor or cultivator requires much less fuel and improves net energy yield (fig. 20.30).

An even better biofuel crop may be *Miscanthus x giganteus,* a perennial grass from Asia. Often called elephant grass (although this name is also used for other species), *Miscanthus* is a sterile, hybrid grass that grows 3 or 4 meters in a single season (fig. 20.31). Europeans have been experimenting with this species for several decades, but it has only recently been introduced to the United States. *Miscanthus* can produce at least five times as much dry biomass per hectare as corn. Part of the reason for this is that *Miscanthus* starts growing four to six weeks earlier in the spring than corn, and it stays green a month or more longer in the fall. This longer growing season, coupled with the nutrients and energy stored in underground rhizomes, gives this giant grass a huge advantage compared to annual crops, such as corn. Its perennial growth and long-lasting canopy also protect the soil from erosion, and it requires much less fuel for cultivation.

Using corn or switchgrass to produce enough ethanol to replace 20 percent of U.S. gasoline consumption would take

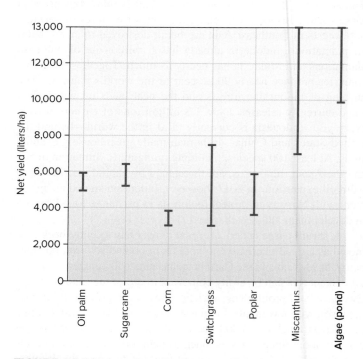

FIGURE 20.30 Proven biofuel sources include oil palms, sugarcane, and corn grain (maize). Other experimental sources may produce better yields, however.
Source: Data from E. Marris, 2006. Nature 444: 670–678.

FIGURE 20.31 *Miscanthus x giganteus* is a perennial grass that can grow 3 or 4 meters in a single season. It thrives on marginal land with little fertilizer or water and can produce five times as much biomass as corn.

Source: Courtesy S. Raghu, Susan Post, Illinois Natural History Survey

about one-quarter of all current U.S. cropland out of food production. *Miscanthus* could produce the same amount on less than half that much area. And it wouldn't need to be prime farm fields. *Miscanthus* can grow on marginal soil with far less fertilizer than corn needs. In the fall, *Miscanthus* moves nutrients into underground rhizomes. This means that the standing stalks are almost entirely cellulose and next year's crop needs very little fertilizer. And because it stores far more carbon in the soil than other crops, *Miscanthus* may be eligible for climate offset credits.

Currently there are no known diseases or pests for *Miscanthus*. However, Professor S. Raghu and his colleagues at the University of Illinois point out that the characteristics that make it an attractive energy crop—rapid growth, highly efficient photosynthesis, low need for nutrients, no known pests, high water-use efficiency—also make it a good candidate to become an invasive pest. The fact that the variety being tested for growth in the United States is a sterile hybrid may make it less likely to spread, but there are cases of invasives that spread vegetatively.

Harvesting, storing, and shipping biomass crops remains a problem. The low energy content of straw or wood chips, compared to oils or sugars, makes it prohibitively expensive to ship them more than about 50 km to a refinery. We might need to have a very large number of small refineries, if we depend on cellulosic ethanol. Alternatively, some studies find that you could drive a hybrid automobile about twice as far on the electricity generated by burning a ton of dry biomass than you could on the ethanol fermented from that same ton. So, burning biomass may still be a better solution than fermentation, if we move to hybrid engines.

Could algae be an efficient energy source?

Algae are aquatic, photosynthetic plants that produce lipids (oils), the basic substances that led to ancient oil deposits we use today. In theory, all algae need to grow are nutrients (such as waste from a sewage plant), carbon dioxide (such as effluent from a power plant), sunlight, and water. In principle, algae could be an extremely productive biofuel source. While *Miscanthus* can yield up to 13,000 liters (3,500 gal) of ethanol per hectare, some algal species growing in a photobioreactor might theoretically produce 30 times as much high-quality oil. This is partly because single-celled algae can grow 30 times as fast as higher plants. Furthermore, some algae species store up to half their total mass as oil. The challenges have been in developing algae varieties that both produce oil efficiently grow well, and in developing growth systems that scale up to produce reliably in quantity. So far, the actual yield from algal test facilities is about the same as for *Miscanthus*.

One of the most intriguing benefits of algae is that it could be grown in facilities next to conventional power plants, where CO_2 from burning either fossil fuels or biomass could be captured and used for algal growth. Thus, they could be carbon negative, providing a net reduction in atmospheric carbon while also creating useful fuel. In 2009, Japan Airlines made a test flight using a combination of jet fuel and algal oils, and in 2013, Australia's New South Wales government began construction on an algal carbon capture and biofuels production facility next to the 2,640 MW Bayswater coal-fired power plant near Sydney. A number of U.S. companies, including Solix Biofuels, Sapphire Energy, OriginOil, PetroAlgae, and Shell Oil, are exploring algal biofuels. Although the potential remains unrealized, the promise remains tantalizing.

Section Review

1. What is a fuel cell and how does it work?
2. How is methane produced from waste material?
3. How do we make biofuels out of plant material?

20.5 HYDROPOWER, TIDAL, AND GEOTHERMAL ENERGY

- *A few countries dominate hydropower development.*
- *Massive hydropower dams produce abundant electricity but also environmental concerns.*
- *Geothermal heat, tides, and waves could be valuable resources.*
- *Our energy future will involve renewable sources.*

Falling water was our first industrial water source, powering the factories of the early Industrial Revolution. The invention of hydroelectric turbines in the nineteenth century, which produced electricity by spinning generators, greatly increased the efficiency of hydropower. By 1925, falling water generated 40 percent of the world's electric power. Since then, hydroelectric production capacity has grown 15-fold, but fossil fuel use has risen so rapidly that water power is now only 20 percent of total electrical generation. Still, many countries produce most of their electricity from falling water. Norway, for instance, depends on hydropower for 99 percent of its electricity. Currently, total world hydroelectric production is about 3,000 terawatt-hours. Six countries—Canada, Brazil, the United States, China, Russia, and Norway—account for more than half that total. In fact, of the approximately 50,000 dams in the world taller than 15 m (45 ft), roughly half are in China. Untapped hydropower resources still exist in Latin America, Central Africa, India, and China, but most large rivers have been dammed or are in the process of hydropower development.

Most hydroelectricity comes from large dams

Much of the hydropower development in recent decades has been in enormous dams. There is a certain efficiency of scale in giant dams, and they bring pride and prestige to the countries that build them, but as we discussed in chapter 17, they can have unwanted social and environmental effects. The largest hydroelectric dam in the world at present is the Three Gorges Dam on China's Yangzi River, which spans 2 km (1.2 mi) and is 185 m (600 ft) tall. Designed to generate 25,000 MW of power, this dam produces as much energy as 25 large nuclear power plants (fig. 20.32). The reservoir behind the dam displaced at least 1.5 million people and submerged 5,000 archaeological sites. China is now in the process of constructing a sequence of dams on the Mekong River, which runs from Tibet to Cambodia and Viet Nam. Chinese dams now block the seasonal floods that have always sustained the fisheries and farms of the river's lower reaches.

There are other problems with big dams, besides human displacement, ecosystem destruction, and wildlife losses. Dam failure can cause catastrophic floods and thousands of deaths. Sedimentation often fills reservoirs rapidly and reduces the usefulness of the dam for either irrigation or hydropower. In China, the Sanmenxia Reservoir silted up in only two years, and the Laoying Reservoir filled with sediment before the dam was even finished. Schistosomiasis, caused by parasitic flatworms called blood flukes (chapter 8), is transmitted to humans by snails that thrive in slow-moving, weedy tropical waters behind these dams. It is thought that 14 million Brazilians suffer from this debilitating disease.

In tropical regions, where vegetation is abundant, rotting trees and other organic material in reservoirs can have disastrous effects on water quality. When Lake Brokopondo in Suriname flooded a large region of uncut rainforest, underwater decomposition of the submerged vegetation produced hydrogen sulfide that killed fish and drove out villagers over a wide area. Acidified water from this reservoir ruined the turbine blades, making the dam useless for power generation. Decaying vegetation also creates methane, an important greenhouse gas. A recent study of one reservoir in Brazil suggested that decaying vegetation produced more greenhouse gases (carbon dioxide and methane) than would have come from generating an equal amount of energy by burning fossil fuels.

Dams in arid regions, meanwhile, can evaporate much of a river's flow, reducing resources for farms and cities. This has been an especially serious problem on the Nile in Egypt and on the Colorado River in the southwestern United States (see chapter 17).

If big dams—our traditional approach to hydropower—have so many problems, how can we continue to exploit the great potential

FIGURE 20.32 Very large dams, such as China's Three Gorges Dam, produce abundant electricity but have tremendous environmental and social costs.
© menabrea/Getty Images RF

of hydropower? Fortunately there is an alternative to gigantic dams and destructive impoundment reservoirs. Small-scale, **low-head hydropower** technology can extract energy from small headwater dams that cause much less damage than larger projects. Some modern, high-efficiency turbines can even operate on **run-of-the-river flow.** Submerged directly in the stream and small enough not to impede navigation in most cases, these turbines don't require a dam or diversion structure and can generate useful power with a current of only a few kilometers per hour. They also cause minimal environmental damage and don't interfere with fish movements, including spawning migration. **Micro-hydro generators** operate on similar principles but are small enough to provide economical power for a single home. If you live close to a small stream or river that runs year-round and you have sufficient water pressure and flow, hydropower is probably a cheaper source of electricity for you than solar or wind power (fig. 20.33).

However, small-scale hydropower systems also can cause abuses of water resources. The Public Utility Regulatory Policies Act of 1978 included economic incentives to encourage small-scale energy projects. As a result, thousands of applications were made to dam or divert small streams in the United States. Many of these projects have little merit. All too often, fish populations, aquatic habitat, recreational opportunities, and the scenic beauty of free-flowing streams and rivers are destroyed to produce a small amount of power or economic gain.

Geothermal energy is everywhere

The earth's internal temperature can provide a useful source of energy in many cases. High-pressure, high-temperature energy exists just below the earth's surface. Around the edges of continental plates or where the earth's crust overlays pools of magma (molten rock) close to the surface, this **geothermal energy** is experienced in the form of hot springs, geysers, and fumaroles.

FIGURE 20.33 Solar collectors capture power only when the sun shines, but hydropower is available 24 hours a day. Small turbines such as this one can generate enough power for a single-family house with only 15 m (50 ft) of head (elevation change) and 200 liters (50 g) per minute flow. The turbine can have up to four nozzles to handle greater water flow and generate more power.
Source: Courtesy of Burkhardt Turbines

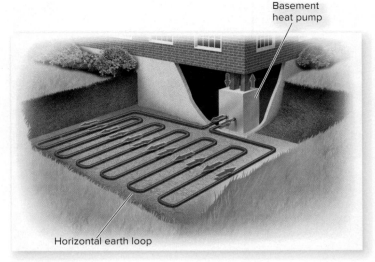

Basement heat pump

Horizontal earth loop

FIGURE 20.34 Geothermal energy can cut heating and cooling costs by half in many areas. In summer (shown here), warm water is pumped through buried tubing (earth loops), where it is cooled by constant underground temperatures. In winter, the system reverses and the relatively warm soil helps warm the house.

Yellowstone National Park is the largest geothermal region in the United States. Iceland, Japan, and New Zealand also have high concentrations of geothermal springs and vents. Depending on the shape, heat content, and access to groundwater, these sources produce wet steam, dry steam, or hot water. Iceland gets most of its heat and electricity from volcanic geothermal heat. Costa Rica is also rich in volcanic geothermal heat, and most of its electricity now comes from geothermal sources.

Although few places have geothermal steam, the earth's warmth can help reduce energy costs nearly everywhere. Pumping water through buried pipes can extract enough heat so that a heat pump (which extracts and concentrates heat) can provide space heating or hot water even in cold regions. In summer, the reverse process can provide air conditioning. If enough space is available, the horizontal tubes can be laid in shallow trenches. Where space is limited, tubes can be dropped into vertical wells. Similarly, the relatively uniform temperature of the ground can be used to augment air conditioning in the summer (fig. 20.34). This can cut home heating costs by half in many areas and pay for itself in five years.

Engineers are now exploring deep wells for community geothermal systems. Drilling 2,000 m (6,000 ft) in the American West gets you into rocks above 100°C. Fracturing them to expose more surface area, and pumping water in can produce enough steam to run

an electrical generator at a cost significantly lower than conventional fossil fuel or nuclear power. The well is no more expensive than most oil wells, and the resource is essentially perpetual. Currently, about 60 new geothermal energy projects are being developed in the United States. However, there are cautions about this technology. As we saw in chapter 19, fracturing (or fracking) can contaminate groundwater aquifers. And in 2010, two large geothermal projects (one in California and another in Switzerland) were abruptly canceled over concerns that they seemed to be triggering earthquakes.

Tides and waves contain significant energy

Ocean tides and waves contain enormous amounts of energy that can be harnessed to do useful work. A **tidal station** works like a hydropower dam, with its turbines spinning as the tide flows through them. A high-tide/low-tide differential of several meters is required to spin the turbines. Unfortunately, variable tidal periods often cause problems in integrating this energy source into the electric utility grid. Nevertheless, demand has kept some plants running for many decades.

Ocean wave energy can easily be seen and felt on any seashore. The energy that waves expend as millions of tons of water are picked up and hurled against the land, over and over, day after day, can far exceed the combined energy budget for both insolation (solar energy) and wind power in localized areas. Captured and turned into useful forms, that energy could make a substantial contribution to meeting local energy needs.

Dutch researchers estimate that 20,000 km of ocean coastline are suitable for harnessing wave power. Among the best places in the world for doing this are the west coasts of Scotland, Canada, the United States (including Hawaii), South Africa, and Australia (fig. 20.35). Wave energy specialists rate these areas at 40 to 70 kW

FIGURE 20.35 Thousands of miles of ocean coastline around the world are suitable for harvesting wave power. If suitable technology can be developed, this could provide a significant amount of electrical power. © William P. Cunningham

per meter of shoreline. Altogether, it's calculated, if the technologies being studied today become widely used, wave power could amount to as much as 16 percent of the world's current electrical output.

Some of the designs being explored include oscillating water columns that push or pull air through a turbine, and a variety of floating buoys, barges, and cylinders that bob up and down as waves pass, using a generator to convert mechanical motion into electricity. It's difficult to design a mechanism that can survive the worst storms.

Ocean thermal electric conversion might be useful

Temperature differentials between upper and lower layers of the ocean's water also are a potential source of renewable energy. In a closed-cycle **ocean thermal electric conversion (OTEC)** system, heat from sun-warmed upper ocean layers is used to evaporate a working fluid, such as ammonia or Freon, which has a low boiling point. The pressure of the gas produced is high enough to spin turbines to generate electricity. Cold water then is pumped from the ocean depths to condense the gas. As long as a temperature difference of about 20°C (36°F) exists between the warm upper layers and cooling water, useful amounts of net power can, in principle, be generated with one of these systems. This differential corresponds, generally, to a depth of about 1,000 m in tropical seas. The places where this much temperature difference is likely to be found close to shore are islands that are the tops of volcanic seamounts, such as Hawaii, or the edges of continental plates along subduction zones (chapter 14) where deep trenches lie just offshore. The west coast of Africa, the south coast of Java, and a number of South Pacific islands, such as Tahiti, have usable temperature differentials for OTEC power. In 2013, Lockheed Martin

announced plans to build a 10 MW closed-cycle OTEC system to provide energy for a luxury resort on Hainan Island in southern China. The company hopes this may be the first of many such plants around the world.

The U.S. needs a supergrid

Many of the places with the greatest potential for renewable energy development are far from the urban centers where power is needed. This means we'll need a vastly increased network of power lines if we're going to depend on wind or solar for a much greater proportion of our energy. In introducing his plans to expand renewable energy, President Obama said, "Today, the electricity we use is carried along a grid of lines and wires that dates back to Thomas Edison—a grid that can't support the demands of clean energy." He designated $4.5 billion to modernize and expand the transmission grid as part of the $86 billion in clean-energy investments in the economic recovery bill (fig. 20.36).

Fortunately, as we've seen earlier in this chapter, high-voltage direct-current lines make it possible to transmit electricity over long distances with relatively minor losses. Interestingly, studies in California show that integration of renewable resources can smooth out daily variations (fig. 20.37). The wind blows more strongly at night, and the sun shines (obviously) during the day. And because hydropower can start up quickly, it easily fills in gaps. Even though the wind doesn't blow every day in most locations, linking together wind farms even a few hundred kilometers apart can give a steadier electrical supply than does a single site. A supergrid, such as the one proposed for Desertech, could make our entire energy supply more robust, reliable, and sustainable.

What will our energy future be?

Former vice president Al Gore described our situation succinctly: Currently "we're borrowing money from China to buy oil from the Persian Gulf to burn in ways that destroy the planet." He urged

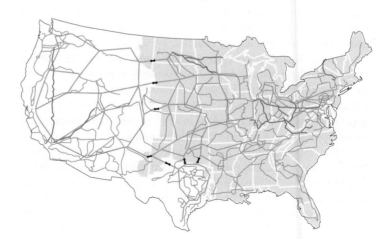

FIGURE 20.36 New high-voltage power lines (orange) will be needed if the United States is to make effective use of its renewable energy potential. The pink area served by the eastern electrical grid needs to be connected to the West by interlinks (black dots) for maximum efficiency.

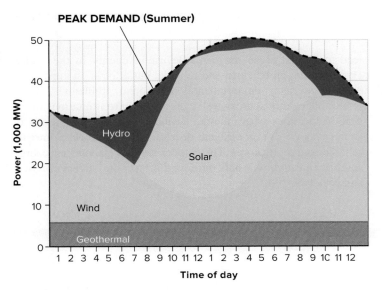

FIGURE 20.37 The wind doesn't blow all the time, nor is sunshine always available, but a mix of renewable resources could supply all the energy we need, especially if distant facilities are linked together. This graph shows hypothetical energy supplies for California on a typical July day.

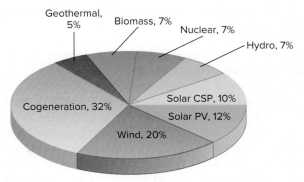

FIGURE 20.38 A renewable energy scenario for 2050. Cogeneration would mostly burn natural gas to generate both electricity and space heating.
Source: 2008 Worldwatch Report

America to repower itself with 100 percent carbon-free electricity within a decade. Doing so, he proposed, would solve the three biggest crises we face—environmental, economic, and security—simultaneously. This ambitious project could create millions of jobs, spur economic development, and eliminate our addiction to imported fossil fuels.

But could we realistically get all our electricity from renewable, environmentally friendly sources in such a short time? Mark Jacobson from Stanford University and Mark Delucchi from the University of California–Davis believe we can. Moreover, they calculate that currently available wind, water, and solar technologies could supply 100 percent of the world's energy by 2030 and completely eliminate all our use of fossil fuels. They calculate that it would take 3.8 million large wind turbines (each rated at 5 MW), 1.7 billion rooftop photovoltaic systems, 720,000 wave converters, half a million tidal turbines, 89,000 concentrated solar power plants and industrial-sized photovoltaic arrays, 5,350 geothermal plants, and 900 hydroelectric plants worldwide.

Wouldn't it be an overwhelming job to build and install all that technology? It would be a huge effort, but it's not impossible. Jacobson and Delucchi point out that society has achieved massive transformations before. In 1956, the United States began building the Interstate Highway System, which now extends 47,000 mi (75,600 km) and has changed commerce, landscapes, and society. And every year roughly 60 million new cars and trucks are added to the world's highways.

Is there enough clean energy to meet our needs? Yes, there is. As we've already seen, the readily available wind, solar, and water power sources are at least 100 times larger than our current power consumption. Even allowing for growth as people in the developing world improve their standard of living, there's more than enough environmentally friendly energy for everyone.

The World Energy Council projects that renewables could provide about 60 percent of world cumulative energy consumption in 2030, assuming that political leaders take global warming seriously and pass taxes to encourage conservation and protect the environment (fig. 20.38). This idealized "ecological scenario" also envisions measures to shift wealth from the north to south, and to enhance economic equity. By the end of the twenty-first century, renewable sources could provide all our energy needs if we take the necessary steps to make this happen. After the meltdown of four nuclear reactors in Japan in 2011, Germany, Japan, Switzerland, Sweden, and several other countries announced intentions to move away from both nuclear and fossil fuels and to emphasize renewable energy sources in the future.

Interestingly, it would take about 30 percent less total energy to meet our needs with sun, wind, and water than to continue using fossil fuels. That's because electricity is a more efficient way to use energy than burning dead plants and animals. For example, only about 20 percent of the energy in gasoline is used to move a vehicle (the rest is wasted as heat). An electric vehicle, on the other hand, uses about three-quarters of the energy in electricity for motion. Furthermore, much of the energy from renewable sources could often be produced closer to where it's used, so there are fewer losses in transmission and processing.

Won't it be expensive to install so much new technology? Yes it will be, but the costs of continuing our current dependence on fossil fuels would be much higher. It's estimated that investing $700 billion per year now in clean energy will avoid twenty times that much in a few decades from the damages of climate change.

Section Review

1. What are some advantages and disadvantages of large hydroelectric dams?
2. How can geothermal energy be used for home heating?
3. Describe how tidal power or ocean wave power generate electricity.
4. What are likely to be some major energy sources of the future?

Conclusion

Conservation is a key factor in a sustainable energy future. New designs in housing, office buildings, industrial production, and transportation can all save huge amounts of energy. There's more than enough solar and wind energy to supply all our needs. Rapid progress in technology has made the cost of these renewable sources competitive with fossil fuels in many situations. But policies that encourage development of clean energy will be needed, to offset the entrenched position of fossil fuels and nuclear power. Biofuels, including ethanol and oil (biodiesel), vary greatly in their net energy yield and environmental effects, but cellulosic feedstocks and algae may provide useful energy in the future. Hydropower can be clean and reliable, but a focus on huge dams has led to many environmental and social problems. All the renewable energy sources we've discussed have limitations in the time or locations where they're available. We will need a mix of sources, as well as an efficient distribution grid, to get a smooth, reliable source of power. The choices we make about our energy sources and uses will have profound effects on our environment and society.

Reviewing Key Terms

Can you define the following terms in environmental science?

active solar system 20.2
biofuel 20.4
biomass 20.4
cellulosic ethanol 20.4
cogeneration 20.1
concentrating solar power (CSP) system 20.2
feed-in tariffs 20.2
flex-fuel boilers 20.4

fuel cells 20.4
geothermal energy 20.5
green pricing 20.2
light-emitting diodes (LEDs) 20.1
low-head hydropower 20.5
methane 20.4
micro-hydro generators 20.5
negawatts 20.1

ocean thermal electric conversion (OTEC) 20.5
ocean wave energy 20.5
passive heat absorption
passive houses
photovoltaic (PV) cell 20.2
power tower 20.2
property assessed clean energy (PACE) 20.2

reformer 20.4
renewables portfolio 20.2
run-of-the-river flow 20.5
smart meter 20.2
solar garden
tidal station 20.5
wind turbines 20.3

Critical Thinking and Discussion Questions

1. What alternative energy sources are most useful in your region and climate? Why?

2. What can you do to conserve energy where you live? In personal habits? In your home, dormitory, or workplace?

3. Do you think building wind farms in remote places, parks, or scenic wilderness areas would be damaging or unsightly?

4. If you were the energy czar of your state, where would you invest your budget?

5. What could (or should) we do to help developing countries move toward energy conservation and renewable energy sources? How can we ask them to conserve when we live so wastefully?

Data Analysis

Energy Calculations

Suppose that you were debating between a high-mileage car, such as a Nissan leaf (rated at the equivalent of 114 mpg), or a Ford F150 4WD E85 (rated at 12 mpg). How do the energy requirements of these two purchases measure up? To put it another way, how long could you run a computer on the energy you would save by buying a Leaf rather than an F150?

Go to Connect to find information to help you calculate energy savings from various automobile models, and then compare that to the savings from running your computer for a year.

TO ACCESS ADDITIONAL RESOURCES FOR THIS CHAPTER, PLEASE VISIT CONNECT AT
www.connect.mheducation.com

You will find LearnSmart, an adaptive learning system, Google Earth™ exercises, additional Case Studies, Data Analysis exercises, and an interactive ebook.

▲An endangered Hawaiian monk seal is disentangled from abandoned fishing nets in the Papahānaumokuākea Marine National Monument.
Source: NOAA

Learning Outcomes

After studying this chapter, you should be able to:

21.1 Identify the components of solid waste, and describe our different methods of waste management.

21.2 Explain how we might shrink the waste stream.

21.3 Investigate hazardous and toxic wastes.

"We have no knowledge, so we have stuff; but stuff without knowledge is never enough."

– Greg Brown

Plastic Seas

The Papahānaumokuākea Marine National Monument is the largest marine reserve in the world, larger than all U.S. national parks combined. Established by President George W. Bush in 2006, the sanctuary was expanded by President Obama in 2016 to encompass a chain of islands, atolls, and reefs extending across 583,000 mi² northwest of Hawaii. The monument protects some of the most pristine and diverse deep coral reefs and over 7,000 marine species, including rare and endangered species such as the Laysan albatross and the Hawaiian monk seal. The string of isolated islets and coral atolls makes up the world's largest tropical seabird rookery, supporting 14 million nesting seabirds. The preserve is also home to a wealth of cultural and historic heritage sites, including shipwrecks and World Heritage cultural sites for native Hawaiians.

Despite its remote location, Papahānaumokuākea,* remains vulnerable to the flotsam and jetsam of modern life. The islands and reefs lie within the vast circulating currents known as the Pacific gyre. These swirling currents, driven by winds and the Coriolis effect (chapter 15), concentrate nutrients, organic debris, and, in recent decades, an ocean of plastic trash. Often called the Great Pacific Garbage Patch or the Pacific Garbage Gyre, this region of floating plastic debris is really a drifting cloud of plastic particles, soda bottles and caps, disposable shopping bags, packaging, discarded fishing nets, and other debris. Much of it consists of tiny fragments floating just below the surface, but some pieces are large and recognizable, and some float 20–30 m deep. The greatest concentrations of plastic debris occur in the eastern Pacific, between California and Hawaii, and in the western Pacific near Japan. But the trash field extends across the ocean, with lesser aggregations near the Papahānaumokuākea preserve. Similar garbage patches have been identified in the Atlantic and elsewhere in the world's oceans, but the Pacific cases are the best studied.

The Pacific Garbage Gyre is thought to contain more than 100 million tons of plastic. In some areas, this debris outweighs the living biomass. Fish have been found with stomachs full of plastic fragments. Seabirds gulp down plastic fragments, then regurgitate them for their chicks. With stomachs blocked by indigestible bottle caps, disposable lighters, and other items, chicks starve to death. In one study of Laysan albatrosses, 90 percent of the carcasses of dead albatross chicks contained plastic fragments (fig. 21.1). Seals, turtles, porpoises, and seabirds become ensnared in ghost fishing nets and drown, or they die from ingesting indigestible materials. Oceanographers worry that this debris is slowly starving ocean ecosystems.

Surveys at sea and on beaches indicate that 50–80 percent of the floating material originates onshore. The rest is discarded or lost at sea. Stray shopping bags, drink containers, fast-food boxes, and other refuse fall from dumpsters, wash away from landfills, or are discarded on the street, then wash into storm sewers and streams. Eventually these items travel to the sea, where they gradually break into smaller pieces as they join the great global masses of ocean plastic.

The problem has been extraordinarily difficult to address because it is widespread, diffuse, abundant, and constantly replenished by careless or incomplete disposal of waste onshore and at sea.

FIGURE 21.1 A Laysan albatross chick, which died after being fed plastic debris rather than fish. Starvation after plastic ingestion is a leading cause of death for these albatross chicks.
Source: National Marine Sanctuary, photographer Claire Fackler/NOAA

But growing awareness is starting to make a difference. Cleanup cruises in Papahānaumokuākea have collected more than 700 metric tons of discarded fishing gear that had clogged reefs.

In Papahānaumokuākea and elsewhere, marine debris has also caught the public's attention, and widespread beach cleanups are having a visible effect. And people are becoming active worldwide. According to the Ocean Conservancy, an organization that conducts regular coastal cleanups around the world and then measures and weighs the trash, nearly 9 million volunteers had collected almost 66,000 tons of debris between 1990 and 2015.

Increasing awareness is also encouraging many fishing boats to reduce disposal of plastic garbage at sea. Because all this material fouls fishing gear, costing time and money, it is in their best interest to bring in the garbage they produce or collect in their nets. This is important because discarded fishing gear is the top accidental killer of marine life, followed by plastic shopping bags, balloons, and plastic bottle caps.

You can help protect the oceans, no matter how far from them you are. Next time you see plastic debris that might wash into a storm sewer, remember that everything ends up eventually in the ocean. Pick it up if you can, and try to prevent your own plastic

*Pronounced Pa-pa-ha-nao-Mo-kua-kea. To hear the pronunciation, visit the monument's website, www.papahanaumokuakea.gov.

bottles, caps, and packaging from escaping into waterways. (And never release helium balloons into the sky.) You can also try to reduce the amount of disposable containers, bottles, and packaging you buy. Containing and minimizing loose garbage is one of the best ways to reduce marine debris.

The remote atolls of northwestern Hawaii show us that no place is too remote to be affected by our waste production and disposal. The materials we buy and the ways we manage our garbage can have dramatic impacts on living systems at home and far away. At the same time, responses to the problem have shown that people everywhere have an interest in taking care of the land and oceans and in keeping them beautiful. Often, the obstacles and volumes of waste seem insurmountable, but cleanup efforts in Hawaii have shown that progress can be real if we keep at it. In this chapter, we'll examine the waste we produce, our methods to dispose of it, and strategies to reduce, reuse, and recycle it.

21.1 WHAT DO WE DO WITH WASTE?

- *Municipal solid waste is the portion of the waste stream for which we are most directly responsible.*
- *Much of the world's waste continues to be disposed of inappropriately, especially e-waste and hazardous waste.*
- *Sanitary landfills dominate U.S. waste management, but a number of other options exist.*

Waste is everyone's business. We all produce wastes in nearly everything we do. Are there ways we can produce less and handle waste better? As you read this chapter, keep in mind the "three Rs:" reduce, reuse, and recycle. Where in our production and management of waste can we improve our performance on the three Rs?

Municipal waste—a combination of household and commercial refuse—amounts to roughly 250 million metric tons per year in the United States (fig. 21.2). That's approximately two-thirds of a ton for each of us every year, twice as much per capita as Europe or Japan and five to ten times as much as most developing countries.

Industrial waste, other than mining and mineral production, amounts to some 400 million metric tons per year in the United States. Most of this material is recycled, converted to other forms, destroyed, or disposed of in private landfills or deep injection wells. About 60 million metric tons of industrial waste falls into a special category of hazardous and toxic waste, which we will discuss later in this chapter.

More than one-third of the 11 billion tons of solid waste we produce each year results from mining—mine tailings, overburden from strip mines, smelter slag, and other residues produced by mining and primary metal processing. Most of the remainder is agricultural waste, such as crop residue and animal manure, which provide valuable nutrients if kept on fields. Mining and agricultural waste is usually stored at or near its source of production. Improper disposal practices, however, can result in serious and widespread pollution. Runoff from uncontrolled agricultural waste, for example, is our single largest source of nonpoint air and water pollution.

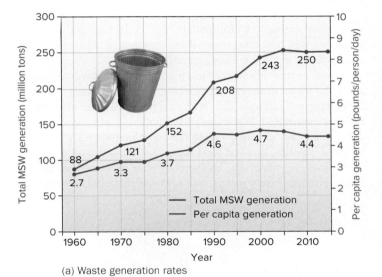

(a) Waste generation rates

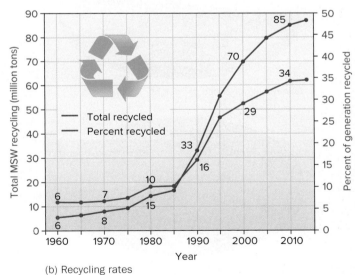

(b) Recycling rates

FIGURE 21.2 Bad news and good news in solid waste production in the U.S. Per capita waste has risen to more than 2 kg/person/day, but it shows signs of declining (a). Recycling rates are rising (b), although they remain lower than in other developed economies.

The waste stream is everything we throw away

Think for a moment about how much we discard every year. There are organic materials, such as yard and garden wastes, food wastes, and sewage sludge from treatment plants; junked cars; worn-out furniture; and consumer products of all types. Newspapers, magazines, catalogs, and office refuse make paper one of our major wastes (fig. 21.3). In spite of recent progress in recycling, many of the 200 *billion* metal, glass, and plastic food and beverage containers used every year in the United States end up in the trash. Wood, concrete, bricks, and glass come from construction and demolition sites, dust and rubble from landscaping and road building. All of this varied and voluminous waste has to arrive at a final resting place somewhere.

The **waste stream** is the steady flow of varied wastes that we all produce, from domestic garbage and yard wastes to industrial, commercial, and construction refuse. Many of the materials in our waste stream would be valuable resources if they were not mixed with other garbage. Unfortunately, our collecting and dumping processes mix and crush everything together, making separation an expensive and sometimes impossible task.

Another problem with refuse mixing is that hazardous materials in the waste stream get dispersed through thousands of tons of miscellaneous garbage. This is especially a problem when mixed waste is incinerated. When pesticides, batteries (zinc, lead, or mercury), cleaning solvents, smoke detectors containing radioactive material, plastics, and wet food scraps are burned with paper and other nontoxic materials, incineration can produce dioxins and PCBs, airborne heavy metals, and acidic particulates and aerosols.

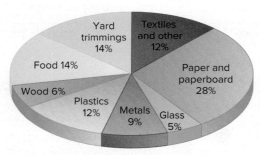

(a) Amount generated, by weight

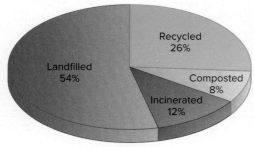

(b) Disposal methods

FIGURE 21.3 Composition of municipal solid waste in the United States by weight, before recycling, and disposal methods.
Source: Data from U.S. Environmental Protection Agency, Office of Solid Waste Management, 2015

Quantitative Reasoning

Figure 21.2 shows a continuing increase in waste production per capita. What is the percentage increase per capita from 1960 to 2005? (*Hint:* Calculate (4.5 − 2.7) ÷ 2.7.) What might account for this increase? Is there a relationship between waste production and our quality of life?

Open dumps pollute air and water

Open, unregulated dumps were historically the way nearly everyone dealt with waste, and these are still the main disposal method in most developing countries (fig. 21.4). There, fast-growing megacities have enormous garbage problems. Mexico City, one of the largest cities in the world, generates some 10,000 tons of trash *each day.* Until recently, most of this torrent of waste was left in giant piles, exposed to the wind and rain, as well as rats and flies. Manila, in the Philippines, generates a similar amount of waste, half of which goes to a giant, constantly smoldering dump called "Smoky Mountain." Over 20,000 people live and work on this mountain of refuse, scavenging for recyclable items or edible food scraps. The government would like to close these dumps, but finding another disposal method has been a challenge.

Most developed countries now forbid open dumping, at least in metropolitan areas, but illegal dumping is still a problem. You may have seen trash accumulating along roadsides and in vacant, weedy lots in the poorer sections of cities. This is not just a question of aesthetics. Plastic and other waste products are long-lasting and sometimes dangerous. Discarded oil and solvents from cars, paints, and household chemicals are also toxic. An estimated 200 million liters of waste motor oil are poured into the sewers or allowed to soak into the ground every year in the United States. This is about five times as much as was spilled by the *Exxon Valdez* in Alaska in 1989. No one knows the volume of solvents and other chemicals disposed of by similar methods.

FIGURE 21.4 Trash disposal has become a crisis in the developing world, where people have adopted cheap plastic goods and packaging but lack good recycling or disposal options.
© William P. Cunningham

(a) Marine litter

(b) Threats to marine life

FIGURE 21.5 Plastic trash dumped on land and at sea ends up on remote beaches (a) and kills unknown numbers of marine organisms (b). This sea turtle is tangled in abandoned fishing nets.
© T. O'Keefe/PhotoLink/Getty Images RF (a); J. Asher, NOAA PIFSC CRED (b)

Dumping is uncontrollable when it's out of sight

When nobody is looking, we often dispose of waste improperly. This is part of the reason that waste is often poorly managed in developing areas, and in low-income areas of wealthy countries, where agencies in charge of waste management are often underfunded or nonexistent.

The oceans have long been a free-for-all and a universal dumping ground, as noted in the opening case study. An estimated 20 million tons of plastic debris ends up in the ocean each year. This includes some 25,000 metric tons (55 million lbs) of packaging, including half a million bottles, cans, and plastic containers, which are dumped at sea. Beaches, even in remote regions, are littered with the nondegradable flotsam and jetsam of industrial society (fig. 21.5a). About 150,000 tons (330 million lbs) of fishing gear—including more than 1,000 km (660 mi) of nets—are lost or discarded at sea each year (fig. 21.5b). Wildlife advocates estimate that 50,000 northern fur seals are entangled in this refuse and drown or starve to death every year in the North Pacific alone.

Recent studies have found similar concentrations of floating plastic debris in the Great Lakes. In some cases, concentrations of garbage patches in these landlocked lakes are even worse than those in the gyres of the global oceans.

We often export e-waste and toxic waste to countries ill-equipped to handle it

Electronic waste, or **e-waste,** is a rapidly growing world problem, because it includes a bewildering variety of mixed and often toxic materials. The United States disposes of about 47 million computers and 1 million cell phones every year, each containing a complex mix of often-toxic metals and plastics. Since 1989, it has been illegal to export this e-waste to developing countries, but we continue to do so. About 80 percent of our e-waste is shipped overseas, initially to China, and more recently to developing countries in Asia and Africa. There, villagers, including young children, break it apart to retrieve valuable metals. Often this scrap recovery is done under primitive conditions, where workers have little or no protective gear (fig. 21.6a) and residue goes into open dumps. Health risks in this work are severe, especially for growing children. Soil, groundwater, and surface water contamination at these sites has been found to be as much as 200 times the World Health Organization's standards. An estimated 100,000 workers handle e-waste in China alone. With increasing regulation in China, however, the trade is shifting to India, Ghana, and other impoverished areas.

E-waste generation is increasing, and soon developing countries themselves will be the leading producers of these toxic materials (fig. 21.6b). Outdated electronic devices are one of the greatest sources of toxic material currently going to developing countries. There are at least 2 billion television sets and personal computers in use globally. Televisions often are discarded after only about five years, while computers, playstations, cell phones, and other electronics become obsolete even faster. As many as 600 million computers are in use in the United States (twice as many as there are residents), and most will be discarded in the next few years. Only about 10 percent of the components are currently recycled. These computers contain at least 2.5 billion kg of lead (as well as mercury, gallium, germanium, nickel, palladium, beryllium, selenium, arsenic), and valuable metals, such as gold, silver, copper, and steel.

Exporting toxic waste is a chronic problem even though it is banned in most countries. In 2006, for example, 400 tons of toxic waste were illegally dumped at 14 open dumps in Abidjan, the capital of the Ivory Coast. The black sludge—petroleum wastes containing hydrogen sulfide and volatile hydrocarbons—killed ten people

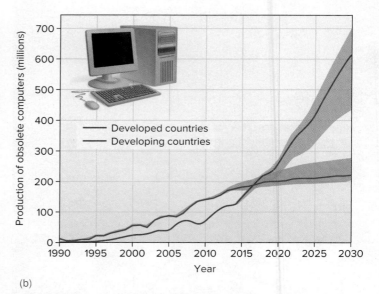

(b)

FIGURE 21.6 A Chinese woman breaks up e-waste to extract valuable metals (a). This kind of unprotected demanufacturing is hazardous to workers and the environment, but production of e-waste is rising in both developed and developing areas (b). Most waste will be produced in developing areas after about 2015.

Source: Modified from Yu et al., 2010, Environmental Science and Technology; © Basel Action Network 2006

and injured many others. At least 100,000 city residents sought medical treatment for vomiting, stomach pains, nausea, breathing difficulties, nosebleeds, and headaches. The sludge—which had been refused entry at European ports—was transported by an Amsterdam-based multinational company on a Panamanian-registered ship and handed over to an Ivorian firm (thought to be connected to corrupt government officials) to be dumped in the Ivory Coast. The Dutch company agreed to clean up the waste and pay the equivalent of (U.S.) $198 million to settle claims.

Another notorious form of waste export is the disposal of obsolete ocean-going ships. Often filled with toxic or hazardous oils, fuels, asbestos, or heavy metals, retired ships are generally sold to ship breakers in India, Bangladesh, Pakistan, or China. There they are dismantled, often under lax labor and environmental safety rules. On India's Alang Beach, for example, more than 40,000 workers tear apart outdated vessels using crowbars, cutting torches, and even their bare hands. Metal is dragged away and sold for recycling. Organic waste is often simply burned on the beach, where ashes and oily residue wash back into the water.

Landfills receive most of our waste

Over the past 50 years most American and European cities have recognized the health and environmental hazards of open dumps. Instead we have **sanitary landfills,** where solid waste is contained more effectively. To decrease smells and litter and to discourage insect and rodent populations, landfill operators are required to compact the refuse and cover it every day with a layer of dirt. This method helps

control pollution, but the dirt fill also takes up as much as 20 percent of landfill space. Since 1994, all operating landfills in the United States have been required to control such hazardous substances as oil, chemical compounds, toxic metals, and contaminated rainwater that seeps through piles of waste. An impermeable clay and/or plastic lining underlies and encloses the storage area (fig. 21.7). Drainage systems are installed in and around the liner to catch drainage and to help monitor chemicals that may be leaking. Modern municipal solid-waste landfills now have many of the safeguards of hazardous waste repositories (described later in this chapter).

Landfill space near population centers is becoming scarce and expensive. Just 25 years ago the United States had 8,000 landfills; today we have fewer than 2,000. Fresh Kills Landfill on Staten Island, New York, was the world's largest until it closed in 2001. New York now sends its garbage to Pennsylvania and Ohio. Many other cities are running out of local landfill space and must export trash, at enormous expense, to neighboring states. More than half the solid waste from New Jersey goes out of state, some of it up to 800 km (500 mi) away.

Quantitative Reasoning

Examine figure 21.2 closely. How has waste production changed from 1960 to today, in pounds per person? What are some of the reasons for this change? Think of someone you know who lived in the United States in 1960. Ask them if they can explain the difference in production per person. Ask these same questions about recycling, in terms of percentage recycled.

FIGURE 21.7 A plastic liner being installed in a sanitary landfill. This liner and a bentonite clay layer below it prevent leakage to groundwater. Trash is also compacted and covered with earth fill every day.
© Doug Sherman/Geofile

More careful attention is now paid to the siting of new landfills. Sites located on highly permeable or faulted rock formations are passed over in favor of sites with less leaky geologic foundations. Landfills are being built away from rivers, lakes, floodplains, and aquifer recharge zones rather than near them, as was often done in the past. More care is being given to a landfill's long-term effects so that costly cleanups and rehabilitation can be avoided.

Historically, landfills have been a convenient and relatively inexpensive waste-disposal option in most places, but this situation is changing rapidly. Rising land prices and shipping costs, as well as increasingly demanding landfill construction and maintenance requirements, are making this a more expensive disposal method. The cost of disposing of a ton of solid waste in Philadelphia went from $20 in 1980 to more than $100 in 2010. Union County, New York, experienced an even steeper price rise. In 1987, it paid $70 to get rid of a ton of waste; a year later that same ton cost $420, or about $10 for a typical garbage bag. In the past decades, costs have continued to rise steadily, though not as sharply. The United States now spends over $10 billion per year to dispose of trash.

Suitable places for waste disposal are becoming scarce in many areas. Other uses compete for open space. Communities have become more concerned and vocal about health hazards, as well as aesthetics. It is difficult to find a neighborhood willing to accept a new landfill. Since 1984, when stricter financial and environmental protection requirements for landfills took effect, thousands of landfills have closed.

A positive trend in landfill management is methane recovery. Methane, the main component of natural gas, is a natural product of decomposing garbage deep in a landfill. Methane is also a potent greenhouse gas. Normally, methane seeps up to the landfill surface and escapes. At 300 U.S. landfills, the methane is being collected and used for fuel. Cumulatively, these landfills could provide enough electricity for a city of a million people. Three times as many landfills could be recovering methane. Tax incentives could be developed to encourage this kind of resource recovery.

Incineration produces energy but also pollutes

Landfilling is still the disposal method for the majority of municipal waste in the United States (fig. 21.8), but incineration is our second most common method. Another term commonly used for this technology is **energy recovery,** or waste-to-energy, because the incinerators often run boilers to produce electricity. Internationally, well over 1,000 waste-to-energy plants in western Europe, Japan, and elsewhere generate much-needed energy while also reducing the amount that needs to be landfilled. In the United States, more than 110 waste incinerators burn nearly 80,000 tons of garbage daily. Most of these incinerators produce energy.

In some municipal incinerators, refuse is sorted on arrival to remove unburnable or recyclable materials before combustion. This is called **refuse-derived fuel** because the enriched burnable fraction has a higher energy content than the raw trash. Another approach, called **mass burn,** is to dump nearly everything into a giant furnace and burn as much as possible (fig. 21.9). This technique avoids the expensive job of sorting through the garbage for

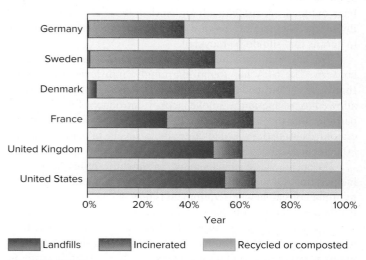

Landfills   Incinerated   Recycled or composted

FIGURE 21.8 Reliance on landfills, recycling/composting, and incineration varies considerably among countries.
Source: Data Source Eurostat; U.S. EPA 2014

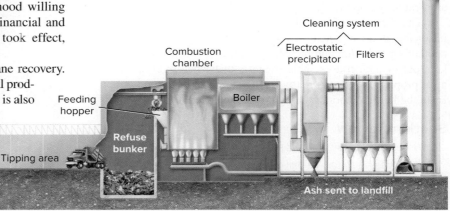

FIGURE 21.9 A municipal garbage incinerator burns waste material, ideally sorted to remove unburnables. Steam produced in the boiler can be used to generate electricity or heat nearby buildings.

nonburnable materials, but it produces much more air pollution than sorted waste does.

Residual ash and unburnable residues, representing 10 to 20 percent of the original volume, are usually taken to a landfill for disposal. Because the volume of burned garbage is reduced by 80 to 90 percent, disposal is a smaller task. However, the residual ash may contain a variety of toxic components that make it an environmental hazard if not disposed of properly. Ironically, one worry about incinerators is whether enough garbage will be available to feed them. Some communities in which recycling has been really successful have had to buy garbage from neighbors to meet contractual obligations to waste-to-energy facilities. In other places, fears that this might happen have discouraged recycling efforts.

Well-run incinerators can be clean

Waste incinerators are often the subject of heated debates. Initial construction costs are high—usually between $100 million and $300 million for a typical municipal facility. Tipping fees at an incinerator, the fee charged to haulers for each ton of garbage dumped, are often much higher than those at a landfill. Most important, incinerators require a waste stream that may include materials that should be recycled.

The environmental hazards of incinerators is another point of concern. The EPA has found alarmingly high levels of dioxins, furans, lead, and cadmium in incinerator ash. If this material is allowed to escape in the smoke stack, it introduces light, airborne particles into the air we breathe. In one EPA study, all of the incinerators examined exceeded acceptable levels of cadmium, and 80 percent exceeded lead standards.

European waste management, meanwhile, relies heavily on incineration. German studies have found that incinerators produce minimal emissions when equipment is new, when waste separation is reliable (no cadmium batteries, for example), and when regulatory agencies monitor facilities. In these conditions, studies have found that incinerators produce far lower air emissions than other industries, urban air pollution sources, and home heating systems. Between 1990 and 2000, German emissions of dioxins and furans, important airborne carcinogens, dropped from 400 to 0.5 toxicity units. Improved waste sorting and emissions control equipment are generally responsible for these improvements.

The high cost of landfills in Europe makes it easier to invest in sorting, improved incineration, and recycling. Plastics, batteries containing cadmium and nickel, and wet food waste are generally kept out of incinerators. This suggests that when built and managed properly, waste-to-energy incinerators can be a good option. Even the best incinerator, though, is less efficient than recycling.

Section Review

1. Why is e-waste difficult to dispose of properly?
2. How does a sanitary landfill differ from an open dump?
3. Why is it good to divert yard waste and organic material from the general waste stream?
4. Why do many northern European countries incinerate much of their waste?

21.2 SHRINKING THE WASTE STREAM

- *Reducing waste is the most important of the three Rs.*
- *Recycling and reuse save materials and energy.*
- *Composting is an important component of recycling.*

Reducing waste in the first place can save money right through the waste disposal chain. For the waste we produce, the best way to handle it may be to re-imagine it as a resource. In nature, everything is reused, and the waste produced by one organism is a resource for another. Can we do the same with our own wastes?

In terms of solid waste management, **recycling** is the reprocessing of discarded materials into new, useful products (fig. 21.10). This contrasts with *reusing* materials, such as refillable beverage containers, which are returned to a factory to be refilled. Some recycling processes reuse materials for the same purposes; for instance, old aluminum cans and glass bottles are usually melted and recast into new cans and bottles. Other recycling processes turn old materials into entirely new products. Old tires, for instance, are shredded and turned into rubberized road surfacing. Newspapers become cellulose insulation, kitchen wastes become a valuable soil amendment, and steel cans become new automobiles and construction materials.

How much we recycle usually depends on the value of material. Aluminum scrap has a high value and is relatively light to transport. It's also expensive to produce from raw materials. A large percentage of aluminum is recycled nearly everywhere (fig. 21.11). About half of all aluminum beverage cans are now recycled, up from only 15 percent in 1970. Aluminum can be recycled rapidly: Half of the cans now on grocery shelves will be made into another can within two months. Even so, we throw away staggering amounts of materials. Every three months Americans landfill enough aluminum drink cans to rebuild the entire commercial airline fleet.

FIGURE 21.10 Creating a stable, economically viable market for recycled products is essential for recycling success. Consumers can help by buying recycled products.
© David Trevor/Alamy RF

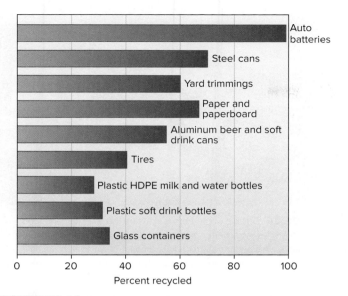

FIGURE 21.11 Recycling rates for selected materials in the United States. Car battery recycling is required by law, to keep lead out of the waste stream.
Source: U.S. Environmental Protection Agency 2015

Recycling has multiple benefits

Recycling saves money, energy, raw materials, and landfill space, while also reducing pollution. Recycling also encourages individual awareness and responsibility for the refuse produced. Household sorting is the bedrock of many recycling systems, but many cities now have recycling facilities with mechanized sorting systems, so that recyclables can be put into a single container.

Curbside pickup of recyclables costs around $35 per ton, as opposed to the $80 paid to dispose of them at an average metropolitan landfill. Many recycling programs cover their own expenses with materials sales and may even bring revenue to the community. Although landfills continue to dominate American waste disposal, recycling (including composting) has quadrupled since 1980 (fig. 21.12).

Recycling cuts our waste volumes drastically. Philadelphia has invested in neighborhood collection centers that recycle enough to eliminate the need for a previously planned, high-priced incinerator. New York, which exports its daily 11,000 tons of waste by truck, train, and barge, to New Jersey, Pennsylvania, Virginia, South Carolina, and Ohio, has set ambitious recycling goals of 50 percent waste reduction, although the city still recycles less than 30 percent of its household and office waste. In contrast, Minneapolis and Seattle recycle nearly 60 percent of domestic waste, Los Angeles and Chicago over 40 percent. In 2002, New York Mayor Michael Bloomberg raised a national outcry by canceling most of the city's recycling program. He argued that the program didn't pay for itself and the money should be spent to balance the city's budget. A year later, Bloomberg relented after realizing that it cost more to ship garbage to Ohio than to recycle. Recycling was reinstated for nearly all recyclable materials.

Japan is probably the world's leader in recycling. Short of land for landfills, Japan recycles about half its municipal waste and incinerates about 30 percent. The country has begun a push to increase recycling, because incineration costs almost as much. Some communities have raised recycling rates to 80 percent, and others aim to reduce waste altogether by 2020. This level of recycling is most successful when waste is well sorted. In Yokohama, a city of 3.5 million, there are now 10 categories of recyclables, including used clothing and sorted plastics. Some communities have 30 or 40 categories for sorting recyclables.

Perhaps most important, recycling lowers our demands for raw resources. In the United States, we cut down 2 million trees every day to produce newsprint and paper products, a heavy drain on our forests. Recycling the print run of a single Sunday issue of *The New York Times* would spare 75,000 trees. Every piece of plastic we make reduces the reserves supply of petroleum and makes us more dependent on foreign oil. Recycling 1 ton of aluminum saves 4 tons of bauxite (aluminum ore) and 700 kg (1,540 lbs) of petroleum coke and pitch, as well as keeping 35 kg (77 lbs) of aluminum fluoride out of the air.

Recycling also reduces energy consumption and air pollution. Plastic bottle recycling can save 50 to 60 percent of the energy needed to make new ones. Making new steel from old scrap offers up to 75 percent energy savings. Producing aluminum from scrap instead of bauxite ore cuts energy use by 95 percent, yet we still throw away more than a million tons of aluminum every year. If aluminum recovery were doubled worldwide, more than a million tons of air pollutants would be eliminated every year.

Recycling plastic is especially difficult

Much of the plastic ocean debris (opening case study) results from carelessness, but another part of the problem is that plastic is tricky to reuse and recycle. Contamination is a major reason

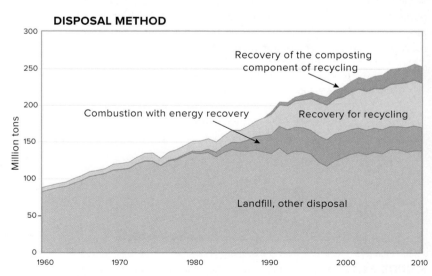

FIGURE 21.12 Disposal of municipal solid waste, 1960–2010. Landfills remain the dominant destination, but recycling and composting are increasing.
Source: U.S. Environmental Protection Agency 2012

for this difficulty. Most of the 24 billion plastic soft drink bottles sold every year in the United States are made of PET (polyethylene terephthalate), which can be melted and remanufactured into carpet, fleece clothing, plastic strapping, and nonfood packaging. However, even a trace of vinyl—a single PVC (polyvinyl chloride) bottle in a truckload, for example—can make PET useless.

Although most bottles are now marked with a recycling number, it's hard for consumers to remember which is which. Another obstacle is that many soft drink bottles are sold and consumed on the go, and never make it into recycling bins. As a consequence, Americans have an extremely low recovery rate for plastics (fig. 21.13).

Reducing litter is an important benefit of recycling. Ever since disposable paper, glass, metal, foam, and plastic packaging began to accompany nearly everything we buy, these discarded wrappings have collected on our roadsides and in our lakes, rivers, and oceans. Without incentives to properly dispose of beverage cans, bottles, and papers, it often seems easier to just toss them aside when we have finished using them. Litter is a costly as well as unsightly problem. The United States pays an estimated 32 cents for each piece of litter picked up by crews along state highways, which adds up to $500 million every year. "Bottle bills" requiring deposits on bottles and cans have reduced littering in many states.

Our present public policies often tend to favor extraction of new raw materials: Energy, water, and raw materials are often sold to industries below their real cost, in order to create jobs and stimulate the economy. A pound of recycled clear PET, the material in most soft drink bottles, is worth about 40¢, while a pound of virgin PET costs about 25¢. Setting the prices of natural resources at their real cost would tend to encourage efficiency and recycling.

Price fluctuations are a constant challenge for businesses trying to make an income on recycling. As with any primary materials, prices can vary dramatically. Some years, copper is so valuable that copper pipes, wires, and flashing are stolen from houses; other years, prices and demand are low. States and cities have often helped to stabilize markets by requiring government agencies to purchase a minimum amount of recycled materials. Each of us can play a role in creating markets, as well. If we buy things made from recycled materials—or ask for them if they aren't available—we

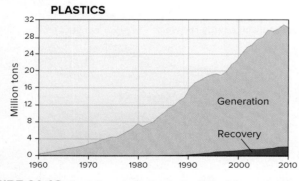

FIGURE 21.13 Recycling of plastics in the United States is improving but remains extremely low. This helps explain why so much plastic ends up in the ocean.
Source: U.S. Environmental Protection Agency 2012

FIGURE 21.14 Organic waste from agriculture, households, and food-processing factories is converted to methane (biogas) in these enclosed digesting tanks. This system diverts a vast amount of waste from landfills while producing gas that is used to power city buses and to heat city buildings.
© Ashley Cooper/Corbis

will help make it possible for recycling programs to succeed. Growing world demand, with expanding consumer economies around the world, is also likely to increase the value of recycled materials.

Compost and biogas are useful products

Composting is often considered a type of recycling, because it converts organic waste to soil-enriching organic fertilizer. Many U.S. cities have municipal compost facilities, which keep yard clippings and leaves out of landfills and provide gardeners with a valuable soil amendment. Composting uses natural aerobic (oxygen-rich) decomposition to reduce organic debris to a nutrient-rich soil amendment. Home gardeners have always composted yard and garden waste in their backyards, but cities and towns increasingly see this as a way to reduce waste disposal costs. Organic debris such as yard waste makes up 13 percent of the waste we generate (fig. 21.3). Almost two-thirds of our yard waste is composted.

While compost is a useful material, its market value is modest. Another option is biogas, or methane, which is produced by fermenting organic waste in air-tight, liquid-filled containers (fig. 21.14). There is tremendous potential in these systems because about 30 percent of municipal waste is biodegradable organic material. Methane "biogas" plants are increasingly common ways for European cities to dispose of organic waste. Over 14,500 biogas digesters convert organic waste to gas that can be burned for heat and electricity. Biogas now supplies about 3 percent of German electricity consumption. Increasingly, U.S. dairies are also starting to run operations on methane produced from manure.

These plants are easiest to operate where organic waste is already separated from other waste. Increasingly, cities are requiring separation in order to reduce greenhouse gas emissions (methane) from landfills and to ensure good incinerator operation. Efficient production of a useful fuel source is a great co-benefit.

Demolition and construction debris is another major source of waste. Every year, thousands of tons of debris from building sites heads to landfills, but recycling facilities are beginning to collect, sort, and resell increasing portions of this debris. Taylor Recycling, in Newburgh, New York, recycles and sells 97 percent of the mixed demolition debris it receives, well above the industry average of 30 to 50 percent. Trees are ground up and converted to mulch for landscaping. Dirt from stumps is screened and sold as clean garden soil. Mixed materials are sorted into recyclable glass, metals, and plastics. Construction debris is sorted and ground: Broken drywall is ground to fresh gypsum, which is sold to drywall producers; wood is composted or burned; and bricks are crushed for fill and construction material. Organic waste that can't be separated, such as food-soaked paper, is sent to a gasifier. The gasifier is like an enclosed, oxygen-free pressure cooker, which converts biomass to natural gas. The gas runs electric generators for the plant, and any extra gas can be sold. Waste heat warms the recycling facility. The 3 percent of incoming waste that doesn't get recycled is mainly mixed plastics, which currently are landfilled.

Appliances and e-waste must be demanufactured

Demanufacturing is the disassembly and recycling of obsolete products, such as TV sets, computers, refrigerators, and air conditioners. As we mentioned earlier, electronics and appliances are among the fastest-growing components of the global waste stream. Americans throw away about 54 million household appliances, such as stoves and refrigerators, 12 million computers, and uncounted cell phones each year. Most office computers are used only 3 years; televisions last 5 years or so; refrigerators last longer, an average of 12 years. In the United States, an estimated 300 million computers await disposal in storage rooms and garages.

Demanufacturing is key to reducing the environmental costs of e-waste and appliances. A single personal computer can contain 700 different chemical compounds, including toxic materials (mercury, lead, and gallium), and valuable metals (gold, silver, copper), as well as brominated fire retardants and plastics. A typical personal computer has about $6 worth of gold, $5 worth of copper, and $1 of silver. Approximately 40 percent of lead entering U.S. landfills, and 70 percent of heavy metals, comes from e-waste. Batteries and switches in toys and electronics make up another 10 to 20 percent of heavy metals in our waste stream. These contaminants can enter groundwater if computers are landfilled, or the air if they are incinerated. When collected, these materials can become a valuable resource—and an alternative to newly mined materials.

To reduce these environmental hazards, the European Union now requires cradle-to-grave responsibility for electronic products. Manufacturers now have to accept used products or fund independent collectors. An extra $20 (less than 1 percent of the price of most computers) is added to the purchase price to pay for collection and demanufacturing. Manufacturers selling computers, televisions, refrigerators, and other appliances in Europe must also phase out many of the toxic compounds used in production. Japan is rapidly adopting European environmental standards, and some U.S. companies are following suit, in order to maintain their international markets. In the United States, at least 29 states have passed, or are considering, legislation to control disposal of appliances and computers, in order to protect groundwater and air quality.

Reuse is more efficient than recycling

Even better than recycling or composting is cleaning and reusing materials in their present form, thus saving the cost and energy of remaking them into something else. We do this already with some specialized items. Auto parts are regularly sold from scrap yards, especially for older car models. In some areas, stained glass windows, brass fittings, fine woodwork, and bricks salvaged from old houses bring high prices. Some communities sort and reuse a variety of materials received in their dumps (fig. 21.15).

In many cities, glass and plastic bottles are routinely returned to beverage producers for washing and refilling. The reusable, refillable bottle is the most efficient beverage container we have. This is better for the environment than remelting and more profitable for local communities. A reusable glass container makes an average of 15 round-trips between factory and customer before it becomes so scratched and chipped that it has to be recycled. Reusable containers also favor local bottling companies and help preserve regional differences.

FIGURE 21.15 Reusing discarded products is a creative and efficient way to reduce wastes. This recycling center in Berkeley, California, is a valuable source of used building supplies and a money saver for the whole community. *Source: Courtesy of Urban Ore, Inc., Berkeley, California*

Since the advent of cheap, lightweight, disposable food and beverage containers, many small, local breweries, canneries, and bottling companies have been forced out of business by huge national conglomerates. These big companies can afford to ship food and beverages great distances as long as it is a one-way trip. If they had to collect their containers and reuse them, canning and bottling factories serving large regions would be uneconomical. Consequently, the national companies favor recycling rather than refilling, because they prefer fewer, larger plants and don't want to be responsible for collecting and reusing containers. In some circumstances, life-cycle assessment shows that washing and decontaminating containers takes as much energy and produces as much air and water pollution as manufacturing new ones.

Reducing waste is the best option

Most of our attention in waste management focuses on recycling. But slowing the consumption of throw-away products is by far the most effective way to save energy, materials, and money. In the 3R waste hierarchy—reduce, reuse, recycle—the most important strategy is reuse. Industries are increasingly finding that reducing saves money. Soft drink makers use less aluminum per can than they did 20 years ago, and plastic bottles use less plastic. Individual action is essential too (What Can You Do?).

Excess packaging of food and consumer products is one of our greatest sources of unnecessary waste. Paper, plastic, glass, and metal packaging material make up 50 percent of our domestic trash by volume. Much of that packaging is primarily for marketing and has little to do with product protection (fig. 21.16). Manufacturers and retailers might be persuaded to reduce these wasteful practices if consumers ask for products without excess packaging. Canada's National Packaging Protocol (NPP) recommends that packaging minimize depletion of virgin resources and production of toxins in manufacturing. The preferred hierarchy is (1) no packaging, (2) minimal packaging, (3) reusable packaging, and (4) recyclable packaging.

FIGURE 21.16 How much more do we need? Where will we put what we already have?
JIM BORGMAN © Cincinnati Enquirer. Reprinted with permission of UNIVERSAL UCLICK. All rights reserved.

What Can You Do?

Reducing Waste

1. Help your school establish zero-waste (or low-waste) policies.
2. Buy foods that come with less packaging; shop at farmers' markets or co-ops, using your own containers.
3. Carry and use your own refillable beverage container, instead of buying new ones.
4. Avoid single-use plastic shopping bags. Try to buy products in recyclable containers.
5. Minimize use of hard-to-recycle products such as aluminum foil.
6. Separate your cans, bottles, papers, and plastics for recycling.
7. Compost yard and garden wastes, leaves, and grass clippings.
8. Write to your elected representatives and urge them to vote for container deposits, recycling, and safe incinerators or landfills.

Source: Minnesota Pollution Control Agency

We can reduce the volume of waste in landfills somewhat by using materials that are compostable or degradable. **Photodegradable plastics** break down when exposed to ultraviolet radiation in sunlight. **Biodegradable plastics** incorporate materials such as cornstarch that can be decomposed by microorganisms. These degradable plastics are an imperfect solution, though. Many don't decompose completely; they only break down to small particles that remain in the environment. In doing so, they can release toxic chemicals into the environment. In modern, lined landfills, they cannot decompose at all. Furthermore, they make recycling less feasible and may lead people to believe that littering is okay.

Section Review

1. Describe some general trends in landfilling, recycling, and incineration of waste.
2. Why are plastics difficult to recycle?
3. What is demanufacturing, and why is it used for some types of waste?

21.3 HAZARDOUS AND TOXIC WASTES

- *Hazardous waste is dangerous in small doses, easily combustible, or corrosive.*
- *Handling and cleaning up hazardous waste is very expensive, so federal legislation enforces cleanup and safe handling.*
- *Hazardous waste can be used, recycled, decontaminated, or stored permanently.*

The most dangerous aspect of the waste stream we have described is that it often contains highly toxic and hazardous materials that are injurious to both human health and environmental quality. We now produce and use a vast array of flammable, explosive, caustic, acidic, and highly toxic chemical substances for industrial, agricultural, and domestic purposes (fig. 21.17). According to the EPA,

FIGURE 21.17 According to the U.S. Environmental Protection Agency, industries produce about one ton of hazardous waste per year for every person in the United States. Responsible handling and disposal is essential.
© Michael Greenlar/The Image Works

industries in the United States generate about 265 million metric tons of *officially* classified hazardous wastes each year, slightly more than 1 ton for each person in the country. In addition, considerably more toxic and hazardous waste material is generated by industries or processes not regulated by the EPA. At least 40 million metric tons (22 billion lbs) of toxic and hazardous wastes are released into the air, water, and land in the United States each year. The biggest source of these toxins are the chemical and petroleum industries (fig. 21.18).

Legally, a **hazardous waste** is any discarded material, liquid or solid, that contains substances known to be (1) fatal to humans or laboratory animals in low doses, (2) toxic, carcinogenic, mutagenic, or teratogenic to humans or other life-forms, (3) ignitable with a flash point less than 60°C, (4) corrosive, or (5) explosive or highly reactive (undergoes violent chemical reactions either by itself or when mixed with other materials). Notice that this definition includes both toxic and hazardous materials as defined in chapter 8. Certain compounds are exempt from regulation as hazardous waste if they are accumulated in less than 1 kg (2.2 lbs) of commercial chemicals or 100 kg of contaminated soil, water,

or debris. Even larger amounts (up to 1,000 kg) are exempt when stored at an approved waste treatment facility for the purpose of being beneficially used, recycled, reclaimed, detoxified, or destroyed.

Hazardous waste must be recycled, contained, or detoxified

To protect the public from hazardous wastes, most is recycled, converted to nonhazardous forms, stored, or disposed of on-site by the generators. Most producers of these wastes are chemical companies, petroleum refiners, and other large industrial facilities. Still, the hazardous waste that does enter the waste stream or the environment represents a serious environmental problem.

Orphan wastes, abandoned by industries that have gone out of business, are an especially tricky threat to both environmental quality and human health. For years, little attention was paid to this material. Forgotten wastes stored on private property, buried, or allowed to soak into the ground were considered of little concern to the public. An estimated 5 billion metric tons of highly poisonous chemicals were improperly disposed of in the United States between 1950 and 1975, before regulatory controls became more stringent.

Federal legislation requires waste management

Two important federal laws regulate hazardous waste management and disposal in the United States. The Resource Conservation and Recovery Act (RCRA, pronounced "rickra") of 1976 is a comprehensive program that requires rigorous testing and management of toxic and hazardous substances. A complex set of rules requires generators, shippers, users, and disposers of these materials to keep meticulous account of everything they handle and what happens to it from generation (cradle) to ultimate disposal (grave) (fig. 21.19).

The Comprehensive Environmental Response, Compensation and Liability Act (CERCLA or Superfund Act), passed in 1980 and modified in 1984 by the Superfund Amendments and Reauthorization Act (SARA), is aimed at rapid containment, cleanup, or remediation of abandoned toxic waste sites. This statute authorizes the Environmental Protection Agency to undertake emergency actions when a threat exists that toxic material will leak into the environment. The agency is empowered to bring suit for the recovery of its costs from potentially responsible parties such as site owners, operators, waste generators, or transporters.

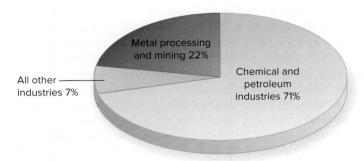

FIGURE 21.18 Producers of hazardous wastes in the United States.
Source: U.S. EPA

> ### Quantitative Reasoning
>
> What proportion of hazardous waste producers are chemical and petroleum industries (fig. 21.18)? What proportion are metal processing and mining? What kinds of contaminants are most likely to occur at these different types of sites? Can you think of two or three locations where chemical and petroleum industries might be concentrated?

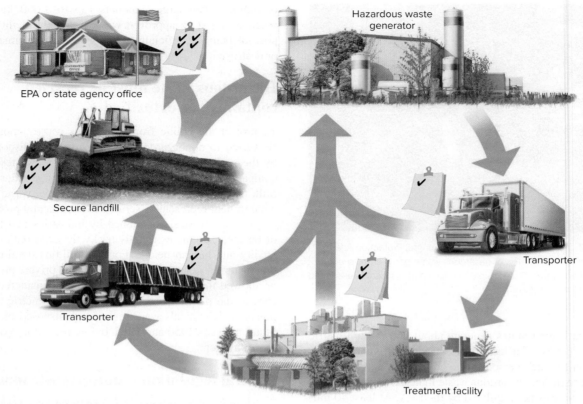

FIGURE 21.19 Toxic and hazardous wastes must be tracked from "cradle to grave" by detailed shipping manifests.

SARA also established that communities have a right to know about toxic substances that are produced or used nearby, which could be released into air or water. To give the public access to this information, SARA established a reporting system, the **Toxic Release Inventory.** More than 20,000 manufacturing facilities are required to report annually on the use, release, or transfer of toxic substances. This inventory is not always maintained completely, but it is the best available source of public information on exposure risk. You can find it on the EPA website and see what potential sources are in your neighborhood.

The government does not have to prove that anyone violated a law or what role they played in a Superfund site. Rather, liability under CERCLA is "strict, joint, and several," meaning that anyone associated with a site can be held responsible for the entire cost of cleaning it up no matter how much of the mess they made. In some cases, property owners have been assessed millions of dollars for removal of wastes left there years earlier by previous owners. This strict liability has been a headache for the real estate and insurance business, but it also allows for protection of public health.

Superfund sites are listed for federal cleanup

The EPA estimates that there are at least 36,000 seriously contaminated sites in the United States. The General Accounting Office (GAO) places the number much higher, perhaps more than 400,000 when all are identified. By 2016, some 1,170 sites were on the National Priority List (NPL) for cleanup with financing

from the federal Superfund program. This number was down from 1,680 sites eight years earlier, The decline represents removal of sites from the list, mostly due to the completion of cleanup projects. Fifty-two new sites had been proposed for NPL listing.

The **Superfund** is a revolving pool designed to (1) provide an immediate response to emergency situations that pose imminent hazards, and (2) to clean up or remediate abandoned or inactive sites. Without this fund, sites would languish for years or decades while the courts decided who was responsible to pay for the cleanup. Originally, a $1.6 billion pool, the fund peaked at $3.6 billion. From its inception, the fund was financed by taxes on producers of toxic and hazardous wastes. Industries opposed this "polluter pays" tax, because current manufacturers are often not the ones responsible for the original contamination. In 1995, Congress agreed to let the tax expire. Since then, the Superfund has dwindled, and the public has picked up an increasing share of the bill. In the 1980s, the public covered less than 20 percent of the Superfund; now public tax dollars from the general fund cover most of the cost of toxic waste cleanup, in combination with fines collected from polluters (fig. 21.20).

Reliance on general funds makes cleanup progress vulnerable to political winds in Congress. In some years, Congress is able and willing to cover the costs needed to protect the public from exposure to hazardous substances; in other years, it can't or won't provide sufficient funding.

Total costs for hazardous waste cleanup in the United States are estimated between $370 billion and $1.7 trillion, depending on

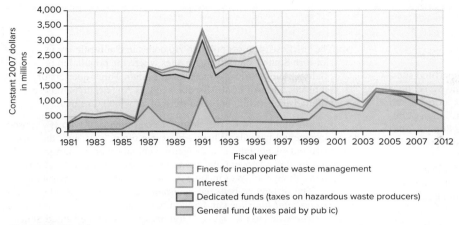

FIGURE 21.20 Congress has gradually shifted the funding sources for the Superfund, used to clean up toxic and hazardous waste sites, from industrial producers of hazardous waste to taxpayers.

how clean sites must be and what methods are used. For years, Superfund money was spent mostly on lawyers and consultants, and cleanup efforts were often bogged down in disputes over liability and the best cleanup methods. Progress depends largely on the priorities of the president and his aids: During the Reagan and the first Bush administrations, there was little investment in cleanup; the Clinton administration significantly accelerated work: From 1993 to 2000, the number of completed NPL cleanups jumped from 155 to 757, almost half the 1,680 sites on the list at that time. From 2000 to 2008, progress slowed again, due to underfunding and a lack of interest from the federal administration. After 2008, the Obama administration accelerated cleanup activity again.

What qualifies a site for the NPL? These sites are considered to be especially hazardous to human health and environmental quality because they are known to be leaking or have a potential for leaking supertoxic, carcinogenic, teratogenic (causing birth defects), or mutagenic (cancer causing) substances (see chapter 8, Environmental Health and Toxicology for discussion of these terms). The ten substances of greatest concern or most commonly detected at Superfund sites are lead, trichloroethylene, toluene, benzene, PCBs, chloroform, phenol, arsenic, cadmium, and chromium. These and other hazardous or toxic materials are known to have contaminated groundwater at 75 percent of the sites now on the NPL. In addition, 56 percent of these sites have contaminated surface waters, and airborne materials are found at 20 percent of the sites.

Where are these thousands of hazardous waste sites, and how did they get contaminated? Old industrial facilities such as smelters, mills, petroleum refineries, and chemical manufacturing plants are highly likely to have been sources of toxic wastes. Regions of the country with high concentrations of aging factories such as the "rust belt" around the Great Lakes or the Gulf Coast petrochemical centers have large numbers of Superfund sites (fig. 21.21). Mining districts also are prime sources of toxic and hazardous waste. Within cities, factories and places such as railroad yards, bus repair barns, and filling stations where solvents, gasoline, oil, and other petrochemicals were spilled or dumped on the ground often are highly contaminated.

Some of the most infamous toxic waste sites were old dumps where many different materials were mixed together indiscriminately.

For instance, Love Canal in Niagara Falls, New York, was an open dump used by both the city and nearby chemical factories as a disposal site. More than 20,000 tons of toxic chemical waste was buried under what later became a housing development. Another infamous example occurred in Hardeman County, Tennessee, where about a quarter of a million barrels of chemical waste were buried in shallow pits that leaked toxins into the groundwater. In other sites, liquid wastes were pumped into open lagoons or abandoned in warehouses.

Studies of populations living closest to Superfund and toxic release inventory sites reveal that minorities often are overrepresented in neighborhoods near waste sites. Charges of environmental racism have been made, but they are difficult to prove conclusively (What Do You Think?, Environmental Justice).

Brownfields present both liability and opportunity

Many cities have unused or abandoned industrial sites, with soil or water contaminated with toxic and hazardous substances. Because these contaminants are health risks, it is unsafe—and legally risky—to rebuild on these **brownfields** without expensive cleanup projects. As a result, brownfields remain unused eyesores, and a drain on a city's productive tax base, for years or decades. Brownfield redevelopment has become a priority for many cities, and in recent decades federal funding has become available to help pay for the millions of dollars of remediation before new building can even begin.

One of the thorny problems with brownfield redevelopment is the question of how clean it needs to be. Must a site be cleaned to the point that young children could play in the dirt safely, and food could be grown in the soil? Is it possible to clean a site somewhat

- ● Hazardous waste site
- ▨ Aquifers

FIGURE 21.21 Hazardous waste sites are often located on aquifer recharge zones, making groundwater contamination a common risk.
Source: U.S. EPA

Environmental Justice

Who do you suppose lives closest to toxic waste dumps, Superfund sites, or other polluted areas in your city or country? If you answered poor people and minorities, you are probably right. Everyday experiences tell us that minority neighborhoods are much more likely to have high pollution levels and unpopular industrial facilities such as waste dumps, landfills, smelters, refineries, and incinerators than are middle- or upper-class, white neighborhoods.

One of the first systematic studies showing this inequitable distribution of environmental hazards based on race in the United States was conducted by Robert D. Bullard in 1978. Asked for help by a predominantly black community in Houston that was slated for a waste incinerator, Bullard discovered that all five of the city's existing landfills and six of the eight incinerators were located in African-American neighborhoods. In a book entitled *Dumping on Dixie,* Bullard showed that this pattern of risk exposure in minority communities is common throughout the United States (fig. 1).

In 1987, the Commission for Racial Justice of the United Church of Christ published an extensive study of environmental racism. Its conclusion was that race is the most significant variable in determining the location of toxic waste sites in the United States. Among the findings of this study are:

- Three of the five largest commercial hazardous waste landfills, accounting for about 40 percent of all hazardous waste disposal in the United States, are located in predominantly black or Hispanic communities.

- 60 percent of African Americans and Latinos and nearly half of all Asians, Pacific Islanders, and Native Americans live in communities with uncontrolled toxic waste sites.

- The average percentage of the population made up by minorities in communities without a hazardous waste facility is 12 percent. By contrast, communities with one hazardous waste facility have, on average, twice as high (24 percent) a minority population, while those with two or more such facilities average three times as high a minority population (38 percent) as those without one.

- The "dirtiest" or most polluted zip codes in California are in riot-torn South Central Los Angeles where the population is predominantly African American or Latino. Three-quarters of all blacks and half of all Hispanics in Los Angeles live in these polluted areas, while only one-third of all whites live there.

Race is claimed to be the strongest determinant of who is exposed to environmental hazards. Where whites can often "vote with their feet" and move out of polluted and dangerous neighborhoods, minorities are restricted by color barriers and prejudice to less desirable locations. In some areas, though, class or income is also associated with environmental hazards. The difference between *environmental racism* and other kinds of *environmental injustice* can be hard to define. Economic opportunity is often closely tied to race and cultural background in the United States.

Racial inequities also are revealed in the way the government cleans up toxic waste sites and punishes polluters (fig. 2). White communities see faster responses and get better results once toxic wastes are discovered than do minority communities. Penalties assessed against polluters of white communities average six times higher than those against polluters of minority communities. Cleanup is more thorough in white communities as well. Most toxic wastes in white communities are removed or destroyed. By contrast, waste sites in minority neighborhoods are generally only "contained" by putting a cap over them, leaving contaminants in place to potentially resurface or leak into groundwater at a later date. The growing environmental justice movement works to combine civil rights and social justice with environmental concerns to call for a decent, livable environment and equal environmental protection for everyone.

Ethical Considerations

What are the ethical considerations in waste disposal? Does everyone have a right to live in a clean environment or only a right to buy one if they can afford it? What would be a fair way to distribute the risks of toxic wastes? If you had to choose between an incinerator, a secure landfill, or a composting facility for your neighborhood, which would you take?

FIGURE 1 Native Americans protest toxic waste dumping on tribal lands.
© Barbara Gauntt

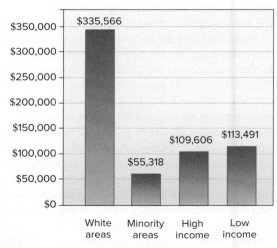

FIGURE 2 Hazardous waste law enforcement. The average fines or penalties per site for violation of the Resource Conservation and Recovery Act vary dramatically with racial composition of the communities where waste was dumped.

Alternatives to Hazardous Household Chemicals

Most household cleaning solutions can be replaced with inexpensive, non-toxic, safe alternatives. Here are some examples.

Chrome cleaner: Use vinegar and nonmetallic scouring pad.

Floor cleaner: Mop floors with 1 cup vinegar mixed with 2 gallons of water.

Silver polish: Rub with toothpaste on a soft cloth.

Furniture polish: Rub with olive, almond, or lemon oil.

Drain opener: Use a plumber's snake; pour boiling water in drain; keep fats and hair out of drains.

Upholstery cleaner: Clean stains with club soda.

Window cleaner: Mix 1/3 cup ammonia, 1/4 cup white vinegar in 1 quart warm water. Spray on window. Wipe with soft cloth.

Toilet cleaner: Pour 1/2 cup vinegar or liquid chlorine bleach into toilet bowl. Let stand for 30 minutes, scrub with brush, flush.

Garden pest control: Spray plants with soap-and-water solution for aphids, mealybugs, and other insects. Plant pest repellent plants such as marigolds, coriander, thyme, yarrow, rue, and tansy. Introduce natural predators such as ladybugs or lacewings.

without planning for every unexpected use far in the future? Might rules become more strict in the future?

These questions were tested in a brownfield site in Columbia, Mississippi. For many years, the 35-ha (81-acre) site in Columbia was used for turpentine and pine tar manufacturing. Soil tests showed concentrations of phenols and other toxic organic compounds exceeding federal safety standards. The site was added to the Superfund National Priority List, and remediation was ordered. Some experts argued that the best solution was to simply cover the surface with clean soil and enclose the property with a fence to keep people out. The total costs would have been about $1 million.

Instead, the EPA ordered Reichhold Chemical, the last known property owner, to excavate more than 12,500 tons of soil and haul it to a commercial hazardous waste dump in Louisiana at a cost of some $4 million. The intention was to make the site safe enough to be used for any purpose, and to remove risk from exposure to anybody, even children, who might be exposed to soil on the site.

Similarly, in places where contaminants have seeped into groundwater, the EPA generally demands that cleanup be carried to drinking water standards, because we can't be sure that people won't be drinking the water sometime in the future. Many critics believe that these pristine standards are unreasonable. Former congressman Jim Florio, a principal author of the original Superfund Act, says, "It doesn't make any sense to clean up a rail yard in downtown Newark so it can be used as a drinking water reservoir."

Over time, brownfield cleanup policies have found compromise positions. Some brownfields are designated specifically as business sites, not for residential use. For organic chemicals, soil might be dug up and put into a mobile incinerator on site,

to break down contaminants, rather than trucked away for storage. Sometimes a visible, impermeable membrane can be laid over a contaminated site, then covered by clean soil (fig. 21.22). The membrane prevents access to contaminated soil, alerts future excavators to the unsafe deeper layers, and allows rapid rebuilding on the site. These steps don't really clean a site, but they do contain the hazards and prevent further contamination. They also allow cities to redevelop near the urban core, and to slow sprawl, with its many environmental costs, at the urban margin.

Hazardous waste can be recycled or contained

We have four general strategies for managing hazardous waste: reduce, recycle, convert to less hazardous forms, and contain safely in perpetuity. Some of these options apply to our personal uses, as well as to public and industrial uses of substances. At home, you can reduce the amount of toxic materials you buy, and you can buy just as much as you need, not extra. You can replace some toxic substances with safer and cheaper alternatives (What Can You Do?). You can't recycle household hazardous materials, but you can take them to a collection center for proper, safe disposal (table 21.1).

In public and industrial contexts, too, the safest and cheapest way to minimize hazardous waste problems is to avoid creating the wastes in the first place. Manufacturing processes can be modified to reduce or eliminate waste production. In Minnesota, the 3M Company reformulated products and redesigned manufacturing processes to eliminate more than 140,000 metric tons of solid and hazardous wastes, 4 billion liters (1 billion gal) of wastewater, and 80,000 metric tons of air pollution each year. Often, these new processes also saved money by using less energy and fewer raw materials.

Recycling and reusing materials can also eliminate hazardous wastes and pollution. Many waste products of one process

FIGURE 21.22 This brownfield site on the Hudson River has soil contaminated with arsenic from lumber treatment. To make the site safe for reuse, contaminated soil is being covered with an orange, impermeable (waterproof) sheet, which is then being covered by several feet of clean soil.
© Mary Ann Cunningham

Table 21.1	How Should You Dispose of Household Hazardous Waste?
Flush to sewer system (drain or toilet)	Cleaning agents with ammonia or bleach, disinfectants, glass cleaner, toilet cleaner
Put dried solids in household trash	Cosmetics, putty, grout, caulking, empty solvent containers, water-based glue, fertilizer (without weed killer)
Save and deliver to a waste collection center	*Solvents:* cleaning agents (drain cleaner, floor wax-stripper, furniture polish, metal cleaner, oven cleaner), paint thinner and other solvents, glue with solvents, varnish, nail polish remover
	Metals: mercury thermometers, button batteries, NiCad batteries, auto batteries, paints with lead or mercury, fluorescent light bulbs/tubes/ballasts, electronics and appliances
	Poisons: bug spray, pesticides, weed killers, rat poison, insect poison, mothballs. Other chemicals: antifreeze, gasoline, fuel oil, brake fluid, transmission fluid, paint, rust remover, hair spray, photo chemicals

Source: U.S. EPA

FIGURE 21.23 Actor Martin Sheen joins activists protesting a hazardous waste incinerator in East Liverpool, Ohio. Concerns over hazardous emissions worry residents near incinerators.
© Piet van Lier

or industry are valuable commodities in another. Already, about 10 percent of the wastes that would otherwise enter the waste stream in the United States are sent to surplus material exchanges where they are sold as raw materials for use by other industries. This figure could probably be raised substantially with better waste management. In Europe, at least one-third of all industrial wastes are exchanged through clearinghouses where beneficial uses are found. This represents a double savings: The generator doesn't have to pay for disposal, and the recipient pays less for raw materials.

Substances can be converted to safer forms

Several processes are available to convert hazardous materials to less dangerous forms. *Physical treatments* capture or isolate substances. Charcoal or resin filters absorb toxins. Distillation separates hazardous components from aqueous solutions. Precipitation and immobilization in ceramics, glass, or cement isolate toxins from the environment so they become essentially nonhazardous. One of the few ways to dispose of metals and radioactive substances is to fuse them in silica at high temperatures to make a stable, impermeable glass that is suitable for long-term storage.

Incineration is a quick way to dispose of many kinds of hazardous waste. Incineration is not necessarily cheap—nor always clean—unless it is done correctly. Wastes must be heated to over 1,000°C (2,000°F) for a sufficient length of time to complete destruction. The ash resulting from thorough incineration is reduced in volume up to 90 percent and often is safer to store in a landfill or other disposal site than the original wastes. Incineration remains highly controversial, though, if only because it has a long history of releasing toxic contaminants to the air (fig. 21.23).

Several sophisticated features of modern incinerators improve their effectiveness. Liquid injection nozzles atomize liquids and mix air into the wastes so they burn thoroughly. Fluidized bed burners force air up from the bottom of the furnace, separating and aerating waste, to allow for rapid and complete combustion. Before heading to the smoke stack, gases are passed by electrically charged plates, which capture particulates in the flue gases.

Chemical processing can transform materials so they become nontoxic. Included in this category are neutralization, removal of metals or halogens (chlorine, bromine, etc.), and oxidation. The Sunohio Corporation of Canton, Ohio, for instance, has developed a process called PCBx in which chlorine in such molecules as PCBs is replaced with other ions that render the compounds less toxic. A portable unit can be moved to the location of the hazardous waste, eliminating the need for shipping them.

Biological waste treatment or **bioremediation** taps the great capacity of microorganisms to absorb, accumulate, and detoxify a variety of toxic compounds (see the Exploring Science box later in this section). Bacteria in activated sludge basins, aquatic plants (such as water hyacinths or cattails), soil microorganisms, and other species remove toxic materials and purify effluents. Recent experiments have produced bacteria that can decontaminate organic waste metals by converting them to harmless substances. After the *Deepwater Horizon* oil spill in the Gulf of Mexico in 2010, most of the waste remediation was performed by naturally occurring bacteria, which were adapted to metabolizing oil compounds. Bioremediation holds exciting possibilities for addressing many organic pollutants, including oils, PCBs, and other toxic compounds.

Permanent storage is often needed

Inevitably, there will be some materials we can't destroy, make into something else, or otherwise cause to vanish. We will have to store them out of harm's way. There are differing opinions about how best to do this.

Retrievable Storage allows for access later, in case we need to recover wastes or use new treatment methods. Often this means **permanent retrievable storage.** Waste storage containers are placed in a secure building, salt mine, or bedrock cavern where

Phytoremediation: Cleaning Up Toxic Waste with Plants

Getting contaminants out of soil and groundwater is one of the most widespread and persistent problems in waste cleanup. Once leaked into the ground, solvents, metals, radioactive elements, and other contaminants are dispersed and difficult to collect and treat. The main method of cleaning up contaminated soil is to dig it up, then decontaminate it or haul it away and store it in a landfill in perpetuity. At a single site, thousands of tons of tainted dirt and rock may require incineration or other treatment. Cleaning up contaminated groundwater usually entails pumping vast amounts of water out of the ground—hopefully extracting the contaminated water faster than it can spread through the water table or aquifer. In the United States alone, there are tens of thousands of contaminated sites on factories, farms, gas stations, military facilities, sewage treatment plants, landfills, chemical warehouses, and other types of facilities. Cleaning up these sites is expected to cost at least $700 billion.

Recently, a number of promising alternatives have been developed using plants, fungi, and bacteria to clean up our messes. *Phytoremediation* (remediation, or cleanup, using plants) can include a variety of strategies for absorbing, extracting, or neutralizing toxic compounds. Certain types of mustards and sunflowers can extract lead, arsenic, zinc, and other metals (*phytoextraction*). Poplar trees can absorb and break down toxic organic chemicals (*phytodegradation*). Reeds and other water-loving plants can filter water tainted with sewage, metals, or other contaminants. Natural bacteria in groundwater, when provided with plenty of oxygen, can neutralize contaminants in aquifers, minimizing or even eliminating the need to extract and treat water deep in the ground. Radioactive strontium and cesium have been extracted from soil near the Chernobyl nuclear power plant using common sunflowers.

How do the plants, bacteria, and fungi do all this? Many of the biophysical details are poorly understood, but in general, plant roots are designed to efficiently extract nutrients, water, and minerals from soil and groundwater. The mechanisms involved may aid extraction of metallic and organic contaminants. Some plants also use toxic elements as a defense against herbivores—locoweed, for example, selectively absorbs elements such as selenium, concentrating toxic levels in its leaves. Absorption can be extremely effective. Braken fern growing in Florida was found to contain arsenic at concentrations more than 200 times higher than the soil in which it was growing.

Genetically modified plants are also being developed to process toxins. Poplars have been grown with a gene borrowed from bacteria that transform a toxic compound of mercury into a safer form. In another experiment, a gene for producing mammalian liver enzymes, which specialize in breaking down toxic organic compounds, was inserted into tobacco plants. The plants succeeded in producing the liver enzymes and breaking down toxins absorbed through their roots.

These remediation methods are not without risks. As plants take up toxins, insects could consume leaves, allowing contaminants to enter the food web. Some absorbed contaminants are volatilized, or emitted in gaseous form, through pores in plant leaves. Once toxic contaminants are absorbed into plants, the plants themselves are usually toxic and must be landfilled. But the cost of phytoremediation can be less than half the cost of landfilling or treating toxic soil, and the volume of plant material requiring secure storage ends up being a fraction of a percent of the volume of the contaminated dirt.

Cleaning up hazardous and toxic waste sites will be a big business for the foreseeable future, both in the United States and around the world. Innovations such as phytoremediation offer promising prospects for business growth as well as for environmental health and saving taxpayers' money.

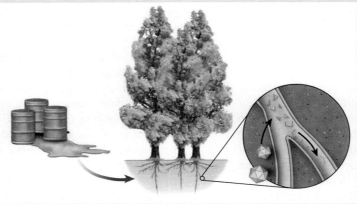

Plants can absorb, concentrate, and even decompose toxic contaminants in soil and groundwater.

they can be inspected periodically and retrieved, if necessary, for repacking or for transfer if a better means of disposal is developed. This technique is more expensive than burial in a landfill because the storage area must be guarded and monitored continuously to prevent leakage, vandalism, or other dispersal of toxic materials.

Remedial measures are much cheaper with this technique, however, and it may be the best system in the long run.

Secure landfills are one of the most popular solutions for hazardous waste disposal. Although, as we saw earlier in this chapter, many such landfills have been environmental disasters; newer techniques make it possible to create safe, modern **secure landfills** that are acceptable for disposing of many hazardous wastes. The first line of defense in a secure landfill is a thick bottom cushion of compacted clay that surrounds the pit like a bathtub (fig. 21.24). Moist clay is flexible and resists cracking if the ground shifts. It is impermeable to groundwater and will safely contain wastes. A layer of gravel is spread over the clay liner, and perforated drain pipes are laid in a grid, so as to collect any seepage that escapes from the stored material. A thick polyethylene liner, protected from punctures by soft padding materials, covers the gravel bed. A layer

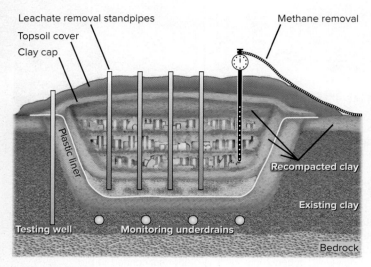

Leachate removal standpipes
Topsoil cover
Clay cap
Methane removal
Plastic liner
Recompacted clay
Existing clay
Testing well
Monitoring underdrains
Bedrock

FIGURE 21.24 A secure landfill for toxic waste. A thick plastic liner and two or more layers of impervious compacted clay enclose the landfill. A gravel bed between the clay layers collects any leachate, which can then be pumped out and treated. Well samples are tested for escaping contaminants, and methane is collected for combustion.

of soil or absorbent sand cushions the inner liner and the wastes are packed in drums, which then are placed into the pit, separated into small units by thick berms of soil or packing material.

When the landfill has reached its maximum capacity, a cover much like the bottom sandwich of clay, plastic, and soil—in that order—caps the site. Vegetation stabilizes the surface and improves its appearance. Sump pumps collect any liquids that filter through the landfill, either from rainwater or leaking drums. This leachate is treated and purified before being released. Monitoring wells check groundwater around the site to ensure no toxins have escaped.

Transportation of hazardous wastes to disposal sites is of concern because of the risk of accidents. Emergency preparedness officials conclude that the greatest risk in most urban areas is not nuclear war or natural disaster but crashes involving trucks or trains carrying hazardous chemicals through densely occupied urban corridors. Another worry is who will bear financial responsibility for abandoned waste sites and for landfills over the long term. As is the case with nuclear wastes (see chapter 19, Conventional Energy), we may need new institutions to ensure perpetual care of these wastes.

Section Review

1. What is a legal definition of hazardous waste?
2. What are brownfields? Why are they both important and difficult to clean up?
3. How might hazardous wastes be converted to safer forms?
4. What is bioremediation?

Conclusion

Waste is a global problem. Each year, we consume more materials and produce more waste. Finding ways to dispose of all our garbage and hazardous substances is a constant challenge. In the United States, recycling rates are improving, but we still landfill more than half our municipal waste. Many other countries, especially those short on landfill space, recycle over half their waste. Modern landfills seek to keep trash from contaminating air and groundwater. These sites are a great improvement over the past, but they are often remote from major cities, which must transport garbage long distances for disposal. Incineration is a costly but widely used alternative to landfilling.

Waste disposal is expensive, but our policies are often better set up for landfilling or incinerating waste than for recycling. Government policies and economies of scale make it cheaper and more convenient to extract virgin raw materials than to reuse or recycle. But the increasing toxicity of modern products, including e-waste,

makes it more urgent that we reduce, reuse, and recycle materials worldwide. Strategies and opportunities for recycling are expanding, including reuse markets, bioenergy generation, deconstructing demolition debris, and composting yard waste. Creating gold from garbage—making money from biogas, compost, and efficient waste disposal—is a growing industry, especially in Europe.

Hazardous and toxic waste remain a serious health threat and environmental risk. Many abandoned and derelict sites must be cleaned up by the Superfund, although that fund has dwindled since it lost its main source of support, industry contributions, in 1995. These sites threaten public health, especially for minority groups, and are a serious problem. At the same time, we are producing new hazardous substances that require safe disposal. We can all help by thinking carefully about what we buy, use, and dispose of in our own communities.

Reviewing Key Terms

Can you define the following terms in environmental science?

biodegradable plastics 21.2
bioremediation 21.3
brownfields 21.3
composting 21.2

demanufacturing 21.2
energy recovery 21.1
e-waste 21.1
hazardous waste 21.3
mass burn 21.1

permanent retrievable storage 21.3
photodegradable plastics 21.2
recycling 21.2
refuse-derived fuel 21.1

sanitary landfills 21.1
secure landfills 21.3
Superfund 21.3
Toxic Release Inventory 21.3

Critical Thinking and Discussion Questions

1. A toxic waste disposal site has been proposed for the Pine Ridge Indian Reservation in South Dakota. Many tribal members oppose this plan, but some favor it because of the jobs and income it will bring to an area with 70 percent unemployment. If local people choose immediate survival over long-term health, should we object or intervene?

2. There is often a tension between getting your personal life in order and working for larger structural changes in society. Evaluate the trade-offs between spending time and energy sorting recyclables at home compared to working in the public arena on a bill to ban excess packaging.

3. Should industry officials be held responsible for dumping chemicals that were legal to dump when they did it but are now known to be extremely dangerous? When can we argue that they should have known about the hazards involved?

4. Look at the discussion of recycling or incineration presented in this chapter. List the premises (implicit or explicit) that underlie the presentation, as well as the conclusions (stated or not) that seem to be drawn from them. Do the conclusions necessarily follow from these premises?

5. The Netherlands incinerates much of its toxic waste at sea by a shipborne incinerator. Would you support this as a way to dispose of our wastes as well? What are the critical considerations for or against this approach?

Data Analysis

How Much Do You Know about Recycling?

As people become aware of waste disposal problems in their communities, more people are recycling more materials. Some things are easy to recycle, such as newsprint, office paper, or aluminum drink cans. Other things are harder to classify. Most of us give up pretty quickly and throw things in the trash if we have to think too hard about how to recycle them.

1. Take a poll to find out how many people in your class know how to recycle the items in the table shown at the end of these questions. Once you have taken your poll, convert the numbers to percentages: Divide the number who know how to recycle each item by the number of students in your class, and then multiply by 100.

2. Now find someone on your campus who works on waste management. This might be someone in your university or college administration, or it might be someone who actually empties trash containers. (You might get more interesting and straightforward answers from the latter.) Ask the following questions: (1) Can this person fill in the items your class didn't know about? (2) Is there a college/university policy about recycling? What are some of the points on that policy? (3) How much does the college spend each year on waste disposal?

How many tuition payments does that total? (4) What are the biggest parts of the campus waste stream? (5) Does the school have a plan for reducing that largest component?

Item	Percentage Who Know How to Recycle
Newspapers	
Paperboard (cereal boxes)	
Cardboard boxes	
Cardboard boxes with tape	
Plastic drink bottles	
Other plastic bottles	
Styrofoam food containers	
Food waste	
Plastic shopping bags	
Plastic packaging materials	
Furniture	
Last year's course books	
Leftover paint	

TO ACCESS ADDITIONAL RESOURCES FOR THIS CHAPTER, PLEASE VISIT CONNECT AT www.connect.mheducation.com.

You will find LearnSmart, an adaptive learning system, Google Earth™ exercises, additional Case Studies, Data Analysis exercises, and an interactive ebook.

▲Car-free roads provide a cleaner, safer, healthier environment for residents of the Vauban district in Freiburg, Germany.
© Martin Specht/The New York Times/Redux Pictures

Learning Outcomes

After studying this chapter, you should be able to:

22.1 Describe trends in urban growth, and reasons for it.
22.2 Understand urban challenges in the developing world.
22.3 Identify urban challenges in the developed world.
22.4 Explain smart growth and green urbanism.

"What kind of world do you want to live in? Demand that your teachers teach you what you need to know to build it."

– Peter Kropotkin

Vauban: A Car-Free Suburb

To live without a car would be unthinkable for most of us. But residents of Vauban, a district in the city of Freiburg, Germany, have a lifestyle that suggests it might be both enjoyable and economical. In Vauban, it's so easy to get around by tram, bicycle, and on foot that there is little need for cars. The community is designed using "smart growth" principles, with stores, banks, schools, and restaurants in easy walking distance of homes. Jobs and office space are available nearby, and trams to the city center run every few minutes through the center and around the edges of Vauban. Residential streets are narrow and vehicle-free, making a great place for bicycles and playing children.

Residents who own cars can park them in a municipal ramp, but cars aren't encouraged: Buying a space in the ramp costs $40,000. (For comparison, most cities cover the cost of on-street parking, at a rate of around $10,000 per parking spot.) Nearly three-quarters of Vauban's families don't own a car, and more than half sold their car when they moved there.

Fewer vehicles means less energy consumption and less pollution. But most families don't move to Vauban for environmental reasons. They move because they find a car-free lifestyle is healthier for children and for themselves (fig. 22.1). The neighborhoods are quiet, the air is clean. Schools, child-care services, playgrounds, and sports facilities are within easy biking or walking distance. Children can play outside and can walk or bike to school without having to cross busy streets.

In most American cities, one-third of the land area is dedicated to cars, mainly for parking and roads. In Vauban, reducing car dependence has saved so much space that neighborhoods have abundant green space, gardens, and play areas while still being small enough for easy walking.

A highly successful and growing car-sharing program makes it still easier to live without owning a car. The city's car-sharing program began in about 1992 and now has more than 2,500 members, who save money and parking space by using shared cars. The cars are available all around town, and they can be reserved online or by mobile phone. In addition, a single monthly bus ticket covers all regional trains and buses, making it especially easy to get around by public transportation.

FIGURE 22.1 Narrow streets in Vauban are designed for children and bicycles first, with limited car use.
© Mary Ann Cunningham

A car-free lifestyle makes economic sense. Owning and operating a vehicle in Germany is even more expensive than it is in the United States, where car ownership costs about $9,000 per year in purchase costs, maintenance, insurance, and fuel. Residents of car-free neighborhoods can put that money to other uses.

Vauban's comfortable row houses, with balconies and private gardens, are built with **passive house standards** that conserve energy but maximize quality of life. Clever use of space, beautiful woodwork, large balconies, and large, triple-pane windows make the small-footprint homes feel spacious. Many houses are so efficient that they don't need a heating system at all. In addition, a highly efficient wood-burning, district heat and power plant provides much of the space heating and electricity for the area, and rooftop solar collectors and photovoltaic panels provide hot water and power for individual homes. Many of the houses produce more energy than they consume.

Vauban's example provides an attractive alternative to suburban sprawl. In most places, decades of advertising, as well as government policies such as easy and abundant public funding for road building and underfunding of public transportation, and low tax rates on the edge of town, have encouraged people to desire a single-family residence on a spacious lot in the suburbs. We disregard many costs associated with suburban expansion: energy use, lack of easy access to schools, shopping, or jobs, especially for people too young or old or poor to have a car of their own; the loss of farmland and woodlands on the edge of town; and the costs of extending fire, water, and sewer service to outlying areas. Cities are increasingly recognizing and accounting for some of these costs, and car-free urban development is being promoted across Europe. There are proposed car-free developments in China and the United States as well. Many cities are also reinvesting in older central districts, which were built from the start on the principle of car-free access to jobs, shopping, and entertainment.

In this chapter, we examine the ways cities grow and change, sustainably and otherwise. We examine the forces that encourage people to migrate to cities, and some of the ways cities might help make our lifestyles more sustainable.

22.1 Urbanization

- *The world is becoming more urban, and cities are growing larger.*
- *The largest cities are now in developing countries.*
- *Economic policies can push people toward cities.*

Cities are growing larger and more numerous than in the past. More than half the world's population now lives in urban areas. Many cities are burdened by pollution and poverty, but cities are also places of opportunity and creativity. Most young people today plan to live in cities, because of their educational, employment, and cultural resources. You may or may not plan to live in a city, but understanding and building better cities, and making them more just and socially equitable, is essential if we are to live sustainably, and if we are to protect the environmental quality, biodiversity, and ecological services on which we depend.

For most of human history, the vast majority of people have lived in rural areas, where they engaged in hunting and gathering, farming, fishing, or other natural resource–based occupations. Since

Table 22.1	Urban Share of Total Population (Percentage)		
	1950	**2000**	**2050***
World	29.4	46.7	67.2
North America	63.9	79.1	88.6
Latin America	41.4	75.5	86.6
Africa	14.4	35.6	57.7
Asia	17.5	37.4	64.4
Europe	51.3	70.8	82.2
Oceania	62.4	70.4	73.0

*Projected

Source: United Nations Population Division, 2012

the beginning of the Industrial Revolution about 300 years ago, however, cities have grown rapidly in both size and power (fig. 22.2). By 2050, over two-thirds of us will live in urban areas (table 22.1)—in fact, many of the world's regions already are 70 to 80 percent urban.

Demographers predict that 90 percent of the human population growth in this century will occur in developing countries, and that almost all of that growth will occur in cities (fig. 22.3). Already, huge **urban agglomerations** (multiple expanding and converging municipalities) are forming throughout the world. Some 370 million people live in urban areas with populations of 10–30 million people, which are often called **megacities.** This explosive growth is a challenge in developing areas, where cities often fail to keep up with rapidly growing demand for housing, fresh water, sanitation, and other services.

Cities are also engines of economic progress and social reform. Some of the greatest promise for innovation comes from places like Vauban, where innovative leaders can share knowledge and resources to resolve common problems. In wealthy countries, cities are far more efficient than suburbs and rural areas in terms of resource use and environmental impact per person. Living together in apartment buildings saves energy. Cities can offer public transportation, and they can provide multiple amenities within walking or biking distance.

FIGURE 22.2 In less than 20 years, Shanghai, China, has built Pudong, a new district of 5 million residents and 500 skyscrapers—located on former marshy farmland. This is a dramatic example of the many new world-class cities in regions that were largely rural only a few decades ago.
© Royalty-Free/Corbis

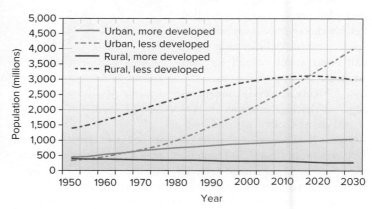

FIGURE 22.3 Growth of urban and rural populations in more-developed regions and in less-developed regions.
Source: United Nations Population Division

Concentrating people in urban areas leaves open space available for farming and biodiversity. But cities can also have concentrations of poverty, pollution, and displaced members of society. Providing food, housing, transportation, jobs, clean water, and sanitation to the two or three billion additional urban residents expected to move into cities in this century may be one of the preeminent challenges of this century.

As in Vauban, there is much we can do to make our cities both livable and sustainable. Of course, Vauban is in Germany, a wealthy country that can afford major transformations to its cities and transportation systems. Can there also be sustainable transformations in developing areas? What will make those changes possible?

Cities have specialized functions

Since their earliest origins, cities have been centers of education, religion, commerce, recordkeeping, communication, and political power. As cradles of civilization, cities have influenced culture and society far beyond their proportion of the total population (fig. 22.4). Until about 1900, only a small percentage of the world's people lived permanently in urban areas, and even the greatest cities of antiquity were small by modern standards.

Just what makes up an urban area or a city? Definitions differ. The U.S. Census Bureau considers any incorporated community to be a city, regardless of size, and defines any city with more than 2,500 residents as urban. More meaningful definitions are based on *functions*. In a **rural area**, most residents depend on agriculture or other ways of harvesting natural resources for their livelihood. In an **urban area**, by contrast, a majority of the people are not directly dependent on natural resource–based occupations.

A **village** is a collection of rural households linked by culture, custom, family ties, and association with the land (fig. 22.5). A **city**, by contrast, is a differentiated community with a population and resource base large enough to allow residents to engage in crafts, services, or professions other than natural resource–based occupations. While the rural village often gives a sense of security

and connection, small communities may also resist changes that might lead to economic or social progress. A city offers more freedom to experiment, to be upwardly mobile, and to break from restrictive traditions, but it can be harsh and impersonal.

Large cities are expanding rapidly

Very large cities are becoming more numerous, and they are shifting to developing areas. In 1900, only 13 cities in the world had populations over 1 million (table 22.2). All of those cities, except Tokyo and Beijing, were in more developed countries in Europe and North America. London was the only city in the world with more than 5 million residents. By 2016, there were at over 500 cities—100 of them in China alone—with more than 1 million residents. Of the 13 largest of these metropolitan areas, all had more than 20 million residents, and none were in Europe. Only New York City and Tokyo are in a developed country. It's expected that by 2030 at least 93 cities will have populations over 5 million, and three-fourths of those cities will be in developing countries (fig. 22.6). In just the next 25 years, Mumbai and Delhi (India), Karachi (Pakistan), Manila (Philippines), and Jakarta (Indonesia) are all expected to grow by at least 50 percent.

Large, regional urban agglomerations, including tens of millions of people, are increasingly common. In the United States, a region stretching from Boston to Washington, D.C., has merged into a more or less continuous urbanized area. This "megalopolis," sometimes called Bos-Wash, contains about 35 million people. Because these agglomerations have expanded beyond what we normally think of as a city, some geographers prefer to think of them as urbanized **core regions** that dominate the social, political, and economic life of most countries (fig. 22.7). Similar urban agglomerations can be found around the world. The Tokyo-Yokohama-Osaka-Kobe megalopolis, for example, has about 50 million residents, while the greater Yangtze River Delta region, which combines Shanghai, Suzhou, Nanjing, and dozens of other once independent cities, now contains roughly 90 million people.

FIGURE 22.4 Since their earliest origins, cities have been centers of education, religion, commerce, politics, and culture. Unfortunately, they have also been sources of pollution, crowding, disease, and misery.
© Royalty-Free/Corbis

FIGURE 22.5 This village in Chiapas, Mexico, is closely tied to the land through culture, economics, and family relationships. While the timeless pattern of life here gives a great sense of identity, it can also be stifling and repressive.
© William P. Cunningham

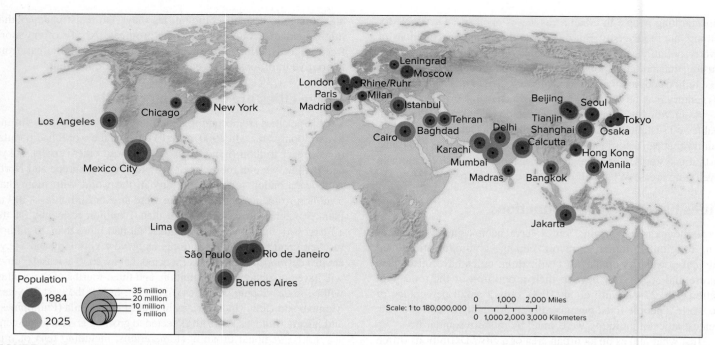

FIGURE 22.6 By 2025, most of the world's largest cities will be in developing countries that already have trouble housing, feeding, and employing their people.

China represents the largest demographic shift in human history. Since the end of Chinese collectivized farming and factory work in 1986, around 250 million people have moved from rural areas to cities. And in the next dozen years an equal number are expected to join this vast exodus. In addition to expanding existing cities, China plans to build 400 new urban centers with populations of at least 500,000 over the next 20 years. Already at least half of the concrete and one-third of the steel used in construction around the world each year is consumed in China.

FIGURE 22.7 Urban core agglomerations (*lavender areas*) are forming megalopolises in many areas. While open space remains in these areas, the flow of information, capital, labor, goods, and services links each into an interacting system.
Source: U.S. Census Bureau

Table 22.2	The World's Largest Urban Areas (Populations in Millions)		
1900		**2016**	
London, England	6.6	Tokyo, Japan	38.8
New York, USA	4.2	Shanghai, China	35.5
Paris, France	3.3	Jakarta, Indonesia	31.5
Berlin, Germany	2.4	Seoul, Korea	25.6
Chicago, USA	1.7	Beijing, China	25.0
Vienna, Austria	1.6	Guangzho, China	24.9
Tokyo, Japan	1.5	Karachi, Pakistan	24.3
St. Petersburg, Russia	1.4	New York City, USA	23.6
Philadelphia, USA	1.4	Mexico City, Mexico	22.2
Manchester, England	1.3	Delhi, India	21.7
Birmingham, England	1.2	São Paulo, Brazil	21.5
Moscow, Russia	1.1	Lagos, Nigeria	21.0
Beijing, China	1.1	Mumbai, India	20.7

Source: World Population Review, 2016

Quantitative Reasoning

How many of the large cities in table 22.2 were in developing countries in 1900? How many cities on the list in 1900 were still on the list in 2016? How much did Tokyo grow between 1900 and 2016? New York? Beijing? How might quality of life in Beijing have changed in this interval?

Consider Shanghai, for example. In 1985, the city had a population of about 10 million. It's now over 35 million—including millions of migrant laborers—and the surrounding megacity is nearly three times as large. In the past decade, Shanghai has built 4,000 skyscrapers (buildings with more than 25 floors). The city already has twice as many tall buildings as Manhattan, and proposals have been made for 1,000 more. Unfortunately, most of this growth has taken place in a swampy area called Pudong, across the Huangpu River from the historic city center (see fig. 22.2). Pudong is now sinking about 1.5 cm per year due to groundwater drainage and the weight of so many buildings.

Other Chinese cities have plans for similar massive building projects to revitalize blighted urban areas. Harbin, an urban complex of about 9 million people and the capital of Heilongjiang Province, for example, recently announced plans to relocate the entire city across the Songhua River on 740 km^2 (285 mi^2) of former farmland. Residents hope these new towns will be both more livable and more ecologically sustainable than the old cities they're replacing. In recent years, the Chinese government contracted with several design firms to build brand-new "ecocities." They hope to show how cities can grow in wealth and size while still having a modest ecological footprint. Plans call for these cities to be self-sufficient in energy, water, and most food products, with the aim of zero emissions of greenhouse gases from transportation.

Push and pull factors motivate people to move to cities

Urban populations grow in two ways: by natural increase (more births than deaths) and by immigration. Natural increase is fueled by improved food supplies, better sanitation, and advances in medical care that reduce death rates and cause populations to grow both within cities and in the rural areas around them (chapter 7). In Latin America and East Asia, natural increase is responsible for two-thirds of urban population growth. In Africa and West Asia, immigration is the largest source of urban growth.

Immigration to cities can be caused both by **push factors** that force people out of the country and by **pull factors** that draw them into the city. Common push factors are population growth and unemployment in rural areas, which drive people to cities in search of jobs, food, and housing. Economic forces can also be push factors: Trade policies might make farming unprofitable, or production of export crops can displace food production and deplete rural food supplies. Political, racial, or religious conflicts also drive people from their homes and villages. These events can actually depopulate the countryside. The United Nations estimates that in 2015 a record 60 million people either fled their native country or were internal refugees within their own country, displaced by political, economic, or social instability. Many of these refugees end up in the already overcrowded megacities of the developing world.

Land tenure patterns and changes in agriculture also play a role in pushing people into cities. The same pattern of agricultural mechanization that eliminated most farm labor in the United States is spreading now to developing countries. In many areas, the cost of mechanization is leading to the increasingly concentrated ownership of land by fewer and wealthier landowners. These changes force subsistence farmers off the land so it can be converted to grazing lands or monoculture cash crops.

Pull factors also draw people to even the largest and most hectic cities. Young people are drawn by the excitement, vitality, and opportunity to meet others like themselves. Cities offer jobs, housing, entertainment, and freedom from the constraints of village traditions. The city provides opportunities for upward social mobility, prestige, and power not ordinarily available in the country. Cities support specialization in arts, crafts, and professions for which markets don't exist elsewhere. Attractive images of urban life on television also draw people to the city. And even for the desperately poor and homeless, cities often provide social services that are sometimes preferable to what rural areas can provide.

Government policies can drive urban growth

Government policies can create both push and pull factors by favoring urban over rural areas. Developing countries commonly spend most of their budgets on improving urban areas, especially around the capital city where leaders live, even when a small percentage of the population lives there or benefits directly from the investment. This gives the major cities a virtual monopoly on new jobs, housing, education, and opportunities, all of which bring in rural people searching for a better life. In Peru, for example, Lima accounts for 20 percent of the country's population but has 50 percent of the national wealth, 60 percent of the manufacturing, 65 percent of the retail trade, 73 percent of the industrial wages, and 90 percent of all banking in the country. Similar statistics pertain to São Paulo, Mexico City, Manila, Cairo, Lagos, Bogotá, and a host of other cities.

Governments often manipulate exchange rates and food prices for the benefit of more politically powerful urban populations but at the expense of rural people. Importing lower-priced food pleases city residents, but local farmers then find it uneconomical to grow crops. As a result, an increased number of people leave rural areas to become part of a large urban workforce, keeping wages down and industrial production high. Zambia, for instance, sets maize prices below the cost of local production to discourage farming and to maintain a large pool of workers for the mines. Keeping the currency exchange rate high stimulates export trade but makes it difficult for small farmers to buy the fuels, machinery, fertilizers, and seeds that they need. This depresses rural employment and rural income while stimulating the urban economy. The effect is to transfer wealth from the country to the city.

Section Review

1. What proportion of the world's population is now urban?
2. List 5 of the largest 13 cities in 2016.
3. Give an example of how government policy can promote migration to cities.

22.2 URBAN CHALLENGES IN THE DEVELOPING WORLD

- *Low-income cities cannot build infrastructure to keep up with rapid urban growth.*
- *Poor water and air quality are especially important concerns.*
- *Housing and waste management are also widespread problems.*

Large cities in both developed and developing countries face similar challenges in accommodating the needs and by-products of dense populations. The problems are most intense, however, in rapidly growing cities of developing nations.

Cities in developing nations are also where most population growth will occur in coming decades (see fig. 22.7). These cities already struggle to supply food, water, housing, jobs, and basic services for their residents. The unplanned and uncontrollable growth of those cities causes tragic urban environmental problems. What responsibilities might we in richer countries have to help people in these developing areas?

Traffic congestion and air quality are growing problems

A first-time visitor to a supercity—particularly in a less-developed country—is often overwhelmed by the immense crush of pedestrians and vehicles of all sorts that clog the streets. The noise, congestion, and confusion of traffic make it seem suicidal to venture onto the street. Delhi, for instance, is one of the most congested cities in the world (fig. 22.8). Traffic is chaotic almost all the time. People commonly spend three or four hours each way commuting

to work from outlying areas. Jakarta, Bangkok, Beijing, and many other cities suffer chronic congestion. The average resident of Bangkok, for example, spends the equivalent of 44 days a year sitting in traffic jams. About 20 percent of all fuel is consumed by vehicles standing still. Hours of work lost each year are worth at least $3 billion.

Traffic congestion is expected to worsen in many developing countries as the number of vehicles increases but road construction fails to keep pace (fig. 22.9). All this traffic, much of it involving old, poorly maintained vehicles, combines with smoky factories and use of wood or coal fires for cooking and heating to create a thick pall of air pollution in the world's supercities. Lenient pollution laws, corrupt officials, inadequate testing equipment, ignorance about the sources and effects of pollution, and lack of funds to correct dangerous situations usually exacerbate the problem.

What is its human toll? An estimated 60 percent of Kolkata's residents are thought to suffer from respiratory diseases linked to air pollution. Lung cancer mortality in Shanghai is reported to be four to seven times higher than rates in the countryside.

Insufficient sewage treatment causes water pollution

Few cities in developing countries can afford to build modern waste treatment systems for their rapidly growing populations. The World Bank estimates that only 35 percent of urban residents in developing countries have satisfactory sanitation services. The situation is especially desperate in Latin America, where only 2 percent of urban sewage receives any treatment. In Egypt, Cairo's sewer system was built about 50 years ago to serve a population

FIGURE 22.8 Motorized rickshaws, cycle rickshaws, scooters, bicycles, and buses compete for space on this crowded street in Delhi. Providing infrastructure to meet growing demand is a challenge.
© Reuters/Corbis

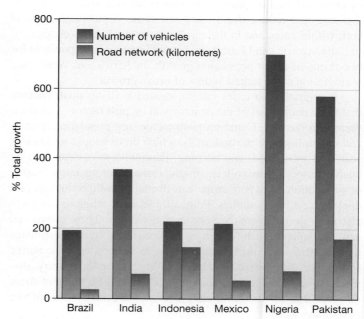

FIGURE 22.9 Transport growth in selected developing countries, 1980–2000.
Source: Earth Trends, 2006

of 2 million people. It is now being overwhelmed by more than 10 million people. Less than one-tenth of India's 3,000 towns and cities have even partial sewage systems or water treatment facilities. Some 150 million of India's urban residents lack access to sanitary sewer systems.

Figure 22.10 shows one of many tidal canals that crisscross Jakarta and serve as the sewage disposal system for many of the 30 million urban residents. In 2007, unusually heavy rain backed up these canals and flooded about half the city. Health officials warned about possible health effects.

Nearly a billion people, or about one-third of the population in developing world cities, do not have safe drinking water, according to the World Bank. Although city dwellers are somewhat more likely than rural people to have clean water, this still represents a large problem. Where people have to buy water from merchants,

it often costs 100 times as much as piped city water and may not be any safer to drink. Many rivers and streams in developing countries are little more than open sewers, and yet they are all that poor people have for washing clothes, bathing, cooking, and—in the worst cases—for drinking. Diarrhea, dysentery, typhoid, and cholera are widespread diseases in these countries, and infant mortality is tragically high (chapter 8).

Many cities lack adequate housing

The United Nations estimates that at least 1 billion people— nearly 15 percent of the world's population—live in crowded, unsanitary slums of the central cities and in the vast shantytowns and squatter settlements that ring the outskirts of most developing world cities. Around 100 million people have no home at all. In Mumbai, India, for example, it is thought that a million people sleep on the streets, sidewalks, and traffic circles because they can find no other place to live. In Brazil, perhaps a million "street kids," who have run away from home or been abandoned by their parents, live however and wherever they can. This is surely a symptom of a tragic failure of social systems.

Slums are generally legal but inadequate multifamily tenements or rooming houses, either custom-built to rent to poor people or converted from some other use. The chals of Mumbai, India, for example, are high-rise tenements built in the 1950s to house immigrant workers. Never very safe or sturdy, these dingy, airless buildings are already crumbling and often collapse without warning. Eighty-four percent of the families in these tenements live in a single room; half of those families consist of six or more people. Typically, they have less than 2 m^2 of floor space per person and only one or two beds for the whole family. They may share kitchen and bathroom facilities down the hall with 50 to 75 other people. Because of this crowding, household accidents are a common cause of injuries and deaths in developing world cities, especially to children. Charcoal braziers or kerosene stoves used in crowded homes are a routine source of fires and injuries.

Shantytowns are settlements created when people move onto undeveloped lands and build their own houses. Shacks are built of corrugated metal, discarded packing crates, brush, plastic sheets, or whatever building materials people can scavenge. Some shantytowns are simply illegal subdivisions where the landowner rents land without city approval. Others are spontaneous or popular settlements or **squatter towns** where people occupy land without the owner's permission. Sometimes this occupation involves thousands of people who move onto unused land in a highly organized, overnight land invasion, building huts and laying out streets, markets, and schools before authorities can root them out. In other cases, shantytowns just gradually "happen."

Called *barriads, barrios,* or *favelas* in Latin America, *bidonvillas* in Africa, or *bustees* in India, shantytowns surround every megacity in the developing world (fig. 22.11). They are not an exclusive feature of the poorest countries, however. Some 250,000 immigrants and impoverished residents live in the *colonias* along the southern Rio Grande in Texas.

FIGURE 22.10 This tidal canal in Jakarta serves as an open sewer. Although most of the city has modernized in recent years, millions of people still lack access to modern sanitation systems.
© William P. Cunningham

FIGURE 22.11 Homeless people have built shacks along this busy railroad track in Jakarta. It's a dangerous place to live, with many trains per day using the tracks, but for the urban poor, there are few other choices.
Source: Courtesy Dr. Helga Leitner

FIGURE 22.12 São Paulo, Brazil, is South America's largest urban area. Traffic congestion is legendary, and the wealthy residents increasingly depend on helicopters to avoid crime and traffic jams. With more than 193 heliports, and more than 700 flights daily, the city has the world's largest helicopter fleet.
© *William P. Cunningham*

These populous but unauthorized settlements usually lack sewers, clean water supplies, electricity, and roads. Often, the land on which they are built was not previously used because it is unsafe or unsuitable for habitation. In Rio de Janeiro, La Paz (Bolivia), Guatemala City, and Caracas (Venezuela), shanty towns are perched on landslide-prone hills. São Paulo, Brazil, is the country's wealthiest city, and the hub of finance and business, but still has at least 2 million people living in favelas or slums (fig. 22.12).

As desperate and inhumane as conditions are in these slums and shantytowns, many people do more than merely survive there. They keep themselves clean, raise families, educate their children, find jobs, and save a little money to send home to their parents. They learn to live in a dangerous, confusing, and rapidly changing world and have hope for the future. The people have parties; they sing and laugh and cry. They are amazingly adaptable able and resilient. In many ways, their lives are no worse than those in the early industrial cities of Europe and America a century ago. Perhaps continuing development will bring better conditions to cities of the developing world—as it has for many in the developed world.

Section Review

1. Why are traffic congestion and water pollution problems in most fast-growing cities in developing countries?
2. What are shantytowns, and why are they common?
3. List several costs of polluted air and water in cities.

22.3 URBAN CHALLENGES IN THE DEVELOPED WORLD

- *Cities in developed countries suffer from pollution and poverty.*
- *Urban sprawl undercuts urban sustainability.*

For the most part, the rapid growth of central cities that accompanied industrialization in nineteenth-and early twentieth-century Europe and North America has now slowed or even reversed. London, for instance, once the most populous city in the world, has lost nearly 2 million people, dropping from its peak of 8.6 million in 1939 to about 6.7 million today. While the greater metropolitan area surrounding London has been expanding to about 10 million inhabitants, the city itself is now only a modest-sized urban area.

Many of the worst urban environmental problems of the more developed countries have been substantially reduced in recent years. Minority groups in inner cities, however, remain vulnerable to legacies of environmental degradation in industrial cities (see the What Do You Think? section titled "People for Community Recovery" below).

In most developed countries, improved sanitation and medical care have reduced or eliminated many of the communicable diseases that once afflicted urban residents. Air and water quality have improved dramatically as heavy industry such as steel smelting and

People for Community Recovery

The Lake Calumet Industrial District on Chicago's far South Side is an environmental disaster area. A heavily industrialized center of steel mills, oil refineries, railroad yards, coke ovens, factories, and waste disposal facilities, much of the site is now a marshy wasteland of landfills, toxic waste lagoons, and slag dumps, around a system of artificial ship channels.

At the southwest corner of this degraded district sits Altgeld Gardens, a low-income public housing project built in the late 1940s by the Chicago Housing Authority. The 2,000 units of "The Gardens" or "The Projects," as they are called by the largely minority residents, are low-rise row houses, many of which are vacant or in poor repair. But residents of Altgeld Gardens are doing something about their neighborhood. People for Community Recovery (PCR) is a grassroots citizen's group organized to work for a clean environment, better schools, decent housing, and job opportunities for the Lake Calumet neighborhood.

PCR was founded in 1982 by Mrs. Hazel Johnson, an Altgeld Gardens resident whose husband died from cancer that may have been pollution-related. PCR has worked to clean up more than two dozen waste sites and contaminated properties in their immediate vicinity. Often this means challenging authorities to follow established rules and enforce existing statutes. Public protests, leafleting, and community meetings have been effective in public education about the dangers of toxic wastes and have helped gain public support for cleanup projects.

PCR's efforts successfully blocked construction of new garbage and hazardous waste landfills, transfer stations, and incinerators in the Lake Calumet district. Pollution prevention programs have been established at plants still in operation. And PCR helped set up a community monitoring program to stop illegal dumping and to review toxic inventory data from local companies.

Education is an important priority for PCR. An environmental education center administered by community members organizes workshops, seminars, fact sheets, and outreach for citizens and local businesses. A public health education and screening program has been set up to improve community health. Partnerships have been established with nearby Chicago State University to provide technical assistance and training in environmental issues.

PCR also works on economic development. Environmentally responsible products and services are now available to residents. Jobs that are being created as green businesses are brought into the community. Wherever possible, local people and minority contractors from the area are hired to clean up waste sites and restore abandoned buildings. Job training for youth and adults, as well as retraining for displaced workers, is a high priority.

In the 1980s, a young community organizer named Barack Obama worked with PCR on jobs creation, housing issues, and education. He credits the lessons he learned there for much of his subsequent political successes. In his best-selling memoir *Dreams from My Father,* Obama devotes more than 100 pages to his formative experiences at Altgeld Gardens and other nearby neighborhoods.

PCR and Mrs. Johnson have received many awards for their fight against environmental racism and despair. In 1992, PCR was the recipient of the President's Environmental and Conservation Challenge Award. PCR is the only African-American grassroots organization in the country to receive this prestigious award.

Although Altgeld Gardens is far from clean, much progress has been made. Perhaps the most important accomplishment is community education and empowerment. Residents have learned how and why they need to work together to improve their living conditions. Could these same lessons be useful in your city or community? What could you do to help improve urban environments where you live?

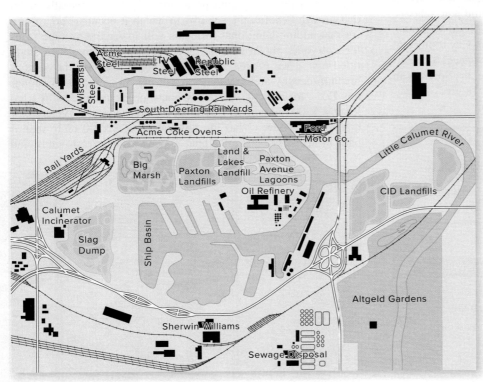

The Calumet industrial district in South Chicago.

chemical manufacturing have moved to developing countries. In consumer and information economies, workers no longer need to be concentrated in central cities. They can live and work in dispersed sites. Automobiles now make it possible for much of the working class to enjoy amenities such as single-family homes, yards, and access to recreation that once were available only to the elite.

In the United States, old, dense manufacturing cities such as Philadelphia and Detroit have lost population as industry has moved to developing countries. In a major demographic shift, both businesses and workers have moved west and south. Some of the most rapidly growing metropolitan areas like Phoenix (Arizona), Boulder (Colorado), Austin (Texas), and San Jose

(California), are centers for high-tech companies located in landscaped suburban office parks. These cities often lack a recognizable downtown, being organized instead around low-density housing developments, national-chain shopping malls, and extensive freeway networks. For many high-tech companies, being located near industrial centers and shipping is less important than a good climate, ready access to air travel, and amenities such as natural beauty and open space.

Urban sprawl consumes land and resources

While the move to suburbs and rural areas has brought many benefits to the average citizen, it also has caused numerous urban problems. Cities that once were compact now spread over the landscape, consuming open space and wasting resources. This pattern of urban growth is known as **sprawl.** While there is no universally accepted definition of the term, sprawl generally includes the characteristics outlined in table 22.3. As former Maryland governor Parris N. Glendening said, "In its path, sprawl consumes thousands of acres of forests and farmland, woodlands and wetlands. It requires government to spend millions extra to build new schools, streets, and water and sewer lines." And Christine Todd Whitman, former New Jersey governor and head of the Environmental Protection Agency, said, "Sprawl eats up our open space. It creates traffic jams that boggle the mind and pollute the air. Sprawl can make one feel downright claustrophobic about our future."

Quantitative Reasoning

Compare the photos in fig. 22.13 and 22.20a. How many families would fit in a city block in these two developments? How would your daily time budget differ in getting to school, work, shopping, and socializing in these two contexts?

Table 22.3   Characteristics of Urban Sprawl

1. Unlimited outward extension.
2. Low-density residential and commercial development.
3. Leapfrog development that consumes farmland and natural areas.
4. Fragmentation of power among many small units of government.
5. Dominance of freeways and private automobiles.
6. No centralized planning or control of land uses.
7. Widespread strip malls and "big-box" shopping centers.
8. Great fiscal disparities among localities.
9. Reliance on deteriorating older neighborhoods for low-income housing.
10. Decaying city centers as new development occurs in previously rural areas.

Source: Excerpt from a speech by Anthony Downs at the CTS Transportation Research Conference, as appeared on the website by Planners Web, Burlington, VT, 2001

In most American metropolitan areas, the bulk of new housing is in large, tract developments that leapfrog out beyond the edge of the city in a search for inexpensive rural land with few restrictions on land use or building practices (fig. 22.13). The U.S. Department of Housing and Urban Development estimates that urban sprawl consumes some 200,000 ha (roughly 500,000 acres) of farmland each year. Because cities often are located in fertile river valleys or shorelines, much of that land would be especially valuable for producing crops for local consumption. But with planning authority divided among many small, local jurisdictions, metropolitan areas have no way to regulate growth or provide for rational, efficient resource use. Small towns and township or county officials generally welcome this growth because it profits local landowners and businesspeople. Although the initial price of tract homes often is less than comparable urban property, there are external costs in the form of new roads, sewers, water mains, power lines, schools, shopping centers, and other extra infrastructure required by this low-density development.

Landowners, builders, real estate agents, and others who profit from this crazy-quilt development pattern generally claim that growth benefits the suburbs in which it occurs. They promise that adding additional residents will lower the average taxes for everyone, but in fact the opposite often is true. In a study titled *Better Not Bigger,* author Eben Fodor analyzed the costs of medium-density and low-density housing. In suburban Washington, D.C., for instance, each new house on a quarter acre (0.1 ha) lot cost $700 more in services than it paid in taxes. A typical new house on a 5-acre (2-ha) lot, however, cost $2,200 more than it paid in taxes, because of higher expenses for infrastructure and

FIGURE 22.13 Huge houses on sprawling lots consume land, alienate us from our neighbors, and make us ever more dependent on automobiles. They also require a lot of lawn mowing!

services. Ironically, people who move out to rural areas to escape from urban problems such as congestion, crime, and pollution often find they have simply brought those problems with them. A neighborhood that seemed tranquil and remote when they first moved in soon becomes just as crowded, noisy, and difficult as the city they left behind, as more people join them in their rural retreat.

In a study of 58 large American urban areas, David Rusk, an author and a former mayor of Albuquerque, found that between 1950 and 1990, populations grew 80 percent, while land area grew 305 percent. In Atlanta, Georgia, the population grew 32 percent between 1990 and 2000, while the total metropolitan area increased by 300 percent. The city is now more than 175 km across. Atlanta loses an estimated $6 million to traffic delays every day. By far the fastest-growing metropolitan region in the United States is Las Vegas, Nevada, which doubled its population but quadrupled its size in the 1990s (fig. 22.14a and b).

Transportation is crucial in city development

Getting people around within a large urban area has become one of the most difficult problems that many city officials face. A century ago, most American cities were organized around transportation corridors. First horse-drawn carriages, then electric streetcars provided a way for people to get to work, school, and shops. Everyone, rich or poor, wanted to live as close to the city center as possible.

The introduction of automobiles allowed people to move to suburbs, and cities began to spread over the landscape. The U.S. Interstate Highway System was the largest construction project in human history. Originally justified as necessary for national defense, it was really a huge subsidy for the oil, rubber, automobile, and construction industries. Its 72,000 km (45,000 mi) of freeways probably did more than anything to encourage sprawl and change America into an auto-centered society.

Because many Americans now live far from where they work, shop, and recreate, most consider it essential to own a private automobile. The average U.S. driver spends about 443 hours per year behind a steering wheel. This means that for most people, the equivalent of one full 8-hour day per week is spent sitting in an automobile. Of the 5.8 billion barrels of oil consumed each year in the United States (60 percent of which is imported), about two-thirds is burned in cars and trucks. As chapter 16 shows, about two-thirds of all carbon monoxide, one-third of all nitrogen oxides, and one-quarter of all volatile organic compounds emitted each year from human-caused sources in the United States are released by automobiles, trucks, and buses.

Building the roads, parking lots, filling stations, and other facilities needed for an automobile-centered society takes a vast amount of space and resources (fig. 22.15). In some metropolitan areas, it is estimated that one-third of all land is devoted to the automobile. To make it easier for suburban residents to get from their homes to jobs and shopping, we provide an amazing network of freeways and highways. At a cost of several trillion dollars to build, the interstate highway system was designed to allow us to drive at high speeds from source to destination without ever having to stop. As more and more drivers clog the highways, however, the reality is far different. In Los Angeles, for example, which has the worst congestion in the United States, the average speed in 1982 was 58 mph (93 km/hr), and the average driver spent less than 4 hours per year in traffic jams. In 2000, the average highway speed in Los Angeles was only 35.6 mph (57.3 km/hr), and the average driver

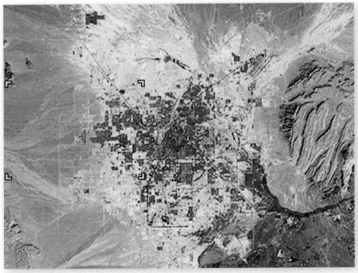
(a) 1972

(b) 2002

FIGURE 22.14 Satellite images of Las Vegas, Nevada, in 1972 (a) and 2002 (b). The metropolitan area quadrupled in three decades.
Source: (a) and (b): USGS Earth Resources Observation and Science (EROS) Center

FIGURE 22.15 Freeways give us the illusion of speed and privacy, but they consume land, encourage sprawl, and create congestion as people move farther from the city to get away from traffic and then have to drive to get anywhere.
© Charles Smith/Corbis RF

spent 82 hours per year waiting for traffic. Although new automobiles are much more efficient and cleaner operating than those of a few decades ago, the fact that we drive so much farther today and spend so much more time idling in stalled traffic means that we burn more fuel and produce more pollution than ever before.

Altogether, it is estimated that traffic congestion costs the United States $78 billion per year in wasted time and fuel. Some people argue that the existence of traffic jams in cities shows that more freeways are needed. Often, however, building more traffic lanes simply encourages more people to drive farther than before. Rather than ease congestion and save fuel, more freeways can exacerbate the problem.

Sprawl impoverishes central cities from which residents and businesses have fled. With a reduced tax base and fewer civic leaders living or working in downtown areas, the city is unable to maintain its infrastructure. Streets, parks, schools, and civic buildings fall

into disrepair at the same time that these facilities are being built at great expense in new suburbs. The poor who are left behind when the upper and middle classes abandon the city center often can't find jobs where they live and have no way to commute to the suburbs where jobs are now located. About one-third of Americans are too young, too old, or too poor to drive. For these people, car-oriented development causes isolation and makes daily tasks like grocery shopping very difficult. Parents, especially mothers, spend long hours transporting young children. Teenagers and aging grandparents are forced to drive, often presenting a hazard on public roads.

Sprawl also is bad for your health. By encouraging driving and discouraging walking, sprawl promotes a sedentary lifestyle that contributes to heart attacks and diabetes, among other problems. In Atlanta, for example, the lowest-density suburbs tend to have significantly higher rates of overweight residents than the highest-density neighborhoods.

Finally, sprawl fosters uniformity and alienation from local history and natural environment. Housing developments often are based on only a few standard housing styles, while shopping centers and strip malls everywhere feature the same national chains. You could drive off the freeway in the outskirts of almost any big city in America and see exactly the same brands of fast-food restaurants, motels, stores, filling stations, and big-box shopping centers.

Public transit can make cities more livable

Many American cities are now rebuilding the public transportation systems that were abandoned in the 1950s (fig. 22.16). Consider how different your life might be if you lived an automobile-free life in a city with good mass transit.

FIGURE 22.16 Many American cities are now rebuilding light rail systems that were abandoned in the 1950s when freeways were built. Light rail is energy-efficient and popular, but it can cost up to $100 million per mile ($60 million per kilometer) to install.
© William P. Cunningham

A famous example of successful mass transit is found in Curitiba, Brazil. High-speed, bi-articulated buses, each of which can carry 270 passengers, travel on dedicated roadways closed to all other vehicles. These bus-trains are linked to 340 feeder routes extending throughout the city. Everyone in the city is within walking distance of a bus stop that has frequent, convenient, affordable service. Curitiba's buses carry some 1.9 million passengers per day, or about three-quarters of all personal trips within the city. Working with existing roadways for the most part, the city was able to construct this system for one-tenth the cost of a light rail system or freeway system, and one-hundredth the cost of a subway.

But many developing countries are adopting the automobile-centered model of the United States rather than Curitiba's model. Traffic accidents have become the third-largest cause of years of lost life worldwide. The number of vehicles increased eightfold in Nigeria and sixfold in Pakistan between 1980 and 2000, while the road networks in those countries expanded by only 10 to 20 percent in the same time.

The recent introduction of the Tata Nano in India raises nightmares for both urban planners and energy experts. Costing less than $2,000 brand new, these tiny vehicles put car ownership within reach for millions who could never afford it before. But they will probably increase gasoline consumption greatly and result in huge traffic jams as inexperienced drivers take to the road for the first time.

In 2012, more than 19.3 million motor vehicles were sold in China, making it both the world's largest manufacturer and the largest market for automobiles. The number of cars, buses, vans, and trucks on the road in China is expected to surpass the number in the United States by 2050. How those vehicles will be powered is of vital importance to our global ecosystem. Already Chinese efficiency standards are higher than those in the United States, and Chinese companies are making rapid progress in developing hybrid and all-electric vehicles.

Section Review

1. Describe urban sprawl, and explain some reasons for it.
2. Identify several costs associated with expansive urban growth.
3. How did Curitiba, Brazil, make public transit work?

22.4 SMART GROWTH

- *Better environmental planning involves mixed uses, alternative transportation, and urban growth boundaries.*
- *Garden cities and new urbanism aimed to make cities appealing.*
- *Green urbanism emphasizes alternative energy and transportation and compact growth.*

Smart growth involves strategies for well-planned developments that make efficient and effective use of land resources and existing infrastructure. An alternative to haphazard, poorly planned sprawling developments, smart growth involves thinking ahead to develop pleasant neighborhoods while minimizing the wasteful use of space and tax dollars for new roads and extended sewer and water lines.

Smart growth aims to make land-use planning democratic. Public discussions allow communities to guide planners. Mixing land uses, rather than zoning exclusive residential areas well away from commercial areas, makes living in neighborhoods more enjoyable. By planning a range of housing styles and costs, smart growth allows people of all income levels, including young families and aging grandparents, to find housing they can afford. Open communication between planners and the community helps make urban expansion fair, predictable, and cost-effective.

Smart growth approaches acknowledge that urban growth is inevitable; the aim is to direct growth, to make pleasant spaces for us to live, and to preserve some accessible, natural spaces for all to enjoy (table 22.4). It strives to promote the safety, livability, and revitalization of existing urban and rural communities.

Smart growth protects environmental quality. It attempts to reduce traffic and to conserve farmlands, wetlands, and open space. This may mean restricting land use, but it also means finding economically sound ways to reuse polluted industrial areas within the city (fig. 22.17). As cities grow and transportation and communications enable communities to interact more, the need for regional planning becomes both more possible and more pressing. Community and business leaders need to make decisions based on a clear understanding of regional growth needs and how infrastructure can be built most efficiently and for the greatest good.

One of the best examples of successful urban land-use planning in the United States is Portland, Oregon, which has rigorously enforced a boundary on its outward expansion, requiring instead that development be focused on "in-filling" unused space within the city limits. Because of its many urban amenities, Portland is considered one of the most livable cities in America. Between 1970 and 1990, the Portland population grew by 50 percent but its total land area grew only 2 percent. During this time, Portland property taxes decreased 29 percent and vehicle miles traveled increased only 2 percent. By contrast, Atlanta, which had similar population growth, experienced an explosion of urban sprawl that increased its land area threefold, drove up property taxes 22 percent, and increased traffic miles by 17 percent. A result of this expanding traffic and increasing congestion was that Atlanta's air pollution increased by 5 percent, while Portland's, which has one of the best public transit systems in the nation, decreased by 86 percent.

Table 22.4 Goals for Smart Growth
1. Create a positive self-image for the community.
2. Make the downtown vital and livable.
3. Alleviate substandard housing.
4. Solve problems with air, water, toxic waste, and noise pollution.
5. Improve communication between groups.
6. Improve community member access to the arts.

Source: Vision 2000, Chattanooga, TN

FIGURE 22.17 Many cities have large amounts of unused open space that could be used to grow food. Residents often need help decontaminating soil and gaining access to the land.
© William P. Cunningham

Garden cities and new towns were early examples of smart growth

The twentieth century saw numerous experiments in building "new towns" for society at large that try to combine the best features of the rural village and the modern city. One of the most influential of all urban planners was Ebenezer Howard (1850–1929), who not only wrote about ideal urban environments but also built real cities to test his theories. In *Garden Cities of Tomorrow,* written in 1898, Howard proposed that the congestion of London could be relieved by moving whole neighborhoods to **garden cities** separated from the central city by a greenbelt of forests and fields.

In the early 1900s, Howard worked with architect Raymond Unwin to build two garden cities outside of London: Letchworth and Welwyn Garden. Interurban rail transportation provided access to these cities. Houses were clustered in "superblocks" surrounded by parks, gardens, and sports grounds. Streets were curved. Safe and convenient walking paths and overpasses protected pedestrians from traffic. Businesses and industries were screened from housing areas by vegetation. Each city was limited to about 30,000 people to facilitate social interaction. Housing and jobs were designed to create a mix of different kinds of people and to integrate work, social activities, and civic life. Trees and natural amenities were carefully preserved, and the towns were laid out to maximize social interactions and healthful living. Care was taken to meet residents' psychological needs for security, identity, and stimulation.

Today, Letchworth and Welwyn Garden each have 70 to 100 people per acre. This is a true urban density, about the same as

New York City in the early 1800s and five times as many people as most suburbs today. By planning the ultimate size in advance and choosing the optimum locations for housing, shopping centers, industry, transportation, and recreation, Howard believed he could create a hospitable and satisfying urban setting while protecting open space and the natural environment. He intended to create parklike surroundings that would preserve small-town values and encourage community spirit in neighborhoods.

Planned communities also have been built in the United States following the theories of Ebenezer Howard, but most plans have been based on personal automobiles rather than public transit. Radburn, New Jersey, was designed in the 1920s, and two highly regarded new towns of the 1960s are Reston, Virginia, and Columbia, Maryland. More recent examples, such as Seaside in northern Florida, represent a modern movement in new towns known as "new urbanism."

New urbanism promoted smart growth

New towns and garden cities included many important ideas, but they still left cities behind. More recently, a movement among architects and planners known as "new urbanism" has sought to make urban environments more appealing, efficient, and livable. Andres Duany and Peter Calthorpe were among the early promoters of this idea. Their idea was to recapture a traditional, small-town neighborhood feel in new developments. The goal of new urbanism has been to rekindle Americans' enthusiasm for cities by building charming, integrated, walkable developments. Sidewalks, porches, and small front yards encourage people to get outside and be sociable. A mix of apartments, townhouses, and detached houses in a variety of price ranges ensures that neighborhoods will include a diversity of ages and income levels.

These are some design principles of new urbanism:

- Limit city size or organize them in modules of 30,000 to 50,000 people, large enough to be a complete city but small enough to be a community. A greenbelt of agricultural and recreational land around the city limits growth while promoting efficient land use. By careful planning and cooperation with neighboring regions, ideally a city of 50,000 people can have real urban amenities such as museums, performing arts centers, schools, and hospitals.

- Identify sites and land uses carefully, to prevent chaotic development in which the lowest uses drive out the better ones. Plan ahead to protect valuable wetlands, agricultural soils, or aesthetically and ecologically valuable sites.

- Locate everyday shopping and services so people can meet daily needs with less automobile dependency. Provide accessible, sociable public spaces (fig. 22.18).

- Increase jobs in the community by locating offices, light industry, and commercial centers in or near suburbs, or by enabling work at home via computer.

- Encourage walking or the use of small, low-speed, energy-efficient vehicles, such as smaller cars, for local trips now performed by full-size automobiles.

FIGURE 22.18 This walking street in Queenstown, New Zealand, provides opportunities for shopping, dining, and socializing in a pleasant outdoor setting.
© William P. Cunningham

- Promote more diverse, flexible housing as alternatives to conventional, detached single-family houses. "In-fill" building between existing houses saves energy, reduces land costs, and might help provide a variety of living arrangements.

- Narrow the streets to slow traffic so children can play more safely. The land released from streets can be used for gardens, linear parks, playgrounds, and other public areas that will foster community spirit and encourage people to get out and walk.

Green urbanism aims for more sustainable cities

New urbanism has helped turn around Americans' enthusiasm for cities. Critics point out, though, that new urbanist developments, much like garden cities and new towns, are usually **greenfield developments,** projects built on previously undeveloped farmlands or forests on the outskirts of cities. Greenfield developments are inherently unsustainable because they contribute to sprawl and require most residents to commute to work by private car, which undermines efforts to reduce car dependence. The architect-designed houses rarely fall into middle- or low-income price ranges, so they remain elite, rather than mixed income, and small size and remote locations can limit the viability of mixed commercial uses.

"Green urbanism" strategies have been developed to improve the environmental profile of urban development. Green urbanism usually promotes redevelopment and in-fill near the urban center, where built infrastructure and mixed uses already are available. Walkable-scale urban density encourages mixed commercial and residential land uses. Energy-efficient housing, like that in Vauban, is aided by the efficiency of apartments versus separated houses. Integrated green space, including public recreational space, gathering places, and "pocket parks" are a focus. Walkable cities, bikable cities, and public transportation are encouraged by

building near transportation hubs (see the What Do You Think? section titled "The Architecture of Hope" below).

"Right-sizing" streets is one way to encourage livable cities. This means building streets smaller, to slow traffic, and easing automobile congestion by planning better turning lanes, traffic circles, and better traffic management. Converting some lanes to bicycle lanes and designated bus lanes can encourage alternative transportation. Improving sidewalks and crosswalks and other infrastructure for pedestrians makes a city more friendly and comfortable for people, as well as for cars.

European cities have been especially innovative in green planning. Stockholm, Sweden, has expanded by building small satellite suburbs linked to the central city by commuter rails and by bicycle routes that pass through a network of green spaces that reach far into the city. Copenhagen, Denmark, has rebuilt most of its transportation infrastructure since the 1960s, including more than 300 km of well-marked bike lanes and separated bike trails. Thirty percent of all trips through central Copenhagen are made using public transportation, and 14 percent of trips are made by bicycle.

Green building strategies are encouraged in many European cities. French cities have new, stringent energy standards for all public buildings. In 2016, San Francisco, California passed an ordinance that all new buildings less than 10 stories tall must have solar PV panels or solar water heaters on their roofs. Many German cities now require that roofs of most new buildings must have "green roofs"—growing grass or other vegetation (fig. 22.19). Green roofs absorb up to 70 percent of rain water, provide bird and butterfly habitat, insulate homes, and, contrary to old mythology, are structurally sturdy and long-lasting.

FIGURE 22.19 This award-winning green roof on the Chicago City Hall is functional as well as beautiful. It reduces rain runoff by about 50 percent, and keeps the surface as much as 30°F cooler than a conventional roof on hot summer days.
© Roofscapes, Inc. Used by permission; all rights reserved.

The Architecture of Hope

How sustainable and self-sufficient can urban areas be? An exciting experiment in minimal impact in London gives us an image of what our future may be. BedZED, short for the Beddington Zero Energy Development, is an integrated urban project built on the grounds of an old sewage plant in South London. BedZED's green strategies begin with recycling the ground on which it stands. Designed by architect Bill Dunster and his colleagues, the complex demonstrates dozens of energy-saving and water-saving ideas. BedZED has been occupied, and winning awards, since it was completed in 2003.

Like Vauban (opening case study), most of BedZED's innovations involve simple, even conventional ideas. Expansive, south-facing, triple-glazed windows provide abundant light, minimize the use of electric lamps, and provide passive solar heat in the winter. Thick, superinsulated walls keep interiors warm in winter and cool in summer. Rotating "wind cowls" on roofs turn to catch fresh breezes, which cool spaces in summer. In winter, heat exchangers warm incoming fresh air with the heat from stale, outgoing air. Energy used in space heating is nearly eliminated. Building materials are recycled, reclaimed, or renewable, which reduces the "embodied energy" invested in producing and transporting them.

BedZED does use energy, but the complex generates its own heat and electricity with a small, on-site, superefficient plant that uses local tree trimmings for fuel. Thus, BedZED uses no *fossil* fuels, and it is "carbon-neutral" because the carbon dioxide released by burning wood was recently captured from the air by trees. In addition, photovoltaic cells on roofs provide enough free energy to power 40 solar cars. Fuel bills for BedZED residents can be as little as 10 percent of what other Londoners pay for similar-sized homes.

Water-efficient appliances and toilets reduce water use. Rainwater collection systems provide "green water" for watering gardens, flushing toilets, and other nonconsumptive uses. Reed-bed filtration systems purify used water without chemicals. Water meters allow residents to see how much water they use. Just knowing about consumption rates helps encourage conservation. Residents use about half as much water per person as other Londoners.

BedZED residents can save money and time by not using, or even owning, a car. Office space is available on-site, so some residents can work where they live, and the commuter rail station is just a ten-minute walk away. The site is also linked to bicycle trails that facilitate bicycle commuting. Car pools and rent-by-the-hour auto memberships allow many residents to avoid owning (and parking) a vehicle altogether.

Building interiors are flooded with natural light, ceilings are high, and most residences have rooftop gardens. Community events and common spaces encourage humane, healthy lifestyles and community ties. Child-care services, shops, entertainment, and sports facilities are built into the project. The approximately 100 housing units are designed for a range of income levels ensuring a racially, ethnically, and age-diverse community.

South-facing windows heat homes, and colorful, rotating "wind cowls" ventilate rooms at BedZED, an ecological housing complex in South London, UK.
© Bill Dunster

Prices are lower than many similar-sized London homes, and few in this price range or inner-city location have abundant sunlight or gardens.

Similar projects are being built across Europe and even in some developing countries, such as China. Architect Dunster says that BedZED-like developments on cleaned-up brownfields could provide all the 3 million homes that the UK expects to need in the next decade with no sacrifice of open space. And as green building techniques, designs, and materials become standard, he argues, they will cost no more than conventional, energy-wasting structures.

What do you think? Would you enjoy living in a neighborhoodlike BedZED? Would it involve a lower or higher standard of living than you now have? How much would it be worth to avoid spending eight to ten hours per week not fighting bumper-to-bumper traffic while commuting to school or work? The average cost of owning and driving a car in the United States is about $9,000 per year. What might you do with that money if owning a vehicle were unnecessary? Try to imagine what urban life might be like if most private automobiles were to vanish. If you live in a typical American city, how much time do you have to enjoy the open space that the suburbs and freeways were supposed to provide? Perhaps, most importantly, how will we provide enough water, energy, and space for the 3 billion people expected to crowd into cities worldwide over the next few decades if we don't adopt some of the sustainable practices and approaches represented by BedZED or Vauban?

Principles of green urbanist planning focus on efficiency and reinvesting in cities:

- Focus on in-fill development—filling in the inner city so as to help preserve green space in and around cities. Where possible, focus on **brownfield developments,** building on abandoned, reclaimed industrial sites. Brownfields have been eyesores and environmental liabilities in cities for decades, but as urban growth proceeds, they are becoming an increasingly valuable land resource.

- Build high-density, attractive, low-rise, mixed-income housing near the center of cities or near public transportation routes

(fig. 22.20a). Densely packed housing saves energy as well as reduces the infrastructure costs per person.

- Provide incentives for alternative transportation, such as reserved parking for shared cars (fig. 22.20b) or bicycle routes and bicycle parking spaces. Figure 22.20c shows an 8,000-bicycle parking garage at the train station in Leiden, the Netherlands. An 8,000-car parking garage at the station would cut out the heart of the city. Discourage car use by minimizing the amount of space devoted to driving and parking cars, or by charging for parking space, once realistic alternatives are available.

(a)

(b)

(c)

FIGURE 22.20 Green urbanism includes (a) concentrated, low-rise housing, (b) car-sharing clubs that receive special parking allowances, and (c) alternative transportation methods. These examples are from Amsterdam and Leiden, the Netherlands.
(a), (b), and (c): © Mary Ann Cunningham

- Encourage ecological building techniques, including green roofs, passive solar energy use, water conservation systems, solar water heating, wind turbines, and appliances that conserve water and electricity.

- Encourage co-housing—groups of households clustered around a common green space that share child care, gardening, maintenance, and other activities. Co-housing can reduce consumption of space, resources, and time while supporting a sense of community.

- Provide facilities for recycling organic waste, building materials, appliances, and plastics, as well as metals, glass, and paper.

- Invite public participation in decision making. Emphasize local history, culture, and environment to create a sense of community and identity. Coordinate regional planning through metropolitan boards that cooperate with but do not supplant local governments.

Open-space design preserves landscapes

Traditional suburban development typically divides land into a checkerboard layout of nearly identical 1- to 5-ha parcels with no designated open space (fig. 22.21, *top*). The result is a sterile landscape consisting entirely of house lots and streets. This style of development, which is permitted—or even required—by local zoning and ordinances, consumes agricultural land and fragments wildlife habitat. Many of the characteristics that people move to the country to find—space, opportunities for outdoor recreation, access to wild nature, a rural ambiance—are destroyed by dividing every acre into lots that are "too large to mow but too small to plow."

An interesting alternative known as **conservation development,** cluster development, or open-space zoning preserves at least half of a subdivision as natural areas, farmland, or other forms of open space. Among the leaders in this design movement are landscape architects Ian McHarg, Frederick Steiner, and Randall Arendt. They have shown that people who move to the country don't necessarily want to own a vast acreage or live miles from their nearest neighbor. What they most desire is long views across an interesting landscape, an opportunity to see wildlife, and access to walking paths through woods or across wildflower meadows.

By carefully clustering houses on smaller lots, a conservation subdivision can provide the same number of buildable lots as a conventional subdivision and still preserve 50 to 70 percent of the land as open space (fig. 22.21, *bottom*). This not only

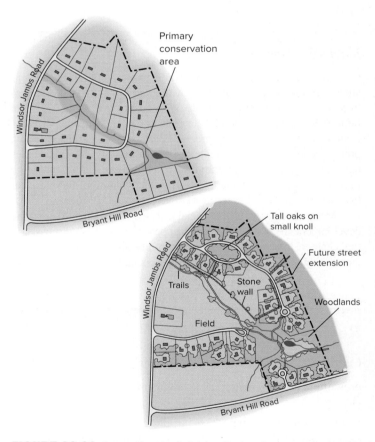

FIGURE 22.21 Conventional subdivision (*top*) and an open space plan (*bottom*). Although both plans provide 36 home sites, the conventional development allows for no public space. Cluster housing on smaller lots in the open space design preserves at least half the area as woods, prairie, wetlands, farms, or other conservation lands, while providing residents with more attractive vistas and recreational opportunities than a checkerboard development.

reduces development costs (less distance to build roads, lay telephone lines, sewers, power cables, and so on) but also helps foster a greater sense of community among new residents. Walking paths and recreation areas get people out of their houses to meet their neighbors. Home owners have smaller lots to care for, and yet everyone has an attractive vista and a feeling of spaciousness (fig. 22.22).

Urban habitat can make a significant contribution toward saving biodiversity. In a groundbreaking series of habitat conservation plans triggered by the need to protect the endangered California gnatcatcher, some 85,000 ha (210,000 acres) of coastal scrub near San Diego was protected as open space within the rapidly expanding urban area. This is an area larger than Yosemite Valley, and will benefit many other species, as well as humans.

Section Review

1. List several reasons for smart growth.
2. Identify some priorities for green urbanism.
3. What are some benefits of clustering houses in a development?

FIGURE 22.22 Jackson Meadows, an award-winning cluster development near Stillwater, Minnesota, groups houses at sociable distances and preserves surrounding open space for walking, gardening, and scenic views from all houses.
© William P. Cunningham

Conclusion

Cities are growing rapidly, many of them to incredible sizes and densities. Demographers predict that by the middle of this century, two-thirds of us will live in urban areas, so ensuring that our cities are built for sustainability should be a global priority. Vauban, Germany, is an outstanding example of green design to improve transportation, protect central cities, and create a sense of civic pride. Other cities have far to go, however, before they reach this standard. Among the immediate needs are housing, clean water, sanitation, food, education, health care, and basic transportation for their residents.

Many planners argue that social justice and sustainable economic development are answers to the urban problems we have discussed in this chapter. If people have the opportunity and money to buy better housing, adequate food, clean water, sanitation, and other things

they need for a decent life, they will do so. Democracy, security, and improved economic conditions help in slowing population growth and reducing rural-to-city movement. An even more important measure of progress may be institution of a social welfare safety net guaranteeing that old or sick people will not be abandoned and alone.

Some countries have accomplished these goals even without industrialization and high incomes. Sri Lanka, for instance, has lessened the disparity between the core and periphery of the country. Giving all people equal access to food, shelter, education, and health care eliminates many incentives for interregional migration. Both population growth and city growth have been stabilized, even though the per capita income is only $800 per year. So, what do you think? Could we help other countries do something similar?

Reviewing Key Terms

Can you define the following terms in environmental science?

brownfield development 22.4	greenfield developments 22.4	shantytowns 22.2	squatter towns 22.2
city 22.1	megacity 22.1	slums 22.2	urban agglomerations 22.1
conservation development 22.4	pull factors 22.1	smart growth 22.4	urban area 22.1
core region 22.1	push factors 22.1	sprawl 22.3	village 22.1
garden city 22.4	rural area 22.1		

Critical Thinking and Discussion Questions

1. Picture yourself living in a rural village or a developing world city. What aspects of life there would you enjoy? What would be the most difficult for you to accept?

2. Why would people move to one of the megacities of the developing world, if conditions are so difficult there?

3. A city could be considered an ecosystem. Using what you learned in chapters 3 and 4, describe the structure and function of a city in ecological terms.

4. Look at the major urban area(s) in your state. Why were they built where they are? Are those features now a benefit or drawback?

5. Weigh the costs and benefits of automobiles in modern American life. Is there a way to have the freedom and convenience of a private automobile without its negative aspects?

6. Boulder, Colorado, has been a leader in controlling urban growth. One consequence is that the city has stayed small and charming, but housing prices have skyrocketed and poor people have been driven out. If you lived in Boulder, what solutions might you suggest? What do you think is an optimum city size?

Data Analysis

Plotting Urban and Economic Indicators

Urbanization and economic growth are two closely related changes going on in societies today. How are these two processes related? How do they compare in different parts of the world? Is the United States more urbanized or less so than other regions? How do economic growth rates compare in different regions?

Gapminder.org is a rich source of data on global population, health, and development that you may have already examined in the Data Analysis exercise for chapter 7. The site includes animated graphs showing change over time. Go to Connect to find and analyze this animated Gapminder graph, and to answer questions about what they tell you.

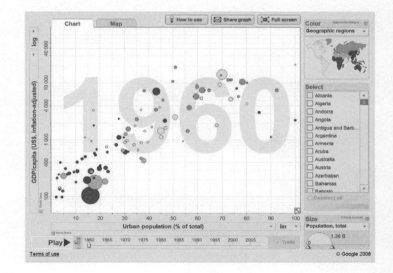

23

Ecological Economics

▲British Columbians are justly proud of their beautiful environment, and worried about the effects of global climate change. The province is using tax policy to shift the economy to a carbon-constrained future.
© Pierre Leclerc Photography/Moment/Getty Images

Learning Outcomes

After studying this chapter, you should be able to:

23.1 Identify some assumptions of classical and neoclassical economics.

23.2 Explain key ideas of ecological economics.

23.3 Describe relationships among population, technology, and scarcity.

23.4 Understand ways we measure growth.

23.5 Summarize how market mechanisms can reduce pollution.

23.6 Discuss the importance of green development and business.

"Underground nuclear testing, destruction of rain forests, toxic waste . . . Let's put it this way: if the world were a big apartment, we wouldn't get our deposit back."

– John Ross

Using Economics to Fight Climate Change

British Columbians are justifiably proud of their spectacular environment, and they want to protect it. One of the greatest challenges they—and all of us—face is global climate change. How can we move our economy to a low-carbon future?

The province has taken a bold and important step in that direction by adopting a tax on carbon fuels. In 2008, British Columbia (BC) passed the first broad-based carbon tax in North America. Conservatives claimed the tax would "destroy jobs and growth," but the opposite has been true. The latest numbers from Statistics Canada show that the carbon tax has been a tremendous environmental and economic success. Since the tax came into effect, fossil fuel use has dropped by 16 percent in BC compared to a 3.5 percent increase in the rest of Canada. At the same time, the BC economy has grown faster and unemployment has been lower than any other province.

The BC carbon tax is designed to be revenue neutral. That is, every penny taken in by the tax is used to reduce other taxes. Roughly 65 percent of the income is used to decrease corporate taxes. About 35 percent goes to lower personal taxes. Half that amount is targeted to low-income families and individuals for whom the carbon tax creates a particular disadvantage. And about half the corporate tax reductions are targeted for industries, such as clean energy, digital media and film, international businesses, and investment capital that the government wants to encourage. The result is that BC has the lowest personal income tax rate in Canada, and one of the lowest corporate rates in North America. Tax reductions have actually been greater than the amount taken in by the carbon tax. The difference is made up by increased taxes in sectors stimulated by tax shifts (fig. 23.1).

Passing this carbon tax was easier in BC than in other places because approximately 95 percent of the province's electricity currently comes from hydroelectric dams, which don't need carbon fuels. Still, most people heat their homes with coal or natural gas, and fuel their vehicles with petroleum-based fuels. The carbon tax started out low (C$10 per metric ton of CO_2) to ease people into the new system, and then rose by C$5 to the current C$30 per ton. In 2012, the tax was frozen so that BC wouldn't be at a disadvantage compared to other provinces.

Nearly 60 percent of BC's residents approved of the carbon tax when it was first proposed. That support has slipped somewhat since then, but most people are still keenly aware of how the insect infestations caused by warmer winters have devastated forests, or how droughts, heat waves, and violent storms impact their lives. Furthermore, air pollution caused by burning fossil fuels costs thousands of lives and adds more than $8 billion per year to Canada's health expenses. The idea of using tax policy to discourage harmful actions (pollution, climate change) and to reward beneficial sectors (such as renewable energy) is a classic in economic theory. This fiscal reform is especially appealing when it may bring a double dividend: reducing pollution and stimulating economic growth in renewable energy and conservation.

In 2016, British Columbia released the results of a Climate Leadership Plan for reducing greenhouse gas emissions while supporting a strong economy. Thousands of organizations, families, and individuals contributed to the plan. As a result of this effort, more than 150 corporate leaders called for a $10 per ton increase in the carbon tax. They argued that the current price of $30 per ton is far below the real social, economic, and environmental costs of global climate change.

There still are many people who hate new taxes of any sort. They rail against the bureaucracy needed to collect the tax—although this is just a shift in which taxes the bureaucracy collects. In fact, most people find their total tax bill has gone down significantly with this measure. And poorer people, as well as those in remote areas in particular, have benefited from programs supported by the tax.

There are other ways to put a price on carbon, or to use economic policy to achieve social goals. In this chapter, we'll look at ecological economics and how we value both natural resources and human well-being. Finally, we'll examine other ways that market mechanisms can help us solve environmental problems, and how businesses can contribute to sustainability.

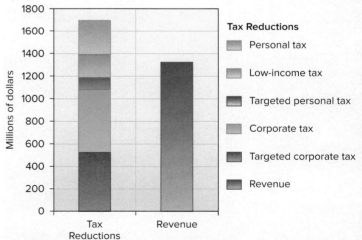

FIGURE 23.1 Revenues from British Columbia's carbon tax and personal and corporate tax reductions made possible by this revenue. (Data from Duke/Nicholas Institute)

23.1 PERSPECTIVES ON THE ECONOMY

- *Economics involves understanding how we evaluate and use resources.*
- *Resources can be nonrenewable, renewable, or intangible.*
- *Classical and neoclassical economics differ in how they define and analyze resources, growth, and price setting.*

Economy is the management of resources, ideally to meet our needs as efficiently as possible. The terms *ecology* and *economy* share a common root, *oikos* (ecos), the Greek word for "household." Economics is the *nomos,* or counting, of the household resources. Ecology is the *logos,* or logic, of how the household works.

Much of our economy involves using natural resources—such as minerals, soil, water, or solar energy—to produce goods (fig. 23.2). Some resources are renewable, others are not. Knowledge, experience, and creativity, for example, are valuable resources that aren't exhausted by use. Economics has traditionally been involved with choices and trade-offs, between goods and services that are nonrenewable. Understanding the balance of costs and benefits of these choices is a major part of classical economics. Environmental economics, like environmental science, tends to ask questions about long-term resource use: Are we using resources efficiently? Are the costs of our resource use reflected in the prices we pay for goods? Are there alternative strategies that could help us produce goods and services with fewer resources?

Can development be sustainable?

One of the most important questions in environmental science is how we can continue to improve human welfare within the limits of the earth's natural resources and biological systems.

FIGURE 23.2 Our economy depends on using resources to produce marketable goods. How we understand and value "resources" is a central question in environmental economics.
© Bruce Heinemann/Getty Images RF

Does our use of resources limit the opportunities of others—either future generations or people in other regions—to lead healthy and productive lives? As we saw in Chapter 1, *development* means improving people's lives, usually through increased access to goods (such as food) or services (such as education). *Sustainability* means living on the earth's renewable resources without damaging the ecological processes that support us all (table 23.1). **Sustainable development** is an effort to marry these two ideas. A definition developed by the World Commission on Environment and Development in 1987 is that "sustainable development is development that meets the needs of the present without compromising the ability of future generations to meet their own needs."

Is this possible? Not at our present population and rates of consumption. Some observers insist that there is no way more people can live at a high standard of living without irreversibly degrading our environment. Others say that as natural resources become scarce, we will simply find alternatives. Still others argue that there's enough for everyone if we can just share equitably and consume less. Much of this debate depends on how we define resources and economic growth.

Resources can be renewable or nonrenewable

A **resource** is anything with potential use in creating wealth or giving satisfaction. Natural resources can be either renewable or nonrenewable. In general, **nonrenewable resources** are materials present in fixed amounts in the environment, especially earth resources such as minerals, metals, and fossil fuels (fig. 23.3). Many of these resources, such as oil and coal, are renewed or recycled over geological time, but on a human time scale they are not renewable. Predictions abound that we are in imminent danger of running out of one or another of these exhaustible resources. Supplies of metals and other commodities, however, have frequently been extended by more efficient use, recycling, substitution of one material for another, or new technologies that can extract resources from dilute or remote sources.

Renewable resources are things that can be replenished or replaced. These include living organisms, fresh water from rain and snow, and sunlight—our ultimate energy source. These systems also provide essential ecological services on which we

Table 23.1	Goals for Sustainable Natural Resource Use

- Harvest rates for renewable resources (those like organisms that regrow, or those like fresh water that are replenished by natural processes) should not exceed regeneration rates.
- Waste emissions should not exceed the ability of nature to assimilate or recycle those wastes.
- Nonrenewable resources (such as minerals) may be exploited by humans, but only at rates equal to the creation of renewable substitutes.

FIGURE 23.3 Nonrenewable resources, such as the oil from this forest of derricks in Huntington Beach, California, are irreplaceable. Once they're exhausted (as this oil field was half a century ago), they will never be restored on a human time scale.
© William P. Cunningham

depend, although most of us don't think of these resources very often (fig. 23.4). We discuss these ideas further in section 23.2.

Because biological organisms and ecological processes are self-renewing, we often can harvest surplus organisms or take advantage of ecological services without diminishing future availability, if we do so carefully. Unfortunately, our stewardship of these resources often is less than ideal. Even once-vast biological populations, such as passenger pigeons, American bison, and Atlantic cod, were exhausted by overharvesting in only a few years. Similarly, we are now reducing renewable water resources (from rainfall and snowpack) in many regions by modifying the climate system. This modification of a renewable resource is leading to drought and reduced crop production in dry regions (chapter 15). Mismanagement of renewable resources, then, often makes them more ephemeral and limited than nonrenewable resources.

We also depend on **intangible resources,** such as open space, beauty, serenity, wisdom, and diversity (fig. 23.5). Paradoxically, these resources can be both infinite *and* exhaustible. There is no upper limit to the amount of beauty, knowledge, or compassion that can exist in the world, yet they can be easily destroyed. A single piece of trash can ruin a beautiful vista, or a single cruel remark can spoil an otherwise perfect day. On the other hand, unlike tangible resources that usually are reduced by use or sharing, intangible resources often are increased by use and multiplied by being shared. Nonmaterial assets can be important economically. Information management and tourism—both based on intangible resources—have become two of the largest and most powerful industries in the world.

Another term used to describe resources is **capital,** or wealth that can be used to produce more wealth. Usually, capital refers to something that has been built up or accumulated over time. There

can be many forms of capital. Banks provide *financial capital* (money) that small businesses need to start or grow. Economists also consider *manufactured* or *built capital* (tools, infrastructure, and technology), *natural capital* (goods and services provided by nature), *human* or *cultural capital* (knowledge, experience, and ideas about how to make or do things), and even *social capital* (shared values, trust, cooperative spirit, and community organization). All these kinds of capital may be needed to produce marketable goods and services.

Classical economics examines supply and demand

Originally, **classical economics** was a branch of moral philosophy concerned with how individual interest and values intersect with larger social goals. We trace many of our ideas about economy to Adam Smith (1723–1790), a moral philosopher concerned with individual freedom of choice. Smith's landmark book *Inquiry into the Nature and Causes of the Wealth of Nations,* published in 1776, argued:

> Every individual endeavors to employ his capital so that its produce may be of the greatest value. He generally neither intends to promote the public interest, nor knows how much he is promoting it. He intends only his own security, only his own gain. And he is in this led by an *invisible hand* to promote an end which was no part of his intention. By pursuing his own interests he frequently promotes that of society more effectually than when he really intends to.

This statement often is taken as a foundation for the capitalist system, in which willing sellers and fully informed buyers agree on a fair price for goods in the market. Smith proposed that this agreement would bring about the greatest efficiency of

FIGURE 23.4 Renewable resources, such as clean water, fresh air, plants, and animals, replenish themselves—unless we overuse and exhaust them. Often, we think of renewable resources in terms of ecosystem services. For instance, the freshwater marsh shown here provides purified water and air, freshwater storage, and biodiversity.
© William P. Cunningham

FIGURE 23.5 Intangible resources, such as the scenic beauty or solitude of this Colorado wilderness area, are widely treasured but are hard to evaluate in economic terms.
© William P. Cunningham

FIGURE 23.6 Informal markets such as this one in Bali, Indonesia, may be the purest example of willing sellers and buyers setting prices based on supply and demand.
© William P. Cunningham

resource use, because efficiency is necessary to produce goods at an acceptably low price. Assuming that all buyers and sellers are free to make any choice, this system also ensures individual liberty (fig. 23.6). The British economist John Maynard Keynes summarized faith in free markets and pricing this way: "Capitalism is the astounding belief that the most wickedest of men will do the most wickedest of things for the greatest good of everyone."

In a real market, producing goods at a low cost often requires that some costs are **externalized,** or passed off to someone else. Environmental costs and social costs are often supported by communities. For example, producing electricity at a power plant requires a stable and educated work force to run the plant, and the cost of educating workers normally is borne by society in general. Transportation networks, mining and oil drilling, and insurance costs are often supported by the public. Tax breaks also represent a subsidy because they transfer costs from one taxpayer to other sectors of society. Producing electricity also usually involves some pollution, and a power company allows the public to absorb the cost of that pollution (such as health care cost or reduced crop production). All these costs are involved in producing and selling electricity, but they are external to the company's cost calculations. Many economists, both conservative and liberal, argue that subsidies are a problem because they mask the real costs of production and lead to inefficiency in our economy.

David Ricardo (1772–1823), a contemporary of Adam Smith, introduced a description of the relationship between supply and demand in economics. **Demand** is the amount of a product or service that consumers are willing and able to buy at various possible prices, assuming they are free to express their preferences. **Supply** is

the quantity of that product being offered for sale at various prices, other things being equal. Classical economics proposes that there is a direct, inverse relationship between supply and demand (fig. 23.7).

With increasing quantity of production, supply increases and prices fall. With decreasing quantity, prices rise and demand falls. The difference between the cost of production and the price buyers are willing to pay, Ricardo called "rent." Today we call it profit.

In a free market of independent and rational buyers and sellers, an optimal price is achieved at the intersection of the supply and demand curves (fig. 23.7). This intersection is known as

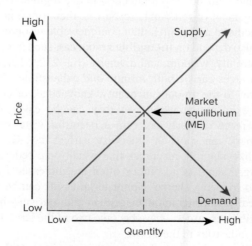

FIGURE 23.7 Classic supply/demand curves. When price is low, supply is low and demand is high. As prices rise, supply increases but demand falls. Market equilibrium is the price at which supply and demand are equal.

the **market equilibrium.** In real life, prices are not determined strictly by total supply and demand as much as by what economists call **marginal costs and benefits.** Sellers ask themselves, "What would it cost to produce one more unit of this product or service? Suppose I add one more worker or buy an extra supply of raw materials, how much more profit could I make?" Buyers ask themselves similar questions, "How much would I benefit and what would it cost if I bought one more widget?" If both buyer and seller find the marginal costs and benefits attractive, a sale is made.

There are exceptions to this theory of supply and demand. Consumers will buy some things regardless of cost. Raising the price of cigarettes, for instance, doesn't necessarily reduce demand. We call this price inelasticity. Other items have **price elasticity:** Price always changes as supply changes. When price goes up, demand falls and vice versa.

Neoclassical economics emphasizes growth

Toward the end of the nineteenth century, the field of economics divided into two broad camps. **Political economy** continued the tradition of moral philosophy and concerned itself with social structures and relationships among the classes. This group included reformers such as Karl Marx and later E. F. Schumacher, who argued that unfettered capital accumulation inevitably leads to inequity, which leads to instability in society. The other camp, called **neoclassical economics,** strove to adapt methods of modern science, and to be mathematically rigorous, noncontextual, abstract, and predictive. Neoclassical economists claim to be objective and value-free, leaving social concerns to other disciplines. Like their classical predecessors, they retain an emphasis on scarcity and the interaction of supply and demand in determining prices and resource allocation (fig. 23.8).

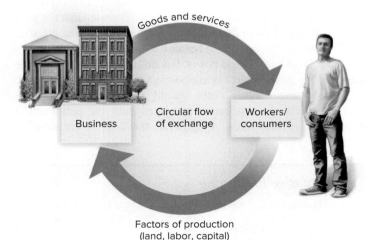

FIGURE 23.8 The neoclassical model of the economy focuses on the flow of goods, services, and factors of production (land, labor, capital) between business and individual workers and consumers. The social and environmental consequences of these relationships are irrelevant in this view.

Constant economic growth is considered necessary and desirable, in the neoclassical view. Growth keeps people happy by always offering more income or goods than people had last year. In a growing population, economic growth is seen as the only way to maintain full employment and avoid class conflict arising from inequitable distribution of wealth. Growth is also essential because businesses borrow resources to operate and grow. Few lenders are willing to share their money without a promise of greater returns later. Thus, businesses must continue to expand in order to increase profits and maintain the confidence of shareholders, whose money they are using to run their operations.

John Stuart Mill (1806–1873), a classical economist and philosopher, argued that perpetual growth in material well-being is neither possible nor desirable. Economies naturally mature to a steady state, he proposed, leaving people free to pursue nonmaterialistic goals. He didn't regard this equilibrium state to be necessarily one of stagnation or poverty. Instead, he wrote that in a stable economy, "there would be as much scope as ever for all kinds of mental culture, and moral and social progress; as much room for improving the art of living, and much more likelihood of its being improved when minds cease to be engrossed by the art of getting on."

Some neoclassical economists point out that not all growth involves increased resource consumption and pollution. Growth based on education, entertainment, and nonconsumptive activities, as suggested by Mill, can still contribute to economic expansion.

Neoclassical economics tends to view natural resources as interchangeable. As one resource becomes scarce, neoclassical economists believe, substitutes will be found. Labor is also substitutable. Because materials and labor are substitutable, they are not considered indispensable. Debates about the nature of growth, consumption, and resource scarcity have become increasingly active lately, in response to recent developments in economics, including environmental economics, the concept of a steady-state economy, and sustainable development.

Section Review

1. Define nonrenewable, renewable, and intangible resources.
2. What does it mean to externalize costs?
3. Why does neoclassical economics emphasize growth?

23.2 ECOLOGICAL ECONOMICS

- *Ecological economics incorporates ecological principles.*
- *Accounting for externalized costs is an important challenge.*
- *Ecosystem services regulate, supply, provide, or produce resources we need. Often these ideas overlap somewhat.*

Classical and neoclassical economics shape most of our economic activities, but the externalized costs and social inequity of these approaches have led to an interest in new ways to evaluate economic progress. **Ecological economics** has emerged as a way to understand the relationship between our economy and the ecological systems that support it. These economists point out that even Adam Smith wrote that the profit motive is good at guiding decision making in the short term, but it is ill-suited to long-term decision making. Ecological economists argue that we need to improve our long-term decision making if we intend to be here for the long term.

Ecological economists have followed two main approaches to resolving the short-term view of conventional economics: One approach is to question the necessity of constant growth. Another is to identify externalities and calculate their costs. If we know these costs, we can make the price reflect the total costs of production, or we can reduce those costs by changing the ways we make things.

Ecological economics draws on ecological concepts such as systems (chapter 2), thermodynamics, and material cycles (chapter 3). In a system, all components are interdependent. Disrupting one component (such as climate conditions) risks destabilizing other components (such as agricultural production) in unpredictable and possibly catastrophic ways. Thermodynamics and material cycles teach that energy and materials are continually reused. One organism's waste is another's nutrient or energy source. This perspective allows us to consider relationships among the many parts of our environment, our economy, and our own well-being (fig. 23.9).

Systems analysis also proposes that the earth has a limited carrying capacity for human populations. A limited carrying capacity means that unlimited economic growth is not possible. Many ecological economists, such as Herman Daly, argue for a **steady-state economy,** characterized by low human birth rates and death rates, the use of renewable energy sources, material recycling, and an emphasis on durability, efficiency, and stability. **Throughput,** the volume of materials and energy consumed and of waste produced, should be minimized. Most neoclassical growth models have assumed constantly increasing consumption and waste production, with resources being substituted as they run out. The steady-state idea, in contrast, reflects the notion of an ecosystem in equilibrium, or a population below its carrying capacity, where overall conditions remain generally stable over time.

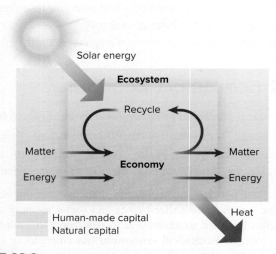

FIGURE 23.9 Ecological economics includes services such as recycling and resource provision and economic accounting. An effort is made to internalize, rather than externalize, natural resources and services.

Ecological economics assigns value to ecosystems

Ecological economics seeks to draw natural systems and ecological processes into economic accounting. Classical and neoclassical economics usually have focused on the value of human capital and built capital and resources, such as buildings, roads, or labor. Natural systems are considered free of cost—in part because in fact they have been free, abundant, or unregulated. Rivers absorb wastewater, bacteria decompose waste, and winds carry away smoke, but until recently there were few mechanisms for monitoring these processes or assigning them a value.

Consider energy production again, in which producers weigh their profits against the costs of coal, labor, buildings, and furnaces. These are **internal costs,** part of the normal expenses of doing business. Costs such as the climate's absorption of carbon dioxide, sulfur dioxide, or mercury are external costs, outside the accounting system. Similarly, if populations downwind suffer from air pollution, incurring lost workdays or illness, traditional economics has argued that society should bear the responsibility.

Can we internalize these costs? Identifying and calculating values takes considerable effort because it is a new process, but a study by Paul Epstein and his colleagues in 2011 produced an unusually complete estimate of the externalized costs of coal power. Epstein and his colleagues examined the costs of land disturbance, methane emissions from mines, fatalities during coal transportation, climate impacts, downwind costs to health, child development, cancers, direct federal subsidies to coal miners and energy producers, the costs of restoring abandoned mines, and other costs. Their best estimate of externalized costs was approximately 18¢/kilowatt-hour (kWh). The average cost of electric power in 2010 was about 9.5¢/kWh, or just half of the externalized costs. In other words, the real cost was about triple the price paid for electricity (fig. 23.10).

High and low estimates were also calculated in this study, to account for uncertainties in the data. These suggested that the public absorbs about 9¢ to 28¢ for every kWh of coal-based electricity. This study focused on coal because it has long been the world's dominant source of electric power, but similar accounting could be done for any power source or economic activity.

Accounting for all costs should make production more efficient, because an accurate price can help the public make more informed decisions. In general, the cost of cleaning up

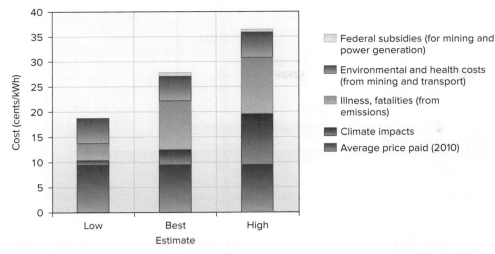

FIGURE 23.10 Estimates of externalized costs of coal power, in ¢/kilowatt-hour (kWh). Externalized costs (red colors) are shown added to the average price paid in 2010 (blue).

a power plant usually is lower than the cost of health care and lost productivity. In economic terms, the extra costs of illness and environmental damage are "market inefficiencies": They represent inefficient overall use of resources (money, time, energy, materials) because of incomplete accounting of costs and benefits.

Ecosystem services include provisioning, regulating, and aesthetic values

Ecosystem services is a general term for the resources provided and waste absorbed by our environment. These services are often grouped into four general classes (table 23.2): regulation (of climate, water supplies, and other factors), provision (of foods and other resources), supporting or preserving (of crop pollinators, nutrient cycling), and aesthetic or cultural benefits (fig. 23.11).

Although many ecological processes have no direct market value, we can estimate replacement costs, contingent values, shadow prices, and other methods of indirect assessment to determine a rough value. For instance, we now dispose of much of our wastes by letting nature detoxify them. How much would it cost if we had to do this ourselves?

Estimates of the annual value of all ecological goods and services provided by nature range from $16 trillion to $54 trillion, with a median worth of $33 trillion, or about three-fourths the combined annual GNP of all countries in the world (table 23.3). These estimates are lower than the real value because they omit ecosystem services from several biomes, such as deserts and tundra, that are poorly understood in terms of their economic contributions.

Accounting for ecosystem services is a focus of several global initiatives on sustainable development. A UN program called The Economics of Ecosystems and Biodiversity (TEEB) has been working to improve estimates of the value of ecosystem services. TEEB studies have shown that preserving ecosystems is far more cost-effective than using up their resources. Even restoring already-damaged ecosystems has enormous paybacks (fig. 23.12). Calculating a price for carbon storage in natural ecosystems has been the aim of REDD (Reducing Emissions from Deforestation and Forest Degradation) programs. These efforts are discussed in chapter 12. A 2003 economic study from Cambridge University (UK) estimated that protecting a series of nature reserves representing samples of all major biomes would cost $45 billion per year, but would preserve ecological services worth 100 times that cost—$4.4 trillion to $5.2 trillion annually.

Table 23.2   Important Ecological Services
1. *Regulate* global energy balance; chemical composition of the atmosphere and oceans; local and global climate; water catchment and groundwater recharge; production, storage, and recycling of organic and inorganic materials; maintenance of biological diversity.
2. *Provide* space and suitable substrates for human habitation, crop cultivation, energy conversion, recreation, and nature protection.
3. *Produce* oxygen, fresh water, food, medicine, fuel, fodder, fertilizer, building materials, and industrial inputs.
4. *Supply* aesthetic, spiritual, historic, cultural, artistic, scientific, and educational opportunities and information.

Source: From *Investing in Natural Capital,* edited by AnnMari Jansson, Monica Hammer, Carl Folke, and Robert Costanza. Washington, DC: Island Press, 1994.

FIGURE 23.11 We rely on ecosystem services to provide resources; they also regulate our environment and support essential biogeochemical processes that support life.
© Author's Image/PunchStock RF

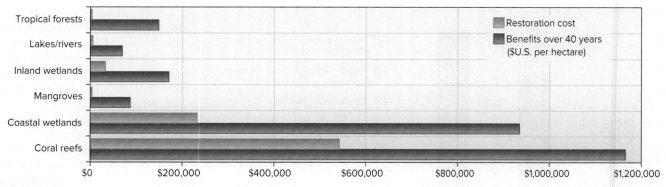

FIGURE 23.12 Can we afford to restore biodiversity? It's harder to find money to restore ecosystems than to destroy them, but the benefits over time greatly exceed the average cost of restoration.

Table 23.3	Estimated Annual Value of Ecological Services
Ecosystem Services	**Value (Trillion $U.S.)**
Soil formation	17.1
Recreation	3.0
Nutrient cycling	2.3
Water regulation and supply	2.3
Climate regulation (temperature and precipitation)	1.8
Habitat	1.4
Flood and storm protection	1.1
Food and raw materials production	0.8
Genetic resources	0.8
Atmospheric gas balance	0.7
Pollination	0.4
All other services	1.6
Total value of ecosystem services	**33.3**

Source: Nature, May 15, 1997, R. Costanza et al., "The Value of the World's Ecosystem Services and Natural Capital," vol. 387, pp 253–260.

Section Review

1. What is ecological about ecological economics?
2. What are ecosystem services? Give several examples.
3. List several externalized costs of power production with coal.
4. What is TEEB, and what are its aims?

23.3 POPULATION, SCARCITY, AND TECHNOLOGY

- *The question of how to avoid depleting finite resources, such as oil, has long worried ecologists and economists.*
- *The tragedy of the commons is one way to understand scarce resources; communal resource management is another.*
- *There is disagreement on whether resources are fixed.*

Despite changing perspectives on resources, many analysts continue to ask the question that worried Adam Smith and Thomas Malthus (chapter 1): Are we about to run out of essential natural resources? It stands to reason that if we consume a fixed supply of nonrenewable resources at a constant rate, eventually we'll use up all the economically recoverable reserves. The dismal Malthusian prospects of diminishing resources, starvation, and social decay inspire many people to call for immediate changes in our consumption rates. Others respond by saying that we simply need to redefine how we think about resources. In this section, we consider some of the ways we might understand resource limits and growth.

Does scarcity lead to new technologies?

One of the most widely known predictions of resource depletion is shown in the **Hubbert curve,** devised in 1956 by a Shell Oil geologist, M. King Hubbert (fig. 23.13). This graph has fairly

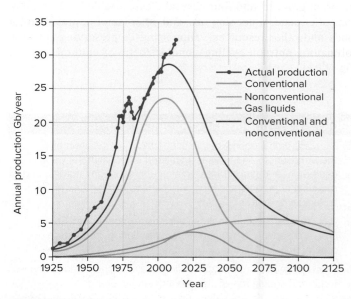

FIGURE 23.13 Hubbert's curve describes petroleum production. Dots show actual production. The area under the curve represents the amount of economically recoverable oil (Gb = billion barrels).
Source: Jean Laherrère, www.hubbertpeak.org

accurately described the peak and decline of U.S. oil supplies. Many energy economists argue that it also predicts our energy future, at least in general shape.

Other economists contend that resource supplies and demand curves can vary for reasons other than extent or location of physical stocks. Transitions to new technologies, such as wind or solar energy, could allow us to reduce our dependence on traditional fuels before supplies are entirely depleted. A common saying is that the stone age didn't end because we ran out of stones. And the fossil fuel age may not end because we're out of fossil fuels, but because we choose to not use them all.

Recently, the challenge of global climate change has introduced a new argument supporting the technology-transition view of the Hubbert curve. Most of the trillions of dollars on the books of major oil and gas companies represents resources still in the ground. But as the international community begins to take impending climate change seriously, and as alternative energy sources become more available, it will become more obvious that the vast wealth represented by untapped fossil fuels simply cannot be extracted and burned. Some economists argue that if it is unburnable, because of international climate accords and new technologies, all that oil wealth will become as inaccessible as if it were on the moon, and thus its value—and the value of the companies and the shares it supports—will crumble.

Common property resources are a classic problem in ecological economics

In 1968, biologist Garret Hardin wrote an influential article entitled "The Tragedy of the Commons." He argued that any commonly held resource inevitably is degraded or destroyed because the narrow self-interest of individuals tends to outweigh public interest. Hardin offered as a metaphor the common woodlands and pastures held by most colonial New England villages. In deciding how many cattle to put on the commons, Hardin explained, each villager would attempt to maximize his or her own personal gain. Adding one more cow to the commons could mean a substantially increased income for an individual farmer. The damage done by overgrazing, however, would be shared among all the farmers (fig. 23.14). This is known as the "free rider" problem, where individuals take more than their fair share of commonly held resources. Hardin concluded that the only solutions would be either to give coercive power to the government or to privatize the resource.

Hardin intended this parable to warn about human overpopulation and overexploitation of resources. Other authors have used his metaphor to explain such diverse problems as famines, air pollution, or collapsing fisheries.

What Hardin was really describing, however, was an **open-access system** in which there are no rules to manage resource use. In fact, many communal resources have been successfully managed for centuries by cooperative arrangements among users. Some examples include Native American management of wild rice beds and hunting grounds; Swiss village-owned mountain forests and pastures; Maine lobster fisheries; communal irrigation

FIGURE 23.14 Adding more cattle to the Brazilian Cerrado (savanna) increases profits for individual ranchers, but is bad for biodiversity and environmental quality.
© William P. Cunningham

systems in Spain, Bali, and Laos; and nearshore fisheries almost everywhere in the world.

A large body of literature has developed to describe how **common property management** works. Features of these systems, also known as communal resource management systems, include the following: (1) community members have lived on the land or used the resource for a long time and anticipate that their children and grandchildren will as well, thus giving them a strong interest in sustaining the resource and maintaining bonds with their neighbors; (2) the resource has clearly defined boundaries; (3) the community group size is known and enforced; (4) the resource is relatively scarce and highly variable so that the community is forced to be interdependent; (5) management strategies appropriate for local conditions have evolved over time and are collectively enforced; that is, those affected by the rules have a say in them; (6) the resource and its use are actively monitored, discouraging anyone from cheating or taking too much; (7) conflict resolution mechanisms reduce discord; and (8) incentives encourage compliance with rules, while sanctions for noncompliance keep community members in line.

In some cases, privatization leads to degradation of common pool resources. Where small villages have owned and managed jointly held forests or fishing grounds for generations, privatization has led to shortsighted decision making, leading to rapid destruction of both society and ecosystems. A tragic example of this was the forced privatization of Indian reservations in the United States. Where communal systems once enforced restraint over harvesting, privatization encouraged narrow self-interest. With individuals making decisions for personal, near-term benefit, many people chose to sell their resources to outsiders, who could easily take advantage of the weakest members of the community. Failing to recognize or value local knowledge and forcing local people to participate in a market economy allowed outsiders to disenfranchise native people and resulted in disastrous resource exploitation. Distinguishing between open-access systems and communal property regimes is important in understanding how best to manage natural resources.

Scarcity can lead to innovation

In a pioneer or frontier economy, methods for harvesting resources and turning them into useful goods and services tend to be inefficient and wasteful. The history of logging in the United States, for example, is a classic case of inefficient resource exploitation. Between about 1860 and 1930, the supply of American forests was vast and unregulated. Logging companies cleared eastern states, then swarmed across the Great Lakes forests. As prime timber in the northern forests was depleted, the companies simply shifted to the Rocky Mountains and the Pacific Northwest. At each stage, logging wasted a vast amount of wood, but this inefficiency seemed unimportant because the supply of trees was so great. Labor, financial capital, and transportation to market were the scarce resources.

Today our use of forest products is considerably more efficient. We have smaller supplies and greater demand, and we have developed better technology and methods that allow us to create the same amount of goods and services using fewer resources. Instead of using giant old-growth timbers for building, we use laminated beams, chipboard, and other products that can be produced from what once was scrap wood. We also require that logging companies replant new forests after logging, because they can no longer simply move on to a new region.

Scarcity often is a catalyst for innovation and change (fig. 23.15). As materials become more expensive and difficult to obtain, it becomes cost-effective to discover new supplies or to use available ones more efficiently. Several important factors play a role in this cycle of technological development:

- Technical inventions can increase the efficiency of extraction, processing, use, and recovery of materials.
- Substitution of new materials or commodities for scarce ones can extend existing supplies or create new ones.
- Trade makes remote supplies of resources available and may also bring unintended benefits in information exchange and cultural awakening.
- Discovery of new reserves through better exploration techniques, more investment, and looking in new areas becomes rewarding as supplies become limited and prices rise.
- Recycling becomes feasible and accepted as resources become more valuable. Recycling now provides about 37 percent of the iron and lead, 20 percent of the copper, 10 percent of the aluminum, and 60 percent of the antimony consumed each year in the United States.

Increasing technological efficiency can dramatically shift supply-and-demand relationships. As technology makes goods and services cheaper to produce, the quantity available at a given price can increase greatly. The market equilibrium, or the point at which supply and demand equilibrate, will shift to lower prices and higher quantities as a market matures (fig. 23.16).

Carrying capacity is not necessarily fixed

Despite repeated warnings that rapidly growing populations and increasing affluence are bound to exhaust natural resources and result in rapid price increases, technological developments have resulted in price decreases for most raw materials over the last

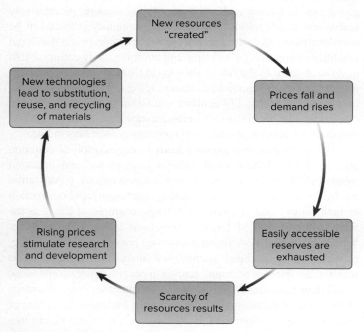

FIGURE 23.15 Scarcity/development cycle. Paradoxically, resource use and depletion of reserves can stimulate research and development, the substitution of new materials, and the effective creation of new resources.

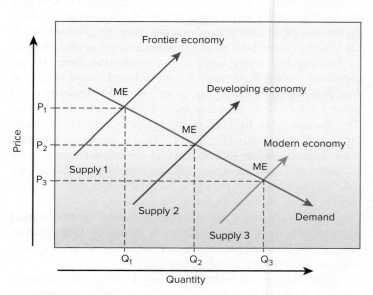

FIGURE 23.16 Supply and demand curves at three different stages of economic development. At each stage, there is a market equilibrium point at which supply and demand are in balance. As the economy becomes more efficient, the equilibrium shifts so there is a larger quantity available at a lower price than before. (P = price, Q = quantity, ME = market equilibrium)

hundred years. Consider copper, for example. Twenty years ago, worries about impending shortages led the United States to buy copper and store it in strategic stockpiles. Estimated demand for this important metal, essential for electric motors, telephone lines, transoceanic cables, and other uses, far exceeded known reserves. It looked as if severe shortages and astronomical price increases were inevitable. But then aluminum power lines, satellites, fiber optic cables, integrated circuits, microwave transmission, and other inventions greatly diminished the need for copper. Although prices are highly variable, the general trend for most materials was downward in the twentieth century.

Recent reports have warned that increasing demand for consumer goods and infrastructure in China will raise demand, and prices, to previously unimagined levels. Will this increase lead to shortages and crisis? Or will it lead to innovation and resource substitution? Economists generally believe that substitutability and technological development will help us avoid catastrophe. Ecologists generally argue that there are bound to be limits to how much we can consume.

A noted example of this debate occurred in 1980. Ecologist Paul Ehrlich bet economist Julian Simon that increasing human populations and growing levels of material consumption would inevitably lead to price increases for natural resources. They chose a package of five metals—chrome, copper, nickel, tin, and tungsten—priced at the time at $1,000. If, in ten years, the combined price (corrected for inflation) was higher than $1,000, Simon would pay the difference. If the combined price had fallen, Ehrlich would pay. In 1990, Ehrlich sent Simon a check for $576.07; the price for these five metals had fallen 47.6 percent.

Does this prove that resource abundance will continue indefinitely? Hardly. Ehrlich argued that the timing and set of commodities chosen simply were the wrong ones. The fact that we haven't yet run out of raw materials doesn't mean that it will never happen.

Many ecological economists now believe that some nonmarket resources such as ecological processes may be more irreplaceable than tangible commodities like metals. What do you think? Are we approaching limits to consumption? Which resources, if any, do you think are most likely to be limiting in the future?

Economic models compare growth scenarios

In the early 1970s, an influential study of resource limitations was funded by the Club of Rome, an organization of wealthy business owners and influential politicians. The study was carried out by a team of scientists from the Massachusetts Institute of Technology headed by the late Donnela Meadows. The results of this study were published in the 1972 book *Limits to Growth.* A complex computer model of the world economy was used to examine various scenarios of different resource depletion rates, growing population, pollution, and industrial output.

Given the Malthusian assumptions built into this model, catastrophic social and environmental collapse seemed inescapable (fig. 23.17*a*). Food supplies and industrial output rise as population grows and resources are consumed. Once past the carrying capacity of the environment, however, a crash occurs as population, food production, and industrial output all decline precipitously. Pollution continues to grow as society decays and people die, but, eventually, it also falls. Notice the similarity between this set of curves and the "boom and bust" population cycles described in chapter 6.

Many economists criticized these results because the models discount technological development and factors that might mitigate the effects of scarcity. In 1992, the Meadows group published updated computer models in *Beyond the Limits* that include technological progress, pollution abatement, population stabilization, and new public policies that work for a sustainable future. If we adopt these changes sooner rather than later, all factors in the model stabilize sometime in this century at an improved standard of living for everyone (fig. 23.17*b*). Of course, neither of these computer models shows what will happen, only what some possible outcomes *might* be, depending on the choices we make.

Section Review

1. Describe Hubbert's curve and its prediction about oil resources.

2. Explain the tragedy of the commons.

3. What is common property management?

4. How do the two "limits to growth" graphs (fig. 23.17) differ and why?

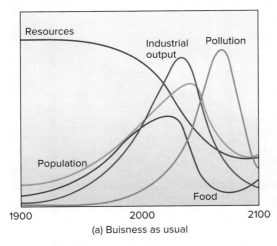

(a) Buisness as usual

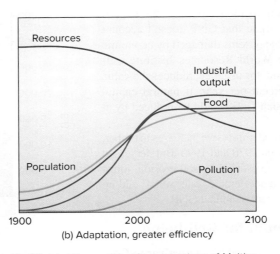

(b) Adaptation, greater efficiency

FIGURE 23.17 Models of resource consumption and scarcity. Running the model with assumptions of Malthusian limits and high consumption causes food, productivity, and populations to crash, while pollution increases (left). Running the same model with assumptions of slowing population growth and consumption, with better technologies, produces stable output and population (right). Which of these models will we follow?

- *GNP and GDP are common, though imperfect, measures of productivity.*
- *GPI, EPI, HDI, and WHI are some alternative measures of progress.*
- *Cost–benefit analysis is used to assess policies, although assigning costs accurately is difficult.*

How do we monitor our economic progress? In order to know if conditions in general are getting better or worse, economists have developed indices that countries or regions can use to monitor change over time. These indices track a variety of activities and values, to produce an overall picture of the economy. Which factors we choose to monitor, though, reflect judgments about what is important in society, and those judgments can vary substantially.

GNP is our dominant growth measure

The most common way to measure a nation's output is **gross national product (GNP).** GNP can be calculated in two ways. One is the money flow from households to businesses in the form of goods and services purchased. The other is to add up all the costs of production in the form of wages, rent, interest, taxes, and profit. In either case, a subtraction is made for capital depreciation, the wear and tear on machines, vehicles, and buildings used in production. Some economists prefer **gross domestic product (GDP),** which includes only the economic activity within national boundaries. Thus, the vehicles made and sold by Ford in Europe don't count in American GDP.

Both GNP and GDP have been criticized as measures of real progress or well-being because they don't attempt to distinguish between beneficial activities and harmful activities. A huge oil spill that pollutes beaches and kills wildlife, for example, shows up as a positive addition to GNP because it generates economic activity in the costs of cleanup. Or if many people die from cancer or AIDS, there is a great deal of economic activity for doctors and undertakers.

Ecological economists also argue that GNP doesn't account for natural resources used up or ecosystems damaged by economic activities. Robert Repeto of the World Resources Institute estimates that soil erosion in Indonesia, for instance, reduces the value of crop production about 40 percent per year. If natural capital were taken into account, Indonesian GNP would be reduced by at least 20 percent annually.

Similarly, Costa Rica experienced impressive increases in timber, beef, and banana production between 1970 and 1990. But decreased natural capital during this period represented by soil erosion, forest destruction, biodiversity losses, and accelerated water runoff add up to at least $4 billion or about 25 percent of annual GNP.

Alternate measures account for well-being

A number of systems have been proposed as alternatives to GNP that reflect genuine progress and social welfare. In their 1989 book, Herman Daly and John Cobb proposed a genuine progress index (GPI) that takes into account real per capita income,

quality of life, distributional equity, natural resource depletion, environmental damage, and the value of unpaid labor. They point out that while per capita GDP in the United States nearly doubled between 1970 and 2000, per capita GPI increased only 4 percent (fig. 23.18). Some social service organizations would add to this index the costs of social breakdown and crime, which would decrease real progress even further over this time span.

A newer measure is the **Environmental performance index (EPI)** created by researchers at Yale and Columbia Universities to evaluate national sustainability and progress toward achievement of the United Nations Millennium Development Goals. The EPI is based on 16 indicators tracked in six categories: environmental health, air quality, water resources, productive natural resources, biodiversity and habitat, and sustainable energy. The top-ranked countries—New Zealand, Sweden, Finland, the Czech Republic, and the United Kingdom—all commit significant resources and effort to environmental protection. In 2006, the United States ranked 28th in the EPI, or lower than Malaysia, Costa Rica, Columbia, and Chile, all of which have a GDP that is between 6 and 15 times lower than the United States.

The United Nations Development Programme (UNDP) uses a benchmark called the **human development index (HDI)** to track social progress. HDI incorporates life expectancy, educational attainment, and standard of living as critical measures of development. Gender issues are accounted for in the gender development index (GDI), which is simply HDI adjusted or discounted for inequality or achievement between men and women.

Although poverty remains widespread in many places, encouraging news also can be found in development statistics. Poverty has fallen more in the past 50 years, the UNDP reports, than in the previous 500 years. Child death rates in developing countries as a whole have been more than halved. Average life expectancy has increased by 30 percent while malnutrition rates have declined

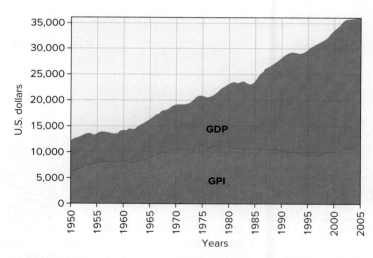

FIGURE 23.18 Although per capita GDP in the United States nearly doubled between 1970 and 2000 in inflation-adjusted dollars, a genuine progress index that takes into account natural resource depletion, environmental damage, and options for future generations hardly increased at all.

by almost a third. The proportion of children who lack primary school has fallen from more than half to less than a quarter. And the share of rural families without access to safe water has fallen from nine-tenths to about one-quarter.

Some of the greatest progress has been made in Asia. China and a dozen other countries, with populations that add up to more than 1.6 billion, have decreased the proportion of their people living below the poverty line by half. Still, in the 1990s, the number of people with incomes less than (U.S.) $1 per day increased by almost 100 million to 1.3 billion—and the number appears to be growing in every region except Southeast Asia and the Pacific. Even in industrial countries, more than 100 million people live below the poverty line and 37 million are chronically unemployed.

The **World Happiness Index (WHI)** is a relatively new measure of human well-being. The concept was first introduced by the King of Bhutan in 1972 as an alternative to GNP. His Gross National Happiness index looks at satisfaction rather than just economic progress. In 2011, the United Nations borrowed this idea to evaluate progress. Every year, the Sustainable Development Division of the UN publishes the WHI, a survey of global contentment and well-being, as a measure of achievement.

In all of the first five WHI reports (2012–2016), Denmark ranked as the happiest nation in the world. Half the top ten countries were in Scandinavia. The index is calculated from six factors that are found to explain most difference among countries in life satisfaction: (1) a comfortable per capita income compared to the cost of living, (2) long life expectancy and excellent health care, (3) lack of corruption in society, (4) strong social networks, (5) freedom to make life choices, and (6) a culture of generosity. In addition, Denmark has several other characteristics that contribute to contentment. The country supports parents with generous family leaves, free or low-cost childcare and early childhood education, universal health care, gender equity, a healthy lifestyle, a generally positive outlook, and dignity with meaningful activities for all ages. Cultural heritage and religion also can create a sense of purpose and contribute to happiness in many societies (fig 23.19).

Many countries in recent years have achieved economic growth at the cost of sharply rising inequality, entrenched social exclusion, and grave damage to the natural environment. Paying attention to economic, social, and environmental objectives advances happiness and general satisfaction for both present and future generations. These are all features of the sustainable development goals discussed in chapter 1. By contrast, the Economic Freedom Index, proposed by libertarians, shows no correlation with widespread human satisfaction.

Cost–benefit analysis aims to optimize benefits

One way to evaluate public projects is to analyze the costs and benefits they generate in a **cost–benefit analysis (CBA).** This process attempts to assign values to resources, as well as to the social and environmental effects of carrying out or not carrying out a given undertaking. It tries to find the optimal efficiency point at which the marginal cost of pollution control equals the marginal benefits (fig. 23.20).

FIGURE 23.19 Preserving cultural heritage and strong social networks contributes to national happiness.
© William P. Cunningham

CBA is one of the main conceptual frameworks of resource economics and is used by decision makers around the world as a way of justifying the building of dams, roads, and airports, as well as in considering what to do about biodiversity loss, air pollution, and global climate change. Deeply entrenched in bureaucratic practice and administrative culture, this technique has become much more widespread in American public affairs since the Reagan administration's executive orders in the 1980s calling for the application of CBA to all regulatory decisions and legislative proposals. Many conservatives see CBA as a way of eliminating what they consider to be unnecessary and burdensome requirements to protect clean air, clear water, human health, or biodiversity. They would like to add a requirement that all regulations be shown to be cost-effective.

The first step in CBA is to identify who or what might be affected by a particular plan. What are the potential outcomes and results? What alternative actions might be considered? After identifying and quantifying all effects, an attempt is made to assign monetary costs and benefits to each one. Usually, the direct

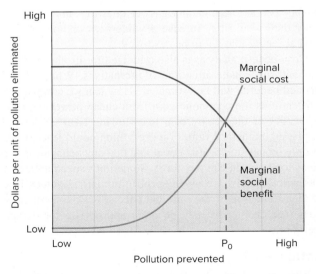

FIGURE 23.20 To achieve maximum economic efficiency, regulations should require pollution prevention up to the optimum point (P_0) at which the costs of eliminating pollution just equal the social benefits of doing so.

FIGURE 23.21 What is the value of solitude or beauty? How would you assign costs and benefits to a scene such as this?
© William P. Cunningham

expenses of a project are easy to ascertain. How much will you have to pay for land, materials, and labor? The monetary value of lost opportunities—to swim or fish in a river, or to see birds in a forest—is much harder to appraise (fig. 23.21). How would you put a price on good health or a long life? It's also important to ask who will bear the costs and who will reap the benefits of any proposal. Are there external costs that should be accounted for? Eventually, the decision maker compares all the costs and benefits to see whether the project is justified or whether some alternative action might bring more benefits at less cost.

Because of the difficulty of assigning monetary prices to intangible or public resources we value, many people object to CBA. In analyzing the costs and benefits of a hydroelectric dam, for example, economists often cannot assign suitable values to land, forests, streams, fisheries and livelihoods, and community. Ordinary people often cannot answer questions about how much money they would pay to save a wilderness or how much they would accept to allow it to be destroyed.

Cost–benefit analysis is also criticized for its absence of standards and inadequate attention to alternatives. Who judges how costs and benefits will be estimated? How can we compare things as different as the economic gain from cheap power with loss of biodiversity or the beauty of a free-flowing river? Critics claim that placing monetary values on everything could lead to a belief that only money and profits count and that any behavior is acceptable as long as you can pay for it. Sometimes speculative or even hypothetical results are given specific numerical values in CBA and then treated as if they are hard facts. Risk-assessment techniques (see chapter 8) may be more appropriate for comparing uncertainties.

Section Review

1. What are GNP and GPI? How do the two differ?
2. List several factors that contribute to the WHI.
3. What is the purpose of cost–benefit analysis?

23.5 CAN MARKETS REDUCE POLLUTION?

- *Setting a price for emissions can create incentives for reducing them.*
- *Some pollutants are easier to price effectively than others, and setting the right price can be difficult.*

We are becoming increasingly aware that our environment and economy are mutually interconnected. Natural resources and ecological services are essential for a healthy economy, and a vigorous economy can provide the means to solve environmental problems. In this section, we'll explore some of these links.

Most scientists regard global climate change as the most serious environmental problem we face. In 2006, the business world got a harsh warning about this problem from British economist, Sir Nicolas Stern. Commissioned by the British treasury department to assess the threat of global warming, Sir Nicolas, who formerly was chief economist at the World Bank, issued a 700-page study that concluded that if we don't act to control greenhouse gases, the damage caused by climate change could be equivalent to losing as much as 20 percent of the global GDP every year. This could have an impact on our lives and environment greater than the worldwide Depression or the great wars of the twentieth century.

The cost of climate change will be far greater than steps we could take now to reduce climate change. Stern calculates that it will take about $500 billion per year (1 percent of global GDP) to avoid the worst impacts of climate change if we act now. That is a lot of money, but it's a bargain compared to his estimates of $10 trillion in annual losses and costs of climate change in 50 years if we don't change our practices. And the longer we wait, the more expensive carbon reduction and adaptation are going to be.

On the other hand, reducing greenhouse gas emissions and adapting to climate change will create significant business opportunities as new markets are created in low-carbon energy technologies and services (fig. 23.22). These markets could create millions of jobs and be worth hundreds of billions of dollars every year. Already, Europe has more than 5 million jobs in renewable energy, and the annual savings from solar, wind, and hydro power are saving the European Union about $10 billion per year in avoided oil and natural gas imports. Being leaders in the fields of renewable energy and carbon reduction gives pioneering countries a tremendous business advantage in the global marketplace. Markets for low-carbon energy could be worth $500 billion per year by 2050, according to the Stern report.

What's the best way to regulate pollutants such as greenhouse gases? Economists sometimes describe authoritarian societies as having **command and control economies.** In a communist country, for example, where the government controls both the means of production and price for consumers, it can just say "stop doing that." In 2015, for example, China cut its coal consumption by more than 30 percent. No capitalist economy could accomplish that.

But individualistic societies aren't willing to give up that much personal freedom. Conservatives generally prefer to let **market forces** determine where, when, and by whom pollution reductions should be made. This approach is based on a belief that competition in a free market will always find the most efficient balance between costs and benefits.

FIGURE 23.22 Markets for low-carbon energy could be worth $500 billion per year by 2050, and could create millions of high-paying jobs. © jorgenjacobsen/E+/Getty Images

Sulfur trading offers a good model

The 1990 U.S. Clean Air Act created one of the first market-based systems for reducing air pollution. It mandated a decrease in acid-rain-causing sulfur dioxide (SO_2) from power plants and other industrial facilities. An SO_2 targeted reduction was set at 10 million tons per year, leaving it to industry to find the most efficient way to do this. The government expected that meeting this goal would cost companies up to $15 billion per year, but the actual cost has been less than one-tenth of that. Prices on the sulfur exchange have varied from $60 to $800 per ton depending on the availability and price of new technology, but most observers agree that the market has found much more cost-effective ways to achieve the desired goal than rigid rules would have required.

This program is regarded as a shining example of the benefits of market-based approaches. There are complaints, however, that while nationwide emissions have come down, "hot spots" remain where local utilities have paid for credits rather than install pollution abatement equipment. If you're living in one of these hot spots and continuing to breathe polluted air, it's not much comfort to know that nationwide average air quality has improved. Currently, credits and allowances of more than 30 different air pollutants are traded in international markets.

Is emissions trading the answer?

The Kyoto Protocol, which was negotiated in 1997, and has been ratified by every industrialized nation in the world except the United States and Australia (oil-rich Canada also withdrew in 2012), sets up a mechanism called **emissions trading** to control greenhouse gases. This is also called a **cap-and-trade** approach. The first step is to mandate upper limits (the cap) on how much each country, sector, or specific industry is allowed to emit.

Companies that can cut pollution by more than they're required to can sell the **carbon credits** to other companies that have more difficulty meeting their mandated levels.

Suppose you've just built a state-of-the-art power plant that allows you to capture and store CO_2 for about $20 per ton, and that allows you to cut your CO_2 emissions far below the amount you are permitted to produce. Suppose, further, that your neighboring utility has a dirty, old coal-fired power plant for which it would cost $60 per ton to reduce CO_2 emissions. You might strike a deal with your neighbor. You reduce your CO_2 emissions, and he pays you $40 for each ton you reduce, so he doesn't have to reduce. You make $20 per ton, and your neighbor saves $20 per ton. Both of you benefit. On the other hand, if your neighbor can find an even cheaper way to offset his carbon emissions (that is, to pay for someone else to reduce emissions), he's free to do so. This creates an incentive to continually search for ever more cost-effective ways to reduce emissions.

Opportunities are increasing for all of us to buy personal carbon offsets. When you buy an airplane ticket, for example, some airlines offer you the chance to pay a few extra dollars, which will be used to pay for projects to reduce greenhouse gas emissions. You can also buy carbon offsets if you have an old, inefficient car: For about $20 per ton (or about $100 per year for the average American car), you can pay for someone elsewhere to plant trees, build a windmill, or provide solar lights to a village in a developing country to compensate for your emissions. You can take pride in being carbon-neutral at a far lower price than buying a new automobile.

Already about 20 countries have some sort of emissions trading markets. Most of the European Union is covered by these agreements, and other countries, such as the United States and Canada, are considering joining them. In 2010, about US$120 billion worth of climate credits, equivalent to 5 billion tons of CO_2, traded hands on international markets. Since then, unfortunately, the carbon market has been in free-fall. Prices have dropped from roughly US$25 per ton to around US$5 per ton in the European Union in 2016. The most disappointing collapse in the carbon market is in China, which economists thought might be the key for control of greenhouse gas emissions. In 2016, carbon credits were selling for only the equivalent of US$1.37 on the Shanghai market. Economists consider about US$30 per ton to be the threshold to trigger a shift to renewable energy. And the real social cost of carbon emissions (taking all the damage from climate change into account) is probably well over US$100 per ton (chapter 15). This sort of weakness in carbon markets occurs when regulators issue too many pollution permits, so that supply exceeds demand. Economic downturns, like that after 2008, and the rapid decrease in renewable energy prices, also help a carbon market collapse.

In the early stages of emissions marketing, more than 80 percent of the international emissions payments went to just four countries, and nearly two-thirds of those payments were for reductions of the refrigerant HFC-23 (fig. 23.23). Critics of our current emissions markets point out that (so far) carbon trading has produced a great deal of wealth for banks and stock brokers, but has

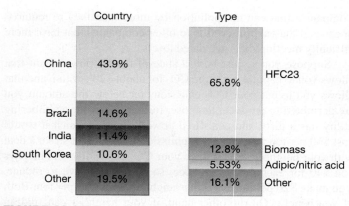

FIGURE 23.23 Distribution of payments for emissions reduction, deriving from emissions trading markets. Four countries collect 80 percent of payments, and two-thirds of payments reward the incineration of relatively cheap HFC23, making this a bonanza for some corporations.

done little to reduce actual carbon emissions or help reduce inequity. Rather than finance renewable energy in developing countries, marketing emission credits is primarily benefiting bankers, consultants, and factory owners so far and is leading to short-term fixes rather than fundamental, long-term solutions.

Are carbon taxes a better answer?

A criticism of cap-and-trade is that it has no mechanism to evaluate or price external costs of complex pollutants, such as greenhouse gases. Corporations calculate how much it costs to capture pollutants, not how much it may cost future generations to live in a hotter, more unstable world.

As the opening case study for this chapter shows, a carbon tax can be an effective way to achieve the goals of reducing GHG emissions, as well as financing other social programs. The BC carbon tax covers a wide base. It started low to ease the transition, and rose to a more substantive level, roughly in line with estimates of the damage per ton or the "social cost of carbon." Rather than have the profits from pollution reduction go into corporate coffers, as they do in emissions trading, the tax revenue can be used to promote social goals or to finance clean industries. The revenues in British Columbia, for example, are used to lower income taxes on businesses and households, as well as to benefit disadvantaged low-income households and residents of remote, rural areas. Reporting of the sources and uses of carbon tax funds is subject to a highly transparent process, in which politicians and the public can track how the revenues are used.

Sweden also has used a carbon tax to reduce greenhouse gas emissions since 1991. Although Sweden has employed a suite of other policies to reduce emissions, its environment ministry estimates that the carbon tax has led to cuts in emissions of 20 to 25 percent above cuts due to regulations. Sweden's carbon emissions have decreased more than seven percent since 1990, while its economy has grown to be one of the most robust in the world. Its carbon tax has been credited with spurring innovation and development in the field of low-carbon energy technologies such as

green-heating technologies that have led to significant reductions in the use of oil for home heating. Sweden's carbon tax has also been credited in part for putting the country on target to achieve and honor its commitment under the Kyoto Protocol.

Section Review

1. How did the cap-and-trade approach work to control sulfur?
2. List some arguments for and against carbon trading.
3. Why do some economists believe a carbon tax is a better policy?

23.6 GREEN DEVELOPMENT AND BUSINESS

- *Trade increases the value of local products, but producers often get only a small share of that value.*
- *The WTO, IMF, World Bank, and NAFTA are some of the many organizations promoting large-scale development and trade.*
- *Green business models emulate ecological ideas of efficiency, diversity, and life-cycle thinking.*

Trade can be a powerful tool in making resources available and raising standards of living. Think of the things you now enjoy that might not be available if you had to live exclusively on the resources available in your immediate neighborhood. Too often, the poorest, least powerful people suffer in this global marketplace. To balance out these inequities, nations can deliberately invest in economic development projects. In this section, we'll look at some aspects of trade, development, business, and jobs that have impacts on our environment and welfare.

International trade brings benefits but also intensifies inequities

The banking and trading systems that regulate credit, currency exchange, shipping rates, and commodity prices were set up by the richer and more powerful nations in their own self-interest. The General Agreement on Tariffs and Trade (GATT) and World Trade Organization (WTO) agreements, for example, negotiated primarily between the largest industrial nations, regulate 90 percent of all international trade.

These systems tend to keep the less-developed countries in a perpetual role of resource suppliers to the more-developed countries. The producers of raw materials, such as mineral ores or agricultural products, get very little of the income generated from international trade (fig. 23.24).

Policies of the WTO and the IMF have provoked criticism and resistance in many countries. As a prerequisite for international development loans, the IMF frequently requires debtor nations to adopt harsh "structural adjustment" plans that slash welfare programs and impose cruel hardships on poor people. The WTO has issued numerous rulings that favor international trade over pollution prevention or protection of endangered species. Trade conventions such as the North American Free Trade Agreement (NAFTA) have been accused of encouraging a "race to the bottom" in which companies can play one country against another and

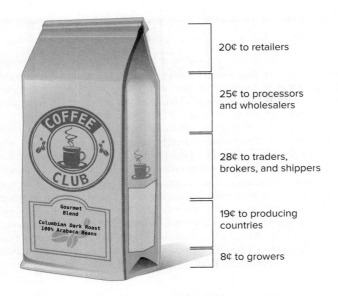

20¢ to retailers

25¢ to processors
and wholesalers

28¢ to traders,
brokers, and shippers

19¢ to producing
countries

8¢ to growers

FIGURE 23.24 What do we really pay for when we purchase a dollar's worth of coffee?

move across borders to find the most lax labor and environmental protection standards.

No single institution has more influence on financing and policies of developing countries than the World Bank. Of some $25 billion loaned each year for development projects by international agencies, about two-thirds comes from the World Bank. Founded in 1945 to fund the reconstruction of Europe and Japan, the World Bank shifted its emphasis to aid developing countries in the 1950s. Many of its projects have had adverse environmental and social effects, however. Its loans often go to corrupt governments and fund ventures such as nuclear power plants, huge dams, and giant water diversion schemes. Former U.S. treasury secretary Paul O'Neill said that these loans have driven poor countries "into a ditch" by loading them with unpayable debt. He said that funds should not be loans, but rather grants to fight poverty.

Microlending helps the poorest of the poor

Global aid from the WTO usually aids banks and industries more than it helps the impoverished populations who most need assistance. Often, structural adjustment leads the poorest to pay back loans negotiated by their governments and industries. These concerns led Dr. Muhammad Yunus of Bangladesh to initiate the micro-loan plan of the Grameen Bank (see the What Do You Think? section below titled "Loans that Change Lives").

One of the most important innovations of the Grameen Bank is that borrowers take out loans in small groups. Everyone in the group is responsible for each other's performance. The group not only guarantees loan repayment, it helps businesses succeed by offering support, encouragement, and advice. Where banks depend on the threat of foreclosure and a low credit rating to ensure debt repayment, the Grameen Bank has something at least as powerful

for poor villagers—the threat of letting down your neighbors and relatives. Even in the United States, organizations assist micro-enterprises with loans, grants, and training. The Women's Self-Employment Project in Chicago, for instance, teaches job skills to single mothers in housing projects. Similarly, "tribal circle" banks on Native American reservations successfully finance microscale economic development ventures.

The most recent venture for the Grameen Bank is providing mobile phone service to rural villages. Supplying mobile phones to poor women not only allows them to communicate, it provides another business opportunity. They rent out their phone to neighbors, giving the owner additional income, and linking the whole village to the outside world. Suddenly, people who had no access to communication can talk with their relatives, order supplies from the city, check on prices at the regional market, and decide when and where to sell their goods and services. This is a great example of "bottom-up development." Founded in 1996, Grameen Phone now has 2.5 million subscribers and is Bangladesh's largest mobile phone company.

Quantitative Reasoning

Suppose the World Bank has $25 billion to lend for development aid and that there are about 5 billion residents of developing countries. How would you distribute that aid? In what ways is that $25 billion more or less effective than the $71 million distributed by Kiva to 171,000 smaller entrepreneurs?

Green business involves efficiency and creative solutions

At home, as well as abroad, business leaders are increasingly discovering that they can save money and protect our environment by greening up their business practices. They can save money through fuel efficiency and reducing electricity consumption. These steps also cut greenhouse gases. Recycling waste and minimizing the use of hazardous materials saves on disposal costs. In addition, these companies win public praise and new customers by demonstrating an interest in our shared environment. By conserving resources, they also help ensure the long-term survival at their own corporations.

The green business movement has had some remarkable successes and presents an encouraging pathway for how we might achieve both environmental protection and social welfare. Some of the leaders in this new approach to business include Paul Hawken, William McDonough, Ray Anderson, Amory Lovins, David Crockett, and John and Nancy Todd.

These individuals and others have shown that operating in a socially responsible manner that is consistent with the principles of sustainable development and environmental protection can be good for employee morale, public relations, and the bottom line simultaneously. Environmentally conscious or "green" companies such as the Body Shop, Patagonia, Aveda, Malden Mills, Johnson and Johnson, and Interface, Inc., consistently earn high marks from

Loans that Save Lives

Ni Made is a young mother of two children who lives in a small Indonesian village. Her husband is a day laborer who makes only a few dollars per day—when he can find work. To supplement their income, Made goes to the village market every morning to sell a drink she makes out of boiled pandanus leaves, coconut milk, and pink tapioca. A small loan would allow her to rent a covered stall during the rainy season and to offer other foods for sale. The extra money she could make could change her life. But traditional banks consider Ni Made too risky to lend to and the amounts she needs too small to bother with.

Around the world, billions of poor people find themselves in the same position as Made—they're eager to work to build a better life for themselves and their families, but lack the resources to succeed. Now, however, a financial revolution is sweeping the world. Small loans are becoming available to the poorest of the poor. This new approach was invented by Dr. Muhammad Yunus, professor of rural economics at Chittagong University in Bangladesh. Talking to village women, Dr. Yunus learned that money-lenders consumed nearly all the profits of their small businesses.

In 1983, Dr. Yunus started the Grameen (village) Bank to show that "given the support of financial capital, however small, the poor are fully capable of improving their lives." His experiment has been tremendously successful. By 2009, the Grameen Bank had nearly 2 billion customers, 97 percent of them women. It had loaned more than $8 billion with 98 percent repayment, nearly twice the collection rate of commercial Bangladesh banks. In 2006, Dr. Yunus was awarded the Nobel Peace Prize for his work in microlending.

The Grameen Bank provides credit to poor people in rural Bangladesh without the need for collateral. It depends, instead, on mutual trust, accountability, participation, and the creativity of the borrowers themselves. Microcredit is now being offered by hundreds of organizations in 43 other countries. Would you like to be part of this movement? Well, now you can. You don't have to own a bank to help someone in need.

A brilliant way to connect entrepreneurs in developing countries with lenders in wealthy countries is offered by Kiva, a San Francisco–based technology startup. The idea for Kiva, which means "unity" or "cooperation" in Swahili, came from Matt and Jessica Flannery. Jessica had worked in East Africa with the Village Enterprise Fund, a California nonprofit that provides training, capital, and mentoring to small businesses in developing countries. Jessica and Matt wanted to help some of the people she had met, but they weren't wealthy enough to get into microfinancing on their own. Joining with four other young people with technology experience, they created Kiva, which uses the power of the Internet to help the poor.

Kiva partners with about two dozen nonprofits with staff in developing countries. The partners identify hardworking entrepreneurs who deserve help. They then post a photo and brief introduction to each one on the Kiva web page. You can browse the collection to find someone whose story touches you. The minimum loan is generally $25. Your loan is bundled with others until it reaches the amount needed by the borrower. You make your loan using your credit card (through PayPal, so it's safe and easy). The loan is generally repaid within 12 to 18 months (although without interest). At that point, you can either withdraw the money or use it to make another loan. It's easy to take part in this innovative human development project. Check out www.kiva.org for more information.

A small amount of seed money will allow Ni Made to expand her business and help provide for her family.
© William P. Cunningham

---

community and environmental groups. Conserving resources, reducing pollution, and treating employees and customers fairly may cost a little more initially but can save money and build a loyal following in the long run.

New business models adopt concepts of ecology

Paul Hawken's 1993 book, *The Ecology of Commerce,* was a seminal influence in convincing many people to reexamine the role of business and economics in environmental and social welfare. Basing his model for a new industrial revolution on the principles of ecology, Hawken points out that almost nothing is discarded or unused in nature. The wastes from one organism become the food of another. Industrial processes, he argues, should be designed on a similar principle (table 23.4). Rather than a linear pattern in which we try to maximize the throughput of material and minimize labor, products and processes should be designed to:

- *be energy efficient;*
- *use renewable materials;*
- *be durable and reusable or easily dismantled for repair and remanufacture, nonpolluting throughout their entire life cycle;*
- *provide meaningful and sustainable livelihoods for as many people as possible;*
- *protect biological and social diversity;*
- *use minimum and appropriate packaging made of reusable or recyclable materials.*

Table 23.4 Goals for an Eco-Efficient Economy

- Introduce no hazardous materials into the air, water, or soil.
- Measure prosperity by how much natural capital we can accrue in productive ways.
- Measure productivity by how many people are gainfully and meaningfully employed.
- Measure progress by how many buildings have no smokestacks or dangerous effluents.
- Make the thousands of complex governmental rules unnecessary that now regulate toxic or hazardous materials.
- Produce nothing that will require constant vigilance from future generations.
- Celebrate the abundance of biological and cultural diversity.
- Live on renewable solar income rather than fossil fuels.

We can do all this and at the same time increase profits, reduce taxes, shrink government, increase social spending, and restore our environment, Hawken claims. Recently, Hawken has served as chairperson for The Natural Step in America, a movement started in Sweden by Dr. K. H. Robert, a physician concerned about the increase in environmentally related cancers. Through a consensus process, a group of 50 leading scientists endorsed a description of the living systems on which our economy and lives depend. More than 60 major European corporations and 55 municipalities have incorporated sustainability principles (table 23.5) into their operations.

Another approach to corporate responsibility is called the **triple bottom line.** Rather than reporting only net profits as a measure of success, ethically sensitive corporations include environmental effects and social justice programs as indications of genuine progress.

Corporations committed to eco-efficiency and clean production include such big names as Monsanto, 3M, DuPont, Duracell, and Johnson & Johnson. Following the famous three Rs—reduce, reuse, recycle—these firms have saved money and gotten welcome publicity. Savings can be substantial. Slashing energy use and redesigning production to use less raw material and to produce less waste is reported to have saved DuPont $3 billion over the past decade, while also reducing its greenhouse emissions 72 percent.

Table 23.5 The Natural Step: System Conditions for Sustainability

1. Minerals and metals from the earth's crust must not systematically increase in nature.
2. Materials produced by human society must not systematically increase in nature.
3. The physical basis for biological productivity must not be systematically diminished.
4. The use of resources must be efficient and just with respect to meeting human needs.

Efficiency starts with product design

Our current manufacturing system often is incredibly wasteful. On average, for every truckload of products delivered in the United States, 32 truckloads of waste are produced along the way. The automobile is a typical example. Industrial ecologist, Amory Lovins, calculates that for every 100 gallons (380 liters) of gasoline burned in your car engine, only one percent (1 gal or 3.8 liters) actually moves passengers. All the rest is used to move the vehicle itself. The wastes produced—carbon dioxide, nitrogen oxides, unburned hydrocarbons, rubber dust, heat—are spread through the environment where they pollute air, water, and soil.

Architect William McDonough urges us to rethink design approaches (table 23.6). In the first place, he says, we should question whether the product is really needed. Could we provide the same service in a more eco-efficient manner? According to McDonough, products should be divided into three categories:

1. *Consumables* are products like food, natural fabrics or paper that can harmlessly go back to the soil as compost.

2. *Service products* are durables such as cars, TVs, and refrigerators. These products should be leased to the customer to provide their intended service, but would always belong to the manufacturer. Eventually, they would be returned to the maker, who would be responsible for recycling or remanufacturing the product. Knowing that they will have to dismantle the product at the end of its life will encourage manufacturers to design for easy disassembly and repair.

3. *Unmarketables* are compounds like radioactive isotopes, persistent toxins, and bioaccumulative chemicals. Ideally, no one would make or use these products. But because eliminating their use will take time, McDonough suggests that in the meantime these materials should belong to the manufacturer and be molecularly tagged with the maker's mark. If they are discovered to be discarded illegally, the manufacturer would be held liable.

Table 23.6 McDonough Design Principles

Inspired by the way living systems actually work, Bill McDonough offers three simple principles for redesigning processes and products:

1. *Waste equals food.* This principle encourages elimination of the concept of waste in industrial design. Every process should be designed so that the products themselves, as well as leftover chemicals, materials, and effluents, can become "food" for other processes.

2. *Rely on current solar income.* This principle has two benefits: First, it diminishes, and may eventually eliminate, our reliance on hydrocarbon fuels. Second, it means designing systems that sip energy rather than gulping it down.

3. *Respect diversity.* Evaluate every design for its impact on plant, animal, and human life. What effects do products and processes have on identity, independence, and integrity of humans and natural systems? Every project should respect the regional, cultural, and material uniqueness of its particular place.

Following these principles, McDonough Bungart Design Chemistry has created nontoxic, easily recyclable materials to use in buildings and for consumer goods. Among some important and innovative "green office" projects designed by the McDonough and Partners architectural firm are the Environmental Defense Fund headquarters in New York City, the Environmental Studies Center at Oberlin College in Ohio (see fig. 20.11), the European Headquarters for Nike in Hilversum, the Netherlands, and the Gap Corporate Offices in San Bruno, California (fig. 23.25). Intended to promote employee well-being and productivity, as well as eco-efficiency, the Gap building has high ceilings, abundant skylights, windows that open, a full-service fitness center (including a pool), and a landscaped atrium for each office bay that brings the outside in. The roof is covered with native grasses. Warm interior tones and natural wood surfaces (all wood used in the building was harvested by certified sustainable methods) give a friendly feeling. Paints, adhesives, and floor coverings are low toxicity and the building is one-third more energy efficient than strict California laws require. A pleasant place to work, the offices help recruit top employees and improve both effectiveness and retention. As for the bottom line, Gap, Inc. estimates that the increased energy and operational efficiency will have a four- to eight-year payback.

Green consumerism gives the public a voice

Consumer choice can play an important role in persuading businesses to produce eco-friendly goods and services (see the What Can You Do? section below titled "Personally Responsible

Economy"). Increasing interest in environmental and social sustainability has caused an explosive growth of green products. You can find eco-travel agencies, telephone companies that donate profits to environmental groups, entrepreneurs selling organic foods, shade-grown coffee, efficient houses, paint thinner made from orange peels, sandals made from recycled auto tires, earthworms for composting, sustainable clothing, shoes, rugs, balm, shampoo, and insect repellent. Although these eco-entrepreneurs represent a tiny sliver of the $7-trillion-per-year U.S. economy, they often serve as pioneers in developing new technologies and offering innovative services.

In some industries, eco-entrepreneurs have found profitable niches within a larger market. In other cases, once a consumer demand has built up, major companies add green products or services to their inventory. Natural foods, for instance, have grown from the domain of a few funky, local co-ops to a $7 billion market segment. Most supermarket chains now carry some organic

FIGURE 23.25 The award-winning Gap, Inc. corporate offices in San Bruno, California, demonstrate some of the best features of environmental design. A roof covered with native grasses provides insulation and reduces runoff. Natural lighting, an open design, and careful relation to its surroundings all make this a pleasant place to work.
© Proehl Studios/Corbis

What Can You Do?

Personally Responsible Economy

There are many things that each of us can do to lower our ecological impacts and support green businesses through responsible consumerism and ecological economics.

- Practice living simply. Ask yourself if you really need more material goods to make your life happy and fulfilled.
- Minimize consumption of resources, to save personal and global costs of electricity, gas, metals, plastics, and other resources. Recycle or reuse products, and avoid excessive packaging.
- Look at the amount of your garbage on trash day. Is that the amount of throughput you would like to produce?
- Support environmentally friendly businesses. Consider spending a little more for high-quality, fairly produced goods, at least some of the time. Ask companies what they are doing about environmental protection and human rights.
- Buy green products. Look for efficient, high-quality materials that will last and that are produced in the most environmentally friendly manner possible. Subscribe to clean-energy programs if they are available in your area.
- Think about the total life-cycle costs of the things you buy, including environmental impacts, service charges, energy use, and disposal costs, as well as initial purchase price.
- Invest in socially and environmentally responsible mutual funds or green businesses when you have money for investment.
- Try making a Kiva or similar micro-loan. You may find that it's fun and educational, and it can feel good to help others.
- Vote thoughtfully. Think carefully about the long-term versus short-term social and environmental impacts of economic policies, and work with others in your community to push elected representatives to act in ways that safeguard resources for the generations to come.

food choices. Similarly, natural-care health and beauty products are now more than 10 percent of a $33 billion industry. By supporting these products, you can ensure that they will continue to be available and, perhaps, even help expand their penetration into the market.

Walmart, the large-volume discount price chain, has established a name as the world's largest seller of organic products and energy-efficient lightbulbs, among other green products. The incentive for Walmart has been that green products and green production processes are often less wasteful, and thus cheaper (when produced in volume), than other products and processes.

Does this mean that Walmart has internalized all its costs and produced sustainable relationships with suppliers? Some critics say no, and that Walmart has diluted the idea of sustainable and organic production. But supporters point out that the chain has also helped legitimize these ideas for the public. Evidently, Walmart shoppers also are enthusiastic about contributing to environmental solutions while they shop, because they eagerly buy the organic and energy-efficient products offered.

Environmental protection creates jobs

For years, business leaders and politicians have portrayed environmental protection and jobs as mutually exclusive. Pollution control, protection of natural areas and endangered species, and limits on use of nonrenewable resources, they claim, will strangle the economy and throw people out of work. Ecological economists dispute this claim, however. Their studies show that only 0.1 percent of all large-scale layoffs in the United States in recent years were due to government regulations (fig. 23.26). Environmental protection, they argue, not only is necessary for a healthy economic system, it actually creates jobs and stimulates business.

Recycling, for instance, makes more new jobs than extracting virgin raw materials. This doesn't necessarily mean that recycled goods are more expensive than those from virgin resources. We're simply substituting labor in the recycling center for energy and huge machines used to extract new materials in remote places.

Japan, already a leader in efficiency and environmental technology, has recognized the multibillion-dollar economic potential of green business. The Japanese government is investing US$4 billion per year on research and development that targets seven areas, ranging from utilitarian projects such as biodegradable plastics and heat-pump refrigerants to exotic schemes such as carbon-dioxide-fixing algae and hydrogen-producing microbes.

Increasingly, people argue that the United States needs a new Apollo Project (like the one that sent men to the moon, but this time focusing on saving planet Earth) to develop renewable energy, break our dependence on fossil fuels, create green jobs, and reinvigorate the economy. The global recession of 2008–2009 strengthened this idea. In 2009, President Barack Obama signed an economic-recovery bill with at least $62 billion in direct spending on green initiatives and $20 billion in green tax incentives. Among the provisions in this bill are $19 billion for renewable energy and upgrading the

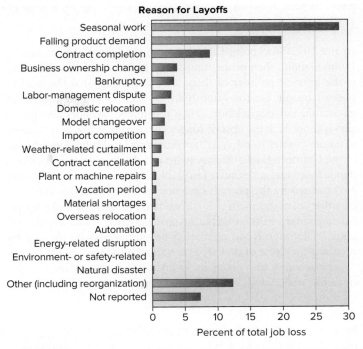

Reason for Layoffs

Percent of total job loss

FIGURE 23.26 Although opponents of environmental regulation often claim that protecting the environment costs jobs, studies by economist E. S. Goodstein show that only 0.1 percent of all large-scale layoffs in the United States were the result of environmental laws.
Source: E. S. Goodstein, Economic Policy Institute, Washington, D.C.

electrical transmission grid; $20 billion for energy conservation, including weatherizing building and providing efficient appliances; $17 billion for mass transit and advanced automobiles; and $500 million for green jobs programs. More than a million "green collar" jobs could result from these investments. Check out the Apollo Alliance for current news about a new green economy.

Economists report that the renewable energy sector already employs more than 2 million workers worldwide. If we were to get half our energy from sustainable sources, it would probably sustain nearly 10 million jobs. Even more people could be employed in energy conservation, ecosystem restoration, and climate remediation programs. Morgan Stanley, a global financial services firm, estimates that global sales from clean energy alone could grow to as much as $1 trillion per year by 2030. Already authors are rushing books to publication giving advice on how to make a fortune investing in green corporations and renewable energy technologies. For students contemplating career choices, clean energy and conservation could be good areas to explore.

Section Review

1. What is the definition and the purpose of NAFTA?
2. Why might microlending produce more widespread results than World Bank lending?
3. What does it mean that green business adopts ideas from ecology?
4. List several benefits and weaknesses of green consumerism.

Conclusion

At the 1972 Stockholm Conference on the Human Environment, Indira Gandhi, then prime minister of India, stated that, "Poverty is the greatest polluter of them all." She meant that the world's poorest people are too often both the victims and the agents of environmental degradation. They are forced to meet short-term survival needs at the cost of long-term sustainability. But "charity is not an answer to poverty," according to Dr. Muhammad Yunus of the Grameen Bank, "It only helps poverty to continue. It creates dependency and takes away [an] individual's initiative . . . Poverty isn't created by the poor, it's created by the institutions and policies that surround them . . . All we need to do is to make appropriate changes in the institutions and policies, and/or create new ones." The microcredit revolution he started may be the key for breaking the cycle of poverty and changing the lives of the poor.

Economics has given us many tools to understand development, trade, and strategies to improve human well-being.

Ecological economics is increasingly finding ways to include natural services, including regulation, provisioning, and aesthetic and cultural factors in that accounting. Because the poorest populations often depend directly on environmental services, this new approach could provide real assistance to needy people in developing regions. Carbon taxes, emissions trading, green business, fair trade, and other strategies are also being used to aid underprivileged individuals and countries.

These strategies also promise to aid wealthier countries by improving efficiency, lowering externalized costs to society, and encouraging the spread of renewable energy and nonpolluting technologies worldwide. Although economists remain divided about the necessity of constant growth, steady-state economies, and the degree to which resources are fixed or flexible, it is increasingly accepted that greater internalization of environmental and social costs are important in any effort toward sustainable development.

Reviewing Key Terms

Can you define the following terms in environmental science?

cap-and-trade 23.5
capital 23.1
carbon credits 23.5
classical economics 23.1
command and control
   economies 23.5
common property
   management 23.3
cost–benefit analysis
   (CBA) 23.4
demand 23.1

ecological
   economics 23.2
ecosystem services 23.2
emissions trading 23.5
Environmental
   Performance Index
   (EPI) 23.4
externalized costs 23.1
gross domestic product
   (GDP) 23.4

gross national product
   (GNP) 23.4
Hubbert curve 23.3
human development
   index (HDI) 23.4
intangible resources 23.1
internal costs 23.2
marginal costs and
   benefits 23.1
market equilibrium 23.1
market forces 23.5

neoclassical
   economics 23.1
nonrenewable
   resources 23.1
open-access system 23.3
political economy 23.1
price elasticity 23.1
renewable resources 23.1
resource 23.1
steady-state
   economy 23.2

supply 23.1
sustainable
   development 23.1
throughput 23.2
triple bottom line 23.6
world happiness index
   (WHI) 23.4

Critical Thinking and Discussion Questions

1. When ecologists warn that we are using up irreplaceable natural resources, and economists respond that ingenuity and enterprise will find substitutes for most resources, what underlying premises and definitions shape their arguments?

2. How can intangible resources be infinite and exhaustible at the same time? Isn't this a contradiction in terms? Can you find other similar paradoxes in this chapter?

3. What would be the effect on the developing countries of the world if we were to change to a steady-state economic system? How could we achieve a just distribution of resource benefits while still protecting environmental quality and future resource use?

4. Resource-use policies bring up questions of intergenerational justice. Suppose you were asked: "What has posterity ever done for me?" How would you answer?

5. If you were doing a cost–benefit study, how would you assign a value to the opportunity for good health or the existence of rare and endangered species in faraway places? Is there a danger or cost in simply saying some things are immeasurable and priceless and therefore off limits to discussion?

6. If natural capitalism or eco-efficiency has been so good for some entrepreneurs, why haven't all businesses moved in this direction?

Data Analysis

Evaluating the Limits to Growth

The graphs shown in figure 23.17 contrast two views of resource supply and scarcity. Models like these are very influential in both describing and informing our view of resources, scarcity, consumption patterns, and the kind of society we may live in, in the future. They also represent specific economic ideas and assumptions. Review the text for this figure, then go to Connect to demonstrate your understanding of these ideas.

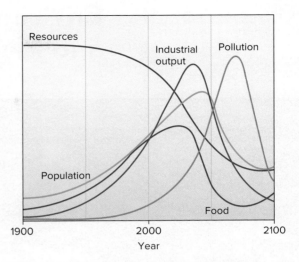

Models of resource consumption and scarcity are used both to describe our economic system and to predict future resource scarcity.

TO ACCESS ADDITIONAL RESOURCES FOR THIS CHAPTER, PLEASE VISIT CONNECT AT
www.connect.mheducation.com
You will find LearnSmart, an adaptive learning system, Google Earth™ exercises, additional Case Studies, Data Analysis exercises, and an interactive ebook

24

Environmental Policy, Law, and Planning

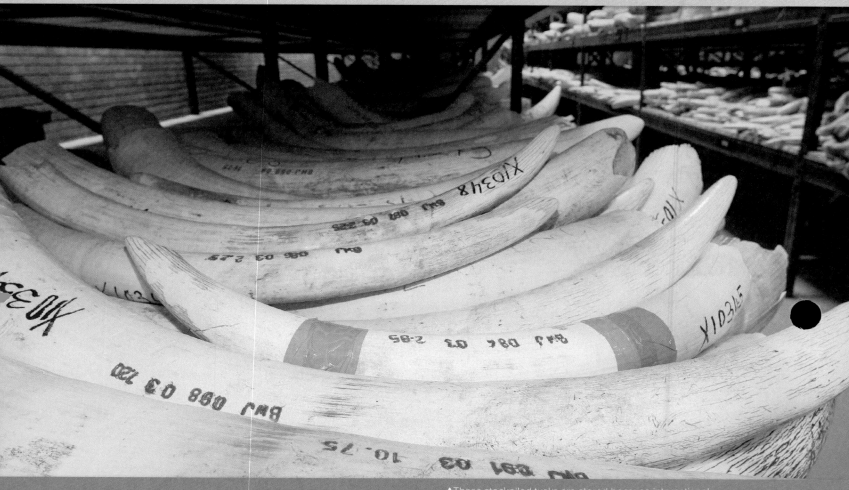

▲These stockpiled tusks are stored because international agreements make most sales illegal.
© Dave Hamman/Getty Images

Learning Outcomes

After studying this chapter, you should be able to:

24.1 List several basic concepts in policy.

24.2 Describe some major environmental laws.

24.3 Identify ways that executive, judicial, and legislative bodies shape policy.

24.4 Explain the purposes of international treaties and conventions.

24.5 Outline dispute resolution and planning.

"The power to command frequently causes failure to think."

– Barbara Tuchman

Can Policy Protect Elephants?

Elephants are among the most charismatic and fascinating of African animals. Powerful national symbols in many parts of Africa, they also draw millions of tourists and support essential tourism economies. Ecologists prize the elephant as a keystone species, a critical part of its ecosystem. Elephants are also considered an umbrella species: Habitat saved for elephants also supports countless other species.

Elephants are also valued for their ivory. Ivory is a luxury commodity that has been traded globally for centuries, but the trade has transformed dramatically with modern guns, growing trade networks, and increasing global wealth. The 1960s–1980s saw unprecedented slaughter of Africa's elephants, mainly for ivory but increasingly for bushmeat, skin, feet, and other parts, as well. In the face of this lucrative global trade, individual countries have been powerless to protect elephant populations. Already by the 1960s, this symbolic species was dwindling and disappearing in many parts of the continent.

African elephants are just one of many species falling to increasingly efficient poaching in the late twentieth century. Indian elephants, South American parrots, and Asian turtles have diminished sharply. Legal but uncontrolled slaughter also threatened most of the world's whale species, as well as sea turtles and many species of fish. By the 1950s, people around the world were questioning whether new policies were needed to prevent extinctions. It was clear that international trade was the primary incentive for most wildlife collectors and poachers. So international trade became the focus of efforts to save these species.

The world's first widely adopted biodiversity policy, the Convention on International Trade in Endangered Species (CITES), was drafted at a 1963 meeting of the International Union for Conservation of Nature (IUCN). Because this was a convention (agreement) among countries, representatives of each participating state had to debate the policies and decide whether to ratify (agree to and sign) the convention. The convention was finally adopted, with 80 countries signing on, at a 1973 IUCN meeting, a full decade after its initial drafting.

What exactly does CITES do? The convention marries science and policy to protect biodiversity. It establishes the principle that exports and imports of endangered species and their parts should be restricted, that scientific evidence should be used in deciding if a species is endangered, that participating states should establish scientific agencies with the power to monitor threats to species, and that an export permit can be granted only if the scientific agency certifies that the export will not further endanger the species. Export of an individual animal for zoos, for example, is often permitted because just one or two individuals can bring a large sum to an impoverished country. Participating countries also agree to penalize anyone found breaking the rules, and to confiscate illegally traded species or body parts (fig. 24.1).

CITES also establishes a list of species that require monitoring. There are three levels of protection. Species of urgent concern can be traded only in exceptional circumstances and require a permit from both importing and exporting countries (the Appendix I species list). Others require export permits but can be traded relatively freely (Appendix II). A few require permits in only a few countries (Appendix III). These lists make it possible to monitor all these species and to follow standard procedures for protecting them.

Establishing and enforcing a new policy is a contentious process. Conflicting economic interests pit ivory marketers against wildlife conservationists and the general public, who tend to be enthusiastic about protecting wildlife. Details require years of bitter negotiation. Establishing independent monitoring and policing agencies supports national pride, but it's also expensive. Enforcing new rules is difficult where established tradition is an unfettered free-for-all.

Despite these inevitable challenges, CITES has been extremely popular. It protects the charismatic and symbolically powerful species that people cherish worldwide, such as India's tigers and China's pandas, as well as Africa's elephants. Most of these species are worth far more as tourist attractions than as body parts, so local support for protection is often strong. There are now 175 signatory nations, all but 20 of the world's independent states. Enforcement varies tremendously, and countries dispute whether or not a species should be added to the endangered list, but CITES now provides some protection to about 30,000 species. Some 900 of these are on the high-concern Appendix I list.

FIGURE 24.1 African elephants (*Elphas maximus* and *Loxodonta africana*) are among the many species CITES is designed to protect.
© Royalty-Free/Corbis

The African elephant was put on the Appendix II list in 1976. That listing proved insufficient to restrict poaching: Ivory is too valuable, and importing countries bore no legal responsibility to help restrict trade. The population plummeted by half between 1979 and 1989, from about 1.2 million to 600,000. Finally, in 1989, the African elephant was upgraded to Appendix I. International trade in ivory is allowed only in exceptional circumstances. The hope is that if ivory cannot be sold, it will no longer be profitable to kill elephants for their tusks. Perhaps most important, it has raised international awareness of the precarious state of Africa's elephant populations. Protected populations in many areas have stabilized. In some national parks in southern Africa, populations have rebounded to the extent that parks must periodically cull some animals, which cannot migrate outside park boundaries and which threaten to exceed local carrying capacity.

As with any policy that threatens lucrative trade, CITES's protection for elephants has been controversial. Debates have been long and bitter. Special exemptions to the Appendix I listing have been granted to four African countries (Botswana, Namibia, South Africa, and Zimbabwe) that insist their populations are stable and that they should be permitted to sell some of the ivory stockpiled during culling operations. Nonetheless, CITES is an important and largely effective agreement. CITES also set global precedents: Among the first broadly accepted environmental policies, the convention predates the U.S. Endangered Species Act by a decade, and many other countries have adopted internal laws following the principles set by CITES.

In this chapter, we'll examine how environmental policies are formed, and will discuss the structures involved in shaping, refining, and enforcing laws, at national, local, and international scales.

24.1 BASIC CONCEPTS IN POLICY

- *Environmental policy includes rules and regulations as well as public opinion.*
- *Money influences policy.*
- *Public awareness and action shape policy.*

The basic idea of **policy** is a statement of intentions and rules that outline acceptable behaviors or accomplish some end. In the case of CITES, these rules were formally stated and voted on, but policy can also be less formally defined. You might have an informal policy to always do your homework; a church might have an open-door policy for visitors; most countries have policies to ensure the welfare of citizens. Laws, such as the U.S. Clean Water Act (chapter 18), are formal statements of national policy. Like the rules that define how you play a board game, policies define agreed-upon limits of behavior. Because it sets the rules we live by, policy making is a contentious and extremely important process.

We will take **environmental policy** to mean official rules and regulations concerning the environment, as well as public opinion that helps shape environmental policy. Ideally, environmental policy serves the needs of human health, economic stability, and ecosystem health. Often, these interests have been seen as contradictory, leading to bitter disputes in policy making. Increasingly, however, the public and policymakers acknowledge that these interests overlap. Protecting species, for example, supports the economy and is often a source of pride for communities. Controlling pollution protects human health, saves money in health care, and preserves essential resources.

The drafting of CITES in 1963 was part of a wave of environmental protections that started in many countries in the 1960s and 1970s. Increasing evidence of the damage caused by unconstrained resource use and pollution, following the industrial expansions

of World War II, helped foster widespread interest in pollution control, species protection, health and human safety, and other safeguards. Since then, global interest in environmental quality has remained high and has even strengthened in recent years, as more people become aware of our dependence on a healthy environment. In a 2007 BBC poll of 22,000 residents of 21 countries, 70 percent said they were personally ready to make sacrifices to protect the environment (fig. 24.2). Overall, 83 percent agreed that individuals would definitely or probably have to make lifestyle changes to reduce the amount of climate-changing gases they produce. Concern about environmental quality varied by country: Just over 40 percent of Russians polled were willing to change their

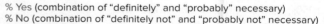

% Yes (combination of "definitely" and "probably" necessary)
% No (combination of "definitely not" and "probably not" necessary)

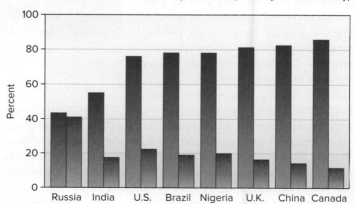

FIGURE 24.2 About 70 percent of the 22,000 people in 21 countries polled by the BBC in 2007 agreed with the statement, "I am ready to make significant changes to the way I live to help prevent global warming or climate change."

lifestyle to prevent global warming, compared to nearly 90 percent of Canadians. The Chinese were the most enthusiastic about energy taxes to prevent climate change. Eighty-five percent of the Chinese polled agreed that such taxes are necessary.

Basic principles guide environmental policy

Protection of fundamental rights is usually one aim of policy making. For many groups, these rights include a right to a safe, sustainable environment (fig. 24.3). The 1987 World Commission on Environment and Development, for example, stated, "All human beings have the fundamental right to an environment adequate for their health and well-being." Of the 195 independent nations in the world, 109 now have constitutional provisions for protection of the environment and natural resources. Often such policies are not legally enforceable, but they provide a statement of intentions that can guide subsequent laws.

In theory, democratic societies try to establish policies that are fair to everyone, at least in principle, and that defend everyone's basic needs. Ideally, the majority interests are served, but interests of the minority are also defended. Fairness in policy making requires that rules are transparent and decided by public debate and input from many groups in society.

Cost–benefit analysis is often involved in policy making. The aim is to assign standard values, such as dollar value, to competing concerns, then compare the costs and benefits of a plan. This way, we can decide if benefits outweigh the costs of a policy. In principle, this is a utilitarian approach to ensuring an objective, methodical decision process. In practice, there are many difficulties in implementing rational cost–benefit analysis:

- Many conflicting values and needs cannot be compared: They are fundamentally different or we lack complete information.

- We often lack agreed-upon broad societal goals; benefits to different groups and individuals often conflict.

FIGURE 24.3 How do we ensure a safe and healthy environment for everyone? Policies of individuals, communities, and governing institutions are all responsible.
© Digital Vision/PunchStock RF

- Policymakers may be motivated less by societal goals than by their own interests in power, status, money, or reelection.

- Past investments in existing programs and policies create "path dependence" (a tendency to follow familiar ways) and hidden costs that prevent policymakers from considering good alternatives foreclosed by previous decisions.

- Policymakers often lack the information or models to calculate costs and benefits accurately. Some values, such as health, freedom, or scenery, are very hard to price in dollars.

Another consideration is the **precautionary principle:** When an activity threatens to harm health or the environment, we should fully understand risks before initiating that activity. According to this principle, for example, we shouldn't mass-market new chemicals, new cars, or new children's toys until we're sure they are safe. There are four widely accepted tenets of this principle:

- People have a duty to take steps to prevent harm. If you suspect something bad might happen, you have an obligation to try to stop it.

- The burden of proof of carelessness of a new technology, process, activity, or chemical lies with the proponents, not with the general public.

- Before using a new technology, process, or chemical, or starting a new activity, people have an obligation to examine a full range of alternatives, including the alternative of not using it.

- Decisions using the precautionary principle must be open and democratic and must include the affected parties.

The European Union has adopted this precautionary principle as the basis of its environmental policy. In the United States, opponents of this approach claim that it threatens productivity and innovation. However, many American firms that do business in Europe—including virtually all of the largest corporations—are having to change their manufacturing processes to adapt to more careful E.U. standards. For example, lead, mercury, and other hazardous materials must be eliminated from electronics, toys, cosmetics, clothing, and a variety of other consumer products. A proposal being debated by the E.U. would require testing of thousands of chemicals, would cost industry billions of dollars, and lead to many more products and compounds being banned as they are shown to be unsafe to the public. What would you do about this? Is this proposal just common sense, or is it an invitation to decision paralysis?

Money influences policy

Politics is often seen as a struggle for power among competing interest groups, which strive to shape public policy to suit their own agendas. Usually, money is a key to political power, so the issue of money in politics is a matter of perennial and contentious debate. Even in a democratic society, money is an important factor in what rules are made, or not made. Often those with the most money can advertise to promote their point of view, draw voters to their side, and win the most legislative seats. Wealthy individuals and corporations can buy influence and friends in government agencies. Large donations are now essential to success in a political campaign, so wealthy donors can influence policymakers with generous funding, or by threatening to withhold funds.

Debates about power and money in politics in the United States have increased since 2010, when the U.S. Supreme Court decided in the Citizen's United case that corporations can spend unlimited amounts of money in political advertising (section 24.3). Because this decision opens elections to potentially vast new infusions of money, many observers expect this decision to shape politics for decades to come. What do you think? Does it matter if an oil company can spend millions on political advertising? Would it matter if the company were a foreign-owned corporation working to shape U.S. laws?

Public awareness and action shape policy

Although power and money are important forces, they can't account for our many policies that serve the public interest. Public participation by scientists and citizens is also an essential force in policy formation. Take the example of CITES, an almost universally adopted set of rules protecting species, mostly for noneconomic purposes. Concerned scientists and communities, as well as policymakers with a strong conscience who objected to excessive poaching, led to the establishment of this convention. Within the United States, civic action has led to many of our strongest environmental and social protections, such as the Clean Water Act, the Clean Air Act, and the Voting Rights Act, which defend the interests of all citizens. Many people consider public citizenship the most important force in a democratic government.

Policy changes often start with protests over environmental contamination, pollution, or resource waste. Media attention helps publicize these protests and the problems they are challenging. One of the most lauded U.S. environmental rules, the Clean Water Act, followed shocking and widely broadcast images in 1969 of the Cuyahoga River in Cleveland burning and the Santa Barbara oil spill. The oil spill resulted from an oil well blowout near the California coast. An estimated 100,000 barrels (16,000 m³) of black, gooey crude oil was spilled, and much of it washed up onto beautiful Southern California beaches. Dead birds, dolphins, and sea lions caught public attention, and cleanup crews rallied to try to protect the beaches (fig. 24.4). For weeks, images were broadcast nationwide on TV news. Daily updates were broadcast nationwide, using live satellite feeds, for which the technology had just become available.

These images helped to build national concern for environmental protections. Images of young volunteers, smudged with oil, trying vainly to sweep gooey oil off a beautiful beach, were ideal for TV. Like the Cuyahoga fires, the Santa Barbara oil spill played an important role in mobilizing public opinion and was a major factor in passage of the 1972 U.S. Clean Water Act.

Opportunities for public awareness and participation have increased in recent years. There is a growing diversity of news outlets. YouTube, Twitter, and other social media let anyone post images and news. In addition, there are many groups committed to encouraging public involvement in environmental policy. One of these is the Environmental Working Group, which monitors

FIGURE 24.4 Beach cleanup efforts after the Santa Barbara oil spill in 1969 made excellent media material and had an important role in U.S. environmental policy.
© Esther Henderson/Science Source

policy and budget issues in Congress. Another is the League of Conservation Voters, which publishes a scorecard for all members of Congress, recording how they voted on environmental laws. You can see how your own representatives are doing by looking at their website: www.lcv.org/scorecard/.

Section Review

1. What is the policy cycle, and how does it work?
2. What is the precautionary principle?
3. How did the 1969 Santa Barbara oil spill help pass environmental legislation?

24.2 MAJOR ENVIRONMENTAL LAWS

- *NEPA (1969) establishes public oversight.*
- *The Clean Air Act (1970) regulates air emissions.*
- *The Clean Water Act (1972) protects surface water.*
- *The Endangered Species Act (1973) protects wildlife.*
- *The Superfund Act (1980) lists hazardous sites.*

We depend on many different laws to protect resources such as clean water, clean air, safe food, and biodiversity (table 24.1). Here we'll review a few of these laws, and in the sections that follow we'll examine how laws are created. We will focus mainly on U.S. laws in this section. Many other countries have followed the early lead of the United States in environmental policy formation. In recent years, other countries and the international community

Table 24.1   Major U.S. Environmental Laws

Legislation	Provisions
Wilderness Act of 1964	Established the national wilderness preservation system.
National Environmental Policy Act of 1969	Declared national environmental policy, required Environmental Impact Statements, created Council on Environmental Quality.
Clean Air Act of 1970	Established national primary and secondary air quality standards. Required states to develop implementation plans. Major amendments in 1977 and 1990.
Clean Water Act of 1972	Set national water quality goals and created pollutant discharge permits. Major amendments in 1977 and 1996.
Federal Pesticides Control Act of 1972	Required registration of all pesticides in U.S. commerce. Major modifications in 1996.
Marine Protection Act of 1972	Regulated dumping of waste into oceans and coastal waters.
Coastal Zone Management Act of 1972	Provided funds for state planning and management of coastal areas.
Endangered Species Act of 1973	Protected threatened and endangered species, directed FWS to prepare recovery plans.
Safe Drinking Water Act of 1974	Set standards for safety of public drinking-water supplies and to safeguard groundwater. Major changes made in 1986 and 1996.
Toxic Substances Control Act of 1976	Authorized EPA to ban or regulate chemicals deemed a risk to health or the environment.
Federal Land Policy and Management Act of 1976	Charged the BLM with long-term management of public lands. Ended homesteading and most sales of public lands.
Resource Conservation and Recovery Act of 1976	Regulated hazardous waste storage, treatment, transportation, and disposal. Major amendments in 1984.
National Forest Management Act of 1976	Gave statutory permanence to national forests. Directed USFS to manage forests for "multiple use."
Surface Mining Control and Reclamation Act of 1977	Limited strip mining on farmland and steep slopes. Required restoration of land to original contours.
Alaska National Interest Lands Act of 1980	Protected 40 million ha (100 million acres) of parks, wilderness, and wildlife refuges.
Comprehensive Environmental Response, Compensation and Liability Act of 1980	Created $1.6 billion "Superfund" for emergency response, spill prevention, and site remediation for toxic wastes. Established liability for cleanup costs.
Superfund Amendments and Reauthorization Act of 1994	Increased Superfund to $8.5 billion. Shares responsibility for cleanup among potentially responsible parties. Emphasizes remediation and public "right to know."

Source: N. Vig and M. Kraft, Environmental Policy in the 1990s, 3rd Congressional Quarterly Press

have increasingly led the way, as global concern for environmental policy has expanded.

Most of the laws we take for granted now are relatively recent. For most of its history, U.S. policy has had a laissez-faire or hands-off attitude toward business and private property. Pollution and environmental degradation were regarded as the unfortunate but necessary costs of doing business. There were some early laws forbidding gross interference with another person's property or rights—the Rivers and Harbors Act of 1899, for example, made it illegal to dump so much refuse in waterways that navigation was blocked. But in general there were few rules regarding actions on private property, even when those actions impaired the health or resources of neighbors.

Many of these attitudes and rules changed in the 1960s and 1970s, which marked a dramatic turning point in our understanding of the dangerous consequences of pollution. Rachel Carson's *Silent Spring* (1962) and Barry Commoner's *Closing Circle* (1971) alerted the public to the ecological and health risks of pesticides, hazardous wastes, and toxic industrial effluents. Public activism in the civil rights movement and protests against the war in Vietnam carried over to environmental protests and demands for environmental protection.

Rising public concern and activism about environmental issues, such as DDT-poisoned birds (chapter 8), water pollution, and rising smog levels, led to more than 27 major federal laws for environmental protection and hundreds of administrative regulations established in the environmental decade of the 1970s. Among the most important were the establishment of the Environmental Protection Agency (EPA) and the National Environmental Policy Act (NEPA), which requires environmental impact statements for all major federal projects. Because of their power, the EPA and NEPA have both been the targets of repeated attacks by groups that dislike regulations imposed on their pollution emissions or resource uses.

In the initial phase of this environmental revolution, the main focus was on direct regulation and lawsuits to force violators to obey the law. In recent years, attention has shifted to pollution prevention and collaborative methods that can provide win–win solutions for all stakeholders. At the same time, many corporations now recognize that unregulated pollution is unacceptable, and that cleaner, more efficient practices are widely understood to be good for profits as well as for the environment.

Environmental, health, and public safety laws, like other rules, impose burdens for some people and provide protections for others. The Clean Water Act, for example, requires that industries take responsibility for treating waste, rather than discharging it into public waters for free. Cities have had to build and maintain expensive sewage treatment plants, instead of discharging sewage into lakes and rivers. These steps internalize costs that previously had been left to the public to deal with. These national laws are intended to protect public health and shared resources for all areas and all citizens. In fact, enforcement varies, but the existence of national, legally enforceable laws allows some recourse for victims when environmental laws are broken.

You can see the text of these laws, together with some explanation, on the EPA's website: www.epa.gov/lawsregs/laws/index .html#env. If you have never examined the text of a law, you should take a look at these.

NEPA (1969) establishes public oversight

Signed into law by President Nixon in 1970, the **National Environmental Policy Act (NEPA)** is the cornerstone of U.S. environmental policy.

NEPA does three important things: (1) it authorizes the Council on Environmental Quality (CEQ), the oversight board for general environmental conditions; (2) it directs federal agencies to take environmental consequences into account in decision making; and (3) it requires an **environmental impact statement (EIS)** be published for every major federal project likely to have an important impact on environmental quality (fig. 24.5). NEPA doesn't forbid environmentally destructive activities if they comply otherwise with relevant laws, but it demands that agencies admit publicly what they plan to do. Once embarrassing information is revealed, however, few agencies will bulldoze ahead, ignoring public opinion. And an EIS can provide valuable information about government actions to public interest groups, which wouldn't otherwise have access to this information.

What kinds of projects require an EIS? The activity must be federal and it must be major, with a significant environmental impact. Evaluations are always subjective as to whether specific activities meet these characteristics. Each case is unique and depends on context, geography, the balance of beneficial versus harmful effects, and whether any areas of special cultural, scientific, or historical importance might be affected. A complete EIS for a project is usually time-consuming and costly. The final document is often hundreds of pages long and generally takes six to nine months to prepare. Sometimes just requesting an EIS is enough to sideline a questionable project. In other cases, the EIS process gives adversaries time to rally public opposition and information with which to criticize what's being proposed. If agencies don't agree to prepare an EIS voluntarily, citizens can petition the courts to force them to do so.

Every EIS must contain the following elements: (1) purpose and need for the project, (2) alternatives to the proposed action (including taking no action), (3) a statement of positive and negative environmental impacts of the proposed activities. In addition,

FIGURE 24.5 Every major federal project in the United States must be preceded by an Environmental Impact Statement.
© Russell Illig/Getty Images RF

an EIS should make clear the relationship between short-term resources and long-term productivity, as well as any irreversible commitment of resources resulting from project implementation.

Many lawmakers in recent years have tried to ignore or limit NEPA in forest policy, energy exploration, and marine wildlife protection (chapter 12). The "Healthy Forest Initiative," for example, eliminated public oversight of many logging projects by bypassing EIS reviews and prohibiting citizen appeals of forest management plans. Similarly, when the Bureau of Land Management proposed 140,000 coal-bed methane wells in Wyoming and Montana, promoters claimed that water pollution and aquifer depletion associated with this technology didn't require environmental review (chapter 19). And in the 2005 Energy Bill, Congress inserted a clause that exempts energy companies from NEPA requirements in a number of situations, with the aim of speeding energy development on federal land.

The Clean Air Act (1970) regulates air emissions

The first major environmental legislation to follow NEPA was the Clean Air Act (CAA) of 1970. Air quality has been a public concern at least since the beginning of the industrial revolution, when coal smoke, airborne sulfuric acid, and airborne metals such as mercury became common in urban and industrial areas around the world (fig. 24.6). Sometimes these conditions produced public health crises: One infamous event was the 1952 Great Smog of London—several days of cold, still weather that trapped coal smoke in the city and killed some 4,000 people from infections and asphyxiation. Another 8,000 died from respiratory illnesses in the months that followed (see section 16.2).

Although crises of this magnitude have been rare, chronic exposure to bad air has long been a leading cause of illness in many areas. The Clean Air Act provided the first nationally standardized rules in the United States to identify, monitor, and reduce air contaminants. The core of the act is an identification and regulation of seven major "criteria pollutants," also known as "conventional pollutants." These seven include sulfur oxides,

FIGURE 24.6 The Clean Air Act has greatly reduced the health and economic losses associated with air pollution.
© William P. Cunningham

lead, carbon monoxide, nitrogen oxides (NO_x), particulates (dust), volatile organic compounds, and metals and halogens (such as mercury and bromine compounds). Recent revisions to the Clean Air Act require the EPA to monitor carbon dioxide, our dominant greenhouse gas, which endangers the public through heat stress, drought, and other mechanisms (chapter 15).

Two significant regulations under the Clean Air Act are the Mercury and Air Toxics Standards and the Clean Power Plant Rule. In 2016, after 20 years of study and discussion, the EPA issued final regulations for mercury and air toxics from power plants. These standards aimed to cut releases of these toxins from about 600 power plants by 75 percent each year. They will force utilities to install and operate equipment that removes mercury and fine particulate material from their emissions. The EPA estimated the rules would cost utilities about $9.6 billion per year, but would save the public between $37 billion and $90 billion annually, mostly by avoiding premature deaths and reducing health problems.

President Obama's Clean Power Plan, proposed in 2014, was designed to improve air quality and also help the U.S. meet its greenhouse gas reduction targets promised at the Paris climate meetings in 2015. The plan aimed to cut carbon emissions from power plants by 30 percent compared to 2005 levels, while creating new jobs in the clean power sector, where employment has been growing rapidly. It also would reduce soot and smog levels by at least 25 percent by 2030. The Environmental Protection Agency estimates that this plan will result in climate and health benefits worth between $55 billion and $93 billion by 2030. It should also avoid around 5,000 premature deaths and 150,000 asthma attacks. In addition, the plan is expected to shrink business and residential electric bills by roughly 8 percent.

The Clean Power Plan sets state-by-state emissions reduction targets, but it lets the states to decide how to achieve those reductions. In 2015, Republican attorneys general from 27 states sued the government to block the plan. Proponents of the plan point out that those attorneys general took a combined $20 million in campaign contributions from the fossil fuel industry before filing their suits. In 2016, the Supreme Court issued a stay that bars implementation of the plan. The plan is widely considered a badly needed strategy for the economy, energy futures, and our climate, but its future remains unclear.

The Clean Water Act (1972) protects surface water

Water protection has been a goal with wide public support, in part because clean water is both healthy and an aesthetic amenity. The act aimed to make the nation's waters "fishable and swimmable"; that is, healthy enough to support the propagation of fish and shellfish that could be consumed by humans, and low enough in contaminants that they were safe for children to swim and play in them.

The first goal of the Clean Water Act (CWA) was to identify and control point source pollutants, end-of-the-pipe discharges from factories, municipal sewage treatment plants, and other sources. Discharges are not eliminated, but water at pipe outfalls must be tested, and permits are issued that allow moderate discharges of low-risk contaminants such as nutrients or salts. Metals, solvents, oil, high counts of fecal bacteria, and other more serious contaminants must be captured before water is discharged from a plant.

By the late 1980s, point sources were increasingly under control, and the CWA was used to address nonpoint sources, such as runoff from urban storm sewers. The act has also been used to promote watershed-based planning, in which communities and agencies collaborate to reduce contaminants in their surface waters. As with the CAA, the CWA provides funding to aid pollution-control projects. Those funds have declined in recent years, however, leaving many municipalities struggling to pay for aging and deteriorating sewage treatment facilities. For more details on the CWA and water pollution control, see chapter 18.

The Endangered Species Act (1973) protects wildlife

While CITES (opening case study) aims to provide international protections for species, the Endangered Species Act (ESA) regulates species within the United States. The ESA provides a structure for identifying and listing species that are vulnerable, threatened, or endangered. Once a species is listed as endangered, the ESA provides rules for protecting it and its habitat, ideally to make recovery possible (fig. 24.7). Listing of a species can be a controversial process because habitat conservation can get in the way of land development. Many ESA controversies arise when developers want to put new housing developments in scenic areas where the last remnants of a species reside. In many cases, however, disputes have been resolved by negotiation and the more creative design of developments, which can sometimes allow both for development and for species protection.

There is considerable collaboration between CITES and the ESA in listing and monitoring endangered species. The ESA maintains

FIGURE 24.7 The Endangered Species Act is charged with protecting species and their habitat. The black-footed ferret was declared extinct in 1979, but a remnant population was discovered in 1981, and captive breeding programs have restored the species to more than 1,000 mature, wild-born animals in eight states.
© Royalty-Free/Corbis

a worldwide list of endangered species, as well as a U.S. list. The ESA also provides for grants and programs to help land owners protect species. The responsibility for studying and attempting to restore threatened and endangered species lies mainly with the Fish and Wildlife Service and the National Oceanic and Atmospheric Administration. You can read more about endangered species, biodiversity, and the ESA in chapter 11.

The Superfund Act (1980) lists hazardous sites

Most people know this law as the Superfund Act because it created a giant fund to help remediate abandoned toxic sites. The proper name of this law is informative, though: the Comprehensive Environmental Response, Compensation, and Liability Act (CERCLA). The act aims to be comprehensive, addressing abandoned sites, emergency spills, or uncontrolled contamination, and it allows the EPA to try to establish liability, so that polluters help to pay for cleanup. It's much cheaper to make toxic waste than to clean it up, so we have thousands of chemical plants, gas stations, and other sites that have been abandoned because they were too expensive to clean properly. The EPA is responsible for finding a contractor to do cleanup, and the Superfund was established to cover the costs, which can be in the billions of dollars. Until recently, the fund was financed mainly by contributions from industrial producers of hazardous wastes. In the 1990s, however, Congress voted to end

that source, and the Superfund was allowed to dwindle to negligible levels. Site cleanup is now funded by taxpayer dollars.

According to the EPA, one in four Americans lives within 3 miles of a hazardous waste site. The Superfund program has identified more than 47,000 sites that may require cleanup. The most serious of these (or the most serious for which proponents have been sufficiently vigorous) have been put on a National Priorities List. Some 1,680 sites have been put on the list, and about half have been cleaned up. The total cost of remediation is thought to be somewhere between $370 billion and $1.7 trillion. To read more, see chapter 21.

Section Review

1. Describe the major provisions of NEPA, the Clean Air Act, the Clean Water Act, the Endangered Species Act, and the Superfund Act.

24.3 HOW ARE POLICIES MADE?

- *Congress and legislatures vote on statutory laws.*
- *Judges decide case law.*
- *Executive agencies make rules and enforce laws.*

The general process by which policies are developed is often called the *policy cycle,* because rules are developed, enacted, and revised repeatedly (fig. 24.8). The cycle starts when a problem is identified as a priority. In the case of endangered species protections (both CITES and the ESA), for example, the public became concerned about accelerating damage to species and ecosystems. Citizen groups and wildlife advocates initiated public debates on the issue. Communities, students, and environmental groups worked to organize stakeholders, choose tactics, and aggregate related issues into a case for species preservation that was of interest to a broad range of groups and communities.

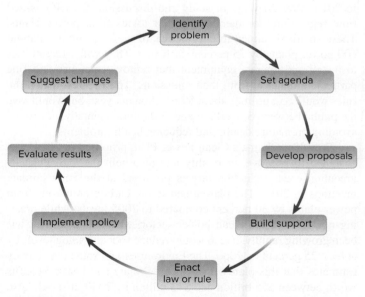

FIGURE 24.8 The policy cycle is a path through which rules are developed, enacted, tested, and revised.

FIGURE 24.9 Marches, sit-ins, and other peaceful demonstrations can help call attention to your cause and encourage your neighbors to join the effort.
© William P. Cunningham

Citizens often draw attention to their cause by organizing protests, marches, street theater, or other kinds of public events (fig. 24.9). Getting involved in local election campaigns, writing letters, or making telephone calls to legislators also influences the decision process. You'd be surprised at how few letters or calls legislators receive from voters, even on important national issues. Your voice can have an important impact (see chapter 25).

Once an issue is defined and support has been gathered, the next stage in the policy cycle is to propose a new law or rule. The rule is debated and negotiated. Media campaigns, public education, and personal lobbying of decision makers are needed to build support for a new policy. Getting a proposal enacted as a law or a rule takes persistence, negotiation, and usually years of effort.

After a rule or law is enacted, continued public oversight and monitoring are usually needed to ensure that government agencies faithfully carry out policies. Measuring impacts and results are also essential, so that amendments can be proposed, if necessary, to make the policy fairer or more effective.

Each branch of government plays a role in establishing laws. **Statute law** consists of documents or laws ("statutes") voted on and enacted by the legislative branch of government. **Case law** is derived from lawsuits (legal cases) in court. **Administrative law** arises from executive orders, administrative rules, and regulations. Because every country has different legislative and legal processes, this discussion focuses on the U.S. system, which shares many general similarities with other democratic systems.

Congress and legislatures vote on statutory laws

Elected legislative bodies, such as Congress, state legislatures, or town councils, debate and vote on policies that become legally enforceable laws. Federal laws (statutes) are enacted by Congress and signed by the president. They originate as legislative proposals called "bills," which are usually drafted by the congressional staff, often in consultation with representatives of various interest groups.

Thousands of bills are introduced every year in Congress. Some are very narrow, providing funds to build a specific road or bridge or to help a particular person, for instance. Others are extremely broad, perhaps overhauling the social security system or changing the entire tax code.

Bills Move Slowly Through Congress

After a bill is introduced, it goes to a committee, where it is discussed and debated. Most hearings take place in Washington, but if the bill is controversial or legislators want to attract publicity for themselves or the issue, they may conduct field hearings closer to the site of the controversy. The public often has an opportunity to give testimony at field hearings (fig. 24.10). Elected officials may be swayed by public opinion, and they need public support in policy making.

The language of a bill is debated, revised, and negotiated in a committee until it is considered widely acceptable enough to send to the full House of Representatives or Senate. Compromises are necessary to make the bill acceptable to different parties. The House has one version of the bill, which is debated on the floor of the House. The Senate has another version that it debates. Often there are further amendments at these stages.

By the time an issue has passed through both the House and Senate, the versions approved by the two bodies are likely to be different. They then go to conference committee to iron out any differences between them. After going back to the House and Senate for confirmation, the final bill goes to the president, who may either sign it into law or veto it. If the president vetoes the bill, it may still become law if two-thirds of the House and Senate vote to override the veto. If the president takes no action within ten days of receiving a bill from Congress, the bill becomes law without his signature.

Each step of this convoluted process is published in print and online in the *Congressional Quarterly Weekly,* which you

FIGURE 24.10 Nevada college students testify against higher education cuts at a Congressional field hearing. Positive participation in an event, such as this, can influence both legislators and the general public.
© AP Photo/Cathleen Allison

can access at any time through the official congressional website, www.thomas.gov. It can take a little practice to find the legislation you want, but it's a rich repository of public records.

Legislative Riders

There are two types of legislation: authorizing bills become laws, while appropriation bills provide the money for federal agencies and programs. Eliminating funding for an agency in an appropriations bill is often an effective way to prevent laws from being implemented. Appropriation bills are not supposed to make policy, but merely fund existing plans and projects. Legislators who can't muster enough votes to pass pet projects through regular channels often will try to add authorizing amendments called **riders** into completely unrelated funding bills. Even if they oppose the riders, other members of Congress have a difficult time voting against an appropriation package for disaster relief or to fund programs that benefit their districts. This often happens in conference committee because when the conference report goes back to the House and Senate, the vote is either to accept or reject with no opportunity to debate or amend further.

Starting with the 104th Congress (in 1995) industry groups began using this tactic to roll back environmental protections and gain access to natural resources. For instance, the 2016 Appropriations bills had more than 113 anti-environmental riders that attempted among other things to block climate change research; require the government to burn more fossil fuels; prohibit pesticide regulations; discourage the use of efficient light bulbs; defund clean water protection; restrict endangered salmon and sea otter protection; reduce public participation in land-use decisions; forbid dam removal on U.S. rivers; open millions of acres of wilderness to mining, drilling, and off-road motor use; and exempt grazing on public lands from environmental laws. Often, these amendments are added after all other debate on the particular bill is finished so that other legislators don't even know what they're voting for or against.

Lobbying Influences Government

Groups or individuals with an interest in pending legislation can often have a great deal of influence through lobbying, or visiting congressional offices, talking directly with representatives, and using personal contacts to persuade elected representatives to vote in their favor. The term *lobbying* derives from the habit of people waiting in hallways and lobbies of Congress to catch the elbow of a passing legislator and plead their case.

Citizens often make trips to Washington—or to state capitals, county seats, or city halls—to try to personally persuade elected officials on upcoming votes. This direct contact is a basic part of the democratic process, but it can sometimes work unevenly because most people can't abandon work or school and fly to Washington to lobby.

Not everyone can go to Washington, but many people join organizations that can collectively send representatives, or hire professional lobbyists, to make sure their message is heard. Most major organizations now have lobbyists in Washington. The biggest single citizen lobbying group is probably the American Association of Retired Persons (AARP), which actively lobbies for issues considered of interest to senior citizens. The National Rifle Association (NRA), Union of Concerned Scientists, and many other groups participate in lobbying. Environmental organizations such as the Natural Resources Defense Council, Audubon, and the Sierra Club lobby on many environmental bills. Lobbyists and volunteer activists attend hearings, draft proposed legislation, and meet with officials. The range of interests involved in lobbying is astounding. Business organizations, workers, property owners, religious and ethnic groups, are all there. Walking the halls of Congress, you see an amazing mixture of people attempting to be heard.

Lobbying firms often hire former legislators, military officers, and agency staff. These people are valuable because they have personal ties that can be a great aid in catching the ear of voting members of Congress. Industry groups have especially large rewards to reap through lobbying. *The Washington Post* reported on a case in which a group of corporations invested $1.6 million lobbying for a special low overseas tax rate, and the effort saved them over $100 billion in tax payments. In another case, the Carmen Group, a lobbying firm, charged $500,000 to lobby for insurance claims following the September 11, 2001, attacks on the World Trade Center, and as a result the government agreed to cover $1 billion in insurance premiums for its clients.

Often, lobbyists write the bills that a legislator introduces. For example, in 2013 Congress passed a continuing resolution to fund the federal government for a few additional months. Tucked into the 587-page bill was a brief provision worth millions of dollars for genetic engineering firms. It exempted firms like Monsanto from judicial review, allowing them to sell and plant genetically engineered crops even if a court of law orders them to stop. It turned out that Monsanto lobbyists wrote this section of the bill.

The number of lobbyists registered in Washington more than doubled between 2000 and 2005, from 16,000 to almost 35,000. The biggest industry lobbying firms, such as The Federalist Group, can charge $20,000 to $40,000 per month for their services. Lobbying is, by its nature, about tipping the tables in the favor of an interest group. But lobbying is also something that many people see as necessary, as part of getting voices heard in a democratic process. What do you think? Is corporate lobbying important? Is it necessary? How would you distinguish the actions of an oil industry's lobbyists from those of an environmental or community group's lobbyists?

Judges decide case law

Often, environmental policies are established when groups bring complaints to the courts, involving damage to property or health, failure to enforce existing laws, or infringement of rights. Judges or juries decide these cases by determining whether written law (statutes) or customary law (common law) has been violated. The body of legal opinions built up by many court cases is called case law.

The United States is divided into 96 federal court districts, each of which has at least one trial court. Disputes about procedural issues and interpretations of the law in district courts are sent on to one of 12 regional appeals courts. Cases might involve criminal prosecutions, claims against the federal government, or complaints in which parties come from multiple states. Each state has its own courts that generally parallel the federal system. These courts decide cases involving state laws.

Legislation is often written in vague and general terms, which help make a bill widely enough accepted to gain passage. Congress often leaves it to the courts to "fill in the gaps," especially in environmental laws. As one senator said when Congress was about to pass the Superfund legislation, "All we know is that the American people want these hazardous waste sites cleaned up . . . Let the courts worry about the details."

The Supreme Court Decides Major Cases

Lawsuits with very far-reaching implications are decided by the United States Supreme Court, a group of nine justices whose job is to judge whether a law is consistent with the U.S. Constitution, or whether a policy is consistent with a law as written by Congress. States also have Supreme Courts for deciding cases at the state level. The Supreme Court can only study and judge a few cases a year, and for many people the Court and its actions are little known and poorly understood. But this is a body that makes pivotal and far-reaching decisions, many of them affecting our national environmental policies (fig. 24.11). In 2016, Justice Antonin Scalia died suddenly leaving the Court with a four to four conservative/liberal deadlock. Replacing him became a hot political contest that will determine the direction of the court for decades.

Perhaps the most sweeping rule change the Supreme Court has made in recent years was the 2010 decision in *Citizens United v. Federal Election Commission.* For nearly a century, anticorruption laws had limited corporate and union spending on political campaigns, generally on the grounds that corporations have more money than individuals, which gives them unfair influence in elections. In a hotly debated, 5–4 split vote, the Court decided that these laws limited free speech (in the form of political advertising), which is protected under the Constitution. The five-person majority argued that because U.S. law gives corporations the same rights and protections as individual people, corporate advertising could not be limited. Within months of the decision, a coalition of oil, gas, and coal corporations announced that they would substantially increase their campaign contributions in the next congressional races.

FIGURE 24.11 The Supreme Court decides pivotal cases, many of them bearing on resources or environmental health.
© Roger L. Wollenberg/Pool/Corbis

Many commentators, including dissenting members of the Supreme Court, argue that the *Citizens United* decision protects corporations from a wide range of public oversight and legal restrictions. Among other concerns, this decision opens the door to new levels of corporate influence in justice, as well as politics. For example, a coal company could easily fund a judge's election campaign. If a community filed a suit against the coal company for polluting streams, or for unsafe working conditions, could they be sure to receive a fair judgment?

A dissenting opinion from the four justices who voted against the *Citizens United* decision stated that "corporations have no consciences, no beliefs, no feelings, no thoughts, no desires . . . and their 'personhood' often serves as a useful legal fiction. But they are not themselves members of 'We the People' by whom and for whom our Constitution was established." The dissent states further, "The fact that corporations are different from human beings might seem to need no elaboration, except that the majority opinion almost completely elides it."

Legal Standing

Before a trial can start, the litigants must establish that they have **standing,** or a right to stand before the bar and be heard. The main criteria for standing is a valid interest in the case. Plaintiffs must show that they are materially affected by the situation they petition the court to redress. This is an important point in environmental cases. Groups or individuals often want to sue a person or corporation for degrading the environment. But unless they can show that they personally suffer from the degradation, courts are likely to deny standing.

In a landmark 1969 case, *Sierra Club v. Morten,* the Sierra Club challenged a decision of the Forest Service and the Department of the Interior to lease public land in California to Walt Disney Enterprises for a ski resort. The land in question was a beautiful valley that cut into the southern boundary of Sequoia National Park. Building a road into the valley would have required cutting down a grove of giant redwood trees within the park (fig. 24.12). The Sierra Club argued that it should have standing in the case to represent the trees, animals, rocks, and mountains that couldn't defend their own interests in court. After all, the club pointed out, corporations—such as Disney—are treated as persons and represented by attorneys in the courts. Why not grant trees the same representation? An influential legal brief by Christopher D. Stone titled *Should Trees Have Standing?* proposed that organisms as well as ecological systems and processes should have legally recognized rights.

The case went all the way to the Supreme Court, which ruled that the Sierra Club failed to show that its members would be materially affected by the development. However, the Court established a key precedent in stating that "aesthetic and environmental wellbeing, like economic wellbeing, are important ingredients of the quality of life" and are "deserving of legal protection."

Criminal Law Prosecutes Lawbreakers

Violation of many environmental statutes constitutes criminal offenses. In 1975, the U.S. Supreme Court ruled that corporate officers can be held criminally liable for violations of environmental laws if they were grossly negligent, or the illegal actions can

FIGURE 24.12 In 1969, the Sierra Club sued the U.S. Department of the Interior on behalf of the giant redwood trees in Sequoia National Park. The U.S. Supreme Court ruled that the Sierra Club didn't have legal standing in this case, but agreed that "aesthetic and environmental well-being" deserve protection.
© Royalty-Free/Corbis

be considered willful and knowing violations. In 1982, the EPA created an Office for Criminal Investigation. Under the Clinton administration, prosecutions for environmental crimes rose to nearly 600 per year. They fell by 75 percent under George W. Bush, however. The Obama administration has again increased new prosecutions to about 400 per year.

Civil law regulates relations between individuals or between individuals and corporations (which have the rights of individuals under U.S. law). Property rights and personal dignity and freedom are protected by civil law. Sometimes legislative statutes, such as the Civil Rights Act, establish specific aspects of civil law. Custom and previous court decisions, collectively called **common law,** can also establish precedents that constitute a working definition of individual rights and responsibilities.

Criminal offenses can lead to jail, while civil cases lead only to fines. Civil judgments can be costly, however. In 2000, the Koch oil company, one of the largest pipeline and refinery operators in the United States, agreed to pay $35 million in fines and penalties to state and federal authorities for negligence in more than 300 oil spills in Texas, Oklahoma, Kansas, Alabama, Louisiana, and Missouri between 1990 and 1997. Koch also agreed to spend more than $1 billion on cleanup and improved operations. In the aftermath of the 2010 Gulf oil spill, BP (formerly British Petroleum) agreed to pay a total of $20.8 billion to settle civil lawsuits, criminal penalties, and fines for violations of the U.S. Clean Water Act. The clean-up costs added another $40 billion to this amount.

Sometimes the purpose of a civil suit is to prevent harmful actions. You might ask the courts, for example, to order the government to cease and desist from activities that are in violation of either the spirit or the letter of the law. Public interest groups have often asked courts to stop logging and mining operations, and to enforce implementation of laws regarding endangered species, public health, and air and water pollution laws.

Lawsuits can also be used to stop public interest groups from challenges to industry. Because defending a lawsuit is so expensive, the mere threat of litigation can be a chilling deterrent. Increasingly environmental activists are being harassed with **strategic lawsuits against public participation (SLAPP).** Citizens who criticize businesses that pollute or government agencies that neglect their responsibility to protect the public are often sued in retaliation.

Most of these preemptive strikes are groundless and ultimately dismissed, but defending yourself against them can be cripplingly expensive and they can halt progress on the issue in question. Public interest groups and individual activists—many of whom have little money to defend themselves—often are intimidated from taking on polluters. For example, when a West Virginia farmer wrote an article about a coal company's pollution of the Buckhannon River, the company sued him for $200,000 for defamation. Similarly, citizen groups fighting a proposed incinerator in upstate New York were sued for $1.5 million by their own county governments. A Texas woman called a nearby landfill a dump—and her husband was named in a $5 million suit for failing to "control his wife."

Executive agencies make rules and enforce laws

More than 100 federal agencies and thousands of state and local boards and commissions oversee environmental policies. They set rules, adjudicate disputes, and investigate misconduct and operate parks and refuges (fig. 24.13). In the federal government, the president's cabinet (a group of department heads in charge of various tasks) includes the heads of major departments such as Agriculture, Interior, or Justice. Rules that these agencies make and enforce decide many of our most important environmental, resource, and health issues.

The public can influence agency rule-making by giving comments on a proposed rule. Comments can be made at regional hearings, which agencies are required to offer, and by mail or email during a "public comment period," usually a few months, that precedes the

FIGURE 24.13 Federal agencies administer public lands and enforce environmental regulations in places, such as Glacier National Park in Montana.
© William P. Cunningham

acceptance of a new rule or action. Often, thousands of comments are collected, and these can substantially influence whether or not a proposed rule or plan is enacted.

Executive orders also can be powerful agents for change. President Barack Obama, for example, used the Antiquities Act more than any other President, establishing 23 new national monuments and expanding dozens of existing national parks and wildlife refuges. Altogether, Obama ordered protection for 107 million ha (265 million acres) of land and water.

Rules and policies made by executive decree in one administration can be quickly undone in the next one. In his first day in office, President George W. Bush ordered all federal agencies to suspend or ignore more than 60 rules and regulations from the Clinton administration. In addition, President Bush called for a sweeping overhaul of environmental laws to ease restrictions on businesses and to speed decisions on development projects, and he prevented implementation of many other environmental rules. Because most of this agenda was pursued through agency regulations and executive orders, most Americans, distracted by fears of terrorism and lingering wars in Afghanistan and Iraq, were unaware of the magnitude and implications of this abrupt policy shift. Barack Obama, in turn, reversed many of President Bush's executive rules, restoring environmental and social protections.

An important administrative rule that has been reversed repeatedly by these presidents is the moratorium (ban) on building new logging roads in nearly 24 million ha of roadless public land. This rule, also established under Clinton, defends wild areas that have no other legal protection (*de facto* wilderness). The pro-industry Bush administration reversed Clinton's rules; the pro-conservation Obama administration then overturned Bush's reversals.

Regulatory Agencies

The EPA is the primary agency with responsibility for protecting environmental quality. It was created in 1970, at the same time as NEPA, and its head is appointed by the president as part of his cabinet, the small body of top administrators who work directly with the president. Like other cabinet positions, the EPA is strongly influenced by political winds. EPA funding, enforcement activity, and rule-making strongly reflect the views of the president's political advisers. Under the Nixon and Carter administrations, the EPA grew rapidly and enforced air and water quality standards vigorously. EPA activity and funding declined sharply during the Reagan administration, recovered under Bill Clinton, declined again during the Bush administrations, and rose again under Obama.

The Departments of the Interior and Agriculture manage natural resources. The National Park Service, which is responsible for more than 376 national parks, monuments, historic sites, and recreational areas, is part of the Interior Department. Other Interior agencies are the Bureau of Land Management (BLM), which administers some 140 million ha of land, mostly in the western United States, and the Fish and Wildlife Service, which operates more than 500 national wildlife refuges and administers endangered species protection.

The Department of Agriculture is home to the U.S. Forest Service, which manages about 175 national forests and grasslands,

FIGURE 24.14 Smokey Bear is the public face of an executive agency, the U.S. Forest Service.
© William P. Cunningham

totaling some 78 million ha (fig. 24.14). With 39,000 employees, the Forest Service is nearly twice as large as the EPA. The Department of Labor houses the Occupational Safety and Health Agency (OSHA), which oversees workplace safety, including exposure to toxic substances. In addition, several independent agencies that are not tied to any specific department also play a role in environmental protection and public health. The Consumer Products Safety Commission passes and enforces regulations to protect consumers, and the Food and Drug Administration is responsible for the purity and wholesomeness of food and drugs.

All of these agencies have a tendency to be "captured" by the industries they are supposed to be regulating. Many of the people with expertise to regulate specific areas came from the industry or sector of society that their agency oversees. Furthermore, the people they work most closely with and often develop friendships with are those they are supposed to watch. And when they leave the agency to return to private life—as many do when the administration changes—they are likely to go back to the same industry or sector where their experience and expertise lies. The effect is often what's called a "revolving door," where workers move back and forth between industry and government. As a result, regulators often become overly sympathetic with and protective of the industry they should be overseeing.

How much government do we want?

In his 1981 inaugural address, President Ronald Reagan famously said, "Government is not the solution to our problem; government is the problem." In this, he invoked a perennial debate in American politics: Is government a power that undermines personal liberties? Or is government a representative of the people and a defender of personal liberties against bullies?

The answer sometimes depends on when you ask. During political campaigns, many of us decry the size and cost of government agencies. But in a crisis most of us assume the government will be there to help us out, as it did after a swarm of tornadoes swept across the American South in early 2016. Within hours, affected states welcomed federal rescue teams, emergency aid, and federal funding for reconstruction (fig. 24.15).

FIGURE 24.15 When tornadoes, floods, hurricanes, or other natural disasters afflict us, we expect government agencies to help us.
© Julie Denesha/Getty Images

Debates about the proper size and role of government are common. We value self-reliance and rugged individualism. Yet we also want someone to protect us from contaminated food and drugs, to educate our children, and to provide roads, bridges, and safe drinking water. President Reagan was among those who favor "free market" capitalism, with businesses unfettered by rules such as the Clean Air Act or the Clean Water Act. Political strategist Grover Norquist, president of the Americans for Tax Reform, famously said he'd like to "shrink the government down to the size where we can drown it in the bathtub." Other observers note that while Reagan's language focused on freedom for individuals and small business owners, the elimination of public health and safety regulations tends to undermine the interests of individuals and small businesses. Advantages are often given to the biggest players in a game without rules.

Part of the reason these disputes persist may be that both views are partially correct. Regulations, such as those imposing expensive pollution abatement technologies for polluters, require private businesses to bear the cost of protecting public resources. Businesses are squeezed between shareholders' demands for ever-higher profits and agency demands for safer, sometimes costly, operating standards. Viewed another way, regulations require businesses to clean up their own messes. Opponents of agency regulations point out that corporations produce the economic vitality on which our prosperity depends. Proponents of regulations point out that corporations couldn't prosper without subsidies, tax breaks, transportation infrastructure, and a healthy and educated workforce. These costs are necessary to doing business but are normally external to the accounting of business costs and profits.

Since about 1981, the small-government philosophy has dismantled much of the regulatory structure set up during President Nixon's term in office. Often, this has been done by agency heads, who have been appointed despite openly opposing the existence of those agencies and their laws. For example, President George W. Bush appointed Christopher Cox, a proponent of bank deregulation, to chair the Securities and Exchange Commission, which oversees Wall Street trading. Subsequent dismantling of trading rules led to risky behavior by banks, which culminated in the Wall Street collapse in 2007–2008. Business failures and high unemployment

spread nationwide and have lasted for years. President Bush also oversaw dramatic reductions in USDA food safety inspections, on the grounds that they represented unnecessary interference in the private business of the food industry. Increasingly, frequent food contamination scares have made many Americans rethink the importance of government inspectors in the food system.

These debates probably will always be with us. Which view is most correct depends on many factors: your interest group, life experience, philosophical perspective, economic position, the time frame you analyze, and other priorities. What factors influence your view on these issues? Do you think there is room for compromise? If not, why not? If yes, where?

Section Review

1. Describe the path of a bill through Congress. When are riders and amendments attached?
2. What is legal standing, and why is it important?
3. What regulatory agencies have major environmental oversight in the United States?

24.4 INTERNATIONAL CONVENTIONS

- *Many international agreements protect our environment.*
- *International enforcement often depends on national honor.*

Growing interconnections in our global environment and economy have made nations increasingly interested in signing on to international agreements (conventions) for environmental protection. A principal motivation for participating in these treaties is the recognition that countries can no longer act alone to protect their resources and interests. Water resources, the atmosphere, trade in endangered species, and many other concerns cross international boundaries. Over time, the number of parties taking part in negotiations has grown sharply (fig. 24.16). The speed at which agreements take force has also increased. The Convention on International Trade in Endangered Species (CITES), for example, was not enforced until 14 years after its ratification in 1973, but the Convention on Biological Diversity (1991) was enforceable after just one year, with 160 nations signing the agreement just four years after its introduction.

Over the past 25 years, more than 170 treaties and conventions have been negotiated to protect our global environment. These agreements have focused on concerns ranging from intercontinental shipping of hazardous waste, to deforestation, overfishing, trade in endangered species, global warming, and wetlands protection.

Major international agreements

International accords and conventions have emerged slowly but fairly steadily from meetings such as those in Stockholm and Rio (table 24.2). A few of the important benchmark agreements are discussed here.

The **Convention on International Trade in Endangered Species** (CITES, 1973) declared that wild flora and fauna are

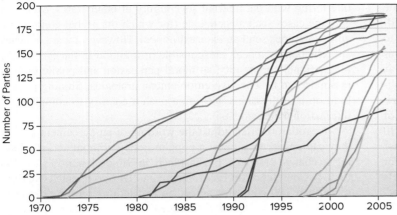

Ramsar Convention, wetland protection
World Heritage Program, protecting cultural sites
Stockholm Declaration
CITES
Convention on Migratory Species
UN Convention on Law of the Sea
Montreal Protocol, ozone protection
Basel Convention on hazardous waste
Convention on Biological Diversity
UN Convention on Combat Desertification
UN Framework Convention on Climate Change
Kyoto Protocol on climate change
Rotterdam Convention, restricting industrial chemicals
Cartagena Protocol on biotechnology

FIGURE 24.16 Major international environmental agreements, listed in order of ratification dates.

valuable, irreplaceable, and threatened by human activities. To protect disappearing species, CITES maintains a list of threatened and endangered species that may be affected by trade. As with most international agreements, this one takes no position on movement or loss of species within national boundaries, but it establishes rules to restrict unauthorized or illegal trade across boundaries. In particular, an export permit is required specifying that a state expert declares an export is legal, that it is not cruel, and that it will not threaten a wild population.

The **Montreal Protocol** (1987) protects stratospheric ozone. This treaty committed signatories to phase out the production and use of several chemicals that break down ozone in the atmosphere. The ozone "hole," a declining concentration of ozone (O_3) molecules over the South Pole, threatened living things: Ozone high in the atmosphere blocks cancer-causing ultraviolet radiation, keeping it from reaching the earth's surface. The stable chlorine- and fluorine-based chemicals at fault for reducing ozone are used mainly as refrigerants. Alternative refrigerants have since been developed, and the use of chlorofluorocarbons (CFCs) and related molecules has plummeted. Although the ozone "hole" has not disappeared, it has declined as predicted by atmospheric scientists since the phase-out of CFCs. The Montreal Protocol is often held up as an example of a highly successful and effective international environmental agreement.

The Montreal Protocol was effective because it bound signatory nations not to purchase CFCs or products made using them

from countries that refused to ratify the treaty. This trade restriction put substantial pressure on producing countries. Initially, the protocol called for only a 50 percent reduction in CFC production, but subsequent research showed that ozone was being depleted faster than previously thought (chapter 16). The protocol was strengthened to an outright ban on CFC production, in spite of the objection of a few countries.

The **Basel Convention** (1992) restricts shipment of hazardous waste across boundaries. The aim of this convention, which has 172 signatories, is to protect health and the environment, especially in developing areas, by stating that hazardous substances should be disposed of in the states that generated them.

Signatories are required to prohibit the export of hazardous wastes unless the receiving state gives prior informed consent, in writing, that a shipment is allowable. Parties are also required to minimize production of hazardous materials and to ensure that there are safe disposal facilities within their own boundaries. This convention establishes that it is the responsibility of states to make sure that their own corporations comply with international laws. The Basel Convention was enhanced by the Rotterdam Convention (1997), which places similar restrictions on unauthorized transboundary shipment of industrial chemicals and pesticides.

The 1994 **UN Framework Convention on Climate Change (UNFCCC)** directs governments to share data on climate change, to develop national plans for reducing greenhouse gas (GHG)

Table 24.2	Some Important International Environmental Treaties
CBD: Convention on Biological Diversity 1992 (1993)	
CITES: Convention on International Trade on Endangered Species of Wild Fauna and Flora 1973 (1987)	
CMS: Convention on the Conservation of Migratory Species of Wild Animals 1979 (1983)	
Basel: Basel Convention on the Transboundary Movements of Hazardous Wastes and their Disposal 1989 (1992)	
Ozone: Vienna Convention for the Protection of the Ozone Layer and Montreal Protocol on Substances that Deplete the Ozone Layer 1985 (1988)	
UNFCCC: United Nations Framework Convention on Climate Change 1992 (1994)	
CCD: United Nations Convention to Combat Desertification in those Countries Experiencing Serious Drought and/or Desertification, Particularly in Africa 1994 (1996)	
Ramsar: Convention on Wetlands of International Importance Especially as Waterfowl Habitat 1971 (1975)	
Heritage: Convention Concerning the Protection of the World Cultural and Natural Heritage 1972 (1975)	
UNCLOS: United Nations Convention on the Law of the Sea 1982 (1994)	

emissions, and to cooperate in planning for adaptation to climate change. Under the UNFCCC, the Kyoto Protocol (1997) set binding targets for signatories to reduce greenhouse gas emissions to less than 1990 levels by 2012. The idea of binding targets was strong, but few countries achieved their goals, and some, such as the United States, declined to be subjected to international policies, particularly because Kyoto set tighter restrictions on industrialized countries, which are responsible for roughly 90 percent of GHG emissions up to the present, than on developing countries. In 2012, Canada also withdrew from the agreement, in part because energy-intensive development of oil shale makes it impossible for Canada to meet its targets. In 2015, the UNFCCC led to the Paris Agreement, which set a climate change target of "well below" 2 degrees C or less, and in which the 195 signatory countries agreed to establish their own national determined contributions to this goal. As with Kyoto, plans and reporting are to be transparent, with regular updates as technology improves (chapter 15, section 15.6).

Some observers consider Paris to be a historic turning point in the effort to reduce climate change, while others criticize the fact that promises are not enforceable. Particularly important questions are whether the United States, where Congress has blocked clean power legislation, and China, which continues to build coal power plants, will be able to achieve intended goals (fig. 24.17).

Enforcement often depends on national pride

In the Paris Agreement and many other international accords, there is no clear means of forcing parties to make good on promises, so enforcement often depends on the fact that countries care about their international reputation. In general, states are wary of interfering with the internal sovereignty of other states, except in extreme cases such as genocide (for which the global community will sometimes send an external police). However, most countries are reluctant to appear irresponsible or immoral in the eyes of the international community, so moral persuasion and public embarrassment can be effective enforcement strategies. Shining a spotlight on transgressions will often push a country to comply with international agreements.

Negotiators often adjust the wording of an agreement so that they can appear to support a popular policy without actually having to change practices. For instance, in Rio de Janeiro in 1992 at the UN Conference on Environment and Development (UNCED), more than 100 countries agreed to restrictions on the release of greenhouse gases. At the insistence of U.S. negotiators, however, the climate convention was reworded so it only urged—and did not require—nations to stabilize their emissions. Similarly, in 2010 negotiations over CITES species protection, Japan almost single-handedly derailed global protections for bluefin tuna, a huge, long-lived fish whose populations have dropped more than 98 percent in some areas (see chapter 6).

When strong accords with meaningful sanctions cannot be passed, sometimes the pressure of world opinion generated by revealing the sources of pollution can be effective. Activists can use this information to expose violators. For example, the environmental group Greenpeace discovered monitoring data in 1990 showing that Britain was disposing of coal ash in the North Sea. Although not explicitly forbidden by the Oslo Convention on ocean dumping, this evidence proved to be an embarrassment, and the practice was halted.

Trade sanctions can be an effective tool to compel compliance with international treaties. The Montreal Protocol used the threat of trade sanctions very effectively to cut CFC production dramatically. On the other hand, trade agreements also can work against environmental protection. The World Trade Organization (WTO) was established in 1995 to promote free international trade and to encourage economic development. The WTO's emphasis on unfettered trade, however, has led to weakening of local environmental rules. For example, in 1990 the United States banned the import of tuna caught using methods that kill thousands of dolphins each year. Shrimp caught with nets that kill endangered sea turtles were also banned. Mexico filed a complaint with the WTO, contending that dolphin-safe tuna laws represented an illegal barrier to trade. Thailand, Malaysia, India, and Pakistan filed a similar suit against turtle-friendly shrimp laws. The WTO ordered the United States to allow the import of both tuna and shrimp from countries that allow fisheries to kill dolphins and turtles. Environmental advocates point out that the WTO has never ruled against a corporation because it is composed of industry leaders. As such, the WTO mainly defends the interests of the business community, not the broader public interest.

FIGURE 24.17 President Barack Obama met with Chinese premier Xi Jinping at the United Nations Climate Change Conference in Paris in 2015. Their two countries are the largest emitters of GHGs.
© Jim Watson/AFP/Getty Images

Section Review

1. Describe five important international environmental treaties.
2. Why has most international environmental protection been based on an honor system?

24.5 NEW APPROACHES TO POLICY

- *Community-based planning uses local knowledge.*
- *Green plans outline goals for sustainability.*
- *Bolivia set a world precedent with its Law of Mother Earth.*

As we gain experience with environmental governance, new policy strategies are being developed. These approaches are growing in part because environmental protection has remained a high priority for the public (fig. 24.18). One of the key changes has been to seek win–win compromises in environmental debates. Dispute resolution and mediation are strategies for reaching agreements without the mutual suspicions and hostility inherent in a lawsuit. Dispute resolution can avoid the time, expense, and winner-take-all confrontation inherent in lawsuits, these techniques encourage compromise and workable solutions with which everyone can live.

Arbitration is a formal process of dispute resolution somewhat like a trial. There are stringent rules of evidence, cross-examination of witnesses, and the process results in a legally binding decision. The arbitrator can actively work to find creative resolutions to the dispute. **Mediation** is generally less formal. Disputants are encouraged to sit down and talk to see if they can come up with a solution. In face-to-face meetings, people are often more willing to see their opponent's viewpoint and seek solutions.

Less rigid strategies for rule making are also being developed. One of these is **adaptive management,** or "learning by doing." This approach proposes that management should be experimental. Environmental policies should be designed from the outset to test clearly formulated hypotheses about the ecological, social, and economic impacts of the actions being undertaken (fig. 24.19). What initially seemed to be the best policy may not always be best, so we need to carefully monitor how conditions are changing. And we need to be able and willing to revise plans if our initial assumptions don't hold up over time.

Ecological principles also suggest that policymakers should plan for resilience—that is, for changes and recovery in a system (table 24.3). This means, for example, that protected forests must be large enough to allow for disturbances (such as fires or pest outbreaks) and recovery, or that policymakers should anticipate the possibility that the climate or species abundance may change over time.

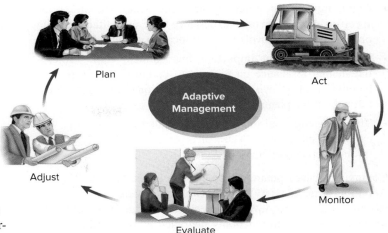

FIGURE 24.19 Adaptive management recognizes that we need to treat management plans for ecosystems as scientific experiments in which we monitor, evaluate, and adjust our policies to fit changing conditions and knowledge.

Community-based planning uses local knowledge

Over the past several decades, natural resource managers have come to recognize the value of holistic planning that acknowledges multiple users and perspectives. Involving all stakeholders and interest groups early in the planning process can help avoid the "train wrecks" in which adversaries become entrenched in non-negotiable positions. Working with local communities can tap into traditional knowledge and gain acceptance for management plans that finally emerge from policy planning. There are many reasons to use collaborative approaches:

- Incorporating a variety of perspectives early in the process is more likely to lead to the development of acceptable solutions in the end. Public buy-in to an idea is likely to be better if many people have a voice from the start.

- Two heads are better than one. Involving multiple stakeholders and multiple sources of information enriches the process.

Table 24.3 Planning for Resilience

1. Interdisciplinary, integrated modes of inquiry are needed for adaptive management of wicked problems.

2. We must recognize that these problems are fundamentally nonlinear and that we need nonlinear approaches to them.

3. We must attend to interactions between long-term processes, such as climate change or soil erosion in the American Corn Belt, and rapid events, such as the collapse of Antarctic ice sheets or the appearance of a dead zone in the Gulf of Mexico.

4. The spatial and temporal scales of our concerns are widening. We must consider global interconnections in our planning.

5. We need adaptive management policies that focus on building resilience and the capacity of renewal both in ecosystems and in human institutions.

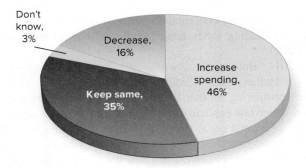

FIGURE 24.18 Americans persistently favor maintaining or increasing spending on environmental protections.

- Community-based planning provides access to situation-specific information and experience that can often only be obtained by active involvement of local residents.

- Participation is an important management tool. Project-threatening resistance on the part of certain stakeholders can be minimized by inviting active cooperation of all stakeholders throughout the planning process.

- The knowledge and understanding needed by those who will carry out subsequent phases of a project can only be gained through active participation.

Community-based planning can be seen in the Atlantic Coastal Action Programme (ACAP) in eastern Canada. The purpose of this project is to develop blueprints for the restoration and maintenance of environmentally degraded harbors and estuaries in ways that are both biologically and socially sustainable. Officially established under Canada's Green Plan and supported by Environment Canada, this program created 13 community groups, some rural and some urban, with membership in each dominated by local residents. Federal and provincial government agencies are represented primarily as nonvoting observers and resource people. Each community group is provided with core funding for full-time staff who operate an office in the community and facilitate meetings.

Four of the 13 ACAP sites are in the Bay of Fundy, an important and unique estuary lying between New Brunswick and Nova Scotia. Approximately 270 km long, and with an area of more than 12,000 km^2, the bay, together with the nearby Georges Bank and the Gulf of Maine once formed one of the richest fisheries in the world (fig. 24.20a). With the world's highest recorded tidal range (up to 16 m at maximum spring tide), the bay still sustains a great variety of fishery and wildlife resources, and provides habitat for a number of rare or endangered species. Now home to more than 1 million people, the coastal region is an important agricultural, lumbering, and paper-producing region. Pollution and sediment damage harbors and biological communities. Overfishing and introduction of exotic species have resulted in endemic species declines. The collapse of cod, halibut, and haddock fishing has had devastating economic effects on the regional economy and the livelihoods of local residents. To cope with these complex, intertwined social and biological problems, ACAP is bringing together different stakeholders from around the bay to create comprehensive plans for ecological, economic, and social sustainability, including citizen monitoring and adaptive management.

Green plans outline goals for sustainability

Several national governments have undertaken integrated environmental planning that incorporates community round-tables for vision development. Canada, New Zealand, Sweden, and Denmark all have so-called **green plans** or comprehensive, long-range national environmental strategies. The best of these plans weave together complex systems, such as water, air, soil, and energy, and mesh them with human factors such as economics, health, and carrying capacity. Perhaps the most thorough and well-thought-out green plan in the world is that of the Netherlands (fig. 24.20b).

Developed in the 1980s through a complex process involving the public, industry, and government, the 400-page Dutch plan contains 223 policy changes aimed at reducing pollution and establishing economic stability. Three important mechanisms have been adopted for achieving these goals: integrated life-cycle management, energy conservation, and improved product quality. These measures should make consumer goods last longer and be more easily recycled or safely disposed of when no longer needed. For example, auto manufacturers are now required to design cars so they can be repaired or recycled rather than being discarded.

Among the guiding principles of the Dutch green plan are: (1) the "stand-still" principle that says environmental quality will not deteriorate, (2) abatement at the source rather than cleaning up afterward, (3) the "polluter pays" principle that says users of a resource pay for the negative effects of that use, (4) the prevention of unnecessary pollution, (5) application of the best practicable means for pollution control, (6) carefully controlled waste disposal, and (7) motivating people to behave responsibly.

The Netherlands have invested billions of euros in implementing this comprehensive plan. Some striking successes already have been accomplished. Between 1980 and 1995, emissions of sulfur dioxide, nitrogen oxides, ammonia, and volatile organic compounds were reduced more than 30 percent; pesticide use had been reduced 25 percent; chlorofluorocarbon use had been virtually eliminated; and industrial wastewater discharge into the Rhine River was down 70 percent. Some 250,000 ha (more than 600,000 acres) of former wetlands that had been drained for agriculture are being restored as nature preserves and 40,000 ha (99,000 acres) of forest are being replanted. This is remarkably generous and foresighted for Europe's most densely populated country.

Bolivia's Law of Mother Earth

The small, impoverished country of Bolivia took remarkably strong stands on many environmental issues with the election of President Evo Morales in 2005. Most of Bolivia's population lives in poor farming communities, directly dependent on natural resources, including water, healthy soil, and natural biodiversity (fig. 24.20c).

In 2011, Bolivia set a world precedent by proposing the "Law of Mother Earth." Following indigenous Andean traditions of considering Mother Earth, or *Pachamama,* to be a living being, the new law explicitly aims to protect life and biodiversity. It grants all people equal rights to a clean environment, including safe water, protection of biodiversity, clean air, and essential ecological functions. Specific terms of the law include requiring

(a)

(b)

(c)

FIGURE 24.20 Innovations in environmental policy and planning can be found worldwide. Examples include community planning in Canada's Bay of Fundy (a), the Dutch green plan that restores ecosystems (b), and the Law of Mother Earth, protecting rights of all Bolivians to a healthy environment (c).
(a): © Rolf Hicker/All Canada Photos/Getty Images (b): © Royalty-Free/Corbis (c): © AFP/ Getty Images

the government to transition toward renewable energy, to develop new economic indicators that account for environmental costs of economic activities, to focus on food sovereignty, and to invest in energy efficiency.

None of these steps will be easy, but like all policies, the first step is to identify a goal. The law also sets a standard by which later policy decisions can be judged. Figuring out how to reach that goal may take years. But without a statement of policy intent, progress might never happen. Can other nations do the same thing? What factors might support or discourage other places from developing their own Mother Earth Laws?

Section Review

1. What is community-based planning?
2. What are green plans?
3. Describe Bolivia's Law of Mother Earth.

Conclusion

Otto von Bismark once said, "Laws are like sausages, it is better not to see them being made." Still, if you hope to improve your environmental quality, it's helpful to understand how policies and laws are made and enforced. Laws, such as the Clean Water Act have been among the most effective tools that conservationists have had to protect biodiversity and habitat. But there is a constant struggle between those who want to strengthen environmental laws and those who want to reduce or eliminate them.

The legislative, administrative, and judicial branches of government all contribute to stages in the policy cycle. Though ordinary individuals might feel powerless in these various stages, you might be surprised at how much impact you can have if you get involved. Probably the best way to participate in environmental policy formation or passage of environmental laws is to join a group that shares

your concerns. Being part of a group amplifies your influence. But even as an individual, you can make an impression. Write to or call your legislator. They do pay attention to constituents.

Global cooperation has also emerged as a key part of environmental protection. Dozens of laws protect resources, biodiversity, and environmental quality. Mechanisms for enforcement are not as obvious as they are within a single country, but creative strategies are evolving, and most nations see it to be in their interest to cooperate with their neighbors, most of the time.

On a smaller scale, community planning and community knowledge have also become key parts of policy formation. Understanding all these aspects of policy is a first step toward empowering yourself to influence the health of the environment in which you live.

Reviewing Key Terms

Can you define the following terms in environmental science?

adaptive management
    24.5

administrative law 24.3

arbitration 24.5

Basel Convention 24.4

case law 24.3

civil law 24.3

common law 24.3

Convention on International
    Trade in Endangered
    Species (CITES) 24.4

cost–benefit analysis
    (CBA) 24.1

environmental impact
    statement (EIS)

environmental policy 24.1

green plans 24.5

mediation 24.5

Montreal Protocol 24.4

National Environmental
    Policy Act (NEPA) 24.2

policy 24.1

precautionary principle 24.1

riders 24.3

standing 24.3

statute law 24.3

strategic lawsuits against
    public participation
    (SLAPPs) 24.3

UN Framework Convention
    on Climate Change
    (UNFCC) 24.4

Critical Thinking and Discussion Questions

1. In your opinion, how much environmental protection is too much? Think of a practical example in which some stakeholders may feel oppressed by government regulations. How would you justify or criticize these regulations?

2. Among the steps in the policy cycle, where would you put your efforts if you wanted influence in establishing policy?

3. Do you believe that trees, wild animals, rocks, or mountains should have legal rights and standing in the courts? Why or why not? Are there other forms of protection you would favor for nature?

4. It's sometimes difficult to determine whether a lawsuit is retaliatory or based on valid reason. How would you define a SLAPP suit and differentiate it from a legitimate case?

5. Create a list of arguments for and against an international body with power to enforce global environmental laws. Can you see a way to create a body that could satisfy both reasons for and against this power?

6. Identify a current environmental problem, and outline some policy approaches that could be used to address it. What strengths and weaknesses would different approaches have?

Data Analysis

Examine Your Environmental Laws

The federal government publicizes the text of laws in multiple locations on the Internet. Reading about these laws is a good way to get a sense of the structures of environmental regulation, and to understand some of the compromises and the complexity of making rules that apply to thousands of different cases across the country. The primary way to access government rules and laws is through www.thomas.gov. A more direct source for environmental legislation is to go to the EPA website: www.epa.gov/lawsregs/laws/index.html#env.

Go to the EPA website, and then log in to Connect to find suggestions about how to examine a specific law.

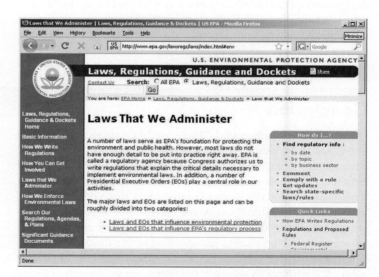

The web address listed above gives you direct access to federal laws that define how the U.S. environment and resources are managed.

25

What Then Shall We Do?

▲400,000 people march in New York City in 2014 to protest climate change. This was the largest climate march in history.
© Robert van Waarden/Survival Media Agency

Learning Outcomes

After studying this chapter, you should be able to:

25.1 Explain why environmental education is important.
25.2 Evaluate what individuals can do.
25.3 Review how we can work together.
25.4 Investigate campus greening.
25.5 Define the challenge of sustainability.

"When spiders unite, their web can tie down an elephant."

– African proverb

350.org: Building a Climate Movement

Could a handful of students at a small liberal arts college in Vermont mobilize an international campaign to tackle the most important environmental challenge we face today? They already have. And it's an example from which all of us could learn. It all started a few years ago when author Bill McKibben, who is scholar in residence at Middlebury College, asked six students to help organize an event focused on global climate change.

They realized they didn't have the money or connections to fight directly against the fossil fuel industry or its minions in Congress. So they decided to use the power of the Internet and social networks to mobilize activists to do something meaningful and newsworthy to call attention to the issue. By acting together on the same day in different places, and publishing photos and news releases to show the interconnections between actions, activists could create a meta-event with greater power to influence local citizens and decision makers than isolated gatherings could have done.

Their first event was called "Step It Up." In 2007, with a minuscule budget and little previous experience, the fledgling group inspired tens of thousands of citizens to participate on a single day in more than 1,400 events in iconic places in all 50 of the United States. These creative actions—from skiers descending a melting glacier to protest global warming, to activists planting endangered chestnut trees to absorb CO_2, to people flying thousands of handmade kites with environmental messages—were designed both to attract attention and to educate the public about the need to cut carbon emissions.

Building on this success, the students decided to broaden their campaign to the international stage. Renaming themselves 350.org, they expanded their team to include young people from all over the world. They chose 350 because it's the number (in parts per million) that climate scientists say is the safe upper limit for CO_2 in the atmosphere. We're already past 400 ppm, and politicians are debating whether we might hold emissions to 450 ppm, so the organization has chosen an ambitious goal. But why not dream big?

Over the past decade, 350.org has organized thousands of other grassroots events in more than 188 countries. CNN called

FIGURE 25.1 Canadian First Nations people and native American tribes join with 350.org in 2015 to protest the Keystone XL pipeline and tar sands extraction.
© Madeleine Desrochers

their International Day of Action in 2009 "the most widespread day of political action in the planet's history." Many events have featured blockades or protest marches, but others have been based on positive steps, such as putting up solar panels, starting community gardens, or planting trees to help reduce carbon emissions. Working together on pragmatic local projects empowers people, gives them hope, and helps build grassroots networks.

Some actions have included civil disobedience. In a 2015 rally in Washington D.C., for example, 1,250 volunteers, politicians, and media personalities were arrested peacefully as they encircled the White House to show their opposition to the Keystone XL pipeline. And in 2016, hundreds of people in Vancouver, BC, used canoes and kayaks to protest oil pipelines and terminals that would trespass on First Nations lands and contribute to climate change. Both of these actions are part of a larger project to keep carbon in the ground. Some other major goals are to build a new, more equitable low-carbon economy, and to limit fossil fuel emissions.

The group's largest event (so far) was the 2014 people's climate march in New York City. More than 400,000 people representing 1,574 organizations marched through Manhattan to demand international action to address this problem. Simultaneously, 2,646 events in 162 countries expressed solidarity with the marchers. The event generated 630,000 social media posts, and 5,200 printed articles. As part of this struggle, 350.org is encouraging churches, colleges, and retirement plans to divest from coal, oil, and gas corporations, and to help build a fossil-free future.

Among the core principles of 350.org are support of climate justice, a belief that we're stronger when we collaborate, and a conviction that mass mobilizations can bring about change. They have captured media attention and demonstrated to political leaders our widespread desire for environmental protection. They also unleashed creativity, motivated a wide diversity of people, and offered a hopeful way to express opinions about the future of our world. Wouldn't you like to get involved? The group has chapters in many places. Go to www.350.org to get suggestions for how to plan an event, create a press release, invite elected officials and media, follow up, and get other useful resources.

In this chapter, we'll look at how other individuals and groups also are working to build a sustainable future.

25.1 Making a Difference

- *Environmental stewardship is everyone's responsibility.*
- *Environmental literacy is a national goal and a lifelong process.*

We are surrounded by environmental challenges, from climate change to biodiversity to energy policy debates. Biodiversity is disappearing at the fastest rate in human history; major ocean fisheries have collapsed; within 50 years, it is expected that two-thirds of countries will experience water shortages, and 3 billion people may live in slums. As we have come to understand these problems, many exciting innovations have been developed to deal with them. New irrigation methods reduce agricultural water use; bioremediation provides inexpensive methods to treat hazardous waste; new energy sources, including wind, solar, geothermal, and even pressure-cooked garbage, offer strategies for weening our society from its dependence on oil and gas. Growth of green consumerism has developed markets for recycled materials, low-energy appliances, and organic foods. Population growth continues, but its rate has plummeted from a couple of generations ago.

Stewardship for our shared resources is increasingly understood to be everybody's business. The environmental justice movement (chapter 21) has shown that minority groups and the poor frequently suffer more from pollution than wealthy white people. African Americans, Latinos, Native Americans, and other minority groups have a clear interest in pursuing environmental solutions. Religious groups are voicing new concerns about preserving our environment (chapter 1). Farmers are seeking ways to save soil and water resources (chapter 10). Loggers are learning about sustainable harvest methods (chapter 12). Business leaders are discovering new ways to do well by doing good work for society and the environment (chapter 23). These changes are exciting, but many challenges remain.

Whatever your skills and interests, you can contribute to understanding and protecting our common environment. If you enjoy science, many disciplines contribute to this cause. As you know by now, biology, chemistry, geology, ecology, climatology, geography, demography, and other sciences all provide ideas and data that are essential to understanding our environment. Environmental scientists usually focus on one of these disciplines, but their work also serves the others. An environmental chemist, for example, might study contaminants in a stream system, and this work might help an aquatic ecologist understand changes in a stream's food web.

You can also help seek environmental solutions if you prefer writing, art, working with children (fig. 25.2), history, politics, economics, or other areas of study. As you have read, environmental science depends on communication, education, good policies, and economics in addition to science.

In this chapter, we discuss some of the steps you can take to help find solutions to environmental dilemmas. You have already taken the most important step, educating yourself. When you understand how environmental systems function—from nutrient cycles and energy flows to ecosystems, climate systems, population dynamics, agriculture, and economies—you can develop well-informed opinions and help find useful answers.

FIGURE 25.2 Helping children develop a sense of wonder is a first step in protecting nature. As the Senegalese poet Baba Dioum said, "In the end, we conserve only what we love. We will love only what we understand. We will understand only what we are taught."
© William P. Cunningham

Environmental literacy has lasting importance

In 1990, Congress recognized the importance of environmental education by passing the National Environmental Education Act. The act established two broad goals: (1) to improve understanding among the general public of the natural and built environment and the relationships between humans and their environment, including global aspects of environmental problems, and (2) to encourage postsecondary students to pursue careers related to the environment. Objectives include developing an awareness and appreciation of our natural and social/cultural environment, knowledge of basic ecological concepts, acquaintance with a broad range of current environmental issues, and experience in using investigative, critical-thinking, and problem-solving skills in solving environmental problems (fig. 25.3). Several states, including Arizona, Florida, Maryland, Minnesota, Pennsylvania, and Wisconsin, have incorporated these goals into their curricula (table 25.1).

A number of organizations have been established to teach ecology and environmental ethics to elementary and secondary school students, as well as to get them involved in active projects to clean up their local community. Groups such as Kids Saving the Earth or Eco-Kids Corps have helped reach this vital audience. Family education results from these efforts as well. In a World Wildlife Fund survey, 63 percent of young people said they "lobby" their parents to recycle and to buy environmentally responsible products.

The importance of environmental education was established in the National Environmental Education Act. Former EPA administrator William K. Reilly, speaking in favor of this act, called for broad **environmental literacy** in which every citizen is fluent in the principles of ecology and has a "working knowledge of the basic grammar and underlying syntax of environmental wisdom." Environmental literacy, according to Reilly, can help establish a

FIGURE 25.3 Environmental education helps develop awareness and appreciation of ecological systems and how they work.
© William P. Cunningham

stewardship ethic—a sense of responsibility to care for and manage wisely our natural endowment and our productive resources for the long haul.

Pursuing your own environmental literacy is a lifelong process. Some of the most widely influential environmental books suggest solutions (table 25.2). To this list we'd add some personal favorites: *The Singing Wilderness* by Sigurd F. Olson, *My First Summer in the Sierra* by John Muir, *Encounters with the Archdruid* by John McPhee, and *The Life and Death of Great American Cities* by Jane Jacobs.

Table 25.1	Outcomes from Environmental Education

The natural context: An environmentally educated person understands the scientific concepts and facts that underlie environmental issues and the interrelationships that shape nature.

The social context: An environmentally educated person understands how human society is influencing the environment, as well as the economic, legal, and political mechanisms that provide avenues for addressing issues and situations.

The valuing context: An environmentally educated person explores his or her values in relation to environmental issues; from an understanding of the natural and social contexts, the person decides whether to keep or change those values.

The action context: An environmentally educated person becomes involved in activities to improve, maintain, or restore natural resources and environmental quality for all.

Source: A Greenprint for Minnesota, Minnesota Office of Environmental Education, 1993.

Table 25.2	The Environmental Scientist's Bookshelf

What are some of the most influential and popular environmental books? In a survey of environmental experts and leaders around the world, votes for the top 12 books on nature and the environment were as follows:

A Sand County Almanac by Aldo Leopold (100)[1]

Silent Spring by Rachel Carson (81)

State of the World by Lester Brown and the Worldwatch Institute (31)

The Population Bomb by Paul Ehrlich (28)

Walden by Henry David Thoreau (28)

Wilderness and the American Mind by Roderick Nash (21)

Small Is Beautiful: Economics as if People Mattered by E. F. Schumacher (21)

Desert Solitaire: A Season in the Wilderness by Edward Abbey (20)

The Closing Circle: Nature, Man, and Technology by Barry Commoner (18)

The Limits to Growth: A Report for the Club of Rome's Project on the Predicament of Mankind by Donella H. Meadows et al. (17)

The Unsettling of America: Culture and Agriculture by Wendell Berry (16)

Man and Nature by George Perkins Marsh (16)

[1]Indicates number of votes for each book. Because this was a survey of American readers, the preponderance of respondents were from the United States (82 percent), and American books are overrepresented.

Source: Robert Merideth, The Environmentalist's Bookshelf: A Guide to the Best Books, 1993, by G. K. Hall, an imprint of Macmillan, Inc.

An important part of environmental education that you shouldn't forget is experience in enjoying your environment. Get outside, have fun, explore, and observe for yourself. You can do this anywhere—in an urban park or waterway, in suburban roadsides and yards, or in remote wilderness areas. The more places you explore, the better. There's always something to see and learn. As author Edward Abbey wrote,

> It is not enough to fight for the land; it is even more important to enjoy it. While it is still there. So get out there and mess around with your friends, ramble out yonder and explore the forests, encounter the grizz, climb the mountains. Run the rivers, breathe deep of that yet sweet and lucid air, sit quietly for a while and contemplate the precious stillness, that lovely mysterious and awesome space.

Citizen science lets everyone participate

While university classes often are theoretical and abstract, many students are discovering they can make authentic contributions to scientific knowledge through active learning and undergraduate research programs. Internships in agencies or environmental organizations are one way of doing this. Another is to get involved in organized **citizen science** projects in which ordinary people join with established scientists to answer real scientific questions.

EXPLORING SCIENCE

Doing Citizen Science with eBird

eBird is a collaborative venture between the Cornell Lab of Ornithology and the National Audubon Society. It provides wonderful tools for amateur birders, but it also provides a rich database about bird abundance and distribution at a variety of spatial and temporal scales for researchers. Every observation, whether made by a recreational birder or highly trained professionals, adds to the resource. Collectively, the thousands of participants in this project are amassing one of the largest and fastest-growing biodiversity data resources in existence. In May 2015, for example, more than 9.5 million bird observations from around the world were added to the record.

How do you get involved? Don't worry—you don't need to be a bird expert, or to go out all day long. Simply go birding and then submit your data to eBird. Even half an hour can make a difference. What if you don't know how to identify birds? A number of online bird identification guides can help you learn to differentiate species. In the beginning, it helps to go out with experienced birders. Your school may have a bird-watching club, or your city recreation department may sponsor bird outings. Nearly every state has an Audubon Club chapter that offers free local trips. Don't worry that you'll be laughed at if you're a beginner. Most birders are happy to share their passion with others. And an extra set of eyes is always welcome. When a group works together, it always sees more than a single person would.

Most birders are happy to share their interest and expertise with others. It's a great way to get outdoors and learn about your environment.
© William P. Cunningham

To get your sightings included in the database, they have to be entered in eBird as one or more checklists. Go to the eBird web page (ebird.org) to learn more. Don't forget, every bird counts. Whether you only saw some pigeons and starlings eating spilled grain along the railroad tracks or spied a critically endangered species on a remote island, eBird and the world want to know about your observations. The more data in the collection, the richer and more meaningful it becomes.

There are many tools at ebird.org to help you have a good outing and report your results. You can make notes about your sightings with a free eBird mobile app. Take photos and upload them directly to your personal checklist on the app. This helps others know what you saw. You can make your checklist a work of art and become a part of the Macaulay Bird Library, too.

What if you don't know where to go to see birds? You can find the best places in your neighborhood, city, state, or country to see birds on the "Trip Planning" tab at http://ebird .org/ebird/places. An interactive map of local "hot spots" can help you decide where to go. Species maps tell you both what birds you can expect to see in a given location as well as when during the year they're most likely to be there.

If you're interested in longer trips, you can explore sightings by state or country. For example, Texas has the highest total number of species (378) in the United States, but Colombia has recorded a total of 1706 species in about the same-sized land area. You can compare the results of the top birders in your neighborhood, or if you're keeping a life list for yourself, you can find suggestions on where to see specific birds needed for your own records.

Perhaps best of all, you can feel good about contributing to real science. Researchers compile and analyze results from the database to follow bird movements and population dynamics. In this way, contributions made to eBird increase our understanding of the distribution, richness, and uniqueness of the biodiversity of our planet.

Community-based research was pioneered in the Netherlands, where several dozen research centers now study environmental issues ranging from water quality in the Rhine River, to cancer rates by geographic area, to substitutes for harmful organic solvents. In each project, students and neighborhood groups team with scientists and university personnel to collect data. Their results have been incorporated into official government policies.

You can participate in a collaboration between Audubon and the Cornell Ornithology Lab, for example (see the Exploring Science box below). Groups such as Earthwatch offer more organized and focused opportunities to take part in research. Every year, hundreds of Earthwatch projects each field a team of a dozen or so volunteers who spend a week or two working on issues ranging from loon nesting behavior to archaeological digs. The American River Watch organizes teams of students to measure water quality. You might be able to get academic credit, as well as helpful practical experience, in one of these research experiences.

Environmental careers range from engineering to education

The need for both environmental educators and environmental professionals opens up many job opportunities in environmental fields. Scientists are needed to understand the natural world and the effects of human activity on the environment. Lawyers and other specialists are needed to develop government and industry policy, laws, and regulations to protect the environment. Engineers are needed to develop technologies and products to clean up pollution and to prevent its production in the first place. Economists, geographers, and social scientists are needed to evaluate the costs of pollution and resource depletion and to develop solutions that are socially, culturally, politically, and economically appropriate for different parts of the world. Businesses increasingly seek environmentally literate and responsible leaders who appreciate how products sold, and services rendered, affect our environment.

Trained people are essential in these professions at every level, from technical and clerical support staff to top managers. Perhaps the biggest national demand over the next few years will be for environmental educators to help train an environmentally literate populace.

We urgently need teachers at every level who are trained in environmental education and environmental science. In the classroom and outdoors, experience in natural sciences and environmental issues are employable. These topics can be incorporated into reading, writing, arithmetic, and every other part of education.

Green business and technology are growing fast

Businesses increasingly consider environmental protection and sustainability to be a strategic advantage. Most large corporations now have an environmental department. Those that don't have many opportunities to reduce waste and improve efficiency by planning for sustainability. There is growing interest in designing products and manufacturing processes to minimize environmental impacts right from the start. In the long run, this can save money and make their businesses more competitive in future markets.

Businesses save money in waste disposal and litigation, too, when they use good environmental design. The market for pollution-control technology and know-how is also growing. Germany and Japan appear to be ahead of America in the pollution-control field because they have had more stringent laws for many years, giving them more experience in reducing effluents, but corporations everywhere are interested in improved efficiency.

The rush to "green up" business is good news for those looking for jobs in environmentally related fields, which are predicted to be among the fastest-growing areas of employment during the next few years. The federal government alone projects a need to hire some 10,000 people per year in a variety of environmental disciplines (fig. 25.4).

How can you prepare yourself to enter this market? The best bet is to get some technical training: Environmental engineering, analytical chemistry, microbiology, ecology, limnology, groundwater hydrology, or computer science all have great potential. Currently, a chemical engineer with a graduate degree and some experience in an environmental field can practically name his or

FIGURE 25.4 Many interesting, well-paying jobs are opening up in environmental fields. Here an environmental technician takes a sample from a monitoring well for chemical analysis.
© William P. Cunningham

her salary. Some other very good possibilities are environmental law and business administration, both rapidly expanding fields.

Good communicators, artists, and especially journalists are also needed. A liberal arts education will help you develop skills such as communication, critical thinking, balance, vision, flexibility, and caring that should serve you well. Employers need a wide variety of people; small companies need a few people who can do many things well. Society as a whole, too, needs planners, health professionals, writers, teachers, and policymakers who understand environmental science and sustainability.

Section Review

1. What is environmental literacy?
2. Why are corporations interested in hiring people with environmental science training?
3. How can artists and writers, as well as chemists and biologists, work for environmental quality and environmental health?

25.2 WHAT CAN INDIVIDUALS DO?

- *Green consumerism is one way to practice sustainability.*
- *Active citizenship includes participation in your community.*

In addition to educating yourself, you can work to reduce overconsumption and promote sustainability in your community. These efforts have become increasingly important, as technology has made consumer goods and services cheap and readily available in the richer countries of the world, because these goods also greatly expand our global footprint.

The average American consumes twice as many goods and services as in 1950. The average house is now more than twice as big as it was 50 years ago, even though the typical family has half as many people. We need more space to hold all the stuff we buy. One study found that three-fourths of American garages couldn't hold a car because they were too full of stuff. Shopping has become the way many people define themselves. A century ago, economist and social critic Thorstein Veblen, in his book *The Theory of the Leisure Class,* coined the term **conspicuous consumption** to describe buying things we don't want or need just to impress others. Does that idea remain relevant today?

Sometimes it becomes unclear whether we control our things, or if our things control us. Some critics argue that consumerism, and earning money to buy all these consumer goods, has eclipsed family, ethnicity, even religion in our lives. With so much attention on earning and spending money, we don't have time to have real friends, to cook real food, to have creative hobbies, or to do work that makes us feel we have accomplished something with our lives. Social critics have called this drive to possess stuff **"affluenza."** We find ourselves stuck in a vicious circle: We work frantically at a job we don't like to buy things we don't need, so we can save time to work even longer hours (fig. 25.5). Seeking a measure of balance in their lives, some opt out of the rat race and adopt simpler, less-consumptive lifestyles. As Thoreau wrote in *Walden,* "Our life is frittered away by detail. . . . simplify, simplify."

All choices are environmental choices

Everything we do has consequences, so it's worth practicing awareness of the impacts of our choices, both positive and negative. Choosing to live near public transit or far from it, for example, determines a great deal of your environmental impact and carbon footprint. Buying efficient appliances or lightbulbs makes you more environmentally responsible without thinking about it. Buying inefficient appliances, or more than you need, makes it extremely hard to reduce your environmental impacts, no matter how good your intentions. Turning off lights, appliances, and

computers, and turning down furnaces and air conditioners all reduce your environmental costs.

A study at the University of Chicago found that switching from a diet based on red meat to a vegetarian one can reduce your carbon footprint as much as trading in a normal-size sedan for a hybrid Prius. But it's not necessary to give up meat altogether: "Meatless Mondays" are an easy way to reduce your impact, and it could lead you to discover healthy food choices, too. You can also reduce your impact by buying chickens or farm-raised fish (preferably vegetarian ones, such as tilapia or catfish) more often than beef. This is because beef takes about 10 times as much energy to produce as a chicken or fish does. Eating locally grown, grass-fed beef has far less environmental impact than eating beef from confined feeding operations. This choice also has a positive impact on your local economy (see the What Can You Do? box below).

FIGURE 25.5 Is this our highest purpose? Used with permission from www.CartoonStock.com

What Can You Do?

Reducing Your Impact

Purchase Less

Ask yourself whether you really need more stuff, and avoid buying things you don't need or won't use.

Use items as long as possible (and don't replace them just because a new product becomes available).

Make gifts, or give nonmaterial gifts.

Reduce Excess Packaging

Carry reusable bags when shopping and refuse bags for small purchases.

Buy items in bulk or with minimal packaging; avoid single-serving foods. Choose packaging that can be recycled or reused.

Avoid Disposable Items

Use cloth napkins, handkerchiefs, and towels.

Bring a washable cup to meetings; use washable plates and utensils rather than single-use items.

Choose items built to last and have them repaired; you will save materials and energy while providing jobs in your community.

Conserve Energy and Water

Walk, bicycle, or use public transportation. Carpool and combine trips to reduce car mileage.

Turn off or turn down lights, water, heat, and air conditioning.

Save money with clotheslines or racks instead of a clothes dryer.

Use water-saving devices and fewer flushes with toilets.

Turn off running water when washing hands, food, dishes, and teeth.

Vote

Participation in local policy making gives you a voice and influence that can have widespread impacts.

Join civic organizations for fun and to influence policy.

Green consumerism encourages corporations to have an environmental conscience

Much of our global impact involves things we choose to buy. As long as corporations know that consumers want sustainably or fairly produced goods, they will take steps to use "green" production methods—or at least they will try to appear to do so. To help consumers make informed choices, several national programs have been set up to independently and scientifically analyze the environmental impacts of major products. Germany's Blue Angel, begun in 1978, is the oldest of these programs. Endorsement is highly sought after by producers because environmentally conscious shoppers have shown that they are willing to pay more for products they know have minimum environmental impacts. Similar programs are being proposed in every Western European country, as well as in Japan, China, and North America. The best of these organizations attempt "cradle-to-grave" **life-cycle analysis** (fig. 25.6) that evaluates material and energy inputs and outputs at each stage of manufacture, use, and disposal of the product.

Other familiar certification systems include fair trade and organic certification programs. These trace production chains for foods, coffee, and other products to ensure that they meet established standards of environmentally or socially acceptable methods. The U.S. EPA certifies appliances, such as refrigerators, as "Energy Star" to indicate accepted standards of efficiency.

Certification can be a powerful force, but sometimes it is also weak, with low or poorly enforced standards. USDA organic certification, for example, allows for a wide variety of chemicals on crops. Fair trade programs are generally superior to conventional trade, but they often fail to provide living wages to the workers they are meant to protect. The EPA Energy Star program can have modest efficiency standards.

Advertisers also use "greenwashing," false or meaningless claims, to exaggerate environmental benefits. The following are some types of greenwashing you might encounter:

- "Nontoxic" suggests that a product has no harmful effects on humans, but the term has many meanings and no legal definition. Substances safe for humans also can be harmful to other organisms.

- "Biodegradable," "recyclable," or "compostable" claims depend on how you dispose of a product. In a landfill, nothing is recycled or biodegraded.

- "Natural" and "organic" can connote different things in different places. Loopholes in standards allow many synthetic chemicals to be included in "organics," especially in shampoos and skin-care products. Some cigarette brands advertise that they're organic, but they're still toxic.

- "Environmentally friendly," "environmentally safe," and "won't harm the ozone layer" are often empty claims. Because there are no standards to define these terms, anyone can use them. How much energy and nonrenewable material are used in manufacturing, shipping, use, or excess packaging?

Sometimes it's hard to be sure we're making good choices as consumers. Is it better to buy organic foods that are produced on an industrial scale and shipped thousands of miles across the country, or is it better to eat only locally grown products? Maybe some of both strategies are needed. Should you buy shampoo and skin-care products that contain environmentally harmful components, such as antibacterial triclosan or environmental estrogens (see chapter 18)? Can you find alternatives that are safer?

You are a citizen, as well as a consumer

We often frame our options as individuals in terms of consumption, and in public discourse we often describe ourselves as consumers. But you also participate, one way or another, in a community. People around you, possibly including yourself, take action to shape policies about energy consumption, urban development, educational policy, and many other matters of global importance.

Participating in political campaigns, or running for office yourself, is an obvious way you can influence environmental policy and sustainability. You can also participate in petition drives, lobbying, and public meetings and hearings. Policy makers watch these events carefully as they gauge public opinion on policy matters about energy, water treatment, development policy, and countless other issues that bear on environmental questions in your area.

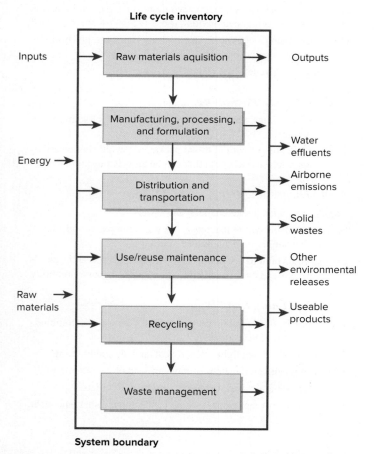

Life cycle inventory

Inputs → | Raw materials aquisition | → Outputs

Energy →

Raw materials →

- Raw materials aquisition
- Manufacturing, processing, and formulation
- Distribution and transportation
- Use/reuse maintenance
- Recycling
- Waste management

Outputs:
- Water effluents
- Airborne emissions
- Solid wastes
- Other environmental releases
- Useable products

System boundary

FIGURE 25.6 At each stage in its life cycle, a product receives inputs of materials and energy, produces outputs of materials or energy that move to subsequent phases, and releases wastes into the environment.

Politics aren't the only way to participate, though. Joining a club, a civic organization, a library board, a church, a school organization, or a neighborhood group also gives you a chance to learn about your area and the issues going on. One of the aims of these activities can be to get to know your neighbors and, ideally, have some fun together. This social networking is one of the keys to strengthening the fabric of civil society. That fabric is necessary to influence public policy. Nobel Prize–winning sociologist and economist Elinor Ostrom found that healthy social networks were an essential element in good management of common resources. Everyone can participate in that networking, and it can be enriching, too.

Voting is something every member of a democracy should do. Freedom and democracy take constant tending to remain strong. It may not always be easy, but getting out to vote, even for small and local elections, can give you pride in your community and yourself.

You can learn leadership

All of the environmental progress you have read about has occurred because individuals decided to organize and lead groups in order to promote change. Changes to environmental policy, such as the Clean Water Act or the Clean Air Act, happened because citizen groups pushed their legislators to back new rules. These changes also happened because people with an interest in public health and environmental quality decided to run for office, and others stood with them and helped their campaigns. These activities take energy and commitment. They are not easy. But they are energizing and often exciting for those who choose to take action.

Leadership is something most of us learn only through practice. How do you organize a group or set goals and priorities? How do you learn to stand up and speak to a group, to get others motivated and directed toward useful action? These are lessons you can practice in student groups, environmental organizations, and civic organizations wherever you are.

Student organizations are an especially important place to learn leadership. You have peers who share your viewpoint, and who can help you plan. You have other peers who disagree but who can challenge you to debate and to sharpen your arguments. You can publish your arguments in student newspapers and explore them in classes. And all these experiences can help you work toward goals you care about—and even jobs you hope to get after you graduate.

You can own this class

It may seem self-evident, but one of the most important things you can do is to think carefully about ways your classes can help you become better at bringing about the world you hope to see. Many students take classes because they are required as part of their major, or because they seem unusual and amusing. But just about any class has something that contributes to your larger goals, and identifying those contributions can make you a more purposeful and successful student and citizen.

Some classes push you to write better. Some encourage you to read more closely and to examine arguments and form arguments of your own. Some teach artistic communication, which can capture the imagination of people around you. Some explain fundamental ideas of chemistry, biology, and other sciences, without which it is impossible for us to fully understand problems of water pollution, pesticides, food production, climate change, and other issues in which environmental science and policy intersect.

Section Review

1. What is "affluenza"?
2. Describe the benefits and challenges of green consumerism.
3. Why is it important to vote?
4. What are some ways you can learn leadership, and why is this important?

25.3 How Can We Work Together?

- *National groups can be effective in promoting resource conservation.*
- *New groups emerge as new actors and issues become important.*
- *International non-governmental organizations pursue diverse causes.*

Collective action, such as that mobilized by 350.org, multiplies the power of individuals (fig. 25.7). As that group's work shows, collective action can help change public and governmental perceptions. 350.org's theory of change is simple: If an international

FIGURE 25.7 Working with others can give you energy, inspiration, and a sense of accomplishment.
© William P. Cunningham

grassroots movement holds our leaders accountable to the latest climate science, we can start the global transformation we do desperately need. By carrying out local projects linked in an international movement, they hope to encourage their neighbors to take positive steps to protect our environment, while also giving decision makers a sense of urgency and possibility for our planet.

Working together gives you encouragement, energy, and useful information from meeting regularly with others who share your interests. It's easy to get discouraged by the slow pace of change, so having a support group is important. In this section, we will look at some environmental organizations that have been influential in environmental policy change.

National organizations influence policy

Public organizations often arise in response to environmental and social challenges. Many of the groups we consider establishment institutions today started as associations of young radicals fighting powerful political and economic forces. Although the largest organizations have sometimes become less radical, they also have important influence in policy circles. In the United States, these include the National Wildlife Federation, the World Wildlife Fund, the Audubon Society, the Sierra Club, the Izaak Walton League, Friends of the Earth, Greenpeace, Ducks Unlimited, the Natural Resources Defense Council, and The Wilderness Society. The Audubon Society organized to protect egrets and other birds, which were being slaughtered for their plumes to decorate ladies' hats. The Sierra Club organized to protect the giant redwood trees of California and the valleys where they grew, which were rapidly being logged for timber (fig. 25.8). It went on to fight to preserve public access to other public lands in the West. Ducks Unlimited formed during the Dust Bowl years to protect ducks and

FIGURE 25.8 The Sierra Club grew to national prominence in the fight to protect the giant sequoias of California, which loggers coveted for lumber.
© Anthony R. Ambrose, UC Berkeley

their habitat, in a time when farmers were draining wetlands as fast as they could and converting them to croplands. The Wilderness Society organized at about the same time to protect open space for all citizens, not just the wealthy elite who could afford private retreats in the mountains.

Sometimes known as the "group of 10," these large organizations have become more established and have lost much of their radical flavor. They have professional staff that work with lawmakers in state and federal governments. Most members are passive and only occasionally involved in organizational activities. As a consequence, these large organizations have been criticized for not being run by volunteer activists, and activities such as litigation (lawsuits), lobbying, and research are often distant from grassroots membership. On the other hand, volunteers rarely have the time or expertise to give effective leadership in policymaking, which is often the field of highly-paid corporate lawyers. As lawyers representing corporations write and lobby for more and more legislation, citizens' groups need larger and more professional representation. Effective work often can't be done without dedicated, paid, and experienced staff.

Despite the activity of major environmental groups in Washington, most of these groups rely on a foundation of local chapters, which gather for social outdoor activities (as with Audubon and the Sierra Club), volunteer work (such as restoration at Nature Conservancy preserves), or political projects. Anybody can join these groups to learn more and meet people.

Established groups are thus influential in policy because their mass membership, professional staffs, and long history give them a degree of respectability and influence not found in newer, smaller groups. The Sierra Club, for instance, with more than 2.5 million members in 64 chapters, has a national staff of about 200, an annual budget of about $100 million, and up to 20 full-time professional lobbyists in Washington, D.C. These national groups have become a potent force in Congress, especially when they band together to push important legislation, such as the Endangered Species Act or the Clean Air Act.

In a survey that asked congressional staff and officials of government agencies to rate the effectiveness of groups that attempt to influence federal policy on pollution control, the top five were national environmental organizations. In spite of their large budgets and important connections, the American Petroleum Institute, the Chemical Manufacturers Association, and the Edison Electric Institute ranked far behind these environmental groups in terms of influence.

Some large environmental groups employ a large professional staff to carry out the goals of the organization through research and land conservation. The Nature Conservancy, for example, buys and preserves land of high ecological value. With more than 3,200 employees and assets around $3 billion, this group manages 7 million acres in what it describes as the world's largest private sanctuary system (fig. 25.9). The Conservancy is controversial for some of its management decisions, such as allowing gas and oil drilling in some reserves, and for including executives from some questionable companies on its governing board and advisory council. The Conservancy replies that it is trying to work with these companies to bring about change rather than just criticize them.

FIGURE 25.9 The Nature Conservancy buys land with high biodiversity or unique natural values to protect it from misuse and development.
© David L. Hansen, University of Minnesota Agricultural Experiment station

FIGURE 25.10. 350.org has inspired thousands of people around the world to work to stop climate change. This isn't just a campaign for our own interests, but for our children and grandchildren.

New players bring energy to policy making

There's a new energy today in environmental groups. In 2004, Michael Shellenberger and Ted Nordhaus, consultants for the Environmental Grant-making Foundation, proclaimed the "Death of Environmentalism." The major environmental organizations, they claimed, had become so embedded in Washington politics and concerned about their own jobs that they had become largely irrelevant. The greatest evidence of this impotence, they charged, was the failure to influence policy on global climate change despite years of work and hundreds of millions of dollars spent on lobbying. Even political candidates, such as John Kerry and Al Gore, who had stellar records in Congress as environmental advocates, barely mentioned it when running for office.

But just when the prospects for environmental progress looked darkest, we seemed to reach a tipping point. The evidence of global climate change became impossible to ignore, and new leaders emerged. Among these was Al Gore, who lost the U.S. presidency by just one vote in the Supreme Court, and who then used his public visibility to reinvent himself as an environmental champion. In a single year, in 2007, he won a Grammy, an Emmy, and an Oscar, wrote a best-selling book, and shared the Nobel Peace Prize with the Intergovernmental Panel on Climate Change (IPCC).

Now a broader, diverse, savvy, and passionate movement is taking shape. The 350.org network, for example, has inspired tens of thousands of citizens to work to stop climate change. As the opening case study shows, this global, grassroots movement uses social media to organize hundreds of student groups and allies around the world. Their campaign is for social justice for the poorest people in the world, who will suffer the most. It's also a matter of intergenerational equity (fig. 25.10). They're working to leave carbon in the ground, pressure governments to limit greenhouse gas emissions, encourage investors to divest from fossil fuels, stop offshore drilling, and to find people-centered solutions to climate change. The divestment campaign has resulted in a larger and faster shifting of money than any similar movement in history. Although Washington insiders considered approval of the Keystone Pipeline to be a foregone conclusion, this scrappy, student-led movement denied some of the largest, richest corporations a project they wanted badly.

350.org's web page offers a wide variety of resources and suggestions for getting involved. You can explore an interactive map to find groups or projects close to you. You can also find sample posters, signs, videos, campaign plans, and DIY templates, stencils, and sample press-releases. Get inspired and "skilled up" at workshops, conferences, and seminars. Attend an action to meet people and get ideas. There's a world of ways to take part.

Many new advocacy and research organizations have formed over time, in response to new needs and opportunities. Some groups are large, national, and professional. Others are local and volunteer. There is always room for improvement in these groups, and there is always room for more energy and participation, which can be rewarding and educational for anyone who can spare some time.

International NGOs mobilize many people

International **nongovernmental organizations (NGOs)** can be vital in the struggle to protect areas of outstanding biological value. Without this help, most local groups could never mobilize the public interest or financial support for major projects.

The rise in international NGOs in recent years has been phenomenal. At the Stockholm Conference in 1972, only a handful of environmental groups attended, almost all from fully developed countries. Twenty years later, at the Rio Earth Summit, more than 30,000 individuals representing several thousand environmental groups, many from developing countries, held a global Ecoforum to debate issues and form alliances for a better world.

FIGURE 25.11 Developing areas produce marketable goods that provide income without natural resource destruction. Often international development projects try to target these alternative income sources, to aid both local economies and environments.
© William P. Cunningham

Some NGOs are located primarily in the more highly developed countries of the north and work mainly on local issues. Others are headquartered in the north but focus their attention on the problems of developing countries in the south. Still others are truly global, with active groups in many different countries. A few are highly professional, combining private individuals with representatives of government agencies on quasi-government boards or standing committees with considerable power. Others are on the fringes of society, sometimes literally voices crying in the wilderness. Many work for political change, more specialize in gathering and disseminating information, and some undertake direct action to protect a specific resource.

Public education and raising consciousness—through protest marches, demonstrations, civil disobedience, and other participatory public actions and media events—are generally important tactics for these groups. Greenpeace, for instance, carries out well-publicized confrontations with whalers, seal hunters, toxic waste dumpers, and others who threaten very specific and visible resources. Greenpeace may well be the largest environmental organization in the world, claiming some 2.8 million contributing members.

In contrast to these highly visible groups, others choose to work behind the scenes, but their impact may be equally important. Conservation International has been a leader in debt-for-nature swaps to protect areas particularly rich in biodiversity. It also has some interesting initiatives in economic development: seeking products made by local people that will provide livelihoods along with environmental protection (fig. 25.11).

25.4 CAMPUS GREENING

- *Many student organizations work to organize, promote change, and teach leadership.*
- *Students have considerable power to push their institutions toward more sustainable practices.*

Colleges and universities can be powerful catalysts for change. Across North America, and around the world, students and faculty study sustainability and exchange ideas about improving both technology and policy for environmental and social progress. Organizations for secondary and college students often are among our most active and effective groups for environmental policy. One of the largest student environmental groups in North America has been the Student Environmental Action Coalition (SEAC). Formed in 1988 by students at the University of North Carolina at Chapel Hill, SEAC has grown rapidly and has chapter organizations on hundreds of campuses. SEAC has since joined other student networks to form the Campus Climate Challenge, which promotes climate action on campus, and Power Shift, which gathers students to learn about influencing energy and climate policy. Members of these groups organize and work for energy and environment policies on campus, at local and state levels, and nationally. National conferences bring together thousands of activists who share tactics and inspiration while also having fun. Students on your campus may participate in these or similar networks. Student groups are one of the most important places to learn about organizing for change (table 25.3).

A long-established and extremely influential student organizing group is the network of Public Interest Research Groups (PIRGs), which are active across the United States. Addressing a range of environmental and social policy issues, the PIRGs have been important forces for public education and policy influence. Joining a PIRG or similar organization is an important way to learn how many others share your concerns and want to work with you to bring about change. Joining an environmental campaign is also an important way to learn the power of working together to publicize policy issues (fig. 25.12). Participating in an organization also helps you learn essential techniques, such as using media to get your message out (table 25.4).

Schools provide environmental leadership

Colleges and universities are important sources of ideas, information, and experimentation. Faculty have knowledge and experience to share, and students have the energy, enthusiasm, curiosity, and creativity to research new ideas. Colleges and universities also have a clear interest in the future of our shared resources, because students are actively preparing to influence the world they expect to inhabit in the future.

Table 25.3	Organizing an Environmental Campaign

1. *What do you want to change?* Can you specify why your goals are focused and reasonable?

2. *What and who will be needed to get the job done?* What resources do you have; who can help you?

3. *Who are the stakeholders in this issue?* Who are your allies and constituents? How can you work with them?

4. *How will your group make decisions and set priorities?* Will you operate by consensus, majority vote, or informal agreement?

5. *Have others already worked on this issue?* What can you learn from their successes or failures?

6. *Who has the power to resolve your issue?* Which individuals, organizations, corporations, or elected officials should be targeted by your campaign, and what arguments will they listen to?

7. *What tactics will be effective?* Finding the right tactics will avoid alienating decision makers and convince people your ideas are sound.

8. *How should you make yourself presentable to your audience?* Gaining trust involves dressing or speaking, in a way your audience understands and respects. Personal presentation can be important in delivering a message.

Table 25.4	Using the Media to Influence Public Opinion

Shaping opinion, reaching consensus, electing public officials, and mobilizing action are accomplished primarily through the use of the communications media. Here are some ways to use these resources:

1. *Write letters to the editor and news releases.* Submit these to local newspapers, blogs, and radio shows.

2. *Collect data on campus resource use or environmental attitudes.* Include this evidence in letters to papers and administrators.

3. *Write feature stories for local news sources.* These outlets may be happy to publish public interest items they don't have to pay for.

4. *Produce a public service announcement for local radio and TV stations.* Publicly supported stations are required to air public service announcements, so why not yours?

5. *Find celebrities to support your position.* Ask them to give a concert or performance to attract attention to the issue. They might like to be associated with your cause.

6. *Hold a media event that is photogenic and newsworthy.* Clean up your local river and invite photographers to accompany you. Hold peaceful or amusing demonstrations. Avoid violence or conflict.

7. *If you hear negative remarks about your issue on TV or radio, ask for free time under the Fairness Doctrine to respond.* Stations must do a certain amount of public service and may be happy to accommodate you.

8. *Ask your local TV or newspaper to do a feature story about your issue or about your organization.* Articles can give valuable publicity and inspire others to participate.

Students have led innovation on countless aspects of sustainability. At many schools, students have persuaded administration to buy locally produced food and to provide organic, vegetarian, and fair trade options in campus cafeterias. Increasing vegetarian options is a way to reduce energy consumption, since meat production involves a great deal of energy. Developing local and fair trade food sources also supports communities

FIGURE 25.12 Protests, marches, and public demonstrations can be an effective way to message to policy makers and administrators, and to influence legislators.
© Tom Finkle

while teaching about alternative practices. Student-led fossil fuel divestment campaigns have transformed how we think about the impacts of investments (see What Do You Think?, below). Students also develop career paths that have an impact: At more than 100 universities and colleges across the United States, graduating students have taken a pledge for social and environmental responsibility. Could you introduce something similar at your school?

> I pledge to explore and take into account the social and environmental consequences of any job I consider and will try to improve these aspects of any organization for which I work.

The Association for the Advancement of Sustainability in Higher Education (AASHE) is one of the leading groups that offer resources to college and university students, faculty, and staff. The AASHE web site offers information about greenhouse gas monitoring, ideas for efficiency, and reports on projects at other campuses. You can find examples of curriculum innovation, green building projects, purchasing standards for fuel-efficient vehicles, food service efforts, recycling, and many other projects.

Campus building projects are important for modeling sustainability research and development. Buildings are visible, they

Divestment: Environmental Science, Policy, and Economics

In recent years, a major push of 350.org has been to encourage colleges and universities to divest their endowments from the most polluting fossil-fuel-producing corporations. Oil, gas, and coal companies have long been part of the backbone of many national economies: They are huge corporations, worth a great deal of money. Their market is locked in, so their profits are more or less guaranteed, as long as manufacturing and transportation economies are strong, and as long as we have limited alternative sources of energy.

Bill McKibben and 350.org have pointed out that educational institutions must be committed to the future well-being of their students. It seems hypocritical, they argue, to be simultaneously educating students and undermining the future they will inhabit, as global climate change threatens cities, farmland, water supplies, public health, biodiversity, and ecosystem services. The 350 organizers have encouraged students to press their administrations to wean their endowments—the investments that colleges rely on to support operations—from this investment sector.

Divestment proponents argue that this strategy helped topple South Africa's racist apartheid regime in the 1980s. Similarly, divestment from Sudan, with its predatory military government, helped to shift policy positions toward this regime in the 2000s. Divestment from tobacco companies was also a way that colleges made statements about reducing political tolerance of smoking-related illness.

Divestment has been a highly contentious issue. Opponents of divestment argue that it is essential to remain engaged, and to own shares in companies, in order to influence corporate practices. They also point out that most educational institutions invest in large, mixed funds, in which it is impossible to tell precisely where investment dollars lie. Unlike personal investments, which can be placed in socially responsible funds or indexes, the large funds of institutions cannot choose socially and environmentally responsible funds. Further, because the size of educational institutions' investments is minuscule compared to the size of the fossil fuel industry, divestment will do little practical harm to those companies. In addition, many colleges and universities are already on delicate ground financially. Risking lower returns is something many administrators are unwilling to do.

Pro-divestment student organizations counter that investment in fossil fuels is no longer financially safe, that divesting from direct holdings makes an important policy statement, and that if funds are mixed, then colleges and universities have no power to influence corporate practices. Acting as engaged shareholders is therefore not possible. Moreover, for companies whose business is to sell oil, gas, and coal, it is hard to imagine ways to push them to adjust corporate policy to ease their climate impacts. Divestment proponents also argue that even if selling shares has little impact on share value directly, the goal is to force public examination of our dependence on fossil fuels, and to question their influence on public policy, from public subsidies and tax breaks to international support for corrupt oil-producing regimes.

The number of institutions agreeing to divest is growing slowly. Union College, in Maine, a liberal arts college focused on environmental science and sustainability, was the first to commit to divestment. This was an easy step for Union College because the school had invested its funds with socially and environmentally committed investment firms. Increasingly, cities and pension funds are developing plans to divest. The growing availability of affordable alternatives is an important part of this economic decision.

The idea remains controversial, challenging, energizing, and uncertain. Divestment challenges colleges and universities to unite theory with practice—often a hard thing to do. What do you think? What arguments draw you to one side of the debate or the other? What information would you need to form a clearer opinion? Where could you get that information?

---

consume most energy on most campuses, and they offer teaching opportunities for students and the public. Many colleges and universities have environmental building standards or have committed to following standards of the U.S. Green Building Council, the organization that does Leadership in Energy and Environmental Design (LEED) certification. Often building standards are adopted because students ask administrators what they are doing to improve building performance. Examples of prize-winning sustainable designs can be found at Stanford University, Oberlin College in Ohio, and the University of California at Santa Barbara, and many other schools. Stanford students worked with the administration to develop *Guidelines for Sustainable Buildings,* a booklet that covers everything from energy-efficient lighting to native landscaping. Hampshire College, in western Massachusetts, has one of the first college buildings built to meet "living building" standards in its Kern Center: With solar panels, excellent insulation, and composting toilets, the center is designed to consume no energy or water not produced within its own footprint. The center is an important part of the college's strategy for achieving campus-wide carbon neutrality.

Sustainable buildings don't need to be especially expensive. UC Santa Barbara's Bren School of Environmental Science and Management looks like a normal institutional building, but it was the first academic laboratory building in the United States to get LEED Platinum certification, the highest level (fig. 25.13). It wasn't originally funded as an experimental green building, but planners found that some simple features like large windows for natural light and ventilation make the building both more functional and more appealing. Motion detectors control light levels, and sensors monitor and refresh the air when there is too much CO_2 putting students to sleep. More than 30 percent of interior materials are recycled. Solar panels supply 10 percent of the electricity, and the building exceeds federal efficiency standards by 30 percent. And all this cost only 2 percent more than a conventional design, an amount that was rapidly recovered in lower energy costs. "The overriding and very powerful message is

FIGURE 25.13 The University of California at Santa Barbara's Bren School of Environmental Science and Management doesn't look radically different, but design features make it extremely sustainable and comfortable.
© *William P. Cunningham*

it really doesn't cost any more to do these things," says Dennis Aigner, dean of Bren School.

A green campus is an educational opportunity

These facilities offer important educational opportunities. At Carnegie Mellon University in Pittsburgh, students helped design a green roof for Hamershlag Hall. They now monitor how the living roof is reducing storm water drainage and improving water quality. A kiosk inside the dorm shows daily energy use and compares it to long-term averages. Classrooms within the dorm offer environmental science classes in which students can see sustainability in action. Green dorms are popular with students, who appreciate natural lighting, clean air, lack of allergens in building materials, and other features of LEED-certified buildings.

Concordia University in Austin, Texas, was the first college or university in the country to purchase all of its energy from renewable sources. The 5.5 million kilowatt-hours of "green power" it uses each year, mainly wind power, eliminate about 8 million pounds of CO_2 emissions annually, the equivalent of planting 1,000 acres of trees or taking 700 cars off the roads. Emory University in Atlanta, Georgia, is a leader in green building standards, with 11 buildings that are or could become LEED certified. Emory's Whitehead Biomedical Research Building was the first facility in the Southeast to be LEED certified. Like a number of other colleges, Carleton College in Northfield, Minnesota, has built its own wind turbines, which generate about two-thirds of the college's electricity. These turbines are expected to pay for themselves in about a decade.

A recent study by the Sustainable Endowments Institute evaluated more than 100 of the leading colleges and universities in the United States on their green building policies, food and recycling programs, climate change impacts, and energy consumption. The report card ranked Dartmouth, Harvard, Stanford, and Williams as the top of the "A list" of 23 greenest campuses. Berea College in Kentucky got special commendation as a small school with a strong commitment to sustainability. Berea's "ecovillage" has a student-designed house that produces its own electricity and treats wastewater in a living system. The college has a full-time sustainability coordinator to provide support to campus programs, community outreach, and teaching.

Section Review

1. Why are colleges and universities good places for exploring new strategies for sustainability?
2. What is fossil fuel divestment, and what are some arguments for and against it?
3. What are some ways students can promote sustainability on campus?

25.5 SUSTAINABILITY IS A GLOBAL CHALLENGE

- *Global goals of sustainability involve widespread commitment to development and to justice.*
- *The Millennium Development Goals are a set of quantifiable measures of sustainability and development.*

As the developing countries of the world become more affluent, they are adopting the resource-intensive consumption patterns of richer countries. Automobile production in China, for example, is increasing at about 19 percent per year, or doubling every 3.7 years. Fuel consumption has risen in response, and Beijing, Shanghai, and other major cities have some of the worst peak levels of urban air pollution on record. Throughout the developing world, increasing ownership of refrigerators, air conditioners, televisions, and other appliances is driving up demand for electric power. Between 1995 and 2015, per capita emissions of CO_2 more than doubled in India and nearly tripled in China. More widely alarming, diseases associated with affluent lifestyles—such as obesity, diabetes, heart attacks, depression, and traffic accidents—are becoming the leading causes of morbidity and mortality worldwide.

On the other hand, there appears to be a dramatic worldwide shift in public attitudes toward environmental protection. In a BBC poll of 22,000 residents of 21 countries, 83 percent agreed that individuals would definitely or probably have to make lifestyle changes to reduce the amount of climate-changing gases they produce. Overall, 70 percent said they were personally ready to make sacrifices to protect the environment. In the U.S., where some policymakers argue that the public opposes EPA pollution standards and greenhouse gas regulation, a 2016 bipartisan poll found that 77 percent of respondents supported stricter CO_2 standards, and 69 percent said that EPA scientists, not Congress, should set pollution standards.

We would all benefit by helping developing countries access more efficient, less-polluting technologies. Education, democracy, and access to information are essential for sustainability. In the 2015 Paris climate negotiations, in which 195 countries agreed on plans to reduce greenhouse gas emissions, one of the central priorities was building an investment fund through which wealthy countries would invest $100 billion in clean technology in developing countries. This agreement was about fairness, but it was also about improving our global environment and supporting regional stability and security. Overall, the costs to control population growth, develop renewable energy sources, stop soil erosion, protect ecosystems, and provide a decent standard of living for the world's poor are estimated at $350 billion per year. This is a great deal of money, but it is small compared to over $1 trillion per year spent on wars and military equipment.

Sustainable development means social, environmental, and economic goals

Sustainable development has been a theme running through much of this book. The idea emphasizes the use of renewable resources in harmony with ecological systems (fig. 25.14); the aim is to seek a combination of economic, well-being, and environmental priorities. A widely used definition of sustainable development, from the 1992 report *Our Common Future,* is "meeting the needs of the present without compromising the ability of future generations to meet their own needs" (see sections 1.2 and 1.3). These are some goals of sustainable development:

- A demographic transition to a stable world population of low birth and death rates.

- An energy transition to high efficiency in production and use, coupled with increasing reliance on renewable resources.

- A resource transition to reliance on nature's "income" without depleting its "capital."

- An economic transition to sustainable development and a broader sharing of its benefits.

- A political transition to global negotiation grounded in complementary interests between North and South, East and West.

- An ethical or spiritual transition to attitudes that do not separate us from nature or each other.

Sustainability is an even more important goal for those who live in wealthy countries than for those in developing countries (fig. 25.15). Because of our resource-intensive lifestyles, we have a much greater impact on our environment than the billions of people in poorer countries. Finding ways for all of us to live more lightly on the earth is a critically important goal.

In 2000, United Nations Secretary-General Kofi Annan called for a **millennium assessment** of the consequences of ecosystem change on human well-being. The assessment also explored the scientific basis for actions to enhance the conservation and sustainable use of environmental systems. More than 1,360 experts from around the world worked on technical reports about the conditions and trends of ecosystems, and scenarios for the future. As a result of this assessment, the United Nations developed a set

FIGURE 25.14 A model for integrating ecosystem health, human needs, and sustainable economic growth.
© Jerry Alexander/Getty Images

of Millennium Development Goals for sustainable development, with targets for poverty reduction and environmental quality that countries were to achieve between 2000 and 2015.

Contrary to many expectations, considerable progress was made on these goals. Hunger, poverty, and child mortality dropped

FIGURE 25.15 Sustainable development to ensure a healthy environment for all the world's people is the aim of the millennium development goals.
© Jerry Alexander/Getty Images

sharply. Access to education rose, as did access to clean water and sanitation. There were some improvements in democracy and women's rights, although this progress was mixed. Building on lessons learned from both successes and failures in the millennium goals, the UN launched a set of Sustainable Development Goals for 2016–2030 (see chapter 1). These goals represent major challenges, but they are also opportunities to improve the world we share. Findings from the millennium assessment serve as a good summary for this book:

- All of us depend on nature and ecosystem services to provide the conditions for a decent, healthy, and secure life.

- We have made unprecedented changes to ecosystems in recent decades to meet growing demands for food, fresh water, fiber, and energy.

- These changes have helped improve the lives of billions, but at the same time they weakened nature's ability to deliver other key services, such as purification of air and water, protection from disasters, and the provision of medicine.

- Among the outstanding problems we face are the dire state of many of the world's fish stocks, the intense vulnerability of the 2 billion people living in dry regions, and the growing threat to ecosystems from climate change and pollution.

- Human actions have taken the planet to the edge of a massive wave of species extinctions, further threatening our own well-being.

- The loss of services derived from ecosystems is a significant barrier to reducing poverty, hunger, and disease.

- The pressures on ecosystems will increase globally unless human attitudes and actions change.

- Measures to conserve natural resources are more likely to succeed if local communities are given ownership of them, share the benefits, and are involved in decisions.

- Even today's technology and knowledge can reduce considerably the human impact on ecosystems. They are unlikely to be deployed fully, however, until ecosystem services cease to be perceived as free and limitless.

- Better protection of natural assets will require coordinated efforts across all sections of governments, businesses, and international institutions.

From what you've learned in this book, how do you think we could work—individually and collectively—to accomplish these goals? What are the ways in which you would like to contribute?

Section Review

1. List several goals or aspects of sustainability.
2. What is the source of the Millennium Development Goals?
3. Identify eight major areas in which quantifiable goals have been set for the Millennium Development Goals.

Conclusion

All through this book you've seen evidence of environmental degradation and resource depletion, but there are also many cases in which individuals and organizations are finding ways to stop pollution, use renewable rather than irreplaceable resources, and even restore biodiversity and habitat. Sometimes all it takes is the catalyst of a pilot project to show people how things can be done differently to change attitudes and habits. In this chapter, you've learned some practical approaches to living more lightly on the world individually, as well as working collectively to create a better world.

Public attention to issues in the United States seems to run in cycles. Concern builds about some set of problems, and people are willing to take action to find solutions, but then interest wanes and other topics come to the forefront. For the past decade, the American public has consistently said that the environment is very important and that government should pay more attention to environmental quality. Nevertheless, people haven't shown this concern for the environment to be a very high priority, either in personal behavior or in how they vote.

Recently, however, approaches to the environment in much of the world seem to have reached a tipping point. Countries, cities, companies, and campuses all are vying to be the most green. This may be a very good time to work on social change and sustainable living. As the famous anthropologist Margaret Mead said, "Never doubt that a small group of thoughtful, committed people can change the world. Indeed, it is the only thing that ever has."

Reviewing Key Terms

Can you define the following terms in environmental science?

affluenza 25.2
citizen science 25.1

conspicuous consumption 25.2
environmental literacy 25.1

life-cycle analysis 25.2
millennium assessment 25.5

nongovernmental organizations (NGOs) 25.4

Critical Thinking and Discussion Questions

1. What lessons do you derive from the case study about 350.org? If you were interested in bringing about change in your neighborhood or in the wider world, which of the tactics used in this effort might you use for your campaign?

2. Reflect on how you learned about environmental issues. What have been the most important formative experiences or persuasive arguments in shaping your own attitudes? If you were designing an environmental education program for youth, what elements would you include?

3. How might it change your life if you were to minimize your consumption of materials and resources? Which aspects could you give up, and what is absolutely essential to your happiness and well-being? Does your list differ from your friends' and classmates' lists?

4. Have you ever been involved in charitable or environmental work? What were the best and worst aspects of that experience? If you haven't yet done anything of this sort, what activities seem appealing and worthwhile to you?

5. What green activities are now occurring at your school? How might you get involved?

6. Identify two key messages from the millennium assessment and two goals and objectives that you believe are most important for environmental science. Why did you choose these messages and goals? How might we accomplish them?

Data Analysis

Campus Environmental Audit

If you want to understand how sustainable your campus is, the first step is to gather data. A campus environmental audit can be a huge, professionally executed task, but you can make a reasonable approximation if you work with your class to gather some basic information.

Among the many places to find established listings of factors to consider, one widely used audit is that of the Association for the Advancement of Sustainability in Higher Education. You can find details of their full audit system here: https://sta rs.aashe.org/pages/about/technical-manual.html. Below is a set of 30 questions, in 9 groups, that approximate the AASHE guidelines. Give one point for each Yes answer. Then, see what your score is out of 30 possible points. Note that, as in all audits, qualifying decisions are often needed to distill complex practices into a simple "yes" or "no" answer.

Which No answers would be easiest or most important to change? Which Yes answers still have room for improvement? What can student groups do to advocate for those improvements?

1. Education: Does your school offer courses focused on sustainability? Are there programs or facilities that encourage student research, monitoring, or data collection on sustainability?

2. Buildings and energy: Are any campus buildings LEED certified? Are LED lighting, automated switches, or other energy-saving devices widely used across campus? Do most buildings allow for use of natural lighting? Are most windows single-paned or double-paned? Are most roofs insulated and well sealed? Do dorms or classroom buildings have energy meters, so that energy can be monitored separately? Does your school purchase or use energy from renewable sources?

3. Climate: Has your campus conducted an inventory of greenhouse gas emissions?

4. Dining: Do dining services compost food waste? Is locally grown or vegan food available? Do disposable products contain recycled content, or are they recyclable? Are reusable take-out containers in use?

5. Grounds and purchasing: Does landscaping include native plantings or wildlife habitat? Is there a policy of minimizing pesticide and herbicide use? Does your school purchase environmentally safe cleaning products? Does office paper contain recycled content?

6. Transportation: Does your school own fuel-efficient vehicles? Are there programs or policies to encourage public transit or ride sharing? Are safe facilities provided for bicyclists (bicycle routes, sufficient bike racks)? Do most faculty and administrators live within a short commuting distance of campus?

7. Waste: Does your school provide convenient recycling facilities? Is there a mechanism for exchanging household goods when students move in and out of dorms? Is disposal of e-waste (computers, phones, etc.) tracked to recycling facilities?

8. Water: Does your school practice xeriscaping, or has it eliminated daytime lawn watering? Are waterless urinals and low-flush toilets widely available?

9. Administration: Does your administration actively promote and coordinate sustainability planning? Are energy conservation and sustainability integrated into campus management and building plans? Does your campus promote farmers' markets, public education, or other public engagement?

TO ACCESS ADDITIONAL RESOURCES FOR THIS CHAPTER, PLEASE VISIT CONNECT AT www.connect.mheducation.com.

You will find LearnSmart, an adaptive learning system, Google Earth™ exercises, additional Case Studies, Data Analysis exercises, and an interactive ebook.

Glossary

A

abiotic Nonliving.

abundance The number or amount of something.

acid deposition Acidic rain, snow, fog, or dry particulates deposited from the air due to an increased release of acids by anthropogenic or natural resources.

acids Substances that release hydrogen ions (protons) in water.

active solar system A mechanical system that uses moving parts or pumped fluids to collect, concentrate, distribute, or store solar energy.

acute effects Sudden, severe effects.

adaptation The acquisition of traits that allow a species to survive and thrive in its environment.

adaptive management A management plan designed from the outset to "learn by doing" and to actively test hypotheses and adjust treatments as new information becomes available.

administrative law Executive orders, administrative rules and regulations, and enforcement decisions by administrative agencies and special administrative courts.

aerosols Minute particles or liquid droplets suspended in the air.

aesthetic degradation Changes in environmental quality that offend our aesthetic senses.

affluenza An addiction to spending and consuming beyond one's needs.

albedo A description of a surface's reflective properties.

allergens Substances that activate the immune system.

allopatric speciation Speciation deriving from a common ancestor due to geographic barriers that cause reproductive isolation; *also called* geographic isolation.

ambient air The air immediately around us.

analytical thinking Thinking that breaks a problem down into its constituent parts.

anemia Low levels of hemoglobin due to iron deficiency or lack of red blood cells.

antigens Chemical compounds to which antibodies bind.

aquaculture Growing aquatic species in net pens or tanks.

aquifers Porous, water-bearing layers of sand, gravel, and rock below the earth's surface; reservoirs for groundwater.

arbitration A formal process of dispute resolution in which there are stringent rules of evidence, cross-examination of witnesses, and a legally binding decision made by the arbitrator that all parties must obey.

artesian well A pressurized aquifer from which water gushes without being pumped, due to the aquifer's intersecting the surface or being penetrated by a pipe or conduit; *also called* a spring.

atmospheric deposition Sedimentation of solids, liquids, or gaseous materials from the air.

atom The smallest unit of matter that has the characteristics of an element; consists of three main types of subatomic particles: protons, neutrons, and electrons.

atomic number The characteristic number of protons per atom of an element; used as an identifying attribute.

B

barrier islands Low, narrow, sandy islands that form offshore from a coastline. They protect the shore from storms.

Basel Convention Restricts shipment of hazardous waste across international boundaries.

bases Substances that bond readily with hydrogen ions.

Batesian mimicry Evolution by one species to resemble the coloration, body shape, or behavior of another species that is protected from predators by a venomous stinger, bad taste, or some other defensive adaptation.

benthic The bottom of a sea or lake.

best available, economically achievable technology (BAT) The best pollution control available.

best practicable control technology (BPT) The best technology for pollution control available at reasonable cost and operable under normal conditions.

binomials Two-part names (genus and species, usually in Latin) invented by Carl Linneaus to show taxonomic relationships.

bioaccumulation The selective absorption and concentration of molecules by cells.

biocentric preservation A philosophy that emphasizes the fundamental right of living organisms to exist and to pursue their own goods.

biochemical oxygen demand (BOD) A standard test of water pollution, measured by the amount of dissolved oxygen consumed by aquatic organisms over a given period.

biocide A broad-spectrum poison that kills a wide range of organisms.

biodegradable plastics Plastics that can be decomposed by microorganisms.

biodiversity The genetic, species, and ecological diversity of the organisms in a given area.

biodiversity hot spots Areas with exceptionally high numbers of endemic species.

biofuel Fuels such as ethanol, methanol, biodiesel, or vegetable oils from crops.

biological community The populations of plants, animals, and microorganisms living and interacting in a certain area at a given time.

biological controls Use of natural predators, pathogens, or competitors to regulate pest populations.

biomagnification An increase in the concentration of certain stable chemicals (for example, heavy metals or fat-soluble pesticides) in successively higher trophic levels of a food chain or web.

biomass The total mass or weight of all the living organisms in a given population or area; also a general term for material produced by living things, such as wood, leaves, and other material.

biome A broad, regional type of ecosystem characterized by distinctive climate and soil conditions and a distinctive kind of biological community adapted to those conditions.

bioremediation The use of biological organisms to remove or detoxify pollutants from a contaminated area.

biosphere reserves World heritage sites identified by the IUCN as worthy of national park or wildlife refuge status because of high biological diversity or unique ecological features.

biotic Pertaining to life; environmental factors created by living organisms.

birth control Any method used to reduce births, including abstinence, delayed marriage, contraception, devices or medications that prevent implantation of fertilized zygotes, and induced abortions.

bitumen A sticky, semi-solid hydrocarbon found in tar sands.

blind experiment Experiment in which those carrying out the experiment don't know, until after the gathering and analysis of data, which was the experimental treatment and which was the control.

body burden The sum total of all persistent toxins in our body that we accumulate from our air, water, diet, and surroundings.

bog An area of waterlogged soil that tends to be peaty; bogs are fed mainly by precipitation and have low productivity, and some are acidic.

boreal forest A broad band of mixed coniferous and deciduous trees that stretches across northern North America (and also Europe and Asia); its northernmost edge, the taiga, intergrades with the arctic tundra.

breeder reactor A nuclear reactor that produces fuel by bombarding isotopes of uranium and thorium with high-energy neutrons that convert inert atoms to fissionable ones.

bronchitis A persistent inflammation of bronchi and bronchioles (large and small airways in the lungs).

brownfield development Building on abandoned or reclaimed polluted industrial sites.

brownfields Abandoned or underused urban areas in which redevelopment is blocked by liability or financing issues related to toxic contamination.

buffalo commons A large open area proposed for the Great Plains in which wildlife and native people could live as they once did without interference by industrialized society.

C

cancer Invasive, out-of-control cell growth that results in malignant tumors.

cap-and-trade An approach to controlling pollution by mandating upper limits (the cap) on how much each country, sector, or specific industry is allowed to emit. Companies that can cut pollution by more than they're required to can sell the credit to other companies that have more difficulty meeting their mandated levels.

capital Any form of wealth, resources, or knowledge available for use in the production of more wealth.

carbon capture and storage (CCS) Carbon dioxide (generally from fuel combustion) is captured and stored in geological formations.

carbon credits Permits for pollution emissions that can be bought and sold. In theory, the trading of carbon credits produces pollution reductions for the lowest cost.

carbon cycle The circulation and reutilization of carbon atoms, especially via the processes of photosynthesis and respiration.

carbon monoxide (CO) A colorless, odorless, non-irritating, but highly toxic gas produced by incomplete combustion of fuel, incineration of biomass or solid waste, or partially anaerobic decomposition of organic material.

carbon neutral A system or process that doesn't release more carbon to the atmosphere than it consumes.

carbon sink Places of carbon accumulation, such as in large forests (organic compounds) or ocean sediments (calcium carbonate); carbon is thus removed from the carbon cycle for moderately long to very long periods of time.

carcinogens Substances that cause cancer.

carnivores Organisms that prey mainly upon animals.

carrying capacity The maximum number of individuals of any species that can be supported by a particular ecosystem on a long-term basis.

case law Precedents from both civil and criminal court cases.

cell Minute biological compartments within which the processes of life are carried out.

cellular respiration The process in which a cell breaks down sugar or other organic compounds to release energy used for cellular work; may be anaerobic or aerobic, depending on the availability of oxygen.

cellulosic ethanol Ethanol made from (plant) material.

chain reaction A self-sustaining reaction in which the fission of nuclei produces subatomic particles that cause the fission of other nuclei.

chaparral Thick, dense, thorny, evergreen scrub found in Mediterranean climates.

chemical energy Potential energy stored in chemical bonds of molecules.

chemosynthesis The process in which inorganic chemicals, such as hydrogen sulfide (HS) or hydrogen gas (H^2), serve as an energy source for synthesis of organic molecules.

chlorinated hydrocarbons Hydrocarbon molecules to which chlorine atoms are attached.

chlorofluorocarbons (CFCs) Chemical compounds with a carbon skeleton and one or more attached chlorine and fluorine atoms. Commonly used as refrigerants, solvents, fire retardants, and blowing agents.

chronic effects Long-lasting results of exposure to a toxin; can be caused by a single, acute exposure or a continuous, low-level exposure.

chronic obstructive lung disease Irreversible damage to the linings of the lungs caused by irritants.

chronically undernourished Those people whose diet doesn't provide the 2,200 kcal per day, on average, considered necessary for a healthy, productive life.

citizen science Projects in which trained volunteers work with scientific researchers to answer real-world questions.

city A differentiated community with a sufficient population and resource base to allow residents to specialize in arts, crafts, services, and professional occupations.

civil law A body of laws regulating relations between individuals, or between individuals and corporations, concerning property rights, personal dignity and freedom, and personal injury.

classical economics Modern, Western economic theories of the effects of resource scarcity, monetary policy, and competition on supply of and demand for goods and services in the marketplace. This is the basis for the capitalist market system.

climate A description of the long-term pattern of weather in a particular area.

climax community A relatively stable, long-lasting community reached in a successional series; usually determined by climate and soil type.

closed system An ecosystem that neither receives nor emits energy or matter to the surrounding environment.

closed-canopy forest A forest where tree crowns spread over 20 percent of the ground; has the potential for commercial timber harvests.

cloud forests High mountain forests where temperatures are uniformly cool and fog or mist keeps vegetation wet all the time.

coarse woody debris Logs, branches, and other woody debris in a stream.

coevolution The process in which species exert selective pressure on each other and gradually evolve new features or behaviors as a result of those pressures.

cogeneration The simultaneous production of electricity and steam or hot water in the same plant.

cold front A moving boundary of cooler air displacing warmer air.

coliform bacteria Bacteria that live in the intestines (including the colon) of humans and other animals; used as a measure of the presence of feces in water or soil.

command and control economies Economies in which government-set rules and standards guide production and consumption, rather than market forces.

commensalism A symbiotic relationship in which one member is benefited and the second is neither harmed nor benefited.

common law The body of court decisions that constitute a working definition of individual rights and responsibilities where no formal statutes define these issues.

common property management Resources managed cooperatively by a community for long-term sustainability. Also known as communal resource management systems.

complexity The number of species at each trophic level and the number of trophic levels in a community.

composting The biological degradation of organic material under aerobic (oxygen-rich) conditions to produce compost, a nutrient-rich soil amendment and conditioner.

compound A molecule made up of two or more kinds of atoms held together by chemical bonds.

concentrating solar power (CSP) system A facility that captures solar energy to generate steam, which turns a turbine to produce electricity.

conclusion A statement that follows logically from a set of premises.

confidence limits A statistical measure of the quality of data that tells you how close the sample's average probably is to the average for the entire population of that species.

confined animal feeding operation (CAFO) Facilities in which large numbers of animals spend most or all of their life in confinement.

conifers Needle-bearing trees that produce seeds in cones.

conservation development Consideration of landscape history, human culture, topography, and ecological values in subdivision design. Using cluster housing, zoning, covenants, and other design features, at least half of a subdivision can be preserved as open space, farmland, or natural areas.

conservation medicine A medical field that attempts to understand how environmental changes threaten our own health, as well as that of the natural communities on which we depend for ecological services.

conservation of matter In any chemical reaction, matter changes form; it is neither created nor destroyed.

conspicuous consumption A term coined by economist and social critic Thorstein Veblen to describe buying things we don't want or need in order to impress others.

constructed wetlands Artificially constructed wetlands used to purify water.

consumer An organism that obtains energy and nutrients by feeding on other organisms or their remains. *See also* heterotroph.

consumption The fraction of withdrawn water that is lost in transmission or that is evaporated, absorbed, chemically transformed, or otherwise made unavailable for other purposes as a result of human use.

contour plowing Plowing along hill contours; reduces erosion.

control rods Neutron-absorbing material inserted into spaces between fuel assemblies in nuclear reactors to regulate fission reactions.

controlled study Study in which comparisons are made between experimental and control populations that are (as far as possible) identical in every factor except the one variable being studied.

convection currents Rising or sinking air currents that stir the atmosphere and transport heat from one area to another. Convection currents also occur in water. *See* spring overturn.

Convention on International Trade in Endangered Species (CITES) The 1963 convention that regulates international trade in endangered species.

conventional or criteria pollutants The seven major pollutants (sulfur dioxide, carbon monoxide, particulates, hydrocarbons, nitrogen oxides, photochemical oxidants, and lead) identified and regulated by the U.S. Clean Air Act.

coral bleaching Whitening of corals caused by expulsion of symbiotic algae—often resulting from high water temperatures, pollution, or disease.

coral reefs Prominent oceanic features composed of hard, limy skeletons produced by coral animals; usually formed along edges of shallow, submerged ocean banks or along shelves in warm, shallow, tropical seas.

core (earth's) The dense, intensely hot mass of molten metal, mostly iron and nickel, thousands of kilometers in diameter at the earth's center.

core habitat Essential habitat for a species.

core region The primary industrial region of a country; usually located around the capital or largest port; has both the greatest population density and the greatest economic activity of the country.

Coriolis effect The influence of friction and drag on air layers near the earth; deflects air currents to the direction of the earth's rotation.

corridor A strip of natural habitat that connects two adjacent nature preserves to allow migration of organisms from one place to another.

cost–benefit analysis (CBA) An evaluation of large-scale public projects by comparing the costs and benefits that accrue from them.

cover crops Plants, such as rye, alfalfa, or clover, that can be planted immediately after harvest to hold and protect the soil.

creative thinking Asks, how could I do this differently.

criteria pollutants *See* conventional pollutants.

critical factor The environmental factor closest to a tolerance limit for a given species, which is the most important in determining distribution at a given time.

critical thinking An ability to evaluate information and opinions in a systematic, purposeful, efficient manner.

crude birth rate The number of births in a year divided by the midyear population.

crude death rate The number of deaths per thousand persons in a given year; *also called* crude mortality rate.

crust The cool, lightweight, outermost layer of the earth's surface that floats on the soft, pliable, underlying layers; similar to the "skin" on a bowl of warm pudding.

cultural eutrophication An increase in biological productivity and ecosystem succession caused by human activities.

cyclonic storms Storms with winds that spiral clockwise out of an area of high pressure in the Northern Hemisphere and counterclockwise into a low-pressure zone.

D

debt-for-nature swap Forgiveness of international debt in exchange for nature protection in developing countries.

deciduous Trees and shrubs that shed their leaves at the end of the growing season.

decomposer Fungi and bacteria that break down complex organic material into smaller molecules.

deductive reasoning Deriving testable predictions about specific cases from general principles.

demand The amount of a product that consumers are willing and able to buy at various possible prices, assuming they are free to express their preferences.

demanufacturing Disassembly of products so components can be reused or recycled.

demographic bottleneck A population founded when just a few members of a species survive a catastrophic event or colonize a new habitat that is geographically isolated from other members of the same species.

demographic transition A pattern of falling death rates and birthrates in response to improved living conditions; could be reversed in deteriorating conditions.

demography Vital statistics about people: births, marriages, deaths, etc.; the statistical study of human populations relating to growth rate, age structure, geographic distribution, etc., and their effects on social, economic, and environmental conditions.

density-dependent Factors affecting population growth that change as population size changes.

density-independent Factors that affect population growth, but that do not vary as population size changes.

dependency ratio The number of nonworking members compared to working members for a given population.

dependent variable A variable that is affected by the condition being altered in a manipulative experiment.

desert A biome characterized by low moisture levels and infrequent and unpredictable precipitation. Daily and seasonal temperatures fluctuate widely.

desertification Conversion of productive lands to desert.

detritivore Organisms that consume organic litter, debris, and dung.

disability-adjusted life years (DALYs) A measure of premature deaths and losses due to illnesses and disabilities in a population.

discharge The amount of water that passes a fixed point in a given amount of time; usually expressed as liters or cubic feet of water per second.

disease A deleterious change in the body's condition in response to destabilizing factors such as nutrition, chemicals, or biological agents.

dissolved oxygen (DO) content The amount of oxygen dissolved in a given volume of water at a given temperature and atmospheric pressure; usually expressed in parts per million (ppm).

disturbance Periodic destructive events such as fires or floods; changes in an ecosystem that affect (positively or negatively) the organisms living there.

disturbance-adapted species Species that depend on disturbances to succeed.

diversity The number of species present in a community (species richness), as well as the relative abundance of each species.

DNA (deoxyribonucleic acid) A giant molecule composed of millions or billions of nucleotides (sugars and bases called purines and pyramidines held together by phosphate bonds) that form a double helix and store genetic information in all living cells.

double-blind experiment A design in which neither the experimenter nor the subjects know, until after the gathering and analysis of data, which was the experimental treatment and which was the control.

E

e-waste Discarded electronic equipment, such as computers, cell phones, and television sets.

earthquake A sudden, violent movement of the earth's crust.

ecological diseases Emergent diseases (new or rarely seen diseases) that cause devastating epidemics among wildlife and domestic animals.

ecological economics A relatively new field that brings the insights of ecology to economic analysis.

ecological footprint A measure that computes the demands placed on nature by individuals and nations.

ecological niche The functional role and position of a species (population) within a community or an ecosystem, including what resources it uses, how and when it uses the resources, and how it interacts with other populations.

ecological restoration Bringing a landscape back to a former condition. Ecological restoration involves active manipulation of nature to re-create conditions that existed before human disturbance.

ecology The scientific study of relationships between organisms and their environment. It is concerned with the life histories, distribution, and behavior of individual species, as well as the structure and function of natural systems at the level of populations, communities, and ecosystems.

ecosystem management An integration of ecological, economic, and social goals in a unified systems approach to resource management.

ecosystem services Resources provided by nature, such as food, water, energy, fiber, waste disposal, and other conditions on which life depends.

ecotone A transition zone or boundary between different biomes, ecosystems, or ecological communities.

ecotourism A combination of adventure travel, cultural exploration, and nature appreciation in wild settings.

edge effects A change in species composition, physical conditions, or other ecological factors at the boundary between two ecosystems.

effluent sewerage A low-cost alternative sewage treatment for cities in poor countries that combines some features of septic systems and centralized municipal treatment systems.

El Niño A climatic change marked by the shifting of a large warm water pool from the western Pacific Ocean toward the east. Wind direction and precipitation patterns are changed over much of the Pacific and perhaps around the world.

electrostatic precipitators The most common particulate controls in power plants; fly ash particles pick up an electrostatic surface charge as they pass between large electrodes in the effluent stream, causing particles to migrate to the oppositely charged plate.

element A molecule composed of one kind of atom; cannot be broken into simpler units by chemical reactions.

emergent diseases A new disease or one that has been absent for at least 20 years.

emergent properties Characteristics of whole, functioning systems that are quantitatively or qualitatively greater than the sum of the system's parts.

emissions trading Programs in which companies that have cut pollution more than they're required to can sell "credits" to other companies that still exceed allowed levels.

endangered species A species considered to be in imminent danger of extinction.

endemic species Those found only in a particular place.

endocrine disrupters Chemicals that disrupt normal hormone functions.

energy The capacity to do work (that is, to change the physical state or motion of an object).

energy recovery Incineration of solid waste to produce useful energy.

entropy Disorder in a system.

environment The circumstances or conditions that surround an organism or group of organisms, as well as the complex of social or cultural conditions that affect an individual or community.

environmental health The science of external factors that cause disease, including elements of the natural, social, cultural, and technological worlds in which we live.

environmental impact statement (EIS) An analysis, required by provisions in the National Environmental Policy Act of 1970, of the effects of any major program a federal agency plans to undertake.

environmental justice A recognition that access to a clean, healthy environment is a fundamental right of all human beings.

environmental literacy Fluency in the principles of ecology that gives us a working knowledge of the basic grammar and underlying syntax of environmental wisdom.

Environmental Performance Index (EPI) A measure that evaluates national sustainability and progress toward achievement of the United Nations Millennium Development Goals.

environmental policy The official rules or regulations concerning the environment adopted, implemented, and enforced by some governmental agency.

environmental racism Decisions that restrict certain people or groups of people to polluted or degraded environments on the basis of race.

environmental science The systematic, scientific study of our environment, as well as our role in it.

environmentalism Active participation in attempts to solve environmental pollution and resource problems.

enzymes Molecules, usually proteins or nucleic acids, that act as catalysts in biochemical reactions.

epigenetic effects Temporary changes in genetic material that affect gene expression

epigenetic effects Temporary changes in genetic material that affect gene expression

estuary A bay or drowned valley where a river empties into the sea.

ethics A branch of philosophy concerned with right and wrong.

eutrophic Rivers and lakes rich in organisms and organic material (*eu* = truly; *trophic* = nutritious).

evolution A theory that explains how random changes in genetic material and competition for scarce resources cause species to change gradually.

existence value The importance we place on just knowing that a particular species or a specific organism exists.

exponential growth Growth at a constant rate of increase per unit of time; can be expressed as a constant fraction or exponent. *See* geometric growth.

externalized costs Expenses, monetary or otherwise, borne by someone other than the individuals or groups who use a resource.

extinction The irrevocable elimination of species; can be a normal process of the natural world as species outcompete or kill off others or as environmental conditions change.

F

family planning Controlling reproduction; planning the timing of birth and having the number of babies that are wanted and can be supported.

famines Acute food shortages characterized by large-scale loss of life, social disruption, and economic chaos.

feed-in tariffs Require utilities to buy surplus power from small producers at a fair price. These tariffs are generally essential to individual solar installations.

fen An area of waterlogged soil that tends to be peaty; fed mainly by ground water; low in productivity.

fetal alcohol syndrome A tragic set of permanent physical, mental, and behavioral birth defects that result from the mother's drinking alcohol during pregnancy.

first law of thermodynamics Energy can be transformed and transferred, but cannot be destroyed or created; i.e., energy is *conserved.*

flagship species Especially interesting or attractive organisms, such as giant pandas or killer whales, to which people react emotionally. These species can motivate the public to preserve biodiversity and contribute to conservation.

flex-fuel boilers Boilers that can use a variety of different fuels: coal, oil, biomass, etc.

food chain A linked feeding series; in an ecosystem, the sequence of organisms through which energy and materials are transferred, in the form of food, from one trophic level to another.

food security The ability of individuals to obtain sufficient food on a day-to-day basis.

food web A complex, interlocking series of individual food chains in an ecosystem.

forest Any area where trees cover more than 10 percent of the land.

fossil fuels Petroleum, natural gas, and coal created by geological forces from organic wastes and the dead bodies of formerly living biological organisms.

founder effect The effect on a population founded when just a few members of a species survive a catastrophic event or colonize a new habitat that is geographically isolated from other members of the same species.

fuel assembly A bundle of hollow metal rods containing uranium oxide pellets; used to fuel a nuclear reactor.

fuel cells Mechanical devices that use hydrogen or hydrogen-containing fuels such as methane to produce an electric current. Fuel cells are clean, quiet, and highly efficient sources of electricity.

fugitive emissions Substances that enter the air without going through a smokestack, such as dust from soil erosion, strip mining, rock crushing, construction, and building demolition.

fumigants Toxic gases such as methyl bromine that are used to kill pests.

fungicide A chemical that kills fungi.

G

gap analysis A biogeographical technique of mapping biological diversity and endemic species to find gaps between protected areas that leave endangered habitats vulnerable to disruption.

garden city A new town with special emphasis on landscaping and rural ambience.

generalist Species that can tolerate a wide range of conditions or exploit a wide range of resources.

genetic drift The gradual changes in gene frequencies in a population due to random events.

genetically modified organisms (GMOs) Organisms whose genetic code has been altered by artificial means such as interspecies gene transfer.

geographic isolation Speciation deriving from a common ancestor due to geographic barriers that cause reproductive isolation; *also called* geographic isolation.

geothermal energy Energy drawn from the internal heat of the earth, either through geysers, fumaroles, hot springs, or other natural geothermal features, or through deep wells that pump heated groundwater.

global environmentalism A concern for, and action to help solve, global environmental problems.

grasslands A biome dominated by grasses and associated herbaceous plants.

green plans Integrated national environmental plans for reducing pollution and resource consumption while achieving sustainable development and environmental restoration.

green pricing Setting prices to encourage conservation or renewable energy; plans that invite customers to pay a premium for energy from renewable sources.

green revolution Dramatically increased agricultural production brought about by "miracle" strains of grain; usually requires high inputs of water, plant nutrients, and pesticides.

greenfield developments Housing projects built on previously undeveloped farmlands or forests on the outskirts of large cities.

greenhouse effect Gases in the atmosphere are transparent to visible light but absorb infrared (heat) waves that are re-radiated from the earth's surface.

greenhouse gases Chemical compounds that trap heat in the atmosphere. The principal anthropogenic greenhouse gases are carbon dioxide, methane, chlorofluorocarbons, nitrous oxide, and sulfur hexafluoride.

gross domestic product (GDP) The total economic activity within national boundaries. Gross domestic product (GDP) is used to distinguish economic activity within a country from that of offshore corporations.

gross national product (GNP) The sum total of all goods and services produced in a national economy.

groundwater Water held in gravel deposits or porous rock below the earth's surface; does not include water or crystallization held by chemical bonds in rocks or moisture in upper soil layers.

gully erosion Removal of layers of soil, creating channels or ravines too large to be removed by normal tillage operations.

H

habitat The place or set of environmental conditions in which a particular organism lives.

habitat conservation plans (HCP) Agreements under which property owners are allowed to harvest resources or develop land as long as habitat is conserved or replaced in ways that benefit resident endangered or threatened species in the long run. Some incidental "taking" or loss of endangered species is generally allowed in such plans.

hazardous air pollutants (HAPs) Especially dangerous air pollutants, including carcinogens, neurotoxins, mutagens, teratogens, endocrine system disrupters, and other highly toxic compounds.

hazardous waste Any discarded material containing substances known to be toxic, mutagenic, carcinogenic, or teratogenic to humans or other life-forms; or ignitable, corrosive, explosive, or highly reactive alone or with other materials.

health A state of physical and emotional well-being; the absence of disease or ailment.

heat A form of energy transferred from one body to another because of a difference in temperatures.

herbicide A chemical that kills plants.

herbivore An organism that eats only plants.

HIPPO *H*abitat destruction, *I*nvasive species, *P*ollution, *P*opulation (human), and *O*verharvesting— the leading causes of extinction.

homeostasis The maintenance of a dynamic, steady state in a living system through opposing, compensating adjustments.

Hubbert curve A curve describing a peak and decline in production of natural resources, especially oil production, defined by M. King Hubbert in 1956.

human development index (HDI) A measure of quality of life using life expectancy, child survival, adult literacy, childhood education, gender equity, and access to clean water and sanitation as well as income.

hurricanes Large cyclonic oceanic storms with heavy rain and winds exceeding 119 km/hr (74 mph).

hydraulic fracturing ("fracking") A mixture of water, sand, and toxic chemicals is pumped into rock formations at extremely high pressure to fracture sediments and release oil or gas.

hydrologic cycle The natural process by which water is purified and made fresh through evaporation and precipitation. This cycle provides all the freshwater available for biological life.

hypothesis A provisional explanation that can be tested scientifically.

I

I = PAT formula A formula that says our environmental impacts (I) are the product of our population size (P) times our affluence (A) and the technology (T) used to produce the goods and services we consume.

igneous rocks Crystalline minerals solidified from molten magma from deep in the earth's interior; basalt, rhyolite, andesite, lava, and granite are examples.

independent variable A factor not affected by the condition being altered in a manipulative experiment.

indicator species Species whose critical tolerance limits can be used to judge environmental conditions.

indicators Species whose critical tolerance limits can be used to judge environmental conditions.

inductive reasoning Inferring general principles from specific examples.

infiltration The process of water percolation into the soil and the pores and hollows of permeable rocks.

inherent value Ethical values or rights that exist as an intrinsic or essential characteristic of a particular thing or class of things simply by the fact of their existence.

inorganic pesticides Inorganic chemicals such as metals, acids, or bases used as pesticides.

insecticide A chemical that kills insects.

instrumental value The value or worth of objects that satisfy the needs and wants of moral agents. Objects that can be used as a means to some desirable end.

intangible resources Factors such as open space, beauty, serenity, wisdom, diversity, and satisfaction that cannot be grasped or contained. Ironically, these resources can be both infinite and exhaustible.

integrated pest management (IPM) An ecologically based pest-control strategy that relies on natural mortality factors, such as natural enemies, weather, cultural control methods, and carefully applied doses of pesticides.

Intergovernmental Panel on Climate Change (IPCC) An international organization formed to assess global climate change and its impacts. The IPCC is concerned with social, economic, and environmental impacts of climate change, and it was established by the United Nations Environment Programme and the World Meteorological Organization.

internal costs The expenses, monetary or otherwise, borne by those who use a resource.

interspecific competition In a community, competition for resources between members of *different* species.

intervention Techniques to discourage or reduce undesired organisms and favor or promote desired species.

intraspecific competition In a community, competition for resources among members of the *same* species.

invasive species Organisms that thrive in new territory where they are free of predators, diseases, or resource limitations that may have controlled their population in their native habitat.

ions Electrically charged atoms that have gained or lost electrons.

island biogeography The study of rates of colonization and extinction of species on islands or other isolated areas, based on size, shape, and distance from other inhabited regions.

isotopes Forms of a single element that differ in atomic mass due to having a different number of neutrons in the nucleus.

J

jet streams Powerful winds or currents of air that circulate in shifting flows; similar to oceanic currents in their extent and effect on climate.

joule A unit of energy. One joule is the energy expended in 1 second by a current of 1 amp flowing through a resistance of 1 ohm.

K

K-selected species Species that reproduce more slowly, occupy higher trophic levels, have fewer offspring, longer life spans, and greater intrinsic control of population growth than *r*-selected species.

kerogen A solid, energy-rich organic material similar to bitumen.

keystone species A species whose impacts on its community or ecosystem are much larger and more influential than would be expected from mere abundance.

kinetic energy Energy contained in moving objects, such as a rock rolling down a hill, the wind blowing through the trees, or water flowing over a dam.

kwashiorkor A widespread human protein-deficiency disease resulting from a starchy diet low in protein and essential amino acids.

Kyoto Protocol An international agreement to reduce greenhouse gas emissions.

L

La Niña The part of a large-scale oscillation in the Pacific (and, perhaps, other oceans) in which trade winds hold warm surface waters in the western part of the basin and cause upwelling of cold, nutrient-rich, deep water in the eastern part of the ocean.

landscape ecology The study of the reciprocal effects of spatial pattern on ecological processes. The study of how landscape history shapes the features of the land and the organisms that inhabit it, as well as our reaction to, and interpretation of, the land.

latent heat Stored energy in a form that is not sensible (cannot be detected by ordinary senses).

LD50 A chemical dose lethal to 50 percent of a test population.

life expectancy The average age that individuals born in a particular time and place can be expected to attain.

life-cycle analysis The evaluation of material and energy inputs and outputs at each stage of manufacture, use, and disposal of a product.

light-emitting diodes (LEDs) A semiconductor light source. When a hole recombines with an electron (from an electric current), light is emitted in an effect called electroluminescence. In a sense, this is the reverse of a photovoltaic panel. The color of the light is determined by the energy band gap of the semiconductor. These lights can be very small, but powerful. They take only a fraction of the energy of an incandescent bulb or a fluorescent light, and are said to last for years.

liquefied natural gas (LNG) When natural gas (mostly methane) is cooled to approximately $-162°C$ ($-260°F$) it becomes a colorless, odorless liquid that takes up about 1/600th the volume of the gaseous state. This makes it much more convenient to ship and store.

locavore Someone who eats locally grown, seasonal food.

logical thinking Asks, can the rules of logic help understand this.

logistic growth Growth rates regulated by internal and external factors that establish an equilibrium with environmental resources.

low-head hydropower Small-scale hydro technology that can extract energy from small headwater dams; causes much less ecological damage.

LULUs *L*ocally *U*nwanted *L*and *U*ses, such as toxic waste dumps, incinerators, smelters, airports, freeways, and other sources of environmental, economic, or social degradation.

M

magma Molten rock from deep in the earth's interior; called lava when it spews from volcanic vents.

malnourishment A nutritional imbalance caused by a lack of specific dietary components or an inability to absorb or utilize essential nutrients.

Man and Biosphere (MAB) program A design for nature preserves that divides protected areas into zones with different purposes. A highly protected core is surrounded by a buffer zone and peripheral regions in which multiple-use resource harvesting is permitted.

managing the commons Systems for managing common resources.

mangroves Trees from a number of genera that live in shallow salt water.

manipulative experiment An experiment in which some conditions are deliberately altered while others are held constant to study cause-and-effect relationships.

mantle A hot, pliable layer of rock that surrounds the earth's core and underlies the cool, outer crust.

marasmus A widespread human protein-deficiency disease caused by a diet low in calories and protein or imbalanced in essential amino acids.

marginal costs and benefits The costs and benefits of producing one additional unit of a good or service.

market equilibrium The dynamic balance between supply and demand under a given set of conditions in a "free" market (one with no monopolies or government interventions).

market forces Dynamics by which prices or wages are set by negotiation between buyers and sellers.

marsh Wetland without trees; in North America, this type of land is characterized by cattails and rushes.

mass burn Incineration of unsorted solid waste.

mass wasting Mass movement of geologic materials downhill caused by rockslides, avalanches, or simple slumping.

matter Anything that takes up space and has mass.

mean Average; a representation of the middle of a group; commonly calculated as the sum of values divided by the number of observations.

mediation An informal dispute resolution process in which parties are encouraged to discuss issues openly but in which all decisions are reached by consensus and any participant can withdraw at any time.

megacity *See* megalopolis.

metabolism All the energy and matter exchanges that occur within a living cell or organism; collectively, the life processes.

metamorphic rock Igneous and sedimentary rocks modified by heat, pressure, and chemical reactions.

metapopulation A collection of populations that have regular or intermittent gene flow between geographically separate units.

methane CH_4; the simplest hydrocarbon and the main component of natural gas.

methane hydrate Small bubbles or individual molecules of methane (natural gas) trapped in a crystalline matrix of frozen water.

micro-hydro generators Small power generators that can be used in low-level rivers to provide economical power for four to six homes, freeing them from dependence on large utilities and foreign energy supplies.

microbial agents Beneficial microbes (bacteria, fungi) that can be used to suppress or control pests; *also called* biological controls.

microbiome The complex community of microorganisms that live in and on your body. The microbiome is essential to digestion and other processes.

mid-ocean ridges Mountain ranges on the ocean floor created where molten magma is forced up through cracks in the planet's crust.

Milankovitch cycles Periodic variations in tilt, eccentricity, and wobble in the earth's orbit; Milutin Milankovitch suggested that these are responsible for cyclic weather changes.

millennium assessment A set of ambitious environmental and human development goals established by the United Nations in 2000.

Millennium Development Goals A set of goals established in 2000 by the United Nations that include ending poverty and hunger, universal education, gender equity, child health, maternal health, combating HIV/AIDS, environmental sustainability, and global cooperation in development efforts.

mineral A naturally occurring, inorganic, crystalline solid with definite chemical composition and characteristic physical properties.

minimum viable population size The number of individuals needed for long-term survival of rare and endangered species.

mitigation Repairing or rehabilitating a damaged ecosystem or compensating for damage by providing a substitute or replacement area.

models Simple representations of more complex systems.

molecule A combination of two or more atoms.

monoculture agroforestry Intensive planting of a single species; an efficient wood production approach, but one that encourages pests and disease infestations and conflicts with wildlife habitat or recreational uses.

monsoon A seasonal reversal of wind patterns caused by the different heating and cooling rates of the oceans and continents.

Montreal Protocol An international convention to eliminate chlorofluorocarbon production.

moral extensionism Expansion of our understanding of the inherent value or rights to persons, organisms, or things that might not be considered worthy of value or rights under some ethical philosophies.

moral value The value or worth of something based on moral principles.

morbidity Illness or disease.

mortality Death rate in a population; the probability of dying.

Müllerian mimicry Evolution of two species, both of which are unpalatable and have poisonous stingers, or some other defense mechanism, to resemble each other.

mutagens Agents, such as chemicals or radiation, that damage or alter genetic material (DNA) in cells.

mutation A change, either spontaneous or by external factors, in the genetic material of a cell. Mutations in the gametes (sex cells) can be inherited by future generations of organisms.

mutualism A symbiotic relationship between individuals of two different species in which both species benefit from the association.

mycorrhizal symbiosis An association between the roots of most plant species and certain fungi. The plant provides organic compounds to the fungus, while the fungus provides water and nutrients to the plant.

N

National Environmental Policy Act (NEPA) *National Environmental Policy Act*, the cornerstone of U.S. environmental policy. Authorizes the Council on Environmental Quality, directs federal agencies to take environmental consequences into account when making decisions, and requires an environmental impact statement for every major federal project likely to have adverse environmental effects.

natural experiment A study of events that have already happened.

natural increase Crude death rate subtracted from crude birthrate.

natural organic pesticides "Botanicals" or organic compounds naturally occurring in plants, animals, or microbes that serve as pesticides.

natural selection The mechanism for evolutionary change in which environmental pressures cause certain genetic combinations in a population to become more abundant. Genetic combinations best adapted for present environmental conditions tend to become predominant.

negative feedback loop A situation in which a factor or condition causes changes that reduce that factor or condition.

negawatts Energy saved through conservation.

neoclassical economics A branch of economics that attempts to apply the principles of modern science to economic analysis in a mathematically rigorous, noncontextual, abstract, predictive manner.

neonicotinoid Pesticides with a chemical structure similar to nicotine. Thought to be a cause of massive bee deaths.

neurotoxins Toxic substances, such as lead or mercury, that specifically poison nerve cells.

nitric oxide NO.

nitrogen cycle The circulation and reutilization of nitrogen in both inorganic and organic phases.

nitrogen dioxide NO_2.

nitrogen oxides Highly reactive gases formed when nitrogen in fuel or combustion air is heated to over 650°C (1,200°F) in the presence of oxygen, or when bacteria in soil or water oxidize nitrogen-containing compounds.

nitrous oxide N_2O.

nongovernmental organizations (NGOs) A term referring collectively to pressure and research groups, advisory agencies, political parties, professional societies, and other groups concerned about environmental quality, resource use, and many other issues.

nonpoint sources Scattered, diffuse sources of pollutants, such as runoff from farm fields, golf courses, and construction sites.

nonrenewable resources Minerals, fossil fuels, and other materials present in essentially fixed amounts (within human time scales) in our environment.

novel ecosystem A specific biological community and its physical environment interacting in an exchange of matter and energy.

nuclear fission The radioactive decay process in which isotopes split apart to create two smaller atoms.

nuclear fusion A process in which two smaller atomic nuclei fuse into one larger nucleus and release energy; the source of power in a hydrogen bomb.

O

oak savanna Open grasslands with sparse tree cover.

obese Generally considered to be a body mass greater than 30 kg/m^2, or roughly 30 pounds above normal for an average person.

ocean thermal electric conversion (OTEC) Energy derived from temperature differentials between warm ocean surface waters and cold deep waters. This differential can be used to drive turbines attached to electric generators.

ocean wave energy Electricity generated by ocean wave power.

oil shale A fine-grained sedimentary rock rich in solid organic material called kerogen. When heated, the kerogen liquefies to produce a fluid petroleum fuel.

oligotrophic The condition of rivers and lakes that have clear water and low biological productivity (*oligo* = little; *trophic* = nutrition); usually clear, cold, infertile headwater lakes and streams.

omnivore An organism that eats both plants and animals.

open system A system that exchanges energy and matter with its environment.

open-access system A commonly held resource for which there are no management rules.

organic compounds Complex molecules organized around skeletons of carbon atoms arranged in rings or chains; includes biomolecules, molecules synthesized by living organisms.

organophosphates Organic molecules to which one or more phosphate groups are attached.

overgrazing Allowing livestock to eat so much forage that the ecological health of the habitat is damaged.

overharvesting Harvesting so much of a resource that its existence is threatened.

oxygen sag Oxygen decline downstream from a pollution source that introduces materials with high biological oxygen demands.

ozone A highly reactive molecule containing three oxygen atoms; a dangerous pollutant in ambient air. In the stratosphere, however, ozone forms an ultraviolet-absorbing shield that protects us from mutagenic radiation.

P

paradigm shift Dramatic changes in models that provide frameworks for interpreting observations.

parasitism An organism that lives in or on another organism, deriving nourishment at the expense of its host, usually without killing it.

particulate matter Atmospheric aerosols, such as dust, ash, soot, lint, smoke, pollen, spores, algal cells, and other suspended materials. The term originally was applied only to solid particles but now is extended to droplets of liquid.

passive heat absorption The use of natural materials or absorptive structures without moving parts to gather and hold heat; the simplest and oldest use of solar energy.

passive houses Highly insulated, virtually airtight buildings heated primarily by passive solar gain and by the energy released by people, electrical appliances, cooking, and other ordinary activities.

pastoralists People who make a living by herding domestic livestock.

peak oil A prediction, made about 1940 by Dr. M. King Hubbert, that oil production in the United States would peak in the 1970s and then decline.

pelagic Zones in the vertical water column of a water body.

permanent retrievable storage Placing waste storage containers in a secure building, salt mine, or bedrock cavern where they can be inspected periodically and retrieved if necessary.

persistent organic pollutants (POPs) Chemical compounds that persist in the environment and retain biological activity for a long time.

pest resurgence Rebound of pest populations due to acquired resistance to chemicals and nonspecific destruction of natural predators and competitors by broadscale pesticides.

pesticide Any chemical that kills, controls, drives away, or modifies the behavior of a pest.

pesticide treadmill A need for constantly increasing doses or new pesticides to prevent pest resurgence.

pH A value that indicates the acidity or alkalinity of a solution on a scale of 0 to 14, based on the proportion of H1 ions present.

phosphorus cycle The movement of phosphorus atoms from rocks through the biosphere and hydrosphere and back to rocks.

photochemical oxidants Products of secondary atmospheric reactions. *See* smog.

photodegradable plastics Plastics that break down to smaller particles when exposed to sunlight or to a specific wavelength of light.

photosynthesis The biochemical process by which green plants and some bacteria capture light energy and use it to produce chemical bonds. Carbon dioxide and water are consumed while oxygen and simple sugars are produced.

photovoltaic (PV) cell An energy-conversion device that captures solar energy and directly converts it to electrical current.

phytoplankton Microscopic, free-floating, autotrophic organisms that function as producers in aquatic ecosystems.

pioneer species In primary succession on a terrestrial site, the plants, lichens, and microbes that first colonize the site.

point sources Specific locations of highly concentrated pollution discharge, such as factories, power plants, sewage treatment plants, underground coal mines, and oil wells.

policy A societal plan or statement of intentions intended to accomplish some social good.

political economy The branch of economics concerned with modes of production, distribution of benefits, social institutions, and class relationships.

population A group of individuals of the same species occupying a given area.

population crash A sudden population decline caused by predation, waste accumulation, or resource depletion; *also called* a dieback.

population momentum A potential for increased population growth as young members reach reproductive age.

positive feedback loop A situation in which a factor or condition causes changes that further enhance that factor or condition.

potential energy Stored energy that is latent but available for use; for example, a rock poised at the top of a hill, or water stored behind a dam.

power The rate of energy delivery; measured in horsepower or watts.

power tower A high-temperature concentrating solar power system in which thousands of mirrors arranged in concentric rings around a tall central tower track the sun and focus light on a heat absorber to generate steam, which drives an electrical generator.

prairie potholes Small ponds on the prairie.

precautionary principle The decision to leave a margin of safety for unexpected developments.

predator An organism that feeds directly on other organisms in order to survive; live-feeders, such as herbivores and carnivores.

predator-mediated competition A situation in which predation reduces prey populations and gives an advantage to competitors that might not otherwise be successful.

premises Introductory statements that set up or define a problem. Those things taken as given.

prescribed burning Periodic fires that clean out brush and trees to maintain a prairie.

price elasticity A situation in which supply and demand of a commodity respond to price.

primary forest Forest composed primarily of native species, where there are no clearly visible indications of human activity and ecological processes are not significantly disturbed.

primary pollutants Chemicals released directly into the air in a harmful form.

primary productivity Synthesis of organic materials (biomass) by green plants using the energy captured in photosynthesis.

primary succession An ecological succession that begins in an area where no biotic community previously existed.

primary treatment A process that removes solids from sewage before it is discharged or treated further.

principle of competitive exclusion No two species can occupy the same ecological niche for long.

producer An organism that synthesizes food molecules from inorganic compounds by using an external energy source; most producers are photosynthetic.

productivity The synthesis of new organic material; synthesis done by green plants using solar energy is called primary productivity.

pronatalist pressures Influences that encourage people to have children.

property assessed clean energy (PACE) A program that uses city bonds to pay for renewable energy and conservation expenses, such as solar power and insulation.

pull factors (in urbanization) Conditions that draw people from the country into the city.

push factors (in urbanization) Conditions that force people out of the country and into the city.

Q

quality-of-life indicators Factors, such as infant mortality, life expectancy, income, sanitation, and education, that indicate quality of life in a country.

R

r-selected species Species that tend to have rapid reproduction and high offspring mortality. They frequently overshoot the carrying capacity of their environment and display boom-and-bust life cycles. They lack intrinsic population controls and tend to occupy lower trophic levels in food webs than do *k*-selected species.

rain shadow A dry area on the downwind side of a mountain.

re-creation (in ecology) Construction of an entirely new biological community to replace one that has been destroyed on that or another site.

reallocation To use a site (and its resources) to create a new and different kind of biological community rather than the existing one.

recharge zone An area where water infiltrates an aquifer.

reclamation Chemical, biological, or physical cleanup and reconstruction of severely contaminated or degraded sites to return them to something like their original topography and vegetation.

recycling Reprocessing of discarded materials into new, useful products; not the same as reuse of materials for their original purpose, but the terms are often used interchangeably.

red tide A population explosion or bloom of minute, single-celled marine organisms called dinoflagellates. Billions of these cells can accumulate in protected bays, where the toxins they contain can poison other marine life.

reflective thinking Asks, what does this all mean.

reformer A device that strips hydrogen from fuels such as natural gas, methanol, ammonia, gasoline, or vegetable oil so they can be used in a fuel cell.

refuse-derived fuel Processing of solid waste to remove metal, glass, and other unburnable materials; organic residue is shredded, formed into pellets, and dried to make fuel for power plants.

rehabilitate To rebuild elements of structure or function in an ecological system without necessarily achieving complete restoration to its original condition.

remediation Cleaning up chemical contaminants from a polluted area.

renewable portfolio standard A mandate that a utility company's generating capacity, or purchasing practices, include a specified amount of renewable energy. This policy regulation is designed to establish a market for alternative energy sources.

renewable resources Resources normally replaced or replenished by natural processes; resources not depleted by moderate use; examples include solar energy, biological resources such as forests and fisheries, biological organisms, and some biogeochemical cycles.

renewable water supplies Annual freshwater surface runoff plus annual infiltration into underground freshwater aquifers that are accessible for human use.

replication Repeating studies or tests to verify reliability.

reproducibility The capacity for a particular result to be observed or obtained more than once.

residence time The length of time a component, such as an individual water molecule, spends in a specific compartment or location before it moves on through a particular process or cycle.

resilience The ability of a community or ecosystem to recover from disturbances.

resource In economic terms, anything with potential use in creating wealth or giving satisfaction.

resource partitioning In a biological community, various populations sharing environmental resources through specialization, thereby reducing direct competition.

riders Amendments attached to bills in conference committee, often completely unrelated to the bill to which they are added.

rill erosion The removal of thin layers of soil by little rivulets of running water that gather and cut small channels in the soil.

risk The probability that something undesirable will happen as a consequence of exposure to a hazard.

risk assessment Evaluation of the short-term and long-term risks associated with a particular activity or hazard; usually compared to benefits in a cost–benefit analysis.

Roadless rule A Clinton-era ban on logging, road building, and other development on the lands identified as deserving of wilderness protection in the Roadless Area Review and Evaluations (RARE).

rock A solid, cohesive aggregate of one or more crystalline minerals.

rock cycle The process whereby rocks are broken down by chemical and physical forces; sediments are moved by wind, water, and gravity, sedimented and reformed into rock, and then crushed, folded, melted, and recrystallized into new forms.

rotational grazing Confining animals to a small area for a short time (often only a day or two) before shifting them to a new location.

run-of-the-river flow Ordinary river flow not accelerated by dams, flumes, etc. Some small, modern, high-efficiency turbines can generate useful power with run-of-the-river flow or with a current of only a few kilometers per hour.

rural area An area in which most residents depend on agriculture or the harvesting of natural resources for their livelihood.

S

salinization A process in which mineral salts accumulate in the soil, killing plants; occurs when soils in dry climates are irrigated profusely.

salt marsh Shallow wetlands along coastlines that are flooded regularly or occasionally with seawater.

saltwater intrusion Movement of saltwater into freshwater aquifers in coastal areas where groundwater is withdrawn faster than it is replenished.

sample Noun: A small portion of a population, used to approximate characteristics of the population. Verb: to collect observations or a small portion of a population, to allow characterization of a population.

sanitary landfills A landfill in which garbage and municipal waste are buried every day under enough soil or fill to eliminate odors, vermin, and litter.

scavenger An organism that feeds on the dead bodies of other organisms.

science A process for producing knowledge methodically and logically.

scientific consensus A general agreement among informed scholars.

scientific theory An explanation supported by many tests and accepted by a general consensus of scientists.

second law of thermodynamics With each successive energy transfer or transformation, less energy is available to do work.

secondary pollutants Chemicals that acquire a hazardous form after entering the air or that are formed by chemical reactions as components of the air interact.

secondary succession Succession on a site where an existing community has been disrupted.

secondary treatment Bacterial decomposition of suspended particulates and dissolved organic compounds that remain after primary sewage treatment.

secure landfill A solid-waste disposal site lined and capped with an impermeable barrier to prevent leakage or leaching. Drain tiles, sampling wells, and vent systems provide monitoring and pollution control.

sedimentary rock Deposited material that remains in place long enough, or is covered with enough material, to compact into stone; examples include shale, sandstone, breccia, and conglomerates.

sedimentation The deposition of organic materials or minerals by chemical, physical, or biological processes.

selection pressures Factors in the environment that favor the successful reproduction of individuals possessing certain heritable traits and that reduce the viability and fertility of individuals that do not possess those traits.

shantytowns Settlements created when people move onto undeveloped lands and build their own shelters with cheap or discarded materials. Some are simply illegal subdivisions where a landowner rents land without city approval; others are land invasions.

sheet erosion The peeling off of thin layers of soil from the land surface; accomplished primarily by wind and water.

sick building syndrome Headaches, allergies, chronic fatigue, and other symptoms caused by poorly vented indoor air contaminated by pathogens or toxins.

significant figure Also known as significant digits, a meaningful degree of precision.

slums Legal but inadequate multifamily tenements or rooming houses; some are custom-built for rent to poor people, others are converted from some other use.

smart growth Efficient use of land resources and existing infrastructure.

smart meter A meter records energy use data and communicates with the utility company, so that you can be charged less for off-peak energy and more for peak energy.

social justice Equitable access to resources and the benefits derived from them: a system that recognizes inalienable rights and adheres to what is fair, honest, and moral.

solar garden A solar array shared by multiple owners, who help finance the facility and gain from the electricity produced. Also called a solar farm.

solar garden Energy derived from temperature differentials between warm ocean surface waters and cold deep waters. This differential can be used to drive turbines attached to electric generators.

species A population of morphologically similar organisms that can reproduce sexually among themselves but that cannot produce fertile offspring when mated with other organisms.

sprawl Unlimited outward extension of city boundaries that lowers population density, consumes open space, generates freeway congestion, and causes decay in central cities.

squatter towns Shantytowns that occupy land without owner's permission. Some are highly organized movements in defiance of authorities; others grow gradually.

standing The right to take part in legal proceedings.

state shift A permanent or long-lasting change in a system to a new set of conditions and relations in response to a disturbance.

statistics Numbers that describe observations or groups of observations; also a field of study that provides methods of comparing descriptive numbers.

statute law Formal documents or decrees enacted by the legislative branch of government.

steady-state economy Characterized by low birth and death rates, use of renewable energy sources, recycling of materials, and emphasis on durability, efficiency, and stability.

stewardship A philosophy that holds that humans have a unique responsibility to manage, care for, and improve nature.

strategic lawsuits against public participation (SLAPPs) Lawsuits that have no merit but are brought merely to intimidate and harass private citizens who act in the public interest.

stratosphere The zone in the atmosphere extending from the tropopause to about 50 km (30 mi) above the earth's surface; temperatures are stable or rise slightly with altitude; has very little water vapor but is rich in ozone.

stratospheric ozone The ozone (O_3) occurring in the stratosphere 10 to 50 km above the earth's surface.

stress-related diseases Diseases caused or accentuated by social stresses such as crowding.

subduction The process by which one tectonic plate is pushed down below another as plates crash into each other.

subsidence A settling of the ground surface caused by the collapse of porous formations that result from a withdrawal of large amounts of groundwater, oil, or other underground materials.

subsoil A layer of soil beneath the topsoil that has lower organic content and higher concentrations of fine mineral particles; often contains soluble compounds and clay particles carried down by percolating water.

sulfur dioxide A colorless, corrosive gas directly damaging to both plants and animals.

Superfund A fund established by Congress to pay for the containment, cleanup, or remediation of abandoned toxic waste sites. The fund is financed by fees paid by toxic waste generators and by cost recovery from cleanup projects.

supply The quantity of a product being offered for sale at various prices, other things being equal.

sustainable development Improving well-being and the standard of life over the long-term; that is, meeting the needs of the present without compromising the ability of future generations to meet their own needs.

Sustainable Development Goals (SDGs) A set of 17 goals adopted by the United Nations in 2015, which aim to end poverty, protect the planet, and ensure prosperity for all as part of a new sustainable development agenda. Each goal has specific targets to be achieved between 2015 and 2030.

swamp A wetland with trees, such as the extensive swamp forests of the southern United States.

symbiosis The intimate living together of members of two different species; includes mutualism, commensalism, and, in some classifications, parasitism.

sympatric speciation Species that arise from a common ancestor due to biological or behavioral barriers that cause reproductive isolation even though the organisms live in the same place.

synergistic effect An interaction in which one substance exacerbates the effects of another. The sum of the interaction is greater than the parts.

systems Networks of interactions among many interdependent factors.

T

taiga The northernmost edge of the boreal forest, including species-poor woodland and peat deposits; intergrading with the arctic tundra.

tar sands Sand deposits containing petroleum or tar.

technological optimists Those who believe that technology and human enterprise will find cures for all our problems; *also called* Promethean environmentalism.

tectonic plates Huge blocks of the earth's crust that slide around slowly, pulling apart to open new ocean basins or crashing ponderously into each other to create new, larger landmasses.

temperate rainforest The cool, dense, rainy forest of the northern Pacific coast; enshrouded in fog much of the time; dominated by large conifers.

temperature inversion A stable layer of warm air overlying cooler air, trapping pollutants near ground level.

teratogens Chemicals or other factors that cause abnormalities during embryonic growth and development.

terracing Shaping the land to create level shelves of earth to hold water and soil; requires extensive hand labor or expensive machinery, but enables farmers to farm very steep hillsides.

tertiary treatment The removal of inorganic minerals and plant nutrients after primary and secondary treatment of sewage.

thermal plume A plume of hot water discharged into a stream or lake by a heat source, such as a power plant.

thermocline In water, a distinctive temperature transition zone that separates an upper layer that is mixed by the wind (the epilimnion) and a colder, deep layer that is not mixed (the hypolimnion).

threatened species A species that is still abundant in parts of its territorial range but that has declined significantly in total numbers and may be on the verge of extinction in certain regions or localities.

throughput The flow of energy and matter into, through, and out of a system.

tidal station A dam built across a narrow bay or estuary that traps tide water flowing both in and out of the bay. Water flowing through the dam spins turbines attached to electric generators.

tide pool Depressions in a rocky shoreline that are flooded at high tide but cut off from the ocean at low tide. They often have a rich collection of marine life.

tolerance limits Chemical or physical factors that limit the existence, growth, abundance, or distribution of an organism.

topsoil The uppermost layer of a soil, including the "O" and "A" layers, which are usually rich in organic material.

tornado A violent storm characterized by strong swirling winds and updrafts. Tornadoes form when a strong cold front pushes under a warm, moist air mass over the land.

total fertility rate The number of children born to an average woman in a population during her entire reproductive life.

total growth rate The net rate of population growth resulting from births, deaths, immigration, and emigration.

total maximum daily load (TMDL) The amount of particular pollutants that a water body can receive from both point and nonpoint sources and still meet water quality standards.

toxic colonialism Shipping toxic wastes to a weaker or poorer nation.

Toxic Release Inventory (TRI) A program created by the Superfund Amendments and Reauthorization Act of 1984 that requires manufacturing facilities and waste-handling and disposal sites to report annually on releases of more than 300 toxic materials.

toxic substances Poisonous chemicals that react with specific cellular components to kill cells or to alter growth or development in undesirable ways; often harmful, even in diluted concentrations.

tragedy of the commons An inexorable process of degradation of communal resources due to the selfishness of "free riders" who use or destroy more than their fair share of common property. *See* open access system.

triple bottom line Corporate accounting that reports social and environmental costs and benefits, as well as merely economic ones.

trophic level A step in the movement of energy through an ecosystem; an organism's feeding status in an ecosystem.

tropical rainforests Forests in which rainfall is abundant—more than 200 cm (80 in.) per year—and temperatures are warm to hot year-round.

tropical seasonal forest Semi-evergreen or partly deciduous forests tending toward open woodlands and grassy savannas dotted with scattered, drought-resistant tree species; distinct wet and dry seasons, hot year-round.

troposphere The layer of air nearest to the earth's surface; both temperature and pressure usually decrease with increasing altitude.

tsunami A giant seismic sea swell that moves rapidly from the center of a submarine earthquake; can be 10 to 20 meters high when it reaches shorelines hundreds or even thousands of kilometers from its source.

tundra Treeless arctic or alpine biome characterized by cold, harsh winters, a short growing season, and a potential for frost any month of the year; vegetation includes low-growing perennial plants, mosses, and lichens.

U

umbrella species Require large blocks of relatively undisturbed habitat to maintain viable populations. Saving this habitat also benefits other species.

UN Framework Convention on Climate Change (UNFCCC) Directs governments to share data on climate change, to develop national plans for controlling greenhouse gases, and to cooperate in planning for adaptation to climate change.

unburnable carbon Carbon fuels that we need to leave in the ground if we want to avoid disastrous climate change.

urban agglomerations An aggregation of many cities into a large metropolitan area.

urban area An area in which a majority of the people are not directly dependent on natural-resource-based occupations.

utilitarian conservation A philosophy that resources should be used for the greatest good for the greatest number for the longest time.

V

vertical zonation Terrestrial vegetation zones determined by altitude.

village A collection of rural households linked by culture, custom, and association with the land.

volatile organic compounds (VOCs) Organic chemicals that evaporate readily and exist as gases in the air.

volcano A vent in the earth's surface through which gases, ash, or molten lava are ejected. Also a mountain formed by this ejecta.

vulnerable species Naturally rare organisms or species whose numbers have been so reduced by human activities that they are susceptible to actions that could push them into threatened or endangered status.

W

warm front A long, wedge-shaped boundary caused when a warmer advancing air mass slides over neighboring cooler air parcels.

waste stream The steady flow of varied wastes, from domestic garbage and yard wastes to industrial, commercial, and construction refuse.

water pollution Anything that degrades water quality.

water scarcity When annual available freshwater supplies are less than 1,000 m³ per person.

water stress A situation when residents of a country don't have enough accessible, high-quality water to meet their everyday needs.

water table The top layer of the zone of saturation; undulates according to the surface topography and subsurface structure.

waterlogging Water saturation of soil that fills all air spaces and causes plant roots to die from lack of oxygen; a result of overirrigation.

watt (W) The force exerted by 1 joule, or the equivalent of a current of 1 amp per second flowing through a resistance of 1 ohm.

weather Description of the physical conditions of the atmosphere (moisture, temperature, pressure, and wind).

weathering Changes in rocks brought about by exposure to air, water, changing temperatures, and reactive chemical agents.

wedge analysis Policy options proposed by R. Socolow and S. Pacala for reducing greenhouse gas emissions using existing technologies. Each wedge represents a cumulative reduction of the equivalent of 1 billion tons of carbon over the next 50 years.

wetland mitigation Replacing a wetland damaged by development (roads, buildings, etc.) with a new or refurbished wetland.

wetlands Ecosystems of several types in which rooted vegetation is surrounded by standing water during part of the year. *See also* swamp, marsh, bog, fen.

wind turbines Large windmills that produce electricity.

withdrawal A description of the total amount of water taken from a lake, river, or aquifer.

work The application of force through a distance; requires energy input.

world conservation strategy A proposal for maintaining essential ecological processes, preserving genetic diversity, and ensuring that the utilization of species and ecosystems is sustainable.

World Happiness Index (WHI) A measure of life-satisfaction rather than mere economic success.

Z

zero population growth (ZPG) The number of births at which people are just replacing themselves; *also called* the replacement level of fertility.

zone of aeration Upper soil layers that hold both air and water.

zone of saturation Lower soil layers where all spaces are filled with water.

California continued
  honeybee shortage, 215
  Klamath River, water reallocation, 394
  Los Angeles, climate change and water supply, 12
  Los Angeles, temperature inversions, 361
  Los Angeles, traffic congestion, 509–510
  Los Angeles, water diversion reversal, 388
  Mojave Desert, CSP systems, 457–458
  Mt. Whitney, vertical zonation in, 99
  pesticide residue in food, study, 217
  pests, resistance, 215
  photovoltaic energy production, 460
  redwood trees, legal standing, 553, 572
  San Francisco earthquakes (1906 and 1989),
    315–316
  Santa Barbara Channel, oil spill, 546
  wetland disturbances, 113
  wind power use, 462, 465–466
Calment, Jeanne Louise, 140
calorie, 57
Calthorpe, Peter, 512
Camel's Hump Mountain (Vermont), 369
Cameroon, family size, 144
campus buildings, LEED certified, 576–577
Campus Climate Challenge, 574
campus greening, 574–577
Canada
  Atlantic Coastal Action Programme, 560
  Eastern pipeline proposal, 436
  energy consumption, per capita, 430
  forest management, 257–258
  Gateway pipeline proposal, 436
  grasslands, 260
  Great Bear Rainforest, 264, 268, 436
  Green Plan, 560
  hydropower use, 472
  indigenous title, 26
  irrigation methods, 206
  lynx population, 124
  Montreal Protocol, 556–557, 558
  National Packaging Protocol, 488
  natural gas supply, for U.S., 439
  oil sands, 434–435
  old-growth forests, 257–258
  parks and preserves, 264
  persistent pollutants, 216
  placer mining, 311
  pothole restoration, prairie provinces, 293
  proven petroleum reserves, 434
  Quittinirpaaq National Park, 264
  snowshoe hare population, 124
  surface mining for oil, 435
  tar sands, 434–435
  Transmountain pipeline proposal, 436
  water use, 385
  wood and paper pulp production, 252
Canadian gold miners, on U.S. public lands, 314
canals, water redistribution using, 389
Canary Islands, insects introduced from, 86
cancer
  anti-cancer drugs from periwinkle, 230–231
  cancer rates, 155, 360
  environmental causes, from, 163
  epigenic changes, 172
cap-and-trade program
  carbon, 359–360
  Kyoto Protocol, 533
  mercury pollution, 358
  wind energy, 463

capillary action, 53
capital (economics), 521
captive breeding, 245–246, 278
carbamates, neurotoxin, 162–163
carbaryl, neurotoxin, 163
carbohydrate, 54–55
carbon
  atoms, bonding, 52
  bubble, 527
  capture and storage, 344–346
  cycle, 66–67
  emissions, 15, 246
  organic compounds, 54
  sinks, 67
  trading in, 533–534
carbon capture and storage (CCS), 344–346, 433
carbon converter, algae, 471
carbon dioxide
  air pollutant, 352, 355, 358–359
  atmospheric, 12, 21, 326, 333
  current levels, 337, 371
  global climate change, 335–338, 358–359
  Kyoto Protocol, 342
  molecular structure, 52
  oil shale vs conventional oil, 436
  photosynthesis, 58–59, 66
  reducing, 343–347
  sources, 13, 58, 346, 432, 453
carbon monoxide, air pollutant, 352–356,
    367–368, 372
carbon neutral crops, 467
carbon neutrality, Kyoto, 533–534
carbon neutrality pledge, NZ, 345
carbon tax
  case study, 519
  results, 534
carbon tetrachloride, 360
carbon trading, results, 342–347
carcinogens, 163, 167, 169
careers, environmental, 567, 568
car-free city, case study, 499
carnivores, 62–63, 82, 187
carrying capacity, factors, 118–119, 124, 135,
    528–529
cars. See automobiles
Carson, Rachel, 18, 154, 211, 547
Carter, Jimmy, 573
Carter, Majorca, 349
case law, 551–554
case studies
  air pollution, Beijing, 350–351
  alternatives to coal, 427
  BedZED, 514
  carbon tax, 519
  car-free city, 499
  CITES, 543–544
  classroom achievement, 2
  coral reef restoration, 226–227
  Elwha River dam, 275–276
  endangered species, 543–544
  environmental activism, 564
  farming the Cerrado, 198–199
  fishing to extinction, 117
  Galápagos Islands, 73
  Ganges river, 401–402
  global warming, forests, 34
  Great London Smog (1952), 363
  Greenbelt Movement, Kenya, 113
  Lake Mead, 377

locavores, 178–179
MTR mining, 301–302
Pacific Garbage Gyre, 478–479
palm oil, 250
population stabilization, 132
sustainable development, 9–10
wedge analysis, 323–324
cations, 53
Catskill Mountains, runoff, 416
CDC. See Centers for Disease Control and Prevention
cells, 55–56
cellular respiration, 59, 66
cellulose-based biofuels, 469–471
cellulosic ethanol, 469–471
Census Bureau, U.S.
  commuter data, 453
  urban areas, defined, 501
Center for Journalistic Excellence, 6
Center for Naval Analyses, 359
Center for Public Integrity, 437
Center for Rural Affairs (Nebraska), 222
Centers for Disease Control and Prevention (CDC)
  antibiotic resistant infections, data, 187
  deaths, antibiotic resistance, from, 79
  diabetes data, 163
  flu vaccines and mutation, 80
  lead levels in children, 166
  Zika virus, 157
Central America. See individual countries
Central Arizona Project, water supply for, 377
cereal use, livestock, 185
Cerrado, case study, 198–199, 203
CFCs (chlorofluorocarbons), 364, 557
chain reaction, nuclear reactor, 442
charcoal
  indoor pollution from, 360
  Terra Preta, use in making, 209
charismatic species, protection, 543–544
cheetahs, genetic diversity, 127, 128
chemical bonds, 52
chemical defenses, species with, 83
chemical energy, 57
Chemical Manufacturers Association, 572
chemical weathering, rock, 306
chemosynthesis, 58
Chesapeake Bay
  case study, 50–51
  dead zones, 407
  ecosystem, 61–65
  fracking, 439
  pollutants, 67–69, 207, 416, 417
  restoration efforts, 109, 293
Chicago, Women's Self-Employment Project, 535
children
  lead poisoning in, 166
  survival rates, 154–155
Chile
  air pollution, 373
  climate, 103, 378, 379
chimps of Gombe, case sturdy, 269
China
  air pollution in, 13, 351–352, 372–373
  automobile production, 577
  biogas used for food, 468
  birth dearth, 141
  carbon dioxide, deaths, 432
  cities, demographic shift, 501, 502–503
  coal, 430, 432–433
  coal mine fires, 311

M

Maathai, Wangari, 8, 18, 19, 98, 283
MacArthur, Robert H., 39, 77–78, 90, 126
Macaulay Bird Library, 567
Madagascar
    biodiversity hot spot, 229, 230
    periwinkle, medicines derived from, 230
Maginnis, Stewart, 282
magma, 303, 305
maize
    ethanol base, 184
    genetic engineering of, 191
    ideal growth conditions, 87
    major food crop, 184
malaria
    avian, 236
    drug resistance, 158, 159, 160
    water, role of, 404
malathion, 162, 213
Malaysia, palm oil plantations, 469
Mali, freshwater shortage, 388
malnourishment, 182
Malpai Borderlands Group, 262
Malthus, Thomas, 73, 135, 136, 147, 526
managing the commons, strategies, 25–26
Man and Biosphere (MAB) program
    (UNESCO), 267
Man and Nature (Marsh), 16
manatees (endangered), Florida, 410
maneb (neurotoxin), 163
mangroves, 108–109
Manila, slums of, 9–10
manipulative experiment
    B4Warmed, 39
    defined, 40
mantle, Earth's, 302
manufactured capital, 521
manure, as fuel, 467–468
Marasmus, 183
marginal costs and benefits, 523
marine ecosystems
    biodiversity of, 106–110
    coastal zones, 108–110, 414–416
    coral reefs (*see* coral reefs)
    deep sea organisms, 58
    open-ocean communities, 107
    parks and preserves, 266–267
    predation in, 83
    productivity in, 87
marine sanctuaries and fish, 187
marine snow, 105–106
marine species, global climate change effects, 235
market equilibrium, 522–523
market inefficiencies, economic, 525
Marsh, George Perkins, 16–17
Marshall, Robert, 18
marshes, 112, 289
Marx, Karl, 135, 523
Maryland, Columbia, planned community, 512
mass burn, 483–484
mass media, 6
mass wasting, 318
Matador ranch (Montana), 287
Mather, Stephen, 17, 265
matter, 51
Mauritius (1769), 16

maximum sustained yield, 120
McClintock, Barbara, 36
McDonough, William, 535
McDonough Design Principles, 537
McHarg, Ian, 515
McKibben, Bill, 564, 576
Meadows, Donnela, 529
mean, statistical, 38
mean global temperature increase, 12
meat
    feed requirements per kg, 185
    key food source, 184–185
mechanical weathering, 305–306
Médecins Sans Frontières (MSF), 160
mediation, 559
medicines, biodiversity and, 230–231
megacities, 500
megadrought projection, Colorado River, 377
megalopolises, 502
megawatts, 452
meltdown, nuclear, 442
Mencken, H. L., 24, 131
Merck, 231
mercury
    air pollutant, 13, 357–358
    American children, levels in, 406–408
    cost-benefit analysis of regulation, 432
    poisoning, 237
    power plant emissions, regulation, 358
    water pollutant, 406, 412
Mercury and Air Toxics Standards, 432, 549
mesolimnion, 111
mesopelagic zones, 107
mesosphere, 325
metabolic degradation of toxins, 167
metabolism, 58
metal contaminants, as water pollutants, 406
metals
    air pollutants, 357–358
    economic resource, 306
    importance of, 307
    mining, 310, 312–313, 314
    new materials substituted for old, 310
    processing, 312–313
    water pollutants, 408–409
metamorphic rocks, 305
metapopulations, 128
meteor (asteroid), extinction scenario, 233, 315
methane (CH$_4$)
    air pollutant, 352
    China, fuel from biomass, 468
    climate change, role in, 438–439
    coal formation and, 309
    extraction by fracking, 439
    global warming, contribution to, 336–337
    landfills, recovery from, 483
    microbes, methane-eating, 58
    molecular structure of, 52
    natural gas, 437, 438–439
    production, anaerobic digestion process, 468
    wells, escape from, 439
methane hydrates, source and effect, 440
methylating, epigenome, 172
Mexico
    biodiesel, *Jatropha curcas* conversion, 469
    dependency ratios, 142–143
    fertility rates, 139
    land degradation, 204

Mexico City, air pollution, 373
Mexico City, garbage problems, 480
Mexico City, subsidence, 389
Sian Ka'an Reserve, 268
Michigan, clear-cut logging in Kingston Plains, 93
microbes, 58–59
microbial agents, 214
microbiome, human, 85
microcephaly, 156–157
micro-hydro generators, 472
microlending, native American, 535
microlending, origin and extent of, 536
Middle East
    natural gas reserves, 437
    population growth rates, 138
    proven petroleum supplies, 434
mid-ocean ridges, 302–303
Migratory Bird Act of 1918, 240
Migratory Bird Hunting Stamp (1934), wetland
    conservation, 293
Milankovitch, Milutin, 332
Milankovitch cycles, 332, 337
Mill, John Stuart, 23, 523
Millennium Development Goals, 23–24
millennium assessment, 578–579
Millennium Development Project (UN), 413
Minamata Convention on Mercury (2009), 358
Mineral Policy Center, 312
minerals
    corruption, supporting, 313–314
    defined, 304
    economic resource, 306–310
    evaporite, 306
    high-value, 313–314
    new materials substituted for old, 310
    rare earth, 307, 308
minimills, 310
minimum viable population size, 128
mining
    General Mining Law (1872), application, 314
    methods, 301, 310–314
    mountaintop removal (case study), 301–302
    oil, surface mining of, 435–436
    pollutants from, 312, 406
    reclamation, 297
    toxic and hazardous wastes from, 491
Minnesota
    forest experiment, 34
    Haubenschild farm, manure powering, 468
    reforestation (Greening the Green River),
        282–283
    St. Paul heating plant, biomass use, 467
    U of M gasification plant, 468
    Zoo, 246, 247
Miscanthus x giganteus, biofuel crop, 470–471
Mississippi, Columbia, Superfund site remediation, 493
Mississippi River
    Army Corps of Engineers reclamation, 290
    dead zones, 407
Missouri Botanical Garden, 245
Missouri Breaks (ranchland), restoration, 287–288
mitigation, 277
Mittermeier, Russell, 229, 263
mobility, of toxins, 163
modeling in science, importance of, 40
molecular structure, 52
molecular taxonomy, 228
mollisol, USDA soil order, characteristics, 202

reflective thinking, 5
reforestation
    in Kenya, 19
    monoculture agroforestry, 253–254
reformers, in fuel cells, 465
refugees, climate change, 314, 394
refuse-derived fuel, 483
Regional Greenhouse Gas Initiative (RGGI), carbon
    trading, 343
regulations, layoffs caused by, data, 539
regulatory agencies, environmental, 555
rehabilitation, 277
Reich, Peter, 34
Reichhold Chemical, brownfield remediation, 493
Reilly, William K., 565–566
reintroduction, 277
remediation
    defined, 277
    water, of, 420–422
remote sensing, 66
remote sensing surveys, 266
renewable energy
    advances in, 13
    biofuel, 469–471
    biomass, 466–469, 471
    fuel cells, 465–466
    global self sufficiency (2030) calculation, 475
    investment in, 9–10
    price reduction, 13
    solar, 457–462
    supergrid for, 474–475
    water, 471–474
    wind, 462–465
renewable portfolio standards, 460
renewable resources, 520
Repeto, Robert, 530
reproducibility, science and, 35
reproductive isolation, 227
reserves, of coal, 430
reservoirs, sedimentation problems, 392–393
residence time, molecule in water compartment, 380
resilience
    biological communities, in, 90–91
    economic policy, in, 559
    systems, in, 43
Resource Conservation and Recovery
    Act (1976), 489
resource extraction, environment, effect on,
    311–314
resource management, 17
resource partitioning, 77, 78, 82
resources
    intangible, 521
    nonrenewable, 520, 521
    overharvesting, 521
    partitioning, 77–78, 82
    renewable, 520–521
    waste of and pollution by, effect of, 16–20
respiration, cellular, 59, 66
restoration ecology
    benefits of, 281–283
    bison, value in, 288–289
    common components of, 278
    Elwha River, case study, 275–276
    goals, 477–478
    measuring progress, central criteria, 291
    native species, of, 280–281
    natural, 279
    neighborhood DIY, 283

origin of, 278–279
    pragmatic side of, 277
    prairie restoration, 285–288
    prescribed burning, 283–286
    reforestation, 282–283
    reintroduction of native birds, 281
    restoration glossary, 277
    restoring wetlands and streams, 288–297
retrievable storage, nuclear waste, 445
reverse osmosis desalination, 395
RGGI (Regional Greenhouse Gas Initiative), carbon
    trading, 343
rhinos, white and Javanese, 245
ribonucleic acid (RNA), 55
Ricardo, David, 522
rice
    genetically modified, 191
    key global food source, 184
ricin, toxin, 169
riders, legislative, 552
right-sizing streets, 513
rill erosion, 204
Rio de Janeiro, slums of, 9–10
risk
    acceptance, 171
    assessment, 170–171
    management, 173–174
    perception, 173
    precautionary principle, 174
rivers, as water compartments, 383
Rivers and Harbors Act (1899), 547
RNA, guide, 56
Roadless Area Review and Evaluation
    (RARE), 258
Roadless Rule, wilderness guideline, 258
Robert, K. H., 537
Robinson, Frances, 4
rock cycle, geological process, 304–305
rocks
    composition, 304
    evaporite minerals, 306
    igneous, 305
    metamorphic, 305
    minerals, 304, 306–307
    sedimentary, 305–306
    weathering and sedimentation, 305–306
Rogers, Will, 177
Roosevelt, Theodore, 17, 18, 279
Ross, John, 518
rotational grazing, 262–263
Rotterdam Convention (1997), 557
Rousseff, Dilma, 132
r-selected species, 120–121
run-of-the-river, power generation, 472
rural areas
    communications issues, 535
    deer population issues, 122
    functional definition, Census Bureau, 501
    groundwater quality, 414
    microlending, 535–536
    urbanization issues, 132, 500, 502–503,
        508–509
    waste treatment, natural processes, 417
    water treatment, 413
Rural Electrification Act (1935), 462
Rusk, David, 509
Russia. *See also* Soviet Union (former)
    birth dearth in, 145
    Chelyabinsk, nuclear waste site explosion, 445

Chernobyl, bioremediation, 296
Chernobyl nuclear accident (1985), 443, 445
coal deposits in, 430
drug-resistant tuberculosis in, 158
hydropower use in, 472
ice core drilling of Vostok ice sheet, 333
Norilsk, toxic air pollution, 373
population growth rate, 138
wood and paper pulp production in, 252

S

saccharin, 169
Sachs, Jeffrey, 24, 160
safe drinking water, access to, 410–414
Safe Drinking Water Act (1986)
    Haliburton clause (exemption for fracking), 439
    preventive measures, 413
    Water Quality Legislation, 423
Safe Harbor Policy, 243
Sagan, Carl, 45, 300
saguaro cactus, adaptation and specialization, 74–76
Sahara Desert (Africa), overgrazing, 103
salinization, in soil, 206, 392–393
salmon
    anti-environmental legislative riders, 552
    aquaculture issues, 187–188
    dams, effects on, 392
    Elwha River dam, 275
    endangered, 240, 242
    genetically modified, 192
salt cedar, 236–237, 277
salt marshes
    biodiversity, 109
    productivity, 87
    restoration, 293
saltwater intrusion, freshwater aquifers, into, 388
Salween dams, Burma and Thailand, 19
sample, statistical, 38
Sand County Almanac, A (Leopold), 17–18
sanitary landfills, 482
Sargasso Sea, 108
satellites,
    air quality monitoring, 13
    deforestation monitoring, 256
    digital mapping, 270
    drought monitoring, 198
    fishery monitoring, 186
    water measuring, 390–391
Saudi Arabia
    desalination in, 395
    Empty Quarter, preserve, 264
    oil price, effect of fluctuating, 435
savanna, oak
    defined, 284
    historic destruction of, 284
    restoration problems, 284–285
savannas, 102–103, 124, 251
Savory, Allan, 262
scarcity
    conflict, 359, 394
    natural selection, and, 73
    population and technology, 526–529
    water, 383–386
scavenger organisms, 62, 82
schistosomiasis, dams, relationship to, 472
Schneider, Steve, 345
Schumacher, E. F., 523

science
  accuracy and precision, 35–36
  basic principles, 35
  consensus and conflict in, 43–44
  deductive and inductive reasoning in, 36
  defined, 35
  environmental, nature of, 10
  experimental design and, 37, 39–40
  gene editing, 56
  hypotheses and theories, 36–37
  models, 40
  probability in, 37
  pseudoscience, detecting, 44–45
  statistics, basic ideas in, 38–39
  systems, nature of, 41–43
Science, 25, 27
scientific consensus, 43
scientific method, 36
scientific theory, 37
Scott, J. Michael, 244
seafood, key food source, 186
sea ice
  Arctic, extent of, 322
  case study, 322–323
  loss of, 339–340
  oil, effect on exploration, 434–435
  polar bears, and, 234
sea otters, threatened, 241
seasonal rains, 330
sedimentary rocks, 305–306, 332, 436
sedimentation
  dam reservoirs, effect on, 392–393
  rock formation by, 306
  stream and river, degradation and restoration, 294–296
sediments, as water pollution, 410
selection pressures, 75, 78, 83
selective breeding, 56
semen quality, decline in, 166–167
Sen, Amartya K., 181
Seneca, 152
septic tanks, operation of, 417
Sequoia National Park (California), 285
service products, 537
sewage treatment
  absence of, consequences, 486, 504–505
  biological, 418
  Clean Water Act, effectiveness, 411, 422–423, 548, 549
  developing countries, in, 412–413, 504–505
  infectious agents in human waste, 403–404
  low-cost treatment, 420
  mechanical method, 418
  municipal, 417–420
  pet waste problem, 412
  polluted discharge from, 50
  primary method, 418–419
  secondary method, 418–419
  tertiary method, 419
  wetland method, 419
  worldwide, data on, 412
shantytowns, 505–506
sharks
  controlling fishing of, 244–245
  pesticides in, 363
  population data, 244
Sheen, Martin, 494
sheet erosion, 204
Shelford, Victor, 75

Shellenberger, Michael, 573
Shell Oil, arctic drilling effort (2010), 434
shelterwood harvesting, 253
Shiva, Vandana, 191
shortgrass prairie restoration, 287–288
Should Trees Have Standing? (Stone), 27, 553
shrublands, 104
Siberia
  deciduous forests in, 105
  deforestation rate in, 104
  Norilsk, effect of toxic air pollution, 373
  Norilsk, smelters in, 312
sick building syndrome, 161
Sierra Club, 17, 18, 553, 572
Sierra Club v. Morten (1969), 27, 553
sigmoidal growth curve, 120
significant number, appropriate level of detail, 35–36
Silent Spring (Carson), 18, 154, 211, 547
silicon collectors, 460
Simon, Julian, 137, 529
Singapore
  drinking water from sewage, 396
  population growth rate in, 142–143
  wealth in, 20
single large or several small reserves (SLOSS), 268–271
single tree selective harvesting, 253
sink and source habitats, 128
skepticism, scientific need for, 35
skills, critical-thinking, 2, 5–7, 44
Skinner, Michael, 172
S'Klallam Nation fishery, 275
slash and burn agriculture, 209
SLOSS debate (single large or several small reserves)
  Brazilian rainforest, 270–271
  corridors, 268–271
slums
  defined, 505
  factors contributing to, 9–10
small-scale hydropower, issues with, 472
smart growth, 511–516
Smart Growth pollution goals, 411
smart growth principles, cities, 499
smart metering, 453, 456
smelting, 312–313
Smith, Adam, 521, 526
Smith, Robert Angus, 368
Smithsonian Institute, oil dumping study, 409
Smithsonian Institution, 270
smog
  Asian Brown Cloud, 362
  China, city levels, 351–352
  Clean Air Act, effect, 371–372
  Cubatao, Brazil, 373–374
  deaths from, 156, 352, 362
  London's Great Smog (1952), 351, 353, 363, 548–549
  nitrogen oxides (NO_2), 354
  photochemical, 355, 362
  temperature inversions and, 361
  U.K. royal response (1273), 18
  U.S. legislative response (1970s), 547–549
  U.S. levels, 14, 362, 371–372
  visibility reduction from, 266, 366–367, 373
Smoky Mountain, Manila open dump, 480
snail darter, 242
snowfields, 12, 381–382
social capital, 521

social justice, 18, 98, 133, 146–147, 450, 461, 537, 573
social networks, 5, 26, 531, 564, 571
social progress, 18–19, 23, 501, 523, 530, 574
Socolow, Robert, 323, 338, 344
sodium chloride, 52, 306
soil
  acid rain, effect, 367–369
  aeration, zone of, 381
  arable land, distribution of, 203
  brownfield reclamation, 491–493
  carbon cycle, 66–67
  Cerrado, cultivation, effect, 198
  city farming techniques, 212
  components of, 199–200
  conservation, 207–210
  Conservation Reserve Program (CRP), and, 191–192
  dark soils, creating, 209
  deaths from contaminated, 156
  desert, 103
  desertification, 206
  dust storms, 356
  ecosystem, 199–202
  emergent properties, effect of, 202
  erosion, 203–206
  fertility, factors, 201–202
  fertilization, issues with, 207
  GMO crops for poor soil, 191–192
  grazing, issues, 260–263
  greenhouse gas production, 336
  horizons, composition, 201–202
  hydrological cycle, role in, 65–66
  irrigation, damage from, 207–208
  land degradation, 203–205
  losses, causes of, 203–210
  metal extraction, plants used for, 495
  night soil fertilizer, 417
  orders, USDA classification, 202
  organic chemical extraction, plants used for, 495
  organisms in, 201
  phytoremediation, 495–496
  pollution, racism and, 202
  prairie restoration, 285–286
  profile, 202
  saturation, zone of, 381
  shade-grown coffee, effect, 189
  structure of, 202
  succession, role in, 93
  temperate grasslands, 103–104
  Terra Preta, creation of, 209
  texture, 199–200
  tillage techniques for improving, 210
  topsoil, 'A' horizon characteristics, 202
  tropical moist rainforest, 100
  types of, 199–200
  use and abuse of, 203–208
  water compartment, residence time, 380
  zones of aeration, saturation, 381
solar energy
  absorption and reflection of, 326–327
  active solar systems, 457
  advances in, 13
  atmospheric absorption of, 327
  atmospheric zones created by, 325
  average solar radiation received, 457
  case study of transition, 450–451
  cost comparison, 463
  decentralized power, 460–462

water use, 505
world's largest cities, 502
urbanization
brownfield developments, 514
core regions, 501
developed world, 506–511
global concentrations, 502
governmental policies and, 503
greenfield developments, 513
green urbanism, 513–515
land tenure patterns and, 503
new urbanism, 512–513
population shift toward, 502–503
process of, 500–503
pull factors in, 503
push factors in, 503
rate of growth, 501
smart growth, 511–516
sprawl, 508
transportation in city development, 509–510
Uruguay, renewable power and, 10
utilitarian, 16–17
utilitarian conservation, 16–17
UV water treatment, 419–420

V

value, intrinsic or instrumental, 26
value added food products, 182
variables, in experimental design, 39
Vassar College, New York, 178
Vauban, Germany, car-free, case study, 499
Veblen, Thorstein, 569
Venezuela, parks and preserves in, 264
Vermont
350.org, 564
Biedler family, sustainable farm of, 221
forests, damage from acid rain, 369
Middlebury College gasification plant, 468
reforestation, 279
vertical axis wind turbines (VAWT), 463
vertical stratification, 107
vertical zonation, vegetation, 99
vertical zones in freshwater ecosystems, 111
vertisol, USDA soil order, characteristics, 202
Vidaza, epigenic change-reversing drug, 172–173
village
functional definition, 501, 506–511
land use planning, 269
Village Enterprise Fund, 536
Village Forest Reserves, 269
vinblastine, plant derived anticancer drug, 230
vincristine, plant derived anticancer drug, 230
viral DNA, 56
Virginia, Reston, planned community, 512
viruses, behavior of flu, 80
vitamin A, deficiency, 183
volatile organic compounds (VOCs), 355, 360
volcanoes
air pollution from, 351
emissions from, 317, 353
Mount Vesuvius (Italy), 317
Mt. Merapi (Indonesia), 317
pyroclastic clouds, 317
tectonic movement, role in, 303–304
Voting Rights Act, 546
vulnerable species, 241

W

Walden (Thoreau), 569
Wallace, Alfred, 83
Walmart, green products, debate about, 539
Ward, Barbara, 10
Ward, William Arthur, 426
warm fronts, weather patterns, 331
Warming, J. E. B., 92
Warren, Karen J., 5
Washington, D.C.
Mineral Policy Center, 312
taxes vs service costs, 509
Washington, Olympic National Park, Elwha River
restoration, 275
Washington Post, 552
waste disposal
exporting waste, 481–482
hazardous wastes (*see* hazardous wastes)
incineration (*see* incineration)
landfills (*see* landfills)
ocean dumping, 481
open dumps, 480
suitable places for, 483
waste (human and animal) disposal, 417–420
wastes
demanufacturing, 487
e-waste, 487
hazardous, 488–496
reducing, 488
reusing, 487–488
shrinking the waste stream, 484–487
solid, 479
waste hierarchy, 488
waste stream, 480
waste-to-energy, 483
wastewater treatment, wealth as a factor, 404
water
accessible, renewable supply of, 383
access to safe, 14, 410–414
as agent of soil erosion, 204
agricultural use, 206, 207, 385–386
availability and use, 383–386
bottled, 415
clean water, access to, 383
climate change, effect of, 14
compartments, 380
consumption, 385
control, Everglades National Park, 292–293
desalination, 394–395
distribution, uneven, 378–379
domestic conservation, 395–396
domestic use, 386
ecological services, 289
energy from, 57
erosion, 204
evaporation, 327–328
fossil water defined, 388
glaciers, 381
groundwater (*see* groundwater)
hydrologic cycle, 65–66, 378
illness and death from, 14
improved water sources, 413
increase in use of, 385
industrial use, 386
infectious disease reduction, 14
loss by evaporation, 391–392

major rivers of the world, 383
molecule, 52, 53
as most critical resource, 14
oceans (*see* ocean)
in photosynthesis, 59–60
policies, 397
prices, 397
properties of, 53
recycling, 396–397
redistribution of, 378
renewable supplies of, 383
saving, 395–396
shortages, freshwater, 387–394
states of, 53
streams, 293–295
units of water measurement, 378
use of, rate, 385
water, "mining," 385
water, source of, 381
waterborne diseases, deaths from, 404
water conservation, DIY methods, 396
water content, aquifers vs surface water (U.S.), 382
water flow projections, 377
Water Health International, 420
water level, glaciers and, 340
waterlogging, 206
water mining, 385
water pollution. *See also* Clean Water Act (1972)
acids and bases, 408
atmospheric deposition. defined, 403
bacteria, testing for, 403–404
benefits of removing, 402
categories of pollution, 403
China, problems in, 412–413
coal burning, from, 432
containment, 420
eutrophication, 405–406
historical issues, 402
improvement in developing countries, 413
infectious agents, 403–404
inorganic pollutants, 406–408
legislation, U.S. and International, 423
major categories of, 403
measuring oxygen levels for, 403–404
metals, 406–408
nonmetallic salts, 406–408
nonpoint sources of, 402
nutrient enrichment, 405–406
oceans, 414–415
organic pollutants, 408–409
pigs, diseased and dead in Chinese river, 413
point sources of, 402
quality, improving, 413, 416–424
sediment, 410
thermal pollution, 410
urbanization and, 506
water pollution control
lead in gasoline, banning of, 416
nonpoint sources of, 416–417
source reduction, 416
water remediation, 420–422
water purification, 420–421
water quality, data on, 411
water remediation, 420–422
water scarcity, 383–384
water shortage warning, 393
water shortage warning, Intergovernmental Panel on
Climate Change (IPCC), 393

Periodic Table of the Elements

Period	IA (1) Alkali Metals	IIA (2) Alkaline Earth Metals	IIIB (3)	IVB (4)	VB (5)	VIB (6)	VIIB (7)	VIIIB (8)	VIIIB (9)	VIIIB (10)	IB (11)	IIB (12)	IIIA (13)	IVA (14)	VA (15)	VIA (16)	VIIA (17) Halogens	VIIIA (18) Noble Gases
1	Hydrogen 1 **H** 1.008																	Helium 2 **He** 4.003
2	Lithium 3 **Li** 6.941	Beryllium 4 **Be** 9.012											Boron 5 **B** 10.81	Carbon 6 **C** 12.01	Nitrogen 7 **N** 14.01	Oxygen 8 **O** 16.00	Fluorine 9 **F** 19.00	Neon 10 **Ne** 20.18
3	Sodium 11 **Na** 22.99	Magnesium 12 **Mg** 24.31											Aluminum 13 **Al** 26.98	Silicon 14 **Si** 28.09	Phosphorus 15 **P** 30.97	Sulfur 16 **S** 32.07	Chlorine 17 **Cl** 35.45	Argon 18 **Ar** 39.95
4	Potassium 19 **K** 39.10	Calcium 20 **Ca** 40.08	Scandium 21 **Sc** 44.96	Titanium 22 **Ti** 47.88	Vanadium 23 **V** 50.94	Chromium 24 **Cr** 52.00	Manganese 25 **Mn** 54.94	Iron 26 **Fe** 55.85	Cobalt 27 **Co** 58.93	Nickel 28 **Ni** 58.69	Copper 29 **Cu** 63.55	Zinc 30 **Zn** 65.39	Gallium 31 **Ga** 69.72	Germanium 32 **Ge** 72.61	Arsenic 33 **As** 74.92	Selenium 34 **Se** 78.96	Bromine 35 **Br** 79.90	Krypton 36 **Kr** 83.80
5	Rubidium 37 **Rb** 85.47	Strontium 38 **Sr** 87.62	Yttrium 39 **Y** 88.91	Zirconium 40 **Zr** 91.22	Niobium 41 **Nb** 92.91	Molybdenum 42 **Mo** 95.94	Technetium 43 **Tc** (98)	Ruthenium 44 **Ru** 101.1	Rhodium 45 **Rh** 102.9	Palladium 46 **Pd** 106.4	Silver 47 **Ag** 107.9	Cadmium 48 **Cd** 112.4	Indium 49 **In** 114.8	Tin 50 **Sn** 118.7	Antimony 51 **Sb** 121.8	Tellurium 52 **Te** 127.6	Iodine 53 **I** 126.9	Xenon 54 **Xe** 131.3
6	Cesium 55 **Cs** 132.9	Barium 56 **Ba** 137.3	Lanthanum 57 **La** 138.9 †	Hafnium 72 **Hf** 178.5	Tantalum 73 **Ta** 180.9	Tungsten 74 **W** 183.8	Rhenium 75 **Re** 186.2	Osmium 76 **Os** 190.2	Iridium 77 **Ir** 192.2	Platinum 78 **Pt** 195.1	Gold 79 **Au** 197.0	Mercury 80 **Hg** 200.6	Thallium 81 **Tl** 204.4	Lead 82 **Pb** 207.2	Bismuth 83 **Bi** 209.0	Polonium 84 **Po** (209)	Astatine 85 **At** (210)	Radon 86 **Rn** (222)
7	Francium 87 **Fr** (223)	Radium 88 **Ra** (226)	Actinium 89 **Ac** (227) ‡	Rutherfordium 104 **Rf** (261)	Dubnium 105 **Db** (262)	Seaborgium 106 **Sg** (266)	Bohrium 107 **Bh** (264)	Hassium 108 **Hs** (277)	Meitnerium 109 **Mt** (268)	Darmstadtium 110 **Ds** (281)	Roentgenium 111 **Rg** (280)	Copernicium 112 **Cn** (285)	Ununtrium 113 **Uut** (284)	Flerovium 114 **Fl** (289)	Ununpentium 115 **Uup** (288)	Livermorium 116 **Lv** (293)	Ununseptium 117 **Uus** (294)	Ununoctium 118 **Uuo** (294)

Transition Elements: groups IIIB (3) through IIB (12)

Legend: Metals; Semiconductors; Nonmetals

Inner Transition Elements

†Lanthanides 6	Cerium 58 **Ce** 140.1	Praseodymium 59 **Pr** 140.9	Neodymium 60 **Nd** 144.2	Promethium 61 **Pm** (145)	Samarium 62 **Sm** 150.4	Europium 63 **Eu** 152.0	Gadolinium 64 **Gd** 157.3	Terbium 65 **Tb** 158.9	Dysprosium 66 **Dy** 162.5	Holmium 67 **Ho** 164.9	Erbium 68 **Er** 167.3	Thulium 69 **Tm** 168.9	Ytterbium 70 **Yb** 173.0	Lutetium 71 **Lu** 175.0
‡Actinides 7	Thorium 90 **Th** 232.0	Protactinium 91 **Pa** 231.0	Uranium 92 **U** 238.0	Neptunium 93 **Np** (237)	Plutonium 94 **Pu** (244)	Americium 95 **Am** (243)	Curium 96 **Cm** (247)	Berkelium 97 **Bk** (247)	Californium 98 **Cf** (251)	Einsteinium 99 **Es** (252)	Fermium 100 **Fm** (257)	Mendelevium 101 **Md** (258)	Nobelium 102 **No** (259)	Lawrencium 103 **Lr** (262)

Key: element name — Hydrogen; atomic number — 1; symbol of element — **H**; atomic weight — 1.008

Values in parentheses are the mass numbers of the most stable or best-known isotopes.

UNITS OF MEASUREMENT METRIC/ENGLISH CONVERSIONS

Length

1 meter = 39.4 inches = 3.28 feet = 1.09 yard
1 foot = 0.305 meters = 12 inches = 0.33 yard
1 inch = 2.54 centimeters
1 centimeter = 10 millimeters = 0.394 inch
1 millimeter = 0.001 meter = 0.01 centimeter = 0.039 inch
1 fathom = 6 feet = 1.83 meters
1 rod = 16.5 feet = 5 meters
1 chain = 4 rods = 66 feet = 20 meters
1 furlong = 10 chains = 40 rods = 660 feet = 200 meters
1 kilometer = 1,000 meters = 0.621 miles = 0.54 nautical miles
1 mile = 5,280 feet = 8 furlongs = 1.61 kilometers
1 nautical mile = 1.15 mile

Area

1 square centimeter = 0.155 square inch
1 square foot = 144 square inches = 929 square centimeters
1 square yard = 9 square feet = 0.836 square meters
1 square meter = 10.76 square feet = 1.196 square yards = 1 million square millimeters
1 hectare = 10,000 square meters = 0.01 square kilometers = 2.47 acres
1 acre = 43,560 square feet = 0.405 hectares
1 square kilometer = 100 hectares = 1 million square meters = 0.386 square miles = 247 acres
1 square mile = 640 acres = 2.59 square kilometers

Volume

1 cubic centimeter = 1 milliliter = 0.001 liter
1 cubic meter = 1 million cubic centimeters = 1,000 liters
1 cubic meter = 35.3 cubic feet = 1.307 cubic yards = 264 US gallons
1 cubic yard = 27 cubic feet = 0.765 cubic meters = 202 US gallons
1 cubic kilometer = 1 million cubic meters = 0.24 cubic mile = 264 billion gallons
1 cubic mile = 4.166 cubic kilometers
1 liter = 1,000 milliliters = 1.06 quarts = 0.265 US gallons = 0.035 cubic feet
1 US gallon = 4 quarts = 3.79 liters = 231 cubic inches = 0.83 imperial (British) gallons
1 quart = 2 pints = 4 cups = 0.94 liters
1 acre foot = 325,851 US gallons = 1,234,975 liters = 1,234 cubic meters
1 barrel (of oil) = 42 US gallons = 159 liters

Mass

1 microgram = 0.001 milligram = 0.000001 gram
1 gram = 1,000 milligrams = 0.035 ounce
1 kilogram = 1,000 grams = 2.205 pounds
1 pound = 16 ounces = 454 grams
1 short ton = 2,000 pounds = 909 kilograms
1 metric ton = 1,000 kilograms = 2,200 pounds

Temperature

Celsius to Fahrenheit $°F = (°C \times 1.8) + 32$
Fahrenheit to Celsius $°C = (°F - 32) \div 1.8$

Energy and Power

1 erg = 1 dyne per square centimeter
1 joule = 10 million ergs
1 calorie = 4.184 joules
1 kilojoule = 1,000 joules = 0.949 British Thermal Units (BTU)
1 megajoule = MJ = 1,000,000 joules
1 kilocalorie = 1,000 calories = 3.97 BTU = 0.00116 kilowatt-hour
1 BTU = 0.293 watt-hour
1 kilowatt-hour = 1,000 watt-hours = 860 kilocalories = 3,400 BTU
1 horsepower = 640 kilocalories
1 quad = 1 quadrillion kilojoules = 2.93 trillion kilowatt-hours

Quantitative Prefixes

Large Numbers	Description	Small Numbers
exa 10^{18}	quintillion	alto 10^{-18}
peta 10^{15}	quadrillion	femto 10^{-15}
tera 10^{12}	trillion	pico 10^{-12}
giga 10^{9}	billion	nano 10^{-9}
mega 10^{6}	million	micro 10^{-6}
kilo 10^{3}	thousand	milli 10^{-3}

(e.g., a kilogram = 1,000 gm; a milligram = one-thousandth of a gram)